The Industrial Electronics Handbook
SECOND EDITION

INDUSTRIAL COMMUNICATION SYSTEMS

The Industrial Electronics Handbook
SECOND EDITION

FUNDAMENTALS OF INDUSTRIAL ELECTRONICS

POWER ELECTRONICS AND MOTOR DRIVES

CONTROL AND MECHATRONICS

INDUSTRIAL COMMUNICATION SYSTEMS

INTELLIGENT SYSTEMS

The Electrical Engineering Handbook Series

Series Editor
Richard C. Dorf
University of California, Davis

Titles Included in the Series

The Avionics Handbook, Second Edition, Cary R. Spitzer
The Biomedical Engineering Handbook, Third Edition, Joseph D. Bronzino
The Circuits and Filters Handbook, Third Edition, Wai-Kai Chen
The Communications Handbook, Second Edition, Jerry Gibson
The Computer Engineering Handbook, Vojin G. Oklobdzija
The Control Handbook, Second Edition, William S. Levine
CRC Handbook of Engineering Tables, Richard C. Dorf
Digital Avionics Handbook, Second Edition, Cary R. Spitzer
The Digital Signal Processing Handbook, Vijay K. Madisetti and Douglas Williams
The Electric Power Engineering Handbook, Second Edition, Leonard L. Grigsby
The Electrical Engineering Handbook, Third Edition, Richard C. Dorf
The Electronics Handbook, Second Edition, Jerry C. Whitaker
The Engineering Handbook, Third Edition, Richard C. Dorf
The Handbook of Ad Hoc Wireless Networks, Mohammad Ilyas
The Handbook of Formulas and Tables for Signal Processing, Alexander D. Poularikas
Handbook of Nanoscience, Engineering, and Technology, Second Edition, William A. Goddard, III, Donald W. Brenner, Sergey E. Lyshevski, and Gerald J. Iafrate
The Handbook of Optical Communication Networks, Mohammad Ilyas and Hussein T. Mouftah
The Industrial Electronics Handbook, Second Edition, Bogdan M. Wilamowski and J. David Irwin
The Measurement, Instrumentation, and Sensors Handbook, John G. Webster
The Mechanical Systems Design Handbook, Osita D.I. Nwokah and Yidirim Hurmuzlu
The Mechatronics Handbook, Second Edition, Robert H. Bishop
The Mobile Communications Handbook, Second Edition, Jerry D. Gibson
The Ocean Engineering Handbook, Ferial El-Hawary
The RF and Microwave Handbook, Second Edition, Mike Golio
The Technology Management Handbook, Richard C. Dorf
Transforms and Applications Handbook, Third Edition, Alexander D. Poularikas
The VLSI Handbook, Second Edition, Wai-Kai Chen

The Industrial Electronics Handbook
SECOND EDITION

INDUSTRIAL COMMUNICATION SYSTEMS

Edited by

Bogdan M. Wilamowski

J. David Irwin

CRC Press is an imprint of the
Taylor & Francis Group, an informa business

CRC Press
Taylor & Francis Group
6000 Broken Sound Parkway NW, Suite 300
Boca Raton, FL 33487-2742

First issued in paperback 2017

ISBN-13: 978-1-4398-0281-6 (hbk)
ISBN-13: 978-1-138-07180-3 (pbk)

Library of Congress Cataloging-in-Publication Data

Industrial communication systems / editors, Bogdan M. Wilamowski and J. David Irwin.
p. cm.
"A CRC title."
Includes bibliographical references and index.
ISBN 978-1-4398-0281-6 (alk. paper)
1. Computer networks. 2. Data transmission systems. 3. Telecommunication systems. I. Wilamowski, Bogdan M. II. Irwin, J. David, 1939- III. Title.

TK5105.5.I477 2010
004.6--dc22 2010020567

Visit the Taylor & Francis Web site at
http://www.taylorandfrancis.com

and the CRC Press Web site at
http://www.crcpress.com

Contents

Preface...xiii
Preambles...xv
Acknowledgments xxiii
Editorial Board xxv
Editors...xxvii
Contributors xxxi

PART I Technical Principles

1 ISO/OSI Model 1-1
Gerhard Zucker and Dietmar Dietrich

2 Media 2-1
Herbert Schweinzer, Saleem Farooq Shaukat, and Holger Arthaber

3 Media Access Methods..3-1
Herbert Haas and Manfred Lindner

4 Routing in Wireless Networks 4-1
Teresa Albero-Albero and Víctor-M. Sempere-Payá

5 Profiles and Interoperability 5-1
Gerhard Zucker and Heinz Frank

6 Industrial Wireless Sensor Networks 6-1
Vehbi Cagri Gungor and Gerhard P. Hancke

7 Ad Hoc Networks 7-1
Sajjad Ahmad Madani, Shahid Khattak, Tariq Jadoon, and Shahzad Sarwar

8 Radio Frequency Identification 8-1
Edward Kai-Ning Yung, Pui-Yi Lau, and Chi-Wai Leung

9 RFID Technology and Its Industrial Applications...9-1
Vidyasagar Potdar, Atif Sharif, and Elizabeth Chang

10 Ultralow-Power Wireless Communication...10-1
Joern Ploennigs, Volodymyr Vasyutynskyy, and Klaus Kabitzsch

11 Industrial Strength Wireless Multimedia Sensor Network Technology 11-1
Vidyasagar Potdar, Atif Sharif, and Elizabeth Chang

12 A Survey of Wireless Sensor Networks for Industrial Applications 12-1
Stig Petersen and Simon Carlsen

13 Vertical Integration 13-1
Thilo Sauter, Stefan Soucek, and Martin Wollschlaeger

14 Multimedia Service Convergence 14-1
Alex Talevski

15 Virtual Automation Networks 15-1
Peter Neumann and Ralf Messerschmidt

16 Industrial Agent Technology 16-1
Aleksey Bratukhin, Yoseba Peña Landaburu, Paulo Leitão, and Rainer Unland

17 Real-Time Systems 17-1
Lucia Lo Bello, José Alberto Fonseca, and Wilfried Elmenreich

18 Clock Synchronization in Distributed Systems 18-1
Georg Gaderer and Patrick Loschmidt

19 Quality of Service 19-1
Gabriel Diaz Orueta, Elio San Cristobal Ruiz, Nuria Oliva Alonso, and Manuel Castro Gil

20 Network-Based Control 20-1
Josep M. Fuertes, Mo-Yuen Chow, Ricard Villà, Rachana Gupta, and Jordi Ayza

21 Functional Safety 21-1
Thomas Novak and Andreas Gerstinger

22 Security in Industrial Communication Systems 22-1
Wolfgang Granzer and Albert Treytl

23 Secure Communication Using Chaos Synchronization 23-1
Yan-Wu Wang and Changyun Wen

PART II Application-Specific Areas

24 Embedded Networks in Civilian Aircraft Avionics Systems 24-1
Christian Fraboul, Fabrice Frances, and Jean-Luc Scharbarg

25 Process Automation 25-1
Alois Zoitl and Wilfried Lepuschitz

26 Building and Home Automation 26-1
Wolfgang Kastner, Stefan Soucek, Christian Reinisch, and Alexander Klapproth

27 Industrial Multimedia 27-1
Javier Silvestre-Blanes, Manfred Weihs, and Víctor-M. Sempere-Payá

28 Industrial Wireless Communications Security (IWCS)/C42........................28-1
Milos Manic and Kurt Derr

29 Protocols in Power Generation........................29-1
Tuan Dang and Gaëlle Marsal

30 Communications in Medical Applications........................30-1
Paulo Bartolomeu, José Alberto Fonseca, Nelson Rocha, and Filipe Basto

PART III Technologies

31 Controller Area Network........................31-1
Joaquim Ferreira and José Alberto Fonseca

32 Profibus........................32-1
Max Felser and Ron Mitchell

33 INTERBUS........................33-1
Juergen Jasperneite and Orazio Mirabella

34 WorldFip........................34-1
Francisco Vasques and Orazio Mirabella

35 Foundation Fieldbus........................35-1
Carlos Eduardo Pereira, Augusto Pereira, and Ian Verhappen

36 Modbus........................36-1
Mário de Sousa and Paulo Portugal

37 Industrial Ethernet........................37-1
Gaëlle Marsal and Denis Trognon

38 EtherCAT........................38-1
Gianluca Cena, Adriano Valenzano, and Claudio Zunino

39 Ethernet POWERLINK........................39-1
Paulo Pedreiras, Stefan Schoenegger, Lucia Seno, and Stefano Vitturi

40 PROFINET........................40-1
Max Felser, Paolo Ferrari, and Alessandra Flammini

41 LonWorks........................41-1
Uwe Ryssel, Henrik Dibowski, Heinz Frank, and Klaus Kabitzsch

42 KNX........................42-1
Wolfgang Kastner, Fritz Praus, Georg Neugschwandtner, and Wolfgang Granzer

43 Protocols of the Time-Triggered Architecture: TTP, TTEthernet, TTP/A........................43-1
Wilfried Elmenreich and Christian El-Salloum

44 FlexRay........................44-1
Martin Horauer and Peter Rössler

45 LIN-Bus........................45-1
Andreas Grzemba, Donal Heffernan, and Thomas Lindner

46 Profisafe 46-1
Ron Mitchell, Max Felser, and Paulo Portugal

47 SafetyLon 47-1
Thomas Novak, Thomas Tamandl, and Peter Preininger

48 Wireless Local Area Networks 48-1
Henning Trsek, Juergen Jasperneite, Lucia Lo Bello, and Milos Manic

49 Bluetooth 49-1
Stefan Mahlknecht, Milos Manic, and Sajjad Ahmad Madani

50 ZigBee 50-1
Stefan Mahlknecht, Tuan Dang, Milos Manic, and Sajjad Ahmad Madani

51 6LoWPAN: IP for Wireless Sensor Networks and Smart Cooperating Objects 51-1
Guido Moritz and Frank Golatowski

52 WiMAX in Industry 52-1
Milos Manic, Sergiu-Dan Stan, and Strahinja Stankovic

53 WirelessHART, ISA100.11a, and OCARI 53-1
Tuan Dang and Emiliano Sisinni

54 Wireless Communication Standards 54-1
Tuan Dang

55 Communication Aspects of IEC 61499 Architecture 55-1
Valeriy Vyatkin, Mário de Sousa, and Alois Zoitl

56 Industrial Internet 56-1
Martin Wollschlaeger and Thilo Sauter

57 OPC UA 57-1
Tuan Dang and Renaud Aubin

58 DNP3 and IEC 60870-5 58-1
Andrew C. West

59 IEC 61850 for Distributed Energy Resources 59-1
Sidonia Mesentean, Heinz Frank, and Karlheinz Schwarz

PART IV Internet Programming

60 User Datagram Protocol—UDP 60-1
Aleksander Malinowski and Bogdan M. Wilamowski

61 Transmission Control Protocol—TCP 61-1
Aleksander Malinowski and Bogdan M. Wilamowski

62 Development of Interactive Web Pages 62-1
Pradeep Dandamudi

63 Interactive Web Site Design Using Python Script 63-1
Hao Yu and Michael Carroll

64 Running Software over Internet 64-1
Nam Pham, Bogdan M. Wilamowski, and Aleksander Malinowski

65 Semantic Web Services for Manufacturing Industry 65-1
Chen Wu and Tharam S. Dillon

66 Automatic Data Mining on Internet by Using PERL Scripting Language 66-1
Nam Pham and Bogdan M. Wilamowski

PART V Outlook

67 Trends and Challenges for Industrial Communication Systems 67-1
Peter Palensky

68 Processing Data in Complex Communication Systems 68-1
Gerhard Zucker, Dietmar Bruckner, and Dietmar Dietrich

Index **Index-1**

Preface

The field of industrial electronics covers a plethora of problems that must be solved in industrial practice. Electronic systems control many processes that begin with the control of relatively simple devices like electric motors, through more complicated devices such as robots, to the control of entire fabrication processes. An industrial electronics engineer deals with many physical phenomena as well as the sensors that are used to measure them. Thus, the knowledge required by this type of engineer is not only traditional electronics but also specialized electronics, for example, that required for high-power applications. The importance of electronic circuits extends well beyond their use as a final product in that they are also important building blocks in large systems, and thus, the industrial electronics engineer must also possess a knowledge of the areas of control and mechatronics. Since most fabrication processes are relatively complex, there is an inherent requirement for the use of communication systems that not only link the various elements of the industrial process but are also tailor-made for the specific industrial environment. Finally, the efficient control and supervision of factories requires the application of intelligent systems in a hierarchical structure to address the needs of all components employed in the production process. This need is accomplished through the use of intelligent systems such as neural networks, fuzzy systems, and evolutionary methods. The Industrial Electronics Handbook addresses all these issues and does so in five books outlined as follows:

1. *Fundamentals of Industrial Electronics*
2. *Power Electronics and Motor Drives*
3. *Control and Mechatronics*
4. *Industrial Communication Systems*
5. *Intelligent Systems*

The editors have gone to great lengths to ensure that this handbook is as current and up to date as possible. Thus, this book closely follows the current research and trends in applications that can be found in *IEEE Transactions on Industrial Electronics*. This journal is not only one of the largest engineering publications of its type in the world but also one of the most respected. In all technical categories in which this journal is evaluated, it is ranked either number 1 or number 2 in the world. As a result, we believe that this handbook, which is written by the world's leading researchers in the field, presents the global trends in the ubiquitous area commonly known as industrial electronics.

Clearly, the successful operation of any production process is dependent on a well-designed and reliable communication system. Modern communication systems that are employed within a factory use a variety of means for sending and receiving information. With time, these systems have become more and more sophisticated. This book is the most voluminous of the five that comprise the Industrial Electronics Handbook, and spans the full gamut of topics that are needed for engineers working with industrial communication systems. A description of the numerous topics covered in this book is outlined in the Preambles, and the readers are directed to the relevant parts for further details.

Preambles

Dietmar Dietrich, Dietmar Bruckner, Gerhard Zucker, and Peter Palensky
Institute of Computer Technology
Vienna University of Technology
Vienna, Austria

Process control requires control units and while in the past these were stand-alone elements, they have now become more and more interconnected. Today, we have networks on multiple layers; for example, we have networks of processes with their attendant control units as well as networks of process components. These communication systems have different requirements, not only on different layers but also in vastly different areas of automation. It is fascinating to see that automation permeates essentially every area of our lives. As a result, we are today able to reach any electrical component, wherever it may be.

This book provides an overview of the many facets of communication that are relevant to industrial systems. Part I deals with the technical principles that are necessary for communication, including both wired and wireless communication, the integration of diverse systems, and quality of service aspects. Part II focuses on the application of communication systems to different domains such as process and building automation, energy distribution, and medical applications.

Part III describes what appear to be the most important communication technologies. Although the list is not exhaustive, it does address the most important areas, including wireless communication, fieldbus systems, and the industrial Ethernet and industrial Internet for building automation and automotive applications. Part IV covers topics related to general integration of Internet technologies into industrial automation. Finally, Part V peers into the future in an attempt to describe possible upcoming developments.

Preamble to Part I: Technical Principles

Friederich Kupzog
Institute of Computer Technology
Vienna University of Technology
Vienna, Austria

Jürgen Jasperneite
Institute Industrial IT
Lemgo, Germany

Thilo Sauter
Institute for Integrated Sensor Systems
Austrian Academy of Sciences
Wiener Neustadt, Austria

Communication is a prerequisite for distributed systems. Such systems can be loosely defined as a group of individual computer systems that appear to the user as a single coherent system. The spatially dispersed nature of industrial processes, on the scale of a factory floor or electric power grid, is actually often used as a guide for the design and layout of automation systems. This can be observed, for example, in network-based control, where the control loop can actually be distributed over different processors in a network. In this environment, the basic principles of distributed systems apply. However, while the classic theory of distributed systems has been developed keeping mainly general-purpose computer systems in mind, industrial automation focuses on dedicated systems with highly specialized hardware and software.

Therefore, in what follows, the relevant aspects of distributed systems are revisited from the viewpoint of industrial communication systems. It begins with a discussion of the classic ISO/OSI model. Although the basic principle of communication layering is very significant for communication in automation processes, not all layers defined in the reference model are of equal importance.

Furthermore, special attention is given to three different aspects: wireless, integration, and quality of service. Wireless communication today has a fixed place in many of the application areas of automation. Wireless-related topics such as wireless sensor networks, low-power wireless communication nodes, and RFID are discussed in detail. The integration of heterogeneous systems into a coherent application environment is another crucial issue that is addressed. Finally, quality of service is revisited for industrial communication systems, ranging from real-time communication for safety and security to network-based control.

Group 1.1: Layers

Thilo Sauter
Institute for Integrated Sensor Systems
Austrian Academy of Sciences
Wiener Neustadt, Austria

The design of complex communication systems is not possible without a structured approach. Therefore, a layered structure is commonly adopted. A landmark for communication system development was the definition of the open system interconnection (OSI) model, a generic framework that is presented in the first chapter of this group (Chapters 1 through 3). Within this model, it is primarily the lower layers that are important for industrial communications in order to guarantee the performance needed for a given application domain. Special attention will therefore be given to the large variety of wired electrical, optical, and wireless communication media, as well as the many methods devised for access control. For larger networks that are gaining importance in distributed systems, flat network structures are not adequate for electrical and logical reasons. Therefore, the information flow through the network has to be controlled by appropriate routing strategies, which has been a topic of interest for researchers and developers for a long time. Experience with the first industrial communication systems, however, showed that the OSI model was insufficient to ensure the interoperability that was a major requirement of industry. A substantial amount of work has been, and still is, devoted to the definition of high-level profiles that, depending on device type or application domain, further constrain the degrees of freedom for system developers and implementers.

Group 1.2: Wireless

Jürgen Jasperneite
Institute Industrial IT
Lemgo, Germany

The integration of wireless technologies in industrial automation systems is the next step in the evolution of industrial networking. Wireless technologies have the potential to reduce the life-cycle costs of machines and plants as well as support future adaptive production concepts, either as an extension to, or as a replacement for, existing wired networks.

In addition to wireless networking, new applications can be enabled such as wireless monitoring and control, or asset and personnel tracking. Most of the wireless technologies are standardized in the IEEE802 family and are driven by consumer market requirements. As a result, they are not designed to meet the automation-specific requirements such as low latencies or the demand for high service reliability. As a consequence, the basic concepts must be reviewed and sometimes revised. Therefore, it is in this group (Chapters 4 through 12) that the principles and architectures of wireless sensor networks are presented. Furthermore, some key functions of wireless networks, such as self-configuration, routing, energy efficiency, and data security are introduced.

Group 1.3: Integration

Thilo Sauter
Institute for Integrated Sensor Systems
Austrian Academy of Sciences
Wiener Neustadt, Austria

Industrial communication systems, and automation solutions in general, were initially developed primarily as islands. In recent years, however, the integration of automation systems into a wider scope has increased its importance and thus requires a technological basis. This group of chapters (Chapters 13 through 18) is therefore focused on selected integration aspects. The first chapter deals with vertical integration aimed at providing a transparent data exchange across all levels of the automation hierarchy, both from a networking and application point of view. A topic receiving increasing attention is the integration of multimedia technologies in automation, which requires the convergence of telecommunications and data services and poses new challenges for both network and application design. The issue of complex heterogeneous networks comprised of wired and wireless as well as automation and office domains is being addressed by the concept of virtual automation networks. Finally, software agent technology will be discussed as one way of achieving integration in automation systems by means of distributing functionalities among a group of autonomous, loosely coupled entities that may interact to accomplish a task that is difficult to solve in a centralized manner.

Group 1.4: Quality of Service

Friederich Kupzog
Institute of Computer Technology
Vienna University of Technology
Vienna, Austria

In many cases, special requirements have to be fulfilled in industrial communication systems. While requirements differ to some extent from those in the consumer products domain, the basic principles are the same. Communication in an industrial environment normally has to be highly reliable, and often has to fulfill special demands in terms of delay, bandwidth, or integrity. This fact, also referred to as quality of service, is therefore revisited in the following group of chapters (Chapters 19 through 23) from the viewpoint of industrial communication, which ranges from real time over safety and security to network-based control. The discussion centers around the manner in which these systems have to be designed in order to fulfill the minimum requirements that guarantee the different properties for these communication areas.

Preamble to Part II: Application-Specific Areas

Peter Palensky
Energy Department
Austrian Institute of Technology
Vienna, Austria

Thomas Novak
SWARCO Futurit Verkehrssignalssysteme GmbH
Perchtoldsdorf, Austria

The plethora of applications for industrial communication systems (ICS) leads to a large variety of technologies and standards. This part gives an overview of the important applications and their specialties—and peculiarities—in ICS. The spectrum of topics ranges from embedded networks in avionics to building and home automation to medical applications. The applications can differ in a number of aspects that are important for designing or selecting an ICS technology, some of which are

- Number of nodes
- Requested latency
- Requested bandwidth
- Real-time requirements
- Cost per node
- Reliability and availability
- Functional safety
- Electromagnetic compatibility
- Physical topology
- Length of network segments
- Scalability and extensibility
- Allowed physical media
- Network management
- Interoperability
- Information security
- Explosion protection

It is therefore no wonder that there is no "universal" network for everybody and everything, but a set of specialized networks that are applicable to one area but probably not to another. Knowing the details and differences of application-specific ICS helps to understand their strengths and weaknesses and greatly helps in design decisions. There is an increasing trend that encompasses technological convergence (runs everything over Ethernet) and semantic convergence (runs everything over Web services), and the following chapters will explain why this has yet to be realized. There are reasons for this phenomenon, and it is important to know them.

Preamble to Part III: Technologies

Stefan Mahlknecht
Institute of Computer Technology
Vienna University of Technology
Vienna, Austria

Gianluca Cena
Istituto di Elettronica e di Ingegneria dell'Informazione e delle Telecomunicazioni
Consiglio Nazionale delle Ricerche
Turin, Italy

Martin Wollschlaeger
Institute of Applied Computer Science
Dresden University of Technology
Dresden, Germany

Introduction

This part describes the technologies for industrial communications. It has been organized in seven different groups by technology family and by application areas.

Group 3.1: Classical Fieldbus Systems

Fieldbus systems date back to the 1980s and represent the first successful attempt to bring concepts related to local area networks to factory automation environments. Thanks to digital serial communication, an unprecedented degree of flexibility was achieved when compared with analogue point-to-point links, allowing remote configuration and diagnostics to be carried out easily. Moreover, noticeable savings were made in both cabling and deployment costs because of the shared communication support. Needless to say, these advantages made fieldbus technology more and more adoptable in industrial plants throughout the 1990s.

One of the main drawbacks of fieldbuses is the lack, among manufacturers, of a unique, standard solution. Instead, a large number (on the order of about 100) of different and incompatible solutions were developed, some of which are still in use. Noticeable examples are PROFIBUS, INTERBUS, MODBUS, as well as CAN-based solutions such as Devicenet and CANopen. In the following chapters (Chapters 31 through 36), some of the most popular fieldbus solutions are described.

Group 3.2: Industrial Ethernet

Ethernet is currently the "de facto" standard networking solution for office automation environments. Since its introduction in the 1970s, it has managed to keep pace with the ever-increasing bandwidth requirements of distributed information systems and has been able to offer increased performance over the years without losing compatibility with the original protocol and equipment.

While Ethernet was initially deemed unsuitable for use in distributed control systems, due to its random access scheme, the extensive improvements that were made to this network made people change their minds by the end of the 1990s. The availability of high-speed (100 Mb/s and beyond) full-duplex connections, VLANs with traffic prioritization, and non-blocking switches made it possible to achieve increased levels of determinism, often suitable for most factory automation systems. Solutions such as EtherNet/IP are based on unmodified Ethernet equipment and the conventional TCP/IP communication stack.

In order to cater to highly demanding control applications with tight timing constraints, such as motion control, a number of modifications have been proposed in the past few years that are aimed at further enhancing the real-time behavior of Ethernet. While relying on the same transceivers, frame format, and access scheme as the original protocol, changes were added to the original Ethernet hardware or communication stack. This is the case, for example, of EtherCAT, Ethernet Powerlink, PROFINET IRT, and so on. The following chapters (Chapters 37 through 40) focus on some of these solutions.

Group 3.3: Building Automation Networks

Modern building automation networks are based on distributed networks where network topologies are flexible enough to reflect the building structure. They are primarily based on wired technologies although wireless extensions also exist. Installation and maintenance are key issues, as large networks may comprise thousands of nodes. Two widely adopted technologies, namely, LonWorks and KNX, have been on the market for many years and occupy different market segments. LonWokrs, due to its flexibility, is applied more in large buildings and industries, while KNX is used more in private homes. In many large buildings, a heterogeneous network with LonWorks-, KNX-, and IP-based networks are implemented. The following chapters (Chapters 41 through 43) present the main building automation networks standardized under ISO.

Group 3.4: Automotive Networks

Automotive networks have the same advantages that fieldbuses bring to industrial automation environments, in particular for in-vehicle control systems such as powertrain, body electronics, or infotainment. There is no doubt that the most popular solution so far has been the controller area network (CAN) protocol introduced by Bosch in the mid-1980s in order to reduce cable clutter in cars and trucks.

Despite being perfectly suitable for most of today's vehicles, CAN has some drawbacks that will likely rule it out for next-generation automotive systems. In particular, when taking steer-by-wire systems into account, a much higher degree of determinism, performance, and, mostly, fault tolerance has to be ensured. This has led to the introduction of the time-triggered architecture (TTA) and, in particular, the TTP/C protocol. In order to reduce design and production costs, high flexibility is required as well. To this extent, the FlexRay protocol has been defined, which combines the dependability and determinism of TTP/C with the ability to carry out data exchanges on demand, through a flexible time division multiple access scheme. The chapters that follow (Chapters 44 and 45) describe the basic principles behind the new high-performance solutions as well as low-cost in-vehicle networks such as LIN.

Group 3.5: Safety

Safety is one of the most important requirements in industrial applications. The guaranteed transmission of secured data in a reliable time frame, order, integrity, and sequence is an evident task in systems where man and equipment are at risk of being harmed. Thus, safety integrity levels (SIL) have been defined that must be met by technical systems (Chapters 46 and 47). Typically, safety-related functions are not originally embedded in industrial communication systems. In order to meet the required criteria, add-ons to existing protocols and systems have been defined. Thus, interoperability with existing protocols and applications can be ensured.

Group 3.6: Wireless Networks

Wireless networks have experienced tremendous growth in the last decade, driven by mobile phones and the computer industry. Most of us are familiar with the widely used technologies in consumer products, such as GSM/3G, WLAN, and Bluetooth. In contrast, wireless automation networks or sensor networks are still a topic of research, and products are either available only in certain segments or are slowly entering the market. To name a few of the candidates presented in the following chapters (Chapters 48 through 55), ZigBee, 6LoWPAN, and WirelessHart are some of the wireless technologies that are capable of replacing many wired fieldbus applications. These technologies allow for added flexibility by placing nodes freely on moving machines and by reducing the installation effort. All wireless networks have to specifically address the issue of security and power consumption for nodes that are battery powered.

Group 3.7: Industrial Internet

The application of Internet- and IT-based protocols and technologies is undoubtedly a promising and up-to-date development (Chapters 56 through 59). Besides acceptance by the users, the adoption of existing, proven technologies in the automation domain reduces efforts by reusing existing concepts, functions, and software components. However, different time frames in technology development cycles—compared to the rather long-term application in industry—are critical issues in the selection of appropriate technologies. The technologies described in Chapter 55 address different application areas and thus use different technologies. Starting with function blocks concepts according to IEC 61499, a generic, function-related approach is described. The concepts allow a network-independent synthesis of application functions, which is a prerequisite for distributed industrial applications. The application of typical IT protocols and system structures can be investigated perfectly in Industrial Internet, and the adoption, specialization, and application of protocols from the Internet is a global trend. Originally developed together with software companies, including Microsoft, OPC has become the de facto standard for providing access from higher-level applications to automation applications. With OPC UA, major enhancements in this technology have been made, including support for Web services and complex information models. Web technology and the Industrial Internet have enabled the application of multimedia technologies as integral parts of automation systems. Advances in machine vision document this fact. Finally, energy production and distribution are important tasks supported by various technologies. These technologies use Ethernet as one of the underlying protocols and thus their development follows that of IT systems in general.

Preamble to Part IV: Internet Programming

J. David Irwin and Bogdan M. Wilamowski
Auburn University
Auburn, Alabama

The rapidly growing Internet is also expanding into the industrial environment. Many of the protocols, techniques, and hardware developed for the public Internet can also be used in closed industrial networks, while enjoying the benefits of reduced component cost due to their mass production. There is also the possibility of using the Internet to watch, supervise, and control industrial environments remotely from any place in the world, assuming that a proper security cover is provided. This part introduces two commonly used Internet protocols, TCP and UDP, and illustrates typical API interfaces and their sample use in simple proof-of-concept client–server applications. Both protocols belong to transport layer protocols and use an underlying IP network layer and a communication media–specific data link layer. UDP is a packet-based connectionless protocol with little overhead for unicast, multicast, and broadcast communication (Chapter 60), while the TCP protocol provides reliable, best-effort delivery of data streams (Chapter 61). The development of interactive Web sites can be done in many languages, the most common being HTML, Javascript, PHP, PERL, and Python. It is of course possible to develop such Web sites using general languages such as Java or C++, but specialized languages are usually preferred. This part shows how interactive Web sites can be programmed using PHP (Chapter 62), Python (Chapter 63), and PERL (Chapter 64). Chapter 65 describes how to run remote applications over the Internet.

In this manner, it is possible to remotely observe and control any equipment or process in industry. Chapter 66 focuses on methods that permit the handling of multiple processes with the ability of easy reconfiguration. Chapter 67 shows how to develop Internet robots that are capable of performing autonomous processes inclosing a search in the Internet. Examples in this chapter were developed in PERL, but this can also be developed in PHP and Python. This part also illustrates the philosophy of distributed programming, in which the software need not be executed locally.

Preamble to Part V: Outlook

Dietmar Dietrich and Dietmar Bruckner
Institute of Computer Technology
Vienna University of Technology
Vienna, Austria

Gerhard Zucker
Energy Department
Austrian Institute of Technology
Vienna, Austria

The final part of this book examines the future of industrial communication systems. With the rapidly increasing capabilities of computation and communication systems, we are able to create systems that have long been only concepts or even dreams. Processes become more complex because more data are available, and these data can be processed by more sophisticated algorithms. New ideas are necessary to control complexity. On the one hand, these are new paradigms for communication structures such as extended vertical integration of systems, service-oriented architecture, and hybrid local networks. On the other hand, industrial electronics as a whole can profit from knowledge that has been gathered in other disciplines such as artificial intelligence using statistical methods to process huge amounts of data with the aim of understanding the human mind, the most sophisticated control and communication device that we have at hand. This part briefly touches upon these topics and glances through ongoing developments that may in the future contribute to new generations of industrial communication systems.

Acknowledgments

The editors wish to express their heartfelt thanks to their wives Barbara Wilamowski and Edie Irwin for their help and support during the execution of this project.

Editorial Board

Editors

Bogdan M. Wilamowski received his MS in computer engineering in 1966, his PhD in neural computing in 1970, and Dr. habil. in integrated circuit design in 1977. He received the title of full professor from the president of Poland in 1987. He was the director of the Institute of Electronics (1979–1981) and the chair of the solid state electronics department (1987–1989) at the Technical University of Gdansk, Poland. He was a professor at the University of Wyoming, Laramie, from 1989 to 2000. From 2000 to 2003, he served as an associate director at the Microelectronics Research and Telecommunication Institute, University of Idaho, Moscow, and as a professor in the electrical and computer engineering department and in the computer science department at the same university. Currently, he is the director of ANMSTC—Alabama Nano/Micro Science and Technology Center, Auburn, and an alumna professor in the electrical and computer engineering department at Auburn University, Alabama. Dr. Wilamowski was with the Communication Institute at Tohoku University, Japan (1968–1970), and spent one year at the Semiconductor Research Institute, Sendai, Japan, as a JSPS fellow (1975–1976). He was also a visiting scholar at Auburn University (1981–1982 and 1995–1996) and a visiting professor at the University of Arizona, Tucson (1982–1984). He is the author of 4 textbooks, more than 300 refereed publications, and has 27 patents. He was the principal professor for about 130 graduate students. His main areas of interest include semiconductor devices and sensors, mixed signal and analog signal processing, and computational intelligence.

Dr. Wilamowski was the vice president of the IEEE Computational Intelligence Society (2000–2004) and the president of the IEEE Industrial Electronics Society (2004–2005). He served as an associate editor of *IEEE Transactions on Neural Networks*, *IEEE Transactions on Education*, *IEEE Transactions on Industrial Electronics*, the *Journal of Intelligent and Fuzzy Systems*, the *Journal of Computing*, and the *International Journal of Circuit Systems and IES Newsletter*. He is currently serving as the editor in chief of *IEEE Transactions on Industrial Electronics*.

Professor Wilamowski is an IEEE fellow and an honorary member of the Hungarian Academy of Science. In 2008, he was awarded the Commander Cross of the Order of Merit of the Republic of Poland for outstanding service in the proliferation of international scientific collaborations and for achievements in the areas of microelectronics and computer science by the president of Poland.

J. David Irwin received his BEE from Auburn University, Alabama, in 1961, and his MS and PhD from the University of Tennessee, Knoxville, in 1962 and 1967, respectively.

In 1967, he joined Bell Telephone Laboratories, Inc., Holmdel, New Jersey, as a member of the technical staff and was made a supervisor in 1968. He then joined Auburn University in 1969 as an assistant professor of electrical engineering. He was made an associate professor in 1972, associate professor and head of department in 1973, and professor and head in 1976. He served as head of the Department of Electrical and Computer Engineering from 1973 to 2009. In 1993, he was named Earle C. Williams Eminent Scholar and Head. From 1982 to 1984, he was also head of the Department of Computer Science and Engineering. He is currently the Earle C. Williams Eminent Scholar in Electrical and Computer Engineering at Auburn.

Dr. Irwin has served the Institute of Electrical and Electronic Engineers, Inc. (IEEE) Computer Society as a member of the Education Committee and as education editor of *Computer*. He has served as chairman of the Southeastern Association of Electrical Engineering Department Heads and the National Association of Electrical Engineering Department Heads and is past president of both the IEEE Industrial Electronics Society and the IEEE Education Society. He is a life member of the IEEE Industrial Electronics Society AdCom and has served as a member of the Oceanic Engineering Society AdCom. He served for two years as editor of *IEEE Transactions on Industrial Electronics*. He has served on the Executive Committee of the Southeastern Center for Electrical Engineering Education, Inc., and was president of the organization in 1983–1984. He has served as an IEEE Adhoc Visitor for ABET Accreditation teams. He has also served as a member of the IEEE Educational Activities Board, and was the accreditation coordinator for IEEE in 1989. He has served as a member of numerous IEEE committees, including the Lamme Medal Award Committee, the Fellow Committee, the Nominations and Appointments Committee, and the Admission and Advancement Committee. He has served as a member of the board of directors of IEEE Press. He has also served as a member of the Secretary of the Army's Advisory Panel for ROTC Affairs, as a nominations chairman for the National Electrical Engineering Department Heads Association, and as a member of the IEEE Education Society's McGraw-Hill/Jacob Millman Award Committee. He has also served as chair of the IEEE Undergraduate and Graduate Teaching Award Committee. He is a member of the board of governors and past president of Eta Kappa Nu, the ECE Honor Society. He has been and continues to be involved in the management of several international conferences sponsored by the IEEE Industrial Electronics Society, and served as general cochair for IECON'05.

Dr. Irwin is the author and coauthor of numerous publications, papers, patent applications, and presentations, including *Basic Engineering Circuit Analysis*, 9th edition, published by John Wiley & Sons, which is one among his 16 textbooks. His textbooks, which span a wide spectrum of engineering subjects, have been published by Macmillan Publishing Company, Prentice Hall Book Company, John Wiley & Sons Book Company, and IEEE Press. He is also the editor in chief of a large handbook published by CRC Press, and is the series editor for Industrial Electronics Handbook for CRC Press.

Dr. Irwin is a fellow of the American Association for the Advancement of Science, the American Society for Engineering Education, and the Institute of Electrical and Electronic Engineers. He received an IEEE Centennial Medal in 1984, and was awarded the Bliss Medal by the Society of American Military Engineers in 1985. He received the IEEE Industrial Electronics Society's Anthony J. Hornfeck Outstanding Service Award in 1986, and was named IEEE Region III (U.S. Southeastern Region) Outstanding Engineering Educator in 1989. In 1991, he received a Meritorious Service Citation from the IEEE Educational Activities Board, the 1991 Eugene Mittelmann Achievement Award from the IEEE Industrial Electronics Society, and the 1991 Achievement Award from the IEEE Education Society. In 1992, he was named a Distinguished Auburn Engineer. In 1993, he received the IEEE Education Society's McGraw-Hill/Jacob Millman Award, and in 1998 he was the recipient of the

IEEE Undergraduate Teaching Award. In 2000, he received an IEEE Third Millennium Medal and the IEEE Richard M. Emberson Award. In 2001, he received the American Society for Engineering Education's (ASEE) ECE Distinguished Educator Award. Dr. Irwin was made an honorary professor, Institute for Semiconductors, Chinese Academy of Science, Beijing, China, in 2004. In 2005, he received the IEEE Education Society's Meritorious Service Award, and in 2006, he received the IEEE Educational Activities Board Vice President's Recognition Award. He received the Diplome of Honor from the University of Patras, Greece, in 2007, and in 2008 he was awarded the IEEE IES Technical Committee on Factory Automation's Lifetime Achievement Award. In 2010, he was awarded the electrical and computer engineering department head's Robert M. Janowiak Outstanding Leadership and Service Award. In addition, he is a member of the following honor societies: Sigma Xi, Phi Kappa Phi, Tau Beta Pi, Eta Kappa Nu, Pi Mu Epsilon, and Omicron Delta Kappa.

Contributors

Teresa Albero-Albero
Escuela Politécnica Superior de Alcoy
Universidad Politécnica de Valencia
Alcoy, Spain

Nuria Oliva Alonso
Department of Electrical, Electronics, and Control Engineering
Spanish University of Distance Education, UNED
Madrid, Spain

Holger Arthaber
Institute of Electrodynamics, Microwave and Circuit Engineering
Vienna University of Technology
Vienna, Austria

Renaud Aubin
Department of Simulation and Information Technologies for Power Generation Systems
EDF Research and Development
Chatou, France

Jordi Ayza
Department of Automatic Control and Industrial Informatics
Universitat Politècnica de Catalunya
Barcelona, Spain

Paulo Bartolomeu
Institute of Telecommunication
University of Aveiro
Aveiro, Portugal

Filipe Basto
Gabinete de Saúde Internacional
Hospital de Sao João
Porto, Portugal

Lucia Lo Bello
Department of Computer and Telecommunications Engineering
University of Catania
Catania, Italy

Aleksey Bratukhin
Institute for Integrated Sensor Systems
Austrian Academy of Sciences
Wiener Neustadt, Austria

Dietmar Bruckner
Institute of Computer Technology
Vienna University of Technology
Vienna, Austria

Simon Carlsen
Statoil ASA
Harstad, Norway

Michael Carroll
Department of Electrical and Computer Engineering
Auburn University
Auburn, Alabama

Gianluca Cena
Istituto di Elettronica e di Ingegneria dell'Informazione e delle Telecomunicazioni
Italian National Research Council
Torino, Italy

Elizabeth Chang
Digital Ecosystems and Business Intelligence Institute
Curtin University of Technology
Perth, Western Australia, Australia

Mo-Yuen Chow
Department of Electrical and Computer Engineering
North Carolina State University
Raleigh, North Carolina

Pradeep Dandamudi
Department of Electrical and Computer Engineering
Auburn University
Auburn, Alabama

Tuan Dang
Department of Simulation and Information Technologies for Power Generation Systems
EDF Research and Development
Chatou, France

Kurt Derr
Idaho National Laboratory
Idaho Falls, Idaho

Henrik Dibowski
Faculty of Computer Science
Institute of Applied Computer Science
Dresden University of Technology
Dresden, Germany

Dietmar Dietrich
Institute of Computer Technology
Vienna University of Technology
Vienna, Austria

Tharam S. Dillon
Digital Ecosystems and Business Intelligence Institute
Curtin University of Technology
Perth, Western Australia, Australia

Wilfried Elmenreich
Institute of Networked and Embedded Systems
University of Klagenfurt
Klagenfurt, Austria

Christian El-Salloum
Institute of Computer Engineering
Vienna University of Technology
Vienna, Austria

Max Felser
Department of Engineering and Information Technology
Bern University of Applied Sciences
Burgdorf, Switzerland

Paolo Ferrari
Department of Information Engineering
University of Brescia
Brescia, Italy

Joaquim Ferreira
Institute of Telecommunications
University of Aveiro
Aveiro, Portugal

Alessandra Flammini
Department of Information Engineering
University of Brescia
Brescia, Italy

José Alberto Fonseca
Department of Electronics, Telecommunications and Informatics
Universidade of Aveiro
Aveiro, Portugal

Christian Fraboul
IRIT INPT-ENSEEIHT
Université de Toulouse
Toulouse, France

Fabrice Frances
ISAE
Université de Toulouse
Toulouse, France

Heinz Frank
Institute of Fast Mechatronic Systems
Reinhold-Würth-University
Künzelsau, Germany

Josep M. Fuertes
Department of Automatic Control and Industrial Informatics
Universitat Politècnica de Catalunya
Barcelona, Spain

Georg Gaderer
Institute for Integrated Sensor Systems
Austrian Academy of Sciences
Wiener Neustadt, Austria

Andreas Gerstinger
Institute of Computer Technology
Vienna University of Technology
Vienna, Austria

Manuel Castro Gil
Department of Electrical, Electronics, and Control Engineering
Spanish University of Distance Education, UNED
Madrid, Spain

Frank Golatowski
Institute of Applied Microelectronics and Computer Engineering
University of Rostock
Rostock, Germany

Wolfgang Granzer
Automation Systems Group
Vienna University of Technology
Vienna, Austria

Andreas Grzemba
Department of Electrical Engineering
University of Applied Sciences-Deggendorf
Deggendorf, Germany

Vehbi Cagri Gungor
Computer Engineering Department
Bahcesehir University
Istanbul, Turkey

Rachana Gupta
Department of Electrical and Computer Engineering
North Carolina State University
Raleigh, North Carolina

Herbert Haas
Institute of Computer Technology
Vienna University of Technology
Vienna, Austria

Gerhard P. Hancke
Department of Electrical, Electronic, and Computer Engineering
University of Pretoria
Pretoria, South Africa

Donal Heffernan
Department of Electronic and Computer Engineering
University of Limerick
Limerick, Ireland

Martin Horauer
Department of Embedded Systems
University of Applied Sciences Technikum Wien
Vienna, Austria

Tariq Jadoon
Department of Computer Science
Lahore University of Management Sciences
Lahore, Pakistan

Juergen Jasperneite
Institut Industrial IT
Ostwestfalen-Lippe University of Applied Sciences
Lemgo, Germany

Klaus Kabitzsch
Faculty of Computer Science
Institute of Applied Computer Science
Dresden University of Technology
Dresden, Germany

Wolfgang Kastner
Automation Systems Group
Vienna University of Technology
Vienna, Austria

Shahid Khattak
Department of Electrical Engineering
COMSATS Institute of Information Technology
Abbotabad, Pakistan

Alexander Klapproth
CEESAR-iHomeLab
Lucerne University of Applied Sciences and Arts
Lucerne, Switzerland

Yoseba Peña Landaburu
Faculty of Economics and Business Administration
University of Deusto
San Sebastian, Spain

Pui-Yi Lau
Department of Electronic Engineering
City University of Hong Kong
Kowloon, Hong Kong

Paulo Leitão
Polytechnic Institute of Bragança
Bragança, Portugal

Wilfried Lepuschitz
Automation and Control Institute
Vienna University of Technology
Vienna, Austria

Chi-Wai Leung
Department of Electronic Engineering
City University of Hong Kong
Kowloon, Hong Kong

Manfred Lindner
Institute of Computer Technology
Vienna University of Technology
Vienna, Austria

Thomas Lindner
BMW Group
Munich, Germany

Patrick Loschmidt
Institute for Integrated Sensor Systems
Austrian Academy of Sciences
Wiener Neustadt, Austria

Sajjad Ahmad Madani
Department of Computer Science
COMSATS Institute of Information Technology
Abbotabad, Pakistan

Stefan Mahlknecht
Department of Electrical Engineering
Vienna University of Technology
Vienna, Austria

Aleksander Malinowski
Department of Electrical and Computer Engineering
Bradley University
Peoria, Illinois

Milos Manic
Department of Computer Science
University of Idaho–Idaho Falls
Idaho Falls, Idaho

Gaëlle Marsal
Department of Simulation and Information Technologies for Power Generation Systems
EDF Research and Development
Chatou, France

Sidonia Mesentean
Institute of Fast Mechatronic Systems
Reinhold-Würth-University
Künzelsau, Germany

Ralf Messerschmidt
Institute for Automation and Communication
Magdeburg, Germany

Orazio Mirabella
Department of Computer Engineering and Telecommunications
University of Catania
Catania, Italy

Ron Mitchell
RC Systems
Johnson City, Tennessee

Guido Moritz
Institute of Applied Microelectronics and Computer Engineering
University of Rostock
Rostock, Germany

Georg Neugschwandtner
Automation Systems Group
Vienna University of Technology
Vienna, Austria

Peter Neumann
Institute for Automation and Communication
Magdeburg, Germany

Thomas Novak
SWARCO Futurit Verkehrssignalssysteme GmbH
Perchtoldsdorf, Austria

Mirabella Orazio
Department of Computer Engineering
and Telecommunications
University of Catania
Catania, Italy

Gabriel Diaz Orueta
Department of Electrical, Electronics, and
Control Engineering
Spanish University of Distance Education, UNED
Madrid, Spain

Peter Palensky
Energy Department
Austrian Institute of Technology
Vienna, Austria

Paulo Pedreiras
University of Aveiro
Aveiro, Portugal

Augusto Pereira
Pepperl-Fuchs
Sao Paulo, Brazil

Carlos Eduardo Pereira
Department of Electrical Engineering
Federal University of Rio Grande do Sul
Porto Alegre, Brazil

Stig Petersen
SINTEF Information and Communication
Technology
Trondheim, Norway

Nam Pham
Department of Electrical and Computer
Engineering
Auburn University
Auburn, Alabama

Joern Ploennigs
Faculty of Computer Science
Institute of Applied Computer Science
Dresden University of Technology
Dresden, Germany

Paulo Portugal
Department of Electrical and Computer
Engineering
University of Porto
Porto, Portugal

Vidyasagar Potdar
Digital Ecosystems and Business Intelligence
Institute
Curtin University of Technology
Perth, Western Australia, Australia

Fritz Praus
Automation Systems Group
Vienna University of Technology
Vienna, Austria

Peter Preininger
LOYTEC Electronics GmbH
Vienna, Austria

Christian Reinisch
Automation Systems Group
Vienna University of Technology
Vienna, Austria

Nelson Rocha
Secção Autónoma de Ciências da Saúde
University of Aveiro
Aveiro, Portugal

Peter Rössler
Department of Embedded Systems
University of Applied Sciences
Technikum Wien
Vienna, Austria

Elio San Cristobal Ruiz
Department of Electrical, Electronics, and
Control Engineering
Spanish University of Distance Education, UNED
Madrid, Spain

Uwe Ryssel
Faculty of Computer Science
Institute of Applied Computer Science
Dresden University of Technology
Dresden, Germany

Shahzad Sarwar
Punjab University College of Information Technology
University of the Punjab
Lahore, Pakistan

Thilo Sauter
Institute for Integrated Sensor Systems
Austrian Academy of Sciences
Wiener Neustadt, Austria

Jean-Luc Scharbarg
IRIT INPT-ENSEEIHT
Université de Toulouse
Toulouse, France

Stefan Schoenegger
B&R Industrial Automation
Eggelsberg, Austria

Karlheinz Schwarz
Schwarz Consultancy Company
Karlsruhe, Germany

Herbert Schweinzer
Institute of Electrodynamics, Microwave and Circuit Engineering
Vienna University of Technology
Vienna, Austria

Víctor-M. Sempere-Payá
Escuela Técnica Superior de Ingenieros de Telecomunicación
Universidad Politécnica de Valencia
Valencia, Spain

Lucia Seno
Istituto di Elettronica e di Ingegneria dell'Informazione e delle Telecomunicazioni
Italian National Research Council
Padova, Italy

Atif Sharif
Digital Ecosystems and Business Intelligence Institute
Curtin University of Technology
Perth, Western Australia, Australia

Saleem Farooq Shaukat
Department of Electrical Engineering
COMSATS Institute of Information Technology
Lahore, Pakistan

Javier Silvestre-Blanes
Instituto Technológico de Informática
Universidad Politécnica de Valencia
Alcoy, Spain

Emiliano Sisinni
Department of Information Engineering
University of Brescia
Brescia, Italy

Stefan Soucek
LOYTEC Electronics GmbH
Vienna, Austria

Mário de Sousa
Department of Electrical and Computer Engineering
University of Porto
Porto, Portugal

Sergiu-Dan Stan
Department of Mechanisms, Precision Mechanics and Mechatronics
Technical University of Cluj-Napoca
Cluj-Napoca, Romania

Strahinja Stankovic
Ninet Company Wireless ISP
Nis, Serbia

Alex Talevski
Digital Ecosystems and Business Intelligence Institute
Curtin University of Technology
Perth, Western Australia, Australia

Thomas Tamandl
SWARCO Futurit Verkehrssignalssysteme GmbH
Perchtoldsdorf, Austria

Albert Treytl
Institute for Integrated Sensor Systems
Austrian Academy of Sciences
Wiener Neustadt, Austria

Denis Trognon
EDF Research and Development
Chatou, France

Henning Trsek
Institut Industrial IT
Ostwestfalen-Lippe University of Applied Sciences
Lemgo, Germany

Rainer Unland
Institute for Computer Science and Business Information Systems
University of Duisburg-Essen
Essen, Germany

Adriano Valenzano
Istituto di Elettronica e di Ingegneria dell'Informazione e delle Telecomunicazioni
Italian National Research Council
Torino, Italy

Francisco Vasques
Mechanical Engineering Department
University of Porto
Porto, Portugal

Volodymyr Vasyutynskyy
Faculty of Computer Science
Institute of Applied Computer Science
Dresden University of Technology
Dresden, Germany

Ian Verhappen
Industrial Automation Networks Inc.
Wainwright, Alberta, Canada

Ricard Villà
Department of Automatic Control and Industrial Informatics
Universitat Politècnica de Catalunya
Barcelona, Spain

Stefano Vitturi
Istituto di Elettronica e di Ingegneria dell'Informazione e delle Telecomunicazioni
Italian National Research Council
Padova, Italy

Valeriy Vyatkin
Department of Electrical and Computer Engineering
University of Auckland
Auckland, New Zealand

Yan-Wu Wang
Department of Control Science and Engineering
Huazhong University of Science and Technology
Hubei, China

Manfred Weihs
TTTech Computertechnik AG
Vienna, Austria

Changyun Wen
School of Electrical and Electronic Engineering
Nanyang Technological University
Singapore, Singapore

Andrew C. West
Invensys Operations Management
Eight Mile Plains, Queensland, Australia

Bogdan M. Wilamowski
Department of Electrical and Computer Engineering
Auburn University
Auburn, Alabama

Martin Wollschlaeger
Faculty of Computer Science
Institute of Applied Computer Science
Dresden University of Technology
Dresden, Germany

Chen Wu
Digital Ecosystems and Business Intelligence Institute
Curtin University of Technology
Perth, Western Australia, Australia

Hao Yu
Department of Electrical and Computer Engineering
Auburn University
Auburn, Alabama

Edward Kai-Ning Yung
Department of Electronic Engineering
City University of Hong Kong
Kowloon, Hong Kong

Alois Zoitl
Automation and Control Institute
Vienna University of Technology
Vienna, Austria

Gerhard Zucker
Institute of Computer Technology
Vienna University of Technology
Vienna, Austria

Claudio Zunino
Istituto di Elettronica e di Ingegneria dell'Informazione e delle Telecomunicazioni
Italian National Research Council
Torino, Italy

I

Technical Principles

1 **ISO/OSI Model** *Gerhard Zucker and Dietmar Dietrich* 1-1
Introduction • Open Standard • Vertical and Horizontal Communication • Dynamic Behavior of Services and Protocols • Extensions, Benefits, and Discussion • References

2 **Media** *Herbert Schweinzer, Saleem Farooq Shaukat, and Holger Arthaber* 2-1
Introduction • Wired Links • Optical Links • Wireless Links • References

3 **Media Access Methods** *Herbert Haas and Manfred Lindner* 3-1
Introduction • Full-Duplex Media Access • Synchronous Access Arbitration Concepts • Statistic Access Arbitration Concepts • Carrier Sense Mechanisms with Exponential Backoff • Other Media Access Issues • References

4 **Routing in Wireless Networks** *Teresa Albero-Albero and Víctor-M. Sempere-Payá* 4-1
Introduction • Routing Protocols and Classification • Routing Protocol Families for Ad Hoc Networks • Routing Protocol Families for Wireless Sensor Networks • Summary of the Main Routing Protocols in Wireless Networks • Conclusions • Acknowledgment • Abbreviations • References

5 **Profiles and Interoperability** *Gerhard Zucker and Heinz Frank* 5-1
Interoperating Components • Application of Profiles • Achieving Interoperability • References

6 **Industrial Wireless Sensor Networks** *Vehbi Cagri Gungor and Gerhard P. Hancke* 6-1
Applications • Standardization Activities • Technical Challenges • Design Goals • Design Principles and Technical Approaches • Conclusions and Future Work • References

7 **Ad Hoc Networks** *Sajjad Ahmad Madani, Shahid Khattak, Tariq Jadoon, and Shahzad Sarwar* 7-1
Introduction • Protocol Stack • Performance Evaluation • Challenges and Issues • References

8 **Radio Frequency Identification** *Edward Kai-Ning Yung, Pui-Yi Lau, and Chi-Wai Leung* 8-1
Prologue • Bar Code System • Magnetic Stripes • Smart Card • Proximity Card • HF RFID • Electronic Cash • Personal Identity • Innovation verus Hi-Tech • Active RFID • Wake-Up Technology • Semi-Active RFID • Backscattering •

Initialization • Vicinity Card • Frequency Selection • UHF RFID • Supply Chain Management • International Standard • Promiscuity • National Standards • Hands-Free Bar Code System • Bar Code Mentality • Affordable Tag • Ubiquity of RFID • Role Reversal • Historical Development • Privacy Infringement • Recent Developments • Dual Authentication • Trace-and-Track • Innovative Applications • Nonionization Radiation • Era of Artificial Perception • Abbreviations • References

9 **RFID Technology and Its Industrial Applications** *Vidyasagar Potdar, Atif Sharif, and Elizabeth Chang* **9**-1
Introduction • RFID Architecture • Item Tracking and Tracing • Access Control • Anticounterfeiting • Conclusion • References

10 **Ultralow-Power Wireless Communication** *Joern Ploennigs, Volodymyr Vasyutynskyy, and Klaus Kabitzsch* **10**-1
Introduction • Hardware Approaches • Communication Protocol Approaches • Application Layer Approaches • Conclusion and Open Topics • References

11 **Industrial Strength Wireless Multimedia Sensor Network Technology** *Vidyasagar Potdar, Atif Sharif, and Elizabeth Chang* **11**-1
Introduction • Wireless Sensor Network • WMSN Architecture • WMSN Hardware • Applications of WMSNs • WMSNs' Technical Challenges • Conclusion • References

12 **A Survey of Wireless Sensor Networks for Industrial Applications** *Stig Petersen and Simon Carlsen* **12**-1
Introduction • Wireless Sensor Network Basics • Motivation and Drivers for Wireless Instrumentation • Industrial Applications and Requirements • Technology Survey and Evaluation • Conclusion • Abbreviations • References

13 **Vertical Integration** *Thilo Sauter, Stefan Soucek, and Martin Wollschlaeger* **13**-1
Introduction • Historical Background • Network Interconnections • Application View • Security Aspects in Vertical Integration • Trends in Vertical Integration • Abbreviations • References

14 **Multimedia Service Convergence** *Alex Talevski* **14**-1
Introduction • Background • Service-Oriented Architecture • Tailorability • Multimedia Convergence Using Service Architecture • Conclusion • References

15 **Virtual Automation Networks** *Peter Neumann and Ralf Messerschmidt* **15**-1
Introduction • Virtual Automation Network: Basics • Name-Based Addressing and Routing, Runtime Tunnel Establishment • Maintenance of the Runtime Tunnel Based on Quality-of-Service Monitoring and Provider Switching • VAN Telecontrol Profile • Abbreviations • References

16 **Industrial Agent Technology** *Aleksey Bratukhin, Yoseba Peña Landaburu, Paulo Leitão, and Rainer Unland* **16**-1
Introduction • Agents and Multi-Agent Systems • Agents and Multi-Agent Systems in Industry • Application Areas • Agents and Multi-Agent Systems in Industry: Conclusions • Abbreviations • References

17 **Real-Time Systems** *Lucia Lo Bello, José Alberto Fonseca, and Wilfried Elmenreich* **17**-1
Introduction on Real-Time Systems • Real-Time Communication • Design Paradigms for Real-Time Systems • Design Challenges in Real-Time Industrial Communication Systems • References

18 Clock Synchronization in Distributed Systems *Georg Gaderer and Patrick Loschmidt*........ **18**-1
Introduction • Precision Time Protocol • IEEE 1588 System Model • Service Access Points • Ordinary Clocks • Boundary Clocks • Precision Time Protocol, IEEE 1588–2008 (PTPv2) • Network Time Protocol • Network Time Protocol Strata • Architecture, Protocol, and Algorithms • NTP Clock Synchronization Hardware Requirements • Synchronization Algorithms of NTP • References

19 Quality of Service *Gabriel Diaz Orueta, Elio San Cristobal Ruiz, Nuria Oliva Alonso, and Manuel Castro Gil*........ **19**-1
Introduction • Relationship with Information Security Topics • Quality of Service for IP Networks • Special Considerations for Managing the Quality of Service • References

20 Network-Based Control *Josep M. Fuertes, Mo-Yuen Chow, Ricard Villà, Rachana Gupta, and Jordi Ayza*........ **20**-1
Introduction • Mutual Concepts in Control and in Communications • Architecture of Networked-Based Control • Network Effects in Control Performance • Design in NBC • Summary • References

21 Functional Safety *Thomas Novak and Andreas Gerstinger*........ **21**-1
Introduction • The Meaning of Safety • Safety Standards • The Safety Lifecycle and Safety Methods • Safety Approach for Industrial Communication System • Acronyms • References

22 Security in Industrial Communication Systems *Wolfgang Granzer and Albert Treytl*........ **22**-1
Introduction to Security in Industrial Communication • Planned Approach to Security: Defense in Depth • Security Measures to Counteract Network Attacks • Security Measures to Counteract Device Attacks • State of the Art in Automation Systems • Outlook and Conclusion • Abbreviations • References

23 Secure Communication Using Chaos Synchronization *Yan-Wu Wang and Changyun Wen*........ **23**-1
Introduction • Chaos Synchronization • Secure Communication Using Chaos Synchronization • References

1

ISO/OSI Model

Gerhard Zucker
Vienna University of Technology

Dietmar Dietrich
Vienna University of Technology

1.1 Introduction 1-1
1.2 Open Standard 1-3
Layer Functionalities
1.3 Vertical and Horizontal Communication 1-5
1.4 Dynamic Behavior of Services and Protocols 1-6
1.5 Extensions, Benefits, and Discussion 1-9
References 1-9

1.1 Introduction

The ISO/OSI model was developed and standardized in the late 1970s by the International Organization for Standardization as the standard ISO IS 7498. It supports designers by easing the definition of communication protocols in a way that they operate correctly and are easy to maintain [Hay 88]. The name OSI originates from open systems interconnection. The model is not intended as strict implementation rules, because a real system will always have to adapt to requirements of price, economy, and flexibility.

Instead, the ISO/OSI model represents an abstract definition (independent from hardware or software implementation) consisting of hierarchical layers. Related functions are grouped together in layers with strict separation between horizontal and vertical communication. The standard defines services, protocols, and interfaces. The original intention was to define a model for connecting computers for data transfer, logging into remote computers, and so on. Requirements like real-time, protocols for embedded microcontrollers or protocols for field buses were of no concern, which has to be considered when applying the ISO/OSI model to such systems.

In a first step, all subfunctions that contributed to communications were collected as shown in Figure 1.1 in the top right cloud, where each subfunction is represented by a circle. These subfunctions were then assigned to separate layers. The fact that the model consists of seven layers does not have technical reasons, but represents the common agreement between the participants of the workgroup that was responsible for the definition of the model. In this way, each layer was assigned to have a dedicated function, which consists of subfunctions.

The protocol stacks (i.e., the systems that are designed based on this model) shall contain evenly distributed subfunctions over all layers. Subfunctions that are too different shall not be included in the same layer and the interfaces shall be designed toward a low amount of information that has to be exchanged between the layers (in order to keep the overhead low).

The standard defines many more rules that shall encourage a uniform and logical structure. The use of specific description language (SDL) is required to avoid endless sequences of unstructured code as well as flow diagram, which in the end lead to the SDL and its corresponding tools [Ols 92] that are used in telecommunications.

The ISO/OSI model was a great leap forward in the design of protocols. It created orderliness and a uniform structure that builds a commonly approved base for the standardization of protocols and

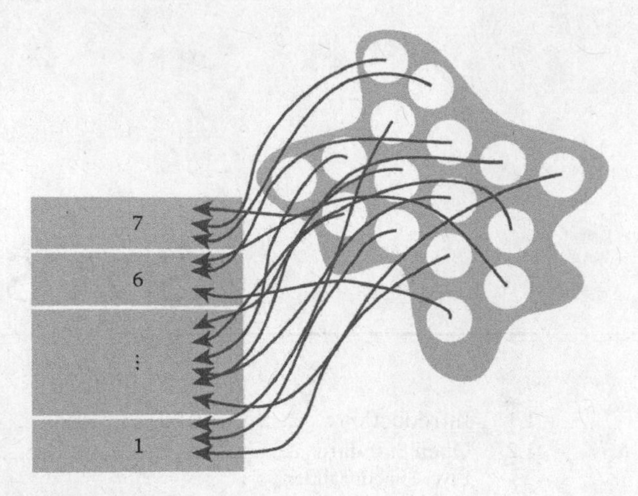

FIGURE 1.1 Development of the model.

communication interfaces. The model defines functions, which are expected by the components that contribute to communication. It explains how these functions can be based on each other and how they cooperate. Before the ISO/OSI model was created, different experts from different fields of communication (e.g., telecommunications, computer communication, automation, process engineering, or equipment technology) created completely different solutions. They defined proprietary sets of functions and often forgot other important functions, they aligned the different functions individually to meet the requirements at hand, and they based the design on different communication principles. The ISO/OSI model defined a layer for each subfunction and specified when which layer should be accessed. Thus, the model helps to simplify the decision, which subfunctions are relevant for a certain kind of communication and how they correlate. It builds the base for new protocol standards (e.g., the various fieldbus standards in [Zur 05]) and eases linking different types of networks.

Some literature states that the ISO/OSI model has finished the language confusion that existed in technical communication. This is a bit of exaggeration, since the model does not claim to define a uniform language for all communication tasks. If that would have been the goal, the model would only contain descriptions for these tasks; instead, it describes the principles of communication. Considering the vastly different communication tasks in, e.g., an airplane and in a washing machine, this appears to be the only feasible approach.

Where shall the communication system be located within a system? There is no common opinion to this question, especially since at the time the ISO/OSI model was defined, the only separation that was done was between application and communication system; an operating system was not considered at all.* Today, the operating system is well established as being located underneath the application, and sometimes the communication system is included into the operating system, sometimes it is not included. Automation prefers a design according to Figure 1.2, where the communication system is a separate unit underneath the operating system and establishes the connection to different networks.

Consistent with the hierarchical model, the interface of the communication unit provides its services to the module above it—the operating system (given that it exists in the system), which again provides its services to the applications. According to the top-down design, the developer shall specify this interface only after the application and the operating system have been specified. However, reality shows that this is often not the case. Communication has to follow a standard, since it generally connects systems of different vendors. The dilemma can be resolved partly by allowing the definition of different

* The fact that the operating system did not play an important role is also shown in the name "application layer," which would more accurately be named "operating system layer."

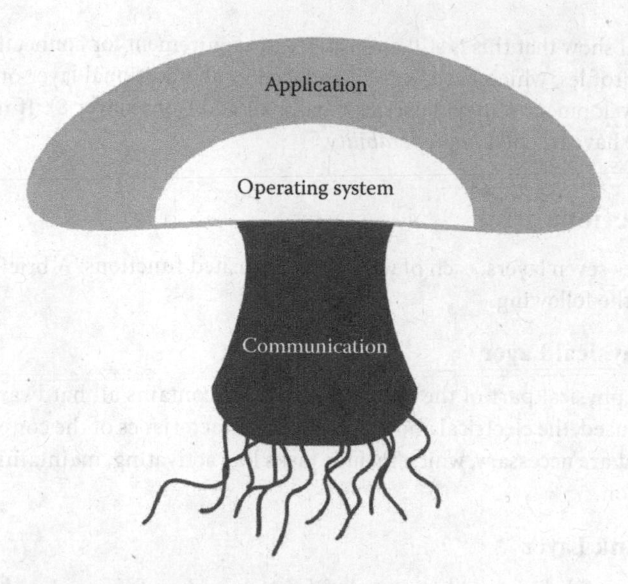

FIGURE 1.2 Location of the communication system in a device.

communication protocols. While on one hand, there should be only a few different protocols to meet the requirements of the standard (i.e., little variation, lots of common properties), on the other hand, we need a lot of different protocols to meet the different requirements.

Today, we see that most fieldbus protocols in industrial automation consist of at most three layers; building automation protocols usually have more layers. The LonTalk protocol [Loy 05] (ANSI/CEA 709.1 and ISO/IEC 14908-1) even implements all seven layers.

1.2 Open Standard

The term "open system" is well defined in ISO: a "system" is in this context a complete facility (unit), that is, computers (like process computers) and data processing machines with peripheral devices like storage, front-end computers, data stations, or application software. Such a system contains communication partners and (parts of) a communication system.

The term "open" has a stricter meaning than as it is used in common language, and cannot be used freely in the area of communications (even if marketing wants to interpret it differently). A system is "open," if it meets certain requirements. The first authoritative requirement is that the protocol is officially standardized (nationally or internationally). A quasi-standard or a special standard defined by a company is not valid. In such a way, the distribution (publication) is guaranteed—which shall ensure equal opportunities between companies.

In 1997, the different European committees struggled hard to define the requirement how the term may be interpreted. CEN and CENELEC agree that—in accordance with the reference model—a system is open, if the protocols have been opened (i.e., standardized) and are not protected by a patent that prevents competitors to design the same protocol. Licenses have to be available for everyone at "reasonable" prices.

The ISO/OSI reference model describes functions, not hardware or software. It does not dictate manufacturers which technology to use in their products. It only prescribes how the technology has to behave, seen from the outside.

Unfortunately, the standard for a protocol cannot be compared with a physical unit like a screw. A standard is complex and thus never completely error free. This was considered regarding different terms. Two systems are called *interconnectable*, if they are subject to the same standard. This does, however, not mean that they cooperate. If cooperation shall be guaranteed, *interworkability* is required, which means that the system can in principle exchange data according to the protocol. However, the following sections will show that this is still an insufficient requirement for connecting devices. This can only be achieved by profiles, which have been introduced as an additional layer on top of the ISO/OSI model during the development of field buses (virtually an additional layer 8). If it can be proven that devices cooperate, we have reached *interoperability*.

1.2.1 Layer Functionalities

The OSI model defines seven layers, each of which has dedicated functions. A brief description of these functions is given in the following.

1.2.1.1 Layer 1: Physical Layer

This layer covers the physical part of the communication. It contains all hardware specification data, including the signals used, the electrical and mechanical characteristics of the connection, and all functional parameters that are necessary, which include tasks like activating, maintaining, and terminating the physical connection.

1.2.1.2 Layer 2: Link Layer

The link layer is responsible for providing an error-free connection from one node to another node in the same network segment (point-to-point communication). It has to correct errors that occur during the physical transmission by using, for example, error-correction codes. For that, it needs error-correction algorithms and redundant information in the received data. It also adds source and destination address to the packets that are transmitted.

1.2.1.3 Layer 3: Network Layer

The network layer defines the path that packets take on their way through the network. A packet that is addressed to a destination address will not always be transmitted directly to its receiver but will rather be passed from one part of the network to the other until it reaches its destination. This is done by routing the packets, an algorithm that can be implemented in different ways, depending on the capabilities of the components. Layer 3 defines addresses, which are not related to the addresses on layer 2 (if they are implemented). The network layer also is responsible for establishing and terminating network connections and reestablishing broken network connections.

1.2.1.4 Layer 4: Transport Layer

The transport layer is responsible for the flow control of data that is sent from one end user to the other (end-to-end connection) and for assigning logical addresses to the physical addresses that are used by the network layer. It uses the network layers' ability to establish network connections in order to guarantee that messages really reach their end users, which also includes retransmission of lost packets.

1.2.1.5 Layer 5: Session Layer

In order to establish a session, the session layer has to make sure that all the end users agree on the same session protocol; therefore, the participants first have to negotiate a common protocol, which is then used throughout the session. The session layer defines how a session is started and terminated, describes how data-exchange is established, and is responsible for end-user identification (e.g., by password).

1.2.1.6 Layer 6: Presentation Layer

The presentation layer defines how the information shall be formatted in order to make it understandable for the end user. If, for example, an integer number is transmitted, the presentation layer knows how to interpret the bytes that make up the number and is able to provide a mathematical value to the application layer (e.g., by first converting big endian to little endian). Conversion of data is covered here as well as optional encryption of information.

1.2.1.7 Layer 7: Application Layer

The application layer provides an interface that can be used by the application. It contains services for the application, which can, for example, provide access to distributed databases or other high-level services. The application layer strongly depends on what the applications (or the operating system) above it need, and is therefore usually designed to meet the requirements of these applications.

As stated earlier, the seven layers cannot provide interoperability by themselves. Profiles, which create a layer on top of the ISO/OSI model, can help to reach the level of interworkability.

1.3 Vertical and Horizontal Communication

Each communication node (i.e., a device that participates in communication) contains at least one communication stack, which is defined by the function layers 1–7. Therefore, we distinguish between horizontal and vertical communication as shown in Figure 1.3.

The layers communicate using a logical connection (horizontal communication) or the different layer protocols (e.g., application protocol or session protocol), respectively. The term "protocol" is ambiguously used for both the complete communication stack and the separate (seven) protocols of the stack. The protocol contains not only data definitions that define the kind of information that is exchanged, but it also defines the set of rules that communicating entities have to use, in terms of appropriate reactions to incoming information, handling of error situations, or timing constraints. Information flows vertically (vertical communication) through the layers, which offer services to higher layers. The following section will go into more detail on this.

If we look at the protocols of the separate layers—on which we want to focus in the following—we see a set of rules that are assigned to the same layer. They are defined by the type of control information, the procedure, and its behavior; this is done in three definitions: (a) definition of the states, (b) definition of the transitions, and (c) the definition of the timing. They should be realized in a way that they can each be replaced separately in the stack.

The seven layers of the ISO/OSI model can be assigned to different block functions. The left side of Figure 1.4 shows the separation between point-to-point connections, i.e., the connection between two units without another unit (e.g., a router) is connected in between, and end-to-end connections. This means that on a connecting line between two end devices, the three lower layers have to be processed by each device on the way from end device to end device, while the upper four layers are only processed by the end devices.

Figure 1.4 on the right side separates between transport-oriented protocols and the upper three application-oriented protocols. The transport-oriented protocols do not process payload data, while the application-oriented protocols depend on the according application.

This indicates the importance of the transport layer: layer four is the first that shows the end-to-end property of the communications. This layer shall guarantee that all data, which are sent, do receive the other side completely and in correct order. On the other side, it is the lowest layer that can still be defined independently from the application

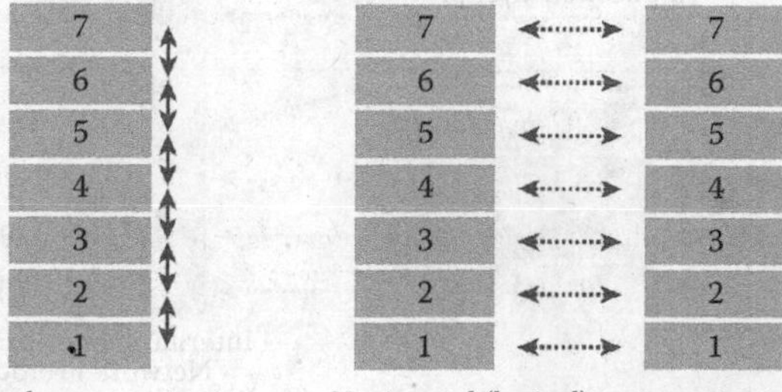

FIGURE 1.3 Layer architecture.

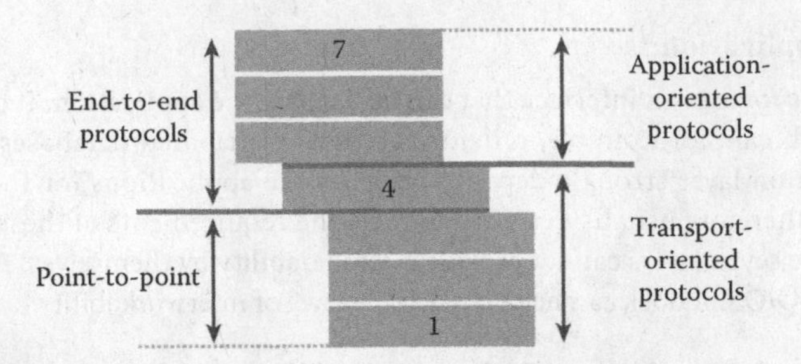

FIGURE 1.4 Layer build-up as blocks.

1.4 Dynamic Behavior of Services and Protocols

Figure 1.5 shows a common example of communication that runs over an intermediate station, e.g., a router. One side, e.g., the left side, starts transmitting a message, which is done by the application using the application layer. Communication runs vertically down to layer 1, horizontally through the router to the receiving device, and vertically up to the application in the receiving device. This communication can also be seen as parallel horizontal communications that run between each layer as peer-to-peer communications.

Due to the modularization that the OSI model requires, all layers operate simultaneously. Consequently, the layers can operate independently, the interfaces can be defined easily and systematically, and the different tasks can be clearly assigned to the different layers.

Each layer in the OSI model contains a definition of the protocol and the services. The protocol specification contains the exact number of functions and messages (protocol data units, PDUs)—so to say the "language" of the layer. The layer itself is characterized by the procedures that operate on the PDUs.

The service definitions can be explained as follows: two opposing entities (the service users, SU) exchange information using protocols (peer-to-peer protocols) by using the service providers (SP). Thus, a service provider offers services to a service user, so that the service user can fulfill its task. Service communication runs vertically, where lower layers offer services to higher layers (with the exception of the lowest layer).

Seen from a logical perspective, the layers of entities communicate with each other (i.e., layer n of one entity communicates with layer n' of the other entity). In the vertical direction, service primitives are passed on using the service access points (SAP) as shown in Figure 1.6. There is no communication defined between layer $n - 1$ and layer $n + 1$. The communication between two adjacent layers can be compared to master–slave communication, because a master does not communicate with a slave,

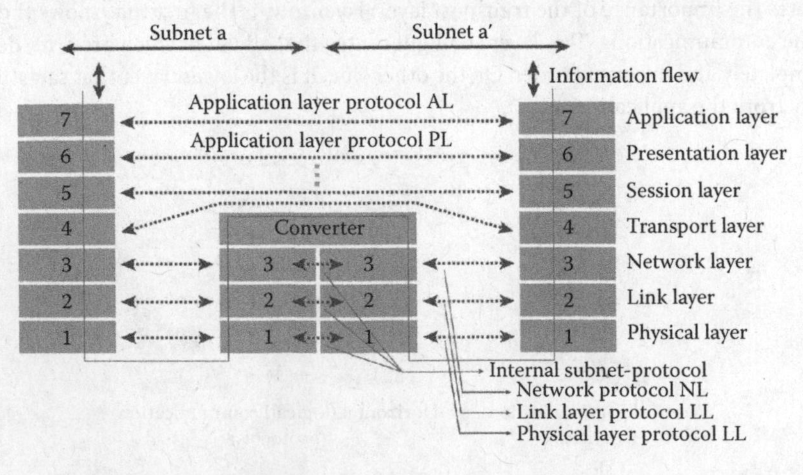

FIGURE 1.5 Communication over subnets.

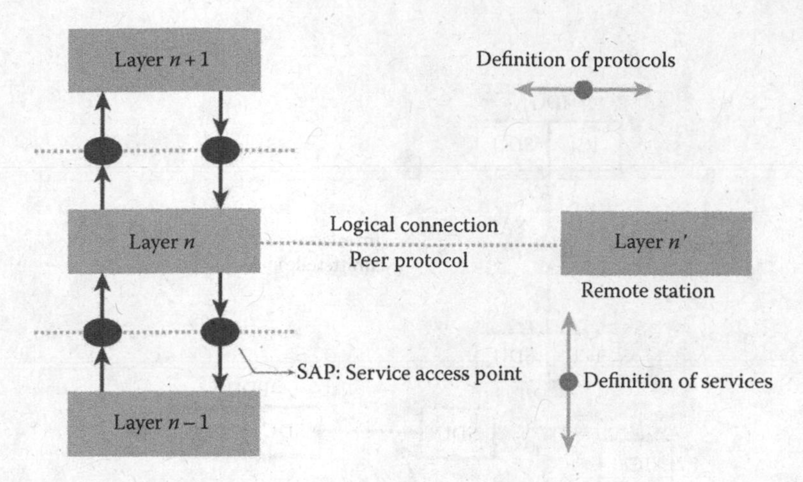

FIGURE 1.6 Services, protocols, and service access points.

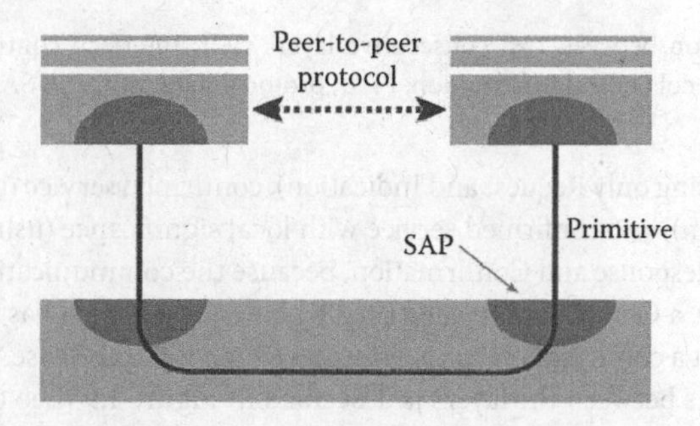

FIGURE 1.7 The SAP is a logical interface between the service user (SU) and the service provider.

it rather requests a service from the slave; the slave itself can again be the master of underlying slaves (Figure 1.7). This is also called a cascaded system. The service user can also be seen as an instance, which runs a process with the opposing instance (i.e., the peer user).

A communication stack according to ISO/OSI is therefore a parallel working system, which does not have priority control (at least in the first definition of the model). This has been softened later by adding a management, which will be described later.

The SAP can be implemented in software (e.g., as a subroutine) or in hardware (e.g., as memory). The decision is up to the developer.

The vertical communication uses predefined primitives to unify the sequence of communication in all systems (Figure 1.8). The sequence of primitives is as follows: Request, Indication, Response, and Confirmation. Based on these primitives, we can derive different communication relations. Three examples

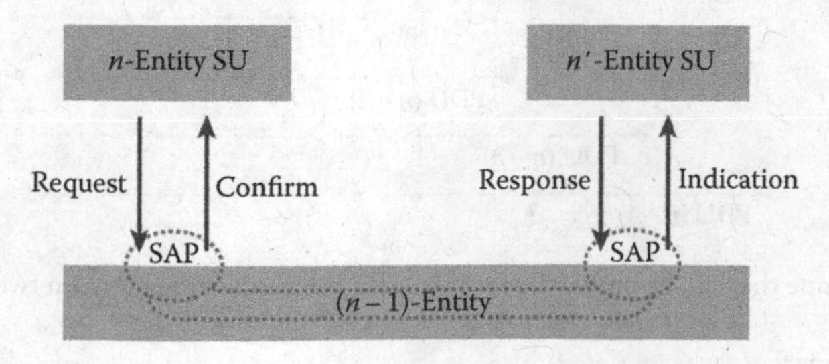

FIGURE 1.8 Predefined service primitives.

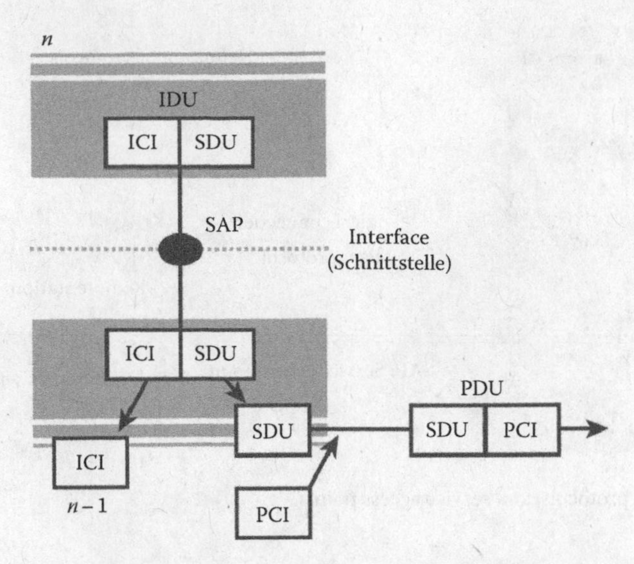

FIGURE 1.9 Communication between two consecutive layers (ICI, interface control unit information; IDU, interface data unit; PCI, protocol control information; PDU, protocol data unit; and SAP, service access point).

are unconfirmed service (using only Request and Indication), confirmed service (using Request, Indication, Response, and Confirmation), and confirmed service with local significance (using all four primitives, but without being able to link Response and Confirmation, because the communication channel is only unidirectional). In the latter case, a Confirmation is initiated by a Request, which has the big disadvantage that the user is led to believe that a confirmation has been sent—which is not the case.*

The process that happens between the layers is described in Figure 1.9 [Leo 00]. Layer n of a communication stack shall send a piece of information to an opposing layer. This information is called service data unit (SDU). The interface process in layer n attaches an interface control unit information (ICI) to the head of the SDU, which together builds the interface data unit (IDU) that is transferred over the SAP. The layer below the ICI is decoded and a new process is triggered. The following sequence (i.e., the execution of a protocol) is to be seen as a logical process, with the exception of the lowest layer, the physical layer. For all other layers, we can say that layer $n - 1$ packs the SDU into a PDU by following a procedure defined by the ICI by prefixing a protocol control information (PCI) to the SDU. The PDU is now transmitted, which means that the layer $n - 1$ uses its layers underneath and acts as a service user. Finally, layer 1 physically transmits information as shown in Figure 1.10. The SDUs can be physically measured as well as all header, i.e., the PCI of all layers.

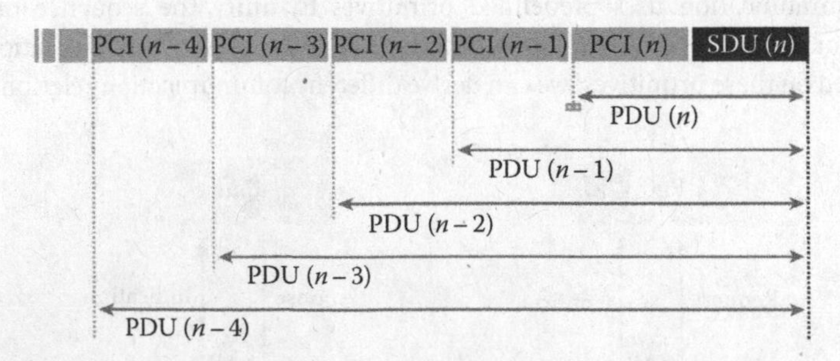

FIGURE 1.10 Data frame that can be physically measured on the connection between two end devices.

* For some implementations this is sufficient, assuming that errors occur rarely and can be caught on higher layers.

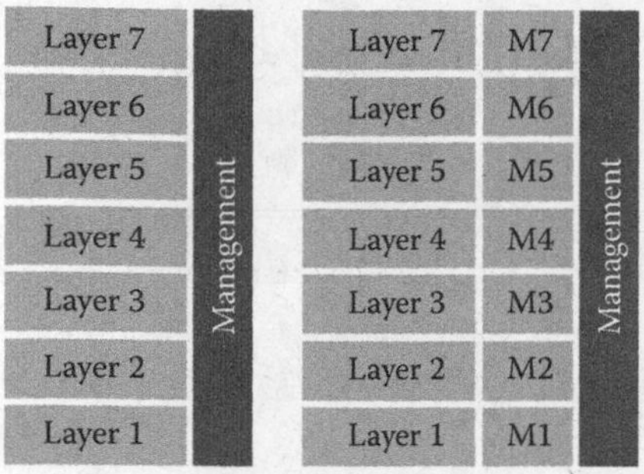

FIGURE 1.11 Adding management units. Mn: management *n* {*n*: 1 to *n*}.

1.5 Extensions, Benefits, and Discussion

The ISO/OSI model is the first model of its kind. It provides a template for the design of protocols and has proven useful over decades. Since its definition, many different protocols have been defined using the ISO/OSI model. Telecommunication relies on it, just like fieldbus systems in automation use it. It has been refined over the years, for example, by introducing a management for automatic handling of different protocol layers as shown in Figure 1.11.

The developer has to decide how to add management systems. For smaller stacks, one management for the whole stack will be sufficient, for bigger stacks each layer will get its own management, and additionally, the whole stack will have an integral management. In such a system, it is necessary to introduce priorities, since error handling has to be considered.

The main advantages as we see them today are modularity, a clear top-down design, independence of the layers, the possibility to handle errors in each layer, the functional description, and the resulting possibility to implement in either hardware or software. Disadvantages of a design according to the ISO/OSI model are the overhead, which can become considerable, and the fulfillment of real-time requirements.

References

[Hay 88] Hayes, J. P., *Computer Architecture and Organization, McGraw-Hill Series in Computer Organization and Architecture*, 2nd edn, McGraw-Hill, New York, 1988, p. 500.

[Leo 00] Leon-Garcia, A. and Widjaja, I., *Communication Networks*, McGraw-Hill, New York, 2000, p. 51.

[Loy 05] Loy, D., Fundamentals of LonWorks/EIA-709, ANSI/EIA-709 protocol standard (LonTalk), in: *The Industrial Communication Technology Handbook*, CRC Press, Boca Raton, FL, Zurawski, R. (Ed.), 2005, p. 35-1.

[Ols 92] Olsen, A., Færgrgemand, O., Møller-Pedersen, B., Reed, R., and Smith, J. R. W., *Systems Engineering Using SDL 92*, Elsevier Science B. V., Amsterdam, the Netherlands, 1994, ISBN 0-444-89872-7.

[Zur 05] Zurawski, R. (Ed.), *The Industrial Information Technology Handbook*, CRC Press, Boca Raton, FL, 2005, p. 37-1.

2

Media

2.1 Introduction 2-1
2.2 Wired Links 2-1
Physical Properties • Cable Types and Operational Characteristics • Single-Ended and Differential Transmission • Simplex and Duplex Communication • Bit Encoding • Standards • Data Transmission Utilizing Existing Cable Infrastructure
2.3 Optical Links 2-7
Physical Properties • Types and Media Access • Transmitters and Receivers • Multiplexing • Implementations and Standards
2.4 Wireless Links 2-11
Physical Properties • Types and Media Access • Modulation • Bit Coding, Multiplexing • Realizations and Standards
References 2-17

Herbert Schweinzer
Vienna University of Technology

Saleem Farooq Shaukat
COMSATS Institute of Information Technology

Holger Arthaber
Vienna University of Technology

2.1 Introduction

This chapter discusses physical mechanisms for transferring digital data. The three dominant media types and their applications are presented: data transmission over cable, over optical links, and over wireless radio frequency (RF) links. Properties of each medium are concisely introduced with respect to physical characteristics, alternatives of realization, topologies, bit coding, and standards. Media for special local data exchange, e.g., infrared transmission, are not contained in this chapter.

2.2 Wired Links

2.2.1 Physical Properties

Data transmission of digital signals over cables is based on electromagnetic wave propagation, which is used for moving data from one location to another. Data bits are coded as short pulses or fast changing modulations of wave properties (amplitude, frequency, phase). The fundamental relation of wave propagation is

$$v = f \cdot \lambda \tag{2.1}$$

where

v is the phase velocity of a sinusoidal electromagnetic wave
f is the frequency
λ is the wavelength

Each cable has a *characteristic impedance* Z_W and has to be matched at both ends with equal terminating resistors R_E. Mismatch at an end leads to a reflection where the wave front, e.g., a pulse, is traveling back

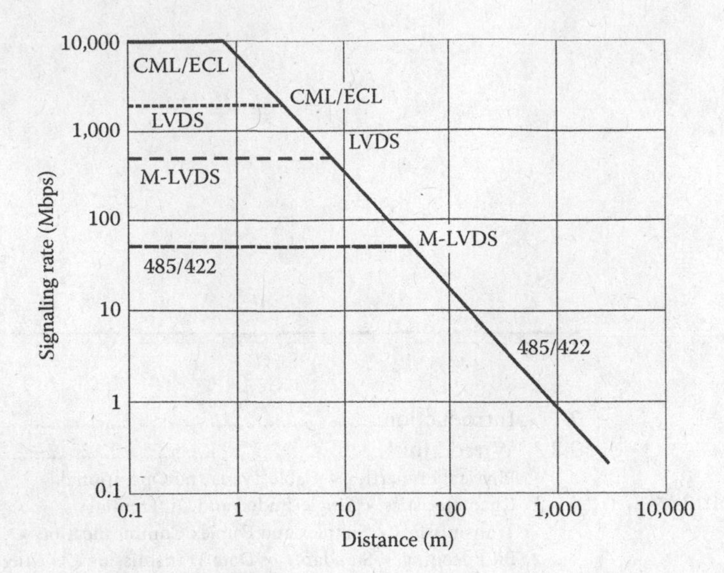

FIGURE 2.1 Signaling rate versus cable length.

to the sender with an amplitude and sign depending on the mismatch's impedance. Cable mismatch mostly reduces the achievable bit rate and, therefore, should be avoided.

The form of a rectangular pulse remains equal as long as the pulse frequency components are transmitted with the same attenuation and velocity. Frequency dependence of phase velocity is called *dispersion*. Mostly, cable attenuation (in dB/m) increases with the frequency. This and increasing cable length are limiting the maximum rate of transmitted bits considering a maximal bit error rate (BER). Defining the distance as length of the signal transmission path between the sending and the receiving system and the signaling rate as the bit rate at which data have to be passed to the receiving device, a characteristic relation can be stated as defined by Figure 2.1.

Different transmission standards, such as TIA/EIA–485, low-voltage differential signaling (LVDS), multiplex LVDS (M-LVDS), and current-mode logic (CML) provide solutions for various needs in terms of bit rate and line length (see Table 2.1 for details of the standards). As shown in Figure 2.1, as the cable length increases, the speed at which the information is transmitted must be lowered in order to keep the BER down. More details with regard to physical properties, transmission types and standards, and devices can be found in [TI04,NS08].

2.2.2 Cable Types and Operational Characteristics

Besides of cable data transmission characteristics, the main important aspect is cable protection against electromagnetic influences of the environment, which gets growing importance with increasing cable length and strong and high-frequent electromagnetic fields near the cable. Two parameters influence protecting the cable against influences: cable type and construction.

Basically, a cable has two lines. This *unshielded* type can be additionally protected by *shielding* where the two lines are placed within a conductive hose. Mostly connected with ground potential at one end, this hose reduces the impact of electromagnetic noise.

Several cable constructions are well suited for supporting high bit rates: coaxial cable and one or multiple twisted pairs. The coaxial cable is set up with a line in the center of a conductive hose that delivers a shielding of the inner signal line by the hose at ground potential. Primarily used for unbalanced signal transmission as described in the following, a coaxial cable can be additionally shielded by an outer hose.

TABLE 2.1 Overview over Standards for Digital Data Transmission and Their Characteristics

Standard	Product Family	Max. data Rate per Line	Max. Data Rate per Device	Max. Distance	Topology
TIA/EIA-232 (ITU-T V.28)	TIA/EIA-232	512 kbps		20 m	Point-to-point (simplex)
CAN (ISO 11898)	CAN	1 Mbps		40 m	Multipoint (Multiplex)
TIA/EIA 422 (ITU-T V.11)	TIA/EIA-422	10 Mbps		10 m (1,200 m)	Multidrop (distributed simplex)
TIA/EIA 485 (ISO8482)	TIA/EIA 485	35 Mbps		10 m (1,200 m)	Multipoint (Multiplex)
USB 1.1 USB 2.0	USB	12 Mbps 480 Mbps		5 m	Multipoint (Multiplex)
IEEE1394-1995 IEEE1394a-2000 IEEE1394b-2002	Firewire	100-400/ 800 Mbps 800 Mbps		4.5 m > 100 m	Multipoint (Multiplex)
TIA/EIA-899 (Multiplex LVDS)	M-LVDS	500 Mbps		0.5 m (~30 m)	Multipoint (Multiplex)
TIA/EIA-644 (LVDS)	LVDS	2 Gbps	4 ch: 1,600/800 Mbps	1 m (~30 m)	Point-to-point (simplex)
IEEE P802.3z	Gigabit Ethernet	1.25 Gbps	1.25 Gpbs full duplex	<10 m	Point-to-point (simplex)
IEEE P802.3ae	10 Gigabit Ethernet	2.5 Gbps	4 ch: 10.0 Gbps full duplex	<10 m	Point-to-point (simplex)

Twisted pair cables are based on two twisted lines in an unshielded (UTP) or shielded cable (STP). Due to twisting, the two lines are similarly influenced by electromagnetic noise. Especially, if operated with differential signal transmission as described in the following, this influence affects only the common mode (CM).

2.2.3 Single-Ended and Differential Transmission

Two forms of operation of electrical interface circuits are used for data transmission via cables: single-ended (or unbalanced, SE) and differential (or balanced) transmission.

Figure 2.2 shows the principle of SE transmission. A single signal line carries the data coding signal referring at both ends to a common ground line. Advantages of SE transmission are simplicity and low cost of implementation, e.g., used with the frequently applied standard TIA/EIA-232 serial communication with many handshaking lines. Cabling costs can be kept to a minimum with short distance communication, depending on data throughput. For longer distances and/or noisy environments, shielding and additional ground lines are essential. Twisted pair cables are recommended for line lengths of

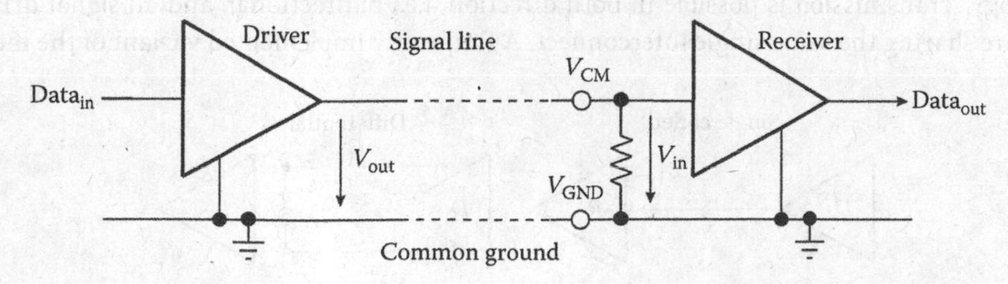

FIGURE 2.2 SE transmission with terminating resistor at the end.

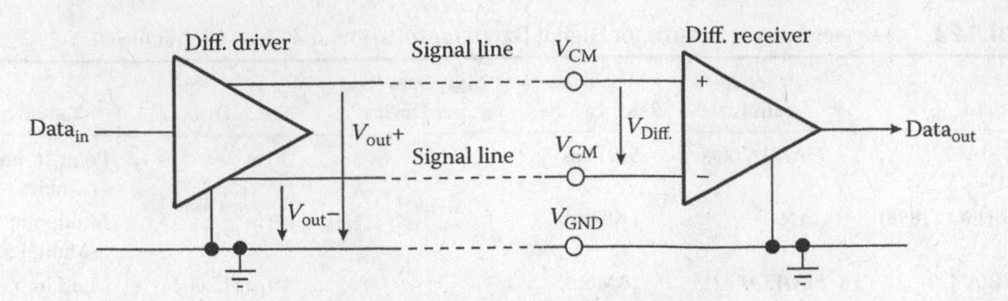

FIGURE 2.3 Differential transmission.

more than 1 m. These additional efforts significantly reduce the cost advantages. Disadvantage of SE transmission is the poor noise immunity originated by the way of ground wiring.

For balanced or differential transmission, a pair of signal lines is necessary together with an additional ground connection (see Figure 2.3). The signal is transmitted by a sender with differential outputs in a way that the signal is placed on one line and the inverted signal on the second line. The receiver detects the voltage difference between its inputs. Good noise performance comes from the use of a twisted pair whereby noise is coupled into both signal lines in the same way. Due to the CM rejection capability of a differential amplifier, this noise is without effect. Twisted-pair cables with correct termination and differential signaling allow very high data rates up to 10 Gbps. Only cost is the primary disadvantage of differential transmission. Furthermore, high data rates require a very well-defined cable impedance and correct termination to avoid reflections.

2.2.4 Simplex and Duplex Communication

Media access for data transmission can be performed in different topologies in respect of directly realizable communication links. These topologies are point-to-point, multidrop, and multipoint.

Point-to-point or *simplex* communication is characterized by one sender and one receiver per signal line or line pair of a cable (Figure 2.4). Data transmission is possible only in one direction, i.e., unidirectional. Being the most elementary topology, it has the advantage of enabling well-controlled cable impedances necessary for very high signaling rates. All signaling technologies may be used for point-to-point links. However, very fast differential signaling technologies are primarily designed for point-to-point signal transmission.

A topology with one sender and multiple receivers is called *multidrop* or *distributed simplex* (Figure 2.5). Only unidirectional transfer is possible. Terminating the cable on the far receiver side is advisable only when the signal driver is on the opposite end of the cable from the terminated receiver. In all other cases (e.g., driver connected to the middle of the bus), the bus needs to be terminated at both ends of the bus.

The *multipoint* or *multiplex* topology is implemented with many transmitters and many receivers per line (Figure 2.6). In practice, this solution is often realized with combined transmitter–receiver pairs called *transceivers.* Any combination of receivers, transmitters, and transceivers is possible for this topology. Transmission is possible in both direction, i.e., bidirectional, and all signal drivers and receivers are sharing the same single interconnect. A frequently implemented variant of the multipoint

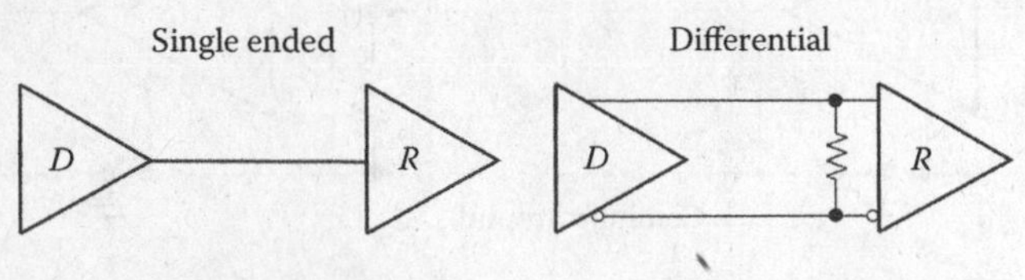

FIGURE 2.4 Point-to-point communication link.

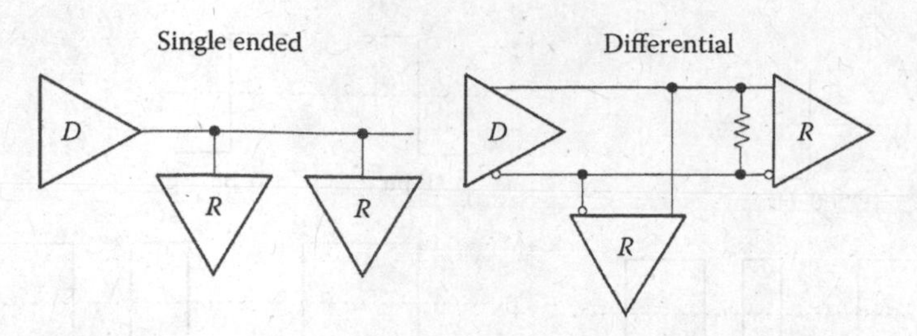

FIGURE 2.5 Multidrop topology.

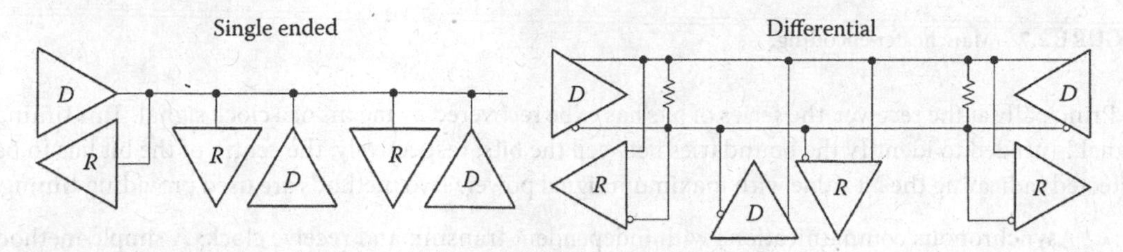

FIGURE 2.6 Multipoint topology.

topology is the *half-duplex* topology, which consists of two driver/receiver pairs that transmit and receive signals between two points over a single interconnect. Since sharing of a single interconnect requires that only one transmitter is active at any time, a *full-duplex* topology implying concurrent transmission in both directions has to be realized with two lines in parallel.

Physical connection of multiple drivers and receivers to a common signal line raises a design challenge mainly encountered by impedance discontinuities that device loading and device connections (stubs) introduce on the common bus.

The topologies presented above can be used with single line or multiple lines in parallel, which correspondingly increases the rate of transferred bits. On the other hand, parallel lines increase the cost because of the need of parallel transmitters and receivers and of cable cost according to the cable length. Parallel data transmission moreover requires provisions for synchronous data clocking, which is basically implemented by a separate clock line. These aspects are responsible for mostly using single or very few parallel lines in communication networks with larger line lengths. In this case, a separate clock line can be saved and replaced by coding techniques allowing for synchronizing the receiver clock as described as follows.

2.2.5 Bit Encoding

For bit transmission, a frame is serialized and sent across a communications link to the destination. As defined by the open systems interconnection (OSI) model, the physical layer (Layer 1) defines the representation of each bit as a voltage, current, phase, or frequency. The following basic schemes are used:

- Return to zero, RZ (pulse signaling)
- Non–return to zero transmission, NRZ (level signaling)
- Manchester encoding or Manchester phase encoding (edge/phase signaling)

In earlier transmission techniques, pulses were used to represent bits, e.g., RZ in which a logic 1 is represented by a pulse and a logic 0 by the absence of a pulse. In NRZ transmission, each data bit is represented by a level. A high level may represent a logic 1 and a low level a logic 0, or vice versa.

Manchester encoding (Figure 2.7) uses a still different scheme where a logic 1 is represented by a transition in a particular direction (usually a rising edge) in the centre of each bit. A transition in the opposite direction (downward in this case) is used to represent a logic 0. Compared with NRZ, advantage of this coding is that the bit timing is intrinsically indicated by the edges characterizing the bits.

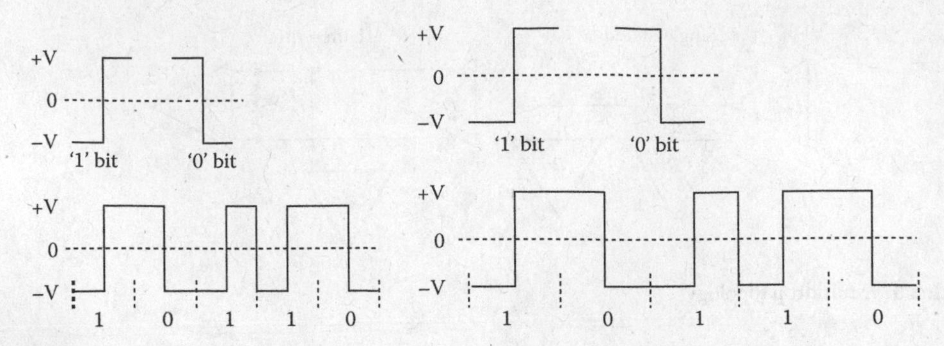

FIGURE 2.7 Manchester encoding.

Principally at the receiver, the series of bits has to be recovered by means of a clock signal. This timing signal is needed to identify the boundaries between the bits; respectively, the centre of the bit has to be detected indicating the bit value with maximum signal power. Two methods are used providing timing:

- Asynchronous communications with independent transmit and receive clock: A simple method where transmit and receive clock are independently set to the same clock rate. Bit synchronization within a single byte is attained by a start bit providing the timing for detecting the data bits (Figure 2.8). Synchronization ends with a stop bit. The data rate is limited mainly caused by increasing synchronization loss, e.g., because of noise disturbing the start bit. A parity bit is often employed to validate correct reception. This type of communication is primarily used for connecting printer, terminal, and modem.
- Synchronous communication using synchronized transmit and receive clocks: This more complex method supports also high data rates (~Gbps). Timing is derived from edges in the bit stream that are used to synchronize the receive clock. Mostly, cyclic redundancy check is used to validate correct reception.
- A lot of additional coding and modulation techniques are used, which are only mentioned shortly.
- Digital frequency modulation (FM) and modified frequency modulation (MFM) related to Manchester coding and primarily used for magnetic recording.
- Differential Manchester encoding, where an additional transition at the beginning of the bit time is only included when coding a zero bit.
- Non–return to zero invert (NRZ-I), where the coding signal level depends on the previous one. With that, frequent level changes allow better synchronizing.
- Block codes 4B/5B, 5B/6B, and 8B/10B use a greater number of bits (5, 6, or 10 instead of 4, 5, or 8) for coding the bit group with the goal of using code combination for reliable clock synchronization. 5B/6B and 8B/10B additionally allow DC balance.
- Block code 8B/6T uses ternary coding (three voltage levels –V, 0, +V).
- Pulse amplitude modulation 5 (PAM-5) uses five voltage levels (e.g., –2, –1, 0, 1, 2 V) where one level (0 V) is used for trellis error correction. PAM-5 is utilized by Ethernet 100BASE-T2 over two wire pairs and 1000BASE-T over four wire pairs. One type of 10 Gbps Ethernet over wire (Standard IEEE 802.3an) makes use of a pulse-amplitude modulation with 16 discrete levels (PAM-16) over UTP or STP cable.

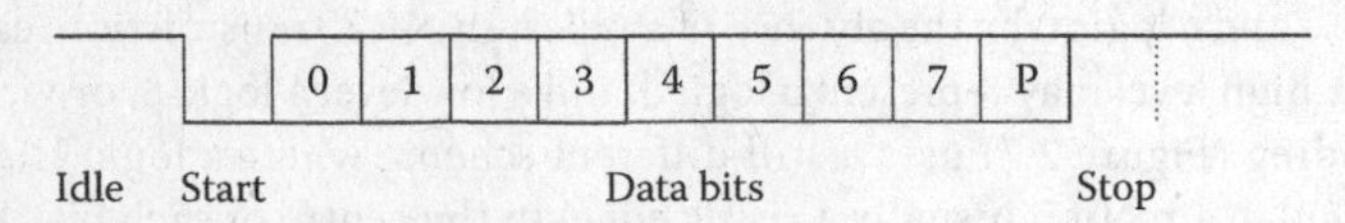

FIGURE 2.8 Asynchronous communication: example with 8 data bits (0–7), a parity bit P and a single stop bit with length of 1 data bit.

2.2.6 Standards

Numerous standards in different versions were developed over the years which are used in the area of industrial communication. Table 2.1 gives an overview of some often used systems.

2.2.7 Data Transmission Utilizing Existing Cable Infrastructure

In special cases, e.g., within buildings, existing cable infrastructure primarily not intended for data transmission can be employed for. Especially used for home applications, *power line communication* or *power line networking* allows data transfer via AC power supply lines. Applications range from narrowband home control up to broadband home networking and internet access with data rates ≥1 Mbps. Power line data transmission is based on medium to high carrier frequencies (narrowband 9–500 kHz, broadband 1.6–80 MHz) modulated by different modulation formats, e.g., OFDM (see Section 2.4.3). Induced by device on and off switching and by nonlinearities causing harmonics, power lines are inherently a noisy environment. Power lines are unshielded and, therefore, radiate signals they carry which may interfere with nearby equipment. Power line technology can also be used for in-vehicle networks. Other cables which can be employed for data transmissions are telephone wires using Digital Subscriber Line (DSL) technology.

2.3 Optical Links

2.3.1 Physical Properties

The core and cladding are the two key elements of an optical fiber. The core is the inner part of the fiber, through which light is guided. The cladding surrounds it completely. The refractive index of the core is higher than that of the cladding, so light in the core strikes the boundary with the cladding at a glancing angle, confined in the core by the principle of *total internal reflection*.

Optical communication is the fastest communication. The tendency of the light to travel in packets through the narrow core is not because of photon size but of speed; photons are many orders of magnitude smaller than the core. Transmission of light by optical fibers is not 100% efficient. Some light is lost, causing attenuation of the signal. Attenuation measures the reduction in signal strength by comparing output power with input power. Measurements are made in decibels (dB). Optical fibers are unique in allowing high-speed signal transmission at low attenuation. The degree of attenuation depends on the wavelength of light transmitted. This makes operating wavelength an important feature of a fiber system. Light intensity attenuation has no direct effect on the bandwidth of the electrical signals being transported. There is a direct correlation, between the S/N of the fiber receiver electronic circuits and the usable recovered optical signal. For commercially available fibers, attenuation ranges from approximately 0.15–0.5 dB/km for single-mode fibers. The gradual reduction in attenuation losses in all the four generations of optical communication systems (OCSs) have been described in Table 2.2.

Dispersion is the spreading or broadening of light pulses as they propagate through fiber. Dispersion limits fiber transmission capacity. Dispersion increases with distance, so the maximum transmission rate decreases with the distance. There are different forms of dispersion, including material and waveguide. The materials used to create the fiber cable use different refractive indices; therefore, each wavelength moves at a different speed inside the fiber cable. This means that some wavelengths arrive prior to others and a signal pulse disperses over a broader range. This is also called *smearing*. The center core creates the waveguide. The shape and the refractive index inside the core can create the dispersion or spreading of the pulse. Information-carrying capacity (band width/bit rate) is very important in all types of communications. The more bits that can pass through a system in a given time, the more information it can carry. The band width enhancement has been the major area of research since 1974. The developments in bandwidth are shown in chronological order in Table 2.2.

TABLE 2.2 Comparison of the Generations of Optical Communication

Serial #	Parameter	First Generation	Second Generation	Third Generation	Fourth Generation
1	Deploying year	1974	1978	1982	1994
2	Operating Wavelength	820 nm	1,330 nm	1,550 nm	1,550 nm
3	Attenuation	20 dB/km reduced to 5 dB/km	0.5 dB/km	0.25 dB/km	0.15 dB/km
4	Max. Bandwidth for Communication	2 Mbps	140 Mbps	1 Gbps	10 Gbps expandable to 50 Tbps
5	Repeater Distance	8–10 km	10 km	50 km	100 km EDFA (erbium doped fiber amplifier)

2.3.2 Types and Media Access

Total internal reflection of rays is a first approximation of light guiding in fibers. Core-cladding structure and material composition are the key factors in determining fiber properties. Optical fibers are classified into single and multimode fibers. Furthermore, the fibers are categorized in terms of their refractive index profile. The most important types of fibers are single-mode (step index), multimode step index, and multimode graded index.

The thickness of a single-mode fiber today is approximately 8.3–10 μm at the center. The outer cladding is approximately 125 μm thick. The single-mode fiber is the focus of most of the activities today. Multimode fiber has given way to single-mode fiber throughout the public carrier networks. Thus, single-mode fiber has an advantage over all other types since it can handle multiple signals without any attenuation for long distances. A single-mode step index fiber between the two ends enables the developers to speed up the input because there is no concern about varying path lengths. If the glass could be made very thin and very pure in the center, the light would have no choice but to follow the same path every time.

Multimode fiber, as the name implies, has multiple paths where the light can reach the end of the fiber. In multimode step index fiber, the core has a uniform index with a sharp change at the boundary of the cladding. The thickness of the glass is crucial to the passage of the light and the path used to propagate from one end to the other. This fiber is the thickest form of fiber, having a core of 120–400 μm thick. The light is both refracted and reflected inside the encased fiber. Light beams in the fiber propagate by using paths of different lengths causing different delays. These fibers are less in practice because of low carrying capacity. They have a relatively wide core, but since they are not graded, the light bounces widely through the fiber exhibiting high levels of modal dispersion.

In a multimode graded-index fiber, the core index is not uniform; it is highest at the center and decreases until it matches the cladding. By using a grading of the density of the glass from the center core out, the capacity of the fiber can be increased. The path in the outer part of the glass is longer than the path in the center of the glass. This means that light arrives at different times because the path lengths are different. Grading the center core to have a higher level of refraction and the outer parts of the glass to be thinner (and thus less refractive) can exploit the characteristics of the glass to get approximately the same length of a wave on the cable and therefore increase the speed of throughput. The better the grading of the index, the more throughput one can expect. Currently, the two forms of graded-index fiber either a 62.5 or a 50 μm center core with a 125 μm outer cladding of glass are being used.

2.3.3 Transmitters and Receivers

Optical transmitters generate the signals carried by fiber-optic communication systems. Light sources compatible to the properties of optical fibers are used in optical communication. Visible red light-emitting diodes (LEDs) are used, that transmit wavelengths better than the near-infrared. Near-infrared LEDs and semiconductor laser made on gallium arsenide and gallium aluminum arsenide emitting at 750–900 nm are used with glass optical fibers for relatively short links and moderate speed systems. These light sources have been used in the first generation of optical communication. The second and third generations of OCSs have been using the semiconductor laser. Its operating wavelength is 1300 nm through glass fibers and has a loss of 0.35–0.5 dB/km at this wavelength. The transmission window for the fourth generation of optical communication is 1550 nm and the attenuation is 0.15–0.5 dB/km. Optical fibers doped with the rare earth erbium also generate light near 1550 nm, but they are used more often to amplify an optical signal that has traveled a long distance. Erbium-doped fiber amplifier (EDFA) has revolutionized optical communication. It has wide bandwidth (20–70 nm), high gain (20–40 dB), high output power (>200 mW), modulation format, wavelength insensitive, low distortion, and low noise.

Photomultipliers respond to incident light by delivering charge to the anode. Photoelectrons are accelerated toward a series of electrodes (dynodes). These are maintained at successively higher potential with respect to the cathode. On striking a dynode surface, each electron causes the emission of several secondary electrons, which in turn are accelerated toward next dynode, and multiplication continues. Silicon photodiodes are one of the most popular radiation detectors. They have small size, high quantum efficiency, wavelength range from 0.4 to 1 μm, good linearity response, large bandwidth, simple biasing requirements, and relatively low cost. Most fast photodiodes require internal amplification. This useful amplification of photocurrent is achieved in the avalanche photodiode. A basic p–n junction is operated under very high reverse bias. Carriers traversing the depletion region gain sufficient energy and enable further carriers to be excited across energy gap by impact excitation.

Optimum performance of the OCS is achieved by the proper alignment. The generic representation of an OCS is shown in Figure 2.9. It is composed of three major blocks: optical transmitter, fiber link, and optical receiver. An optical source, which is generally a LASER, signal is fed to a modulator's input and the second input in the form of electrical signal is fed to the modulator's other input. This input in case of optical modulator can be used to directly modulate the optical signal up to 5 Gbps. Several kinds of modulators are being used for communication purposes, the Mach–Zender modulator is the most common one. The modulated signal is transmitted through the fiber link, which can span over hundreds to thousands of kilometers. The fiber link, in addition to regenerative repeaters and optical amplifiers, also employs splitters to distribute or drop optical signal on any particular link. The splitters can be thought of as toll points at the main highway where the routing of the traffic is to be decided.

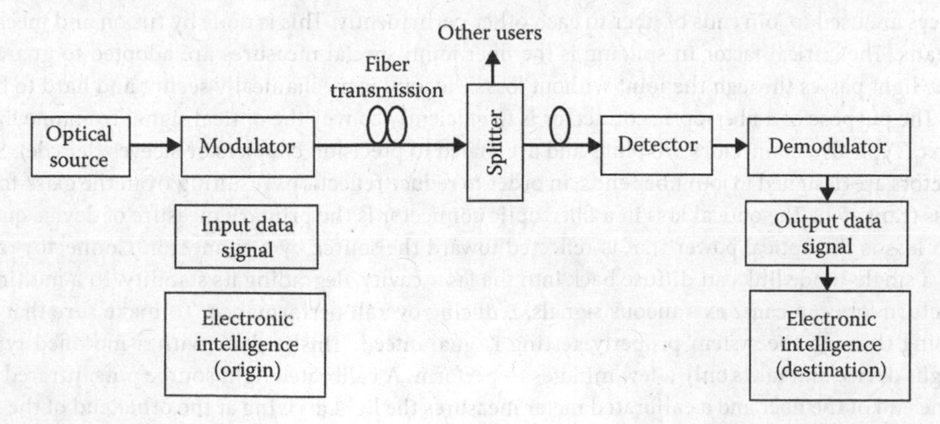

FIGURE 2.9 Generic representation of an OCS.

The optical detector detects the modulated optical signal and feeds that detected signal to a demodulator. The demodulator usually compares the incoming signal with threshold values and generates the electrical signal. This electrical signal is further modified to proper data sequence if necessary, and finally it is reproduced at the destination.

2.3.4 Multiplexing

Wavelength division multiplexing (WDM) technology uses multiple wavelengths to transmit information over a single fiber. Coarse WDM (CWDM) has wider channel spacing (20 nm) and is of low cost. Dense WDM (DWDM) has high capacity and dense channel spacing (0.8 nm) that allows simultaneous transmission of 16+ wavelengths. First WDM networks used just two wavelengths, 1310 and 1550 nm. Today's DWDM systems utilize 16, 32, 64, 128, or more wavelengths in the 1550 nm window. Each of these wavelengths provides an independent channel. The range of standardized channel grids includes 50, 100, 200, and 1000 GHz spacing. Wavelength spacing practically depends on laser line width and optical filter bandwidth. Each optical channel can carry any transmission format (different asynchronous bit rates, analog or digital). Wavelength is used as another dimension to time and space. Multiplexing is used throughout communications because it greatly increases the transmission capacity and reduces system costs.

2.3.5 Implementations and Standards

The fundamental requirements for making optical fibers sound deceptively simple. The material that is transparent can be drawn into thin fibers with a distinct core-cladding structure. It should be uniform along the length of the fibers and can survive in the desired working environment. Glass is the most common material used in optical fibers. Ordinary glass is a noncrystalline compound of silica. Many different variations of glass fibers have been developed. Simple glass-clad fibers are made by collapsing a low-index tube onto a higher-index rod. Refractive indices of most optical glasses are between 1.44 and 1.8. Impurities limit transmission of standard glasses. Fibers are drawn from the bottom of a hot preform. The preform has a physical make up identical to the final fiber, except that it is much wider and shorter. The preform is heated very precisely, and a thin strand of glass is pulled off on one end. The diameter of this strand is controlled very carefully through variances in heating and pulling tension. This strand is the optical fiber, containing both the core and the cladding. Once the fiber is pulled off the preform, it must be carefully and slowly cooled, covered with the final coating, and wound on to reels. All silica fibers are used for communications. Hard-clad silica fibers are used for illumination and beam delivery. Fibers made of fluoride and chalcogenide glasses transmit infrared wavelength, which silica fiber absorbs.

Splices are used to join ends of fiber to each other permanently. This is done by fusion and mechanical means. The critical factor in splicing is the fiber joint; special measures are adopted to guarantee that the light passes through the joint without loss. The joint is mechanically secure and hard to break easily. The purpose of a fiber-optic connector is to efficiently convey the optical signal from one link to the next. Typically, connectors are plugs and are mated to precision couplers or sleeves (female). Some connectors are designed to join fiber ends, in order to reduce reflections resulting from the glass-to-air-to-glass transition. The optical loss in a fiber-optic connector is the primary measure of device quality. Return loss is the optical power that is reflected toward the source by a connector. Connector return loss in a single-mode link can diffuse back into the laser cavity, degrading its stability. In a multimode link, return loss can cause extraneous signals, reducing overall performance. To make sure that light is passing through the system properly, testing is guaranteed. This is done with a modified type of flashlight device and takes only a few minutes to perform. A calibrated light source puts infrared light into one end of the fiber and a calibrated meter measures the light arriving at the other end of the fiber. The equipment used for testing is called an optical time domain reflectometer (OTDR). This device uses light backscattering to analyze fibers. OTDR takes a snapshot of the fiber's optical characteristics. It sends a high-powered pulse into the fiber and measures the light scattered back toward the instrument. OTDR can be used to locate fiber breaks, splices, and connectors, as well as to measure loss. In order to completely specify a fiber-optic cable, four primary performance categories must be quantified: installation specifications, environmental specifications, fiber specifications, and optical specifications.

2.4 Wireless Links

Although a wireless radio link has many advantages with respect to mobility, the list of technical hurdles to be tackled is impressive. The wireless link depends on many parameters like distance, reflectors, interferers, weather condition, and much more. In order to achieve a desired link quality, large safety margins are need to be used and suitable modulation formats have to be chosen. A link will never guarantee 100% perfect transmission of the information; higher BERs than for wired links have to be expected.

2.4.1 Physical Properties

2.4.1.1 Wavelength

Solving the wave equation (based on Maxwell's formulas) turns out that electromagnetic waves propagate with the speed $c = 1/\sqrt{\mu\varepsilon}$ with μ denoting the permeability and ε denoting the permittivity of the medium the wave is propagating in. For air, which is the typical transport medium, the free-space permeability and permittivity μ_0 and ε_0, respectively, can be used as a first approximation, resulting in a propagation speed of about the speed of light, $c \approx c_0$ (the error due to approximation is $<10^{-3}$). Hence, the wavelength in air can be written as

$$\lambda \approx \frac{c_0}{f} \tag{2.2}$$

with f denoting the radio wave's frequency.

Because of having no boundary conditions like in an electromagnetic waveguide, the propagation speed c reflects the phase velocity. As this is a constant value, no dispersion occurs.

2.4.1.2 Thermal Noise

The range of wireless communication systems is limited by several reasons; one of them being a too small signal and, therefore, a too low signal-to-noise ratio (SNR). The power spectral density of thermal noise of a resistor can be calculated by $\overline{V_n^2} = 4K_BTR$ with k_B denoting Boltzmann's constant, T denoting the temperature, and R is the resistor value. As antennas of wireless communication systems on earth are typically operated at or around room temperature, the noise power for a given bandwidth B can be expressed as $P_n = k_BTB$, or with the power in dBm by

$$P_{n(\text{dBm})} = -174 + 10\log(B). \tag{2.3}$$

The very commonly used dBm defines decibels relative to 1 mW, meaning 0 dBm = 1 mW. For example, a wireless system with bandwidth B = 2 MHz receives P_n = 7.9 fW (−111 dBm) noise power.

2.4.1.3 Channel Capacity

In order to determine the maximum data rate that can be achieved over a wireless link, the Shannon–Hartley theorem can be used. It gives an upper bound for the channel capacity for an error-free transmission rate of random data. The channel capacity C (bits/s) is given by

$$C = B\log_2(1+\text{SNR}) \tag{2.4}$$

where B is the channel bandwidth and the signal to noise ratio (SNR) = P_s/P_n (signal power P_s and noise power P_n).

2.4.1.4 Free-Space Path Loss, Fresnel Zone

For determination of the required transmit power levels and/or receiver sensitivities, it is important to know the attenuation of the air-link between transmit and receive antenna. In the most simple case of two antennas that are in line of sight (e.g., point-to-point radio systems), the so-called free-space path loss L_0 equals

$$L_0 = \left(\frac{4\pi d}{\lambda}\right)^2 \tag{2.5}$$

with d as the distance between the transmitter's and receiver's antenna. Note that the loss depends on the frequency: Doubling the transmission frequency increases the path loss by four.

Having a direct line of sight path between both antennas is not the only prerequisite for applying formula 2.5. Radio waves do not propagate only at a direct line from the transmitter to the receiver. They occupy a certain volume defined by the concentric Fresnel ellipsoids. By ensuring no or almost no obstacles (20% obstruction as a rule of thumb) to be within the first Fresnel ellipsoid/first Fresnel zone, the wireless link can be assumed not to be influenced (no additional attenuation). This is shown in Figure 2.10.

The first Fresnel zone's radius, F_1, at a given point can be calculated by

$$F_1 = \sqrt{\frac{\lambda d_1 d_2}{d_1 + d_2}}. \tag{2.6}$$

For example, if someone plans to run a point-to-point radio link at 2.45 GHz over a distance $d = d_1 + d_2 = 1$ km, the maximum cross-section radius r (see Figure 2.10) becomes $r = 5.5$ m. This means that the antennas need to be mounted 5.5 m higher than an obstacle in the middle of the radio link.

2.4.1.5 Antennas

Each wireless transmission needs antennas for reception and transmission. An antenna is characterized by lots of parameters like size, bandwidth, gain, return loss (matching to system impedance), and many

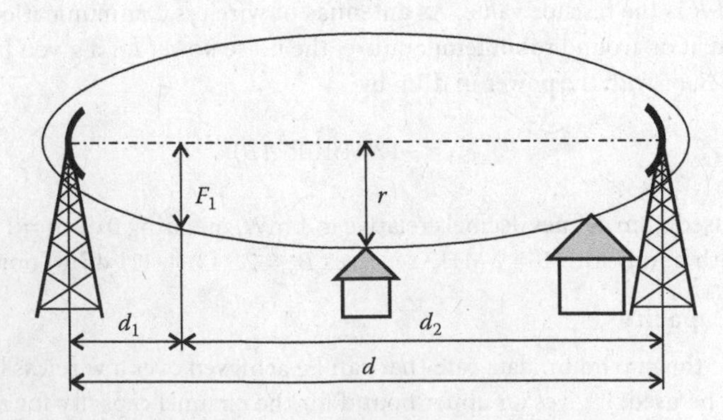

FIGURE 2.10 First Fresnel zone of a line of sight radio link.

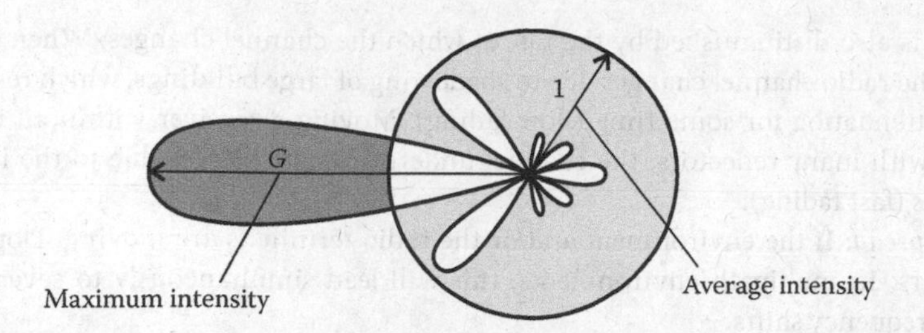

FIGURE 2.11 Antenna directional characteristics (2D equivalent).

others. For planning a radio link, the most important parameter is the antenna's gain: The gain is the ratio of the radiation intensity $P_{rad}(\phi, \vartheta)$ in the angular direction ϕ, ϑ to the antenna input power P_n:

$$G(\phi,\vartheta) = 4\pi \frac{P_{rad}(\phi,\vartheta)}{P_{in}} \quad (2.7)$$

If the atenna gain is specified without an angular dependency, the maximum value $G = \max(G(\phi, \vartheta))$ is meant. The gain G represents the maximum intensity compared to the average intensity (Figure 2.11).

A theoretical antenna that emits the entire input power provided uniformly into all angular directions, a so called isotropic antenna, has a gain of 1. The physically feasible Hertzian dipole or short dipole has a gain of 1.76 dBi (dB relative to isotropic antenna). It consists of two wires with a total length $\ll\lambda$ and has the feed point in the center of the antenna. Sometimes, also gain measures in dBd units are found: these refer to the short dipole. For example, 4.2 dBd = (4.2 + 1.76) dBi = 5.96 dBi.

2.4.1.6 Non–Line of Sight Channels and Moving Antennas/Objects

There are much more problems with generic wireless links. The most important impairments are as follows:

- *Interference*: The assumption that the range of a communication link is limited by thermal noise is only true when operating at an exclusive frequency that is not reused by someone else within reception range of a wireless receiver. It is much more typical that the link is limited by interference from other users. For example, a cellular phone system where the same frequency is reused some radio cells away can be mentioned. Another example is a license-free industrial, scientific, and medical (ISM) radio frequency band. Because everyone can operate radio equipment within those bands, the range of a wireless local area network (LAN) is often limited due to other access points or other devices operating at the same frequency: For 802.11b/g wireless LANs this could be a Bluetooth radio as well as a microwave oven.
- *Multipath, Delay Spread, Fading*: For the majority of wireless links, there is not just a single propagation path from the transmitter to the receiver. For a cellular system, e.g., the received signal might be the line of sight signal from the base station, plus reflections/diffractions from buildings, plus multiple reflections from parked cars, and so on. Hence, the received field strength is determined by the superposition of all paths at the point of the antenna. Due to the different phase shifts of the waves, this results in constructive and destructive interference as well, making the instantaneous received power level dependent on the receiver's position and the positions of all reflectors.

  The variation of the channel's attenuation over time is called fading. Depending on the transmission bandwidth B, different behavior occurs: For narrowband channels, fading can be represented by a time-varying gain and phase (flat fading). But in case of a broadband, transmission fading needs to be represented as a frequency-dependent function (frequency selective fading). Hence, the estimation of the channel and equalization might be required on the receiver side, making the wireless receiver much more complicated.

Fading is also distinguished by the rate at which the channel changes: When moving a radio receiver, the radio channel changes due to shadowing of large buildings, which results in a higher channel attenuation for some time (slow fading). Moving a receiver within an industrial environment with many reflectors, the channel undergoes fast changes due to the high number of multipaths (fast fading).

- *Doppler Spread*: If the environment and/or the radio terminals are moving, Doppler frequency shift occurs. In multipath environments, this will lead simultaneously to several positive and negative frequency shifts.

2.4.1.7 Link Budget

For planning a radio system, a link budget calculation needs to be done. This can be either used for determining a wireless link's range under a given transmit power or, in the other way round, it can be used for calculating the required transmit power for a certain link range. For a line of sight radio link, this can be done by (all values in decibels)

$$P_{rx} = \underbrace{P_{tx} + G_{tx} - L_{tx}}_{P_{EIRP,tx}} - L_0 - L_m + G_{rx} - L_{rx} \qquad (2.8)$$

with

the received and transmitted power P_{rx}, P_{tx} in dBm
the transmit and receive antenna gain G_{tx}, G_{rx} in dBi
the cable/connector losses on transmitter and receiver side L_{tx}, L_{rx} in dB
the free-space and miscellaneous losses L_0, L_m in dB

The effectively radiated power $P_{EIRP,tx}$ denotes the feed power of an isotropic antenna generating the same maximum field strength as the antenna used for link budget calculation. Therefore, almost all radio regulations limit $P_{EIRP,tx}$ instead of the transmitter power P_{tx} as it provides a fair comparison of maximum power densities.

The miscellaneous losses, L_m, are very important for the reliability of the link: They include additional losses due to changing weather conditions, component variations, imperfect alignment of the antennas to each other, and so on. In case of multipath environments, they also account for losses due to fading.

2.4.2 Types and Media Access

Because of the electromagnetic spectrum being a limited resource, access to frequencies is regulated by national authorities. Deploying a radio system, therefore, requires getting a local license. Typically, licensing a system with a wide bandwidth and high transmit power is not as easy as for a low-power narrowband system. The more bandwidth and transmit power is required, the more difficult it is to get a license and/or the more expensive the license will be.

Only a small fraction of the bandwidth used nowadays is available for unlicensed use. Furthermore, frequencies are not equivalent: Depending on the application, a higher or lower frequency is suitable.

2.4.2.1 Frequencies for Wireless Links

Covering all frequency bands in detail would go beyond the scope of this book. Generally, it can be said that lower frequencies allow longer ranges due to less free-space attenuation show better penetration into buildings, but need larger antennas, and allow only low data rates. In contrary, higher frequencies have higher attenuation, show less penetration-performance, but allow high data-rate applications and small/integrated antennas.

As an example, the three most popular license-free ISM frequency bands shall be listed:

- *433.05 ... 434.79 MHz* (*ITU-Region 1 only: Europe, Africa, Middle East*): Used for speech applications (baby-phones, two-way radios) and low-speed data communications (remotes); distances of 1 km and more can be achieved; good penetration of walls; high number of users.
- *2.4 ... 2.5 GHz* (*worldwide*): Mainly used for wireless LANs (IEEE 802.11b/g/n), Bluetooth radios, video transmitters, and also by microwave ovens; high channel bandwidths allow for high-speed data communications; typical ranges are below 200 m; poor penetration of walls; very high number of users.
- *5.725 ... 5.875 GHz* (*worldwide*): Mainly used for wireless LANs (IEEE 802.11a), high transmit powers allow compensation of higher free-space attenuation; high channel bandwidths allow for high-speed data communications; very poor wall penetration; still low number of users.

2.4.2.2 Media Access

While cellular systems access the wireless channel coordinated by a master (e.g., a base station), the opposite is true for ISM bands. As several systems coexist (typically not knowing from each other), a coordinated access is not possible. Also, for many ISM bands, a system is allowed to transmit a radio signal, even if the channel is in use by someone else. To prevent ISM bands from becoming useless due to too much interference, several requirements have to be fulfilled:

- *Maximum transmit power*: Limiting the transmit power results in a maximum range for the communication system and, thus, limits the distance where interference can occur.
- *Maximum bandwidth*: By limiting a system's radio bandwidth, a single system is not able to block the entire ISM band and, therefore, collisions are reduced.
- *Fixed channel list*: Defining a set of center frequencies reduces collisions.
- *Maximum duty cycle and maximum transmit time*: Both requirements reduce the time a system is allowed to use the radio channel. As a consequence, the probability for a collision reduces.
- *Listen before talk*: Allowing a system to use a channel only in case it is not in use gives a high reduction of collisions but increases access times and latency.

2.4.3 Modulation

Typical modulation formats of present communication systems are as follows (detailed information is given by [JP95] and [SH88]):

- *Gaussian minimum shift keying* (*GMSK*): This nonlinear modulation format has very poor spectral efficiency (bit rate per bandwidth) and can, therefore, be used for low data rates only. Typical applications are low power, low data-rate sensor transmitters. GMSK is also used by the cellular global system for mobile communication (GSM) system.
- *Single carrier pulse-amplitude modulation* (*PAM*) *with high bit rates*: Here, the carrier frequency is multiplied by a complex value representing the transmission symbol. This factor remains constant until the next symbol is transmitted. The complex values representing the symbols are spread across the complex plane but centered around the origin. Therefore, the resulting transmit signal is altered in magnitude and phase depending around the transmit symbol. Also, several bits are packed into a single symbol. This can be one bit/symbol for binary phase-shift keying (BPSK), two bits/symbol for quadrature phase-shift keying (QPSK), or even more bits/symbol for quadrature amplitude modulation (QAM) formats. A high data rate is achieved by using a large symbol alphabet (many bits per transmit symbol) and high symbol rates.
- *Orthogonal frequency division multiplex* (*OFDM*): Because classic PAM uses a large symbol alphabet (QAM-16, QAM-64) and high symbol rates, each bit is "spread" across the entire system's bandwidth. In case of a frequency selective channel (e.g., due to multipath), the channel response

needs to be equalized by the receiver that requires estimation of the channel. If the channel response has multiple zeros in the frequency domain, it might happen that equalization is not possible any longer, resulting in a high BER.

The OFDM approach is different to that: the single, high bit rate carrier is split into multiple lower bit rate carriers (multicarrier modulation). This has some benefits: the sub-channels have low bandwidth and flat fading can be assumed for each sub-channel (even if the entire bandwidth across all sub-channels shows frequency selective fading). Hence, now channel estimation/equalization is necessary. Furthermore, if the channel's frequency response has nulls on some sub-carriers, only those show bit errors. By appropriate coding, the overall BER can be kept low.

- *Ultra wideband* (*UWB*): This very new technology spreads the data bits over bandwidths in excess of 500 MHz. This can be done by the generation of extremely short pulses in the range of some 100 ps for example. This results in high immunity to multipath fading while having the potential to provide very high data rates (>500 MBit/s). UWB reuses frequency bands already occupied by other communication systems. Due to UWB's broadband nature, the spectral power density is very low; existing communication systems are hardly affected by UWB.

2.4.4 Bit Coding, Multiplexing

2.4.4.1 Bit Coding

Because of the relatively high BERs of a wireless channel (due to fading), much effort has to be spent on forward error correction. Typically, two different types of bit errors need to be considered:

- *Single bit errors*: Due to very small received signal strength the SNR decreases, leading to higher BERs. By using error correcting codes, it is possible to dramatically reduce the BER by sacrificing the data rate. The IEEE 802.11x standard uses coding rates down to 0.5, meaning each transmitted bit is accompanied by an error-correction bit and, thus, reducing the usable bit rate by 50%.
- *Burst errors*: Because of the time-varying fading properties of a wireless link, a complete loss of the reception often occurs for short time intervals. Because error-correction codes cannot deal with burst errors, interleaving will be applied. This means that the data bits are spread over time: If a short burst error occurs, the burst error is spread over the bit stream by the receiver's de-interleaver. As a consequence, the block of erroneous bits is spread into multiple single bit errors. Now, the error-correction code can be used for calculating the correct bits.

2.4.4.2 Multiplexing

Sharing a wireless frequency band can be done by one or a combination of the following three ways:

- *Frequency division multiple access* (*FDMA*): Each communication link uses its own frequency. Hence, no coordination between the different links is required.
- *Time division multiple access* (*TDMA*): Multiple users share one frequency on the basis of time slots. This access scheme requires coordination of all transmitters.
- *Code domain multiple access* (*CDMA*): Here, all users use the same frequency at the same time (continuously). To distinguish between multiple users, each transmitter encodes the data bits by a spreading sequence (direct sequence spread spectrum, DSSS). By using different spreading sequences with low cross-correlation, it is possible to separately extract a single transmitter's bits at the receiver side for each user.
- *Frequency hopping spread spectrum* (*FHSS*): By rapidly changing the center frequency in a pseudo-random sequence (known by the transmitter and receiver), the transmission is spread in bandwidth. This is not a real multiplexing technique as transmissions on the same frequency can occur but helps to reduce BER in multiuser applications.

TABLE 2.3 Overview of Different Communication Standards

Standard	$P_{EIRP,tx}$ (dBm)	Required P_{rx} (dBm)	Net Bit Rate (Mbit/s)	Modulation	Frequency (GHz)	Bandwidth (MHz)
FM radio	+80	−116	n.a.	FM	~0.1	0.2
GSM mobile	+33	−110[a]	0.384	GMSK	.../0.9/1.8/...	0.2
UMTS mobile	+33	−124[a]/−113[b]	14.4	W-CDMA	1.9...2.2	4.7
GPS	+46	−127.5[c]	0.00005	DSSS	1.575	2.0
Bluetooth 1.2	+20	−70	2.1	FHSS	ISM 2.45	80.0
IEEE 802.11a	+30	−82...−65[d]	54	OFDM	5.2...5.8	16.6
IEEE 802.11g	+20	−82...−65[d]	54	OFDM	ISM 2.45	16.6
IEEE 802.11p[e]	+44.8[f]	−85...−68[d]	27	OFDM	5.85...5.92	8.3

[a] For speech transmission (values for base stations, mobiles do not perform that good).
[b] For 384 kbit/s data transmission (values for base stations, mobiles do not perform that good).
[c] SNR = −16 dB (received power is 16 dB below thermal noise power).
[d] Depending on coding rate and sub-carrier modulation.
[e] Wireless access in vehicular environments (WAVE), e.g., car-to-car communication.
[f] +44.8 dBm EIRP is for the United States, additional limitations apply per FCC 47CFR [B8], 90.375.

The cellular GSM system uses a combination of FDMA and TDMA (with eight timeslots). On the contrary, UMTS is based on a combination of FDMA and CDMA.

2.4.5 Realizations and Standards

Table 2.3 presents a summary of several radio communication standards.

References

[JP95] J. G. Proakis, *Digital Communications,* McGraw-Hill, New York (1995).

[NS08] National Semiconductor, *LVDS Owner's Manual,* National Semiconductor, Santa Clara, California (2008).

[SH88] S. Haykin, *Digital Communications,* John Wiley & Sons, New York (1988).

[TI04] Texas Instruments, Application Report SLLA067A, Texas Instruments, Dallas, Texas (2004).

3

Media Access Methods

Herbert Haas
Vienna University of Technology

Manfred Lindner
Vienna University of Technology

3.1 Introduction 3-1
3.2 Full-Duplex Media Access 3-2
3.3 Synchronous Access Arbitration Concepts 3-2
Static Timeslot Mechanisms • Dynamic Timeslot Mechanisms
3.4 Statistic Access Arbitration Concepts 3-3
Aloha Mechanisms • Pure Aloha • Slotted Aloha
3.5 Carrier Sense Mechanisms with Exponential Backoff 3-4
Using Collision Detection • Using Collision Avoidance • Token Mechanisms • Polling Mechanisms
3.6 Other Media Access Issues 3-6
Access Fairness • Access Priorities • Quality of Service • Hidden Stations
References 3-7

3.1 Introduction

The technical details of media access are generally implemented in OSI layer 1 and 2. Usually the nature of OSI layer 1 determines the operation conditions and feasible access mechanisms that can be implemented in layer 2.

Media access on full-duplex media is often a simple task because each station can send data independently of each other. On the other hand, media access on half-duplex media is typically nontrivial and requires solutions for the following issues:

- Access arbitration
- Collision treatment
- Access fairness
- Access priorities for selected stations or messages
- Quality of service (QoS) requirements (queuing)
- Real-time requirements

Furthermore, most interesting for each access mechanism is the question about media utilization. Typically, there is a natural trade-off between real-time support and maximum media utilization, which might be solved only by increasing the complexity of the system (for example, by introducing a base station which is typically used in wireless data/voice communications).

The subsequent sections provide a detailed problem description, especially considering half-duplex media access in more detail, and explain possible solutions. Historical and current examples are also given.

3.2 Full-Duplex Media Access

If the communication medium supports full-duplex access, most communication problems are inherently solved. Full duplex can be achieved using the following methods:

- Separated wire pairs within cables, one for transmitter (TX), the other for receiver (RX). This is the most important solution in practical implementations (e.g., used in 100BaseT).
- Separated TX and RX fibers in optical communication environments.
- Separated frequencies, one frequency for TX, the other for RX.
- Separated wavelengths, one for TX, one for RX, within fibers. This method is typically more expensive than using separate fibers.
- Echo cancellation—both stations send and receive at the same time over the same wires. This scheme has been implemented, for example, on Gigabit Ethernet on twisted pair media (1000BaseT).
- Alternating TX and RX timeslots—this method is, for example, supported on ISDN (Basic Rate Interfaces, BRI) lines.

When full-duplex communication is supported, a station willing to transmit a frame does not need to consider the transmission states of the other station. Instead, the current station can immediately initiate the transmission. Any issues regarding QoS, blocking, or real-time requirements must be solved by the actual network node (e.g., an Ethernet switch) using queuing and scheduling to decide what should be forwarded first to the next link.

Consequently, half-duplex media deserve closer attention, especially because several modern systems such as wireless networks are typically half duplex. Furthermore, the half duplex mode is preferred in many real-time communication systems (e.g., Ethernet Powerlink) in order to avoid bridges which would introduce some additional latency.

3.3 Synchronous Access Arbitration Concepts

Access on half-duplex media requires algorithms that select exactly one station at each time for transmission. For a particular station, the problem arises *when* to send data. The algorithms can be divided in deterministic and statistical approaches.

Obviously, access arbitration is the central mechanism that must be tuned and controlled to achieve or improve fairness, precedence, QoS, and real-time.

3.3.1 Static Timeslot Mechanisms

A technically simple approach to handle half-duplex media access is to assign a dedicated timeslot of fixed size to each station. If there are N timeslots within a *periodic frame*, then the media can support up to N stations (however, usually $N - 1$ because one timeslot is often used for management purposes).

This mode is also known as time division multiple access (TDMA) and is generally best suited for isochronous services such as telephony, where the communication patterns can be described by *calls* or *circuits* (steady data rates, blocking, non-bursty). TDMA is implemented, for example, in GSM, UMTS, DECT, or TETRA.

In some cases (e.g., TETRA), multiple stations that wish to communicate with each other can statistically share a particular timeslot. The major disadvantage is that one station cannot utilize unused timeslots easily (without significant increase of the overall management complexity; an example is GSM). Therefore, TDMA is not a good choice to transmit bursty data traffic.

3.3.2 Dynamic Timeslot Mechanisms

In order to achieve better media utilization, TDMA can be implemented with dynamically negotiated timeslot sizes. Every additional station registering at the central base station will be assigned a dedicated timeslot at the expense of the timeslot sizes of other existing stations. That is, assuming a fair environment, the base station would create new timeslots by dividing the whole time period (typically called a superframe) by the total number of stations.

Alternatively, timeslot sizes may remain constant but the base station may assign a set of D timeslots for particular stations. That is, a single station may be assigned a single timeslot or D timeslots, assuming the superframe consists of $N > D$ timeslots.

This solution achieves a mix of both real-time behavior and good trunk utilization and is therefore common in modern wireless cellular network solutions, including WiMAX, GSM, and UMTS.

3.4 Statistic Access Arbitration Concepts

3.4.1 Aloha Mechanisms

The DARPA-founded Aloha media access protocol has been invented in 1971 to interconnect several minicomputers distributed over the islands of Hawaii via a single wireless channel [Abr70]. Although today Aloha is seldom used practically, it is fundamental to understand the Aloha mechanisms (especially the drawbacks) before studying modern concepts such as carrier sense multiple access (CSMA).

All classic Aloha methods described below used fixed frame sizes (cells) and did not support carrier sense (CS).

3.4.2 Pure Aloha

The basic idea of Aloha is implemented in *Pure Aloha*:

1. Every station may transmit whenever it wants
2. Upon collision with another station (no acknowledgment received on a dedicated broadcast channel), the involved station may retransmit the message after a random time

Due to this rather anarchistic access method, the collision probability is relatively high. Let T denote the fixed message length in seconds and R the rate of both messages and retransmissions. Assuming that the starts of messages and retransmissions can be described via a Poisson process, the probability of zero messages within time T is $\exp(-RT)$. However, a collision occurs if a particular message is sent at time t and another message appears within $[t - T, t + T]$. That is, a collision occurs if another (or more) message(s) appear within a time interval of $2T$ s. Therefore, the collision probability is $1 - \exp(-2TR)$. Hence the average number of retransmission per second is $R[1 - \exp(-2TR)]$ and adding the message rate r leads to $R = r + R(1 - \exp(-2TR))$, or equivalently, $r = R\exp(-2TR)$.

Multiplying both sides by T gives a relation between the channel utilization rT and total channel traffic. The channel utilization reaches a maximum value of $1/2e = 0.184$ when RT reaches 0.5. That is, the channel is saturated as soon as two messages per message duration time T are sent on average. At maximum, only 18.4% of the channel bandwidth can be utilized.

Since the message rate r is the product of the number of stations k times the average single-station message rate λ, the maximum number of stations is determined solely by λ and T, that is, $k_{max} = 1/2e\lambda T$.

3.4.3 Slotted Aloha

The throughput of Aloha can be doubled if all transmissions must take place within fixed timeslots of size T, which is usually equal to the frame size. That is, all stations maintain a synchronized clock and transmissions must start at the beginning of a timeslot. In this situation, collisions can only occur within a single timeslot T and not within $2T$ as in the case of Pure Aloha. Therefore, using the same considerations as above, the maximum channel utilization is 36.8%.

Today, Slotted Aloha is still used, for example, in RFID networks because the algorithm can be implemented on simple ASIC structures and the throughput requirements are relatively low.

3.5 Carrier Sense Mechanisms with Exponential Backoff

3.5.1 Using Collision Detection

The most important achievement was made by Robert Metcalfe in 1972 when he created Ethernet-based on Carrier Sense Multiple Access with Collision Detection (CSMA/CD). Actually, three important improvements [Cha00] were made compared to the Aloha principle:

1. Before starting a transmission, a station must sense for an existing carrier to avoid immediate collisions. This is the CS part of the protocol.
2. Even during a transmission, the transmitting station must sense for a collision. This is the CD part of the protocol. Ten Megabit IEEE 802.3 and Ethernet use Manchester coding with a negative DC level; during collisions, each station measures an increase of the average DC level. When the collision is detected, the sender must continue the transmission for the duration of 32 bits. This *jam* signal ensures that all involved senders back off for a random time.
3. If the network diameter is relatively large so that the signal propagation time t_p becomes significant compared to the transmission delay t_d, frames "detach" from the senders and collisions may appear somewhere on the cable.
4. To guarantee that all collisions are detectable, it is required that frames have a minimum size so that any collision signal can propagate back to the sender. In general, the dimensions of the network size and the minimum frame size must satisfy $2t_p \leq t_d$.
5. The random backoff time is actually the most important innovation. After detection of a collision each sender calculates a backoff time, which is the product of the *slot time* (a fundamental base time unit) times a randomly chosen number from the interval $[0 \ldots 2^k]$, with $k = \min\{\text{number of retransmission attempts}, 10\}$. The slot time is defined as the transmission delay of a 64 byte frame.

This *truncated binary exponential backoff* algorithm ensures that the probability of repeated collisions is exponentially reduced. The backoff is truncated by $k = 10$, so up to 1024 potential timeslots can be occupied by stations (theoretically).

These improvements enable Ethernet to reach throughputs close to the physical bandwidth, in particular 95% with small frames and close to 100% with large frames [Bog88]. It is important to understand that collisions in CSMA/CD-based network technologies are part of the arbitration process; the occurrence of collisions does not indicate a bad design.

3.5.2 Using Collision Avoidance

In certain environments, most importantly wireless media within the RF range, the concept of CD is not implementable because

1. The ratio between TX and RX powers is typically "dramatic" (e.g., 110 dBm), so that "listen while talk" would fully saturate (if not damage) the RX circuits.
2. Due to the long distances, the transmission time is often smaller than twice the signal propagation time; hence, collision signals (i.e., the sum of two or more frames) might not reach each sender.

Therefore, wireless technologies, such as IEEE 802.11 WLANs, utilize a collision avoidance (CA) method with the following properties:

- Collisions should be avoided if possible. Therefore, a sender willing to transmit a frame, must first listen whether the medium is non-occupied, and additionally start a random timer. Only after expiration of that timer, the frame can be sent. This *initial backoff* method reduces the collision probability if multiple stations wait for the meduim to become idle.

- Each frame must be acknowledged by the sender. CSMA/CA interprets the absence of acknowledgment frames as collision. Obviously, acknowledgment frames have a higher priority than data frames and should not suffer from the initial backoff.
- In order to give certain (management) frames precedence over data frames, data frames must not only await the initial backoff delay but also a constant inter-frame space (the DIFS in IEEE 802.11). Management frames, for example, acknowledgments, are sent immediately after the medium is idle and after a short interframe space (SIFS in IEEE 802.11, where SIFS < DIFS) is over.

In case of collisions, again a *truncated binary exponential backoff* algorithm ensures that the probability of repeated collisions is exponentially reduced.

Generally, the throughput of CSMA/CA transmissions is approximately 50% lower than the nominal bit rate because each packet must await an acknowledgment.

3.5.3 Token Mechanisms

Statistic but collision-free medium access methods can be implemented via a token mechanism. The principle is simple: A single "token" (e.g., a bit flag) is handed over from one station to the other, allowing each station that possesses the token to initiate a data transmission. After a transmission, the station must forward the token to the next station.

Logically, the network must follow a ring topology because the single token must "rotate" through all stations. This method has been implemented in IEEE 802.4 (token bus, the physical medium is a bus) or IEEE 802.5 (token ring, also the physical medium is a ring).

Obviously token-based access arbitration provides statistic access to the medium because idle stations simply forward the token without any frame being sent. However, note that the arbitration itself is deterministic, i.e., the order is foreseeable.

In general, token mechanisms prevent *channel occupation*, i.e., other than with CSMA/CD-based networks, it is not possible that particular stations are preferred by a random backoff algorithm during specific network conditions.

Moreover, the rotating token provides each station periodic transmission opportunities with guaranteed maximum interval duration. This maximum token rotation time (MTRT) depends on the number of stations and the maximum transmission delay per station.

Since the token is always forwarded deterministically and the MTRT is a known constant, token-based access arbitration is well suited in real-time environments. Today, token mechanisms are mainly used in field bus systems, such as Profibus (IEC 61158/IEC 61784).

3.5.4 Polling Mechanisms

Polling mechanisms logically resemble token passing; the only difference is that a dedicated central polling station is required (while token mechanisms do not need dedicated stations, except for token monitoring and reestablishment tasks).

Instead of a token that rotates through all stations, a central polling instance polls each station whether it wants to transmit data. If the station does not want to transmit, the central station receives some sort of negative acknowledgment and immediately polls the next station.

The polling arbitration mechanism has the following distinguishing properties:

- There is no token that can be lost or that needs to be monitored and maintained. Therefore, polling methods are usually simpler to implement than token schemes.
- This simplicity allows for scalability: The central polling instance may not only poll normal stations but also other polling instances that are responsible to poll a subset of stations. This way, a hierarchical tree of polling realms can be implemented.
- The central polling instance is a single point of failure.

Polling-based access arbitration has been used with IEEE 802.12 and is still implemented in several real-time systems. The standard IEEE 802.11 defines an optional polling mode for real-time services though it is only seldom implemented because queue-based CSMA/CA is practically sufficient. Another more recent example is Ethernet POWERLINK, an Ethernet-based real-time protocol.

3.6 Other Media Access Issues

Although the problem of access arbitration is the most important issue, this short section shall highlight other issues that must be considered with each implementation.

3.6.1 Access Fairness

Regarding the issue of fairness, the two main questions are

- Can the arbitration mechanism prevent single stations from dominating the medium?
- Are there scenarios where single stations are not scheduled for a long duration ("starvation")?

Obviously, only statistic arbitration methods may result in unfair situations. One example for a fairness-enforcing mechanism is to introduce a *post-backoff* in CSMA/CA-based systems. This concept requires that every station that successfully occupied the channel must start a post-backoff timer after transmission, so that unsuccessful stations are luckier in the next arbitration cycle.

3.6.2 Access Priorities

In many applications, several stations or frames (especially control frames) must be given a higher access priority. If statistical access methods are used (e.g., CSMA), priorities can be implemented by tuning backoff times. In deterministic access schemes (e.g., Token) priority information (flags) might be signaled within the token. Obviously, TDM methods provide the same access priority to each station.

3.6.3 Quality of Service

Again, the problem of QoS is basically an issue with statistical access schemes because token or polling principles inherently provide constraints on delay. Obviously, QoS is not an issue with (synchronous) TDM techniques.

With statistical access schemes, prioritized queuing and consequently tuned backoff is the only method to implement QoS. For example, IEEE 802.11e defines up to eight queues per station (practically, WiFi MultiMedia only requires four queues) to differentiate higher or lower priority packets. Each queue can be considered as a virtual station with its own backoff constraints. That is, the truncated binary exponential backoff algorithm is implemented for each queue separately while the range of possible random numbers (the *contention window*, CW) has a different size in each queue, depending on the priority.

This QoS mechanism does *not* guarantee that high-priority frames are always preferred when low-priority queues are also filled; rather, this mechanism provides *statistical QoS*—which turned out to be sufficient for practical interactive voice and video services.

3.6.4 Hidden Stations

In wireless environments, some stations may interfere with sessions between other pairs of stations. This is especially an important issue when the disturbing stations cannot detect the peers of existing sessions.

For example, assume that station A communicates with station B. Another station C can detect A but not B. Therefore, CSMA/CA on C will not take care about B and station A will suffer from collisions caused by the simultaneous transmissions of B and C. In this example, stations B and C are "hidden" from each other.

One (optional) solution defined in IEEE 802.11 (WLAN) is to announce a transmission via a dedicated request-to-send (RTS) frame, which are confirmed by the receiver by a clear-to-send (CTS) frame. Both frames, RTS and CTS, carry an estimation of the total handshake duration. Hidden stations will receive either the RTS or the CTS frame and shall be silent during the announced handshake duration.

That is, the total transaction consists of one RTS, one CTS, one data, and one acknowledgment frame. Clearly, this four-way handshake decreases the throughput of plain CSMA/CA by approximately 50%; only 25% of the bandwidth remains for data throughput.

In reality, the RTS/CTS mechanism is typically only used in point-to-multipoint wireless bridging scenarios, where the concurrent non-headend bridges use directional antennas (pointing to the single head-end bridge) and therefore cannot see each other.

References

[Abr70] N. Abramson, The Aloha system—Another alternative for computer communications, SRMA Aerospace Research, Technical Report, April 1970.

[Bog88] D. R. Boggs, J. C. Mogul, and C. A. Kent, Measured capacity of an Ethernet: Myths and reality, in *Proceedings of the SIGCOMM'88 Symposium on Communications Architectures and Protocols*, Stanford, CA, pp. 222–234, ACM SIGCOMM, August 1988.

[Cha00] C. E. Spurgeon, Ethernet, in *The Definitive Guide*, O'Reilly, Sebastopol, CA, 2000.

4

Routing in Wireless Networks

4.1 Introduction 4-1
4.2 Routing Protocols and Classification 4-4
4.3 Routing Protocol Families for Ad Hoc Networks 4-5
Proactive Routing Protocols • Reactive Routing Protocols
4.4 Routing Protocol Families for Wireless Sensor Networks 4-6
Flat Routing Protocols • Hierarchical Routing Protocols • Location-Based Routing Protocols
4.5 Summary of the Main Routing Protocols in Wireless Networks 4-7
Optimized Link-State Routing Protocol • Topology Dissemination Based on Reverse Path Forwarding • Dynamic Source Routing Protocol • Ad Hoc On-Demand Distance Vector Routing Protocol • Dynamic MANET On-Demand Routing Protocol • Sensor Protocols for Information via Negotiation • Low Energy Adaptive Clustering Hierarchy • Geographic Adaptive Fidelity
4.6 Conclusions 4-13
Acknowledgment 4-14
Abbreviations 4-14
References 4-14

Teresa Albero-Albero
Universidad Politécnica de Valencia

Víctor-M. Sempere-Payá
Universidad Politécnica de Valencia

4.1 Introduction

Wireless networks allow communication between devices over a wireless medium using electromagnetic waves. They offer flexibility and speed when setting up the network, need less maintenance than cabled networks, and allow mobility. There are different ways of classifying these networks, the most common of which is based on their purpose and radio range. Depending on these aspects, wireless networks can be separated into four groups: wireless personal area networks (WPAN), wireless local area networks (WLAN), wireless metropolitan area networks (WMAN), and wireless wide area networks (WWAN), see Figure 4.1.

- *Wireless personal area networks* interconnect devices in a small area. The standard used is IEEE 802.15.1 based on the Bluetooth specifications. The IEEE 802.15.4 standard [WMAC03], approved in 2004 and promoted by the ZigBee Alliance, has been developed to enable applications with relaxed bandwidth and delay requirements, where the emphasis is device battery lifetime maximization. These applications will be run on platforms such as sensors.
- *Wireless local area networks* are technologies based on High Performance Radio LAN (HiperLAN), a group of the standard group European Telecommunications Standards Institute (ETSI), and Wi-Fi standardized under IEEE 802.11 series.

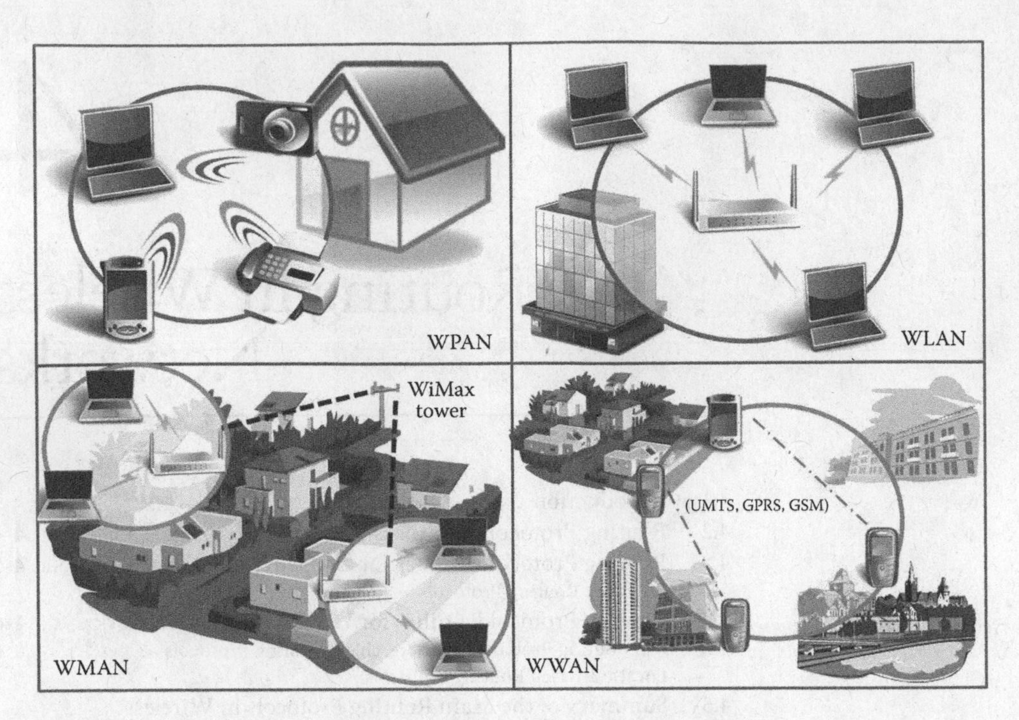

FIGURE 4.1 Wireless network classification (WPAN, WLAN, WMAN, WWAN).

- *Wireless metropolitan area networks* are a type of network that connects several Wireless LANs. WiMAX is the term used to refer to WMAN and is covered in IEEE 802.16d/802.16e.
- *Wireless wide area networks* with technologies such as universal mobile telecommunication system (UMTS), general packet radio service (GPRS), and global system for mobile communication (GSM).

To further increase radio coverage for clients and facilitate ease of movement, wireless mesh networks (WMN) are used. WMNs span all segments and are a completely flat and nonhierarchical organization formed to cover all areas, and for this reason it is outside the previous classification. A WMN has been defined as a network composed of mesh routers and mesh clients. Mesh routers have minimal mobility and form the backbone of the network, and the mesh clients may be stationary or mobile, and can form a client mesh network among themselves and with mesh routers. In this scheme, each node operates not only as a host but also as a router, forwarding packets on behalf of other nodes that may not be within direct wireless transmission range of their destinations [AWW05]. In mesh networks, full physical layer connectivity is not required; a mesh network employs one of two connection arrangements, full mesh topology or partial mesh topology. As long as a node is connected to at least one other node in a mesh network, it will have full connectivity to the entire network because each mesh node forwards packets to other nodes in the network as required.

There are many different types of mesh networks. Mesh networks can be wired or wireless. For wireless networks, there are ad hoc mobile mesh networks or wireless sensor networks,* and permanent infrastructure mesh networks. Regarding radio transmission, there is another classification: single radio mesh networks, dual-radio mesh networks, and multi-radio mesh networks.

WMNs can widely find applications in WLANs, WMANs, WPANs, and wireless sensor networks (WSNs). Major industrial organizations are actively working on introducing multi-hop mesh elements in their next generation standards. For example, IEEE has created a working group to define how the mesh works in 802.11 networks (WLAN). 802.11s [AESSM] networks also include two routing

* In ad hoc networks and sensor networks every node functions as a router.

mechanisms, through an Ad hoc On-Demand Distance Vector (AODV) and Optimized Link-State Routing Protocol (OLSR) hybrid solution [MZKD04]. The standard 802.16 (WMAN) supports a mesh form of working [LMAN04], although it is not compatible with the standard IEEE 802.16e for wireless metropolitan networks. The IEEE 802.15 standard defines the physical layer and media access control (MAC) of the WPANs and the working group IEEE 802.15.5 [TG5] is studying how to establish a mesh architecture in this type of network. Figure 4.2 shows how the four types of wireless networks (WPAN, WLAN, WMAN, WWAN) and the WMNs live side by side. In the case of WMNs, mesh routers are distinguished from mesh clients.

If we use the term WMN in a wide sense to denote any kind of multi-hop network, regardless of the wireless technology used or the hardware features of the related devices, an ad hoc network can be understood as a subset of WMNs, since no infrastructure exists in an ad hoc network [AWW05]. A mobile ad hoc network (MANET) is a collection of wireless mobile hosts forming a temporary network without the aid of any centralized administration or standard support services. In such an environment, it may be necessary for one mobile host to enlist the aid of others in forwarding a packet to its destination due to the limited propagation range of each mobile host's wireless transmissions. Routing in a MANET is challenging because of the dynamic topology and the lack of an existing fixed infrastructure. For this reason, conventional routing protocols are not useful in ad hoc networks. In this chapter, there is a brief description of the routing protocols in ad hoc networks, which are currently undergoing a standardization process.

WSNs can be understood as a subset of WMNs. These share similarities with ad hoc wireless networks. The dominant communication method in both is multi-hop networking, but several important distinctions can be drawn between the two. These differences result from both the technological structure of sensor nodes and from the intended application scenarios. Ad hoc networks typically support routing between any pair of nodes, whereas sensor networks have a more specialized communication pattern [AY05]; for this reason, new protocols have been designed to be used in WSNs. Three of these protocols are introduced in this chapter.

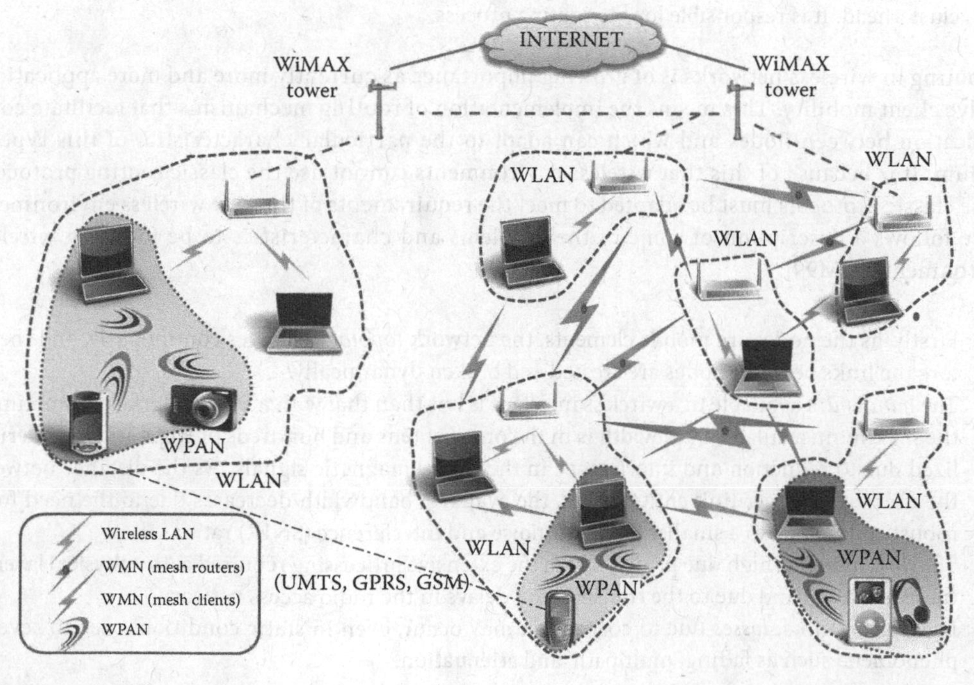

FIGURE 4.2 Wireless networks.

4.2 Routing Protocols and Classification

The purpose of routing protocols is to determine the best path from source to destination. The routing protocols have to enable routers to communicate with each other exchanging topology and state information. There is a diverse range of routing protocols, which may be classified according to several different characteristics.

Depending on the type of routing algorithms used, routing protocols can be divided into three families: distance vector, link state, and source routing:

- Protocols that use the *distance vector* technique maintain a table for communication and employ diffusion (not flooding) for information exchanged between their neighbors.
- *Link state* technique maintains a table with the entire topology network. The topology is built by finding the shortest path in terms of link cost. This cost is periodically exchanged among all the nodes through a flooding technique. Each node updates its routing table using the link cost information. This technique can cause loops in networks that change the topology quickly [M95].
- In the *source routing* technique, all the data packets have the routing information on the headers. The source node makes the route decision. With this technique, loops can be avoided, but the protocol overhead is quite significant. This technique is not efficient for fast moving topologies due to route invalidation around the path of a packet.

However, there are other classifications that take into account other characteristics.

- *Dynamic* or *static*. A dynamic or adaptative routing protocol changes its behavior according to the network state, while a static one does not.
- *Centralized* or *distributed*. Dynamic routing protocols can also be classified as centralized or distributed. The classification is based on which node makes the routing decisions. In a distributed routing protocol, all nodes are responsible for making their own routing decisions. In a centralized routing protocol, the decisions take place at a central node. One example of centralized routing protocols is the cluster-based routing protocols, where the central node is the so-called clusterhead. It is responsible for the routing process.

Routing in wireless networks is of growing importance as currently more and more applications involve client mobility. This means the implementation of routing mechanisms that facilitate communication between nodes and which can adapt to the particular characteristics of this type of medium. It is because of this that wireless environments cannot use the classic routing protocols. These classic protocols must be adapted to meet the requirements of the new wireless environment. There follows a description of some of the problems and characteristics to be found in wireless environments [CM99]:

- Firstly, as the nodes are mobile elements, the network *topology* changes continuously, and therefore the links between nodes are created and broken dynamically.
- The *bandwidth* available in a wireless interface is less than that with a wired interface. Sometimes, the maximum available bandwidth is in the order of tens and hundreds of Kbps and is underutilized due to reduction and interference in the electromagnetic signals. As the distance between the user and its peer link entity grows, the available bandwidth decreases due to the need for a robust coding due to a smaller signal-to-noise and interference (SNIR) ratio.
- The *link latency* is high due principally to the extensive processing required at the physical layer of these networks and due to the transmission delays in the radio access network.
- In wireless links, *losses* due to corruption may occur, even in static conditions, due to several phenomena such as fading, multipath, and attenuation.

- On some occasions, a mobile user may experience a *link outage* in some situations, such as driving through a tunnel or being in an elevator. During these periods, the user has no radio connectivity, and if such periods have a duration greater than application timers, ongoing communications may be aborted.
- *Delay jitter* may be caused by a number of reasons when using wireless technologies.
- Some or all of the nodes will be powered by batteries, which means that *energy* saving is an important factor when dealing with this type of network. As energy consumption is directly proportional to the distance between hosts (proportional to radius squared), single-hop transmissions between two hosts require a lot of power, causing interference with other hosts. To avoid this problem in routing, two hosts can use a multi-hop transmission to communicate with each other via other nodes in the network.
- Problems related to *security* are accentuated in wireless networks. As it is a shared environment to which anybody can have access, confidentiality of data is an important consideration.
- When we work with wireless networks, we must consider some problems such as *hidden terminals*, which occur when there are two nodes within communication range of a third node but not of each other. Both may try to communicate with the third node simultaneously and might not detect any interference in the wireless medium. Thus, the signals collide at the third node, which will not be able to receive the transmissions from either node [P99]. The typical solution is that the terminals coordinate their transmissions, by means of request to send/clear to send (RTS/CTS).
- Supposing node B communicates with node A, and node C wants to transmit a packet to node D. During the transmission between node B and node A, node C perceives the channel as busy. Node C falsely concludes that it may not send to node D, even though both transmissions would succeed. Bad reception would only occur in the zone between node B and node C, where neither of the receivers are located. That is the problem of the *exposed terminals*, and this together with the hidden terminal cause significant reduction of network throughput when the traffic load is high.

4.3 Routing Protocol Families for Ad Hoc Networks

MANET routing protocols are traditionally divided into three main categories [AWD04] depending on the type of information being exchanged and the frequency with which it is done: proactive or global protocols, reactive or on-demand protocols, and hybrid protocols. The first group aims to maintain up-to-date routing information in the nodes through periodic control message exchange, while the second attempts to find routes on demand. There are also hybrid routing protocols, which combine features from both proactive and reactive approaches.

4.3.1 Proactive Routing Protocols

In proactive routing protocols, each node keeps the routing tables constantly up-to-date. The routing tables contain the routing information for every other node in the network, which is updated periodically and when there are changes in the topology of the network. The routing information can be kept in different routing tables depending on the routing protocol. Proactive techniques typically use algorithms such as distance vector and link state.

This type of protocol will operate in networks in which it is necessary for the route discovery procedure to not have excessive latency, and this type of protocol is appropriate in terms of consumption of resources such as bandwidth and energy. In most of the proactive routing protocols, the control message overhead grows as $O(N^2)$, where N represents the number of nodes in the network.

The difference between these protocols is in the manner that the routing information is updated, detected, and the type of information that is stored in each routing table.

4.3.2 Reactive Routing Protocols

Reactive or on-demand* routing protocols were designed to reduce the overhead of the proactive routing protocols; they maintain information on active routes only. When one node wants to communicate with another node, one route to the destination is demanded. Although network resources such as energy and bandwidth are used more efficiently than in the proactive protocols, the delay in the routing discovery procedure increases. The route discovery procedure is usually made by flooding a route request packet through the network. If a node with a route to the destination (or the destination itself) receives the route request packet, it sends back a route reply packet to the source node using the link reversal if the packet has traveled through bidirectional links. The route replay packet contains the desired route. The routing algorithms found among the reactive routing protocols are distance vector and source routing.

Reactive routing protocols are divided into two groups, source-based and hop-by-hop or point-to-point.

- In source-based on-demand protocols, each data packet transports in its header the complete source to the destination, that is, the information of every neighboring node from the source to the destination. Every intermediate packet consults the header packet to know where to forward it to. Therefore, there is no need for the intermediate nodes to keep the routing information continuously up to date by means of forwarding routing messages periodically as in proactive routing protocols. In contrast, in large ad hoc networks, the probability of link failures occurring increases with the number of nodes. As the number of intermediate nodes increases, the size of the header packet increases too. This kind of protocol is not recommended in large networks with a lot of hops and high mobility because they do not scale well.
- In the hop-by-hop routing protocols, the packet only carries the destination address and the next hop address; in this way, every intermediate node that participates in the route would consult its routing table to decide which way to send the packet. The advantage is that every intermediate node keeps its routing table up to date continuously and independently, so that when a packet reaches an intermediate node, it can decide how to route it according to the current state of the network; the routes can therefore adapt to the dynamic topology of this kind of route more easily. The drawback is that every intermediate node along the route must constantly update its routing table by means of beaconing messages.

The route discovery overhead can grow $O(N + M)$ if a link reversal is possible, where N represents the number of nodes in the network and M the number of nodes in the reply path; for unidirectional links, the discovery overhead is $O(2N)$.

4.3.2.1 Hybrid Routing Protocols

Hybrid routing protocols are a new generation of protocols that use the characteristics of reactive and proactive protocols. These are designed to increase scalability and reduce the overload involved in route discovery. This is possible by means of proactive route maintenance for the closer nodes and a determination of routes for the more distant nodes by means of a route discovery strategy.

Most of the hybrid protocols are zone based, that is, the network is divided into a number of zones, each containing one node that is selected to be in charge of all the mobile nodes within their transmission range, or a subset of them. Other hybrid routing protocols group the nodes into trees or clusters.

4.4 Routing Protocol Families for Wireless Sensor Networks

Routing in sensor networks is very challenging due to a number of characteristics that distinguish them from contemporary communication and wireless ad hoc networks.

* Note that the terms reactive and on demand are used interchangeably.

Firstly, due to the relatively large number of sensor nodes, it is not possible to build a global addressing scheme for the deployment of a large number of sensor nodes as the overhead of ID maintenance is high. Therefore, classic IP-based protocols cannot be applied to sensor networks.

Secondly, in contrast to typical communication networks, almost all applications of sensor networks require the flow of data from multiple sources to a particular base station.

Thirdly, sensor nodes are tightly constrained in terms of energy, processing, and storage capacities and this must be considered when designing a routing protocol. The next consideration is that in most application scenarios, nodes in WSNs are generally stationary after deployment except for a few mobile nodes. The next characteristic is that sensor networks are application specific (the requirements change according to the application). Position awareness of sensor nodes is also important since data collection is normally based on the location. Finally, data collected by many sensors in WSNs is typically based on common phenomena, so there is a high probability that this data has some redundancy. This characteristic must be used by the routing protocols to improve energy and bandwidth utilization [AK04].

Due to such differences, many new algorithms have been proposed for the problem of routing data in sensor networks. The three main categories are flat, hierarchical, and location based, although there are several others based on network flow or quality of service (QoS) awareness.

4.4.1 Flat Routing Protocols

In flat networks, each node normally plays the same role and sensor nodes collaborate to perform the sensing task. Due to the large number of nodes, it is not feasible to assign a global identifier to each node. This consideration has led to data-centric routing, where the base station sends queries to certain regions and waits for data from the sensors located in the selected regions. Since data are being requested through queries, attribute-based naming is necessary to specify the properties of data.

4.4.2 Hierarchical Routing Protocols

The hierarchical routing protocols use methods originally proposed in wireline networks. These are well-known techniques with special advantages related to scalability and efficient communication. The concept of hierarchical or cluster-based routing is also utilized to perform energy-efficient routing in WSNs. In this type of architecture, higher-energy nodes are used to process and send the information, while low-energy nodes can be used to perform the sensing in the proximity of the target. The creation of clusters and assigning special tasks to cluster heads can greatly contribute to overall system scalability, lifetime, and energy efficiency.

4.4.3 Location-Based Routing Protocols

Sensor nodes that use this routing protocol are addressed by means of their location. The distance between neighboring nodes is estimated according to signal strengths. Coordinates of the neighboring nodes can be obtained by exchanging information between neighbors. Some protocols use GPS to find the location of neighboring nodes. In order to save energy in some location-based schemes, the nodes that are not active go into sleep mode.

4.5 Summary of the Main Routing Protocols in Wireless Networks

The most common MANET routing protocols are described in this chapter. These are promoted by the IETF MANET Working Group by publishing them as experimental RFC, and are OLSR, Topology dissemination Based on Reverse Path Forwarding (TBRPF), AODV, and Dynamic Source Routing (DSR).

After this, a new one is explained—Dynamic MANET On-Demand (DYMO), which has been published as Standard Track RFC.

Along similar lines, three more routing protocols used in WSNs are explained. These are Sensor Protocols for Information via Negotiation (SPIN), which are classified as flat routing protocols, and Low Energy Adaptive Clustering Hierarchy (LEACH) protocol, and Geographic Adaptive Fidelity (GAF), which is a location-based routing protocol.

4.5.1 Optimized Link-State Routing Protocol

OLSR [JMC01] is a proactive routing protocol based on the link state algorithm. It is in experimental state in the IETF Working Group [CJ03].

Each node maintains a route to the rest of the nodes in the ad hoc network. The nodes forming the network exchange messages about the link state periodically, but they use the multipoint relaying (MPR) strategy to minimize both, the routing message size and the number of nodes that resend the routing messages in broadcast mode. In the MPR [BCGS04] strategy, each node uses "Hello" messages to find out which nodes are placed at one hop distance and a list is created, see Figure 4.3.

Each node selects in that list a subset of neighbor nodes able to reach every node at a distance of two hops from the node that is making the selection. Only these selected neighbor nodes will be used to retransmit the routing messages, and these are called multipoint relays. The rest of the neighboring nodes will process the routing messages they receive, but they will be unable to retransmit them. Each node determines the best route (in number of hops) for each destination, using the information stored (in the routing table of the topology and in that of their neighbors) [AWD04], and stores that information in a routing table, so that it is available at the precise moment when a node wants to start sending data. This protocol selects bidirectional links to send messages [ChCL03], dispensing with unidirectional links. OLSR is well suited to large, dense mobile networks. Because of the use of MPRs, larger and denser networks link state routing offers optimum performance.

4.5.2 Topology Dissemination Based on Reverse Path Forwarding

Topology Dissemination Based on Reverse-Path Forwarding [OTL04] is a proactive, link-state routing protocol designed for mobile ad hoc networks, which provides hop-by-hop routing along shortest paths to each destination. Each node running TBRPF computes a source tree based on partial topology information stored in its topology table, using a modification of Dijkstra's algorithm. To minimize overhead, each node reports only part of its source tree to neighbors. TBRPF uses a combination of periodic and differential updates to keep all neighbors informed of the reported part of its source tree. Each node also has the option to report additional topology information, to provide improved robustness

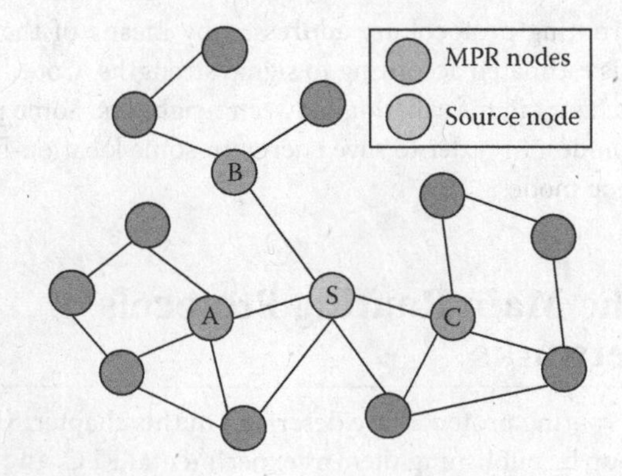

FIGURE 4.3 OLSR multipoint relay selection.

in highly mobile networks. TBRPF performs neighbor discovery using "differential" HELLO messages that report only changes in the status of neighbors. This results in HELLO messages that are much smaller than those of other link-state routing protocols.

4.5.3 Dynamic Source Routing Protocol

This protocol was created in 1996 [JM96,JM01], and like other protocols, it has been widely studied and also modified [DRL03]. Currently, its state is experimental [JHM07], like that of OLSR. DSR is a reactive routing protocol. An important characteristic to consider is that this protocol requires that each message sent contains the complete address from the source node to the destination node (source based). When a node needs to know the path to the destination, it sends a Route Request (RREQ) message and the route discovery procedure is initiated. Each node adds its own identifier when the RREQ is resent, and in this way, the route record is created.

To detect RREQ duplicates, each node in the network maintains a list with the pair [source address, request_id]. When a node receives a RREQ message, it follows the following steps: (1) If the pair [source address, request_id] for this route discovery are in the recent search list of this node, it dismisses it and does not process the RREQ message. (2) If the address of this node is in the route record of the RREQ message, it dismisses it and does not process it. (3) If the search objective is the node address, the route record already contains the route from source to destination and the route reply (RREP) message is sent. (4) Otherwise, this node adds its own address in the route record of the RREQ message and resends the search.

If the node that receives the RREQ is the destination or it contains information about the destination, it replies with a RREP message. The RREP message is sent through the same path that has been discovered, and this same message holds the route from the source node to the destination. If the link is not bidirectional, a new route discovery is carried out to send the RREP message. When the source node receives the RREP, it stores the route from source to destination included in the RREP in the route cache. While RREQ and RREP messages pass through intermediate nodes, they store the information in their route cache. While waiting for the route discovery to finish, the node can continue sending and receiving messages to/from other nodes. When the source node receives the RREP message, it can send data packets to the destination node, and the header of those packets includes the route they have to follow. This way, intermediate nodes use the route included in the packet to decide which node they have to send the packet to. When the packet reaches its destination, it is delivered to the network layer.

Each node has a route cache where the path to a destination node is stored. Each entry in the route cache has an associated lifetime after which the entry is deleted from the cache. The use of this route cache in each node offers two advantages: Firstly, more speed in the route discovery—if a node that receives a RREQ for a path to destination that is already known (because it already has it in its cache), it can reply with a RREP, using the local routing cache. Secondly, the transmission of RREQ is reduced, because if the path to the destination is already known because a node has it stored in its route cache, there is no need to continue transmitting RREQ messages. If a route breaks, the source node can consider another route that it has in its cache for the same address. If it does not have a route stored, it starts route discovery again. Figure 4.4 shows an example of the route cache stored in the nodes. However, using the route cache can cause two problems. When two nodes receive a RREQ at the same time and both of them reply based on the routes stored in their caches, sending their responses at the same time, this can cause message collisions and network congestion. However, simultaneous replies can be avoided by adding a delay before replying with the information in the cache. The second problem is that formation of a loop may occur in a route sent in a RREP message. Therefore, to solve this problem, if a node receives a RREQ and is not the objective of the search but can reply with information in its cache, the node dismisses the reply if the route contains a loop. The node will reply only with its cache with a route in which it appears at the end of the route stored in the RREQ message and at the beginning of the path obtained in the route cache.

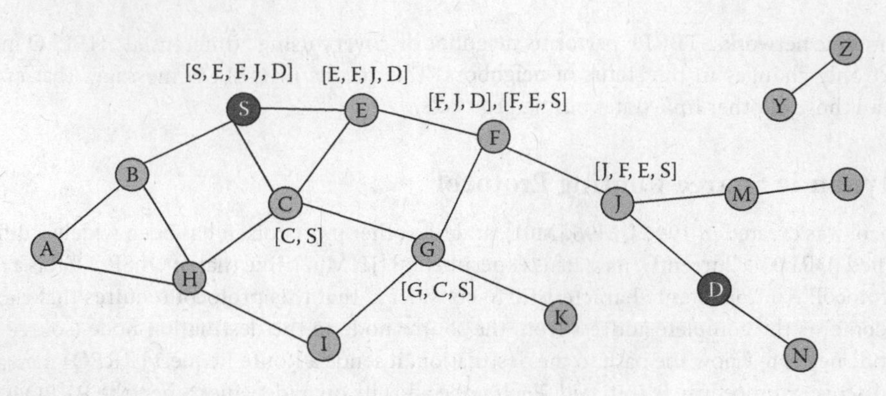

FIGURE 4.4 Example of the route cache in the DSR protocol.

4.5.4 Ad Hoc On-Demand Distance Vector Routing Protocol

The AODV protocol is built on the destination-sequenced distance-vector (DSDV) protocol. The improvement here is to minimize the number of transmissions required to create routes, because as they are on demand, the nodes that are not placed in the chosen path do not have to maintain the route or participate in table interchanges. Several studies on this protocol have been carried out [PR99,RP99,RP00], and it has been modified [SWL03] several times. Currently, it is a protocol in experimental state [PBD03].

When a node wants to transmit to a destination and it does not have a valid route, it must start the path discovery process. To do this, in the first place, it broadcasts a RREQ message to its neighbors, which in turn sends it to their own neighbors and so on, until it gets to the destination or to an intermediate node that has a route to the destination. Like DSDV, it uses sequence numbers to identify the most recent routes and delete loops. Each node maintains two counters: the node's sequence number (to avoid loops) and Broadcast_ID that is increased when a transmission is initiated in the node.

To identify a single RREQ, it uses the Broadcast_ID and the IP address of the source node. The RREQ contains the following fields: source_addr, number_sequence_#, broadcast_id, dest_addr, dest_sequence_#, and hop_cnt. Intermediate nodes only reply to RREQ messages if they have a route to the destination with a sequence number greater or equal to the one stored in the RREQ, in other words, only if they have equal routes (in age) or more recent. While the RREQ is sent, intermediate nodes increase the "hop_cnt" field, and they also register in their routing table the address of the neighbor from which they first received the message, to establish the Reverse Path, where RREQs are still sent, and at the same time, the Reverse Path is being established. The copies of the same RREQ received after that coming from other neighbors are dismissed.

Once the "destination node/intermediate node with recent route" has been found, it replies with a unicast packet known as RREP to the neighbor from which it received the first RREQ. The RREP uses the links that had been established before as reverse path. The RREP contains the following fields: source_addr, dest_addr, dest_sequence_#, hop_cnt, and lifetime or expiration time for the Reverse Path. The RREP uses the reverse path established to the source node. In its route, all the nodes the RREP passes through write the Reverse Path as the most recent route to the destination node. From that we can conclude that the AODV holds only bidirectional links. In Figure 4.5, the AODV route discovery procedure is shown.

If a source node moves, it is able to restart the discovery protocol to find a new route to the destination. If an intermediate node moves, its previous neighbor (in the source-destination direction) broadcasts (until the source node is reached) a not-requested RREP with a "recent" sequence number (in other words, greater than the sequence number known) and with a number of hops to the destination equal to infinite. That way, the source node restarts the route discovery process in the case that it still needs the route.

When the link breakage happens, the node must invalidate the existing route in the routing table entry. The node must list the affected destinations and determine which neighbors can be affected by

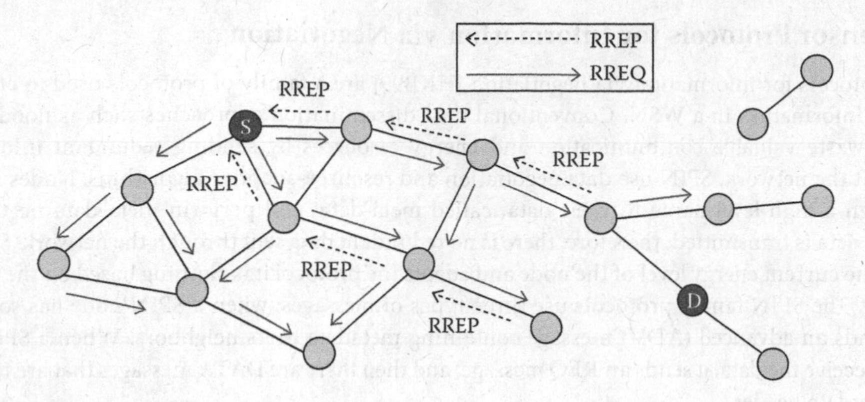

FIGURE 4.5 ADOV route discovery procedure.

this breakage. Finally, the node must send the route error (RERR) message to the corresponding neighbors. The RERR message can be broadcasted if there are many neighbors that need that information or unicasted if there is only one neighbor. The sequence numbers for the entries in the routing table for the unreachable destinations must be set to invalid lifetime. When the link breakage happens, then the host can try to locally repair the link if the destination is no further than a specified amount of hops. If the RREP message is not received, then it changes the routing table status for the entry to "invalid." If the host receives the RREP message, then the hop count measurement is compared. If the hop measurement from the message is greater than the previous one, then the RERR with the *N* field that is set up is broadcasted. The *N* field in the RERR message indicates that the host has been locally repaired, and the link and the entry in the table should not be deleted.

An additional feature is the use of "Hello" messages (periodical broadcast) to inform a mobile node about every neighboring node. These are a special kind of not-requested RREPs, whose sequence number is equal to the last RREP sent (the sequence number is not increased) and that have a time-to-live (TTL) = 1 to not flood the net. They can be used for the network route maintenance.

4.5.5 Dynamic MANET On-Demand Routing Protocol

DYMO [ChP08] is a reactive protocol that inherits most of its functionality from its predecessors, AODV and DSR. It is intended for use by mobile nodes in wireless multi-hop networks. It offers adaptation to changing network topology and determines unicast routes between nodes within the network. One of the main features of this protocol is path accumulation, which consists of the following mechanism. During the RREQ dissemination process, each intermediate node records a route to the originating node. When the target node receives the RREQ, it responds with a RREP unicast toward the originating node. It is expected that since intermediate nodes learn routes gratuitously from RREQs and RREPs, protocol overhead will be small since part of the route discovery processes is avoided.

In Table 4.1, there is a comparison of the five routing protocols for ad hoc networks presented in this section. This table also summarizes the routing algorithm used and the family protocol of every routing protocol presented.

TABLE 4.1 Routing Protocols in Wireless Sensor Networks

	Routing Protocols in Wireless Sensor Networks		
	SPIN	LEACH	GAF
Routing algorithm	Resource adaptative	Cluster based	Energy aware location based
Routing protocol family	Flat	Hierarchical	Location based

4.5.6 Sensor Protocols for Information via Negotiation

Sensor protocols for information via negotiation [HKB99] are a family of protocols used to efficiently distribute information in a WSN. Conventional data dissemination approaches such as flooding and gossiping waste valuable communication and energy resources by sending redundant information throughout the network. SPIN use data negotiation and resource-adaptive algorithms. Nodes running SPIN assign a high-level name to their data, called meta-data, and perform meta-data negotiations before any data is transmitted; therefore, there is no redundant data sent through the network. SPIN has access to the current energy level of the node and adapts the protocol it is running based on the remaining energy. The SPIN family protocols use three types of messages: when a SPIN node has some new data, it sends an advanced (ADV) message containing metadata to its neighbors. When a SPIN node wishes to receive the data, it sends an REQ message, and then there are DATA messages that are messages with a metadata header.

The SPIN family of protocols is made up of different protocols, SPIN-1, and SPIN-2; they incorporate negotiation before transmitting data in order to ensure that only useful information will be transferred, SPIN-BC (for broadcast channels), SPIN-PP (designed for point-to-point communications), SPIN-EC (with low energy threshold), and SPIN-RL.

4.5.7 Low Energy Adaptive Clustering Hierarchy

LEACH [HChB00] is a cluster-based protocol, which includes distributed cluster information. This protocol randomly selects a few sensor nodes as cluster heads (CHs) and rotates this role to evenly distribute the energy load among the sensors in the network. Figure 4.6 shows firstly a typical network based on clusters, and secondly how a WSN is applied.

LEACH has two phases, the setup phase and the steady state phase. In the setup phase, the clusters are organized and CHs are selected. In the steady state phase, the actual data transfer to the base station (BS) takes place. The duration of the steady state phase is longer than the duration of the setup phase in order to minimize overhead. All elected CHs broadcast an advertisement message to the rest of the nodes in the network that they are the new CHs, and the rest of nodes decide to which they want to belong to and inform the appropriate CHs. This decision is based on the signal strength of the advertisement.

During the steady state phase, the sensor nodes can begin sensing and transmitting data to the CHs. The CH node, after receiving all the data, aggregates it before sending it to the BS. Each cluster communicates using different code division multiple access (CDMA) codes to reduce interference from nodes belonging to other clusters.

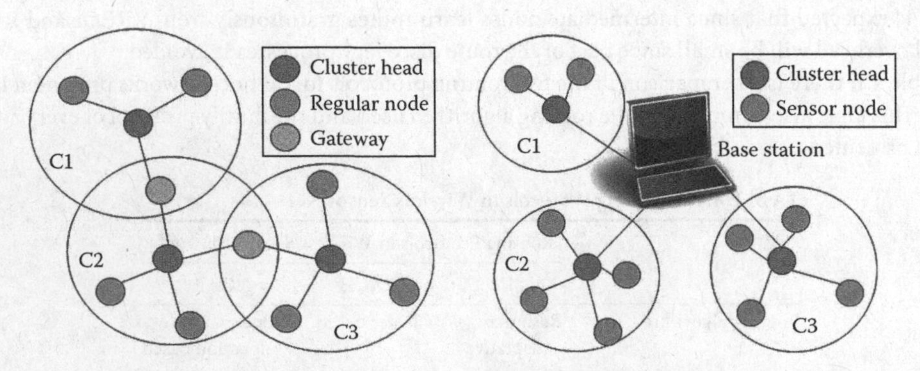

FIGURE 4.6 Example of a typical network based on clusters and application in a sensor network.

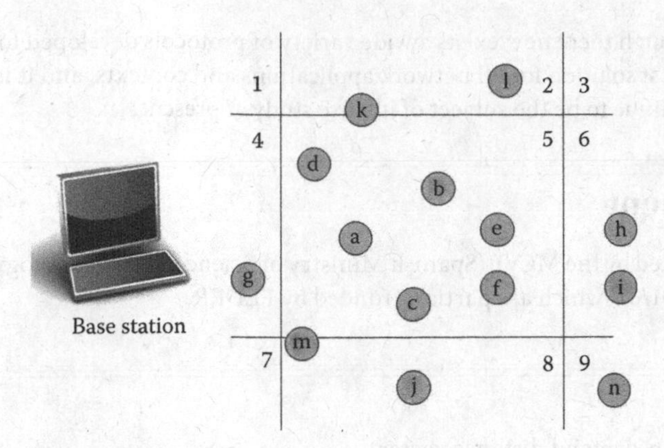

FIGURE 4.7 Example of virtual grid in GAF.

TABLE 4.2 Routing Protocols in Wireless Ad Hoc Networks

	Routing Protocols in Wireless Ad Hoc Networks				
	OLSR	TBRPF	DSR	AODV	DYMO
Routing algorithm	Link state	Link state	Source routing based	Distance vector	Source routing based
Routing protocol family	Proactive	Proactive	Reactive	Reactive	Reactive

4.5.8 Geographic Adaptive Fidelity

GAF [XHE01] is an energy-aware location-based routing algorithm designed primarily for mobile ad hoc networks, but may be applicable to sensor networks as well. The network area is first divided into fixed zones and forms a virtual grid, see Figure 4.7. Inside each zone, nodes collaborate with each other to play different roles. GAF conserves energy by turning off unnecessary nodes in the network without affecting the level of routing fidelity. Each node uses its GPS-indicated location to associate itself with a point in the virtual grid. Nodes associated with the same point on the grid are considered equivalent in terms of the cost of packet routing. GAF can substantially increase the network's lifetime as the number of nodes increases.

There are three states defined in GAF: discovery, for determining the neighbors in the grid; active, reflecting participation in routing; and sleep, when the radio is turned off. In order to handle mobility, each node in the grid estimates its leaving time and sends this to its neighbors. The sleeping neighbors adjust their sleeping time accordingly in order to keep the routing fidelity. Before the leaving time of the active node expires, sleeping nodes wake up and one of them becomes active. GAF is implemented both for nonmobility (GAF basic) and mobility (GAF-mobility adaptation) of nodes.

In Table 4.2, there is a comparison of the three routing protocols for WSNs presented in this section. The routing algorithm used and the family protocol of every routing protocol is summarized.

4.6 Conclusions

As a general rule, we can say that while proactive protocols are very efficient when the network is small with high traffic and low mobility, the constant propagation of routing information generally has a negative effect on the efficiency of these protocols. On the other hand, reactive protocols create a greater latency due to delays in route discovery to the destination. On-demand routing protocols are suitable for low traffic load and/or moderate mobility. With high mobility, flooding of data packets may be the only option, and it is because of this that new solutions such as hybrid protocols are being sought in the hope of achieving the best of both systems.

In conclusion, although there now exists a wide variety of protocols developed for wireless networks, none represents the best solution for all network applications and contexts, and it is for this reason that routing protocols continue to be the subject of intense study at present.

Acknowledgment

This work was supported by the MCyT (Spanish Ministry of Science and Technology) under the projects TSI2007-66637-C02-01/02, which are partially funded by FEDER.

Abbreviations

AODV	Ad hoc on-demand distance vector
BS	Base station
CH	Cluster head
DSDV	Destination-sequenced distance-vector
DSR	Dynamic source routing
DYMO	Dynamic MANET on-demand
ETSI	European Telecommunications Standards Institute
GAF	Geographic adaptive fidelity
GPRS	General packet radio service
GSM	Global system for mobile communication
HiperLAN	High performance radio LAN
IETF	Internet engineering task force
LEACH	Low energy adaptive clustering hierarchy
MANET	Mobile ad hoc networks
MPR	Multipoint relaying
OLSR	Optimized link-state routing protocol
QoS	Quality of service
RREP	Route reply
RREQ	Route request
RTS/CTS	Request to send/clear to send
SNIR	Signal to noise and interference ratio
SPIN	Sensor protocols for information via negotiation
TBRPF	Topology dissemination based on reverse path forwarding
TTL	Time-to-live
UMTS	Universal mobile telecommunications system
WLAN	Wireless local area networks
WMAN	Wireless metropolitan area networks
WMN	Wireless mesh network
WPAN	Personal area networks
WSN	Wireless sensor network
WWAN	Wireless wide area networks

References

[AESSM] IEEE 802.11s Task Group, Draft Amendment to Standard for Information Technology—Telecommunications and Information Exchange between Systems-LAN/MAN Specific Requirements. Part 11: Wireless Medium Access Control (MAC) and physical layer (PHY) specifications. Amendment: ESS Mesh Networking, IEEE P802.11s/D3.O, March 2009.

[AK04] Al-Karaki, J.N. and Kamal, A.E., Routing techniques in wireless sensor networks: A survey. *IEEE Wireless Communications*, 11(6), 6–28, 2004.

[AWD04] Abolhasan, M., Wysocki, T., and Dutkiewicz, E., A review of routing protocols for mobile ad hoc networks. *Journal Ad Hoc Networks*, 2, 1–22, 2004.

[AWW05] Akyildiz, I.F., Wang, X., and Wang, W., Wireless mesh networks: A survey. *Computer Networks and ISDN Systems*, 47(4), 445–487, March 2005.

[AY05] Akkaya, K. and Younis, M., A survey on routing protocols for wireless sensor networks. *Journal Ad Hoc Networks*, 3(3), 325–349, 2005.

[BCGS04] Basagni, S., Conti, M., Giordando, S., and Stojmenovic, I., *Mobile Ad Hoc Networking*, IEEE Press & Wiley Inter-Science, New York, 2004.

[ChCL03] Chlamtac, I., Conti, M., and Liu, J., Mobile ad hoc networking: Imperatives and challenges, *Ad Hoc Network Journal*, 1(1), January–March 2003.

[ChP08] Chakeres, I.D. and Perkins, C.E., Dynamic MANET on-demand (DYMO) routing protocol. Internet Draft version 21, IETF, July 26, 2010.

[CJ03] Clausen, T. and Jacquet, P., Optimized Link State Routing Protocol. RFC 3626, October 2003.

[CM99] Corson, S. and Macker, J., Mobile Ad Hoc Networking (MANET): Routing Protocol Performance. Issues and Evaluation Considerations. RFC 2501, June 1999.

[DRL03] Domingo, M.C., Remondo, D., and Leon, O., A simple routing scheme for improving ad hoc network survivability. *GLOBECOM 2003-IEEE Global Telecommunications Conference*, 22(1), 718–723, December 2003.

[HChB00] Heinzelman, W., Chandrakasan, A., and Balakrishnan, H., Energy-efficient communication protocol for wireless microsensor networks. In *Proceeding 33rd Hawaii International Conference of System Science*, Maui, HI, January 4–7, 2000.

[HKB99] Heinzelman, W.R., Kulik, J., and Balakrishnan, H., Adaptive protocols for information dissemination in wireless sensor networks. In *Mobicom 1999*, Seattle, WA, August 15–20, 1999, pp. 174–185.

[JHM07] Johnson, D., Hu, Y., and Maltz, D., The Dynamic Source Routing Protocol (DSR) for Mobile Ad Hoc Networks for IPv4. RFC 4728, February 2007.

[JM01] Johnson, D.B., Maltz, D.A., and Broch, J., DSR: The dynamic source routing protocol for multi-hop wireless ad hoc networks. In C.E. Perkins (Ed.), *Ad Hoc Networking*, Addison-Wesley, Reading, MA, 2001.

[JM96] Johnson, D.B. and Maltz, D.A., Dynamic source routing in ad hoc wireless networks. In T. Imielinski and H. Korth (Eds.), *Mobile Computing*, Kluwer Academic Publishers, Dordrecht, the Netherlands, 1996.

[JMC01] Jacquet, P., Muhlethaler, P., Clausen, T., Laouiti, A., Qayyum, A., and Viennot, L., Optimized link state routing protocol for ad hoc networks, In *IEEE INMIC 01*, Lahore, Pakistan, December 28–30, 2001.

[LMAN04] IEEE Std 802.16-2004 (Revision of IEEE Std 802.16-2001), IEEE Standard for Local and Metropolitan Area Networks Part 16: Air Interface for Fixed Broadband Wireless Access Systems. IEEE Computer Society and IEEE Microwave Theory and Techniques Society. Sponsored by the LAN/MAN Standards Comittee, New York, October 2004.

[M95] Moy, J., Link-state routing. In Steenstrup, M.E. (Ed.), *Routing in Communications Networks*, Prentice Hall, Englewood Cliffs, NJ, 1995, pp. 135–157.

[MZKD04] Lee, M.J., Zheng, J., Ko, Y-B, Shrestha, D.M., Emerging standards for wireless mesh technology. *Wireless Communications, IEEE*, 13(2), 56–63, 2004.

[OTL04] Ogier, R., Templin, F., and Lewis, M., Topology Dissemination Based on Reverse-Path Forwarding (TBRPF). RFC 3684. February 2004.

[P99] Perkins, C.E., Mobile networking in the internet. *Mobile Networks and Applications*, 3(4), 319–334, 1999.

[PBD03] Perkins, C., Belding-Royer, E., Das, S., Ad hoc On-Demand Distance Vector (AODV) Routing. RFC 3561, July 2003.

[PR99] Perkins, C.E. and Royer, E.M., Ad hoc on demand distance vector routing. In *Proceedings of the 2nd IEEE Workshop on Mobile Computing Systems and Applications*, New Orleans, LA, February 1999, pp. 90–100.

[RP00] Royer, E.M. and Perkins, C.E., An implementation study of the AODV routing protocol. In *Proceedings of the IEEE Wireless Communications and Networking Conference*, Chicago, IL, September 2000.

[RP99] Royer, E.M. and Perkins, C.E., Multicast operation of the ad hoc on-demand distance vector routing protocol. In *Proceedings of MobiCom '99*, Seattle, WA, August 1999, pp. 207–218.

[SWL03] Song, J.-H., Wong, V.W.S., and Leung, V.C.M., Efficient on-demand routing for mobile ad hoc wireless access networks. *GLOBECOM 2003-IEEE Global Telecommunications Conference,* 22(1), 558–563, December 2003.

[TG5] IEEE 802.15 WPANTM Task Group 5 (TG5). http://ieee802.org/15/pub/TG5.html

[WMAC03] IEEE 802.15.4, Wireless Medium Access Control (MAC) and Physical Layer (PHY) Specifications for Low-Rate Wireless Personal Area Networks (LRWPANs), IEEE, New York, October 2003.

[XHE01] Xu, Y., Heidemann, J., and Estrin, D., Geography-informed energy conservation for ad-hoc routing. In *Proceeding 7th Annual ACM/IEEE International Conference Mobile Computers and Networks*, Rome, Italy, July 2001, pp. 70–84.

5 Profiles and Interoperability

Gerhard Zucker
Vienna University of Technology

Heinz Frank
Reinhold-Würth-University

5.1 Interoperating Components 5-1
5.2 Application of Profiles 5-4
Function Blocks of IEC 61499 • Functional Profiles in LON • Logical Nodes of the IEC 61850
5.3 Achieving Interoperability 5-6
References 5-7

5.1 Interoperating Components

In distributed automation systems, automation devices like sensors, actuators, and controllers are connected through an industrial communication system (Figure 5.1). Automation systems of the past were freely configurable, allowing the system to be adapted specifically to one given installation. Although this could already be done in software (and not by connecting wires directly), it was still necessary that every system had to be programmed individually, which resulted in a unique and nonreusable installation. Costs for integrating multiple industries, maintenance, and extensions of existing systems were considerable and required well-educated experts. When automation systems became more sophisticated and consisted of a considerable amount of communicating components, this effort became too high and a new solution had to be found. Instead of programming each component individually, an existing definition has since been used as a template and reproduced for as many components (or nodes, as they are called from communication point of view) as needed. The abilities (i.e., the functions) of each component are standardized, the nodes do not need to be programmed, they merely need to be configured. These functions can be coupled, thus creating the functionality of the whole system.

The users do not need to know about the internal design of a node; they only have to know about the functions that a node offers. Well-known functions are, for example, actuators, sensors, and controllers, each of which can be offered by a separate node (or all integrated into one complex node). Instead of knowing the whole component, the user only has to know the interface of the component, that is, the variables that it offers and their behavior.

When configuring such a system, the user has to connect the outputs of one function block with the inputs of other function blocks. Physical data transmission does not require dedicated wires, but can be done on a shared bus, where messages are transported between nodes using distinct addresses for sender to receiver in the message.

The goal is to get a system that can easily be put into operation, preferably with little (or even no) commissioning. While the advantages are obvious, one also has to consider how to achieve cooperating components. This is an issue on multiple layers. While before it was sufficient to check for the correct physical parameters like voltage before connecting two components by a dedicated wire, we now have

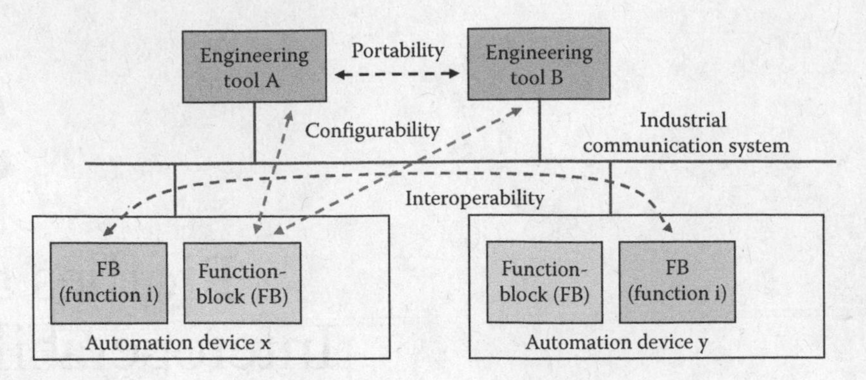

FIGURE 5.1 Distributed automation system.

to consider not only the physical connection but also different other levels. This ability to cooperate can be described by different terms.

Two components need to have identical properties to reach a certain level of cooperation. These five properties are

1. Identical communication protocol on layer 1–7
2. Usage of services by the application
3. Definition of the interface variables regarding, for example, data type, resolution, or measuring unit
4. Semantics of the application (algorithms, amount, and meaning of interface variables)
5. Dynamic behavior regarding, for example, control parameters or filter constants

Depending on the number of properties that a system fulfills, we can assign the following terms (Table 5.1):

Compatible and interconnectable require loosely defined communication agreements; the components must not interfere with each other, which translate to requirements on the physical channel, for example, voltage level or bitrates. Interworkable systems need to have an agreement on the meaning of transmitted information in terms of data types (e.g., how many bytes make up a number?, is it an integer or a floating point?) and also have to define error values in case a value cannot be transmitted (e.g., too big or sensor is broken). An example for this level of cooperation can be found in local operating network (LON) [1], where standard network variable types (SNVTs) have been defined. For example, the type snvt_temp defines temperature, which can be used for transmitting common temperatures with the following properties:

- Data type: unsigned long integer
- Total length: two bytes
- Measuring unit and range: degree Celsius ranging from −274°C to +6279.5°C
- Resolution: 0.1°C

On this level, the user of the system can be sure that the components do not interfere with each other, that they can exchange data, and also have the same understanding of the interpretation of

TABLE 5.1 Term Definitions by Properties 1–5 Given above

Incompatible	None				
Compatible	1				
Interconnectable	1	2			
Interworkable	1	2	3		
Interoperable	1	2	3	4	
Interchangeable	1	2	3	4	5

these data. However, this does not ensure that the components can also cooperate seamlessly. In addition to an understanding of values, the components need to be designed toward other requirements. For one, there is the direction from where communication is triggered: A node can either send its information regularly to any component that is interested in the value or it can wait until another component queries it for its value. A controller component on the other side has to implement error behavior in case a required value cannot be retrieved, for example, due to communication failure or failure of a sensor component. In some cases, it may be necessary to calibrate a sensor component. A controller component has to consider the fact that during calibration, the component is unable to deliver a reliable measurement. These issues are beyond mere communication; they relate to the functionality of a component.

In an interoperable system, the devices from different suppliers have to be able to exchange information and to use the information that has been exchanged [1]. If a system is capable of communicating and exchanging data, it is syntactically interoperable. When a device is able to process such data with useful results, it is semantically interoperable. To achieve interoperable or interchangeable components, the functions of the components have to be standardized. This includes defining groups of components, for example, sensor, actuator or controller and their properties, for example, their time response. In LON, this level of cooperation is achieved by defining functional profiles. On this level, it is possible to achieve true interoperability, which means that components (e.g., from different manufacturers) can be combined to one system to cooperatively provide the system functionality. Components can be replaced (also by components of other manufacturers) without affecting functionality.

The function blocks shown in Figure 5.1 consist of programs, data, and communication services (Figure 5.2). To ensure interoperability between distributed function blocks for one control function, we have to standardize the functions, data, and communication services of the function blocks in a functional profile.

Users benefit from standardized components by being able to use components, which can be manufactured by different companies. Manufacturers on the other side can extend their market segment by offering only parts or single components of a system and do not have to offer all components of a system. Generally, standardized distributed systems that are interoperable have the following advantages:

- Automation systems can be built up with autonomous subsystems.
- Autonomous subsystems can be manufactured and tested independently from the complete system.
- Subsystems from different manufacturers can be integrated.
- Existing systems can be easily extended with new automation devices.
- Integration of automation devices can be done by configuration.

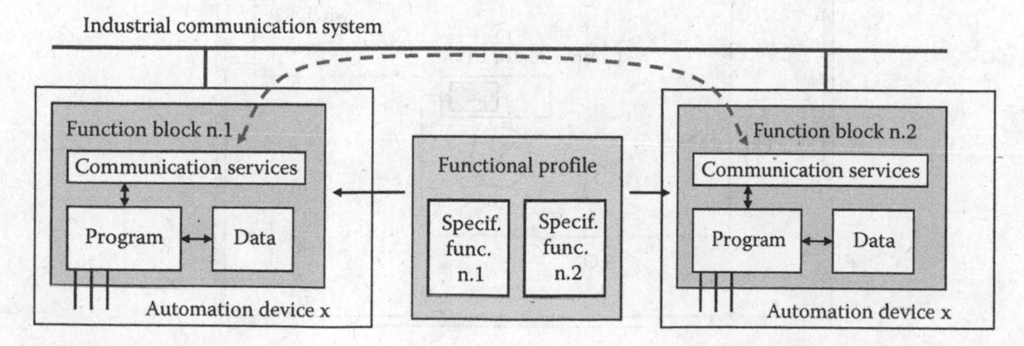

FIGURE 5.2 Distributed automation function.

5.2 Application of Profiles

This section describes some examples for functional profiles as they are covered by different standards.

5.2.1 Function Blocks of IEC 61499

Distributed systems used for automation have been standardized in the IEC 61499 standard. This standard describes an architecture for communication networks and processes that can be used for designing system applications. Interoperability is a central concept that has to be achieved. The core component in IEC 61499 is a Function Block, which is a module that has a certain function and provides the output of this function based on its input to other components using an interface. This interface consists of both Event Inputs/Outputs (I/Os) and Data I/Os. Internally, the Function Block executes an algorithm, which processes input data and produces output data, but this algorithms is not visible from the outside. Output data can be transferred to other Function Blocks and become input data for the other block. Basic Function Blocks are the most elementary blocks; Composite Function Blocks can be composed of multiple Basic Function Blocks. Using these Function Blocks, it is possible to design a system with a high degree of modularity. IEC 61499 is intended to describe a generic modelling approach for control applications that are distributed over multiple components and thus have to be built in a modular way. Another aspect of modularity is the fact that information flow and control flow are separated by means of Event I/O and Data I/O.

5.2.2 Functional Profiles in LON

LON was developed for building automation. It makes it possible to distribute the automation functions in, for example, light control, sunblind control, heating, ventilation, and air conditioning all over the buildings [2].

Figure 5.3 shows a simple example for a light control. It is possible to switch a lamp on and off from two different switches. Both sensors (i.e., the switches) and the actuator have their own microcomputer (Neuron-Chip). The function for the light control is distributed on these three microcomputers.

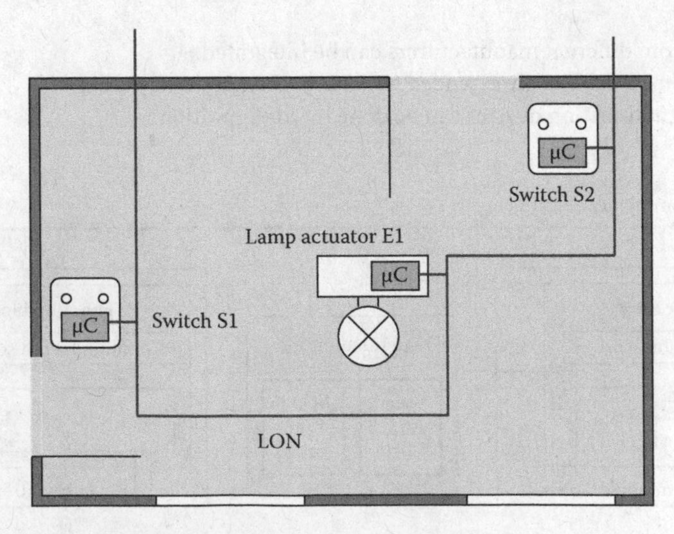

FIGURE 5.3 Example for a simple LON network.

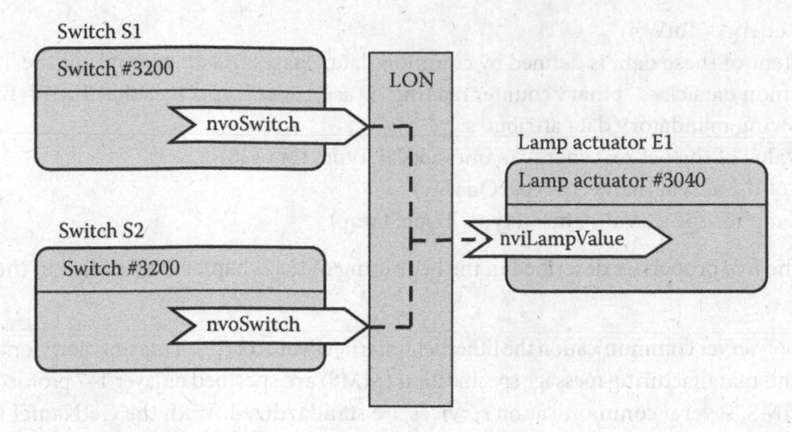

FIGURE 5.4 Nodes and functional profiles.

The standardization organization for LON (LonMark) has specified "functional profiles" for such a light control [3]. For the application according to Figure 5.3, two functional profiles have to be considered:

- The switch profile (standardized as No. 3200) defines the function, the configuration parameters, and the network variables (the types of which are called SNVTs, see below) that are required in a switch (Figure 5.4). If a user presses the left button on the switch, the microcontroller writes an "ON"-information into the network-variable nvoSwitch; if the right button is pressed, it writes an "OFF"-information into this variable. The representation of these values in the variable is standardized with so-called SNVTs. The nvoSwitch-variable is an output variable and can be connected to other input variables at other nodes. So each time this variable is changed, its value is transmitted over the LON network to one or more other automation devices.

  Figure 5.4 shows the implementation of the switch profile on two automation devices, which are switch S1 and switch S2. In this network, the value from the nvo-Switch values must be transmitted over the LON to the lamp actuator.
- The lamp-actuator-profile (standardized as No. 3040) defines the function, the configuration parameters, and the network variables that are required in a lamp actuator (Figure 5.4). A network variable nviLampValue, which is an input variable, is required. If an "ON"-information is received in this input value from a switch, the lamp actuator switches the lamp on; accordingly, the lamp is switched off, if an "OFF"-information is received. Figure 5.4 shows the implementation of one lamp-actuator-profile on one automation device, which is the lamp actuator E1.

In LON, the integration of the automation devices is done with integration tools. With such tools, the structure of the network and the so-called bindings must be configured. A binding defines the connection of input variables to output variables.

5.2.3 Logical Nodes of the IEC 61850

The IEC 61850 standard was developed for the communication of devices in substations of electrical power grids and for distributed energy resources [4–6].

For generic functions like supervision and control of transformers, switches, circuit breakers, and metering devices, so-called logical node classes were standardized.

For example, the logical node class for a metering device (abbrivated as MMTR according to IEC 61850) includes the following data:

- Health status of the metering device (EEHealth)
- Net apparent energy (TotVAh)

- Net real energy (TotWh)
  The content of these data is defined by common data classes. As an example of the TotWh-data, the common data class "binary counter reading" is assigned. It specifies that TotWh must include the following mandatory data attributes:
- Actual value of the net real energy (name: actVal, type: INT128)
- Quality of the actVal (name: q, type: Quality)
- Time stamp for the actVal(name: t, type: TimeStamp)

Therewith the five properties described in the beginning of this chapter are fulfilled in the IEC 61850 as follows:

1. For a client-server communication the Ethernet, Internet Protocol (IP), Transmission Control Protocol (TCP), and manufacturing message specification (MMS) are specified as layer 1–7 protocols [7,8].
2. In the MMS, several communication services are standardized. With the GetNameList-service a server can retrieve the names of all interface variables from a client. Afterwards, it can retrieve the data types for these variables by using the GetVarAccessAttributes service. With the read and the write services, it can access the values of the interface variables.
3. The data types and the resolution of the interface variables are defined by standardized data attributes.
4. The semantics of the applications are standardized by the logical nodes.
5. IEC 61850 does not consider the dynamic behavior of communication.

Therefore, we can say that the IEC 61850 standard enables an "interoperability" of devices.

5.3 Achieving Interoperability

Looking at the examples in the previous sections, we see that the key to successful interoperability is a proper standardization of all levels of functionality. By ensuring that components can cooperate on physical and logical levels, a lot can be achieved. For simple applications like switching lights in an office, the process of thoroughly describing the application in written form will suffice. For more complex applications, it will most likely not be possible to describe every single detail. It may happen that a written standard contains ambiguities that result in incompatibilities of components that have been built by different manufacturers. The next step has to be to run tests for all components and test the achieved level of interoperability. Only by examining the components during their operation, it is possible to ensure that full interoperability is given.

Software interoperability for complex systems such as distributed automation systems is achieved by following five approaches that depend on each other:

1. Product testing
   Given decent clarity of the written standard, products can be produced to meet this standard or a subprofile of it. System testing and unit testing can reveal incompatibilities, but do not suffice. Conformance-based product testing ensures only conformance to a standard, but does not always create interoperability with other products that have also been tested for conformance. Differences in product implementations can only be detected in a production scenario; such a test in a realistic environment can ensure interoperability.
2. Product engineering
   Product engineering has the task to create implementations that meet a common standard (or a subprofile of the standard). It has to achieve interoperability with other software implementations that also follow the same standard (or a subprofile of the standard).
3. Industry/community partnership
   Industry/community partnerships are the driving force toward standardization. They can be organized nationally or internationally and create standardization workgroups. Such a workgroup

defines a standard (i.e., creates a written description) based upon the discussions and negotiations in the workgroup. This standard is then used to build software systems, which meet the standard and are thus able to communicate with other software systems following the same standard. Standardization organizations cooperate by adopting existing standards from other organizations; this reduces the workload, makes national standards internationally available, and reduces the number of available options when selecting a standard. This again reduces costs for achieving interoperability.

4. Common technology and Intellectual Property
   Sometimes, it is better to use existing technology instead of creating a new system from scratch. Additionally, the number of options can be reduced by using common technology or intellectual property. Therefore, it is sometimes feasible to use third party technology instead of own developments.
5. Standard implementation
   A common agreement to implement a standard, may it be national or international, which opens freely or is only accessible by paying a fee.

Each of these has an important role in reducing variability in intercommunication software and enhancing a common understanding of the end goal to be achieved.

We shall finally take a look at the important factors characterizing interoperability testing. Two components set the boundaries and are the core components for testing: the Equipment Under Test (EUT) and the Qualified Equipment (QE). Ideally, both the EUT and QE are manufactured by different suppliers; if this is not possible they shall be manufactured in different product lines. If this is not the case, there is a chance that devices from the same supplier interoperate with each other, but not with devices from other suppliers. The tests for interoperability have to be done from a user perspective: they are done only at interfaces that are used for normal user control and observation; furthermore, the interoperability tests can only be based on functionality that is accessible by a user and only as experienced by a user (i.e., they are not specified at the protocol level). Note that in this context a user need not be human, it can also be a software application. This is because the interoperability tests are performed and evaluated at functional interfaces such as Man-Machine Interfaces, Application Programming Interfaces and protocol service interfaces. The fact that interoperability tests are performed at the end points and at functional interfaces means that interoperability test cases can only specify functional behavior. They cannot explicitly create protocol error behavior or test this error behavior.

References

1. D. Dietrich, D. Loy, and H.J. Schweinzer, editors. *Open Control Networks LonWorks/EIA-709 Technology*. Kluwer Academic Publishers, Norwell, MA, 2001.
2. LON Nutzer Organisation e.V. *LonWorks Installation Handbook*. VDE-Verlag, Berlin, Germany, 2005.
3. LonMark International. LonMark application-layer interoperability guidelines Version 3.4, 2005. Available at http://www.lonmark.org/technical_resources/guidelines/developer
4. IEC 61850. Communication networks and systems in substations, IEC Standard, 14 parts, Beuth-Verlag, Berlin, Wien, Zurich, 2002–2004.
5. IEC 61850. Part 7-420 DER logical nodes. Beuth-Verlag, Berlin, Wien, Zurich, 2008.
6. K.H. Schwarz. *An introduction to IEC 61850. Basics and User-Oriented Project-Examples for the IEC 61850 Series for Substation Automation*. Vogel Verlag, Würzburg, Germany, 2005.
7. ISO 9506. Industrial automation systems—Manufacturing message specification, 2 Parts, Beuth-Verlag, Berlin, Wien, Zurich, 2003.
8. J.T. Sorenson and M.G. Jaatun. *An Analysis of the Manufacturing Message Specification Protocol*. Springer, Berlin, Germany, 2008.

6

Industrial Wireless Sensor Networks

Vehbi Cagri Gungor
Bahcesehir University

Gerhard P. Hancke
University of Pretoria

6.1 Applications....................6-2
Factory Automation • Building Automation • Industrial Process Automation • Inventory Management • Utility Automation • Automatic Meter Reading
6.2 Standardization Activities....................6-3
ZigBee • Wireless Hart • IETF 6loWPAN • Bluetooth and Bluetooth Low Energy • Ultra-Wideband
6.3 Technical Challenges....................6-5
6.4 Design Goals....................6-6
6.5 Design Principles and Technical Approaches....................6-7
Hardware Development • Software Development • System Architecture and Protocol Design
6.6 Conclusions and Future Work....................6-12
References....................6-13

Given the increasing age of many industrial systems and the dynamic industrial manufacturing market, industrial monitoring and control systems are critical to maintain safety, reliability, efficiency, and uptime. Industrial wireless sensor networks (IWSNs), with their unique characteristics such as flexibility, self-organization, low cost, and rapid deployment, have become a promising technology for intelligent and low-cost industrial automation systems [3,9,47–49]. In these systems, wireless tiny sensor nodes are installed on industrial equipment and monitor the parameters critical to each equipment's efficiency based on a combination of measurements such as vibration, temperature, pressure, and power quality. These data are then wirelessly transmitted to a sink node that analyzes the data from each sensor. Any potential problems are notified to the plant personnel as an advanced warning system. This enables plant personnel to repair or replace equipment, before their efficiency drops or they fail entirely [9,29]. In this way, catastrophic equipment failures and the associated repair and replacement costs can be prevented while complying with strict environmental regulations.

The collaborative nature of IWSNs brings several advantages over traditional wired industrial monitoring and control systems, including self-organization, rapid deployment, flexibility, and inherent intelligent processing capability. In this regard, IWSN plays a vital role in creating a highly reliable and self-healing industrial system that rapidly responds to real-time events with appropriate actions. However, to realize the envisioned industrial applications and hence take the advantages of the potential gains of IWSN, effective communication protocols, which can address the unique challenges posed by such systems, are required.

Recently, many researchers have been engaged in developing schemes that address the unique challenges of IWSNs. In this chapter, first a short review about the emerging and already employed IWSN applications and technologies is presented. In addition, IWSN standards are presented for the system owners, who plan to utilize new IWSN technologies for industrial automation applications. After a short review of IWSN applications and standardization activities, technical challenges and design principles are introduced in terms of hardware development, system architectures and protocols, and software development. Specifically, radio technologies, energy-harvesting techniques, and cross-layer design for IWSNs are discussed. Here, our aim is to provide a contemporary look at the current state of the art in IWSNs and discuss the still-open research issues in this field, and hence, to make the decision-making process more effective and direct. Note that some parts of this chapter have also been presented in [9].

The remainder of the chapter is organized as follows. Section 6.1 presents the IWSN applications. In Section 6.2, the IWSN standardization activities are reviewed. Sections 6.3 and 6.4 review the technical challenges and corresponding design directions, respectively. In Section 6.5, design principles of IWSNs are discussed. Finally, the chapter is concluded in Section 6.6.

6.1 Applications

IWSNs can be used for various applications in which deployment environment and technical challenges can greatly differ and thus, the design of IWSNs is application oriented. In addition, they can be deployed to harsh industrial environments, in which installation and maintenance of conventional wired sensor systems may not be cost-effective. Some of the key IWSN applications are briefly described below.

6.1.1 Factory Automation

Modern factory facilities are characterized by highly flexible manufacturing plants and highly dynamic processes, where clusters of fixed or moving sensors and actuators have to be controlled in a limited space under stringent real-time and reliability constraints [8,11,25,35]. In such demanding industrial environments, wireless sensor networks can also be beneficial by improving flexibility, cutting cables, and enabling solutions, which are cumbersome or even not possible to realize with wireline systems, especially in controlling moving or rotating parts. The wireless way of communicating makes plant setup and modification easier, cheaper, and more flexible.

6.1.2 Building Automation

The primary focus of building automation is the reduction of energy consumption in building installations through automated mechanisms to lower total energy costs and comply with governmental regulations while maximizing the comfort [10]. Building automation comprises a set of different functionalities including energy conservation, environment control, safety, and security. The interest in using wireless sensor networking in building automation applications is based on the need to lower installation cost, which comes in the form of cabling, labor, materials, testing, and verification.

6.1.3 Industrial Process Automation

Wireless sensor networks can be used to enable condition-based maintenance and remote management of industrial equipment and processes by continuously monitoring time-critical process information, such as temperature, pressure, humidity, vibration, and energy usage. In addition, the preventive maintenance information can be collected by IWSNs to support a manufacturing process such as those used in oil tankers [27], conveyor belts [14], automobiles [14], electric motors [29,46] and pumps [1], and food or pharmaceutical products.

6.1.4 Inventory Management

Inventory management systems based on manual processes can cause out-of-stocks, expedited shipments or billing delays. With IWSN technology, the inventory and assets could be monitored in real time, and time-critical information such as the arrival of raw materials can be communicated to the remote control center. For example, General Motors has implemented a real-time tracking system [14], in which the tracking process starts from the component suppliers to the assembled cars in the factory, and to the car buyers. This system improves the visibility of materials, its location, and asset utilization, leading to improved supply-chain efficiency.

6.1.5 Utility Automation

Due to several reasons such as equipment failure, lightning strikes, accidents, and natural catastrophes, power disturbances and outages in electric systems occur and often result in long service interruptions [13]. Thus, electric systems need to be properly controlled and monitored to take the necessary precautions accurately and timely. In this respect, IWSNs provide cost-effective, real-time, and reliable monitoring system for the electric utilities. Efficient monitoring systems constructed by smart sensor nodes reduce the time for detection of faults and resumption of electric supply service in distribution and transmission networks.

6.1.6 Automatic Meter Reading

IWSNs can be used in remote reading of utility meters, such as water, gas, or electricity, and then transmit the readings through wireless connections [13]. Wireless collection of electric utility meter data is a very cost-efficient way of gathering energy consumption data to the billing system, and it adds value in terms of new services such as remote deactivation of a customer's service, real-time price signals, and control of customers' applications and demand response applications. The present demand for more data in order to make cost-effective decisions and to provide improved customer service has played a major role in the move toward wireless automatic meter reading (WAMR) systems.

6.2 Standardization Activities

Several standards for IWSNs are currently either ratified or under development. In this section, major standardization efforts related to IWSNs, such as ZigBee, Wireless Hart, IETF 6lowPAN, Bluetooth, Bluetooth low energy, and ultra-wideband (UWB), are briefly described.

6.2.1 ZigBee

ZigBee is a mesh-networking standard based on IEEE 802.15.4 radio technology targeted at industrial control and monitoring, building and home automation, embedded sensing, and energy system automation. Zigbee is promoted by a large consortium of industry players. The advantages of ZigBee are extremely low energy consumption and support for several different topologies, which makes it a good candidate for several sensor network applications [1]. However, in [7], it is reported that ZigBee cannot meet all the requirements for at least some industrial applications. For example, it cannot serve the high number of nodes within the specified cycle time. Note that in 2007, the ZigBee Alliance has approved the ZigBee PRO profile stack, which adds advanced features and greater flexibility to the original specification, particularly related to ease-of-use and support for larger networks [54].

6.2.2 Wireless Hart

Wireless HART is an extension of the HART Protocol and is specifically designed for process monitoring and control. Wireless HART was added to the overall HART protocol suite as part of the HART 7 specification, which was approved by the HART Communication Foundation in June 2007 [20].

The technology employs IEEE 802.15.4-based radio, frequency hopping, redundant data paths, and retry mechanisms. Devices that comply with the wireless HART specification are interoperable and support tools that work with both wired and wireless HART equipment. Wireless HART networks utilize mesh-networking, in which each device is able to transmit its own data as well as relay information from other devices in the network. Each field device has two routes to send data to the network gateway; the alternative route is used when the primary route is blocked either physically or by interference. Additionally, a new route is established if a route is seen permanently blocked [20].

The standard uses time division multiple access (TDMA) to schedule transmissions over the network. In TDMA, a series of time slots (10 ms for Wireless HART) are used to coordinate transmissions. The media access control header is designed to support the coexistence of other networks, such as ZigBee. Wireless HART is aimed for new applications (e.g., asset monitoring) rather than replacing the wired solutions.

6.2.3 IETF 6loWPAN

6loWPAN aims for standard Internet Protocol (IP) communication over low-power, wireless IEEE 802.15.4 networks utilizing Internet Protocol version 6 (IPv6) [32]. The advantages of 6loWPAN from the industrial point of view are its ability to communicate directly with other IP devices locally or via an IP network (e.g., Internet, Ethernet), existing architecture and security, established application level data model and services (e.g., HTTP, HTML, XML), established network management tools, transport protocols, and existing support for an IP option in most industrial wireless standards.

6.2.4 Bluetooth and Bluetooth Low Energy

Bluetooth operates at the 2.4 GHz ISM band and employs a frequency hopping spread spectrum (FHSS) modulation technique. Bluetooth has been considered as one possible alternative for WSN implementation. However, due to its high complexity and inadequate power characteristics for sensors, the interest toward Bluetooth-based WSN applications has decreased [1]. Additionally, Bluetooth is designed for high-throughput applications between small numbers of terminals [7]. The IEEE 802.15.1 specification covers physical and link layers from Bluetooth technology.

The Bluetooth low energy specification is part of the Bluetooth aimed at addressing devices with very-low battery capacity. This extension to Bluetooth allows for data rates of up to 1 Mbit/s over distances of 5–10 m in the 2.45 GHz band. Though Bluetooth low energy is similar to Bluetooth and can employ the same chips and antennas, it has some important differences. Bluetooth low energy has a variable-length packet structure, compared to Bluetooth's fixed length. It also employs a different modulation scheme. Additionally, the implementation of the security algorithm Advanced Encryption Standard (AES) has been taken into account from the start. In [30], it has been pointed out that Bluetooth low energy also may not be applicable to industrial environments due to short range (<10 m), lack of mesh-networking capabilities, and limited number of radios in the network.

6.2.5 Ultra-Wideband

UWB is a short-range wireless communication technology based on the transmission of very short impulses emitted in periodic sequences [16,53]. The initial applications of UWB include multimedia and personal area networking [37]. Recently, UWB-based industrial applications have gained attention [16,22,53]. In [22], it has been demonstrated that UWB technology is applicable for automated registration and localization of industrial equipment and people, remote control, the rapid commissioning of the process control equipment, industrial plant wide communications not related to process control, and secondary sensor and control networks like gauging in tanks. On the other hand, UWB is not a viable approach for communication over longer distances or measuring data from unsafe zones because of the high-peak energy of pulses.

The advantages of UWB are good localization capabilities [53], possibility to share previously allocated radio frequency bands by hiding signals under the noise floor, ability to transmit high data rates with low power, good security characteristics due to the unique mode of operation, and ability to cope with multipath environments. The existing challenges lie, e.g., in hardware development, dealing with medium access control (MAC) and multipath interference, and understanding propagation characteristics [16].

6.3 Technical Challenges

The design of IWSNs is influenced by several technical challenges, which are outlined as follows:

- *Resource constraints*: The design and implementation of IWSNs are constrained by three types of resources: (i) energy, (ii) memory, and (iii) processing. Constrained by the limited physical size and low-cost nature, sensor nodes have limited battery energy supply [13]. At the same time, their memories are limited and have restricted computational capabilities.
- *Dynamic topologies and harsh environmental conditions*: In industrial environments, the topology and connectivity of the network may vary due to link and sensor node failures. Furthermore, sensors may also be subject to RF interference, highly caustic or corrosive environments, high humidity levels, vibrations, dirt and dust, or other conditions that challenge performance [47]. These harsh environmental conditions and dynamic network topologies may cause a portion of industrial sensor nodes to malfunction or render the information they gather obsolete [12,9].
- *Quality of service requirements*: The wide variety of applications envisaged on IWSNs will have different quality of service (QoS) requirements and specifications. In IWSNs, the sensor data represent the physical condition of industrial equipment, which is valid only for a limited time duration. Hence, the sensed data should be transported to the control center within a certain delay bound, which depends on the criticality of the sensed data. For example, the temperature information of equipment that is in the range of normal operating conditions can be transported to the control center within a long delay bound. On the other hand, the critical sensor data that contain an abnormally high temperature of the industrial equipment should be delivered to the control center within a short delay bound, since it can be a sign of a system failure.
- *Data redundancy*: Because of the high density in the network topology, sensor observations are highly correlated in the space domain. In addition, the nature of the physical phenomenon constitutes the temporal correlation between each consecutive observation of the sensor node.
- *Packet errors and variable link capacity*: Compared to wired networks, in IWSNs, the attainable capacity of each wireless link depends on the interference level perceived at the receiver and high-bit error rates (BER = 10^{-2} to 10^{-6}) are observed in communication. Also, wireless links exhibit widely varying characteristics over time and space due to obstructions and noisy environment [51]. Thus, capacity and delay attainable at each link are location dependent and vary continuously, making QoS provisioning a challenging task.
- *Security*: Security should be an essential feature in the design of IWSNs to make the communication safe from external denial of service (DoS) attacks and intrusion. IWSNs have special characteristics that enable new ways of security attacks [15,44]. Passive attacks are carried out by eavesdropping on transmissions including traffic analysis or disclosure of message contents. Active attacks consist of modification, fabrication, and interruption, which in IWSN cases may include node capturing, routing attacks, or flooding.
- *Large-scale deployment and ad hoc architecture*: Most IWSNs contain a large number of sensor nodes (hundreds to thousands or even more), which might be spread randomly over the deployment field. Moreover, the lack of predetermined network infrastructure necessitates the IWSNs to establish connections and maintain network connectivity autonomously.

- *Integration with Internet and other wireless networks*: It is of fundamental importance for the commercial development of IWSNs to provide services that allow querying the network to retrieve useful information from anywhere and at any time. For this reason, the IWSNs should be remotely accessible from the Internet and hence need to be integrated with the IP architecture. The current sensor network platforms use gateways for integration between IWSNs and the Internet [2,43]. Note that although today's sensor networks use gateways for integration between IWSNs and the Internet, the sensor nodes may have IP connectivity in the future [32].

6.4 Design Goals

The IWSNs have the potential to improve productivity of industrial systems by providing greater awareness, control, and integration of business processes. To deal with the technical challenges and meet the diverse IWSN application requirements, the following design goals need to be followed:

- *Low-cost and small sensor nodes*: Compact and low-cost sensor devices are essential to accomplish large-scale deployments of IWSNs. In addition, the smaller the sensor is, the easier the industrial deployment would be. Note that the system owner should consider the cost of ownership (packaging requirements, modifications, maintainability, etc.), implementation costs, replacement and logistics costs, and training and servicing costs as well as the per unit costs all together [18].
- *Scalable architectures and efficient protocols*: The IWSNs support heterogeneous industrial applications with different requirements. It is necessary to develop flexible and scalable architectures that can accommodate the requirements of all these applications in the same infrastructure. A modular and hierarchical systems enhance the system flexibility, robustness, and reliability [39]. Also, interoperability with existing legacy solutions, such as fieldbus and Ethernet-based systems, is required. This can be achieved by using standard-based communication protocols and open network architectures.
- *Data fusion and localized processing*: Instead of sending the raw data to the sink node directly, sensor nodes can locally filter the sensed data based on the application requirements and transmit only the processed data, i.e., in-network processing. Thus, only necessary information is transported to the end-user and communication overheads can be significantly reduced.
- *Resource efficient design*: In IWSNs, energy efficiency is important to maximize the network lifetime while providing the QoS required by the application. Energy saving can be accomplished in every component of the network by integrating network functionalities with energy efficient protocols, e.g., energy-aware routing on network layer, energy-saving mode on MAC layer, etc.
- *Self-configuration and organization*: In IWSNs, the dynamic topologies caused by node failure, mobility, temporary power-down, and large-scale node deployments necessitate self-organizing architectures and protocols. Note that with the use of self-configurable IWSNs, new sensor nodes can be added to replace failed sensor nodes in the deployment field and existing nodes can also be removed from the system without affecting the general objective of the application.
- *Adaptive network operation*: The adaptability of IWSNs is extremely crucial, since it enables end-users to cope with dynamic/varying wireless channel conditions in industrial environments and new connectivity requirements driven by new industrial processes. To balance the trade-offs among resources, accuracy, latency, and time-synchronization requirements, adaptive signal processing algorithms and communication protocols are required.
- *Time synchronization*: In IWSNs, large numbers of sensor nodes need to collaborate to perform the sensing task and the collected data are usually delay sensitive [2,18]. Thus, time synchronization is one of the key design goals for communication protocol design to meet the deadlines of the application. However, due to resource and size limitations and lack of a fixed infrastructure, as

TABLE 6.1 Challenges vs. Design Goals in IWSNs

Challenges	Design Goals
Resource constraints	Resource efficient design
Dynamic topologies and harsh environmental conditions	Adaptive network operation
Quality of service requirements	Application-specific design and time synchronization
Data redundancy	Data fusion and localized processing
Packet errors and variable link capacity	Fault tolerance and reliability
Security	Secure design
Large-scale deployment and ad hoc architecture	Low-cost sensor nodes and self-organization
Integration with Internet and other wireless networks	Scalable architectures and efficient protocols

well as the dynamic topologies in IWSNs, existing time-synchronization strategies designed for other traditional wired and wireless networks may not be appropriate for IWSNs. Adaptive and scalable time-synchronization protocols are required for IWSNs.

- *Fault tolerance and reliability*: In IWSNs, based on the application requirements, the sensed data should be reliably transferred to the sink node [11]. Similarly, the programming/retasking data for sensor operation, command, and queries should be reliably delivered to the target sensor nodes to assure proper functioning of the IWSN. However, for many IWSN applications, the sensed data are exchanged over a time-varying and error-prone wireless medium. Thus, data verification and correction on each communication layer and self-recovery procedures are extremely critical to provide accurate results to the end-user.
- *Application-specific design*: In IWSNs, there exists no one-size-fits-all solution; instead, the alternative designs and techniques should be developed based on the application-specific QoS requirements and constraints.
- *Secure design*: When designing the security mechanisms for IWSNs, both low-level (key establishment and trust control, secrecy and authentication, privacy, robustness to communication DoS, secure routing, resilience to node capture) and high-level (secure group management, intrusion detection, secure data aggregation) security primitives should be addressed [34]. Also, because of resource limitations in IWSNs, the overhead associated with security protocols should be balanced against other QoS performance requirements.

It is very challenging to meet all the above-mentioned design goals simultaneously. Fortunately, most IWSN designs have different requirements and priorities on design objectives. Therefore, the network designers and application developers should balance the trade-offs among the different parameters when designing protocols and architectures for IWSNs. Table 6.1 summarizes the technical challenges and corresponding design goals for IWSNs.

6.5 Design Principles and Technical Approaches

In this section, the design principles and technical approaches in IWSNs are broadly classified into three categories: (1) hardware development; (2) software development; and (3) system architecture and protocol design.

6.5.1 Hardware Development

6.5.1.1 Low-Power and Low-Cost Sensor Node Development

An IWSN node integrates sensing, data collection and processing, and wireless communications along with an attached power supply on a single chip. The hardware architecture of a typical industrial sensor node is composed of four basic components: (1) sensor, (2) processor, (3) transceiver, and (4) power source:

- *Sensor*: Sensors are hardware devices that produce measurable response to a change in a physical condition, e.g., temperature, pressure, voltage, current, etc. The analog signals produced by the sensors based on the observed phenomenon are converted to digital signals by the analog-to-digital converter (ADC) and sent to the processor for further processing. Several sources of power consumption in sensors are (1) signal sampling and conversion of physical signals to electrical ones, (2) signal conditioning, and (3) analog-to-digital conversion.
- *Processor*: The processing unit, which is generally associated with a small storage unit, performs tasks, processes data, and controls the functionality of other components in the sensor node.
- *Transceiver*: A transceiver unit connects the node to the network. Generally, radios used in the transceivers of industrial sensor nodes operate in four different modes: (i) transmit, (ii) receive, (iii) idle, and (iv) sleep. Radios operating in idle mode result in power consumption almost equal to power consumed in receive mode. Hence, it is better to completely shut down the radios rather than run it in the idle mode when it is not transmitting or receiving. A significant amount of power is consumed when switching from sleep mode to transmit mode for transmitting a packet.
- *Power source*: One of the most important components of an industrial sensor node is the power source. In sensor networks, power consumption is generally divided into three domains: sensing, data processing, and communication. Compared to sensing and data processing, much more energy is required for data communication in a typical sensor node. For example, the energy cost of transmitting 1 kb of data over a distance of 100 m is approximately the same as that for executing 3 million instructions by a 100 million instructions per second/W processor [38]. Hence, local data processing is crucial in minimizing power consumption in IWSNs.

Generally, the lifetime of IWSNs shows a strong dependence on battery capacity. Furthermore, in a multi-hop ad hoc sensor network, each node plays the dual role of data originator and data router. The failure of a few nodes can cause topological changes and might require rerouting of data packets and reorganization of the network. In this regard, power conservation and management take on additional significance. Due to these reasons, technical approaches for prolonging the lifetime of battery-powered sensors have been the focus of a vast amount of literature in sensor networks. These approaches include energy-aware protocol development for sensor network communications and hardware optimizations, such as sleeping schedules to keep electronics inactive most of the time, dynamic optimization of voltage, and clock rate.

All these research and development efforts in both academy and industry lead to several commercially available industrial sensor hardware platforms and components. In the recent sensor radio chips, such as CC2430 and EM250, the *system-on-chip* (SOC) technology has been used for low power consumption by integrating a complete system on a single chip. For example, EM250 from Ember provides a ZigBee SOC that combines a 2.4 GHz IEEE 802.15.4 compliant radio transceiver with a 16 bit microprocessor to extend battery lifetime. Specifically, SOC solutions provide significant amount of power consumption improvement in the sleep mode, but modest improvement in transmission (Tx) and receiving (Tx) modes.

In a multi-hop IWSN, communicating nodes are linked by a wireless medium, which can be formed by radio, infrared, or optical media [3]. To enable global operation of these networks, the chosen transmission medium must be available worldwide. An overview of commercially available radio chips and sensor hardware platforms for IWSNs is given in Tables 6.2 and 6.3, respectively.

6.5.1.2 Radio Technologies

In industrial environments, the coverage area of WSN as well as the reliability of the data may suffer from noise, co-channel interferences, multipath propagation, and other interferers [28]. For example, the signal strength may be severely affected by the reflections from the walls (multipath propagation),

TABLE 6.2 Comparison of Commercial Off-the-Shelf Radio Chips for IWSNs

Features	CC2520	CC2430	AT86RF230	JN5139	MC1321	EM250
Manufacturer	TI	TI	Atmel	Jennic	Freescale	Ember
Frequency (GHz)	2.4	2.4	2.4	2.4	2.4	2.4
Bit rate (kbps)	250	250	250	250	250	250
Supply voltage (V)	1.8–3.8	2.0–3.6	1.8–3.6	2.2–3.6	2.0–3.4	2.1–3.6
Sleep (μA)	1	0.5	0.02	0.2	1	1
Rx (mA)	18.5	27	15.5	34	37	29
Tx minimum (mA)	16.2 (−18 dBm)	—	9.5 (−17 dBm)	—	20.9 (−28 dBm)	19 (−32 dBm)
TX maximum (mA)	33.6 (+5 dBm)	27 (0 dBm)	16.5 (+3 dBm)	35 (+3 dBm)	30 (0 dBm)	33 (+5 dBm)

TABLE 6.3 Comparison of Commercial Off-the-Shelf Sensor Platforms

Features	XBee	M1030 Mote	M2135 Mote	MicaZ	Mica2
Manufacturer	Digi	Dust networks	Dust networks	Crossbow	Crossbow
Radio frequency	2.4 GHz	900 MHz	2.4 GHz	2.4 GHz	900 MHz
Bandwidth (kbps)	250	76.8	250	250	40
Current consumption Listening/Rx/Tx (mA)	–/40/40	–/14/28	–/22/50	8/20/18	08/10/17
Power sleep (μA)	1	8	10	27	19
CPU type @(MHz)	—	—	—	8 bit Atmel @8	8 bit Atmel @8
Memory (SRAM [kB])	—	—	—	4	4

by interferences from other devices using ISM bands, and by the noise generated from equipment or heavy machinery [28]. In these conditions, it is important that data reliability and integrity are maintained for operation-critical data, for example, alarm conditions.

Specifically, interference signals can be classified in two different categories, broadband and narrowband [28]. Broadband interference signals have a constant energy spectrum over all frequencies and high energy. They are usually emitted unintentionally from radiating sources, whereas narrowband interference signals are intentional and have less energy. Both interferences have a varying type of degradation effect on wireless link reliability. In an industrial environment, broadband interference can be caused by motors, SCR circuits, inverters, computers, electric switch contacts, voltage regulators, pulse generators, thermostats, and welding equipment. On the other hand, narrowband interference can be caused by UPS systems, power-line hum, electronic ballasts, test equipment, cellular networks, radio-TV transmitters, signal generators, and microwave equipment [28]. In industrial environments, other types of interference sources can be wide operating temperatures, strong vibrations, and airborne contaminants. In this respect, it is important to study and understand the radio channel characteristics to predict the communication performance in these operating conditions.

Two main classes of mechanisms are traditionally employed to combat the unreliability of the wireless channel at the physical and data link layer, namely forward error correction (FEC) and automatic repeat request (ARQ), along with hybrid schemes. Compared to FEC techniques, ARQ mechanisms use bandwidth efficiently at the cost of additional latency. Hence, while carefully designed selective repeat schemes may be of some interest, naive use of ARQ techniques is clearly infeasible for applications requiring real-time delivery [2]. In [50], FEC schemes are shown to improve the error resiliency compared to ARQ. In a multi-hop network, this improvement can be exploited by reducing the transmit power (transmit power control) or by constructing longer hops (hop length extension) through channel-aware routing protocols. The analysis reveals that, for certain FEC codes, hop length extension decreases both the energy consumption and the end-to-end latency subject to a target packet error rate compared to ARQ. Thus, FEC codes can be preferred for delay-sensitive traffic in IWSNs [2].

Moreover, radio modulation techniques can be applied to reduce the interference and improve wireless communication reliability in an industrial facility. In this respect, to reduce the interference in IWSNs, spread spectrum radio modulation techniques can be applicable because of their multiple access, anti-multipath fading, and antijamming capabilities. The two main spread spectrum techniques employed are direct sequence spread spectrum (DSSS) and frequency hopping spread spectrum (FHSS). These have different physical mechanisms and thus react differently in industrial settings. The choice between radio techniques is dependent on application requirements and the industrial environment characteristics.

In IWSNs, interference to the mission-critical data can have costly consequences in terms of money, manpower, and even lives of employees and public [28]. Therefore, the coexistence of WSN with other systems operating in the same band should be examined. The coexistence should be considered in both directions, "from system" and "to system" points of views, i.e., how the system reduces its effects to other systems or how it can reduce the effects from other systems and interference sources to itself. An example about coexistence impact analysis between IEEE 802.15.4 and 802.11b is presented in [4,17,19,36].

6.5.1.3 Energy-Harvesting Techniques

In IWSNs, the use of batteries as power source for the sensor nodes can be troublesome due to their limited lifetime, making periodic replacements unavoidable [5]. In this respect, *energy-harvesting* (also referred to as energy scavenging) techniques, which extract energy from the environment where the sensor itself resides, offer another important way to prolong the lifetime of sensor devices.

Systems able to perpetually power sensors based on simple COTS photovoltaic cells coupled with rechargeable batteries and supercapacitors have already been demonstrated [21]. In [33], the state of the art in more unconventional techniques for energy harvesting is surveyed. Technologies to generate energy from background radio signals, thermoelectric conversion, vibrational excitation, and the human body are investigated. As far as collecting energy from background radio signals is concerned, unfortunately, an electric field of 1 V/m yields only 0.26 W/cm^2, as opposed to 100 W/cm^2 produced by a crystalline silicon solar cell exposed to bright sunlight [2]. Electric fields of intensity of a few volts per meter are only encountered close to strong transmitters.

Another practice, which consists in broadcasting RF energy deliberately to power electronic devices, is severely limited by legal limits set due to health and safety concerns. Recently, it has been also demonstrated that wireless power transfer using resonant magnetic coupling between two copper coils is possible [23]. In this experiment, it was shown that a 60 W light bulb was lighted at an efficiency of 40%, with the distance between the transmitter and receiver being 2 m. The efficiencies demonstrated were almost a million times larger than nonresonant magnetic induction. Wireless energy transfer via magnetic coupling is important, since biological organisms only weakly interact with magnetic fields.

While thermoelectric conversion may not be suitable for wireless devices, harvesting energy from vibrations in the surrounding environment provides another useful source of energy. Vibrational magnetic power generators based on moving magnets or coils could yield power that range from tens of microwatts when based on micro-electromechanical system (MEMS) technologies to over a milliwatt for larger devices. Other vibrational microgenerators are based on charged capacitors with moving plates and, depending on their excitation and power conditioning, yield power in the order of 10 μW. In [33], it is also reported that recent analysis [31] suggested that 1 cm^3 vibrational microgenerators can be expected to yield up to 800 W/cm^3 from machine-induced stimuli, which is orders of magnitude higher than what is provided by currently available microgenerators. Hence, this is a promising area of research for small battery-powered devices.

Other energy-scavenging approaches employ piezoelectric materials. In [5] and [33], it is reported that these materials can generate power between 100 and 330 μW/cm^3. Please note that while energy-harvesting techniques provide an additional source of energy and help prolong the lifetime of sensor devices, they yield power that is several orders of magnitude lower as compared to the power consumption of state-of-the-art industrial multimedia devices. Hence, they may currently be suitable only for very-low duty cycle devices. An overview of different energy-harvesting techniques is presented in Table 6.4 [5,2,25,33].

TABLE 6.4 Energy-Harvesting Techniques in Wireless Sensor Networks

Energy Source	Performance	Secondary Storage	Commercially Available	Dimension
Primary battery	2880 J/cm³	—	Yes	—
Secondary battery	1080 J/cm³	—	Yes	—
Light (indoor)	10–100 μW/cm²	Yes	Yes	59–590 cm²
Airflow	0.4–1 mW/cm³	Yes	No	6–15 cm³
Vibrations	200–380 μW/cm³	Yes	Yes	16–30 cm³
Thermoelectric	40–60 μW/cm²	Yes	Yes	98–148 cm²
Electromagnetic radiation	0.2–1 mW/cm²	Yes	Yes	6–30 cm²
Piezoelectric	100–330 μW/cm³	Yes	Yes	

6.5.2 Software Development

6.5.2.1 Application Programming Interface

In IWSNs, the application software should be accessible through a simple application programming interface (API) customized for both standards-based and customer-specific requirements. This also enables rapid developments and network deployments [6,42]. With a proper API, the underlying network complexity can be transparent to the end-users who are experts in their specific application domain, but not necessarily experts in networking and wireless communications. Moreover, the deployed sensor network should be able to integrate seamlessly with the legacy fieldbus systems existing in most of the industrial facilities [52].

6.5.2.2 Operating System and Middleware Design

In IWSNs, the design of operating system is very critical to balance the trade-off between energy and QoS requirements. In this regard, TinyOS is one of the earliest operating systems dedicated for tiny sensor nodes [45]. It incorporates a component-based architecture, which minimizes the code size and provides a flexible platform for implementing new communication protocols. It fits in 178 bytes of memory and supports communication, multitasking, and code modularity.

Furthermore, in IWSNs, the design of a proper middleware is required for an efficient network and system management. In this regard, the middleware abstracts the system as a collection of massively distributed objects and enables industrial sensor applications to originate queries and tasks, gather responses and results, and monitor the changes within the network. For example, the sensor information networking architecture provides a middleware implementation of the general abstraction and describes sensor query and tasking language to implement such middleware architecture [41].

6.5.2.3 System Installation and Commissioning

During installation of IWSNs, the system owners must be able to indicate to the system what a sensor is monitoring and where it is. After deployment in the field, network management and commissioning tools are essential. For example, a graphical user display could display network connectivity and help the system owner to set the operational parameters of the sensor nodes. Network management tools can also provide whole-network performance analysis and other management features, such as detecting failed nodes (e.g., for replacement), assigning sensing tasks, monitoring network health, upgrading firmware, and providing QoS provisioning [42].

6.5.3 System Architecture and Protocol Design

6.5.3.1 Network Architecture

In IWSNs, designing a scalable network architecture is of primary importance. One of the design approaches is to deploy homogeneous sensors and program each sensor to perform all possible application tasks. Such an approach yields a flat, single-tier network of homogeneous sensor nodes. An alternative, multitier approach is to utilize heterogeneous elements. In this approach, resource-constrained, low-power elements are in charge of performing simpler tasks, such as detecting scalar physical measurements, while resource-rich, high-power devices (such as gateways) perform more complex tasks [2]. In multitier approaches, the system partitioning/clustering is applied to reduce power dissipation in the sensor nodes by spreading some of the complex energy-consuming computation among resource-rich nodes that are not energy constrained. Generally, IWSNs support several heterogeneous and independent applications with different requirements. Therefore, it is necessary to develop flexible and hierarchical architectures that can accommodate the requirements of all these applications in the same infrastructure [2].

In harsh industrial conditions, device failures can occur because of energy depletion or destruction. In industrial applications, it is also possible to have sensor networks with highly mobile nodes. Furthermore, sensor nodes and the network experience varying task dynamics, and they may be a target for deliberate jamming. Therefore, sensor network topologies are prone to frequent changes after deployment. In this respect, additional sensor nodes can be redeployed at any time to replace the malfunctioning nodes or due to changes in task dynamics. Addition of new nodes poses a need to reorganize the network. Coping with frequent topology changes in an ad hoc network that has myriads of nodes and very stringent power consumption constraints requires special communication protocols.

6.5.3.2 Data Aggregation and Fusion

In IWSNs, local processing of raw data before directly forwarding reduces the amount of communication and improve the communication efficiency (information per bit transmitted). Data aggregation and fusion are typical localized mechanisms for the purpose of in-network data processing in IWSNs. These mechanisms minimize traffic load (in terms of number and/or length of packets) through eliminating redundancy. Specifically, when an intermediate node receives data from multiple source nodes, instead of forwarding all of them directly, it checks the contents of incoming data and then combines them by eliminating redundant information under some accuracy constraints. In this way, dense spatial sampling of events and optimized processing and communication through data fusion can be achieved.

6.5.3.3 Cross-Layer Design

In multi-hop IWSNs, there is an interdependence among functions handled at all layers of the communication protocol stack [2,12,26,40]. Functionalities handled at different layers are inherently coupled due to the shared nature of the wireless communication channel. The physical, MAC, routing, and transport layers together affect the contention for available network resources. The physical layer has a direct effect on multiple access of nodes in wireless channels by changing the interference levels at the receivers. The MAC layer determines the network bandwidth allocated to each node, which naturally influences the performance of the physical layer in terms of successfully detecting the desired signals. On the other hand, as a result of transmission schedules, high packet delays and low bandwidth can occur, forcing the routing layer to modify its route decisions. Different routing decisions change the set of nodes to be scheduled, and thereby impact the performance of the MAC layer. Moreover, congestion control and transmission power control are also inherently coupled, as the capacity available on each link depends on the transmission power [2]. Therefore, technical challenges caused by harsh industrial conditions and application-specific QoS requirements in IWSNs call for new research on cross-layer optimization and design methodologies to leverage potential improvements in exchanging information between different layers of the communication stack. However, it is still important to keep some form of logical separation of these functionalities to preserve modularity, and ease of design and testing [24].

6.6 Conclusions and Future Work

The collaborative nature of IWSNs brings several advantages over traditional wired industrial monitoring and control systems, including self-organization, rapid deployment, flexibility, and inherent intelligent processing capability. Despite the great progress on development of IWSNs, quite a few issues still need to be explored in the future. For example, because of the diverse industrial application requirements and large scale of the network, several technical problems still remain to be solved in analytical IWSN models in terms of communication latency and reliability, and energy efficiency. Other open issues include optimal sensor node deployment, localization, security, and interoperability between different IWSN manufacturers. Finally, to cope with RF interference and dynamic/varying wireless channel conditions in industrial environments, porting a cognitive radio paradigm to a low-power industrial sensor node and developing controlling mechanisms for channel hand-off is another challenging area yet to be explored.

References

1. N. Aakvaag, M. Mathiesen, and G. Thonet, Timing and power issues in wireless sensor networks an industrial test case, in *Proceedings of International Conference on Parallel Processing Workshops,* Oslo, Norway, June 2005.
2. I.F. Akyildiz, T. Melodia, and K. Chowdhury, A survey on wireless multimedia sensor networks, *Computer Networks Journal,* 51(4), 921–960, March 2007.
3. I.F. Akyildiz, W. Su, Y. Sankarasubramaniam, and E. Cayirci, Wireless sensor networks: A survey, *Computer Networks Journal,* 38(4), 393–422, March 2002.
4. L. Angrisani et al., Experimental study of coexistence issues between IEEE 802.11b and IEEE 802.15.4 wireless networks, *IEEE Transactions on Instrumentation and Measurement,* 57(8), 1514–1523, August 2008.
5. S.R. Anton and H.A. Sodano, A review of power harvesting using piezoelectric materials (2003–2006), *Smart Materials and Structures,* 16(3), 1–21, 2007.
6. L.L. Bello et al., Design and implementation of an educational testbed for experiencing with industrial communication networks, *IEEE Transactions on Industrial Electronics,* 54(6), 3122–3133, December 2007.
7. D. Dzung, C. Apneseth, J. Endersen, and J.E. Frey, Design and implementation of a real-time wireless sensor/actuator communication system, in *IEEE Conference on Emerging Technologies and Factory Automation,* Catania, Italy, December 19–22, 2005.
8. J. Garcia et al., Reconfigurable distributed network control system for industrial plant automation, *IEEE Transactions on Industrial Electronics,* 51(6), 1168–1180, December 2004.
9. V.C. Gungor and G.P. Hancke, Industrial wireless sensor networks: Challenges, design principles, and technical approaches, *IEEE Transactions on Industrial Electronics,* 56(10), 4258–4265, October 2009.
10. J.A. Gutierrez, On the use of IEEE Std. 802.15.4 to enable wireless sensor networks in building automation, *International Journal of Wireless Information Networks,* 14(4), December 2007.
11. V.C. Gungor, O.B. Akan, and I.F. Akyildiz, A real-time and reliable transport protocol for wireless sensor and actor networks, *IEEE/ACM Transactions on Networking,* 16(2), 359–370, April 2008.
12. V.C. Gungor, M.C. Vuran, and O.B. Akan, On the cross-layer interactions between congestion and contention in wireless sensor and actor networks, *Ad Hoc Networks Journal,* 5(6), 897–910, August 2007.
13. V.C. Gungor and F.C. Lambert, A survey on communication networks for electric system automation, *Computer Networks Journal,* 50, 877–897, May 2006.
14. P. Hochmuth, Case study: GM cuts the cords to cut costs. *Techworld,* June 2005 Available: http://howto.techworld.com/mobile-wireless/1530/case-study-gm-cuts-the-cords-to-cut-costs/.
15. P. Hamalainen et al., Security in wireless sensor networks: Considerations and experiments, *Embedded Computer Systems: Architectures, Modeling and Simulation,* Samos, Greece, pp. 167–177, July 17–20, 2006.
16. G.P. Hancke and B. Allen, UWB as an industrial wireless solution, *IEEE Pervasive Computing,* 5(4), 78–85, October–December 2006.

17. J. Hauer, V. Handziski, and A. Wolisz, Experimental study of the impact of WLAN interference on IEEE 802.15.4 body area networks, in *Proceedings of EWSN*, Cork, Ireland, February 11–13, 2009.
18. I. Howitt et al., Wireless industrial sensor networks: Framework for QoS assessment and QoS management, *ISA Transactions*, 45, 347–359, 2006.
19. I. Howitt and J.A. Gutierrez, IEEE 802.15.4 low rate wireless personal area network coexistence issues, *IEEE Wireless Communications and Networking*, 3, 1481–1486, 2003.
20. IML Group Plc, Wireless hart specification released, *Control Engineering Europe*, June/July 2007.
21. X. Jiang, J. Polastre, and D. Culler, Perpetual environmentally powered sensor networks, in *Proceedings of IEEE Workshop on Sensor Platform, Tools and Design Methods for Networked Embedded Systems (SPOTS)*, Los Angeles, CA, April 25–27, 2005.
22. E.V. Kar, Z. Lukszo, and G. Leus, Wireless networks in the process industry: Opportunities for ultra-wideband applications, in *Proceedings of IEEE International Conference on Networking, Sensing and Control*, Ft. Lauderdale, FL, April 23–25, 2006.
23. A. Karalisa, J.D. Joannopoulosb, and M. Soljacic, Efficient wireless non-radiative mid-range energy transfer, *Annals of Physics*, 323(1), 34–48, 2008.
24. V. Kawadia and P.R. Kumar, A cautionary perspective on cross layer design, *IEEE Wireless Communication*, 12(1), 3–11, 2005.
25. H. Korber, H. Wattar, and G. Scholl, Modular wireless real-time sensor/actuator network for factory automation applications, *IEEE Transactions on Industrial Informatics*, 3(2), May 2007.
26. U.C. Kozat, I. Koutsopoulos, and L. Tassiulas, A framework for cross-layer design of energy-efficient communication with Qos provisioning in multi-hop wireless networks, *Proceedings of IEEE INFOCOM*, Hong Kong, China, Vol. 2, pp. 1446–1456, March 2004.
27. L. Krishnamurthy et al., Design and deployment of industrial sensor networks: Experiences from a semiconductor plant and the north sea, in *Proceedings of Sensys*, San Diego, CA, November 2005.
28. K.S. Low, W.N.N. Win, and J.E. Meng, Wireless sensor networks for industrial environments, in *Proceedings of International Conference on Computational Modelling, Control and Automation*, Vienna, Austria, pp. 271–276, 2005.
29. B. Lu and V.C. Gungor, Online and remote energy monitoring and fault diagnostics for industrial motor systems using wireless sensor networks, in *IEEE Transactions on Industrial Electronics*, 56(11), 4651–4659, November 2009.
30. P. Mannion, Handset forms sensor gateway, *Electronic Engineering Times*, October 2007.
31. P.D. Mitcheson, T.C. Green, E.M. Yeatman, and A.S. Holmes, Architectures for vibration-driven micropower generators, *Journal of Microelectromechanical Systems*, 13(3), 429–440, 2004.
32. G. Montenegro, N. Kushalnagar, J. Hui, and D. Culler, Transmission of IPv6 Packets over IEEE 802.15.4 networks, *IETF RFC 4944*, September 2007.
33. J. Paradiso and T. Starner, Energy scavenging for mobile and wireless electronics, *Proceedings of IEEE Pervasive Computing*, 4(1), 18–27, 2005.
34. A. Perrig, J. Stankovich, and D. Wagner, Security in wireless sensor networks, *Communications of the ACM*, 47, 53–57, 2004.
35. F. Pellegrini, D. Miorandi, S. Vitturi, and A. Zanella, On the use of wireless networks at low level of factory automation systems, *IEEE Transactions on Industrial Informatics*, 2, 129–143, 2006.
36. S. Pollin et al., Harmful coexistence between 802.15.4 and 802.11: A measurement-based study, in *Proceeding of IEEE CrownCom*, Piscataway, NJ, May 2008.
37. D. Porcino and W. Hirt, Ultra-wideband radio technology: Potential and challenges ahead, *IEEE Communications Magazine*, 41, 66–74, 2003.
38. G.J. Pottie and W.J. Kaiser, Wireless integrated network sensors, *Communications of the ACM*, 43(5), 551–558, May 2000.
39. H. Ramamurthy, Wireless industrial monitoring and control using a smart sensor platform, *IEEE Sensors Journal*, 7(5), 611–618, May 2007.

40. S. Shakkottai, T.S. Rappaport, and P.C. Karlsson, Cross layer design for wireless networks, *IEEE Communications Magazine*, 41, 74–80, October 2003.
41. C-C. Shen, C. Srisathapornphat, and C. Jaikaeo, Sensor information networking architecture and applications, *IEEE Personal Communications,* 52–59, August 2001.
42. X. Shen, Z. Wang, and Y. Sun, Wireless sensor networks for industrial applications, in *Proceedings of Fifth World Congress on Intelligent Control and Automation*, Hangzhou, China, June 15–19, 2004.
43. S. Soucek and T. Sauter, Quality of service concerns in IP-based control systems, *IEEE Transactions on Industrial Electronics,* 51(6), 1249–1258, December 2004.
44. W. Stallings, *Network and Internetwork Security: Principles and Practice*, Prentice-Hall, Englewood Cliffs, NJ, 1995.
45. TinyOS, [Online] Available: http://www.tinyos.net
46. U.S. Department of Energy, Sensors and automation eaton wireless sensor network for advanced energy management solutions, June 2004.
47. U.S. Department of Energy, Industrial wireless technology for the 21st century, Office of Energy and Renewable Energy Report, 2002.
48. U.S. Department of Energy, Sensors and automation annual report, Office of Energy and Renewable Energy Report, 2004.
49. U.S. Department of Energy, Assessment study on sensors and automation in the industries of the future, Office of Energy and Renewable Energy Report, 2004.
50. M.C. Vuran and I.F. Akyildiz, Cross-layer analysis of error control in wireless sensor networks, in *Proceedings of IEEE SECON*, Reston, VA, September 2006.
51. A. Willig et al., Measurements of a wireless link in an industrial environment using an IEEE 802.11-compliant physical layer, *IEEE Transactions on Industrial Electronics,* 49(6), 1265–1282, December 2002.
52. A. Willig, K. Matheus, and A. Wolisz, Wireless technology in industrial networks, *Proceedings of the IEEE*, 93, 1130–1151, 2005.
53. W. Zeng, H. Wang, H. Yu, and A. Xu, The research and application of UWB based industrial network, in *Proceedings of Industrial Conference on Ultrawideband and Ultrashort Impulse Signals,* Sevastopol, Ukraine, pp. 153–155, 2006.
54. ZigBee Alliance Home Page, [Online] Available: www.zigbee.org

7

Ad Hoc Networks

Sajjad Ahmad Madani
COMSATS Institute of Information Technology

Shahid Khattak
COMSATS Institute of Information Technology

Tariq Jadoon
Lahore University of Management Sciences

Shahzad Sarwar
University of the Punjab

7.1 Introduction 7-1
Principles and Benefits • Applications • Ad Hoc Network Characteristic • Enabling Technologies
7.2 Protocol Stack 7-4
Transport Layer • Network Layer • MAC Layer • Physical Layer
7.3 Performance Evaluation 7-8
7.4 Challenges and Issues 7-9
Quality-of-Service • Energy Management • Topology and Connectivity • Security
References 7-11

7.1 Introduction

7.1.1 Principles and Benefits

Ad hoc network is a collection of mobile nodes that dynamically organize themselves to form arbitrary and temporary networks. These mobile nodes can get connected without any preexisting infrastructure or centralized administration where connection is facilitated through peer-to-peer (P2P) communication which may involve multihop structures. The mobile nodes are free to move and organize themselves in arbitrary fashion while communicating with each other. There can be multiple paths employing heterogeneous radio between any pair of nodes. Ad hoc networks range from small single-hop personal area networks (PANs) to large-scale multihop networks involving thousands of nodes. Based on the coverage area, the ad hoc networks are classified into four main categories: body, personal, local, and wide-area networks. These either connect to Internet through fixed infrastructure or may operate as an isolated P2P network in a stand-alone fashion as shown in Figure 7.1a and b, respectively.

Although the idea of ad hoc network was first conceived in early 1970s, it did not generate wider interest earlier as the original application scenarios were not directed to mass users. Recently, new concepts of opportunistic ad hoc networks have emerged which have commercial applications as well. Here, ad hoc network is used opportunistically to extend home or campus networks to previously inaccessible areas. The development of sensor networks, which can be treated as specialized ad hoc networks, has also contributed to an increased interest in this area.

Unlike the classical cellular network that requires prelocated cell sites and base stations, an ad hoc network does not require any detailed infrastructure planning as it relies on highly dynamic network topologies to deal with hostile environment and irregular connectivity. Since an ad hoc network automatically establishes the communication channels (links) and adopts to change, the setup time is short and cost is low. It can, therefore, be used to complement overloaded fixed wireless infrastructures (hot spots) and can

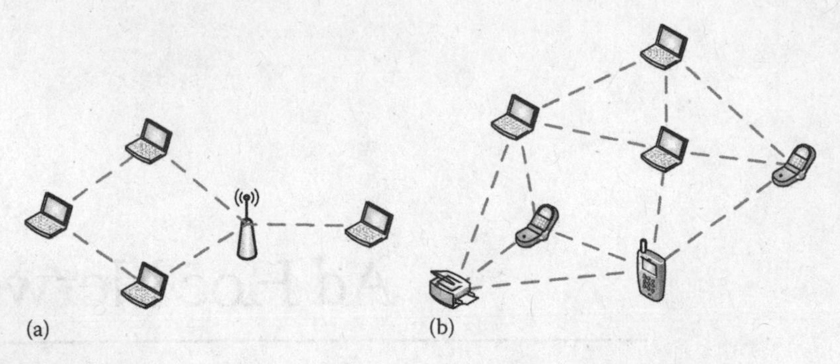

FIGURE 7.1 Ad hoc networks: (a) fixed infrastructure network and (b) peer-to-peer network.

TABLE 7.1 Differences between Ad Hoc Networks and Conventional Cellular Systems

Conventional Systems	Ad Hoc Networks
Requires infrastructure	No infrastructure required
Requires careful planning before setting up base stations	Adapts to changing network conditions
Static backbone network topology	Dynamic network topologies
Predictable network conditions	Highly unpredictable network conditions
Relatively more secure	More susceptible to malicious attacks

also serve as an emergency backup in case of massive fixed infrastructure failure due to war, natural, or industrial disaster. It can extend the service area of the access networks and provide wireless connectivity into areas with limited or no coverage. Although ad hoc architecture offers several benefits such as self-reconfiguration and adaptability to highly variable mobile characteristics, it poses several new challenges primarily resulting from the unpredictability of the network topology due to mobility and count of nodes. The key differences of ad hoc networks with conventional cellular systems are summarized in Table 7.1.

7.1.2 Applications

The evolution toward a seamless connectivity through a heterogeneous network is made possible due to progress in wireless communication standards, radio access technologies, and network architecture. Although still in its infancy, developments in these areas are already resulting in several novel business models.

Wireless home networks are now offering to move broadband data around the house without wires through P2P routing and without requiring each device to be in close proximity of a centralized server. Military applications are another key area, where P2P technology is implemented as "mesh networks." These networks enable instant and reliable communication among troops across the battlefield without requiring predeployed large structures, antennas, or traditional backhaul connections. Similarly, this concept is applicable to establish communication among the public safety workers (of police, fire-fighting, and rescue department), allowing them to establish/extend networks by simply bringing mesh-enabled devices to the site of an incident. Ad hoc networks can be deployed in mining companies and other enterprises, which operate in extremely difficult environment, in order to improve communication and safety of workers. Also, ad hoc technology can be used to extend wireless connectivity to hard-to-reach areas such as old buildings that are difficult to wire, or to buildings that are too distant to reach with Wi-Fi. Another interesting application of the ad hoc networks is in creating an intelligent transportation system that improves safety and reduces congestion on roads. As shown in Figure 7.2, each vehicle incorporates smart devices that monitor and communicate directly with each other in order to optimize timing, warning of danger, and conveys this information to the traffic control server for strategic planning. Ad hoc technology enables

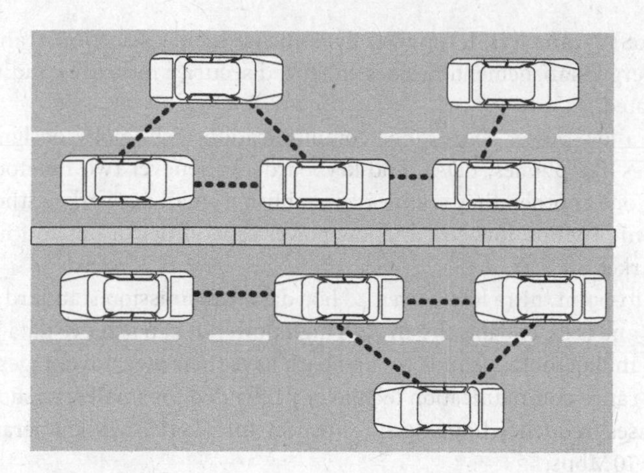

FIGURE 7.2 Vehicular ad hoc networks.

rapid deployment of these systems at a lower cost, without having to lay cables to each sensor. To sum up, ad hoc networks and alternative technologies are enabling direct intelligent communication between entities. This new paradigm is poised to change the way we interact with the world and perform different tasks.

7.1.3 Ad Hoc Network Characteristic

One of the key characteristics of ad hoc networks is mobility due to rapid repositioning of nodes. These nodes can either move independently or are grouped together while their movement can be random or along preplanned routes. These mobility models can have major impact on selecting an appropriate routing scheme, which may influence the performance of the network. Another characteristic of ad hoc networks is multihopping where the path from the source to destination includes several other intermediate nodes. Wide-area ad hoc networks often exhibit multiple hops for obstacle negotiation, spectrum reuse, and energy conservation. Besides, an ad hoc network must effectively handle problems pertaining to addressing, routing, clustering, position identification, and power control, just to name a few.

Another important feature of the ad hoc networks is energy conservation. Designing energy-efficient protocol is critical for prolonged operation since most ad hoc nodes have limited power. The ad hoc networks can sometimes grow up to several thousand nodes, all moving in an unpredictable manner. Although it may be possible to find an ad hoc solution for a few fast nodes or for a very large number of static nodes, the problem arises when a large number of heterogeneous nodes move in random direction with different speed over an unpredictable terrain. Thus, multihop, mobility, and large network size combined with varying device characteristics (heterogeneity, bandwidth, and battery power constraints) make the design of adequate routing protocol a major challenge. Besides, ad hoc network should be able to prevent any attempt from intruders to eavesdrop and jam the channel. Due to open P2P network architecture, the ad hoc networks are more vulnerable against malicious attacks than the infrastructured counterparts. These attacks can be active or passive; either an attacker actively disrupts network operations or simply monitors data, controls traffic patterns, and relays this information to the enemy headquarters. The security for ad hoc networks, therefore, involves key establishment, trust setup, secure routing, authentication, data aggression, etc., and is therefore a lot more challenging than in conventional networks.

7.1.4 Enabling Technologies

IEEE 802.11 WLAN [802.11] is the most popular technology for ad hoc networks. There are various working groups for the standards offering highest data rate up to 54 Mbps and enhancing quality-of-service (QoS) parameters. It supports dynamic frequency selection, transmit power control, and spectrum and energy management. Besides enhanced security measures, radio resource management is also incorporated.

Bluetooth [B04] is a low-power short-range communication technology designed to connect portable electronic devices like phones, PDAs, and keyboards. Whenever two Bluetooth-enabled devices come within range of one another, they seamlessly establish a small network. Although the maximum possible data rate is only 1 Mbps, low-cost and low-power capabilities of Bluetooth have stimulated its penetration in the market.

Infrared is a point-to-point, ultra low-power, ad hoc data transmission standard designed to operate over a distance of 1 m and is extendable to 2 m with high power. It can achieve data rates up to 16 Mbps. Infrared is cheaper than Bluetooth technology, and both have their own advantages and disadvantages.

HomeRF is a short-range communication technology intended for small area such as homes or small buildings. HomeRF uses frequency hopping spread spectrum (FHSS) while operating at 2.4 GHz and offers data rates up to 10 Mbps.

ZigBee standard is a short-range wireless communication standard for PANs. It offers data rate up to 250 kbps at 2.4 GHz, 40 kbps at 915 MHz, and 20 kbps at 868 MHz with a range of 10–100 m. ZigBee uses IEEE 802.15.4 as physical and medium access control (MAC) layers. Upper layers as well as ZigBee security architecture are defined by the ZigBee standard.

7.2 Protocol Stack

7.2.1 Transport Layer

Transport layer protocols are responsible for end-to-end delivery of data. Transmission control protocol (TCP) is a dominant transport layer protocol for wired networks. TCP is responsible for congestion and flow control, and reliable and in-order delivery of packets. The unique characteristics of wireless ad hoc networks such as lack of infrastructure, mobility, shared bandwidth [SAHS03], contention, and high bit error rate (BER) as well as the design principles of TCP motivate to design customized transport layer protocols.

7.2.1.1 Why TCP Does Not Suit Ad Hoc Networks?

The failure of TCP to work well in ad hoc networks is because of the following known problems:

- In layered architectures, TCP implicitly assumes that packet loss is caused due to collisions. Whereas, collisions in ad hoc wireless networks are one of the possible causes of packet loss, although there are also other potential causes such as fading, varying link quality [CZWF04], interference, and noise.
- Two communicating nodes as well as other nodes sharing the same medium contend for the medium.
- The packet loss in wireless networks is comparatively higher, and loss of retransmitted packet further degrades the performance.

7.2.1.2 Transport Layer Protocols for Ad Hoc Networks

Transport layer protocols for ad hoc networks can broadly be classified into two major approaches [MK04]; TCP variants and non-TCP variants.

The basic idea of TCP variants is to retain TCP as a transport layer protocol, because of its global existence, and to suggest modifications in order to overcome the problems associated with wireless links and mobility. TCP feedback (TCP-f) [F94] is customized in such a way that the sender is able to differentiate between congestion and a lost link. In this way, the invocation of congestion control algorithm is restricted, which in turn stops performance degradation. In case of link failure, the sender node is explicitly notified by the network layer of the neighbor node. The sender stops sending further packets, while all other nodes listening to these notifications invalidate these routes to avoid packet loss. Explicit link failure notification (ELFN) is a technique similar to TCP-f where route failures are notified to downstream nodes in ELFN message sent by upstream nodes of the failed link. Another TCP variant is ad hoc TCP [LS01] which requires network layer feedback and split TCP [KKFT02] to deal with congestion control and end-to-end reliability separately.

In the non-TCP variant approach, the transport layer is built from scratch while considering the limitations of wireless ad hoc networks. Although this approach outperforms TCP variants approach in a stand-alone environment [MK04, p. 130], it poses many challenges when the machines with such transport layer have to talk to the global Internet. Ad hoc transport protocol (ATP), one of the non-TCP variants, is specifically built for ad hoc networks. ATP exploits synergies between different layers to enhance performance. As mentioned earlier, one of the disadvantages of ATP is its incompatibility with systems that are running plain TCP.

A reliable transport protocol for ad hoc networks is still needed. While designing a transport layer protocol, the key trade-off is between compatibility and performance of the network.

7.2.2 Network Layer

Routing is used to send the data from source to destination. The easiest way is to broadcast the data in the network that is wasteful of the bandwidth resource. A more efficient way is to compute route from source to destination and send data in a multihop fashion. There are different ways to classify routing protocols. The routing protocols can be classified on the basis of (a) the way the routing information is updated, (b) network structure (flat and hierarchical), and (c) position and nonposition-based routing protocols. The different categories are discussed in the following sections.

7.2.2.1 Proactive, Reactive, and Hybrid Routing Protocols

Proactive routing protocols are also known as table-driven protocols. The consistent and up-to-date routing tables are computed ahead of transmission time. Whenever topology changes, update messages are triggered to assure consistency of network map being maintained at each node. Destination-sequenced distance vector (DSDV) [PP94] is a proactive routing protocol, which uses Bellman Ford algorithm to compute the shortest path and ensures loop-free routing tables. Wireless routing protocols (WRP) [MSGJ96], based on improved Bellman Ford, restrict route updates to immediate neighbors only. The cluster gateway switch routing (CGSR) reduces routing table size and routing information exchange by employing a clustering hierarchy where only the cluster heads communicate among themselves. Optimized link state routing (OLSR) is a distributed, table-driven, and proactive routing protocol for mobile ad hoc networks that work on the principle of "multipoint relays," which means selecting a set of nodes from within one-hop neighbor to provide ability to reach all the two-hop neighbors. OLSR works well in large-scale and high-density network but incurs large routing overhead.

Overhead of periodic updates, slow route repair, and maintenance of unused routing information are few of the problems of the proactive routing protocols. To overcome these issues, reactive routing protocols have been designed.

Unlike the proactive routing protocols, the reactive routing protocols do not maintain routing tables and compute routes whenever required. These are also known as "on-demand" routing protocols. Route discovery is accomplished by flooding request in the network. Although, less control overhead makes them scalable, larger delays are incurred while discovering route whenever needed. Dynamic source routing (DSR) [DD96] and ad hoc on-demand distance vector (AODV) [PR99] are two examples of reactive protocols. AODV is based on DSDV but is reactive and loop free. Temporally ordered routing algorithm (TORA) [PC97] is a reactive scheme and is uniquely featured by maintaining multiple paths between given pair of source and destination nodes.

There are different routing protocols which exhibit a hybrid approach for routing, for example, zone-based routing (ZBR) [DHY03]. ZBR is more suitable for large networks where the network is divided into clusters. Intracluster routing is proactive while intercluster routing is reactive in ZBR. Other protocols in this category comprise core extraction distributed ad hoc routing (CEDAR) protocol [SSB99] and zone-based hierarchical link state (ZHLS) routing protocol.

7.2.2.2 Flat and Hierarchical Routing Protocols

Hierarchical routing protocols arrange nodes in the form of clusters or trees where every cluster has a cluster head. Cluster heads may also aggregate data from cluster nodes to reduce the number of packet transmissions and hence conserve energy [CPF05] or may provide gateway to external networks. Advantages of hierarchical routing protocols include scalability and efficient communication [AK04]. AODV, DSDV, and DSR are examples of flat routing protocols, while CGSR is an example of hierarchical routing scheme.

7.2.2.3 Position- and Nonposition-Based Routing Protocols

It is experimentally verified that reactive and proactive routing protocols, including AODV, DSDV, and DSR that do not use location information in routing decisions, face scalability issues as opposed to location-based routing strategies [S02]. Location-based routing depends upon the physical position of the nodes in the network to take routing decisions [MWH01]. To get location/position information, the node may use a low-power global positioning system (GPS) if the nodes are outdoor or may rely on relative positioning techniques based on the signal strength or manual registration process. Several position-based routing schemes [SL01] have already been presented by the research community. Position-based routing algorithms are classified into greedy and restricted directional flooding. In greedy approaches, the distance toward the sink node is either maximized or minimized depending upon forwarding strategy, while in directional flooding, the data are always flooded toward the nodes which are in the direction of the destination node. Greedy perimeter stateless routing (GPSR) [KK00] and geographic distance routing (GEDIR) [SL01] are examples of position-based routing protocols working on the principle of greedy forwarding approach.

7.2.3 MAC Layer

MAC schemes are used to define policy for accessing the shared medium in ad hoc networks. Limited spectrum, multiple access, node mobility, error-prone environment, and characteristics of wireless medium like noise, fading, and interference compel to design MAC schemes customized for wireless environments. The MAC schemes for ad hoc networks can be classified into contention-free and contention-based protocols. Contention-free protocols work better in infrastructure-driven networks with a master node controlling the medium access. Contention-based protocols work in a decentralized fashion and can be divided into random access and controlled access (scheduling- and reservation-based) protocols. The performance of such protocols is deteriorated by hidden and exposed terminal problems and requires special attention.

Let us assume that node-1 is sending data to node-2, as shown in Figure 7.3. Furthermore, assume that node-3 is also about to send data to node-2. As node-3 is out of the transmission range of node-1,

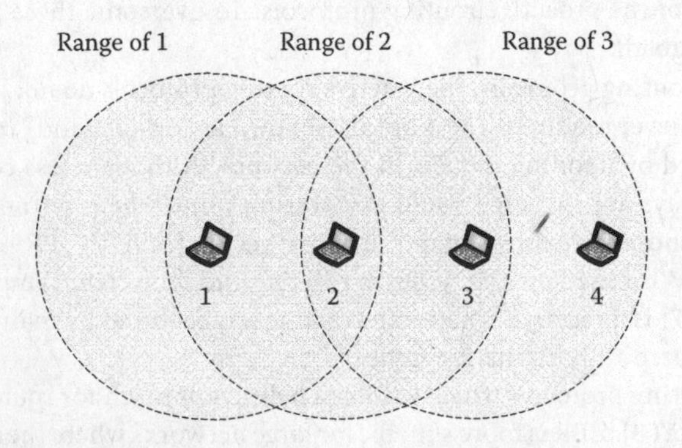

FIGURE 7.3 Hidden and exposed terminal problem.

it will sense the medium as free and sends data to node-2 which will result in a collision. This is known as hidden terminal problem; node-1 and node-3 are hidden from each other. This issue is solved by request-to-send/clear-to-send (RTS/CTS) mechanism. Now, assume that node-3 knows by RTS/CTS mechanism that node-1 and node-2 are communicating; it will refrain from communicating with node-4 as well, though node-4 is beyond the range of node-2 and thus channel capacity will be unnecessarily wasted. This is known as exposed terminal problem.

7.2.3.1 Random Access Schemes

In random access schemes (also know as contention-based scheme), nodes contend for the shared channel with the neighboring nodes (nodes within interference range) and the winning nodes get hold of the medium. Such schemes cannot guarantee QoS because access to the channel is not guaranteed and is random. In ALOHA [N70], a node simply sends data as soon as it is available. Slotted ALOHA introduces time division multiple access (TDMA)–like time-slots to reduce the number of collisions. Carrier sense multiple access (CSMA)–based schemes further enhance the performance by carrier sense mechanism. The performance of all such schemes is adversely affected by hidden and exposed terminal problem.

7.2.3.2 Reservation-Driven Contention-Based Access

In order to overcome the problems of hidden and exposed terminal problems, dynamic reservation-based protocols are designed, which operates on the basis of RTS/CTS mechanism. In such protocols, a node initiates the reservation process if it has some data to send. The sending node sends an RTS packet to the next hop. The next hop in turn replies by a CTS message. Such a CTS is heard by other nodes in the vicinity, and they keep quiet unless the data are actually sent by the initiating node. MAC layer defined by IEEE 802.11 standard includes distributed coordination function (DCF) and point coordination function (PCF). DCF is based on CSMA/CA and uses RTS-CTS-DATA-ACK mechanism for data transmission, while PCF is designed for infrastructure-based networks. Other important protocols in this category includes multiple access collision avoidance (MACA) [K90], MACA for wireless LANs (MACAW) [BDSZ94], and floor acquisition multiple access (FAMA) [FG95], which are all based on CSMA/CA.

7.2.3.3 Scheduling-Driven Contention-Based Access

Such protocols are designed to provide a certain level of QoS. Scheduling can be a function of different parameters, for example, remaining node energy, traffic loads, and/or bound on packet delays. Some of the scheduling-driven MAC schemes include distributed priority scheduling and medium access [KLSSK02] and distributed laxity-based priority scheduling scheme [KMS05].

7.2.4 Physical Layer

Physical layer translates communication requests from the upper layers into hardware-specific operations to enable transmission or reception of signals. These operations normally involve time and frequency synchronization while also dealing with harsh time variant and frequency-selective wireless channels. Besides, the performances can be severely compromised by interference originating from other terminals operating on nonorthogonal resources, thereby requiring efficient receive strategies which give good complexity performance trade-off. Here, we give an overview of physical layer specification for only two main standards supporting ad hoc wireless networks, i.e., the IEEE 802.11 standard for WLANs [802.11] and the Bluetooth specifications for short-range wireless communications [B01,FBP98]. The first one not only allows single-hop WLAN ad hoc network but can also be extended to multihop networks covering areas of several square kilometers, while the second one can only be used to build smaller scale ad hoc wireless body and PANs.

IEEE 802.11 is the first wireless local area network standard with data rates of up to 2 Mbps [802.11]. The original standard has then been extended to 802.11b, 802.11a, and 802.11g which operate in the

2.4 and 5 GHz unlicensed band. The peak data rates of 11 Mbps are supported by 802.11b and 54 Mbps for 802.11a and 802.11g. Different transmission technologies such as infrared, orthogonal frequency domain multiplexing (OFDM), FHSS, and direct sequence spread spectrum (DSSS) can be employed. For any given bandwidth, these standards support different data rates by selecting different modulation schemes, i.e., DBPSK, DQPSK, 16-QAM, etc. Besides, half rate convolution codes are used with puncturing to ensure reliable data transfer for a given QoS. Synchronization is achieved by sending a predefined sequence which alerts the receiver that a signal is present. This is followed by start frame delimiter which defines the beginning of a frame. The next generation for 802.11 is 802.11n, which is designed to effectively replace all the previous 802.11 standards and enables speed up to 540 Mbps.

The Bluetooth technology is the commonly adopted standard for low-cost, short-range radio links between different portable devices [MB00]. Also, based on the portions of the Bluetooth specification, the IEEE 802.15 has been put forward as a WPAN standard. Like 802.11, the Bluetooth system is operating in the 2.4 GHz industrial, scientific, and medicine band (ISM). The modulation uses Gaussian frequency shift keying (GFSK) with a maximum frequency deviation of 140–175 kHz. The transmission technology adopted is FHSS, where hopping at up to 1600 hops/s among 79 channels, which are spaced at 1 MHz separation. The maximum transmission power depends upon which of the three power classes (as defined by the standard) are supported and ranges from 1 to 100 mW. A maximum base-band data rate of 723.2 kbps is supported for each link, with options for 1/3 rate repetition and 2/3 rate Hamming forward error correction codes.

7.3 Performance Evaluation

Traditionally, the performance evaluation of ad hoc networks has been done on a layer-by-layer basis. New protocols are proposed at a certain layer, and performance studies compare them with one another for a number of scenarios and varying conditions. The key issues in the design and performance of MANETs are comprehensively presented in the authoritative reference [CCL03]. Scores of studies compare MAC protocols for both random access and reservation-based schemes for use in ad hoc wireless networks presenting novel ideas. Similarly at the network layer, the performance evaluation of a number of routing protocols for a variety of workload scenarios has been simulated [B04]. More recent and comprehensive simulation results using OPNET Modeler 10.5 [IR1] have been carried out for the most commonly used routing protocols, for instance AODV, DSR, and TORA [AA06]. Similarly, the performance of transport layer protocols over MANETs has been considered in a number of studies and schemes such as delaying ACKs [CGLS08] or using a fractional window increment (FeW) have been recently presented [WHX09]. In the same vein, the impact of P2P traffic on various routing protocols in MANETS has been evaluated in [OSL05]. A consideration that cuts across all studies is the choice of the mobility model. Most studies assume a random way point mobility model which may not be the most appropriate model for ad hoc networks. A comprehensive framework for evaluating the impact of mobility in ad hoc networks is presented in [BSH03], and more recent studies have evaluated the impact of the mobility model on routing protocols [TBD07]. A further consideration especially for sensor networks is energy efficiency, and a number of studies present energy-efficient schemes both at the MAC and at the network layer in order to increase node life.

The performance modeling of ad hoc networks is a complex problem, primarily due to the fact that the research community does not have a unified framework for understanding the interaction of MAC layer, congestion, interference, network coding, and reliability [KHP08]. The performance evaluation of ad hoc networks is further compounded by the choice of the mobility model as well as by the traffic patterns [PDH07], wherein it has been demonstrated that the choice of more representative traffic patterns might produce results that are different than traditional choices of traffic sources. Cross-layer designs have been a major theme, and novel stacks have been proposed to improve the performance of ad hoc and sensor networks. Understanding the role and interactions between each subcomponent will improve our understanding of these complex distributed systems.

7.4 Challenges and Issues

7.4.1 Quality-of-Service

With the increase in portable computing devices and demand for multimedia applications, the importance of providing QoS in ad hoc networks will certainly grow. There are two key challenges in providing QoS over ad hoc networks: the fickle nature of the radio channel and the mobility of the nodes. The varying quality of the transmission channel due to interference and fading results in dynamically changing BERs and bandwidth throughput as well as problems for the network layer in distinguishing between congestion or link-layer loss. Furthermore, it is difficult to apply conventional techniques that estimate the effective bandwidth at a node for providing QoS in order to ensure bandwidth reservation, admission control, policing, and flow control in ad hoc networks. The survey papers by [KRD06] and [RKMM06] provide an excellent overview to the issues and challenges by providing a layer-wise classification and QoS frameworks for ad hoc networks. QoS over multihop networks poses further challenges as each node may be forwarding multiple flows of varying demands [MRV09]. Recent advances in multiple input multiple output (MIMO) technology have led to a growing interest in multiantenna wireless ad hoc networks where each node has more than one antenna. The advances in MIMO technology result in significant gains in link capacity while combating multiple path interference and multiple-user interference, making cross-layer approaches to providing QoS attractive [GLB09].

7.4.2 Energy Management

Since ad hoc wireless networks are mostly collection of battery-operated portable devices, efficient energy management is critical. These networks offer communication using intermediate mobile hosts to route information, severely straining their already limited power. Additionally, in wireless sensor networks, many devices are operated by batteries that cannot be replaced at all. Therefore, energy-efficient designs are required for these devices to enable them to operate for prolonged period of time without replacing batteries. The power-conservative schemes are classified into transmitter power-control mechanisms and power-management algorithms.

Power control involves controlling the host transmission powers in accordance with distances between the terminals so that any required criterion for the received power can be satisfied. In [RR00], two centralized algorithms are proposed which output the transmission power of each host by constructing either a uni-connected network or a bi-connected network. Although these algorithms are designed for static networks where location of all hosts is known, they allow a continuous adjustment of the host transmission powers to maintain a certain QoS. These algorithms, however, require a guaranteed knowledge of global connectivity without which they may lead to a disconnected network. This global network connectivity is ensured in [WLBW01], which assumes that each receiver is able to determine the direction of the sender when receiving a message.

Power-management strategies are applied to MAC or higher layers of OSI model. The power consumption of hosts in ad hoc networks strongly depends on the MAC policy [WESW98]. In [BCD99,BCD01], the authors propose a distributed contention control (DCC) protocol, which preemptively estimates the probability of successful transmission before transmitting each frame and defers transmission if the probability of success is too low. Thus, battery energy is saved by avoiding unnecessary retransmissions, which consume significant power. In [RS98], a power-aware multiaccess protocol with signaling (PAMAS) is introduced for power conservation in ad hoc networks, which proposes separate data and signaling channels. The signaling channel is used for exchanging RTS/CTS packets which enables a host to determine when and for how long to power its antenna off. Similarly, several power-saving techniques are proposed [CT00,SL01] for the network layer for ad hoc networks. For route selection, these protocols not only take into account the traditional routing protocol metrics such as hop-count, link quality, and congestion but also consider battery power levels. Power-management protocols are also implemented for transport layer where the TCP is modified to respond appropriately to unreliable wireless environment by avoiding retransmission and effective congestion control [KK98]. In addition, power-efficient strategies are also suggested for the application layer where the operating system (OS) is modified to handle the hardware and software more efficiently [LS98].

7.4.3 Topology and Connectivity

In most of the ad hoc networks, P2P communication topology is required, which means each node in the network must be able to communicate with any other node in the same network. To attain this, connectivity between the entire networks must be provided all the times. But the limited resources, uncontrolled mobility of nodes, and the characteristics of wireless communication channels make it a challenging issue. Topology and connectivity control can have a significant impact on the performance of wireless networks. Such algorithms are used to enhance energy efficiency, minimize interference, and increase effective rate by adjusting transmission power.

Topology control algorithms can be categorized into centralized and distributed [SBC03]. Centralized approach need to have the global knowledge of the network and works well in static networks. Distributed approach uses localized information only and works well in mobile networks. There are many issues with topology control algorithms. Optimal transmit power computation is one of the main challenges. If high transmit power is used, interference /contention range is increased. If lower power is used, it may result in a disconnected network. Furthermore, adjusting power level may introduce additional latency. In centralized approach, where all the information is sent to a central authority, reducing processing and communication overhead is a challenging task. Efficient topology control algorithms which take node mobility into account still require a great deal of attention.

7.4.4 Security

Security is required for secure communication and to maintain desired network performance in ad hoc network [YLYLZ04]. Security issues in ad hoc networks are more challenging than in conventional networks because of open access mechanism and P2P architecture, no infrastructure, highly dynamic network topology, publicly known protocols, stringent resource constraints, and unattended operation.

Security for ad hoc networks entails key establishment and trust setup, secrecy and authentication, privacy, secure routing, intrusion, and secure data aggregation [PSW04]. Three main goals of security as discussed in [DPT03] are to maintain confidentiality (data protection), integrity (unauthorized data tampering), and availability (service availability anywhere and anytime). Common security threats outlined in [PSS00] and discussed in [DPT03] include "stealing confidential information, tampering information, resource stealing, and denial-of- service attacks."

Following is the summary of the open research opportunities and challenges in the security implementation for ad hoc networks, which are discussed in [ZFZ08,R96,RFLW96]. Because of the wireless environment and lack of firewall-like devices, the eavesdroppers can spoof the ongoing communication that compromises confidentiality, privacy, and integrity of data. On the network layer, malicious nodes can introduce denial-of-service attacks, holes, and tamper information compromising basic security primitives, for example, availability and integrity of data. Therefore, designing secure routing protocols is one of the challenging areas in ad hoc networks. [T97] and [SMG97] are a few efforts in this direction. The major challenge is the authentication of the public keys, which can be solved by using identity-based cryptography. As the security operations are very frequent in ad hoc networks, it requires an efficient symmetric key algorithm for authentication and encryption. Key revocation should happen when a node is declared as malicious or when it is under compromise node attack. Although different key revocation schemes are working, still there is a demand for universal key revocation scheme [ZFZ08].

References

[802.11] *IEEE standard for Wireless LAN Meduim Access Control: MAC and PHY Specifications*, ISO/IEC 8802-11, 1999(E), 1999.

[AA06] S. Ahmed and M. S. Alam, Performance evaluation of important ad-hoc network protocols, *EURASIP Journal on Wireless Communications and Networking*, 1–11, 2006.

[AK04] J. N. Al-Karaki and A. E. Kamal, Routing techniques in wireless sensor networks: A survey, *IEEE Wireless Communications*, 11, 6–28, 2004.

[B01] C. Bisdikian, An overview of the bluetooth wireless technology, *IEEE Communication Magazine*, December 2001.

[B04] A. Boukerche, Performance evaluation of routing protocols for ad-hoc wireless networks, *Journal of Mobile Networks and Applications*, 9, 333–342, 2004.

[BCD01] L. Bononi, M. Conti, and L. Donatiello, A distributed mechanism for power saving in IEEE 802.11 Wireless LAN, *ACM Journal of Mobile Networks and Applications* (*MONET*), 6, 211–222, 2001.

[BCD99] L. Bononi, M. Conti, and L. Donatiello, A distributed contention control mechanism for power saving in random-access ad-hoc wireless local area networks, in *IEEE International Workshop on Mobile Multimedia Communications* (*MoMuC*), San Diego, CA, pp. 114–123, November 15–17, 1999.

[BDSZ94] V. Bhargavan, A. Demers, S. Shenker, and L. Zhang, MACAW: A media access protocol for wireless LANs, *ACM SIGCOMM Computer Communication Review*, 24(4), 212–225, 1994.

[BSH03] F. Bai, N. Sadagopan, and A. Helmy, The IMPORTANT framework for analyzing the Impact of Mobility on Performance Of RouTing protocols for Adhoc NeTworks, *Journal of Ad-Hoc Networks*, 1, 383–403, 2003.

[CCL03] I. Chlamtac, M. Conti, and J. J.-N. Liu, Mobile ad-hoc networking: Imperatives and challenges, *Journal of Ad-Hoc Networks*, 1, 13–64, 2003.

[CGLS08] J. Chen, M. Gerla, Y. Z. Lee, and M. Y. Sanadidi, TCP with delayed ACK for wireless networks, *Journal of Ad-Hoc Networks*, 6, 1098–1116, 2008.

[CPF05] K. S. Chan, H. Pishro-Nik, and F. Fekri, Analysis of hierarchical algorithms for wireless sensor network routing protocols, *IEEE Wireless Communications and Networking Conference*, 3, 13–17, March 2005.

[CT00] J.-H. Chang and L. Tassiulas, Energy conserving routing in wireless ad-hoc networks, in *19th Annual Joint Conference of the IEEE Computer and Communications Societies (INFOCOM)*, vol. 1, Tel-Aviv, Israel, pp. 22–31, March 2000.

[CZWF04] X. Chen, H. Zhai, J. Wang, and Y. Fang, TCP performance over mobile ad-hoc networks, *Canadian Journal of Electrical and Computer Engineering*, 29, 129–134, 2004.

[DD96] D. B. Johnson and D. A. Maltz, Dynamic source routing in ad hoc wireless networks, in *Mobile Computing*, Kluwer, Boston, MA, 1996.

[DPT03] D. Dietrich, P. Palensky, and A. Treytl, Communication in automation with the emphasis on security, in *EUSAS (European Society for Automatic Alarm Systems) Workshop*, Rome, Italy, pp. 89–104, June 16–17, 2003.

[F94] S. Floyd, TCP and explicit congestion notification, *ACM Journal of SIGCOMM Computer Communication Review*, 24, 8–23, 1994.

[FBP98] M. Faloutsos, A. Banerjea, and R. Pankaj, QoSMIC: Quality of service sensitive multicast Internet Protocol, *SIGCOMM Computer Communication Rev.* 28(4), 144–153, 1998.

[FG95] C. L. Fullmer and J. J. Garcia-Luna-Aceves, Floor acquisition multiple access (FAMA) for packet-radio networks, *ACM SIGCOMM Computer Communication Review*, 25, 262–273, 1995.

[GLB09] R. Guimarães, L. Cerdà, J. M. Barceló, J. García, M. Voorhaen, and C. Blondia, Quality of service through bandwidth reservation on multirate ad-hoc wireless networks, *Ad-hoc Networks*, 7(2), 388–400, March 2009.

[IR1] OPNET Modeler version 10.5, OPNET Technologies, Inc., available at http://www.opnet.com, accessed on December 13, 2009.

[KHP08] D. Koutsonikolas1, Y. C. Hu, and K. Papagiannaki, How to evaluate exotic wireless routing protocols? in *Proceedings ACM HotNets*, Calgary, Canada, October 2008.

[KK00] B. Karp and H. T. Kung, GPSR: Greedy perimeter stateless routing for wireless networks, in *Proceedings of the 6th Annual international Conference on Mobile Computing and Networking*, Boston, MA, August 2000.

[KK98] R. Kravets and P. Krishnan, Power management techniques for mobile communication, in *ACM International Conference on Mobile Computing and Networking (MobiCom)*, New York, pp. 157–168, October 1998.

[KKFT02] S. Kopparty, S. Krishnamurthy, M. Faloutsos, and S. K. Tripathi, Split TCP for mobile ad-hoc networks, *IEEE Global Telecommunications Conference GLOBECOM '02*, vol. 1, Taipei, Taiwan, pp. 138–142, November 2002.

[KLSSK02] V. Kanodia, C. Li, A. Sabharwal, B. Sadeghi, and E. Knightly, Distributed priority scheduling and medium access in ad-hoc networks, *ACM Journal of Wireless Networks*, 8, 455–466, 2002.

[KMS05] I. Karthigeyan, B. S. Manoj, and C. Siva Ram Murthy, A distributed laxity-based priority scheduling scheme for time-sensitive traffic in mobile ad-hoc networks, *Journal of Ad-hoc Networks*, 3, 27–50, 2005.

[KRD06] S. Kumar, V. S. Raghavan, and J. Deng, Medium access control protocols for ad-hoc wireless networks: A survey, *Ad Hoc Networks*, 4(3), 326–358, May 2006.

[LS01] J. Liu and S. Singh, ATCP: TCP for mobile ad-hoc networks, *IEEE Journal on Selected Areas in Communications*, 19, 1300–1315, 2001.

[LS98] J. R. Lorch and A. J. Smith, Software strategies for portable computer energy management, *IEEE Journal of Personal Communications*, 5, 60–73, 1998.

[MB00] B. A. Miller and C. Bisdikian, *Bluetooth Revealed*, Prentice Hall, New York, 2000.

[MK04] P. Mohapatra and S. Krishnamurthy, *Ad-hoc Networks: Technologies and Protocols*, Springer, New York, 2005.

[MRV09] J. C. Mundarath, P. Ramanathan, and B. D. Van Veen, A quality of service aware cross-layer approach for wireless ad-hoc networks with smart antennas, *Ad Hoc Networks*, 7(5), 891–903, July 2009.

[MSGJ96] S. Murthy and J. J. Garcia-Luna-Aceves, An efficient routing protocol for wireless networks, *Journal of Mobile Networks and Applications*, 1, 183–197, 1996. http://dx.doi.org/10.1007/BF01193336

[MWH01] M. Mauve, J. Widmer, and H. Hartenstein, A survey on position-based routing in mobile ad-hoc networks, *IEEE Network Magazine*, 15, 30–39, 2001.

[N70] N. Abramson, The ALOHA system: Another alternative for computer communications, *AFIPS Joint Computer Conferences*, Vol. 37, Houston, TX, pp. 281–285, November 17–19, 1970.

[OSL05] L. B. Oliveira, I. G. Siqueira, and A. A. F. Loureiro, On the performance of ad-hoc routing protocols under a peer-to-peer application, *Journal of Parallel and Distributed Computing*, 65, 1337–1347, 2005.

[PC97] V. Park and S. Corson, Temporally-ordered routing algorithm (TORA), *version 1 functional specification, ETF Internet draft, IETF MANET Working Group*, Flarion Technologies, Inc. Bridgewater, NJ, 1997.

[PDH07] H. Pucha, S. M. Das, and Y. C. Hu, The performance impact of traffic patterns on routing protocols in mobile ad-hoc networks, *Computer Networks: The International Journal of Computer and Telecommunications Networking*, 51, 3595–3616, 2007.

[PP94] C. E. Perkins and P. Bhagwat, Highly dynamic destination sequenced distance vector routing (DSDV) for mobile computers, *Computer Communication Review*, 234–244, October 1994.

[PR99] C. Perkins and E. Royer, Ad-hoc on demand distance vector routing, in *Proceedings of the 2nd IEEE Workshop on Mobile Computing Systems & Applications*, New Orleans, LA, February 1999.

[PSS00] P. Palensky, T. Sauter, and C. Schwaiger, Security und Feldbusse—ein Widerspruch, *Journal of Information Technology*, 42, 31–37, 2000.

[PSW04] A. Perrig, J. Stankovic, and D. Wagner, Security in wireless sensor networks, *Communications of the ACM*, 47, 53–57, 2004.

[R96] M. K. Reiter, Distributing trust with the Rampart toolkit, *Communications of the ACM*, 39(4), 71–74, 1996.

[RFLW96] M. K. Reiter, M. K. Franklin, J. B. Lacy, and R. N. Wright, The key management service, *Journal of Computer Security*, 4, 267–297, 1996.

[RKMM06] T. B. Reddy, I. Karthigeyan, B. S. Manoj, and C. S. R. Murthy, Quality of service provisioning in ad-hoc wireless networks: A survey of issues and solutions, *Ad Hoc Networks*, 4(1), 83–124, January 2006.

[RR00] R. Ramanathan and R. Rosales-Hain, Topology control of multihop wireless networks using transmit power adjustment, in *19th Annual Joint Conference of the IEEE Computer and Communications Societies (INFOCOM)*, Vol. 2, Tel Aviv, Israel, pp. 404–413, March 2000.

[RS98] C. S. Raghavendra and S. Singh, PAMAS-power-aware multi-access protocol with signaling for ad-hoc networks, *ACM Journal of SIGCOMM Computer Communication Review*, 28, 5–26, 1998.

[S02] I. Stojmenovic, Position-based routing in ad-hoc networks, *IEEE Communication Magazine*, 40, 128–134, 2002.

[SAHS03] K. Sundaresan, V. Anantharaman, H. Hsieh, and R. Sivakumar, A reliable transport protocol for ad-hoc networks, *IEEE Journal on Mobile computing*, 4, 588–603, 2005.

[SBC03] G. Srivastava, P. Boustead, and J. F. Chicharo, A Comparison of topology control ad-hoc networks, in *Proceedings of the 2003 Australian Telecommunications, Networks and Applications Conference*, Melbourne, Australia, December 8–10, 2003.

[SL01] I. Stojmenovic and X. Lin, Power-aware localized routing in wireless networks, *IEEE Transactions on Parallel and Distributed Systems*, 12(11), 1122–1133, 2001.

[SMG97] B. R. Smith, S. Murphy, and J. J. Garcia-Luna-aceves, Securing distance-vector routing protocols, in *Proceedings of Symposium on Network and Distributed System Security*, The Internet Society, IEEE Computer Society Press. Los Alamitos, CA, pp. 85–92, February 1997.

[SSB99] P. Sinha, R. Sivakumar, and V. Bharghaven, CEDAR: A core-extraction distributed ad-hoc routing algorithm, *IEEE Transaction on Selected Areas in Communications*, 17, 1454–1465, 1999.

[T97] C.-K. Toh, Associativity-based routing for ad-hoc mobile networks, *Wireless Personal Communications Journal, Special Issue on Mobile Networking and Computing Systems*, 4, 103–139, 1997.

[TBD07] H. D. Trung, W. Benjapolakul, and P. M. Duc, Performance evaluation and comparison of different ad-hoc routing protocols, *Journal of Computer Communications*, 30, 2478–2496, 2007.

[WESW98] H. Woesner, J.-P. Ebert, M. Schlager, and A. Wolisz, Power-saving mechanisms in emerging standards for wireless LANS: The MAC level perspective, *IEEE Journal of Personal Communications*, 5, 40–48, 1998.

[WHX09] X. Wang, Y. Han, and Y. Xu, APS-FeW: Improving TCP throughput over multihop adhoc networks, *Journal of Computer Communications*, 32, 19–24, 2009.

[WLBW01] R. Wattenhofer, L. Li, P. Bahl, and Y.-M. Wang, Distributed topology control for power efficient operation in multihop wireless ad-hoc networks, in *19th Annual Joint Conference of the IEEE Computer and Communications Societies (INFOCOM)*, vol. 3, Anchorage, AK, pp. 1388–1397, April 2001.

[YLYLZ04] H. Yang, H. Luo, F. Ye, S. Lu, and L. Zhang, Security in mobile ad-hoc networks: Challenges and solutions, *IEEE Journal on Wireless Communications*, 11, 38–47, 2004.

[ZFZ08] Y. Zhou, Y. Fang, and Y. Zhang, Securing wireless sensor networks: A survey, *IEEE Communications Surveys and Tutorials*, 10, 6–28, 2008.

8

Radio Frequency Identification

8.1 Prologue........8-2
8.2 Bar Code System........8-3
8.3 Magnetic Stripes........8-4
8.4 Smart Card........8-5
8.5 Proximity Card........8-6
8.6 HF RFID........8-6
8.7 Electronic Cash........8-7
8.8 Personal Identity........8-7
8.9 Innovation verus Hi-Tech........8-8
8.10 Active RFID........8-8
8.11 Wake-Up Technology........8-8
8.12 Semi-Active RFID........8-10
8.13 Backscattering........8-11
8.14 Initialization........8-11
8.15 Vicinity Card........8-12
8.16 Frequency Selection........8-12
8.17 UHF RFID........8-13
8.18 Supply Chain Management........8-13
8.19 International Standard........8-14
8.20 Promiscuity........8-14
8.21 National Standards........8-15
8.22 Hands-Free Bar Code System........8-16
8.23 Bar Code Mentality........8-16
8.24 Affordable Tag........8-18
8.25 Ubiquity of RFID........8-18
8.26 Role Reversal........8-20
8.27 Historical Development........8-20
8.28 Privacy Infringement........8-20
8.29 Recent Developments........8-21
8.30 Dual Authentication........8-22
8.31 Trace-and-Track........8-23
8.32 Innovative Applications........8-23
8.33 Nonionization Radiation........8-26
8.34 Era of Artificial Perception........8-27
Abbreviations........8-28
References........8-29

Edward Kai-Ning Yung
City University of Hong Kong

Pui-Yi Lau
City University of Hong Kong

Chi-Wai Leung
City University of Hong Kong

8.1 Prologue

In prehistorical years, our ancestors tied knots on a string to mark episodes of importance, so that these memorable moments could be recalled afterward. This apparently simple maneuver has, in fact, distinguished human beings from other living things, as it was a preamble of engraved symbols and written words, the icon of civilization. In ancient China, circa 1000 BC in the Shang Dynasty, a series of smoke towers on mountain peaks was used to transmit messages to and fro the imperial court, a simple yet effective link of communications covering thousands of kilometers. It is the first documented telecommunications system in history.

In recent history, one of the most memorable moments is the invention of the electromagnetic telegraph by Baron Schilling in 1832 [1] as it symbolized the genesis of the era of electrical telecommunications. Of comparable importance is the postulation of electromagnetism disseminated by James Maxwell in 1865 [2], because it laid a solid theoretical ground for the development of wireless communications. Equally remarkable are the developments of the Z1, the first programmable electromechanical binary machine by Konrad Zuse in 1936 [3], and the ABC, the first electronic computer, by John Atanasoff and Cliff Berry in 1942 [4]. Since these machines enabled electronic automation, many men and women have been emancipated from mundane repetitive daily chores.

In fact, electronic automation was inherited from an electromechanical one. Under a contract for the Census Office of the Federal Government of the United States, the punch card was invented by Herman Hollerith (Figure 8.1), and the patent was eventually approved in 1884 [5]. Since punch cards could be counted or sorted mechanically, statistics compilation was made easier. From another point of view, this invention marked the beginning of a new era of information processing because the punch card is a reliable method of data input and data storage.

However, the association of an object with the datum being input was only made possible with the advent of the bull's eye by Joseph Woodland and Bernard Silver in 1949 [6]; therefore, it is credited as being the father of automatic identity. Ever since then many automatic identification and data capature (AIDC) schemes have been developed, such as the magnetic stripe by Forrest Parry in 1966 [7] and the reflective color stripes by David Collin in 1967.

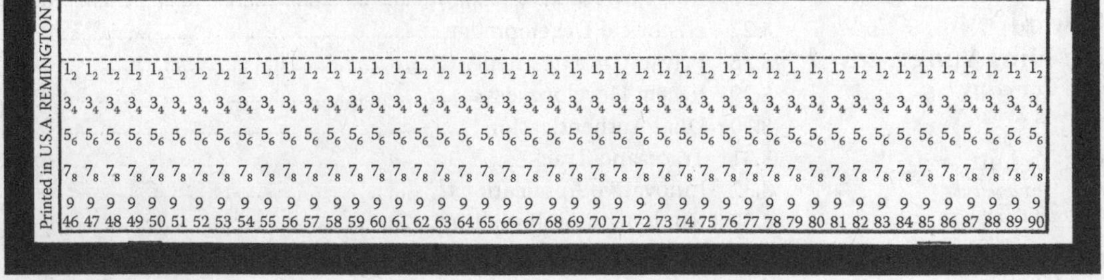

FIGURE 8.1 Hollerith's punch card.

8.2 Bar Code System

Inspired by the dots and dashes in a Morse code (Figure 8.2) [8], Joe Woodland drew the first linear bar code of narrow and wide lines on sand when he was vacationing in Florida. Using a converted movie projector with a 500 W bulb as a reader, Joe found that a bull's eye pattern with circular lines was more reliable. Unlike punch cards, the circular or the linear bar code is a noncontact method of data entry, but it is vulnerable to smear and smash. Competing with the time-tested punch cards, little progress was seen until the light bulb was replaced by a helium–neon laser in 1969. A linear bar code of 11 digits was reinvented in 1972. Together with a laser scanner, the system regained competitiveness and prevailed eventually. Other than minor augmentations, the revised version is virtually identical to the bar code used today, such as Code 128 shown in Figure 8.3, because there has been no major change thereafter.

In the linear or the circular form, bar code reading is a slow process because the pattern must be properly aligned with the scanner on the one hand, and on the other, it is read one by one, manually. Nonetheless, it was a winner because printing a bar code on a box costs nothing and the helium–neon laser scanner was affordable by most merchants in western countries at that time. Moreover, a bar code provides a primitive method of authentication because it cannot be changed once printed on paper. It was welcomed by manufacturers, distributors, retailers, and other merchandise handlers. The logistics management industry was especially enthusiastic because tracking-and-tracing of lost cargo was made easier. With the arrival of the low-cost solid-state laser scanner in the 1990s, the dominance of the bar code was established.

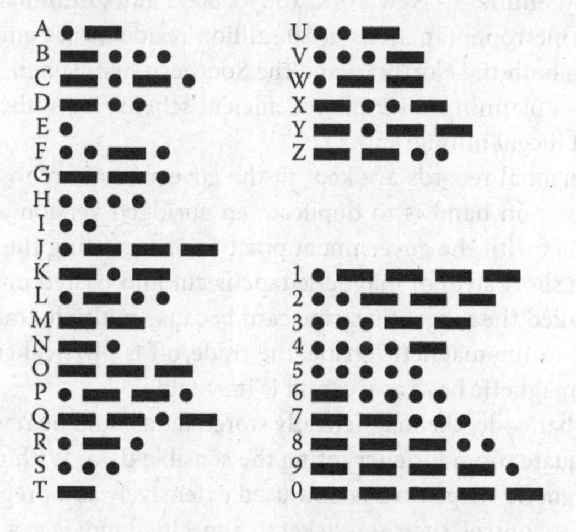

FIGURE 8.2 Morse code.

FIGURE 8.3 Code 128 bar code.

To facilitate a common platform for national trade in the United States, the Universal Product Code (UPC) was adopted in 1973 [9]. On the other side of the Atlantic Ocean, the UPC code was augmented to form the European Article Numbering Scheme (EAN-13). It was accepted by most European nations as a common standard in 1974. By adding an extra code to the European scheme, an international standard (EAN-14) was born.

8.3 Magnetic Stripes

Besides coding a product, a bar code is applicable in marking a pack of fresh produce and identifying a person. Strictly speaking, a bar code cannot be treated as an identity because a code is usually assigned to a line of products, not an individual item. In the existing format, it has no information on its owner, not to mention when and where the product was made. Should bar code be used to specify trillions of items sold annually all over the world uniquely, it must be substantially elongated. Consequently, at least one side of the box must be reserved for accommodating its code of more than a hundred digits. In fact, a bar code is hardly adequate to code 6.7 billion men and women on earth. However, for a simple stand-alone case, it is feasible to assign every person in a specific group an exclusive bar code, such as the boarding pass for a given air flight.

However, with industrialization and urbanization come the inevitable socioeconomic changes, structurally and nonstructurally. Similar to New York, Tokyo, Sao Paulo, Mumbai, Lagos, Mexico City, and Shanghai, more and more metropolitan areas of 20 million residents are emerging fast in the Eastern and the Western worlds, in both the Northern and the Southern hemispheres. For an effective population control and better town planning, a simple yet efficient scheme for authenticating personal identities is needed to screen out illegal immigrants.

It is understood that personal records are kept in the government archive in both the written and electronic formats. The issue on hand is to duplicate an abridged version and store it on a portable device, inexpensively. In line with the government practice of recording the data semipermanently in decks of magnetic tapes, a short strip of magnetic tape is cut and pasted on a name-card sized paper or polymer card. This is called the magnetic stripe card because multiple tracks of record are found in a strip. From the bar code to the magnetic stripe, the trade-off is physical contact due to swiping the magnetic tape over a ferromagnetic head as wear off is inevitable.

Unlike data shown in a bar code, the magnetically stored data could be rewritten easily. Fortunately, a magnetic stripe has adequate room for encrypting the sensible data. With data encryption and other security measures, the magnetic stripe card is now used extensively as a prepaid calling card or a debit card. It is also used in access control, such as a ticket to a specified movie at a given time or a bus ride at a given time (Figure 8.4).

(a)

(b)

FIGURE 8.4 (a) A cable car ticket and (b) magnetic stripe card in Hong Kong.

8.4 Smart Card

Every sword has two blades. The ability to write and rewrite implies that the personal record of its bearer is vulnerable to unauthorized alterations. Addressing this issue, a microprocessor chip with a sizable memory is embedded in a polyvinyl chloride (PVC) card because the said polymer is cheap, durable, and easy to assemble at a relatively low temperature. Using the old-fashioned write-once memory, top security could be obtained because the data stored in this type of memory cannot be altered. Alternatively, specially designed software could be used to protect the data from external hacking. Short of absolute security, the latter method is tamper resistant.

The microprocessor-controlled memory could be used in conjunction with the magnetic stripe. With or without a magnetic stripe, it is called a smart card. Although the first version was unveiled by Helmut Gröttrup and Jürgen Dethloff for SEL (Standard Elektrik Lorenz) in 1968, the smart card remains very popular 4 decades later as most credit cards in our wallets belong to this category.

With significant progress made in the mass production of semiconductor devices in recent years, the price of EPROMs (erasable programmable read-only memory), E-EPROMs (rewritable EPROM), or flash memories is dropping very fast. Thus, a memory of 1024 kB or more could be found in many smart cards; therefore, many new features are made possible. For example, a high-resolution image of its bearer could be included for authentication. One may also choose to download one's resume, medical record, and other information, such as the name of the physician and the date of the last visit in an eye-catching card such that erroneous blood transfusion and the prescription of allergic medicine could be reduced, especially when the bearer is knocked unconscious after an accident.

By incorporating a proprietary hardware onboard, cryptographic smart cards could be used in top-security transactions by providing a digital signature. For example, it is adopted for interbank fund transfers, as in the electronic payment system (EPS) in Hong Kong. It could also be used in money withdrawals and cash deposits via an autoteller machine (ATM). With additional reconfirmation through various biofeatures, such as fingerprint and face recognition, smart cards could be used at border control. For example, a resident in Hong Kong could clear the immigration process between the Hong Kong Special Administrative Region (SAR) and Mainland China through the "e-way" in less than 7 s or 5 s in the upgraded version.

A boom in smart card use came in 1990 after it was packaged into the form of a subscriber identity module (SIM) card for the GSM mobile phone in Europe (Figure 8.5). Besides mobile phones, SIM cards are extensively used in video games, software licenses, and for watching premium movies on a high-definition television (HDTV) format at home.

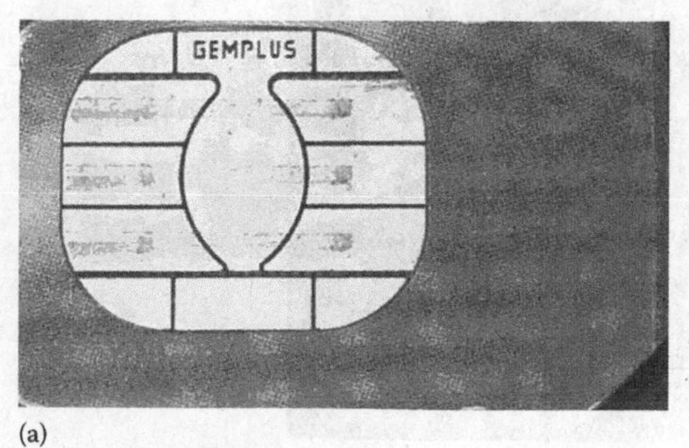

(a)

(b)

FIGURE 8.5 (a) SIM card and (b) GSM telephone card.

8.5 Proximity Card

Since a smart card draws current via direct contact, it is not contactless. With the unprecedented success of cellular phones, the spoilt public wants that every electronic device be wireless. As the use of smart cards for daily transactions is ever increasing, more and more people are annoyed by the need to look for an appropriate one out of numerous cards in a wallet or a purse, pull it out, and insert it into a reader for every business deal.

Addressing public demand, current loops are added in both reader and smart card, such that power could be drawn from the reader by aligning the coils in series without physical contact, as if it were a mutual inductor. Via energy stored in a capacitor, the built-in microprocessor of a smart card could feed back signals through the mutual inductor. Unlike the 50 Hz power supply at home (60 Hz in North America), the frequency used is 125 kHz for a stronger coupling. Due to the lack of space, the number of coils in a smart card is very small; thus, the effectiveness of this inductive coupling is very weak and the coil separation is correspondingly small, in millimeters. Reflecting that a contactless smart card must be very close to a reader, it is called a proximity card. Other than the reduction in wear off, the millimeter separation does not bring us significant changes because the card must be pulled out of our wallet for every transaction. Having said that, the proximity card finds extensive applications in hotels, gasoline stations, dry cleaners, and toys (Figure 8.6).

8.6 HF RFID

Based on fundamental electromagnetic theories, the coupling coefficient of a mutual inductor could be enhanced via a frequency increase. However, it cannot be increased indefinitely because a higher coupling is obtained at the expense of higher energy dissipation. In fact, the frequency cannot be chosen arbitrarily as it must fall within one of the (industrial, scientific, and medical) ISM frequency bands [10], the spectrum open for public use without license. Incidentally, the frequency adopted for enhancing the performance of a proximity card is 13.56 MHz, one of the frequencies used in radio broadcast in the high-frequency band. Thus, it is called a radio frequency identification card with the acronym HF RFID. The name is actually attributable to Charles Walton who is perhaps the first person who filed a US patent using a "radio frequency" and an "identifier" in 1980 [11].

Another gain obtained from using a higher frequency is size reduction because it makes placing a ferrimagnetic disk inside the primary coil economically feasible. With the magnetic coupling, the range of operation could be increased more than tenfold to 200 mm.

FIGURE 8.6 HF RFID pair for toys.

FIGURE 8.7 Smart card, the Octopus card in Hong Kong.

8.7 Electronic Cash

At 13.56 MHz, most materials encountered in our daily lives are RF lucent [12]. In other words, an RF wave could pass through the material without excessive attenuation. Simply put, a lady need not turn her purse inside out to search for her RFID card; instead, she could gain access by placing her purse on top of a reader. This is not a trivial gain because it protects the bearer from exposing valuable items in a public area. Extra expediency could be observed by hiding the RF device inside a watch or a pendant. Probably due to this reason, the HF RFID has been extensively used in mass public transport systems under various brand names, such as Octopus in Hong Kong (Figure 8.7) and Compass in San Diego.

Besides gains in efficient passenger entry and better inventory control, the company involved could use less manpower in retrieving and counting millions of coins collected daily, and therefore the subsequent saving in bank charge. No wonder Octopus was adopted by convenient stores, super markets, and fast food outlets as cash payment. It is also used for attendance checks in schools, entry to private properties, fee payment in public garages, and circulation control in libraries.

An added convenience could be obtained by authorizing one's bank to transfer a finite sum to the Octopus account automatically, whenever the latter balance is below a prescribed threshold. In short, Octopus is treated as electronic cash (known as the electronic purse in Europe). With an Octopus card, we need not carry any coins, paper bills, staff IDs, garage keys, and other miscellaneous cards on our way to school, to work, or when shopping, dining, and for other activities.

8.8 Personal Identity

After years of successful operation, even government bureaucrats were convinced. Parking meters were first modified to accept Octopus, and now it can be used to pay miscellaneous fees, including fines. To be fair, some governments are very proactive. For example, a smart driver license system was adopted in Argentina as early as 1995. Commencing in 2001, 22 million Malaysians are now benefited by the extra features included in MyKad [13], the first smart national identity card in the world. For better authentication, an electronic signature has been added in the national identity cards in Spain and Belgium in 2009.

As demonstrated in Malaysia, the benefits provided by the RFID-ID card are welcome by the users and the issuer. Unfortunately, it is only applicable within the national boundary. The prerequisite for roaming across this man-made demarcation line is an international standard, similar to EAN-14 mentioned previously. After years of endless domestic arguments and international negotiations, the said standard was finally ready for deployment in 2006 and it is managed by the International Civil

Aviation Organization [14]. Using the ISO 14443 RFID air interface, e-passports were adopted in the same year in many countries, including Japan, Germany, Norway, Poland, Ireland, and Pakistan. In fact, the e-passport was first used in Malaysia in 1998, but it was meant for internal consumption only. In addition to a digital picture of its bearer for authentication, one's track record of entries and exits from Malaysia is also registered.

8.9 Innovation verus Hi-Tech

Surprisingly, the apparently trivial range enhancement in the proximity card provides us an unforeseen convenience. The hands-free RFID profoundly affected our daily lives and induced undue changes in our habits. Equally surprising is the way how RFID was developed. From a feasibility study, to product development, to system design, to an in situ field test, and the final deployment, no expensive new hi-tech equipment was needed. It is simply an ingenious application of the existing technologies and uses of the direct-off-the-shelf components and materials. Technologically, hands-free RFID is merely a step forward, definitely not a major revolution, but sociologically, it is unquestionably a paramount breakthrough. This explains why developing countries such as Argentina and Malaysia could claim some firsts in the world. In short, RFID development exemplifies that "innovative technology is not necessarily hi-tech."

8.10 Active RFID

Greed is an intrinsic trait of human beings, yet it is also the primary motive force for developments in our civilization. Furthermore, success breeds demand; thus, our frustration is understandable. Today, we are no longer contented with the limited range offered by HF RFID. Accustomed to the freedom provided by the tether-less mobile phone, we are convinced that RFID should reach beyond meters and, hopefully, kilometers.

As the strength of a traveling electromagnetic wave is inversely proportional to distance-square, the wave incident on a tag due to a tenfold range extension is subject to an additional 60 dB attenuation. Hence, a sizable energy source is needed in a tag to generate a response of 60 dB stronger because the returning wave will attenuate correspondingly.

For efficient wave reception and radiation, half-wave resonant antennas with a gain of over 6 dB are preferred at both ends of the communications link. The trade off is an anisotropic reception. Thus, circularly polarized antennas are used in setting up the reader-tag linkage. For tags with a limited surface area, two orthogonal short dipoles could be used instead.

With a wavelength of 22 m, a half-wave 13.56 MHz RFID tag could be excessively bulky to carry around. Hence, a higher ISM frequency must be chosen for making active RFID. For a long-range operation in kilometers, two frequencies at the lower end of the ultrahigh frequency band, 315 and 433 MHz, are preferred because the resultant waves could better circumvent a metallic obstacle of the size of an automobile. As a result, an active 433 MHz RFID system has been adopted by the logistics industry for locating containers in international trade and for managing truck movements near a container port or airport. Similar systems are being used by transport authorities for monitoring their fleets of buses. For easy reference, it is called active RFID.

8.11 Wake-Up Technology

Since the power needed for driving a vehicle-based RFID device could be drawn from the vehicle, energy efficiency is not a major issue of concern. Power consumption is, however, a major one in a handheld device, such as a smart bracelet used in keeping a detainee under house arrest inside a designated place.

Given that an active RFID tag has been registered in a given setting, significant power could be conserved by instructing the tag to listen for signals from the nearby interrogators as usual, but to do nothing unless its number is called, the so-called listen before talk (LBT) mode. In cases when the person involved can only move slowly, such as an elderly patient in a sanatorium, additional energy could be

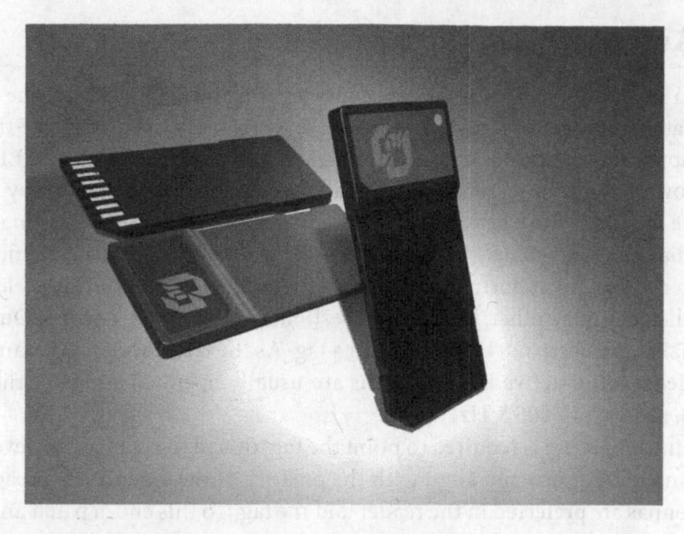

FIGURE 8.8 Active RFID reader to be inserted in a PDA for the tourism industry.

saved by directing the tag to sleep for an extended period, wake up momentarily, listen to the broadcast, and go back to sleep unless it is summoned. Once woken up, a reader-tag linkage is established via energy stored in the battery. A fast response is observed because there is no need to accumulate energy from the incoming wave. Using pager technology, a tag with one CR2025 battery could run for months without battery replacement, such as the RFID system for a tourist group (Figure 8.8) and the RFID tag for mountain hikers with GPS and GPRS capabilities (Figure 8.9), both of which are prototype systems developed at the City University of Hong Kong.

Sometimes, waking up periodically remains unnecessary because there is no reader in the vicinity. This scenario is often found in a vehicle-based RFID tag, such as the toll-payment label on the windshield glass. Addressing this scenario, an application-specific integrated circuit (ASIC) has been tailor-made to wake up under the illumination of a strong electromagnetic field. Surprisingly, most ASICs available in the market use a low frequency 125 kHz signal in the wake-up process, and the sensitivity ranges from 100 μV_{rms} to 1 mV_{rms}. Other than the wake-up mechanism, the instruction set is similar to that of a pager-tag; therefore, an equally fast response is observed.

FIGURE 8.9 RFID locator for mountain hikers with GPS and GPRS capabilities.

8.12 Semi-Active RFID

A wake-up chip is a passive device that wakes up under a strong incident field, yet the responses of a tag are powered by a battery. Hence, it is called a semi-active RFID or a semi-passive RFID.

Dependent on applications, an incident electromagnetic wave as strong as 4 W EIRP (effective isotropic radiation power) could be used to activate a wake-up module meters away from the source. Compared with the milliwatt wave generated by an active RFID reader, it is a strong radiation. At the UHF band, the radiation is a low-energy nonionizing one whose effects or lack of effects on our health is a heated controversy in many industrialized nations. In order to excite a strong field on the one hand and to protect our health on the other, high-gain directional antennas are needed. Due to physical constrains, a directional antenna cannot be installed in a tag. As the last resort, a resonant half-wavelength antenna is used. Hence, semi-active RFID systems are usually operated at one of the microwave ISM frequencies; that is, at 2450 or 5800 MHz.

For an efficient linkage, a user is required to point the tag toward the reader. However, the orientation of the dipole antenna in a tag may not align with the polarity of the antenna in a reader. Hence, circularly polarized antennas are preferred in the reader and the tag. To this end, a patch antenna is preferred in a tag because it is a low profile and the cost of production is low. At 2450 MHz, the length of a patch antenna without a dielectric substrate is 60 mm. The length could be reduced to 20 mm by printing the patch antenna on aluminum dioxide ($\varepsilon_r = 10$) or any substrate with a high dielectric constant. With or without a dense substrate, a 60 mm tag is small enough to serve as a handheld mobile unit in most applications.

As a traveling plane wave, wave propagation is vulnerable to many obstacles because it is not supposed to circumvent an RF-opaque object whose size is as large as its wavelength, and in this case, 122 mm or 52 mm. Consequently, the effectiveness of a microwave RFID is restricted to the light-of-sight tags.

In addition to the above blockage, microwave RFID is subject to higher losses. Besides a high rate of attenuation in moist air, some of the RF-lucent materials at HF and UHF bands would become RF absorbent at microwave frequencies. The notable examples are dry wood, wet wood, water, and some liquids. In general, microwave RFID is designed to operate within a small cell, a terminology used in cellular phones.

One of the exceptions is traffic control on a highway because the line-of-sight requirement is not an issue of concern there. Hence, a semi-active microwave RFID is often used in toll payments, such as the Autotoll system in Hong Kong (Figure 8.10). It is also used in electronic road pricing schemes, such as 407 ETR in Toronto. Powered by a CR2025 button battery, an Autotoll tag could run for several

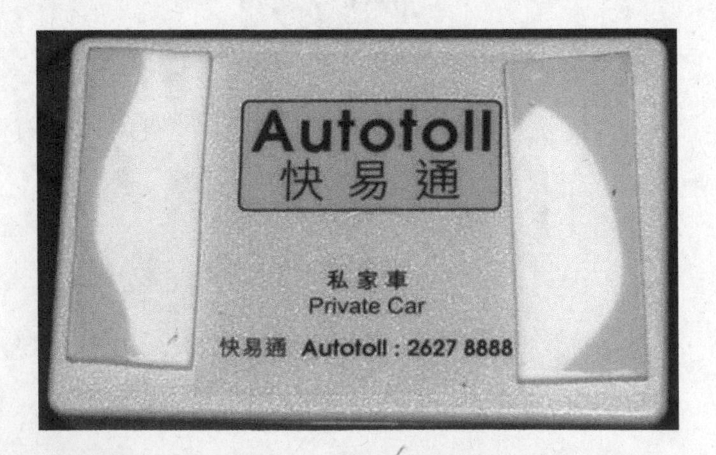

FIGURE 8.10 Semi-active RFID tag, Autotoll system for tariff payment in Hong Kong.

years without battery replacement. In other words, no battery replacement is needed before the car is replaced by a new one. Compared with the most energy-efficient active RFID, it has extremely long life expectancy.

8.13 Backscattering

Although semi-active RFID is very versatile and reliable, it finds limited utilizations because its tag is expensive in absolute and relative terms. Of course, the finite life span and the bulkiness of its tag are also factors of concern.

Based on the successful development of the passive HF RFID, attempts to develop a long-range system with inexpensive, low profile, and passive tags have been reported. However, little or marginal success was observed. The culprit is fixation, as engineers have been fascinated and captivated by the advances developed for mobile communications; few have tried to solve the problem intuitively from the basic principles.

To this end, attention is focused on the antenna in a passive tag. Illuminated by an incident wave, current is induced on the antenna. As an antenna, current is drawn from its feed, and as a result, a significant dip is found in the current distribution there. Energy is drawn by the tag for internal consumption, but it represents only a small portion of the incident energy. At equilibrium, the remaining portion of energy is "lost" through the secondary radiation, the so-called scattered field. As the scattered electromagnetic wave could be determined by treating the current induced on an antenna as an impressed source, the radiation pattern is dependent on the current distribution. Following a reciprocal path, part of the backscattered wave will reach the reader and part of the energy carried by the backscattered wave picked up by the reader.

In case the feed is shorted, the tag antenna would behave as a dummy metallic scatterer with the dip at the original feed having disappeared. As the backscattered field picked up by the reader will honestly reflect this phenomenal change, a message is sent. Simply put, by opening and shorting the antenna feed in accordance with the data stream to be sent, the backscattered field is digitally modulated. In fact, the modulation technique used is more complex than the simple open-and-short, but it is beyond the scope of this chapter. Although the currents on the reader antenna due to the backscattered fields in both cases are very weak, a radio-frequency circuit with high sensitivity is able to differentiate the subtle changes. In doing so, the digitally modulated signal is demodulated and the data stream recovered.

It is obvious that the antenna current generated by an energy storage device in a passive tag is much weaker than the current induced by an incoming wave. It then follows that the backscattered field picked up by the reader is stronger than that due to energy accumulation by several orders. Hence, the range of coverage provided by backscattering is longer than that of the traditional techniques. Moreover, a fast response is seen as the energy needed to turn the antenna feed on-and-off is much less than that of setting up a conventional wireless communications channel.

Backscattering communications is not a new concept. The pioneering study conducted by Harry Stockman had been summarized in his classical paper titled "Communication by means of reflected power" in a Proceedings of IRE in 1948 [15]. It was overlooked for a quarter of a century until Steven Depp and his colleagues demonstrated at the Los Alamos National Laboratory in 1973 [16] that reflective power is deployable.

8.14 Initialization

To a certain extent, the technique adopted in setting up a link of communications in a passive RFID system is similar to the LBT process used in semi-active RFID systems. After sufficient energy has been accumulated in a passive tag, a short message is broadcast to alert the reader. In the case when many tags are found in the near vicinity at a given time, all tags would be woken up simultaneously and all initializations would fail. Every tag will then wait for a randomly determined duration in accordance with the predetermined anticollision scheme before trying again. The most common standard used in the air interface at 860–960 MHz is ISO/IEC 18000 Part 6.

8.15 Vicinity Card

In layman's terminology, a communications link in backscattering is established by modifying the environment for wave scattering. A longer range results because the current responsible for the backscattered wave is as strong as the current induced by the incident wave. In principle, a similar technique could be used to extend the range of coverage due to inductive coupling. Using energy drawn from inductive coupling, the microprocessor in a passive HF RFID tag could be instructed to load or short the secondary coil in accordance with the data streams to be sent [17]. As a result of the physical change in the secondary inductor, the coupling coefficient is affected correspondingly. It then follows that the driving source of the primary coil would detect a small yet apprehensible change in the input impedance, and as a result, a communications link is established. Based on the ISO 15693 standard, the effective range of a HD RFID could be extended beyond 1 meter. To distinguish it from the conventional HF RFID, it is called a vicinity card.

Although mutual inductors are used in both cases, migration from HF RFID to vicinity card is not simple. Of course, the computer network and the associated software could be reused with minor modifications, but the readers and tags must be replaced. Thus, existing users of HF RFID systems will be confronted with the dilemma on choosing between a familiar HF system or a versatile one at a different frequency.

8.16 Frequency Selection

Since backscattering is frequency independent, the next step in the design of a long-range passive RFID system is the selection of a frequency for operation. It is a tall order for the spectrum regulator because finding a vacant frequency slot in the one-and-only-one spectrum for an approved wireless usage requires a comprehensive knowledge of the technology itself, the trend of development, and an appreciation of the urgent socioeconomic needs on the one hand, and on the other, a regulator needs to command an excellent political skill in balancing the technical and socioeconomic factors subtly with little discretionary leverage.

While socioeconomic issues will be discussed later, the easier one is first addressed. Technically, the issues of concern are range of coverage, physical constraints, scope of application, speed of data transmission, electromagnetic compatibility, environmental friendliness, intersystem and intrasystem interferences, product safety, and cost of production.

For long-range operations, an efficient antenna is needed in a tag. Hence, microwave RFID is preferred as room is available for accommodating the 30 mm half-wave antenna on the covering box of most products. Having said that, UHF RFID is a strong alternative because its antenna could be shortened to less than 30 mm by slow-wave techniques, and further shrinkage could be achieved by fabricating the patch antenna on a dense substrate. Nonetheless, it is a choice between a UHF and a microwave.

On the scope of applications, UHF prevails because dry wood, an RF-lucent material at ultrahigh frequencies, would become RF absorbent at microwave frequencies. This change has a steep consequence as wood is one of the key materials used in making pallets for the logistics industry. Moreover, UHF wins again on cost-effectiveness because the technologies used in making radio frequency–integrated circuits (RFIC) are more mature; therefore, a 900 MHz die is much cheaper than its counterpart at 2450 MHz.

Most important of all, UHF has a clear edge on intersystem electromagnetic interference. As a result of the migration of mobile phone subscribers to the 2¾ generation at 1800 MHz and the third generation at 2000 MHz, a relatively clean environment is found in the UHF band as some of the slots allocated for the second generation at 900 MHz have been underutilized for a long time. On the other hand, traffic congestions are often seen in the microwave ISM frequency because 2450 MHz is currently used in microwave ovens, active RFIDs, wireless local area networks, and various wireless interconnect gadgets. The selection of 5800 MHz is not advisable because the relevant technologies for mass production of low-cost tags are far from mature.

In summary, regulators of the major trading blocks of nations have reached a consensus in assigning some of the ISM slots in the ultrahigh frequency band for the license-free passive RFID.

8.17 UHF RFID

Using the method of backscattering, a long-range passive RFID system with a low-cost compact tag could be designed. Since it is operating at one of the empty UHF frequency slots in the ISM band, it is called the UHF RFID or simply the RFID.

Recall that one of the major objectives in developing the RFID is to replace the existing bar code with a long-range, hands-free, rewritable, tamper resistant, immune of electromagnetic interference, efficient, and versatile one, yet the advantages of the bar code, such as low-cost, low-profile, long-life, smallness, and robustness are kept intact. To be honest, no electronic device could be cheaper than the bar code. Similarly, no wireless device could be immunized from electromagnetic interference but its vulnerability (EMI/EMC) could be assessed for comparison. Hence, it is a choice between the vicinity card and the UHF RFID, as both have met most requirements, apparently.

Since the microelectronic technologies used in making HF components are more mature than that of the UHF ones, chips for vicinity cards are less expensive. As the HF band is less utilized, the vicinity card is subject to less intersystem interference. Moreover, due to its principle of operation, the range of coverage of the vicinity card is very small, and therefore it is not vulnerable to intrasystem interference. In other words, on the EMI/EMC, the vicinity card wins. However, due to its limited range of coverage, the vicinity card is rejected because it does not satisfy the requirement on a long-range operation.

For various technical, economical, medical, and political reasons, most of the active and passive RFID systems mentioned under the preceding headings will not be replaced by the UHF RFID in the foreseeable future. However, the importance of these RFID technologies will fade with the gradual eminence of the UHF RFID. Hence, attention will be focused on the UHF RFID hereafter, unless stated otherwise.

8.18 Supply Chain Management

As a bar code replacement, an RFID tag is embedded in every pallet or product for shipment. Due to globalization, the shipment may have to travel thousands of kilometers on land, on sea, and in air. At a national border, it has to endure the long waiting line for custom control, quarantine check, and other inspections. Then, the cargos are stored in one of the warehouses of the distributor before they are sent to the point-of-sale. It is a journey of months that involves tens of handlers speaking different languages and dialects.

Equipped with an active RFID tag, significant saving in waiting time inside and outside a container port or an airport could be achieved by dispatching the container trucks via a WiFi wireless local area network to move in an orderly fashion. Assisted by the same active RFID system, the loading and the unloading of the containers could be carried out more efficiently. On the other hand, clearance of custom control and other paper work could be speeded up for merchandise equipped with a passive RFID tag. The passive RFID system is also applicable in accelerating inventory control at various warehouses and the eventual retail outlet.

Unfortunately, a container port or an airport is only one of the bottlenecks along a 1000-km journey. Instead of waiting helplessly at the receiving end, a retailer would like to know the location of the merchandise selected for the upcoming sales promotion. In fact, the said cargo could be sitting idly in stock or traveling. For the former case, the network of readers in a warehouse could be used to locate the misplaced item through its RFID tag. For a cargo on its way, a global trace-and-track could be conducted via GPRS (general public radio service) provided by most terrestrial mobile phone networks because the position of a truck could be extracted from the GPS (global positioning system) map. Similarly, the GPS reading of a vessel could be sent back through Inmarsat, a satellite communications system operated by the International Mobile Satellite Organization. Other than the GPS system, with or without a local map on every truck, a real time global track-and-trace could be conducted through the existing communications network with no auxiliary equipment.

Simply put, the supplementary capital investment needed for deploying the RFID in the supply chain management is relatively small compared with the total cost spent in equipping every product with an RFID tag. Nevertheless, it is a worthwhile investment because the tangible gains will surpass the cost of a billion tags, not to mention the intangible ones. Hence, the industry of logistics management has reached a consensus in equipping every product with a RFID tag of a unique ID number when the price of a tag is right. Baring unforeseeable political and economical obstacles, this fantasy will become a reality in the near future.

8.19 International Standard

Unlike the bar code or the HF RFID, there is no common platform for a UHF RFID tag to roam seamlessly across a man-made political boundary. The outstanding issues to be solved are, however, more political than technical.

In the past, the national standard where a technology was invented could easily be steamrollered as the de facto international standard as illustrated in bar code development. However, in the fast changing world of electronics, changes are not just found in technologies, they are also observed in places where the technologies were first developed. As a result, revolutionary changes are noted on the map of technology three decades after promulgation of the universal product codes (UPC). Today, there are at least three international standards for consumer electronics as exemplified by CDMA 2000, W-CDMA, and TD-SCDMA in the third generation mobile phone. Analogous confusions are also seen in high-definition televisions.

The RFID differs from the above examples. After all, the RFID tag is not a product, but an electronic component in every electronic or nonelectronic product. In competing with the bar code, the RFID tag must be made cost effectively. Ironically, the production of this tiny and inexpensive tag involves a super clean ambiance and an extremely stable environment for the growth of a large silicon–germanium ingot, a state-of-the-art foundry for the fabrication of a microelectronic disk, and sophisticated equipment for packaging a die into a tag. Furthermore, an effective engineering management team and cheap skilled labor are required for embedding a tag into every product for shipment. Very likely, these processes are carried out in different countries. As a result, no nation could claim a monopolistic control of all RFID technologies. Finally, but not the least, as the RFID has been destined to play a key role in the vertically integrated industry of supply chain management comprising raw material suppliers, component makers, product manufacturers, tariff controllers, shipping companies, warehouses, distributors, and retailers in different countries, an international standard is not desirable, but mandatory.

8.20 Promiscuity

Even though printing a bar code on paper is costless, the overall turnover of the related markets on hardware, software, computer network, and bar code management is more than a billion US dollars. Similarly, the expenditure consumed in the fabrication of RFID tags, the production of RFID equipment, the development of specific software, and the management of the relevant network represents only part of the total revenue generated by the RFID. Together with industries, due to the predictable and the unpredictable applications of RFID technologies in affluent nations as well as in the emerging ones, the annual turnover of all RFID-related industries will be measured in hundreds of billion dollars, and eventually in trillion dollars.

That is why every nation wants to play a dominant role in drafting an international standard that may allow the drafter to keep one's strengths intact yet sacrifice nothing. In all fairness, most nations have solid grounds for standing firm because they have dissimilar sociopolitical structures, unequal levels of industrialization, distinct socioeconomic problems to be solved, and therefore different expectations from the RFID.

Short of a unique standard for all nations, political entities of similar socioeconomic backgrounds and of comparable levels of industrialization will settle for a multinational standard, so that the economy of scale will remain applicable in keeping the unit cost of the special equipment and the tag at a low level. As depicted in the third generation mobile phone, three to four standards are expected. Instead of full compliance, promiscuity or multisystem and multistandard operations will be supported. In other words, the multinational standards will share some common features, such as compatible air interface, similar initialization procedures, accessible identification number, and preferably, the ability to write in some common areas.

8.21 National Standards

While regulatory authorities from various trading blocks of nations are working hard in drafting a multinational standard that is acceptable to all parties concerned; some of the national standards are first examined.

Subsequent to the immense success of mobile phones, many wired systems are being converted to wireless, including local area networks and the last mile system in the plain old telephone network. In addition to the conversion of the existing systems, many new wireless gadgets at the consumer level have been launched under the aegis of various wireless interconnect formats such as WiFi, WiMax, ZigBee, Bluetooth, and ultra wideband. Due to the proliferation of wireless devices, the power rating of any proposal must be rigorously scrutinized.

Probably due to its projected role in the logistics industry, the RFID has been favorably treated by many regulators. For example, the permissible output power in the United States is 4 W EIRP. With special approval from the Federal Commission of Communications (FCC), the rating could be increased to 20 W. Under the normal condition without license, the effective range is 8 m and the rate of reading is 200 tags per second, provided that the tags are moving at a speed of less than 3 m/s. Compared with the HF RFID, significant improvements are observed.

Although it is the right of a sovereign country to draft a national RFID standard, more and more countries are convinced of the advantages in adopting one of the multinational standards. Today, there are two major multinational standards led by the European Union and the United States as listed in Table 8.1. Unfortunately, there is none in Asia where most products are made. For easy reference, national standards in selected countries are listed in Table 8.2 [18].

TABLE 8.1 Major Multinational Standards

Lead Nation	Nations
European Union	Armenia, Austria, Azerbaijan, Belgium, Bosnia, Bulgaria, Croatia, Cyprus, Czech Republic, Denmark, Estonia, Finland, France, Germany, Greece, Herzegovina, Hungary, Iceland, Ireland, Italy, Jordan, Latvia, Lithuania, Luxembourg, Macedonia, Malta, Moldova, the Netherlands, Norway, Poland, Portugal, Romania, Russia, Serbia and Montenegro, Slovak, Slovenia, Spain, Sweden, Switzerland, Tunisia, Turkey, the United Arab Emirates, the United Kingdom.
United States	Argentina, Canada, Chile, Costa Rica, the Dominican Republic, Mexico, Peru, Puerto Rico, Uruguay, the United States.

TABLE 8.2 National Standards in Selected Countries

Authority	Frequency	Maximum Power	Characteristics
European Union	865.6–867.6 MHz	2.0 W ERP[a]	LBT
United States of America	902.0–928.0 MHz	4.0 W EIRP	FHSS
China	840.5–844.5 MHz	2.0 W ERP	FHSS
	920.5–924.5 MHz		
Hong Kong	865.0–868.0 MHz	2.0 W ERP	LBT
	920.0–925.0 MHz	4.0 W EIRP	FHSS
Taiwan	922.0–928.0 MHz	1.0 W ERP, indoor	FHSS
		0.5 W ERP, outdoor	
Singapore	866.0–869.0 MHz	0.5 W ERP	
	920.0–925.0 MHz	2.0 W ERP	
Japan	952.0–954.0 MHz	4.0 W EIRP, with license	LBT
	952.0–955.0 MHz	20 mW EIRP, w/o license	
Korea	908.5–910.0 MHz	4.0 W EIRP	LBT
	910.0–914.0 MHz		
India	865.0–867.0 MHz	4.0 W EIRP	

[a] 1 ERP = 1.64 EIRP.

8.22 Hands-Free Bar Code System

Apparently, the primary objective in developing the UHF RFID is to extend the range of coverage of the HF RFID. Having a longer range is definitely one of the means of achieving the aim, but it is not the ultimate one. The principal aim appears to be the development of a hands-free bar code-scanning technique for automation processes.

The development began in 2003. As the military forces of the United States were preparing for a swift and decisive invasion of Iraq, it was deemed that all military supplies and other provisions for the allied force of 300,000 for the blitzkrieg must be ready for immediate use in Saudi Arabia before the first bomb was dropped. It was also decided to use the UHF RFID for speeding up the counting of military and nonmilitary supplies, in hundreds of million. Partly due to the RFID, the 6-month task in the First Gulf War in 1990 was completed in 2 weeks [19].

By 2004, critical components, specially designed hardware, and tailor-made software were ready for development of a pilot RFID inventory system for civilian use in peace time. In early 2005, the RFID was used by the U.S. armed forces for receiving and dispatching various supplies bundled on a pallet. For the enhanced version, every case was required to possess a unique RFID tag. It was scheduled that every individual item must be equipped with an RFID label in 2009. The Pentagon claimed that the pilot RFID inventory system saves over 80 million dollars for its government annually, not to mention the intangible gains.

An equally pioneering project was conducted on the other side of the Pacific Ocean in 2006. By replacing the bar code on a baggage label with an RFID tag (Figure 8.11), the luggage accepted by a check-in counter in the Hong Kong Airport could be automatically sent through a maze of RFID-monitored conveyor belts to the cart earmarked for the designated aircraft (Figure 8.12). Besides gain in speed, the rate of erroneous delivery has been significantly reduced. With tens of strategically located interrogators, a baggage dropped from the kilometers of conveyor belts or mishandled in whatever way could be located on time before flight departure.

8.23 Bar Code Mentality

As a hands-free bar code, the development of the UHF RFID is practically completed and the remaining tasks are simply fine-tuning the microchip, miniaturizing the hardware, enhancing the antenna, and debugging the control system and other software. In general, these chores should not arouse the interest

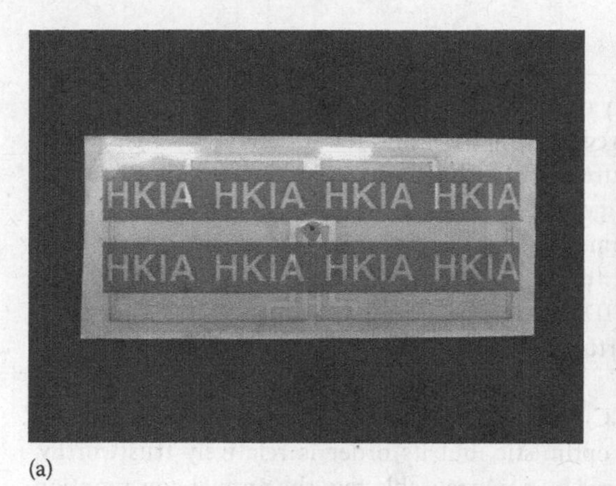

(a)

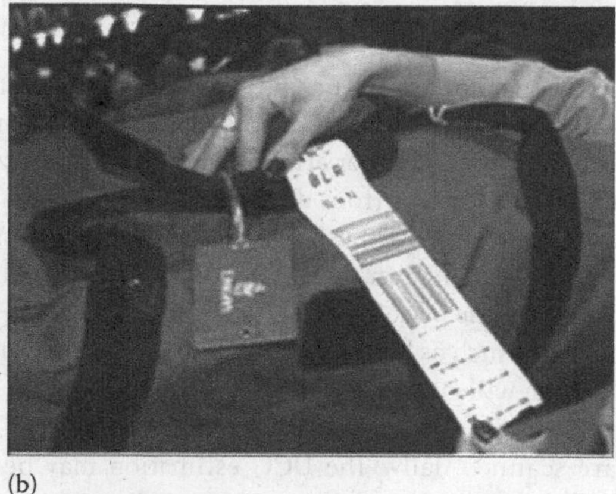

(b)

FIGURE 8.11 UHF RFID baggage label, Hong Kong Air.

FIGURE 8.12 UHF RFID antenna in the Hong Kong Airport.

of researchers in a university or a national research center. On the contrary, the RFID has been gold-fingered by the national authorities as an area of intensive development in many countries, including the United States and the Peoples Republic of China.

On closer scrutiny, it is found that the UHF RFID is a facilitator. Simply put, the RFID could enable us to realize our fantasies, and in fact, it could give us something beyond our imagination. To a certain extent, the scenario is similar to the invention of the electricity generator for lighting up light bulbs. At least, it is comparable to the development of the Electronic Numerical Integrator and Computer (ENIAC) [20] in 1946 for handling the massive paper load due to the discharge of millions of veterans after World War II. No one at that time realized that a necessity for human beings had been created. Ironically, we have lived for millions of years without this "necessity."

It is understandable that the passive UHF RFID has been treated as an enhanced bar code at its infancy. However, it is unfortunate that the bar code mentality is now shared among professionals and nonprofessionals in technical and nontechnical communities, especially among the old-timers. Echoing this mentality, the translation of its Chinese name in Hong Kong is "Radio Frequency Bar Code," while it is called "Electronic Bar Code" in Mainland China. Fixation and obstinacy could hinder or curtail the further development of the RFID; thus, its intrinsic capabilities and potential abilities would never be fully utilized.

8.24 Affordable Tag

From the aforementioned pilot projects, the ability of RFID in detecting a tag at a distance has been fully demonstrated. Reasonable saving was observed in manpower and fewer penalties were paid for late delivery and other errors. However, it is not sure whether the saving is good enough to cover the extra cost arising from dumping the free-of-charge bar codes. Moreover, could the capital investment be recovered? Both systems were highly commended because accuracy and efficiency have been heavily weighted in the system assessment, while the cost of the implementation was played down. Based on the price of an RFID tag in 2006, the RFID can only be afforded by shops with a high profit margin, such as boutiques and gourmet stores. Fortunately, due to the economy of scale, the unit cost of a tag will go down.

According to the Uniform Code Council (UCC), the issuer of bar codes, five billion bar codes are scanned daily. The UCC estimation may be optimistic, but its order is relatively trustworthy. Supposing the bar code on every product is replaced by a nonreusable tag, the annual consumption would be measured in hundreds of billions. With the projected modernization of populous nations, such as China, India, and Brazil, this number could easily be doubled or tripled. Hence, it is foreseen that the direct-from-foundry price of a microchip could be reduced to 5 ¢ in the coming years, if not months. Besides the die, making a tag involves printing an antenna on a piece of paper, bonding a die onto the antenna, downloading a unique ID number and other information into the memory, performing a critical functional test, packaging the tags in a box, and sending it to the end user. Then, a tag is embedded in a product for shipment. Using efficient manufacturing technologies and proven logistics management in China, the total cost of adding an RFID tag in a made-in-China product is approximately one renminbi (RMB), equal to 12 ¢ US. At this price, the RFID is affordable for most products sold in department stores. In fact, another factor that drives the tag price down is competition. Today, there are hundreds of UHF RFID tags of various sizes available in the market and the prices on the forms shown in Figure 8.13 range from 8 ¢ to 50 ¢ US.

Since the technologies used in making 900 MHz RFIC is rather mature, the price of a microchip will not fall substantially. On the other hand, the cost of supplementation will probably stay firm because the gain in technological improvement, if any, may not offset the rises in overheads such as salary, rental, financial charge, and code administration. Together with the projected devaluation of the U.S. dollar, the ultimate price of a tag will probably hang around 10 ¢, twice as expensive as the usual prediction of 5 ¢.

At 10 ¢ a piece, the RFID is ready for use in supermarkets, convenient stores, and corner shops. For small-value items, such as candies and lollipops, the RFID is also applicable provided that they are bundled together into a family pack. However, use of disposable RFID tags for fresh produce at the retail level may not be ready in the coming years. Thus, it is foreseen that the RFID and the bar code will coexist in a computerized point-of-sale; while cash, electronic cash, credit card, and debit cards will be used elsewhere. RFID is not just used in affluent countries; it will be extensively used in the emergent markets too.

8.25 Ubiquity of RFID

At 900 MHz, most packaging materials are RF-lucent; thus, there is no need to place an RFID tag at a prominent position on the covering box as would be the case if it were a bar code. For better security, it is advisable to hide the tag inside the box, especially for expensive items, such as jewelry, clothes, purses, bonnets, watches, mobile phones, cameras, and digital organizers. In fact, the number of tags inside a carton is not necessarily one. By tagging each module individually, the completeness of an electronic or a nonelectronic system, such as an audio-visual system, a personal computer, or a business suit, could be double-checked before the final packaging process.

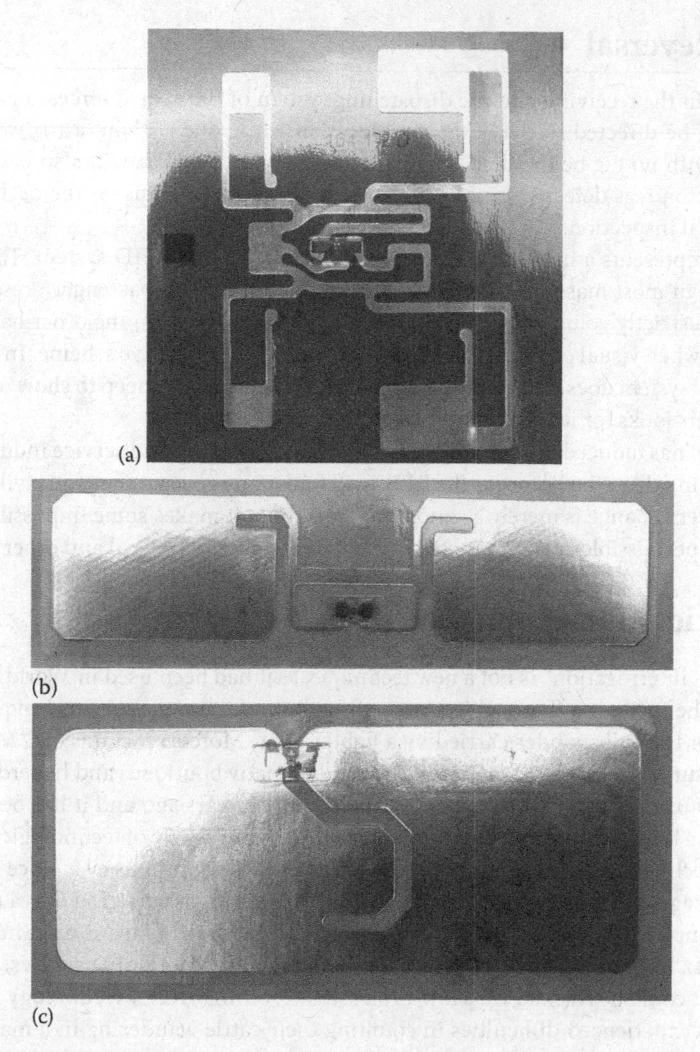

FIGURE 8.13 UHF RFID tags.

Experience acquired in the successful deployment of an RFID for productivity enhancement in the manufacturing sector is certainly pertinent in the service sector, including those provided by governmental and the nongovernmental organizations. To this end, the existing applications of the HF RFID or the vicinity card in the service industry are especially vulnerable for replacement.

The use of a long-range RFID in e-IDs, e-passports, and tickets for access control is of particular interest because it would induce profound changes in our daily lives. However, many groups of special interest have claimed that the RFID would infringe on their privacy; therefore, many RFID applications are experiencing undue oppositions, especially in liberal western countries. This issue will be further examined under a subsequent heading. Even though many hurdles must be scaled, it is foreseen that many vicinity card systems will be replaced by UHF RFID schemes, eventually.

Although a tag could be killed at a point-of-sale, it is advisable to keep it alive in a different mode. In fact, keeping its tag accessible in post-sales interrogations could distinguish the UHF RFID from other AIDC schemes. However, it also implies that every man or woman will carry many tags at all times, and they will be interrogated numerous times by equally numerous readers hidden everywhere, everyday. In short, the ubiquity of the RFID will create many opportunities on the one hand and equally many technical and nontechnical problems on the other.

8.26 Role Reversal

As demonstrated in the receiving and the dispatching system of the armed forces, a pallet with a valid RFID label would be directed to the designated location while one without a tag would be rejected. Should a carton with no tag be bundled on a pallet with a valid label, it will also be stopped because the number of carton tags detected is not equal to the number of cartons on the declaration and that confirmed by visual inspection.

This sequence represents a marked difference from that of an HF RFID system. The access control currently adopted in most mass transit systems works, if and only if a passenger presents a valid card for examination, a strictly volunteering process. A UHF RFID reader, on the other hand, detects a tag at a distance even when visual contact is blocked by an obstacle, a proactive scheme. In other words, the gatekeeper in a HF system does nothing but wait passively for the black sheep to show up, while its UHF counterpart actively looks for it and acts accordingly.

The role reversal has induced drastic changes in the manufacturing and service industries. In fact, its impact will be felt in other sectors too, as the RFID will find many government and civilian applications. Although the present change is merely a simple augmentation, it makes some impossible tasks feasible; while some of the permissible current practices would be forbidden for legal and other reasons.

8.27 Historical Development

Strictly speaking, "interrogation" is not a new technique, as it had been used in World War II for determining whether the incoming flying object was a friend or a foe. Analogous techniques are still used today to search for the transponder carried by a fighter pilot. Moreover, clumsy 8.2 MHz 1 bit passive electronic article surveillance (EAS) tags are still found in many boutiques and haberdashers for deterring shoplifting. Surprisingly, this technique was launched 46 years ago and it has been working continuously without a break, an extremely rare phenomenon in the history of technological developments.

In fact, RFID technology has been used in registering domestic animals ever since the 1980s. Using sophisticated packaging technologies, a passive 134.2 kHz HDX tag is squeezed into a rice-size capsule. Registration is completed by implanting the capsule beneath the skin of an animal. Similarly, by attaching a large 125 kHz FDX transponder on the back of an endangered wildlife species, its movement in the natural habitat could be tracked via a low-orbit satellite. Similar RFID technology has been used by farmers who have experienced difficulties in counting their cattle wandering in a mammoth range in Canada. The latter tags are also used in shipping animals for slaughter and other business transactions.

In fact, the true ancestor of the modern passive RFID with memory was invented by Mario Cardullo in 1971 and a U.S. patent was approved in 1973 [21]. The idea was sound but the technologies at that time were not ready. Ironically, the needed technologies were ready after the expiry of Cardullo's patent.

8.28 Privacy Infringement

After years of passionate debates, fierce quarrels, and occasional fights in parliamentary chambers and at informal gatherings on the probable privacy infringement, most people in liberal countries have finally accepted that the current trend of RFID developments cannot be stopped. The claimed privacy infringement may be valid but it is not due to the use of HF RFID technologies in making a reliable and a versatile identity card or passport. Today, the heated debate has been reopened as a result of the planned upgrading of the HF RFID to the UHF RFID, probably due to role reversal.

Claims of privacy infringement are a complete fallacy because we have already deliberately chosen to inform the service provider of our exact locations by turning on our mobile phones. The network operator needs to keep track of our movement, otherwise we could not receive any incoming call. Using the latest technologies, our locations could be accurately assessed to within a few meters. For accurate billing, the operator has a complete record of our locations in months. It is totally illogical to claim that a private company could be trusted to protect our personal records from illegal use, when we cannot entrust the government with it. It is also argued that a mobile phone is different because we could return to anonymity by turning our mobile phones off. Similarly, we need to carry our e-ID if and only if we want to enter some controlled areas; otherwise it could be safely kept at home. After entry, we could keep our privacy by hiding the e-ID in a metallic bag, such as a bag for potato chips.

Concern about the loss of privacy goes hand-in-hand with the development of the AIDC from the issuance of the first EPC bar code. Today, a personal trail could be tracked via one's use of the credit card, the vicinity card, the prepaid card, the electronic key, the electronic ticket, the mobile phone, the iPod, the iPhone, the fixed-line or the wireless network connection, the e-ID, and the e-passport. The RFID is simply another variation of the same theme. The association of a product brought with a buyer may leave a faint trace of the whereabouts of the person involved, but it is certainly not a reliable one. However, it may be the last cue for locating a missing person for rescue. From the technical point of view, the assured benefits obtained from having a long-range hands-free ID outweigh the potential loss of privacy in absolute terms [22]. At least, it is a view of an old-fashioned Asian living in a conservative Asian country.

8.29 Recent Developments

At the present stage of development, there exist many compatible and incompatible protocols in the market. To better understand the subsequent discussions, the ISO 18000-6C, an air interface for RFID at 860–960 MHz, is used as an illustration.

As usual, the memory of an RFID tag is divided into many fields. In addition to the 8 bit header and the 36 bit serial number, rooms have been reserved for the company (28 bit) and the object class of the tag (24 bit). With the ever-increasing popularity of the RFID, the memory has been expanded from 64 to 96 bit as illustrated in Table 8.3. For the EPCglobal Class 1 Generation 2 (Gen 2) tag, a 32 bit error correction was augmented to the 96 bit standard and a kill command was added.

The memory of Gen 2 was later amended to 256 bit [23]. The extra space of 128 bit is undefined, a room reserved for any nation to define special commands to be used within its jurisdiction for catering to its specific needs due to its unique cultural heritage, religious background, and level of industrialization.

Nowadays, the trend of development is to further expand the memory while keeping the Gen 2 data structure intact. Together with a 48 bit overhead, a 304 bit EPC bank is created. It is very likely that the EPC bank would be used as a common platform for multisystem and multistandard operations. In addition to the serial number in Gen 2, a bank of 64 bit is added on top of the EPC bank for a unique identification of the tag (TID) according to the ISO 15963 recommendations. To distinguish this TID from the ID found in the EPC bank, they are called ISO-TID and EPC-TID, respectively. To address diverse national needs, an extra 64 bit is included for storage of four passwords for accessing special commands and the "kill" commands as shown in Table 8.4.

Compared with the 32 bit EPC-TID in the 36 bit serial number, 42 bit in the TID bank are reserved for the ISO-TID. Hence, the maximum number of identifications has been increased from 4.295×10^9 to 4.398×10^{12}. Assuming that one trillion tags will be consumed annually, the ISO-TID alone could last for several years. However, if it is used in conjunction with the EPC-TID, the tag could be used almost indefinitely.

The "kill" command in Gen 2 is a noteworthy new feature that warrants in-depth discussion. This command enables any operator to disable any tag permanently. It was perhaps created for lessening our worry on the loss of privacy as it dissociates a product from its new owner. In doing so, it could also relieve the projected traffic congestion due to the simultaneous arrival of many tags.

TABLE 8.3 The 96 bit EPC Bank

Header 8 Bit	Manager Number 28 Bit	Object Class 24 Bit	Serial Number 36 Bit
A0	01FBA3	2EF970	7345A351BE

TABLE 8.4 The 432 bit ST XRAG 2 Memory Structure

User	TID	EPC					Reserved	
Bank 11	Bank 10	Bank 01					Bank 00	
128 Bit	64 Bit	176 Bit					64 Bit	
00_h to $7F_h$	00_h to $3F_h$	00_h to $0F_h$	10_h to $1F_h$	20_h to $9F_h$	$A0_h$ to AF_h		00_h to $1F_h$	20_h to $3F_h$
User (128 bit)	TID (64 bit)	CRC16 (16 bit)	PC + AFI/NSI (16 bit)	EPC (128 bit)	RFU (16 bit)		Kill password (16 bit)	Access password (16 bit)

It is definitely an overkill as it would make tags vulnerable to intentional sabotage. Preferably, a "kill" command should be used to downgrade the mode of operation to the "killed" level. In the modified initialization process, the reader would be alerted that "killed" tags are found in the near vicinity. In case of failed initialization, a "killed" tag will go to sleep, wake up after a relatively long period, go back to sleep again unless a broadcast from the interrogator is heard announcing that all tags operating in the first mode have been cleared. Then all remaining tags will compete for connection on equal ground.

8.30 Dual Authentication

Since sophisticated equipment is needed to modify the data stored in the protected area, it is beyond the reach of most makers of counterfeit products. Hence, both the EPC-TID and the ISO-TID are relatively safe from alteration. Unfortunately, the tag on a brand name product could be removed and pasted on a counterfeit one. To this end, it is proposed to register a sales transaction with the manufacturer via a real-time communications channel. Once confirmed, the tag could be killed by the retailer by checking one of the undefined bits in the EPC bank. By erasing one of the kill passwords, the killing process cannot be reversed. As a result of this dual-authentication process, a product with a "killed" tag could only be sold as a second handed item.

This process is similar to the registration of a new car with the government and the subsequent sale of a used car. Moreover, the registration ensures that a second hand product could only be sold by the registered owner. In other words, by filing a complaint of an item lost in burglary with the police, the chance of recovering it from a pawnshop could be increased. In theory, a stolen valuable could also be detected via a matrix of interrogators installed at the major meeting points, such as shopping centers, subway stations, and airports. In practice, it is not feasible because extremely fast interrogators would be required to handle the large database of stolen goods in a metropolitan area at the real time mode, not to mention the manpower needed. The database in a small town may be manageable, but unfortunately the latter's budget is also very small for installing a network of interrogators. Nonetheless, as the tag of authentication is sold with the product, its presence or absence would provide a law enforcement officer an almighty legal tool in identifying and confiscating a counterfeit product.

Information obtained from the manufacturer is not necessarily limited to the 1-bit confirmation. Pharmaceutical companies and processed food producers could return the date of expiry and other relevant information with the confirmation. At the discretion of the buyer, processed food could still be sold shortly after expiry, but both parties of concern must be informed. With the real-time dual-authentication assurance, consumers in China will regain confidence in buying brand-named liquor.

8.31 Trace-and-Track

Fully convinced of the potential gains offered by the RFID, a smart R&D system has been installed in Walmart for more efficient inventory control and better market intelligence. Since line-of-sight is not required for tag detection, misplaced pallets and cartons could be found via a network of stationary interrogators and supplemented by handheld readers. Significant saving could be achieved by locating an overstocked item in one warehouse for replenishing the understocked condition in another store. In conjunction with terrestrial and satellite communications links, an accurate real-time trace-and-track of merchandise throughout the world could be conducted for upcoming promotional activities.

Recall that information obtained from the manufacturer is more than a simple confirmation, the reverse is also valid. Dependent on the object class of a tag, the IP address of the reader and the ID number of the operator could be sent to the manufacturer with the TID. Together with the date and the time of transaction, these ID numbers could be stored in the database of the given tag. This proposal is welcome in China where careless contamination and deliberate poisoning in processed food is a major issue of concern. Based on the information provided by the manufacturer, the origin of error could be located and cleared. Moreover, as the handlers' identities in all nodes of the logistics chain have been recorded, the person responsible for the error could be identified. The ability to trace is a powerful tool in deterring a frustrated employee or any other person from committing this type of criminal act. Finally, but not the least, it must be emphasized that this apparently tedious dual-authentication process is costless because sending a 1-bit confirmation takes as much time as sending it with other ID numbers because the overheads needed to set up a communication channel involves the sending of thousands of bits to and fro between both ends of the link.

8.32 Innovative Applications

Through examples cited in the previous headings, the intrinsic characteristics of the UHF RFID have been demonstrated. Its potentials have also been explored via some propositions. Probably due to the bar code mentality, developments of RFID technologies have outpaced conceptions of ingeniously new RFID applications, not to mention system deployments.

Although the RFID is treated as a wireless bar code in Walmart's inventory system, substantial values have been added to justify the finite cost of RFID tags. Certainly, deployment of an RFID inventory system in a hazardous environment is justified. For the closely packed tanks of inflammable gas in an open space or a warehouse, a real-time check of tags via a network of interrogators on lamp posts or mounted on the ceiling is definitely preferred to the close up reading of vicinity cards one-by-one by an operator using a handheld device. The extra expenditure is justified in some high-profile events, such as the New York Marathon and the Tour de France where thousands of participants must be counted in no time. Due to role reversal, the RFID system could also be used to detect cheating, an invaluable value added.

As previously mentioned, HF RFID systems are vulnerable for replacement. One of the likely casualties is the inventory system for libraries. The principal reason for this change is the adoption of the UHF RFID by the logistics industry. It is foreseen that every new book will have a UHF RFID tag embedded in an invisible location before shipment; therefore, significant manpower could be saved from attaching an HF RFID tag (Figure 8.14) on a new book manually, the most time-consuming and expensive process in the deployment of a smart library system. Besides the multibook self-service check out and the multibook easy-return systems (Figure 8.15), library utilization could be enhanced through the smart book sorting machine (Figure 8.16) and the smart bookshelves equipped with multiple antennas for a real-time inventory check. For example, the library of the City University of Hong Kong was originally designed for 10,000 students, but it now serves more than 20,000 students with more than a million volumes. Hence, one half of the books are stored off campus. For access to these books, a student has to browse through the Web and order it as if it were an interlibrary loan. Usually, the book will be

FIGURE 8.14 HF RFID tag for libraries.

FIGURE 8.15 Multibook self-check out machine for the library of the City University of Hong Kong.

available for collection in less than 24 h. Instead of getting it from a library assistant, a student has to browse through the Web again and then fetch it from one of the smart bookshelves. Moreover, better book utilization could be obtained by converting the cards for temporarily holding books put back by students after reading it in the library. Usually, it takes days before these books are returned to the designated locations by a library assistant. With smart cards (Figure 8.17), these books could be located electronically during the blackout period. For better cost effectiveness, it is not advisable to convert all bookshelves in a library into smart ones because an occasional inventory could be conducted by operators using handheld readers.

Since many tags could be read at the same time, the RFID is ideal for cases where simultaneous tag matching is preferred. Today, the issue of most concern is the dispensing of medicine. In fact, the process could be made automatic provided that pills in standard sizes are delivered in tapes as if they were electronic components. Similar to the placement of electronic components onto a PCB board, different pills for a given patient are orderly transferred into another tape in accordance with the pattern of

FIGURE 8.16 RFID-enabled book sorting machine.

FIGURE 8.17 RFID-enabled book return bin.

daily consumption for the entire hospital stay. A tag is then attached at the end of the tape (Figure 8.18). Erroneous dispensing could be prevented by matching the patient's ID and that of the tape, simultaneously. Using the same system, in the prescription of allegoric medicine, over- or underdosage could also be avoided by comparing the patient's medical history and the medicine to be dispensed. This lifesaving system will be welcomed by the Hospital Authority in Hong Kong at any cost because the practically free public medical services in Hong Kong has been grossly overloaded during the worst global economic recession triggered by the CDO (collateralized debt obligation) and the CDS (credit default swap) debacles on Wall Street.

To a certain extent, the existing RFID technology resembles a large man-made lake of the size of Lake Superior. Better lake utilization does not depend on the development of bigger machines for enlarging the lake. Instead, the community could feel the impact of a new lake better, if many piers were built at strategically selected locations using conventional technologies, and if necessary, supplemented by new ones. Simply put, auxiliary facilities with and without auxiliary technologies are needed to strengthen the impact of the RFID in our daily lives. The situation is similar to the earlier days of the Internet;

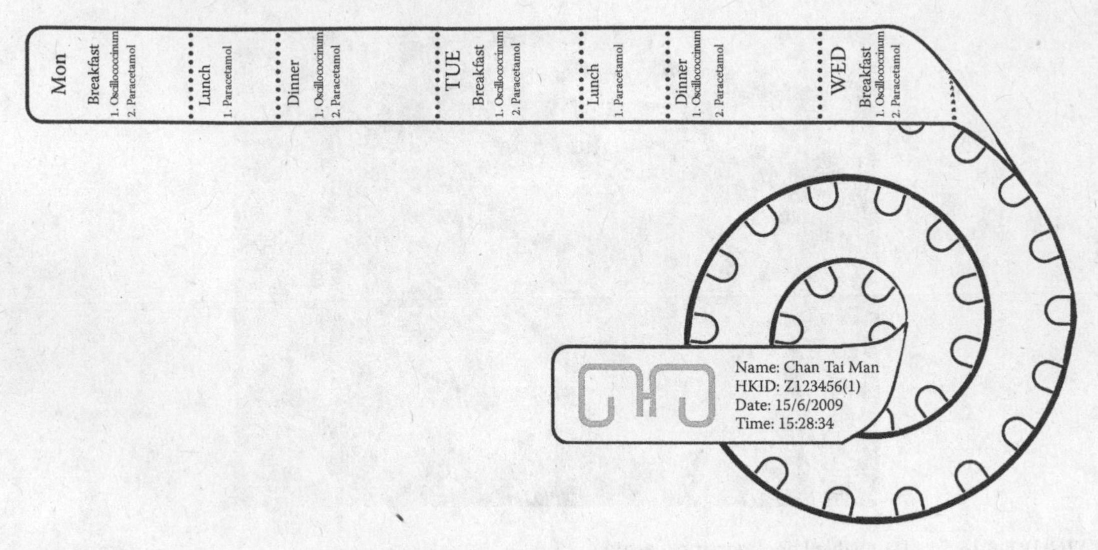

FIGURE 8.18 Illustration of the RFID-medicine dispensary tape.

therefore the RFID versions of Cisco and Yahoo are desperately needed. In fact, technology alone cannot make the Internet great. The joint force of novel technologies and time-honored ones has facilitated a platform on the one hand, and on the other, the collective wisdom of innovative semitechnical applications and ingenious nontechnical usages has contributed equally, if not more, in building the almighty global system. In other words, an RFID styled e-Bay, Facebook, and MySpace are also needed.

8.33 Nonionization Radiation

In the evolution of the AIDC from Hollerith's punch card to the UHF RFID, a new scheme was invented because the existing one might have lost its competitiveness due to the emergence of a challenging new technique. Sometimes, a system might have reached its capacity or capability, yet further expansion was ruled out for spectral reasons and other physical constraints. From time to time, the deployment of a new system was objected to by various activist groups on grounds of environmental protection, potential cultural conflicts, and other socioeconomic issues, such as loss of privacy.

Today, however, one of the issues of great concern is nonionizing electromagnetic radiation. So far, there is no solid scientific evidence proving that illumination of the electromagnetic field is hazardous to our health. However, there is no proof showing that it is harmless. On this life and death issue, we must be very cautious.

Unlike burning or poisoning, there exists no proven case indicating that prolonged exposure to a strong electromagnetic field is fatal. Statistical analyses with questionable inductions and inconclusive remarks are available. Hence, it is very difficult to determine the critical field strength and the threshold duration. Based on disputable studies, admissible limits on field strengths have been promulgated by many national authorities. Nonetheless, it is better to be conservative than aggressive on this issue, especially when the aggregated field intensity due to a large number of interrogators is difficult to estimate in advance.

For convenience, an RFID reader is usually hidden beneath the counter with its antenna pointing upward, so that the head of a tag carrier is subject to a strong nonionization radiation. Moreover, the 900 MHz electromagnetic field and its harmonics will interfere with the operations of many wireless communications systems, especially cellular phones working at 900 and 1800 MHz. What is more, as a proactive device, an RFID interrogator is turned on all the time even when no tag is found within its turf for a long time. For a person working inside the same office, it is an absolutely unnecessary exposure,

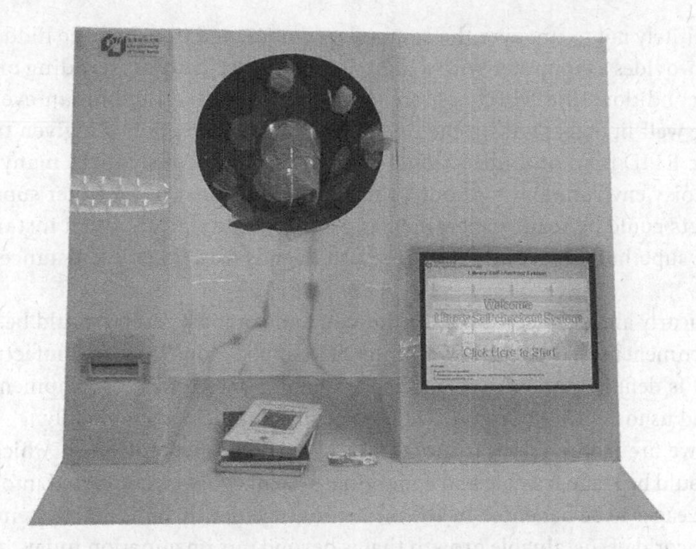

FIGURE 8.19 A multibook check out system with an accordion helical antenna camouflaged as a decorative lighting that points downward to the ferrite sheet hidden under the desk.

irrespective of whether the field exposure is harmful or not. To this end, it is highly recommended that the high-gain antenna be turned upside down and a sheet of ferrite or some other field-absorbent material be placed on the spot where the tag is supposed to be read to absorb the excessive electromagnetic energy (Figure 8.19). In case of a turnstile for access control, the ferrite sheet could be hung on the other side of the interrogator. In addition, a motion detector could be used to turn on the interrogator, especially at night.

As the UHF RFID operates at one of the ISM frequencies, its uses in hospitals must be carefully monitored because an imperfectly shielded interrogator could interfere with the normal operation of sophisticated medical apparatuses. Due to the unpredictable consequences of electromagnetic interference, a patient with a pacemaker should never come close to an RFID reader. In other words, a vicinity card is preferred to an UHF RFID in dispensing medicine in wards because its range is limited to a meter and the frequency of operation is much lower than those used in medical equipment. The UHF RFID loses again in the long-range monitoring system for kindergarten children because most parents would not allow their kids to be exposed to a strong electromagnetic field. In this connection, an active RFID operating in tailor-made protocols will prevail because the radiation power can be scaled down to milliwatts.

8.34 Era of Artificial Perception

Technically, the RFID is simply another enhancement of the bar code system developed in 1949. The modification may be minor; its impact on our daily lives is tremendous. Instead of waiting passively for a tag to present itself one by one for recognition, the UHF RFID differs from other AIDC schemes in its proactive search for all tags within its range of coverage. Via a tiny tag embedded in every product, or a small capsule implanted in an animal, or a card carried by a person, an interrogator detects everything at a distance even when it is blocked by obstacles.

The role reversal is not just new in the AIDC; it is also revolutionary in a computer. Today, a computer may feel the effects due to various stimulants, such as light, sound, smell, temperature, humidity, and pressure via appropriate transducers. These inputs could then be used to adjust the ambient conditions, to respond to the imminent collision, to fine-tune the predetermined path of motion, and for other passive reactions. However, a computer is not equipped to look for the sources of stimulants preemptively, definitely not to uncover the camouflaged ones, and to detect the hidden ones. In other words, the RFID provides a computer with a torch light for a better understanding of its environment in a totally dark condition. The RFID is more than an eye. Unlike the human eye that sees only a one-and-only-one well-lit object within the line of sight in one direction at a given time, the eyesight empowered by the RFID is an omnidirectional one that simultaneously locks many exposed or hidden objects in a noisy environment without a third party illumination or other supports. Moreover, the detected objects could be readily identified and appropriate actions taken instantly. With auxiliary supports, the superhuman eye could distinguish friends from foes at a distance well beyond the radar horizon.

Armed with a nearly almighty sensor, many human and computer errors could be reduced. Besides military and government uses, the RFID finds immediate applications in the manufacturing and service sectors. The RFID is definitely the driver for technical and nontechnical developments in the coming decades. It will lead us to another horizon, technologically and socioeconomically.

In conclusion, we are looking forward to an **era of artificial perception** in which an implausible operation today could be made feasible and a marginal system could be converted into an economically viable one. We are eager to be benefited by the numerous ingeniously built new systems to enhance our quality of life in a world of sustainable growth that is beyond our imagination today.

Abbreviations

AIDC	Automatic identification and data capture
ASIC	Application specific integrated circuit
ATM	Auto teller machine
CDMA	Code division multiple access
EAN	European article numbering scheme
EAS	Electronic article surveillance
EIRP	Effective isotropic radiation power
EMC	Electromagnetic compatibility
ENIAC	Electronic Numerical Integrator and Computer
EPC	Electronic product code
EEPROM	Electronic erasable programmable read only memory
EPS	Electronic payment system
FCC	Federal Commission of Communications
FDX	Frequency division multiplexing
FHSS	Frequency hopping spread spectrum
GPRS	General public radio service
GPS	Global positioning system
GSM	Groupe spécial mobile
HDTV	High definition television
HDX	Hopping duration multiplexing
HF	High frequency
ICAO	International Civil Aviation Organization
IMSO	International Mobile Satellite Organization
IP	Internet protocol
ISM	Industrial, scientific and medical
ISO	International Organization for Standardization
LBT	Listen before talk
PVC	Polyvinyl chloride
RF	Radio frequency
RFIC	Radio frequency integrated circuit

RFID	Radio frequency identification
SAR	Special administrative region
SIM	Subscriber identity module
TID	Tag identification
UCC	Uniform Code Council
UHF	Ultrahigh frequency
UPC	Universal Product Code

References

1. G. Schilling, *The Memoirs of a Baron*, Lightning Source, La Vergne, TN, March 2007.
2. J. Maxwell, A dynamic theory of the electromagnetic field, *Philosophical Transactions of the Royal Society of London*, 155, 1865, 459–512.
3. H. Dorsch, *Der erste computer, Konard Zuses Z1-Berlin 1936, beginn und entwicklung einer technischen revolution*, Berlin, Germany: Museum für Verkehr und Technik, 1989.
4. C. Mollenhoff, *Atanasoff: Forgotten Father of the Computer*, Ames, IA: Iowa State University Press, 1988, pp. 47–48.
5. H. Hollerith, Art of compiling statistics, US Patent 395,782, January 8, 1889.
6. J. Woodland, B. Silver, and J. Johanson, Classifying apparatus and method, US Patent 2,612,994, October 7, 1952.
7. F. Parry, Identification card, *IBM Technical Disclosure Bulletin*, 3(6), November 1960, 8.
8. S. Morse, Improvement in the mode of communicating information by signals by the application of electro-magnetism, US Patent 1,647, June 20, 1840.
9. R. Adams and J. Lane, *The Black and White Solution, Bar Code and the IBM PC*, Peterborough, NH: Helmers Publishing, 1987.
10. Radio regulations ITU-R in 5.138, 5.150, and 5.280, International Telecommunication Union, p. 4, April 17, 2008.
11. C. Walton, Portable radio frequency emitting identifier, US Patent 4,384,288, May 1983.
12. S. Lahiri, *RFID Sourcebook*, Westford, MA: IBM Press, January 2006.
13. W. Knight, Malaysia pioneers smart cards with fingerprint data, *New Scientist*, September 21, 2001.
14. *ICAO Document 9303*, Part 1, Vols. 1 and 2, 6th edition, International Civil Aviation Organization, 2006.
15. H. Stockman, Communication by means of reflected power, *Proceedings of the IRE*, 36(10), 1196–1204, October 1948.
16. H. Baldwin, S. Depp, A. Koelle, and R. Freyman, US Patent 4,075,632, February 1978.
17. M. L. Beigel, Identification device, US Patent 4,333,072, June 1, 1982.
18. EPCglobal Regulatory status for using RFIF in the UHF spectrum, GS1, October 2008.
19. S. Garfinkel and B. Rosenberg, *RFID Applications, Security, and Privacy*, Upper Saddle River, NJ: Addison Wesley, July 2005.
20. E. Berkeley, *Giant Brains or Machines that Think*, New York: Wiley, 1949.
21. M. Gardullo and W. Parks III, Transponder apparatus and system, US Patent 3,713,148, January 1973.
22. S. Garfinkel and B. Rosenberg, Preface, in *RFID Applications, Security, and Privacy*, Upper Saddle River, NJ: Addison Wesley, July 2005, p. xxx.
23. ST XRAG2 432-bit contactless memory chip, STMicroelectronics, p. 1, April 2008.

9

RFID Technology and Its Industrial Applications

Vidyasagar Potdar
Curtin University of Technology

Atif Sharif
Curtin University of Technology

Elizabeth Chang
Curtin University of Technology

9.1 Introduction 9-1
9.2 RFID Architecture 9-2
RFID Tags • RFID Readers • RFID Antenna • RFID Middleware
9.3 Item Tracking and Tracing 9-4
Baggage Tracking • Library Book Tracking • Animal Tracking • Hospital Equipment Tracking • Patient Tracking • Newborn Baby Tracking • Tracking Children • Golf Ball Tracking • Crowd Control
9.4 Access Control 9-8
Family Access to Babies in Neonatal Care • Vehicle Identification
9.5 Anticounterfeiting 9-10
Electronic Drug Pedigree • Bank Notes • Secure Passports and Visas • Automobile Parts
9.6 Conclusion 9-12
References 9-12

9.1 Introduction

In this chapter, we give a detailed introduction of radio frequency identification (RFID) technology and 15 different applications for different industries. We introduce RFID transponders, readers, middleware, and labels. A detailed classification and explanation of each of these components is provided. This is, then, followed by outlining the key applications of this technology in the real world as well as industrial scenario. The following applications are covered in this chapter: RFID for item tracking and tracing, for access control, and for anticounterfeit. The benefits that can be achieved by adopting this technology are outlined in detail. The following applications are explained in detail:

- Baggage tracking
- Library book tracking
- Animal tracking
- Hospital equipment tracking
- Patient tracking
- Newborn baby tracking
- Tracking children
- Golf ball tracking
- Crowd control
- Family access to babies in neonatal care
- Vehicle identification

- Electronic drug pedigree
- Anticounterfeit bank notes
- Secure passports and visas
- Anticounterfeit automobile parts

9.2 RFID Architecture

RFID system is composed of the following main elements: tags, readers, antennas, and middleware [EAS05, MS08]. A typical RFID system is shown in Figure 9.1. When the RFID tag comes in the range of the reader, the reader activates the tag to transmit its unique information. This information is propagated to the RFID middleware, which appropriately processes the gathered information and then updates the backend database.

9.2.1 RFID Tags

An RFID tag is a microchip attached with an antenna, which is attached to a product that needs to be tracked. The tag picks up signals from the reader and reflects back the information to the reader. The tag usually contains a unique serial number, which may represent information such as a customers' name, address, etc. [RFID06]. A detailed classification is discussed next. RFID tags can be classified using three schemes. First, the tags can be classified based on their ability to perform radio communication: active, semiactive (semipassive), and passive tags. Second, the tags can be classified based on their memory: read only, read/write or write once, and read many. Finally, the tags can also be classified based on the frequency in which they operate: LF, HF, or UHF.

9.2.2 RFID Readers

The RFID readers send radio waves to the RFID tags to enquire about their data contents. The tags then respond by sending back the requested data. The readers may have some processing and storage capabilities. The reader is linked via the RFID middleware with the backend database to do any other computationally intensive data processing. There are two different types of RFID readers [GLOE05]. RFID readers can be classified using two different schemes. First, the readers can be classified based on their location: handheld readers and fixed readers. Second, the tags can be classified based on the frequency in which they operate: single frequency and multiple frequency.

9.2.3 RFID Antenna

In RFID, antennas are classified into RFID tag's antenna and RFID reader's antenna [POT07].

RFID Tag Antenna: This type of antenna is the conductive element that enables the transmission of data between the tag and the reader [RFGA06]. Antennas play a major role in deciding the communication

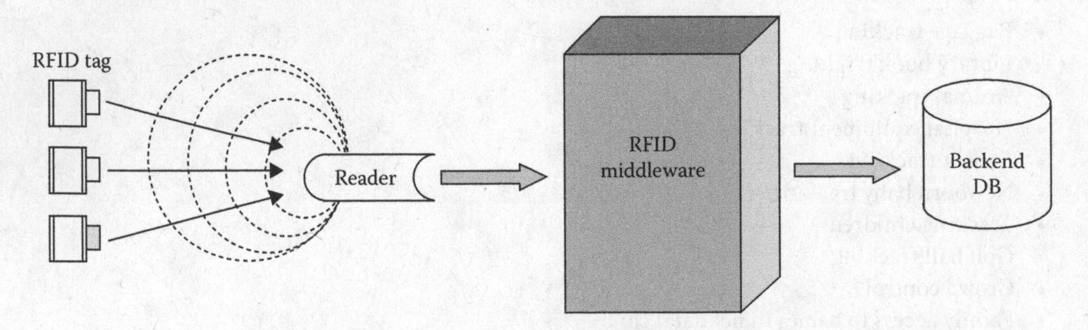

FIGURE 9.1 RFID architecture. (From Potdar, V. et al., *Automated Data Capture Technologies: RFID*, Zhang, Q. (Ed.), Hershey, London, U.K., pp. 112–141, Information Science Reference.)

distance; normally, a larger antenna offers more area to capture electromagnetic energy from the reader and, hence, provides a greater communication distance. There are several kinds of antennas like rectangular planar spiral antenna, fractal antennas, microstrip patch antenna (monopole, dipole), etc. Different types of tags have different kinds of antennas, e.g., low frequency and high frequency tags usually have a coiled antenna that couples with the coiled antenna of the reader to form a magnetic field [RFGA06]. UHF tag antennas look more like old radio or television antennas because UHF frequency is more electric in nature [SWEE05]. Recent advanced in technology have even facilitated the deployment of printed antennas to achieve similar functionality like the traditional antennas. One possible way of printing antennas is to use silver conductive inks on plastic substrates or papers [TECS05]. The main advantage of printed antennas is that they are cheap.

RFID Reader Antenna: Every RFID reader is equipped with one or more antennas. These antennas generate the required electromagnetic field to sense the RFID tags. There are many different kinds of antennas like linearly polarized, circularly polarized, or ferrite stick antennas. Several antennas can connect a single reader at the same time. There are different kinds of antennas that are equipped with standard readers, e.g., the IP3 Intellitag Portable Reader (UHF) [INTE06] comes with an integrated circular polarized antennas. The advantage of such antennas is that it can read tags in any orientation.

9.2.4 RFID Middleware

In general, the RFID middleware manages the readers and extracts electronic product code (EPC) data from the readers, performs tag data filtering, aggregating, and counting, and sends the data to the enterprise warehouse management systems (WMS), the backend database, and the information exchange broker [CHA07]. Figure 9.1 shows the relationship between tag, reader, RFID middleware, and backend database. An RFID middleware works within the organization, moving information (i.e., EPC data) from the RFID tag to the integration point of high-level supply-chain management systems through a series of data related services. From the architectural perspective, RFID middleware has four layers of functionality: reader API, data management, security, and integration management. The reader Application Programming Interface (API) provides upper layer the interface interacting with the reader. Meanwhile, it supports flexible interaction patterns (e.g., asynchronous subscription) and active "context-ware" strategy to sense the reader. The data management layer mainly deals with filtering redundant data, aggregating duplicate data, and routing data to appropriate destination based on the content. Integration layer provides data connectivity to legacy data source and supporting systems at different integration levels and, thus, can be further divided into three sublayers as specified in [LEAV05]: application integration, partner integration, and process integration. The application integration provides varieties of reliable connection mechanisms (e.g., messaging, adaptor, or the driver) that connect the RFID data with the existing enterprise systems such as Enterprise Resource Planning (ERP), or WMS. The partner integration enables the RFID middleware to share the RFID data with other RFID systems via other system communication components (e.g., the data exchange broker in Figure 9.2). The process integration provides capability to orchestrate the RFID-enabled business process. The security layer obtains input data from the data management layer and detects data tampering, which might occur either in the tag by wicked RFID reader during the transportation or in the backend internal database by malicious attacks. The overall architecture of RFID middleware and its related information systems in an organization are depicted in Figure 9.2.

Backend database component stores the complete record of RFID items. It maintains the detailed item information as well as tag data, which has to be coherent with those read from the RFID. It is worth noting that the backend database is one of the data tampering sources, where malicious attacks might occur to change the nature of RFID item data by circumventing the protection of organization firewall. The WMS integrates mechanical and human activities with an information system to effectively manage warehouse business processes and direct warehouse activities. WMS automates receiving, put-away, picking, and shipping in warehouses and prompts workers to do inventory cycle counts. The RFID middleware employs the integration layer to allow real-time data transfer toward WMS. The data exchange

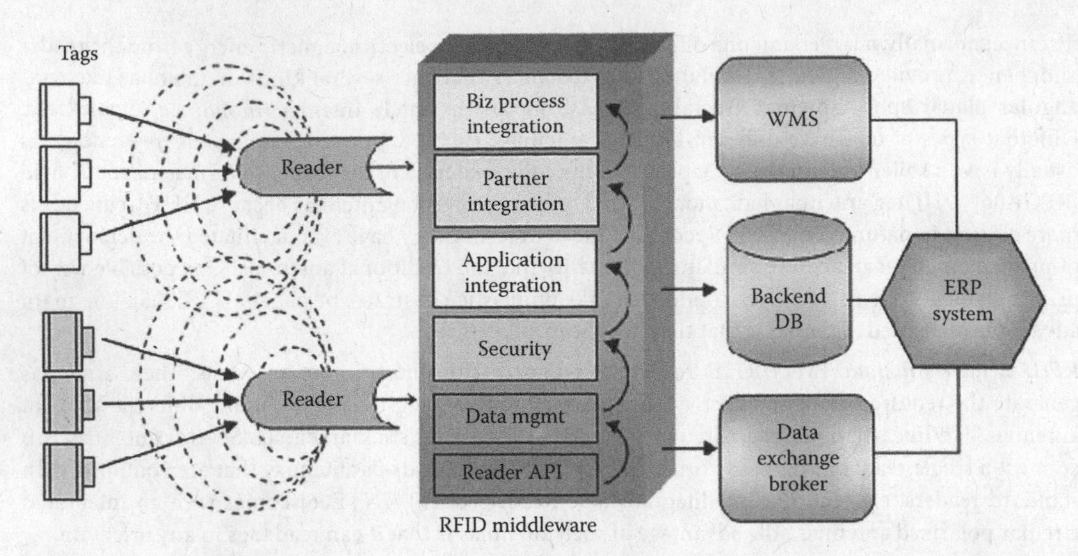

FIGURE 9.2 RFID middleware architecture. (From Potdar, V. et al., in: Zhang, Q. (Ed.), *Automated Data Capture Technologies: RFID*, Hershey, London, U.K., pp. 112–141, Information Science Reference.)

broker is employed in this architecture to share, query, and update the public data structure and schema of RFID tag data by exchanging XML document. Any update of the data structure will be reflected and propagate to all involved RFID data items stored in the backend database. From the standardization view, it enables users to exchange RFID-related data with trading partners through the Internet. From the implementation angle, it might be a virtual web services consumer and provider running as peers in the distributed logistics network.

9.3 Item Tracking and Tracing

Item tracking and tracing applications aim to track the movement of items (goods or people) at different points at different times for a specific reason [HE08], for example,

- Tracking baggage at airports for transferring bags to correct destination
- Tracking patients in hospitals to provide correct treatment
- Tracking passports in the airline terminal to easily find passports in case they are lost
- Tracking newborn babies to distinguish one from another
- Tracking animals in farms to prevent spread of diseases
- Tracking books in library
- Tracking golf ball in the golf court

The advantages of item tracking and tracing applications are very clear: It can provide you with detailed information about an item's movement during its entire lifecycle and can help you answer a number of questions that would assist you to make informed decisions in the future. Consider the scenario of manufacturing consumer goods like shampoo. The normal process employed to manufacture shampoo would be

- Picking up the raw materials from the warehouse
- Moving it to the production line
- Transferring it to finished goods section
- Packaging
- Preparing for dispatch

At every stage during the manufacturing process, organizations can capture a lot of information, but whether it is captured or not is a different question [IA07]. One reason why all this information may not be captured would be insufficient human resources. But what if one wants to know

- When was the raw material picked for a batch number "XYZ"?
- Which personnel picked the raw material from the warehouse?
- What time was it picked?
- When did it go to the production line?
- Who authorized the production?
- How long did it take to finish the batch "XYZ"?
- Who authorized the packaging?
- What time did the packaging begin and finish?

If tracking is not implemented, answers to these questions would be difficult to get. However, if RFID-enabled tracking solution is deployed, it would be quite easy to have access to this information, which can prove to be very useful [ZHA06]. Thus, we see item tracking and tracing can provide organizations with that information, which would otherwise not be accessible. How the organizations use this information to their competitive advantage is up to them, but RFID can get them the required data that can be turned to their advantage. We, now, discuss few item tracking and tracing applications deployed in different industries.

9.3.1 Baggage Tracking

The increasing growth in the number of air passengers and the resulting increase in checked baggage are straining the world's baggage-handling infrastructure, as evidenced by major increases in reports of mishandled baggage across the globe [BAG08]. In Europe, the Association of European Airlines reported that the incidence of mishandled bags has increased 14.6%—representing an additional 1.2 million bags reported missing on arrival. In the United States, baggage complaints have been on the rise, since 2002 [BR07]. The U.S. Department of Transportation has reported an increase from 3.84 mishandled bags per 1000 enplaned passengers in 2002 to 6.73 in 2006—an effective 75% increase over a 4 year period, representing an additional 2.3 million mishandled bags each year. The existing barcode systems are used for baggage tracking, however, using barcode reading system up to 85%–90% [RBC07,BR07,PM09] read rate is possible. Also, optical read rates are very dependent on line of sight; thus, it is necessary to make sure that the bar code has to be seen by an automated device.

The problem can be solved by employing RFID technology, provides 99% read rate and is not constrained by sight lines, for this purpose passenger and baggage reconciliation system has been designed that employs nonprogrammable, passive transponder, RFID tags attached to checked baggage, which enables continuity of reconciliation through any number of intra-air transfer points until the destination of the air travel is reached. And if any separation between baggage and the passenger occurs, the result would be an alarm notification to the airline. The passenger's and baggage enrollment occur upon payment for travel, although checked baggage enrollment can occur variously outside the airport building (curbside), at the ticket counter, at the departure gate, on board the airplane, etc. The interface to the airline carrier's computer reservation system to obtain passenger and schedule data permits continuous tracking of passengers and baggage for the reconciliation process. RFID has numerous advantages over traditional barcode systems, and they are

- Reduction in tagging errors
- Decrease in the number of left behind/short-shipped baggage
- Drastic reduction in load discrepancies
- Timely arrival and departure of the vessel avoiding airport congestions
- Improved turn around time of interline transfers
- Avoid baggage damages and misplace of items from the bags
- Tags are self-adhesive and easy to install on totes and do not interfere with x-ray inspections

9.3.2 Library Book Tracking

Institutional (academic) libraries and public libraries face growing challenges in managing the assets of their collection and maintaining or improving service levels to patrons. Over the past 20 years, libraries have grown their services to their patrons from simply offering books and periodicals to now adding multimedia items such as videos, CDs, and DVDs. Add to this the fact that many libraries provide patrons computers to access the Internet as another source for information. Throughout this growth in material resources offered by libraries, there are also growing concerns on the security of the collection assets and maintaining (or improving) service levels to the patrons [GAO09]. Throughout this time, many municipalities and institutions who manage our library systems have sought ways to reduce operating budgets (cut staff) to offset the growing capital costs of the multimedia resources and computer hardware. For the concerns that library managers face each day, RFID technology brings solutions that magnetic stripe or barcode technology simply cannot deliver so effectively. The RFID tag is assigned and attached to the asset in similar ways, depending on the material, paper, plastics, etc. The advantages of RFID tags over barcode or magnetic stripe are as follows:

- Faster scanning of the data stored on the RFID tag
- Simple and easy way for patrons to self-checking for material loans and returns
- Superior detection rates
- Significantly reduce the number of false alarms at the exit sensors (up to 75% less)
- High-speed inventory—reducing time to staff for "shelf-reading" and other inventory activities
- Automated return of materials that speeds up sorting of materials and reshelving for the next patron to access
- A longer lifecycle than a barcode

9.3.3 Animal Tracking

Animal shipment from one country to another country is a major cause of disease transmission. The number of cases due to this disease spread keeps on growing in various parts of the world. To limit this disease spread, some sort of identification or record mechanism is generally required. This growing number of countries exporting animals to other countries has led to the establishment of record keeping standards that use RFID as the main identification [LIV09]. The countries like the United States, Canada, Australia, etc. have already been using some RFID supported International Standards Organization record keeping standards like 11784/5 and 14223-1. This technology adoption led other countries that wish to export animal to these countries to have proper tracking and tracing records, which mandate the use of RFID. RFID carries a unique information code for each animal species that could be used to access the database, which stores very vital information related to that animal, which includes its blood group, data related to its birth, inoculations, and much more.

9.3.4 Hospital Equipment Tracking

RFID also has a significant importance in modern health care. Every year health-care facilities invest large amount of capital in upgrading and maintaining expensive medicinal equipment for diagnostics, treatment, and monitoring of patients [HEA09]. In many cases, it is evident that the medical equipment is mistakenly misplaced, thus causing serious problems in times of emergency. This in turn led to the following undesirable scenarios:

- Unwanted time wastage in equipment surfing by medical staff, so a number of patient's are being ignored by their absence and they also have to wait for long.
- An institution acquires additional equipment to compensate for constantly misplaced equipment.
- In critical scenarios, especially when the patient is being operated, it would be a big compromise when a medical device in untraceable.

Again, in this case, RFID technology resolves the above-mentioned problem. Tagging each medicinal equipment helps in keeping an updated record of equipment. RFID reader is generally used for reading the RFID tags, and for real-time equipment tracking, the facility should have an RFID reader network. In this way, any device can be located when needed. The staff can know where in the facility a certain piece of equipment is located, when they really need to have it. The result is less time wasted, better utilization of existing assets, and improvement of patient care. However, care should be taken when implementing an RFID zone in hospitals so that it does not interfere with other medical equipments operating in similar frequencies.

9.3.5 Patient Tracking

Patient tracking is quite a big issue nowadays in hospitals and health-care facilities. Presently, the patient's record management is either done manually or electronically in various parts of the world. But to track the medical records of an individual patient, especially during intercountry or intracountry travel, needs some quicker tracking ways to manage the information much more efficiently [HEA09]. As we have seen that RFID did its best in medicinal equipment tracking, it can also be used for patient tracking in hospital. The system can be designed to track hospital assets, curb excess expenditures, increase safety, and ensure security and access control by placing small RFID devices, or "RFID tags" on people and objects. In hospitals, emergency medical services find difficulty in tracking the patients, especially during the time of responding to emergencies. There must be some system that enables the emergency units to efficiently track the patients and medicinal assets in order to fully benefit the patients. As the institution is responsible for patient's health, sometimes patients are found missing from their beds, and medical staff find it difficult in locating them. In their search, they spend much time and, in the process, the patient may sometimes even miss his or her medicinal dose. RFID makes tracking and finding patients easier. Administering bedside care is safer, faster, because of real-time and accurate identification and verification of patient.

- RFID-embedded bracelets can be read through bed linens, so patients do not have to be disturbed when sleeping.
- RFID-embedded bracelets communicates with the reader network so this equipment helps in patient's tracking and ensures patients do not stray beyond a predetermined perimeter.
- RFID bracelets contains the patient's secret ID, which is linked to the database for getting the complete patient's record.

9.3.6 Newborn Baby Tracking

Another area where RFID possibly show its miracle is its use in maternity ward. Maternity ward is a place of happiness where the newborn babies are welcomed. Not only the families but also staff members are celebrating the arrival of the newborn [HEA09,DA03]. But, if the maternity ward is handling many cases at one time, the possible mix-up of mother and newborn might happen accidently; the wrong baby being sent to the wrong mother for the crucial early bonding period, breast-feeding, etc. In this situation as well, RFID technology can play a vital role in keeping record of mother and her newborn. To the mother, an RFID tag, with a unique ID, is attached during admission and this ID will be used for tracking all the processes relevant to that labor. And, once the child is born, an ankle RFID tag (unique ID referenced to child's mother) is attached so as to identify the child's mother. Thus, in this way, RFID prevents the child and mother mix-up in a simple manner.

9.3.7 Tracking Children

Fear of children abduction is a serious issue nowadays and is the biggest threat to parents as well. To track abducted children, the only way is a phone call from the abductors. A fashion designer in California has launched a sleepwear for small children that contain RFID tags (which when passed from the reading device alarms), providing some peace of mind to parents, who might fear that their young ones may be abducted while they are asleep [CLO05].

9.3.8 Golf Ball Tracking

RFID can also help in tracing golf balls. When a player plays a shot off the fairway into the unknown place and if no RFID tag is embedded, this may result in time wasting in surfing the ball or possibly the ball gets lost. Importantly, when multiple players are playing in the same area this possibly results in ball mix-up if all or some are using the same brand [GBT04]. This situation can be prevented by tagging the ball with an RFID tag with a unique ID. Not only does it save money but also speeds up the game. The reader, a handheld device with an LCD screen and a pulsing audio tone, is generally used to pinpoint the right ball when the ball is in its effective range (roughly 10–30 m). It would not be long before a new golf will come embedded with an RFID reader to detect RFID-embedded golf balls.

9.3.9 Crowd Control

A visitor control and tracking system for a venue having various access points and multiple destination points include a monitoring system that tracks location and movement of visitors to the venue. In accordance with the invention, a monitoring system tracks a location and movement of individuals or crowds in the venue and works in combination with a distributed network of screening units or kiosks to provide effective crowd control and monitoring. In one embodiment, the monitoring system employs RFID tags or devices, which are distributed at kiosks to visitors entering the venue. A plurality of RFID readers or receivers is arranged about the venue and function to scan for the RFID devices. A central control, operatively connected to each of the RFID receivers, tracks location and movement of each of the RFID devices to determine a visitor-associated metric for each of the plurality of destination points [VCC07]. In another embodiment, the visitor-associated metric is simply established by visual crowd monitoring [ESM06]. For example, RFID will help control crowd trouble at the World Cup, says FIFA. More than 3.5 million tickets will be sold with an embedded RFID chip containing identification information, to be checked against a database as fans pass through entrance gates at all 12 stadiums, in June. It is the first World Cup tournament to use RFID technology to identify cardholders. But, if one senior FIFA official has his way, the amount of personal information required of fans all of which is quickly identifiable with the help of RFID will be kept to a minimum at future events.

9.4 Access Control

Access control is another area that needs key attention so as to avoid the risks of terrorism and theft. In the past manual, procedure was used for accessing the building premises, such act of assessing the building had increased the chances of theft and terrorism [AC07]. The efficient access control system is generally demanded by many civil and military organizations, because of many reasons, which include

- Enhanced security to limit access to restricted areas
- Tracking employee activity
- Efficient and secure entry and exit
- Improve loss prevention

The various solutions were proposed like barcode, magnetic stripes, and proximity readers, all rely on the user to either make contact or place the badge very close to the reader. Also, for example, the barcode scheme has another problem of reading once at a time in addition to the line of sight issue. To address the above listed issues and the drawbacks of the proposed access control schemes like barcode, etc., researchers proposed RFID access control system, which can provide an easy and efficient solution. Using RFID for access control has the following advantages:

- The embedded electronic information for each badge can be overwritten repeatedly.
- RFID badges can be read from much further distances than other traditional technologies.
- Multiple RFID badges can be read all at the same time.
- The increased reading distance thus enables other tracking technologies like surveillance cameras to be activated in conjunction with an employee being in their vicinity.
- Information about employee access, attendance, and duties performed can be easily and efficiently monitored and stored in a database.
- Access information can also be tied to a Windows Active Directory or Lightweight Directory Access Protocol (LDAP) for user authentication and, therefore, can be synchronized to an authorized access scheme.
- Entry and exit doors are operated reliably by RFID reader/tag network without showing or making line of sight position of the card in front of reader.

9.4.1 Family Access to Babies in Neonatal Care

In intensive care units (ICUs), the patient's relatives find difficulty in having access to the patients, e.g., take the case of a newborn needing neonatal care; family members are almost as anxious as the new mother, to greet and bond with the new family member. Due to restricted access in the ICU, this would be very difficult for them to see the condition of the newborn. To address this problem RFID is used. The newborn is assigned a unique ID, embedded in its RFID tag, which correlates to its mother's ID. RFID solves this problem by making available to families the ability to view on a computer monitor outside of the ICU the key data about the baby, a photograph, height, weight, skin color, and temperature after keying in, for example, the mother's family name and first name [HEA09].

9.4.2 Vehicle Identification

In many major cities of the world like New York in the United States, Sydney in Australia, and London in the United Kingdom, car parking is a big issue nowadays. This is because of the growing number of traffic and the misuse of the existing parking lot resources [VI07]. Existing parking lot operators, whether public or private, face a number of challenges such as:

- Accurately identify and authorize vehicle movement.
- Collect and record vehicle movement data.
- Analyze traffic patterns to maximize facility utilization.
- Increase security within the parking facility.
- Manage staffing for peak traffic periods.
- Relieve congestion.
- Improve customer service.

To meet the challenges of the fully automated parking lot, RFID technology gives operators the ability to enhance parking control management systems in the following ways [APS06]:

- The use of RFID system makes it easier to automate the "in and out" privileges of parking subscribers.
- Existing private parking policies can be seamlessly integrated with public parking areas without increased manpower. This allows automated entry and exit; automated payment and shorter queues equate to satisfied customers.
- Using RFID network security can be improved.
- Occupancy rates and dwell times can be instantly calculated.
- RFID-based intelligent parking lot system helps in studying the existing resource occupancy for future expansion.
- Above all, these benefits will drive higher revenues.

9.5 Anticounterfeiting

Counterfeiting is the process of fraudulently manufacturing, altering, or distributing a product that is of lesser value than the genuine product. Counterfeiters now have the ability to replicate brand logos and product quality criteria nearly perfectly—making any attempt to disprove their authenticity both expensive and time-consuming. And the fight against gray-market trade has been made even more difficult because counterfeiters now frequently use genuine parts in their fake products. Counterfeiting is now a global problem. Seven percent (7%) [BGK07] of all world trade is in counterfeit goods and that over the past 10 years counterfeiting has destroyed 120,000 jobs each year in the United States, and 100,000 in Europe [ICCC06]. With the counterfeit industry growing at a rate of 6%–8% annually, it is estimated to be a US$30 billion industry in the United States and US$50 billion worldwide. The German advocacy group APM, which fights product and brand piracy, calculates annual losses for the German economy due to counterfeiting at €25 billion. Counterfeit goods do not only target famous brand names but anything that can buy. Virtually every country in the world suffers from counterfeiting which results in—lost tax revenue, job losses, health and safety problems, and business losses.

RFID technology has been proposed to address the above concerns since RFID tags can be attached to individual items encoded with product specific information which can prove its authenticity and originality. A number of RFID anticounterfeiting mechanisms have recently been proposed. These systems are aimed at relatively high-end consumer products, and helps protect genuine products by maintaining the product pedigree and the supply chain integrity. The main application areas for RFID anticounterfeit technology include electronic drug pedigree, tracking genuine automobile parts, document authentication, and detecting fake currency and passports. We now discuss each of these applications.

9.5.1 Electronic Drug Pedigree

One of the most important applications of RFID anticounterfeit technology is detecting genuine drugs from counterfeit drugs. It is estimated that between 5% and 8% of the 500 billion USD in medicines sold worldwide are counterfeit [STF05, BASC04] and for developing countries the percentage of counterfeit drugs account for up to 60% of all drugs [IP05,KSCB03] and can lead to injury and, in some cases, fatality. It is projected that 95% of counterfeit pharmaceuticals provide little or no therapeutic value. According to the World Health Organization, 43% of them contain no active ingredient, and another 21% are subpotent, often caused by counterfeiters expanding volume through dilution. As many as 25,000 cancer patients may have received subpotent medicine that was 1/20th the strength prescribed by their physicians when Procrit® was relabeled by U.S. counterfeiters in 2002. Of 110,000 vials labeled as the highest-dose 40,000-unit product, but actually containing the lowest-dose 2,000-unit product, only 8,000 were recovered; the counterfeiters gained approximately $46 million [GFL03]. With the global nature of modern supply chains, it has never been easy to cope with counterfeiters. Seeking solutions to counterfeiting, the Food and Drug Administration (FDA) recommended, in February 2004, that each drug have an electronic pedigree, which is a "secure record documenting the drug was manufactured and distributed under safe and secure conditions" [CCD04]. The FDA's approach to cost-effectively tracking individual drug products in the supply chain is to serialize each drug product, adopt electronic pedigrees, and automate the tracking process using RFID. The Authenticated RFID model [SA05] enhances item-level product security in real-time, independent of a connection to a host network, by creating strong authentication between the tag and an authenticated RFID reader. By initially deploying the model at the point of manufacturing and the point of dispensing, the pharmaceutical industry immediately provides a higher level of item-level authentication against counterfeit products [POT06,EDP09]. After initial introduction of RFID and Public Key Infrastructure (PKI) end-to-end item-level authentication, the integration of other points in the chain of custody provide ever increasing levels of confidence in the supply chain. While bar code solutions may cost less in the short term, there are a number of shortcomings, compared with RFID, that limit their effectiveness over time. For example,

- RFID has the capacity to store larger amounts of information and can be read more faster than bar codes (40-plus reads per second, compared with one to two for bar codes), and requires far less human involvement.
- Bar codes require a direct line of sight to be read, while RFID tags do not.
- Bar code must be able to survive on multiple types of printed media in harsh conditions, sometimes over long periods of time.
- Barcodes are easy to duplicate by using a simple inkjet printer.
- Barcodes cannot store the lifecycle information about any product but RFID can.

Thus, RFID technology has seen to be very effective in managing counterfeit pharmaceutical products and preventing the entry of fake drugs in the supply chain.

9.5.2 Bank Notes

Other than pharmaceutical drugs, another areas where RFID anticounterfeiting technology is making significant inroads is in securing currency or bills. Counterfeit euro bills were assumed to be coming from Greece and other new member states in the European Union. In 2003, Greek authorities dealt with 2411 counterfeiting cases and seized 4776 counterfeit banknotes, whereas Polish authorities arrested a gang that circulated more than 1 million fake euros in the market [BIL06,JUN01]. To address this problem of counterfeit currency, European central bank started looking at the possibility of implementing RFID technology in euro bills and rolling out such bills by 2005. The main question then was to decide on which bills to tag as it would not be cost effective to tag €5 or €10 euro notes. So the most likely option was to tag high denomination notes like the €200 or €500 to deter money laundering activities [AVO04,JUE03]. Some other challenges that need to be met would be the privacy concerns; an RFID-embedded bank note can be used to track people who are carrying lot of cash and would prove to be targets of robbery. Other than that, retailers could misuse this feature by estimating the buying power of the customer and can device strategies to reduce the bargaining power of the customer. These are some of the issues to be considered before implementing an RFID-enabled bank note.

9.5.3 Secure Passports and Visas

Other than bank notes, travel documents like passports are now already using RFID chips that are making the passports machine readable often termed as machine readable travel documents. Many countries like Australia, United States, and European nations are already issuing passports with RFID tag embedded within them. There is also been talks about implementing RFID visas to increase the authenticity of the visas issued. In the past, many people have been fraudulently issuing visas for developed nations. These activities can now be controlled by having an RFID-enabled visa [DIM07].

9.5.4 Automobile Parts

Counterfeiting of automobile parts is also a big industry. It is estimated that on an average about 5%–10% of all spare parts are counterfeits [LALK03]. Counterfeit products pose a big threat to the automobile industry, since fake products not only eat up the revenue of the original manufacturer but it can also increase the risk of warranty issues for the automobile manufacturer. Hence, the automobile industry is now looking at the possibility of using RFID technology for tracking the automobile parts throughout the supply chain to ensure that counterfeits do not enter the market.

9.6 Conclusion

In this chapter, we introduced RFID technology by explaining the main RFID architecture, which included RFID transponders, readers, antennas, and middleware. We, then, outlined the main application areas of RFID, which included item tracking, access control, and anticounterfeiting. Within each application umbrella, we covered the most popular applications like baggage tracking, library book tracking, animal tracking, hospital equipment tracking, patient tracking, newborn baby tracking, tracking children, golf ball tracking, crowd control, vehicle identification, electronic drug pedigree, anticounterfeit bank notes, secure passports and visas, and anticounterfeit automobile parts. There are many other applications of RFID technology, and it can be said that the applications can only be limited by imagination.

References

[AC07] GAO RFID Inc., RFID Solutions for ID Badges and Access Control, available online at http://www.findwhitepapers.com/whitepaper844/, accessed on Friday, January 9, 2009.

[APS06] Unipart Logistics Deploys RFID-Based Information Service From Savi Networks To Track Jaguar Car Parts From U.K. to U.S., available online at http://www.savi.com/about/press-releases/2006–09–18.html, accessed on Friday, January 9, 2009.

[AVO04] G. Avoine, Privacy issues in RFID banknotes protection schemes, in J.-J. Quisquater, P. Paradinas, Y. Deswarte, and A. Abou El Kalam (Eds.), *Proceedings of the Sixth International Conference on Smart Card Research and Advanced Applications (CARDIS 2004)*, Toulouse, France, Kluwer, August 2004, pp. 33–48.

[BAG08] Motorola, Industry Brief-Baggage Tracking RFID Solutions, available online at http://www.motorola.com/staticfiles/Business/_Documents/static%20files/Baggage_Tracking_RFID_Solutions%209-08.pdf

[BASC04] Business Action to Stop Counterfeiting and Piracy Fact Sheet, Technical report, ICC.

[BGK07] L. Batina, J. Guajardo, T. Kerins, N. Mentens, P. Tuyls, and I. Verbauwhede, Public-key cryptography for RFID-tags, in *Fifth Annual IEEE International Conference on Pervasive Computing and Communications PerCom 2007*, New York, 2007.

[BIL06] RFID banknotes, available online at http://www.fleur-de-coin.com/eurocoins/rfid.asp, accessed on Tuesday, January 13, 2009.

[BR07] The Baggage Report 2007, SITA, www.sita.aero.

[CCD04] Combating Counterfeit Drugs, A Report of the Food and Drug Administration, FDA, February 2004, p. 3.

[CHA07] S. Chalasani and R. V. Boppana, Data architectures for RFID transactions, *IEEE Transactions on Industrial Informatics*, 3(3), August 2007, 246–259.

[CLO05] PHYSORG, Anti-theft RFID clothing, available online at http://www.physorg.com/news5537.html, accessed on Friday, January 9, 2009.

[DA03] Birth and other legal documents having an RFID device and method of use for certification and authentication, U.S. Patent 7170391, available online at http://www.freepatentsonline.com/7170391.html, accessed on Friday, January 9, 2009.

[DIM07] D. Lekkas and D. Gritzalis, e-Passports as a means towards the first world-wide Public Key Infrastructure, in *Proceedings: Fourth European PKI Workshop (EuroPKI 2007), Lecture Notes in Computer Science*, Vol. 4582, Mallorca, Spain, June 28–30, 2007, Springer, 2007.

[EAS05] Electronic Article Surveillance (EAS), available online at http://www.technovelgy.com/ct/Technology-Article.asp?ArtNum=33, accessed on Friday, January 9, 2009.

[EDP09] RFID in the Pharmaceutical Supply Chain, available online at http://www.ascet.com/documents.asp?d_ID=3435, accessed on Friday, January 9, 2009.

[ESM06] RFID tag with visual environmental condition monitor, available online at http://www.wipo.int/pctdb/en/wo.jsp?IA=US2005045945&DISPLAY=STATUS, accessed on Friday, January 9, 2009.

[GAO09] RFID Solutions for Library Systems, available online at http://library.gaorfid.com, accessed on Friday, January 9, 2009.

[GBT04] RFIDNews.org, Golf Ball Tracking System, available online at http://www.rfidnews.org/2004/05/28/golf-balltracking-system/, accessed on Friday, January 9, 2009.

[GFL03] G.M. Gaul and M.P. Flaherty, Lax system allows criminals to invade the supply chain, Washington Post, October 22, 2003.

[GLOE05] D. Gloeckler, What is RFID? Retrieved September 1, 2005, from http://www.controlelectric.com/RFID/What_is_RFID.html

[HE08] W. He, N. Zhang, P.S. Tan, E.W Lee, T.Y. Li, and T.L. Lim, A secure RFID-based track and trace solution in supply chains, in *Proceedings of the Sixth IEEE International Conference on Industrial Informatics*, Daejeon, South Korea, July 13–16, 2008, pp. 1364–1369.

[HEA09] GAO RFID Inc., RFID Solutions for Healthcare Industry, available online at http://healthcare.gaorfid.com/, accessed on Friday, January 9, 2009.

[IA07] RFID for Manufacturing, available online at http://manufacturing.gaorfid.com/, accessed on Friday, January 9, 2009.

[ICCC06] International Chamber of Commerce Commercial Crime Services, International Guide to IP Rights Enforcement First Edition 2006, International Chamber of Commerce Counterfeiting Intelligence Bureau, 2006.

[INTE06] Intermec, IP3 Intelllitag Portable Reader (UHF), June 14, 2006, from http://www.pointofsaleinc.com/pdf/Intermec/ip3.pdf

[IP05] Intellectual property: Source of innovation, creativity, growth and progress, Technical report, ICC, August 2005.

[JUE03] A. Juels and R. Pappu, Squealing euros: Privacy protection in RFID-enabled banknotes, in R.N. Wright (Ed.), *Proceedings of the Seventh International Conference on Financial Cryptography* (*FC 2003*), *Lecture Notes in Computer Science*, Vol. 2742, Springer-Verlag, Guadeloupe, France, January 2003, pp. 103–121.

[JUN01] J. Yoshida, Euro bank notes to embed RFID chips by 2005, EETimes, December 19, 2001, available online at http://www.eetimes.com/story/OEG20011219S0016, accessed on Tuesday, January 13, 2009.

[KSCB03] R. Koh, E.W. Schuster, I. Chackrabarti, and A. Bellman, *Securing the Pharmaceutical Supply Chain*, White Paper MIT-AUTOID-WH-021, Auto-Id Center MIT, Cambridge, MA, September 1, 2003, available online at http://www.mitdatacenter.org/MIT-AUTOID-WH021.pdf

[LALK03] B. Lalk-Menzel and D. Chrysler, available online at http://www.autoid.org/sc31/clr/200305_3826_Automotive%20Prpsl.pdf, accessed on Tuesday, January 13, 2009.

[LEAV05] S. Leaver, Evaluating RFID Middleware, 2005, retrieved August 07, 2005, from http://www.forrester.com/Research/Document/Excerpt/0,7211,34390,00.html

[LIV09] GAORFID Solutions for Livestock and Other Animal Tracking, available online at http://livestock.gaorfid.com/, accessed on Friday, January 9, 2009.

[MS08] RFID for Postal and Courier Services 2008–2018, available online at http://www.researchandmarkets.com/reportinfo.asp?report_id=615186&t=t&cat_id=, accessed on Friday, January 9, 2009.

[PM09] Axcess International Inc., RFID in the Security Industry, available online at http://www.rfidsb.com/index.php?page=rfidsb&c_ID=145, accessed on Friday, January 9, 2009.

[POT06] M. Potdar, E. Chang, and V. Potdar, Applications of RFID in pharmaceutical industry, in *Proceedings of the IEEE International Conference on Industrial Technology*, Mumbai, India, December 15–17, 2006.

[POT07] V. Potdar, C. Wu, and E. Chang, *Automated Data Capture Technologies: RFID*, Q. Zhang (Ed.). Hershey, London, U.K. Information Science Reference, pp. 112–141.

[RBC07] *RFID Business Case for Baggage Tagging*, International Air Transport Association (IATA), 2007, pp. 27, 36.

[RFGA06] RFidGazzete Tag shapes, retrieved June 14, 2006, from http://www.rfidgazette.org/2005/10/tag_shapes.html

[RFID06] *RFID Journal*, 2006a, retrieved June 14, 2006, from http://www.rfidjournal.com/article/articleview/208#Anchor-scanners-5989

[SA05] Proven Security for the Supply Chain and Anti-Counterfeiting Prevention, available online at http://www.certicom.com/index.php/rfid, accessed on Friday, January 9, 2009.

[STF05] T. Staake, F. Thiesse, and E. Fleisch, Extending the EPC network: The potential of RFID in anti-counterfeiting, in A. Omicini, H. Haddad, L.M. Liebrock, and R.L. Wainwright (Eds.), *ACM Symposium on Applied Computing, SAC 2005*, ACM Press, New York, March 13–17, 2005, pp. 1607–1612.

[SWEE05] P.J. Sweeney, RFID for Dummies. Symbol Technologies, Business Benefits from Radio Frequency Identification (RFID), Wiley Publishing Inc., Indianapolis, IN, 2004, retrieved June 14, 2006, from http://www.symbol.com/products/rfid/rfid.html

[TECS05] Tecstra RFID, retrieved September 1, 2005, from http://glossary.ippaper.com/

[VCC07] R.E. Dugan, T.E. McVeigh, J. Kotowicz, and L. Carin. Visitor control and tracking system, USPTO Application #: 20070109134, available online at http://www.freshpatents.com/Visitor-control-and-tracking-system-dt20070517ptan20070109134.php, accessed on Friday, January 9, 2009.

[VI07] GAO RFID Inc., RFID Automated Parking Control Systems, available online at http://parking.gaorfid.com/, accessed on Friday, January 9, 2009.

[ZHA06] Y.Z. Zhao and O.P. Gan, Distributed design of RFID network for large-scale RFID deployment, in *Proceedings of the Fourth International IEEE Conference on Industrial Informatics* (*INDIN 2006*), Singapore, August 16–18, 2006.

10

Ultralow-Power Wireless Communication

Joern Ploennigs
Dresden University of Technology

Volodymyr Vasyutynskyy
Dresden University of Technology

Klaus Kabitzsch
Dresden University of Technology

10.1 Introduction 10-1
10.2 Hardware Approaches 10-2
Overview • Energy Harvesting
10.3 Communication Protocol Approaches 10-4
10.4 Application Layer Approaches 10-6
10.5 Conclusion and Open Topics 10-8
References 10-9

10.1 Introduction

Wireless networks permit quick and easy installation, but depend on independent energy sources such as batteries or on energy harvesting. Energy efficiency is therefore a common requirement. Nevertheless, in some application domains, the solutions available in the market are too energy hungry as they are designed for general purpose appliances. Especially when wireless devices have to run independently for a long time and a high miniaturization allows only small batteries or energy-harvesting modules, ultralow-power wireless devices are needed.

An example is the structural health monitoring of new composite materials for airplanes. Composite materials are strong, light, and can significantly reduce the fuel consumption of the planes, but tiny cracks in the fabric reduce the integrity of the materials. These cracks can be detected by acoustic structural health monitoring [FFH05]. The necessary sensors are ingrained within the composite material and need to communicate wirelessly to avoid the weight of wires and the structural impact of cable ducts. The wireless sensors have to survive on their own energy sources for the lifetime of the plane of 10–20 years and provide a reliable structural health monitoring. This requires specialized ultralow-power, energy-autonomous wireless sensor networks that are investigated as part of the German cutting-edge research cluster CoolSilicon. Other scenarios for ultralow-power communication are, for example, wireless switches that harvest energy from a single press on a switch by a human. Node and network need to be optimized to exploit the short period of time the node has energy for communication. Rain sensors in windshields of modern cars are another example for which industry requests miniaturized wireless sensors for easy assembling.

Ultralow-Power Wireless Communication requires a harmonized design of the nodes from hardware to application software covering the topics shown in Figure 10.1. The present chapter follows this outline and discusses the state of the art in research to develop a guideline to design ultralow-power wireless communication systems. The next section focuses on the hardware approaches with an emphasis on energy harvesting for a long-lasting energy supply. Then, communication protocols and topologies are

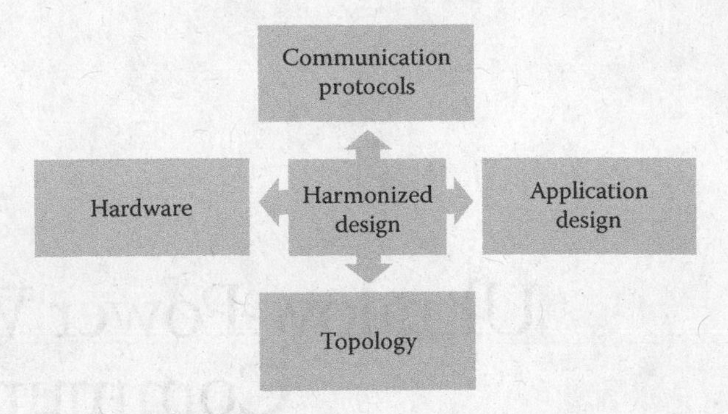

FIGURE 10.1 Aspects of ultralow-power wireless communication design.

introduced that utilize energy sources efficiently. The final section looks at application design concepts for ultralow-power wireless communication.

10.2 Hardware Approaches

10.2.1 Overview

Figure 10.2 shows the common architecture of an ultralow-power node. It consists of four basic parts plus two optional parts. The basic parts, namely communication, processing, sensor and actuation, as well as energy storage, are known from the introduction of energy-efficient WSN in Chapter 6. For ultralow-power applications, the energy consumption of each component has to be minimized to provide only the core functionality necessary. This may require developing customized hardware.

A new addition to this basic design for ultralow-power wireless nodes is the *energy supply*, which allows recharging the energy storage with energy-harvesting approaches from the node environment. Basic energy-harvesting approaches will be compared in the next section.

Ultralow energy consumption is reached by the option to deactivate individual hardware parts or whole nodes. This is scheduled by the *energy management* that is usually running on the microcontroller or separately on a low-power circuit depending on the power consumption of the microcontroller and the computing requirements of the task. For instance, the aforementioned example of acoustic structural health monitoring for airplanes is a task with high computation demands that requires a dedicated digital signal processor (DSP) during measurements. Separating the energy

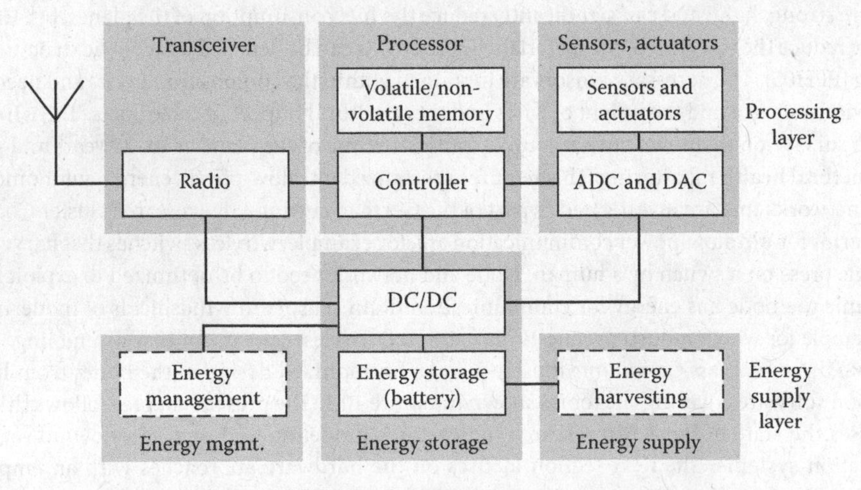

FIGURE 10.2 Common architecture of ultralow-power nodes.

management and basic communication tasks on a second low-power microcontroller and activating the DSP only for measurements does save energy.

10.2.2 Energy Harvesting

The power supply is a bottleneck for many ultralow-power scenarios. Battery size is the most limiting factor in miniaturizing nodes as battery capacities double only every 10 years in contrast to circuits, which need only 2 years due to Moore's law. As a result, batteries need to be replaced regularly, which is laborious if it is difficult to access nodes. Energy-harvesting approaches provide an important alternative to batteries and allow extending the node lifetime to its hardware limits. The obtained energy is usually temporarily available depending on various conditions. The following energy sources can be used for harvesting:

- *Solar*: Several aspects should be considered if solar cells are used. First, the energy density for outdoor and indoor usage varies strongly as can be seen in Figure 10.3. Second, the sun as a power source is not available during the night. This can be in principle compensated by adding matching energy storages such as batteries or ultracapacitors, but this is not feasible in Nordic countries with polar nights. Third, the device needs access to light and cannot be integrated in nontransparent materials or placed in dark areas.
- *Radio-Frequency Power*: Scheible et al. [SDE07] demonstrated how a wireless network is supplied by an electromagnetic field created around an industrial production cell. This commercially available product works well for static medium-sized environments that are less frequented by humans.
- *Thermalelectric*: The temperature difference between two adjoining materials can be used as an energy source through the Seebeck effect. A first low-power thermalelectric generator was demonstrated by Stordeur and Stark [SS97]. Its commercial successor by Thermo Life produces 30 μW/cm² from 5°C temperature difference on a 5 cm² module.
- *Vibration*: Ambient motions or vibrations are promising energy sources especially in industrial, automotive, and avionic domains where vibrations are common. Mitcheson presents in [MGY04] a good introduction of the different approaches from electromagnetic, electrostatic, to piezoelectric technologies. The potential of energy harvesting for wireless communication from vibrations was demonstrated in [R03].
- *Human Body*: The human body is a steady energy source emitting heat, movement, and vibration. Thus, it can be scavenged by thermalelectric or vibration generators. Paradiso and Starner outline the current approaches in [PS05]. Practical examples are switches of the EnOceans technology [EO08] that use only the energy created by the human body for actuating.

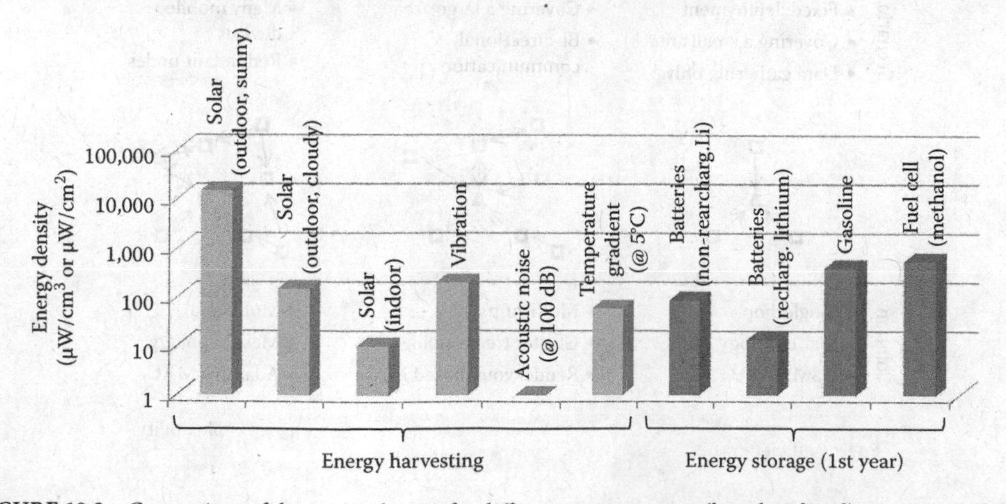

FIGURE 10.3 Comparison of the energy density for different energy sources (based on [R03]).

Figure 10.3 compares different energy sources for ultralow-power devices from the viewpoint of their energy density, which defines the amount of energy that can be obtained from an element of a comparable size of one cubic or square (for flat elements) centimeter. The displayed energy density presents averaged values for the first year of operation based on published results. The energy capacity of batteries and fuel cells falls with the consumption of resources. The comparison shows that the energy densities of batteries and energy harvesting have the same dimension, implying that energy-harvesting modules are difficult to miniaturize further. This emphasizes the necessity for designing ultralow-power wireless devices that minimize energy consumption of such modules.

10.3 Communication Protocol Approaches

Energy efficiency is reached by turning hardware parts off or to a *sleep mode* with minimal energy consumption as often as possible. Therefore, nodes are unavailable for processing and communication tasks for most of their lifetime, which requires specially designed communication protocols and applications. This principle is commonly known as *duty-cycling*, while the *duty-cycle* is defined as the fraction of time the node is not in sleep mode. The duty-cycle should be as low as possible for ultralow-power nodes. Nevertheless, certain communication and application performance parameters need to be assured like transmission delay or sampling accuracy. Hence, communication and application layer requirements and approaches need to be harmonized to efficiently use wake-ups.

Communication aspects such as medium access control (MAC), routing, and topology control have strong influence on the duty-cycle and should be selected adequately. Figure 10.4 illustrates three example scenarios that result in adequate combinations of approaches to be discussed in the following paragraphs. The following design questions are relevant: Does the network consist of a fixed deployment of nodes or mobile nodes? How large is the area to cover? Is redundancy needed for functional safety? Is the network used only for data gathering or additionally for bidirectional communication (like decentralized control with actuators)?

For a fixed deployment covering a small area, the simplest and most energy-efficient solution is a single-hop star architecture with a hub or coordinator that has a wired energy supply and is always on. This permits the usage of simple CSMA protocols for the sensor nodes that can wake-up at any time in order to perform their task, wait until the communication channel is free, and send their message to the coordinator before going to sleep again. IEEE 802.15.4 is one example that uses this approach in non-beacon mode. As collision detection is not easily handled in wireless networks due to the

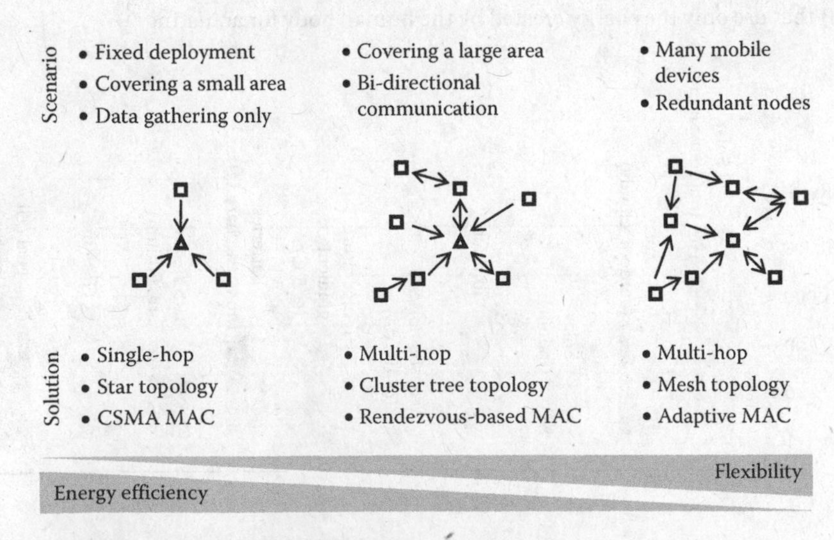

FIGURE 10.4 Different scenarios and energy-efficient solutions.

hidden-terminal problem, some protocols also neglect it and use direct channel access, assuming that collisions are unlikely due to rare wake-ups [EO08]. The resulting network protocols are quite simple from a WSN perspective, but very energy efficient for the sensor nodes, as long as the always-on coordinator is not taken into account.

If the coverage area is large, then multi-hop approaches are more energy efficient than single-hop approaches [NPK06], because the signal strength falls in a logarithmic manner making short communication distances significantly more energy efficient. These multi-hop approaches require specific routing approaches and coordinated wake-ups of the nodes forwarding messages. Simple CSMA protocols are not sufficient for that and more sophisticated MAC protocols need to be used, which are discussed later on in this section.

Full flexibility of the network is usually reached with mesh networks. They add a topology control protocol layer allowing easy integration of mobile nodes. Other protocols provide a self-healing capability of the network often taking advantage of redundant nodes. Some approaches also consider enhancing the network lifetime through redundant nodes that mainly sleep for the first part of their life to take over the job of nodes that have used up their energy resources [TG02]. These redundancy approaches improve network reliability, but require higher hardware costs. Nonetheless, all these protocols add an overhead of management messages to the communication that require energy for transmission and keep nodes awake.

A low duty-cycle is most important for ultralow-power devices. The communication needs to be coordinated in a way that still allows most of the network to be asleep. Several issues should to be avoided during communication like *idle listening* to the channel if no other node intents to send a message or *overhearing* messages not addressed to the node. Also, nodes should avoid transmitting messages to sleeping receivers (*overemitting*). Two or more senders should avoid sending messages at the same time on the wireless channel that causes *collisions* corrupting messages. As a result, an ideal communication protocol ensures that sender and receiver are only awake for message transmission and sleep the rest of the time. This is manageable as long as only two nodes are involved and both work on a periodic schedule with synchronized clocks. In multi-hop networks, where a node may need to forward messages of other nodes, this requires suitable protocols that are discussed in the following paragraphs.

Depending on how the wake-up intervals of nodes are coordinated in the network, the protocols can be distinguished in on-demand, asynchronous, and rendezvous-based approaches (Figure 10.5). *On*-demand approaches often use a second low-power, low-rate radio to wake-up a node for communication [STS02]. This approach is limited in practice, due to the cost for the second radio and its range difference to the main radio as less energy is used.

In *asynchronous* approaches, the sleep cycles of nodes are not synchronized. The common approach is that a sender waits until a receiver is awake to transmit the message. Therefore, nodes periodically wake-up and check for messages. B-MAC is the most commonly used protocol in which a sender starts with a long preamble [PHC04]. Other nodes waking up on their periodical schedule recognize this

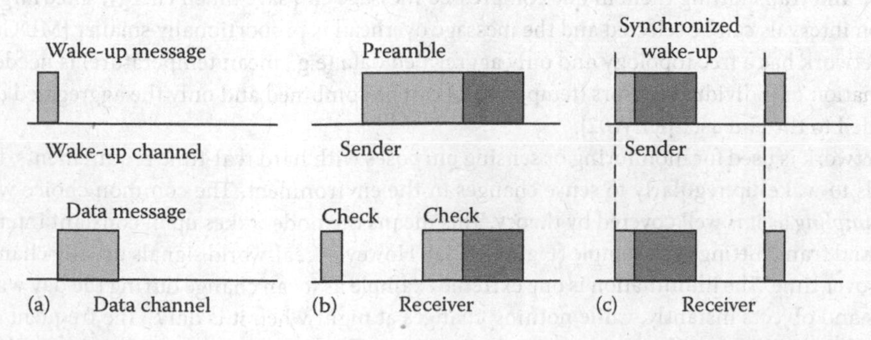

FIGURE 10.5 Illustration of typical wake-up approaches for MAC. (a) On-demand, (b) asynchronous, and (c) rendezvous based.

preamble and stay awake to receive this message. The drawbacks of this approach are overemitting for the preamble and overhearing at nodes not involved (addressed) in the communication as they need to stay awake until the preamble ends to check the message. More energy efficient are X-MAC [BYAH03] and CSMA-MPS [MB04], where multiple short messages, with the relevant node address, are repeated instead of a single long preamble. This not only avoids the overhearing of nodes that are not involved, but also allows interested nodes to indicate when they are awake and ready to receive. The downside of these asynchronous approaches is that the energy problem is mainly shifted to the sender, who needs to send the preamble. The length of the preamble depends on the check interval of the receiver and the energy efficiency becomes a trade-off between overemitting at the sender and idle listening at the receiver.

Rendezvous-based approaches synchronize the wake-up periods of a group of nodes. The first approach was the S-MAC protocol [YHW04], in which all nodes awake periodically for a fixed time to exchange messages. IEEE 802.15.4 uses a comparable approach in its beacon-enabled mode. As in both approaches, nodes not involved in communication also stay awake for the active time; this results in overhearing and idle-listening again. T-MAC [DL03] improves this issue by setting uninvolved nodes to sleep. The downside of all rendezvous-based approaches is that all nodes wake-up at least for the synchronization, and the energy consumption depends strongly on the wake-up cycle, which is usually constant. Especially in scenarios with a dynamic communication load, this leads to idle listening, when no node has a message to transmit, and collisions, when multiple nodes intend to send messages. The adaption of the wake-up cycles to the traffic load of the nodes is therefore one approach to reduce energy-consumption while providing a high throughput [NPK05].

Rendezvous-based approaches are usually more energy efficient than asynchronous approaches. But, application requirements should also be considered when deciding about communication protocols. For example, if a temperature sensor samples a new value every minute, then the node should not have to synchronize in a rendezvous-based MAC with a node sampling the illumination every 100 ms. In this case, an *asynchronous* approach may be a better choice, but then the preamble for the temperature sensor might be very long. Hence, network topology, communication protocols, and application design need to be harmonized for ultralow-power communication.

10.4 Application Layer Approaches

An appropriate application design is essential for ultralow-power communication. Again, the application requirements of the node need to be considered to choose the most energy-aware design. Two general design principles should be used: first, allow the node to sleep as often and as long as possible; and second, process information locally, as the data processing consumes significantly less power than transmitting data [RSS02]. Thus, the *number of transmitted messages*, the *message size* and the *number of wake-ups* should be reduced, while still providing the required functional quality. For example, if the network is used for monitoring applications with no hard real-time requirements, then gathering several samples in the node and transmitting them in one compressed message can save much energy, since larger communication intervals can be selected and the message overhead is proportionally smaller [MDG06].

If the network has a tree topology and only aggregated data (e.g., mean temperature) is needed, then the information of individual sensors (temperature) can be combined and only the aggregated data can be forwarded to the end user [KEW02].

If the network is used for monitoring or sensing purposes with hard real-time requirements, then the node needs to wake-up regularly to sense changes in the environment. The common choice would be *periodic sampling* as it is well covered by theory. This means the node wakes up in constant intervals for sampling and transmitting each sample (Figure 10.6a). However, real-world signals usually change their dynamics over time. The illumination is one extreme example as it can change during the day with moving clouds and objects instantly, while nothing changes at night when it is dark. The frequent changes during the day require a small sampling period, which is inefficient in the night with only small changes and sampled values that are very similar.

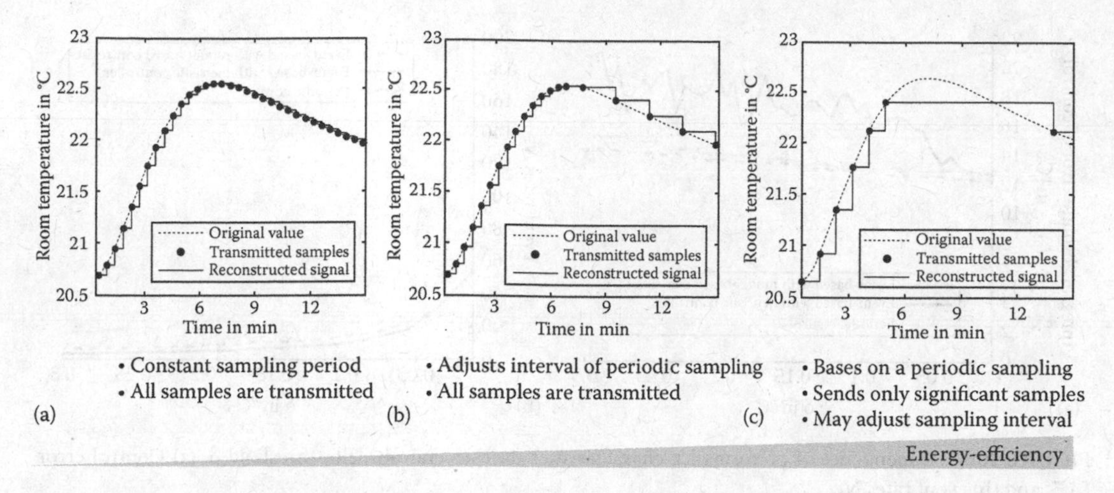

FIGURE 10.6 Step responses for different sampling approaches in a temperature control loop: (a) periodic sampling every 30 s, (b) adaptive periodic sampling every 30 s for the first 3 min and 90 s afterward, and (c) event-based (send-on-delta) sampling using a sampling interval of 30 s and a delta of 0.25°C.

Adaptive sampling approaches adjust the sampling intervals based on different criteria, e.g., network load [GDD04], round trip time, signal dynamics, sampling error, or operation mode. They can be generally distinguished in adaptive periodic sampling and event-based sampling approaches.

Adaptive periodic sampling adjusts the sampling interval of periodic sampling, for example, in Figure 10.6b, where the sampling period is increased after the step response passed its peak after 6 min. The benefit is that approaches for periodic systems can still be used, e.g., for signal reconstruction or fault-tolerance mechanisms. The downside is that reaction times are limited by the constant sampling period.

Event-based sampling transmits only significant changes in sampling values. Different decision criteria may be used like the difference between the actual and last transmitted value (send-on delta sampling [NK04]) or the integral of this difference [VK07b]. Send-on-delta, also referred as absolute deadband sampling [OMT02], is the most common approach. It samples with a constant interval T_A, but sends a sample $y(t)$ only if it differs more than a threshold δ from the last transmitted sample $y(t_L)$, hence $|y(t) - y(t_L)| > \delta$. The benefit of these approaches is that the number of transmitted messages is reduced based on the signal dynamic. As sampling and processing usually cost less energy than transmission, event-based sampling is generally more energy efficient. Recent event-based sampling approaches adjust also the period of wake-ups to the signal dynamics [PVK09] to remove the overhead of periodic sampling and improve energy efficiency.

Model-based reconstruction may increase the energy efficiency of adaptive sampling approaches even further by permitting larger sampling intervals, because they use a process model in the sender and receiver to reconstruct the signal transition between samples. They were applied on adaptive periodic sampling [MA02] and event-based sampling [LYT06]. However, they need precise process models—otherwise the sampling quality decreases dramatically.

Also, closed-loop controls can be run in an ultralow-power network using adaptive sampling approaches. However, the usage of standard, nonadapted control algorithms like proportional–integral–derivative (PID) lead to a significant degradation of control loop performance [VK07b]. The reason is that these control algorithms are based on periodic sampling and are not adapted to the rare, nonperiodic update events. A uniform theory for such *event-based controls* does not exist and their properties were analyzed only for some scenarios [A07].

The degradation of control loop performance is also visible in Figure 10.6. The maximum of the step response for the event-based sampling in Figure 10.6c is slightly higher with 22.7°C than in Figure 10.6a and b with 22.55°C. Note that the sampling period of the event-based sampling is identical

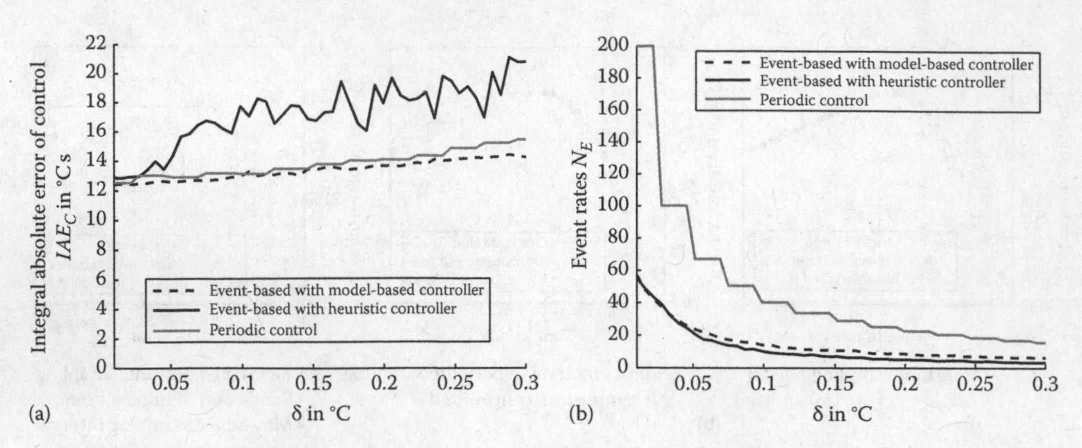

FIGURE 10.7 Dependence of performance characteristics on the send-on-delta threshold δ. (a) Control error IAE_C and (b) event rates N_E.

to the periodic sampling in Figure 10.6a, only less messages are transmitted to the PID controller. Similar simulations can be used during design time to analyze specific scenarios [LYT06], but they are not always feasible in practice due to time constraints. Model-based control approaches are more robust as they estimate the signal transition between samples [LYT06], [A07], but they require system models that are often unknown. Heuristic algorithms do not require such explicit models and still allow an acceptable quality-of-control level [VK07a].

Figure 10.7 compares periodic sampling with event-based control using a model-based and heuristic algorithm for the step response shown in Figure 10.6. The control loop performance (quality of control) is measured by the integral absolute error of control IAE_C, which equals the integral of the difference between the set-point and actual value.

The comparison in Figure 10.7 shows that with increasing threshold δ, the error IAE_C raises and the event rates N_E decrease. For larger threshold values (δ > 0.1), the event rates do not decrease significantly, but the error IAE_C continues to grow. The model-based algorithm is less sensitive to the threshold value due to a more precise estimation of the system behavior. However, it requires a detailed system model, contrary to the heuristic approach. Both algorithms perform effectively, and reduce the number of events to up to one third of that of the equivalent periodic loop, while retaining a similar quality of control.

10.5 Conclusion and Open Topics

Ultralow-power wireless communication devices require a holistic design from hardware to application. The different aspects that need to be considered during design and common solutions were introduced in individual sections of this chapter. A general best practice solution for ultralow-power communication devices cannot be given, as minimizing the energy consumption requires manual tuning of the device, protocols, topology, and application to the requirements of each scenario. Design tools that support engineers in this task do not exist yet. Model-driven design approaches for device applications can support the creation of optimized, energy-efficient applications and communication protocols. Other open research questions cover the synchronization of application and communication protocols, for example, the synchronization of sampling intervals of device applications with rendezvous-based MAC. Especially, cross-layer protocols that consider the requirements of event-based, real-time actuation and combine a fast delivery of messages to devices with a low duty-cycle are rarely researched. Current research also covers the extended usage and miniaturization of energy-harvesting for ultralow-power devices.

References

[A07] K. J. Aström, Event based control, in *Analysis and Design of Nonlinear Control Systems: In Honor of Alberto Isidori*, Springer Verlag, pp. 127–147, New York, 2007.

[BYAH03] M. Buettner, G. V. Yee, E. Anderson, and R. Han, X-MAC: A short preamble MAC protocol for duty-cycled wireless sensor networks, in *Proceedings of Fourth International Conference on Embedded Networked Sensor Systems*, pp. 307–320, Boulder, CO, October–November 2006.

[DL03] T. V. Dam and K. Langendoen, An adaptive energy-efficient MAC protocol for wireless sensor networks, in *Proceedings of First ACM Conference on Embedded Networked Sensor Systems*, Los Angeles, CA, November 2003.

[EO08] EnOcean GmbH, Energy for free—wireless technology without batteries, EnOcean Alliance, White Paper, 2008.

[FFH05] B. Frankenstein, K. J. Fröhlich, D. Hentschel, and G. Reppe, Microsystem for signal processing applications, in *Proceedings of 10th SPIE International Symposium on Nondestructive Evaluation for Health Monitoring and Diagnostics*, Vol. 5770, pp. 152–159, San Diego, CA, 2005.

[GDD04] I. A. Gravagne, J. M. Davis, J. J. DaCunha, and R. J. Marks II, Bandwidth reduction for controller area networks using adaptive sampling, in *Proceedings of International Conference on Robotics and Automation*, pp. 5250–5255, New Orleans, LA, April 2004.

[KEW02] L. Krishnamachari, D. Estrin, and S. Wicker, The impact of data aggregation in wireless sensor networks, in *Proceedings of 22nd International Conference on Distributed Computing Systems Workshops*, pp. 575–578, Vienna, Austria, July 2–5, 2002.

[LYT06] F. L. Lian, J. K. Yook, D. M. Tilbury, and J. Moyne, Network architecture and communication modules for guaranteeing acceptable control and communication performance for networked multi-agent systems, *IEEE Transactions on Industrial Informatics*, 2(1), 12–24, 2006.

[MA02] L. A. Montestruque and P. J. Antsaklis, Model-based networked control systems: Necessary and sufficient conditions for stability, in *Proceedings of 10th Mediterranean Conf. on Control and Automation*, Lisbon, Portugal, July 2002.

[MB04] S. Mahlknecht and M. Bock, CSMA-MPS: A minimum preamble sampling MAC protocol for low power wireless sensor networks, in *Proceedings of the Fifth International Workshop on Factory Communication Systems*, pp. 73–80, Vienna, Austria, September 22–24, 2004.

[MDG06] G. Mathur, P. Desnoyers, D. Ganesan, and P. Shenoy, Ultra-low power data storage for sensor networks, in *Proceedings of Fifth International Conference on Information Processing in Sensor Networks*, pp. 374–381, ACM, New York, 2006.

[MGY04] P. D. Mitcheson, T. C. Green, E. M. Yeatman, and A. S. Holmes, Architectures for vibration-driven micropower generators, *Journal of Microelectromechanical Systems*, 13(3), 429–440, 2004.

[NK04] M. Neugebauer and K. Kabitzsch, A new protocol for a low power sensor network, in *Proceedings of IEEE International Conference on Performance, Computing, and Communications*, pp. 393–399, Phoenix, AZ, April 2004.

[NPK05] M. Neugebauer, J. Plönnigs, and K. Kabitzsch, A new beacon order adaptation algorithm for IEEE 802.15.4 networks, in *Proceedings of Second European Workshop on Wireless Sensor Networks*, pp. 302–311, Istanbul, Turkey, January–February 2005.

[NPK06] M. Neugebauer, J. Ploennigs, and K. Kabitzsch, Evaluation of energy costs for single hop vs. multi hop with respect to topology parameters, in *Proceedings of IEEE International Workshop on Factory Communication Systems*, pp. 175–182, Torino, Italy, June 2006.

[OMT02] P. Otanez, J. Moyne, and D. Tilbury, Using deadbands to reduce communication in networked control systems, in *Proceedings of American Control Conference*, Vol. 4, pp. 3015–3020, Anchorage, AK, May 2002.

[PHC04] J. Polastre, J. Hill, and D. Culler, Versatile low power media access for wireless sensor networks, in *Proceedings of Second International Conference on Embedded Networked Sensor Systems*, pp. 95–107, ACM Press, New York, 2004.

[PS05] J. A. Paradiso and T. Starner, Energy scavenging for mobile and wireless electronics, *IEEE Pervasive Computing*, 4(1), 18–27, 2005.

[PVK09] J. Ploennigs, V. Vasyutynskyy, and K. Kabitzsch, Comparison of energy-efficient sampling methods for WSNs in building automation scenarios, in *Proceedings of 14th IEEE International Conference on Emerging Technologies and Factory Automation*, pp. 1053–1060, Palma de Mallorca, Spain, September 22–25, 2009.

[R03] S. J. Roundy, Energy scavenging for wireless sensor nodes with a focus on vibration to electricity conversion, PhD thesis, University of California, Berkeley, CA, 2003.

[RSS02] V. Raghunathan, C. Schurgers, S. Park, and M. B. Srivastava, Energy-aware wireless microsensor networks, *IEEE Signal Processing Magazine*, 19(2), 40–50, 2002.

[SDE07] G. Scheible, D. Dzung, J. Endresen, and J.-E. Frey, Unplugged but connected—Design and implementation of a truly wireless real-time sensor/actuator interface, *IEEE Industrial Electronics Magazine*, 1(2), 25–34, 2007.

[SS97] M. Stordeur and I. Stark, Low power thermoelectric generator-self-sufficient energy supply for micro systems, in *Proceedings of 16th International Conference on Thermoelectrics*, pp. 575–577, Dresden, Germany, August 26–29, 1997.

[STS02] C. Schurgers, V. Tsiatsis, and M. B. Srivastava, STEM: Topology management for energy efficient sensor networks, in *Proceedings of IEEE Aerospace Conference*, Vol. 3, pp. 1099–1108, Big Sky, MT, March 2002.

[TG02] D. Tian and N. D. Georganas, A coverage-preserving node scheduling scheme for large wireless sensor networks, in *Proceedings of First ACM International Workshop on Wireless Sensor Networks and Applications*, pp. 32–41, New York, September 2002.

[VK07a] V. Vasyutynskyy and K. Kabitzsch, Simple PID control algorithm adapted to deadband sampling, in *Proceedings of 12th IEEE International Conference on Emerging Technologies and Factory Automation*, pp. 932–940, Patras, Greece, September 2007.

[VK07b] V. Vasyutynskyy and K. Kabitzsch, Towards comparison of deadband sampling types, in *Proceedings of IEEE International Symposium on Industrial Electronics*, pp. 2899–2904, Vigo, Spain, June 4–7, 2007.

[YHW04] W. Ye, J. Heidemann, and D. Estrin, Medium access control with coordinated adaptive sleeping for wireless sensor networks, *IEEE/ACM Transactions on Networking*, 12(3), 493–506, 2004.

11

Industrial Strength Wireless Multimedia Sensor Network Technology

11.1 Introduction 11-1
11.2 Wireless Sensor Network 11-2
WSN System Requirements • Wireless Multimedia Sensor Networks
11.3 WMSN Architecture 11-4
Single-Tier Flat Architecture • Single-Tier Clustered Architecture • Multitier Architecture
11.4 WMSN Hardware 11-6
Low-Resolution WMSN Motes • Medium-Resolution WMSN Motes • High-Resolution WMSN Motes
11.5 Applications of WMSNs 11-8
Surveillance • Traffic Monitoring • Personal and Health Care • Habitat Monitoring • Target Tracking
11.6 WMSNs' Technical Challenges 11-10
WMSN Application-Specific QoS Requirement • Scalable and Flexible Architectures and Protocols to Support Heterogeneous Applications • High Bandwidth • Localized Processing and Data Fusion • Energy-Efficient Design • Reliability and Fault Tolerance • Multimedia Coverage • Integration with IP and Various Other Wireless Technologies
11.7 Conclusion 11-12
References 11-12

Vidyasagar Potdar
Curtin University of Technology

Atif Sharif
Curtin University of Technology

Elizabeth Chang
Curtin University of Technology

11.1 Introduction

This chapter provides an introduction to wireless multimedia sensor network (WMSN). We introduce the WMSN technology by discussing the WMSN architecture and WMSN hardware. WMSN architecture outlines the different network configurations that can be deployed when setting up a WMSN like single-tier flat, single-tier clustered, and multitiered architecture. The next section on WMSN hardware (Section 11.4) covers the wireless motes that are used in deploying the WMSN itself, which include low-resolution motes, medium-resolution motes, and high-resolution motes. Each mote has additional processing and sensing capability, which makes it suitable for different applications. The following section (Section 11.5) then describes potential WMSN applications. These applications include surveillance, traffic monitoring, personal and health care, habitat monitoring, and target tracking. However, these applications bring with them a new set of technical challenges like QoS, high bandwidth, localized processing and data fusion, energy efficient design, multimedia coverage, etc., which are then discussed in depth to give the reader complete understanding of these needs since any WMSN application would need to keep these challenges in mind when developing new motes or deploying WMSNs. This concludes the chapter. Chapter 12 will focus on WMSN communication protocol stack, which forms the key basis of the wireless network communication.

11.2 Wireless Sensor Network

Wireless sensor networks (WSN) as shown in Figure 11.1 have attracted wide attention of research community during the past few years in applications ranging from environmental monitoring, heath care, to mission-critical industrial and military applications. This growing interest can be largely attributed to new applications enabled by large-scale networks of small devices capable of harvesting information from the physical environment [AMC07], performing simple processing on the extracted data and transmitting it to remote locations.

WSN is composed of intelligent embedded nodes capable of communicating over wireless link with base station and other nodes. WSN node as shown in Figure 11.2 comprises low-power sensing devices,

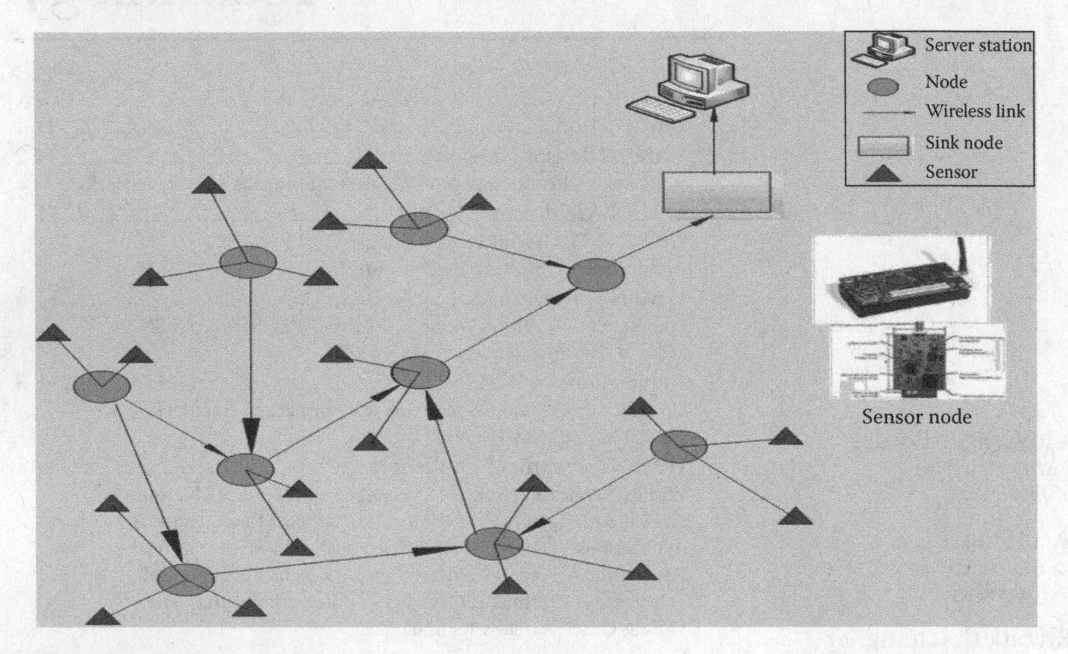

FIGURE 11.1 WSN architecture.

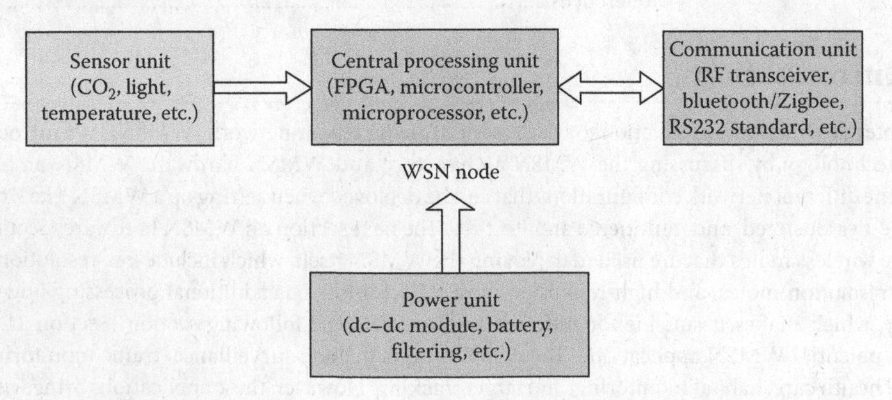

FIGURE 11.2 WSN node.

embedded processor, communication channel, and power module. The embedded processor is generally used for collecting and processing the signal data taken from the sensors. Sensor element produces a measurable response to a change in the physical condition like temperature, humidity, measuring carbon dioxide, etc. The wireless communication channel provides a medium to transfer the information extracted from the sensor node to the exterior world, which may be a computer network, and inter-node communication. Finally, power unit mostly comprises AA-size batteries and dc–dc module for feeding the entire WSN node.

11.2.1 WSN System Requirements

Here, we discuss some of the characteristic requirements of a system comprising wireless sensor nodes. The system should be

1. *Fault tolerant*: The system should be robust against node failure (running out of energy, physical destruction, H/W, S/W issues, etc.).
2. *Scalable*: The system should support a large number of sensor nodes to cater for different applications.
3. *Long life*: The node's lifetime entirely defines the network's lifetime and it should be high enough. The sensor node should be power efficient against the limited power resource that it has since it is difficult to replace or recharge thousands of nodes. The node's communication, computing, sensing, and actuating operations should be energy efficient too.
4. *Programmable*: The reprogramming of sensor nodes in the field might be necessary to improve flexibility.
5. *Secure*: The system should have the following:
   a. *Access control*: To prevent unauthorized attempts.
   b. *Message integrity*: To detect and prevent unauthorized changes in the message.
   c. *Confidentiality*: To assure that sensor node should encrypt messages so that only those nodes that have the secret key would listen.
   d. *Replay protection*: To assure that sensor node should provide protection against an adversary reusing an authentic packet for gaining confidence/network access; man-in-the-middle attack can be prevented by time-stamping the data packets.
6. *Affordable*: The system should use low-cost devices since the network comprises thousands of sensor nodes. The installation and maintenance of system elements should also be significantly low to make its deployment realistic.

11.2.2 Wireless Multimedia Sensor Networks

Wireless sensor network node, if equipped with audio- or video-sensing device, forms a WMSN node as shown in Figure 11.3. And the network formed by such nodes (WMSN) is defined as a network of wirelessly interconnected devices that are able to ubiquitously retrieve multimedia content such as video and audio streams, still images, and scalar sensor data from the environment. The key application in which WMSN is widely used includes surveillance systems, intelligent traffic management systems, habitat monitoring, and military applications. All these applications are heterogeneous by nature as they not only require scalar information but also multimedia information as well. Multitier WMSN architectures are generally used for these applications to have a complete understanding of the environment

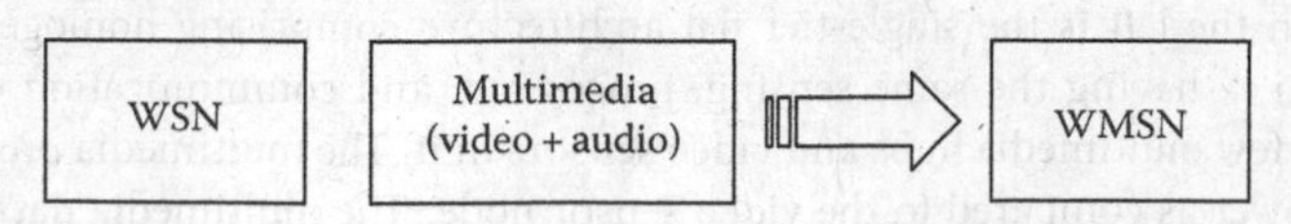

FIGURE 11.3 Wireless multimedia sensor network evolution from WSN.

behavior. For such heterogeneous applications, WSN node does not fulfill the application requirements because of the following key aspects that WMSN nodes have:

- *Processing power*: Since the multimedia information processing is computationally intensive, high-end processor or field-programmable gate array (FPGA)/ASIC is generally used for WMSN node as opposed to simple microcontrollers as in the case of WSN.
- The *amount of data* in WMSN is more in magnitude than that in WSN.
- *Storage memory*: Multimedia information local processing does require temporary storage/buffering of the data during manipulation and during sensing. However, there is no need of input buffer and high memory requirement in case of WSN that deals with only scalar information.
- *Communication standard which it uses for sending and receiving the multimedia and scalar information*: UWB is generally used as wireless communication standard for multimedia information communication; whereas in the case of WSN, Zigbee is generally used as a wireless communication standard having a data rate of 250 kbps.
- *Network configuration*: As opposed to star, cluster, or Mesh topologies in WSN, in WMSN, the basic network topology, although having the same configuration standard, is broadly classified as single-tier or multitier, based on the nature of application (details can be found in the following Section 11.3).
- *Homogeneous vs. heterogeneous*: WSN is capable of sensing only scalar information (flat homogeneous architecture; each node comprising same sensor and having same computational and communication power); however, WMSN, having multitier configuration, has heterogeneous sensing elements (scalar sensors, audio sensors, low/medium/high resolution video sensors, etc.).
- Because of the difference between the application environments and data features of stream data in WMSN and scalar data in WSN, traditional secure *Routing* for scalar data is not fit for stream data.
- The traditional WSN transport protocol is not used in WMSN because it is unable to meet the multimedia application-specific QoS parameter. In contrast to traditional WSN transport protocols are the WMSN protocols like congestion control, end-to-end reliability of communication, and packet reordering due to multipath.
- The existing WSN motes, like Telos/TelosB, Micaz mote family, did not support the *in-network processing feature*, which is generally desirable in multimedia applications due to the large size of the multimedia information (audio/video stream or snapshot). If such motes are used in multimedia applications, they will rapidly deplete their energy because of enormous power consumption due to the multimedia processing and radio communication (raw multimedia data).

11.3 WMSN Architecture

Multimedia sensor networks, in particular, exploit multilevel tier architectures for communication and network management. This is because, in addition to sensor compression, data aggregation among nodes in a common neighborhood is necessary for scalable decision-making and network operation. Figure 11.3 [AMC07,BFNPV07] provides an example in which hierarchical three-tier architecture [BP06] is used for a video-based sensor network. The architecture in Figure 11.4 depicts the three different types of sensor clouds based on the nature of application.

11.3.1 Single-Tier Flat Architecture

The sensor cloud on the left is the single-tier flat architecture comprising homogeneous sensors (in fact, same sensors, i.e., having the same sensing, processing, and communication capability). As we have seen, there are few multimedia hubs and video sensors in it. The multimedia processing hubs have more computing power as compared to the video sensor node. The multimedia data, in this network, are transferred from the parent node to sink/storage device via the gateway following the hop-by-hop

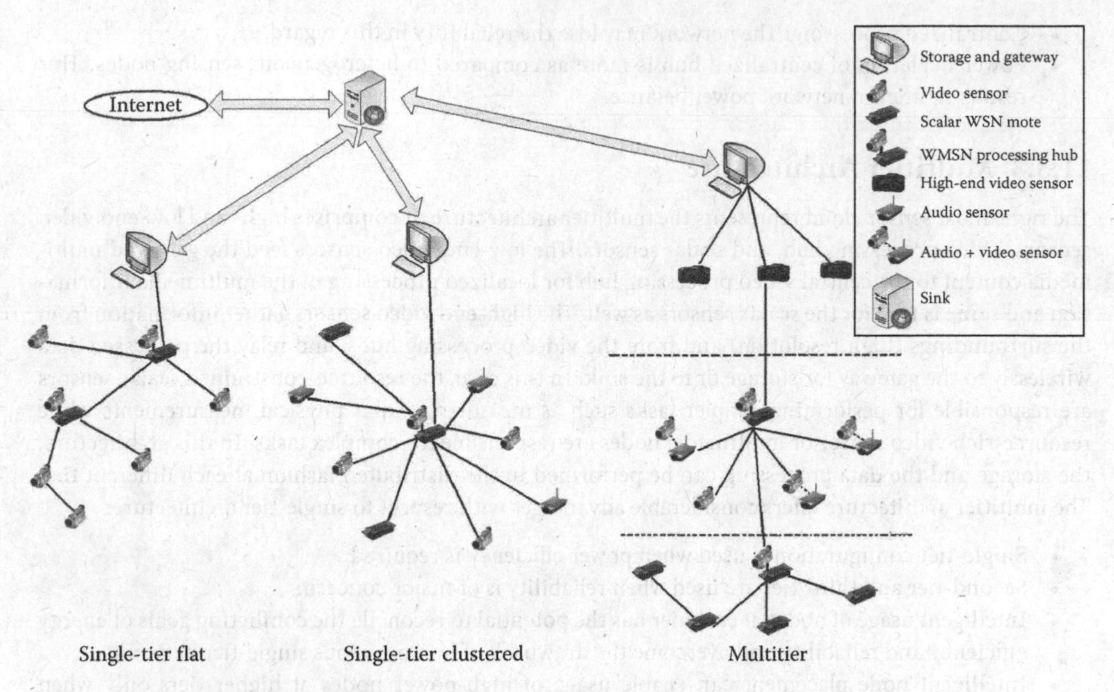

FIGURE 11.4 Architecture of WMSN.

network data routing topology (because of energy constraints as imposed by the WMSN). The advantages of single-tier flat network architecture are as follows:

- Flat homogeneous network configuration of low-power sensor nodes having low-resolution camera and scalar sensor.
- Low-power nodes result in long network lifetime.
- Easy to manage because of the homogeneous nature of nodes.
- Distributed processing.

Disadvantages include the following:

- Low-power consumption nodes have lower reliability and functionality.
- Not suitable for a range of applications involving scalar, variable resolution multimedia information.
- Also, low-power consumption nodes although having high energy efficiency have a high latency of detection.

11.3.2 Single-Tier Clustered Architecture

The second sensor cloud (in the middle) represents a single-tier clustered architecture comprising heterogeneous sensors (video, audio, scalar sensors, etc.). The clustered architecture has a central processing hub that acts as an in-charge of all the data coming from multiple sensors belonging to the same cluster. This node also performs computationally intensive multimedia processing on the gathered data and transfers them to the gateway on wireless link for storage or to sink node. The advantage of single-tier clustered network architecture is

- Heterogeneous sensing elements are used to address a range of application scenarios.

The disadvantages include the following:

- Since single high-end processing hub is used as an in-charge of the network, in case of its failure due to malfunction or power depletion, it results in network failure.

- Centralized processing: The network may lose the reliability in this regard.
- Power depletion of centralized hub is more as compared to heterogeneous sensing nodes. This results in uneven network power balance.

11.3.3 Multitier Architecture

The right-most sensor cloud represents the multitier architecture. It comprises high- and low-end video sensors, video processing hub, and scalar sensors. The low-end video sensors feed the gathered multimedia content to the central video processing hub for localized processing of the multimedia information and same is true for the scalar sensors as well. The high-end video sensors gather information from the surroundings (high resolution) and from the video processing hubs, and relay the processed data wirelessly to the gateway for storage or to the sink. In this case, the resource-constrained scalar sensors are responsible for performing simpler tasks such as measuring scalar physical measurements while resource-rich video sensor or multimedia nodes are responsible for complex tasks. In this architecture, the storage and the data processing can be performed in the distributed fashion at each different tier. The multitier architecture offers considerable advantages with respect to single-tier architecture:

- Single-tier configuration is used when power efficiency is required.
- Second-tier and third-tier are used when reliability is of major concern.
- Intelligent usage of nodes at each tier has the potential to reconcile the conflicting goals of energy efficiency and reliability and overcome the drawback of homogeneous single-tier networks.
- Intelligent node placement can enable usage of high-power nodes at higher tiers only when required with a wake-up mechanism, resulting in energy benefits.
- Coverage of a region with nodes from multiple tiers leads to availability of extra sensing and computation resources, resulting in redundancy benefits.
- Better scalability, lower cost, and higher functionality.

The disadvantages include the following:

- Special consideration should be put for inter-tier and intra-tier communication.
- Synchronization among various tiered nodes is a very difficult task.

11.4 WMSN Hardware

The recent advances in complementary metal-oxide-semiconductor (CMOS) camera technology and audio-capturing devices have taken WSN to the next level, i.e., WMSNs. In this section, we will discuss some of the existing video sensor nodes. The video sensor nodes, or more specifically WMSN nodes, are broadly classified into three different classes, based on the video-sensing node's capability:

1. Low-resolution WMSN motes
2. Medium-resolution WMSN motes
3. High-resolution WMSN motes

11.4.1 Low-Resolution WMSN Motes

Low-resolution WMSN motes comprise CMOS camera and processing module (also named CMOS imaging sensor), thus eliminating the need of several processing chips required by the traditional charge coupled device (CCD) technology. Cyclops [RIGWES05] is an electronic interface between a CMOS camera module (CMOS Agilent ADCM-1700 common intermediate format (CIF) camera) and a wireless mote such as MICA2 [CM08] or MICAz [CMZ08] and contains programmable logic and memory for high-speed data communication. The MCU controls the imager, configures its parameters, and performs local processing on the image to produce an inference. Since image capture requires faster data transfer, a complex programmable logic device (CPLD) is used to provide access to the high-speed clock. However, it can be better addressed by using single FPGA.

Researchers at Carnegie Mellon University are developing the CMUcam 3 [CMU08], which is an embedded camera endowed with a CIF Resolution (352 × 288) red green blue (RGB) color sensor that can load images into memory at 26 frames per second. The CMUcam 3 is not equipped with onboard networking capabilities, but an external mote can be attached via a serial communication channel. CMUcam 3 has software JPEG compression and has a basic image manipulation library, and can be interfaced with an 802.15.4 compliant TelosB mote [TB08].

Another integrated imaging mote was proposed [DRA06] that comprised ARM7 32 bit CPU clocked at 48 MHz with external flash storage, 802.15.4 complaint CC2420 radio chip, ADCM-1670 (CMOS sensor), and CIF (low-resolution 30 × 30 pixel optical sensor). The 32 bit ARM7 is used in this case for reasons of energy constrain as it would take a shorter time to process various image processing algorithms as compared to the traditionally used 8 bit architecture (Atmel's ATmega128 in WSN).

11.4.2 Medium-Resolution WMSN Motes

The Stargate board [SP08], designed by Intel and produced by Crossbow, is an example of medium-resolution imaging (WMSN) mote when combined with a Logitech webcam. The Stargate board is a high-performance processing platform despite being designed for sensor, signal processing, control, robotics, etc. and can also be used in sensor network applications as an independent mote (WMSN mote when combined with webcam). Stargate is based on Intel's PXA-255 XScale 400 MHz RISC processor. It has 32 MB of Flash memory, 64 MB of SDRAM, and an onboard connector for Crossbow's MICA2 or MICAz motes as well as PCMCIA Bluetooth or IEEE 802.11 cards. Hence, it can work as a wireless gateway and as a computational hub for in-network processing algorithms. WMSN mote based on Stargate platform consumes more energy and is not good from the energy consumption point of view [MPOM06]. Besides, it has high energy consumption; another drawback of the PXA-255 XScale processor is that it supports fixed-point computations only. But since the video application demands floating-point operations, we can use efficient software implementations needed to efficiently perform multimedia processing algorithms.

Intel has also developed two prototypal generations of wireless sensors, known as Imote and Imote2. Imote is built around an integrated wireless microcontroller consisting of an 8 bit 12 MHz ARM7 processor, a Bluetooth radio, 64 KB RAM, and 32 KB FLASH memory, as well as several I/O options. The software architecture is based on an ARM port of TinyOS. The second generation of Intel motes has a common core to the next-generation Stargate 2 platform and is built around a new low-power 32 bit PXA271 XScale processor at 320/416/520 MHz, which enables performing DSP operations for storage or compression, and an IEEE 802.15.4 ChipCon CC2420 radio. It has a large onboard RAM and Flash memories (32 MB), additional support for alternate radios, and a variety of high-speed I/O to connect digital sensors or cameras. Its size is also very limited, 48 × 33 mm, and it can run the Linux operating system and Java applications.

11.4.3 High-Resolution WMSN Motes

Researchers at BWN laboratory have deployed an experimental testbed [GT08] based on currently-off-the-shelf advanced devices to demonstrate the efficiency of their newly developed algorithms and protocols for multimedia communications through wireless sensor networks. The testbed includes three different types of multimedia sensors support, i.e., low-end imaging sensors, medium-quality webcam-based multimedia sensors, and high-end pan-tilt cameras. The high-resolution WMSN motes consist of pan-tilt cameras installed on a robotic platform. The objective is to develop a high-quality mobile platform that can perform adaptive sampling based on event features detected by low-end motes. The mobile actor can then redirect high-resolution cameras to a region of interest when events are detected by lower tier (densely deployed low-resolution video sensors). Another development was presented in [WOL02], a first generation smart camera prototype for detecting people and analyzing their movement in real time. For the implementation, they equipped a standard PC with additional PCI-boards featuring a TriMedia TM-1300 VLIW processor. A Hi8 video camera is connected to each PCI board for image acquisition. Now that we have outlined the basic WMSN architecture and the WMSN hardware, we would like to discuss the main applications that are making WMSN a very interesting field of research.

11.5 Applications of WMSNs

With the availability of audio and video sensors, the WMSN has enabled various new applications that WSN failed to target. The key applications that WMSN targets, some of them are shown in Figure 11.2, are broadly classified into surveillance, traffic monitoring, personal and health care, habitat monitoring, and target tracking.

11.5.1 Surveillance

Surveillance is defined as the monitoring of behavior. So, systems surveillance is the process of monitoring the behavior of people, objects, or processes within systems for conformity to expected or desired norms in trusted systems for security or social control. The importance of surveillance is increasing day by day with the increasing violence and terroristic activities around the globe. Traditional surveillance, using CCTV cameras, has come a long way, but these technologies are limited in terms of intelligence. Surveillance using WMSN has added advantages: WMSN can provide more detail and precise information, reducing the cost for network installation, extending the monitoring target areas from indoors of houses or buildings to the whole residential districts, and easily adapting upon changes of utilizations, and because a large number of wireless cameras with constrained resources and low costs replace a few traditional powerful, expensive wired cameras, a wider range of the region can be monitored in more detail.

Audio and video sensors are used to monitor a targeted place, building, or border for surveillance. The streaming multimedia content and the advanced signal processing of the captured data enable the law-enforcement agencies to keep an eye on events. Multimedia sensor node could infer and record potentially relevant activities (thefts, car accidents, traffic violations) in the form of snapshot or stream and report to control or base station.

11.5.2 Traffic Monitoring

Road traffic monitoring is defined as monitoring the behavior of vehicles (speed, number of vehicles, etc.) on different lanes of the road. Today road traffic forecasting is quite important so as to cut short the journey time. Also this helps in fuel saving. In the past, no such traffic monitoring system existed, but the advent of WMSN technology has opened a new dimension in road traffic measurement and forecasting. WMSN helps reduce travel time on road by intelligently routing traffic through the least-congested areas and results in fuel saving; protecting the environment by reducing pollution; helping to better utilize the existing road infrastructure and assets by diffusing traffic to alternate routes some of which may currently be experiencing very low traffic flows; managing abrupt driving behavior on roads by tracking abrupt drivers, tracking unusual driving patterns, and predicting potential drunk-driving activity; tracking stolen cars and attempted robbery in real time, etc.; and providing driver-assistance technology for accident prevention and evidence gathering.

Managing road traffic congestion has become a major issue for many countries, especially in cities like London and New York [ALG05]; hence, strong infrastructure investments should be made to provide real-time traffic status for intelligent traffic routing [ALG05,DCIT08,KCITS08]. Some countries have implemented peak-hour congestion tax, e.g., London, or others are providing free public transport, e.g., Transperth CAT service. WMSN identifies traffic patterns; chain effects; and scenarios in volume, capacity, and rate of change in networks of roads that will lead to heavy congestion or traffic jams. Sensors could also detect violations and transmit video streams to law-enforcement agencies to identify the violator, or buffer images and streams in case of accidents for subsequent accident scene analysis. The significance of WMSN is twofold: First, it is cost-effective as it eliminates the need to have a dedicated wired infrastructure for communication with the traffic control center and, second, its deployment is highly flexible, i.e., the wireless motes can be easily reconfigured for deploying at a new location.

11.5.3 Personal and Health Care

Health care is the prevention, treatment, and management of illness and the preservation of mental and physical well-being through the services offered by the medical, nursing, and allied health professions. According to World Health Organization health care embraces all the goods and services designed to promote health, including "preventive, curative, and palliative interventions, whether directed to individuals or to populations." The organized provision of such services may constitute a health-care system.

Multimedia sensor networks can also be used in elderly personal and health care. WMSN is used to identify the causes of illnesses that affect the elderly (e.g., dementia [R05]). Networks of wearable or video and audio sensors can infer emergency situations and immediately connect elderly patients with remote assistance services or with relatives. WMSN with telemedicine sensors can be used in monitoring patients parameters such as body temperature, blood pressure, pulse oximetry, electrocardiogram, and breathing activity remotely [HK03].

11.5.4 Habitat Monitoring

An explicitly spatial definition of habitat: The physical and biological environment used by an individual, a population, a species, or perhaps a group of species. Habitat-monitoring schemes evaluate the conservation status of habitats or habitat types by estimating the following sets of habitat attributes: extent, biotic composition, biological structure, and physical structure. Habitat-monitoring schemes can be classified into those with and those without a spatial aspect. Both can monitor one, few, most, or all specific habitat types within a country, region, or landscape.

Several projects on habitat monitoring that use acoustic and video feeds are being envisaged, in which information has to be conveyed in a time-critical fashion. For example, Intel Research Laboratory at Berkeley initiated a collaborative project with the College of the Atlantic in Bar Harbor and the University of California at Berkeley to deploy wireless sensor networks on Great Duck Island, Maine. Their goal was to establish a habitat-monitoring kit for researchers worldwide.

11.5.5 Target Tracking

Tracking concerns setting up a track on other objects momentarily viewed from the observer's own location. Target tracking concerns a process starting with determining the current and past locations and other status of property in transit. Tracking and tracing is the completion of this process with uniformly building a track of such property that are forwarded to, processed for, applied in, or disposed of usage.

WMSN can also be used in target-tracking application, where acoustic sensors are adopted to localize the target. Each sensor node can acquire acoustic signals from the target. Various efficient algorithms are used to perform robust target-position forecasting during target tracking. Then sensor nodes around the target are awakened according to the forecasted target position. With committee decision of sensor nodes, target localization is performed in a distributed manner and the uncertainty of detection is reduced.

These WMSN applications are very useful and interesting; however, they bring with them a lot of challenges that need to be considered when developing any new WMSN motes or deploying a WMSN network. These technical challenges are discussed in the following section.

11.6 WMSNs' Technical Challenges

WMSN is a cross-disciplinary research area that involves communication, signal processing, embedded computing, and control theory. Since most of the information in WMSNs is related to multimedia, there are a number of factors that influence the design of WMSN and are described in this section.

11.6.1 WMSN Application-Specific QoS Requirement

WMSNs must support application-specific QoS requirements. WMSNs are designed to address a range of applications from simple data transfer as in case of scalar networks to multimedia communication. The multimedia information may be a snapshot of an event or a streaming multimedia content. The streaming multimedia content communication requires high bandwidth and is delay sensitive. Also, the streaming multimedia information is generated over longer time periods, as opposed to snapshot multimedia information that contains event-triggered observations obtained over a short period of time, and is to be conveyed reliably. So the high-level WMSN multimedia algorithms and underlying WMSN hardware should have strong linkage so as to support application-specific QoS requirements, keeping in view the delay, energy consumption, reliability, and network lifetime constraints of WMSNs.

11.6.2 Scalable and Flexible Architectures and Protocols to Support Heterogeneous Applications

The WMSN design should be flexible enough for the network expansion and support various independent and heterogeneous applications. So while designing WMSN, it is of key importance that the underlying WMSN-embedded design and protocol suite must support the heterogeneous applications up to their required QoS specifications while ensuring the reliability, privacy, delay, and energy constraints as well.

11.6.3 High Bandwidth

Existing WSN architecture that supports scalar data communication requires less bandwidth as compared to WMSN architecture, where the data is video stream by nature requiring large bandwidth and is highly delay sensitive. In WMSN, hop-by-hop routing nature, the node not only sends its own multimedia information (either snapshot of an event or streaming multimedia contents) but also the multimedia and scalar data information from the child nodes as well. So the parent node must have large throughput to support the large bandwidth requirement of multimedia information. The existing WSN nodes that support 802.15.4 standard (MICAz, TelosB, etc.) have maximum capability of 250 kbps, but in the case of WMSN application, we may require a bandwidth order of magnitude higher than this ultra-wideband (UWB), or impulse radio technologies are considered as promising communication technologies to provide high bandwidth capacity to existing WMSN applications.

11.6.4 Localized Processing and Data Fusion

In WMSN, each node gathers information (multimedia by nature) from surroundings and from child nodes (hop-by-hop network topology). Most of the data are unprocessed or raw in nature [GGC01], carrying too much redundant information, which requires high bandwidth. So the concept of in-network or localized processing was introduced to overcome the effects of raw data transmission. In case of in-network processing, the gathered information is processed locally at the node side prior to transmission to the parent node, e.g., data of the same event taken by multiple video sensors may need to be filtered to reduce the repeated information; in video security application, the information from the interesting scenes can be compressed to a simple scalar value; and so on. However, the traditional video/audio coding techniques generally use predictive encoding, which requires complex encoders, powerful processing algorithms that consume large power and are not suitable for energy-limited WMSN nodes. So an effort to develop distributed source coding that employs simple encoder is discussed in [GARM05]. Performing the localized or in-network processing techniques substantially results in the reduction of the bandwidth consumed. While using these algorithms, it should be kept in mind that WMSN has more strict energy constraints as compared to WSN, so the in-network processing algorithms should be energy efficient.

11.6.5 Energy-Efficient Design

Power consumption is of greater concern than in traditional sensor networks as multimedia applications produce high volumes of data, which require high transmission rates and extensive processing. This challenge claims for energy awareness at all the layers of the protocol stack, in order to maximize the network lifetime, providing the required QoS provisions.

11.6.6 Reliability and Fault Tolerance

In WMSN, the reliability and fault-tolerance behavior of node is of critical importance. A number of sensor nodes in a WMSN may fail or breakdown due to the depletion of battery power, physical damage, or because of the extreme interference caused by the harsh environments in which they operate. The failure of these sensor nodes should not have an effect on the overall functioning and reliability of the WMSN. More importantly, the multimedia sensor nodes operating under strict energy constraints have limited energy budget dedicated to testing and fault tolerance. Moreover, the security and privacy concerns often prevent extensive testing procedures. Fault tolerance is the ability to maintain the functionalities of the WMSN without any disruption caused by sensor node malfunctions. Protocols and algorithms need to be developed to deal with the level of fault tolerance required by the WMSNs. The level of fault tolerance will depend on the specific application scenario. For example, a WMSN deployed for home automation is most likely to have a low fault-tolerance requirement since the chance of sensor nodes being easily damaged or interfered by harsh environmental conditions is extremely low. Alternatively, a WMSN deployed in a battlefield for surveillance would require a very high fault-tolerance level because the sensed data are crucial and the sensor nodes have a higher probability of being damaged by the hostile conditions.

11.6.7 Multimedia Coverage

Video camera is generally used as a video sensor in WMSN node; it has large sensing radii and is sensitive to the direction of acquisition. The coverage radii of the video sensor define the effective coverage of the WMSN node and their collective contribution defines the effective coverage of the WMSN. The efficiency of a WMSN heavily depends on the correct orientation [YHSG06] (i.e., view) of its individual sensory units in the field [NW08]. Efficient energy-aware algorithms to determine node's multimedia coverage and find the sensor orientation that minimizes the negative effect of occlusions and overlapping regions in the sensing field are generally required. This enables multimedia sensor nodes to compute their directional coverage leading to an efficient and self-configurable sensor orientation calculation.

11.6.8 Integration with IP and Various Other Wireless Technologies

Integrating the wireless sensor network to the existing IP or other wireless technologies is today's hot topic of research. It is of fundamental importance for the commercial development of sensor networks to provide services that allow querying the sensor network from anywhere so as to retrieve the desired information. So IP protocol suite is to be incorporated in the existing sensor's communication protocol stack. Also large-scale sensor network may be created by interconnecting the local islands of the sensors through other wireless technologies without sacrificing the operational efficiencies within each individual technology [AMC07,BFNPV07].

11.7 Conclusion

In this chapter, we discussed in general about WMSN. A basic introduction to the WMSN technology was provided by discussing WMSN architecture and the associated hardware, which included low-to-high resolution WMSN motes used in deploying any WMSN network. A detailed explanation was provided on different WMSN architectures and their associated advantages and disadvantages, which can be used by any network designed when designing or deploying a WMSN for any given application like surveillance, traffic monitoring, personal and health care, habitat monitoring, and target tracking. A detailed description was then provided on the technical challenges associated with WMSN and how these could be mitigated by arriving at reasonable trade-offs in design. The next chapter will focus on WMSN communication protocol stack, which includes the following layers: application layer, transport layer, network layer, MAC layer, and physical layer.

References

[ALG05] ALG, An Independent Assessment of the Central London Congestion Charging Scheme, Association of London Government (www.alg.gov.uk), London, U.K., 2005.

[AMC07] I.F. Akyildiz, T. Melodia, and K.R. Chowdhury, A survey on wireless multimedia sensor networks, *Computer Networks*, 51, 921–960, 2007.

[BFNPV07] A. Bonivento, C. Fischione, L. Necchi, F. Pianegiani, and A.S. Vincentelli, System level design for clustered wireless sensor networks, *IEEE Transactions on Industrial Informatics*, 3(3), 202–214, 2007.

[BP06] U. Bilstrup, and P.-A. Wiberg, An implementation of a 3-tier hierarchical wireless sensor network, in *Proceedings of the 2006 IEEE International Conference on Industrial Informatics*, Singapore, pp. 138–143, August 16–18, 2006.

[CM08] Crossbow Technology Inc., Crossbow MICA2 Mote Specifications, http://www.xbow.com, August 21, 2008.

[CMU08] The Robotics Institute, Carnegie Mellon University, CMUcam3: Open Source Programmable Embedded Color Vision Platform, http://www.cmucam.org/, October 12, 2008.

[CMZ08] Crossbow Technology Inc., Crossbow MICAz Mote Specifications, http://www.xbow.com, August 21, 2008.

[DCIT08] SIMBAII, Development Status and Trend of Chinese Intelligent Traffic, http://www.simbaproject.org/en/simba_regions/china/publications/development_status_and_trend_of_chinese_intelligent_traffic.htm, October 20, 2008.

[DRA06] I. Downes, L.B. Rad, and H. Aghajan, Development of a mote for wireless image sensor networks, in *Proceedings of COGnitive systems with Interactive Sensors (COGIS)*, Paris, France, March 2006.

[GARM05] B. Girod, A. Aaron, S. Rane, and D. Monedero, Distributed video coding, *Proceedings of the IEEE*, 93(1), 71–83, 2005.

[GGC01] S. Grgic, M. Grgic, and B.Z. Cihlar, Performance analysis of image compression using wavelets, *IEEE Transactions on Industrial Electronics*, 48(3), 682–695, 2001.

[GT08] Wireless Multimedia Sensor Testbed, Broadband Wireless Networking Lab, School of Electrical and Computer Engineering, Georgia Institute of Technology, http://www.ece.gatech.edu/research/labs/bwn/WMSN/testbed.html, August 22, 2008.

[HK03] F. Hu and S. Kumar, Multimedia query with QoS considerations for wireless sensor networks in telemedicine, in *Proceedings of the Society of Photo-Optical Instrumentation Engineers-International Conference on Internet Multimedia Management Systems*, Orlando, FL, September 2003.

[KCITS08] KANSAS Department of Transportation, Kansas City Scout Intelligent Transportation System, MO, http://www.roadtraffic-technology.com/projects/kansas/, October 20, 2008.

[MPOM06] C.B. Margi, V. Petkov, K. Obraczka, and R. Manduchi, Characterizing energy consumption in a visual sensor network testbed, in *Proceedings of IEEE/Create-Net International Conference on Testbeds and Research Infrastructures for the Development of Networks and Communities (TridentCom)*, Barcelona, Spain, March 2006.

[NW08] T. Nurca, W. WANG, Self-orienting wireless multimedia sensor networks for occlusion-free viewpoints, *Computer Networks*, 52(13), 2558–2567, 2008.

[R05] A.A. Reeves, Remote monitoring of patients suffering from early symptoms of dementia, in *Proceedings of International Workshop Wearable Implantable Body Sensor Network*, London, U.K., April 2005.

[RIGWES05] M. Rahimi, R. Baer, O. Iroezi, J. Garcia, J. Warrior, D. Estrin, and M. Srivastava, Cyclops: In situ image sensing and interpretation in wireless sensor networks, in *Proceedings of the ACM Conference on Embedded Networked Sensor Systems (SenSys)*, San Diego, CA, November 2005.

[SP08] Crossbow Technology Inc., The Stargate Platform, http://www.xbow.com/Products/Xscale.htm, August 21, 2008.

[TB08] Crossbow Technology Inc., Crossbow TelosB Mote Specifications, http://www.xbow.com, August 21, 2008.

[WOL02] W. Wolf, B. Ozer, and T. Lv, Smart cameras as embedded systems, *IEEE Computer*, 35(9), 48–53, September 2002.

[YHSG06] K. Yu, M. Hedley, I. Sharp, and Y.J. Guo, Node positioning in ad hoc wireless sensor networks, in *4th International IEEE Conference on Industrial Informatics*, pp. 641–646, IEEE, Singapore, 2006.

12

A Survey of Wireless Sensor Networks for Industrial Applications

12.1 Introduction 12-1
12.2 Wireless Sensor Network Basics 12-2
Wireless Sensor Node • Wireless Sensor Network Stack
12.3 Motivation and Drivers for Wireless Instrumentation 12-3
12.4 Industrial Applications and Requirements 12-4
Standardized Solutions • Reliable Network Performance • Battery Lifetime • Friendly Coexistence with Wireless Local Area Networks • Security • Operation in Harsh and Hazardous Environments
12.5 Technology Survey and Evaluation 12-6
IEEE Std 802.15.4 • ZigBee • WirelessHART • ISA-100 • Coexistence in the 2.4 GHz Band
12.6 Conclusion 12-9
Abbreviations 12-9
References 12-9

Stig Petersen
SINTEF Information and Communication Technology

Simon Carlsen
Statoil ASA

12.1 Introduction

Sensors are used in industrial plants and facilities to monitor the performance and the operational environment in order to provide better insight into operational requirements and potential safety problems. The monitored parameters can be temperature, vibration, pressure, flow, humidity, valve positions, gas leaks, fire outbreaks, equipment condition, and others. The data collected by the sensors are used to make informed, just-in-time decisions on plant operational performance and conditions. In addition, by using intelligent techniques to analyze historical sensor data in conjunction with historical operational performance, certain characteristics and patterns in typical operational conditions can be identified, providing the possibility to optimize plant safety, production, turnarounds and shutdowns, maintenance operations, and error and fault tolerance.

Recent advances and standardization work within the field of wireless technology have enabled the development of low-power and low-cost solutions capable of robust and reliable wireless communication [ASS02]. The IEEE Std 802.15.4 [802.15.4] defines the physical layer (PHY) and the medium access control (MAC) sublayer for low-rate wireless personal area networks. Features such as low complexity, cost, and power, make it a suitable candidate for wireless sensor networks (WSNs) [YXZ06]. With a growing number of solutions based on the IEEE Std 802.15.4 PHY and MAC, it has become the de facto standard for many WSN solutions, both standardized and proprietary. The multi-hop mesh network topologies and self-configuring and self-healing capabilities offered by these technologies have made them a viable addition to, or replacement of, traditional wired instrument networks for factory automation and communication.

12.2 Wireless Sensor Network Basics

In a WSN, a *wireless sensor node* is an individual device in the network which measures a physical phenomenon and transmits the sensor reading to the network manager. The data transmission used by the wireless sensor nodes in a WSN is specified by a network protocol. A *WSN stack* is traditionally used as a layered and abstract description of the network protocol design.

12.2.1 Wireless Sensor Node

A typical wireless sensor node consists of several elements as shown in Figure 12.1. The *sensing* unit is a sensor element which measures a physical phenomenon (e.g., temperature, pressure, or vibration). An analog-to-digital converter quantifies and converts the sensor data to the digital representation needed for further processing and communication. The *processing* unit analyses and processes the digital representation of the sensor data and is also responsible for handling the local radio-frequency (RF) transceiver. The digital sensor data are encapsulated in data packets according to the communication protocol, before being transmitted to their destination. The *communication* unit provides the wireless interface which handles transmission and reception of data packets. It consists of an antenna and an RF transceiver. The *memory* and *storage* are support units for the processing unit, used for both temporary and permanent storage of firmware, configuration parameters, and sensor data. The *power* unit, which normally consists of a battery and some control circuitry, provides power to all the components of the sensor node. For some applications, a sensor node might also be equipped with a *location engine*, which estimates the spatial location of a sensor node based on received signal strengths from neighboring sensor nodes. The location engine could either be a separate microprocessor or implemented as an algorithm on the main processing unit.

One of the main challenges for many WSN applications is the desire to have a long battery lifetime. Low power consumption is therefore a key parameter for a wireless sensor node. As a result, the available hardware resources on a sensor node are limited. The processing unit is typically a low-power microprocessor capable of entering a deep-sleep mode with very low power consumption when there is no scheduled activity.

12.2.2 Wireless Sensor Network Stack

For communication protocols and standards, stacks are used as a layered and abstract description of the network protocol design. A stack consists of several layers where each layer is a collection of functions related to the specific task of the layer. A layer is responsible for providing information and services to

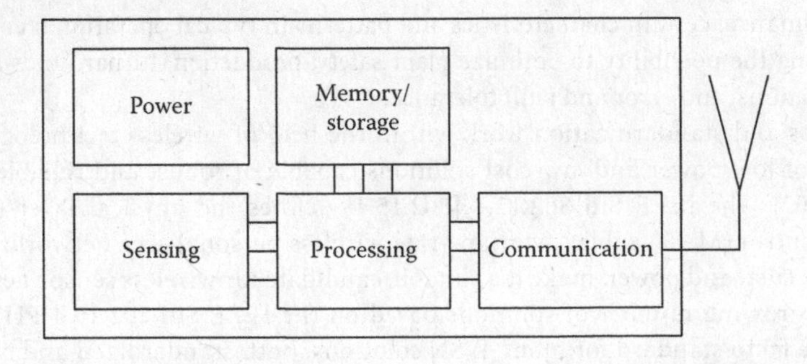

FIGURE 12.1 Wireless sensor node.

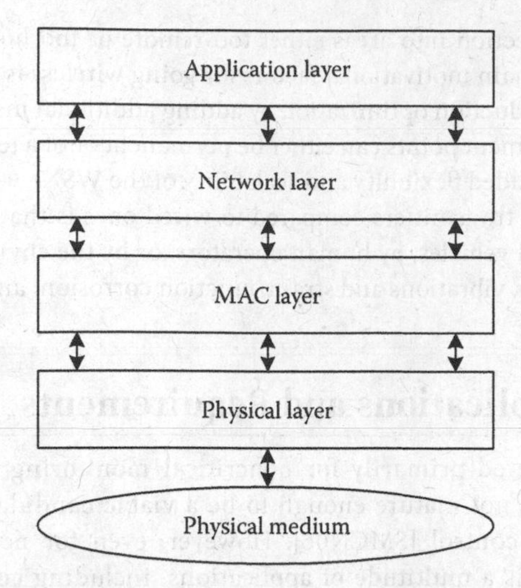

FIGURE 12.2 Wireless sensor network stack.

the layer above it, and it receives information and services from the layer below it. This information and service exchange is performed in a well-defined and standardized message-exchange format.

The most commonly used WSN stack is a simplified version of the seven-layered Open Systems Interconnection (OSI) Basic Reference Model [X.200]. The WSN stack consists of four layers; the PHY layer, the MAC layer, the network layer, and the application layer. The structure of the WSN stack is visualized in Figure 12.2.

The *PHY layer* handles functionality related to the RF transceiver, and it is the interface to the physical medium where the communication occurs. It handles the transmission and reception of data packets and provides control mechanisms for selecting operating channels, performing clear channel assessment, and RF energy detection.

The *MAC layer* provides access to the radio channel and is responsible for radio synchronization. It also handles acknowledgment frames, association/disassociation with other radio devices, and security control. Its main task is to provide a reliable link between two peer MAC entities.

The *network layer* handles functions for how to join and leave a network. It provides mechanisms for end-to-end (source to destination) packet delivery, which for mesh network topologies might involve multiple hops. To achieve multi-hop routing, it is necessary to have updated routing tables. The discovery and maintenance of routing tables is also the task of the network layer.

The *application layer* provides services to user-defined application processes. It handles fragmentation and reassembly of data packets, and it is responsible for defining the network role (end-device, router node, or network coordinator) of the device. The application layer is often referred to as the application programmers interface (API).

12.3 Motivation and Drivers for Wireless Instrumentation

To enable the introduction of any new technology in the industry, a major driver and motivational factor is the potential financial gains, i.e., reduced costs and/or increased revenue. Secondly, if new technology has the potential to benefit other important aspects such as health, safety, and the environment (HSE), they would also be considered interesting.

Wireless sensors/transmitters have many benefits over their traditional wired counterparts. Eliminating the need for cables, combined with the self-configuring capabilities of WSNs, reduces the installation cost compared to wired transmitters, both when it comes to the construction of new plants and facilities and to modification projects on existing facilities. In addition, wireless transmitters allow for the extension of data collection into areas either too remote or too hostile to be viable for wired instrumentation. One of the main motivational factors for going wireless is therefore to provide a cost-efficient way of improving production optimization by adding additional measurement points to a production site. The new measurement points can either be permanent or of a temporary nature, due to the low installation cost and the added flexibility and scalability of the WSN.

Another benefit of wireless transmitters compared to wired ones is that cables can be cut or damaged by moving machines and vehicles, by human operators, or by the environment itself in the shape of extreme weather conditions, vibrations and strain, junction corrosion, and exposure to chemicals or other fluids.

12.4 Industrial Applications and Requirements

WSNs are currently considered primarily for noncritical monitoring and control applications. The WSN technology is still not mature enough to be a viable candidate for critical monitoring applications or closed-loop control [SMCN06]. However, even for noncritical monitoring and control-related usage, there is a multitude of applications, including condition and performance monitoring, surveillance and monitoring, environmental monitoring, process control, emergency management, and asset and personnel tracking and monitoring. For some of these applications, the wireless transmitter will function either as a replacement or extension of the traditional wired fieldbus-based (i.e., FOUNDATION Fieldbus [IEC61158], PROFIBUS [IEC61158], and HART [HART]) transmitters. In this setting, the wireless transmitters can also function as an extension of a wired network.

In order for WSNs to be viable for industrial applications, they have to fulfill a set of requirements. In [PDA08], the general technical requirements for the use of WSN technology in both onshore and offshore facilities in the oil and gas industry have been identified. These requirements should be applicable to other industries as well, since the oil and gas industry can be considered as one of the most demanding industries, representing the strictest requirements regarding technology robustness and reliability. Note that some requirements are application dependent and will vary depending on the individual usage scenarios.

12.4.1 Standardized Solutions

By using standardized, open communication protocols over proprietary protocols, the industry is provided with the flexibility and freedom to choose between multiple suppliers and still have guaranteed interoperability. In addition, a standardized solution usually has a longer life span with respect to component availability and support compared to their proprietary counterparts as well as the added benefit of not being committed to a single vendor. When it comes to replacing a broken transmitter in a network, the new transmitter needs to be of the same standard, but with guaranteed interoperability it does not necessarily have to come from the same provider.

Section 12.5 provides a survey and evaluation of international WSN standards which are suitable for industrial applications.

12.4.2 Reliable Network Performance

The most basic requirement on a WSN is that it should have a reliable and quantifiable network performance. As the performance of a wireless network is more susceptible to changes in the local environment, such as antenna adjustments, moving personnel or equipment, and fluctuations in temperature and humidity, compared to a wired network, it is important to be able to quantify the expected operational performance of the wireless network. For most applications, it is desirable to have the network reliability close to 100%, i.e., there should be practically no loss of sensor data.

It is possible to achieve reliable operation of wireless networks, even in harsh environments, by employing redundant paths, self-healing algorithms, and retransmissions of lost data packets.

12.4.3 Battery Lifetime

One of the main benefits of using WSNs is the fact that they do not require any cabling. However, this means that the power needed to operate the wireless sensor must come from a local power source, in most cases a battery. This makes it important to keep the power consumption of a wireless sensor as low as possible in order to increase the battery lifetime. With the expectation that possibly hundreds or even thousands of wireless sensors could be deployed in a facility, exchanging batteries can be a time-consuming and costly maintenance task, especially if the wireless sensors are installed in hazardous areas. Wireless sensors should therefore be able to provide a battery lifetime in excess of a few years.

Due to the multi-hop mesh network topologies and the self-configuring and self-healing capabilities often present in WSNs, the power consumption of a wireless sensor node is usually nondeterministic. Measurements in [PDA08] have shown that the power consumption of a node in a mesh network depends on the amount of packets it has to forward for other nodes in the network. With constant changes in the RF environment due to climatic changes, RF noise/interference, and moving personnel or equipment, the routing paths in a WSN are dynamic and are subject to change over time. This means that the power consumption of the nodes in a network will change accordingly.

A way to eliminate the need for batteries is to have self-powered wireless sensors. This can be achieved through energy harvesting, where a wireless sensor captures and stores energy from its environment. Energy harvesting can either replace or augment the battery usage of a wireless sensor, depending on the amount of harvested energy. The power generation capability of the harvesting methods are at the time of writing 40 $\mu W/cm^3$ for thermoelectric, 116 $\mu W/cm^3$ for vibration, and 330 $\mu W/cm^3$ for piezoelectric harvesting [CC08].

12.4.4 Friendly Coexistence with Wireless Local Area Networks

Wireless local area networks (WLANs) based on the IEEE 802.11 family of standards [802.11], [802.11n], which provide access to intranet/Internet from the field, are becoming increasingly popular in industrial plants and facilities. To enable the possibility of deploying both WSN and WLAN in the same facility, which represents a highly relevant scenario, it is imperative that the two technologies are able to coexist in a friendly manner. This means no critical degradation of either WSN or WLAN performance when they are operating in the same area.

Coexistence between WSNs and WLAN are discussed in more detail in Section 12.5.5.

12.4.5 Security

The data in a WSN are transmitted over the air, which makes it more susceptible to eavesdropping and other security breaches compared to wired communication. Privacy and access violations are the most common threats to a wireless network.

For WSNs, encryption techniques are employed to protect the privacy of the transmitted data in the network. Transmitter authentication and data consistency tools are used to combat and counteract access threats.

12.4.6 Operation in Harsh and Hazardous Environments

WSNs are a viable solution for noncritical monitoring applications for practically any industrial plant or facility worldwide. The environment in which the wireless sensors can be deployed will thus vary greatly from site to site, ranging from cold winter conditions with snow and ice in arctic regions, to heat and humidity in equatorial regions. In addition, many plants and facilities are defined as hazardous locations, where strict requirements apply to any equipment installed in potentially explosive areas.

National regulations for equipment operating in hazardous areas vary from country to country. In the European Union, the 94/9/EC ATEX (*ATmosphère EXplosible*) [94/9/EC] directive is the governing document, and for the United States and Canada, the North American Hazardous Locations Installation Codes (National Electric Code for the United States and the Canadian Electrical Code for Canada) define rules and regulations on equipment and area classifications' requirements for hazardous locations. Most other countries in the world adhere to the hazardous locations certification documents from the International Electrotechnical Commission (IEC).

Before deployment of a WSN in a plant or facility, it is imperative to verify that the equipment is constructed to withstand local weather conditions, and in the case of deployment in a hazardous location, that the equipment has the required certifications according to national rules and regulations.

12.5 Technology Survey and Evaluation

In Section 12.4, a set of requirements for the operation of wireless instrumentation in industrial application was identified. One of the requirements states that the wireless technology should be based on international standards. A growing number of standards for WSNs are emerging. This chapter contains a survey and evaluation of the standards which are suitable for industrial wireless instrumentation.

12.5.1 IEEE Std 802.15.4

The IEEE Std 802.15.4 defines the PHY and the MAC sublayer for low-rate wireless personal area networks [802.15.4]. The standard specifies four different PHYs, shown in Table 12.1. The two optional high-data-rate PHYs in the 868/915 MHz band were introduced in the 2006 revision of the standard.

The IEEE Std 802.15.4 defines a total of 27 channels, numbered 0–26. Channel 0 is in the 868 MHz band with a center frequency of 868.3 MHz. Channels 1–10 are in the 915 MHz band, with a channel spacing of 2 MHz, and channel 1 having a center frequency of 906 MHz. Channels 11–26 are in the 2.4 GHz band; the channel spacing is 5 MHz; and the center frequency of channel 11 is 2.405 GHz.

12.5.2 ZigBee

The ZigBee specification [ZigBee], initially released in 2004 and updated in 2006, is a low-rate, low-power WSN standard developed by the ZigBee Alliance. The specification defines network and application layers on top of the PHY and MAC layers of the IEEE Std 802.15.4, and it is primarily targeting home automation and consumer electronics applications. Since the ZigBee specification uses the PHY and MAC layers of the IEEE Std 802.15.4, they have the same modulation techniques, bandwidth, and channel configurations. A ZigBee network operates on the same, user-defined channel throughout its

TABLE 12.1 IEEE Std 802.15.4 Frequency Bands and Data Rates

PHY (MHz)	Frequency Band (MHz)	Bit Rate (kb/s)
868/915	868–868.6	20
	902–928	40
868/915 (optional)	868–868.6	250
	902–928	250
868/915 (optional)	868–868.6	100
	902–928	250
2450	2400–2483.5	250

entire lifetime. This makes it susceptible both to interference from other networks operating on the same frequency and to noise from other sources in the environment. As a result, ZigBee has thus not been regarded as robust enough for harsh industrial environments [PDA08]. To combat this challenge, the ZigBee Alliance released the ZigBee PRO specification [ZigBeeP] in 2007. ZigBee PRO is specifically aimed at the industrial market, having enhanced security features and a *frequency agility* concept where the entire network may change its operating channel when faced with large amounts of noise and/or interference.

The ZigBee Alliance announced in April 2009 that it will incorporate standards from the Internet Engineering Task Force (IETF) into future ZigBee releases, thereby opening up for IP-based communication in ZigBee networks. Of special interest for the ZigBee Alliance in the IETF standard portfolio is the 6loWPAN working group that has created a Request for Comments (RFC4944) investigating the transmission of IPv6 packets over IEEE 802.15.4 networks [6loWPAN].

12.5.3 WirelessHART

WirelessHART is a part of the HART Field Communication Specification, Revision 7.0 [HART], which was ratified in September 2007 as the first standard specifically targeting industrial applications. WirelessHART is based on the PHY and MAC layers of the IEEE Std 802.15.4, although the MAC layer has been modified to allow for frequency hopping. Another modification is that WirelessHART operates exclusively in the 2.4 GHz band and to allow for full global availability, channel 26 as defined by the IEEE 802.15.4 is not utilized, since it is, due to national regulations, not legal to use in some countries.

WirelessHART employs a multi-hop full mesh network topology, using time-division multiple access (TDMA) as the channel access method [KHP08]. With TDMA, the network communication is divided into guaranteed time slots (GTS), where each GTS is reserved for a specific communication link. This ensures contention-free utilization of the radio channel. Each time slot is 10 ms long, providing enough time for the transmission of one data packet from the source device and the acknowledgment from the recipient of the data packet. A collection of time slots is called a superframe, and the superframe is repeated periodically throughout the network lifetime. At the beginning of a new superframe, the gateway transmits a beacon which is used for time synchronization of the network.

With capabilities such as self-configuration and self-healing, deploying a WirelessHART network should require minimal detailed understanding of low-level communication and radio propagation aspects.

12.5.4 ISA-100

The goal of the ISA-100 standards committee of the International Society of Automation (ISA) is to deliver a family of standards defining wireless systems for industrial automation and control applications [ISA-100]. The ISA-100.11a, ratified in September 2009, was the first standard to emerge. ISA-100.11a aims to provide secure and reliable wireless communication for noncritical monitoring and control applications, while critical applications are planned to be addressed in later releases of the standard.

The ISA-100.11a is based on the IEEE Std 802.15.4 PHY and MAC, operating in the 2.4 GHz band, defining a frequency hopping, multi-hop mesh network. Like WirelessHART, TDMA is used as the channel access method, along with network self-configuring and self-healing algorithms. The ISA-100.11a also enables a network to carry existing wired fieldbus protocols such as FOUNDATION Fieldbus [IEC61158], PROFIBUS [IEC61158], and HART [HART]. Existing wired installations can thus be conveniently converted to a wireless infrastructure, with a transparent data transfer between systems. ISA-100.11a also supports the IPv6 protocol, having adopted 6loWPAN [60loWPAN] as their network layer.

The ISA-100 has established the ISA-100.12 subcommittee to investigate options for the convergence of WirelessHART and ISA-100.11a. The goal of the committee is to merge the two standards into a single standard, which will then be published as a future release of the ISA-100.11a.

12.5.5 Coexistence in the 2.4 GHz Band

Friendly coexistence between WLANs and WSNs is an absolute requirement for the successful deployment of both solutions in the same area of a plant or facility, as both the IEEE Std 802.11b/g/n and the IEEE Std 802.15.4 define operation in the 2.4 GHz ISM-band.

The IEEE Std 802.11b/g/n divides the 2.4 GHz band into 14 overlapping channels with a bandwidth of 20 MHz [802.11], although to follow national regulations, only channels 1–11 are available in the United States and Canada, and channels 1 through 13 in the rest of the world except Japan.

To avoid interference between neighboring channels in an IEEE Std 802.11b/g/n network and at the same time enable friendly coexistence with IEEE Std 802.15.4 based networks, the nonoverlapping channels 1, 6, and 11 should be used for WLAN installations. As illustrated in Figure 12.3, this allows the IEEE Std 802.15.4 channels 15, 20, 25, and 26 (only channel 15, 20, and 25 in the United States and Canada) to operate in-between the IEEE Std 802.11b/g/n channels without suffering major interference, and vice versa. As the IEEE Std 802.11b/g/n has both higher maximum-allowed transmit power and a wider channel bandwidth compared to the IEEE Std 802.15.4, the risk, and the consequences, of interference between coexisting WLAN and WSN are higher for the WSN than for the WLAN. Experiments performed in [ABFS08] and [PDA08] show that the packet loss for the IEEE 802.15.4 networks increase significantly when coexisting with active IEEE 802.11b/g/n networks.

When it comes to the three international standards described in the previous section, they each employ different mechanisms for coexistence with IEEE 802.11 networks. ISA-100.11a utilizes adaptive frequency hopping and will blacklist channels which suffer from high packet loss due to noise/interference. An ISA-100.11a network will therefore after a period of time stop using the channels which are subject to interference from the WLAN networks, and might, in a worst-case scenario, end up using the interference-free channels 15, 20, 25, and 26 (as illustrated in Figure 12.3).

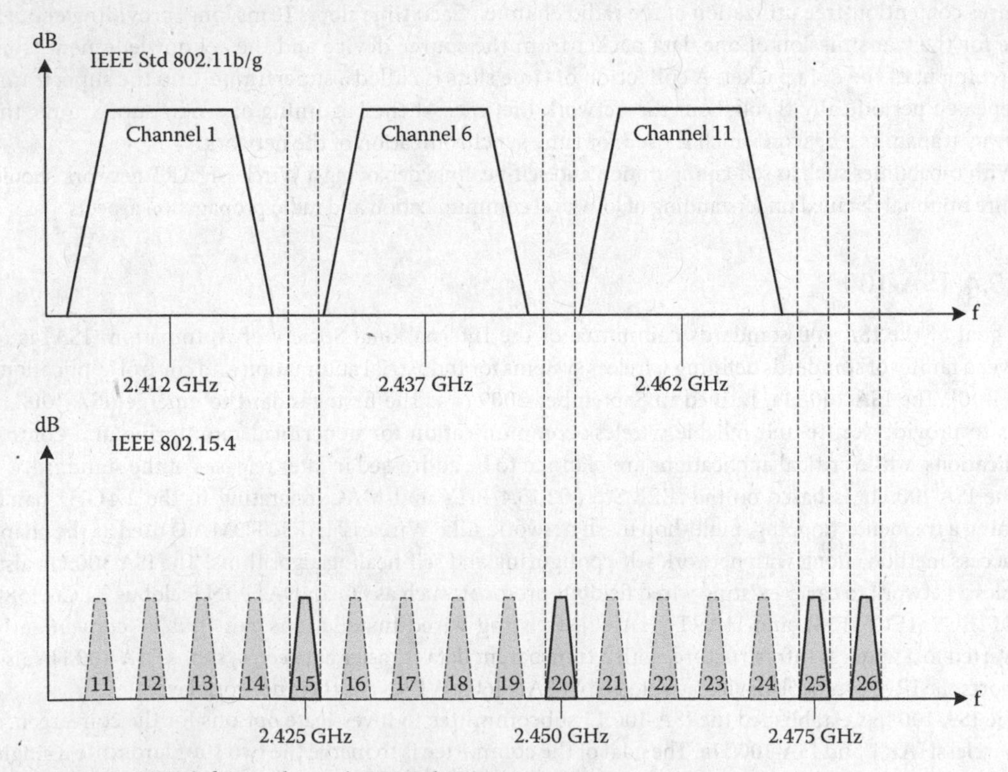

FIGURE 12.3 IEEE Std 802.11b/g and IEEE Std 802.15.4 coexistence.

WirelessHART, on the other hand, employs frequency hopping with no blacklisting, meaning that the network communication is constantly changing between the 15 available channels. The network will in other words take no preventive measures to cope with the interference from IEEE 802.11 networks in the area. As a result, the average packet loss for a WirelessHART network will increase when coexisting with IEEE 802.11 networks [PC09].

For ZigBee, one of the interference-free channels would have to be selected manually by an operator when initializing the network, as a ZigBee network is unable to dynamically switch channels while operational. ZigBee PRO, on the other hand, is capable of changing the active network channel when faced with noise/interference, and thus a ZigBee PRO network might, over time, end up using one of the interference-free channels (15, 20, 25, or 26) regardless of the initial channel configuration.

12.6 Conclusion

Using WSNs as a replacement or addition to the traditional wired industrial networks currently used for factory automation and communication has a number of benefits, both expected and proven. These include reduced cost, increased production, improved flexibility and scalability, and improved HSE. Both laboratory experiments [PDA08,PC09] and installation on an operational offshore platform [CSP08] show that WSN technology is capable of delivering robust and reliable communication in harsh environments.

As a consequence of the industry's demand for open, standardized solutions on these technologies, several international standards for industrial WSNs are emerging. Also, with the industry moving from the wired world into the wireless domain, coming international standards are expected to integrate different wireless technologies, allowing enterprises to take the step into a future with plant-wide wireless infrastructure enabling applications like wireless networking, wireless monitoring and control, and asset and personnel tracking.

Abbreviations

API	Application programmers interface
ATEX	Atmosphère explosible
GTS	Guaranteed time slots
HSE	Health, safety, and the environment
IEC	International Electrotechnical Commission
IETF	Internet Engineering Task Force
ISA	International Society of Automation
MAC	Medium access control sublayer
OSI	Open Systems Interconnection
PHY	Physical layer
RF	Radio frequency
TDMA	Time-division multiple access
WLAN	Wireless local area network
WSN	Wireless sensor network

References

[ABFS08] L. Angrisani, M. Bertocco, D. Fortin, and A. Sona, Experimental study of coexistence issues between IEEE 802.11b and IEEE 802.15.4 wireless networks, *IEEE Transactions on Instrumentation and Measurement*, 57(8), 1514–1523, 2008.

[ASS02] I. F. Akyildiz, W. Su, Y. Sankarasubramaniam, and E. Cayirci, A survey on sensor networks, *IEEE Communications Magazine*, 40(8), 102–114, 2002.

[CC08] S. Chalasani and J. M. Conrad, A survey of energy harvesting sources for embedded systems, *Proceedings of the IEEE Southeastcon*, Huntsville, AL, pp. 442–447, April 3–6, 2008.

[CSP08] S. Carlsen, A. Skavhaug, S. Petersen, and P. Doyle, Using wireless sensor networks to enable increased oil recovery, *Proceedings of the IEEE Conference on Emerging Technologies and Factory Automation*, Hamburg, Germany, pp. 1039–1048, September 15–18, 2008.

[HART] HART Communication Foundation (HCF), HART Field Communication Protocol Specification, Revision 7.0 (HCF_SPEC-13), Austin, TX, 2007.

[IEC61158] International Electrotechnical Commission (IEC), IEC 61158—Industrial communication networks—Fieldbus specifications, Geneva, Switzerland, 2007.

[ISA-100] International Society of Automation (ISA), ISA-100.11a-2009 Standard, Wireless systems for industrial automation: Process control and related applications, 2009.

[KHP08] A. N. Kim, F. Hekland, S. Petersen, and P. Doyle, When HART goes wireless: Understanding and implementing the WirelessHART standard, *Proceedings of the 13th IEEE Conference on Emerging Technologies and Factory Automation ETFA 2008*, Hamburg, Germany, pp. 899–907, September 15–18, 2008.

[PC09] S. Petersen and S. Carlsen, Performance evaluation of WirelessHART for factory automation, *Proceedings of the IEEE Conference on Emerging Technologies and Factory Automation*, Palma, Spain, September 22–25, 2009.

[PDA08] S. Petersen, P. Doyle, S. Vatland et al., Requirements, drivers and analysis of wireless sensor network solutions for the oil & gas industry, *Proceedings of the IEEE Conference on Emerging Technologies and Factory Automation*, Patras, Greece, pp. 219–226, September 25–28, 2007.

[SMCN06] J. Song, A. Mok, D. Chen, and M. Nixon, Challenge of wireless control in process industry, *Workshop on Research Directions for Security and Networking in Critical Real-Time and Embedded Systems in conjunction with the IEEE Real-Time and Embedded Technology and Applications Symposium*, San Jose, CA, April, 7, 2006.

[X.200] ITU-T X.200 (07/94), Information Technology—Open Systems Interconnection—Basic Reference Model: The Basic Model, 1994.

[YXZ06] Q. Yu, J. Xing, and Y. Zhou, Performance research of the IEEE 802.15.4 protocol in wireless sensor networks, *Proceedings of the 2nd IEEE/ASME International Conference on Mechatronic and Embedded Systems and Applications*, Beijing, China, pp. 1–4, August 13–16, 2006.

[ZigBee] ZigBee Alliance, ZigBee-2006 Specification, 2006.

[ZigBeeP] ZigBee Alliance, ZigBee PRO Specification, 2007.

[6loWPAN] The Internet Engineering Task Force (IETF), Request for Comments (RFC) 4944: Transmission of IPv6 Packets over IEEE 802.15.4 Networks, 2007.

[802.11] IEEE Standards Association, IEEE Std 802.11-2007, IEEE Standard for Information Technology—Telecommunications and Information Exchange between Systems—Local and Metropolitan Area Networks—Specific Requirements—Part 11: Wireless LAN Medium Access Control (MAC) and Physical Layer (PHY) Specifications, 2007.

[802.11n] IEEE Standards Association, IEEE Std 802.11n, IEEE Standard for Information Technology—Telecommunications and Information Exchange between Systems—Local and Metropolitan Area Networks—Specific Requirements—Part 11: Wireless LAN Medium Access Control (MAC) and Physical Layer (PHY) Specifications Amendment 5: Enhancements for Higher Throughput, 2009.

[802.15.4] IEEE Standards Association, IEEE Std 802.15.4-2006, IEEE Standard for Information Technology—Telecommunications and Information Exchange between Systems—Local and Metropolitan area Networks—Specific Requirements—Part 15.4: Wireless Medium Access Control (MAC) and Physical Layer (PHY) Specifications for Low-Rate Wireless Personal Area Networks (WPANs), 2006.

[94/9/EC] European Parliament and the Council, Directive 94/9/EC, 1994.

13

Vertical Integration

Thilo Sauter
Austrian Academy of Sciences

Stefan Soucek
LOYTEC electronics GmbH

Martin Wollschlaeger
Dresden University of Technology

13.1 Introduction 13-1
13.2 Historical Background 13-2
13.3 Network Interconnections 13-4
13.4 Application View 13-7
13.5 Security Aspects in Vertical Integration 13-8
13.6 Trends in Vertical Integration 13-10
Abbreviations 13-11
References 13-12

13.1 Introduction

The past two decades have brought tremendous advances in network technology, both in the office world and in all fields of automation. Two obvious indicators for this are on the one hand the stunning success of the Internet, whose technological principles have also widely been adopted in small-size local area networks (LANs). On the other hand, the lengthy and fierce struggle for an acceptable compromise in fieldbus standardization shows that the automation domain is seen as an important market by the big players. Consequently, as networking in both worlds has reached a mature point, the recent years have seen many attempts to bring the two sides together and to finally achieve something that had remained wishful thinking for a long time: the idea of vertical integration, meaning a seamless integration of automation data into a higher level information technology (IT) context beyond mere data acquisition [WV03].

The term "vertical integration" has become trendy in the last years and has been used as a marketing argument for a number of automation solutions. Strictly speaking, however, vertical integration is not a concept peculiar to automation. In fact, it has several connotations in different domains that may be distinguished:

- In an economical context, it denotes the extent to which a company has control of the processes involved in the manufacturing of its products. Full vertical integration in this case means full control over the entire value chain from raw material production until the assembly of the final product [RO03].
- In a business and organizational context, it concerns the information exchange between the various management levels in a company or an agglomerate of companies. The relevant property here is the control of the decision flow across the hierarchies.
- In an automation context, finally, it means the data exchange between all automation levels inside a plant. In particular, it denotes the interconnection of automation and office networks on a technological level, and the linkage of field-level control and planning tools on an application level [S07].

This chapter will focus on the latter aspects of vertical integration, and there again especially on the technological issues. It will shed some light on the evolution of the integration idea from the first attempts to today's situation. A section is devoted to the network interconnection problem that is a cornerstone for integration. The application view of vertical integration will be briefly sketched, and the question of how to achieve security will be discussed. Finally, a few trends of the ongoing evolution will be highlighted.

13.2 Historical Background

Historically speaking, the idea of achieving a data flow from the shop floor and the field level to the office level together with proper network integration is not new. What is new is the current focus on Internet technologies as upper level of the interconnection. However, the integration idea itself was already part of the computer-integrated manufacturing (CIM) concept suggested in the 1970s [H73]. The approach put forward then was to create a transparent, multilevel network to structure the information flow required for factory and process automation [DSP88]. To cope with the expected complexity, a strict subdivision of the information processing into a hierarchical model was devised that became to be known as the automation pyramid. The model exists in various versions with different naming conventions and numbers of levels, but it typically comprises four to six functional levels as depicted in Figure 13.1. Along with the definitions of functionalities—which of course varied according to the application domain—networks were associated to the individual levels. Originally, different networks were intended for these purposes. In recent years, however, the complexity of the pyramid, i.e., the number of levels was reduced chiefly because of the consolidation process that took place in the field of networking technologies. Today, the upper levels are dominated by IP-based networks (mostly on the basis of Ethernet) and, more generally, technologies from the office domain, whereas the lowest level is still home to field-level networks in various forms.

A critical review shows that the CIM approach proposed 30 years ago practically flopped. There are niches where integration became tight and very successful, like in CAD/CAM solutions for tooling machines or (micro)electronic manufacturing. The comprehensive integration of all levels across all possible application areas however failed, and still today, the term CIM has a largely negative connotation in many industries. The reasons for the spectacular failure of a basically good idea are manifold: The overall concept was overloaded and far too ambitious. In particular, it had been designed mostly without a view to the feasibility of implementation. In fact, the concepts did not show much

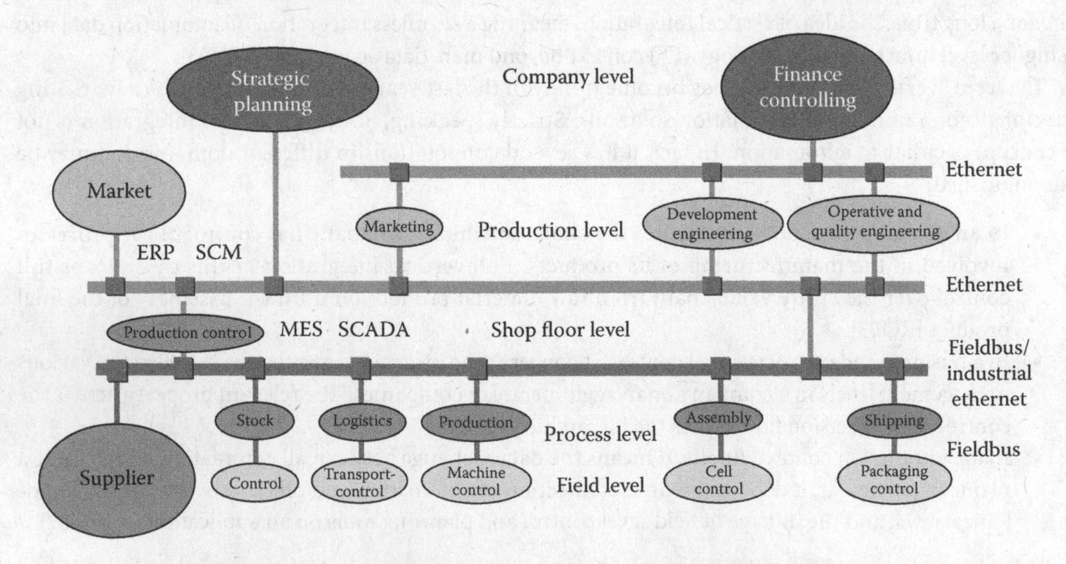

FIGURE 13.1 Typical functional units inside a contemporary company and their interrelation with the automation hierarchy as well as associated network technologies.

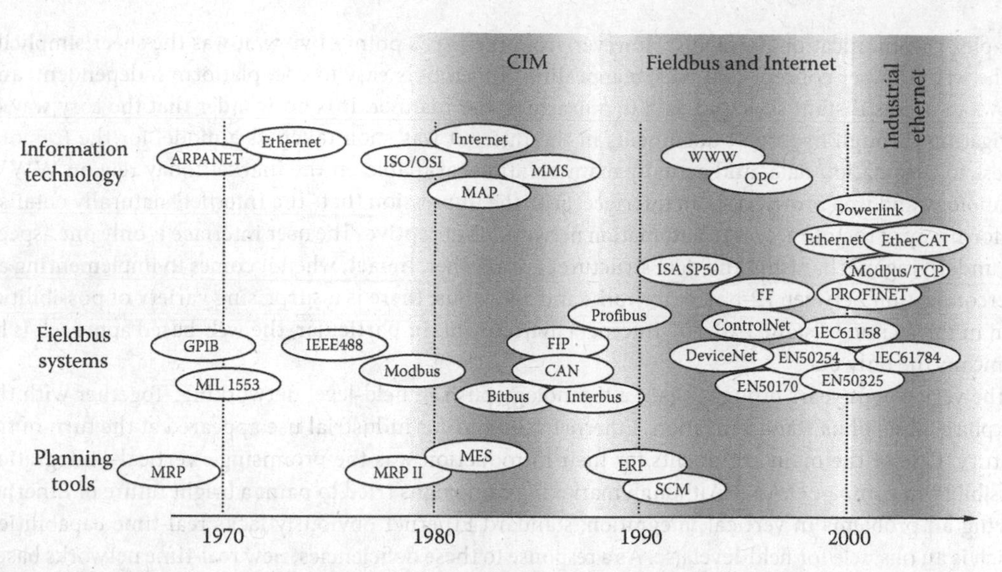

FIGURE 13.2 Selected milestones relevant for the evolution of vertical integration.

flexibility or modularity that would have facilitated their adoption by industry. Rather, it was more of an all-or-nothing decision to be taken. But more important, critical technological elements were simply missing or in a premature stage. Fieldbus systems as low-level networks were not yet widely known; actually the CIM concept was (also) a stimulus for their evolution. Microelectronics had not yet produced high performance and affordable processors and controllers, and IT as a driving force in networking was not yet a determining factor.

The timeline in Figure 13.2 presents a few technological milestones in the evolution of IT and networking concepts and compares them with the introduction of various application concepts relevant for the implementation of vertical integration. It reveals that one of the most decisive factors also for automation networks was the definition if the ISO/OSI model in the late 1970s. This reference model was (and still is) the starting point for the development of many complex communication protocols. The first application of the OSI model to the domain of automation was the definition of the Manufacturing Automation Protocol (MAP), which was the communication technological foundation of CIM. MAP was intended to be a framework for the comprehensive control of industrial processes, and the result of the definition was a powerful and flexible, but overly complex protocol that was not accepted. The tightened standard Mini-MAP did not have the anticipated success either; however, it introduced a reduced three-layer communication model that became the golden model for most fieldbus systems [S05]. What did have success was Manufacturing Message Specification (MMS). It defined the cooperation of various automation components by means of abstract objects and services and was used as a starting point for other fieldbus definitions.

What Figure 13.2 clearly shows is that CIM in its original definition came far too early. The golden era of the fieldbus was the late 1980s and especially the 1990s with lots of different systems and fierce struggles in the international standardization committees. Although fieldbus systems were originally also conceived as a means to bridge the communication gap in the lower levels of the CIM model, the interconnection aspect was gradually left aside for the sake of simplicity. In most cases, developers adopted a pragmatic viewpoint and designed fieldbus systems as isolated networks to get solutions ready for the market. The typical end point for vertical communication was, therefore, some sort of control room, located on the cell level of the automation pyramid.

The great leap forward—and the revival of the integration idea under a new name—came with the success of the Internet and, in particular, with the invention of the World Wide Web (WWW). Of course, advances in microelectronics provided the technological backbone for the processing of increasingly complex communication protocols. However, from the user's point of view, it was the sheer simplicity of the web browser concept that was so appealing. This tool is easy to use, platform independent, and allows to access distant sites and data in a nearly trivial manner. It is no wonder that the easy way of navigating through hypertext documents in the Internet was soon taken as a model for the (remote) access to automation data, and actually many solutions available on the market today rely on WWW technology and web browsers as an interface. Still, the impression that "the Internet" naturally entails a uniform way of remote access to automation networks is deceptive. The user interface is only one aspect, the underlying mechanisms and data structures are another. In fact, when it comes to implementing an interconnection between IP-based networks and a fieldbus, there is a surprising variety of possibilities even in the so much "standardized" Internet environment. In particular, the web-based approach is by no means the only one.

The very recent years brought again a technology push in field-level networking. Together with the last phase of fieldbus standardization, Ethernet solutions for industrial use appeared at the turn of the century. One of the main arguments for their introduction was the promising—vertical—integration possibility in company LANs. Although marketing campaigns tried to paint a bright future of Ethernet solving all problems in vertical integration, standard Ethernet obviously lacks real-time capabilities, which is an obstacle for field-level use. As a response to these deficiencies, new real-time networks based on Ethernet were developed that broke with the original definition and can be expected to replace several of today's fieldbus systems in the long run.

What does all this mean for vertical integration? From a technology point of view, all basic elements needed are finally available—contrary to the situation in the CIM era. What is required today is to put them together in a reasonable way. The technological issues involved are considered in the subsequent sections. A further challenge lurks in the application level. Figure 13.1 also shows how typical functional units are associated with the individual levels. It appears that vertical integration is reasonably advanced in both the upper and lower layers. This is not surprising since the two sections mirror two different worlds inside a company:

- Business-oriented applications like enterprise resource planning (ERP) tools, supply chain management (SCM), or financial tools for accounting and bookkeeping are tightly integrated today. The users as well as the developers of such software tools share an economics background and are focused on a strategic view of the company.
- Production-oriented applications like supervisory control and data acquisition (SCADA), production control, assembly, or to a certain extent also manufacturing execution systems (MES) belong to the operational world. Again, they are mostly well integrated, but the background of users and developers is typically engineering, which stipulates an entirely different view on the overall system.

Figure 13.1 demonstrates that these two worlds are also separated by different networking concepts. In the business context, LANs are predominant whereas the control level is home of the fieldbus. Again, this is not surprising because one of the major design guidelines for fieldbus systems was the recognition that in a (real-time) process control context, completely different networking characteristics are required than in the (strategic) office world. Hence, the gap to bridge today not only concerns the linking of two different network types but also the interconnection of two different mindsets.

13.3 Network Interconnections

The implementation of communication networks has been designed for particular levels of the communication hierarchy, using dedicated I/O buses and fieldbuses at the lowest level. Typically, intermediate levels have been using other, tailored communication systems up to the highest level. Process data has been abstracted and processed from level to level. Network interconnections usually served horizontal communication, i.e., the exchange of data on the same level in the hierarchy, mostly within the same technological domain.

The requirements for vertical access to data throughout the various communications levels require a higher level of interconnection. Here, different communication media, different networking protocols, and different topologies have to be connected. Among other technologies, the use of Ethernet and IP-based protocols serves as the major integration platform of today's automation systems. IP-based systems possess the special appeal that communication can be established using the public Internet infrastructure, avoiding costly and proprietary phone lines or data links. A set of approaches are implemented that lead to tighter vertical integration:

- *Flat network hierarchy*: While staying within the same technology domain, vertical access can still be limited due to different physical media and network partitioning into islands. A flat network hierarchy eliminates those boundaries. Routers and tunneling routers are used today for interconnecting formerly isolated trunks of fieldbus networks.
- *Cross-domain communication*: A number of systems have been designed to be specific to their application domain. It is common that separate systems are found for heating, ventilation, and air-conditioning (HVAC) and emergency lighting in building automation. The former advantage of separating the installation and engineering tasks is now becoming a burden for communication between those domains. For gaining access to different domains, proxies are used.
- *Protocol convergence*: Providing protocol, service, and data translation between different networking technologies is necessary to fully distribute services around the automation system. Vertical access through all levels requires a gateway approach. These gateways translate the high-level services between the different technologies.

There is a tendency to eliminate the separation between the automation and management network, and the implied differences in network media. In some cases, even the fieldbus level is integrated into the same, global network domain. The technology of choice today is the IP transport over Ethernet's structured cabling. It also brings the networks of building automation systems (BAS) and factory automation networks closer to the IT network, which has formerly been separated. For doing so, most of the widely deployed networks have specified and implemented an IP-based transport layer (CEA-852 for LonWorks [CEA852], KNX/IP [KNXIP], BACnet/IP [BACIP], Modbus/IP [ModIP]).

In general, this reduces cabling efforts and increases maintainability. In the field, however, the advantages of special fieldbus media still persist: First, for the ease of installation, where free topology allows to install cabling fitting more closely to the nodes in a room, and second, for better extendibility. The remaining trunks are coupled via control network/IP (CN/IP) routers. These routers are tunneling routers, which encapsulate the native frame formats into IP transport. This architecture is depicted in Figure 13.3. The advantage of tunneling is that fieldbus nodes can be left unmodified, i.e., do not need to implement an IP stack. The tunneling routers are store-and-forward routers, which can additionally perform selective forwarding. This means, the traffic is not broadcast over the IP network but routed precisely to the designated tunneling routers based on CN addressing information. Other high-end nodes such as powerful building controllers or SCADA systems can implement the tunneling protocol

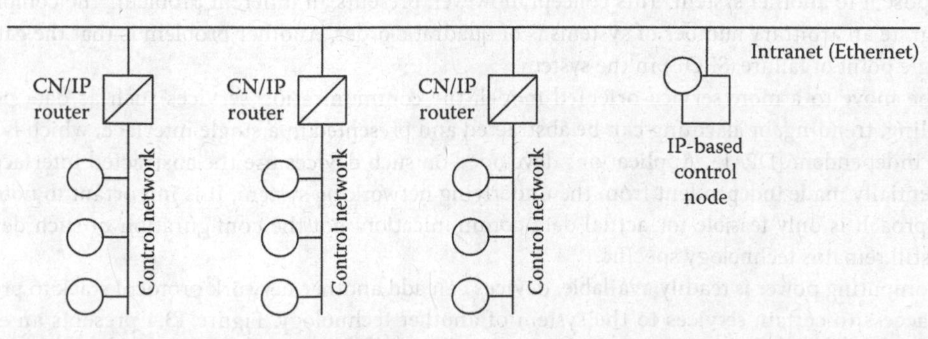

FIGURE 13.3 System with tunneling CN/IP routers.

themselves and process the encapsulated traffic as a node, not as a router. This effectively enables purely IP-based fieldbus nodes, which can communicate directly to legacy (non IP-based) nodes in the system. The tunneling approach has been chosen for LonWorks or KNX/IP.

Other fieldbus systems implement a dedicated IP data link layer for the fieldbus protocol natively. Nodes in these systems can be connected directly onto the Ethernet network without the use of a tunneling technique. Examples are BACnet/IP or Modbus over TCP. To access other, non IP-based CN nodes, routers are used.

Notwithstanding the ease of IP everywhere, a separation of field-level and office-level networks is still recommended for performance and security reasons, even if—as in the case of industrial Ethernet—a direct integration of both networks is possible. Another issue in using IP-based transport systems for fieldbus protocols is that the original, native communication media had partly different properties. For example, delay times, delay variance, and packet ordering may impair the fieldbus protocol state machine itself. These effects are typically combated by using extra sequence numbers and synchronized time-stamping on the tunnel. Once the protocol state machines are maintained, the application on top is still influenced in its design. Application timers or feedback control loops must be aware of the changed timing relations.

Once the problem of merging isolated network islands into a flat network hierarchy is solved, logical boundaries within the same networking technology still exist. These are typically imposed by different application domains to keep the traffic, addressing, and installation tasks separated. While communication would be enabled over the flat network, the domain boundary bars traffic. The use of proxy devices solves this problem. Proxies are basically devices that pass the process data and expose it from one domain into another. An implementation of a proxy is placed as one instance in each domain. Strictly speaking, a proxy can be seen as a gateway between the same technologies. An example for using proxies is LonWorks in automation systems. In this system, address spaces are separated by address domains. The address domains are set up according to different application domains, such as HVAC or emergency lighting. A LonWorks proxy can expose data points of one domain to another domain. Other examples of proxies are HTTP proxies. These are frequently used to isolate local from public IP traffic for security reasons. For instance, the office-level network needs access to Web services in the field-level network. While it is common practice to keep those two IP networks separate, an HTTP proxy can establish the Web service link.

Another area of focus is protocol convergence. Today's automation systems represent highly heterogeneous systems. Different application service domains may use different CN technologies. The goal for vertical access or distribution of communication services is clearly a seamless communication also between the application domains and across all layers of the automation system. The classical approach to let system A communicate with system B is the use of gateways. This concept is being adopted widely. Gateways abstract the network service at the application level (e.g., data point values, trend data, schedules, and alarms) and represent it in different networking technologies. Commonly used gateway concepts today are based on monolithic gateways. These are devices that have the global view of one system and expose it to another system. This concept, however, presents an inherent problem: The complexity to integrate an arbitrary number of systems is of quadratic order. Another problem is that the gateway is a single point of failure (SPOF) in the system.

In the move to a more service-oriented model, the communication services such as data points, scheduling, trending, or alarming can be abstracted and presented in a single interface, which is technology independent [DPAL]. Applications developed on such devices use the abstracted interface and are essentially made independent from the underlying networking system. It is important to note that this approach is only feasible for actual data communication, not the configuration of such devices, which still remains technology specific.

As computing power is readily available, devices can add another network protocol stack to provide native access to certain services to the system of another technology. Figure 13.4 presents an example for this scenario. A controller implements not only the LonWorks technology but also the BACnet

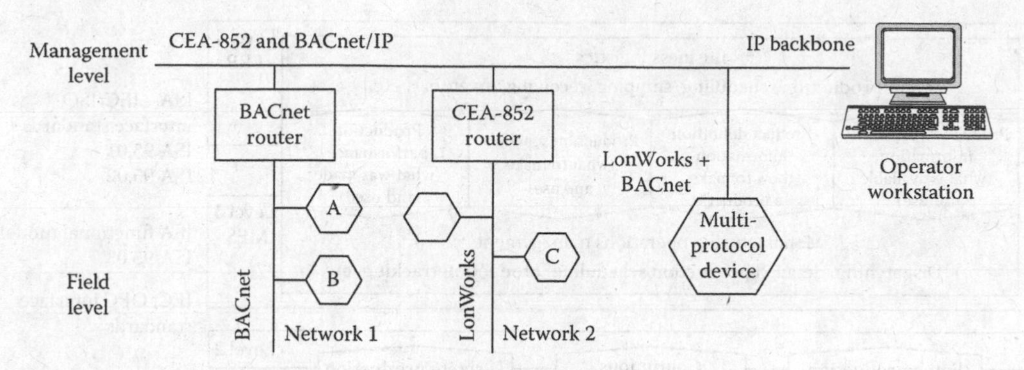

FIGURE 13.4 Protocol convergence using multi-protocol gateways.

technology (multi-protocol device). This way, the fieldbus node A can directly retrieve process data from the controller. Nodes B and C, which do not need access to the respective other network technology, can still be accessed by the operator workstation's native protocol. No additional gateway is needed as the area of overlap does contain the needed technologies. A given number of those multi-protocol devices can be seen as a distributed gateway. This means the SPOF has been eliminated. Also, the complexity of gateway configuration is reduced as the nodes know already about their data points natively. Clearly, not all nodes can be equipped with an arbitrary number of networking technologies.

13.4 Application View

An important aspect of vertical integration is the interconnection of the software application frameworks that are needed for the individual levels of the automation hierarchy. From today's view, the link between the top-level strategic planning and administration tools usually comprised in ERP systems and the distributed control systems on the factory floor is to be seen in the MES concepts that translate the overall, mainly long- and mid-term plans into short-term work and production orders. To achieve this integration, there is still work to be done [V02]. In the context of enterprise engineering, the current focus is on efficient modeling languages and techniques, on analysis methods, and advanced computer-based tools [MPZM03].

An eminent role is played by standardization. The more CIM and its successors found the approval of industry, the more it became apparent that only concerted actions could finally produce results with a high application potential. Standards generally provide the basis for durable solutions and finally products, which is particularly important in such complex areas as the integration of rather diverse functions inside an enterprise. After all, the problems of enterprise integration in actual practice are not so much the functional modules as such (the ERP, MES, SCADA systems in all their varieties) but the interfaces between them. Only standardized interfaces together with appropriate functional models for the application processes and their data objects ultimately allow for interoperability between solutions from different vendors.

Consequently, work being done in various standardization groups covers all aspects ranging from the provision of modeling frameworks and languages through concrete reference and data models for enterprise systems or subsystems down to standards for IT services and infrastructures [CV04]. Among the many standards developed in this context, the two most relevant for this chapter are the Manufacturing Execution Systems Association (MESA) model and, in particular, the ISA S95 Standard for Enterprise-Control System Integration, both of which have already been widely adopted as a reference for the implementation of vertical integration on the higher level of the automation hierarchy [S07].

MESA provided a definition of 11 distinct functional areas of MES that cover the basic information needed to run any type of plant—much in contrast to ERP systems that have no clearly defined functionalities. This model has become an essential starting point for many MES tools currently on the market. The purpose of ISA S95, on the other hand, is to define the interface between control functions

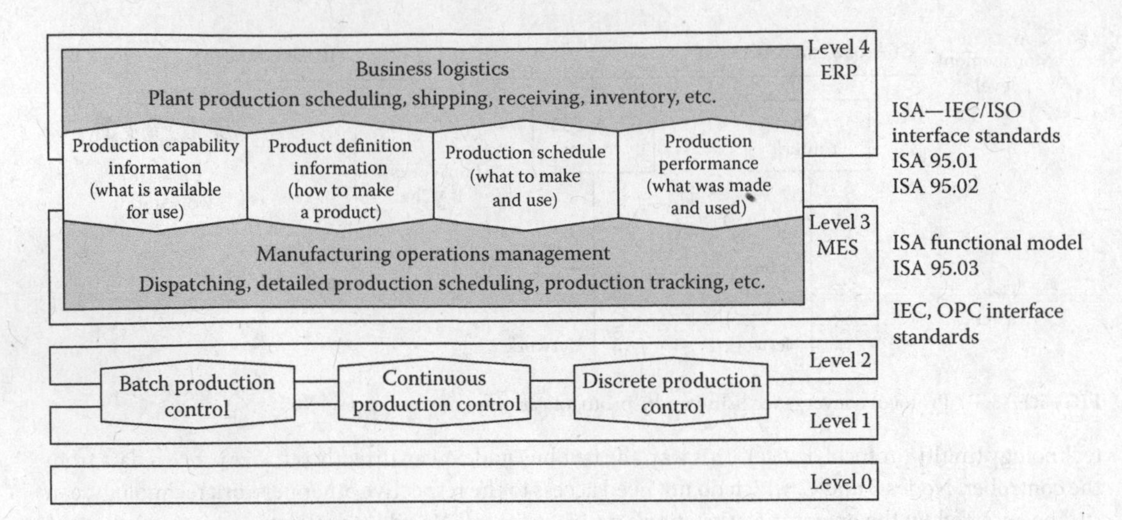

FIGURE 13.5 Scope of ISA 95 and ISA 88.

and other enterprise functions, effectively between the ERP and MES levels of an enterprise. The scope of the standardization work, as seen by the committee, is

- Define in detail an abstract model of the enterprise, including manufacturing control functions and business functions, and its information exchange.
- Establish common terminology for the description and understanding of enterprise, including manufacturing control functions and business process functions, and its information exchange.
- Define electronic information exchange between the manufacturing control functions and other enterprise functions including data models and exchange definitions.

ISA S95 addresses exclusively the information flow between the control domain and the enterprise domain of a company. Consequently, in the model adopted by ISA, the MES functionality is explicitly included in the control domain (Figure 13.5). The information exchanged between the levels can be categorized in information about the product itself (its definition), the availability of production capabilities, and the production of the product itself (scheduling and performance information). Functions concerning only the ERP level or the control level are beyond the scope of the standard and are therefore not covered.

ISA S95 should be regarded in connection with ISA S88 (Batch processing model), which covers the lower levels of the automation model. Both have been developed in parallel, and they are converging in that there are direct mappings between the two standards for models and representations. Likewise, the reference model suggested by MESA has influenced the work, and cross references between the two approaches are given in order to facilitate common understanding.

The standard does not describe an actual implementation; it provides only terminology, functional requirements, and formal models. Yet, inside the standardization group, the common belief was that XML might be a promising approach. Consequently, a working group was formed under the umbrella of the World Batch Forum, and developed the XML schema "Business to Manufacturing Markup Language" (B2MML). This reference implementation was released in a first version in 2002 and is constantly being updated and enhanced.

13.5 Security Aspects in Vertical Integration

Vertical integration in the sense of connecting automation systems to a more open environment naturally raises security questions. It has already been stated that even though network interconnection was an essential goal of the old CIM idea, fieldbus systems were mostly developed as isolated solutions.

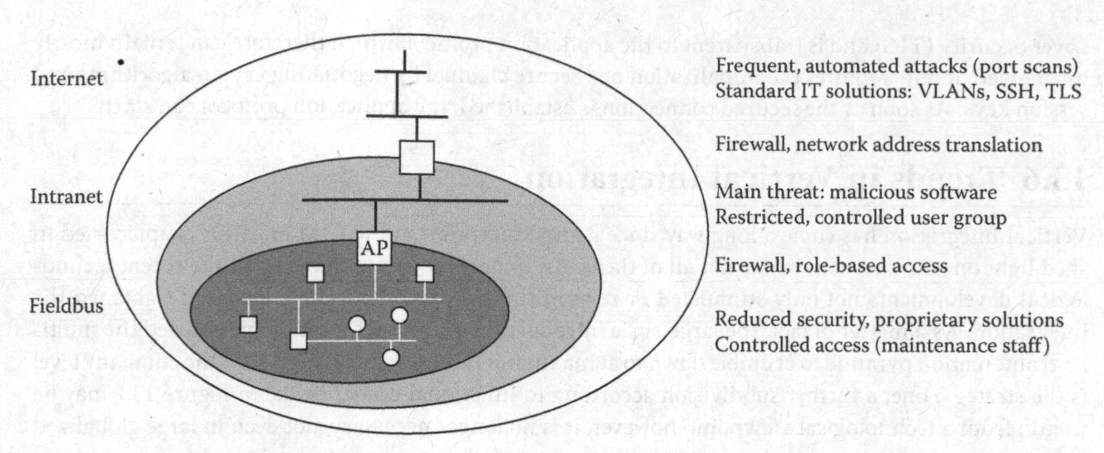

FIGURE 13.6 Defense-in-depth security model for vertically integrated company networks.

As the timeline in Figure 13.2 shows, most of these developments were concluded long before the Internet became known and was widely employed by the broad public. It is thus no wonder that security considerations never played a prominent role in automation before the end of the 1990s [D97,PS00].

The most promising approach to cope with the heterogeneity of the networks is a hierarchical security model following the defense-in-depth approach. Figure 13.6 shows the three typical interconnection zones in the company network. The zones are separated from each other by dedicated network nodes that can be used as anchor points for the security strategy. The inner connection node (access point) to the actual fieldbus is the gateway or proxy already discussed in Section 13.3.

As existing standards cannot be changed, the only viable option for the field level is to use the fieldbus simply as a transport channel and to tunnel secure packets over standard fieldbus protocols and services [SS02]. Such an application-level approach could easily achieve end-to-end security, which is desired in most applications, anyway. Unfortunately, it is likely to cause problems with interoperability unless all nodes in the network adhere to this enhanced standard, i.e., a mixture of secure and insecure devices would normally not be feasible. In addition, the limited messages size in some fieldbuses leaves only little room for the additional data blocks required by efficient security extensions. This problem is similar to the one encountered in IP tunneling discussed in Section 13.3.

The access point itself has so far been the focal point of interest for most researchers. Given the lack of general field-level security, it is the only part of the automation system (apart from the Internet side) where security can easily be applied. One very common approach is to combine the access point with a firewall, which is the most widely employed security measure in the design of network interconnections today even in the field of automation [D97]. Indeed, the term "firewall" is nowadays frequently used as a synonym for network security. In practice, however, their application is not so straightforward. Owing to the inherently asymmetric operation of a firewall (transparent from the private network, opaque from outside), it is difficult to place a standard firewall in front of a fieldbus access point; it is better to tightly integrate the firewall into the access point and use, e.g., port forwarding to control the traffic [PS00]. On top of this, the access point is also the ideal point to control and manage access to the fieldbus zone. A suitable model for handling access rights is role-based access control, where all communication partners (users, devices, tools) are associated with particular roles depending on the context of the data exchange. In the IT world, this concept is widely employed, and it can also be used in an automation context [WB04].

Security in the Internet finally is a well-researched topic with a large number of meanwhile very mature solutions. In fact, for nearly all Internet application layer protocols, secure versions exist, or conventional insecure protocols can be used on top of a secure transport layer. This standard mechanism of first establishing an encrypted channel between communication partners is called Transport

Layer Security (TLS) and is transparent to the application protocol, which therefore can remain mostly unchanged; it only requires the initialization of a secure channel by negotiating crypto algorithms and session keys. As soon as the secured connection is established, the application protocol can start.

13.6 Trends in Vertical Integration

Vertical integration has come a long way since its first inception in the CIM era. This chapter tried to shed light on some aspects, by far not all of them. An important recognition is that the recent technological developments not only stimulated or revived the old idea, they accelerated and tightened the integration. As a matter of fact, the large-scale introduction of Internet technologies caused the multi-level automation pyramid to crumble down to at maximum three levels (Figure 13.7). The company level is the strategic one; a further subdivision according to functional domains like in Figure 13.1 may be useful; from a technological viewpoint, however, it is no longer necessary, not even in large globalized enterprises. Internet has already provided the means to fully integrate remote branches in one single corporate network with one common management or planning framework. From the many operational levels, only two remain: the actual field level with the process control devices and an intermediate level typically for controlling purposes.

From the networking point of view, there are in fact only two levels left today. In the office world and the upper automation level, IP-based networks and Ethernet dominate. Even if fieldbus systems still have their share on the cell level, they are bound to vanish within the next few years. It is the field level that will still belong to them, maybe with some additional specialized and low-level local sensor/actuator systems integrated via controllers acting as simple gateways. Nevertheless, the recent emergence of real-time Ethernet extensions already sparked new competition in the field. In the long run, real-time industrial Ethernet will offer a tempting alternative to the classical fieldbus systems and will most probably make one or the other obsolete.

But, also from the application point of view, there is a trend toward a further reduction of the hierarchy. In plant automation, which was traditionally strictly centralized in the upper two levels, modular concepts are becoming popular. The overall goal is to enhance flexibility of planning systems and to allow for solutions tailored to the needs of the customer. This makes ERP and, especially, manufacturing execution tools attractive also for smaller companies that so far could not benefit from full-fledged solutions. One step ahead are approaches to decentralize essential parts of the planning and control functionality. Ideas like holonic manufacturing systems or agent-based automation practically dissolve the rigid middle layer of the pyramid, implementing its functionality in software agents attached to the outer levels of the hierarchy.

This tendency is supported by complementary trends in the control and enterprise field: In the long run, control devices will resemble standard PCs in terms of computing resources and network interfaces (Ethernet and IP). Thus, they will be able to run more complex planning tasks. On the other hand,

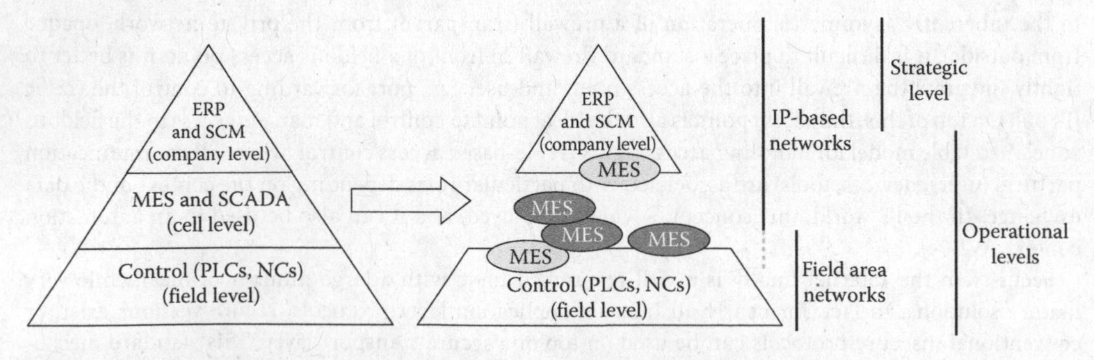

FIGURE 13.7 Decentralization of the middle level in a typical industrial automation context.

ERP frameworks will become even more comprehensive and include functional modules currently belonging to the MES level. Such a mostly distributed approach for the MES level will, however, impose more demanding requirements on the middleware to support better flexibility. Concepts emerging in the IT world like ontologies to formally describe semantic information will have to be used and additional abstraction levels must be introduced. Another emerging middleware concept possibly facilitating the interconnection of heterogeneous software frameworks are service-oriented architectures (SoA) that might ease the actual vertical integration.

The need for abstract interoperability layers is visible already today. Contrary to the beginning of the CIM era, many powerful technological building blocks are available: widely used protocol standards, middleware concepts, distributed application paradigms, databases, and last but not least, high-performance computers. But, the main problem has not much changed over the years: to find proper interfaces between lots of applications that are to a large extent still not interoperable. Nevertheless, interfaces alone are not sufficient to achieve interoperability. The semantic definitions for data exchange will require more effort. This is exactly why there are so many standardization activities in the automation area.

End users may have the impression that it took the emergence of standards like ISA S95 or S88 to stimulate actual product developments. Yet, the need for integration has been there before and has been the driving force behind all standardization efforts. It simply took a long time until the technological environment was mature enough to provide a solid basis. On the other hand, it appears that what has been reached today is only an intermediate step. All standardization activities had a clear focus, either from a technology or application domain point of view. The results are therefore optimized according to the respective goals, but they need to be properly aligned across similar or complementary activities. Major groups like ISA SP95, the OPC Foundation, and Machinery Information Management Open Systems Alliance (MIMOSA) already started joint efforts in this direction.

Throughout the chapter, vertical integration has mostly been treated in the context of actual plant operation. It should be noted, however, that also an automation system has a life cycle with a design phase, installation, and commissioning that need to be finished before the system becomes operational, not to forget maintenance and management during operation. Especially, the first phases of the life cycle are typically still supported by individual design frameworks that are not compatible with the one used later. One future goal of vertical integration will therefore be to include the entire life cycle in one consistent database with just different, phase-specific tools that change with the life cycle. Therefore, vertical integration in its ultimate all-inclusive form also gets a time aspect, and this aspect will gain importance in the future.

Abbreviations

CIM	Computer integrated manufacturing
CN/IP	Control network/IP
ERP	Enterprise resource planning
HTTP	Hypertext transfer protocol
HVAC	Heating, ventilation, air-conditioning
IP	Internet protocol
LAN	Local area network
MAP	Manufacturing automation protocol
MES	Manufacturing execution systems
MIMOSA	Machinery Information Management Open Systems Alliance
OPC	Open process control
SCADA	Supervisory control and data acquisition
SCM	Supply chain management
SoA	Service-oriented architecture
TCP	Transmission control protocol
TLS	Transport layer security

References

[BACIP] BACnet—A Data Communication Protocol for Building Automation and Control Networks, ANSI/ASHRAE Std. 135, American Society of Heating, Refrigerating and Air-Conditioning Engineers, Atlanta, GA, 2004.

[CEA852] Tunneling Component Network Protocols over Internet Protocol Channels, ANSI/EIA/CEA Std. 852, Rev. A, Consumer Electronics Association, 2002.

[CV04] D. Chen and F. Vernadat, Standards on enterprise integration and engineering—State of the art, *International Journal of Computer Integrated Manufacturing*, 17, 235–253, 2004.

[D97] M. S. DePriest, Network security considerations in TCP/IP-based manufacturing automation, *ISA Transactions*, 36, 37–48, 1997.

[DPAL] W. Burgstaller, S. Soucek, and P. Palensky, Current challenges in abstracting datapoints, in *Proceedings of 6th IFAC International Conference on Fieldbus Systems and their Applications*, Puebla, Mexico, pp. 40–47, 2005.

[DSP88] J. N. Daigle, A. Seidmann, and J. R. Pimentel, Communications for manufacturing: An overview, *IEEE Network*, 2, 6–13, 1988.

[H73] J. Harrington, *Computer-Integrated Manufacturing*, Industrial Press, New York, 1973.

[KNXIP] Open Data Communication in Building Automation, Controls and Building Management Systems—Konnex Network Protocol—Part 2: KNXnet/IP, EN 13321-2, European Committee for Standardization, 2006.

[ModIP] Modbus Messaging on TCP/IP Implementation Guide, Version 1.0a, June 2004. Available: http://www.modbus-ida.org

[MPZM03] G. Morel, H. Panetto, M. Zaremba, and F. Mayer, Manufacturing enterprise control and management system engineering: Paradigms and open issues, *Annual Reviews in Control*, 27, 199–209, 2003.

[PS00] P. Palensky and T. Sauter, Security considerations for FAN-Internet connections, *IEEE Workshop on Factory Communication Systems (WFCS)*, Porto, Portugal, pp. 27–35, September 6–8, 2000.

[RO03] M. Rudberg and J. Olhager, Manufacturing networks and supply chains: An operations strategy perspective, *Omega*, 31, 29–39, 2003.

[S05] T. Sauter, Fieldbus systems—History and evolution, in R. Zurawski (editor), *The Industrial Communication Technology Handbook*, CRC Press, Boca Raton, FL, pp. 7-1–7-39, 2005.

[S07] T. Sauter, The continuing evolution of integration in manufacturing automation, *IEEE Industrial Electronics Magazine*, 1, 10–19, 2007.

[SS02] C. Schwaiger and T. Sauter, Security strategies for field area networks, *Annual Conference of the IEEE Industrial Electronics Society (IECON)*, Sevilla, Spain, pp. 2915–2920, November 5–8, 2002.

[V02] F. B. Vernadat, Enterprise modeling and integration (EMI): Current status and research perspectives, *Annual Reviews in Control*, 26, 15–25, 2002.

[WB04] M. Wollschlaeger and T. Bangemann, Maintenance portals in automation networks—Requirements, structures and model for web-based solutions, *IEEE Workshop on Factory Communication Systems (WFCS)*, Vienna, Austria, pp. 193–199, September 22–24, 2004.

[WV03] T. Werner and C. Vetter, From order to production: A distinct view on integration of plant floor and business systems, *IEEE International Conference on Emerging Technologies and Factory Automation (ETFA)*, Lisbon, Portugal, pp. 276–281, September 16–19, 2003.

14

Multimedia Service Convergence

14.1 Introduction 14-1
14.2 Background.......... 14-2
Computer Telephony Integration • Voice over Internet Protocol • Interactive Voice Response Telecom
14.3 Service-Oriented Architecture 14-3
Service Meta Architectures
14.4 Tailorability 14-5
14.5 Multimedia Convergence Using Service Architecture 14-5
Converged Services • MC^2 Interact Reconfigurable Convergence Architecture • Converged Voice Plugins
14.6 Conclusion 14-11
References 14-12

Alex Talevski
Curtin University of Technology

14.1 Introduction

The advent of personal computers (PCs) and the Web have provided distributed mechanisms for information sharing while telephones serve almost exclusively as communication devices [1,2]. Currently, enterprises utilize a heterogeneous mixture of

- **Voice Networks**—Voice telephony networks are used to perform daily telecommunications activities mostly over the public switched telephone network and global system for mobile networks.
- **Data Networks**—Internet protocol (IP) wired and wireless networks provide a data communication medium for access to Internet and intranet data services and applications.
- **Enterprise Applications**—Enterprise software applications allow users to interface with and process information.
- **Data Repositories**—Databases and other repositories provide structured storage facilities for enterprise data.
- **Devices**—Wireless and wired telephones, computing devices, and sensors are used to visualize, vocalize, and process enterprise data, applications, and services.

In the IT&T world, convergence refers to the move toward the use of a single united interaction medium and media as opposed to the many that we use today [3]. Convergence enables telecommunications services that are concurrently coupled with enterprise and Internet data. The ability to visualize a concept via images, graphs, tables, and procedures while communicating over the telephone or a video link greatly enhances interaction. Interaction is more pleasing, meaningful, effective, and efficient. Furthermore, such media-rich services and the devices that support them allow for actions to be taken while on the move with greater precision and faster response to market drivers. Therefore, data and telecommunications convergence promises a wide range of possible solutions that will increase productivity and flexibility, and provide new on-demand opportunities for enterprises.

Converged voice and data services have rapidly emerged as a popular alternative to existing telecommunications networks and computer services. Many sources [4–7] indicate that converged voice and data networks of various forms are rapidly growing across industry in the last 5 years. However, converged telecommunications and data services have been largely isolated to static environments where fixed PCs and network connections are used in conjunction with various software tools that simulate pseudo-converged sessions. Generally, data presented on the Internet and in enterprise applications is not available on voice networks and devices and vice versa.

From an enterprise perspective, converged voice and data services and interaction mediums to its employees, partners, and customers is essential. Due to the diverse nature of this environment, a converged solution that employs the features outlined above is required. The following features are essential:

- **Multimedia**—Converged voice, video and data services, and interaction mediums are required as a consequence of the increased flexibility that businesses demand [8]. Such a collaboration environment is more pleasing, meaningful, effective, and efficient.
- **Accessibility**—The demand for media- and function-rich services on the move is rising. The accessibility of converged services via a variety of mobile devices is essential. Such devices (TV, PC, PDA, mobile phone, and others), voice/data transportation mediums (IP, Wi-Fi, Bluetooth, GPRS, UMTS, etc.), and a variety of protocols and encodings (H323, SIP, IAX, etc.) must be considered.
- **Feature-Rich Services**—Combined telecommunication functions such interactive voice response (IVR) with services that access enterprise and Internet data such as databases, applications, and Web services via multiple interfaces. Access to enterprise repositories, applications (databases, ERP, CRM, and others) and Internet data services (e-mail, news, weather, stocks, and others) is essential.
- **Multi-Interface**—Being able to visualize the context of a conversation greatly enhances interaction. Voice, video, and data interfaces project service data and system/user interaction. Dynamic interaction must be interchangeably available through any available device interface (DTMF, keyboard/mouse, etc.) and/or through voice, video, and/or data-driven commands.
- **Flexible and Adaptive Environment**—A problem faced in developing a feature-rich, flexible, and widely accessible solution is the dynamically evolving system requirements that occur on the disparate layers of numerous telecommunications and computer services. Reconfigurable software architectures promote simplified software evolution in such complex environments.

14.2. Background

14.2.1 Computer Telephony Integration

Telephone and computer systems are two technologies that impact many aspects of our daily lives. These technologies drive the world's economy and are central to the operation of virtually every enterprise. A large number of organizations can only exist with the support of these technologies. Computer telephony integration (CTI) is defined as the integration between computers and telephony systems [1]. CTI technologies bridge the features of computers such as data handling, media processing, and graphical user interface with telephone features such as call handling and routing. Currently, CTI is predominantly used to drive software-based Private Automatic Branch eXchange (PABX) telecommunications systems. However, CTI is heading toward the convergence of both data and voice services over data networks.

14.2.2 Voice over Internet Protocol

Voice over Internet protocol (VoIP) (also termed IP telephony) refers to the transport of voice traffic over a data network. VoIP hardware and software acts as an Internet transmission medium for telephone calls and other converged data. Using VoIP, carrier grade voice communication is digitized and routed in discrete IP packets through a broadband connection. VoIP is particularly useful when there is limited or financially prohibitive access to alternative telephony networks. Telephone calls can be transmitted with little or no loss in functionality, reliability, or voice quality. VoIP has rapidly emerged as a popular alternative to existing telephony networks [3–6]. However, to date, VoIP has been a solution that is mostly used as an alternative medium to carry out cost-effective, long-distance telephone calls. The following services are currently available:

- Telecommunication can be performed on numerous devices (TV, PC, PDA, mobile phone, and others).
- Various voice/data transportation mediums can be used (IP, Wi-Fi, Bluetooth, GPRS, UMTS, etc.).
- A variety of transportation protocols and encodings exist (H323, SIP, IAX, etc.).
- Custom call routing and messaging based on VoIP information can be configured.

As an example, by using caller ID, it is possible to automatically retrieve a caller's details and records from a data repository while the call is routed to an appropriate party. The call can then be transferred to a different network and telephony device. While the user is on hold, he or she can be provided with personalized enterprise and Internet information using a simplified voice interface. A popular CTI solution is an IVR telecomsystem.

14.2.3 Interactive Voice Response Telecom

IVR systems provide computer-controlled telephone answering and routing functions, as well as facilities for the collection and provision of information. Interactive voice and keypad-driven menus allow callers to request information and respond to prompts. IVR devices route calls to the appropriate place or data, based on developer-defined configuration. These systems may use a mixture of human and computer interaction that is provided live or prerecorded by an attendee or digitally synthesized sound to convey data to the caller.

Unfortunately, industry has failed to make the most of CTI, VoIP, and IVR technologies. Current use of these technologies has been limited to purely telephony applications like telephone conversations and conferencing. Existing systems rarely provide a flexible and integrated approach where more than telecommunication services are provided. A flexible approach that integrates the features of CTI, VoIP, and IVR technologies is required. Service-oriented architectures (SOA) are the key to providing flexible, converged solutions for enterprise applications.

14.3 Service-Oriented Architecture

Complex industry systems account for the majority of all software development undertakings [9]. However, the software development industry tends to develop large-scale systems that are monolithic, error prone, and expensive [10–12]. Such systems normally evolve from an uncoordinated build-and-fix pattern. A lack of a systematic approach to software development has resulted in many project failures [10–12].

The ability of software to adapt or be adapted to the typical changes in enterprise environments is referred to as the software's flexibility [13]. A flexible system is needed due to the requirement for a system to be deployed in diverse roles and environments (its diversity) and to be flexible to unclear, lacking, and evolving information regarding these roles and environment (uncertainty) [13]. Versatile systems exhibit generic and function-rich properties. Growing enterprises demand rapid and frequent development, maintenance, customization, and evolution. Such software needs to adapt to changes in requirements in order to support its enterprise.

Service-based software engineering is a way of raising the level of abstraction for software development so that software can be built by easily reusing previously designed, implemented, and refined converged services. Composite component architectures are formed from converged services where components at the lower layers provide component services to the components above them. A SOA is a way of connecting loosely coupled converged application services across a network via common interaction primitives. Services map distinct business processes that can be chained together in order to realize certain collaborative behavior.

14.3.1 Service Meta Architectures

The Meta data model illustrated in Figure 14.1 represents an explicit Meta specification of a single software component and its interface. This explicit model supports software reconfiguration by clearly defining application constructs and their interface. Such data is used to specify construct convergence and interaction. External entities (other applications and constructs) may access this Meta data in order to utilize service constructs and their compositions. A composition is typically formed from many interconnected components that are constructed in a layered and hierarchical manner. Connectors and adaptors provide a level of indirection that reduces dependencies among components. Interconnected compositions are coupled using operation and attribute connectors and adaptors [14].

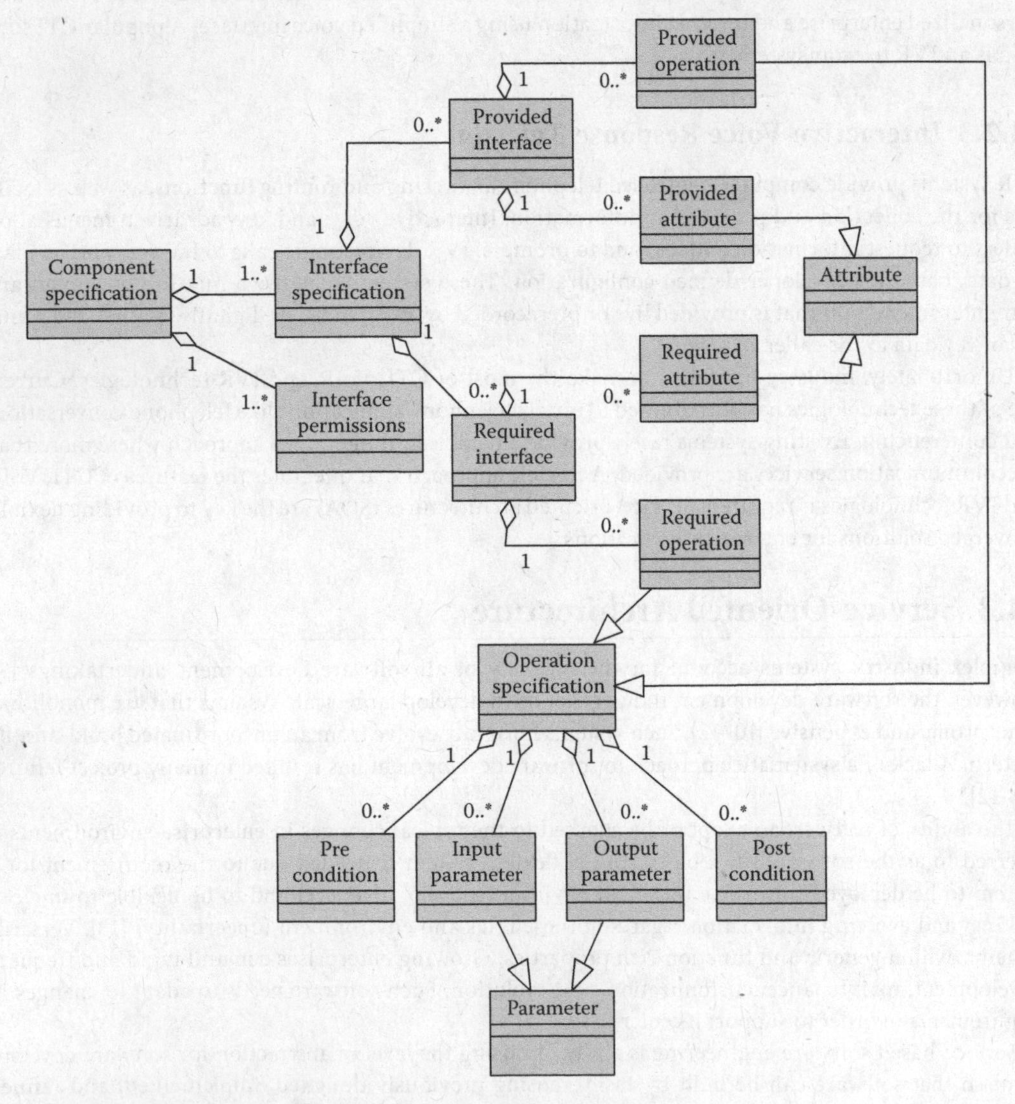

FIGURE 14.1 Conceptual Meta model of a component.

A component may expose a number of interfaces. Each interface either provides or requires services in the form of data attributes and operations. An attribute is a named property value that defines the characteristics and state of a component. An operation is the implementation of a specific service that represents the dynamic behavior of a component.

14.4 Tailorability

Tailorable software allows generic software to be easily constructed and/or modified to satisfy the specialized, rapidly changing, unclear, and/or evolving changes in system requirements. It provides a means for the straightforward creation and modification of software at runtime based on multiple levels of detail and complexity.

Morch [15] has identified three tailoring activities where the system user interface and underlying logic can be modified dynamically:

- **Customization**—Customization refers to user interface modifications.
- **Integration**—The integration level of tailoring refers to changes in application components, compositions, interconnections, and their configuration.
- **Extension**—The extension level of tailoring refers to the addition of new application components and configurations.

Using a reconfigurable plug and play component-based framework as a basis for the creation and modification of software, it is possible to construct, customize, integrate, and evolve convergence solutions.

14.5 Multimedia Convergence Using Service Architecture

Reconfigurable converged applications eliminate traditional barriers by offering a new and novel solution that allows flexible access to enterprise and Internet data. They provide the glue between enterprise business applications and multimodal forms of converged business communication to access a variety of telecommunications and data services Connections may be achieved using plain old telephone service (POTS) telephones, cell phones, PCs, and/or personal digital assistants (PDAs). Multimedia convergence promotes media collaboration using a variety of communication devices and networks anywhere and at any time. It enables individuals and teams to interact in a powerful, simple, and convenient manner. The following approach uses a thin client style that allows system access via a range of devices and connections with no specialized software. Access is provided online via an Internet browser-based client. Calls, conferences, and services are provided over wired and wireless telephony and data networks as routed by a software Private Branch eXchange (PBX). Interaction can be performed using a Web, voice, SMS, video, data, and instant messaging interfaces. Customized switchboard forwarding and voice mail features are used to manage calls as required. Whiteboarding, meeting, and presentation management features are also used to enhance collaboration. Figure 14.2 illustrates a high-level overview of such a convergence approach.

The multimedia convergence center (MC^2) convergence solution is proposed using proven open concepts with a focus on wide compatibility. No specialized hardware devices are deemed to be required. Computer-aided interaction is performed using a Web instant messaging client and VoIP capability that communicates with a central convergence server via an asynchronous approach. The approach illustrated in Figure 14.3 is adopted in order to promote a thin client solution with wide compatibility.

Telecommunications functions are provided by a software PBX solution that couples the features of computers with a telephone interface (Figure 14.4).

14.5.1 Converged Services

Traditionally, enterprise applications allow users to access corporate data in a static location using a PC. However, today's business professionals have a busy schedule and they are frequently on the move.

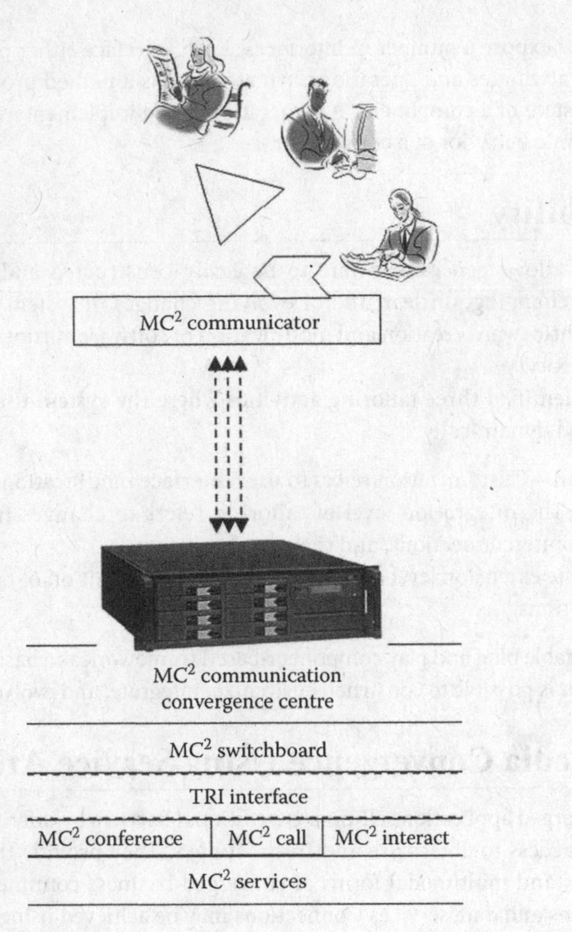

FIGURE 14.2 Multimedia convergence.

Telecommunications and data convergence is required as a consequence of the increased flexibility that businesses demand [8]. An IVR system is programmed using reconfigurable plug and play software architecture [14] to dynamically compose, integrate, and tailor voice access to data services. System interaction is performed using voice commands or touch-tones.

Figure 14.5 illustrates high-level architecture of the Interact portion of the MC² proposal [16]. It provides access to enterprise and Internet data using an IVR system and concurrent Web interface. The plugin services used here interact with the user using voice prompts and/or a Web interface and access other components, services, and repositories as required. Sample vE-mail, vFinance, vStocks, vWeather, and vNews plugin services are illustrated.

14.5.2 MC² Interact Reconfigurable Convergence Architecture

The following framework promotes simplified software construction, customization, integration, and evolution of convergence solutions. The framework allows the solution to easily integrate existing enterprise and Internet applications and newly implemented components as communication/interaction services. Services may be added and removed dynamically as per business requirements. This allows for prompt awareness and response to enterprise triggers.

At the highest level, MC² Interact host behaves as an IVR entrypoint. As illustrated below (Figure 14.6), voice plugin discovery, query, identification, and invocation are used to situate, define, describe, and utilize available converged services. Once a voice plugin has been identified, it is assigned a telephone

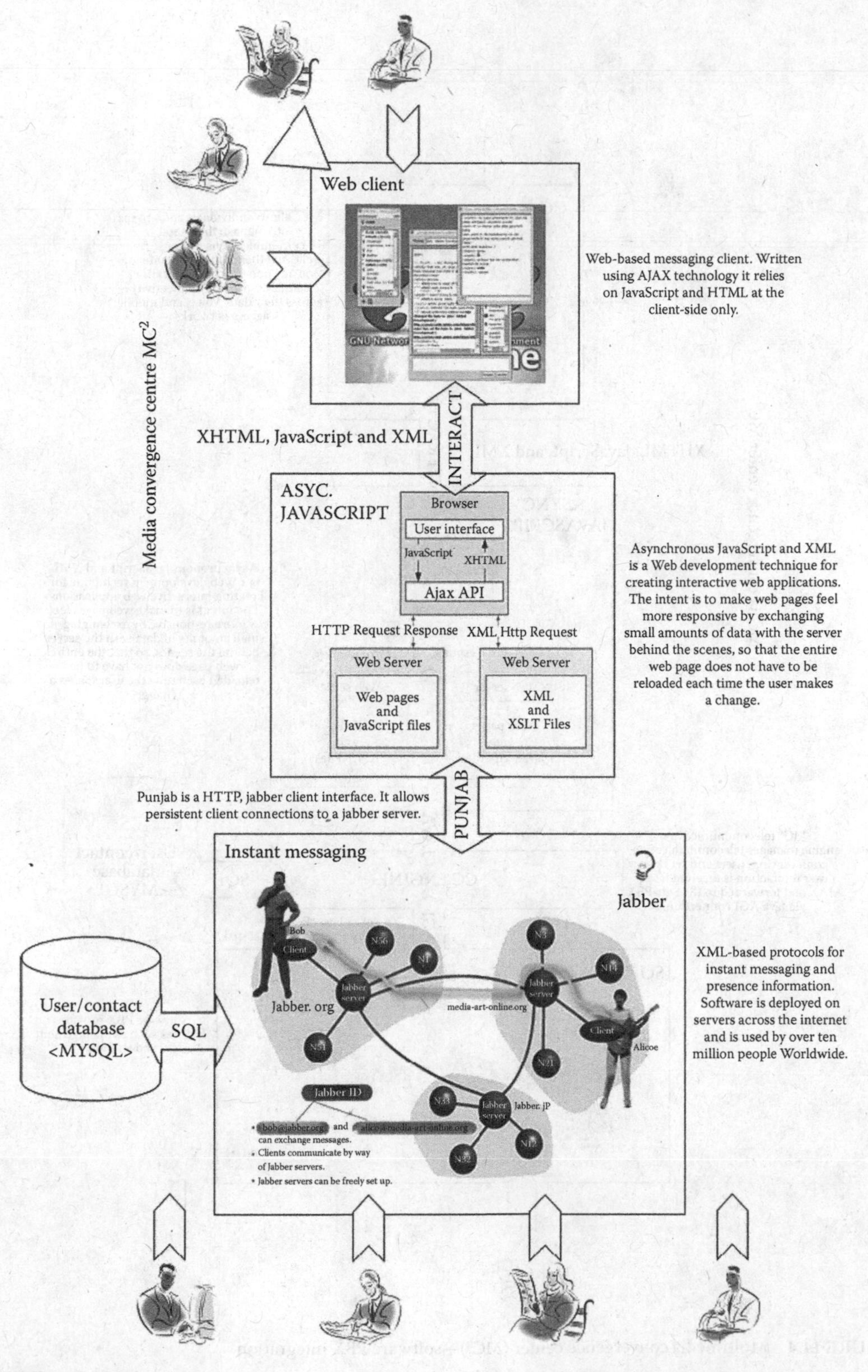

FIGURE 14.3 Multimedia convergence center (MC²)—computer-aided interaction.

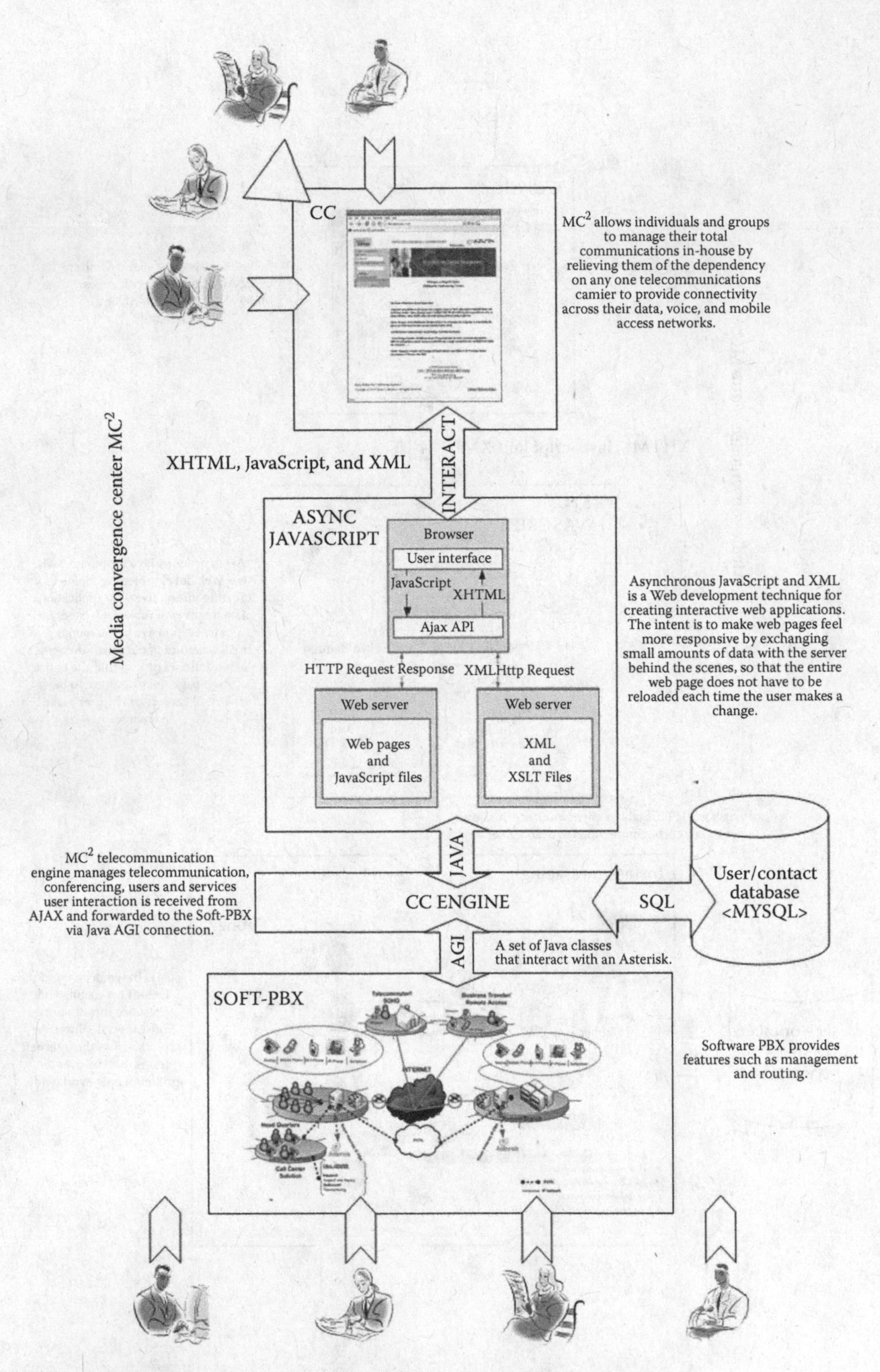

FIGURE 14.4 Multimedia convergence center (MC^2)—software PBX integration.

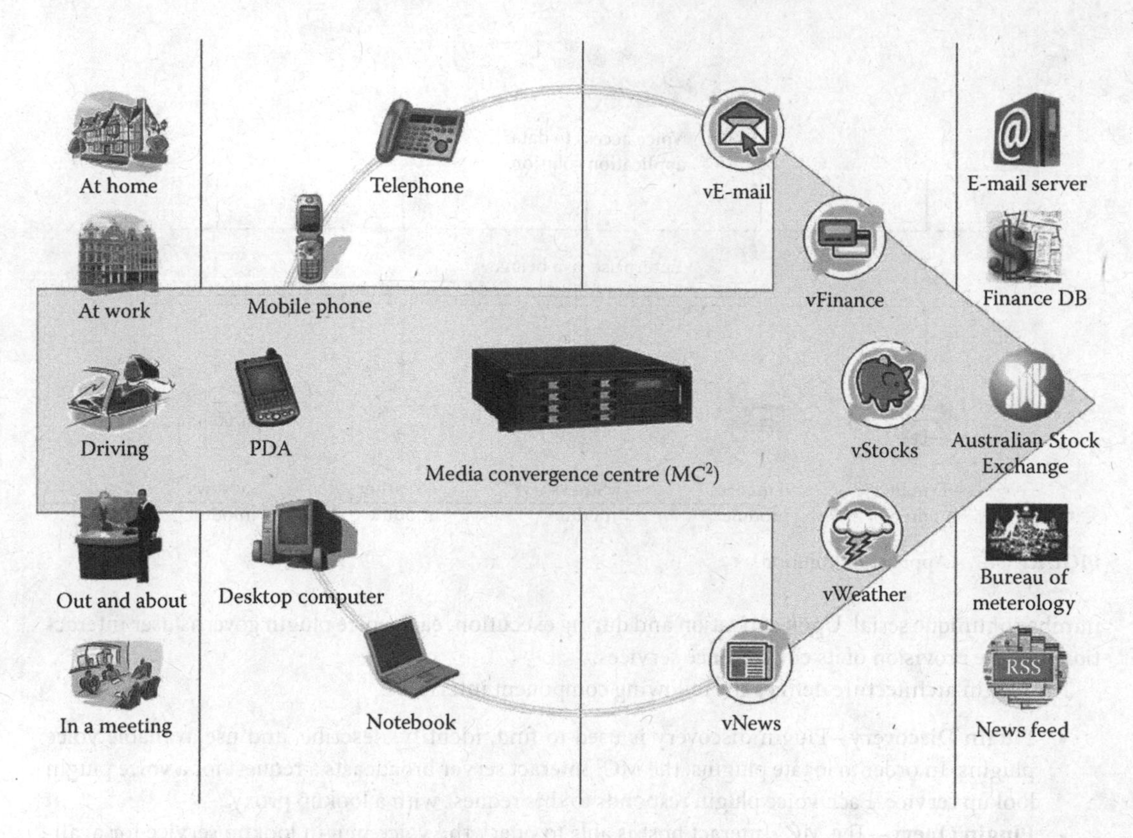

FIGURE 14.5 Converged voice access to data solution.

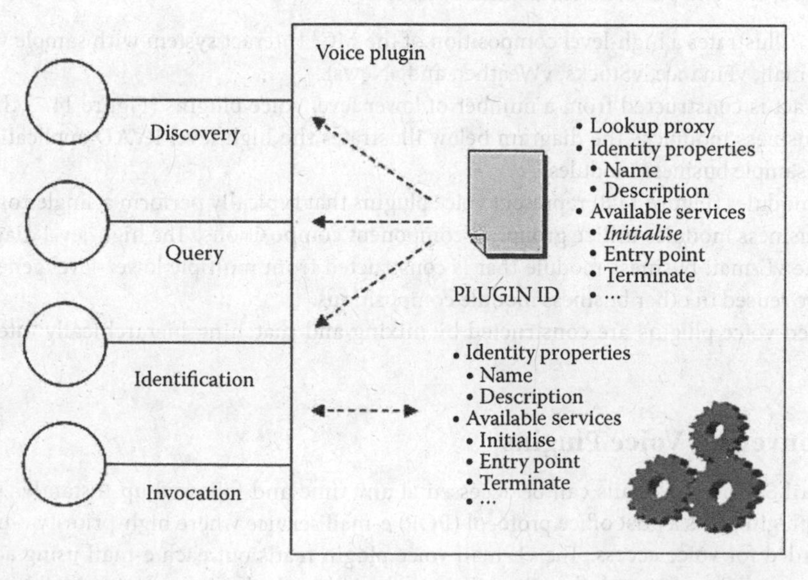

FIGURE 14.6 Voice plugin.

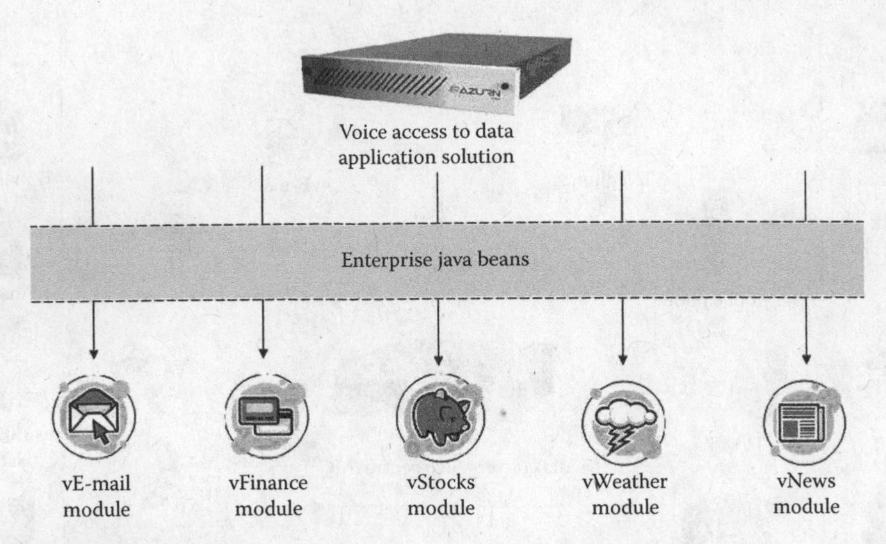

FIGURE 14.7 Application solution.

number or unique serial. Upon activation and during execution, each voice plugin governs user interaction and the provision of its convergence services.

The plugin architecture defines the following component interfaces:

- **Plugin Discovery**—Plugin discovery is used to find, identify, describe, and use available voice plugins. In order to locate plugins, the MC^2 Interact server broadcasts a request for a voice plugin lookup service. Each voice plugin responds to this request with a lookup proxy.
- **Plugin Query**—The MC^2 Interact host is able to query the voice plugin lookup service for available services.
- **Plugin Identification**—The voice plugin lookup service is used to define voice plugin characteristics.
- **Plugin Invocation**—When a voice plugin is selected via the IVR, the MC^2 Interact host dynamically binds to the voice plugin and invokes its entrypoint. The voice plugin then takes over interaction control and performs its identified services.

Figure 14.7 illustrates a high-level composition of the MC^2 Interact system with sample voice plugin services (vE-mail, vFinance, vStocks, vWeather, and vNews).

MC^2 Interact is constructed from a number of lower level voice plugins (Figure 14.7) that are fully composed business modules. The diagram below illustrates the high-level RVAD application solution with a set of sample business modules.

Business modules (Figure 14.8) represent voice plugins that typically perform a single converged service. Such business modules utilize groups of component compositions. The high-level diagram below illustrates the vE-mail business module that is constructed from multiple lower-level generic components that are reused in other business module compositions.

Customized voice plugins are constructed by mixing and matching hierarchically interconnected components.

14.5.3 Converged Voice Plugins

- **vE-Mail Service**—E-mails can be accessed at any time and followed up instantly. The vE-mail voice plugin hosts a post office protocol (POP) e-mail service where high-priority e-mails can be forwarded for voice access. The vE-mail voice plugin reads out each e-mail using a clear voice. The user may interact with the vE-mail voice plugin by telephone key tones. MC^2 Interact allows

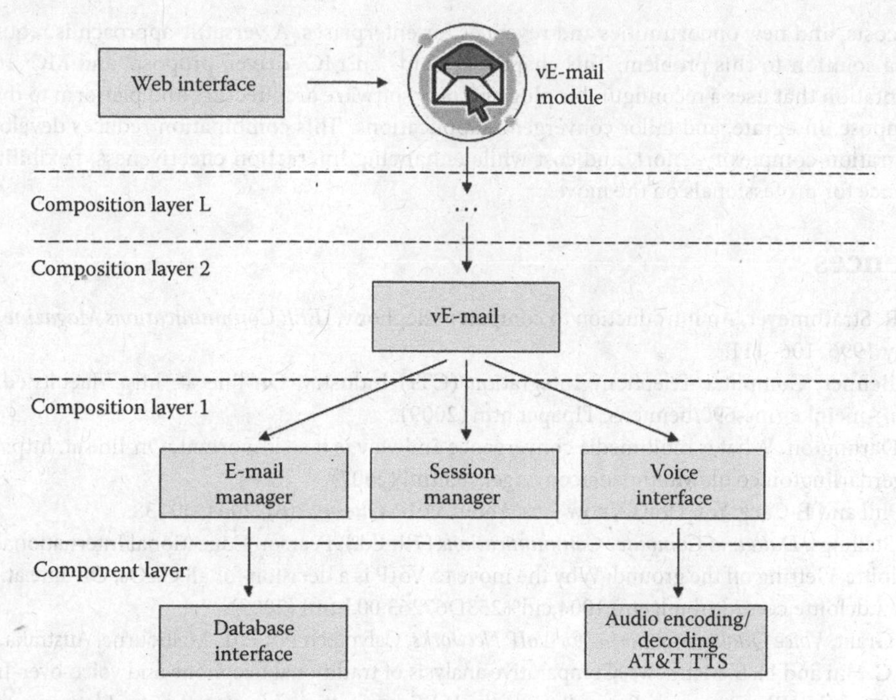

FIGURE 14.8 Business module.

the automatic browsing of e-mails without user intervention. It is possible to reply to e-mails with predefined e-mail templates and forward messages to predefined contacts immediately. Users may manage their e-mail messages by saving, moving, and deleting selected items. It is also possible to customize the way that the system performs and to manage e-mail contacts and template messages via an easy-to-use Web interface.

- **vStocks Service**—Live Australian Stock Exchange (ASX) values can be heard at the user's convenience. The vStocks service reads out detailed information on each user's individually predefined stocks. Stock list navigation is performed using telephone key tones. MC^2 Interact allows the browsing of stock data without user intervention. The vStocks service is able to announce each stock's trade date, time, change, previous close, day high, day low, and volume. Users may customize the stock properties they wish to hear to suit their individual preferences.
- **vWeather Service**—Live Bureau of Meteorology (BOM) weather forecasts can be accessed at any time. The vWeather service reads out detailed weather information for a user's predefined city or town. Weather forecasts are read out for up to one week in advance. Forecast information, days high, and days low are given for each day. Users may customize their city to suit their individual preferences and travel arrangements. MC^2 Interact allows the weekly weather forecast to be read out without user intervention.
- **vNews Service**—Live Rich Site Summary (RSS) news feeds can be heard at a preferred occasion based on a user's preference. The vNews service reads out each news item as requested by the user's telephone key tone interaction or automatically.

14.6 Conclusion

Widespread use of mobile computing devices, telephones along with Internet services has resulted in a broad range of ways to access information and to communicate. However, true voice and data convergence has been isolated to the use of various software tools that simulate a pseudo-converged environment. Converged services that are accessible anywhere and anyhow promise increased productivity, reduced costs, and new opportunities and revenues for enterprises. A versatile approach is required to produce a solution to this problem. This chapter presents an MC^2-driven proposal and MC^2 Interact implementation that uses a reconfigurable plug and play software architecture and platform to dynamically compose, integrate, and tailor convergence applications. This combination reduces development and integration complexity, effort, and cost while enhancing interaction effectiveness, flexibility, and convenience for professionals on the move.

References

1. C. R. Strathmeyer, An introduction to computer telephony, *IEEE Communications Magazine*, 35(5), May 1996, 106–111.
2. B. Benner, Computer Telephony Integration (CTI) Industry, On-line at: http://faculty.ed.umuc.edu/~meinkej/inss690/benner/CTIpaper.htm (2009).
3. R. Darlington, What is multimedia convergence and why is it so important, On-line at: http://www.rogerdarlington.co.uk/Multimediaconvergence.html (2007).
4. S. Phil and F. Cary, You Don't Know Jack About VoIP, *Queue*, 2(6), 2004, 30–38.
5. W. Stallings, *Data and Computer Communications* (7th Ed.), Pearson Educational International, 2004.
6. Deloitte, Getting off the ground: Why the move to VoIP is a decision for all CXOs, On-line at: http://www.deloitte.com/dtt/budget/0,1004,cid%253D67263,00.html (2009).
7. M. Grant, *Voice Quality Monitoring for VoIP Networks*, Calyptech Pty. Ltd., Melbourne, Australia, 2005.
8. M. C. Hui and H. S. Matthews, Comparative analysis of traditional telephone and voice-over-Internet protocol (VoIP) systems, in *Proceedings of the IEEE International Symposium on Electronics and the Environment*, 2004.
9. ATP FOCUSED PROGRAM: Component-Based Software, On-line at: http://www.atp.nist.gov/focus/cbs.htm (2009).
10. M. Doane, The Overwhelming Failure of Go-It-Alone CRM, Meta Group, On-line at: http://www.metagroup.com/us/displayArticle.do?oid=35932 (2002).
11. I. Sommerville, G. Dewsbury, K. Clarke, and M. Rouncefield, Dependability and trust in organisational and domestic computer systems, in *Trust in Technology: A Socio-technical Perspective*, Kluwer, 2004.
12. D. Parnas, The influence of software structure on reliability, in *Current Trends in Programming Methodology: Software Specification and Design*, Prentice-Hall, Englewood Cliffs, NJ, 1977.
13. G. Booch, *Object-Oriented Analysis and Design with Applications* (2nd Ed.), Benjamin/Cummings, Redwood City, CA, 1994.
14. A. Talevski, E. Chang, and T. Dillon, Re-configurable web services for extended logistics enterprise, *IEEE Transaction on Industrial Informatics*, 1(2), 74–84, May 2005.
15. A. Morch, *Three Levels of End-User Tailoring: Customisation, Integration, and Extension, Computers and Design in Context*, The MIT Press, Cambridge, MA, pp. 51–76, 1997.
16. A. Talevski, E. Chang, and T. Dillon, Converged voice access to data (CVAD), in *Proceedings of the International Conference on Convergence Information Technology*, Gyeongju, Korea, November 2007.

15

Virtual Automation Networks

Peter Neumann
Institute for Automation and Communication

Ralf Messerschmidt
Institute for Automation and Communication

15.1 Introduction 15-1
15.2 Virtual Automation Network: Basics 15-3
Domains • Components • VAN System Architecture
15.3 Name-Based Addressing and Routing, Runtime Tunnel Establishment 15-7
15.4 Maintenance of the Runtime Tunnel Based on Quality-of-Service Monitoring and Provider Switching 15-9
15.5 VAN Telecontrol Profile 15-11
Abbreviations 15-13
References 15-14

15.1 Introduction

Various tendencies influence the development of industrial communications:

- Industrial automation projects become more complex
- The offered communication technologies become more comprehensive
- Stronger integration of enterprise (business) and plant automation
- Striving for non-interrupted engineering and tool chains

Future scenarios of distributed automation lead to desired mechanisms for geographically distributed automation functions due to various reasons:

- Centralized supervisory and control of (many) decentralized (small) technological plants
- Remote control, commissioning, parameterization, and maintenance of distributed automation systems
- Inclusion of remote experts or external machine-readable knowledge for the plant operation and maintenance

Since the infrastructure for the office automation as well as for the factory and process automation are merging, all the advantages but also drawbacks of Ethernet-TCP/IP have to be taken under consideration. The communication infrastructure is becoming more and more heterogeneous. Thus, this can be characterized by the expression "heterogeneous networking." The end-to-end communication between geographically distributed automation functions, performed in devices with large distances between them, is becoming increasingly important. The data has to pass through a chain of very different types of communication technologies (circuit or packet switched, wired or wireless, provider oriented or provider less, public or private, etc.). The growing complexity of the necessary industrial communications led in the past to different technical approaches. In the fieldbus technology, different fieldbus networks

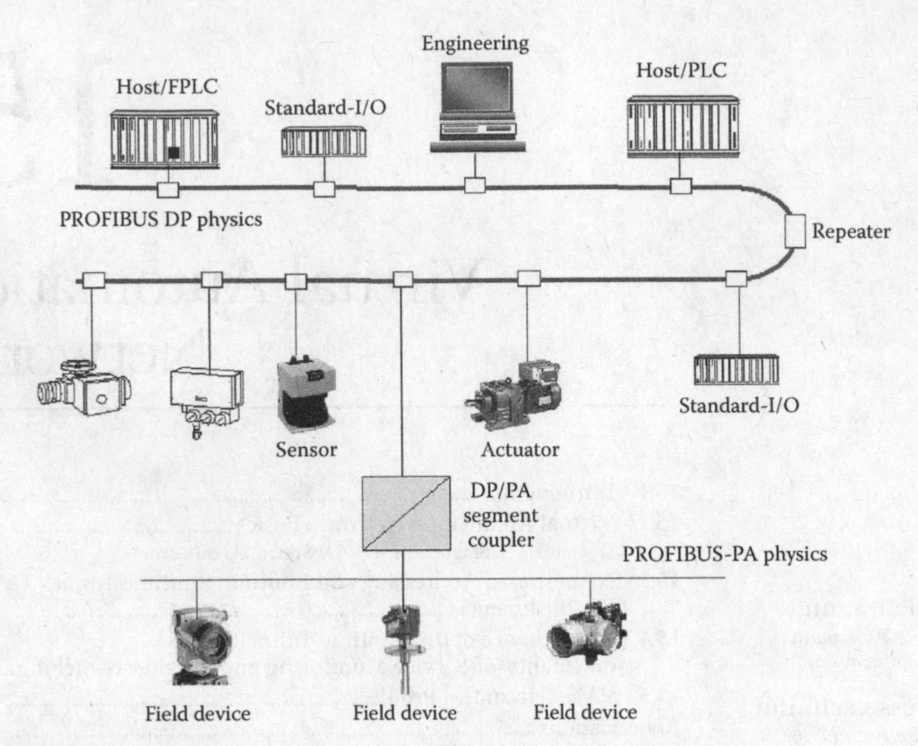

FIGURE 15.1 Coupling of different fieldbus technologies (physics) using link devices.

have been coupled by specific link devices. Thus, different transmission technologies, particularly for various application areas, could be used in a complex application. For example, as shown in Figure 15.1, PROFIBUS-PA using intrinsic-safe and non-intrinsic-safe fieldbus parts, coupled via link devices (DP/PA segment coupler), mapping different stack layers and application object models.

The Ethernet-based fieldbus technology (e.g., PROFINET [IEC61158,IEC61784]) opened the opportunity to integrate real-time Ethernet approaches into Office and Internet applications, and thereby the real-time Ethernet approaches opened the possibility to integrate different fieldbus technologies. As a result, PROFINET network serves as an automation backbone and enables the connection of various fieldbus systems with different fieldbus technologies and the exchange of data between the fieldbus-specific application objects [PIP] (Figure 15.2).

There are various integration concepts (e.g., in terms of PROFINET) and integration mechanisms (mapping of layer functions, proxy concept, mapping of application objects between fieldbus and PROFINET, and architecture of necessary link devices):

- *Transparent integration*: Each fieldbus device (including its data objects) will be mapped to one PROFINET-IO device.
- *Modular integration*: Each fieldbus device (including its data objects) will be mapped to a module of a PROFINET-IO device.
- *Compact integration*: A complete fieldbus system (including its data objects) will be mapped to a module of a PROFINET-IO device.

These integration concepts have been focused on locally concentrated industrial communications and include the opportunity to integrate the plant automation world into the office automation and Internet application world.

The next step to handle the growing complexity of automation systems (geographically distributed applications) requires the use of heterogeneous wide area networks (WANs) [CPP06].

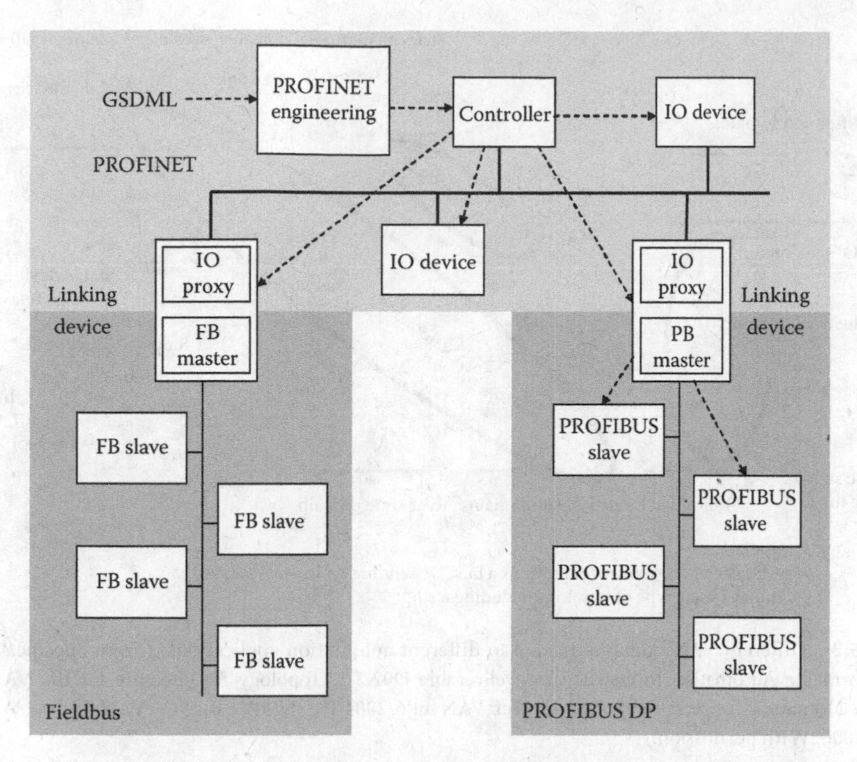

FIGURE 15.2 Integration of fieldbus systems by PROFINET (IO, input/output; FB, fieldbus; PB, PROFIBUS; DP, decentral periphery).

A heterogeneous network has to handle different classes of application data (application classes):

- Decentralized periphery applications
- Real-time applications/control loops (networked control)
- Component-based applications
- Telecontrol (TC) applications
- Teleservices

For these application classes, different requirements and opportunities exist. Limitations are, e.g., regarding isochronous and hard real-time communication caused by the performance of heterogeneous networks. The next sections give an introduction into the problems to be solved and into the developed concept. The main idea is to connect the local industrial communication systems (e.g., PROFINET as automation backbone connected with different fieldbus systems following the mentioned integration concept) via tunnel through the heterogeneous WAN with one or more remote industrial communication system.

15.2 Virtual Automation Network: Basics

In a geographically distributed automation project, various local and remote installations based on fieldbus communication, connected by a heterogeneous WAN, exist. In virtual automation networks (VANs), the distributed automation function (local and remote parts) forms a domain (Figure 15.3), containing all installed equipment to realize these automation functions. Thus, a VAN domain covers all devices that shall be grouped together on a logical or virtual basis to represent a complex application. Depending on the complexity of the automation function, a hierarchical domain approach makes sense. It means that there are also domains and subdomains that can consist of subdomains, etc. Consequently,

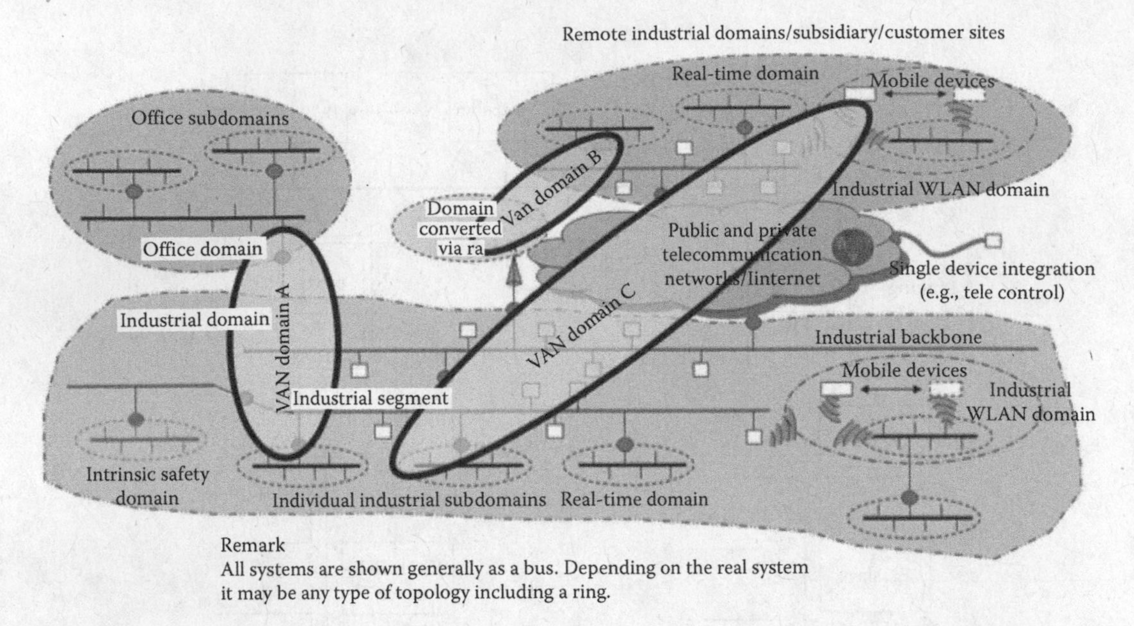

FIGURE 15.3 Different VAN domains related to different automation applications. (From Specification of the Open Platform for Automation Infrastructure, Deliverable D02.2-1. Topology Architecture for the VAN. Virtual Automation Domain. European Integrated Project VAN FP6/2004/IST/NMP/2-016969 VAN Virtual Automation Networks, 2006. With permission.)

within a domain (or subdomain), one or more automation devices (ADs), fieldbus systems, and infrastructure components exist. A domain keeps track of engineered relations between VAN devices that are the end points of a configured communication line between the local domain and the related other domain [DHH08]. The connected end points are linked via a tunnel, which seems to be a (virtual) communication line between the distributed application objects (runtime objects). Once the tunnel has been established, the data exchange between the distributed runtime objects follows the rules of the used fieldbus system (e.g., PROFINET with connected PROFIBUS or other connected fieldbus systems). These rules are well known by the automation engineer.

All devices with the same object model can interact by using the VAN infrastructure. As an example, the standardized runtime object model of PROFINET [IEC61158,IEC61784] can be used as a common object definition. Within the VAN device architecture also, other object models can be used if they are based on the VAN communication stack. Devices with other object models can be connected through a VAN proxy device (VAN-PD).

15.2.1 Domains

To connect distributed application objects, a VAN infrastructure, which offers VAN characteristics, is necessary. The VAN characteristics are defined for domains [PNO05,N07]. The expression "domain" is widely used to address areas and devices with common application purposes (Figure 15.3). A domain is a logical group of VAN devices. The devices related to a VAN domain may reside in a homogeneous network domain (e.g., local industrial domain). However, depending on the application, additional VAN-relevant devices may be only reached by crossing other network types (e.g., by WAN type communication) or they need to use proxy technology to be represented in the VAN domain view of a complex application.

A VAN domain can include given industrial domains (equipped with ADs and connected via industrial communications, e.g., fieldbus), completely or partially. Thus, an industrial domain can consist of segments related to a VAN domain (VAN segment) or of segments that are not related to a VAN domain (Figure 15.3) [N08].

15.2.2 Components

There are two groups of VAN components [NPM08,SOP06]:

- *VAN infrastructure components*: VAN access point (VAN-AP), VAN gateway (VAN-GW), VAN server device (VAN-SVD), VAN security infrastructure device (VAN-SID), VAN engineering station.
- *Automation-specific VAN components*: VAN automation device (VAN-AD), VAN-PD, VAN virtual device (VAN-VD).

The infrastructure components enable the establishment and maintenance of the tunnel between distributed application objects (runtime objects). The automation-specific components are VAN-enabled ADs.

A VAN-AP, the most important infrastructure component, connects VAN network segments. It does not contain an automation function or an automation application process. It can work as a gateway or a router to ADs that are members in a VAN application context (VAN domain). Thus, a VAN-AP may handle different transmission technologies, contains all relevant administration functions (necessary for configuration and parameter setting in connected subnetworks). It enables the switching between available communication paths and finds the best transmission technology and the best route through the different subnetworks of the heterogeneous network.

Within a VAN domain, there are different opportunities how any AD can participate in a VAN network (Figure 15.4):

- By integrating the VAN communication capabilities at the AD itself (VAN-AD)
- By connecting a single device via a separate VAN gateway device
- By connecting the whole network segment (e.g., connecting the ADs using any industrial communication system/fieldbus) through a VAN-PD and mapping its application objects to the VAN object pool that its parts (VAN-VD) can be exchanged within the VAN domain

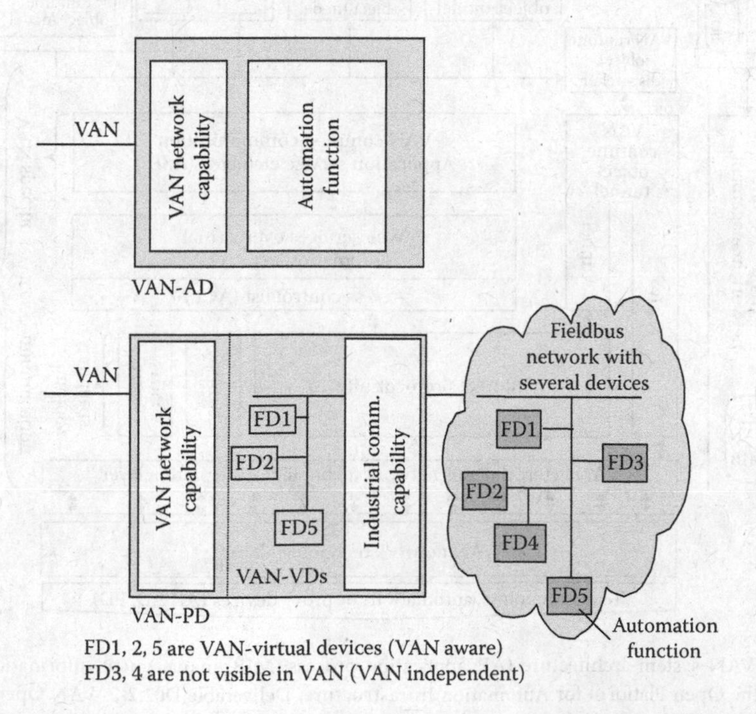

FIGURE 15.4 Automation-specific components (AD, automation device; PD, proxy device; VD, virtual device; FD, field device).

The proxy functionality enables the representation of the field device (FD) application data in context of the object model used in the VAN domain. This mechanism also allows all relevant devices' data of a certain physical device without VAN capabilities to be represented as a VAN-VD. This device is visible within the VAN domain, although it has not been connected to the VAN directly.

It can be seen that the VAN-AD and the VAN-PD offer an integration path to combine different communication technologies in a given automation application, thus to handle the heterogeneity of participating communication networks.

15.2.3 VAN System Architecture

The properties of a VAN can be represented by a block scheme (Figure 15.5). Blocks are logical components offering specific VAN functionality.

Single devices do not offer all possible properties, since they are context specific. Thus, only the necessary blocks have to be allocated to a VAN device, depending on the device type and communication profile.

To classify the block functions, one can distinguish

- Used standards (communication technologies, Internet protocol suite, adopted safety, security, and time synchronization standards)
- VAN-specific functions; the configuration of these functions is stored in application service elements (ASEs)

Since the object-oriented ASEs contain the data for the VAN functionality, they are crucial components of the VAN system architecture. All attributes of an ASE can be accessed via Web services by uniformly defined Get and Set services. The access rights are defined in the access control list (ACL).

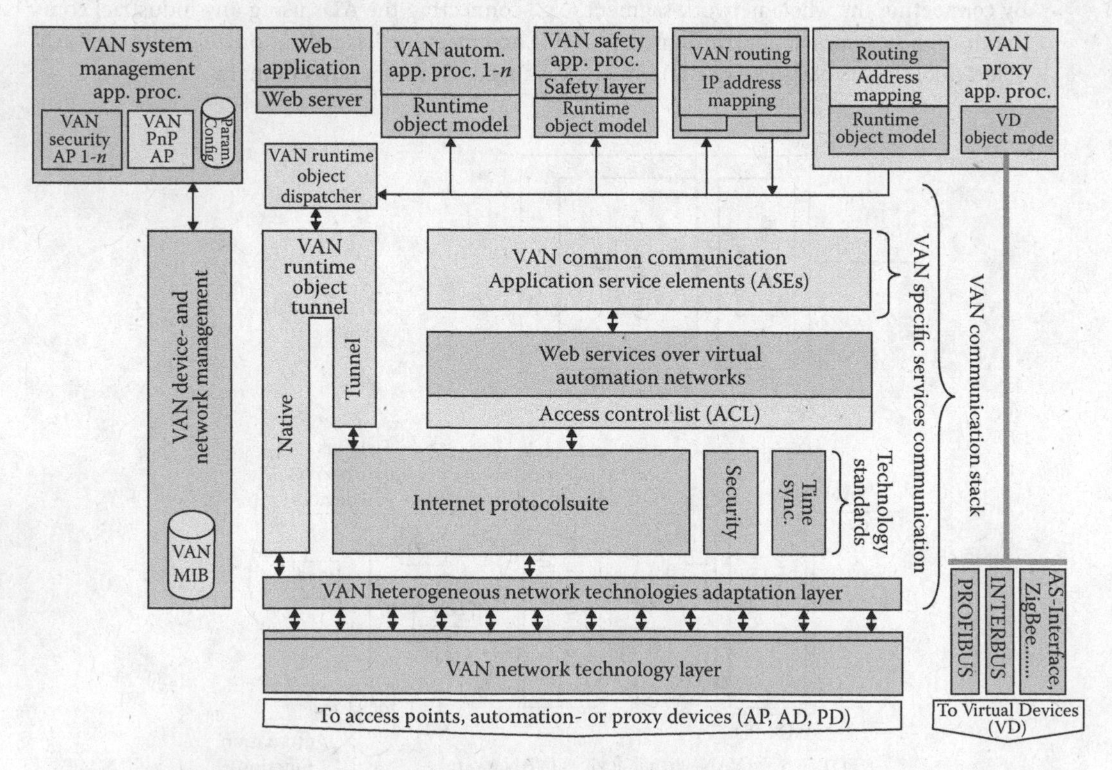

FIGURE 15.5 VAN system architecture (AP, application process; MIB, management information base). (From Specification of the Open Platform for Automation Infrastructure, Deliverable D02.2-2: VAN Open Platform API-Specification, European Integrated Project VAN FP6/2004/IST/NMP/2-016969 VAN Virtual Automation Networks, 2006. With permission.)

Important functionalities are, e.g., the runtime tunnel establishment and maintenance, security (including ACL), device and network management, VAN name-based addressing routing, and VAN (provider) switching.

15.3 Name-Based Addressing and Routing, Runtime Tunnel Establishment

To exchange distributed application data objects (runtime objects), two phases of the connection establishment have to be considered [NPM08]:

The establishment of a runtime tunnel, i.e., to organize the VAN infrastructure, comparable with laying a wired line between the applications processes (e.g., automation application process handling the automation objects to be exchanged). For this, Web services are used. This runtime tunnel is capable of offering the necessary quality of service (QoS) (e.g., availability, real-time capabilities, or security level).

The establishment of the application layer connection between the application objects itself (e.g., PROFINET application objects [IEC61158] to be exchanged) using the established runtime tunnel. That connection establishment follows the rules of the protocols that are used to realize the distributed application (e.g., PROFINET ASE definitions and protocols), and is, therefore, not in the scope of VAN (instead within the scope of e.g., PROFINET). In the following, the important aspects for establishing a runtime tunnel will be described.

A VAN domain usually consists of several subdomains that are interconnected via public or private WANs. Within a subdomain also, further subdomain structures can be realized.

Figure 15.6 depicts a typical VAN scenario [SOPSA07,WM08]: two subdomains, containing VAN-ADs and VAN-PDs, connected via a WAN. The subdomains are parts of industrial automation networks. In a demilitarized zone (DMZ) of the company's network, a VAN-AP is situated as a bastion host. The

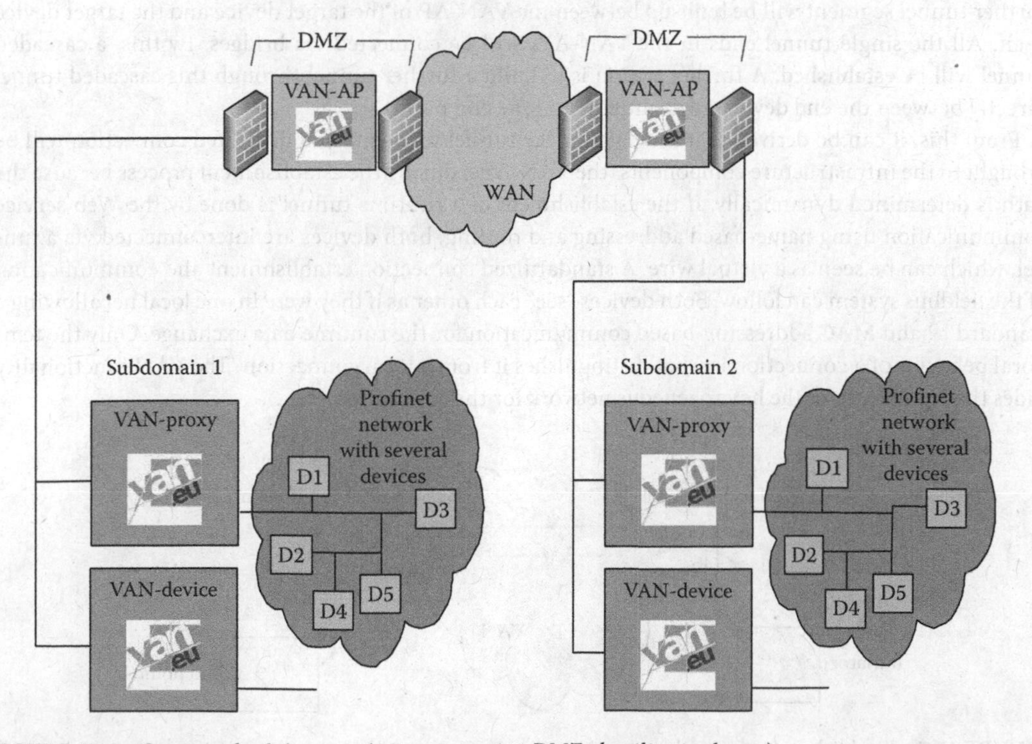

FIGURE 15.6 Connected subdomains (AP, access point; DMZ, demilitarized zone).

DMZ protects the industrial network and by this, the VAN subdomain against unauthorized access from the outside (and related to the VAN-AP also from inside) according to the company's security policy. Via the VAN-AP, all incoming and outgoing VAN messages will be forwarded. A subdomain can also have several VAN-APs.

Standard IT-routing mechanism alone will not work for the interconnection of different industrial domains because a device of an industrial domain will not be visible outside its local domain, neither by its DNS name nor its IP address. Furthermore, it is to be expected that in most cases, the industrial domains are administrated independently, e.g., overlapping IP address spaces may exist or addresses may change over time. To enable communication between devices of those industrial domains, a name-based addressing and routing concept was developed—the VAN name-based routing. Here, the routing decision is not based on network addresses, but on names. A VAN name is a complete fully qualified domain name (FQDN) containing the structure of the devices location within a VAN domain. VAN routing is a proactive next-hop routing mechanism allowing the forwarding of Web services without knowing the entire path also in between different industrial domains and via public networks.

Each VAN end-to-end connection is preconfigured in both involved VAN devices by VAN engineering. So, it is configured who is the initiator and who is the end point of that connection. The initiator at first sends a tunnel setup Web service request to its end point. Since this request is targeted to a device outside its own subdomain, the requester/initiator sends this message to its VAN-AP. The latter knows via which next VAN-AP the subdomain of the targeted device can be reached (either directly or indirectly) and hands over the request respectively. If the request hits its destination, the message will be processed and a response will be issued.

In case of a positive response, the establishment of the tunnel connection starts. This means between all affected VAN devices on the path (the path is determined by the VAN routing), tunnel segments, which belong to that connection, will be built up.

For example, in an easy case as depicted in Figures 15.6 and 15.7, the requestor builds a tunnel to its VAN-AP. The latter builds up a tunnel to the VAN-AP of the subdomain of the target device, and a further tunnel segment will be built-up between the VAN-AP of the target device and the target device itself. All the single tunnel ends in the VAN-APs will be connected via bridges. By this, a cascaded tunnel will be established. A further option is to build a further tunnel through this cascaded tunnel directly between the end devices (the initiator and the end point).

From this, it can be derived that the data for the tunnel segments of a dedicated connection will be brought to the infrastructure components (the VAN-APs) during the establishment process because the path is determined dynamically. If the establishment of a runtime tunnel is done by the Web service communication using name-based addressing and routing, both devices are interconnected via a tunnel, which can be seen as a virtual wire. A standardized connection establishment and communication of the fieldbus system can follow. Both devices "see" each other as if they were in one local net allowing a standard IP and MAC addressing-based communication for the runtime data exchange. Only the temporal behavior of a connection is what distinguishes it from a local connection. The VAN functionality hides the complexity of the heterogeneous network for the application.

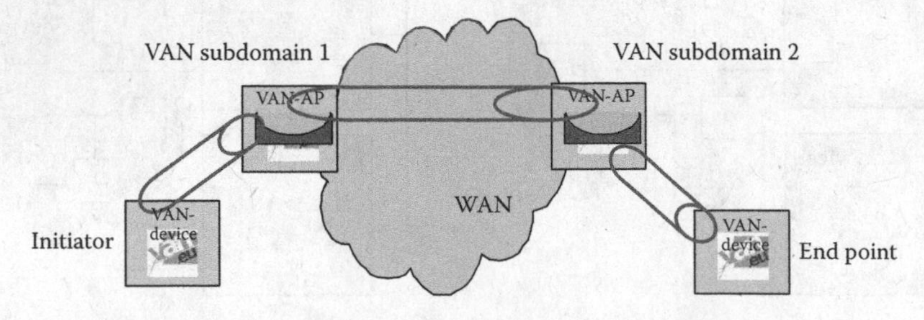

FIGURE 15.7 Establishment of cascaded tunnels.

15.4 Maintenance of the Runtime Tunnel Based on Quality-of-Service Monitoring and Provider Switching

To use a heterogeneous network for the exchange of application data in the automation domain, the following requirements have to be fulfilled:

- Non-interrupted availability
- Guarantee of the desired safety integrity level to protect the human life/health as well as the assets [AW08]
- Automation-specific security measures by the reasons of overall safety [AW08,WA08]
- Suitable real-time behavior for geographically distributed automation applications that can be offered by the used transmission technology [BZ08,RL08]

Since the aspects of safety and security have been described in several chapters of this handbook, only the aspect of availability should be discussed here in the context of using WAN.

Besides a suiting service level agreement (SLA—the contract with the provider), the main VAN mechanisms to guarantee the availability of an end-to-end communication within the heterogeneous network, especially regarding WANs, are

- Monitoring the actual behavior of the selected communication path through the WAN [WM08]; from the monitoring results, a VAN switching can be derived
- VAN switching between providers or suitable transmission technologies that are configured in the VAN device communication properties
- VAN priority mapping (mapping of the priorities of the tunneled packets to the tunnel packets)

In contrast to a local automation network, where the entire transmission behavior can be completely controlled, this is extremely limited in public WANs. Here, the end user has very little technical possibilities to influence the behavior of the transmission channel. The detailed infrastructure composition and therefore the exact communication path cannot be described in end-to-end detail (see Figure 15.8). The communication path is also subject to temporary changes and is often even not known by the service provider himself. The quality of the line is contractually determined by an SLA with the provider. In the SLA, parameters are defined such as bandwidth, delays, fluctuation range, availability, priority classes, data loss rate, etc. Besides those quantitative technical aspects also organizational items as service hours, reaction times on problems, responsibilities, and escalation scenarios. Sometimes, incentive awards for service levels exceeded and usually penalty provisions for unprovided services are defined.

A 100% availability cannot be guaranteed, a usual contract value is about 98%. Also, 99.5% is possible, but at the moment, this is too expensive for a broad use in respective automation applications.

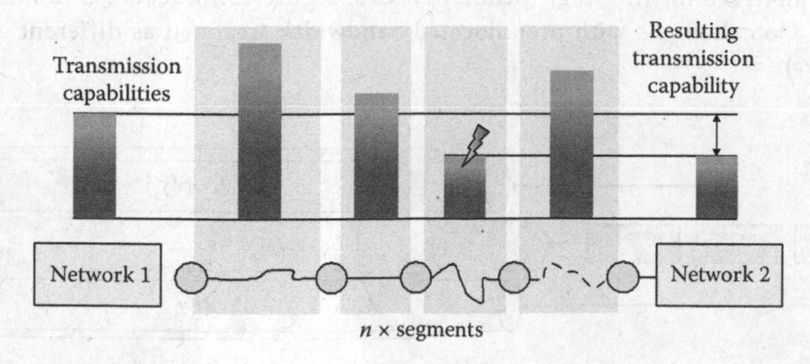

FIGURE 15.8 End-to-end path capabilities in a chain of segments.

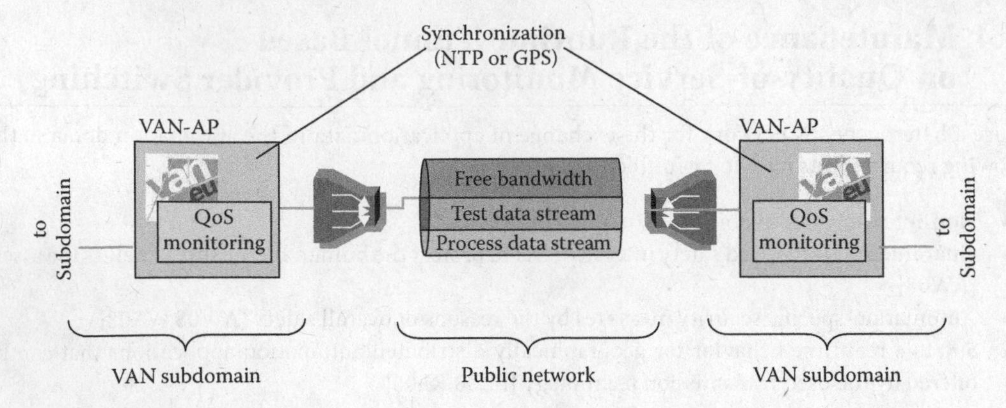

FIGURE 15.9 VAN QoS monitoring.

In both cases, this means the network can be "not available" for several days per year. During runtime, no online measuring data on the current state of the connection is available to the user. The provider does, however, carry out measurements within his network, but these are only made available to the user after the corresponding accounting period at the earliest.

To be able to react on QoS failure scenarios and to have verification against the SLA, VAN defines its own QoS monitoring for the runtime quality of the connection. The VAN QoS monitoring focuses on the public wide network, since it is the most uncertain part of an end-to-end connection. The public network is considered as a black box.

Figure 15.9 gives the general topology structure of the VAN QoS monitoring. The main approach focuses on an active measurement by generating an additional data stream directly between the involved VAN-APs as the entrance points to the public network part of a connection. The data stream will be produced and analyzed by a special application distributed on the producing and receiving VAN-AP. This means the entire public network path is considered as black box. Important is the time synchronization between them to get analyzable data. The time synchronization will be realized using a precise GPS receiver at each location or by using NTP. Each packet of the measurement data stream will be time stamped as base of further calculation (e.g., latency, jitter).

Most important point is that the process data stream will not be disturbed by the measurement. Therefore, it is necessary to know the bandwidth of the line contracted with the provider and the bandwidth needed by the entire process data stream (sum of the bandwidth of all running connections). Since for tunneling, a UDP-based openVPN tunnel is used, the generated stream for monitoring is also a UDP stream.

Also, the behavior and status of the tunnel has to be investigated according to the single defined QoS classes (if more than one is used). Therefore, the monitoring traffic also has to be generated and analyzed for the single priority classes. Figure 15.10 depicts a tunnel with different prioritized QoS channels with pre-allocated bandwidth (realized as different queues in the network routers).

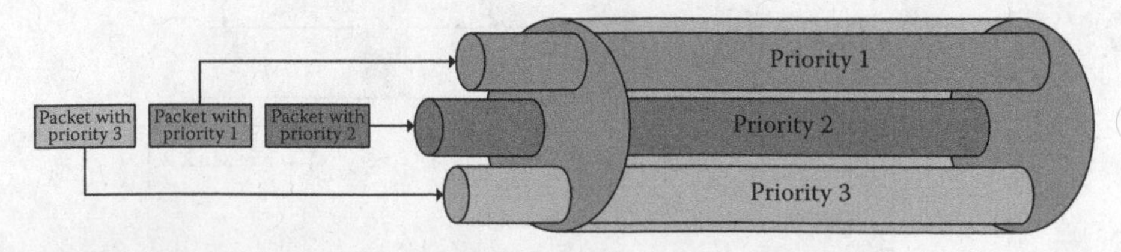

FIGURE 15.10 Different prioritized QoS channels.

There are two monitoring modes defined:

- The stand-alone mode is for testing the link quality between the end points without real automation traffic. So, the test traffic transferred along the path is issued and controlled by the measurement process only.
- In the runtime mode, the measurement traffic runs in parallel with the runtime automation traffic, thus it has to be assured that the measurement does not impact the runtime automation traffic.

The monitoring can be used to measure the current network parameter values and to identify whether the provider network meets the agreed limits contracted in the SLA. The latter also covers logging and processing of the history of parameter values.

The VAN QoS monitoring can be extended to a condition monitoring that will even allow deriving forecasts of the behavior of a transmission path from its current and preceding status.

If a monitored value reaches a predefined threshold, then a VAN switching event can be derived. VAN switching provides alternative transmission paths for different provider accesses or different access technologies. If there are alternative lines between two networks, then a VAN-AP can choose between them. The reasons to switch between alternative communication paths could be

- Bad QoS or degraded QoS: changes in QoS
- Loss of link, line failure (redundancy case; hot standby systems)
- Least cost routing: selecting the cheapest line

15.5 VAN Telecontrol Profile

The so far described aspects refer to the approach of realizing classical fieldbus communication via WANs. Fieldbus standards are primarily based on cyclic transfer of IO data and do not support distributed applications across subnet boundaries. On the other hand, a variety of proprietary and standardized protocols designed to communicate across public networks in the separate automation field of telematic systems do exist. Thus, fieldbus technologies and TC systems today are based on different mechanisms, technologies, and protocols. In VAN, a common open platform and common means to combine both classical fieldbus tasks and TC tasks are strived.

For this, a definition of a TC profile has been done, aiming for a seamless integration of TC functionalities into the international fieldbus standard IEC61158 type 10 [BWM08]. TC takes advantage of the fact that there is no direct feedback from inputs to outputs. The transfer from the data provider to the data consumer is decoupled. The major reasons for the application of TC are minimization of communication costs, minimization of communication volume, delays caused by network transfers, and tolerance to a network being temporarily unavailable. In order to reach the minimization of the data volume, different mechanisms are used: mechanism to generate a transfer object, mechanism to buffer transfer objects, and mechanism to initiate a data transfer (Figure 15.11). A minimized amount of data is the precondition in order to keep a small network load and to minimize the transmission costs. The generation of the data and its transfer occur at different times. This kind of communication is called event-driven communication.

There are different adjustable filters that control whether the input data is copied into a transfer object and stored in the transfer buffer or not. Criteria are event oriented at each value/information change, event-oriented at threshold settings (different strategies), spontaneous (user-controlled, user may be provider or consumer), and cyclic (time driven).

These mechanisms guarantee a minimum quantity of transfer objects to be generated and triggering of a data transfer, when the transfer condition is unconditional/immediately, that means an alarm is identified.

The transfer buffer supports two tasks:

- The objects to be transferred are buffered during communication fault/error. Therefore, no data is lost.
- The amount of data to be transferred and, therefore, also the communication costs can be further reduced.

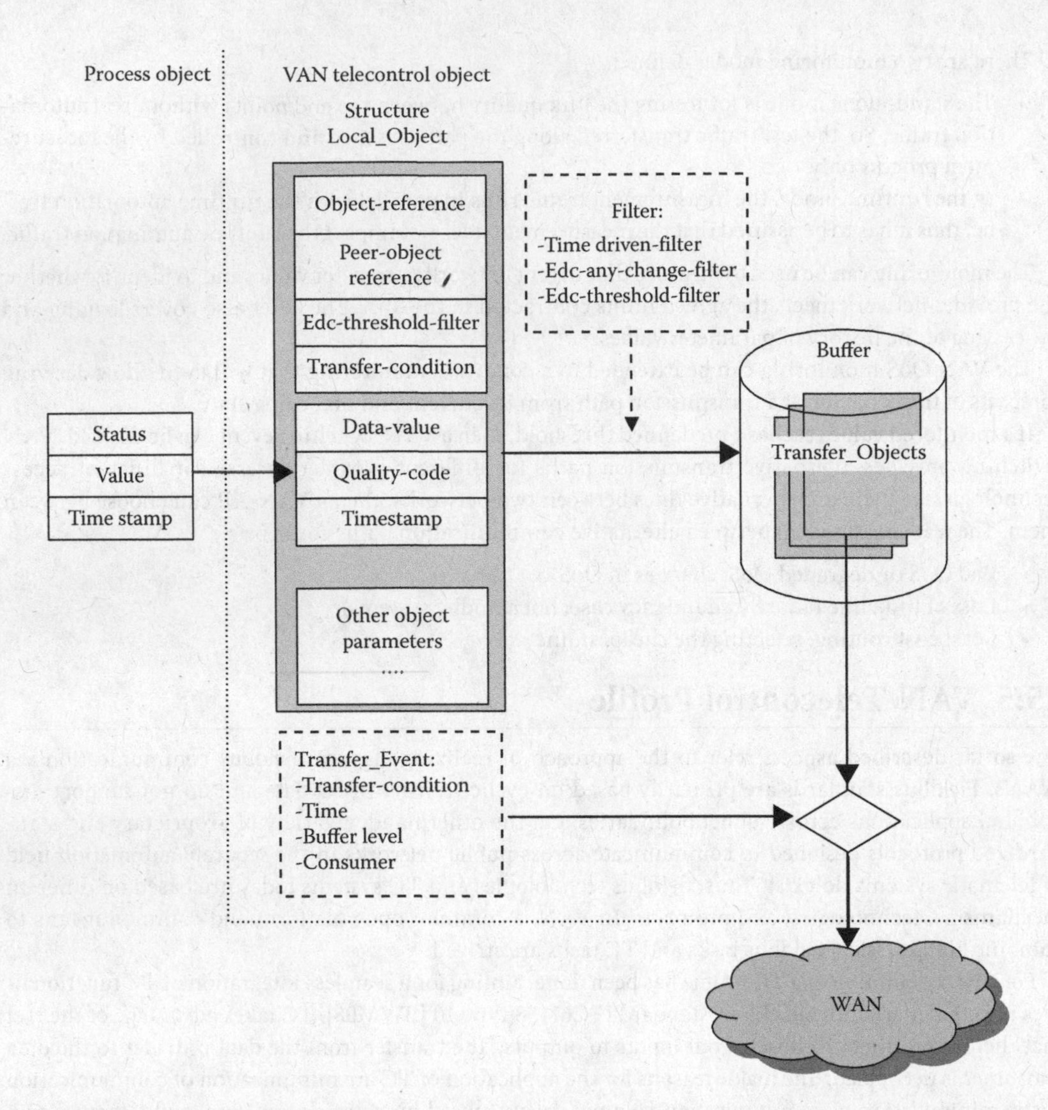

FIGURE 15.11 VAN telecontrol object principle.

The data transfer mechanisms are aligned to the meaning of the data and the reduction of communication costs:

- *Spontaneously*—Data is transmitted immediately (unconditional). The decision for immediate or later transmission is only relevant for dial-up networks. In other networks with permanent connectivity, data is transferred immediately, even if the transfer condition is set to conditionally. Alarms are always transmitted immediately.
- *Conditionally*—Data is transmitted when the send buffer is filled or an alarm (unconditional data) occurs.
- Time driven means that data is transmitted periodically or at given times.

The TC functionality will be realized as a profile, independent from the underlying fieldbus. Figure 15.12 shows this TC add-on put on the IEC61158 type 10 fieldbus (IEC61158) stack architecture of the VAN demonstrator.

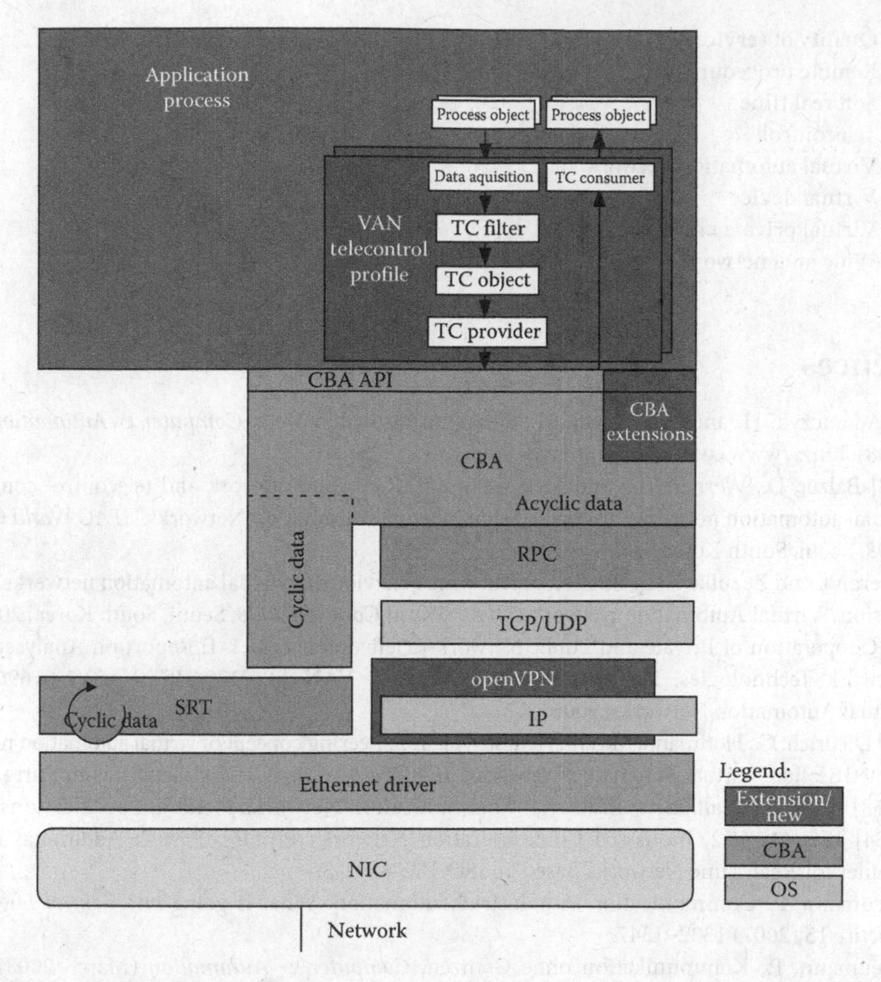

FIGURE 15.12 VAN telecontrol profile approach.

The profile uses the functions provided by the communication stack. In order to enable communication relations via a temporarily unavailable network path, extensions within the communication stack may be necessary.

By the TC profile, the integration of TC functionality into fieldbus technologies has been done, which starts the merging of both classically separated fields.

Abbreviations

ACL	Access control list
AD	Automation device
AP	Access point
CBA	Component-based automation
FD	Field device
FQDN	Fully qualified domain name
GPS	Global positioning system
NIC	Network interface card
NTP	Network time protocol
PD	Proxy device OS

QoS Quality of service
RPC Remote procedure call
SRT Soft real time
TC Telecontrol
VAN Virtual automation network
VD Virtual device
VPN Virtual private network
WAN Wide area network

References

[AW08] Adamczyk, H. and Wolframm, M., Sicher im virtuellen Netz, *Computer & Automation* (April 2008). http://www.computerautomaton.de/

[BWM08] Balzer, D., Werner, Th., and Messerschmidt, R., Public network and telecontrol concepts in virtual automation networks. Invited Session "Virtual Automation Networks." *IFAC World Congress 2008*, Seoul, South Korea, 2008.

[BZ08] Beran J. and Zezulka F., Evaluation of real-time behaviour in virtual automation networks. Invited Session "Virtual Automation Networks." *IFAC World Congress 2008*, Seoul, South Korea, 2008.

[CPP06] Cooperation of Private and Public Networks, Deliverable D07.1-1: Report on Analysed Public Network Technologies, European Integrated Project VAN FP6/2004/IST/NMP/2-016969 VAN Virtual Automation Networks, 2006.

[DHH08] Diedrich, C., Hoffmann, M., and Hengster, H., Engineering concept of virtual automation networks. Invited Session "Virtual Automation Networks." *IFAC World Congress 2008*, Seoul, South Korea, 2008.

[IEC61158] IEC 61158 (all parts) Industrial Communication Networks Fieldbus Specifications.

[IEC61784] IEC 61784-2, Industrial Communication Networks Profiles. Part 2: Additional Fieldbus Profiles for Real-Time Networks Based on ISO/IEC 8802-3.

[N07] Neumann, P., Communication in industrial automation. What is going on? *Control Engineering Practice* 15 (2007) 1332–1347.

[N08] Neumann, P., Kommunikation ohne Grenzen, *Computer & Automation* (March 2008) 54–61. http://www.computerautomation.de/

[NPF07] Neumann, P., Pöschmann, A., and Flaschka, E., Virtual automation networks. Heterogeneous networks for industrial automation, *ATP International* 2 (2007) 36–46.

[NPM08] Neumann, P., Pöschmann, A., and Messerschmidt, R., Architectural concept of virtual automation networks. Invited Session "Virtual Automation Networks." *IFAC World Congress*, Seoul, South Korea, 2008.

[PIP] PROFIBUS Integration in PROFINET IO, Order No: 7.012. PROFIBUS User Organisation.

[PNO05] PNO, PROFIsafe—Profile for Safety Technology on PROFIBUS DP and PROFINET IO, Version 2.0, 09/2005.

[RL08] Rauchhaupt, L. and Lakkundi, V., Wireless network integration into virtual automation networks. Invited Session "Virtual Automation Networks." *IFAC World Congress 2008*, Seoul, South Korea, 2008.

[SOP06] Specification of the Open Platform for Automation Infrastructure, Deliverable D02.2-1. Topology Architecture for the VAN. Virtual Automation Domain. European Integrated Project VAN FP6/2004/IST/NMP/2-016969 VAN Virtual Automation Networks, 2006.

[SOPSA07] Specification of the Open Platform and System Architecture, Prototype Implementation Integration. Deliverable D02.4-1: Software Architecture and Interface Specification. European Integrated Project VAN FP6/2004/IST/NMP/2-016969 VAN Virtual Automation Networks, 2007.

[VANOP06] Specification of the Open Platform for Automation Infrastructure, Deliverable D02.2-2: VAN Open Platform API-Specification. European Integrated Project VAN FP6/2004/IST/NMP/2-016969 VAN Virtual Automation Networks, 2006.

[WA08] Wolframm, M. and Adamczyk, H., Secure virtual automation networks based on generic procedure model. Invited Session "Virtual Automation Networks." *IFAC World Congress 2008*, Seoul, South Korea, 2008.

[WM08] Werner, T. and Messerschmidt, R., Der Telecontrol-Aspekt, *Computer & Automation* (May 2008). http://www.computerautomation.de/

16 Industrial Agent Technology

Aleksey Bratukhin
Austrian Academy of Sciences

Yoseba Peña Landaburu
University of Deusto

Paulo Leitão
Polytechnic Institute of Bragança

Rainer Unland
University of Duisburg-Essen

16.1 Introduction 16-1
16.2 Agents and Multi-Agent Systems 16-2
Intelligent Agents Definition • Multi-Agent Systems • Ontologies • Self-Organization and Emergence • The Holonic Paradigm • Holonic Multi-Agent Systems • How Agents Can Be Implemented
16.3 Agents and Multi-Agent Systems in Industry 16-6
16.4 Application Areas 16-6
Resource Handling • Order Handling • Comparison • Challenges of Industrial Agents' Usage • Other Application Areas in Brief
16.5 Agents and Multi-Agent Systems in Industry: Conclusions 16-12
Abbreviations 16-13
References 16-13

16.1 Introduction

The conventional centralized, rigid information systems cannot serve the demands of modern (manufacturing) industry adequately any longer. In order to stay competitive, industry needs to be highly flexible, with shorter job sizes, variable product portfolios, and always changing shop floors. Although centralized approaches can in the meantime provide highly sophisticated and efficient scheduling solutions, the requirements imposed by novel manufacturing trends renders centralized control systems more and more unfeasible (cf. [8]).

Against this background, multi-agent based industrial information systems seem to be a promising and natural alternative. They provide decentralized architecture, modularity, robustness, and adaptability to changes (cf. [9]). Moreover, the evolution of industry in the previous decades has rearranged the principal goals, especially in manufacturing control. In the past, the main objective was achieving an optimal scheduling algorithm. Nowadays, this aim has been sacrificed for the sake of long-term efficiency: on the one hand, the scheduling problem usually cannot be solved in polynomial time. Modern centralized control system can provide a good solution if they have enough time, but, in real time, this is still unrealistic. On the other hand, possible variations that may occur include demand changes, apparition of new products, changes on the physical layout, disturbances, and errors. Therefore, the stress is currently set upon other desirable features of manufacturing control systems, such as robustness, adaptability, and real-time ability.

Finally, recent or upcoming technologies like service-oriented architectures (SOA) and modern trends in automation (e.g., mass customization), have completely redrawn the problem domain, bringing a bunch of new demands, such as upgradeability, platform independence, scalability, or resilience.

16.2 Agents and Multi-Agent Systems

16.2.1 Intelligent Agents Definition

Intelligent agents can be regarded as autonomous, problem-solving computational entities with social abilities that are capable of effective proactive behavior in open and dynamic environments. There are a number of different definitions of intelligent agents (cf. e.g., [1]); many of them associate the following properties with an agent:

- *Autonomy*: An intelligent agent has control over its behavior, that is, it operates without the direct intervention of human beings or other agents, and has control over its internal state and its goals.
- *Responsiveness/Reactivity*: An intelligent agent perceives its environment and responds in a timely fashion to changes that occur in it in order to satisfy its design objectives/goals.
- *Social ability*: An intelligent agent is capable of interacting with other agents (and humans) and will do so if that helps to achieve its design objectives/goals or organizational or combined goals.
- *Intelligence*: An agent has specific expertise and knowledge. Thus, it is capable of dealing with and maybe solving problems that fall into its domain of expertise.
- *Proactiveness*: An intelligent agent is goal directed, deliberative, intelligent, opportunistic, and initiative. Due to its goal-directed behavior, the agent takes initiative whenever there is an opportunity to satisfy its goals. It especially may react proactively to changes in its environment; that is, it responds to it without being explicitly asked for it from the outside.

The most popular model for cognitive, intentional agents is the belief-desire-intention (BDI) architecture (cf. [36]). Here, agents react to changes in their environment with the help of practical reasoning. Roughly speaking an agent comes with a number of sets. The beliefs reflect its abstracted understanding of that part of the real world it is interested in. This understanding is subjective to the agent. The desires represent the goals of the agent. In order to achieve its goals, an agent has to sense its environment and react to relevant changes, usually by executing plans from its predefined set of plans/actions. Thus, a reasoning process first identifies the affected goals and appropriate plans. Reasoning is enabled by conditions that are annotated to plans and goals. The intensions of an agent are reflected by a data structure that lists those plans/actions in an appropriate/ordered way an agent has chosen to execute as a reaction to the last (and prior) relevant events in its environment.

16.2.2 Multi-Agent Systems

Due to the limited capabilities of a single agent, more complex real-world problems may require the common and cooperative effort of a number of agents in order to get the problem at hand solved. A *multi-agent system* (MAS) is a federation of fully autonomous or semiautonomous agents (problem solvers) that are willing to join forces in order to achieve their individual goals and/or the overall goals of the federation. In order to succeed, they rely on communication, collaboration, negotiation, and responsibility delegation, all of which are based on individual rationality and social intelligence of the involved agents. The global behavior of a MAS is defined by the emergent interactions among its agents, which implies that the capability of a MAS surpasses the capabilities of each individual agent. Reduction of complexity is achieved by (recursively) decomposing a complex task into a number of well-defined subtasks, each of which being solved by a specific agent. However, unlike hard-wired cooperation domains, these coalitions or teams are very flexible. Depending on the organizational structure, agents may autonomously join or leave the coalition whenever they feel like—provided their commitments are fulfilled.

For agents, in order to be able to cooperate, there is a need for communication. The most important communication languages Knowledge Query and Manipulation Language (KQML, standardized by Defense Advanced Research Projects Agency (DARPA) and agent communication language

(ACL, a Foundation for Intelligent Physical Agents [FIPA] standard, see [29]) are based on so-called speech acts that define a set of performatives and their meaning like *agree*, *propose*, *refuse*, *request*, *query-if*. While the performatives of a communication language are standardized, its actual contents are not—it varies from system to system. A popular form of indirect communication between agents can be done via blackboards on which agents can post their messages and other agents can read and react to them. Another kind of communication is contract net protocols, which are similar to procedures that may lead to a contract (task announcement, bidding, awarding, solution providing, and rewarding).

16.2.3 Ontologies

Especially deliberative agents are supposed to exhibit some intelligence. In order to understand what they are doing they usually maintain an internal model of that part of the real world in which they are willing to act. If different agents are meant to work together they need to have a common understanding about the overall problem to be solved. However, since these agents may be specialists in different fields they may not be able to understand each other, mainly because they may not speak the same "language" (different vocabulary) or because there is only an overlap between the expertise of two agents but not a complete one. In such cases, ontologies may help. An *ontology* (cf. [30]) is a formal, preferably machine processable taxonomy of an underlying domain. As such, it contains all relevant entities or concepts within that domain, the underlying axioms and properties, the relationships between them, and constraints on their logically consistent application; that is, it defines a domain of knowledge or discourse by a (shared) vocabulary and, by that, permits to reason about its properties. Ontologies are typically specified in languages that allow abstraction and expression of semantics, e.g., first-order logic languages.

16.2.4 Self-Organization and Emergence

A *self-organizing* system is a system running without central control, which means that it is functioning without external direction, control, manipulation, interference, pressures, or involvement (cf. [31]). It constantly tries to improve its spatial, temporal, and/or functional structure in the course of time by organizing its components in a more suitable way in order to improve its behavior, performance, and/or accuracy. The driving force behind these alignments is to achieve its overall goal(s) as efficiently as possible (cf. [32]). This implies that the combined cooperative behaviors of the individual entities lead to an emergent coherent result.

Emergence (cf. [33–35]) can be seen as an evolving process that leads to the creation of novel coherent structures, patterns of behavior, and properties often but not only during the process of self-organization in complex systems. It is the appearance of a coherent pattern that arises out of interactions among simpler objects (agents) but is more than just their summed behavior. This implies that novel structures and patterns of behavior are neither directly described by nor ingrained in the defining patterns and rules that form the foundation of the system. Thus, the functioning of the system can only be understood by looking at each of the parts in the context of the system as a whole, however, not by simply taking the system apart and looking at the parts. As a consequence emergence can neither be controlled nor predicted nor managed.

Self-organization and emergence have some similarities and some differences. They are both self-sustained systems that can neither be directly controlled nor manipulated in any way from the outside. They both evolve over time; however, only self-organizing systems need to exhibit a goal-directed development. Emergent systems consist of a larger number of low-level (micro-)entities that collaborate in order to exhibit a higher level (macro-)behavior. The unavailability of one or more of those lower level entities does not abrogate the functioning of the system (graceful degradation) although this may be the case in self-organizing systems.

16.2.5 The Holonic Paradigm

One of the first concepts attempting to bring agents and *manufacturing* derived from the holonic paradigm that was proposed by Koestler (cf. [2]). So-called *holons* are the underlying concept of such a design. In social and biological organizations, holons represent basic units. Basic here means observed from the current level of abstraction of an organization. It does not mean that a unit cannot be further subdivided when analyzed on the next lower level of abstraction. Thus, holons may consist of subordinate holons, which, in turn, means that they may be part of a larger whole. Altogether, they may form a tree-like hierarchy, called *holarchy*. Its members cooperate in order to achieve a complex goal or objective. A holarchy may be formed in an ad hoc way and may only exist temporarily until its overall goal is achieved. The holarchy defines the basic rules of cooperation between holons. The stability of holons and holarchies stems from the fact that holons are self-sustained units that have a given degree of independence. They handle circumstances and problems on their particular level of existence without need of support by higher level holons. Depending on the underlying organizational structure, holons may receive instructions from and may be controlled to a certain extent by higher level holons.

A holarchy that is not part of an encompassing holarchy is a *holonic system/organization*. The strength of a holonic organization is that it enables the construction of complex systems that are nonetheless efficient in the use of resources, highly resilient to disturbances (both internal and external), and highly adaptable to changes in their environment.

16.2.6 Holonic Multi-Agent Systems

A Holonic MASS, as a special and important type especially for industrial MASS, combines the concepts of holonic with the concept of MASS. It consists of autonomous agents with *holonic* properties that recursively group together to form holons and, by this, holarchies [3].

In literature, three main architectures for holonic MAS were proposed: the hierarchical, the federation, and the autonomous agents approach. At the one extreme of the spectrum, the *hierarchical* approach, agents completely give up their autonomy. From the outside it may even look like they merge into a new agent. The other side of the spectrum, the *autonomous agents* approach, is defined by a loose coupling of the agents. They share a common goal for some time before separating to continue to work on their own goals. During the collaboration, all agents are equal. Thus, autonomy is not restricted. In between those extremes, all other kinds of associations are possible (*federation* approach). This mainly means that agents give up part of their autonomy, that is, that they may depend on their superordinate agent with respect to some aspects. What aspects exactly are given up is either subject to negotiation between the agents participating in the holon or is a directive from a superordinate agent along the hierarchy. However, it evolves in the interaction process and cannot be predefined by the designer (cf. [4]).

Holonic systems combine top-down organizational structure with decentralized control. The latter realizes a bottom-up perspective. Although it is possible to organize a federated holonic structure in a completely decentralized manner, it is more effective to use an individual agent to represent a holon (cf. [5]). Ref. [6] identifies three possible approaches for federation architectures: facilitators, brokers, and mediators. In the *facilitator* approach, communication between the agents of the federation takes place always through an interface called a facilitator. A facilitator is responsible only for a designated group of agents. The *broker* approach is similar to the facilitator approach. However, here, each agent of the federation may contact each broker in the same federation in order to find appropriate service agents to complete the desired task.

In the *mediator* approach, the mediator agent executes the role of a system coordinator. It promotes and supervises cooperation among intelligent agents. It may even learn by this from the behavior of the involved agents. In every *nonleaf* holon, exactly one (sub-)agent acts as a mediator, that is, it represents the interface between the agents in the holon (*inbound duties*) and those outside of the holon (*outbound duties*). The inbound duties of the mediator can range from pure administrative tasks to the authority to issue directives to the subordinate agents of the holon. Thus, it may broker and/or supervise the interactions between the subholons of that holon. The outbound interface of a mediator implements the interaction of the holon with the rest of the agent society. The mediator can plan and negotiate for the holon on the basis of the common plans and goals of its underlying agents.

All this three kinds of federation architectures permit to coordinate the system activity via facilitation as a means of reducing overheads, ensuring stability, and providing scalability (cf. [6]).

16.2.7 How Agents Can Be Implemented

In order to implement communities of agents, several layers need to be realized. The lowest layers are the network and communication layers that allow agents to abstract from their exact physical location and to exchange messages. On the next level, the actual agent infrastructure needs to be provided, which means that usually a number of different agent types need to be provided (the actual agents, broker agents, white and yellow pages, etc.) as well as ontologies and agent life-cycle management services. These first layers can be seen as syntactical layers in the sense that they do not reflect the actual intelligence of the system—they just provide the necessary foundation for a MAS to function. Especially for these "syntactical" layers, a significant number of commercial and academic agent construction tools are available, providing a variety of services and agent models, the differences reflecting the philosophy and the target problems envisioned by the platform developers. A reference for a broad number of agent construction tools can be found on the AgentBuilder Web site (http://www.agentbuilder.com/AgentTools) and in [28].

Java Agent DEvelopment Framework (JADE; cf. [27]) is perhaps the best-known FIPA (cf. [29]) compliant platform to develop agent-based solutions. It provides the mandatory components defined by FIPA to manage the agent's infrastructure, which are the agent communication channel (ACC), the agent management system (AMS), and the directory facilitator (DF). The AMS agent provides white pages and agent life-cycle management services, maintaining a directory of agent identifiers and states, and the DF provides yellow pages services and the capability of federation within other DFs on other existing platforms. The communication among agents is done via message passing. The messages are encoded using the FIPA-ACL and their content is formatted according to the FIPA-SL (Semantic Language), both specified by FIPA. Ontologies should be used to support a common understanding of the concepts of their knowledge domain, passed in the messages' content. Ontologies can be designed using a knowledge representation tool, such as Protégé (see http://protege.stanford.edu/), and then translated into Java classes according to the JADE guidelines that follow the FIPA Ontology Service Recommendations specifications (cf. [29]).

JADE also provides a set of graphical tools that allows supervising the status of agents and supporting the debugging phase which are usually quite complex tasks in distributed systems. Figure 16.1 illustrates

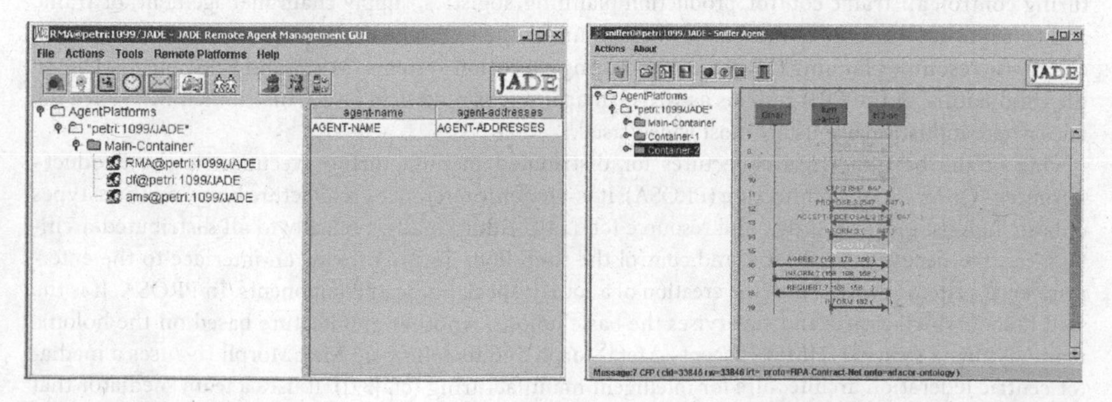

FIGURE 16.1 Graphical user interfaces of the Remote Monitoring and Sniffer agents.

the graphical interface of two tools provided by JADE, namely the Remote Management and the Sniffer agents. The last one is a debugging tool that allows tracking messages exchanged in a JADE environment using a notation similar to Unified Modeling Language (UML) sequence diagrams.

Jadex, as an extension of Jade, is one of the few examples of a platform that also provides support for the reasoning capabilities of agents (cf. http://jadex.informatik.uni-hamburg.de/). Its reasoning engine implements the Belief-Desire-Intention model.

16.3 Agents and Multi-Agent Systems in Industry

Classical centralized solutions perform well in static industrial environments in which neither the physical layout of the plant nor the product portfolio varies. Thus, their rigid control architecture does not have to be modified, for example, to deal with new unknown situations. Modern industries, however, pose different challenges to which the centralized approach cannot respond. The main ones are ever changing, unpredictable order flows, which, in order to be fulfilled, need a dynamic shop floor. Both require more flexible approaches that can be naturally provided by MAS. In fact, MAS are designed as a group of autonomous interacting units that pursue (directly or indirectly) a common goal. This is exactly what modern manufacturing systems need.

Altogether, a MAS-based manufacturing control has to satisfy the following characteristics:

- *Reliability and robustness*: The system must respond properly to disturbances.
- *Adaptability to changes*: A new physical layout, changes on product designs, extension of the product portfolio, demand changes, and other novelties should not prevent the MAS from carrying out its task.
- *Scalability*: It should offer an easy coupling of new parts or systems and should easily allow integrating new automation devices or other information systems (e.g., the scheduling system with the plan execution, the supply chain manager, and so on).
- *Easy upgradeability to new technologies*: In case of a new version of platforms, protocols, or brand new technologies, the system should offer an easy upgrade procedure.
- *Real-time response*: It should be able to react in real time to disturbances or changes.
- *Portability and platform independence*: The MAS should not depend on a certain platform but be able to work under different platforms.

16.4 Application Areas

Responding to the modern trends in automation such as mass customization and dynamic configuration, over the last decade distributed systems found its place in several industrial areas, like manufacturing control, air traffic control, production planning, logistics, supply chain management, or traffic and transportation. Some of them are combined under the umbrella of the plant automation pyramid: enterprise resource planning (ERP), manufacturing execution systems (MES), and field control. Due to the rigid nature of the ERP systems as well as machine-oriented field levels, distributed MESs are the ones where industrial agents are most widely used.

One of the best-known architectures for distributed manufacturing execution is the Product–Resource–Order–Staff Architecture (PROSA). It is a holonic reference architecture based on three types of basic holons: product, order, and resource (cf. [24]). Additionally, typically to all distributed architectures, the necessity to observe and control the shop floor and to provide an interface to the enterprise level systems evolved into the creation of a fourth special type of components. In PROSA, it is the staff holon, which assists and supervises the basic holons. Another architecture based on the holonic manufacturing systems (HMS) concept—MetaMorph and its follow-up MetaMorph II—uses a mediator centric federation architecture for intelligent manufacturing (cf. [37]). It uses a team mediator that provides communication mechanisms to different systems and components and takes over some of the functionalities of the MES. Generally speaking, a mediator is a supervisory agent that temporarily or permanently takes a role of decision making and combines certain resources into a virtual cluster to solve certain problems. However, please keep in mind that MetaMorph is meant to be an integration tool for other systems rather than a complete solution.

Similar agent-based architectures with slightly different implementation focuses are Autonomous Agents at Rock Island Arsenal (AARIA) (cf. [9]), manufacturing control systems capable of managing production change and disturbances (MASCADA) (cf. [26]), and ADAptive holonic COntrol aRchitecture (ADACOR) for distributed manufacturing systems (cf. [38]). Architectures like Advanced Fractal Companies Use Information Supply Chain (ADRENALIN, cf. [25]) consider agent-based resource brokering functionality enabling a decentralized manufacturing order navigation based on local optimization strategies.

In contrast to that, Plant Automation Based on Distributed Systems (PABADIS, cf. http://www.uni-magdeburg.de/iaf/cvs/pabadis/, [8]) and its follow-up PABADIS'PROMISE (PABADIS-based Product-Oriented Manufacturing Systems for Reconfigurable Enterprises, cf. http://www.pabadis-promise.org/) provide agent-based architectures that cover the whole automation pyramid but with the primary focus on distributed MES (cf. [25]).

Due to the fact that different MES solutions usually focus on a specific area of implementation, for example, scheduling, customer support, resource utilization, or system integration, they often cannot be compared. For this article, the focus is set on systems that cover the complete manufacturing process, that is, on systems that cover the vertical integration of all layers of the automation pyramid from the ERP down to the field control level. More specifically, the focus is on the HMS-based system PROSA (and some aspects of MetaMorph) and on the MAS-based systems PABADIS and PABADIS'PROMISE.

Although the above-mentioned solutions differ in their approaches, they exhibit in their concepts common patterns in approaching MES implementations. The main pattern is the division of decision making into two main entities: orders and resources. The first one represents the customers and the upper layers of systems that provide a MES with the orders on what to do (in practice a set of steps how to produce a product) that are then distributed into the agent community. While different solutions differ in the order handling concepts, the second entity—representing the resources on the shop floor that execute particular operations—has similar characteristics in all solutions.

In addition to the main patterns, all solutions provide supervising entities that show how far each concept deviated from conventional centralized approaches (see Figure 16.2). And last but not least, in order to glue the agents together, each architecture provides certain brokering functionality—as a centralized entity or via a hierarchy of objects, either static or dynamic.

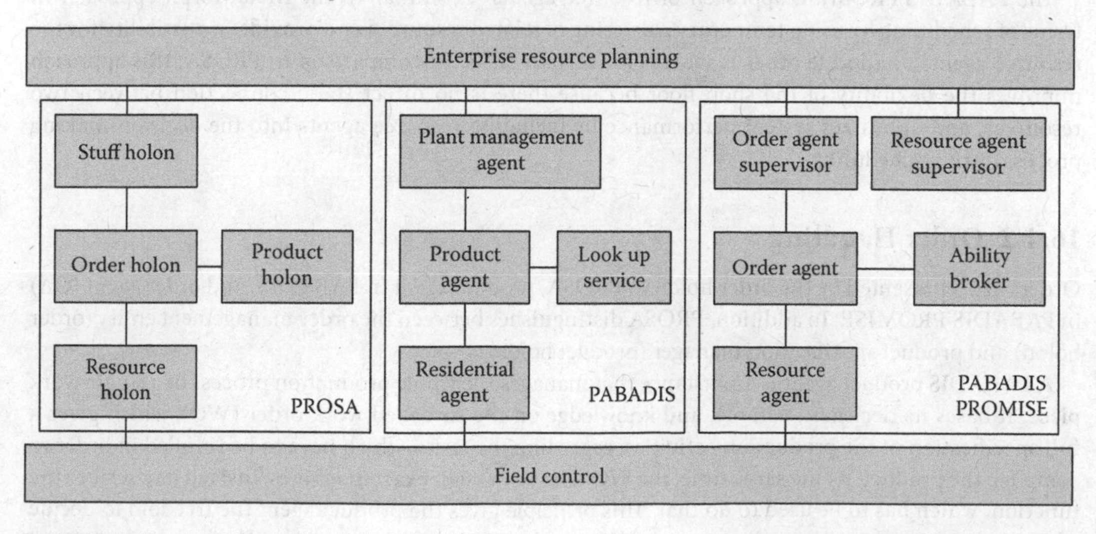

FIGURE 16.2 General patterns in distributed MES.

The following analysis of solutions is mainly based on resource and order handling approaches. PROSA implements the de facto state of the art. The other two are evolutions of the PABADIS architecture that differ from the HMS concept in terms of optimization versus flexibility (cf. [8]).

16.4.1 Resource Handling

Resources in HMS are represented by the resource holons, which are responsible for machine level representations. The resource holon consists of a physical part, namely a production resource in HMS, and an information processing part that controls the resource. This second part holds the methods to allocate the production resources, and the knowledge and procedures to organize, use, and control the physical production resources to drive production.

In terms of PABADIS, the resource holon is a cooperative manufacturing unit (CMU) that is a building unit of the shop floor. The information processing part of the resource holon perfectly fits the concept of the resource agent or a residential agent (RA) as it is called in PABADIS, which represented the CMU and partly controls a certain production or logical operation—so-called function.

The main differences between HMS and PABADIS are (1) in PABADIS the RA is mainly an interface between the agent community and CMU (machine, function unit) and (2) RA has a generic interface enabling the integration of arbitrary machine hardware. In HMS, the resource holon is more than this interface. It is rather a CMU-RA analog, where the whole functionality of the CMU (function, control level, agent communication level) is implemented. PABADIS distinguishes the standard (logical) part of the control level, which can be used for each CMU, from the "machine"-specific part, which has to be implemented by the customer of the system. This makes the system more flexible and encapsulates the MAS from the control level. That makes the adding of new functionality easier, because the customer only needs to add a specific plant dependent resource to the system and does not need to take into account the interoperability of the new component with the existing control system. Incorporation into the system is provided by the generic RA, which is capable of communicating within the system.

Hierarchy of resources is a key distinction point of distributed concepts. The PROSA architecture provides strict control of resources over each other. It improves resource utilization, but makes it complex to implement an actual system and has problems in adapting to a changing environment. Systems like PABADIS, however, treat resources totally independent from each other. PABADIS resource agents do not even communicate with each other. This makes scheduling or order handling independent from the shop floor configuration and hierarchy of resources.

The PABADIS'PROMISE approach of resource agents is similar to the MetaMorph approach in terms of scheduling by using temporary clustering of related resources and provides a possibility for the resource agents to allocate other resources for certain tasks. In comparison to PROSA, this approach improves the flexibility of the shop floor because there is no direct static connection between two resources, and optimizes system performance by including resource agents into the decision-making process during scheduling.

16.4.2 Order Handling

Orders are represented by the order holon in PROSA, product agent in PABADIS, and order agent (OA) in PABADIS'PROMISE. In addition, PROSA distinguishes between the order management entity (order holon) and product specification manager (product holon).

A PABADIS product agent is an instance that manages the whole production process of a single work piece. It bases its decisions, actions, and knowledge on the so-called work order (WO), which gives a full specification of the production activities, regarding the tasks, which have to be fulfilled in order to complete the product. At the same time, the WO does not assign exact machines; instead it describes the function, which has to be used to do that. This principle gives the product agent the freedom to decide what machine to use and introduces a possibility to change the machine during the execution.

An order holon has simpler logic but is more complex in hierarchy, meaning that it can divide a given order into suborders that are controlled by a parent holon of a higher layer. It mainly represents the customer request, where all requirements on the product are defined. It has an advantage in mass production because an order holon is not attached to a single work piece. An order holon does neither deal with scheduling tasks nor resource allocations because it simply does not have knowledge about the physical layout of the plant and the product specification. The latter, instead, is managed by a separate entity—the product holon. It is nothing more than a "recipe" for a particular type of product. A disadvantage of this solution is the lack of customization of products—a major reason for distributed automation systems.

Similar to the other aspects of general patterns there is always a balance between optimization and flexibility and hierarchy of orders and management entities related to them. While PROSA organizes orders in a strong hierarchy and PABADIS minimizes interdependencies between the orders, PABADIS'PROMISE provides what can be called an "implicit hierarchy." There is no strict decision making like in PROSA. The organization is implemented via the structure of the production order that reflects interdependencies between different production steps (called process segments) via so-called node operators. The mechanism of order decomposition developed in PABADIS'PROMISE allows assigning autonomously acting OAs to each process segment. It is similar to the PABADIS approach with the difference that higher level OAs can influence the agents responsible for subtasks execution. For instance, the OA that is responsible for the assembly of a car needs an engine to be produced before performing its tasks. Therefore, it assigns the deadlines for the engine agent, so the engine is delivered in time.

Another issue is the behavior of agents involved in the production. PROSA with its hierarchy does not need to pay attention to this topic, but in more distributed systems such as PABADIS it has higher importance because it can increase efficiency and performance. Product agents in PABADIS are selfish in nature. As a result, the system provides lower optimization of the order execution compared to PROSA, but much more flexibility. PABADIS'PROMISE aims to combine both benefits by introducing benevolent behavior of agents; that is, while agents rely on a certain set of rules that benefit their local goals (e.g., finishing assigned process segments on time) they may, nevertheless, sacrifice their goals in order to help struggling agents to meet their deadlines. This approach improves optimization considerably without reducing flexibility substantially, thus guaranteeing higher overall performance and stability of the system.

16.4.3 Comparison

From a general point of view PABADIS'PROMISE, PROSA, and PABADIS can be summarized as follows:

- *Autonomy and aggregation*: On the one hand PROSA lacks flexibility due to the direct control over aggregated entities and, on the other hand, PABADIS lacks efficiency because of total distribution. PABADIS'PROMISE defines production order decomposition that provides rules for controlling autonomous entities without dramatically reducing flexibility.
- *Cooperation and hierarchy*: Where PROSA has explicit hierarchy that causes rigidness of order changes, PABADIS provides no hierarchy that implies overhead in the management of complex products. PABADIS'PROMISE offers an implicit hierarchy (production order decomposition; flexible structure of orders; dynamic control of resources by resources) that finds the balance between the first approaches.
- *Decision making*: While decision-making in PROSA is centralized, meaning there is one control entity per functionality, PABADIS supports completely distributed decision-making mechanism. PABADIS'PROMISE provides a semi-distributed approach based on clustering of resources at the shop floor and production order decomposition at the MES layer.

- *Data interoperability*: In PROSA, one has to deal with implementation specific data that may cause difficulties during system installation. In PABADIS, data virtually does not have an established connection to the ERP system. In PABADIS'PROMISE, the ontology is spread throughout the entire automation pyramid linking all three layers (ERP–MES–field control) together.
- *Control flow*: PROSA has a strict vertical control flow that lacks feedback to the upper layer of the ERP. PABADIS has a limited feedback to the ERP. It actually approaches it only at the end of the production cycle. PABADIS'PROMISE supports permanent connection with the ERP via planned, periodic, or event-based reports during the production order life cycle.

16.4.4 Challenges of Industrial Agents' Usage

All above-mentioned architectures approach the manufacturing automation from the conceptual point of view and often overlook the application-related aspects that are implied by the distributed nature of the concepts. In particular, security, product identification, and data interoperability are vital aspects for practical implementations.

Distributed systems lack a single point of control with decision-making being spread over multiple entities that communicate with each other. This causes higher security risks, due to the intensive communication that such architectures require and the fact that there is no single entity that controls the system.

Another challenge for distributed systems is to provide a general overview of the processes and components on the shop floor. It is difficult to keep track of the products, work-in-progress, and materials in the plant, which implies that tracking the exact location of pieces is a huge challenge. Therefore, more advanced identification of the work pieces is required in order to guarantee the efficient operation of the system.

Last but not least, distribution of decision-making functionality requires interoperability of information flow over all three layers of the automation pyramid as well as mechanisms of distributed databases. Therefore, a common ontology with mechanisms of data abstraction for different control entities is required to guarantee the coordination within the system.

16.4.5 Other Application Areas in Brief

While agent-based applications have not yet achieved a noteworthy penetration in industry they nevertheless are heavily knocking at the door of several industrial application areas, including manufacturing and supply chain management, logistics, process control, telecommunication, (air) traffic and transportation, or defense (cf., e.g., [7]). In the center of many of those agent-based industrial systems stand aspects like coordination, cooperation, self-organization, timely and reasonable/intelligent reactions to unforeseen developments, planning and decision-making, especially also under real-time constraints and mainly in distributed areas where the individual units are supposed to be autonomous, and in intelligent behavior in general.

The aviation and space control industry is another frontrunner of agent technology. Especially, autonomous missions, such as collision control systems for unmanned, often also referred to as "uninhabited" or "remotely piloted" aerial vehicles (UAV) have attracted a lot of (industrial) research (cf. e.g., [10–13,15,17,18]). In the AGENTFLY project (cf. [15]) developed in the Gerstner Laboratory, Czech Technical University (CTU) in Prague has been working with the Air Force Research Laboratory, New York, on a software prototype supporting the planning and execution of aerial missions of UAVs that includes the free-flight-based collision avoidance of UAVs with other aerial vehicles. Here, UAVs are truly autonomous entities whose autonomy is realized by its own set of intelligent software agents. Each flight mission is tentatively planned before take-off, obviously without being able to consider already likely collisions. During a flight, the agents detect possible collisions and engage in peer-to-peer negotiations aimed at sophisticated replanning in order to avoid any collisions. Ref. [10] describes a similar project, in which two different approaches for adding autonomous control to an existing UAV are explored, tested, and evaluated. The AERIAL project (cf. [13]) is a distributed system that is able to ensure coordination and control of a fleet of UAVs involved in temporally constrained mission. Ref. [12] uses a different technology, namely swarm intelligence, for coordinating the movements of multiple UAVs. The control logic is executed autonomously by a UAV that results in a situation where a single human is able to monitor an entire swarm of UAVs instead of at least one human for each UAV. In Ref. [17], UAVs with small received signal strength indicator sensors cooperatively work together to locate targets emitting radio frequency signals in a large area. Ref. [11] describes a decision making partnership between a human operator and an intelligent UAV. Here, the intelligence is provided by a MAS. The MAS controls the UAV and self-organizes to achieve the tasks set by the operator, however, also through interaction via what they call a variable autonomy interface with the operator. In all these projects, especially the autonomy and intelligence aspects are realized by agent technology. Autonomy, however, requires a lot of communication in order for a set of autonomous units to achieve their goals cooperatively. Thus, communication and decision-making plays a major role in those projects.

Related to supply chain management is logistics in general. In logistics, the information and data required for efficient planning are, for various reasons, typically not available centrally. Quite a number of systems have been deployed in this area from which only two will be presented briefly. The Living Systems/Adaptive Transport Network (LS/ATN, cf. [40]), provided by Whitestein Technologies, for example, to ABX, a European logistics company, is an agent-based system for network planning and transportation optimization for charter business logistics. Based on a simulation study that was performed by Whitestein on the basis of 3500 transportation requests, 11.7% cost savings were achieved. The other very successful commercial deployment is the Magenta i-scheduler, an intelligent, event-driven logistics scheduling system based on Magenta agent technology. It is characterized by a number of unique, advanced features such as real-time, incremental, continuous scheduling and schedule improvement, scalability from small to very large enterprise networks, which also includes balancing of their conflicting interests, multi-criteria schedule analysis, and rich decision support for the user. The system is deployed already in a number of industrial applications. As a scheduling/logistics system for Tankers International, it provides intelligent support in the scheduling of a 46-strong very large crude carrier (VLCC) fleet. I-scheduler implements virtual marketplaces with about 1000 agents running during a typical execution of the system (cf. [19]). I-schedulers were also deployed in several road transportation applications with several U.K. road logistics operators (cf. [21]). The application reported in [20] copes with about 1500 orders daily, in 650 locations, 150 own trucks, and, additionally, 25 third-party carriers. I-Scheduler assigns an autonomous agent to every player in the transportation system and, additionally, tasking agents to obtain the best possible deals for their clients. Players include the transportation enterprise as a whole and all individual transportation demands and resources, like drivers and resource owners. Demands and resources that have common interests self-organize into groups represented by a single agent. Agents may decide to compete or cooperate depending on prevailing circumstances. To construct a schedule, a formal description of business domain knowledge (Ontology), of particular situations (a Scene), of agent goals (e.g., to increase profit), and of real-world events are used.

Another prominent area for agent-based systems is virtual organizations (VO) or enterprises (VEs) (see, e.g., [14] for a nice overview). VOs may deeply integrate all manufacturing and supply chain aspects, sales networks, as well as suppliers, customers, and third-party maintenance and coordination. Here, again, the cooperation and coordination aspect of such conglomerates stays in the center of what agents are supposed to provide.

More recently, research tried to add more semantics to industrial application systems. This usually means that ontology-based concepts are integrated. A nice overview about this trend is given in [22]. An example for a prominent and successful approach is presented in [23]. The OOONEIDA consortium provides a technological infrastructure for a new, open knowledge economy for automation components and automated industrial products. Their framework aims to provide interoperability on the hardware as well as on the software at all levels of the automation components market, that is, from device and machine vendors to system integrators up to the industrial enterprises. In the center of their approach are the so-called searchable repositories of automation objects. Here, each player can deposit its encapsulated intellectual property along with appropriate semantic information to facilitate searching by intelligent repository agents.

16.5 Agents and Multi-Agent Systems in Industry: Conclusions

The use of agent technology in industrial applications provides several important strengths, namely in terms of modularity, adaptability, flexibility, robustness, reusability, and reconfigurability (cf. [7,16,39,41,42]).

In contrast to conventional, centralized, top-down, rigid, and static control architectures, agent-based systems are developed under the fundamental principles of autonomy and cooperation, exploring the distribution and decentralization of entities and functions, over a bottom-up approach. The agent-based solution can lead to more simplicity in the debugging, maintenance, adaptability, and scalability of the system.

The robustness of the control system is essentially achieved since this approach does not consider a central decision element, which means that the loss of one decision component will not cause any fatal failure of any other decision component. In fact, if production is restructured, example, due to the occurrence of disturbances, the same negotiation process continues to be executed for, in spite of the presence of different actors, making the system robust to changes.

Agent-based systems are pluggable systems, allowing changes in production facilities, as the addition, removal, or modification of hardware equipment as well as software modules, without the need to stop, reprogram, and reinitialize the system. This feature is crucial to support the current requirements imposed by customized processing, allowing dynamic system reconfigurability to face the variability of the demand. The migration or update of old technologies or systems by new ones can also be performed in a smooth way without the need to stop the system (cf. [7]).

However, in spite of the huge potential of agent-based solutions, its industrial adoption has yet fallen short of expectations. Several reasons can be identified for this slow industrial adoption, grouped in two categories: conceptual limitations and technical limitations.

In terms of conceptual limitations, the agent-based paradigm is a new way of thinking that requires a bit of a paradigm shift from the way manufacturing systems have been realized in the last 100 years. In fact, as companies have invested a lot of time, effort, and money implementing centralized approaches, they do not want to change them. Additionally, one of the consensuses that people have about agent technology is the missing centralized component for decision making, which also causes some obstacles for the acceptance of these concepts.

In terms of technical limitations, interoperability, robustness, scalability, and reconfiguration mechanisms remain important issues that may inhibit the wider use of agent-based solutions by industry. Taking the scalability, for example, the current reported laboratorial prototypes deal with dozens or hundreds of agents, but industrial applications usually require systems that comprise thousands of agents. Current platforms cannot handle systems of that size with the robustness required by industry (cf. [7,42]).

The previously identified limitations constitute research opportunities and challenges from which especially the following ones can be pointed out: benchmarking, interoperability, and development of engineering frameworks.

In order to proof the maturity and merits of the technology and to convince industry about its applicability in industrial scenarios, a benchmarking framework is required that provides realistic test cases that allow evaluating agent-based solutions and comparing them with traditional ones. Some benchmarking issues remain undefined, namely the selection of proper performance indicators, especially those that allow to evaluate qualitative indicators, the definition of evaluation criteria, the storage and maintenance of the best practices, and an easy access to this service.

Interoperability is probably the main technical problem associated to agent technology that should be addressed. The solution for the interoperability question requires more research on ontologies and Semantic Web domains and opens a door to the use of service-oriented systems (cf. [39]), combining its best features with the agent technology.

The specification of formal, structured, and integrated development engineering frameworks is absolutely needed for the industrial practice in order to support the specification, design, verification, and implementation of agent-based control applications, allowing an easy, modular, and rapid development/reconfiguration of control solutions.

Abbreviations

AARIA	Autonomous Agents at Rock Island Arsenal
ADACOR	ADAptive holonic COntrol aRchitecture for distributed manufacturing systems
ADRENALIN	Advanced Fractal Companies Use Information Supply Chain
CMU	Cooperative manufacturing unit
ERP	Enterprise resource planning
HMS	Holonic manufacturing systems
LS/ATN	Living systems/adaptive transport network
MAS	Multi-agent system
MASCADA	Manufacturing Control Systems Capable of Managing Production Change and Disturbances
MES	Manufacturing execution system
OAS	Order agent supervisor
PABADIS	Plant automation based on distributed systems
PABADIS'PROMISE	PABADIS-based product-oriented manufacturing systems for reconfigurable enterprises
PROSA	Product–resource–order–staff architecture
RA	Residential agent, resource agent
RAS	Resource agent supervisor
SOA	Service oriented architecture
VLCC	Very large crude carrier
VO	Virtual organizations
WO	Work order
UAV	"Uninhabited" or "remotely piloted" aerial vehicles

References

1. Wooldridge, M. and Jennings, N. R., Intelligent agents: Theory and practice. *The Knowledge Engineering Review*, 10(2):115–152, 1995.
2. Koestler, A., *The Ghost in the Machine*, Arkana Press, London, U.K., 1967.
3. Marík, V., Fletcher, M., and Pěchouček, M., Holons and agents: Recent developments and mutual impacts, in *Proceedings of the Ninth ECCAI-ACAI/EASSS 2001, AEMAS 2001, HoloMAS 2001, Lecture Notes in Artificial Intelligence*, Vol. 2322, Springer Publishing Company, New York, pp. 3–43, 2002.
4. Schillo, M., Self-organization and adjustable autonomy: Two sides of the same coin? *Connection Science*, 14(4), 345–359, 2003.
5. Fischer, K., Agent-based design of holonic manufacturing systems. *Journal of Robotics and Autonomous Systems*, 27, 3–13, 1999.
6. Shen, W. and Norrie, D. H., Agent-based approaches for intelligent manufacturing: A state-of-the-art survey, in *Proceedings of DAI'98, Fourth Australian Workshop on Distributed Intelligence*, Brisbane, Australia, July 13, 1998.
7. Marik, V. and McFarlane, D., Industrial adoption of agent-based technologies, *IEEE Intelligent Systems*, 20(1), 27–35, 2005.
8. Luder, A., Klostermeyer, A., Peschke, J., Bratoukhine, A., and Sauter, T., Distributed automation: PABADIS versus HMS, *IEEE Transactions on Industrial Informatics*, 1(1), 31–38, 2005.
9. Parunak, H. V. D., What can agents do in industry, and why? An overview of industrially-oriented R&D at CEC, in Klusch, M. and Weiss, G. (Eds.), *Cooperative Information Agents II—Learning, Mobility and Electronic Commerce for Information Discovery on the Internet, Lecture Notes in Computers Science*, Vol. 1435, Springer, Paris, France, pp. 1–18, 1998.

10. Karim, S. and Heinze, C., Experiences with the design and implementation of an agent-based autonomous UAV controller, in Pěchouček, M., Steiner, D., and Thompson, S. G. (Eds.), *Proceedings of Fourth International Joint Conference on Autonomous Agents and Multiagent Systems (AAMAS 2005)*, Utrecht, the Netherlands, July 25–29, 2005—Special Track for Industrial Applications, ACM Press, New York, 2005.
11. Baxter, J. W. and Horn, G. S., Controlling teams of uninhabited air vehicles, in Pěchouček, M., Steiner, D., and Thompson, S. G. (Eds.), *Proceedings of Fourth International Joint Conference on Autonomous Agents and Multiagent Systems (AAMAS 2005)*, Utrecht, the Netherlands, July 25–29, 2005—Special Track for Industrial Applications, ACM Press, New York, 2005.
12. Parunak, H. V. D., Brueckner, S., and Sauter, J., Digital pheromone mechanisms for coordination of unmanned vehicles, in *Proceedings of First International Joint Conference on Autonomous Agents and MultiAgent Systems (AAMAS): Part 1*, Bologna, Italy, July 15–19, 2002.
13. Marson, P.-E., Soulignac, M., and Taillibert, P., AERIAL: Hypothetical trajectory planning for multi-UAVs coordination and control, in *Proceedings of Seventh International Joint Conference on Autonomous Agents And MultiAgent Systems: Demo Papers*, Estoril, Portugal, May 12–16, 2008.
14. Camarinha-Matos, L. M. and Afsarmanesh, H., Virtual enterprise modeling and support infrastructures: Applying multi-agent system approaches, in *Proceedings of the Ninth ECCAI Advanced Course ACAI 2001 and Agent Link's Third European Agent Systems Summer School, EASSS 2001*, *Lecture Notes in Artificial Intelligence*, Vol. 2086, Prague, Czech Republic, July 2–13, 2001, Selected Tutorial Papers; Springer Publishing Company, New York, 2001.
15. Pěchouček, M., Šišlák, D., Pavlíček, D., and Uller, M., Autonomous agents for air-traffic deconfliction, in *Proceedings of Fifth International Joint Conference on Autonomous Agents and MultiAgent Systems (industry track)*, Hakodate, Japan, May 2006.
16. Belecheanu, S. M., Michael, L., and Terry, P., Commercial applications of agents: Lessons, experiences and challenges, in *Proceedings of Fifth International Joint Conference on Autonomous Agents and MultiAgent Systems (industry track)*, Hakodate, Japan, May 2006.
17. Scerri, P., Von Gonten, T., Fudge, G., Owens, S., and Sycara, K., Transitioning multiagent technology to UAV applications, in *Proceedings of Seventh International Conference on Autonomous Agents and Multiagent Systems (AAMAS)—Industry and Applications Track*, Estoril, Portugal, May 12–16, 2008.
18. Reichel, K., Hochgeschwender, N., and Voos, H., OpCog: An industrial development approach for cognitive agent systems in military UAV applications, in *Proceedings of Seventh International Conference on Autonomous Agents and Multiagent Systems (AAMAS)—Industry and Applications Track*, Estoril, Portugal, May 12–16, 2008.
19. Himoff, J., Skobelev, P., and Wooldridge, M., MAGENTA technology: Multi-agent systems for industrial logistics, in Pěchouček, M., Steiner, D., and Thompson, S. G. (Eds.), *Proceedings of Fourth International Joint Conference on Autonomous Agents and Multiagent Systems (AAMAS 2005)*, Utrecht, the Netherlands, July 25–29, 2005—Special Track for Industrial Applications, ACM Press, New York, 2005.
20. Himoff, J., Rzevski, G., and Skobelev, P., MAGENTA technology: Multi-agent logistics i-scheduler for road transportation, in *Proceedings Fifth International Joint Conference on Autonomous Agents and MultiAgent Systems (industry track)*, Hakodate, Japan, 2006.
21. Skobelev, P., Glaschenko, A., Grachev, I., and Inozemtsev, S., MAGENTA technology case studies of magenta i-scheduler for road transportation, in *Proceedings of Sixth International Joint Conference on Autonomous Agents and MultiAgent Systems*, Honolulu, HI, 2007.
22. Obitko, M., Vrba, P., Mařík, V., and Radakovič, M., Semantics in industrial distributed systems, in *Proceedings of the 17th World Congress*, The International Federation of Automatic Control, Seoul, Korea, July 6–11, 2008.
23. Vyatkin, V., Christensen, J., Lastra, J. L. M., and Auinger, F., OOONEIDA: An open, object-oriented knowledge economy for intelligent industrial automation, *IEEE Transactions on Industrial Informatics*, 1(1), 4–17, 2005.

24. Deen, S. M., *Agent Based Manufacturing—Advances in the Holonic Approach: Advanced Information Processing*, Springer Verlag, Berlin, Germany, 2003.
25. Klostermeyer, A., *Agentengestützte Navigation wandlungsfähiger Produktionssysteme*, Dissertation, Faculty of Mechanical Engineering, Otto-v.-Guericke University, Magdeburg, Germany, ISBN 3-00-011618-4.
26. Heikkila, T., Kollingbaum, M., Valckenaers, P., and Bluemink, G. J., manAge: An agent architecture for manufacturing control, in *Proceedings of the Second International Workshop on Intelligent Manufacturing Systems*, Leuven, Belgium, September 1999.
27. Bellifemine, F., Caire, G., and Greenwood, D., *Developing Multi-Agent Systems with JADE*, Wiley, Chichester, U.K., 2007.
28. Vrba, P., JAVA-based agent platform evaluation, in Marík, V. et al. (Eds.), *Multi-Agent Systems and Applications III, LNAI 2691*, Springer-Verlag, Berlin/Heidelberg, Germany, pp. 47–58, 2003.
29. Foundation for Intelligent Physical Agents (FIPA), 2009, http://www.fipa.org.
30. Gruber, T. R., A translation approach to portable ontology specifications, *Knowledge Acquisition*, 5(2), 199–220, June 1993.
31. Haken, H., *Information and Self-Organization: A Macroscopic Approach to Complex Systems*, Springer-Verlag, Berlin, Germany, 1988.
32. Klir, G., *Facets of System Science*, Plenum Press, New York, 1991.
33. De Wolf, T. and Holvoet, T., Emergence and self-organisation: A statement of similarities and differences, in *Workshop on Engineering Self-Organizing Applications (ESOA); as Part of Third International Joint Conference on Autonomous Agents and MultiAgent Systems (AAMAS)*, New York, July 21–24, 2004.
34. Valckenaers, P. et al., On the design of emergent systems: An investigation of integration and interoperability issues, *Engineering Applications of Artificial Intelligence*, 16, 377–393, 2003.
35. Ulieru, M. and Unland, R., Enabling technologies for the creation and restructuring process of emergent enterprise alliances, *International Journal of Information Technology and Decision Making*, 3(1), 33–60, March 2004.
36. Bratman, M. E., *Intentions, Plans, and Practical Reason*, Harvard University Press, Cambridge, MA, 1987.
37. Maturana, F. and Norrie, D., Multi-agent mediator architecture for distributed manufacturing, *Journal of Intelligent Manufacturing*, 7, 257–270, 1996.
38. Leitão, P. and Restivo, F., ADACOR: A holonic architecture for agile and adaptive manufacturing control, *Computers in Industry*, 57(2), 121–130, February 2006.
39. Leitão, P., Agent-based distributed manufacturing control: A state-of-the-art survey, *Engineering Applications of Artificial Intelligence*, 22(7), 979–991, 2009.
40. Neagu, N., Dorer, K., Greenwood, D., and Calisti, M., LS/ATN: Reporting on a successful agent-based solution for transport logistics optimization, in *Proceedings of the IEEE Workshop on Distributed Intelligent Systems: Collective Intelligence and Its Applications (DIS'06)*, Washington, DC, June 15–16, 2006.
41. Shen, W., Wang, L., and Hao, Q., Agent-based distributed manufacturing process planning and scheduling: A state-of-the-art survey, *IEEE Transactions on Systems, Man, and Cybernetics—Part C: Applications And Reviews*, 36(4), July 2006.
42. Pěchouček, M. and Marík, V., Industrial deployment of multi-agent technologies: Review and selected case studies, *Autonomous Agents and Multi-Agent Systems*, 17(3), 397–431, December 2008.

17

Real-Time Systems

17.1 Introduction on Real-Time Systems 17-1
Definition • Real-Time Constraints Characterization • Typical Application Domains • Real-Time Scheduling and Relevant Metrics

17.2 Real-Time Communication 17-4
Deterministic vs. Statistical Communication • Best Effort vs. Guaranteed Service • Performance Metrics • Analytical Methods to Assess Performance of Real-Time Networks

17.3 Design Paradigms for Real-Time Systems 17-7
Centralized vs. Distributed Architectures • Composability and Scalability • Time-Triggered vs. Event-Triggered Systems • Comparison of the Real-Time Support Provided by Notable Approaches

17.4 Design Challenges in Real-Time Industrial Communication Systems 17-9
Real-Time and Security • Real-Time and Flexibility • Offering Real-Time Support to Wireless Communication

References 17-10

Lucia Lo Bello
University of Catania

José Alberto Fonseca
Universidade of Aveiro

Wilfried Elmenreich
University of Klagenfurt

17.1 Introduction on Real-Time Systems

17.1.1 Definition

A real-time system is a system in which the correctness of the results does not only depend on the logical correctness of the computations/operations performed to obtain them but also on the time the results are obtained [Kop97,Sta88]. Logically correct, but late results can be either not useful or even harmful to the system, depending on the nature of the application considered. Examples of real-time applications are found in industrial plant controls, automotive, avionics, industrial automation, robotics, monitoring systems, multimedia systems, telecommunications, interactive systems, and consumer electronics. Unlike conventional, non-real-time computer systems, real-time computer systems are closely coupled with the physical process or the environment being monitored and controlled.

A *real-time system* (Figure 17.1) consists of a real-time computer system, a controlled object, and may also feature an operator. The real-time computer system must react to stimuli from the controlled object (or the operator) within a time interval specified by the *deadline*. The concept of deadlines must not be confused with fast- or high-performance computing. Real-time computing is different since the objective of fast computing is to minimize the *average* response time of a task, while real-time computing is concerned about the *maximum* or *worst-case* response time and the difference between minimum and maximum response time, the so-called *jitter* [Sta88].

The most important property of a real-time system is not high speed, but *predictability* [Sta90], which means the ability to determine in advance whether the task can be completed within its deadlines or not.

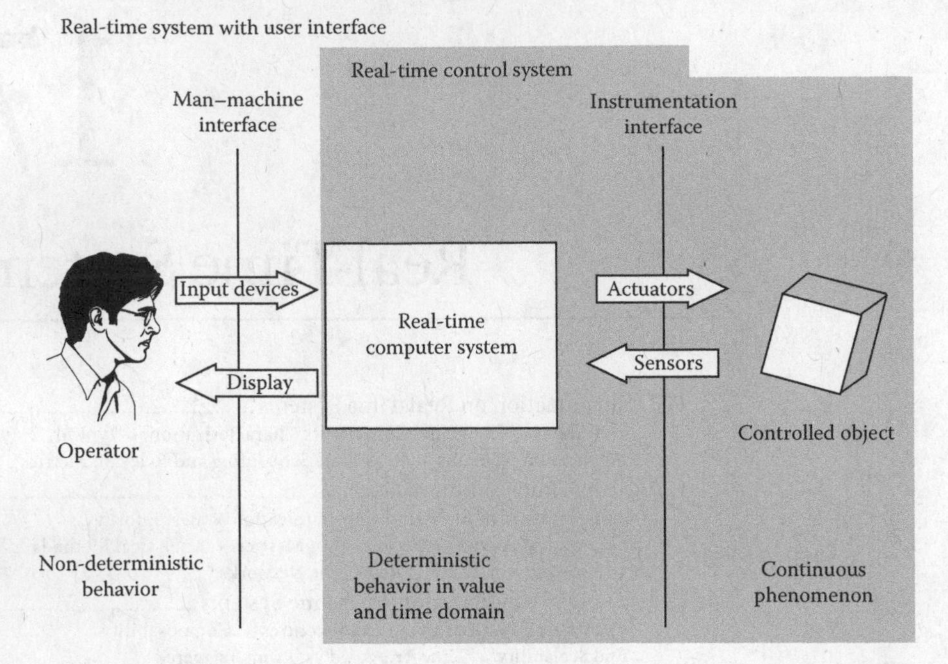

FIGURE 17.1 Elements of a real-time system.

17.1.2 Real-Time Constraints Characterization

A criterion commonly used to classify the timing constraints of real-time tasks is the usefulness of their results (*value*) as functions of the tasks' completion time and, in particular, of their *lateness*. The lateness of a task is defined as the difference between the task's actual completion time and its *deadline*. If the task is on time, its lateness will be zero or negative. If the task is late, a positive lateness will be found. To what extent a deadline miss will compromise the system depends on the real-time nature of the deadlines imposed on the system. When missing a deadline can have catastrophic consequences to the system, the deadline is called *hard*. If a deadline miss only entails a performance degradation, but does not jeopardize the correct system behavior, the deadline is called *soft*.

According to this criterion, three types of tasks can be identified, i.e., *hard*, *soft*, and *firm* real-time. All these kinds of tasks give a positive *value* to the system if they complete on time. The difference between the three categories becomes significant when the tasks are late. If the late task is *hard* real-time, the value of the result produced by the task after its deadline is negative, thus indicating that a damage on the system occurs. As a result, hard real-time tasks have to meet their deadlines under any circumstances. Conversely, the value of the result produced by a *soft* real-time task starts decreasing as the lateness of the task becomes positive and will eventually become null at a given point of time after the deadline. So, the late completion of a soft real-time task can be occasionally tolerated, and it is advisable to minimize the lateness of the task, as the value of a late task decreases with time. Finally, the value of the result produced by a *firm* real-time task drops abruptly to zero as soon as the lateness becomes zero and remains null with increasing lateness values. As it brings no value to the system, a firm real-time task should be either completed on time or dropped if it is late, as there is no point to continue its execution after the deadline.

In the case of hard deadlines, suitable validation techniques are required to demonstrate that the system adheres to the intended timing behavior. If the deadlines are soft, in general no validation is required and it is sufficient to prove that it is possible to meet a timing constraint specified in terms of statistical averages (e.g., that the average deadline miss rate is below a given application-dependent threshold).

17.1.3 Typical Application Domains

Real-time systems can be found in many different application domains. However, unlike personal computers (PCs) and workstations where users are fully aware of (non-real-time) applications such as E-mail, Internet browser, and text processing, real-time applications are often implemented as *embedded systems* providing their service hidden from our view [Elm09]. The typical application domains for real-time systems are

- Digital control systems implementing a feedback loop from one or more sensor readings to one or more actuators. Thus, the controlled parameter is kept at a desired value even when conditions change in the system. Since such control loops are typically very sensitive to jitter and delay, a digital control system has to be implemented in a way that the delay between measurement and actuating is small in comparison to the timing parameters of the controlled system. Jitter is especially critical in such applications. Digital control systems are hard real-time systems. Violation of the timing requirements can result in instabilities of the control loop and, therefore, often in damage of physical systems. An example is the ignition control system in a car's combustion engine. The requirement of this application is basically to make the fuel–air mixture ignite in the cylinder so that the piston is accelerated on its way down. Now imagine a badly designed control system that may eventually delay the ignition instant by a few milliseconds—the fuel–air mixture is now ignited while the piston is in the wrong position. This small time delay changes the forces to piston, connection rod, and crankshaft in a way that may permanently damage the engine.
- Man–machine based real-time systems. The real-time control system incorporates the instrumentation interface consisting of input devices and output devices that interface to a human operator. Across this interface, the real-time computer system is connected to the process environment via sensors and actuators that transform the physical signals of the controlled object into a processable form and vice versa.
- High-level controls involve flight management systems, factory automation, robot control, etc. The real-time tasks require, for example, path and trajectory planning. The real-time control system is required to provide its results within a given maximum response time. Many of these applications, e.g., in transportation or plant automation, are critical requiring a highly dependable implementation of the real-time control system.
- Many signal processing applications come with real-time requirements, needing to provide throughput at a given sampling rate. Voice processing applications for telephony further require that the processing delay does not exceed a given maximum. This kind of application often has soft real-time requirements, since a violation of the timing merely diminishes the quality of the service. Other soft real-time applications include video conference applications, online gaming, and, to some extent, instant messaging.

17.1.4 Real-Time Scheduling and Relevant Metrics

A *schedule* is a mapping of task executions on a resource. For example, if the tasks are processes, the resource will be the processor, while if the tasks to be performed are data packet transmissions, the resource will be the network bandwidth. A schedule is *feasible* if all the tasks complete their execution under a set of specified constraints. A task set is *schedulable* if there exists a feasible schedule for it. In the following, like in [Liu00], we define a *task* as a sequence of *jobs* that jointly provide a system function, while a *job* is the unit of work that is scheduled and executed by the system. This notation allows us to deal with different system activities, such as the execution of a process on a processor or the transmission of a data packet on a network, in a uniform way.

While in conventional non-real-time systems, the commonly used metrics are throughput, fairness, and average response times, in real-time scheduling all the target metrics are relevant to the system timeliness. Typical metrics are therefore

- Response time, defined as the difference between the completion time and the release time of a job
- Absolute jitter, defined as the difference between the minimum and the maximum response time for a job
- Maximum lateness for soft real-time tasks, defined as the maximum difference between the finishing time of a job and its deadline
- Deadline miss ratio for soft real-time tasks, i.e., the ratio between the number of jobs that missed their deadlines and the overall number of soft real-time jobs released
- Throughput on time, defined as the amount of jobs completed by the deadline per time unit
- Cumulative value for a set of soft or firm real-time tasks, defined as the sum of all the values calculated at each job completion times, given their value functions

Since the seminal paper of Liu and Layland [Liu73], a very large amount of works dealt with real-time scheduling, addressing it in multiple scenarios and from different perspectives. Interested readers may refer to [Liu00] and [Sha04].

17.2 Real-Time Communication

17.2.1 Deterministic vs. Statistical Communication

In hard real-time systems, deadline miss should never happen, as deadline misses severely affect the application behavior and determine non-recoverable errors. In soft real-time systems, occasional violations of deadlines can be tolerated, although at the expense of a degradation of the *quality of service* (QoS) provided to the application. Offering real-time support on networks means that a predictable time behavior of communications can be guaranteed, either in a deterministic or in a stochastic way. Predictable time behavior is not necessarily a synonym of constant time behavior. What really counts is the existence of an upper bound on some relevant QoS index, such as end-to-end packet delivery time, packet channel access time, roundtrip time, etc. The existence of such an upper bound makes it possible to assess whether the packet deadlines will be met or not, in either a deterministic or a stochastic way. It is therefore possible to define a

- Deterministic real-time channel as one that provides a priori deterministic guarantees for the timely delivery of packets between two end-points
- Statistical real-time channel as one that guarantees the timely delivery of packets between two end points in statistical terms, i.e., that the probability that a packet misses its deadline is less than a certain loss tolerance Z, i.e., $Pr(packet\ delay > delay\ bound) \leq Z$

Statistical guarantees on deadline meeting satisfy the requirements of real-time non-mission-critical applications, such as periodic control or industrial multimedia, which can tolerate a small violation probability with delay bounds. In an automated manufacturing system, for example, real-time periodic control messages may need to be delivered within 20 ms of their generation with a 98% probability, while voice packets may require delivery within 40 ms with a 94% probability, and so on. Deterministic and statistical analysis techniques for wired industrial networks have been thoroughly investigated in the literature. For instance, in [Kwe03] Shin et al. analytically demonstrate that, in order to statistically bound the medium access time for a real-time packet transmitted over Ethernet networks in the presence of both real-time (RT) and non-real-time (NRT) traffic, it is sufficient to keep the total arrival rate for new packets generated by stations below a threshold called the *network-wide input limit*. To enforce such a limit on a per-station basis, each station is assigned a local threshold, called a *station input limit*, and a middleware called a *traffic smoother* is entrusted with the regulation of the outgoing non-real-time stream on each node. Several traffic smoothers have been proposed in the literature, for either shared or switched Ethernet, which differ in the way the station input limit is enforced. More details can be found in [Kwe04] and [LoB05]. Among recent approaches to obtain real-time performance over an Ethernet, there are the FTT-Ethernet [Ped05], the PEAC protocol [Bon03], and the Time-Triggered Ethernet (TTE) [Kop05].

17.2.2 Best Effort vs. Guaranteed Service

Guaranteed service means that the user is given a *guarantee* on the system timing behavior (for instance, on a message end-to-end delay and/or jitter, medium access time, etc.) and that the meeting of the relevant timing constraints has to be validated, through formal methods or exhaustive simulation and testing. This service is required for hard real-time traffic. On the contrary, *best-effort service* means that, although the user requires the best QoS the system can provide, the system is allowed to deliver a lower quality than expected. This service is suitable for soft real-time traffic.

17.2.3 Performance Metrics

In real-time communication, the primary performance metrics are related to the timeliness of data exchange over the network. As real-time flows have to be provided with a different QoS than non-real-time ones, when different types of traffic have to be transmitted over the same channel, traffic prioritization is required. In particular, end-to-end real-time performance of real-time traffic is very important, as usually networked real-time systems handle monitoring and control applications, which typically require a response within *bounded delay*, often combined with *low jitter* values.

Typically, the end-to-end delay is composed of several stages of data processing and transmission. To be able to provide a bounded delay, the delay of each stage must be bounded. For example, the queuing delay in the network, which in turn depends on the queuing delay in the local queues and on the number of hops in the path from the source to the destination node of the packet in packet switched networks. In each network device traversed by the packet, multiple queues with different priorities are needed to handle different traffic classes and real-time scheduling algorithms, such as weighted fair queuing (WFQ) [Dem90], or non-preemptive versions of rate monotonic or earliest deadline first [Liu73] can be exploited to deal with real-time traffic and calculate delay bounds for it. However, WFQ introduces a significant computational overhead, while the effectiveness of priority queuing depends on the number of priorities, which can be defined (the finer the granularity, the better the resulting traffic differentiation). Even the capacity of the queues of the network devices the packet traverses on its path and the total network load have to be taken into account, as packet dropping may occur if any of these queues fills up. Suitable flow control algorithms, selective discarding, and packet marking policies are needed to support QoS for real-time traffic. Another significant contribution to the end-to-end delay is given by the transmission delay, which is the time span between the emission of the first bit and the emission of the last bit of a frame. The transmission delay depends on the packet size and the network transmission rate. The propagation time, defined as the time between the emission of the first bit on the transmitter side and the reception of such a bit on the receiver side, also contributes to the end-to-end delay, although its impact for local area networks is negligible. Channel errors occurring during the transmissions of a packet may entail the retransmission of the corrupted packet. However, when dealing with real-time traffic, especially with periodic control packets in factory communication, the advantage of retransmissions has to be carefully evaluated, as there is a non-null probability that the retransmitted data becomes obsolete during its way to the destination, due to the generation of a "fresh" value. However, a large part of the end-to-end delay at the application level is due to the *latency* at the *end nodes*. In [Ske02], with reference to switched Ethernet, it was demonstrated that the use of priorities for packets in the communication infrastructure is not on its own sufficient to guarantee that application-to-application deadlines will be met and that the concept of priority has to be extended to the end nodes in the protocol stack. Other works dealt with limitations of switched Ethernet as far as real-time communication is concerned [Jas02]. In [LoB06], a significant reduction in the roundtrip delay for high-priority traffic is obtained thanks to an approach based on both a prioritization mechanism in the protocol stack and multiple reception and transmission queues in the end nodes.

Many networked real-time systems feature applications with further requirements than time constraints only, for instance, mobility, dependability, composability, scalability, flexibility, security, and safety. The way of dealing with real-time constraints in the presence of such requirements depends on the particular scenario.

17.2.4 Analytical Methods to Assess Performance of Real-Time Networks

The analysis of real-time networks can be done with methods that have emerged either in the field of real-time systems or in the field of communication networks. Notable examples are response-time analysis (RTA) and network calculus (NC).

RTA is a method to verify the schedulability of a task set in a processor or of a message set in a communication network, which is based on the response time computation and comparison with the deadline for each task or message in the set. If the deadlines are met for every instance of the tasks or messages that feature real-time constraints, then the schedulability is guaranteed. So, RTA provides an exact schedulability test. RTA has been first presented in [Jos86], but many evolutions of RTA then followed. In [Aud91], the restriction of the deadline being equal to the period was released and RTA was extended to tasks with deadlines less or equal to the period. [Leh90] extended RTA to deadlines greater than the period, thus enabling several active instances of the same messages. One of the most used results in industrial communications is the work where RTA was extended to controller area network (CAN) [Tin94], thus accounting for non-preemptive fixed-priority scheduling. Using RTA, the queuing delay of a message can be obtained using a recursive equation. Following this result, several specific RTA-based analyses have been derived for different real-time networks, in the field of industrial automation. In [Alm02], an RTA for both periodic, using fixed priorities, and aperiodic exchanges of variables (messages) is presented for WorldFIP, leading to a schedulability condition necessary and sufficient for the periodic traffic although just sufficient for the aperiodic one. A recent work by Bril [Bri06] revisited the RTA for ideal CAN network showing by means of examples with a high load (≈98%) that the analysis as presented in [Tin94] is optimistic. They proved that, assuming discrete scheduling, the problem can be resolved by applying the analysis for fixed-priority non-preemptive scheduling presented in [Geo96].

Network calculus was introduced by Cruz in [Cru91a,Cru91b] to perform deterministic analysis in networks where the incoming traffic is not random but unknown, being limited by a known arrival curve. Network calculus differs from queuing theory in the sense that it deals with worst case instead of average cases. The parameters of interest in this methodology are the delay, the buffer allocation requirements, and the throughput. Using the arrival curve, often named α, and a service curve offered by the network, named β, and considering a traffic flow $R(t)$, it is possible to determine upper bounds for the backlog (the number of bits in "transit" in the system [Bou04]) and the delay.

Network calculus has been used in several real-time domains such as industrial automation, avionics, and wireless sensor networks. Considering industrial automation, in [Geo04] network calculus was used to model a switched Ethernet architecture in order to evaluate the maximum end-to-end delays that are critical for industrial applications. In [Rid07], a stochastic network calculus approach to evaluate the distribution of end-to-end delays in the AFDX (avionics full-duplex switched Ethernet, ARINC 664) network used in aircrafts such as Airbus A380 is presented. Network calculus has also been applied to the dimensioning of wireless sensor networks [Sch05].

17.3 Design Paradigms for Real-Time Systems

17.3.1 Centralized vs. Distributed Architectures

A centralized architecture for a real-time system comes with the advantage that no communication between system components has to be considered, which saves cost and design effort. On the other hand, there are several reasons for building distributed real-time systems:

- *Performance*: A centralized system might not be able to provide all the necessary computations within the required time frames. Thus, tasks are to be executed concurrently on networked hardware.
- *Complexity*: Even if it is possible to implement a real-time system with a centralized architecture, a distributed solution might come in with a lower complexity regarding scheduling decisions, separation of concerns, etc. In many cases, a complex monolithic system would be too complex in order to be verified and accepted as a dependable system.
- *Fault tolerance*: Ultra-dependable systems that are to be deployed in safety-critical applications are expected to have a mean time to failure of better than 10–9 h [Sur95]. Single components cannot provide this level of dependability; therefore, a system must be designed as a distributed system with redundant components in order to provide its service despite of particular faults.
- *Instrumentation*: Using a distributed sensor/actuator network allows to perform the instrumentation of sensors and actuators directly at the device. The interface to the particular devices is then implemented as a standardized digital network interface, avoiding possible noise pickup over long analog transmission lines and reducing the complexity in instrumentation [Elm03].

The distributed approach requires a well-suited communication system for connecting its different components. Such a communication system must provide hard real-time guarantees, a high level of dependability, and means for fault isolation and diagnostics.

17.3.2 Composability and Scalability

In a distributed real-time system, the nodes interact via the communication system to execute an overall application. An architecture supports *composability* if it follows from correct verification and testing of the components of the system (i.e., its nodes and the communication system) that the overall system is correct. Think of a complex real-time system integrating components from many manufacturers. Each manufacturer does tests on the provided components. However, without supporting composability, the overall system cannot be proven to work correctly. Thus, the integrated system has to be evaluated again, which comes with high effort and costs for each small change in its components. For real-time systems, composability in the value domain and composability in the time domain need to be ensured. An architecture supporting composability supports then a two-level design approach: The design of the components (possibly done by particular manufacturers) and the overall system design do not influence each other as soon as the composability principle holds [Kop02].

Scalability of a real-time system also heavily depends on composability, since it requires new components not to interact with the stability of already established services. If network resources are managed dynamically, it must be ascertained that after the integration of the new components even at the critical instant, i.e., when all nodes request the network resources at the same instant, the timeliness of all communication requests can be satisfied.

17.3.3 Time-Triggered vs. Event-Triggered Systems

There are two major design paradigms for constructing real-time systems, the *event-triggered* and the *time-triggered* approach. Event-triggered systems follow the principle of reaction on demand. In such systems, the environment enforces temporal control onto the system in an unpredictable manner (interrupts). Conversely, time-triggered systems derive control from the global progression of time; thus, the concept of time that appears in the problem statement is also used as a basic mechanism for the solution.

The event-triggered approach is well-suited for sporadic actions and data, low-power sleep modes, and best-effort soft real-time systems with high utilization of resources. Event-triggered systems can easily adapt to online changes and are flexible. However, they do not ideally cope with the demands for predictability, determinism, and guaranteed latencies as a high analytical effort would be required to prove such properties.

Conversely, the time-triggered approach [Elm07] provides a low jitter for message transmission and task execution. The predictable communication scheme simplifies diagnosis of timing failures and the periodically transmitted messages enable a short and bounded error detection latency for timing and omission errors. The principle of resource adequacy guarantees the nominative message throughput independently of the network load. Moreover, the time-triggered paradigm avoids bus conflicts using a time division multiple access (TDMA) scheme, thus eliminating the need for an explicit bus arbitration. The downside of time-triggered systems is limited flexibility, worse average performance, the permanent full resource utilization, and a higher effort (clock synchronization) in establishing the system. Support to online changes can be provided, but at the expense of efficiency (see Section 17.4.2). Most state-of-the-art safety-critical systems are based on a time-triggered approach. Furthermore, hybrid systems like the FlexRay bus [Fle05] or TTE [Kop05] have gain attention in the last years.

17.3.4 Comparison of the Real-Time Support Provided by Notable Approaches

In order to preserve the real-time capabilities, real-time support in industrial networks has to be provided from the medium access control (MAC) layer up to the application layer. There are several possible techniques to handle a shared medium. Table 17.1 lists some notable examples. A first approach is through controlled-access protocols, which can be further classified as either centralized or distributed ones. The second approach is represented by uncontrolled-access protocols (e.g., those based on the carrier-sense multiple access protocol), which can be extended with additional features to improve their real-time behavior.

In controlled-access centralized methods, there is a central node acting as polling master that grants the stations the right to access to the channel. In this case, the timeliness of the transmissions on the network is up to such a central node, so it turns into a local scheduling problem at that node. Notable examples of these access methods in industrial networks are found in the Ethernet Powerlink [Eth07], Profibus DP [PRO07], and Bluetooth [IEE05] protocols. In the literature, it has been shown that the use of a real-time scheduling algorithm (instead of a non-real-time one such as round-robin or First In, First Out (FIFO) algorithms) combined with a proper timing analysis makes it possible to compute the response time of real-time messages and avoid deadline miss [Tov99a,Col07]. In controlled-access distributed protocols, transmissions are ruled according to a distributed mechanism, involving either the circulation of a physical token, or a virtual token passing, or a TDMA policy that allocates transmission instants to each node. In these cases, to enforce a timely behavior on the network and thus to achieve transmission timeliness, suitable timers and policies [Mal95,Tov99b] or a global synchronization

TABLE 17.1 Classifications of Notable MAC Layer Protocols

Controlled Access		Uncontrolled Access
Centralized	Distributed	CSMA/CD (CSMA/collision detection)
Master/slave	Token passing (virtual or physical)	CSMA/BA (CSMA/collision bitwise arbitration)
	TDMA	CSMA/DCR (CSMA/deterministic collision resolution)
	Timed token	CSMA/CA (CSMA/collision avoidance)

framework [Kop97] are required, respectively. Notable examples of networks provided with these access methods are Profibus [ES96], P-NET [ES96], TTP/C [Kop97], TTE [Kop05]. Conversely, in uncontrolled protocols, such as Carrier Sense Multiple Access (CSMA) based ones, each station decides autonomously when to transmit, obeying only to the protocol-specific rules. When collisions may occur, like in the CSMA/CD protocol, the one used by the Ethernet, achieving predictability becomes an issue, and suitable techniques, such as traffic smoothing, have to be adopted to bound the medium access time [Kwe04,LoB05]. A different situation arises with the CSMA/BA, where the collisions are non-destructive for the highest priority message, which then goes through without delay. CSMA/BA is used in the CAN bus and several studies exist which show that it is possible to calculate the response time of real-time messages over the CAN bus [Tin94,Bri06]. In CSMA/CA collisions may occur, thus the medium access time is non-predictable. The highest probability of a collision exists just after the medium becomes idle following a busy medium because multiple stations could have been waiting for the medium to become available again.

17.4 Design Challenges in Real-Time Industrial Communication Systems

17.4.1 Real-Time and Security

Real-time industrial communication systems can be subject to different security threats. Such attacks can be message integrity, fake messages, intrusion, impersonation, denial of service (DoS), etc. DoS attacks can be difficult to handle due to the constrained resources available in the embedded networks used to support automation applications. DoS attacks can be handled either by prevention or by detection. A common requirement for preventing DoS attacks is the possibility to limit physical access to the network. Detection is typically done by observing the network behavior to detect uncommon situations and, in case of detection, by acting with combat measures. In [Gra08], a solution combining detection and prevention is proposed for automation networks. As industrial communications often deal with safety-critical real-time systems, security services must also be taken into consideration from the point of view of efficient resource utilization. A balanced harmonization among dependability, security, and real-time aspects must then be performed. In a failure situation, when resources become scarce, one must prioritize certain applications/users/services over others to be able to retain the most critical applications [Fal08]. Different dependability, security, and/or real-time requirements will lead to different solutions and, for the most demanding services, the supporting mechanisms could be the most expensive ones, at least in terms of resource usage.

17.4.2 Real-Time and Flexibility

The specification of the requirements of real-time systems is a task that can be hard due to incomplete knowledge of the application and of the dynamic scenarios where the systems will operate. If the requirements are not fully known before runtime or if the operating conditions change during operation or if it is required to add or remove system components during operation, then flexibility is an important issue. For industrial control systems, Zoitl [Zoi09] discusses an approach for reconfigurable real-time applications based on the real-time execution models of the new International Electrotechnical Commission (IEC)'s family of standards (IEC 61499).

At the level of the communication protocol, different approaches for flexibility can be identified. As discussed previously, event-triggered communication systems such as the ones based on CAN can react promptly to communication requests that can be issued at any instant in time, letting the bus available in the absence of events to communicate. On the other side, time-triggered systems such as Time-Triggered Protocol (TTP) [TTA03] are less flexible because communication must take place at predefined instants, being defined at pre-runtime. Protocols such as TTP allow mode changes among predefined static modes or allow reservation of empty slots to include messages whose transmission is decided online; but this wastes bandwidth, as the slots are reserved even if they are not used.

In spite of this division, a number of hybrid protocols has been defined in the last 20 years. The most prominent hybrid protocols are ARINC-629 [AEE94] from the avionics domain, FlexRay [Fle05] from the automotive field, as well as TTE [Kop05]. Both these protocols enable online scheduling using FTDMA that is a flexible TDMA. Another example is FTT-CAN [Alm02], which uses a time-triggered approach where periodic traffic is conveyed with the control of a centralized dispatcher coexisting with "legacy" event-triggered CAN traffic transmitted in a specific window.

While TDMA is in general a static scheme, extensions to make it more flexible have been proposed. For example, P-NET fieldbus [ES96] uses a virtual token-passing approach that enables a node to transmit earlier if the previous slot was not used to transmit a stream. This technique was ported to Ethernet and called VTPE, virtual token-passing Ethernet [Car03].

17.4.3 Offering Real-Time Support to Wireless Communication

In wireless industrial networks, due to the error-prone nature of the wireless channel, the deterministic view on schedulability of a set of real-time streams is no longer applicable and it should be replaced by a probabilistic view, based on the probability of fulfilling some industrial-related QoS figures. Such figures may take packet losses into account as well. Under this perspective, a set of real-time streams is considered to be schedulable (for fixed assumptions on the wireless channel error process) when all of its streams achieve a long-term success probability of, for example, 99% [Wil08]. To assess the schedulability of real-time traffic flows in wireless industrial networks, suitable analysis methodologies are needed. Promising approaches are represented by stochastic models, such as finite-state Markov channels [Ara03,Bab00,Has04], in which the signal strength at a receiver is varied according to a Markov chain with a finite number of states, and stochastic network calculus, in which deterministic end-to-end time bounds are replaced by stochastic ones [Jia06].

References

[AEE94] Airlines Electronic Engineering Committee, ARINC Specification 629-3: IMA Multitransmitter Databus, 1994.

[Alm02] L. Almeida, P. Pedreiras, and J. A. Fonseca, The FTT_CAN protocol: Why and how, *IEEE Transactions on Industrial Electronics*, 49(6), 1189–1201, 2002.

[Ara03] J. Arauz and P. Krishnamurthy, Markov modeling of 802.11 channels, in *Proceedings of 58th IEEE Vehicular Technology Conference (VTC) 2003-Fall*, pp. 771–775, October 2003.

[Aud91] N. C. Audsley, A. Burns, M. F. Richardson, and A. J. Wellings, Real-time scheduling: The deadline monotonic approach, in *Proceedings of Eighth IEEE Workshop on Real-Time Operating Systems and Software*, Atlanta, GA, May 15–17, 1991.

[Bab00] F. Babich, O. E. Kelly, and G. Lombardi, Generalized Markov modeling for flat fading, *IEEE Transactions on Communications*, 48(4), 547–551, April 2000.

[Bon03] A. Bonaccorsi, L. Lo Bello, O. Mirabella, P. Neumann, and A. Pöschmann, A distributed approach to achieve predictable ethernet access control in industrial environments, in *Proceedings of the IFAC International Conference on Fieldbus Systems and their Applications (FET'03)*, pp. 173–176, Aveiro, Portugal, July 2003.

[Bou04] J.-L. Le Boudec and P. Thiran, *Network Calculus*, Lecture Notes in Computer Sciences, Vol. 2050, Springer Verlag, Berlin, Germany, 2001.

[Bri06] R. J. Bril, J. J. Lukkien, R. I. Davis, and A. Burns, Message response time analysis for ideal controller area network (CAN) refuted, in *The Fifth International Workshop on Real-Time Networks, Satellite Workshop of ECRTS 06*, Dresden, Germany, July 4, 2006.

[Car03] F. B. Carreiro, J. A. Gouveia Fonseca, and P. Pedreiras, Virtual token-passing Ethernet—VTPE, in *Proceedings of the Fifth IFAC International Conference on Fieldbus Systems and their Applications (FeT'2003)*, Aveiro, Portugal, July 2003.

[Col07] M. Collotta, L. Lo Bello, and O. Mirabella, Deadline-aware scheduling policies for bluetooth networks in industrial communications, in *Proceedings of the IEEE Second International Symposium on Industrial Embedded Systems—SIES'2007*, pp. 156–163, Lisbon, Portugal, July 4–6, 2007.

[Cru91a] R. L. Cruz, A calculus for network delay, part I: Network elements in isolation, *IEEE Transactions on Information Theory*, 37(1), 114–131, January 1991.

[Cru91b] R. L. Cruz, A calculus for network delay, part II: Network analysis, *IEEE Transactions on Information Theory*, 37(1), 132–141, January 1991.

[Dem90] A. Demers, S. Keshav, and S. Shenker, Analysis and simulation of a fair queuing algorithm, *Journal of Internetwoking Research and Experience*, 1(1), 3–26, October 1990.

[Elm03] W. Elmenreich and S. Pitzek, Smart transducers—Principles, communications, and configuration, in *Proceedings of the Seventh IEEE International Conference on Intelligent Engineering Systems*, Vol. 2, pp. 510–515, Assuit-Luxor, Egypt, March 2003.

[Elm07] W. Elmenreich, Time-triggered transducer networks. Habilitation thesis, Vienna University of Technology, Vienna, Austria, 2007.

[Elm09] W. Elmenreich, editor, Embedded systems engineering. Vienna University of Technology, Vienna, Austria, 2009, ISBN 978-3-902463-08-1.

[ES96] European Standard EN 50170. Fieldbus. Vol. 1: P-Net, Vol. 2: PROFIBUS, Vol. 3: WorldFIP, 1996.

[Eth07] Ethernet Powerlink Standard, Included in IEC 61158-2, Edition 4.0, 2007.

[Fal08] L. Falai et al., Mechanisms to provide strict dependability and real-time requirements, EU FP6 IST Project HIDENETS, Deliverable D3.3, June 2008.

[Fle05] Flexray Consortium, FlexRay Communications System Protocol Specification Version 2.1, 2005. Available at http://www.flexray.com

[Geo04] J.-P. Georges, T. Divoux, and E. Rondeau, A formal method to guarantee a deterministic behaviour of switched Ethernet networks for time-critical applications, in *IEEE International Symposium on Computer Aided Control Systems Design*, Taipei, Taiwan, September 2–4, 2004.

[Geo96] L. George, N. Rivierre, and M. Spuri, Preemptive and nonpreemptive real-time uni-processor scheduling, Technical Report 2966, Institut National de Recherche et Informatique et en Automatique (INRIA), France, September 1996.

[Gra08] W. Granzer, C. Reinisch, and W. Kastner, Denial-of-service in automation systems, in *Proceedings of the IEEE Conference on Emerging Technologies and Factory Automation*, Lisbon, Port, 2008.

[Has04] M. Hassan, M. M. Krunz, and I. Matta, Markov-based channel characterization for tractable performance analysis in wireless packet networks, *IEEE Transactions on Wireless Communications*, 3(3), 821–831, May 2004.

[IEE05] IEEE Standard for Information Technology—Telecommunications and Information Exchange between Systems—Local and Metropolitan Area Networks—Specific Requirements—Part 15.1: Wireless Medium Access Control (MAC) and Physical Layer (PHY) Specifications for Wireless Personal Area Networks (WPANs), IEEE Computer Society—Sponsored by the LAN/MAN Standards Committee, 2005.

[Jas02] J. Jasperneite, P. Neumann, M. Theis, and K. Watson, Deterministic real-time communication with switched Ethernet, in *Proceedings of WFCS'02, Fourth IEEE Workshop on factory Communications Systems*, pp.11–18, Vasteras, Sweden, August 2002.

[Jia06] Y. Jiang, A basic stochastic network calculus, in *Proceedings of ACM SIGCOMM Conference on Network Architectures and Protocols*, Pisa, Italy, September 2006.

[Jos86] M. Joseph and P. Pandya, Finding response times in a real-time system, *The Computer Journal*, 29(5), 390–395, 1986.

[Kop02] H. Kopetz and R. Obermaisser, Temporal Composability, *IEE's Computing and Control Engineering Journal*, 13(4), 156–162, August 2002.

[Kop05] H. Kopetz, A. Ademaj, P. Grillinger, and K. Steinhammer, The time-triggered Ethernet (TTE) design, in *Proceedings of the Eighth International Symposium on Object-Oriented Real-Time Distributed Computing* (*ISORC*), pp. 22–33, Seattle, WA, May 2005.

[Kop97] H. Kopetz, *Real-Time Systems, Design Principles for Distributed Embedded Applications*, Kluwer Academic Publishers, Boston, MA/Dordrecht, the Netherlands/London, U.K., 1997.

[Kwe03] S. K. Kweon and K. G. Shin. Statistical real-time communication over Ethernet, *IEEE Transactions on Parallel and Distributed Systems*, 14(3), 322–335, 2003.

[Kwe04] S. K. Kweon, M. G. Cho, and K. G. Shin, Soft real-time communication over Ethernet with adaptive traffic smoothing, *IEEE Transactions on Parallel and Distributed Systems*, 15(10), 946–959, 2004.

[Leh90] J. Lehoczky, Fixed priority scheduling of periodic task sets with arbitrary deadlines, in *Proceedings of 11th IEEE Real-Time Systems Symposium* (*RTSS*), pp. 201–209, December 1990.

[Liu00] J. W. S. Liu, *Real-Time Systems*, Prentice Hall, Englewood Cliffs, NJ, 2000.

[Liu73] L. Liu and J. Layland, Scheduling algorithms for multiprogramming in a hard real-time environment, *Journal of ACM*, 20(1), 46–61, 1973.

[LoB05] L. Lo Bello, G. Kaczynski, and O. Mirabella, Improving the real-time behaviour of Ethernet networks using traffic smoothing, *IEEE Transactions on Industrial Informatics, Special Section on Factory Communication Systems*, 1(3), 151–161, ISSN: 1551-3203, IEEE Industrial Electronics Society, Piscataway, NJ, August 2005.

[LoB06] L. Lo Bello, G. Kaczynski, F. Sgro', and O. Mirabella, A wireless traffic smoother for soft real-time communications over IEEE 802.11, in *Proceedings of ETFA06, the 11th IEEE International Conference on Emerging Technologies and Factory Automation*, pp. 1073–1079, Prague, Czech Republic, September 2006.

[Mal95] N. Malcom and W. Zhao, Hard real-time communications in multiple access networks, *Real-Time Systems*, 9, 75–107, 1995.

[Ped05] P. Pedreiras, P. Gai, L. Almeida, and G. C. Buttazzo, FTT-ethernet: A flexible real-time communication protocol that supports dynamic QoS management on ethernet-based systems, *IEEE Transactions on Industrial Informatics*, 1(3), 162–172, 2005.

[PRO07] PROFIBUS Standard—DP Specification, Included in IEC 61784-3-3, Edition 1.0, 2007.

[Rid07] F. Ridouard, J.-L. Scharbarg, and C. Fraboul, Stochastic network calculus for end-to-end delays distribution evaluation on an avionics switched ethernet, in *Proceedings of the Fifth International Conference on Industrial Informatics* (*INDIN 2007*), Vienna, Austria.

[Sch05] J. Schmitt and U. Roedig, Sensor network calculus—A framework for worst case analysis, in *Proceedings of the International Conference on Distributed Computing in Sensor Systems (DCOSS05)*, pp. 141–154, Marina del Rey, CA, June 2005.

[Sha04] L. Sha, T. Abdelzaher, K. Arzen, A. Cervin, T. Baker, A. Burns, G. Buttazzo, M. Caccamo, J. Lehoczky, and A. Mok, Real-time scheduling theory: A historical perspective, *Real-Time Systems*, 28(2–3), 101–155, December 2004.

[Ske02] T. Skeie, S. Johannessen, and Ø. Holmeide, The road to an end-to-end deterministic ethernet, in *Proceedings of Fourth IEEE International Workshop on Factory Communication Systems* (*WFCS*), Vasterås, Sweden, September 2002.

[Sta88] J. Stankovic and K. Ramamrithan, editors, *Tutorial on Hard Real-Time Systems*, IEEE Computer Society Press, Washington, DC, 1988.

[Sta90] J. Stankovic and K. Ramamrithan, What is predictability for real-time systems? *Journal of Real-Time Systems*, 2, 247–254, 1990.

[Sur95] N. Suri, C. J. Walter, and M. M. Hugue, editors, *Advances in Ultra-Dependable Systems*, IEEE Press, Los Alamitos, CA, 1995.

[Tin94] K. Tindell, H. Hansson, and A. Wellings, Analysing real time communications: Controller area network (CAN), in *Proceedings of 15th IEEE Real-Time Systems Symposium* (*RTSS*), pp. 259–263, Los Alamitos, CA, December 1994.

[Tov99a] E. Tovar, Supporting real-time communications with standard factory-floor networks, PhD dissertation, Department of Electrical Engineering, University of Porto, Porto, Portugal, 1999.

[Tov99b] E. Tovar and F. Vasques, Cycle time properties of the PROFIBUS timed token protocol, *Computer Communications*, 22, 1206–1216, 1999.

[TTA03] TTAGroup, Specification of the TTP/C Protocol V1.1, 2003. Available at http://www.ttagroup.org

[Wil08] A. Willig, Recent and emerging topics in wireless industrial communications: A selection, *IEEE Transactions on Industrial Informatics*, 4(2), 102–124, May 2008.

[Zoi09] A. Zoitl, Real-time execution for IEC 61499, International Society of Automation (ISA), Noida, India, 2009.

18

Clock Synchronization in Distributed Systems

18.1 Introduction 18-1
18.2 Precision Time Protocol 18-1
18.3 IEEE 1588 System Model 18-2
18.4 Service Access Points 18-3
18.5 Ordinary Clocks 18-3
18.6 Boundary Clocks 18-4
18.7 Precision Time Protocol, IEEE 1588–2008 (PTPv2) 18-5
18.8 Network Time Protocol 18-6
18.9 Network Time Protocol Strata 18-7
18.10 Architecture, Protocol, and Algorithms 18-8
18.11 NTP Clock Synchronization Hardware Requirements 18-8
18.12 Synchronization Algorithms of NTP 18-8
References 18-10

Georg Gaderer
Austrian Academy of Sciences

Patrick Loschmidt
Austrian Academy of Sciences

18.1 Introduction

It is known that technologies to synchronize clocks are not an exclusive research topic of the computer age. As a matter of fact, the construction of highly accurate clocks and the resulting requirements for synchronization date back to the seventeenth century. However, the actual research for the usage in distributed systems began with the introduction of networked computing. The issue was first discussed in depth with the topic of event ordering by Lamport [Lamport1978]. A good, general overview on available algorithms can be found in [Anceaume1997]. Later, especially with the upcoming real-time networks [Kopetz1987,Palumbo1992,Liskov,Patt1994], the problem came again into scientific discussion. This chapter gives an overview of the state of the art of clock synchronization protocols as well as their application in real-time networks.

18.2 Precision Time Protocol

The precision time protocol (PTP or IEEE 1588) is a protocol designed for highly accurate clock synchronization focused especially on test and measurement, as well as power engineering (e.g., substation automation), and factory automation. The standard is IEEE approved since 2002 [IEEE1588v1]. In 2008, a positive ballot on version 2 of the protocol was submitted to the IEEE [IEEE1588v2]. While the test and measurement industry needs synchronized clocks for data collection [Eidson2005,Owen2005, Pleasant2005], other classes of applications, like the factory automation community, need synchronized clocks in order to coordinate the access to the network and thus guarantee

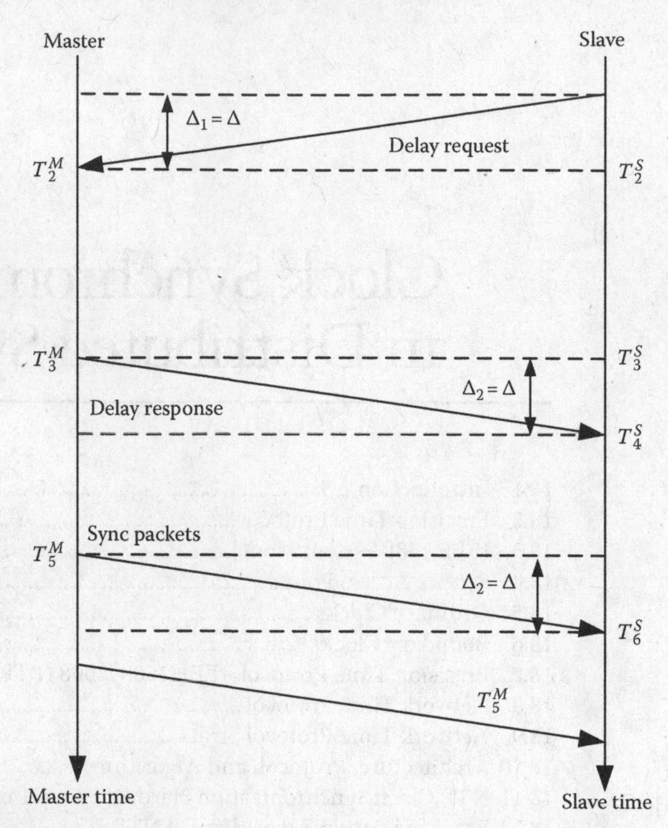

FIGURE 18.1 Synchronization packets and round-trip delay measurement (1588_Sync_DelayRequResp_packets.pdf).

real time for communication [Sauter2005, Brennan2005,Hansson2005]. IEEE 1588 synchronizes clocks in a master/slave fashion, whereas the master is the node with the highest accuracy. The accuracy class (also called stratum in version 1) can of course be the same at every node in the network (e.g., same oscillator type used). However, in the case where the stratum is the same, a unique node identifier (layer 2 MAC address in the Ethernet case) is used to favor one node. The master election follows a certain message-based election algorithm and ends up with the decision which node should be the master for the next synchronization process. After a proper master is elected, synchronization can take place.

The synchronization process in IEEE 1588 networks is organized in two stages. A so-called delay request delay response process measures the round-trip delay, to determine the packet delay between master and slave. This is done upon initiation by the slave with a telegram, which is returned by the master with the local timestamp of reception. In addition, the master regularly sends out synchronization packets. Again, these packets are sent in a two-stage process. The first set of packets is timestamped with an estimation of the sending time. The actual sending time (which obviously can only be known after completing the transmission) is sent with a so-called follow-up packet. Figure 18.1 shows this synchronization principle. A detailed description of this process together with implementation issues can be found in [Eidson2006].

18.3 IEEE 1588 System Model

The starting point of the protocol specified in IEEE 1588 is a distributed system of nodes. The medium as such is not relevant for the protocol. As shown in Figure 18.2 in general, the position of the ingress message timestamp point and the egress timestamp point are not required to be identical, since two different

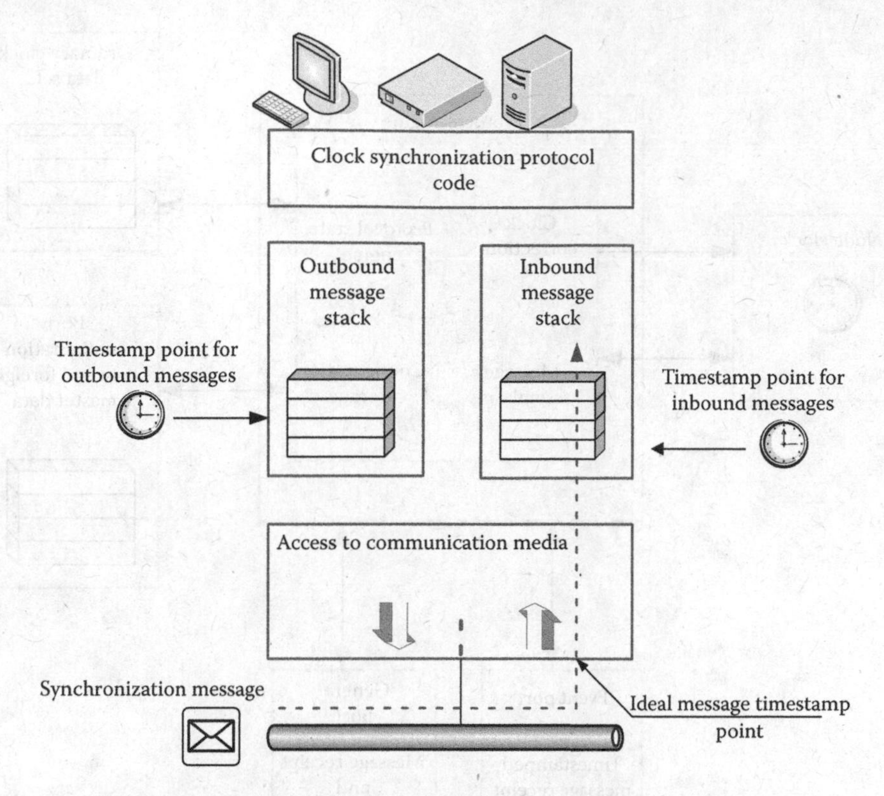

FIGURE 18.2 IEEE 1588 node model (PTPNodeModel.pdf).

data flows handle the data transmission for the two directions. The exact timestamp point for both the DELAY_REQ and SYNC is unimportant for the further processing of the protocol as long as the position is the same for all communication partners.

As soon as a node detects an incoming DELAY_REQ or SYNC message, timestamps for further processing are generated. If the actual implementation is not able to detect the exact time of incoming messages, the timestamp has to be corrected in an appropriate way. The time for a message to travel between the ideal timestamp point and the actual timestamp point for inbound or outbound message point is referred to as *inbound* latency and *outbound latency* for IEEE 1588.

18.4 Service Access Points

Beyond the logical division of subnets, PTP relies on the possibility of distinguishing messages with the identification of the incoming service access point (SAP) mapped to the ISO model. For the example of user datagram protocol (UDP), which is defined in the annex of PTP as a possible transport protocol, this means that synchronization messages (i.e., SYNC and FOLLOW_UP) are sent to another UDP port than all other messages.

18.5 Ordinary Clocks

Figure 18.3 shows a model for an ordinary PTP clock. Each PTP clock uses a communication path for message exchange with other PTP nodes. This path is accessed in PTP via two different SAPs, namely the event and the general port. Both hand over messages to the protocol engine of the PTP clock. In order to model the fact that a clock may not be limited to only one PTP port, the configuration information is

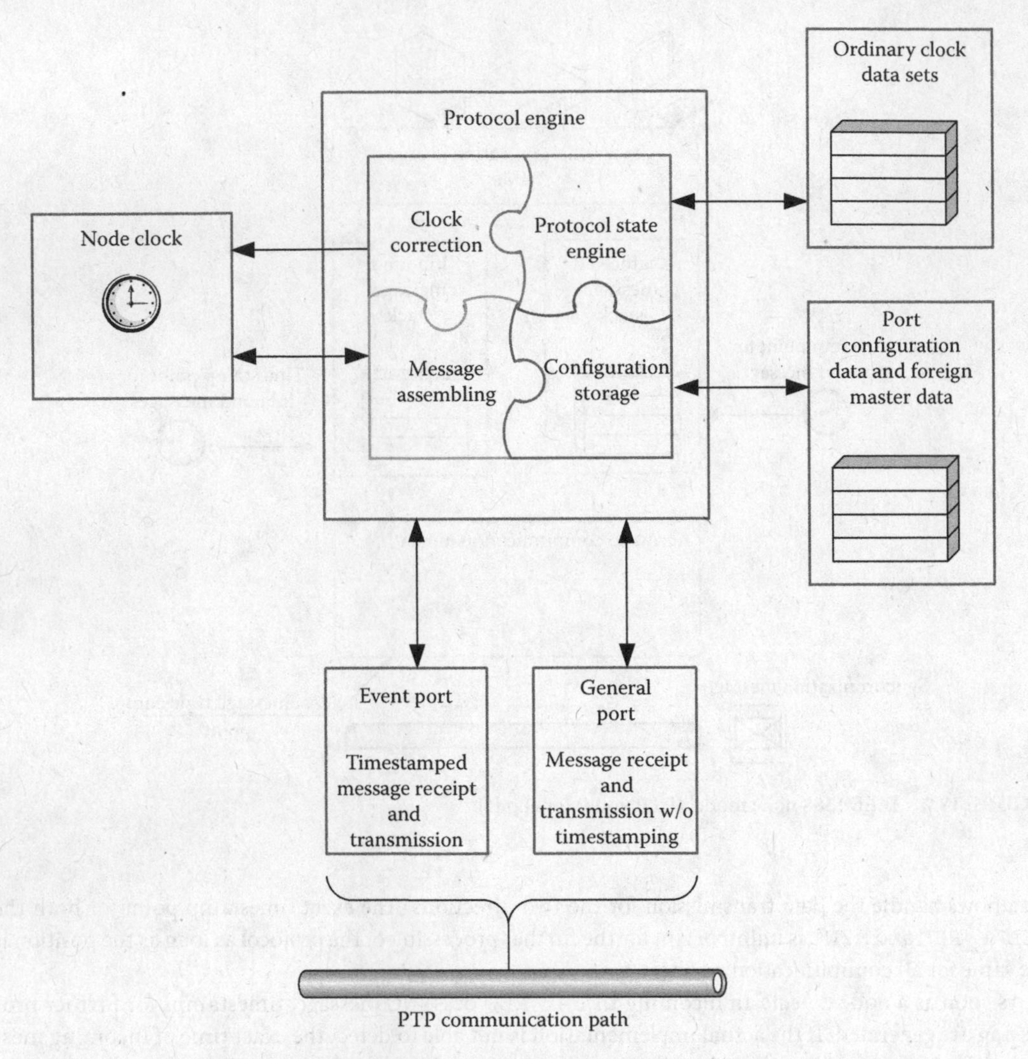

FIGURE 18.3 Model of an ordinary PTP clock (ModelOrdinaryPTPClock.pdf).

split into a port-specific part and general information about the clock. Finally, the protocol engine has access to the timekeeping part—the clock itself. A clock in IEEE 1588 can be adjusted and read out by the protocol engine.

18.6 Boundary Clocks

It is easy to broaden the concept shown for the ordinary clocks by simply duplicating the PTP ports, each having access to a common clock data set and sharing the clock. In order to implement this communication in terms of intra-node time exchange between the PTP ports, ordinary clock data sets are needed. The IEEE 1588 standard calls these generalized clocks *boundary clocks*. A model for a boundary clock with a general number of *n* ports is shown in Figure 18.4. It has to be noted that only one port may act as master node, which implies that the local clock will only be set by one instance.

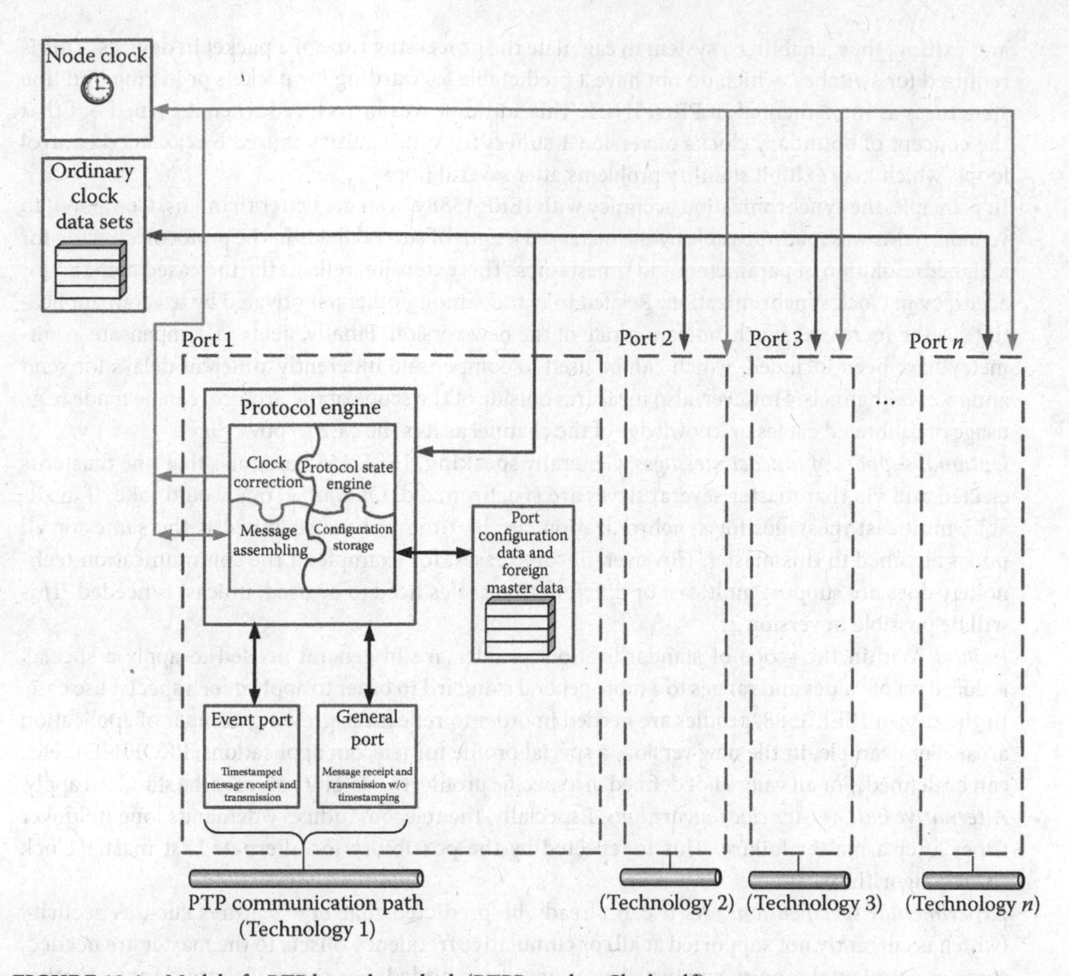

FIGURE 18.4 Model of a PTP boundary clock (PTPBoundaryClock.pdf).

18.7 Precision Time Protocol, IEEE 1588–2008 (PTPv2)

Version 1 of IEEE 1588 was kept as simple as possible, but growing industrial interest in precision clock synchronization soon made clear that an improved version of the standard was needed. By the time of writing, the second ballot on IEEE 1588v2 has started and the final publication of the standard is expected to be in the second half of 2008. The new version of the standard will have several improvements [IEEE 1588v2]:

- *Mappings.* As the main clauses of the standard specify a communication protocol, rather then actual payload positions in packets, mappings to communication protocols such as IP have to be made. As an improvement to version 1, additional mappings such as to UDP in IPv4 and IPv6, DeviceNet, raw Ethernet, PROFINET, and ControlNet will be made.
- *Type length value fields.* Version 1 suffered from the problem that formally, messages could not be extended with arbitrary information. Such extensions have been added via partly user-definable type length value (TLV) fields. Nevertheless, not only proprietary fields are defined via TLVs. Also, standardized functionality, like security updates, is transported via these fields.
- *Transparent clocks.* Under the name "on-the-fly timestamping" [Gaderer2004], version 2 supports transparent clocks, which are basic network elements that timestamp messages when entering

and exiting, thus, enabling a system to calculate the processing time of a packet in devices. This is required for switches, which do not have a predictable forwarding for packets or in ring and line structures as implemented in PROFINET. This addition was introduced when it turned out that the concept of boundary clocks of version 1 suffers from difficulties caused by cascaded control loops, which may exhibit stability problems after several hops.

- In principle, the synchronization accuracy with IEEE 1588v2 can get better than 1 ns. Compared to version 1, this was made possible by the increased length of several fields in the protocol allowing for a higher resolution of parameters and timestamps. This extension reflects the increased demand for accuracy in clock synchronization. Related to it and, among others, motivated by telecom applications is the increased synchronization rate of the new version. Finally, fields to compensate asymmetry have been included, which can be used to compensate inherently, different delays for send and receive channels. However, also measures outside of the scope of the protocol can be made (e.g., usage of calibrated cables or knowledge of the channel as it is the case in powerline).
- *Optional support of unicast messages*. Generally speaking, IEEE 1588 assumes that one master is elected and via that master several slaves are synchronized. Of course, one would take, if available, multicast messages for synchronization, as the time to be distributed is the same for all nodes attached to this master. However, in some cases, for example, if the communication technology does not support multicast or different timescales need to be used, unicast is needed. This will be possible in version 2.
- *Profiles*. Within the scope of standardization, profiles are in general needed to apply a special, reduced set of values and ranges to a more general standard in order to apply it for a special use case. In the case of IEEE 1588, profiles are needed in order to reflect the growing number of application areas. For example, in the new version, a special profile for telecom applications, PROFINET, etc., can be defined. For all values not defined in a specific profile, the default values of the standard apply.
- *Alternative best master clock algorithms*. Especially, the telecom industry demands long holdover times after a master failure. This is reflected by the possibility for alternate best master clock (BMC) algorithms.
- *Experimental specifications*. As it can already be predicted that new features such as security (which is currently not supported at all) or cumulative frequency offsets to the master are needed, some experimental, non-normative clauses have been added. For some of those clauses, further research and case studies are needed; however, the experimental annex eases the start for such investigations [Treytl2007,Gaderer2006a].

18.8 Network Time Protocol

With the introduction of distributed applications in the Internet, a new requirement arose: the synchronization between the participating computers. For example, simple applications like the synchronization of files on two servers, like the well-known rsync application, use timestamps of files to distinguish between *older* and *newer*. This problem was solved with the probably oldest application layer protocol of the Internet still in use: The network time protocol (NTP). The open protocol standard NTP is a specification for a TCP/IP-based clock synchronization scheme for network devices. The main work on NTP was contributed by David L. Mills. The standardization was done via request for comment (RFC) documents—the key contribution can be found in [Mills1991a]. NTP is a very carefully designed protocol almost to every extent investigated. It covers, in opposite to IEEE 1588, all aspects like the control of the clock and respective filter algorithms [Gurewitz2003,Mills1998b,Mills1998a] with a wide range of application studies [Richards2006].

The latest version of NTP is 4; nevertheless, in the meanwhile, a simplified version called SNTP has been specified and is in use.

18.9 Network Time Protocol Strata

A basic principle in NTP is to assign to every clock a so-called *stratum*. This hierarchically organized system of clock strata defines the accuracy of a computer's clock or more precisely the distance to a reference clock. Within NTP, device strata are identified with an integer value starting at stratum-0. Like illustrated in Figure 18.5, these strata are defined as follows:

- *Stratum*-0 devices have highly accurate clocks such as cesium or rubidium time normals. It has to be noted that devices with this stratum are never connected directly to a network—time and frequency transfer happens via serial interfaces or transmission of defined electrical pulses.
- *Stratum*-1 nodes are always directly connected to devices with stratum-0. The usual way is to have these nodes act as time servers for stratum-2 clients.
- *Stratum*-2 nodes communicate via NTP with stratum-1 servers. It is possible for these devices to peer with other stratum-2 nodes in order to obtain a more robust time source for all other devices in the so-called *peer group*. Again, in the usual case, these devices will act as servers for stratum-3 clocks.
- *Stratum*-3 is the class for the next hierarchical layer in NTP. These nodes perform basically the same tasks as NTP devices in the layer below, with the difference that time is obtained from stratum-2 devices. In general, NTP, depending on the version, allows strata higher than stratum-3 (up to 256 depending on the version). Nevertheless, future developments will most likely restrict the NTP strata to 8–16.

A study by Minar in 1999 [Minar1999] revealed that at the time of the study, at least 175,000 hosts use NTP in the Internet. Among those are more then 300 stratum-1 servers and 20,000 and 80,000 stratum-2 and stratum-3 servers, respectively.

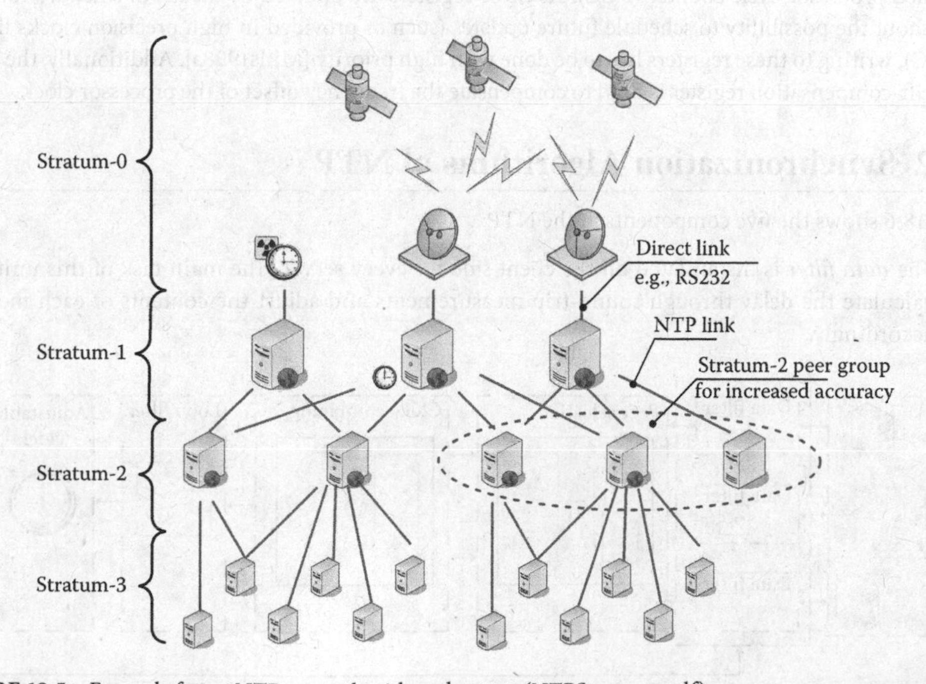

FIGURE 18.5 Example for an NTP network with node strata (NTPStratums.pdf).

18.10 Architecture, Protocol, and Algorithms

The idea behind the NTP is to distribute a common notion of time among all nodes of a network in terms of a reference clock. This goal can be achieved in quite different strategies:

- Minimization of *clock offset*, which tries to minimize the sampled difference between the local and the reference time.
- Alignment of the *frequency*. This is equal to the first strategy, with the exception that the absolute offset of periodically generated clock signals is not considered, but the stability of the offset optimized.
- The alignment of the absolute clock value in terms of minimizing clock offset and optimizing the difference between all nodes. This criterion is equal to internal clock synchronization (alignment of all nodes of a network with no respect to the absolute time scale), with the exception that NTP inherently assumes an alignment to an absolute correct time source of the overall master.

18.11 NTP Clock Synchronization Hardware Requirements

Like any other clock synchronization system, NTP has to make certain assumptions regarding the clocks. As NTP aims to synchronize hosts within the Internet, one assumption is that the architecture of every node is similar to a clock in a PC. One further assumption of NTP is that a network node does not necessarily need to be powered up all time. However, this requires keeping up with the unstoppable progress of time even during power down. In practice, this is usually done with a so-called real-time clock (RTC). This clock is battery-powered and thus, running also during power-down times. For cost and energy efficiency reasons, these clocks are usually equipped with 32,768 kHz crystals. This frequency is very convenient, since 32,768 transitions can be represented as 2^{15}, which means that a 15 bit counter can be used without further fractional parts to represent 1 s. In common PC architectures, the content of this clock is loaded during boot-up into a tick register. Usual clock registers for NTP are 48 bit wide, and updated by adding the 32 bit wide clock-adjust register, with an interrupt typically scheduled every 1–20 ms. The content is also called processor cycle counter (PCC). As these registers are updated by means of software routines and without the possibility to schedule future updates (such as provided in high precision clocks like in SynUTC), writing to these registers has to be done with high priority [Mills1998a]. Additionally, the 16 bit wide drift-compensation register is used to compensate the frequency offset of the processor clock.

18.12 Synchronization Algorithms of NTP

Figure 18.6 shows the five components of the NTP:

- The *data filter* is instantiated on the client side for every server. The main task of this unit is to calculate the delay through round-trip measurements and adjust the contents of each message accordingly.

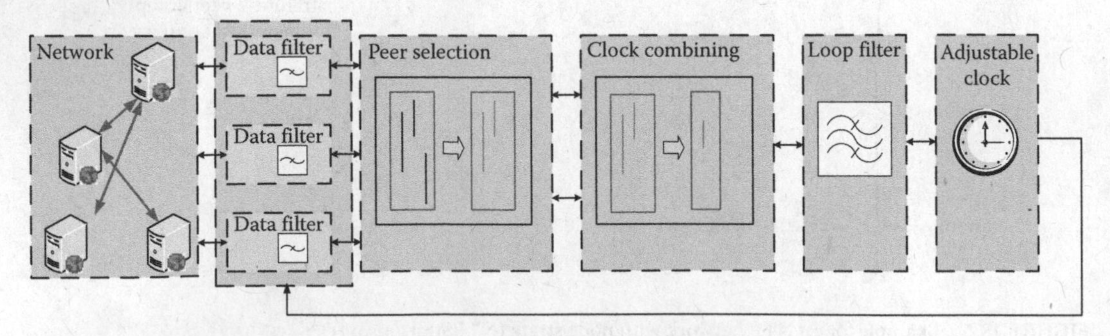

FIGURE 18.6 System concept of a typical NTP node (NTPSystemConcept.pdf).

- The second component in the data flow is the so-called *peer selection*. This unit is included due to the fact that not only delayed messages can arrive from a node, but also mutually wrong clocks can present themselves as masters. Reasons for wrong hosts in NTP networks can be broken server clocks or attacks with the goal to change the time on a network host. The latter can be tackled by authentication of clock servers, which is nevertheless not sufficient for the first issue of accidentally wrong clocks. In order to avoid problems resulting from such broken clocks, a parameter-driven decision of sorting out those clocks is done in peer selection.
- After the peer selection is done—from a message point of view—often contradicting information regarding the offset and drift of the local clock exist. These data have to be merged in the *clock combining* component.
- The last two components, namely the *loop filter* and *voltage-controlled oscillator* (VCO), have the task of conditioning the control signal with respect to control circuit stability and adjusting the local clock, respectively.

The first element of the NTP signal flow is the data filter. This element is provided to eliminate statistical errors as well as byzantine nodes. The filter collects data on a per-peer basis and calculates two measures: the *filter dispersion* ε_σ, and the *select dispersion* ε_ζ. The dispersion is calculated with $\vartheta_i (0 \le i \le n)$ as the offset of the *i*th sample and the sample difference $\varepsilon_{ij} = |\vartheta_i - \vartheta_j|$. The dispersion, ε_j is then defined relative to the *j*th entry as weighted sum

$$\varepsilon_j = \sum_{i=1}^{n-1} \varepsilon_{ij} w^{i+1}$$

Both ε_σ and ε_ζ are defined by protocol parameters, namely `NTP.FILTER` and `NTP.SELECT`, which are usually set to values less than 0.5 and evaluated relative to the first entry ε_0. In that context, w is the weighting factor, which is used to build an exponential average, with a value of $w < 0.5$. The second, functional part, the peer selection is most important for the stability of NTP. As soon as new offsets to a reference clock are estimated, this algorithm decides which peer is used as synchronization source. Of course, also the unavailability of offset measurements (e.g., due to timeouts) are also used to (de-)select peers. The actual peer selection is done by assembling of the candidates in a sorted list. This list represents the available peers sorted by stratum and synchronization dispersion. Also, several sanity checks to detect defective nodes are a knock-out criterion to remove defective nodes. In case that the result of this sanity check is an empty list, the local nodes continue running with their local unadjusted clock frequencies. In the other case, if the list holds more than one node, these statistical equivalent offsets have to be combined. The clock assembly is done by construction of so-called final clock correction, η_n, for all nodes. Moreover, the synchronization distance, λ_n, is also internally required by the previously described algorithms. The value of $1/\lambda_n$ is then used as weight of each clock offset in the list.

The virtual control loop is finally closed by the loop filter and the VCO, respectively. The task of these two parts is to adapt the calculated settings from NTP to the actual clock of the device. This is done by adjusting offset and frequency—both algorithms are implemented in a feedback loop. The general structure of the disciplining is shown in Figure 18.7. In general, this disciplining structure is a subordinate control loop. The set value ϑ_r is taken from the above-described clock selection part of NTP. The value is compared to the control phase of the variable frequency oscillator, which is the actual driver of the node clock. The difference between ϑ_r and ϑ_c is afterward used in the clock filter, which is actually a tapped delay line, whereas the tap is selected by an algorithm described in [Mills2000].

Afterward, the filtered signal is used in the loop filter (Figure 18.7). This filter adjusts the clock via the calculation of three variables. First, the phase offset is calculated and directly fed back to the clock adjustment unit. Second, the frequency offset is calculated via two ways: a phase prediction and a frequency

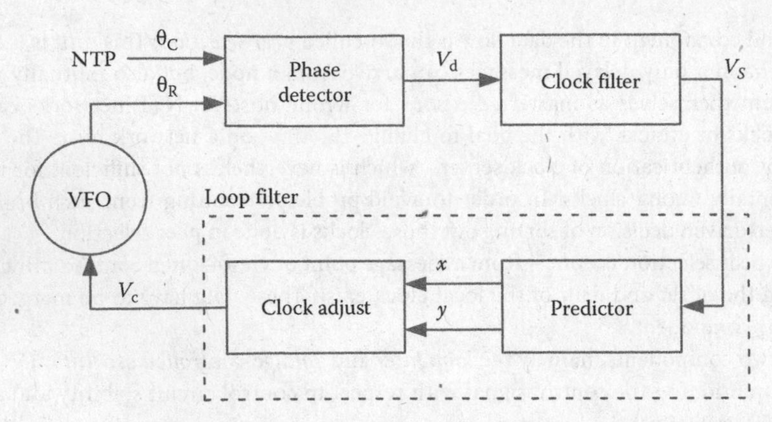

FIGURE 18.7 NTP clock discipline (NTPClockDiscipline.pdf). (Redrawn from Mills, D., *IEEE/ACM Trans. Netw.*, 6(5), 505, 1998.)

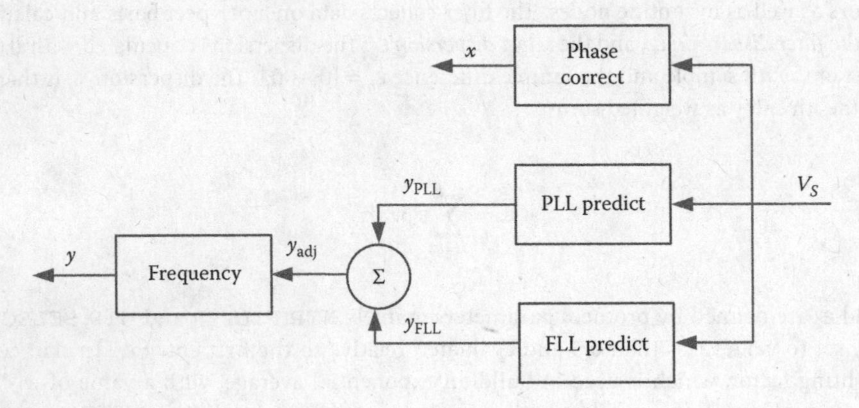

FIGURE 18.8 NTP phase and frequency prediction functions (NTPPredictionFunctions.pdf). (Redrawn from Mills, D., *IEEE/ACM Trans. Netw.*, 6(5), 505, 1998.)

prediction (Figure 18.8). The phase-locked loop (PLL) predicts a frequency offset as integral of $V_S\tau$, whereas the frequency-locked loop (FLL) use an exponential average of V_S/τ. These two values are combined via a simple summarization. It has to be specifically noted that the averaging interval of these two components is chosen with respect to the Allan-intercept, in order to gain an optimal behavior of the structure.

References

[Anceaume1997] Anceaume, E. and Puaut, I.A., Taxonomy of clock synchronization algorithms, *Campus Universitaire de Beaulieu*, Rennes Cedex, France, 1997, 25 pp.

[Brennan2005] Brennan, R.W., Christensen, J.H., Gruver, W.A., Kotak, D.B., Norrie, D.H., and van Leeuwen, E.H., Holonic manufacturing systems: A technical overview, in *The Industrial Information Technology Handbook*, Part II (Industrial Information Technology), Section 7 (Integration Technologies), Chapter 106, CRC Press, Boca Raton, FL, 2005, 106-1–106-15.

[Eidson2005] Eidson, J. C., The application of IEEE 1588 to test & measurement systems, White Paper Agilent, Agilent Laboratories, Palo Alto, CA, December 2005, pp. 9–13.

[Eidson2006] Eidson, J.C., *Measurement, Control, and Communication Using IEEE 1588*, Springer, London, U.K., 2006.

[Gaderer2004] Gaderer, G. and Sauter, T., Strategies for clock synchronization in powerline networks, in *Proceedings of the 3rd International Workshop on Real-Time Networks in Conjunction with the 16th Euromicro International Conference on Real-Time Systems*, Catania, Italy, 2004.

[Gaderer2006a] Gaderer, G., Treytl, A., and Sauter, T., Security aspects for IEEE 1588 based clock synchronization protocols, in *Proceedings of the IEEE International Workshop on Factory Communication Systems*, Turin, Italy, June 27, 2006, pp. 247–250.

[Gurewitz2003] Gurewitz, O., Cidom, I., and Sidi, M., Network time synchronization using clock offset optimization, in *Proceedings of the 2003 International Conference on Network Protocols*, Atlanta, GA, November 4–7, 2003, pp. 212–221.

[Hansson2005] Hansson, H., Nolin, M., and Nolte, T., Real-time systems, Section 6 (Real-time embedded systems), in *The Industrial Information Technology Handbook*, Part II (Industrial Information Technology), Section 6 (Real-Time Embedded Systems), Chapter 81, CRC Press, Boca Raton, FL, 2005, 81-1–81-28.

[IEEE1588v1] Arlt, V., Johnson, R., Lee, K., Munz, H., Powers, E., Read, J., and Schmidt, R., IEEE Standard for a Precision Clock Synchronization Protocol for Networked Measurement and Control Systems TC9, Technical Committee on Sensor Technology of the IEEE Instrumentation and Measurement Society.

[IEEE1588v2] McKay, J. (ed.), *IEEE 1588 (tm) 2.1 Standard for a Precision Clock Synchronization Protocol for Networked Measurement and Control Systems*, Institute of Electrical and Electronics Engineers, Inc., New York, 2006.

[Kopetz1987] Kopetz, H. and Ochsenreiter, W., Clock synchronization in distributed real-time systems, *IEEE Transactions on Computers*, C-36(8), August 1987, 933–940.

[Lamport1978] Lamport, L., Time, clocks, and the ordering of events in a distributed system, *Communications of the ACM*, 1978, 21, 558–565.

[Liskov] Liskov, B., Practical uses of synchronized clocks in distributed systems, *Distributed Computing*, 1993, 6, 211–219.

[Mills1991a] Mills, D.L., Internet time synchronization: The network time protocol, *IEEE Transactions on Communications*, 1991, 39, 1482–1493.

[Mills1998a] Mills, D., RFC 1059: Network Time Protocol (Version 1), University of Delaware, Newark, DE, 1998.

[Mills1998b] Mills, D., Adaptive hybrid clock discipline algorithm for the network time protocol, *IEEE/ACM Transactions on Networking*, 1998, 6(5), 505–514.

[Mills2000] Mills, D., RFC 1119: Network Time Protocol (Version 2), University of Delaware, Newark, DE, 2000.

[Minar1999] Minar, N., A Survey of the NTP Network, Technical Report, Massachusetts Institute of Technology, Cambridge, MA, 1999.

[Owen2005] Owen, D., Wired trigger bus physical aspects 2005, White Paper Agilent, Agilent Laboratories, Palo Alto, CA, 2005.

[Palumbo1992] Palumbo, D.L., The derivation and experimental verification of clock synchronization theory, *IEEE Transactions on Computers*, 1992, 43, 676–686.

[Patt1994] Patt, B., A theory of clock synchronization, PhD thesis, Massachusetts Institute of Technology, Cambridge, MA, 1994.

[Pleasant2005] Pleasant, D., LXI triggering, White Paper Agilent, Agilent Laboratories, Palo Alto, CA, 2005.

[Richards2006] Richards, J.M., Mandell, D.E., Makarand Karanjkar, P., and Pierre-Anthony, S., *Frame Synchronization in a Ethernet NTP Time-Keeping Digital Cinema Playback System*, Dolby Laboratories Licensing Corporation, San Francisco, CA, 2006.

[Sauter2005] Sauter, T., Linking factory floor and the internet, in *The Industrial Information Technology Handbook*, Part II (Industrial Information Technology), Section 3 (Industrial Communication Systems), Chapter 52, CRC Press, Boca Raton, FL, 2005, 52-1–52-15.

[Treytl2007] Treytl, A., Gaderer, G., Hirschler, B., and Cohen, R., Traps and pitfalls in secure clock synchronization, in *Proceedings of the IEEE International Symposium on Precision Clock Synchronization for Measurement, Control and Communication ISPCS 2007*, Vienna, Austria, October 1–3, 2007, pp. 18–24.

19

Quality of Service

Gabriel Diaz Orueta
Spanish University of Distance Education, UNED

Elio San Cristobal Ruiz
Spanish University of Distance Education, UNED

Nuria Oliva Alonso
Spanish University of Distance Education, UNED

Manuel Castro Gil
Spanish University of Distance Education, UNED

19.1 Introduction 19-1
19.2 Relationship with Information Security Topics 19-3
19.3 Quality of Service for IP Networks 19-4
Integrated Services (IntServ) Model • Differentiated Services (DiffServ) Model • Classification and Marking • Queuing and Congestion Management
19.4 Special Considerations for Managing the Quality of Service 19-11
Congestion Avoidance • High Availability Solutions for the Routers
References 19-14

19.1 Introduction

Nowadays, almost any industrial network uses applications based on IP networks [1]. This means that people controlling these networks need to know at least the basics of quality of service (QoS) for IP networks. Every device used in an industrial network is going to send and receive IP packets but, depending on the nature of the industrial application we use in the device, the way to process the IP packets can be decisive to the objectives we want to get through the device. The IP network is the fundamental infrastructure for every industrial network so, although we cannot change the details of the IP implementation in the operating systems of the devices, we need to know how to get the maximum performance and reliability of our network for every industrial application we decide to use. It is also important to note that the devices we can use to manage and implement these QoS characteristics are not the same devices we use to implement our industrial applications but the usual communication devices in every IP network—switches and routers. It is so possible to say that the QoS requirements in industrial networks depend on the industrial applications we want to use and, as a consequence, of the complexity of the network. If the network uses routers, we can apply special features in the routers to improve our QoS; but if the network only needs switches, we can only use a restricted set of QoS features.

Any qualified network professional must know the limits for the networks in his or her organization. It is necessary to know how to get the maximum of the features of the network devices used, being switches, routers, or any network security element that can modify the network performance, as firewalls or intrusion detection systems.

Until recently, there was not much to say about how to improve the services offered by IP networks. It was enough that the network does its job properly 24 h/day. But today, and in almost every organization, it is necessary a network be able to adapt the changing needs for the different kind of applications using it as a means of transmission and communication. There will be applications that need more bandwidth at specific moments and other applications that impose no more than a certain delay for its signal. If these conditions are not reached simply, the client side of these applications will not work properly and we will be losing quality.

Technically speaking, network QoS can be defined as the application of functionality and features to actively manage and satisfy networking requirements of applications sensitive to delay, loss, or delay variation (jitter), guarantying also the availability of bandwidth for critical application flows.

Using QoS methods and technologies [2,3] several benefits are reached:

- *Control over resources*: It can be determined which network resources (bandwidth, equipment, etc.) are being used. For example, critical traffic as voice or video over IP may consume a link and QoS helps control the use of the link by dropping low-priority packets.
- *More efficient use of network resources*: Using network analysis management and accounting tools, it can be determined how traffic is handled, and which traffic shows latency, jitter, or packet loss. QoS tools [1] can be used to tune the handling of the traffic.
- *Coexistence of critical applications*: By using QoS technologies, it is possible that most critical applications receive the most efficient use of the network. It is possible, for example, that time-sensitive multimedia, which requires bandwidth and minimized delay, get them while other applications in the same link receive fair service without interfering with critical applications traffic.

These needs are directly linked with the idea of QoS, seen as a sum of *value* and *warranty*, being only a part of the framework of services management for best practices IT recommendations, as ITIL or CobIT [4]. The *value* of the network is what we search using a particular network, the particular communication between different kind of devices. For the case of an industrial network, the value is going to be what we want to get with communication between servers and clients in our industrial network. The *warranty* is much more linked with the quality, meaning a defined set of features for a concrete network and can be expressed as the level of service of the network for a number of features:

- Availability or how much time is guaranteed that the network is up continuously
- Capacity or how many resources (dedicated bandwidth, delay, etc.) are available for each of the applications and how is the capacity to react to a change in the need of resources
- Network security or the ability to protect any piece of information, following a security policy that include facts like the confidentiality, integrity, or authentication for any kind of access events in the network
- Continuity as a sum of the other features, but specially thought for disaster incidents such as a fire or a flood

Once the network management staff has built a Network Policy for all these features, it is necessary to know the tools for implementing each of the directives in the policy and, fortunately, we have now real technical solutions in place for these kinds of situations, and we have enough reliable standards.

The rest of this chapter presents an introduction on how to manage all these features to get the best QoS for IP industrial networks, although practically every single piece of information in the chapter can be applied to any IP network. The real difference between industrial IP networks and nonindustrial IP networks are the applications we run on them and, for IP QoS, we only need to know some general characteristic of these applications. The next section shows the important connections for QoS, previously defined, with the basics of information security topics and the efforts developed in the last years to build some standards for managing information security topics related with industrial networks. Then, we go in depth of the classical terminology, techniques, and architectures used today for getting a good QoS in any IP network, showing the two most important models: IntServ and DiffServ. Also, we analyze how the communication devices in the network can classify and mark any IP packet and how to manage the possible traffic congestion we can see in the same devices. To end, we analyze two possible situations we must manage only if our IP industrial network uses routers: the congestion avoidance and high availability solutions for routers.

19.2 Relationship with Information Security Topics

As remarked before, some of the topics related with QoS are the information security facts that any network professional must know, at least at an introductory level, related with the need of continuous availability and reliability of the industrial network and data.

As special security problems in industrial networks, it is necessary to emphasize the following:

- *Physical security problems*: The scope of these problems is the range of devices found in a network: PCs or client workstations, servers, routers, switches, or special dedicated equipment. Especially for the most sensible devices, if they are not correctly placed, they faced a possible destruction or an easy unauthorized access to the console of the device, with the consequent risk for the control of the device or for the data being there and, as a consequence, for the value of the network.
- *Software security problems*: It is necessary to include here the operating systems of devices, any IP protocols implementation, and any industrial application installed in the devices. The most common problems are the security bugs due to bad coding (very common), the insecure configuration of the authentication and authorization files in the operating system, the insecure configuration of the permissions for the file systems, or a bad policy for the security copies of the data. Also, it must be taken into account the problems due to incorrect implementation of protocol stacks or the use of dangerous protocols or applications, as the case for Telnet, ftp, tftp, RIP, SMTP, and so on.
- A significant emphasis must be put on the need of having a good antivirus system and special control for the attachments in any mail entered in the network, especially if our industrial network shares the media with the staff network.
- It is also important to take time to analyze the security in the fundamental network devices, the switches and the routers. They are no more than hardware and software; but the consequences of security breaches in them can be extremely dangerous. For example, if one of the switches in our network suffers a denial of service attack, every device running our applications is disconnected and the final result is that we have no network. Sometimes also we can suffer a capture attack in one of our routers and, from the router, using a privilege escalation technique, our servers can be hacked and obliged not to do its usual job.

From the technical point of view [5], we have many tools to face all these security problems, but, in a consistent way with the idea of quality expressed in the previous section, we need a security policy to organize all the implementations of security defenses.

An information security policy can be defined as "a formal statement of the rules by which people, who are given access to an organization's technology and information assets, must abide" [6].

The main goal is to inform all the people in the organization about the obligatory requirements to observe, to protect the technological values and resources and all the organization's information.

The policy must make explicit the mechanisms for us to obey each norm and the methodology to follow for each of the particular cases. If it is complete enough, it can give the hints to select the technologies, including software and hardware to implement the policy.

To build a good network security policy for a concrete organization, a number of key questions must be answered:

- What exactly must be protected? First a complete inventory must be built, taking into account different security standards like ISO/IEC 27001 [7] or ISO/IEC 15408 [8]. Then, a selection (with assigned priorities) of the actives to protect must be built.
- Who could attack our network? It is extremely important to analyze every internal and external people that work with us and decide in which trust. Also, it is a good idea to think about the possible motivations of virtual attackers.

- How are the systems and tools selected to implement the security policy going to be used? There will be cryptographic defenses and no cryptographic ones and it will be important to know the different kinds of attacks and defenses.
- How much money could the organization afford to implement the policy? This data item gives a great input to decide which way can be used to begin to implement the policy.

Once the security policy (or at least the first operating version) is built, it is necessary to start the security process within the information security management system [7]. It can be structured in three continuous phases:

1. *Implementation phase*: Every organizational measure of the policy must be activated, including all the configurations for servers, routers, switches, firewalls, etc.
2. *Monitoring phase*: Audits and continuous monitoring of systems and networks must be done to assure that the security restrictions are working and to investigate possible new security problems, not previously detected or thought of.
3. *Security adulting phase*: Vulnerability audits and security tests must be done, especially to sensible devices and servers, in order to get new possible technical or organizational vulnerabilities.

Adding the results of phases 2 and 3, we get new directions for updating the security policy building, say, the operating version 2, that we must implement putting again in the first phase of the process, and so on.

To finish this section, it is important to point out a number of committees and standards that can help to build a concrete security policy for a concrete industrial network:

- IEEE 1402 [9], a security standard for substations, especially devoted to physical security.
- Process Control Security Requirements Forum (PCSRF) [10], sponsorized by NIST. Its goal is to build a quality and security standards set for building new industrial process control systems. It has developed norms for DCS and SCADA systems, like the SCP-ICS (Industrial Control System Security Capabilities Profile) or the SPP-ICS (Industrial Control System Security Profile).
- *ISA SP99*: The SP99 [11] committee of Instrumentation Systems and Automation (ISA) develops "guide-alike" documents for introducing information security in automatized control systems.
- *IEC TC65*: Technical subcommittee 65C of the IEC is working since 2004 in security standards for field buses and other industrial communication networks.

19.3 Quality of Service for IP Networks

Traditionally, the term "quality of service" refers to a series of techniques [12] and methodologies [2,3] whose targets are to assign different priorities to different kinds of traffic in the IP network. Also refers to implement management policies to handle the latency, delay, and bandwidth. These techniques try to give a distinctive and preferential treatment to the traffic corresponding to the most critical applications in the network. This treatment must be homogeneous, the same treatment in all the areas in the network. The policy must allow the changes, must be dynamic because of the possible changing situations of the services in the network. Nowadays, they are an essential component to assure the correct performance of the applications in the network.

Any network may experience any of these network availability problems:

- *Delay*: Also named latency, is the amount of time that it takes a packet to reach the receiving endpoint after being transmitted from the sending endpoint. This period can be broken into two areas: fixed network delay and variable network delay. The fixed part includes encoding and decoding times (for voice and video, for example) as well as the amount of time required for the electrical and optical pulses to traverse the media to their destination. The variable part generally refers to network conditions, such as congestion, that is going to affect the overall time required for transit. For data networks it is usual to talk about packetization delay, the time that it takes to segment data (if needed), encode signals (if necessary), process data, and turn data into packets; serialization delay, the time that takes to place the bits of a packet, encapsulated in a frame, onto the physical media; the propagation delay or time that it takes to transmit the bits of a frame across the physical wire; the processing delay, the time that takes a network device to take the frame from the input interface, place it into a receive queue, and place it into the output queue of the outgoing interface; and queuing delay, the amount of time that a packet resides in the output queue of an interface.

- *Delay variation*: Also named "jitter" is the difference in the end-to-end delay between packets. If, for example, one packet requires 100 ms to traverse the network from the source endpoint to the destination endpoint, and the following packet requires 130 ms to make the same path, the delay variation between packets is calculated as 30 ms. Many devices in a network have a jitter buffer for smoothing out changes in arrival times of data packets containing, for example, voice or video.
- *Packet loss*: It is a measure of packets transmitted and received compared to the total number that were transmitted and expressed as the percentage of packets that were dropped. Tail drops occur when the output queue is full, being not the unique drops but the most frequent ones, and the origin of the drops is the congestion in the link.

It is important to see that many applications can deliver unpredictable bursts of traffic. It is practically impossible, for example, to predict the usage patterns for Web or file transfer, but network professionals need to be able to support critical applications during peak periods.

QoS technology permits to do the following:

- Predict response times for end-to-end network services
- Manage loss in times of inevitable bursty congestion
- Manage jitter or delay sensitive applications
- Set traffic priorities across the network
- Support dedicated bandwidth
- Avoid or manage network congestion

To achieve these results, it is necessary to do a constant monitoring, real-time monitoring, a careful traffic engineering and, once we get these features, it is possible to create a good planning for every application.

Each of them will require a defined service level. The objectives can be resumed:

- All network traffic must reach its service levels. For example, keeping the latency below a certain value for the voice over IP traffic to get a good quality.
- During the periods of network congestion, the most important traffic must always enjoy the resources it needs.
- As an added benefit, these techniques allow to optimize the use of network resources, delaying the need to spend more money to add resources.

There are a number of processes [2,12] involved in QoS:

- *Classifying the traffic*: A descriptor value must be used to categorize a packet or frame within a specified group. This will make the packet accessible for QoS handling in the network or within a network device. The classification process will allow segmenting the traffic into multiple priority levels or classes of service. Another possible approach is "QoS signaling" in which the classification is done based on application-flow rather than a packet basis.
- *Traffic shaping*: It is used to create a traffic flow for limiting the bandwidth potential of the flow. It determines if a packet is in or out of profile by comparing the traffic rate to the configured police, which limits the bandwidth consumed by a flow of traffic. The traffic exceeding the configured rate is buffered first in an attempt to minimize loss.
- Marking is the process of setting or changing a priority value of a frame or packet.

- Queuing is used to evaluate the priority value and the configured police and to determine in which of the output queues the packet must be put.
- Scheduling is the process that services the transmit queues, based on the sharing and shaping configuration of the transmit port.
- Dropping is used to drop packets, to take advantage of Transmission Control Protocol (TCP) and windowing mechanisms that drop selectively some packets in an attempt to avoid dropping much more packets. This does not work with applications using User Datagram Protocol (UDP) as the transport protocol, only for the TCP ones.

By the other side, the QoS features must be configured throughout the network for providing end-to-end QoS delivery. And there are three components needed to get this consistency:

- QoS components within a single network element, including queuing, scheduling, and traffic shaping.
- QoS signaling techniques to coordinate QoS from end to end between network elements.
- QoS policing and management functions to control and administer end-to-end traffic across a network.

There are nowadays only two QoS architectures for IP networks: the Integrated Services model, (IntServ) [13], and the Differentiated Services model, (DiffServ) [14]. They are different in the way they enable applications to send data and in how the network attempts to deliver the data within the specified level of service.

19.3.1 Integrated Services (IntServ) Model

Integrated Services (IntServ) model is defined in RFC 1633 [13]. It provides multiple services that can accommodate multiple QoS requirements. It can be used through the introduction of Resource reSerVation Protocol (RSVP, RFC 2205 and 3936) [16–18], enabled at both the end points and the networks devices in between. Each RSVP-enabled application requests a specific kind of service from the RSVP-enabled network before sending the data. Explicit signaling via RSVP must occur to facilitate the request. Afterward, the application informs the network of its traffic profile and requests a particular kind of service that can contains its bandwidth and delay needs. The idea is that the application is only going to send data after it gets a confirmation from the network.

This kind of RSVP-enabled network does also admission control, based on information from the requesting host application and available network resources, admitting or rejecting the application request for bandwidth. The network commits to meet the QoS requirements of the application while the specific traffic flow for which the request was made remains within the profile specifications.

It is also possible to centralize admission control by using the Common Open Policy Service (COPS, RFC 2748) [19] protocol at a policy decision point. When used with RSVP, COPS provides several benefits:

- Centralized management of services
- Centralized admission control and authorization of RSVP flows
- Increased scalability of RSVP-based QoS solutions

The two main drawbacks of RSVP and IntServ are the need of continuous signaling due to the stateless architecture and the lack of scalability and, although there is an IETF working group devoted to its study, it is not the most commonly used.

19.3.2 Differentiated Services (DiffServ) Model

The DiffServ model [14,15] is a multiple service-level model that can satisfy differing QoS needs. However, unlike in the IntServ model, an application that uses DiffServ does not signal explicitly the

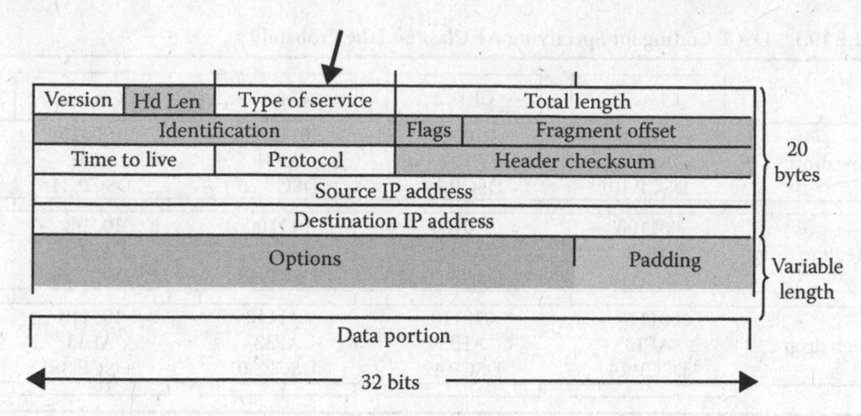

FIGURE 19.1 ToS field in the IP header of any IP packet.

network devices before sending data. DiffServ reassigns bits in the type of service (ToS) (8 bits in ToS, Figure 19.1) field of the IP header of any IP packet [20] and uses differentiated services code points (DSCP) as the QoS descriptor value, which supports 64 possible classifications.

For a differentiated service, the network is trying to deliver a concrete kind of service based on the QoS descriptor specified in each IP packet header on a per-hop basis rather than per-traffic flow as IntServ does.

Usually, this service model is especially appropriate for aggregate flows; complex traffic classification are performed at network edges resulting in a per-packet QoS handling.

The device, typically at the edge of the network, identifies packets based on the IP precedence (the three first bits in the ToS field of the IP header) or on the DSCP fields in the ToS byte of the IP header (Figure 19.2) that are going to be used in a per-hop behavior (PHB), packet by packet.

The six most significant bits of the ToS byte form the DiffServ field and the last two bits are used as early congestion notification (ECN) bits. IP precedence uses three bits, while DSCP, really an extension of IP precedence, uses six bits to select the PHB for the packet at each network node.

There are two typical ways for specifying this PHB: the assured forwarding (AF) PHB [21] and the expedited forwarding (EF) PHB [22].

The AF group allows offering different levels of forwarding assurances for IP packets received from a DiffServ domain. There are four AF classes, named AF1x through AF4x. Within each class, there are three drop probabilities. Depending on the policy of the network, packets can be selected for a PHB based on required throughput, delay, jitter or loss, or simply according to priority of access to network services.

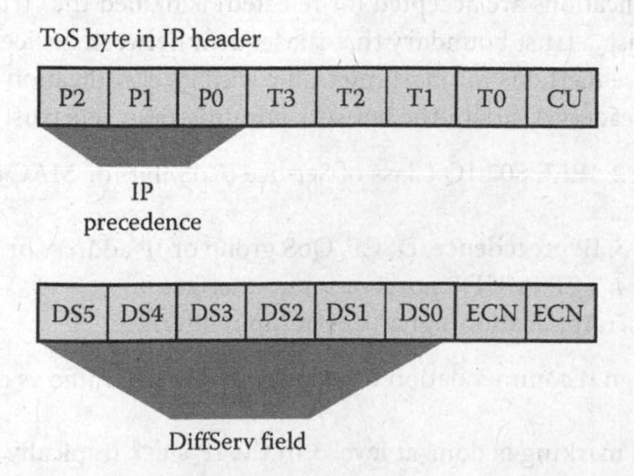

FIGURE 19.2 DiffServ field into the ToS byte of an IP packet.

TABLE 19.1 DSCP Coding for Specifying AF Class and the Probability

	Class 1	Class 2	Class 3	Class 4
Low drop	001010 AF11 DSCP 10	010010 AF21 DSCP 18	011010 AF31 DSCP 26	100010 AF41 DSCP 34
Medium drop	001100 AF12 DSCP 12	010100 AF22 DSCP 20	011100 AF32 DSCP 28	100100 AF42 DSCP 36
High drop	001110 AF13 DSCP 14	010110 AF23 DSCP 42	011110 AF33 DSCP 30	100110 AF43 DSCP 38

The AF PHB model is nearly equivalent to the controlled load service of the IntServ model. It defines a method to give different forwarding assurances to different traffic classes. For example, network traffic can be divided into three classes:

- Gold, 50% of the available bandwidth
- Silver, 30% of the available bandwidth
- Bronze, 20% of the available bandwidth

Table 19.1 shows the DSCP coding for specifying the AF class with the probability: bits 0, 1, and 2 define the class; bits 3 and 4 the drop probability; and bit 5 is always 0.

The EF model can be used to build a low-loss, low-latency, low-jitter, assured bandwidth end-to-end service through different DiffServ domains. This kind of service is also named a premium service.

19.3.3 Classification and Marking

The marking process identifies which frames or packets are processed to meet a specific level of service of QoS policy for end-to-end service.

Using classifications, the network traffic can be classified into multiple priority levels or classes of service. There are different methods of identifying types of traffic, but the two most used are the inclusion of access filters or the network-based application recognition (NBAR) [12] method, a classification engine that can recognize a wide variety of applications, including Web-based applications and client/server applications that dynamically assign TCP or UDP port numbers. After the application is recognized, the network can invoke specific services for that particular application.

The device where classifications are accepted (or rejected) is named the "trust boundary." The trust port configurations establish a trust boundary that subsequent network devices or elements in the network will enforce. There are methods for marking traffic with its classification that allow setting information in layer 2, 3, or 4 headers. Usually, the network administrator sets trust boundaries based on

- *Parameters on layer 2*: IEEE 802.1Q Class of Service (CoS) bits or MAC address in Ethernet networks or input interface
- *Parameters on layer 3*: IP precedence, DSCP, QoS group or IP address or input interface
- *Parameters on layer 4*: TCP or UDP ports or input interface
- *Parameters on layer 7*: Application signatures or input interface

It is a best practice design recommendation to identify and mark traffic as close to the source of the traffic as possible.

If the classification and marking is done at level 3 in the IP stack (typically at a router), the method used is the one cited with DSCP codes; but at level 2 (at a switch), it is typical to use CoS bits in the incoming frame. Then the CoS to DSCP map is used to generate the internal DSCP value. With VLANs in use, the IEEE 802.1Q frame header carries the CoS value in the three most significant bits of the tag control information (TCI) field. The CoS values range from 0, for low priority, to 7, for high priority. If the frame does not contain a CoS value, assign the default port CoS to the incoming frame.

19.3.4 Queuing and Congestion Management

Network devices handle a possible overflow of arriving traffic by using a queuing algorithm to sort the traffic, and, then, determine a method for assigning priorities onto an output interface. Each queuing algorithm solves a specific type of network traffic condition and has a particular effect on network performance.

Figure 19.3 shows the actions that a QoS-enabled network device must take before transmitting a frame. It takes these steps:

- Classification of packets.
- It must be determined if it can put the packet into the queue or it has to drop the packet. By default, queuing mechanisms will drop a packet only if the corresponding queue is full, this is named "tail drop."
- If the packet is queued, it is placed into the queue for that particular class.
- The packets are then taken from the individual per-class software queues and put into a FIFO hardware or transmit queue.

The queuing methods differ in that some of them classify packets automatically while others need manual configuration; but the most important part of every queuing mechanism is the scheduling policy because it says the order in which the packets will leave the network device.

On the other side, the queuing mechanisms directly relate to the DiffServ model and usually in this way:

- *Expedited forwarding PHB Implementations*: Use the priority queuing (PQ) or the IP real-time transport protocol (RTP) [23] prioritization
- *Assured forwarding PHB implementations*: Use the class-based weighted fair queuing (CBWFQ), four classes with weighted random early detection (WRED) within each class, or the custom queuing

The PQ scheme (Figure 19.4) is designed to give strict priority to important traffic and assures that priority traffic is serviced most often and strictly prioritizes according to network protocol (such as IP or IPX), incoming interface, packet size, source, or destination address.

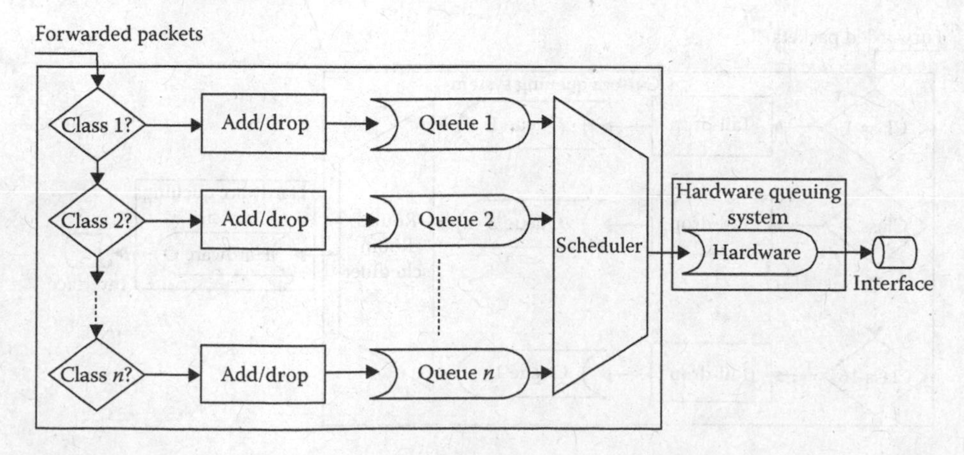

FIGURE 19.3 Queuing components.

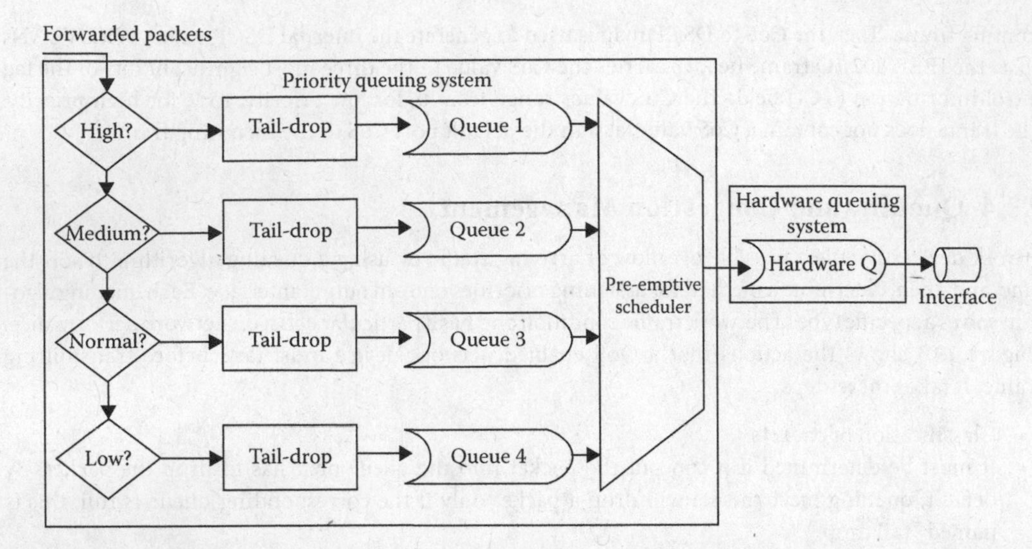

FIGURE 19.4 Priority queuing system.

PQ uses to be statically configured and implemented through the use of the expedite queue. One major drawback is that low queues may never be sampled as long as higher priority traffic is being processed, and this results in queue non-use.

Figure 19.5 shows the custom queuing system, which reserves a percentage of the available bandwidth of an interface for each selected traffic type. If a particular type of traffic is not using the bandwidth reserved for it, other traffic types may use the remaining reserved bandwidth.

Custom queuing must be configured statically and does not provide automatic adaptation for changing network conditions.

The IP RTP Priority model provides a strict PQ scheme in which delay-sensitive data, such as voice, can be dequeued and sent before packets in other queues. It is used typically on serial interfaces in conjunction with CBWFQ on the same outgoing interface. Traffic matching the range of UDP ports specified for the priority queue is guaranteed strict priority over other CBWFQ classes: packets in the priority queues are always serviced first.

The weighted fair queuing (WFQ) approach classifies traffic into different flows based on such characteristics as source and destination address, protocol and port and socket of the session, and is the default

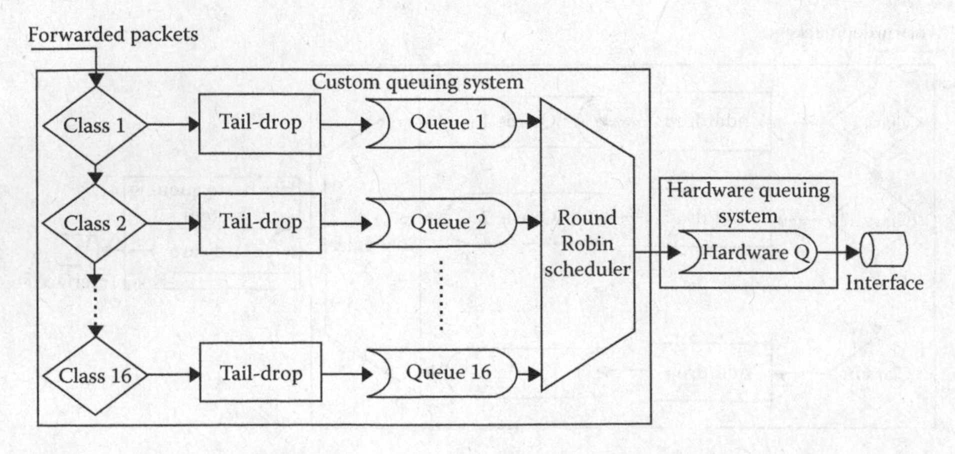

FIGURE 19.5 Custom queuing system.

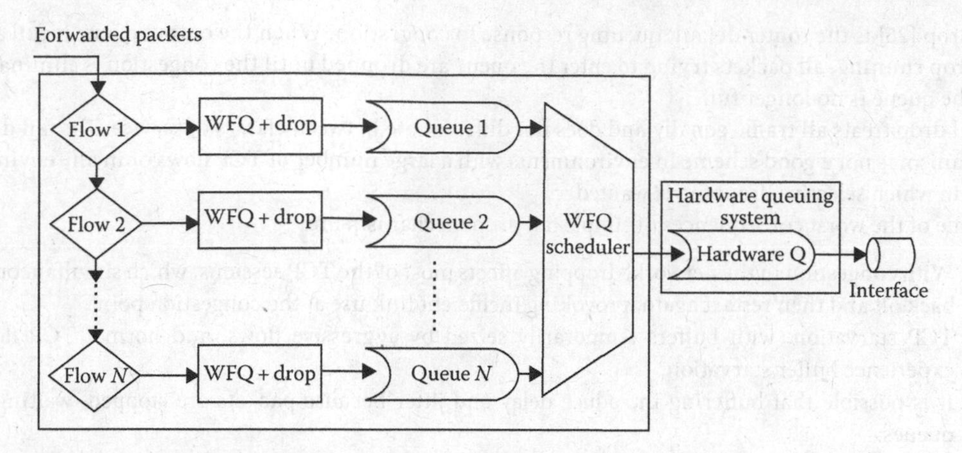

FIGURE 19.6 Class-based weighted fair queuing.

queuing system for Ethernet links. An extension is CBWFQ (Figure 19.6), which provides support for user-defined traffic classes.

The classes allow specifying the exact amount of bandwidth to be allocated for a specific class of traffic. Taking into account available bandwidth on the interface, you can configure up to 64 classes and control distribution among them.

19.4 Special Considerations for Managing the Quality of Service

The way to specify the parameters and rules that control QoS for IP networks is using software tools and devices. It will depend on the type of network, the magnitude and the application running on it.

Many times it is only necessary to apply the QoS mechanisms in each device, case by case. But, in many others, for more complex and big networks, the management is done using different kind of network management tools as HP OpenView or Cisco Works. These tools always use as the preferred management protocol the well-established simple network management protocol (SNMP) [24], to manage many different situations in a network (general network management and network monitoring tasks), some of them being probably QoS tasks.

Although they are not the only devices to implement the QoS techniques, the routers (and any level 3 switch is also a router in this sense) are the most frequently used ones. At the routers, and using the real-time data from the packets they process, the QoS statistics are calculated. A consequence that must be taken into account is that we must size correctly the needed resources for the router with its QoS mechanisms implemented and using them. Also we must be consistent and apply the same mechanisms for every router in the network.

At the routers, there are other special considerations to do about QoS, the congestion avoidance features and the high availability solutions for the routers being two of the most important ways it is handled.

19.4.1 Congestion Avoidance

Congestion avoidance techniques monitor network traffic loads, trying to anticipate and avoid congestion at common network bottleneck points. It is got through packet dropping by using more complex techniques than simple tail drop.

When an interface on a router cannot transmit a packet immediately, the packet is queued. Packets are afterward taken out of the queue and eventually transmitted on the interface.

If the arrival rate of packets to the outgoing interface exceeds the router capability to buffer and forward traffic, the queues increase to their maximum length and the interface becomes congested.

Tail drop [25] is the router default queuing response to congestion. When the output queue is full and tail drop running, all packets trying to enter the queue are dropped until the congestion is eliminated and the queue is no longer full.

Tail drop treats all traffic equally and does not differentiate between classes of service. The tail drop mechanism is not a good scheme in environments with a large number of TCP flows or in any environment in which selective dropping is wanted.

Some of the worst consequences of simple tail drop mechanism are

- With congestion in the network, dropping affects most of the TCP sessions, which simultaneously back off and then restart again, provoking inefficient link use at the congestion point.
- TCP starvation, with buffers temporarily seized by aggressive flows, and normal TCP flows experience buffer starvation.
- It is possible that buffering introduce delay and jitter because packets are stopped, waiting in queues.
- Premium traffic is also dropped because of the inexistence of differentiated drop mechanisms.

One of the approaches for avoiding these problems is the random early detection (RED) feature. RED [25] does not take into account the precedence or CoS, it simply uses one of the single threshold when that threshold value for the buffer fills. RED start to drop packets randomly (not all packets, as with tail drop) until the maximum threshold is got, and then all the packets are dropped. The idea is that the probability of dropping a packet rises linearly with the increase of buffer filling above the threshold.

RED is very efficient only with TCP-based traffic, because it takes advantage of the windowing mechanism used by TCP to manage congestion. As a consequence, it is not very useful when the traffic type is not TCP-based.

Other useful mechanism is Weighted RED (WRED, Figure 19.7) [25]. It is very similar to RED in that both define some threshold, and that threshold starts to randomly drop packets; but WRED is CoS-aware, meaning that a CoS value is added to each of the thresholds.

When the threshold is exceeded, WRED randomly drops packets with the CoS assigned. For example, with two thresholds in the queue

- CoS 0 and 1 are assigned to threshold 1, and the threshold set to 40% of buffer filling.
- CoS 2 and 3 are assigned to threshold 2, and the threshold set to 80% of buffer filling.

At the moment that buffer exceeds 40% of use, packets with CoS 0 and 1 will start to be randomly dropped. More packets will be dropped while the buffer use grows. If 80% is reached, packets with CoS

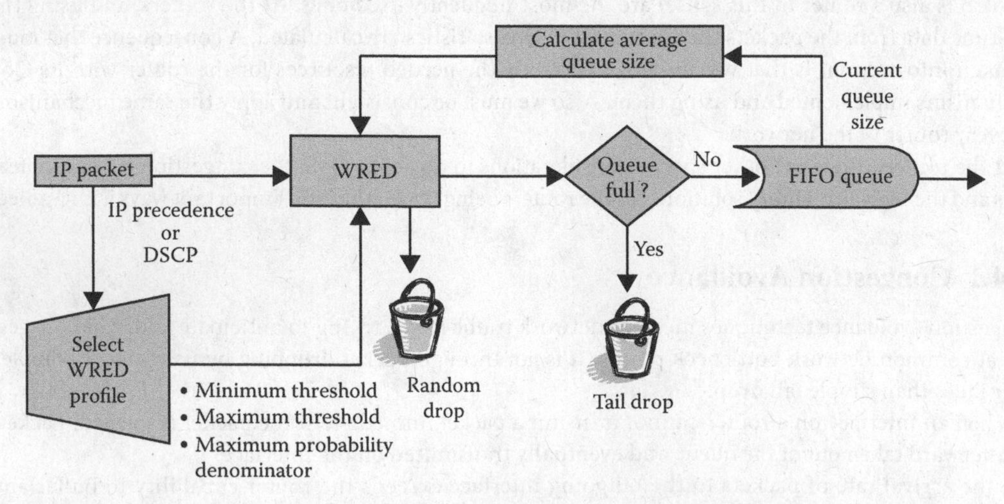

FIGURE 19.7 Weighted random early detection mechanism.

2 and 3 will start to drop, but WRED will also continue to randomly drop packets with CoS 0 and 1. The final result is that there will be more packets dropped with CoS 0 and 1 than with CoS 2 and 3.

19.4.2 High Availability Solutions for the Routers

As mentioned previously, another feature for the "service level" solution is the availability of the network. None of the previous QoS solutions will work if the network devices (switches and/or routers) where must be implemented are not working. This means that any complete network QoS solution must take into account the redundancy for, at least, the most sensible routers in the network.

Nowadays, this is implemented almost universally by using the Virtual Router Redundancy Protocol (VRRP). The VRRP [26] obliges to duplicate routers and links to ensure continuity of service across failures. It introduces the concept of "virtual router" that is addressed by IP clients requiring routing service. The real routing service is provided by physical routers running the VRRP.

There are a number of descriptive terms introduced by VRRP:

- Virtual router, a single router image created through the operation of one or more routers running VRRP.
- Virtual router ID or VRID, a numerical identification of a particular virtual router. It must be unique on a network segment.
- Virtual router IP, an IP address associated with a VRID that the other hosts use to obtain network service. It is managed by the VRRP instances belonging to a VRID.
- Virtual MAC address, a predefined MAC address used for all VRRP actions (for Ethernet) instead of the real adapter MAC address(es). It is derived from the VRID.
- Master, the VRRP instance that performs the routing function for the virtual router at a given time. Only one master is active at a time for a given VRID.
- Backup, other VRRP instances for a VRID that are active, but not in the master state. Backups are ready to take on the role of the master if the current master fails.
- Priority, a value assigned to different VRRP instances, as a way to determine which router will take on the role of the master if the current master fails.

Figure 19.8 shows a simple VRRP configuration, with two routers connecting to a network cloud and VRRP providing a resilient routing function for the client machines in the local area network.

Router rA is the master of virtual router VRID 1 and router rB is at the backup state. When a VRRP instance is in the master state for a VRID, it sends multicast packets to the registered VRRP multicast address advising other VRRP instances that it is the master for the VRID, with a twofold objective:

- If a VRRP instance with a higher priority for that VRID is started, the new VRRP instance can force and election and take on the master role.
- VRRP instances in the backup state for the VRID listen for the master's packets; if an interval elapses without a packet being received, the instances in the backup state take action to elect a new master.

In Figure 19.8, one of the following must be true about router rA:

- It has a higher priority than router rB.
- If both routers have the same priority, its interface IP address is higher than that of router rB.

If router rA fails, after a short interval router rB would notice that no multicast packet has been received and transitions to the master state, taking over the handling of the virtual IP address and sending its own multicast packets.

The time router rB waits before making its state transition is called the master down interval and is based on the length of time between master updates (advertisement interval) and a value called skew time calculated from the priority value.

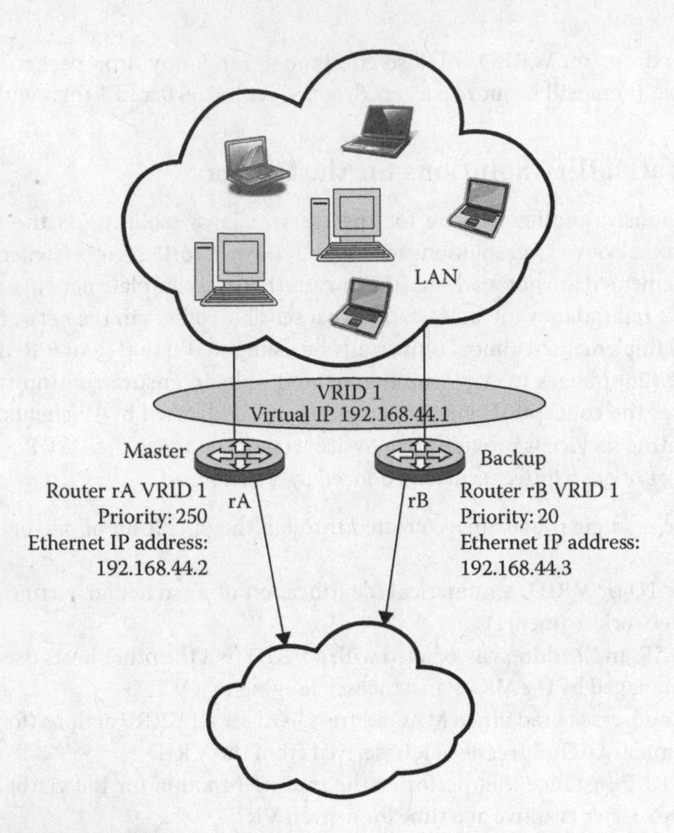

FIGURE 19.8 Simple VRRP example.

When the problem with router rA is resolved, and depending on its configuration, either of two situations would occur:

- If Router rA is configured to start as master, it will force an election immediately by sending its first advertisement as master. Router rB will receive this advertisement and transition to backup state.
- If Router rA is configured to start as backup, it will transition from initialization to backup state. Nothing will happen until it receives an advertisement from router rB. Router rB has a lower priority than router rA, so router rA will start an election by commencing its transition to master state and sending an advertisement. When it receives this advertisement, router rB will transition to backup state because router rA has higher priority.

Although VRRP is based on Cisco's proprietary Hot Standby Router Protocol (HSRP, [26]) concepts, VRRP is actually a standardized version of Cisco's HSRP. Those protocols, while similar in concept, are not compatible. There are a great number of implementations besides the Cisco devices one.

References

1. R. Zurawski, ed., *The Industrial Communication Technology Handbook*, CRC Press, Boca Raton, FL, 2005.
2. W. Stallings, *High Speed Networks and Internets. Performance and Quality of Service*, Prentice Hall, Englewood Cliffs, NJ, 2002.
3. J. García Tomás, J.L. Raya, and V. Rodrigo, *Alta velocidad y calidad de servicio en Redes IP*, Ed. Ra-Ma, Alfaomega, Mexico, 2002.

4. itSMF International, *IT Service Management: An Introduction*, itSMF Library, Van Haren Publishing, Zaltbommel, Holland, 2007.
5. R. Anderson, *Security Engineering*, John Wiley & Sons, New York, 2001.
6. RFC 2196, *Site Security Handbook*, IETF (Internet Engineering Task Force), September 1997, available at http://www.ietf.org/, access in July, 2010.
7. ISO/IEC 27001, Information technology—Security techniques—Information security management systems—Requirements, an ISO (International Organization for Standardization) standard, Geneva, Switzerland, 2005, available at http://www.iso.org/, access in July, 2010.
8. ISO/IEC 15408, Information technology—Security techniques—Evaluation criteria for IT security—Part 1: Introduction and general model, an ISO (International Organization for Standardization) standard, Geneva, Switzerland, 2005, available at http://www.iso.org/, access in July, 2010.
9. IEEE 1402, *Guide for Electric Power Substation Physical and Electronic Security*, The Institute of Electrical and Electronics Engineers, Inc., Piscataway, NJ, 2000.
10. Process Control Security Requirements Forum, sponsorized by National Institute of Standards and Technologies, Gaithersbug, MD, available at http://www.isd.mel.nist.gov/projects/processcontrol/, access on May 2009.
11. ANSI/ISA-TR99.00.01-2007, *Security Technologies for Industrial Automation and Control Systems*, an ANSI/ISA (American National of Standards Institute/International Society of Automation) standard, 2007, Washington D.C., available at http://www.standardsportal.org/usa_en/sdo/isa.aspx, access in July, 2010.
12. R. Froom, B. Sivasubramanian, and E. Frahim, *Building Cisco Multilayer Switched Networks*, Cisco Press, Indianapolis, IN, 2006.
13. RFC 1633, *Integrated Services in the Internet Architecture: An Overview*, IETF (Internet Engineering Task Force), June 1994, available at http://www.ietf.org/, access in July, 2010.
14. RFC 2474, *Definition of the Differentiated Services Field (DS field in the IPv4 and IPv6 headers)*, IETF (Internet Engineering Task Force), December 1998, available at http://www.ietf.org/, access in July, 2010.
15. RFC 2475, *An Architecture for Differentiated Services*, IETF (Internet Engineering Task Force), December 1998, available at http://www.ietf.org/, access in July, 2010.
16. RFC 2205, *Resource ReSerVation Protocol (RSVP)*, Version 1 Functional Specification, IETF (Internet Engineering Task Force), September 1997, available at http://www.ietf.org/, access in July, 2010.
17. RFC 2210, *The Use of RSVP with IETF Integrated Services*, IETF (Internet Engineering Task Force), September 1997, available at http://www.ietf.org/, access in July, 2010.
18. RFC 3936, *Procedures for Modifying the Resource reSerVation Protocol (RSVP)*, IETF (Internet Engineering Task Force), October 2004, available at http://www.ietf.org/, access in July, 2010.
19. RFC 2748, *The COPS (Common Open Policy Service) Protocol*, IETF (Internet Engineering Task Force), January 2000, available at http://www.ietf.org/, access in July, 2010.
21. RFC 2597, *Assured Forwarding PHB Group*, IETF (Internet Engineering Task Force), July 1999, available at http://www.ietf.org/, access in July, 2010.
22. RFC 3246, *An Expedited Forwarding PHB (Per-Hop Behavior)*, IETF (Internet Engineering Task Force), March 2002, available at http://www.ietf.org/, access in July, 2010.
23. RFC 3550, RTP: *A Transport Protocol for Real-Time Applications*, IETF (Internet Engineering Task Force), July 2003, available at http://www.ietf.org/, access in July, 2010.
24. RFC 3413, *Simple Network Management Protocol (SNMP) Applications*, IETF (Internet Engineering Task Force), December 2002, available at http://www.ietf.org/, access in July, 2010.
25. RFC 2309, *Recommendations on Queue Management and Congestion Avoidance in the Internet*, IETF (Internet Engineering Task Force), April 1998, available at http://www.ietf.org/, access in July, 2010.
26. RFC 3768, *Virtual Router Redundancy Protocol (VRRP)*, IETF (Internet Engineering Task Force), April 2004, available at http://www.ietf.org/, access in July, 2010.

20 Network-Based Control

Josep M. Fuertes
Universitat Politècnica de Catalunya

Mo-Yuen Chow
North Carolina State University

Ricard Villà
Universitat Politècnica de Catalunya

Rachana Gupta
North Carolina State University

Jordi Ayza
Universitat Politècnica de Catalunya

20.1 Introduction 20-1
20.2 Mutual Concepts in Control and in Communications 20-2
20.3 Architecture of Networked-Based Control 20-2
Connection Types
20.4 Network Effects in Control Performance 20-6
20.5 Design in NBC 20-6
Design Constraints in the Network Side • Design Constraints in the Control Side • Network and Control Co-Design
20.6 Summary 20-7
References 20-8

20.1 Introduction

Industrial control applications use controllers, sensors, and actuators to close control loops. Sensors and actuators are usually distributed topologically through the controlled plant. As ubiquitous embedded controllers, sensors, and actuators are more and more common, while inexpensive and reliable communication systems are ready for use in the industry, many of the control loops are closed using industrial communication systems. A large part of the industry control loops are discrete in nature, where a sequence of discrete actions are done after a sequence of sensings, e.g., after a displacement has been completed, a motor has to stop. From the control side, a discrete system has as major constraint, the system response time, or the maximum time that can be accepted between occurrence of a stimulus and the reaction on the plant. Previous chapters of the book address how the industrial communication systems attain the goal of sending messages keeping this constraint.

Continuous control loops differ from discrete control loops in that the continuous signal measurement is used by the controller to calculate (using differential equations) the continuous signal to be send to the actuator. Those continuous signals can be properly approximated by time-discretized sequences of values that maintain the dynamic characteristics of the sampled system. A goal when connecting through a communication network sensors, controllers, and actuators in a continuous-time control loop is to maintain the dynamic characteristics of the extracted data from the sequence of messages.

Communication networks for feedback control are in increasing use. A network-based control (NBC) system is composed of distributed nodes sharing feedback control-related information through a communication channel. The basic configuration is given by one sensor node that collects process sensor data, one controller node that receives the sensor data, calculates the control action and sends it to an actuator node, and that actuator node that actuates on the process. At least two feedback control nodes are needed to have a proper NBC system (as one of those nodes can have two functions, i.e., a sensor and controller node or a controller and actuator node). In both cases, there is messaging of sensor to controller or controller to actuator nodes. In general, we should consider that the network is shared with other nodes, they being feedback control or any other applications. The result is that NBC systems can contain a large number of interconnected devices (controllers, actuators, sensors) that use the network for interchanging control messages.

Feedback control imposes several restrictions on the communication networks. The multiple access schemes of common industrial buses serialize the sending of the messages through the shared media. As a result, communication delays are created in the feedback control loop. Delays inside a control loop tend to instabilize the controlled plant. In addition, packets may be lost in a noisy channel, breaking the control loop path and having also a possible bad consequence on the controlled process. This chapter reviews the effects on the control side of the communication technology. It also talks about how a proper use of the communication channel can favor the dynamic operation of the control loop and how the control law is to be adjusted to take into consideration the communication-imposed constraints.

20.2 Mutual Concepts in Control and in Communications

In a networked control system (NCS), the feedback and the command signals use a communication network link. When introducing a network in the control loop, some inconveniences such as band-limited channels, delays, and packet dropouts occur. Such network may also be shared by other applications, resulting in the channel being inaccessible, as sharing a communication channel imposes a wait time until the channel is accessible again. From the point of view of control, it represents an additional delay between the sensor measurement and the control actuation, resulting in a potential loss of control performance. Classical discrete controller design imposes a periodic sampling scheme, which also imposes hard real-time operation. The classical definition of hard real-time system is that the actions or tasks shall be executed within a hard deadline or the result is no longer valid. In a controlled system, the restriction is more severe, as the sampling and actuation is required to be with (not within) a prescribed period. As in many cases strictly periodic operation cannot be assured, due to message loss or communication incurred delays, several schemes can be used.

20.3 Architecture of Networked-Based Control

When a traditional feedback control system is closed via a communication channel (such as a network), which maybe shared with other nodes outside the control system, then the control system is classified as an NCS or NBC system. All definitions found in literature for NBC have one key feature in common. This defining feature is that information (reference input, plant output, control input, etc.) is exchanged among control system components (sensor, controller, actuator, etc.) using a *shared network* (see Figure 20.1).

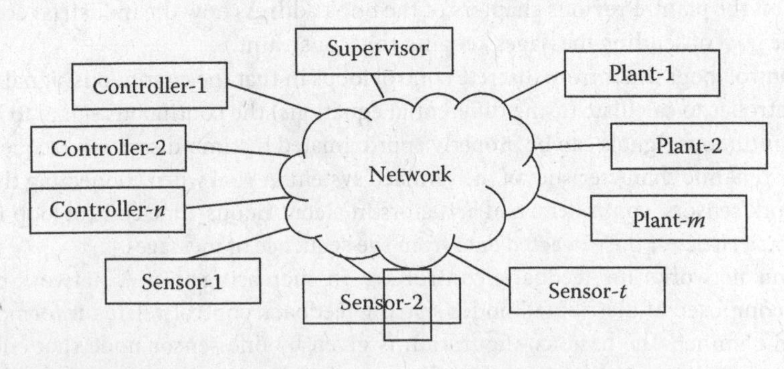

FIGURE 20.1 Typical NBC architecture.

Thus, the basic capabilities of any NBC are information acquisition (sensors/users), command (controllers/users), communication, and network and control (actuators). In broader terms, NCS research and application is categorized into two parts.

1. *Control of network*: Study and research on communications and networks to make them suitable for real-time NCS, e.g., routing control, congestion reduction, efficient data communication, and networking protocol.
2. *Control over network*: This deals more with control strategies and control systems design over the network to minimize the effect of adverse network parameters on NCS performance such as network delay.

This chapter is mainly focused on "control over network."

20.3.1 Connection Types

There are two major types of control systems that utilize communication networks. They are (1) shared-network control systems and (2) remote control systems.

20.3.1.1 Shared-Network Control Systems

Using shared-network resources to transfer measurements, from sensors to controllers and control signals from controllers to actuators can greatly reduce the complexity of connections. This method, as shown in Figure 20.2, is systematic and structured, provides more flexibility in installation, and eases maintenance and troubleshooting. Furthermore, networks enable communication among control loops. This feature is extremely useful when a control loop exchanges information with other control loops to perform more sophisticated controls, such as fault accommodation and control. Similar structures for NBC have been applied to automobiles and industrial plants.

20.3.1.2 Remote Control Systems

On the other hand, a remote control system can be thought of as a system controlled by a controller located far away from it. This is sometimes referred to as teleoperation control. Remote data acquisition systems and remote monitoring systems can also be included in this class of systems.

There are two general approaches to design an NBC system. The first approach is to have several subsystems form a hierarchical structure, in which each of the subsystems contains a sensor, an actuator, and a controller by itself, as depicted in Figure 20.3. These system components are attached to the same control plant. In this case, a subsystem controller receives a set point from the central controller C_M. The subsystem then tries to satisfy this set point by itself. The sensor data or status signal is transmitted back via a network to the central controller.

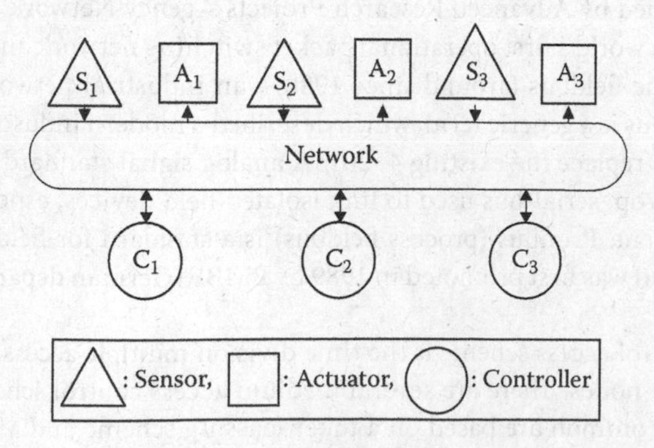

FIGURE 20.2 Shared-network connections.

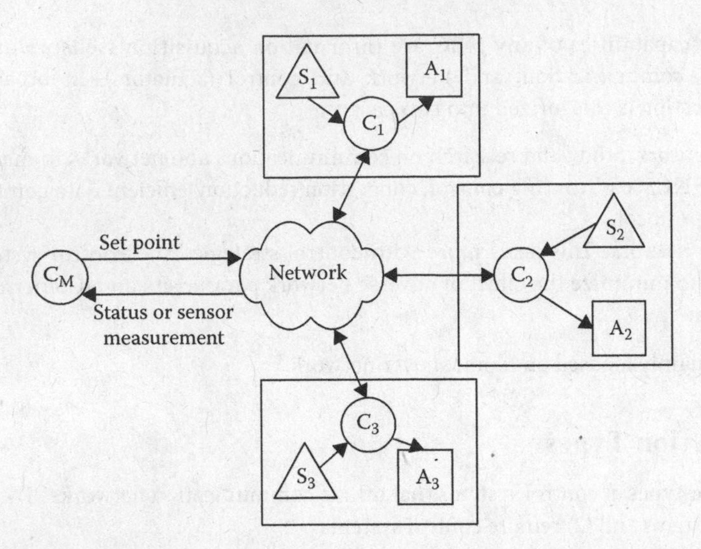

FIGURE 20.3 Data transfers of hierarchical structure.

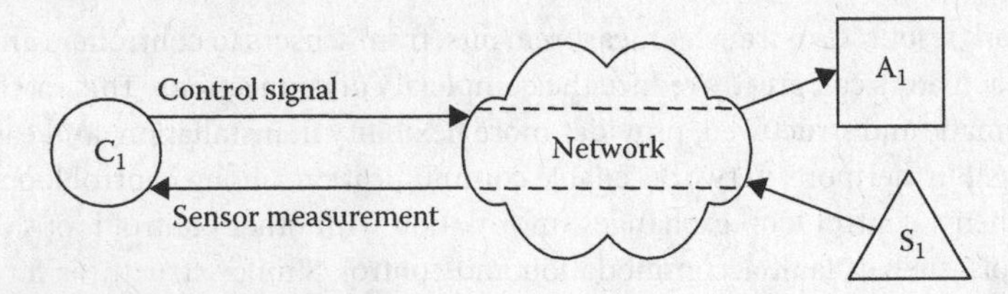

FIGURE 20.4 Data transfers of direct structure.

The second approach of network control is the direct structure, as shown in Figure 20.4. This structure has sensors, actuators, and a controller of a control loop connected directly to a network. Both the hierarchical and direct structures have their own pros and cons. Many NCS include features from both structures, to benefit from their advantages, thus defining the hybrid NCSs.

More details can be found in [1,2].

20.3.1.2.1 Communication Networks

A communication network is the backbone of the NBC. Reliability, security, ease of use, and availability are the main issues while choosing the communication type. A communication network giving support to an NBC should have a deterministic end-to-end message response as a principal property. This is the only way to guarantee the real-time response required by the control algorithm.

The ARPANET developed by Advanced Research Projects Agency Network of the U.S. Department of Defense in 1969 was the world's first operational packet switching network and the predecessor of the global Internet. Later came fieldbus (around since 1988)—an industrial network system for real-time distributed control. Fieldbus is a generic term, which described a modern industrial digital communications network intended to replace the existing 4–20 mA analog signal standard. This network is a digital, bidirectional, multi-drop, serial-bus used to link isolated field devices, especially in the automated manufacturing environment. Profibus (process fieldbus) is a standard for fieldbus communication in automation technology and was first promoted in 1989 by BMBF (German department of education and research) [3].

The basic medium control access scheme is the time division multiple access, where a fraction of the channel is given to all the nodes. There are several medium access control schemes used by industrial networks. The two more common are based on a token-passing scheme and a random access scheme. The token-passing scheme uses a portion of the network bandwidth for transmitting an authorizing message (token) that gives rights to its receiving node to transmit one or more queued messages. The token is usually sent to an ordered and cycled list of nodes. The result is that the network gives an ordered access to all the nodes and each node can know in advance the maximum delay a given message will incur before arriving to its destination.

The random access scheme allows any node to transmit whenever it needs to send a queued message provided not any other node is sending a message. If there is a message being sent through the network, the node waits till the end of this message and then it attempts to send its queued message after an active wait, where a competition between the waiting nodes will give the message access to only one of them. The result is that a random access scheme is generally nondeterministic.

On top of the access scheme is the end-to-end communication relationship between nodes. The most common relationships are the master–slave, producer–consumer, and client–server.

- *Master–Slave*: The network has master nodes and slave nodes. Slave nodes only react to master nodes commands. Master nodes share the channel access by using a round-robin (i.e., token passing) scheme. Profibus is an example of this scheme.
- *Producer–Consumer*: The network has nodes that are producers, consumers, or both. A producer node when it has a message to send looks for the first opportunity to send it in a broadcast way. Consumer nodes'pick up this message contents as soon as they need it. As an example, Control Area Network (CAN) uses this paradigm. CAN is one of the accepted fieldbus standards—a serial, asynchronous, multi-master communication protocol designed for applications needing high-level data integrity and data rates of up to 1 Mbps.
- *Client–Server*: In this case, the network has client nodes and server nodes. The client nodes send requests to the server nodes for services. Based on the best effort, the server provides this service and notifies the completion. This scheme is not necessarily deterministic as the server node takes as much time as it needs.

The result of application of those access and relationship schemes is that a given industrial network may partially use a client–server paradigm and a producer–consumer or master–slave in conjunction with any of the access schemes. How the relationship is to be operated depends on the transaction to be done. For example, in an industrial application, during the commissioning phase, a client application installs an automation program into a Programmable Logic Controller (PLC), which acts as a server, while during the PLC cyclic operation, it connects to sensors and actuators based on a master–slave scheme.

There is a wide variety of concurring fieldbus standards and, therefore, many times interoperability becomes an issue. Another communications network used in NCS—Ethernet—has evolved into the most widely implemented physical and link layer protocol today mainly because of the low cost of the network components and their backward compatibility with the existing Ethernet infrastructure. Now we have fast Ethernet (10–100 Mbps) and Gigabit Ethernet (1000 Mbps) [5]. Recently, switched Ethernet became a very promising alternative for real-time industrial application because it eleminates the collision in traditional Ethernet [4].

Overall, the choice of network depends upon the desired application. Today, with the help of technologies like GPS, electronics atlas (Google maps), we are looking at multi-agent traffic control in urban areas with efficient vehicle communication [7]. Military, surgical, and other emergency medical applications can use dedicated optical networks to ensure fast-speed and reliable data communication. The Internet is the most suitable and inexpensive choice for many applications, where the plant and the controller are far away from each other [6].

20.3.1.2.2 Sampling and Messaging in NBC

Sampling is the activity of measuring the feedback control variable to be sent to the controller. The cyclic control conceptual scheme follows the order: sampling, control law computation, order to actuator. The most usual requirement for this sequence is that it has to be executed based on a fixed clock cycle time. General control law models assume that the time spent between successive samples is a constant, called the sampling period T, while the time spent between the sampling and the actuation instants is negligible or constant and known; but the sampling and actuation messages take some time, which depends on the access scheme and the relationship used, and the control law requires some computation time, so the above assumption does not hold. This effect becomes more important as the network traffic load increases.

Considering an NBC scheme (Figure 20.2), in which the control loops consist of plants G, and controllers C, which exchange through messages the sampled variables $y(kT)$ and the control variables $c(kT)$, where k is the number of the sample. As soon as the messages are sent through the network, they experiment a delay, so the controller receives a delayed value of the sampled signal, $y(kT + t_{sc})$, while the actuator receives a delayed value of the control signal, $c(kT + t_{ca})$, where t_{sc} and t_{ca} are the time delay between sensor to controller and controller to actuator, respectively. In order to ensure system performance, communication-induced time delays have to be considered at the controller design stage.

20.4 Network Effects in Control Performance

The main control performance parameters in a controlled system are stability and the system error. It is clear that stability must be ensured in all controlled systems. The system error, measured on a time window, shows the ability of a closed-loop system to recover after a perturbation. In general, we can state that the smaller the system error, the better the performance.

The performance of an NBC system depends on the control strategy, the message scheduling algorithm, and the reliability of the communication network. We shall consider the case of limited communication resources (sensors, controllers, and actuators), which share and compete with other nodes for accessing the network. The message access scheme and the communication relationship are used as scheduling strategies to assign network access to nodes. The resulting scheduling will cause a shortening or increasing of the message latencies, thus influencing system performance.

A delay in the control loop reduces the system performance [8]. As larger is the delay, worse is the performance. If the delay is known, it can be included into the control design as a parameter and its effect can be minimized. But in a networked control, the network-induced delay usually varies from actuation to actuation as the messages are sent only after competing for and gaining access to the network. The result is that the actual delay varies from cycle to cycle, which not only degrades performance but can drive the system to instability. In particular, the sensor to controller delay can be time-stamped, so the controller can know exactly the time instant of the sampled instance, and from there, calculate the actuation value. But the message from the controller to the actuator can suffer also unpredictable delay not being possible to introduce it into the control calculation. One possible way to overcome this situation and obtaining predictability is using the one-shot model [9] where the actuation is synchronized at the end of the period.

Another important effect in the NCS occurs when a message is lost (there can be several causes for losing a message). It is clear that losing sampling or control messages may also degrade the performance of the controlled system. From the control side, there are different strategies for reducing the effects of lost messages, while from the communication systems side, some kind of redundancy can be used in high-integrity systems. One of the used methods is including some kind of observer or predictor that calculates or predicts the lost value [10].

20.5 Design in NBC

Designing an NBC system requires a careful work within the areas of industrial communications, with reference to the network used and the communication relationship; the control system design areas, with special emphasis on including the induced network delays into the designed control law; and the real-time computation areas, as all the control tasks, together with other tasks that possibly use the network, are to be deployed in computing platforms that shall ensure their continued execution.

20.5.1 Design Constraints in the Network Side

The main design constraints in the network side are the network bandwidth assigned to the controlled system, the resulting sampling to actuation delay, and the probability of message drops. The assigned bandwidth is used to interchange messages between the sensor node, the control node, and the actuator node. On a periodic control model, this bandwidth has to provide enough room for at least one sampling message and one control message in each control period. The network message scheduler has to respect all of the control loops timing constraints, each one with its own sampling period. Depending on the type of access scheme and communication relationship, the assignment of loop bandwidth can become crucial in the resulting network induced delay.

20.5.2 Design Constraints in the Control Side

In the control side, the constraints are mainly given by the dynamic response of the controlled system. The dynamic response, also known as the control system bandwidth, can be expressed as the range of frequencies the control system is able to compensate and is directly related to the characteristic polynomial of the closed-loop system. This control bandwidth, together with the sampling theorem, expresses the range of sampling periods that can be chosen for the discretized control law. The application of the sampling theorem to a discrete control system allows to follow one of the accepted rule of thumb that states that the sampling frequency should be 4–20 times the system's control bandwidth. The result is that there is some flexibility in selecting the control period T. Sampling fast (small T) usually allows better performance, but requires the use of more communication bandwidth. Slow sampling (large T) degrades the control performance, but reduces the network usage. A trade-off must be considered between the common resource usage (network bandwidth) and the control performance.

20.5.3 Network and Control Co-Design

From the above constraints, it naturally appears the consideration of co-design of the network and the control. The assignment of the message scheduling and the control law design is done in parallel as each one of the indicators depends on the others. The control law can include some estimations of the sampling to actuation delays, the control period can be chosen depending on the network conditions.

In addition, other considerations and paradigms can be applied. The preceding triggering scheme for the control loop execution is based on a constant time base, it is time triggered. This scheme has been proved to be quite efficient for continuous signals, so the result is predictable and deterministic, but delays are large. An alternative is to use an event-trigger paradigm. In this case, the control messages are triggered by the occurrence of events, e.g., when a perturbation produces a control error [11]. This scheme is quite efficient for discrete signals as network bandwidth is used when it is necessary, but reduced when the controlled system is stationary. The drawback of this scheme is that, although delays are low in average, it is neither predictable nor deterministic. Guaranteeing the stability and performance of those event-driven systems requires research and applied effort [12,13].

20.6 Summary

NBC systems are control and communication systems where messages are used to close the loops between sensors, controller, and actuator nodes through a shared, band-limited, digital communication network. Designing this kind of system involves the integration of the disciplines of digital communications, control design, and real-time computing. In this chapter, we have shown the architectures for NBC systems, the concepts associated with the network side seen from the control side, the concepts related to the control side seen from the network side, and the principal paradigms and constraints for designing an NBC system. This NBC chapter aims to provide a description not only of the fundamental aspects in this practical area but also where some of the principal research directions lie.

References

1. Chow, M.-Y., Network-based control and application, *Proceedings of the 2007 IEEE International Symposium on Industrial Electronics (ISIE07)*, Vigo, Spain, June 4–7, 2007.
2. Chow, M.-Y., Time sensitive network-based control systems and applications, *IEEE Transactions on Industrial Electronics Newsletter*, 52(2), 13–15, June 2005.
3. Lee, K.C., Lee, S., and Lee, M.H., QoS-based remote control of networked control systems via Profibus token passing protocol, *IEEE Transactions on Industrial Informatics*, 1(3), 183–191, August 2005.
4. Lee, K.C., Lee, S., and Lee, M.H., Worst case communication delay of real-time industrial switched Ethernet with multiple levels, *IEEE Transactions on Industrial Electronics*, 53(5), 1669–1676, October 2006.
5. Daoud, R.M., Elsayed, H.M., and Amer, H.H., Gigabit Ethernet for redundant networked control systems, *IEEE International Conference on Industrial Technology*, 2, 869–873, December 2004.
6. Luo, R.C., Su, K.L., Shen, S.H., and Tsai, K.H., Networked intelligent robots through the Internet: Issues and opportunities, *Proceedings of the IEEE*, 91(3), 371–382, March 2003.
7. Daoud, R.M., Amer, H.H., El-Dakroury, M.A., El-Soudani, M., Elsayed, H.M., and Sallez, Y., Wireless vehicle communication for traffic control in urban areas, *Proceedings of IECON 2006*, Paris, France, pp. 748–753, November 2006.
8. Cervin, A., Henriksson, D., Lincoln, B., Eker, J., and Arzen K.E. How does control timing affect performance? *IEEE Control Systems Magazine*, 23(3), 16–30, 2003.
9. Lozoya, C., Martí, P., Velasco, M., and Fuertes, J.M. Analysis and design of networked control loops with synchronization at the actuation instants, *Proceedings of the 34th Annual Conference of the IEEE Industrial Electronics Society (IECON08)*, Orlando, FL, November 2008.
10. Liu, G.P., Xia, Y., Chen, J., Rees, D., and Hu, W., Networked predictive control of systems with random network delays in both forward and feedback channels, *IEEE Transactions on Industrial Electronics*, 54(3), 1282–1297, 2007.
11. Martí, P., Velasco, M., and Bini, E. The optimal boundary and regulator design problem for event-driven controllers, *Proceedings of the 12th International Conference on Hybrid Systems: Computation and Control (HSCC09)*, San Francisco, CA, April 2009.
12. Zhang, W., Branicky, M., and Phillips, S., Stability of networked control systems, *IEEE Control Systems Magazine*, 21(1), 84–99, February 2001.
13. Hespanha, J., Naghshtabrizi, P., and Xu, Y., A survey of recent results in networked control systems, *Proceedings of IEEE, Special Issue on Technology of Networked Control Systems*, 95(1), 137–162, January 2007.

21 Functional Safety

21.1 Introduction 21-1
21.2 The Meaning of Safety 21-1
21.3 Safety Standards 21-2
Overview of Safety Standards • Basics of IEC61508
21.4 The Safety Lifecycle and Safety Methods 21-4
Generic Lifecycle • HAZOP • FMEA • Fault Tree Analysis • Safety Cases
21.5 Safety Approach for Industrial Communication System 21-8
Overview of Safety-Related Systems • Hazard and Risk Analysis • Failure Mitigation
Acronyms 21-15
References 21-15

Thomas Novak
SWARCO Futurit Verkehrssignalssysteme GmbH

Andreas Gerstinger
Vienna University of Technology

21.1 Introduction

Industrial communication systems take over more and more critical tasks. The criticality of the tasks can be of various types. One of the most common requirements is the achievement of a certain level of availability, in order to reduce times when a system is not productive. Therefore, a certain level of availability is economically necessary. As soon as such systems take over tasks that are critical for the safety of people or the environment, availability alone is not enough, but it is also necessary to achieve a certain level of safety integrity. The following are the typical safety functions in the field of industrial communication systems [WRA07]:

- Emergency stop of an engine: In case of an emergency (e.g., an operator can be hurt), an engine has to stop immediately.
- Unexpected starting of an engine (safe torque off): In case of maintenance activities, it shall not be possible that the engine starts unexpectedly.
- Safe reduced velocity: It shall be guaranteed that the maximum safe level of velocity is not exceeded. Exceeding must be detected and velocity reduced.

All safety functions reduce the likelihood of a person being hurt. The requirements to be met by such safety functions are specified in standards that are either generic, specific for a domain such as industrial communication systems, or application dependent. All of them have in common that they specify a lifecycle model where a hazard and risk analysis, a safety requirements specification, a safety analysis, and safety validation are the crucial activities.

21.2 The Meaning of Safety

Safety is a property that is generally highly desired by the public. Safety consciousness is a property that is continuously increasing, and the demands that society puts on technical systems regarding their safety is increasing at the same rate. The term safety can be interpreted very broadly. Safety is used in various contexts, such as electrical safety, personal safety, safety on the streets, system safety, or functional safety. In this chapter, we are exclusively concerned with the latter.

We use the definitions according to [IEC61508]. Safety is defined as "freedom from unacceptable risk of harm." and functional safety is the part of the overall safety that relates to the correct functioning of some system (e.g., a control system or an industrial communication system) to maintain safety. The functions that need to be performed correctly in order to achieve safety can hence be classified as safety related.

Generally, safety is not an absolute property. To state that the risk related to a specific system has been reduced to zero is generally not a realistic statement. Risk—measured as a combination of probability and severity of an undesired event—is something that we are confronted with continuously. The crucial question is the risk acceptability. What level of risk are we willing to accept. This complex question will not be delved into here, but it should be said that this depends on society, on time, and on the domain in question. Luckily, for industrial communication and control systems, this question has already been posed and answered many times, such that reference can be made to these results.

Safety should not be confused with security: Security is concerned with the protection against intentional attacks, in order to safeguard integrity, confidentiality, and availability of services. Safety is concerned with the avoidance of unintended accidents involving harm to people and the environment. A common denominator is that both disciplines—safety and security—have the goal of avoiding unwanted events. This also makes it hard to determine the success of safety (and security) measures, since they were successful if something unwanted does *not* happen.

A basic concept in safety engineering is the concept of a hazard. A hazard is defined as a "potential source of harm" [IEC61508]. If all hazards are identified and corrected and adequate measures are taken to control and mitigate the hazards, safety is achieved. This is the reason why the identification of hazards has to be done at the very beginning of system development. Typical hazards for industrial communication systems would be the loss or corruption of a message.

The importance of hazards and their identification before they actually cause an accident also highlights the proactive nature of safety. Safety engineering is concerned with predicting unwanted events before they happen, not (only) to learn from failures.

21.3 Safety Standards

In order to guarantee the achievement of a well-defined goal, in our case sufficient safety, standardization plays an important role. Standardization helps to make systems interoperable and comparable. The goal of safety standards is to devise generally applicable rules and recommendations, in order to be able to build, operate, and maintain a safety-related system.

21.3.1 Overview of Safety Standards

The general problem with standards is that they must achieve a compromise between general applicability and practical usefulness for specific systems giving concrete guidance. A high-level standard may be applicable to a wide range of systems, but in order to be applicable to such a large group of systems, it must make several generalizations, which make the standard less understandable for the engineers building a system in a specific domain. Standards that contain a lot of specific information are on the other hand only applicable for a small number of systems and applications. In order to handle this compromise, standards are often classified as generic, domain specific, and application specific.

Standards are abundant, and there is no exception in the domain of safety-critical and safety-related systems. A selection of some standards that are all concerned with safety-related systems can be found in Figure 21.1. Bear in mind that this is surely not a comprehensive figure, but it illustrates the proliferation of safety standards.

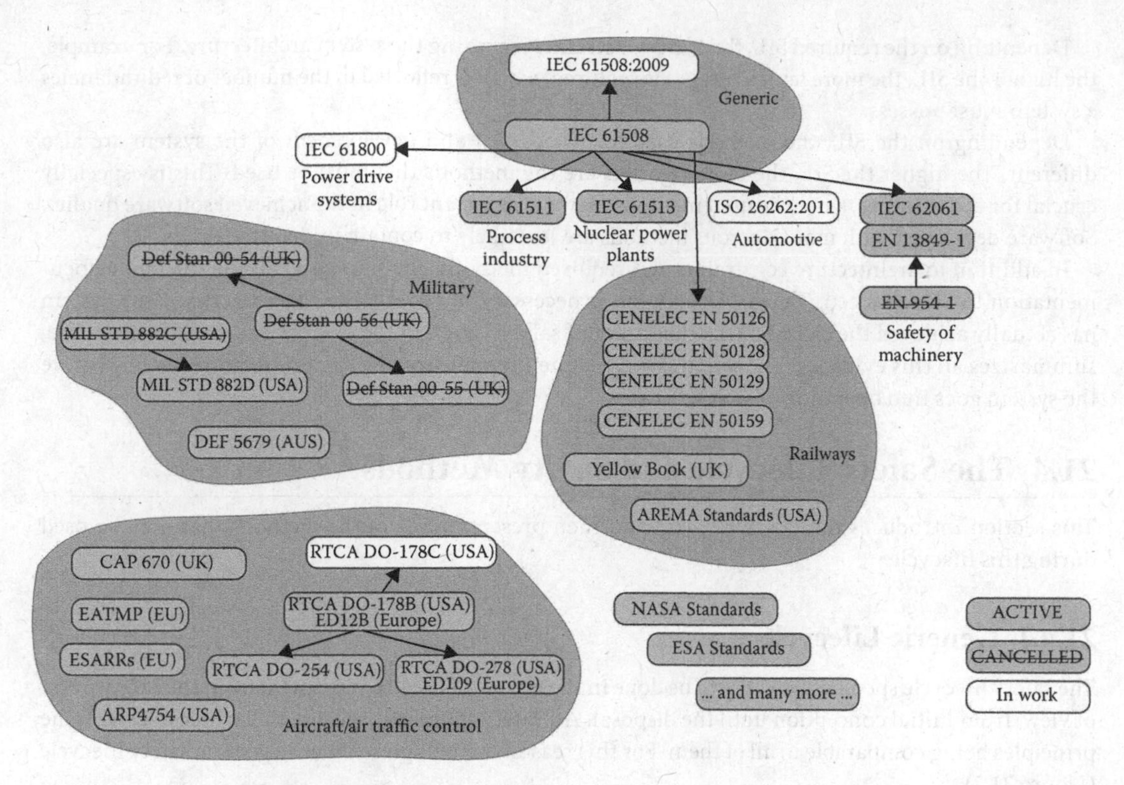

FIGURE 21.1 Safety standards quagmire.

For our purposes, we will consider only one of the standards: IEC61508 [IEC61508]. This standard is generic, i.e., it is not geared toward a specific application domain, but still it contains enough concrete guidance to be useful. Several domains have derived domain-specific versions of this standard, as can be seen in Figure 21.1. The principles of IEC61508 are reflected in all these domain-specific extensions.

21.3.2 Basics of IEC61508

IEC61508 applies to electrical, electronic, programmable electronic safety-related systems. The general assumption is that there is some equipment under control (EUC), which has an inherent risk. In order to reduce this risk, a safety-related system controls the risk of the EUC. For example, in a factory, some heavy machines may be the EUC. The uncontrolled EUC may pose a risk to the operator, and a safety-related industrial communication system can be used to reduce the risk of this EUC. The safety-related system must be functionally safe, i.e., it must function correctly, and shall not fail in a dangerous way.

Depending on the necessary risk reduction, the safety-related system is classified as being in one of four so-called safety integrity levels (SILs). The SILs are defined in terms of accepted failure rates regarding dangerous failures per hour (see Table 21.1).

In a SIL1 system, it is accepted that the system fails in a dangerous way less frequently than every 10^5 h, whereas in a SIL4 system, a dangerous failure is allowed not more often than every 10^8 h. The SIL that a system must achieve is determined by a risk analysis, which is performed at the beginning of the system lifecycle.

TABLE 21.1 Safety Integrity Levels

SIL	Dangerous Failure per Hour
4	$\geq 10^{-9}$ to $<10^{-8}$
3	$\geq 10^{-8}$ to $<10^{-7}$
2	$\geq 10^{-7}$ to $<10^{-6}$
1	$\geq 10^{-6}$ to $<10^{-5}$

Depending on the required SIL, there are constraints regarding the system architecture. For example, the higher the SIL, the more fault tolerance is required, which is reflected in the number of redundancies a system must possess.

Depending on the SIL, the methods used for the design and development of the system are also different. The higher the SIL, the more rigorous are the methods that must be used. This is especially crucial for the software, since the used methods play an important role in the achieved software quality. Software developed with more rigorous methods are less likely to contain undetected faults.

In addition to architecture constraints and required methods, the standard calls for adequate documentation to be produced. The documentation is necessary in order to be able to verify if the system has actually achieved the claimed SIL. The "proof of safety," often in the form of a so-called safety case, summarizes all this evidence. The safety case must then be approved by some regulatory agency before the system goes into operation.

21.4 The Safety Lifecycle and Safety Methods

This section introduces the safety lifecycle and then presents some of the methods that may be used during this lifecycle.

21.4.1 Generic Lifecycle

The safety lifecycle specifies what has to be done in the course of the life of a system from the safety point of view, from initial conception until the disposal. All safety standards specify such a lifecycle, with the principles being comparable in all of them. For this reason, we will present here a generic safety lifecycle (Figure 21.2).

The first step, the hazard and risk analysis starts with an identification of the hazards the system may pose. At this time, there should already be a rough requirements specification of the system, and the major functions should be defined. Based on this information, the potential hazards should be identified. This step is crucial: all theoretical hazards have to be identified, otherwise it is impossible to devise the right methods and countermeasures for them. Hazards that are not identified at this step will not be considered adequately during the remainder of the safety lifecycle and may later pose a risk during system operation. After the identification of the hazards, the hazards must be classified according to their severity and probability, so that the resulting risk can be determined.

The next step in the safety lifecycle of Figure 21.2 is the safety requirements specification. Based on the results of the risk analysis, it is generally necessary to define requirements to ensure that the risks are mitigated to a level so that they are acceptable. The safety requirements can be of various types. They can either be functional requirements, which require an additional implementation of some kind, or they may be nonfunctional requirements, such as the requirement to perform a certain specific analysis, or the requirement to use certain processes during the development. For example, a functional safety requirement in a communication system may specify a checksum with certain failure detection and/or correction properties. A nonfunctional safety requirement may specify the need to perform a common-mode failure analysis on the design of the system. The specified safety requirements should be included in the requirement specification of the complete system.

The third step in the safety lifecycle consists of the performance of all the safety analyses. These analyses can be of various types, and some of them will be described in the following sections. In general, the

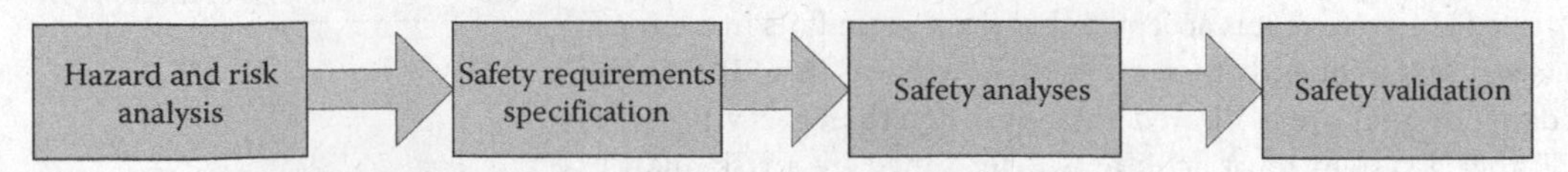

FIGURE 21.2 Generic safety lifecycle.

safety engineer looks at the system in its present stage (this may be only the documented architecture and design, or the fully implemented system, depending on the state of the development) and analyses potential scenarios that lead to hazards. Most analyses in this stage are concerned with faults and failures that may occur. These analyses present a core of the work of a safety engineer.

Safety validation, as the final step, summarizes all the results and evidences. During this phase, it has to be shown that all hazards are mitigated and that all safety requirements have been met. This is usually done in the form of a "safety case." A safety case is a sound and rigorous argument, which clearly demonstrates sufficient safety of a system in a given context. The safety case document references and uses all previously obtained results in order to construct this proof. All these building blocks, which are necessary for this proof, are usually called "evidences." A notation how such a safety case can be constructed is presented later in this section.

The safety lifecycle generally continues until the end of life of the system. The full lifecycle can be found in [IEC61508]. However, for brevity, we have concentrated on the major stages.

All methods that are shown in the following sections are to be used during the safety lifecycle. A reference to the relevant stage of the lifecycle as shown in Figure 21.2 will be made. Generally, the project safety manager is responsible for the conduct of the analyses and executes them together with the project safety engineer and relevant team members. In smaller projects, the role of safety management and safety engineering can be combined. In larger projects, there may be more than one safety engineer.

The methods presented here are only a selection of possible methods. [IEC61508], part 7, annexes A, B, and C contain a comprehensive list of methods with brief descriptions and further references.

21.4.2 HAZOP

Hazard and Operability Study (HAZOP) is a method to identify hazards. It should therefore be used at the very beginning of the safety lifecycle, during the hazard and risk analysis phase. Although this method originates from the process industry, it is very generic and can therefore be applied broadly. For electronic systems, it is standardized in [DS00-58].

The intention of the HAZOP is to analyze all possible functions of a given system and examine all credible malfunctions or deviations from the correct function. In order to identify all potential hazards, this method suggests the use of a given set of guide words that shall be applied to functions (Table 21.2).

All these guide words shall be systematically applied to all functions. It remains to be decided what the "functions" of a system are. Basically, the functions should be defined in a system description or

TABLE 21.2 HAZOP Guide Words

No	This is the complete negation of the design intention. No part of the intention is achieved and nothing else happens.
More	This is a quantitative increase.
Less	This is a quantitative decrease.
As well as	All the design intention is achieved together with additions.
Part of	Only some of the design intention is achieved.
Reverse	The logical opposite of the intention is achieved.
Other than	Complete substitution, where no part of the original intention is achieved but something quite different happens.
Early	Something happens earlier than expected relative to clock time.
Late	Something happens later than expected relative to clock time.
Before	Something happens before it is expected, relating to order or sequence.
After	Something happens after it is expected, relating to order or sequence.

Source: Ministry of Defence, *HAZOP Studies on Systems Containing Programmable Electronics*, U.K. Defence Standard, Def Stan 00-58, Ministry of Defence, Glasgow, U.K., 2000, Part 1, p. 14. With permission.

a requirements document. Therefore, such documentation should be available. It also remains to be decided at which granularity the analysis is carried out. A high-level system function can usually be divided into a number of sub-functions, which can in turn be divided into more sub-functions. The lower the level at which the analysis is conducted, the more effort is required, but the more benefit can be gained from the analysis. The right level of the analysis should be decided at the beginning. It will generally always be a compromise between rigor and efficiency. A good starting point for the analysis is the highest-level requirements description of a system.

HAZOP should also be conducted as a team activity. This means that the team should include not only safety professionals but also engineers, system designers, software engineers, or even end users. One person shall take the lead and moderate the HAZOP sessions. This lead will generally be taken over by the safety manager or engineer. The HAZOP sessions should be planned well in advance, and it is important to keep the focus and not to get sidetracked into long discussions, which do not contribute to the goal. Also, soft factors such as sufficient breaks and an adequate environment are crucial to the success of the sessions.

Finally, the results are documented in a tabular format. The main outputs are the hazards, against which adequate safety requirements have to be specified. In the course of the remainder of the safety lifecycle, it must be shown that these hazards are under control.

21.4.3 FMEA

The Failure Modes and Effects Analysis (FMEA) is a method that systematically analyzes all failure modes of components and determines the associated risk and potential mitigations. It is a bottom-up approach, which starts on the component level and aims to determine the effect of failure modes on the system functions. For a safety engineer, it is essential to determine the safety impact of the analyzed failure modes.

An international standard, which comprehensively describes the method as a technique for system reliability, is available in [IEC60812]. The method consists of analyzing all components or items of a system and filling out of all the columns of a given table. It starts from the component level, and hence is a typical "bottom-up" method. The headings of such a possible FMEA table are given in Table 21.3, an example will be shown in Table 21.6.

Similar to HAZOP, an FMEA is best performed as a group activity, since diverse expertise is needed to record all information necessary.

21.4.4 Fault Tree Analysis

Whereas FMEA has its origins in reliability engineering, fault tree analysis [IEC61025] is a typical safety engineering technique. It is in many ways complementary to an FMEA, and therefore should not be considered as an alternative, but as another necessary method to be used during the safety analysis phase. It is a top-down method, which starts from the undesired events, in our case, it starts with the hazards. It is then analyzed, what the contributing factors to a hazard could be, and this is then presented in the form of logic gates. For example, if the communication system and the backup communication system must fail for a hazard to occur, this is represented by an "and" gate. If one wants to say that the communication system fails when the power supply fails or when the medium fails, this is represented by an "or" gate. This tree is then constructed until one reaches the basic events, which cannot be split up further. The example just described is shown in Figure 21.3.

TABLE 21.3 FMEA Table Header

Item ID	Description and Function	Failure Mode	Possible Causes	Effect	Detection Method	Mitigation	Severity	Probability	Safety Impact

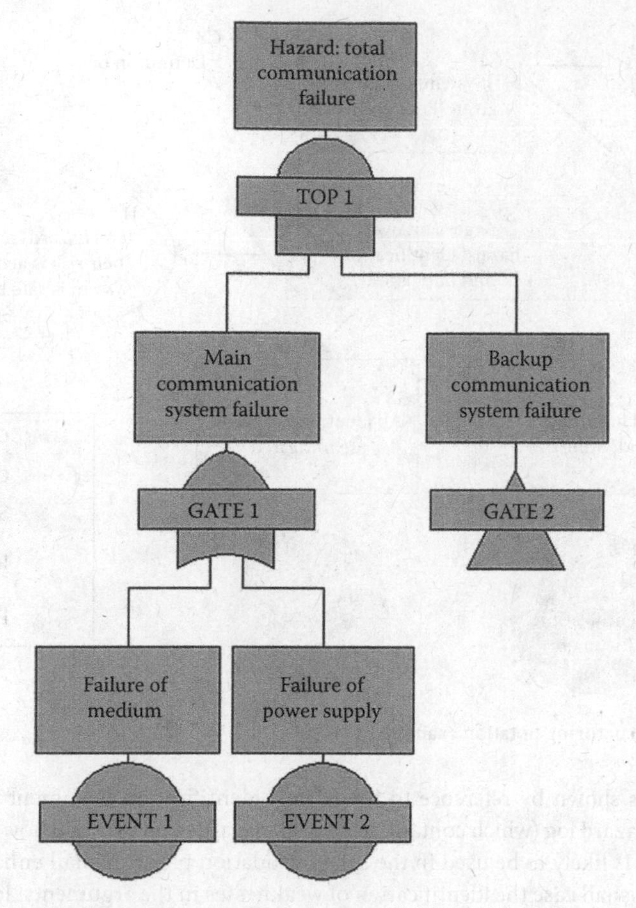

FIGURE 21.3 Fault tree example.

Probabilities can then be assigned to the basic events. If this is done completely, it is possible to calculate the numerical probability for the hazard on the top. Several tools are available, which aid in the construction of such fault trees, and which can perform these calculations automatically.

21.4.5 Safety Cases

A safety case shall show that the safety requirements and objectives are met. Simply said, it shall show that "the system is acceptably safe for its intended use." In order to prove this statement, structured textual arguments can be made up, which show that this claim is true. However, textual arguments can become very long and difficult to read. Especially, it can become hard to verify if the line of argumentation is sound and the safety objectives are really met.

For this reason, [Kel&04] suggests a clearly defined notation for safety cases, called the Goal Structuring Notation (GSN). GSN is already used widely and is likely to be standardized in the near future. The basic elements of the notation are goals, strategies, and solutions. The goals contain some propositional statement that shall be shown to be true. The strategies contain arguments on how this proof can be made. The solutions are the evidences that contain the necessary information to show the truth of the goal. An example is shown in Figure 21.4.

This example contains a claim that a system is safe. The argument is made over the hazard mitigation. It is suggested that if all hazards are found and they are mitigated, the system is safe by definition. Finally, this argument requires two subgoals, namely that all hazards are identified and that all hazards

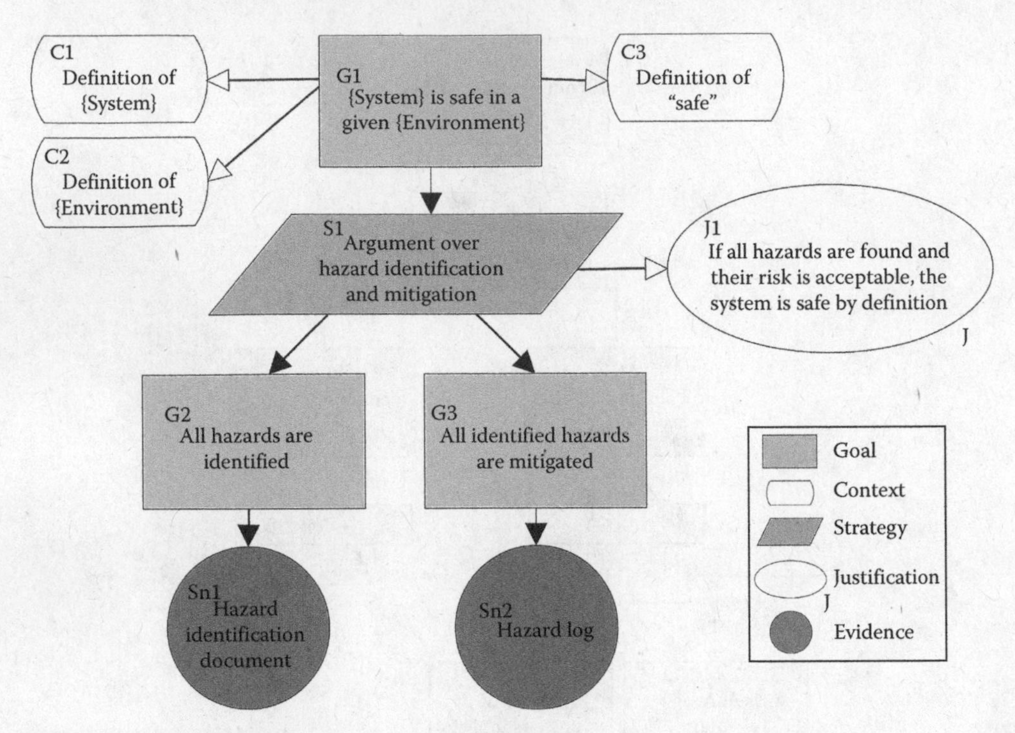

FIGURE 21.4 Goal structuring notation example.

are mitigated. This is shown by reference to the hazard-identification document (such as a HAZOP analysis) and to the hazard log (which contains a log of all hazards and why and how they are mitigated).

The GSN notation is likely to be used in the safety validation phase. It shall enhance the readability of the safety case and shall ease the identification of weaknesses in the arguments. Ideally, the regulator or certifier of a system can read this notation and can then decide if a system may go into operation.

21.5 Safety Approach for Industrial Communication System

Safety-related industrial communication systems are accomplished by enhancing the standard communication protocol and/or adapting the hardware of the standard nodes. Such systems have to be implemented in a way so that systematic failures (failures where the fault can be clearly identified [IEC61508]) and stochastic failures (failures that occur randomly and where various faults can be the reason for a single failure [IEC61508]) are detected during the operation, leading to hazards and resulting from node or network faults.

21.5.1 Overview of Safety-Related Systems

What all safety-related industrial communication systems have in common is that the standard non-safe system is used and safety-related functions are integrated. Most of the communication systems adhere to the requirements of SIL3 specified in [IEC61508]. Predominantly, the communication systems use a two-channel architecture realized by a one-out-of-two (1oo2) architecture, meaning that such architecture consists of two independent channels that execute a given task, but have to agree on the result unanimously. Furthermore, safety-related industrial communication systems use a safety-related message format including a cyclic redundancy check (CRC), a timestamp, or sequence number. That message is embedded into the payload field of the non-safety-related protocol. Finally, the safety-related firmware is very often placed above layer 7 of the ISO/OSI reference model to guarantee compatibility between

TABLE 21.4 Safety-Related Communication Systems

	Compliant to Standard	Safe Node Architecture	Safety-Related Message Format	Safety Firmware in the ISO/OSI Model
CANopen safety	IEC61508, SIL 3	Two channel (1oo2)	NO—standard message sent twice; second time inverted	Integrated into layer 7
DeviceNet safety	IEC61508, SIL 3	Two channel (2× 1oo1, 1oo2)	YES—embedded into standard protocol payload field	Safety layer above layer 7
PROFIsafe	IEC61508, SIL 3	Single channel	YES—embedded into standard protocol payload field	Safety layer above layer 7
SafetyBUS p	EN 954-1, Cat. 4	Two channel (1oo2)	YES—embedded into standard protocol payload field	Safety layer above layer 7
SafetyLon	IEC61508, SIL 3	Two channel (1oo2)	YES—embedded into standard protocol payload field	Safety layer above layer 7

non-safe and safe communication systems. Another reason is that the protocol stack of industrial communication systems is very often standardized and therefore should not to be altered. Table 21.4 gives an overview of the safety features of some safety-related communication systems.

As already mentioned, the main goal of safety is to avoid systematic faults and detect stochastic faults leading to hazards such as a corruption of a message. Safety measures like a CRC are required to mitigate the risk resulting from the hazards. Typically, standard communication systems already include some measures to detect faults (e.g., a CRC or cross-parity mechanism to ensure data integrity). During safety consideration, there are two approaches of how to treat the standard communication system: either take the available fault detection measures into account and add more measures to finally reach the target safety integrity level. Or the standard communication system with its fault detection measures is neglected and the system is treated as a black box referred to as black-channel concept.

Safety of control area network (CAN) relies on the fault detection mechanism of standard CAN. Additionally, to increase integrity and reduce residual failure, probability messages are sent twice and the second one is inverted. The two standard messages comprise a single safety-related message. To distinguish safety- and non-safety-related messages, the identifier range in the standard message was extended. Finally, watchdog functionality is integrated to detect if a safety-related message part or a complete safety-related message was lost. CANopen Safety meets safety requirements of SIL3 given by the international standard IEC61508. Since the protocol enhancement was not sufficient to meet the desired safety level, in addition, a hardware architecture with two channels was chosen.

ProfiSafe or SafetyLon are based on the black-channel concept as shown in Figure 21.5. The network and the standard protocol, and the standard hardware component (non-safety-related part) of the node are treated as a black box. Node A inserts a message into the black-channel and Node B receives it. The idea is that safety measures are required to detect all faults that are possible in the black-channel without paying attention to existing measures. Typical faults are systematic faults on the network or faults in the standard hardware component. Any measures of the standard protocol to detect faults like parity bits to identify message corruption are not considered. The standard protocol is just a means to transport data somehow from Node A to Node B.

The gray part as shown in Figure 21.5 is the safety-related part. It comprises the safety-related hardware and firmware. Both are used to increase the safety integrity and consequently reduce the level of risk to a tolerable target level.

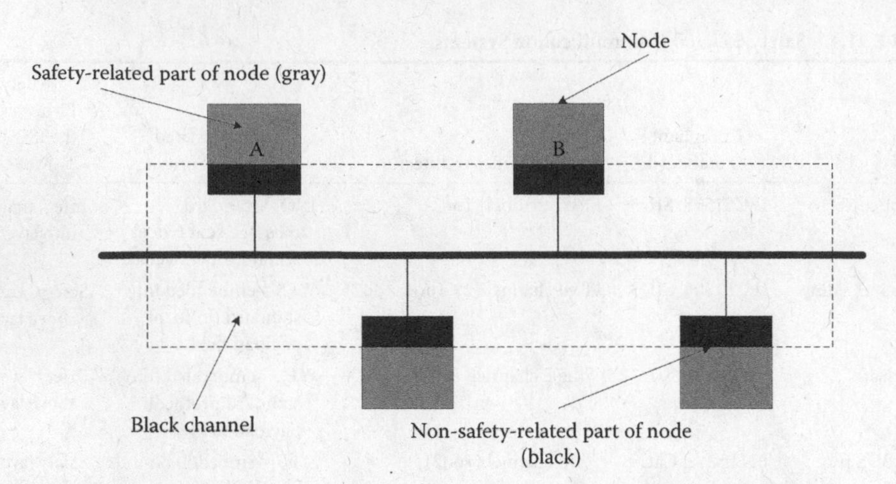

FIGURE 21.5 Black-channel concept.

21.5.2 Hazard and Risk Analysis

Hazard and risk analysis is a crucial step in every safety lifecycle and safety development of a communication system (see, e.g., VDI2184 [VDI2184]). Hazards increase the risk of a fatality and not only must be identified by a hazard and risk analysis but also mitigated by implementing safety measures according to safety requirements. As outlined before, there is a magnitude of possibilities (e.g., HAZOP or FMEA) of how to carry out a hazard analysis.

The scope of the hazard and risk analysis is the node connected to the communication system. First, hazards resulting from failures on the network, which are influencing node safety, must be investigated. It is not intended to ensure network safety in case of a black-channel approach and hence not included into safety considerations. Second, the node hardware and firmware itself has to be examined regarding failures. Moreover, the risk caused by the failures must be assessed and safety measures according to the target SIL have to be specified.

Table 21.5 lists typical network faults. The faults can be categorized in three different groups: faults directly corresponding to human mistakes (8, 9), faults not directly relating to human mistakes (1–7), and faults either directly or indirectly referring to human mistakes (10). The cause of such a grouping is to point out that human mistakes must be considered during investigation of (network) faults.

The next step is to identify the network failures resulting from faults listed in Table 21.5. Network failures can be separated into stochastic (i.e., hardware failures) and systematic failures. Another important topic is to analyze the effect of the failure (i.e., the hazard). Therefore, an FMEA is an appropriate means as shown in Table 21.6.

Risk analysis can be performed by quantifying or qualifying the risk. According to IEC61508, stochastic or hardware failures can be quantified, others only qualified by specifying discrete levels such as "low," "medium," or "high." For example, risk of a "bits being destroyed" can be calculated by the following.

TABLE 21.5 Typical Network Faults

1. Crosstalk	6. Aging
2. Broken cable	7. Temperature
3. EMC failure	8. Human failure
4. Stochastic failure	9. Wiring failure by human
5. Stuck at failure	10. Transmission of non-authorized messages

Source: Reinert, D. and Schaefer, M. (Publisher), *Sichere Bussysteme in der Automation*, Hüthig Verlag, Heidelberg, Germany, 2001, 32 pp. With permission.

TABLE 21.6 FMEA of Network

Network Failure	Failure Mode	Description	Effect (Hazard)
Data corruption	Single bit destroyed	EMC effect destroys bit	Change of message can produce malfunction at receiver side
Loss Incorrect sequence Repetition	Broken wire or gateway problem, wrong routing	Aging of wire or hard software problem in gateway	A safety-related message is not or in the wrong sequence or more often transmitted to the receiver
Delay Insertion	Buffering of messages in a gateway	The gateway cannot process more than a single message at the same time	The delayed or inserted message includes an outdated value
Coupling between safety and non-safety-related messages	Wrong network configuration	A non-authorized message looks like a safety-related message	Receiver performs a non-intended safety-related action being triggered from a non-safety-related sender

According to [PHO97], the probability of a bit being destroyed on shielded twisted pair cables is $p = 10^{-5}$. It is assumed that $v = 10$ safety-related messages are sent every second. The rate of single bit transmission error U is

$$U_1 = p * v = 10^{-4}/\text{s} \tag{21.1}$$

In other words, every 10,000 s, a single transmission error occurs and the risk of a single bit being destroyed per second is 10^{-4}. The probability that two bits are destroyed within 1 s is 10^{-9}, assuming that probabilities are independent and therefore p^2 is inserted into the aforementioned formula instead of p. As a result, the safety integrity of a safety measure must be chosen so that the residual failure probability of a message is at least below 10^{-7} h^{-1} to reach the target SIL3.

Also, failures on node side must be investigated. That is, a hazard analysis and risk analysis has also to be performed to identify failures on the node itself (Table 21.7). In IEC61508, part 2, a detailed list is provided that gives information on the various failures in the components of a microcontroller that must be taken into consideration. Such a hazard analysis is also carried out by means of FMEA.

TABLE 21.7 FMEA of Node Hardware

Hardware Component	Failure Mode	Description	Effect (Hazard)
Controller	Malfunction	Due to high temperature, the ALU of the CPU delivers wrong values	Wrong operation of device
Memory	Wrong values stored	Stuck at fault in the volatile memory results in wrong values stored	Wrong operation of device
Input device	Wrong data read by controller	Stuck at failure or shortcut failure between different channels	Wrong operation of device
Output device to network interface	Corrupted messages sent or received	Stochastic and systematic failures during data transfer	Wrong operation of device
Output device to actuator	Failure of output switch	Stuck at failure or shortcut failure in the output circuit	Unable to switch off

In literature such as [GIE95] or [LIG02], failure rates for standard hardware components are listed. Failure rates are available for resistors, diodes, transistors or capacitors, and the like, and are quantified in failure in time (FIT), which equals a risk of 10^{-9} h^{-1}. They are used to perform the risk analysis. For example, a risk resulting from the input device is a combination of failures rates of different components finally leading to a quantified risk value for the complete circuit. The same is true for the CPU, or non-volatile and volatile memory. See [BOE04] for examples on how to calculate failures rates of circuits.

In conclusion, the hazard and risk analysis is the base for two types of requirements:

- Hazards shall be addressed by adequate *safety function requirements.*
- The risk resulting from the hazards has an impact on the *safety integrity requirements.* The lower the residual risk shall be, the higher and stricter the safety integrity requirements are.

Both requirements have an impact on the type of safety measures used to reach the target SIL. The first defines what measures are required while the latter specifies the safety performance of the measures (e.g., what type of faults can be detected).

21.5.3 Failure Mitigation

Failure mitigation comprises all activities in the safety lifecycle relating to the identification of safety functions and safety integrity requirements as well as derivation of safety measures from the requirements.

As mentioned before, both types of requirements are a result of the hazard and risk analysis. It is separated into two parts: network (cf. Table 21.6) and node hardware (cf. Table 21.7). Typical hazards and measures to detect them are listed in Table 21.8.

Table 21.8 shows that eight safety measures are specified to address the various hazards resulting from network failures. The measures result in a generic safety-related message format: in a specific communication model and a required hardware architecture.

As in most cases, the standard protocol shall not be altered (black-channel approach), the safety-related protocol is embedded into the payload field of the communication protocol. And it shall be different from the standard protocol to reduce the likelihood of coupling between non-safety-related and safety-related messages. As shown in Figure 21.6, the safety-related message has the following structure:

- *ID (Identifier) consists of length field and a version field*: Version denotes the version of the protocol. It is possible that different types of safety-related message structures are used. The identifier is a means of detecting insertion of a message by a faulty node.
- *Safe address*: Safety-critical and non-safety-critical services shall be provided by the system. Therefore, safety- and non-safety-related messages coexist in the same network. To guarantee absence of reaction between them explicitly, every safety-related message includes a safe address. Data sent from a safe output data point (called producer) to a safe input data point (called consumer) is only processed if the safe address of the producer is known to the consumer.
- *Timestamp*: A timestamp is the only means to detect the delay between sending at the sender side and receiving at the receiver. In case of event-triggered communication and devices storing messages temporarily (e.g., routers), congestion on the network, or wrong routing, a message can be delayed and therefore outdated. Additionally, wrong sequence of a message and repetition of the same message is detected. However, a timestamp mechanism requires time synchronization among the different nodes, which is an additional effort to be considered.
- *Safety-related data*: Sensor and actuator data is embedded into that field.
- *Cyclic redundancy check*: The measures mentioned before are implemented to detect systematic faults (faults that cannot be quantified). The CRC is a means against stochastic faults like a bit fault leading to a corruption of the message, and it ensures the integrity of the safety-related message. Depending on the medium of the network (twisted pair, powerline, ...), different CRC polynomials have to be chosen [KOO04].

TABLE 21.8 Network-Related Hazards and Measures to Detect Them

Measures Hazard	Consecutive Number	Time Mark	Time Out (Watchdog)	Acknowledgment (Echo)	Identifier for Sender and Receiver	Data Protection	Redundancy with Cross Comparison	Different Data Protection for Safe/Non-Safe Messages
Repetition	✓	✓					✓	
Loss	✓			✓			✓	
Insertion	✓			✓[a]	✓[b]		✓	
Incorrect sequence	✓	✓					✓	
Data corruption				✓		✓	✓[c]	
Delay		✓	✓[d]					
Coupling between safety and non-safety-related messages				✓	✓			✓

Source: BG-PRÜFZERT, Fachausschuss Elektrotechnik. *GS-ET-26—Bussysteme für die Übertragung sicherheitsrelevanter Nachrichten*. Prüfgrundsätze, Köln, Deutschland, Germany, 2002. With permission.

[a] Application dependent.

[b] Detection of insertion of an invalid source only.

[c] Serial bus only.

[d] Always required.

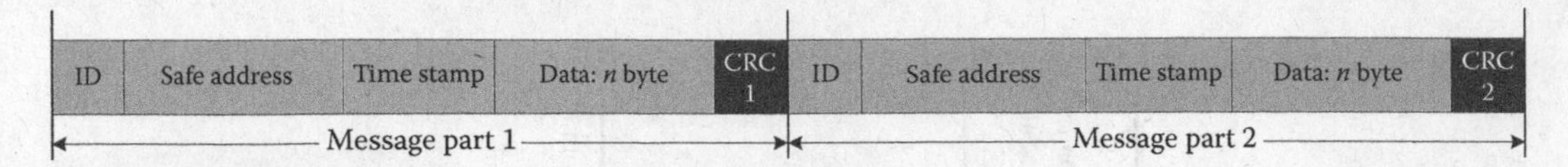

FIGURE 21.6 Generic safety-related message format.

The value of each data point (sensor or actuator value), irrespective of whether it has changed or not, is periodically sent to the receiver by using a safety-related message. That mechanism is called heartbeat. On the receiver side, a watchdog is used, which is reset every time a valid message has been received. As a consequence, a malfunction at the sender side or a fault in the black-channel (e.g., unavailability of the network due to a broke wire or a defect in the standard network interface) can also be detected.

In addition, each safety-related message can be duplicated as shown in Figure 21.6. Just the CRC are different. At the receiver side, not only the CRC are verified but also the duplicated data is compared bit-by-bit. Such a mechanism increases the integrity of the message and reduces the risk of corruption to a minimum.

Finally, every message received by a node is processed by two safe controllers. Each of them verifies the ID, the safe address, the timestamp, and the CRC. After that they compare their results. Only if both safe controllers agree on the same *positive* result, action according to the data point value is taken. Such an approach is called redundancy with cross-comparison.

By using such a safety-related message structure as shown in Figure 21.6, a heartbeat and redundancy with cross-comparison, all faults mentioned in Table 21.8 are addressed and can be detected. Embedding the safety-related message into the data field of the standard industrial communication system message format leaves the standard protocol unchanged. Consequently, non-safe and safe nodes use the same protocol, only the structure of the data field is different.

Node hardware-related safety measures are outlined in detail in [HOE86]. Faults in the hardware components as listed in Table 21.9 are stochastic faults. They cannot be avoided but detected and handled properly. Fault detection is performed by means of online and offline hardware self tests.

The online tests that are executed guarantee a high integrity of the hardware by revealing faults in the different parts of the hardware. Tests are separated into volatile memory (RAM), nonvolatile read-only memory (FLASH), CPU (controller) tests, and test of the input/output unit. Faults of the communication interface to the network interface are tested by sending heartbeats implicitly.

In general, volatile memory test algorithms differ in test effort and diagnostic coverage. A high test effort and a high diagnostic coverage is ensured when using the galloping pattern test, a low one when

TABLE 21.9 Node Hardware-Related Hazards and Measures

Hardware Component	Effect (Hazard)	Safety Measure
Controller	Wrong operation of device	Use of a 1oo2 structure with watchdog and cross-checking; use of cyclic communication via serial interface between both channels; use of CPU test at startup and during operation
Memory	Wrong operation of device	Use of memory test at startup and during operation
Input device	Wrong operation of device	Use of test pulses; use of test pattern with different test pulses
Output device to network interface	Wrong operation of device	Check of message by independent channels and comparison of check results
Output device to actuator	Unable to switch off	Use of test pulses

implementing the marching bit test [HOE86]. The level of the diagnostic coverage depends on the faults revealed by the test. Tests with a high level detect faults according to the DC-fault model, others only detect stuck-at faults [IEC61508].

Tests of the nonvolatile read-only memory rely on safety measures like parity bits, checksums, or CRCs, whereas diagnostic coverage of the first is low and the last is high. The data stored is split up in blocks of equal length and each block is safeguarded with a safety measure. During operation, the CRC or checksum of the block is recalculated and verified with the stored one. In case of equality, the integrity of data is guaranteed.

The CPU test consists of the test of the registers used (equal to the test of the volatile memory), the stack test, the flag test, and the test of the CPU's arithmetic logic unit (ALU):

- *Stack test*: Testing the stack pointer without manipulation of the program flow is not possible. Consequently, the stack test is reduced to the verification of a stack overflow, which is the most common stack fault. Therefore, a defined bit mask is stored at the upper and lower bound of the stack. It is checked periodically, if the bit mask remains unchanged. In case of a changed bit mask, it is obvious that a stack overflow occurred resulting in an undefined and unsafe behavior.
- *Flag test*: Testing the flags is done by setting and resetting the CPU flags and verification of the results. In addition, the conditional execution of instructions has to be verified.
- *Arithmetic logic unit*: The test of the ALU should cover all physical paths within the CPU. Obviously, the problem is that the entire CPU layout is unknown and, more important, too complex for a comprehensive analysis. In order to address this problem, the instruction set of the microcontroller is separated into command classes. Command classes are, e.g., the AND and the MOV command. For every command class, the input values are chosen so that every bit of the output performs a "1–0" and a "0–1" transition.

All the safety requirements and safety measures result in additional safety-related firmware. In most cases, it is logically located above ISO/OSI layer 7 of the standard protocol stack. And another consequence is that mostly a dual channel hardware architecture (see Table 21.4) is required. In other words, the standard hardware of a node is extended with a safety controller to allow cross-comparison of data.

Acronyms

1oo2	One out of two
DC	Direct current
EUC	Equipment under control
FIT	Failure in time
FMEA	Failure mode and effect analysis
FTA	Fault tree analysis
GSN	Goal structuring notation
HAZOP	Hazard and operability study
SIL	Safety integrity level

References

[BOE04] J. Börcsök. *Electronic Safety Systems*. Hüthig Verlag, Heidelberg, Germany, 2004, 107 pp.

[DS00-58] Ministry of Defence. HAZOP studies on systems containing programmable electronics. U.K. Defence Standard. Def Stan 00-58. Ministry of Defence, Glasgow, U.K., 2000.

[GIE95] K. Gieck and R. Gieck. *Technische Formelsammlung*. Gieck Verlag, Germering, Germany, 1995.

[GS-ET26] BG-PRÜFZERT, Fachausschuss Elektrotechnik. *GS-ET-26—Bussysteme für die Übertragung sicherheitsrelevanter Nachrichten*. Prüfgrundsätze, Köln, Deutschland, Germany, 2002.

[HOE86] H. Hölscher and J. Rader, *Microcomputers in Safety Technique, An Aid to Orientation for Developer and Manufacturer*. TÜV Rheinland, Cologne, Germany, 1986.

[IEC60812] Analysis techniques for system reliability—Procedure for failure mode and effects analysis (FMEA). International Electrotechnical Commission, Geneva, Switzerland, 2006.

[IEC61025] Fault tree analysis (FTA). International Electrotechnical Commission, Geneva, Switzerland, 2006.

[IEC61508] Functional safety of electrical/electronic/programmable electronic safety-related systems. International Electrotechnical Commission, Geneva, Switzerland, 2000.

[Kel&04] T. Kelly and R. Weaver. The goal structuring notation—A safety argument notation. *Conference on Dependable Systems and Networks (DSN)*. Florence, Italy, 2004.

[KOO04] Ph. Koopman and T. Chakravarty. Cyclic redundancy code (CRC) polynomial selection for embedded networks. In *Proceedings of the International Conference on Dependable Systems and Networks*, Florence, Italy, pp. 145–154, June 28–July 1, 2004.

[LIG02] P. Liggesmeyer. *Software Qualität*. Spektrum Akademischer Verlag, Heidelberg, Germany, 2002.

[PHO97] Phoenix Contact (Publisher). *Grundkurs Sensor-Aktor-Feldbustechnik*. Vogel-Verlag, Würzburg, Germany, 1997.

[REI01] D. Reinert and M. Schaefer (Publisher). *Sichere Bussysteme in der Automation*. Hüthig Verlag, Heidelberg, Germany, 2001, 32 pp.

[VDI2184] VDI 2184. Reliable operation and maintenance of field bus systems. Verein Deutscher Ingenieure, Germany, 2008.

[WRA07] P. Wratil and M. Kieviet. *Sicherheitstechnik für Komponenten und Systeme*. Hüthig Verlag, Heidelberg, Germany, 2007, 150 pp.

22

Security in Industrial Communication Systems

22.1 Introduction to Security in Industrial Communication 22-1
22.2 Planned Approach to Security: Defense in Depth 22-3
22.3 Security Measures to Counteract Network Attacks 22-4
Virtual Private Networks • Firewalls • Cryptography • DoS Prevention and Detection
22.4 Security Measures to Counteract Device Attacks 22-9
Protected Hardware and Security Token • Secure Software Environments
22.5 State of the Art in Automation Systems 22-12
Security in Building Automation Systems • Security in Industrial Communication • Security in IP-Based Networks • Security in Wireless Communication Systems
22.6 Outlook and Conclusion 22-15
Abbreviations 22-15
References 22-16

Wolfgang Granzer
Vienna University of Technology

Albert Treytl
Austrian Academy of Sciences

22.1 Introduction to Security in Industrial Communication

Modern industrial communication systems go far beyond small automated islands that have been in mind when developing the original communication protocols. Vertical integration and transmission over the Internet are common today, but they require additional security measures to protect the assets.

In literature, a multitude of security definitions that are more or less ambiguous exist [R49,PFL,MEN, BIS,I15]. In the context of industrial communication, security can be defined as measures that protect system resources against *adversaries* that intentionally try to gain unauthorized, malicious access. The aim of such an access can be manifold. In the field of industrial communication, an adversary may be a human or some piece of malicious software (e.g., Trojan horse, virus, worm) with the intention to gain unauthorized access to control functions, i.e., functions that interact with the process under control. Note that this definition of security is contrary to the definition of safety. While security protects the system against *intentional* actions that may result in damage to the system and as a consequence *may* also be harmful to people, safety measures reduce the risk of *unintentional* system states that *do* cause harm to humans.

The actual action that an adversary performs to gain access to the control functions is generally referred to as a *security attack*. A security attack is only possible if the system suffers *vulnerabilities*, i.e., flaws and weaknesses that may be exploited. The existence of vulnerabilities leads to *security threats* that can be seen as the potential for violation of security. To provide measures that avoid a violation of the system's security, these security threats have to be determined. Note that a security attack is different to a security threat. While a security attack is the actual action that tries to violate the security of a system, a security threat is the potential for violation of security that may never be utilized [R49].

Up to now, security has been neglected in the industrial communication domain. However, as mentioned in [DZU], industrial communication systems are already the target of security attacks as reported incidents show. The consequences of a successful security attack on an industrial communication system may be manifold. In addition to a malfunction of safety critical services, which are harmful to humans, security attacks on industrial communication systems can also have massive economic impact. Consider, for example, a power plant being the target of a security attack.

Since security has been a major topic in the Information Technology (IT) world for years, many available security mechanisms exist. However, it is not always possible to trivially map these mechanisms to the industrial communication domain. This is for various reasons. The requirements regarding the used communication protocol(s) may differ. Industrial communication systems often have real-time and safety requirements that cannot be met by protocols and their extensions used in the IT world. Additionally, while Internet protocol (IP)-based networks are getting more popular in industrial communication systems, non-IP-based fieldbus media and protocols are still used at the field level. Since most IT security mechanisms are tailored to IP networks, they are of limited use in non-IP-based fieldbusses. Furthermore, in contrast to devices typically found in the IT domain, industrial communication systems may consist of low-power embedded devices with limited system resources. This is especially true in wireless sensor networks. Since security mechanisms are computationally intensive, their use must not exceed the available resources of these embedded devices.

Industrial communication systems are distributed systems where the control functionality is spread across different devices. To interact with each other, these devices are interconnected by a common network. Therefore, an adversary has two different opportunities to gain unauthorized access to control functions: The adversary may try to maliciously interfere with the data that is exchanged between the devices (*network attacks*) or the adversary may directly attack the devices that implement the control functionality (*device attacks*).

Network attacks can be divided into four classes [PFL]:

- *Interception attacks*: The adversary tries to gain unauthorized access to confidential data exchanged over the network (e.g., network sniffing).
- *Modification attacks*: The adversary tries to change the data while it is transmitted over the network (e.g., modification of network messages).
- *Fabrication attacks*: The adversary tries to insert malicious data (e.g., replay previously sent network messages).
- *Interruption attacks*: The adversary tries to interrupt the communication between devices and thus makes data unavailable (e.g., denial-of-service (DoS) attacks).

Device attacks, on the other hand, can be classified into

- *Software attacks*: An adversary may use regular communication channels to exploit weaknesses in a device's software.
- *Physical attacks*: An adversary may use physical intrusion or manipulation (e.g., probing of bus lines, replacement of ROM chips) to interfere with a device.
- *Side-channel attacks*: Side-channel attacks are based on observing external parameters of a device such as current consumption of EM emissions that are measurable during operation to collect information about its internals.

To counteract these types of security attacks, different *security objectives* have to be guaranteed. According to [DZU], these are integrity, availability, authentication, authorization, confidentiality, and non-repudiation or traceability, whereas the first four have a very high ranking in industrial communication systems.

22.2 Planned Approach to Security: Defense in Depth

Security should be a planned process that comprises the complete system. The *security policy*, also called *security architecture*, describes all security measures but also all relevant organizational procedures and the system environment required to protect a system. It includes the organization's approach to risk, a formal statement of rules through which people are given access to the organization's assets, a definition of business and security goals, and a description of the implemented security measures.

Security is only 20% technology and 80% organization including personnel, process descriptions, etc. Relevant areas can be comprehended in the 4 P's of security: People, Policy, Processes, and Procedures. The setup and maintenance of a security policy includes the following steps:

1. *Asset and risk analysis* identifying the values to be protected (e.g., information or hardware), possible attacks, and the possible damage to the system.
2. *Threat analysis* is based on the risk analysis and is assigning the impact of damage. Whereas the risk analysis is only committed to the identification of risks, the threat analysis is giving a priority to the impact of risks and assigns occurrence probabilities as well as the extent of damage. The resulting risk finally is the product of caused damage and the probability of the threat. In general, for industrial communication systems, a qualitative risk assessment is common since it is hard to precisely determine values for damage and its probability.
3. *Analysis of weaknesses* tracking vulnerabilities of the system. In this step, the system will be carefully analyzed. Based on the previous two steps, actual weaknesses will be identified. Only for these weaknesses countermeasures must be designed.
4. *Design and specification of security measures* dealing with the planning and implementation of appropriate countermeasures. Especially important in this step is to find the trade-off between consequences of an attack and "comfort." For example, consequences could be a monetary loss but also a damage of the image or an environmental damage, whereas "comfort" subsumes many areas beginning from efficient data transmission, low installation costs, easy usability, or social acceptance. Hundred percent security is not possible. Only an optimum within this trade-off can be found on the base of a certain application or application class. Another important part of this phase is the definition of the system boundaries of the security architecture. For example, the best cryptography to protect the transmission channel is useless, if the password requires being so complex that all users maintain a paper copy of the password beneath their keyboard.

This general procedure to design a security policy is a continuous process throughout the life time of a system. In particular, a good security policy is active and not only reactive in the sense that it foresees incidents and avoids being a pure reaction on attacks. Adversaries are favored; they can concentrate on certain vulnerabilities and do not have to protect the whole system. To all these procedures, the principle of Kerckhoff should be applied demanding that the complete cryptographic algorithm must be public and the security should only rely on the secrecy of the keys [KER]. No security by obscurity. Allowing algorithms and procedures to be open allows wide spread security analysis, and therefore reduces the risk of undiscovered security holes. Additionally, this principle also reduces the risk of insider attacks since the knowledge of the system structure offers no advantages unless the proper keys are known.

For industrial automation systems a "defense-in-depth" concept is the favorable approach to protect systems since many industrial communication systems are already structured in a hierarchical way following the computer-integrated manufacturing (CIM) model philosophy and also following the need to structure the network. Defense in depth relies on different security layers to protect valuable assets: Using the famous onion analogon, we see that removing the outmost (security) layer reveals another layer and many more remain to be peeled away before the assets become vulnerable. In the physical world such a procedure is natural (also in industrial automation) having a fence around the factory area,

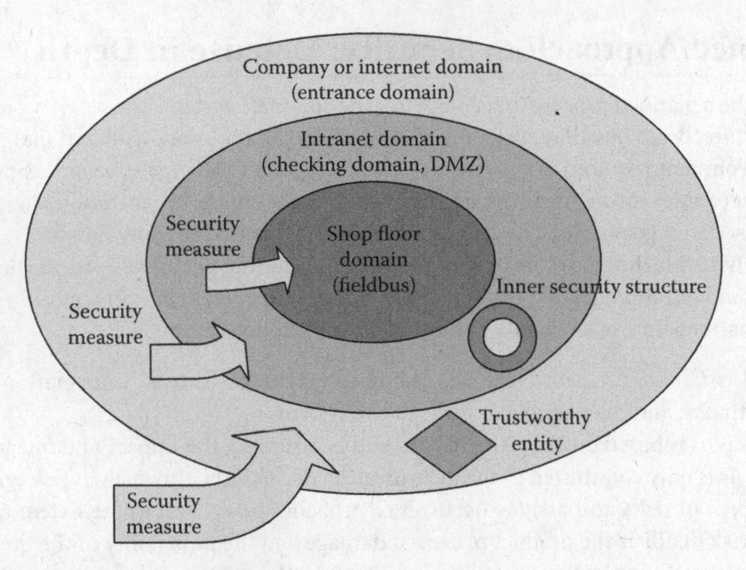

FIGURE 22.1 Three-zone security model.

locked doors to the shop floor, and locked cabinets to the control. Figure 22.1 shows a three-zone security model optimized for industrial communication applications as also introduced by [SC2,TR1,KHA1]. Corresponding to the interconnection zones in the company network three zones are available. The inner fieldbus zone hosting the field-level communication systems mostly located at the shop floor, the intranet zone often build upon IP-based LANs inside the plant (it also includes demilitarized zones (DMZs) and checking domains or inner security structures to strengthen the security), and finally the Internet zone also referred to as company or entrance domain connecting multiple plants, remote maintenance sites, customers, etc. The zones are separated from each other by dedicated security measures often located at dedicated network nodes that can be used as anchor points for the security strategy. Typical examples are firewalls between the company and the intranet domain or application gateways between the intranet domain and the shop floor. As there is only a limited number of those network nodes (between every couple of zones), it is possible to use state-of-the-art components in regard to security without straining available resources too much. In this way, a defense-in-depth approach is installed preventing that an adversary has access to all zones and especially to the most inner zone forming the core of production. Additionally, focusing most of the efforts to those anchor points of the network infrastructure avoids misconfiguration that may happen if the number of security-relevant nodes grows.

22.3 Security Measures to Counteract Network Attacks

To counteract network attacks, two possibilities exist. First, unauthorized access to the network and thus to the data that is transmitted over the network can be avoided. One approach is to limit the physical access to the network medium by, for example, immuring the network cable. Obviously, preventing physical access is not always possible. Consider, for example, the use of wireless or wide area networks where public access cannot be avoided. Therefore, organizational measures that limit the logical access can be used instead. Typical examples are the use of virtual private networks (VPNs) (cf. Section 22.3.1) and firewalls (cf. Section 22.3.2).

Second, the transmitted data itself can be protected in a way that an adversary is not able to maliciously interfere with it even in cases where an adversary has access to the network. According to [GR1], the following security objectives can be guaranteed: *data confidentiality* (against interception attacks), *data integrity and authentication* (against modification attacks), *data freshness* (against fabrication attacks), and *data availability* (against interruption attacks). Depending on the security requirements of the application, it is not always necessary to guarantee all of the security objectives mentioned above. For example, if the nondisclosure of the transmitted data is not a strict requirement, guaranteeing data integrity, availability, and freshness may be sufficient. In general, only those security mechanisms that are absolutely necessary to satisfy the security demands of the application shall be implemented (good enough security). This is especially true for embedded networks that consist of devices with limited system resources that are just sufficient to fulfill the devices' tasks. This concerns primarily processing power (persistent and volatile), memory, power consumption, and network bandwidth. Since security mechanisms are computationally intensive, the realization of security objectives is a critical design step and must not exceed the available device resources.

Guaranteeing data confidentiality, integrity, and freshness can be achieved using cryptographic techniques (cf. Section 22.3.3). However, counteracting interruption attacks like DoS attacks is not possible using a cryptographically secured data transmission. Therefore, additional security measures are required to guarantee data availability (cf. Section 22.3.4).

22.3.1 Virtual Private Networks

A VPN is a logical secure network that is built upon a possibly insecure network. A VPN is transparent to the connected devices. Usually, a device opens a secure unicast connection to a trusted third party (e.g., centralized VPN server) where the whole network traffic to and from the device is tunneled through. In industrial systems, VPNs are most commonly used to connect either two dislocated fieldbus segments or to connect remote maintenance or control centers. Section 22.5.3 describes the open VPN solution and IPsec application. Chapter 15 is further dedicated to Virtual Automation Networks.

22.3.2 Firewalls

A firewall is a network entity that protects a trusted network, host, or service against unauthorized access by inspecting the incoming and outgoing network traffic to decide whether the traffic is allowed or not [PFL]. The decision about allowing or denying network traffic is made based on the firewall's security policy. Such a security policy commonly consists of a default policy and a set of application-specific rules that specify exceptions or amendments to the default policy. Consider, for example, a management interface to an automation controller. The default policy to that interface may be set to "DENY" while a specific rule may be added that allows the system operator's management workstation to access the interface. A security policy is normally predefined according to the system's policy. However, it may also be necessary to dynamically change the rule set of a firewall. For example, if an intrusion detection system (IDS) (cf. Section 22.3.4) identifies a malicious host within a network, the IDS may add a specific rule to the rule set of the firewall that explicitly drops all traffic originated from the identified malicious host (dynamic blacklist).

Depending on the capabilities provided by the firewall, three different types can be distinguished: A *packet filtering firewall* is the simplest form of firewall. It uses part of the header information to decide whether a packet shall be accepted or dropped. For example, to filter IP traffic, the address information (IP address, UDP/TCP port number) as well as the encapsulated application protocol type (e.g., HTTP, FTP) may be considered for identifying valid traffic. However, the state of connections (e.g., has the connection already been closed?) as well as details about the application data (e.g., distinguish between different HTTP methods) are not considered.

A *stateful filtering firewall*, on the other hand, additionally maintains the state of a connection. Using this extra information, the firewall is able to detect illegal connection states and so it is possible to specify a more advanced rule set. For example, to avoid unsolicited connections with a protected entity, the firewall only permits client packets after a dedicated connection has been established to the server.

Finally, an *application proxy* also inspects the application data of network packets. To fully analyze the effects of incoming and outgoing network packets, an application proxy simulates the behavior of the

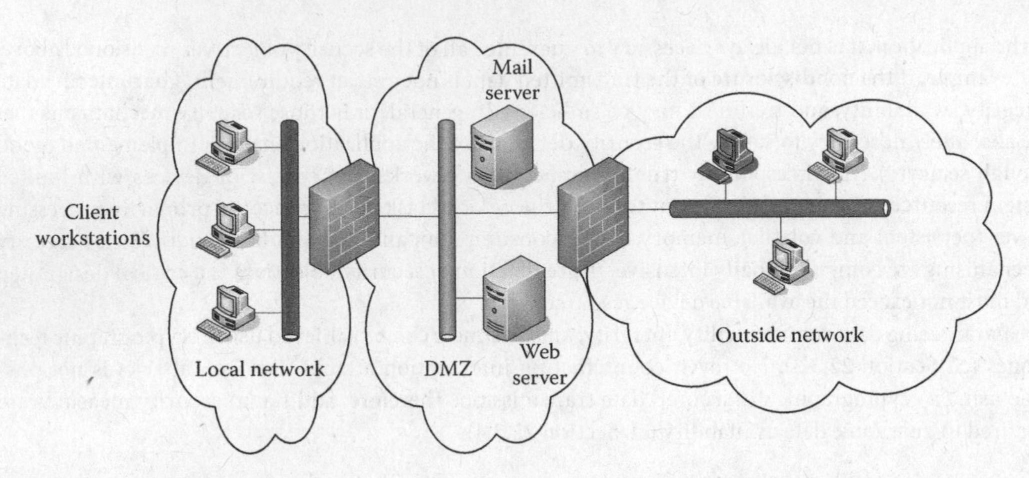

FIGURE 22.2 Demilitarized zone.

entire application. Therefore, a proxy acts as a man-in-the-middle: to the outside network, a proxy behaves like the destination device; to the inside network, the proxy acts as the request origin; and vice versa. Sophisticated application proxies are completely transparent to the involved communication parties.

Due to the importance of firewalls, they have to be reliable and robust against security attacks. To minimize vulnerabilities, firewalls are often isolated, stand-alone devices that are kept as simple as possible. Furthermore, access to firewalls is only permitted to users with special administrator privileges (if at all).

From a security point of view, using a single firewall that acts as a single wall of protection may not be sufficient. If an adversary is able to bypass the firewall, he has full access to the entire network. Therefore, it is more appropriate to separate the network into several zones where each zone is protected with a dedicated firewall and a corresponding security policy. A typical example is shown in Figure 22.2. Besides a separation between an outside and an inside network, the inner network is further divided into a local network that is home for the different client workstations and a so-called DMZ. The firewall between the outside network and the DMZ is responsible for filtering the incoming network traffic that is intended for the servers as well as the client workstations. However, to further protect the clients, a second firewall is located at the boundary between the DMZ and the local network. This firewall is able to additionally filter traffic that is irrelevant for the client workstations (e.g., HTTP traffic to the Web server). The main advantage of this second firewall is that if a server within the DMZ gets compromised, the second firewall acts as an additional wall of protection between the client workstations and the compromised server. Finally, to further secure the client workstations against compromised clients, each client may have its own personal firewall that provides an additional layer of protection.

22.3.3 Cryptography

Cryptographic algorithms use mathematical techniques to guarantee a protection of data against unauthorized interference [MEN]. In industrial communication systems, these cryptographic algorithms can be used to secure data while it is transmitted over a possible insecure network. Depending on the used algorithms, the following security objectives can be guaranteed:

- *Data confidentiality*: To avoid an unauthorized disclosure of confidential data, the producer has to transform it in way that unauthorized entities are not able to interpret the data's meaning while it is transmitted over the network. To achieve this, *encryption algorithms* can be used. The output of the encryption (called cipher text) is transmitted over the network where the consumer receives it. To retrieve the clear-text version again, the consumer applies the inverse operation using the corresponding *decryption algorithm*.

- *Data integrity*: To protect the transmitted data against unauthorized modification, *digital signatures* or *message authentication codes* (MACs) are commonly used. In both cases, the producer calculates a secure tag with a fixed length. The producer transmits it together with the message over the network. The consumer retrieves both, calculates the same secure tag, and verifies whether the received tag corresponds to the calculated one. If the verification process was successful, the consumer has proved the integrity of the message.
- *Data freshness*: Data freshness is guaranteed by including a time-variant parameter. Usually, a *number used only once* (*nonce*) is taken as unique identifier (e.g., random number, monotonically increasing counter, time stamp). However, to avoid a malicious manipulation of the unique token during transmission, it has to be used in combination with an encryption and/or a MAC or digital signature algorithm.

Today, many different cryptographic algorithms that are suitable to guarantee the security objectives mentioned above exist. Most of them are based on the Kerckhoff's principle ([KER], see Section 22.2): While the algorithm itself can be made publicly available, only key parameters must be kept secret (secret keys). Generally, such a cryptographic algorithm works as follows: the producer that wants to securely transmit data via an insecure communication channel (e.g., public network) generates a secured version of the unprotected data by using an algorithm, secret keys, and some other kind of public input parameters. The consumer on the other side of the communication channel receives the secured data and applies the reverse operation to retrieve the plain version of the data.

Depending on the secret keys used at both sides, cryptographic algorithms can be classified into symmetric and asymmetric algorithms. In the case of *symmetric algorithms*, it is relatively easy to derive the secret at the consumer site out of the secret on the producer site. In symmetric algorithms, the secrets on the producer and consumer site are the same. Typical examples of symmetric encryption algorithms are DES [DES], AES [AES], Camellia [R37], and SAFER [SAF]. Popular MAC calculation algorithms that use symmetric algorithms are CMAC [R44] and HMAC [HMA].

In *asymmetric algorithms* (also called public key algorithms), each entity has a so-called public/private key pair where the public key is made publicly available and the private key is only known to the entity itself. In the case of asymmetric encryption algorithms, the public key of the consumer is used to encrypt the data at the producer site. The consumer is able to decrypt the data using its private key. In the case of asymmetric signature calculation and verification algorithms, the producer signs the tag of the data using its private key and the consumer is able to verify the signature using the public key of the producer. Asymmetric algorithms rely on the fact that it is computationally infeasible to derive the private key out of the public key. Popular asymmetric encryption algorithms are RSA [RSA], ElGamal encryption system [ELG], and elliptic curve integrated encryption scheme (ECIES) [HAN]. Typical examples of digital signature generation and verification algorithms based on asymmetric algorithms are the digital signature algorithm (DSA) [DSS], ElGamal signature scheme [ELG], and elliptic curve DSA (ECDSA) [HAN].

In addition to symmetric and asymmetric algorithms that are generally referred to as keyed algorithms, there are cryptographic algorithms that do not need any secret parameters at all. These so-called *unkeyed algorithms* do not directly guarantee any of the security objectives mentioned above. Moreover, they are often used in combination with keyed algorithms. A typical example is a *cryptographic hash function* that may be used in combination with other algorithms to guarantee data integrity. A cryptographic hash function is a computationally efficient one-way function that maps an input of arbitrary length to an output of fixed length. The main property of such a function is collision resistance: It is computationally infeasible to find two distinct inputs that have the same hash value as output. The most commonly used cryptographic hash functions are MD5 [R13] and functions from the SHA family [SHA]. Cryptographic hash functions are used to calculate the tag that is used in MAC and digital signatures. Other examples of unkeyed algorithms are pseudo *random bit generators* where the generated pseudo random number may be used as nonces to guarantee data freshness.

22.3.4 DoS Prevention and Detection

Nevertheless, there are security attacks that cannot be prevented using cryptographic methods. Typical representatives of such attacks are DoS attacks that threaten data availability. DoS attacks can be classified into host-based and network-based DoS attacks. *Host-based DoS attacks* try to waste system resources (e.g., by consuming processor time of a server by sending multiple requests) to prevent the target from performing its expected function. *Network-based DoS attacks*, on the other hand, try to interrupt the communication in a network. A common example is an adversary that tries to consume the network bandwidth (e.g., by flooding the network with unsolicited messages). The situation is further aggravated if multiple sources attack a single victim (distributed DoS attack).

DoS attacks foremost have massive economic impact. Consider, for example, an assembly line that is the target of a DoS attack. A shutdown of the line leads to an economic impact that can be easily compared to the impact of a successful attack on the company Web server. The only difference is that for the Web server elaborate IT security measures are already common practice.

To counteract DoS attacks, two possibilities exist. *DoS prevention*, on the one hand, has the aim to limit the access to system resources in a way that an adversary does not have the opportunity to successfully perform DoS attacks. One opportunity to fully prevent DoS attacks is to limit the physical access to the network medium and to the devices that have an interface to the medium. (e.g., immuring the network cable or by locking the devices into a safe containment.) Obviously, such isolation is not always easy to achieve.

Another possibility to prevent host-based DoS attacks is the use of so-called *client puzzles* [TUO]. The main objective of a client puzzle is to make a DoS attack at least as expensive for the adversary as for the target in terms of computational cost. Consider, for example, a client (e.g., operator workstation) wants to establish a connection to a server (e.g., controller). To set up a connection, the client sends an initial request to the server. If the server is busy (e.g., there are other open connections), the server sends back a client puzzle that the client has to solve. A typical example of such a client puzzle would be a hash value of limited length [WEI]. The client has the objective to find the input value that produces this hash value. To achieve this, the client has to solve this problem by brute force. As it is very easy to verify whether the solution is valid or not, the client must pay more computing costs than the server. After the client has solved the puzzle, it sends the solution to the server. The server verifies the solution, and if it is correct the server accepts further requests. If the solution is not valid, access to the requested service is denied and further requests are temporally blocked from this client.

In situations where prevention methods are inapplicable, DoS attacks shall at least be detected (*DoS detection*). In general, the aim is to make the system intrusion tolerant and limit the affected area. A typical example would be the use of an IDS [CHR]. An IDS tries to detect abnormal system states by comparing the actual behavior with the expected one. If a situation that may lead to a security attack has been detected, countermeasures have to be initiated to minimize the consequences.

An IDS commonly consists of four components (cf. Figure 22.3). The *data gathering component* is responsible for collecting the data by observing the network traffic as well as the behavior of the different network devices. IDS are classified according to the type of data collection (i.e., the location of the data gathering components). A host-based IDS tries to discover abnormal activities on a single host. These methods observe the activities on a single device and compare the behavior pattern with a reference pattern (profiling). Host-based intrusion detection is especially applicable for security-critical devices (e.g., key servers). A network-based IDS observes the entire network traffic. Therefore, these systems are able to discover anomalies that affect more than a single host.

The core unit of an IDS is the *data processing component*. This component processes the collected data and determines whether abnormal behavior is present. Again, different approaches exist. The two most important ones are called misuse-based and anomaly-based. Misuse-based systems use a priori knowledge of activities that form an attack. This knowledge is stored in a database that contains typical patterns of known attacks also called signatures. To determine whether an observation can be classified

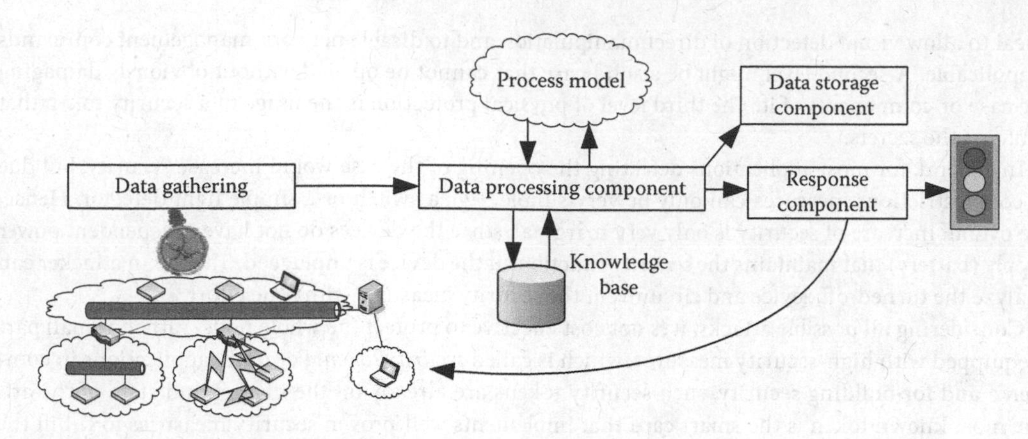

FIGURE 22.3 Intrusion detection system.

as an attack, different techniques such as expert systems as well as signature detection mechanisms can be used. Anomaly-based intrusion detection tries to detect abnormal behavior by comparing the observed behavior with the normal and expected behavior also called reference pattern. To achieve such a comparison, a system model must be specified. This model must define the default reference pattern (i.e., network traffic or device behavior) that represents the expected and normal behavior of the system. Obviously, this default behavior is not static since it can change during the life time of the system. Therefore, self-learning techniques (e.g., neural networks) are usually applied.

Collecting the results as well as the observed data (communication traces) is the task of the *data storage unit*. *The response unit*, finally, is responsible for initiating actions to minimize the consequences of a detected security attack. This can be done by performing a direct feedback to the network. For example, it could decouple the affected network segment(s).

Clearly, DoS prevention is preferable to detection since prevention mechanisms avoid even the occurrence of an attack. Since a full prevention is not always possible, a combination of the advantages of both methods by using a hybrid approach is the most appropriate solution.

22.4 Security Measures to Counteract Device Attacks

So far, only the protection of network traffic has been discussed, yet also attacks to the device itself have to be considered. These attacks can be divided into two categories: physical protection of the device and security software environments.

22.4.1 Protected Hardware and Security Token

Keys used for authentication and encryption of messages typically are confidential information to be stored on a fieldbus node. Yet, confidential information is not limited to this and also includes application data such as application counters, e.g., the power consumption value of a electricity meter, which should not be altered illegitimately. Countermeasures on a first level restrict access to confidential data as offered by most cryptographic units in today's communication chips—keys can be written to the key storage but only be retrieved by the crypto engine, and hence are not visible to the other applications on the node (cf. Section 22.4.2). Yet, some implementations do not honor this fully, and in some fieldbus systems the key data can nevertheless be read by dedicated management commands originally designed to read an arbitrary part of the nodes' memory. For application data, such protected memory areas do not exists in general.

If equipment should be tamper resistant and tamper evident, additionally a protected security token must be added. A trade-off between security and costs is required. A first level measure might already be a seal to allow visual detection of direct manipulation and to disable network management commands if applicable. A second level might be a solid case that cannot be opened without obviously damaging the case or components of it. The third level of physical protection is the usage of a security token that contains the secrets.

In general, for most applications detecting the opening of the case would increase security, but due to cost restrictions measures can only be very simple, e.g., a switch or a simple light detector. Hence, the overall increase of security is only very marginal, since the devices do not have independent power supply (battery) that maintains the security functions if the device is unplugged. That is, an attacker can analyze the turned-off device and circumvent the security measures with little effort.

Considering all possible attacks, it is not cost effective to protect the whole node, rather a small part is equipped with high-security measures, which is called *security token*. For other applications in commerce and for building security, such security tokens are already on the market and state of the art. The most known token is the smart card that implements well-proven security measures to fulfill the requirements defined before. These commercially available security tokens offer an advanced security design and satisfy the requirements of high-security applications (e.g., electronic money, digital signatures). Measures like scrambled RAM placement, power supply monitoring, fail-safe operation, and power consumption ciphering are already included in such devices. Common products often have a certificate to be resistant against malicious scrutiny and manipulation (e.g., common criteria evaluation levels).

In [SC1], the authors already implemented such an approach for the LonWorks [LON] protocol offering two advantages: First the smart card provided the required secure data storage for the node, and second the card also supported state-of-the-art cryptographic functions. The smart card was connected to a LonWorks node using a serial interface and integrated in the application in such a way that each message sent was protected by an HMAC-SHA-1 cryptographic check sum and encrypted using 3-DES overcoming the limitations of the LonWorks security mechanisms with respect to strength, number of security groups (only one is supported by LonWorks), and the support of security services (LonWorks only supports a weak authentication for unicast services).

Yet, a complete integration of the security token in the chip is often favorable compared to the external smart card and the required interconnection circuit. Especially for wireless networks, many chips already have a security token integrated for cryptographic operations reducing the necessity for an additional security token.

22.4.2 Secure Software Environments

A common approach that directly manipulates the behavior of a device is to interfere with the software running on it. To achieve this, an adversary may try to change existing program code or even add new code fragments to existing ones. Thus, modifications that have never been the intention of the software developer can result in malicious software behavior. This changed or newly added malicious code is generally called *malware*, an abbreviation for *malicious software*. Note that unintentional software faults are by definition not malware even if they may result in vulnerabilities that an adversary may utilize.

Depending on the intention of malware and on the damage it may cause, malware can further be categorized. In the literature, different definitions that may differ to each other exist. The most important kinds of malware are the following:

- *Virus*: A virus is a self-replicating piece of software that inserts itself into existing program code or replaces part of it. By definition, a virus needs always a host program that activates it, since a virus cannot run by itself.
- *Worm*: In contrast to a virus, a worm is a malicious stand-alone program that propagates a complete copy of itself usually over the network. The main characteristic is that a worm does not need any user interaction.

- *Trojan horse*: A Trojan horse obscures its malicious intent by pretending to have a useful function. However, after being activated by the user, a hidden malicious function is performed that allows a remote adversary to gain unauthorized access to the host that executes the Trojan horse.
- *Logical bomb*: A logical bomb is malware that only activates itself after certain conditions are met. A special kind of logical bomb is a time bomb where a specific point in time is used as trigger.
- *Rootkit*: A rootkit is a special kind of malware that infects the system in a way that it remains invisible at the host. To achieve this, rootkits typically replace core components of the system software like kernel modules in operating systems.
- *Backdoor or trapdoor*: This kind of malware provides an alternative way to access the infected system that bypasses the normal authentication procedure.
- *Spyware*: Spyware usually does not cause direct harm. It rather collects confidential information on the host being used for further attacks, e.g., passwords or other major assets (e.g., chemical recipes).

Like any other security attack, malware is only able to utilize existing vulnerabilities in the device to infect it. To counteract attacks on software two possibilities exit: First, methods can be used that prevent the existence of vulnerabilities. Second, if a full prevention is not possible, malware or the attempt to insert it into the device's software shall at least be detected. Countermeasures that prevent and/or detect the existence of malware can be categorized as follows.

- *Static software methods* try to avoid the existence of software vulnerabilities a priori during the software development. Typical techniques are static code analysis methods that try to detect programming flaws (e.g., buffer overflows) as well as code-signing methods where the developer signs the executables to confirm the non-modification. Another example is proof-carrying code where the developer provides a proof along with a program that allows checking with certainty that the code can be executed in a secure way.
- *Dynamic software methods* try to identify malware by detecting a malicious behavior during runtime. A common approach is to use a host-based IDS that detects malicious behavior or actions during runtime (cf. Section 22.3.4). A similar technique is called software monitoring, where program execution is observed to check whether the software behaves according to a specified security policy. Another approach is to use self-checking code. Here, the software itself verifies the program code for unauthorized modifications during execution. A further approach that is popular in environments where untrustworthy and thus possible malicious software has to be executed is called sandboxing. Using this technique, the untrustworthy software is running in a so-called sandbox where it is executed in a controlled way with restricted permissions. The main advantage of this scheme is that malicious software is not able to leave the sandbox and infect other parts of the system since an interaction with the rest of the system is only possible using well-defined and protected interfaces.
- *Hardware supported methods* use hardware-specific implementations that try to avoid an insertion of malware. A typical example is the use of microcontrollers that are implementing the Harvard architecture. Due to the physical partitioning of instruction and data memory, code injection is not possible by design. Another recent technique is the no execute (NX) bit in modern CPUs, where memory regions can be designated as being non-executable. Another common approach is the use of a coprocessor that is dedicated to perform security checks during runtime.
- *Human-assisted methods*: In cases where automatic methods cannot fully prevent or detect the existence of malware, human-assisted methods can additionally be used. Typical examples are manual inspection or the use of certificated software. Obviously, human-assisted methods are time consuming and require extensive knowledge. Therefore, it is reasonable to use them only in combination with automatic techniques.

Guaranteeing a secure environment for the execution of software is a typical task of the operation system running on the device. However, industrial communication systems contain embedded devices with limited system resources. Therefore, these devices are often not equipped with an operation system that provides support for advanced security services. Therefore, providing a secure software execution environment is especially challenging in embedded networks.

22.5 State of the Art in Automation Systems

This section will give an overview of the state of the art of security measures in industrial communication systems. These are divided into fieldbus systems originally designed for building automation, but also used in industrial communications, classical industrial fieldbus systems, and IP-based networks. An overview of the supported security mechanisms can be found in Table 22.1 at the end of this section.

22.5.1 Security in Building Automation Systems

Today, many different protocol standards for building automation systems exist. The most important open ones that span more than one application domain are BACnet [BAC], KNX [KNX], LonWorks [LON], and IEEE 802.15.4/ZigBee [IEE,ZIG].

BACnet in its current version—version number 2008—offers an authentication service as well as different security services that guarantee data confidentiality, data integrity, and data freshness. These mechanisms use the symmetric Data Encryption Standard (DES) algorithm and a trusted key server, which is responsible for generating and distributing the secret keys. However, due to several security flaws [GR2], there are investigations underway to replace these security services by new ones. The new security architecture, which is defined in BACnet Addendum g, uses AES and HMAC in combination with a message ID and a time stamp. At the time of writing, BACnet Addendum g has not been declared as final yet.

KNX, on the other hand, only provides a basic access control scheme based on clear-text passwords. Up to 255 different access levels can be defined, each of them associated with a different (otherwise unspecified) set of privileges. For each of these access levels, a 4 byte password (key) can be specified. However, since this rudimentary access control mechanism does not provide strong security [GR2], it is only of limited use in security-critical environments. Furthermore, it is not available for process data exchange, which is called group communication in KNX.

In *LonWorks*, an authentication mechanism that guarantees data integrity and freshness is available. Support for data confidentiality is not provided. This mechanism is based on a four-step challenge–response protocol. The used cryptographic algorithm calculates a 64 bit hash value over the plain

TABLE 22.1 Security Services of Common Fieldbus Systems

Fieldbus	Security Services
Foundation Fieldbus	8 access groups, 8 bit unencrypted password
ControlNet	Connection authentication, unencrypted password
PROFIBUS	Access control for predefined addresses
P-Net	Simple write protection for variables
WorldFIP	8 access groups, 8 bit unencrypted password
Interbus	8 access groups, 8 bit unencrypted password
SwiftNet	None
LonWorks	Challenge–response auth., integrity check, MD5 for IP
KNX	Unencrypted password for management access
BACnet Addendum g	AES for data confidentiality, HMAC for data integrity, time stamp + message ID for data freshness
ZigBee	CCM* algorithm for data integrity, freshness, and confidentiality

message and a random number using a shared secret key. However, due to key length and several protocol flaws, this mechanism cannot be considered as secure [SC1]. In addition to this basic authentication mechanism, a more advanced one based on MD5 is available for LonWorks/IP. However, since MD5 is not collision resistant, it is also regarded as insecure.

IEEE 802.15.4 supports security mechanisms at the data link layer. The current version IEEE 802.15.4-2006 uses a variant of the counter with CBC-MAC (CCM) algorithm called CCM to guarantee data integrity, freshness, and confidentiality. *ZigBee* in its current version ZigBee 2007 utilizes the IEEE 802.15.4-2003 transmission services of the data link layer. However, ZigBee 2007 does not use the security mechanisms provided by IEEE 802.15.4-2003—they are completely replaced by a more advanced security concept. This concept supports the use of different key types and provides advanced key management services. Again, CCM* is used as a cryptographic algorithm.

Beside the security features within the protocol, an important issue is the connection of the building to remote maintenance centers. For such connections, buildings are equipped with gateways that allow remote access to the underlying fieldbus. Typical security measures for these gateways are IP connections protected by transport layer security (TLS) or VPN that securely tunnel the commands from the remote site to the fieldbus (see Section 22.5.3). In rare cases also Web Services Security is used. To the fieldbus site, the gateways usually provide no security or the security protocols mentioned above.

22.5.2 Security in Industrial Communication

Although almost never used, some industrial automation fieldbusses support security. However, these security measures are very limited in scope and capabilities, and therefore cannot be considered as serious protection if a considerable threat level needs to be assumed. The fieldbus systems for industrial and process automation currently standardized by the IEC in the standards IEC 61784-1 [IEC1] and the underlying IEC 61158 series [IEC2] are a heterogeneous collection of solutions. Their security features can be divided into two categories: in systems like *ControlNet*, *P-Net*, and *SwiftNet*, hardly any state-of-the-art security measures are used at all. *Foundation Fieldbus*, *Profibus*, *WorldFIP*, and *Interbus* on the other hand implement very similar application layer services and protocols modeled after their fieldbus ancestor MMS. The simple access protection mechanism of these systems is based on access rights management similar to UNIX-based systems. Every object has a 8 bit long password associated with it and a list of 8 different access groups coded in an 8 bit word. Both password and access groups are transmitted in plain text. If they match for a specific request, the additional access rights parameter defines the allowed operations. Usually the access rights depend on the type of object and allow reading, writing, executing, deletion, etc.

Unfortunately, these security means are not mandatory for implementation, and the password itself, additionally, has no explicit protection. That is, it is transmitted in clear text, and hence should not be used at all. A note in IEC 61158-5 reveals the actual intention of the security mechanisms: "This is not a protection against intentional misuse of the communication facilities of a field device but helps to protect a system for accidental erroneous use of process data." This statement is in fact true for all traditional fieldbus systems in industrial and process automation. Hence, to secure data traffic security measures can only be implemented on top of the protocol stack. That is, the fieldbus is only used as a transparent transport channel to transmit secure messages over standard fieldbus protocols and services. Such an application-level approach could easily achieve end-to-end security, which is the goal in many applications, anyway. Similar approaches have already been implemented for safety applications such as the Profisafe profile of Profibus (IEC 61158). Drawback of adding such an application-level approach for security is to cause problems with interoperability, since in general the interoperability mechanism of the underlying fieldbus system cannot be used. Existing variable types or messages cannot be used since they define no properties for security measures. Hence, the only way is to use general purpose messages and redefine their behavior. Although a parallel usage of secure and insecure devices is possible with this approach, the practical application demands a joint use. Yet connecting secure and unsecure messages sacrifices all security. Another problem is the limited message size in some field-busses that leaves only little room for efficient security extensions. Especially, integrity services such as hash codes or MACs require additional data blocks reducing the actually available payload per packet, and consequently the performance of a secured channel drastically.

22.5.3 Security in IP-Based Networks

IP-based technologies are gaining increased importance in industrial communication systems. Due to the widespread use of IP-based LANs and more recently the Internet, IP-based networks are already widely used at the management level and as backbones to connect remote fieldbus segments. However, due to the decreasing costs for IP cabling and network interface hardware, even small embedded micro-controllers can be equipped with a dedicated Ethernet interface chip. Therefore, IP and LAN technologies have started to penetrate the field level, and their use is no longer limited to the management level where PC-based devices are located.

From the security point of view, IP-based networks are especially prone to security attacks. This is for various reasons. Since IP as well as the underlying data link protocols (e.g., Ethernet) do not provide native security mechanisms, many well-known vulnerabilities exist. Additionally, since IP networks may be shared with other applications (e.g., office LAN) and interconnections to foreign networks for remote access are common, gaining access to the network may be easier.

Due to the widespread use of the Internet, security has been a major research field in the IT world for years. Therefore, many security extensions for IP-based networks are available where each of these mechanisms is suitable for a certain application field. In this section, a small subset of available, state-of-the-art mechanisms that are suitable for industrial communication systems are presented.

Internet Protocol Security (*IPsec*) [IPS] is a security extension to the IP protocol, and thus operates on the network layer. IPsec is a part of IPv6, but since IPv4 is still the predominantly used network protocol in the IT world, it has been ported to extend IPv4. IPsec ensures data integrity, freshness, and confidentiality. To achieve this, various cryptographic algorithms can be selected (e.g., 3-DES, AES, HMAC-SHA1). For key exchange, the Internet Key Exchange (IKE) protocol is used. IKE uses asymmetric algorithms like RSA, ECC, or symmetric algorithms with pre-shared secret keys, alternatively. One of the main concepts of IPsec is the notion of a Security Association (SA). An SA is a one-way connection between a sender and a receiver that specifies the security services to apply to the traffic carried over the connection. Each SA contains the following parameters that may be used to uniquely identify it: a Security Parameter Index (SPI), the IP destination address, and the security protocol identifier. Parameters of every SA are stored in an SA database. SAs either support transport mode for communication between two hosts or tunnel mode for communication between a host and a security gateway or between two security gateways.

Secure Sockets Layer (SSL) and its successor *Transport Layer Security* (TLS) [TLS] is a protocol developed for securing communication between two parties. During the initial handshake, the devices are authenticated, the used cryptographic algorithms are negotiated, and shared secret keys are exchanged. After the initial handshake, these secret keys are used to establish a secured channel between the two entities that provides data confidentiality, integrity, and freshness. Like in IPsec, the initial key exchange is usually done using asymmetric algorithms, while the secured channel is protected using symmetric algorithms exclusively. TLS offers similar algorithms but operates on a higher level in the ISO/OSI layer model. It encrypts data between the transport layer (TCP or UDP) and the application layer. TLS is very flexible with respect to the use of algorithms. So, it is possible to implement secured unicast connections on embedded devices, too. [GUP] shows an implementation of a complete secure Web server, using HTTP and TLS. In this implementation, the asymmetric encryption part of TLS is done with the help of elliptic curve cryptography (ECC) [HAN].

A popular VPN implementation for IP-based networks is *OpenVPN* [OVPN]. In OpenVPN, each device opens a secure unicast connection to a centralized server where the whole network traffic to and from the device is tunneled through. To secure the tunneling connection, TLS is used. OpenVPN provides several methods to ensure authentication: a pre-shared symmetric key, a username and password combination, a TLS certificate, or a combination of these methods. While IPsec manipulates the network protocol, OpenVPN encapsulates the encrypted VPN packets into untouched TCP or UDP packets of the host network. However, the VPN packets themselves are secured down to the IP layer (layer 3) in the routing mode or down to Ethernet (layer 2) in the bridging mode.

22.5.4 Security in Wireless Communication Systems

Like many other sectors, wireless networks are also more and more introduced in industrial and building automation infrastructures. Main concern is the robustness to electromagnetic interference and in general the security goal of availability that can only be solved by proper organizational measures (e.g., redundant transmission paths). Also, the wireless transmission is prone to eavesdropping and message insertion since the media can be accessed by everyone. Chapter 28 introduces the security measures to fulfill the above-mentioned security goals. To combine existing physical security measures (e.g., fences, closed rooms) to increase security, new research is going in the direction of location-based security services limiting the accessibility of a network not only by the signal strength, but also by an active localization of the node. The Austrian Academy of Sciences investigates such security measures based on a localization of COTS Wireless LAN hardware via dedicated access points [TR2].

22.6 Outlook and Conclusion

The lack of state-of-the-art security measures at the field level of many industrial communication systems almost solely allows implementing security measures on top of the automation networks. This usually introduces overhead and undermines interoperability. To omit security is related to high risks. At the intranet/Internet zone, a general awareness of these risks exists and security measures are commonly implemented in terms of secured tunneling. At the field-level zone, this awareness is missing. Looking at threats to industrial communication systems, omitting security measures at the field level must be very carefully considered. Security measures (at all levels) also help to tackle other problems such as accidental errors within remote maintenance or safety issues. Therefore, it is worth the effort to integrate security measures at all levels. Yet, security must not solely be equated to network security or cryptography. Many security threats, e.g. DoS attacks, require more organizational measures and show that security must include the complete system and not only a dedicated part. Setting the appropriate boundaries for security is the most critical part in security design: Defining what is trusted and what is untrusted requires deep insight into the overall applications, especially since industrial communication and industrial process control strongly interact with its environment. In practical implementations the security goals of integrity, availability, and authentication/authorization are the most important ones to protect. Confidentiality and non-repudiation are of minor interest for industrial applications. Additionally, traceability is important during maintenance.

Abbreviations

AES Advanced encryption standard (cryptographic cipher)
BACnet Building automation control network
CMAC Cipher-based message authentication code
CIM Computer-integrated manufacturing
CPU Central processing unit
DMZ Demilitarized zone
DES Data Encryption Standard (cryptographic cipher)
DoS Denial of service (attack)

DSA	Digital signature algorithm
ECC	Elliptic curve cryptography
ECIES	Elliptic curve integrated encryption scheme
ECDSA	Elliptic curve digital signature algorithms
FTP	File transfer protocol
HMAC	Hashed message authentication code
HTTP	Hypertext transfer protocol
IDS	Intrusion detection system
IEC	International Electric Commission
IKE	Internet key exchange
IP	Internet protocol
IPsec	Internet protocol security
ISO/OSI	International Standard Organization/Open System Interface
IT	Information technology
KNX	Standard for home and building automation
LAN	Local area network
LON	Local operation network; LonWorks (fieldbus)
MAC	Message authentication code (cryptographic checksum)
MAC	Media access control
MD5	Message Digest #5
NOUNCE	Number used only once
NX	No execute
PC	Personal computer
RAM	Random access memory
ROM	Read only memory
RSA	Rivest Shamir Adelman (cryptographic cipher)
SA	Security Association
SHA	Secure hash algorithm
SPI	Security Parameter Index
SSL	Secure socket layer
TCP	Transmission control protocol
TLS	Transport layer security
UDP	User datagram protocol
VPN	Virtual private network
3-DES	Triple Data Encryption Standard (cryptographic cipher)

References

[AES] National Institute of Standards and Technology, Advanced Encryption Standard (AES), FIPS PUB 197, Gaithersburg, MD, 2001.

[BAC] BACnet—A Data Communication Protocol for Building Automation and Control Networks, American Society of Heating, Refrigerating, and Air-Conditioning Engineers or ASHRAE SSPC 135, Atlanta, GA, 2008.

[BIS] M. Bishop, *Computer Security: Art and Science*, Addison-Wesley, Boston, MA, 2002.

[CHR] C. Krügel, Network alertness: Towards an adaptive, collaborating intrusion detection system, PhD thesis, Vienna University of Technology, Vienna, Austria, 2002.

[DES] National Institute of Standards and Technology, Data Encryption Standard (DES), FIPS PUB 46-3, Gaithersburg, MD, 1999.

[DSS] National Institute of Standards and Technology, Digital Signature Standard, FIPS PUB 186, Gaithersburg, MD, 1994.

[DZU] D. Dzung, M. Naedele, T. P. von Hoff, and M. Crevatin, Security for industrial communication systems, *Proceedings of the IEEE*, 93(6), 1152–1177, June 2005.

[ELG] T. ElGamal, A public key cryptosystem and a signature scheme based on discrete logarithms, *IEEE Transactions on Information Theory*, IT-31(4), 469–472, 1985.

[GR1] W. Granzer, C. Reinisch, and W. Kastner, Key set management in networked building automation systems using multiple key servers, in *Proceedings of Seventh IEEE International Workshop on Factory Communication Systems*, Dresden, Germany, pp. 205–214, May 21–23, 2008.

[GR2] W. Granzer, W. Kastner, G. Neugschwandtner, and F. Praus, Security in networked building automation systems, in *Proceedings of Sixth IEEE International Workshop on Factory Communication Systems*, Torino, Italy, pp. 283–292, June 2006.

[GUP] V. Gupta, M. Millard, S. Fung, Y. Zhu, N. Gura, H. Eberle, and S. C. Shantz, Sizzle: A standards-based end-to-end security architecture for the embedded Internet, in *Proceedings of the Third IEEE International Conference on Pervasive Computing and Communications*, Hawaii, pp. 247–256, December 2005.

[HAN] D. R. Hankerson, S. A. Vanstine, and A. J. Menezes, *Guide to Elliptic Curve Cryptography*, Springer, New York, 2004.

[HMA] National Institute of Standards and Technology, *The Keyed-Hash Message Authentication Code* (*HMAC*), FIPS PUB 198, Gaithersburg, MD, 2002.

[I15] Information Technology—Security Technique—Evaluation Criteria for IT Security, IEC 15408, International Electrotechnical Commission or IEC, Geneva, Switzerland, 2005.

[IEC1] International Electrotechnical Commission, Digital data communications for measurement and control—Part 1: Profile sets for continuous and discrete manufacturing relative to fieldbus use in industrial control systems, IEC 61784-1, Geneva, Switzerland, 2003.

[IEC2] International Electrotechnical Commission, Digital data communications for measurement and control—Fieldbus for use in industrial control systems, IEC 61158, Geneva, Switzerland, 2003.

[IEE] IEEE 802.15.4, *Wireless Medium Access Control* (*MAC*) *and Physical Layer* (*PHY*) *Specifications for Low-Rate Wireless Personal Area Networks* (*WPANs*), IEEE Computer Society, Washington, DC, 2006.

[IPS] S. Kent and K. Seo, Security Architecture for the Internet Protocol, RFC 4301, Internet Engineering Task Force (IETF) Network Working Group, Fermont, CA, 2005.

[KER] A. Kerckhoffs, La cryptographie militaire, *Journal des Sciences Militaires*, 9, 5–38, Janvier 1883; 161–191, Février 1883.

[KHA1] B.A. Khan, J. Mad, and A. Treytl, Security in agent-based automation systems, in *Proceedings of 11th IEEE Conference on Emerging Technologies and Factory Automation*, Prague, Czech Republic, pp. 768–771, 2007.

[KNX] Information technology—Home Electronic Systems (HES) Architecture, ISO/IEC 14543-3, International Electrotechnical Commission or IEC, Geneva, Switzerland, 2006.

[LON] Control Network Protocol Specification, ANSI/EIA/CEA 709.1, American National Standards Institute or ANSI, Washington, DC, 1999.

[MEN] A. J. Menezes, P. C. van Oorschot, and S. A. Vanstone, *Handbook of Applied Cryptography*, CRC Press, Boca Raton, FL, 2001.

[OVPN] *OpenVPN*, 2010, http://openvpn.net [Online: 25.01.2010].

[PFL] C.P. Pfleeger and S.L. Pfleeger, *Security in Computing*, 4th edn., Prentice Hall, Upper Saddle River, NJ, 2006.

[R13] R. Rivest, The MD5 Message-Digest Algorithm, RFC 1321, Internet Engineering Task Force (IETF) Network Working Group, Fermont, CA, 1992.

[R37] M. Matsui, J. Nakajima, and S. Moriai, A Description of the Camellia Encryption Algorithm, RFC 3713, Internet Engineering Task Force (IETF) Network Working Group, Fermont, CA, 2004.

[R44] J. H. Song, R. Poovendran, J. Lee, and T. Iwata, The AES-CMAC Algorithm, RFC 4493, Internet Engineering Task Force (IETF) Network Working Group, Fermont, CA, 2006.

[R49] R. Shirey, Internet Security Glossary, Version 2, RFC 4949, Internet Engineering Task Force (IETF) Network Working Group, Fermont, CA, 2007.

[RSA] R. L. Rivest, A. Shamir, and K. Adleman, A method for obtaining digital signatures and public-key cryptosystems, *Communications of the ACM*, 21(2), 120–126, 1978.

[SAF] J. Massey, SAFER K-64: A byte-oriented block-ciphering algorithm, *Lecture Notes in Computer Science*, 809, 1–17, 1994.

[SC1] C. Schwaiger and A. Treytl, Smart card based security for fieldbus systems, in *Proceedings of Ninth IEEE Conference on Emerging Technologies and Factory Automation*, Catania, Italy, pp. 398–406, 2003.

[SC2] C. Schwaiger and T. Sauter, Security strategies for field area networks, in *28th Annual Conference of the IEEE Industrial Electronics Society*, Sevilla, Spain, pp. 2915–2920, 2002.

[SHA] National Institute of Standards and Technology, Secure Hash Standard, FIPS PUB 180–2, Gaithersburg, MD, 2002.

[TLS] T. Dierks and K. Resorla, The Transport Layer Security (TLS) Protocol Version 1.2, RFC 5246, 2008.

[TR1] A. Treytl, T. Sauter, and C. Schwaiger, Security measures in automation systems—A practice-oriented approach, in *Proceedings of 11th IEEE Conference of Emerging Technologies and Factory Automation*, Catania, Italy, September 19–22, 2005.

[TR2] A. Treytl, T. Sauter, H. Adamcyk, S. Ivanov, and S. H. Tresk, Security concepts for flexible wireless automation in real-time environments, in *Proceedings of IEEE Conference on Emerging Technologies and Factory Automation 2009*, Palma de Mallorca, Spain, pp. 1–8, 2009.

[TUO] A. Tuomas, N. Pekka, and L. Jussipekka, DOS resistant authentication with client puzzles, in *Proceedings of Eighth International Workshop on Security Protocols*, Cambridge, U.K., pp. 170–177, April 3–5, 2000.

[WEI] B. Weiss, Authentication under denial-of-service attacks, in *Proceedings of IASTED Conference on Communications and Computer Networks*, Boston, MA, pp. 134–139, November 2002.

[ZIG] *ZigBee Specification 2007*, ZigBee Alliance, San Ramon, CA, 2007.

23

Secure Communication Using Chaos Synchronization

Yan-Wu Wang
Huazhong University of Science and Technology

Changyun Wen
Nanyang Technological University

23.1 Introduction 23-1
23.2 Chaos Synchronization 23-2
Feedback Control for Chaos Synchronization via Partial States • Adaptive Control for Chaos Synchronization via Partial States • Impulsive Control for Chaos Synchronization via Partial States • Practical Impulsive Synchronization of Chaotic Systems with Parametric Uncertainty and Mismatch
23.3 Secure Communication Using Chaos Synchronization 23-8
Secure Communication Schema • The Encrypter • The Decrypter • Synchronization Time Estimation • Implementation
References 23-13

23.1 Introduction

Chaos is common in nature. The earliest discovery of chaotic behavior can be casted back to Poincare, a French mathematician in the nineteenth century. In his study of celestial mechanics of conservative system, Poincare found that some solutions of a deterministic dynamic equation were unpredictable. But the research work on chaos is usually said to begin with Lorenz [L63], an American meteorologist. When he was working on the problem of weather prediction using computer, he found that the solution of a deterministic dynamic equation with certain parameters can be random-liked. This phenomenon came to be known as the butterfly effect. After then, Henon [H76] and Rossler [R79] obtained similar conclusions. Since then, chaos theory has become one of the most popular research topics and great development has been made in chaos science.

Chaos has a series of interesting and useful features, such as being extremely sensitive to very minor variations of initial conditions, possessing at least one positive Lyapunov exponent, having bounded trajectories in the phase space, being ergodic in the field of the strange attractor, and so on. Due to its distinguished features, a broad-spectrum and noise-like chaotic signal is particularly appropriate for secure communications. In a secure communication scheme, encryption and decryption are processed at the transmitter and the receiver, respectively. In a typical chaotic secure communication system as shown in Figure 23.1 [LLWS03], which is discussed in Section 23.3, a chaotic cryptosystem is often used. With a suitable cryptographic scheme, the original signal is hidden in the noise-like chaotic signals. The cryptosystem is composed of two main parts, encrypter and decrypter. Each part includes a chaotic system, called driving and response system, respectively. It is essential to synchronize the two chaotic systems in order to recover the original data from encrypted ones in the receiver side. This can be ensured

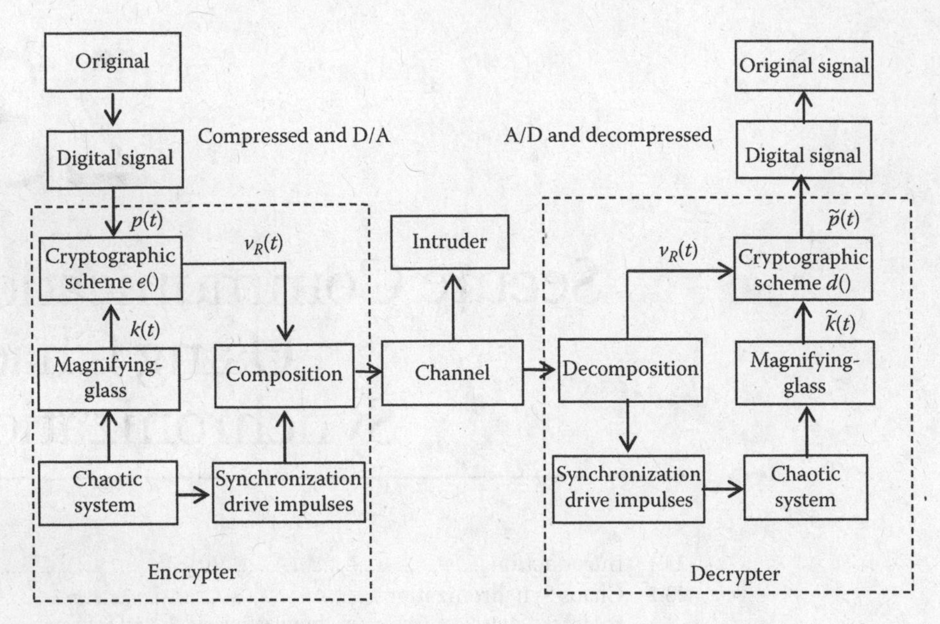

FIGURE 23.1 System block diagram of the chaotic cryptosystem. (Reprinted from Li, Z.G. et al., *IEEE Trans. Commun.*, 51(8), 1306, 2003. Copyright @ 2003 by The Institute of Electrical and Electronics Engineers. With permission.)

by chaos synchronization. But before 1990, chaos synchronization was very difficult and nearly impossible. The ground-breaking scheme, called Pecara and Carroll scheme [PC90], firstly offered a method to control and realize chaos synchronization. From then on, chaos synchronization and its application to secure communication have seen a flurry of research activities for decades.

The main advantage of a chaotic secure communication system over conventional cryptosystems is that chaotic secure communication schemas can often be realized using very simple circuits on a part of a chip [GHGS00] and they can be used in applications that do not require a high level of information security such as remote keyless entry system, video phone, and wireless telephone [GHGS00]. In this chapter, some typical control methods for chaos synchronization will be presented first, and then how they are applied to secure communication will be illustrated.

23.2 Chaos Synchronization

A number of methods have been proposed for the synchronization of chaotic systems, including the linear and nonlinear feedback control, fuzzy control, adaptive control, sliding mode control, and impulsive control. In this section, some typical synchronization control methods are introduced with the Chua's circuit as an illustrated example.

23.2.1 Feedback Control for Chaos Synchronization via Partial States

One of the important aspects to evaluate the effectiveness of the secure communication schema is the transmission efficiency, i.e., the ratio of the data information to the entire transferred information. In order to achieve chaos synchronization, it is unavoidable to occupy the communication channel by synchronization control information of the driving system that needs to be transmitted to the receiver. In the full-order synchronization control schema in which synchronization is achieved by controlling all the system states, information on all the states of the driving system should be sent to the receiver, which leads to a poor transmission efficiency. In addition, the full-order synchronization control schema is not feasible for the case when some of the system states are not available or controllable. In order to overcome the problem for such a case and to improve the transmission efficiency, one may wish to transmit less control information to achieve chaos synchronization. An intuitive idea is to realize the chaos synchronization only by using part of the states information.

Suppose the driving system is given as

$$\begin{cases} \dot{X} = -AX + f(X,Y) + h_1(t), \\ \dot{Y} = g(X,Y) + h_2(t), \end{cases} \tag{23.1}$$

where $X = [x_1 \quad \dots \quad x_n]^T \in R^n$ and $Y = [x_{n+1} \quad \dots \quad x_{n+m}]^T \in R^m$ are the state vectors of the nonlinear system, $A \in R^{n\times n}$. $f: R^n \times R^m \to R^n, g: R^n \times R^m \to R^m$ are nonlinear functions of X and Y, $h_1: R_+ \to R^n$ and $h_2: R_+ \to R^m$ are linear or nonlinear functions of t.

The response system under control using partial system states is given as follows:

$$\begin{cases} \dot{\tilde{X}} = -A\tilde{X} + f(\tilde{X},\tilde{Y}) + h_1(t), \\ \dot{\tilde{Y}} = g(\tilde{X},\tilde{Y}) + h_2(t) + U(t), \end{cases} \tag{23.2}$$

where $\tilde{X} = [\tilde{x}_1 \quad \dots \quad \tilde{x}_n]^T \in R^n$, $\tilde{Y} = [\tilde{x}_{n+1} \quad \dots \quad \tilde{x}_{n+m}]^T \in R^m$, and $U(t)$ is the control action which is only added to states $\tilde{Y}$ of the $n + m$th order system (23.2).

The synchronization errors are defined as

$$E_1 = \tilde{X}(t) - X(t), \quad E_2 = \tilde{Y}(t) - Y(t)$$

and they can be denoted as vectors

$$E_1 = \begin{bmatrix} e_1 & \cdots & e_n \end{bmatrix}^T, \quad E_2 = \begin{bmatrix} e_{n+1} & \cdots & e_{n+m} \end{bmatrix}^T.$$

In the linear feedback control schema, the controller is designed as

$$U(t) = -KE_2(t) \quad \text{with } K = diag(k_j) \quad \text{for } j = 1,\dots,m, \tag{23.3}$$

where k_j is a parameter determined according to the conditions given in [WWSX06].

Obviously, such a feedback controller saves the bandwidth of the communication channel as only m states of the driving system are required to be transmitted, instead of all the $n + m$ states.

Note that the system (23.1) is quite general as $f(X, Y)$, $g(X, Y)$, $h_1(t)$, and $h_2(t)$ are general nonlinear functions. In fact, it includes the majority of typical nonlinear chaotic systems studied so far and thus can be easily implemented. Now, the Chua's circuit given below is taken as an example for illustration:

$$\begin{cases} \dot{x}_1 = -\beta x_2 - \gamma x_1, \\ \dot{x}_2 = y_1 - x_2 + x_1, \\ \dot{y}_1 = -\alpha y_1 + \alpha x_2 - \alpha h(y_1), \end{cases} \tag{23.4}$$

where α, β, and γ are parameters, $h(y_1) = by_1 + 0.5(a - b)(|y_1 + 1| - |y_1 - 1|)$, a and b are constants which satisfying $a < b < 0$. The response system under feedback control is given by

$$\begin{cases} \dot{\tilde{x}}_1 = -\beta\tilde{x}_2 - \gamma\tilde{x}_1, \\ \dot{\tilde{x}}_2 = \tilde{y}_1 - \tilde{x}_2 + \tilde{x}_1, \\ \dot{\tilde{y}}_1 = -\alpha\tilde{y}_1 + \alpha\tilde{x}_2 - \alpha h(\tilde{y}_1) + U(t). \end{cases} \tag{23.5}$$

Clearly,

$$A = \begin{bmatrix} \gamma & 0 \\ 0 & 1 \end{bmatrix}, \quad f(X,Y) = \begin{bmatrix} -\beta x_2 \\ y_1 + x_1 \end{bmatrix}, \quad g(X,Y) = -\alpha y_1 + \alpha x_2 - \alpha h(y_1).$$

In this case, only one state $\tilde{y}_1$ in the response system (23.5) is controlled by using one state y_1 from the driving system, namely

$$U(t) = -k_1\left[\tilde{y}_1(t) - y_1(t)\right].$$

To illustrate the synchronization of the two Chua's circuits, choose $\alpha = 10$, $\beta = 16$, $\gamma = 0.0385$, $a = -8/7$, and $b = -5/7$, and the initial conditions for the driving system and the response system are $(18 \quad -26 \quad -17)^T$ and $(2 \quad 10 \quad 10)^T$, respectively. According to the guidelines in [WWSX06] and based on the above parameters, it is computed that $k_1 > 11.9911$ in order to ensure synchronization of the two chaotic systems. Let $k_1 = 30$ in the simulation. The synchronization errors of two identical Chua's circuits (23.4) and (23.5) under feedback control are shown in Figure 23.2.

Clearly, the errors are almost zero for $t > 10$ s.

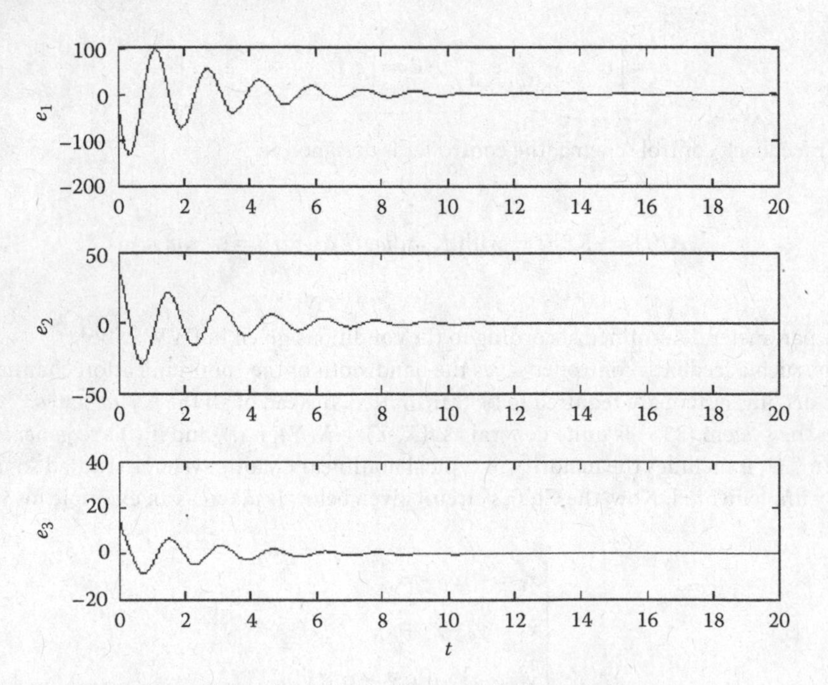

FIGURE 23.2 Synchronization of two Chua's circuits under feedback control.

23.2.2 Adaptive Control for Chaos Synchronization via Partial States

Sometimes, the determination of suitable controller gain K in (23.3) is complicated. So, an adaptive version of the controller can be designed to continuously adjust the gain parameters based on the system responses for synchronization. Based on the results in [WWSX06], the following adaptive controller can be designed:

$$U(t) = -K(t)E_2(t), \quad \text{with } K(t) = diag(k_j(t)) \quad \text{and} \quad \dot{k}_j(t) = \theta_j e_{n+j}^2(t) \quad \text{for } j = 1, \ldots, m \tag{23.6}$$

where θ_j is a positive parameter chosen by users.

Again, the two Chua's circuits (23.4) and (23.5) are considered under the adaptive control (23.6). Let $\theta_1 = 10$, and the initial condition of the adaptive control (23.6) is $k_1(0) = 0$, the synchronization errors of two identical Chua's circuits (23.4) and (23.5) and the changing parameter of the adaptive control (23.6) are shown in Figures 23.3 and 23.4, respectively.

It is observed that the two systems are synchronized.

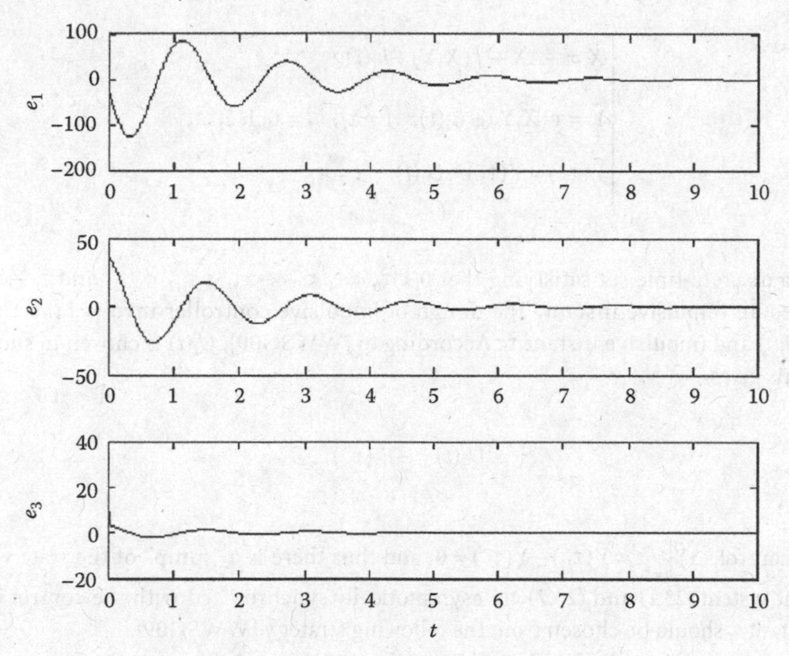

FIGURE 23.3 Synchronization of two Chua's circuits under adaptive control.

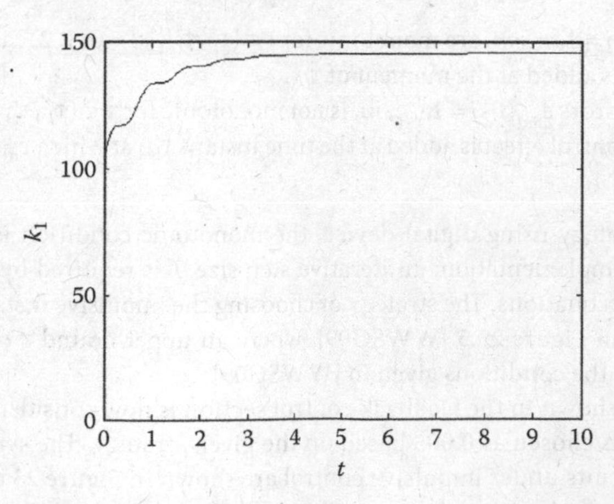

FIGURE 23.4 Changing parameter of the adaptive control.

23.2.3 Impulsive Control for Chaos Synchronization via Partial States

In the continuous synchronization control strategies like the linear feedback control and adaptive control above, continuous transmission of state variables of the driving system is required to obtain synchronization control signals in implementation. Thus, the security level is significantly decreased if the scheme is applied to secure communication. On the other hand, a purely impulsive synchronization scheme is implemented by transferring the states information of the driving system only at discrete impulsive instants, therefore, is much more feasible in secure communication to enhance transmission security and also to improve the transmission efficiency. In addition, a secure communication system based on impulsive synchronization has another advantage of being less sensitive to channel noise than that based on continuous synchronization.

Suppose that the driving system is given by (23.1). The response system with impulsive control added to partial states $\tilde{Y}$ of the system at only certain discrete-time instants τ_l is given as

$$\begin{cases} \dot{\tilde{X}} = -A\tilde{X} + f(\tilde{X},\tilde{Y}) + h_1(t), \\ \dot{\tilde{Y}} = g(\tilde{X},\tilde{Y}) + h_2(t), \quad t \neq \tau_l, \quad l = 0, 1, 2, \ldots, \\ \tilde{Y}(\tau_l^+) = \tilde{Y}(\tau_l^-) + U_l(t), \quad t = \tau_l, \end{cases} \tag{23.7}$$

where $\{\tau_l\}$ is a discrete-time set satisfying that $0 < \tau_0 < \tau_1 < \cdots < \tau_l < \tau_{l+1} < \cdots$, and $\tau_l \to \infty$ as $l \to \infty$ and τ_l is called an impulsive instant. The design of impulsive controller involves the choices of the control law $U_l(t)$ and impulsive instant τ_l. According to [WWSG09], $U_l(t)$ is chosen in such a way that $E_2(\tau_l^+) = 0$. This gives

$$U_l(t) = -E_2(\tau_l^-). \tag{23.8}$$

With such a control, $\Delta\tilde{Y}\big|_{t=\tau_l} = \tilde{Y}(\tau_l^+) - \tilde{Y}(\tau_l^-) \neq 0$, and thus there is a "jump" of the state vector $\tilde{Y}$ at τ_l. To ensure that system (23.1) and (23.7) are asymptotically synchronized with the control in (23.8), the impulsive instant τ_l should be chosen from the following strategy [WWSG09].

Strategy for the choice of impulsive instant τ_l

1. If the states $e_{n+j}(t), j = 1, \ldots, m$, are monotonic for $t \in (\tau_l^+, \tau_{l+1}^-]$, $l = 0, 1, 2, \ldots$, then the $l + 1$th impulsive control effect is added at the moment of τ_{l+1}.
2. If one of the state errors $e_{n+j}(t), j = 1, \ldots, m$, is not monotonic for $t \in (\tau_l^+, \tau_{l+1}^-]$, $l = 0, 1, 2, \ldots$, then the $l + 1$th impulsive control effect is added at the time instant τ'_{l+1} at which $e_{n+j}(t)$ changes monotonicity, i.e., $\tau_{l+1} = \tau'_{l+1}$.

To implement the strategy using digital device, the monotonic condition is approximated in programming. During the implementation, an iterative step size T is required by applying Runge–Kutta method to solve the state equations. The strategy of choosing the impulsive instant τ_l, $l > 0$, is illustrated in the flowchart shown in Figure 23.5 [WWSG09], where an upper bound τ of impulsive intervals is determined according to the conditions given in [WWSG09].

The same example as shown in the feedback control section is now considered. In the implementation, the upper lound τ is chosen as 0.005 based on the given strategy. The synchronization errors of two identical Chua's circuits under impulsive control are shown in Figure 23.6.

Again, the errors converge to zero asymptotically.

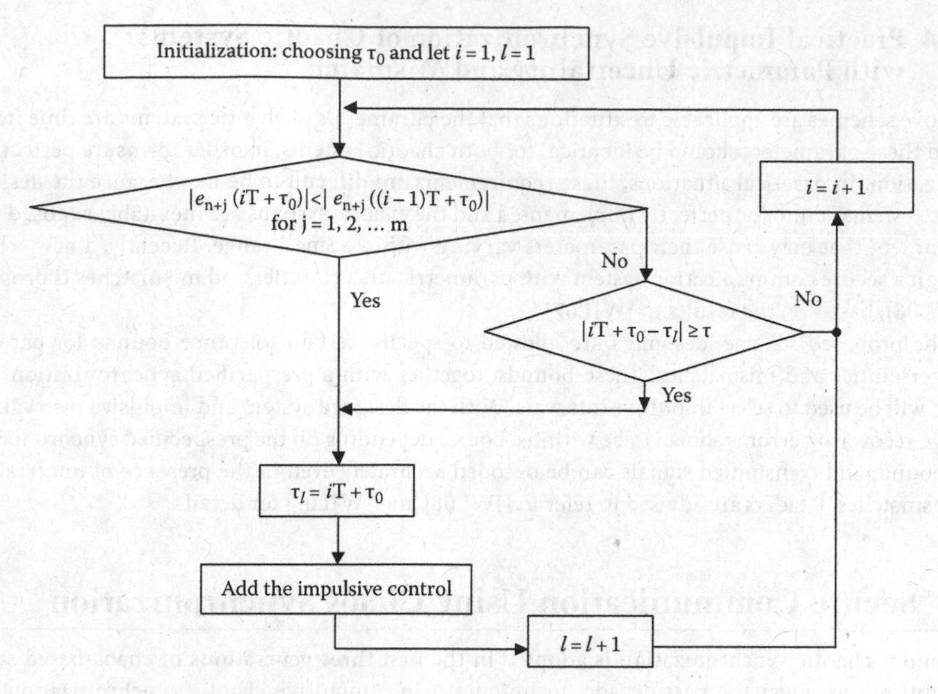

FIGURE 23.5 Choice of impulsive instance τ_l.

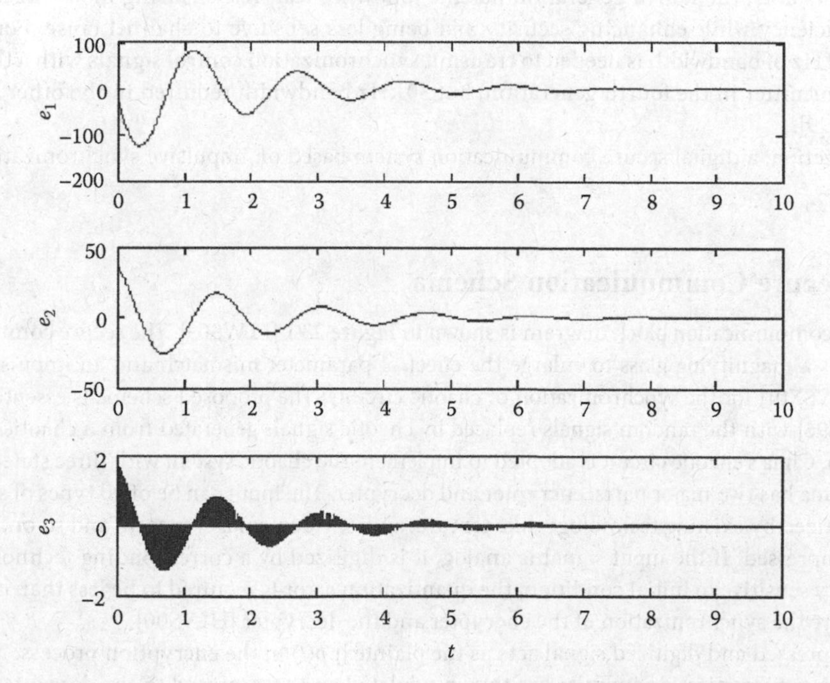

FIGURE 23.6 Synchronization of two Chua's circuits under impulsive control.

23.2.4 Practical Impulsive Synchronization of Chaotic Systems with Parametric Uncertainty and Mismatch

The above schemes are applicable to situations that the parameters of chaotic systems are time invariant and these parameters should be identical for both chaotic systems, in order to ensure perfect synchronization. In practical situations, these requirements are difficult to be met because the designed chaotic systems cannot be perfectly implemented and the chaotic systems are inevitably exposed to an environment that may cause their parameters varying within a small range. Recently, a new scheme to design a secure communication system with parametric uncertainties and mismatches is proposed in [JWL08], based on the results in [WJL07].

In the proposed scheme, designers are allowed to specify certain tolerance bounds for parametric uncertainties and mismatches. These bounds, together with a prespecified synchronization error bound, will be used to select impulsive intervals. With the designed system and impulsive intervals, the message recovering error is shown to be within a bound depending on the prespecified synchronization error bound, and transmitted signals can be decoded accurately even in the presence of uncertainties and mismatches. Readers are advised to refer to [JWL08] and [WJL07] for details.

23.3 Secure Communication Using Chaos Synchronization

Continuous chaotic synchronization is adopted in the first three generations of chaos-based secure communications. Over the past decade, techniques using impulsive chaotic synchronization have been developed and this has started the fourth generation [COS93,WC93,YWC97,ZL97,T99]. Since in impulsive synchronization information on states of the driving system is transmitted only at the impulsive instants, the fourth generation has the following features: ensuring more effective transmission efficiency while enhancing security and being less sensitive to channel noise. For example, less than 94 Hz of bandwidth is needed to transmit synchronization control signals with a third-order chaotic transmitter in the fourth generation, but 30 kHz bandwidth required in the other three generations [T99].

In this section, a digital secure communication system based on impulsive synchronization will be presented.

23.3.1 Secure Communication Schema

The secure communication block diagram is shown in Figure 23.1 [LLWS03]. The secure communication schema uses a magnifying glass to enlarge the effect of parameter mismatch and an impulsive control strategy [LWSX01] for the synchronization of chaotic circuits. The proposed scheme is essentially a one-time pad [S96] with the random signals replaced by chaotic signals generated from a chaotic system. In this schema, Chua's chaotic circuit is adopted to implement the chaotic system with three state variables.

The schema has two major parts: encrypter and decrypter. The input can be of all types of signals that can be digitized by existing technology, like texture, audio, video, image, speech, and so on, which can be first compressed. If the input signal is analog, it is digitized by a corresponding technology. Since chaos is very sensitive to initial condition, the quantization error is required to be less than certain values to ensure the synchronization of the encrypter and the decrypter [HLYS00].

The compressed and digitized signal acts as the plaintext $p(t)$ in the encryption process. The ciphertext and the synchronization impulse are then modulated and transmitted to the decrypter via a network. After demodulating and decrypting, the received plaintext is decompressed and the original message is recovered.

The encrypter consists of a chaotic system and a classical encryption function $e(\cdot)$. The decrypter is composed of an impulsive differential system, a corresponding decryption function $d(\cdot)$, and an impulsive controller. The impulsive differential system is essentially a Chua's circuit, which is the same as that used in the encrypter. The synchronization of the encrypter and the decrypter can be achieved by using an impulsive controller with piecewise constant impulsive intervals. The key signal $k(t)$ is a combination of all three state variables of the chaotic systems. The ciphertext is obtained from an XOR operation on the plaintext and the key sequence bit by bit. The decryption is the same as the encryption, including the XOR operation of the transmitted scrambled signal with the key signal $\tilde{k}(t)$. When the chaotic systems in the decrypter and the encrypter are synchronized, the decrypter can find the same key signal sequence $\tilde{k}(t)$ as that in the encrypter $k(t)$.

23.3.2 The Encrypter

The dimensionless state equations of Chua's circuit are given as

$$\begin{cases} \dot{x}_1 = k\alpha(x_2 - x_1 - f(x_1)), \\ \dot{x}_2 = k(x_1 - x_2 + x_3), \\ \dot{x}_3 = k(-\beta x_2 - \gamma x_3), \end{cases} \tag{23.9}$$

where α, β, and γ are constants, $k \in [-1, 1]$ and $f(x)$ is the nonlinear characteristic of Chua's diode in Chua's circuit given by

$$f(x) = m_1 x + \frac{1}{2}(m_0 - m_1)\{|x+1| - |x-1|\}, \tag{23.10}$$

with m_0 and m_1 being two negative constants. Since the signals are transmitted through a digital channel, the synchronization pulses should be first quantized by a predefined quantizer $Q(\cdot)$, which depends on the amplification factor K used in (23.11). Since chaos is very sensitive to initial condition, the quantization error should be less than certain values to ensure that the encrypter and the decrypter can be synchronized [HLYS00].

To provide the desired key sequence, the concept of a magnifying glass is introduced, which is composed of an amplifier and an observer. They are given in details as follows.

The amplifier:

$$k'(t) = K(x_1^2(t) + x_2^2(t) + x_3^2(t))^{1/2}. \tag{23.11}$$

The observer:

$$k(t) = (\lfloor k'(t) \rfloor + \lambda)\text{mod}(256), \tag{23.12}$$

where

K is a large number chosen to influence the sensitivity of the system

λ is an arbitrary integer

$\lfloor a \rfloor$ is the integer truncation of a

Remark 23.1: One can replace the amplifier function in (23.11) by $k'(t) = Kg(x_1, x_2, x_3)$, where g is a predefined function. Since the function g is a function of the state variables of the chaotic system, the sensitivities to the parameter mismatches will be different for different g.

The scrambled signal v_R is given by

$$v_R(t) = E(p(t)) = p(t) \oplus k(t), \tag{23.13}$$

where

$p(t)$ is the plaintext
$k(t)$ is given in (23.12)
$E(p(t))$ is the ciphertext
$\oplus$ denotes XOR operation

23.3.3 The Decrypter

Both Chua's circuit and the impulsive controller in the decrypter are given by

$$\begin{cases} \dot{\tilde{x}}_1 = k\alpha(\tilde{x}_2 - \tilde{x}_1 - f(\tilde{x}_1)), \\ \dot{\tilde{x}}_2 = k(\tilde{x}_1 - \tilde{x}_2 + \tilde{x}_3), \qquad t \neq \tau_l;\ l = 1, 2, \ldots, \\ \dot{\tilde{x}}_2 = k(-\beta\tilde{x}_2 - \gamma\tilde{x}_3), \end{cases} \tag{23.14}$$

and

$$\begin{bmatrix} \tilde{x}_1(\tau_l) \\ \tilde{x}_2(\tau_l) \\ \tilde{x}_3(\tau_l) \end{bmatrix} = \begin{bmatrix} \tilde{x}_1(\tau_l^-) \\ \tilde{x}_2(\tau_l^-) \\ \tilde{x}_3(\tau_l^-) \end{bmatrix} - B \begin{bmatrix} Q(x_1(\tau_l)) - \tilde{x}_1(\tau_l^-) \\ Q(x_2(\tau_l)) - \tilde{x}_2(\tau_l^-) \\ Q(x_3(\tau_l)) - \tilde{x}_3(\tau_l^-) \end{bmatrix}, \quad l = 1, 2, \ldots, \tag{23.15}$$

where

B is a 3×3 matrix to be designed to satisfy certain inequality for synchronization
$Q(\cdot)$ is a predefined quantizer
τ_l^- are the time instant immediately prior to the time instant τ_l

Let $T_l = (\tau_l - \tau_{l-1})$, ($\tau_0 = 0, l = 1, 2, \ldots$), denote impulsive time intervals.

In the decrypter, the plaintext is recovered via

$$\tilde{k}(t) = \left(\left\lfloor K(\tilde{x}_1^2(t) + \tilde{x}_2^2(t) + \tilde{x}_3^2(t))^{1/2} \right\rfloor + \lambda\right) \text{mod}(256), \tag{23.16}$$

$$\tilde{p}(t) = \tilde{E}(p(t)) = v_R(t) \oplus \tilde{k}(t), \tag{23.17}$$

where

$\tilde{E}(p(t))$ is the recovered encrypted signal
$\tilde{k}(t)$ is recovered in the receiver circuit and should approximate $k(t)$

It can be known from (23.16) and (23.17) that the original signal can be recovered only when two identical chaotic circuits in both the encrypter and the decrypter are synchronized. Similarly, an intruder can know the original message only when he knows the parameters and the structure of the circuits and the synchronization impulses.

Remark 23.2: The magnifying glass is used to transform the chaotic state variables that act as key sequence before XOR with the plaintext. Assuming that there is a small mismatch that results in perturbations $\Delta x_i(t) = \sigma_i$ ($i = 1, 2, 3$), then the signal getting through the amplifier gives $k(t) = \left(\left\lfloor K \left(\sum_{i=1}^{3} (x_i(t) + \sigma_i)^2 \right)^{1/2} \right\rfloor + \lambda \right) \text{mod}(256)$. Since the parameter K is a large number, it can be seen that the signal is enlarged many times, which implies that the parameters mismatch can be enlarged. Thus, even a minor mismatch in the parameters will produce a large decryption error, resulting in a decryption key sequence that is not the same as the encryption key signal. So, one cannot recover the plaintext signal. The security of the chaotic communication system is thus improved.

Before transmitting the ciphertext through a digital channel, it is necessary to packetize the ciphertext. To simplify the operation, a packetization algorithm is designed, in which the packet length is fixed. Since the lengths of the impulsive intervals are piecewise constant, it is difficult to find the synchronization impulse. The security of the system can then be improved. To maintain the simplicity of the presented secure system, the length of the packets is the same as the length of the first impulsive interval. The length of other impulsive interval is determined by the length of the first impulsive interval and the parameters of the Chua's circuits.

23.3.4 Synchronization Time Estimation

It is noticeable that the time-varying impulsive intervals do not affect the synchronization of two identical chaotic circuits embedded in the encrypter and the decrypter, and the time required for synchronization can be estimated. A scheme in [LLWS03] is presented to achieve this objective.

From the estimation, it is remarked that the value of K will influence the synchronization stable time of two identical chaotic systems and the desired precision of the system. A larger K will require longer synchronization time. On the other hand, the security of the system is higher with a larger K. A trade-off should be obtained in practice.

23.3.5 Implementation

Text transmission is taken to illustrate the performance of the proposed chaotic secure communication system. In order to make the presented encryption system widely used in the computer, 256 characters ASCII code is used, which includes 128 standard ASCII codes and other 128 that are known as extended ASCII. Thus, the ciphertext in the simulation has not only the standard characters but also some local symbols and marks. When the ciphertext can be decrypted by the correct key, the message is exactly recovered. Otherwise, the recovered message is spread in the extended ASCII code.

Choose the parameters of Chua's circuit as $k = 1$, $\alpha = 9.35159085$, $\beta = 14.790313805$, $\gamma = 0.016072965$, $m_0 = -1.138411196$, and $m_1 = -0.723451121$. For synchronization to be ensured, matrix B is selected as $B = diag(-1.05, -1, -1)$ and the lengths of impulsive time intervals are chosen as $T_{2h-1} = 5 \times 10^{-3}$ s and $T_{2h} = 2 \times 10^{-3}$ s where $h = 1, 2, \ldots$. Then starting with $\tau_0 = 0$, the impulsive instants can be determined. The initial condition is given by $[x_1(0); x_2(0); x_3(0)] = [-2.12; -0.05; 0.8]$ and $[\tilde{x}_1(0); \tilde{x}_2(0); \tilde{x}_3(0)] = [-0.2; -0.02; 0.1]$, respectively. As the initial error is $e(0) = [-1.92; -0.03; 0.7]$, so the encrypter and decrypter are initially not synchronized. Choose the magnifying glass as $K = 1 \times 10^3$, $\lambda = 12$, and the quantization step is $q = 5 \times 10^{-8}$.

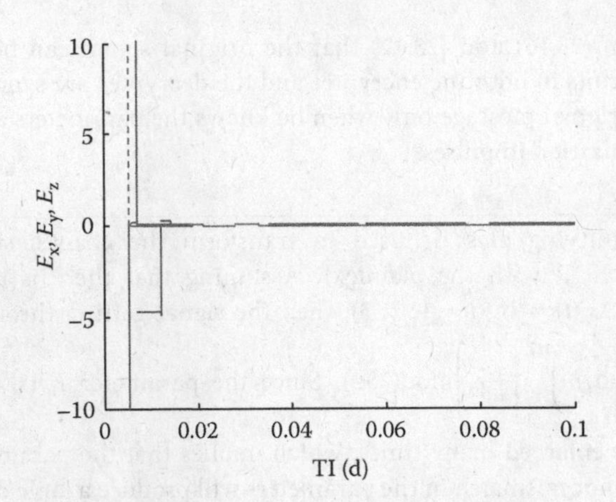

FIGURE 23.7 The time responses of $e^T = (e_1, e_2, e_3)$. (Reprinted from Li, Z.G. et al., *IEEE Trans. Commun.*, 51(8), 1306, 2003. Copyright @ 2003 by The Institute of Electrical and Electronics Engineers. With permission.)

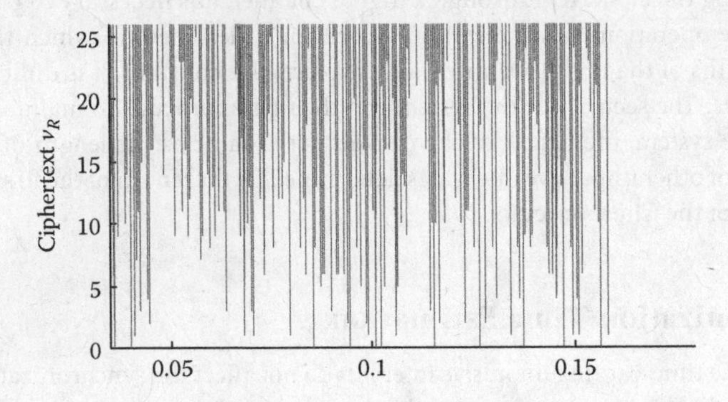

FIGURE 23.8 The transmitted ciphertext v_R. (Reprinted from Li, Z.G. et al., *IEEE Trans. Commun.*, 51(8), 1306, 2003. Copyright @ 2003 by The Institute of Electrical and Electronics Engineers. With permission.)

TABLE 23.1 A Sample of Simulation Result

Message	Chaotic Secure Communication System
Plaintext	=›kž □ ~{ XR □ m“³Ð\E§Í □ u¢`„fÌEm¡o □
Transmitted	□ ¥U£2CG26ci#3W©‰ée|Ÿ • õ1B • V²PùpX • [0
Received	=›kž □ ~{ XR □ m“³Ð\E§Í □ u¢`„fÌEm¡o □
Recovered	Chaotic Secure Communication System

Source: Reprinted from Li, Z.G. et al., *IEEE Trans. Commun.*, 51(8), 1306, 2003. Copyright @ 2003 by The Institute of Electrical and Electronics Engineers. With permission.

It is shown in Figure 23.7 that the synchronization error approaches zero very quickly. After the encrypter and decrypter are synchronized, they will generate the same key sequences.

The ciphertext that is transmitted through the public channel is shown in Figure 23.8. In Table 23.1, one example is shown for the original message, the corresponding arithmetic-coded message, which is the plaintext in encryption, the transmitted ciphertext, the received decrypted text, and the recovered message, respectively. It can be seen from the table that the ciphertext is completely different from the plaintext in transmission while the decrypted text is the exact version of the plaintext.

References

[COS93] K. M. Cuomo, A. V. Oppenheim, and S. H. Strogatz, Synchronization of Lorenz-based chaotic circuits with applications to communications, *IEEE Trans. Circuits Syst. II*, 40, 626–633, 1993.

[GHGS00] O. Gonzales, G. Han, J. Gyvez, and E. Sanchez-Sinencio, Lorenz-based chaotic cryptosystem: A monolithic implementation, *IEEE Trans. CAS-I*, 47, 1243–1247, 2000.

[HLYS00] Z. He, K. Li, L. Yang, and Y. Shi, A robust digital secure communication scheme based on sporadic coupling chaos synchronization, *IEEE Trans. CAS-I*, 47, 397–403, 2000.

[H76] M. N. Henon. A two dimensional mapping with a strange attractor, *Commun. Math. Phys.*, 50, 69–77, 1976.

[JWL08] Y. Ji, C. Y. Wen, and Z. G. Li, Chaotic communication systems in the presence of parametric uncertainty and mismatch, *Int. J. Commun. Syst.*, 21, 1137–1154, 2008.

[LLWS03] Z. G. Li, K. Li, C. Wen, and Y. C. Soh, A new chaotic secure communication system, *IEEE Trans. Commun.*, 51(8), 1306–1312, 2003.

[LWSX01] Z. G. Li, C. Y. Wen, Y. C. Soh, and W. X. Xie, The stabilization and synchronization of Chua's circuits via impulsive control, *IEEE Trans. CAS-I*, 48, 1351–1355, 2001.

[L63] E. N. Lorenz, Deterministic non-periodic flows, *J. Atmos. Sci.*, 20, 130–141, 1963.

[PC90] L. M. Pecora and T. L. Caroll, Synchronization in chaotic system, *Phys. Rev. Lett.*, 64, 821–824, 1990.

[R79] O. Rossler, Continuous chaos: Four prototype equations, *Ann. NY Acad. Sci.*, 316, 376–392, 1979.

[S96] B. Schneier, *Applied Cryptography: Protocols, Algorithms, and Source Code in C* (2nd edn.), J. Wiley, New York, 1996.

[T99] Y. Tao, Chaotic secure communication systems: History and new results, *Telecommun. Rev.*, 9, 597–634, 1999.

[WWSG09] Y.-W. Wang, C. Wen, Y. C. Soh, and Z.-H. Guan, Partial state impulsive synchronization of a class of nonlinear systems, *Int. J. Bifurc. Chaos*, 19(1), 387–393, 2009.

[WWSX06] Y.-W. Wang, C. Wen, Y. C. Soh, and J.-W. Xiao, Adaptive control and synchronization for a class of nonlinear chaotic systems using partial system states, *Phys. Lett. A*, 351, 79–84, 2006.

[WJL07] C. Y. Wen, Y. Ji, and Z. G. Li, Practical impulsive synchronization of chaotic systems with parametric uncertainty and mismatch, *Phys. Lett. A*, 361, 108–114, 2007.

[WC93] C. W. Wu and L. O. Chua, A simple way to synchronize chaotic system with application to secure communication system, *Int. J. Bifurc. Chaos*, 3, 1619–1627, 1993.

[YWC97] T. Yang, C. W. Wu, and L. O. Chua, Cryptography based on chaotic systems, *IEEE Trans. CAS-I*, 44, 469–472, 1997.

[ZL97] H. Zhou and Y. T. Ling, Problems with the chaotic inverse system encryption approach, *IEEE Trans. CAS-I*, 44, 268–271, 1997.

[illegible] 2004 [illegible]
[illegible] quantum [illegible] 1997 [illegible]

II

Application-Specific Areas

24 Embedded Networks in Civilian Aircraft Avionics Systems *Christian Fraboul, Fabrice Frances, and Jean-Luc Scharbarg* **24**-1
Introduction • Avionics Systems Evolution and ARINC Context • Classic Avionics and ARINC 429 • Integrated Modular Avionics • ARINC 629 Multiplexed Data Bus • ARINC 664: Avionics Full-Duplex Ethernet • AFDX End-to-End Delay Analysis • Conclusion • Abbreviations • References

25 Process Automation *Alois Zoitl and Wilfried Lepuschitz* **25**-1
Introduction • Structures and Models of Batch Manufacturing Systems • Currently Applied Communication Systems • Upcoming Requirements of Distributed Process Automation • Industrial Ethernet as the "Silver Bullet" for Future Process Automation Communication Needs • References

26 Building and Home Automation *Wolfgang Kastner, Stefan Soucek, Christian Reinisch, and Alexander Klapproth* **26**-1
Introduction • Building Automation • Home Automation • Outlook and Further Challenges • References

27 Industrial Multimedia *Javier Silvestre-Blanes, Manfred Weihs, and Víctor-M. Sempere-Payá* **27**-1
Introduction • Multimedia Compression: A Review • Industrial Multimedia Applications • Image Transmission • Conclusions • Acknowledgment • References

28 Industrial Wireless Communications Security (IWCS)/C42 *Milos Manic and Kurt Derr* **28**-1
Introduction • Wireless LAN Security • PAN Security • Summary • References

29 Protocols in Power Generation *Tuan Dang and Gaëlle Marsal* **29**-1
Introduction • Power Plant Automation Systems and Intra-Plant Communications • Power Plant Information Systems and Extra-Plant Communications • Conclusions • References

30 Communications in Medical Applications *Paulo Bartolomeu, José Alberto Fonseca, Nelson Rocha, and Filipe Basto* **30**-1
Introduction • Requirements • Localization • Clinical Monitoring • Automation • Issues and Challenges • References

24

Embedded Networks in Civilian Aircraft Avionics Systems

Christian Fraboul
Université de Toulouse

Fabrice Frances
Université de Toulouse

Jean-Luc Scharbarg
Université de Toulouse

24.1 Introduction 24-1
24.2 Avionics Systems Evolution and ARINC Context 24-2
24.3 Classic Avionics and ARINC 429 24-3
24.4 Integrated Modular Avionics 24-4
24.5 ARINC 629 Multiplexed Data Bus 24-5
Basic Protocol • Combined Protocol
24.6 ARINC 664: Avionics Full-Duplex Ethernet 24-7
Full-Duplex Switched Ethernet • The ARINC 664 Standard • Virtual Link Paradigm • Virtual Link Properties • Network Redundancy for Safety and Fault Tolerance
24.7 AFDX End-to-End Delay Analysis 24-13
24.8 Conclusion 24-13
Abbreviations 24-14
References 24-15

24.1 Introduction

The evolution of civilian aircraft avionics systems is mainly due to increasing complexity, which is illustrated by a larger number of integrated functions, a growing volume of exchanged data, and a multiplication of connections between functions. Consequently, the growth in the number of multipoint communication links could not be taken into account by classic avionics mono-emitter data buses (such as ARINC 429 [ARI01]). The first solution proposed for the Boeing 777 led to the design of a new multiplexed data bus based on CSMA-CA medium access control (ARINC 629 standard [ARI99]). The solution adopted by Airbus for the A380 consists in the utilization of a switched Ethernet technology (called avionics full-duplex switched Ethernet [AFDX]). This allows a reuse of development tools as well as of existing communication components while providing better performance; it has been standardized in ARINC 664 [ARI02,ARI03]. This new communication standard represents a major step in the deployment of modular avionics architectures (integrated modular avionics: IMA ARINC 651 [ARI91] and 653 [ARI97]). However, the main problem lies in the indeterminism of the switched Ethernet and a network designer must prove that no frame will be lost by the AFDX (no switch queue overflow) and must evaluate the end-to-end transfer delay through the network (guaranteed upper bound and distribution of delays) according to given avionics applications traffic.

While other means of data communication have been, or could be, used in the context of avionics systems such as automotive field buses (CAN and FlexRay [CAN93,FCS04]) and real-time buses (TTP and real-time Ethernet [KG04,EI03,F05]), this chapter focuses solely on protocols that have already been standardized by the ARINC committee.

In Section 24.2 of this chapter, the main evolution of avionics systems and the standardization undertaken by ARINC are briefly reviewed. The classic avionics context is presented in Section 24.3 and integrated modular avionics concepts are outlined in Section 24.4. ARINC 629 protocols for controlling access to a multiplexed data bus are described in Section 24.5. Section 24.6 is devoted to the presentation of ARINC 664 avionics full-duplex Ethernet mechanisms. End-to-end delay analysis methods are summarized in Section 24.7.

24.2 Avionics Systems Evolution and ARINC Context

Avionics is the generic name given to the electronic systems installed in an aircraft. It includes, for example, the calculators and their software, the sensors and actuators, and all the communication links between these elements. Each function of an aircraft is implemented through a given avionics system: command controls, autopilot, navigation, information display, and cabin management. In the 1950s, avionics were very simple standalone systems in which each function was executed using a single calculator. Modern avionics began in the 1960s with the replacement of analog devices by their digital equivalents. Since then, the complexity of avionics has continually increased: More and more analog devices are becoming outdated and require new digital replacements, and new functions are continuously being added. These functions benefit from improvements in electronics and hardware execution and communication architectures. They can be answers to the new needs inherent in the evolution of civil aviation or simply better answers to existing problems, like the fly-by-wire command system. A few facts clearly illustrate the growth of the role of avionics in civilian aircraft: From 1983 (A310) to 1993 (A340), the number of onboard avionics systems increased by almost 50% (from 77 to 115), while total processing power was multiplied fourfold, from 60 to 250 Mips. Designing and manufacturing new civilian aircraft has led to an increase in the number of embedded systems and functions and consequently, an increase in communication needs [CC93].

Avionics systems are subject to volume and weight limitations and must also operate correctly even in severe conditions: heat, vibrations, electromagnetic interferences, etc. However, the main characteristics lie in the safety level they require. Following ARP 4754 [SAE96], systems are classified according to the effects of their failure: catastrophic, hazardous, major, minor, or no effect. For example, a system is classified as "catastrophic" if its failure can lead to the total loss of an aircraft. The segregation of critical functions leads to a distributed architecture without any centralized control: the systems cooperate through data exchange thanks to data buses. Typically, classic avionics uses pieces of equipment that can be seen as hardware and software black boxes, which are each responsible for one given function. The physical devices adopt a standardized module form, called line replaceable units (LRU). These modules incorporate memory and processing resources and provide a standardized application interface (API) to these resources. This standardization eases maintenance activity, since any piece of equipment can be easily replaced by another LRU, provided the appropriate code has been loaded. New concepts based on better sharing of execution and communication resources were proposed in the 1990s. IMA introduced the definition of a generic platform called "cabinet" composed of standard execution and transmission modules, called line replaceable modules (LRM); this will be presented in Section 24.3. The global IMA architecture is thus composed of modules housed in cabinets distributed throughout the aircraft, leading to new intracabinet and intercabinet communication needs.

Aeronautical Radio Incorporated (ARINC) provided leadership in developing specifications and standards for avionics equipment. Most of the proposed specifications and standards were developed and maintained by the Airlines Electronic Engineering Committee (AEEC) comprising members that represent airlines, governments, and ARINC. ARINC standardized the traditional ARINC 429 mono-emitter data bus [ARI01], which has been widely used in classic avionics architectures context. The ARINC 629 multiplexed data bus [ARI99] was designed with IMA in mind. IMA itself is composed of two

ARINC standards: ARINC 651 for the description of the modular hardware architecture [ARI91] and ARINC 653 for the modular software architecture [ARI97]. More recently, the ARINC 664 has provided an answer to the problem of increased communication by multiplexing huge amounts of communication data over a full-duplex switched Ethernet network [ARI02,ARI03]. It has become the reference communication technology in the context of civilian aircraft avionics since it provides a backbone network for the avionics platform.

24.3 Classic Avionics and ARINC 429

ARINC 429 is the earliest and most commonly used standard for civilian aircraft avionics and has been installed on the majority of Airbus and Boeing aircraft. It employs a unidirectional data bus standard known as the Mark 33 digital information transfer system (DITS). Messages are transmitted over two twisted pairs by the owner of the bus to other system units at a bit rate of either 12.5 or 100 kbps. The specification defines electrical transmission characteristics and protocol data unit features [ARI01].

The ARINC 429 protocol implements a mono-emitter broadcast bus with up to 20 receivers. Messages sent by the sole transmitter consist of a single 32 bits data word. The label field of the word defines the type of data that is contained in the rest of the word. ARINC 429 data words use five primary fields: parity, SSM, data, SDI, and label. As represented in Figure 24.1, the ARINC 429 data word is composed as follows: bit 32 is the parity bit; bits 31 and 30 contain the sign/status matrix (SSM) field that contains equipment conditions, operational modes, or validity of data content; bits 29 through 11 contain the data; bits 10 and 9 provide a source/destination identifier (SDI); and bits 8 through 1 contain a label identifying the data and associated parameters.

The label is an important part of the message. It is used to identify the transmitted data and to determine the data type. The emitter broadcasts the data (with a given periodicity) on the bus and does not know which equipment will receive a given instance of data. Each receiver filters the data according to the label of each data. The SDI is used to identify the receiver to which the data is destined. If a given label can be sent on different data buses, it can also be used to identify the source of the transmission (as units are assigned digital identification numbers called equipment id).

ARINC 429 is confronted with an increasing number of intercommunicating avionics functions. The (point to multi-point and unidirectional) characteristics of the ARINC 429 protocol mean that the avionics system must include an ARINC 429 bus for each communication path as depicted in Figure 24.2.

32	31	30	29 ... 11	10	9	8 ... 1
P	SSM		Data	SDI		Label

FIGURE 24.1 ARINC 429 data word.

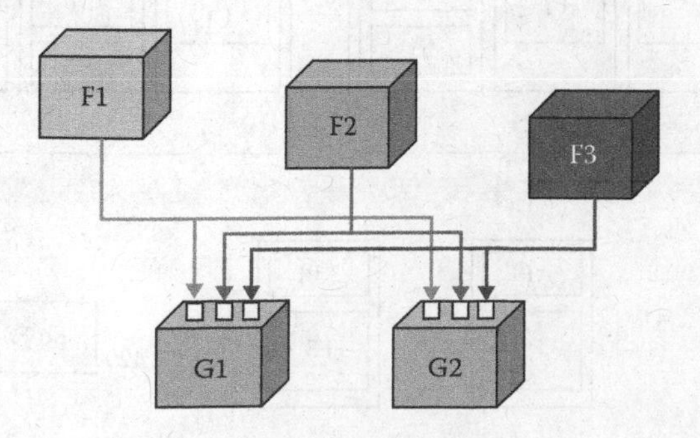

FIGURE 24.2 Classic avionics architecture.

Moreover, an ARINC 429 transmitter can transmit to a maximum of 20 receivers. While the ARINC 429 data bus provides high determinism, point-to-point wiring has become a major problem in systems composed of multiple emitting units.

24.4 Integrated Modular Avionics

The IMA concept has introduced the sharing of execution and communication resources. IMA hardware architecture, as described in the standard ARINC 651 [ARI91], is represented in Figure 24.3. Execution and communication resources are described as standard LRMs that are installed in common cabinets. Three types of modules have been designed: core modules for application execution, input–output modules for communications with non-IMA equipment (existing LRUs), and gateway modules for communicating among cabinets. The ARINC 659 back plane data bus standard has been used for communication between modules hosted in the same cabinet [ARI93]. Communication among cabinets needs new aircraft data buses such as ARINC 629 [ARI99].

ARINC 653 [ARI97] describes how many avionics subsystems can share the execution resources of an IMA architecture. In an ARINC 653 environment, an avionics subsystem is no longer implemented on a dedicated CPU or LRU but is viewed as a partition, to which is assigned a time window to execute its code on a shared core processing module (CPM). Each partition becomes a virtual CPU or LRU. A robust partitioning concept is introduced to guarantee isolation between subsystems running on the same execution computer. Spatial isolation is guaranteed by protecting the address space of each partition (for example, by a memory management unit). Temporal isolation is based on the static allocation of CPU time to each partition. Time partitioning guarantees that each user partition obtains a slice of time for execution as determined by the integrator as a worst-case execution time (easier to evaluate for systems that have mostly periodic processes). The main objective is to guarantee that an avionic subsystem running in a given partition will have no effect on other subsystems running in other partitions. Partitions communicate with each other by exchanging messages through communication ports. Two types of communication ports can be used: sampling ports (periodic data—only the last value of data is stored) and queuing ports (all the values of data are stored). A logical channel concept is used to link communicating ports (multicast scheme—1 emitter and *N* receivers) independently of the underlying data bus or network stack. Mechanisms are needed in order to schedule inputs–outputs according to the needs of each partition (for example, port sizes and depths must be determined on a worst case communication basis). In conclusion, the ARINC 653 interface between application software and the run-time

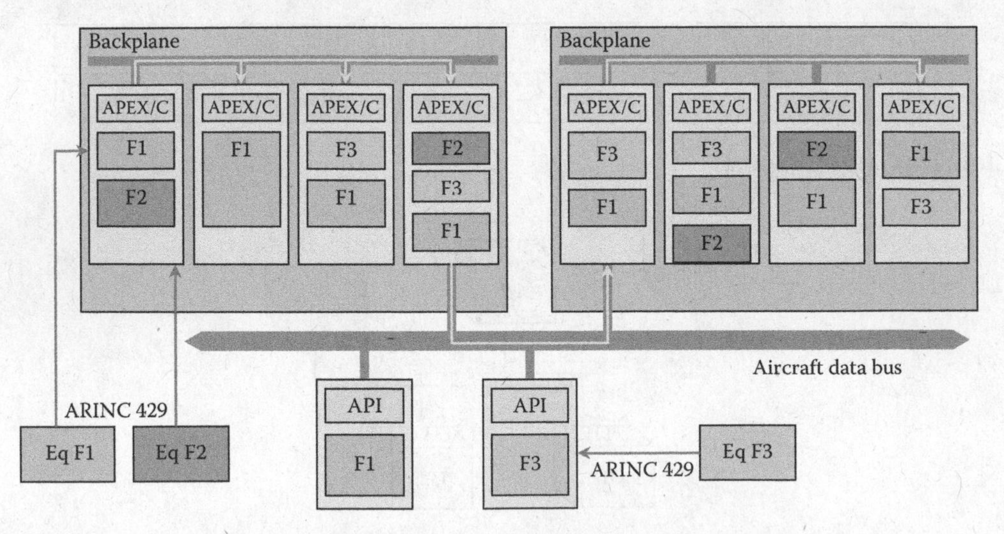

FIGURE 24.3 IMA avionics architecture.

executive called application executive (APEX) is an important step in developing avionics applications without knowing the characteristics of the target processing and communicating architecture.

The first examples of IMA implementation, such as the Boeing 777 airplane information management system (AIMS) show many potential benefits: adaptability (thanks to the modularity of the architecture, it is easy to configure the aircraft for a given mission), maintainability (the standardization of the modules simplifies the maintenance), cost reductions (easier maintenance and hardware evolution cut down costs), and a decrease in weight and volume (fewer devices are needed than with previous architectures). Considering these benefits, Airbus chose the IMA architecture for the A380, and Boeing reused the IMA concepts for the 787 aircraft. Since the architecture is independent of the chosen communication method used, we will present in following sections the ARINC 629 multiplexed data bus, which was chosen for the Boeing 777 AIMS and the ARINC 664 or AFDX, which was chosen as the basis of the IMA architecture of the A380.

24.5 ARINC 629 Multiplexed Data Bus

Development of the ARINC 629 [ARI99] multiplexed data bus began when avionics systems designers realized that the ARINC 429 single transmitter/multiple receiver concept could not cope with the increasing amount of intersystem data transfer required for evolving commercial aircraft. The primary advantages of a multi-transmitter data bus such as the ARINC 629 include the ability to move more data between LRUs at higher rates using fewer wires. Also they are generally more reliable and provide an architecture allowing the integration of complex systems. However, multitransmitter buses have been mainly used in the context of military aircraft. The MIL-STD-1553 data bus standard [MIL78] has been developed for military and space systems. It implements a bus architecture in which all the devices attached to the bus are capable of receiving and transmitting data. However, it uses a command/response (centralized) mechanism that does not satisfy the terminal independence needed for certification of commercial aircraft. Moreover, classic Ethernet architecture and its random CSMA-CD medium access control mechanism cannot guarantee periodic traffic. The digital autonomous terminal access communication (DATAC) project engendered the ARINC 629 [ARI99], an avionics industry digital communications standard, which was chosen as the primary digital communication system on the Boeing 777.

This standard allows the transmission of messages at a 2 Mbps serial data rate on twisted pair conductors. A message has variable length and is comprised of up to 31 word strings. Each word string has variable length and contains a 16 bit label word and up to 256 data words (each 16 bit word is actually transmitted with 3 additional bits of synchronization and a parity bit). Each LRU communicating on the ARINC 629 data bus contains a terminal controller, which implements the bus protocol access logic that determines when the terminal will transmit. The terminal listens to the bus and waits for series of quiet periods, one of unique duration before transmitting (CSMA like mechanism + TDMA like allocation by the definition of preassigned waiting times). Only one terminal is allowed to transmit at a time, and once a terminal has transmitted, it must wait before it transmits again to give time to all other terminals which want to transmit on the bus. The ARINC 629 standard supports two alternative data link level protocols, the basic protocol (BP) and the combined protocol (CP), which cannot coexist on the same bus due to fundamental differences.

24.5.1 Basic Protocol

The BP provides an equal priority access to the bus for each terminal, giving the opportunity to transmit periodic data. However, as shown in Figure 24.4, if the cycle is too short to transmit the data from all terminals, the bus automatically loses this periodic behavior in favor of an aperiodic mode where all messages are transmitted, without data loss.

For a given terminal i (from a set of n connected to the bus), TGi represents the unique time (terminal gap), during which a terminal must wait after bus activity before starting its own transmission. For all n terminals, TI represents a common time interval (transmit interval), separating two successive

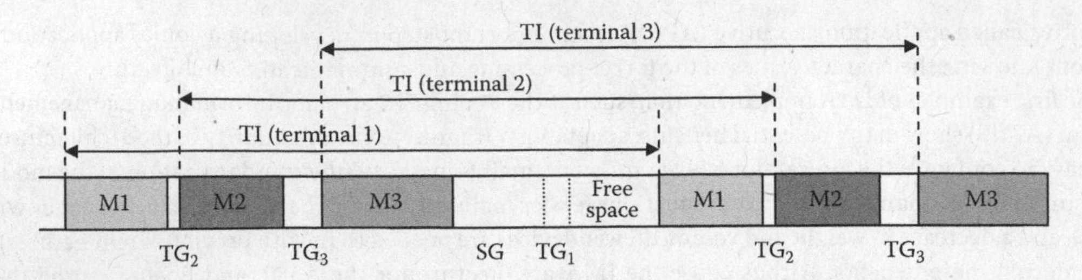

Basic protocol : periodic mode (when TI is sufficiently large)

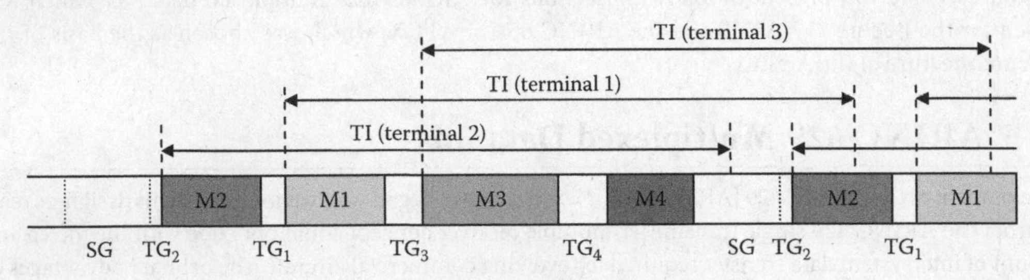

Basic protocol : aperiodic mode (when TI is too small)

FIGURE 24.4 ARINC 629 basic protocol.

transmissions of each terminal. Finally, the synchronization gap (SG) is a common quiet time, longer than any individual TG, which is used to synchronize all the terminals at the beginning of a new cycle.

These three timers are used at each terminal in the following manner:

- The TI timer starts immediately every time the terminal begins transmitting.
- The SG timer starts immediately every time the bus is sensed quiet. This timer may be reset before it has elapsed if any activity is detected on the bus. This timer does not affect periodic mode but in aperiodic mode, after it has elapsed we know that all the terminals have had transmit access to the bus (as SG timer is greater than each TG*i* timer) and so a new cycle can begin.
- A TG*i* timer will start after SG has elapsed. It also starts immediately every time the bus is sensed quiet. This timer may also be reset before it has elapsed if any bus activity is detected.

If all three timers have elapsed, the local terminal begins transmitting. As all TG*i* are different, only one terminal will be allowed to transmit after any quiet time. In aperiodic mode, the terminals are scheduled in ascending order of their TGs.

The assignment of timer values is done on a per-system basis and must address a trade-off between bus cycle time (all terminals transmit exactly once within a cycle) and terminal transmission frequency (all terminals must be allowed to transmit at a frequency suitable for data update rate requirements).

24.5.2 Combined Protocol

The CP has been proposed for combining periodic and aperiodic data transmission. As shown in Figure 24.5, three levels of transmission are handled: level one for periodic transmissions, level two for shorter (and higher priority) aperiodic transmissions, and level three for longer (and lower priority) aperiodic transmissions.

As with BP, each terminal has a unique preassigned TG. The transmit interval (TI) is only applicable during the first periodic transmission in each cycle; for all other terminals, a concatenation event (CE) forces all unelapsed TI timers to be canceled. This has the effect of compressing periodic transmissions at the start of each cycle (separated only by TG delays). Two types of SGs are defined: the periodic synchronization gap (PSG), which is used for cycle level synchronization; and the aperiodic synchronization gap (ASG), which is needed to handle transition in a cycle between levels 1 and (and between levels 2 and 3). In order to guarantee

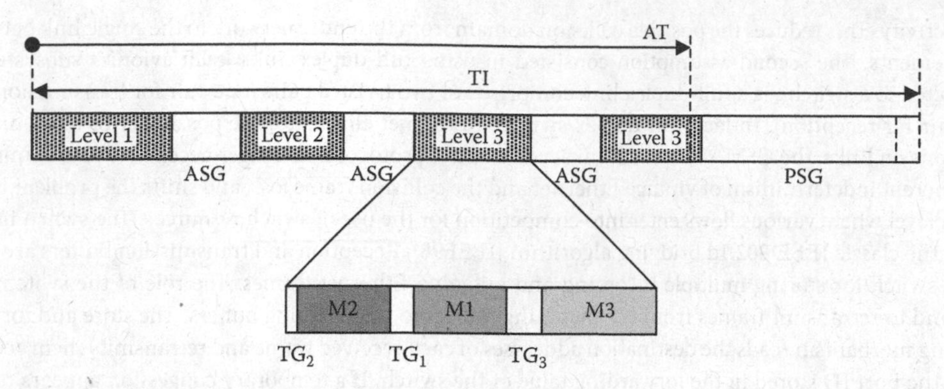

FIGURE 24.5 ARINC 629 combined protocol.

a fixed cycle duration (needed for level 1 periodic transmissions), an aperiodic time-out (AT) is used to avoid transmission of aperiodic data that could make the cycle overflow. Level 3 aperiodic messages are allowed to span multiple cycles (backlog messages have a higher priority than current cycle level three messages). Level 2 aperiodic messages must be transmitted in the current cycle otherwise they are lost; the transmit interval must be calculated so as to give enough space for both levels 1 and 2 messages.

The assignment of timer values is similar to BP mode: ASG is equivalent to SG (greater than each TG*i*) and PSG must be greater than ASG (the standard defines PSG as five times the selected ASG).

In conclusion, implementing an ARINC 629 data bus requires the development of a global avionics system communication scheme in order to guarantee deterministic transmission ordering for critical data. BP delivers true real-time behavior for systems requiring periodic and fixed length messages (periodic mode). Variable length periodic messages or aperiodic messages result in variable length cycles (aperiodic mode). CP protocol offers stable periodic response time for systems, which require some combination of periodic and aperiodic transmissions (with no interference of aperiodic messages on periodic messages). However, taking into account sporadic or aperiodic messages can lead to an under-utilization of the data bus. In fact, the main drawback of the ARINC 629 has been the cost of interfacing an efficient but complex avionics-specific communication protocol.

24.6 ARINC 664: Avionics Full-Duplex Ethernet

Avionics full-duplex Ethernet (AFDX) is part of the ARINC 664 standard and is based on classic switched Ethernet technology but introduces a virtual link (VL) paradigm, which is an important concept for characterizing the incoming traffic of the network.

24.6.1 Full-Duplex Switched Ethernet

Reusing Ethernet technology for avionics presents many advantages: high throughput offered to the connected units (100 Mbps compared to a few Mbps with the ARINC 629 standard), high connectivity given by the network structure, a mature industrial standard, and a significantly lower connection cost than with a proprietary or aeronautical specific protocol. Despite these advantages, Ethernet was not used for critical systems in previous aircraft because of its random physical medium access protocol: CSMA/CD [IEEE98]. An embedded network must have determinism properties such as the bounded transmission delay of any data. The CSMA/CD protocol fails to offer such a guarantee because of potential collisions on the physical medium during the transmission and because of random (binary exponential back-off [BEB]) retransmission algorithms.

To solve this problem, the first assumption was to adopt a switched Ethernet technology; all units are directly connected by a point-to-point link to an Ethernet switch since cascading switches offer the desired connectivity. This reduces the possible collision domain from the entire network to the single link between two elements. The second assumption consisted in using full-duplex links: each avionics subsystem is connected to a switch via a full-duplex link comprised of two twisted pairs (one pair for transmission and one pair for reception). In fact, full-duplex switched Ethernet eliminates the possibility of transmission collisions on links: the CSMA-CD medium access control protocol is no longer necessary. This eliminates the inherent indeterminism of vintage Ethernet and the collision frame loss, and shifts the problem to the switch level where various flows enter into competition for the use of switch resources. The switch implements the classic IEEE 802.1d bridging algorithm [IEEE98]: Reception and transmission buffers are used in the switch for storing multiple incoming and outgoing Ethernet frames. The role of the switch is to filter and to retransmit frames from the incoming buffers to the outgoing buffers. The store and forward bridging mechanism reads the destination addresses of each received frame and retransmits them according to the port ID stored in the forwarding table of the switch. If a temporary congestion appears on the output port of a switch, it can significantly increase end-to-end delays of frames and even lead to frame loss through buffer overflow. This is why dedicated mechanisms have been added to the classic full-duplex Ethernet in order to guarantee the determinism of an AFDX network.

24.6.2 The ARINC 664 Standard

The AFDX has been initiated by Airbus for the evolution of the A380 aircraft toward IMA as represented in Figure 24.6. AFDX concepts have been standardized as in ARINC 664 with the help of many avionics manufacturers: Airbus, Boeing, Rockwell Collins, Honeywell, etc. [ARI02,ARI03]. This standard adapts existing Ethernet standards, describes the global communication system of the aircraft, and focuses on the interconnection of domains with different safety levels. In particular, it explains how the critical avionics domain, responsible for aircraft control, can be connected to the open-world domain.

More precisely, the standard defines two kinds of embedded networks: compliant networks and profiled networks. Compliant networks conform exactly to existing standards: IEEE 802.3, 802.1d [IEEE98] and can be used in the lowest safety-level domains, such as the open-world domain. Profiled networks deviate from existing standards, when specific deviations are needed to achieve required levels of

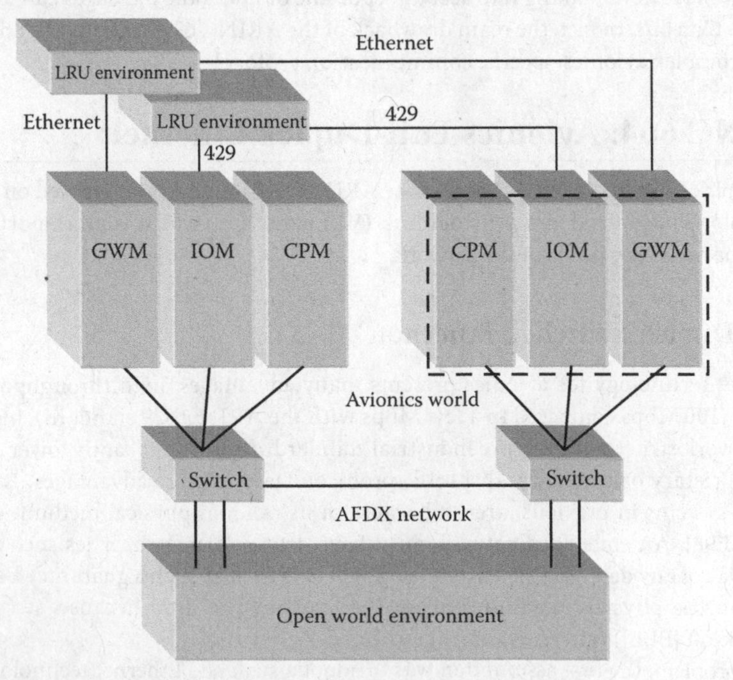

FIGURE 24.6 AFDX IMA avionics architecture.

performance and safety. Among these profiled networks is defined the subset of deterministic networks whose main characteristics can be summarized as follows:

- *Static configuration*: The entire network configuration must be static and precisely defined before any take-off. This includes the network topology, the number of connected units, the switch parameters, and switching tables. Algorithms such as ARP, GMRP, and Spanning-Tree must not be implemented.
- *Flow segregation and error confinement*: The main safety principle is that errors must be locally contained, which means that a local error must not be propagated in the network or cause the malfunction of another system. This led to the recommendation to implement a filtering function in the switches, whose role is to check all incoming frames and discard invalid ones including incorrect length, unknown sender or destination address, and corrupted frames (FCS check). Filtering operations prior to relaying the frames and differentiating the service offered to each flow are mandatory for robust segregation of flows.
- *Controlled traffic*: The main characteristic of deterministic networks is that they guarantee a given quality of service that is provable (bounded delay or jitter, frame loss probability by congestion, and availability rate). Each traffic source must assume a given traffic pattern for each flow (needed regulation is managed by traffic shaping mechanisms at end system level). The network has to provide mechanisms in order to check that no source will transmit more data than allowed (traffic policing mechanisms are needed at the switch level to discard frames if the source has transmitted too many frames for a given flow over a short period).

In conclusion, an ARINC 664 deterministic network is based upon the Ethernet frame definition and IEEE 802.1d switching protocol but includes specific mechanisms in order to guarantee the determinism of avionic communications. The main AFDX specific assumptions deal with the static definition of avionic flows that have to respect a bandwidth envelope (burst and rate) at each network ingress point.

24.6.3 Virtual Link Paradigm

The end-to-end traffic flow characterization is standardized by ARINC 664 in part 7 where the VL paradigm is presented [ARI03]. The idea behind this concept is to segregate the flows for safety reasons. Yet, each VL can also be seen as a virtual ARINC 429 bus dedicated to one emitting end system. This makes it possible to reuse existing avionics components or concepts, which decreases the transition time between ARINC 429 and ARINC 664 technologies. The definition of VLs is based on the concept of virtual communication channels. Thus, it is possible to statically define all flows (VL), which enter a network, as shown in Figure 24.7, which corresponds to an AFDX network test configuration provided by Airbus for industrial research.

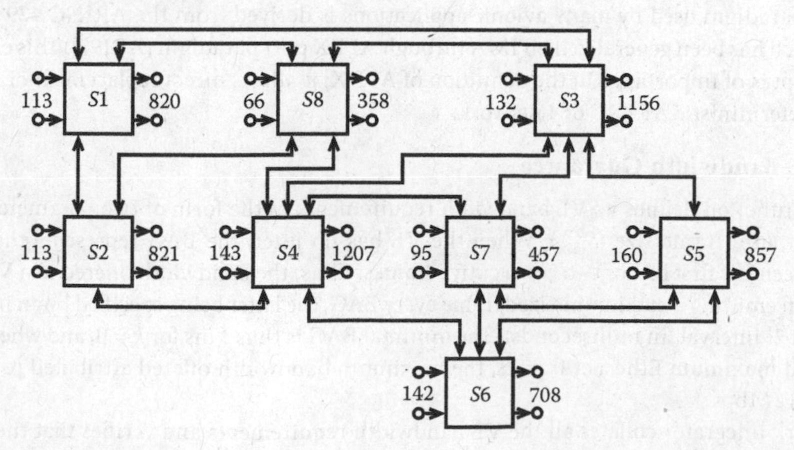

FIGURE 24.7 Example of AFDX architecture configuration.

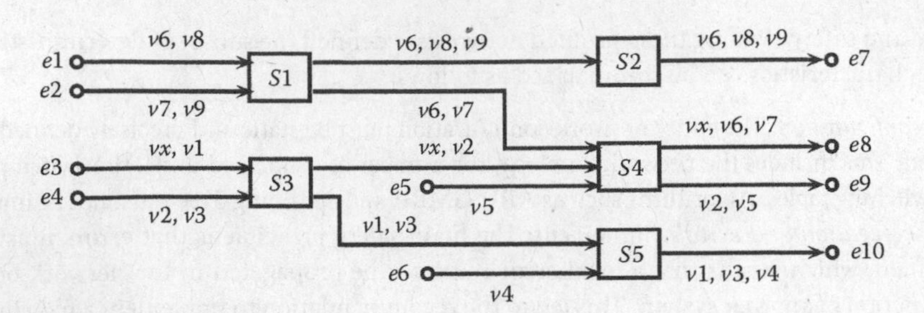

FIGURE 24.8 Example of AFDX VLs configuration.

The AFDX network architecture is composed of several interconnected switches. The inputs and outputs of the network are the end systems (the little circles in Figure 24.7). Each end system is connected to exactly one switch port and each switch port is connected to either an end system or another switch. The links are all full duplex. In Figure 24.7, the values on the end systems indicate the number of VL that are dispatched between the end systems and a given switch. Thus, the VL concept of virtual communication channels has the advantage of statically defining the flows, which enter the network, and associating some performance properties to each flow. Each VL can be statically mapped on the network of interconnected AFDX switches. Transmitting an Ethernet frame from one end system to another is based on a VL identifier, which is used for the deterministic routing of each VL (the switch forwarding tables are statically defined after allocation of all VL on the AFDX network architecture). Each VL defines a logical unidirectional connection from one source end system to one or more destination end systems. For example, Figure 24.8 illustrates different kinds of VL: *vx* is a unicast VL with path {*e*3–*S*3–*S*4–*e*8}, while *v*6 is a multicast VL with paths {*e*1–*S*1–*S*2–*e*7} and {*e*1–*S*1–*S*4–*e*8}.

A VL definition includes the bandwidth allocation gap (BAG) value, the minimum frame size (S_{min}), and the maximum frame size (S_{max}). The BAG is the minimum delay between two consecutive frames of the associated VL (which actually defines a VL as a sporadic flow). BAG and S_{max} values guarantee an allocated bandwidth for each VL. Moreover, a jitter value is associated to each VL to establish an upper bound on the maximum admissible jitter after multiplexing different regulated VL flows.

24.6.4 Virtual Link Properties

From the avionics systems designer's point of view, classic ARINC 429 buses have many interesting features. For example, the single-emitter assumption implies a dedicated link offered to the emitter, and thus guaranteed access to the bus, guaranteed bandwidth, and high determinism. Moreover, the communication paradigm used by many avionic applications is derived from the ARINC 429's properties behavior, which has been generalized on IMA through APEX port paradigm [ARI97]. This explains why the VL concept is of importance in the definition of AFDX; it allows direct replacement of ARINC 429 buses, on a deterministic ARINC 664 network.

24.6.4.1 VL Bandwidth Guarantee

Each avionic function defines its VL bandwidth requirements in the form of two parameters: the BAG and the maximum frame size (S_{max}). When the VL has no jitter, the BAG represents the minimum interval between the first bits of two consecutive frames. Thus, the bandwidth offered to a VL is the one obtained when emitting a maximum-sized frame every BAG, the latter being specified by an integer (k), to give (create) a 2^k interval (in milliseconds). The minimal BAG is thus 1 ms for $k = 0$, and when combined with standard maximum Ethernet frames, the maximum bandwidth offered attributed to a single VL can be up to 12 Mbps.

The network integrator collates all the VL bandwidth requirements and verifies that the sum of VL bandwidths on any physical link of the full-duplex switched Ethernet network does not exceed the

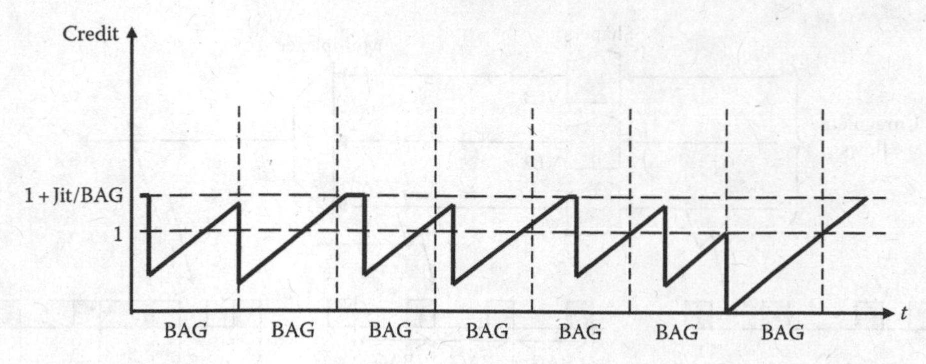

FIGURE 24.9 AFDX VLs shaping and policing.

100 Mbps capacity. Moreover, the network integrator is responsible for VL placement: A multipath VL (a tree originating from the emitter) has to be statically allocated on the network's physical topology and the path can be chosen from among several possibilities.

One must note that guaranteeing bandwidth on the network requires that all VLs remain compliant with their specified BAG and S_{max} parameters; otherwise, the sum of some VL bandwidths could possibly exceed the 100 Mbps capacity. For this reason, every VL is shaped inside the emitter end system, and all incoming VLs are also controlled by policers at the switch level; frames exceeding the VL's contract are deleted by the switches (leaky bucket algorithm) as represented in Figure 24.9.

24.6.4.2 VL Latency Guarantee

The communication latency of an ARINC 429 data bus is very easy to measure. It is simply the time required to emit data plus a negligible propagation time; it is absolutely deterministic.

The picture is completely different for a VL on an AFDX network. Obviously, the latency of a frame cannot be predicted as it depends on whether or not the frame will be delayed by other frames sharing a section of its path. Thus, a strong form of determinism cannot be obtained; a weaker form of determinism is proposed, it is based on VL end-to-end latency upper bounding methods. VLs competing for a given switch output port generate contentions in the output port queue and can be delayed depending on the number of confluent VLs and their characteristics (maximum frame size for example).

Of course, end-to-end latencies have to be compared to the real-time constraints of avionics communications, but in reality, these constraints are very different from one avionics function to another. Thus, in the early prototype specifications, a maximum latency of 1 ms per switch crossing was integrated into the switch specification. When four or five switches are crossed, the maximum obtained latency is comparable with the highest "freshness" required for some critical data. At 100 Mbps, 1 ms of continuous emission of minimum-sized frames represents nearly 150 frames exiting from an output port's queue. Of course, the physical size of the switches' queues is the actual constraint: overflowing this queue capacity necessarily implies lost frames.

24.6.4.3 VL Jitter Issues

Jitter is the third parameter of a VL, after the BAG and S_{max}. As the AFDX definition in the ARINC 664 standard states, jitter is considered null at the output of the end system shapers applied to each VL. Since the VLs are multiplexed in the end system just after having been shaped, as represented in Figure 24.10, jitter is non-null at the output of the multiplexer and is limited by two bounds: the first one comes from the calculation of the multiplexer's latency and the second is an absolute value of 500 μs.

This value has been chosen as half the minimal BAG in order to avoid burst occurrences where a frame pertaining to some VLs would be overtaken by a following frame of the same VL. Of course, avoiding this problem in end systems does not prevent it from occurring later down the network.

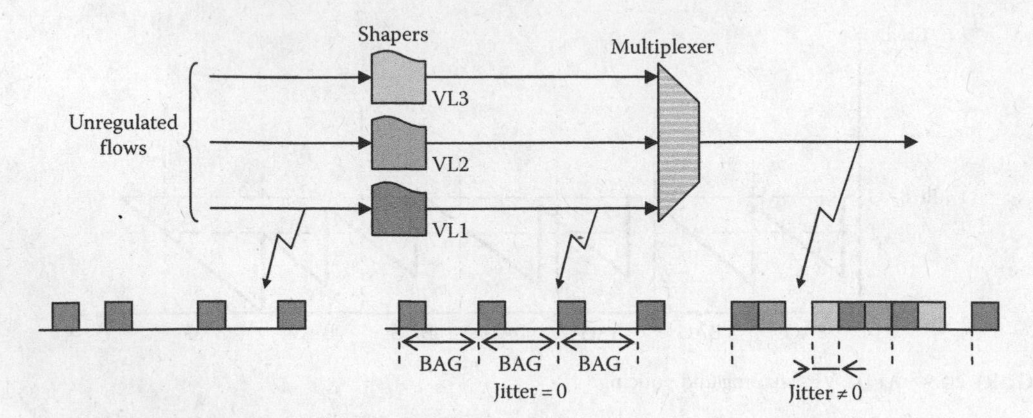

FIGURE 24.10 AFDX VLs scheduling and jitter.

24.6.5 Network Redundancy for Safety and Fault Tolerance

As on any Ethernet network, bit errors may occur. In this case, a frame will be deleted thanks to detection by an invalid Ethernet frame check sequence (FCS), and it will not arrive at the destination. Likewise, faulty equipment (switch or end-system) may send rubbish on the network or simply exceed its VL contracts. In this case, the next switch on the path will reject the extra communication (thanks to policing mechanisms in the switch input ports), and prevent the faulty equipment from polluting the rest of the network. Again some data will be lost in the process.

In order to provide fault tolerance for both types of errors, the AFDX is actually doubled as depicted in Figure 24.11. End systems have two Ethernet network interfaces and the whole topology of switches and links is doubled, leading to a so called "red" network and a so called "blue" network. An additional parameter in the VL definition states/defines whether the VL needs redundancy, i.e., whether frames of this VL are sent on both the red and the blue networks. Nearly all VLs are duplicated on the two networks. Of course, only one frame is returned by the protocol stack on a receiver: the first valid frame to arrive is returned and the second one (if it ever arrives) is deleted. However, this scheme requires proper matching of red and blue frames. For this reason, the standard Ethernet frame format has been modified to include an 8 bits sequence number (protected by the FCS).

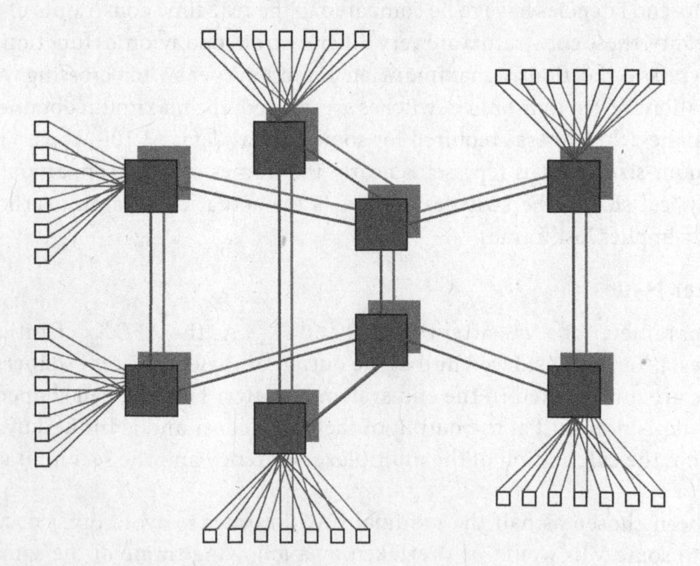

FIGURE 24.11 AFDX redundant networks.

The impact of this redundancy mechanism on guaranteed latencies is limited. Of course, average latency for a single frame improves since Latency = Min ($Latency_{red}$, $Latency_{blue}$), but as we deal with latency bounds and since latency bounds which are mostly the same on both the red and blue networks, the final latency bound is unaffected by the redundant network.

24.7 AFDX End-to-End Delay Analysis

As presented before, the main problem is to guarantee that frames exchanged between end systems satisfy the temporal constraints of transported data. Therefore, the end-to-end delay of each path of each VL has to be studied. The end-to-end delay of each path of VL includes the following characteristics:

- The lower bound for the end-to-end delay corresponds to the case where there is no waiting service time for the frame in queues. This can be easily computed by adding transmission times over the physical links and switching delays.
- The upper bound for the end-to-end delay corresponds to the longest aggregate waiting service time for the frame in queues. This upper bound is mandatory in avionics context for certification reasons.

Different approaches have been proposed in order to analyze networks flow's end-to-end delay and jitter [ACOR99,C98,CSC01,CSEF06,IETF89,IETF98,JNTW02,SCP95,TC94].

Deterministic network calculus (or worst-case network calculus) has been used, for certification reasons, to compute AFDX upper bounds [C91,C95]. Network calculus theory uses envelopes of arrival curves and computes a worst-case scenario on each node visited by a flow, taking into account maximum possible jitter introduced by previously visited nodes [LB98,LBT01]. This approach is obviously pessimistic as it can lead to impossible scenarios, but it has been useful to prove the determinism of an AFDX network as it computes a guaranteed upper bound of end-to-end transmission delay for each VL on the network. This worst-case communication delay analysis also gives intermediate information on latency time in switch output ports, which permits the scaling of the switch memory buffers and avoiding buffer overflows (i.e., frame loss). Moreover, such an approach has been used for the optimization of a given AFDX configuration [FFG06]. In conclusion, a certifiable network (with no switch buffer overflow and bounded latency configuration) often underuses AFDX network capabilities.

Nevertheless, the pessimism of the network calculus approach cannot be evaluated by simply comparing the end-to-end delay measured on a real AFDX configuration with the computed upper bound because of the fact that rare events are difficult to observe on a real configuration in a reasonable time. In order to better understand the real behavior of an AFDX network, a simulation approach has also been proposed. A distribution of observed end-to-end delay on a simulated AFDX network can thereby be obtained for each flow [CSF06]. Stochastic network calculus as proposed in [H63,VB01,VB02] might also help to obtain a probabilistic end-to-end delay analysis [SRF09].

The model-checking approach might help to determine an exact worst-case end-to-end delay and the corresponding scenario since it explores all possible states of the network. Yet up to now model checking has not been able to be used on a real-size industrial network configuration [CSEF06]. In order to better evaluate the pessimism of the end-to-end delay upper bound obtained by deterministic network calculus, the recent trajectory approach seems to be a promising method [MM06]. This approach is based on the analysis of the worst-case scenario experienced by a packet on its trajectory and not on any visited node.

24.8 Conclusion

The traditional mono-emitter ARINC 429 data bus has played a major role in the architecture of classic civilian avionics, but it can no longer cope with the communication needs of new avionics architectures based on IMA. A first solution was to use a multiplexed data bus such as the ARINC 629 data bus, but the ARINC 664 network based on Ethernet full-duplex switching is now becoming the avionics reference communication technology. Its ability to connect avionics world and open-world onboard a civilian aircraft is one of its major advantages. As certified AFDX networks are often underused for avionics applications, the next step will be to use the AFDX network for other applications provided this has absolutely no consequences on critical avionics applications: Quality of service (QoS)-aware AFDX switches could be proposed for deterministic ARINC 664 networks. Such a network could be seen as a federator network, which interconnects, via specific gateways, other communication networks or buses (classic Ethernet, field buses, etc.). In the near future, avionics architectures will have to handle heterogeneous communication flows and heterogeneous communication means.

Abbreviations

AEEC	Airlines Electronic Engineering Committee
AFDX	Avionics Full-DupleX Switched Ethernet
AIMS	Airplane Information Management System
APEX	APplication EXecutive
API	Application Programming Interface
ARINC	Aeronautical Radio INCorporated
ARINC 429	Mark 33 Digital Information Transfer System
ARINC 629	Multi-Transmitter Data Bus
ARINC 651	Design Guidance for Integrated Modular Avionics
ARINC 653	Avionics Application Software Standard Interface
ARINC 659	Backplane Data Bus
ARINC 664	Aircraft Data Network
ARP	Address Resolution Protocol
ARP 4754	Certification Considerations for Highly Integrated or Complex Aircraft Systems
ASG	Aperiodic Synchronization Gap
AT	Aperiodic Timeout
BAG	Bandwidth Access Gap
BEB	Binary Exponential Back Off
BP	Basic Protocol
CE	Concatenation Event
CP	Combined Protocol
CPM	Core Processing Module
CPU	Central Processing Unit
CSMA/CA	Carrier Sense Medium Access/Collision Avoidance
CSMA/CD	Carrier Sense Medium Access/Collision Detection
DATAC	Digital Autonomous Terminal Access Communication
DITS	Digital Information Transfer System
FCS	Frame Check Sequence
GARP	Generic Attribute Registration Protocol
GMRP	Garp Multicast Registration Protocol
IEEE	Institute of Electrical and Electronics Engineers
IEEE 802	Local and Metropolitan Area Networks
IEEE 802.1d	Media Access Control Level Bridging
IEEE 802.3	CSMA/CD Access Method
IMA	Integrated Modular Avionics
LRM	Line Replaceable Module
LRU	Line Replaceable Unit
MIL-STD-1553	Aircraft Internal Time Division Command/Response Multiplex Data Bus

MC	Model Checking
NC	Network Calculus
PSG	Periodic Synchronization Gap
SAE	Society of Automotive Engineers
SDI	Source/Destination Identifier
SG	Synchronization Gap
SSM	Sign/Status Matrix
TDMA	Time Division Multiple Access
TG	Terminal Gap
TI	Transmit Interval
VL	Virtual Link

References

[ACOR99] R. Agrawal, R. L. Cruz, C. Okino, and R. Rajan, Performance bounds for flow control protocols, *IEEE Transactions on Networking*, 7(3), 310–323, June 1999.

[ARI01] ARINC 429 Aeronautical Radio Inc., ARINC specification 429-ALL: Mark 33 digital information transfer system (DITS), Parts 1, 2, 3, Annapolis, MD, 2001.

[ARI02] ARINC 664, Aircraft Data Network, Part 1: Systems Concepts and Overview, 2002; ARINC 664, Aircraft Data Network, Part 2: Ethernet Physical and Data Link Layer Specification, Annapolis, MD, 2002.

[ARI03] ARINC 664, Aircraft Data Network, Part 7: Deterministic Networks, Annapolis, MD, 2003.

[ARI91] ARINC 651 Aeronautical Radio Inc., ARINC specification 651: Design guidance for integrated modular avionics, Annapolis, MD, 1991.

[ARI93] ARINC 659 Aeronautical Radio Inc., ARINC project paper 659: Backplane data bus, Annapolis, MD, 1993.

[ARI97] ARINC 653 Aeronautical Radio Inc., ARINC project 653: Avionics application software standard interface, Annapolis, MD, 1997.

[ARI99] ARINC 629 Aeronautical Radio Inc., ARINC specification 629: Multi-transmitter data bus, Part 1-Technical description, Annapolis, MD, 1999.

[C91] R. Cruz, A calculus for network delay, Part I and II, *IEEE Transactions on Information Theory*, 37(1), 114–131, January 1991.

[C95] R. L. Cruz, Quality of service guarantees in virtual circuit switched networks, *IEEE Journal of Selected Areas in Communication*, special issue on *Advances in the Fundamentals of Networking*, 13(6), August 1995.

[C98] C. S. Chang, On deterministic traffic regulation and service guarantee: A systematic approach by filtering, *IEEE Transactions on Information Theory*, 44, 913–931, May 1998.

[CAN93] CAN Standard Road vehicles Controller Area Network (CAN), ISO Standard 11898, International Organization for Standardization, Geneva, Switzerland, 1993.

[CC93] P. Chanet and V. Cassigneul, How to control the increase in the complexity of civil aircraft on-board systems?, in *AGARD Meeting on Aerospace Software Engineering for Advanced Systems Architectures*, Paris, France, May 1993.

[CSC01] C. Chang, W. Song, and Y. Chiu, On the performance of multiplexing independent regulated inputs, in *Proceedings of Sigmetrics 2001*, Boston, MA, May 2001.

[CSEF06] H. Charara, J. L. Scharbarg, J. Ermont, and C. Fraboul, Methods for bounding end-to-end delays on an AFDX network, in *Proceedings of the 18th ECRTS*, Dresden, Germany, July 2006.

[CSF06] H. Charara, J. L. Scharbarg, and C. Fraboul, Focusing simulation for end-to-end delays analysis on a switched Ethernet, in *Proceedings of the WiP Session of RTSS*, Rio de Janeiro, Brazil, December 2006.

[EI03] W. Elmenreich and R. Ipp, Introduction to TTP/C and TTP/A, in *Workshop on Time-Triggered and Real-Time Communication Systems*, Manno, Switzerland, December 2003.

[F05] M. Felser, Real-time Ethernet—Industry prospective, *Proceedings of IEEE*, 93(6), 1118–1129, June 2005.

[FCS04] FlexRay Communications System—Protocol Specification, Version 2.0, 2004. Available at http://www.flexray.com

[FFG06] F. Frances, C. Fraboul, and J. Grieu, Using network calculus to optimize the AFDX network, in *Proceedings of ERTS*, Toulouse, France, January 25–27, 2006.

[H63] W. Hoeffding, Probability inequalities for sum of bounded random variables, *American Statistical Association Journal*, 58, 13–130, March 1963.

[IEEE98] IEEE 802.1d, 1998 Edition: Local and Metropolitan Area Networks: Media Access Control Level Bridging; IEEE 802.1p: 1998 Edition: LAN Layer 2 QoS/CoS Protocol for Traffic Prioritization; IEEE 802.3, 1998 Edition: Information technology—Telecommunications and information exchange between systems—Local and metropolitan area networks—CSMA/CD access method and physical layer specifications, IEEE publication office Los Alamitos, CA 1998.

[IETF89] IETF Internet Engineering Task Force, RFC 1122: Requirements for Internet Hosts: Communication Layers—Required Internet Standard, Fermont, CA 1989.

[IETF98] S. Blake, D. Black, M. Carlson, E. Davies, Z. Wang, and W. Weiss, An architecture for differentiated services, Internet Draft, IETF Diffserv Working Group, August 1998.

[JNTW02] J. Jasperneite, P. Neumann, M. Theis, and K. Watson, Deterministic real-time communication with switched Ethernet, in *Proceedings of the fourth IEEE International Workshop on Factory Communication Systems*, pp. 11–18, Västeras, Sweden, August 2002, IEEE Press.

[KG04] H. Kopetz and G. Grunsteidl, TTP—A protocol for fault-tolerant real-time systems, *Computer*, 27(1), 14–23, January 1994.

[LB98] J.-Y. Le Boudec, Application of network calculus to guaranteed service networks, *IEEE Transactions on Information Theory*, 44, 1087–1096, May 1998.

[LBT01] J.-Y. Le Boudec and P. Thiran, *Network Calculus: A Theory of Deterministic Queuing Systems for the Internet*, Vol. 2050, *Lecture Notes in Computer Science*, Springer-Verlag, New York, 2001. ISBN: 3-540-42184-X.

[MIL78] MIL-STD-1553B Specification: Aircraft Internal Time Division Command/Response Multiplex Data Bus, U.S. Department of Defense, Aeronautical Systems Division, Washington, DC, 1978.

[MIL78] MIL-STD-1553B Specification: Aircraft Internal Time Division, Command/Response Multiplex Data Bus, U.S. Department of Defense, Aeronautical Systems Division, Washington DC, 1978.

[MM06] S. Martin and P. Minet, Schedulability analysis of flows scheduled with FIFO: Application to the expedited forwarding class, *Parallel and Distributed Processing Symposium, 2006. IPDPS 2006. 20th International*, Miami, FL, 8 pp., April 2006.

[SAE96] SAE ARP4754, Certification considerations for highly-integrated or complex aircraft systems, SAE, Warrendale, PA, November 1996.

[SCP95] H. Sariowan, R. L. Cruz, and G. C. Polyzos, Scheduling for quality of service guarantees via service curves, in *Proceedings of the International Conference on Computer Communications and Networks (ICCCN) 1995*, Las Vegas, NV, pp. 512–520, September 20–23, 1995.

[SRF09] J.-L. Scharbarg, F. Ridouard, and C. Fraboul, A probabilistic analysis of end-to-end delays on an AFDX avionic network, *IEEE Transactions on Industrial Informatics*, 5(1), 38–49, 2009.

[TC94] K. Tindell and J. Clark, Holistic schedulability analysis for distributed hard real-time systems, *Microprocessing and Microprogramming*, 40(2–3), 117–134, 1994.

[VB01] M. Vojnović and J. Y. Le Boudec, Bounds for independent regulated inputs multiplexed in a service curve element, in *Proceedings of Globecom 2001*, San Antonio, TX, November 2001.

[VB02] M. Vojnović and J. Y. Le Boudec, Stochastic analysis of some expedited forwarding networks, in *Proceedings of INFOCOM*, New York, June 2002.

25

Process Automation

25.1 Introduction....25-1
25.2 Structures and Models of Batch Manufacturing Systems....25-3
Process Model • Physical Model • Procedural Control Model • Equipment Entities • Recipes • Classification of Process Cells • Tasks and Functions of a Batch Management and Operation System • Integration of Batch Management and Operation System with Other Information Systems
25.3 Currently Applied Communication Systems....25-8
25.4 Upcoming Requirements of Distributed Process Automation....25-8
25.5 Industrial Ethernet as the "Silver Bullet" for Future Process Automation Communication Needs....25-10
References....25-11

Alois Zoitl
Vienna University of Technology

Wilfried Lepuschitz
Vienna University of Technology

25.1 Introduction

The expression "process automation" confronts us with two meaningful words: process and automation. Usually the following definition is presented in the technical literature, e.g., [POL94], for a process:

> A process represents the entirety of consecutive interacting activities within a system to transform, store or transport material, energy or information.

This definition covers not only processes encountered in technical domains but also nontechnical ones such as business processes. Therefore, the technical literature provides us with a definition for a technical process as follows:

> A technical process is a process, whose physical values are measured and influenced by technical means.

The word automation is derived from the ancient Greek word *automatos* meaning "self moving, self thinking." Automation implicates the usage of artificial instruments to enable the automatic execution of a process. With respect to the technical domain, it means equipping an installation with certain machines and instruments to ensure its automatic job execution.

Process automation therefore refers to the automation of processes, including also nontechnical ones. In the case of technical processes, it involves three coupled kind of systems:

- A technical system (a technical product or installation) with an executed technical process
- A data processing and communication system with an executed information process
- Operator personnel that guide and observe the job execution and react in the case of exceptions or system errors

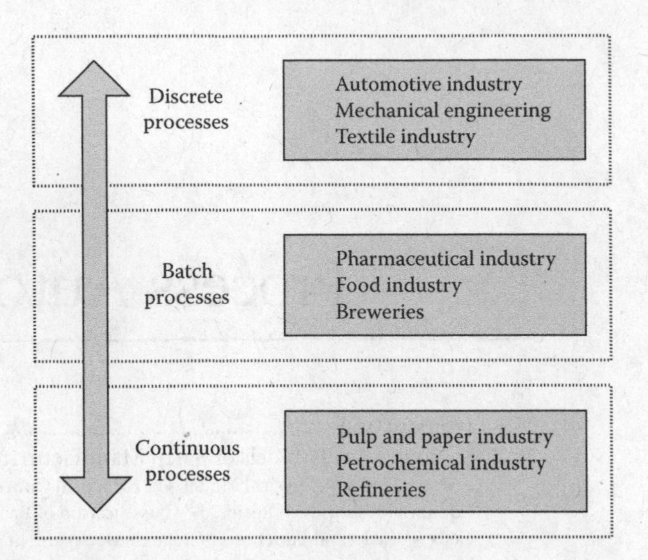

FIGURE 25.1 Types of processes and application domain examples.

Different types of processes can be identified and their application domains are depicted in Figure 25.1. The process types are characterized as follows:

- Discrete processes contain data of fixed points in time and space to follow determined time schedules and determined target locations. Discrete processes can therefore be regarded as nonstationary processes. They usually appear at the production of individual work pieces with a defined shape and determined measurements. Common activities involve the treatment of surfaces, shaping of work pieces, and the assembly of parts to create more complex products such as the engine of a car.
- Continuous processes remain in a constant operational state without down times due to a continuous supply of material and energy. Simultaneously, these processes produce a continuous output. All characteristic values (such as pressure, temperature, or concentration) along the flow path within the process equipment are locally different but temporally constant. Therefore, a continuous process runs stationary. This type of processes usually appears at the production of shapeless materials such as gas, liquid, or powder. It aims at the alteration of the material's nature, structure, property, or chemical composition. Continuous processes are found, for instance, in a petrochemical refinery.
- Batch processes are neither discrete nor continuous. Nevertheless they share similarities with both other process types. On the one hand, batch processes represent a subset treatment of goods normally treated in continuous processes but moreover also follow determined time schedules similar to discrete processes. Regarding an ideal material mixture, all characteristic values within the process equipment are locally equal but change during the dwell period. Hence, a batch process runs nonstationary. The name is derived from the term "batch," which represents the amount of material located in the reactor vessel that is treated as a whole within in the process. Batch processes are found, for instance, in the food and beverage industry as well as the pharmaceutical industry.

As already stated, the term process automation can be used widely for a broad range of domains. Nevertheless, it is commonly deemed to be closely connected with technical domains and more specific often with those domains that treat the production of shapeless materials. In fact, the German word *Verfahrenstechnik*, which represents only one possible translation for process engineering, is more specific and indeed only related to technical engineering.

For our purposes, we adopt the common understanding of process automation in the context of material transforming processes of shapeless materials. Furthermore, our focus lies on batch processes, as they combine attributes of both discrete and continuous processes and are more complex to handle during start-up as well as in operation. Production plants with batch processes need to be more flexible and are often connected to facilities with applied discrete processes such as bottling plants. Nevertheless, the shown communication system architectures and requirements especially from the continuous parts of batch processes can be applied also to continuous processes.

25.2 Structures and Models of Batch Manufacturing Systems

International standards comprise commonly agreed reference points and specifications and therefore are of vital importance for companies to be competitive. With respect to the automation of batch manufacturing plants, the standard ISA/ANSI S88 Batch Control and its counterpart IEC 61512 Batch Control provide a standardized terminology and a consistent set of methods. Both standards are similar and separated into the following four parts:

- ISA/ANSI S88.01 (IEC 61512-1) Batch Control, Part 1: Models and Terminology
- ISA/ANSI S88.02 (IEC 61512-2) Batch Control, Part 2: Data Structures and Guidelines for Languages
- ISA/ANSI S88.03 (IEC 61512-3) Batch Control, Part 3: General and Site Recipe Models
- ISA/ANSI S88.03 (IEC 61512-4) Batch Control, Part 4: Batch Production Records

Especially, Part 1 of this standard, Models and Terminology [IEC97], assists in determining the requirements for communications systems by providing reference models for batch control as applied in the process industries. In general, the utilization of this standard shall lead to

- A shortening of the setup time of production facilities to manufacture new products
- Easier development of tools for implementing batch control
- Enabling the users to better determine their requirements
- Easier development of recipes without the services of a control systems engineer
- A cost reduction of automating batch processes
- A decrease of life-cycle engineering efforts

According to the standard, the described models may be diminished or extended as long as the consistency of each model remains assured. Currently employed batch control systems are based on these models, allowing a certain grade of interoperability between these products. ISA/ANSI S88 introduces a set of structural models that deal with processes, physical equipment, and control software. Figure 25.2 shows the correlations between the procedural control model, the physical model, and the process model.

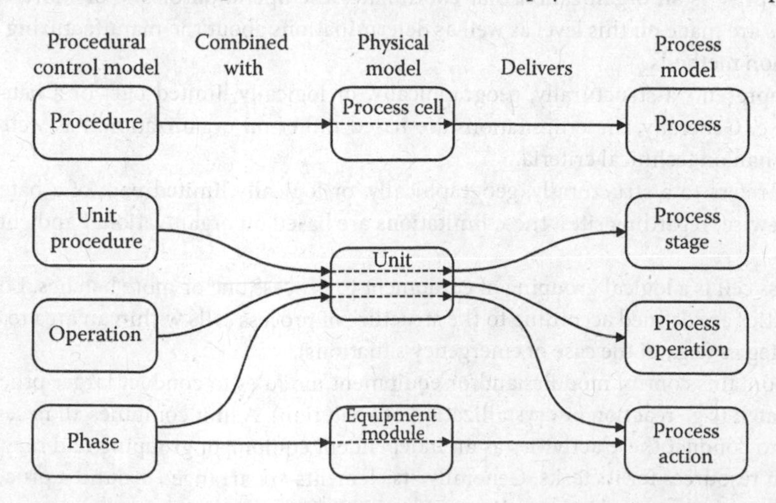

FIGURE 25.2 Procedural control mapped onto equipment to achieve process functionality. (Based on IEC TC65, IEC 61512-1: Batch Control—Part 1: Models and terminology. International Electrotechnical Commission, 1997.)

25.2.1 Process Model

A batch process can be segmented hierarchically according to the process model of the standard into process, process stage, process operation, and process action.

- A process represents the sequence of chemical, physical, or biological activities for the transformation, transport, or storage of material or energy. IEC 61512 presents the production of polyvinyl chloride (PVC) as a batch process example.
- A process consists of one or more process stages that usually run independently from each other in a serial or parallel arbitration to conduct a scheduled sequence of transformations of the processed material. Regarding the PVC process, typical process stages could be the polymerization of vinyl chloride monomer to polyvinyl chloride or the drying of PVC powder.
- A process stage contains a number of process operations. Each process operation refers to a larger process activity, which in fact means usually one chemical or physical transformation of the processed material. Typical process operations of the PVC process could be the preparation of the reactor or the reaction itself.
- Process actions represent smaller process activities that are combined to a process operation. The process operation "reaction" could consist of the following process actions:
  1. Fill a certain amount of catalyst into the reactor.
  2. Fill a certain amount of vinyl chloride monomer into the reactor.
  3. Heat up the reactor to a temperature between 55°C and 60°C.
  4. Hold the reactor temperature between 55°C and 60°C until the reactor pressure drops.

25.2.2 Physical Model

The physical model abstracts the physical organization of a batch manufacturing company into a hierarchical structure. Usually, groupings of lower levels are merged to form a higher level but ISA/ANSI S88 treats the upper three levels—enterprise, site, and area—of the physical model only marginally since these levels are connected closer to entrepreneurial than technical regards. The lower four levels—process cell, unit, equipment module, and control module—represent delimited groupings of equipment. Combining equipment into functional groupings eases operational tasks since each of these groupings can be treated as a larger single device.

- An enterprise is an organization that coordinates the operation of one or more sites. Product decisions are made on this level as well as determinations about the manufacturing location and production methods.
- A site represents a structurally, geographically, or logically limited part of a batch producing enterprise. Generally, these limitations are based rather on organizational as well as entrepreneurial than on technical criteria.
- An area refers to a structurally, geographically, or logically limited part of a batch producing site. Likewise, regarding sites, these limitations are based on organizational and entrepreneurial criteria.
- A process cell is a logical grouping of equipment to process one or more batches. Logical control possibilities are defined according to the structure of process cells within an area to develop control strategies (e.g., in the case of emergency situations).
- A unit contains control modules and/or equipment modules to conduct larger process activities of one batch (e.g., reaction or crystallization of a dilution). A unit combines all necessary process devices to conduct these activities as an independent equipment grouping and may allocate also common resources for its tasks. Generally, its elements are arranged around a processing installation (e.g., a reactor tank). According to ISA/ANSI S88, a unit processes not more than a single batch or one part of it at any point in time.

- An equipment module refers to a functional aggregation of devices, which is generally arranged around a smaller processing installation (e.g., a filter) to conduct smaller process activities (e.g., dosing of material). It may either be part of a unit or act as a common resource if defined as an independent device.
- A control module represents the lowest level of device aggregation and is capable of performing basic control activities. Typically, it contains sensors, actuators, and other control modules to form a functional unit that can be operated as a single entity (for example, a closed-loop control device that consists of a transmitter, a controller, and a control valve).

25.2.3 Procedural Control Model

The procedural control model consists of the four levels procedure, unit procedure, operation, and phase.

- A procedure refers to a strategy to conduct a process (e.g., the production of a batch) and is defined by a sequence of unit procedures. As an equipment procedure, it is part of the equipment control, which enables an entity (i.e., a technical device) to be controllable.
- A unit procedure represents a strategy to execute a process within a unit and contains a sequence of operations as well as start, control, and organization algorithms. These operations cannot be distributed over several units and only one operation may be active within its unit during any point in time. Nevertheless, several unit procedures of one procedure can be executed simultaneously if distributed over several units—one unit procedure per unit. As an equipment unit procedure, it is part of the equipment control.
- An operation is an independent process activity and consists of a sequence of phases as well as start, control, and organization algorithms. Typically, an operation refers to one chemical or physical transformation. As an equipment operation, it is part of the equipment control.
- A phase represents the lowest level of procedural elements within the procedural control model. It refers to a small process-oriented task and can be segmented into steps and transitions as in a sequential function chart. The task of a phase is the execution of a process-oriented action, whereas its steps focus on the technical specifications of the used equipment. For example, a step could be the activation or deactivation of a closed-loop control. A phase that is part of an equipment control is called an equipment phase.

25.2.4 Equipment Entities

Equipment entities are composed by combining the lower four levels of the physical model with the elements of the procedural control model to provide the necessary functionality to produce a batch. Based on this context, the terms process cell, unit, equipment module, and control module still refer to the physical entities but also include the corresponding equipment control.

The correlations between the procedural control model, the physical model, and the process model are depicted in Figure 25.2. Process functionality can be achieved by mapping the procedural control onto the physical equipment. Though all models resemble each other with respect to the number of levels (as already mentioned only the lower four levels of the physical model are treated in this standard), elements of different levels are combined to provide process functionality of a certain level.

25.2.5 Recipes

A "recipe" contains the minimal necessary information and requirements to produce a certain product. It represents a method to describe a product including its corresponding production procedures and process-related information. The standard distinguishes general recipe, site recipe, master recipe, and control recipe.

Control recipes are structured similarly to the procedural control model and contain sufficient information with respect to the attributes of a process cell to be linked with the equipment control. Depending on the control recipe's granularity and the available software elements, this interlinking can be done theoretically on any layer. However, in practice, the interlinking of the control recipe, which resides on a batch server, with the equipment control, which is executed on controllers, is commonly done on the phase layer (see Section 25.2.3) by the so-called phase logic interface [BR05]. This interface is responsible to convert steps of the control recipe into commands for the linked equipment control and to return data and information from the process to the batch server. Most commercially available batch software tools perform this connection on the phase layer. This means that the higher more abstract parts of the control recipe (procedure, unit procedures and operations) are controlled by the batch server while only the phases, which comprise the most precise information about the procedural activities, are actually executed on controllers such as Programmable Logic Controllers (PLCs).

25.2.6 Classification of Process Cells

Process cells can be organized not only according to the physical model described above but also categorized according to either the number of different fabricated products or the physical structure of their installations (i.e., units) [IEC97].

If we categorize process cells according to the number of different fabricated products, we can distinguish

- Single-product process cells that fabricate always the same type of product in each batch though variations of the procedures or material and production parameters are possible to compensate changes of the environment conditions
- Multiproduct process cells that are able to fabricate different types of products by applying different procedures and parameters

A classification according to the physical structure of process cells leads to the following categories:

- Single-path structures represent a group of units that process one or several batches at the same time strictly sequentially.
- Multipath structures consist of several parallel single-path structures excluding material exchanges between the paths. However, material sources as well as storages for the final products may be shared by all paths.
- Network structures may have determined or variable paths. The actually used path for a batch can be scheduled either at the beginning or also during its processing. Flexibility is therefore a key issue as even units may be movable within the process cell. Hence, material sources and storages may need to be accessible for all units.

Evidently, process cells can be either simple in the case of a single-product process cell with a single-path structure or very complex in the case of a multiproduct process cell with a network structure. Especially, the latter rather complex category demands flexible control paradigms.

25.2.7 Tasks and Functions of a Batch Management and Operation System

The realization of certain tasks is necessary to perform the successful production of a batch. Seven tasks are identified, which define the means of controlling the equipment [IEC97]. Figure 25.3a displays these tasks—the connections represent a relation realized by an information flow between the tasks.

- Recipe management comprises activities to create and store general recipes, site recipes, and master recipes. The final master recipe is provided to the process management that uses it to create a control recipe.
- Production planning and scheduling is a high-level activity incorporating as its most important task the creation of a batch plan, which is transmitted to the process control.

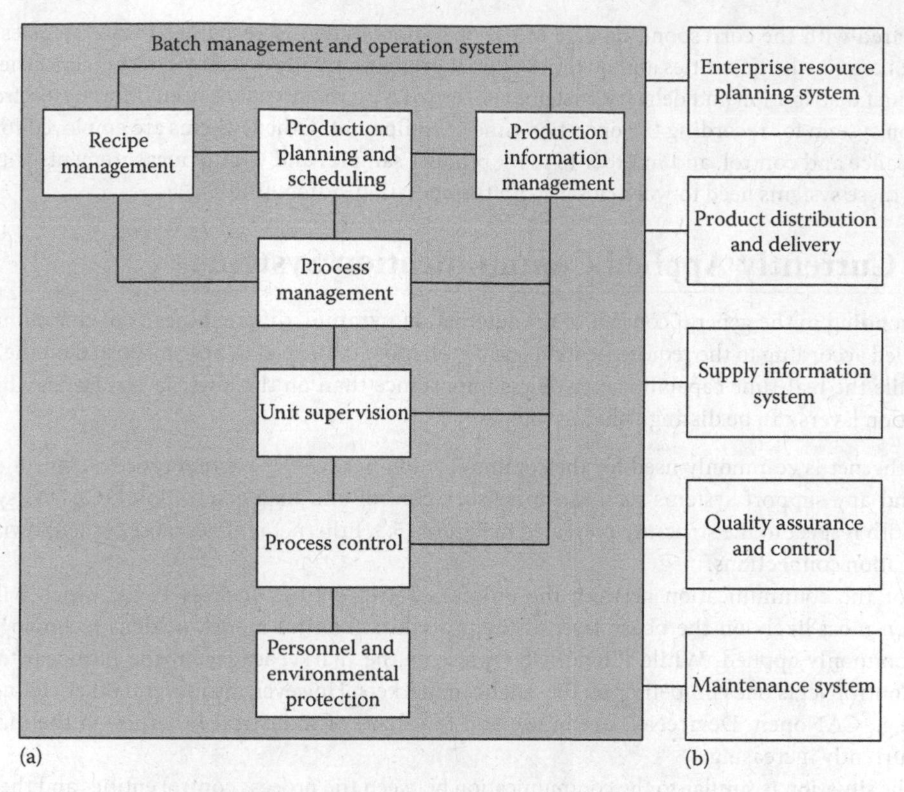

FIGURE 25.3 (a) Batch management and operation system (b) integrated with other information systems.

- Production information management is concerned with gathering, storing, processing, and reporting production information including batch history management. A batch history is a volume of data related to a certain batch to support other tasks such as batch and material tracking.
- Process management manages all batches and equipment entities of a process cell. Control recipes are derived from master recipes, batches are started and observed, process cell equipment is managed, and data related to equipment and batches is gathered.
- Unit supervision executes procedural elements, manages unit equipment, and gathers data related to equipment and batches.
- Process control executes control software on the phase level and gathers data related to equipment modules and control modules.
- Personnel and environmental protection is concerned with safety measurements and shown underneath the process control as no other control activity should interfere between this task and the field hardware.

25.2.8 Integration of Batch Management and Operation System with Other Information Systems

A batch manufacturing company employs not solely the actual batch management and operation system but also several other systems that produce and require certain data and information (see Figure 25.3b) such as enterprise resource planning (ERP) systems, which offer financial reports; order management; and warehouse management. Production data is demanded from the batch operation system and customer orders and material requirements converted into certain data formats to be committed to the operation system. Supply information systems keep track of the quantities and qualities of raw materials. This data is compared with the corresponding data of the operation system to recognize inconsistencies of the values. Likewise, the quantities and qualities of final products are observed by systems concerned with the product distribution and delivery. Systems concerned with maintenance need information from the operation system for recording the operation time of equipment. Other systems are employed for quality assurance and control, and in order to mark product samples and quality measurements with time stamps, these systems need to work closely with the operation system [BR05].

25.3 Currently Applied Communication Systems

Corresponding to the general concept of the automation pyramid, different means of communication are applied according to the requirements at each level. More data need to be transported on the higher level while the real-time capabilities are of less importance than on the lower levels. Hence, the communication layers can be distinguished as follows:

- Ethernet is commonly used for the communication between the actual process control entities and any support systems such as supervisory control and data acquisition (SCADA) systems. With respect to the structure presented in Figure 25.3, Ethernet realizes most of the shown information connections.
- For the communication between the process control entities such as PLCs, which will happen most likely on the phase layer of the procedural control model, fieldbus technologies are commonly applied. While PROFIBUS represents the market leader on the European market, ControlNet is one competitor on the Anglican markets. However, also several other technologies (e.g., CANopen, DeviceNet) are in use and the share of industrial Ethernet on the market is currently increasing.
- The situation is similar to the communication between the process control entities and their allocated peripheral I/O-modules. However, the real-time capabilities are of a vital importance at this level of communication. This type of communication realizes only basic functionality of the automation components without procedural activities and can therefore not be allocated to a layer of the IEC 61512 models.
- With respect to the communication of I/O-modules with their assigned sensor/actuator units, fieldbus technologies such as PROFIBUS-PA, DeviceNet, FOUNDATION fieldbus, or the AS-Interface (AS-i) are commonly applied. Also the HART Protocol represents a common fieldbus for this level of communication. However, even though the number of sensor/actuator units with fieldbus technology is rising, the majority of currently employed sensor/actuator units is still connected via ordinary wires to the I/O-modules. Also this type of communication cannot be allocated to a layer of the IEC 61512 models.

A result of all these different networks is a high integration effort. This is especially a problem if one system has to support different communication technologies. Therefore, attempts are undertaken to unify the interaction of control systems and control equipment. An approach targeting the level of control units and SCADA systems is the new OPC unified architecture (OPC UA) specification from the OPC foundation (http://www.opcfoundation.org). This specification defines at a high level and network independently how control equipment can exchange data. However, in its current implementation, it relies on Ethernet technology and is therefore not suitable for all types of control equipment.

25.4 Upcoming Requirements of Distributed Process Automation

Industrial automation—including process automation—is currently faced with a major paradigm change. Our globalized markets are getting more volatile. Customers are getting more discerning and favors customized special products. This leads to shorter product life cycles, smaller lot sizes, and less predictable sales volume. Especially, for the process automation domain, this new environment is very challenging as process plants incorporate a large investment. Therefore, new means are necessary that allow utilizing production facilities for different products and allow fast switches between products to be produced. For such systems, [GNP94] coined the term "agile manufacturing." In order to provide such manufacturing systems, the dynamic reconfiguration of plant structures as well as the control software is required. Parametric changes are not enough as they lead to unmanageable control software modules. Fifteen years later, recently conducted surveys such as the MANUfuture Strategic Research Agenda [EC06], which is one of the platforms that are integrated in the definition of the frame programs of the European Union, claim the same objectives.

Multipurpose facilities built in the above described network plant structure provide the demanded flexibility and adaptability from the mechanical/physical point of view [K99]. In order to operate such plants effectively and efficiently, new ways for the control structure have to be used. Distributed control with its control nodes locally attached to the plant controlled part represents such a means. Distributed control architectures allow localizing changes and disturbances in the plant. This helps to keep changes manageable for the human as not the whole system has to be investigated. Furthermore, distributed control systems feature an increased scalability allowing to add or remove plant parts. First works on this topic were able to show these benefits, e.g., [LZ08,PCSK07,SSPK06, and TSPK07].

Distributed control systems are an important foundation for flexibility and reconfigurability of production systems. However, in case of changes, disturbances, plant failures, or adaptations, labor-intensive engineering work is necessary. Therefore, more autonomous technologies are needed in order to reduce the human effort incorporated in production changes, as the human effort needs time (i.e., costs) and is typically error prone. A possible approach is to enhance the distributed control modules with autonomous cooperative functionalities. These allow the unit to provide its functionality (e.g., process reactor) to the production system and cooperate with other units like pipes, valves, pumps, or other process reactors in order to fulfill the overall goal production. For assembly automation, several such approaches were investigated and tested (see for example [B06,LR06, or VMM08]). They were able to show that the flexibility and adaptability of the plant is increased. More research is nevertheless still needed in this area before these concepts can be applied in industrial systems, especially for process automation where only little research has been done.

These new paradigms do not change the general production information exchange as described above; however, they change the overall structuring of the control system's architecture and therefore also the ways control components interact. The hierarchical structuring as shown in Figure 25.1 will be broken up into a network oriented one. Therefore, the control architecture resembles the mechanical structure of a flexible (batch) process plant. This leads away from the currently employed request/reply (e.g., client/server) interaction to a peer-to-peer interaction. Future control components will be equal members of the production system and interact with each other on a negotiation basis. However, as process automation systems interact with the real world, there are different kinds of interaction on different levels with different properties. On the higher levels, there are the negotiations between control components in order to fulfill the overall task production. These negotiations feature a higher amount of data but are in general less time critical. On the contrary, on the lower interaction levels—the real-time control levels—the communication's task is to synchronize the control algorithms. For this purpose, we encounter short messages that have to be transmitted and processed in real time (see Figure 25.4). Communication systems for upcoming process automation systems, therefore, have to support the peer-to-peer communication of equal distributed control components. This communication has to be flexible, allowing changeable communication relations and also support real-time interaction as needed by the real-time control elements.

Current communication systems for process automation are developed for the hierarchical command/response or master/slave paradigm as applied in today's process automation systems. These systems are not suitable for the upcoming needs of distributed control and autonomous production resources. Therefore, new communication systems have to be developed in order to support the new paradigms.

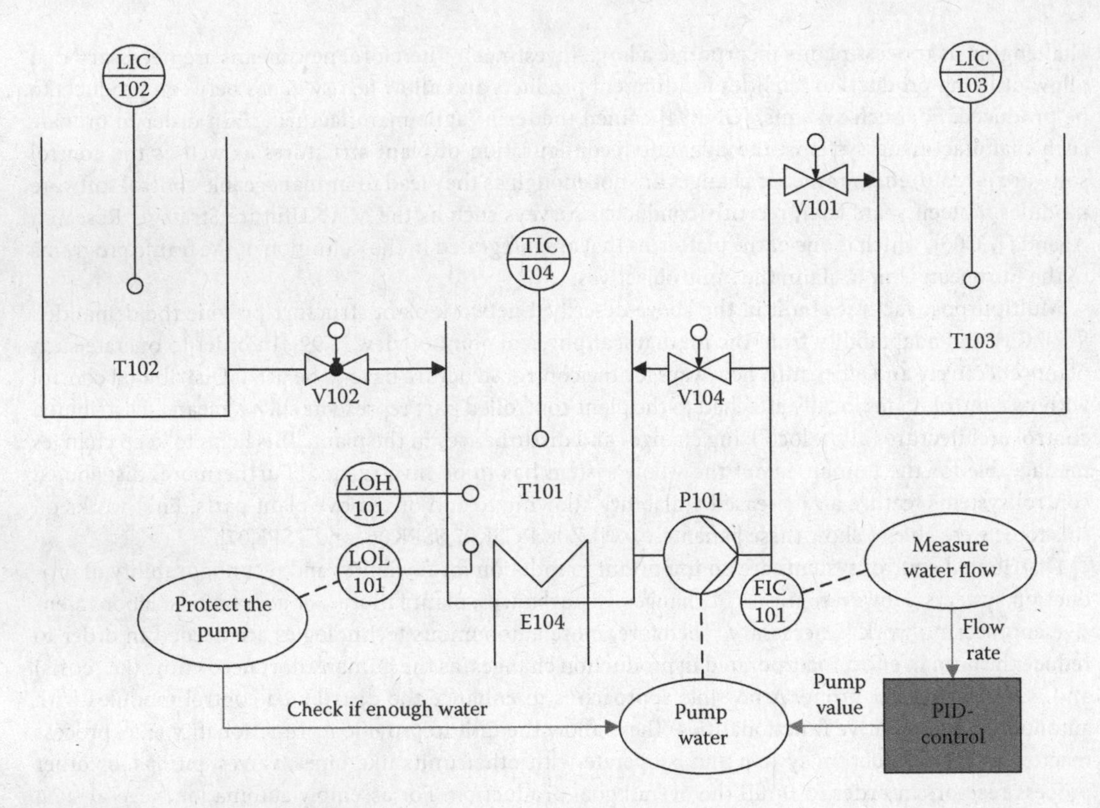

FIGURE 25.4 Example of messages between real-time capable control units.

25.5 Industrial Ethernet as the "Silver Bullet" for Future Process Automation Communication Needs

In the last few years, Ethernet moved more and more from the office/Internet world into the industrial automation world. This trend began with connecting control devices with high-level systems like ERP and SCADA systems, which was followed by replacing controller networks providing less time critical controller-to-controller communication. Currently, Ethernet moves also into the fieldbus world. The advantage of uniform data access and exchange on all levels of a production system and between all components is one of the reasons for this trend. Especially, ERP systems and production planning and scheduling system are systems for which it is believed that they can benefit most of this uniform data access. This should enable such systems to get a more complete view of a plant's state and therefore allow a better planning and scheduling of the production.

However, Ethernet by itself is not capable of fulfilling the needs of typical fieldbus systems. Therefore, different enhancements and changes to Ethernet were developed. Examples for such Ethernet-based fieldbus technologies described in this book are as follows: EtherCAT, Powerlink, Profinet, and ModbusTCP. The main problem of many of these systems is that they apply major changes to the protocols and system behavior of what is commonly known as Ethernet. In the end, many of the described systems have just the wiring of Ethernet (e.g., Profinet IRT). Furthermore, as all of the industrial Ethernet protocols use different methods, they cannot be used in parallel on the same wire or operate together. Therefore, we are again in the same situation the fieldbus diversity has brought us. So, finally the great expectations especially from the higher production system level could not be met. The current state of industrial Ethernet is as such that we have a new bunch of high bandwidth fieldbuses.

Looking at industrial Ethernet from the point of view of distributed autonomous control as described in this section, we are also faced with a similar situation as with conventional fieldbuses. Ethernet with its peer-to-peer, multicast, and broadcast capabilities would fulfill the needs of a distributed control architecture. However, most of the developed industrial Ethernet systems are mimicking the behavior of conventional fieldbus systems. That means they are typically organized in a master/slave hierarchical way and feature a cyclic communication behavior. There are a few that have already a support for multi-master peer-to-peer communication like EtherNet/IP.

The current use of Ethernet in the industrial domain shows that it can be the solution for the fast transmission of large data volumes or for systems where conventional fieldbus systems provide not enough bandwidth. However, for upcoming distributed control systems and also autonomous control equipment current fieldbus and industrial Ethernet technologies are too restrictive. Their applied master/slave communication architecture requires extensive efforts for providing peer-to-peer communication mechanisms as demanded by distributed control systems. In order to support such new control architectures, effective and efficient new communication methods have to be developed, which provide on the one hand the flexibility of Ethernet and on the other hand meet the requirement of industrial control systems as in conventional fieldbus systems.

References

[B06] J. Barata, The Cobasa architecture as an answer to shop floor agility, in *Manufacturing the Future—Concepts, Technologies, Visions*, pp. 31–76, pro literatur Verlag, Germany, 2006.

[BR05] M. Barker and J. Rawtani, *Practical Batch Process Management*, Newnes, Amsterdam, the Netherlands, 2005.

[EC06] European Commission, MANUfuture: Strategic Research Agenda, Assuring the future of Manufacturing in Europe, Report of the High-Level-Group, Luxembourg, ISBN 92-79-01026-3, 2006.

[GNP94] S. L. Goldman, R. N. Nagel, and K. Preiss, *Agile Competitors and Virtual Organizations, Strategies for Enriching the Customer*, Van Nostrand Reinhold ITP, New York, 1994.

[IEC97] IEC TC65, IEC 61512-1: Batch Control—Part 1: Models and terminology. International Electrotechnical Commission, Geneva, Switzerland, 1997.

[K99] S. Kuikka, A batch process management framework: Domain-specific, design pattern and software component based approach, PhD thesis, Helsinki University of Technology, Tapiola, Finland, 1999.

[LR06] P. Leitao and F. Restivo, ADACOR: A holonic architecture for agile and adaptive manufacturing control, *Computers and Industry*, 57(2), 121–130, 2006.

[LZ08] W. Lepuschitz and A. Zoitl, An engineering method for batch process automation using a component oriented design based on IEC 61499, in *Proceedings of the IEEE International Conference on Emerging Technologies and Factory Automation (ETFA'08)*, Hamburg, Germany, 2008, pp. 207–214.

[PCSK07] J. Peltola, J. H. Christensen, S. Sierla, and K. Koskinen, A migration path to IEC 61499 for the batch process industry, in *Proceedings of Fifth IEEE International Conference on Industrial Informatics*, Vienna, Austria, 2007, pp. 811–816.

[POL94] M. Polke, *Process Control Engineering*, VCH, Weinheim, Germany/New York, 1994.

[SSPK06] M. Ströman, S. Sierla, J. Peltola, and K. Koskinen, Professional designers' adaptations of IEC 61499 to their individual work practices, in *Proceedings of the IEEE Conference on Emerging Technologies and Factory Automation (ETFA'06)*, Prague, Czech Republic, 2006, pp. 743–749.

[TSPK07] K. Thramboulidis, S. Sierla, N. Papakonstantinou, and K. Koskinen, An IEC 61499 based approach for distributed batch process control, in *Proceedings of Fifth IEEE International Conference on Industrial Informatics*, Vienna, Austria, 2007, pp. 177–182.

[VMM08] P. Vrba, V. Marík, and M. Merdan, Physical deployment of agent-based industrial control solutions: MAST story, in *Proceedings of IEEE International Conference on Distributed Human-Machine Systems*, Athens, Greece, 2008, pp. 133–139.

26 Building and Home Automation

Wolfgang Kastner
Vienna University of Technology

Stefan Soucek
LOYTEC electronics GmbH

Christian Reinisch
Vienna University of Technology

Alexander Klapproth
Lucerne University of Applied Sciences and Arts

26.1 Introduction 26-1
26.2 Building Automation 26-1
Motivation and Overview • Distributed Functions • Technologies and Integration Aspects • Applications
26.3 Home Automation 26-8
Motivation and Overview • Technologies and Integration Aspects • Applications
26.4 Outlook and Further Challenges 26-13
References 26-14

26.1 Introduction

Building automation systems (BAS) provide automatic control of the conditions of indoor environments, with a focus on large functional buildings. The historical root and still core domain of BAS is the control of heating, ventilation, and air-conditioning (HVAC) systems, where automation helps to yield significant savings in energy (and thus cost). Yet, the reach of BAS has extended toward integrating information from all kinds of building systems, thus paving the way for "intelligent buildings." Since these systems are diverse by tradition, the integration of heterogeneous systems is an important issue. The field of building automation also exposes further special characteristics and differences with respect to industrial automation. This chapter gives an insight into the task of automating large functional buildings and also addresses the necessary systems and required communications infrastructure.

In the second part, the emerging field of home automation is under review. Home automation, which is becoming widely known under the term "smart homes," allows not only the economic benefit already known of building automation to be yielded in the residential sector, but additionally exhibits a huge potential for social benefits. In this chapter, the requirements and applications of home automation as well as a brief overview of common communication protocols will constitute the main topics.

26.2 Building Automation

26.2.1 Motivation and Overview

Building automation is concerned with measurement and control of plants, systems, and installations of (mechanical) building services [27]. The historical roots range back to the traditional core domain of HVAC, where BAS were designed to increase both comfort and energy savings in the building.

They were based on the concept of direct digital control (DDC) [29]. Ever since, BAS have constantly evolved by embracing more building services such as systems used for constant light control, scene management, and shading.

Motivations for the use of BAS are manifold. First and most important are considerable savings in operational cost since highly energy intensive areas such as HVAC can be optimized through advanced control strategies. Also, limiting peak energy consumption by coordinating devices provides high savings potential. In functional buildings, like office buildings, schools, or hospitals, this economic benefit that can be yielded over the entire building life cycle justifies the required investments. Additionally, a number of automation services can lead to positive environmental effects, which are of growing concern as legislation increasingly demands energy-efficient buildings, and sustainability is becoming of concern.

With the employment of BAS, an improved building management also becomes possible. Here, the focus is on central access to all building systems. This allows problems to be detected and probably fixed more easily and quickly. Preventive as well as corrective measures can be planned more efficiently, and thus costs are further decreased. This common way of transparently accessing all systems is supported by tailor-made user controls. They allow changes to be applied during regular operation and additionally do not require specially trained staff, despite the fact that the equipment to be controlled is technically complex. Additionally, remote access to building systems becomes possible, which is particularly useful when building sites are geographically dispersed. Travel time and costs are avoided and faults can, ideally, be resolved without being on-site. Another benefit is the possibility introduced by BAS to collect high quality-data on energy consumption. Last, but not least, increasing comfort for people in the building is a key benefit. Although it cannot be easily quantified, the positive effect is well understood. People who feel comfortable in a building will exhibit higher motivation and therefore also be more productive. The use of modern technology also appeals to tenants concerned with their image and allows building owners and tenants to set themselves apart from the competition by offering higher value. Finally, building security, access control, and people safety are other important topics that can be tackled with the help of BAS.

The technological implementations to the requirements in BAS are described by building services. At the bottom of a building service lies the physical processes that have to be controlled. These processes can be very complex and spatially distributed. Due to the architectural properties of the building, a certain partitioning of the services is required. Typically, buildings are separated into floors. Floors are separated into rooms, and rooms are limited by walls. A common way to approach the problem of high complexity is to define the smallest unit of the building, for example, a window axle, which contains the same control elements as all other units. Individually controlled units are referred to as zones. Zones may cover entire floors or reach down to single window axles. Different zones may exist for different building services. For example, sunblinds might be controlled by window axle while HVAC controls an entire room. Areas such as corridors, elevators, and stairwells pose special zones, as they typically link different building units. For instance, the corridor light may be coupled to room occupancy. These areas often require particular automation functions. Also, certain rooms may exist, such as meeting rooms or executive offices, which pose different demands, for example, different light scenes.

Contractors who engineer BAS are facing different requirements. There can be different contractors for different building services, which need to interact. Systems must be planned beforehand and commissioned at the end of the construction process. Today, a general contractor might need to tie these parts together and coordinate the different technologies and services. The market leaves room for large automation companies that implement an integrated but proprietary set of control equipment and smaller integrators, which rely on open systems to integrate different vendors' systems. Clearly, the open system approach will provide long-term maintenance benefits, while a single-source solution appeals through lower initial cost.

26.2.2 Distributed Functions

Functional buildings often reach considerable spatial dimensions. At the same time, in each building and probably in each building part different requirements are made on building services. New technologies have also enabled new services. The original, centralized approach of DDC stations has reached its limits. The I/O points of a DDC station typically controlled a single zone of a specific building service. The higher number of devices today also requires a considerably higher level of networking between sensors and actuators.

The building control network can be modeled as a directional graph of data points. A data point is a node in that graph and represents a single, physical value of the system, for example, a temperature. The data point is an object that contains this physical value as a present value and a number of metadata describing the context of the value, such as engineering unit, resolution, and minimum and maximum value. The present value is communicated along the links of the graph according to its direction implied by input and output data points. In this model, sensors are data sources, controllers are data processors, and actuators are data sinks. The metadata can be used by engineering and commissioning tools to identify data points in the building and to establish the correct communication links. This process is referred to as *binding*.

The basic purpose of data points is to represent typed values that may be communicated to implement distributed control. Thus, automation functions in BAS operate on data points. Frequently used functions are *trending*, *alarming*, and *scheduling*. Trending refers to the historical logging of data point values over time. Current implementations embrace interval-based change of value or trigger-based logging and record present values or aggregated values such as minimum, average, and maximum. Trend logs can be temporarily instantiated to monitor the effect of specific changes in the BAS. They can also be used to generate reports on energy usage or provide data to detect aging components in order to assist early replacement. Alarming is a function in BAS, which monitors data point values according to predefined rules. A typical alarm generator monitors one or a set of data points and generates an alarm if one of these values exceeds the specified normal range longer than a given time delay. Most alarm systems require an operator to acknowledge the alarm condition for recording that the condition was taken care of. This can be done on an operator panel, via E-mail, or Web-based services. Other alarm systems simply collect and log alarm conditions as they occur and go away for later processing. Scheduling in BAS allows to set and withdraw data point values at given times of a day. This is, for example, commonly used to issue set points of HVAC systems, turn on lights before opening hours, or schedule maintenance tasks at night. The simplest schedule is week based and defines times and values for each day of the week. To cover exceptional days such as holidays, a calendar-based approach is used. In this case, a (central) calendar defines the fixed or recurring dates of special days. The schedule refers to such a calendar and defines exception days, which override the regular weekday schedule.

The standard approach to apply the control network model to a building was to split the system into three layers [8]. This architecture is depicted in Figure 26.1. On the lowest layer, the field layer, control over actuators, such as fans or valves for hot water and coolant is provided, and an adaption of them in response to data point values for sensors, actuators, and set points (which may be provided from a central location or via local operating panels) is performed. The automation level communicates vertically with the field level by exchanging values with the field devices as well as horizontally to peer controllers on the same level. Typical functions in this level are controlling functions, alarm generation, and scheduling. The management layer is the top level and the most abstract layer in the BAS. Here, global parameters are maintained for the automation layer. Typical functions are management, data retrieval and storage with trending, global schedules, global alarm collection and acknowledgement, visualization, and interfaces to other systems.

In today's BAS, the different layers are comprised of certain types of devices. The field layer is typically populated by I/O devices. These can be accessed through direct I/Os linked to a DDC or by devices con-

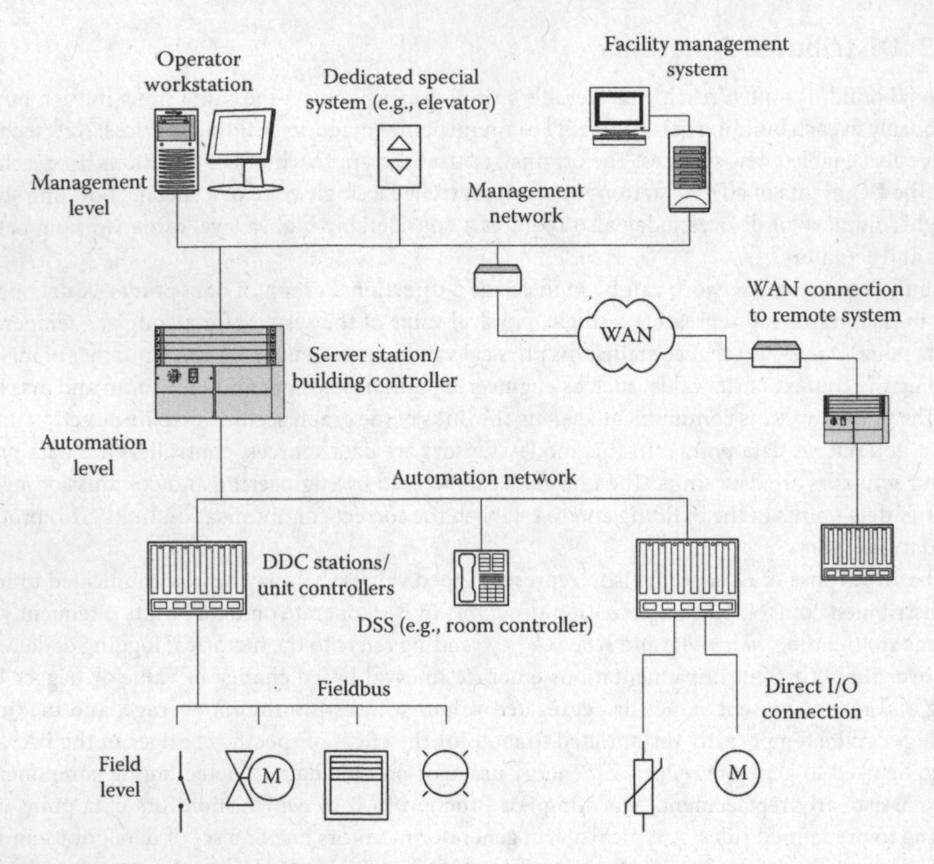

FIGURE 26.1 Field, automation and management level and respective devices.

nected via some fieldbus network. While the first variant requires relative closeness of the devices and is very cheap, the latter approach allows for spatially distributed devices and provides more flexibility.

The data point values provided by the field level (sensor values, subsystem status information) are communicated to a server station (building controller) that also serves as a central point of integration. Process data from the (sub)system is collected as well as integrated with data from other traditionally stand-alone building systems such as safety alarm or access control systems. As a result all those data points are accessible in the building controller. This unified view allows a transparent access to all building operation data, which primarily functions as input for an integrated visualization as well as automation functions. Hence, integration across different application domains is possible and frequently implemented. Also, building automation functions often supervise electrical circuits and provide supervisory control of central plants such as hot water production and air handling units. Especially the latter allows to influence the amount of energy consumed considerably, for example, automatically based on calendar data.

Furthermore, the building controller allows to automate sequences that involve multiple, heterogeneous systems, and also provides the interface to facility management systems. This building management infrastructure is typically known as the building management system (BMS). The user interface is displayed on local workstations, which establish a connection to the server mostly via the office network infrastructure. Using technologies from the IT domain, remote access to the building controllers is provided.

Progress in computer engineering has led to new perspectives in building automation. Until recently, the model described just above had to be considered no more than a rough guide. However, with ever increasing processing power and available memory as well as the overall size and cost of end devices permanently decreasing, so-called intelligent field devices became available. Sensors, actuators, and room units (control panels with sensor function) now take over automation functions that were previously exclusively provided by DDC stations. At the same time, DDC stations are increasingly often equipped with network interfaces and, therefore, can be used for automation and management functions. This leads to the virtual collapse of the automation layer and the reassignment of its functions.

While communication at the field and automation layers still remains geared toward process data exchange, a change in the network model can be observed. Process values can typically be represented in a compact way (considering a single value at a single point in time). Moreover, the control of the building environment has relaxed requirements regarding the frequency of the control loops [33]. Also, the spatial extension of most control loops is limited. Therefore, large systems are divided into network segments, in which low data rates are sufficient even if acceptable response times to selected events are required. The main criteria for the communication infrastructure are cost efficiency, robustness, and easy installation (free topology wiring, link power).

At the management layer, access to data from all segments is mandatory. Therefore, all system data pass through the network segment that the management layer devices are connected to. The amount of traffic accumulated there can be considerably high, especially in larger systems where thousands of data points can be found. Therefore, networks with a higher bandwidth are employed at this layer.

For the backbone, Internet protocol (IP) has found widespread acceptance. Its not entirely deterministic behavior does not restrict its use in building automation due to the moderate requirements present. Because of compatibility issues with installed devices, the transport and network layers of existing fieldbus protocols cannot be changed to take care of the intricacies for IP [32]. This shortcoming can be alleviated by the use of *tunneling* [2,3]. Tunneling routers encapsulate fieldbus data packets and distribute them over any given backbone network. This happens completely transparent to the fieldbus devices. Tunneling routers also provide an access point for remote communication. Nearly all manufacturers offer gateways and interfaces that allow access to data points handled by their DDC stations for the purpose of integrating them into BMS.

26.2.3 Technologies and Integration Aspects

In today's BAS, a number of different technologies can be found [28]. Figure 26.2 displays a brief overview, which technologies are prevalent in the respective layers. By tradition, non-open, proprietary solutions dominate the field of building automation on the automation and management layer. This may be due to the fact that custom solutions are possible at relatively low cost, given the moderate performance requirements involved. For the classic automation field bus, the majority of these solutions follow the EIA-485 standard. In certain cases, also the open standards Profibus, CANOpen

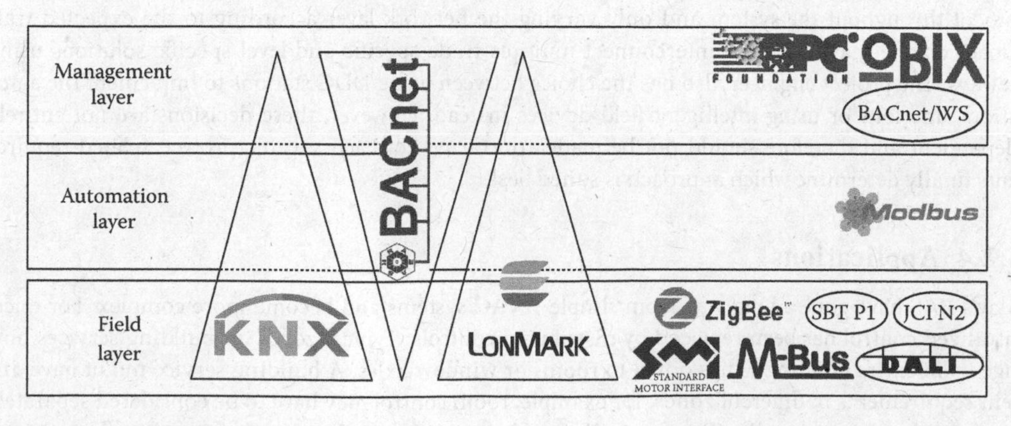

FIGURE 26.2 Networking technologies in BAS.

(and their specific profiles), or Modbus are used at this level. As a software interface into BMS software, OPC [19] has reached importance. Today, also Web technologies are used, for example, for management and configuration of DDC stations. However, these are still of limited use for machine-to-machine communication.

One fact that must not be neglected is the long life cycle of a building and its respective services. When faced with a future extension or change of the system, access to all new data points must be preserved. In this case, only an open system can guarantee that one does not find oneself tied to the contractor that originally installed the system. With BACnet™ [9,11], KNX™ [6,10,19], and LonWorks [1,5,20], three major, open standards exist in the field of building automation. While these standards cover multiple application domains, and hence a broad range of applications, also several domain-specific busses are emerging. These replace the interfaces that were traditionally used to connect field devices to controllers. Since they are designed for the use in a specific application domain, they often provide additional functionality, which makes them especially appealing and cost-effective to install.

A prominent example of these application-specific standards is Digital Addressable Lighting Interface (DALI) [7]. DALI allows control of up to 64 electronic ballasts for fluorescent lamps and intends to replace the classical 1–10 V interface. Another example is M-Bus [4], which has been designed in Europe for (remote) meter reading. The Standard Motor Interface (SMI) bus [22] is mainly used for shutters, protection systems, and sunblinds.

Another trend is the tighter integration of BAS between building service domains. Previously, each building service domain such as HVAC had its own dedicated control system based on a specific technology. Other domains might have used different technologies. A set of arriving new functions requires cross-communication between the BAS domains. It is, for instance, no longer acceptable that presence detectors from the lighting domains are not accessible for intrusion detection systems. Another example is constant light control, which can benefit from integration with shading. If two domains are not based on the same communication technology, the use of gateways is mandatory (e.g., to provide a set of DALI registers to BACnet). If two domains use the same technology, but different address domains, the use of proxies is required (e.g., provide network variables of LonWorks domain A to domain B). The problem with gateways is that the number of translations between systems is of quadratic order. Also, configuring gateways is challenging, as the mapping between technologies can never be a 1:1 mapping, and therefore information is always lost. A new trend is to equip those devices, which shall integrate the singular functions, with network technologies of both domains. For example, the DALI constant light controller may provide both a LonWorks and BACnet interface for transparent access by the respective system. The advantage is no extra configuration as the device already knows about its data points and no loss of information into the other system has to be faced.

It is the task of the project engineer to choose among a variety of applicable protocols as well as various options regarding network topology and function distribution. One may opt to use a common protocol throughout the system and only varying the network layer according to the expected traffic load, or one may choose to interconnect multiple trade-specific and level-specific solutions using gateways. The project engineer also has the choice between using DDC stations to implement the automation functions or using intelligent field devices instead. However, these decisions are not entirely independent and therefore should not be made arbitrarily. For each specific project, related requirements finally determine which approach is suited best.

26.2.4 Applications

Today's BAS have gone a long way from simple HVAC systems and become more complex. For once, centralized control has been replaced by distributed controls. Typical zones for building services now range from entire building wings down to rooms or window axles. A building service might have different requirements in different zones, for example, room control may have to be considered separately from corridors or stairwells. This typically results in different designs for rooms and corridors.

TABLE 26.1 Domains of Building Services

Domain	Typical Building Services
Climate control	HVAC, humidity, air quality
Visual comfort	Artificial lighting, daylighting (motorized blinds/shutters), constant light control
Safety	Fire alarm, gas alarm, water leak detection, emergency sound system, emergency lighting, closed-circuit television (CCTV)
Security	Intrusion alarm, access control, CCTV, audio surveillance
Transportation	Elevators, escalators, conveyor belts
One-way audio	Public address/audio distribution and sound reinforcement systems
Energy management	Peak avoidance
Supply and disposal	Power distribution, waste management, freshwater/domestic hot water, wastewater
Communication and information exchange	IT networks, private branch exchange (PBX), intercom, shared WAN access, wireless access (WLAN)
Miscellaneous special domains	Clock systems, flextime systems, presentation equipment (e.g., video walls), medical gas, pneumatic structure support systems (for airhouses)

Moreover, different building services have been embraced by BAS. A summary of different building service domains is given in Table 26.1.

Climate control is still the most important domain, including the original HVAC applications. The underlying physical process can be measured and influenced using well-known devices. Heating systems, for example, include variable air volume systems to fan-coil units. The timing constants of the process are long and thus make the process simple to control using network-based control systems like BACnet, KNX, and LonWorks. More novel systems also include control loops for air quality and humidity.

Another well-established building service domain is visual comfort. This domain comprises lighting applications (e.g., daylight), constant light controllers, and scene selections for special rooms such as meeting rooms or executive offices. Lighting applications have much more stringent requirements on response time than HVAC. Typical switch commands must be transmitted faster than 400 ms to give the user an instant feedback. Also constant light control requires faster adaption to environment conditions than temperature control, as the human eye responds much faster to changes. Typical installations for lighting rely on KNX or LonWorks systems. The light switches, touch panel units, presence detectors, and luminaires are interconnected over a communication system. The constant light controller units often also have direct I/O ports to light switches. Modern luminaires frequently use the DALI standard for communication, which is especially easy to install, as the same cabling is used for the communication path as for the power cabling. Two installation scenarios are in place: room-based installations are based on KNX or LonWorks, and floor-based cabling is used for DALI luminaires. The DALI light controllers are often installed per floor or floor segment spanning multiple rooms and provide an interface to the BACnet-, KNX-, or LonWorks-based infrastructure installed in the rooms to realize constant light control. Other components in visual comfort comprise shading and sunblinds, which control the admission of daylight into a room.

Two important domains that are still often neglected are security [26] and safety [30]. Security refers to protecting the building premises from unauthorized access (access control) and intrusion detection. Systems in this area use proprietary protocols as well as open technologies, such as BACnet (access door objects) and LonWorks (authorized services). Typical devices are window and door contacts, CCTV devices, and presence detectors. Safety applications refer to life safety and include monitoring and alarm management of numerous building parameters. If a parameter exceeds its normal range, alarms are generated and dissipated through the system. This is a typical function of building controllers. Systems such as BACnet provide standardized objects for life safety and life safety zones. Fire alarm and emergency lighting systems are examples of alarm processing systems. Shutters are frequently found to use Modbus to control them. LonWorks can be found in a number of emergency lighting installations. The main task is to react to a building alarm and constantly monitor the health status of emergency light devices as well as the generation of alarms to the operator as soon as they show failures.

To operate a building, supervisory control and data acquisition (SCADA) systems are frequently used. These high-level management systems collect and visualize operational data from the building and provide parameters to the building manager for configuration of building services. As opposed to low-level control loops, the timing aspects of these systems are very relaxed. Still, SCADA services must exhibit high reliability. To improve reliability, the data acquisition may be distributed and placed as near to the field level as possible. In this case, the subsystems can function autonomously in case the backbone or central SCADA fails to operate. Trending of data is partly already done in building controllers or even lower in the system, and SCADA systems only need to collect and store data in batch cycles. Room control panels are also examples of a small SCADA service in rooms or corridors.

Traditional BAS are often installed by different contractors for different building service domains. This means that in the past, almost no cross-communication existed between service domains. The emergence of new, tighter integrated building services, however, now requires different domains to share data. Examples are data exchange between presence detectors from lighting with HVAC or controlling sunblinds of the lighting domain from the HVAC service. From a communication viewpoint, this requires horizontal integration. Performance contracting and energy monitoring also demand a global view of the building across different building service domains. These are high-level systems that need access to data in the lower levels of the BAS. This is the area of vertical integration.

26.3 Home Automation

26.3.1 Motivation and Overview

Adopting automation technology is equally attractive in the residential domain. Besides efficiency considerations, increased comfort and peace of mind are key motives here. Also, elderly people can live in their own homes longer (ambient assisted living, smart home care). The importance of this aspect continuously increases as life expectancy does.

Although the number of devices involved in a home automation project is by orders of magnitude lower than for large functional buildings, the complexity of such projects must not be underestimated. While many applications and techniques are similar to the ones found in building automation, additional challenges arise in systems tailored to people's needs in private households. Additional application domains (household appliances and consumer electronics) must be integrated. Easy configuration is of particular importance, since the disproportionately high setup cost will otherwise reduce the attractiveness of automation. Acknowledging these specific challenges, one explicitly speaks of "home automation" systems (HAS).

A key challenge is the *integration* of the traditional automation domains (HVAC, lighting/shading) with the many established services in the home such as home appliances, telecommunications, multimedia, as well as energy metering and security devices. Solutions that want to persist in the home market must guarantee a seamless integration of these heterogeneous systems (both in terms of applications and manufacturers) in order to fully activate all inherent benefits. However, it is not sufficient that systems merely provide all functionalities, but they must also become usable for nonprofessionals in everyday life. This ultimately calls for two qualifications. First, installation, setup, configuration, and maintenance of a HAS should come near the famous "plug and play" paradigm and configure itself automatically as far as possible (e.g., zero-configuration networking). Especially, no technician or engineer should be needed to successfully install and use the system. Additionally, integrating any existing equipment of homes (i.e., in case of refurbishment or extension) should get by with a minimum of physical intervention. Second, the system must be usable for the residents during regular operation. This demands that special attention is devoted to the *usability* of the system as such, its user interfaces, and all controls visible to the residents. The design must focus on a broad target group of different ages and (technical) skills, as user friendliness is crucial for the acceptance of any such system. During regular operation, the *user involvement* must also be guaranteed. The users must retain full control of their systems that therefore have to offer sufficient degrees of freedom in controlling. While certain tasks may happen without intervention (i.e., automatically), it is important that the system remains transparent to the user, does not behave unexpectedly, and allows for the user to choose his preferences and—if desired—overrule the system at any time.

During system operation, data is collected, processed, and exchanged over a communication network. These data may consist of personal information on user preferences, activities, and details on the home and its equipment. Clearly, knowledge of this information would provide a global system view and allow backtracking of the behavior of its residents, if misused in such a way. Therefore, systems must be designed with *privacy* in mind and permanently allow the user to decide which information to share and which parts to keep secret. Modern HAS may also be accessible via remote or mobile terminals, (Web) gateways, or even use unsecured communication media for data exchange. This raises the question of *security* in communication that is required to keep sensitive, personal information confidential and that prevents abuse of the system for economic or even life safety reasons. Last, but not least, the *economic dimension* in the residential sector must not be underestimated. The residential market is very sensitive to the price of any equipment, so home automation products need to have a return on investment within a few years after installation time. This return on investment can, for example, be yielded by increased energy efficiency.

26.3.2 Technologies and Integration Aspects

Apart from the big (protocol) players in building automation (i.e., BACnet, KNX, and LonWorks), several other protocols have emerged that specifically target the home automation market. Their common denominator is the support and focus on applications pertaining to the home domain. The special requirements of home automation are targeted and reflected in the protocol design.

With *HomePlug* [17], the HomePlug Powerline Alliance proposes a protocol that enables and promotes rapid availability, adoption, and implementation of cost-effective, interoperable, and standards-based home powerline networks and products. Communication channels are the almost ubiquitous powerlines, which in general are ready to use as basic infrastructure. The underlying powerline communication standard allows for all aspects of powerline communication. Device interconnection with lower data rates is established using the HomePlug 1.0 protocol, while high data rate multimedia streaming and VoIP services are provided by HomePlug-AV. Traditional home automation applications are realized by a low-cost, low data-rate branch called HomePlug Command & Control. With HomePlug Access BPL, a broadband powerline technology intended to provide to-the-home access (i.e., bridging the "last mile") has also been specified. Together with the ZigBee alliance, the HomePlug alliance has formed a working group in order to devise high-level protocol specifications to be used on both powerline and wireless systems.

The approach of the upcoming *digitalSTROM* [13] standard is to control electrical home equipment through core components called "luster terminals." These consist of a high-voltage chip located directly at the end device or appliance. A novelty is the dual use of the chip, which, on the one hand, acts as an actuator (e.g., switching, dimming) and, on the other hand, as a sensor that can at least roughly measure the energy consumption. Installed in the fuse box is the "meter" device that controls the terminals and accurately measures power on a circuit. Additionally, an interface to a Web server is provided that records energy information from sensors, controls terminals through the meter, and provides information visualization. DigitalStrom therefore provides very low-rate powerline communication and features such as energy measurement, plug and play, and remote control.

Wireless systems with all their benefits (e.g., free placement, no additional wires) and challenges (e.g., power consumption and supply, interference) are also on the rise [31]. This is partly not only due to their excellent qualification for refurbishment but also their reduced complexity compared to wired systems. The two most well-known wireless protocols for home automation are *IEEE 802.15.4/ZigBee* [24] and *KNX RF* [19]. While the latter is conceived as the wireless extension of the established open KNX protocol, especially ZigBee was designed with a broad focus on industrial, building, and home automation (in this order). However, ZigBee has not yet achieved a breakthrough in the residential automation market, most obviously due to very limited availability of commercial products. To tackle this shortcoming, in November 2007, the ZigBee alliance published the ZigBee home automation profile. It provides both product manufacturers and end users with standardized device templates, thus guaranteeing specific functionalities and ensuring interoperability of certified devices at the same time.

With *EnOcean* [16], the concept of energy harvesting was commercially exploited for the first time. Energy-autarkic devices employ the EnOcean technology to gather enough energy from the environment (by means of solar cells, piezoelectric elements, thermocouples) to perform computations within the node and set off telegrams over the air. In order to get by with the small amount of energy at disposition, the EnOcean protocol is tailored for lowest possible power consumption. Featuring a comparably high data rate and a small telegram size, EnOcean achieves very short transmission durations, therefore encountering a lower statistical probability for telegram collisions. This robustness, in turn, increases the protocol scalability (more nodes may send simultaneously) and reduces power consumption (less frequent telegram retransmissions). Because of its limited set of functions, EnOcean is often employed in combination with other systems, for example, KNX (RF) or ZigBee, to fully exploit the benefits of both wireless and battery-less technologies.

The *Z-Wave* [25] protocol was developed with an explicit focus on home control applications. With low data rate communication and a device count limited to 232 devices, applications for lighting, home access control, entertainment systems, and household appliances are serviced. Z-Wave devices can be roughly classified into controllers and slaves devices. The latter only act upon messages, while controllers are aware of the network topology and all devices belonging to the network. Therefore, they can perform more complex automation control and provide routing and security services in the network. With the set up of Z-Wave networks being easy and the control of the automation functions intuitive, Z-Wave is often chosen for scenarios involving less technology-experienced users that do not require extensive configuration, management, or visualization possibilities.

Especially in HAS, the interaction among multiple systems, which always has to be seen under the importance of usability, is a key asset. Since these systems are diverse and not only cover the historic core domains HVAC and lighting/shading but also consumer electronics (brown ware) to home appliances (white ware), integration in the home has special requirements. Several technologies that support the transition toward a smart home are discussed in the following paragraphs.

Universal Plug and Play (UPnP) [23] provides a set of open IP-based protocols and mechanisms to easily integrate and control devices in a peer-to-peer network (zero-configuration networking). While UPnP has initially been developed for Windows to enable plug and play for PC peripherals, it has now become a widely accepted, powerful and yet simple approach to connect and control Internet gateways and multimedia devices in homes. UPnP is media and device independent. The standard is hosted by the UPnP forum, an organization with several hundred members across industry. The UPnP device architecture covers IP addressing (DHCP or APIPA), device and service discovery (SSDP), description (XML), control (HTTP/SOAP), eventing (GENA), and presentation (URL/HTML). At the top level, specifications for standardized device classes, the "device control protocols" (DCPs), are specified. The DCPs, which are developed and maintained by the UPnP forum, define a common interface that allows vendor-independent easy handle access to UPnP DCP compliant devices. Currently, DCPs have been defined for media servers and renderers, printers, Internet gateways, HVAC, and lighting devices.

The Chinese standard for *Intelligent Grouping and Resource Sharing* (IGRS) [18] defines a software architecture similar to the one known from UPnP. It builds upon the three goals intelligent grouping, which allows devices to discover available services; resource sharing, which makes resources of all devices available in the whole network; and service collaboration, which enables a realization of intelligent services through device collaboration. Currently, IGRS is predominantly used in China; however, at the moment it is also listed as "under development" standard from ISO (ISO/IEC 14543-5-1).

Also related to the approach taken by UPnP are the *Devices Profile for Web Services* (DPWS) [12]. As in UPnP, services for device and service discovery, messaging, and eventing are specified. However, all services are foreseen to be accessed or exposed via Web services and to be executable on resource constraint devices. For this purpose, DPWS builds on established Web service standards such as WSDL, XML, and SOAP, and specifies a lightweight subset of the overall Web services protocol suite. It adds features that are missing in UPnP, such as a homogeneous security model, standards-based description language, and the ability for a seamless integration into service-oriented enterprise architectures. In a prototype exhibition, the interconnection of a lighting system, audio distribution system, security system, motorized shades, security cameras, thermostats, washers and dryers, and a motorized television mount was already demonstrated.

The basic concept of the *Digital Living Network Association* (DLNA) [14] is the realization of a network of interoperable consumer electronics, personal computers, and mobile devices. The cross-industry organization develops interoperability guidelines based on existing open industry standards such as UPnP, JPG, and MPEG-2 to achieve digital convergence. DLNA has defined a set of media devices such as renderer, server, player, and controller, but also mobile and interoperability devices that guarantee interoperability when being DLNA compliant. The vision is to provide consumers with services to easily acquire, store, manage, access, and share any content at anytime and anywhere.

With *Echonet* (*Energy Conservation and HOmecare Network*) [15], a Japanese standard that primarily targets the two applications its name is derived from was defined in 1997 by the Echonet Consortium. Simple and affordable heterogeneous devices of multiple vendors shall be easily interconnected exploiting plug and play services and realize applications ranging from the reduction of carbon dioxide emissions to the assistance of elderly people. For data exchange, a broad range of media including powerline, RF, Ethernet, infrared, and LonTalk, with a communication middleware handling the communication are available. Access to functionalities (e.g., a gateway) is provided via a so-called service middleware Application Programming Interface (API) that contains common processes in the form of libraries.

OSGi [21] defines a dynamic module system for Java that can function as an open standard for home gateways. Maintained by the OSGi Alliance (formerly known as Open Services Gateway initiative), it is a powerful, Java-based service platform that has been adopted in various domains, for example, home automation, automotive and, more generally, embedded or mobile applications. The main benefit of OSGi core is the definition and management of dynamically loadable modules called OSGi bundles. OSGi provides the user with commands for the complete life cycle management of these bundles that can be handled locally or from remote. The OSGi platform introduces a service concept with a service registry that allows locating and using services of all active bundles. Around the core, OSGi has defined an extensive set of standard services within bundles that, among others, provide support for logging, configuration, user administration, events, and preferences, HTTP, XML, and UPnP.

26.3.3 Applications

In home automation, applications also include the traditional HVAC and lighting/shading use cases. However, the main gain that constitutes a strong argument for the employment of HAS can be found in more advanced use cases incorporating additional domains and devices. Technologies such as sensing the environment and setting user actions "in context" as well as (automatically) inferring user intentions and preferences followed by their execution in control strategies depict the additional value automation systems have to offer particularly in the home.

Ambient Assisted Living (AAL) aims at improving safety, security, entertainment, and general quality of life of the home's inhabitants. Especially when considering demographic developments of the societies in Europe and the United States, elderly people will soon constitute a major part of the population. One main target of AAL is to allow them to live an independent life in their familiar surroundings, which would otherwise not be possible without certain support. This support shall be delivered by a multitude of home automation services that are designed for assisting the inhabitants to live longer and at higher quality in their homes. Because of the target user group of AAL systems, several special requirements emerge. Most important, systems have to become usable also for technically not versed people. Therefore, an intuitive user interface to act as a mediator between system and user is of prime importance. Also, the necessary interaction shall be kept to a minimum. This demands that AAL systems are capable of self-learning. Through the use of a network of sensors and actuators, and with occasional user inputs, the system shall be able to perceive the environment, predict (desired) user actions, and perform actions to alleviate the burdens of daily tasks and provide additional measures of safety, security, and comfort. For example, systems could detect users leaving the home and thereupon automatically turn off miscellaneous electronic equipment, appliances, and lights, reduce temperature set points, and activate the alarm system. Another possible application is to detect if a person has collapsed on the floor by means of motion and pressure sensors as well as built-in logic and set off an emergency call for help.

The "smartness" in home automation needed for modern applications such as AAL benefits greatly from the concepts presented by *ambient intelligence* (AmI) and *context awareness*. Context-aware devices can sense and react based on the state of their environment. Information gained includes location, physical properties (e.g., temperature), and information on ongoing processes and infrastructure, available in the surrounding environment and the device itself. Senses are applied against a rule-set that is based on the user's profile and preferences and generates responses. Therefore, a context-aware device such as a smartphone may "understand" that its bearer has arrived home and, based upon its understanding of the user's location and the environmental settings the user prefers at the time, decide to illuminate the area and turn on the stereo through commands to the home's infrastructure (i.e., in this case the home automation network). The keystone of this automation approach is proper pattern recognition of routine tasks, for example, learning and predicting user tasks. This learning can be accomplished by exploitation of mechanisms originating in artificial intelligence, for example, neural networks or fuzzy control.

Another benefit of home automation is increased *energy efficiency* and energy conservation. Energy consumption throughout the world has been steadily increasing over the past half-century, and now has become an urgent topic for mankind and thus politics. Efficiency of energy use is the primary purchase reason behind new appliances—especially Energy Star certified appliances—and also behind new miscellaneous electrical loads. Conservation of energy is a much broader topic that combines modern building technologies (e.g., insulation, energy-saving lightbulbs) with improvement in the inhabitant's behavior toward more energy-efficient ways of going about daily tasks. In the latter case, support from HAS can be expected. Through automatic metering (*smart metering*) of the energy consumption and visualization of the data, users become aware of their consumption and can now actively change their behavior toward acting more sustainably. The feedback is given in real-time, and may be more or less abstract (e.g., numerically displaying kilowatt-hours and currency compared to feedback in the form of dynamic color changes from green to red), therefore always adapting to the user's needs.

HAS also offer perspectives to use energy more efficiently. On the one hand, this is possible through context awareness (e.g., turn off the lights in currently not occupied rooms). On the other hand, automation systems allow tasks to be executed at times when they are most energy efficient. An example is to turn the heating on as late as possible yet reaching a predefined temperature at a defined point of time. This becomes only possible with the system being aware of inside and outside temperatures, weather, time of the year, and building inertia to name just a few. This information is accumulated and exploited by the automation system to provide better control while guaranteeing optimum comfort for the user. Other energy-aware approaches include "smart thermostats" that can automatically reduce the temperature when energy is expensive to generate, as well as "smart appliances" that wait until energy is inexpensive to start operation (e.g., a dishwasher starting its program when energy costs less).

Finally, through the integration of smart metering devices in the system, additional services such as remote meter read-out, detection and handling of events and alarms in situations involving tampering, vandalism, outages or leaks, and extensive diagnostics, monitoring, and logging functionality are added to the application portfolio of home automation.

26.4 Outlook and Further Challenges

The emergence of BAS is a nonreversible trend in modern buildings. While in the past the main focus was put on the primary goal of automation in certain domains such as HVAC or lighting applications, today's potential lies in the tighter integration of different building domains. This requires the merge of functions as well as technologies and results in more complex systems than before. While it is a primary design goal to aggregate functions into more valuable building functions, the resulting technological implications on how to use gateways, bridges, and open interfaces to previously proprietary systems are manifold. More complex functions also require a steadily increasing number of data points to be configured and maintained. While in the past a number of 1,000–10,000 data points in a building was common, the number tops 100,000 today and keeps growing.

One of the main issues in today's workflow concerns configuration and commissioning. Different technologies, vendors, and even devices of the same vendor require different paradigms, different workflows, and different tools. Connecting 10,000 data points of a LonWorks system into a BACnet BAS can be a daunting task. Also, the growing number of data points per building poses new challenges even within a single technology domain. While more data points mean better control of the process, it also requires a higher planning and configuration effort. Automatisms in the process are needed along with templating common tasks in automated tools that assist the installer.

Another area of development in BAS is flexibility of already installed systems. Changing small parts of a system can incur high costs today. Devices need to be recommissioned, bindings changed, and even cabling restructured. For facility managers, the goal is to reconfigure the office spaces according to tenant's needs while keeping inflicted cost as low as possible. This increases the turnover in renting out space and reduces cost of unrented space. This process should be easy enough so that the facility manager himself or herself can simply move walls and change a limited set of configuration parameters in the BAS software instead of calling in cost-intensive installation personnel to reconfigure the communications systems. This is an area of development, which needs new and innovative solutions such as rule-based commissioning and more advanced self-binding mechanisms to name a few. The ultimate solution is to provide a unified view of the buildings operational state, CAD drawings of its structure, architectural information, configuration parameters, and data mining options from the BAS data into a building information model.

One fact that is still underestimated today is network security in BAS. Early systems used direct I/O and proprietary communications. The introduction of open systems and the move to IP-based backbones opens the BAS network to new levels of security issues not known before. Only now, first solutions are emerging, which are not yet widely accepted.

Moving from the building to the home stresses the above arguments even more. Consumer equipment for home automation needs to be more easily installed and provide valuable solutions to the owner to justify the investment. This is also the main reason why home automation has not taken off recently. Niche markets for the technophile home user or holiday vacation owners pose the exception. Improving comfort and giving peace of mind are main driving factors in that area. Today's solutions with dial-in heating presets, air-conditioning scheduler, or alarm services are just arriving. Companies can gain benefits from giving away price-reduced home automation equipment by generating revenue from advanced Web services such as assisted energy savings or monitored alarm and intrusion detection. While the market in BAS is mostly established, the home automation area still has to be considered being in its infancy.

References

1. ANSI/EIA/CEA-709.1-A, Control network protocol specification, ANSI, Washington DC, 1999.
2. ANSI/EIA/CEA 852, Tunnelling component network protocols over Internet protocol channels, ANSI, Washington DC, 2002.
3. EN 13321-2, Open data communication in building automation, controls and building management - Home and building electronic systems—Part 2: KNXnet/IP communication, CEN, Brussels, Belgium, 2006.
4. EN 1434-3, Heat meters—Data exchange and interfaces, CEN, Brussels, Belgium, 1997.
5. EN 14908, Open data communication in building automation, controls and building management. Control network protocol, CEN, Brussels, Belgium, 2005.
6. EN 50090, Home and building electronic systems (HBES), CENELEC, Brussels, Belgium, 1994–2009.
7. IEC 60929, AC-supplied electronic ballasts for tubular fluorescent lamps, IEC, Geneva, Switzerland.
8. ISO 16484-2, Building automation and control systems (BACS)—Part 2: Hardware, Geneva, Switzerland, 2004.
9. ISO 16484-5, Building automation and control systems—Part 5: Data communication protocol, ISO, Geneva, Switzerland, 2007.
10. ISO/IEC 14543, Information technology—Home electronic system (HES) architecture, ISO, Geneva, Switzerland, 2006-2010.
11. ASHRAE Standing Standard Project Committee 135, www.bacnet.org
12. Profile for Web Services, schemas.xmlsoap.org/ws/2006/02/devprof
13. Digitalstrom.org, www.digitalstrom.org
14. Digital Living Network Alliance, www.dlna.org/home
15. Echonet Consortium, www.echonet.gr.jp
16. EnOcean GmbH, www.enocean.com
17. HomePlug Alliance, www.homeplug.org
18. IGRS, www.igrs.org
19. OPC Foundation, www.opcfoundation.org
20. Konnex Association, www.knx.org
21. LonMark International, www.lonmark.org
22. OSGi Alliance, www.osgi.org
23. The SMI Group, www.smi-group.com
24. UPnP Forum, www.upnp.org
25. ZigBee Alliance, www.zigbee.org
26. Z-Wave Alliance, www.z-wavealliance.org
27. W. Granzer, W. Kastner, G. Neugschwandtner, and F. Praus, Security in networked building automation systems, *Proceedings of the Sixth IEEE International Workshop on Factory Communication Systems*, pp. 283–292, Torino, Italy, June, 2006.
28. W. Kastner, G. Neugschwandtner, S. Soucek, and H. M. Newman, Communication systems for building automation and control, *Proceedings of the IEEE*, 93(6): 1178–1203, 2005.
29. W. Kastner and G. Neugschwandtner, *Data Communication for Distributed Building Automation*, CRC Press, Boca Raton, FL, 2009.
30. H. M. Newman, *Direct Digital Control of Building Systems: Theory and Practice*, John Wiley & Sons, New York, 1994.

31. T. Novak, A. Treytl, and P. Palensky, Common approach to functional safety and system security in building automation and control systems, *12th IEEE Conference on Emerging Technologies and Factory Automation*, pp. 1141–1148, Patras, Greece, September 2007.
32. C. Reinisch, W. Kastner, G. Neugschwandtner, and W. Granzer, Wireless technologies in home and building automation, *Proceedings of the Fifth IEEE International Conference on Industrial Informatics*, pp. 93–98, Prague, Czech Republic, June 23–27, 2007.
33. S. Soucek and T. Sauter, Quality of service concerns in IP-based control systems, *IEEE Transactions on Industrial Electronics*, 51(6): 1249–1258, 2004.

27

Industrial Multimedia

Javier Silvestre-Blanes
Universidad Politécnica de Valencia

Manfred Weihs
TTTech Computertechnik AG

Víctor-M. Sempere-Payá
Universidad Politécnica de Valencia

27.1 Introduction 27-1
27.2 Multimedia Compression: A Review 27-4
Image Compressors • Video Compressors • Quality Evaluation
27.3 Industrial Multimedia Applications 27-8
Monitoring Applications • Computer Vision Applications
27.4 Image Transmission 27-9
IEEE 1394 • IP-Based Networks
27.5 Conclusions 27-11
Acknowledgment 27-11
References 27-11

27.1 Introduction

It is commonly accepted that the multimedia applications that appeared in the mid-1990s are the third generation of computer applications [P98]. The first generation was characterized by its ability to manage data, while the second generation was considered to be principally for communication. This has had an influence on computer instruction set architecture (ISA) in current processors with the introduction of multimedia extensions (e.g., MMX and SSE in Intel) or in the development and definition of LAN and WAN networks that were able to provide quality of service (QoS) to this traffic (such as IEEE 802.11e). The area of industry has not been an exception to this phenomenon and, from the beginning, the development of new industrial applications [P98,RSB99,WIF01] was predicted both for the high degree of interaction between human and industrial process that they allow and for their part in the developments in monitoring and control, quality control, factory automation, and factory communication. Nowadays it is possible to find multimedia applications in nearly all areas of industrial communications—automation, monitoring, image processing, robotics, remote control, etc.

These applications have a wide range of requirements, and there are numerous technologies that support them. Data compression techniques and techniques for transmission of this kind of information are among the most important of these technologies. However, it is necessary to first consider the structure of the data managed in this kind of application. Images or image sequences have a height and width, called resolution, which is measured in pixels. These pixels represent information for which a specific number of bits per pixel (bpp) is required, the most common being 8 bits in grayscale and 15 (5 bits for each color plane), 24, or 32 in color. Resolution, together with the necessary rate of frames per second (fps) provides the bit rate necessary for the application, normally measured in bits per second (bps). The most commonly used format in industrial and monitoring applications is 8 bpp in grayscale, and when working in color the RGB (red, green, blue) format, using 8 bits for each color plane. This means 24 bpp are used for color, as can be seen in Figure 27.1. However, the use of the YUV color format with subsampling is common, and this is generally used in the compression of color images. This format,

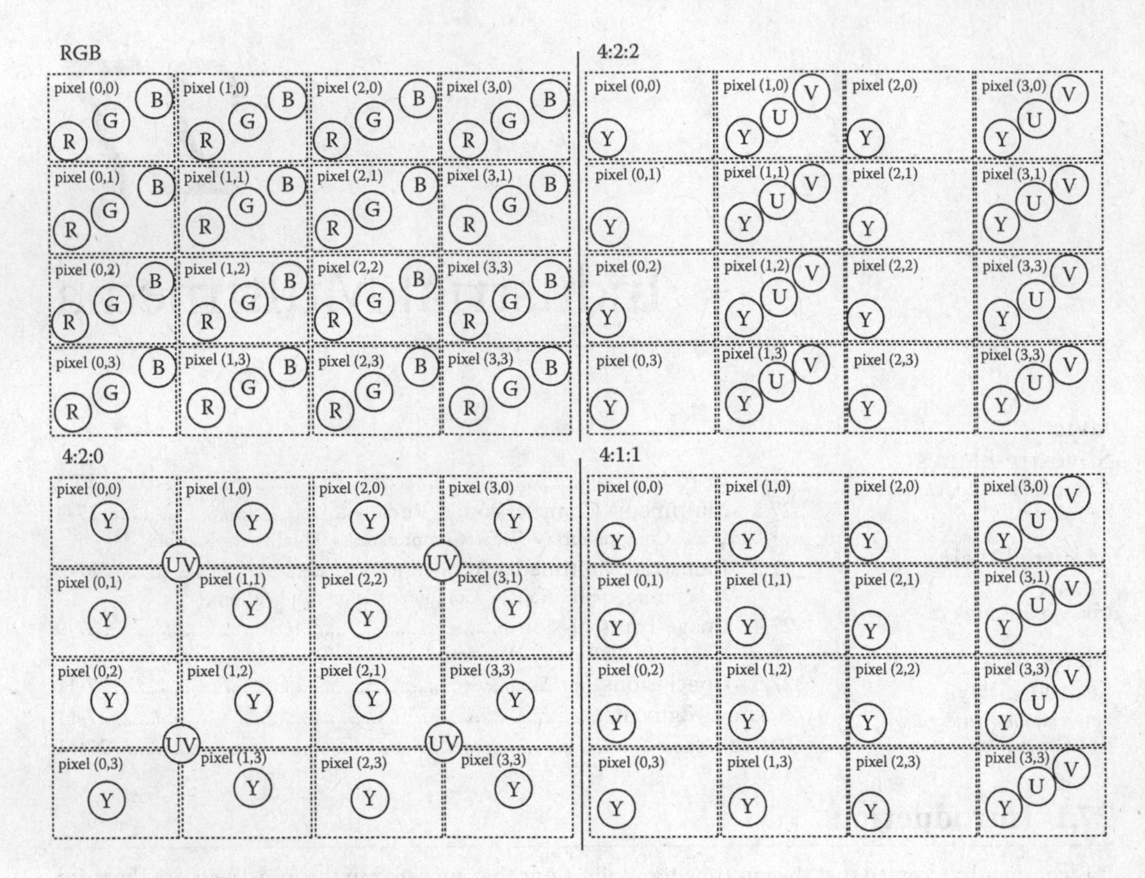

FIGURE 27.1 Multimedia information.

used in analog signals, has one plane used for luminance (Y or brightness) and the other two used by the chrominance component (U and V, the information on color). It is known that the human vision system (HVS) is more sensitive to the quality of the Y plane than the U and V planes, so subsampling is applied to reduce the amount of information in analog signals and can be seen as a first stage of compression.

The resolution and the fps rate are established by the type of application. Once established, these factors influence the requirements of the image's characteristics, the spatial relationship of the equipment participating in the application, and the implications for the temporal requirements that must be satisfied. Although there is a wide range of multimedia applications being used in factory automation, such as virtual manufacturing [WIF01] or online interactive training, we consider [SS07] process monitoring and control through image processing to be the most relevant. In the case of monitoring processes, where an application's temporal requirements are not critical, the sensor and the visualization node can be in the same machine, in the same plant, or in any part of the world. In the case of automated control, where applications do have critical temporal requirements, sensor and processing node are normally in the same machine (personal area network, distances of a few meters), but other solutions are possible as well. Reductions in costs and the increases in processing capacity are giving rise to a new generation of intelligent cameras, where capture and image processing are integrated in the same device (distances in centimeters). On the other hand, this area is beginning to see the influence of the expanding use of Ethernet to areas outside the office, and this is leading to the use of IP technology in image transmission over distances of hundreds meters normally, but also over greater distances, when there are less strict temporal requirements.

In Figure 27.2, the use of different types of communication networks used in multimedia applications depending on communication distances between sensor and processing node can be seen.

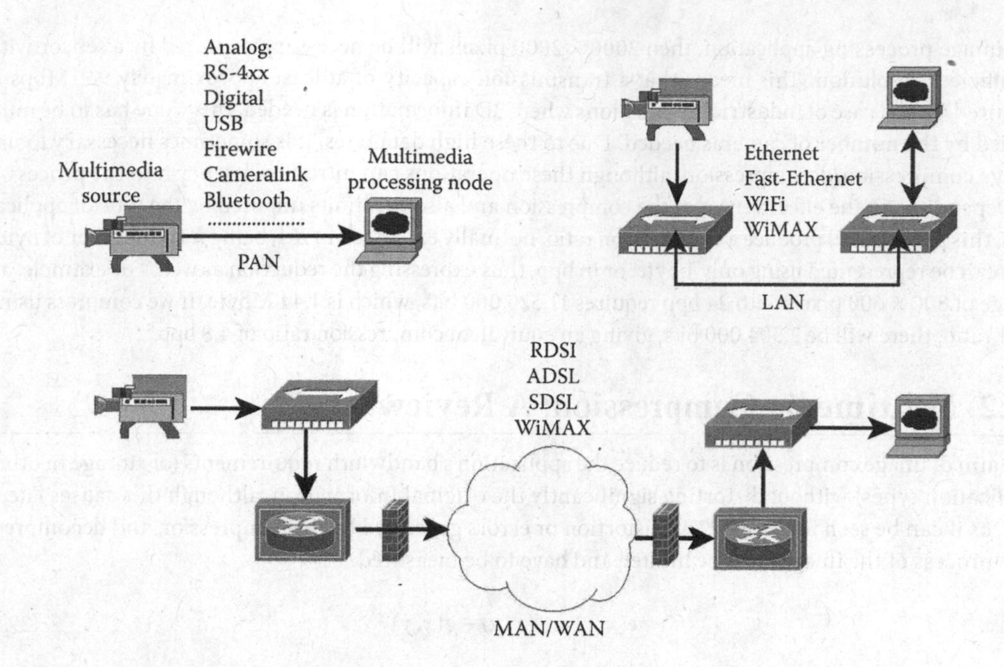

FIGURE 27.2 Multimedia communications in industrial environment.

TABLE 27.1 Image Resolutions in Multimedia Applications

Name	Resolution	Mbps	Name	Resolution	Mbps
SQCIF	128 × 96	8.84	QCIF	176 × 144	18.2
CIF	352 × 288	72.9	4CIF	704 × 576	291.9
QQVGA	160 × 120	13.8	QVGA	320 × 240	55.2
VGA	640 × 480	221.1	SVGA	800 × 600	345.6
XGA	1024 × 768	556.2	UXGA	1600 × 1200	1,382.4
QXGA	2048 × 1536	2264.9	SXGA	1280 × 1024	943.7
QSXGA	2560 × 2048	2831.1	HSXGA	5120 × 4096	15,099.5
WVGA	852 × 480	294.4	WXGA	1366 × 768	755.34
WSXGA	1600 × 1024	1179.6	WUXGA	1920 × 1200	1,658.8
WOXGA	2560 × 1600	2949.1	WQSXGA	3200 × 2048	4,718.6
WQUXGA	3840 × 2400	6635.5	WHSXGA	6400 × 4096	18,874.36
WHUXGA	7680 × 4800	26542.0	HD480	852 × 480	294.4
HD720	1280 × 720	663.5	HD1080	1920 × 1080	1,492.9

These networks have transmission capacities ranging from kbps to Gbps. Image resolutions can also be wide ranging, as can be seen in Table 27.1, where some of the currently existing resolutions are enumerated and where the bandwidth necessary for raw transmission with a rate of 30 images/s, expressed in megabit per second (Mbps), is shown.

In monitoring applications, there is more freedom to choose resolutions and capture rates, but with some restrictions. For example, in a security application in which the movement of people is being monitored, the capture speed must ensure that someone's passing is captured in at least one image. Further, the resolution used will determine if their face can be identified or not. In control applications there is less freedom, and these parameters are strictly defined by the application requirements. If the objects being monitored move at a speed of 600 objects/min, at least 10 images/s must be captured. If the objects have a dimension of 500 × 500 mm and 4 pixels/mm are needed to guarantee the target of the image processing application, then 2000 × 2000 pixels will be necessary, captured by a sensor with a sufficient resolution. This means that a transmission capacity of at least approximately 960 Mbps is required. In the case of industrial applications where 3D information is needed, this value has to be multiplied by the number of cameras needed. Due to these high data rates, it is sometimes necessary to use image compression/decompression, although these operations can introduce latency into the process.

Depending on the effectiveness of the compression and also the limits imposed by the type of application, this process will produce a compression ratio, normally expressed in X:1, being X the number of bytes that can be represented using only 1 byte, or in bpp, thus expressing the reduction as well. For example, an image of 800 × 600 pixels with 24 bpp requires 11,520,000 bits, which is 1.44 Mbyte. If we compress using a 5:1 ratio, there will be 2,304,000 bits, giving an equivalent compression ratio of 4.8 bpp.

27.2 Multimedia Compression: A Review

The aim of image compression is to reduce the application's bandwidth requirements (or storage in other application types) without distorting significantly the original information, although this causes latencies, as it can be seen in Figure 27.3. Distortion or errors generated by the compression and decompression process of the image must be limited and have to be measured:

$$e(x,y) = f(x,y) - g(x,y) \tag{27.1}$$

In Figure 27.3, we can see how the original image must go through a series of steps to achieve a reduction of the information to be transmitted. This will cause a latency T_c (compression time) that will depend on the complexity of the algorithm used. The compressed information must then be transmitted through the network, which means it is necessary to packetize the information and wait a period of time that will depend on the scheduling policy of the source and the priority of the multimedia information with regard to other information that is using the same medium (e.g., an alarm could have more priority). Finally, we have to access the medium and carry out the transmission of the information. This process will introduce a latency T_t (transmission time) that will be smaller the more efficient the compression process has been. There is also an unpack/reception time, but this is almost insignificant, and above all, a decompression time (T_u) depending on the chosen compressor.

There are several classifications of compressors, depending on the distortion that they introduce in the source, the complexity of the transmitter and receiver, and the type of redundancy employed. In the first place, there are compressors with or without losses. Lossless compressors do not distort the image, that is, $f(x, y) = g(x, y)$. These compressors are not used in excess because the rate of compression they can achieve is approximately 2:1, which restricts their use in the industrial area. However, they are used in other types of applications, such as medical applications, where degradation that can distort the medical diagnosis is not tolerated and where transmission latencies are not critical. In other types of

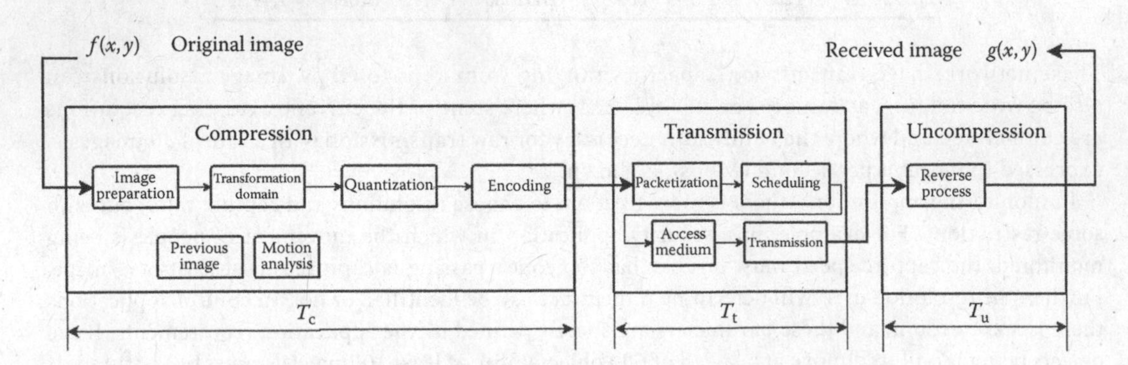

FIGURE 27.3 General multimedia application scheme.

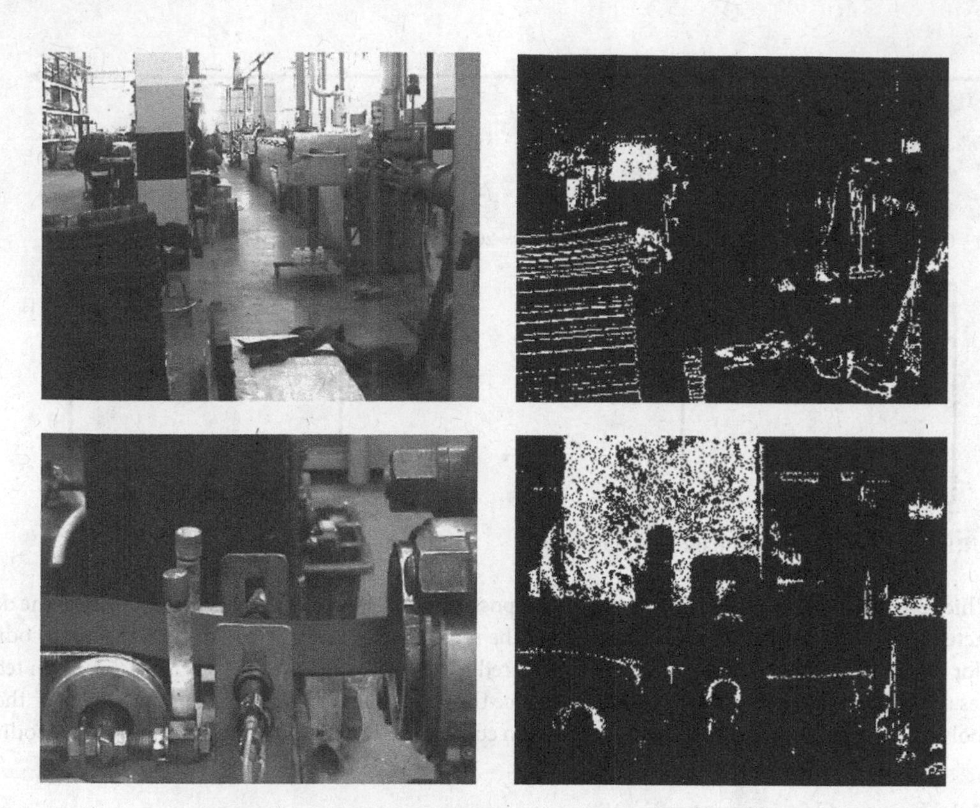

FIGURE 27.4 Spatial redundancy examples.

application where loss of information is tolerated, lossy compressors are used, which allow a significant increase in the compression rate. Concerning the complexity of the process, through an asymmetric approach, it is possible to use quite complex algorithms in the source to optimize the quality and bit rate (this can involve full two passes), while the decompression is performed in real time. However, this approach cannot be used in live real-time applications, which are the more common, so usually symmetric approaches are used, which have a similar complexity in compression and uncompression. Finally, there are compressors that only use the spatial redundancy of images, that is, of the self-contained information in an image, and compressors that also use the temporal redundancy, that is, the redundancy created by similarities between consecutive frames.

Spatial redundancy exploits the proven statistical correlation between the pixels of an image, so it allows us to extract a pixel's value from the neighboring pixels. With this technique, it is not necessary to reproduce each pixel of an image independently. In Figure 27.4, we can see two scenes of industrial monitoring [SSA04]. On the left, there are wide areas in the image that have a similar value, which is used by compressors to reduce the necessary bits to reproduce the information. On the left, all pixels in the interval [28–32] have been represented in white, which gives an idea of the spatial redundancy and its potential to reduce the number of bits to represent the multimedia information.

The temporal redundancy comes from the fact that a correlation between pixels of consecutive images in a sequence exists. Between consecutive frames, less than 10% of pixels change their value in about 1% of the peak signal. In this way, images pixels can be predicted from the pixels of an image nearby in the sequence. On the other hand, the fact that changes between consecutive images are caused by the movement of some objects in the scene has caused the development of motion compensation coding techniques. Figure 27.5 shows this in an industrial monitoring sequence captured at 25 frames/s. As we can see, the difference between consecutive images is very little in this case, and for this reason, in this type of sequence, the use of temporal redundancy allows us to obtain high rates of compression.

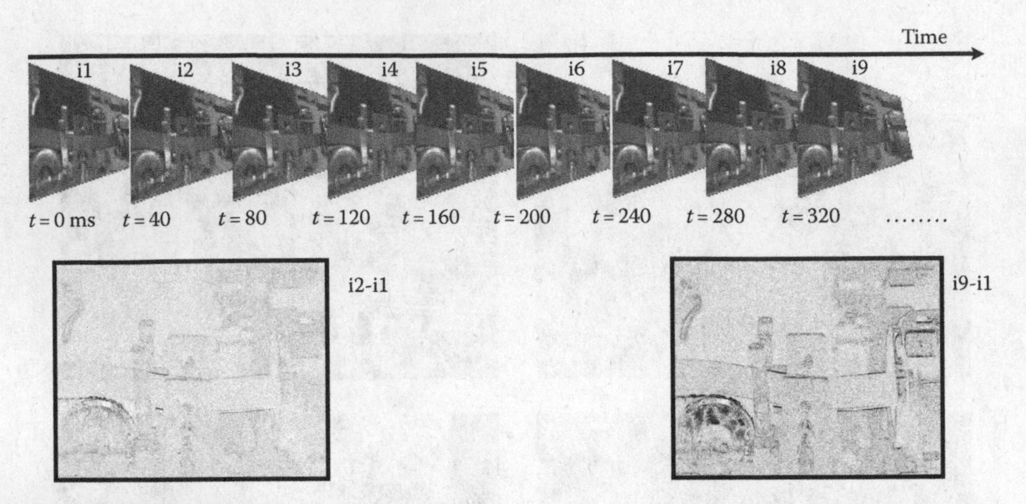

FIGURE 27.5 Temporal redundancy example.

Whichever method the compressor uses, it is still possible to make use of another redundancy of the data extracted—the code redundancy. This is related to the reproduction of information with the same coding and for which there are techniques that allow us to reduce the number of bits per symbol, based on techniques that allow the use of very few bits to reproduce the most common symbols, and more bits for those symbols that appear less frequently, such as Huffman coding, arithmetic code, and variable-length coding.

27.2.1 Image Compressors

Still image compressors were developed in mid-1980s, with International Telecommunication Union (ITU) and International Organization for Standardization (ISO), followed by ISO/IEC international standard JPEG (Joint Photographic Experts Group) in 1992 [ISO94]. This standard takes in four compression modes, although not all of them are always supported. These modes are sequential discrete cosine transformation (DCT)-based mode, progressive DCT-based mode, hierarchical mode, and lossless mode. The first is the baseline of JPEG and the most common one. Progressive and hierarchical modes enable a progressive presentation of the image, and the latter is also useful in applications with multiresolution requirements. The standard JPEG prepares the image by carrying out a color space transformation from RGB to YUV 4:2:2 or 4:1:1, and the data unit is a block of 8 × 8 pixels. These units are transformed through a DCT obtaining 64 DCT coefficients. These coefficients are mapped to integer numbers and, depending on the quality specified, the coefficients lower than a given value are set to 0. These coefficients are ordered in a "zig-zag" form and coded using Huffman followed by a runlength coding. The standard JPEG2000 [CSE00] tries to solve most of JPEG's limitations while at the same time providing a significant quality improvement for the same compression rates. To achieve this objective, it uses more advanced techniques in each stage of the compression. In this way, instead of using the DCT transform, it uses wavelet transformation on the tile components, which gives different decomposition levels, each of them containing a number of subbands with the transformation coefficients. The transformed coefficients of each subband are arranged in code blocks and the symbols produced by a bit-plane coding technique are coded through an adaptative binary arithmetic code. Today, it is generally accepted that this standard only outperforms JPG significantly when very high compression rates have to be used, since the computation complexity is quite high compared to JPEG.

27.2.2 Video Compressors

There are two main families of video compressors, the ITU standards, denominated H.26x, and the ISO Motion Picture Expert Group standards, denominated MPEGx. The first ITU standard was H.261, which works only with QCIF and CIF resolutions (see Table 27.1). It works with blocks of 8 × 8 pixels and in the color space YUV with subsampling 4:1:1. The intraframe coding (denominated I frames using MPEG nomenclature) is based on JPEG technology, and the interframe coding (denominated P, from predictive) is based on motion estimation with respect to previous frames, with a search range of ±15 pixels, and coding the differences between macroblocks when these are higher than a threshold with DCT. The standard H.263 is an extension of H.261 reaching higher bit rates for the same quality. The more relevant differences are that the motion vector may be based also on future frames (if the frame is coded using past and future frames, it is denominated B, from bidirectional), an unlimited search space for motion vector, and the possibility of work with more resolutions such as 4CIF and 16CIF.

The other most important family of compressors are those generated by MPEG (Motion Picture Expert Group), from ISO/IEC. The first of these, MPEG-1 (ISO 11172, 1991) is very similar to H.261 and was developed to send audio/video at about 1.5 Mbps, using a subsampled 4:2:0 and the higher resolution of 786 × 576 pixels. Another difference is that it supports VCR-like operations (fast forward, rewind, direct random access, etc.). MPEG-2 (ISO 13818, 1993) was a standard developed for the broadcasting industry, with a bit rate between 4 and 6 Mbps. MPEG-4 (ISO 14496, 1999) was developed originally for teleconferencing applications but soon it showed adequate properties to be used in distributed multimedia applications. Virtual scenarios can be composed of different objects, which can be managed by the user independently of others. ITU introduced some interesting improvements in the standards that followed H.263. H.263+ presented SNR scalability (spatial and temporal) and H.26L presents some improvements in information coding. This work coincided with the developments made by MPEG, developments carried out in unison with developments on the standard H.264 (JVT, Joint Video Team), better known as MPEG-4 AVC (advanced video coding) (MPEG4 part 10). The target of new compressors was to get a bit rate reduction of 50% with respect to previous standards and achieve a more reliable codec in the presence of errors.

27.2.3 Quality Evaluation

The measurement of distortion produced by the compression/decompression process is still open to debate, and has been approached from various angles. First, the destination of the information is usually the human eye, so the degradation measurement must take this into account. This has produce mechanisms to get subjective measurement, being the mean opinion score (MOS) from experiments with human subjects the most common. However, this evaluation is time-consuming and expensive, so different objective methodologies designed to get quantitative measurement has been developed. These can be classified as full-reference and no-reference, depending on the availability of an original image, although here we are going to talk about full-reference methods, as it is the most common approach.

To measure the error between two signals, the simplest metric is the mean square error (MSE) which in multimedia is calculated by averaging the squared intensity differences of distorted and reference image pixels, and the most used is the peak signal-to-noise ratio (PSNR), measured in decibels. Being MAX the maximum possible pixel value of the image, it can be calculated as

$$\mathrm{PSNR} = 20\log_{10}\left(\frac{\mathrm{MAX}^2}{\mathrm{MSE}}\right) \tag{27.2}$$

These metrics are quite common because they are simple to calculate and have clear physical meanings, but they are not well-matched to the perceived visual quality. This fact has promoted the development of new metrics, such as the structural-similarity (SSIM) based [WBSS04] for a deeper explanation on the formulas that enable their calculation, which is not based on a measurement of the error since it is based on image formation properties: luminance, contrast, and structural information. Other examples of new metrics are UQI, VQM, PEVQ, and CZD, analyzed by the Video Quality Expert Group (VQEG), some of which were standardized as ITU-T Rec. J. 24 g and J. 247 in 2008 [VRZ07].

27.3 Industrial Multimedia Applications

27.3.1 Monitoring Applications

Monitoring applications are very well known in industrial areas, although mainly at the LAN and WAN level. Typically, in all kinds of industries machines are quite large, and because of this different cameras transmit the information to the control point to ensure the correct management of the machine by the worker. Another application is that the images are required from another point of the factory for work supervision from global control centers. These systems are often used in areas that present a certain danger for the worker. Finally, global control centers need to monitor the remote installations' state in order to control their performance (especially to control network utilities [SAS06][SSA07]). In all of these situations, multimedia information is transmitted through networks for a better use of machines and equipments due to the rich context information that multimedia has for the workers.

The requirements of this type of application are not typically too demanding, but they are connected to the final use of the images in the application and the networks available, since it can involve strong limitations. The image resolutions are usually between CIF and 4CIF or VGA since these resolutions are more than adequate to gather the information of interest and do not overload the human capacity for processing this type of information. Greater resolutions can be used if the images, in addition to live human supervision, are stored for a later use (e.g., technical reports in the control of utilities networks, complaints in surveillance applications). The quality of the images is another important parameter, and for monitoring applications, it is considered appropriate to have at least 30 dB. Concerning transmission times, we can consider here two groups. In the applications where only live monitoring is performed (without storage) and the information is used by the worker who is operating the machine, this latency has to be lower than 1000 ms, since the machinery has to be managed with real-time information. In the second group we can consider the applications where a human operator receives and processes the information, usually from several sources, and in which there is no storage. In this case, the latency is connected with the human capacity for processing this type of information.

27.3.2 Computer Vision Applications

These types of applications are characterized by the use of image processing techniques for industrial automation. There are basically three application types: offline inspection applications, online inspection applications, and guidance applications (such as robots). In the first case, they are applications of quality control. Their objective is to know the product's state in order to determine what further processing is necessary, or in the case of finished products, its delivery to the customer. A very high working speed may be required even though there are no hard real-time requirements. For example, quality inspection of 20–80 m/min of fabric, but in which the faults are stored for report purposes, or, in cases where it is necessary to stop the machine, there is some tolerance in the order delay. In the second case, the vision system inspects the product or the process in order to determine what action is necessary. In other words, the image processing result will have consequences on the production process (e.g., regulation on the width of rubber bands [SS07]), and therefore, there are some temporal restrictions that must be adhered to in the transmission and processing of images. In the third application type, the information extracted from the images is used to guide the management of cutting stands, robots, etc., and in which case adherence to temporal requirements is even more important.

A common requirement of the three types of computer vision application is the necessity of high-quality images. There has been some work carried out to analyze the limits of image degradation, but this has been mainly for specific medical applications, and there have been no studies on industrial environments or into a general theory to gain prior knowledge of this. The target of the computer vision application determines the image resolution and the transmission times. Typical applications use 4CIF resolution with transmission between 1 and 60 frames/s. In some cases, more than one camera is necessary, or even in wide network inspection, the use of lineal cameras (lineal cameras are devices with a sensor of only one line of pixels forming the image with the concatenation of consecutive lines).

27.4 Image Transmission

Since processing of image/video data typically does not take place in the camera that captures the data, the data has to be transmitted over some kind of data network. There are transport protocols that define how multimedia data has to be put into data packets that are sent on the network [W06]. This typically also involves some timestamps that are used for synchronization purposes. There are also protocols necessary to set up and start streaming of the data over the network. These protocols are usually tightly coupled with the transport protocols. Each kind of network has different kinds of data formats that are transported by the transport protocol. This depends on the special features (like available bandwidth) of the network. Two kinds of network will be described more in detail here: IEEE 1394 and IP-based networks (typically based on Ethernet) [W05].

27.4.1 IEEE 1394

IEEE 1394 [IEEE1394], also known as FireWire or i.Link, is a serial bus designed to connect up to 63 devices (like camcorders, CD players, satellite tuners, but also personal computers and peripherals like hard disks, scanners, etc.). The original version of this bus supports up to 400 Mbps, has an automatic configuration mechanism allowing hot plugging of devices, and is very well suited for streaming of multimedia content [A99]. It supports 4.5 m links and up to 16 hops between two devices leading to a maximum distance of 72 m between two nodes. This clearly shows that it is suited for small area environments like homes or automobiles, but it is used in industrial environments as well. IEEE 1394a [IEEE1394a] introduced some new features, mainly to increase the bus efficiency (reduction of idle times), but the main characteristics were preserved. The newer version IEEE 1394b supports up to 1600 Mbps and has architectural support for 3200 Mbps [IEEE1394b]. It introduced some new physical media in addition to shielded twisted pair (STP) copper cables and also increased the maximum distance between devices (up to 100 m using optical fibers or CAT-5 UTP). The transmission model and coding of data is completely different to IEEE 1394a; however, compatibility is maintained by introduction of bilingual nodes.

On the physical layer, a FireWire bus is a tree consisting of point-to-point connections between the devices. However, the physical layer transforms this tree into a logical bus, where each node (except the node that is currently transmitting a packet) acts as repeater. The higher layers (link layer and transaction layer of the three-layer design used in IEEE 1394) have a view as a bus, where all packets are received by every node, provided the used speed is supported on all links in between.

Two different data transfer modes are supported: asynchronous and isochronous data transfer. The isochronous mode has been designed to meet the requirements of multimedia streaming. There are 64 isochronous channels. Such channel can be regarded as a multicast group. In fact, the packets are transmitted to each node on the network, but the hardware inside the node usually filters and discards packets of isochronous channels that are not of interest. There is a resource management mechanism: bandwidth and channel can be reserved at the isochronous resource manager (a functionality provided by one of the nodes in the network) and are then guaranteed. This means that quality of service is provided. The requirement for the guarantee of isochronous bandwidth is that all nodes have to obey the rules of IEEE 1394 and reserve bandwidth and channel before they are used. Since this is not enforced by hardware or software, nodes violating these rules can compromise QoS. The source can send one isochronous packet every 125 μs (leading to 8000 isochronous cycles/s). The allowed size of the packet corresponds to the reserved bandwidth. In case of errors, no retransmission occurs, so timing is guaranteed, but not delivery. On the other side, the asynchronous transfer mode is used for delivery of variable-length packets to explicitly addressed nodes. There is an acknowledgment and a retry procedure in case of errors. Therefore, the delivery of integer data can be guaranteed, but timing is not guaranteed.

In the area of streaming multimedia data, asynchronous transactions are not used for the transfer of multimedia, but they are used for control purposes. IEC 61883-1 uses asynchronous transactions for setup and teardown of isochronous connections, which also includes the isochronous resource management [IEC61883-1]. Higher level control protocols like AV/C are based on IEC 61883-1 and use asynchronous transactions to communicate. Real-time transmission of audio and video data is well-standardized by the series of IEC61883. IEC 61883-1 defines general things like a common format for isochronous packets as well as mechanisms for establishing and breaking isochronous connections [IEC61883-1]. This also includes timestamps, which are used for synchronization [WZ03].

Transmission of data in the digital video (DV) format [IEC61834-2] (typically used on handheld digital video cameras) is specified in IEC 61883-2 [IEC61883-2]. An advantage of this format is the high quality and a low complexity of encoding and decoding because it does not use the sophisticated interframe compression techniques known from MPEG-2. On the other hand, it imposes rather high bandwidth requirements (about 25 Mbps for an SD format DV video stream, together with audio and other information it yields about 29 Mbps).

For transmission of audio data, IEC 61883-6 is used [IEC61883-6]. It is used by consumer electronic audio equipment like CD players, amplifiers, speaker systems, etc. Usually, audio is distributed as uncompressed audio samples at 44,100 or 48,000 Hz sample rate, but other data formats are supported as well. But audio distribution is definitely more of interest in home environments than in industrial ones.

Transmission of MPEG-2 transport streams [MPG2] over IEEE 1394 is specified in IEC 61883-4 [IEC61883-4]. This allows the distribution of digital television over IEEE 1394, since both DVB and ATSC use MPEG-2 transport streams. MPEG-2 provides a good trade-off between quality and utilized bandwidth. A typical PAL TV program requires about 6 Mbps.

In addition to the two possibilities based on IEC 61883 mentioned above to transmit video in DV or MPEG-2 data format, there is another protocol that is primarily targeting industrial environments (albeit there are some Web cameras using this protocol as well): the Instrumentation & Industrial Digital Camera (IIDC) protocol [IIDC1394]. This is a very simple protocol (not based on IEC 61883) that transmits uncompressed digital video over isochronous FireWire channels. Therefore, it requires significantly more bandwidth, but does not require encoding and decoding.

27.4.2 IP-Based Networks

IP, the Internet protocol, is a very widely used network protocol. The main reason is that it can be used on top of almost every lower layer networking system (Ethernet, token ring, IEEE 1394, etc.) [AST03]. These networks were not designed for the purpose of multimedia transmission and, therefore, have some weaknesses with regard to real-time data transmission. QoS is very limited. There are mechanisms for QoS in IP networks, but they provide only soft guarantees. The resource reservation protocol (RSVP) is the standard protocol to reserve resources in IP networks [WZ01]. In fact, RSVP is only a signaling protocol for resource reservation requests and does not by itself reserve resources. IP networks only offer best effort services, and therefore, QoS is usually only achieved by oversizing the network and providing much more performance than needed in average to ensure good performance even in worst-case scenarios.

Besides the so-called integrated services approach with per-stream reservations, there is also the differentiated services approach, which does not provide per-stream reservations, but has advantages in scalability [WZ01]. Below IP, on the Ethernet layer, IEEE 802.1p also provides some kind of QoS mechanism. But this is very weak. It just allows defining priorities of individual Ethernet packets. It does not reserve resources and does not provide any guarantees. One problem of IP networks is that transmission of audio and video is not very well-standardized. There is a wide variety of data and transmission formats, open and proprietary ones, which are not compatible with each other.

In IP-based networks, many different protocols are commonly used for real-time transmission of multimedia. The most primitive one is HTTP, the hypertext transfer protocol. There are implementations that use HTTP to stream real-time audio data. But besides some advantages (it is very simple and works well with proxies and firewalls), it also has many disadvantages. It does not support multicast, only unicast. Therefore, if more than one node is listening to the stream, several connections have to be established that use the corresponding multiple of the necessary bandwidth. Furthermore, HTTP is based on TCP, which is a reliable protocol, that is, it does perform retransmission of packets that were not transmitted correctly. This is completely inadequate for real-time transmission because the retransmitted data will often be late and, therefore, useless. In fact, the use of HTTP is just an extension of progressive download (the presentation of a multimedia file starts while downloading), but it was never designed for real streaming.

A much more suitable protocol for transmission of multimedia data is real-time transport protocol (RTP) as proposed in RFC1889, which is usually based on UDP and in principle supports multicast as well as unicast. It also includes timestamps, which are used for synchronization purposes; but it has to deal with the fact that in IP networks there is no global time base that is synchronized automatically. Real-time streaming protocol (RTSP) can be used for setting up the transmission. There are also proprietary protocols like Microsoft media server (MMS) or Real delivery transport (RDT) and others. They have similar features and are used by the tools of the corresponding companies (Microsoft, RealNetworks, etc.).

Many different (open and proprietary) formats for audio and video exist and can be transmitted in an RTP stream [W05]. Concerning audio, the most important open formats are uncompressed audio with 16, 20, or 24 bit resolution and MPEG compressed audio streams. For video, RFC2250 specifies how to transmit MPEG video elementary streams. If audio and video are to be transmitted in one stream, MPEG-2 transport streams can be used, but the concept of RTP is to transport each media in a separate stream rather than using multiplexes. Digital video in the DV format of IEC 61834 [IEC61834-2] can be transported in RTP according to RFC3189 [IEC61883-2]. However, it should be mentioned that this is a format commonly used on IEEE 1394 [IEC61883-2] but not on IP networks. Important proprietary formats are RealAudio and RealVideo.

IPv6, the successor of IP version 4, has some advantages concerning multimedia streaming. Multicast has been integrated into IPv6, so every compliant implementation supports multicast. There are features (especially the flow labels should be mentioned) that make the provision of QoS for streams easier. The functionality of Internet Group Management Protocol (IGMP) is integrated into ICMPv6 (Internet control message protocol). The "multicast listener discovery" (MLD) used for the signaling between routers and hosts is a subprotocol of ICMPv6.

27.5 Conclusions

Multimedia has a wide range of applications in the industrial area. These applications also have a wide range of requirements, like resolution, frame rates, image quality, latency, etc. Industrial communications networks provide the necessary infrastructure for many applications. In this chapter, industrial multimedia applications and their particular requirements are discussed. Compression technology, which is needed in many applications due to bandwidth or storage limitations, is briefly reviewed. Finally, two different examples of network and the associate protocols for transmission of multimedia data are presented.

Acknowledgment

This work was partially supported by de MCYT of Spain under the project TSI2007-66637-C02-02.

References

[A99] D. Anderson. *FireWire System Architecture: IEEE 1394a*, 2nd edn., Addison-Wesley, Reading, MA, 1999. ISBN 0-201-48535-04.

[AST03] A. S. Tanenbaum. *Computer Networks. Paerson Education International*, 4th edn., Upper Saddle River, NJ, 2003. ISBN 0-13-038488-7.

[CSE00] C. Christopoulos, A. Skodras, and T. Ebrahimi. The JPEG2000 still image coding system: an overview, *IEEE Transactions on Consumer Electronics* 46(4), 1103–1127, 2000.

[IEC61883-1] IEC. *Consumer Audio/Video Equipment—Digital Interface—Part 1: General. Standard IEC 61883-1*, 1st edn., IEC, Geneva, Switzerland, February 1998.

[IEC61834-2] IEC. *Recording—Helical-Scan Digital Video Cassette Recording System Using 6,35 mm Magnetic Tape for Consumer Use (525–60, 625–50, 1125–60 and 1250–50 systems)—Part 2: SD Format for 525–60 and 625–50 Systems*. Standard IEC 61834-2, 1st edn., IEC, Geneva, Switzerland, August 1998.

[IEC61883-2] IEC. *Consumer Audio/Video Equipment—Digital Interface—Part 2: SD-DVCR Data Transmission*. Standard IEC 61883-2, 1st edn., IEC, Geneva, Switzerland, February 1998.

[IEC61883-6] IEC. *Consumer Audio/Video Equipment—Digital Interface—Part 6: Audio and Music Data Transmission Protocol*. Standard IEC 61883-6, 1st edn., IEC, Geneva, Switzerland, October 2002.

[IEC61883-4] IEC. *Consumer Audio/Video Equipment—Digital Interface—Part 4: MPEG2-TS Data Transmission*. Standard IEC 61883-4, 1st edn., IEC, Geneva, Switzerland, February 1998.

[IEEE1394] IEEE Computer Society. *IEEE Standard for a High Performance Serial Bus*. Standard IEEE 1394-1995, New York, 1995.

[IEEE1394a] IEEE Computer Society. *IEEE Standard for a High Performance Serial Bus—Amendment 1. Standard IEEE 1394a-2000*, New York, March 2000.

[IEEE1394b] IEEE Computer Society. *IEEE Standard for a High Performance Serial Bus—Amendment 2. Standard IEEE 1394b-2002*, New York, December 2002.

[IIDC1394] 1394 Trade Association. *IIDC 1394-based Digital Camera Specification Ver.1.31, TA Document 2003017*, Grapevine, TX, February 2004.

[ISO94] *Information Technology-Digital Compression and Coding of Continuous Tone Still Images: Requirements and Guidelines*. International Organization for Standardization/International Electrotechnical Comission (ISO/IEC) 10918-1, Geneva, Switzerland, February 1994.

[MPG2] ISO/IEC. *Information Technology—Generic Coding of Moving Pictures and Associated Audio Information: Systems*. International Standard ISO/IEC 13818-1, 2nd edn., Geneva, Switzerland ISO/IEC, 2000.

[P98] J. R. Pimentel. Industrial multimedia system, *IEEE Industrial Electronics Newsletter* 45(2), 1998.

[RSB99] S. M. Rahman, R. Sarker, B. Bignaill. Application of multimedia technology in manufacturing: A review, *Computer in Industry* 38, 43–52, 1999.

[SAS06] V. Sempere, T. Albero, and J. Silvestre. Analysis of communication alternatives in a heterogeneous network for a supervision and control system, *Computer Communications* 29(8), 1133–114, 2006.

[SS07] J. Silvestre and V. Sempere. An architecture for flexible scheduling in Profibus networks, *Computer Standards & Interfaces* 29, 546–560, 2007.

[SSA04] J. Silvestre, V. Sempere, and T. Albero. Industrial video sequences for network performance evaluation, in *Proceedings of International IEEE Factory Communication Systems*, Vienna, Austria, 2004, pp. 343–347.

[SSA07] J. Silvestre, V. Sempere, and T. Albero. Wireless metropolitan area networks for telemonitoring applications, in *Proceedings of the Seventh IFAC International Conference on Fieldbuses & Networks in Industrial & Embedded System*, Toulouse University, France, November 7–9, 2007.

[VRZ07] M. Vranjez, S. Rimac-Drlje, and D. Zagar. Objective video quality metrics, in *Proceedings of the 49th International IEEE Symposium ELMAR*, Zadar, Croatia, September 12–14, 2007, pp. 45–49.

[W05] M. Weihs. Multimedia streaming in home environments, in P. Lorenz and P. Dini, editors, *Networking—ICN 2005: Fourth International Conference on Networking*, Reunion Island, France, April 17–21, 2005; Proceedings, Part I, number 3420 in LNCS, pp. 873–881, Reunion Island, France, April 2005. IEEE, IEE, IARIA, Springer-Verlag. ISBN 3-540-25339-4. doi: 10.1007/b107117.

[W06] M. Weihs. Convergence of real-time audio and video streaming technologies. PhD dissertation, Vienna University of Technology, Vienna, Austria, 2006.

[WBSS04] Z. Wang, C. Bovik, H. R. Sheikh, and E. P. Simoncelli. Image quality assessment: From visibility to structural Similarity, *IEEE Transactions on Image Processing* 13(4), 600–612, 2004.

[WIF01] C.-H. Wu, J. D. Irwin, and F. F. Dai. Enabling multimedia applications for factory automation, *IEEE Transactions on Industrial Electronics* 48(5), 913–919, 2001.

[WZ01] R. Wittmann and M. Zitterbart. *Multicast Communication—Protocols and Applications*, Academic Press, San Diego, CA, 2001. ISBN 1-55860-645-9.

[WZ03] M. Weihs and M. Ziehensack. Convergence between IEEE 1394 and IP for real-time A/V transmission, in D. Dietrich, P. Neumann, and J.-P. Thomesse, editors, *Fieldbus Systems and Their Applications 2003*, Aveiro, Portugal, IFAC, Elsevier, 2003, pp. 299–305. ISBN 0-08-044247-1.

28

Industrial Wireless Communications Security (IWCS)/C42

Milos Manic
University of Idaho
Idaho Falls

Kurt Derr
Idaho National Laboratory

28.1 Introduction 28-1
28.2 Wireless LAN Security 28-2
Security Issues/Attacks on WiFi • Security Mechanisms • Deployment Issues
28.3 PAN Security 28-6
28.4 Summary 28-8
References 28-8

28.1 Introduction

The use of wireless communications in industrial environments poses a different set of challenges than deployment in home and enterprise environments. Significant amounts of electrical noise such as variable frequency drives, competing radio systems, radar and microwave sources, and welders are produced in industrial environments. Signal attenuation and reflects are always present [1].

The placement of wireless access points (APs), or radio nodes, and adjustment of transmitter power to match receiver sensitivities is more crucial in industrial settings where the reliability of the overall system can affect plant safety and security as well as operational cost. Data latency, the time delay experienced when data is sent from one point to another, as well as possible data corruption can affect system performance.

Security is a significant obstacle to wider deployment of wireless technology in enterprise and industrial environments. The security issues and protocols as well as deployment guidelines are presented for wireless local area networks, personal area networks, and mesh networks that may concern industrial users.

The relationship between various wireless networks in terms of range versus data rate (number of bits processed per unit of time) is shown in Figure 28.1. Those are wireless personal area network (WPAN), wireless metropolitan area network (WMAN), wireless local area network (WLAN), and wireless wide area network (WWAN). WWAN typically assume protocols such as IEEE 802.22 or wireless regional area network, while WMAN include IEEE 802.20, or mobile broadband wireless access (MBWA). These wireless networks are in detail specified by IEEE standard specifications [2,3]. The most ubiquitous networks belong to WPAN technologies. These are (lowest to highest data rate) ZigBee, Bluetooth, and 802.15.3 (high data rate WPAN). This chapter presents network security issues and solutions of WLAN, WPAN, and mesh networks, in the order which is representative of their market size and their longevity in the marketplace.

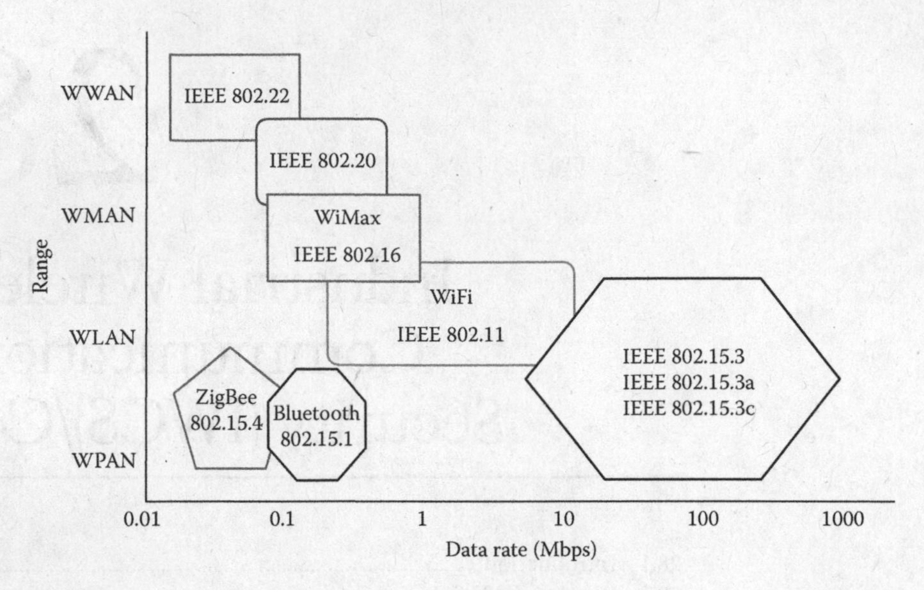

FIGURE 28.1 Wireless technologies: range versus data rates.

28.2 Wireless LAN Security

WLAN standards extend the functionality of a wired LAN providing location freedom (cable-free networking). WLANs are based on the IEEE 802.11 standard. The most widely used WLANs are in the industrial, scientific, and medical (ISM) radio band. Table 28.1 shows the 802.11 Protocol Family. The IEEE 802.11n standard was finalized in September 2009 and published in October 2009. Many "draft IEEE 802.11n" products were available in the marketplace prior to the adoption of this standard.

28.2.1 Security Issues/Attacks on WiFi

The characteristics of the wireless communications channel leave WiFi communications open to attack (authentication, access control, encryption, jamming, interception, and hijacking) [4–6]. These characteristics are discussed further in this section.

Authentication is the process of verifying that someone/something is authentic; i.e., the claims they make are true. Security protocols are used to authenticate computers in possession of a cryptographic key and users. Mobility and ease of connectivity are advantages of wireless versus wired communications.

Access control is necessary to protect against unauthorized use of wireless communications systems. Several approaches to controlling access to a wireless LAN include

1. Use of shared keys
2. Media access control (MAC) address filtering by a wireless AP
3. Link-layer security protocols
4. Firewalls
5. Virtual private network (VPN) encryption in order to provide security

TABLE 28.1 IEEE 802.11 Standards

Standard	802.11	802.11a	802.11b	802.11g	802.11n
Year	1997	1999	1999	2003	2009
Frequency	2.4 GHz	5 GHz	2.4 GHz	2.4 GHz	2.4 GHz, 5 GHz
Max Data Rate	2 Mbps	54 Mbps	11 Mbps	54 Mbps	600 Mbps
Modulation Type	FHSS, DSSS, & IR	Orthogonal FDM	DSSS	Orthogonal FDM	Orthogonal FDM

Encryption transforms data from a readable form to a nonreadable form for humans. The key length is an indicator for the strength of the encryption algorithm. Examples of encryption algorithms include RC4 (Rivest Cypher 4), Data Encryption Standard (DES), Triple-DES, Blowfish, International Data Encryption Algorithm (IDEA), Software-Optimized Encryption Algorithm (SEAL), RSA (Rivest Shamir Adelman), and RC4. Encryption prevents eavesdropping of wirelessly transmitted data.

Radio communications are subject to jamming regardless of the form of wireless signal. APs monitor channel quality and bit rate for other stations, which enables the detection of jamming. However, unless APs have collaborative software for sharing and analyzing this information, the location of the attacker cannot be identified.

Wireless signals radiate in free space and are subject to interception. Wireless laptop computers placed near industrial enterprises can intercept WLAN signals, collect sensitive information, and potentially disrupt the network.

Hijacking a wireless channel is a difficult task because the attacker must ensure that the two parties cannot communicate with one another [7]. The two users must be out of wireless range or be desynchronized to set up a man in the middle (MITM) attack. In an MITM attack, the attacker must eavesdrop on both users and impersonate each user to the other. One MITM attack approach would be to jam the receiver of one user using a directional antenna while receiving the transmitted traffic from another user.

28.2.2 Security Mechanisms

Security mechanisms and protocols are necessary to maintain the secrecy of data transmitted through the air and to ensure that the data is not tampered with. Since the introduction of 802.11 WLAN, new protocols have been developed as insecurities were found in existing deployed protocols (WEP, WPA, WPA2, TKIP, CCMP, and WAPI). A survey of these protocols follows.

28.2.2.1 Wireless Encryption Protocol

The first security mechanism for WiFi is the wireless encryption protocol (WEP), which requires little computational power. WEP is based on the RC4 encryption algorithm and is not as sophisticated as the cryptographic protocols that follow. Researchers demonstrated the insecurity of WEP within the first few years of its deployment [8].

WEP uses a symmetric secret key cipher, k, and an initialization vector, IV, to generate a keystream (a pseudorandom sequence of bits) as shown in Figure 28.2.

The decryption key is identical to the encryption key in symmetric key algorithms. WEP is not scalable due to a lack of automatic key management. The key is either 64 or 128 bits. An integrity checksum is computed on the message/data and then the two are concatenated to create a plaintext. The keystream is mathematically combined (Exclusive OR) with the plaintext to create a ciphertext. The ciphertext and IV are transmitted between a sender and receiver. The receiver uses the identical secret key to recover the message from the ciphertext. RC4 was developed in 1987 by Ron Rivest of RSA Security. The process for the RC4 encryption and decryption is shown in Figure 28.3.

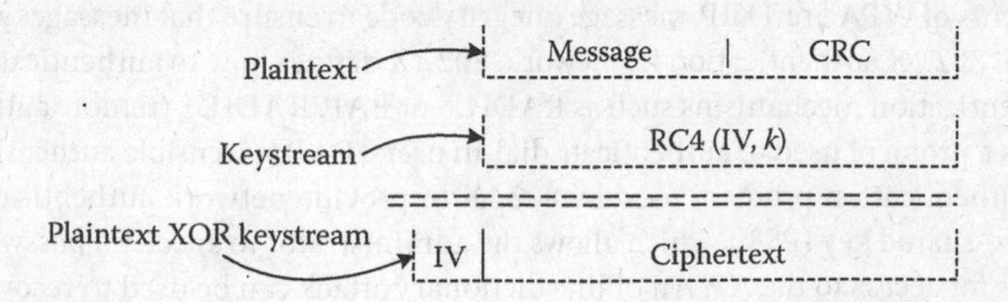

FIGURE 28.2 Encrypted WEP frame.

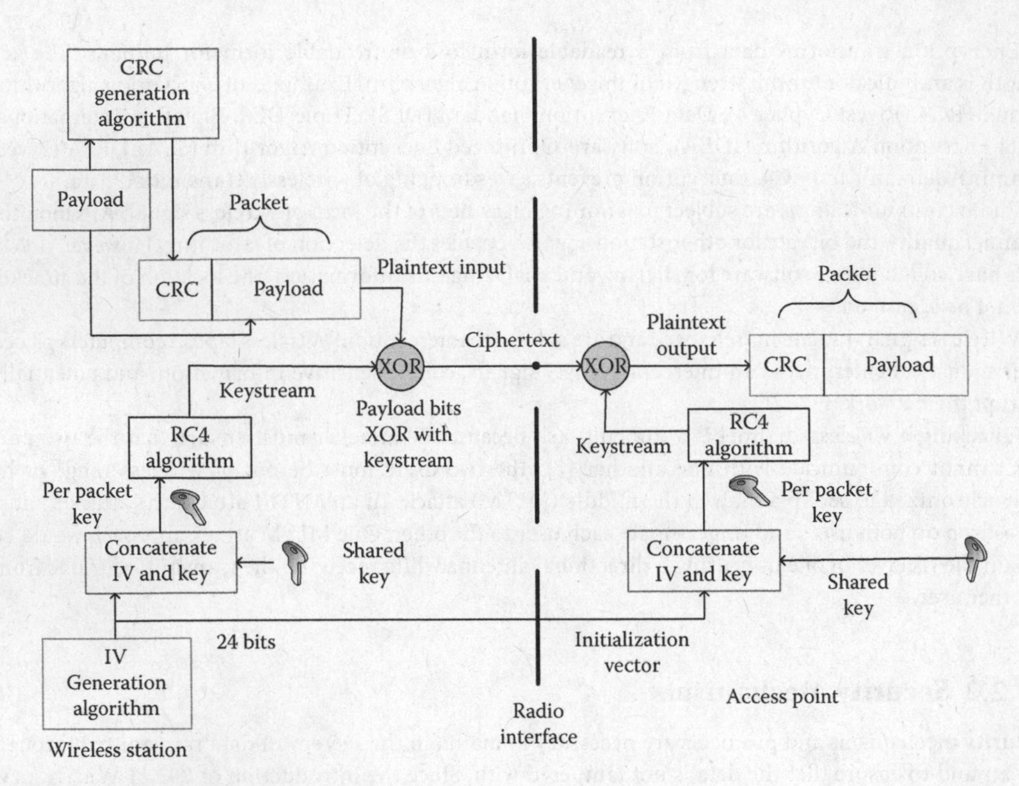

FIGURE 28.3 RC4 encryption/decryption process.

The basic insecurities of RC4 are

- Pseudorandom IV
- Exclusive OR based
- Weak keys
- Keystream reuse

WEP has been cryptographically broken due to its reuse of IVs. If the attacker can capture enough data, the attacker can decrypt the encrypted data without ever learning the encryption key. There are a variety of WEP cracking tools freely available on the Internet, which include, but are not limited to, AirSnort, Wepcrack, WepAttack, and Asleap-imp [4].

28.2.2.2 WiFi Protected Access

WiFi protected access (WPA) is a security standard based on IEEE 802.11i draft 3 with early TKIP implementations. The main design goal of WPA was to make this protocol WEP compatible but more secure and to coordinate vendor solutions for WEP flaws. WPA corrects the problems with WEP by using a larger key length and improving the handling of IVs. WPA distributes different keys to different users with an IEEE 802.1X authentication server. A 48 bit IV and a 128 bit key are used with RC4 encryption.

The main elements of WPA are TKIP, message integrity code to ensure that messages are not tampered with, and the IEEE 802.1X authentication framework. 802.1X defines how to authenticate wired/wireless clients using authentication mechanisms such as RADIUS or EAP. RADIUS (remote authentication dial-in-user services) is a protocol used to authenticate dial-in users. EAP (extensible authentication protocol) is a framework authentication protocol used by 802.1X to provide network authentication. One of the flaws of WPA is pre-shared key (PSK), which allows the administrator to specify a password that must be known by all users for access to the AP. An offline dictionary attack can be used to recover the PSK if the password used is not sufficiently long [9].

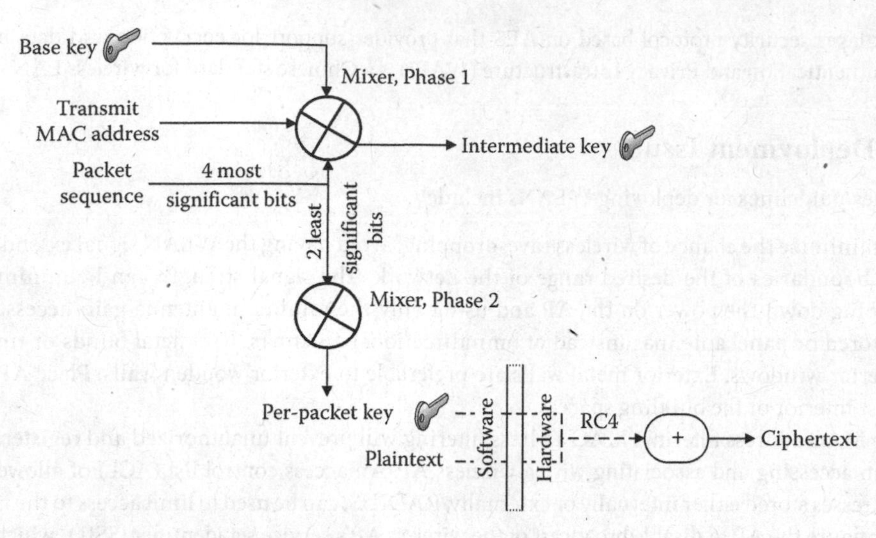

FIGURE 28.4 Temporal key integrity protocol.

28.2.2.3 WiFi Protected Access 2

WiFi protected access 2 (WPA2) is based on the fully ratified version of IEEE 802.11i. WPA2 is WPA with a new advanced encryption standard (AES) based algorithm known as Counter Model with CBC-MAC protocol (CCMP). AES supports multiple encryption key sizes: 128, 196, and 256 bits. WPA2 provides enhanced security over WPA [10].

The temporal key integrity protocol (TKIP) depicted in Figure 28.4 is a link layer encryption protocol initially called WEP2. TKIP has security features that improve upon WEP and is designed as a software upgrade to WEP-based solutions; i.e., a wrapper around WEP. Some of the improvements made by TKIP include

1. Using multiple master keys instead of a single key as used in WEP
2. Deriving a unique RC4 key for each frame generated from a master key
3. Numbering of each frame with a sequence number to mitigate against replay attacks
4. Using a new integrity check hashing algorithm called *Michael* to detect frame forgeries.

28.2.2.4 TKIP, CCMP, WAPI

TKIP adds new keying and message integrity check (MIC) mechanisms to WEP to offer additional security. These new mechanisms are key hierarchy and automatic key management, per-frame keying, sequence counter, new MIC, and countermeasures on MIC failures [11].

Key hierarchy and automatic key management: Keys used to encrypt frames are derived from master keys that are refreshed in a secure manner.

Per-frame keying: A unique RC4 key is derived for each frame from the master key to lessen the likelihood of key attacks.

Sequence counter: Each frame is numbered with a sequence number to eliminate replay attacks (capturing traffic and transmitting the traffic at another time).

New MIC: Michael, a cryptographic integrity check hashing algorithm, is used to defeat forgeries.

Countermeasures on MIC failures: Two failed forgeries in a second cause a station to invoke a rekeying procedure.

TKIP dynamically changes keys as the system is used. A large IV used with TKIP defeats the key recovery attacks that were possible in WEP. TKIP temporarily solves four problems: forgeries with the MIC, replays with IV sequence enforcement, weak key attacks with key mixing, and collision attacks with rekeying. CCMP is the link layer security protocol based on AES that provides support for encryption and data integrity. WLAN Authentication and Privacy Infrastructure (WAPI) is a Chinese standard for wireless LAN.

28.2.3 Deployment Issues

Some issues/guidelines for deploying WLANs include

1. To minimize the chance of wireless eavesdropping, avoid having the WLAN signal extend beyond the boundaries of the desired range of the network. The signal strength can be minimized by turning down the power on the AP and using only the minimum antenna gain necessary. Use sectored or panel antennas instead of omnidirectional antennas. Use metal blinds or tinting on exterior windows. Exterior metal walls are preferable to exterior wooden walls. Place APs in the most interior of the building space.
2. Use MAC address filtering. MAC address filtering will prevent unauthorized and registered users from accessing and associating with a wireless AP. An access control list (ACL) of allowed MAC addresses stored either internally or externally, RADIUS, can be used to limit access to the network.
3. Configure the AP to disable broadcast of the wireless AP's service set identifier (SSID) which is used to identify a LAN that a user may associate with. This limits the exposure of the network to attackers.
4. WPA2 is preferable to WPA which is preferable to WEP.

28.3 PAN Security

A personal area network (PAN) represents a limited number of computer and communications devices that have the ability to form networks and exchange information. WPANs allow computing devices within close range of one another to communicate whether they are stationary or in motion. WPANs may use different technologies to communicate, such as Bluetooth, Infrared Data Association (IrDA), ultra-wideband (UWB), and ZigBee [12,13]. The IEEE 802.15 working group has defined several classes of WPANs: 802.15.1 (Bluetooth), 802.15.3 (high data rate WPAN), and 802.15.4 (low-rate WPAN). When it comes to security aspects of PAN, the entities of interest are Bluetooth Special Interest Group (SIG), the ZigBee Alliance, and the IEEE 802.15.4 working group (wireless sensor networks).

PANs have a wide application scope and may be used in building automation such as access control; smoke detection; heating, ventilation, and air conditioning (HVAC); and lighting. Other applications include industrial monitoring, automatic meter reading, medical sensing, and environmental data collection [14,15]. Security is an important issue in these control and data acquisition applications. These technologies (ZigBee and IEEE 802.15.4) can be linked together into networks of virtually unlimited distance.

PAN security concerns will be addressed with regards to the following aspects: eavesdropping, authentication, location tracking, configuration, jamming, collision, and others.

Eavesdropping—wireless communications in general is subject to eavesdropping and potentially destroying the confidentiality of the data. If encryption is not enabled, then an attacker may attempt to substitute misleading data for the authentic data.

Authentication and encryption, when enabled, occur at the low levels of the communications stack. This makes impersonation of a sending or receiving PAN unit difficult [16,17].

Location tracking—all PAN devices have unique addresses. Many devices are mobile and the radio signal is tractable. If the device can be identified with a specific user, the movements of the user can be tracked [18,19].

No matter how good a design specification is, implementation flaws can cause security holes. The PAN qualification process cannot test every possible permutation of PAN protocol exchanges. Some implementations of the first PAN specifications contained vulnerabilities that granted access to some service on the local device. Additionally, the protocols specified for PAN profiles leave room for manufacturer-specific data.

When PAN devices first appeared in the marketplace the typical configuration of these devices were with mode 1 or no security. Although most PAN devices today ship with some level of security enabled for various profiles/services, users have a tendency to turn PAN security off for ease of use and convenience. "Passkeys" are typically used as an authentication tool. Default or preset passkeys that are well known are standard on some devices and should be changed to protect confidentiality of information.

802.15.4 is also the basis for wireless sensor networks and mesh networking. An example of a mesh network for a factory environment is shown in Figure 28.5. Mesh APs are wirelessly networked with one of the APs possibly providing access to the Internet. Other applications include building automation where electrical devices that need to be controlled and monitored could interface to a mesh AP. A wireless mesh network (WMN) would enable maintenance workers using mobile handheld devices such as smartphones or personal digital assistants to gather data from wherever they are in a building.

Wireless mesh networks are an effective solution for building automation and control. The mesh APs in Figure 28.5 communicate only with their adjacent routers. In larger mesh network deployments, some of the APs would act as intermediate routers or relays between APs (multi-hopping) not within wireless range (non-line-of-sight) of one another. When an intermediate router becomes faulty, other routers/APs adapt and pick up the routing; i.e., the mesh network is self-healing, ensuring constant availability of information.

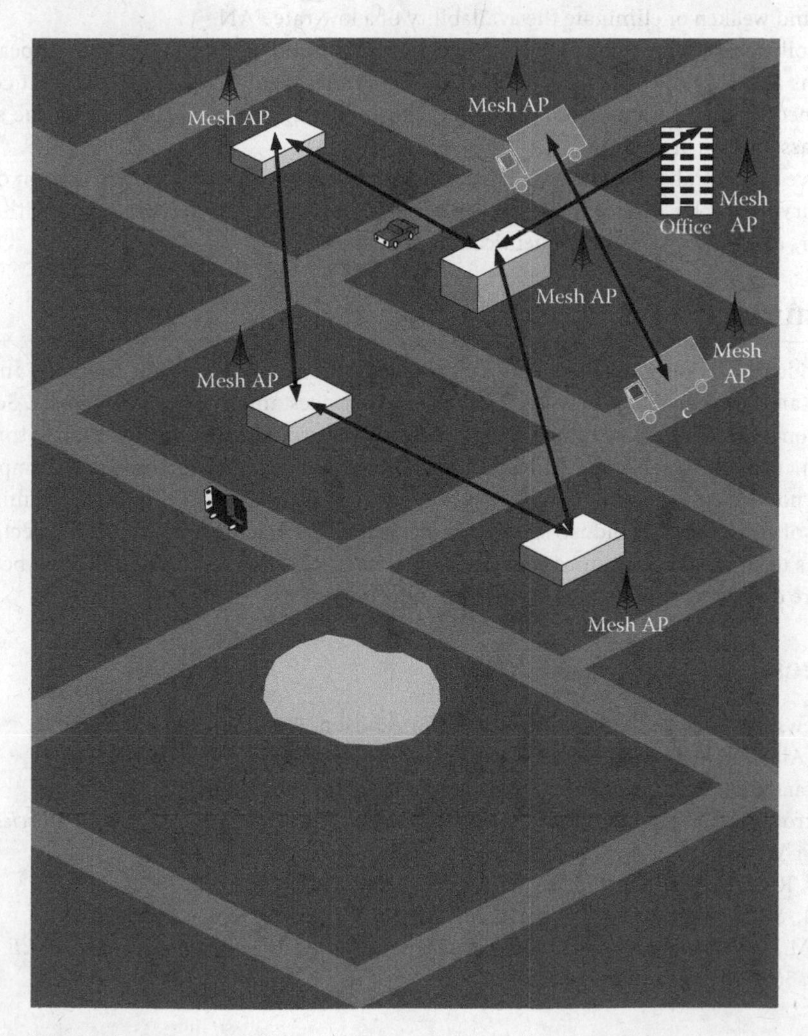

FIGURE 28.5 Large industrial mesh network environment.

The security issues of wireless mesh networks are similar to the security issues of other communications networks: availability, authenticity, integrity, and confidentiality. Attackers can affect the availability of the network by employing signal jamming. A defense against signal jamming is frequency hopping and spread spectrum, which widens the jamming range and changes frequencies periodically. Authenticity refers to nodes in a WMN ensuring the identity of their peer nodes. An authentication mechanism can be implemented on the basis of public key cryptography. Integrity is ensuring that the contents of a communication between sender and receiver is preserved intact. Integrity can be achieved through cryptography and the use of digital signatures. Confidentiality is ensuring that data is protected from breaches by unauthorized persons. Confidentiality can be achieved through the authentication mechanism previously discussed as well as encrypting the communications session.

A WMN has multiple nodes and channels in an open wireless environment that are open to attack. Eavesdropping on wireless channels and injecting fake messages into the channel is a potential threat. The self-healing adaptive change of network topology may make it easier for an attacker to impersonate a legitimate node. The nodes may not reside in physically protected places. Illegitimate nodes may violate the routing protocol and forward packets to destinations inconsistent with the routing protocol.

Low-rate PANs such as ZigBee are highly vulnerable to jamming attacks due to their very low transmission power [20]. A high-power-compliant transmitter could send out radio signals at the physical layer (PHY) and weaken or eliminate the availability of a low-rate PAN.

Selective collision of some control and management frames by an adversary may appear like random collisions and are not easily detectable. An attacker may send frames (packets) that collide with acknowledgment of association response frames, causing an exponential back off by the sender and restart of an association procedure.

WMNs have no central authority or trusted third party responsible for the distribution of keys that are used for cryptography. ZigBee does make use of keys, which can be preconfigured with devices, at different layers as well as challenge-authentication procedures.

28.4 Summary

Wireless in industrial environments is a difficult problem that requires multiple solutions and products while at the same time enabling new applications and services and driving down costs. Security is a paramount concern. Wireless systems provide security features such as confidentiality through AES end-to-end encryption so sensitive data is not interpreted, integrity codes to prevent tampering with information, non-repeating counters to prevent replay attacks, and channel hopping to minimize the risk of denial of service attack. Industrial environments contain machinery and metal objects that present challenges to wireless communications. Reliability and security considerations must be taken into account before deploying applications over wireless networks.

References

1. Goleniewski, L., *Telecommunications Essentials*, Addison-Wesley, Boston, MA, 2007.
2. ZigBee Alliance, http://www.zigbee.org/
3. IEEE Standards, http://standards.ieee.org/getieee802/download/802.15.4-2006.pdf
4. Vlasimirov, A., Gavrilenko, K., and Mikhailovsky, A., *WI-FOO, The Secrets of Wireless Hacking*, Addison-Wesley, Reading, MA, 2004.
5. Nichols, R. and Lekkas, P., *Wireless Security*, McGraw-Hill, New York, 2002.
6. Gast, M., *802.11 Wireless Networks*, O'Reilly Media Inc., Sebastopol, CA, 2005.
7. Aime, M., Calandriello, G., and Lioy, A., Dependability in wireless networks, *IEEE Security & Privacy*, 5(1), 23–29, January/February 2007.

8. Masika, K., *Securing WLANs using 802.11i Draft Recommended Practice Prepared for Idaho National Laboratory Critical Infrastructure Protection Center*, Lawrence Livermore National Laboratory (LLNL), Livermore, CA, February 2007.
9. Berghel, H. and Uecker, J., WiFi attack vectors, *Communications of the ACM*, 48(8), 21–28, August 2005.
10. Zahur, Y. and Yang, T., Wireless LAN security and laboratory designs, *Journal of Computing Sciences in Colleges*, 19(3), 44–60, January 2004.
11. Walker, J., 802.11 Sectuity Series, Part II: The Temporal Key Integrity Protocol (TKIP), *Intel Technical Report*, 2004.
12. Farahani, S., *ZigBee Wireless Networks and Transceivers*, Newnes, Newton, MA, 2008.
13. Holger, K. and Willig, A., *Protocols and Architectures for Wireless Sensor Networks*, Wiley-Interscience, Chichester, U.K., October 2007.
14. Eady, F., *Hands-On ZigBee: Implementing 802.15.4 with Microcontrollers* (*Embedded Technology*), Newnes, Oxford, U.K., March 2007.
15. Gislason, D., *Zigbee Wireless Networking*, Newnes, Oxford, U.K., August 2008.
16. Scarfone, K. and Padgette, J., *Guide to Bluetooth Security, Special Publication 800-121*, Recommendations of the National Institute of Standards and Technology, Gaithersburg, MD, September 2008.
17. Stouffer, K., Falco, J., and Scarfone, K., *Guide to Industrial Control Systems (ICS) Security, Special Publication 800-82*, Recommendations of the National Institute of Standards and Technology, Gaithersburg, MD, September 2008.
18. Scarfone, K., Dicoi, D., Sexton, M., and Tibbs, C., *Guide to Securing Legacy IEEE 802.11 Wireless Networks, Special Publication 800-48*, Recommendations of the National Institute of Standards and Technology, Gaithersburg, MD, July 2008.
19. Frankel, S., Eydt, B., Owens, L., and Scarfone, K., *Establishing Wireless Robust Security Networks: A Guide to IEEE 802.11i, Special Publication 800-97*, Recommendations of the National Institute of Standards and Technology, Gaithersburg, MD, February 2007.
20. Masika, K., *Recommended Practices Guide For Securing ZigBee Wireless Networks in Process Control System Environments Draft Prepared for DHS US CERT Control Systems Security Program* (*CSSP*), Lawrence Livermore National Laboratory (LLNL), Livermore, CA, April 2007.

29

Protocols in Power Generation

Tuan Dang
EDF Research and Development

Gaëlle Marsal
EDF Research and Development

29.1 Introduction 29-1
29.2 Power Plant Automation Systems and Intra-Plant Communications 29-2
Example of Nuclear Power Plant Automation Systems • Safety Requirements and System Classifications (IEC 61226, F1A, F1B, F2)
29.3 Power Plant Information Systems and Extra-Plant Communications 29-4
Common Information Model • Distributed Energy Resource Model
29.4 Conclusions 29-7
References 29-7

29.1 Introduction

Power generation is a complex process in which automation systems play important roles in assisting humans to operate the plant. Safety automation systems are designed, however, to run autonomously in order to protect the process running out of its operating domain or to avoid human errors. A power plant is designed to operate in complex power grids in which human and automation systems interact together to deal with unexpected situations. These situations may concern the imbalance between generation and consumption following hazards or failure of the power plant or the network, or the need of power ancillary services to stabilize the grid's frequency or voltage. Today, most of the power plants control room is connected to the utility economic dispatch systems where power generation optimization is performed in conjunction with trading operation. To improve the efficiency of maintenance operation, modern power plant information systems are more and more connected to a centralized e-monitoring center (Figure 29.1) where plant experts can analyze process parameters for studying the improvement of plant performances or for planning condition-based maintenance.

As shown in the above figure, modern power plant information systems are designed to interface with the world outside the plant: the grid, the economic dispatch center, the e-monitoring expertise center, etc. In the following paragraphs, we will present the most common communication protocols used for intra- and extra-power plant communications.

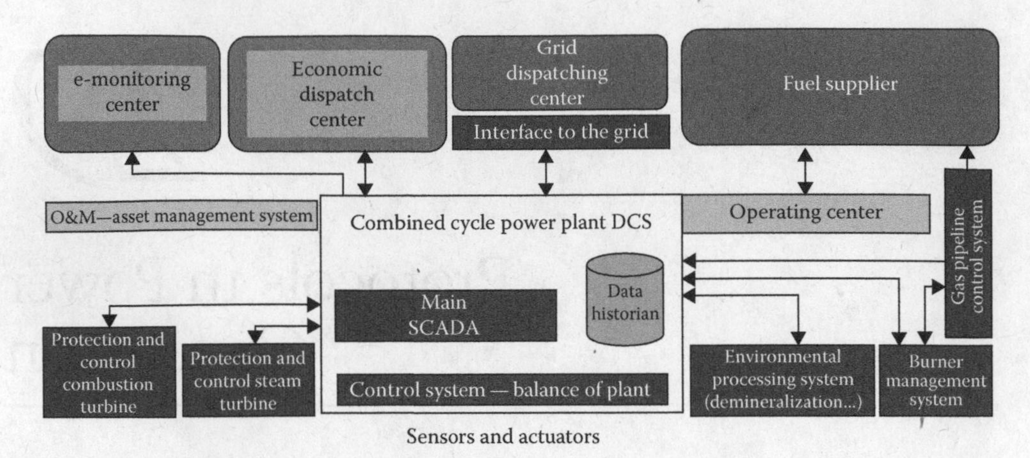

FIGURE 29.1 Example of combined cycle power plant information system.

29.2 Power Plant Automation Systems and Intra-Plant Communications

Power plant automation systems architecture is structured by functional layers as shown below (Figure 29.2):

For example, level 0 deals with instrumentation such as sensors and actuators. Level 1 gathers control and protection functions that are typically implemented in programmable logic controller (PLC). This layer concerns simple and fast control logics. Level 2 gathers more complex supervisory logics including human system interface (HSI). Levels 3 and 4 deal with computer-aided scheduling of generation, maintenance, and refueling if relevant.

These layers may slightly differ from one power generation technology to another, depending on the integration scale of automation systems. In distributed generation such as wind turbine, micro-turbine, fuel cell, and small hydropower plant (up to hundreds of kilowatt), the integration scale is relatively high to reduce the automation cost. Centralized generation and traditional power plant, such as nuclear power plant, are complex processes so that they require many automation systems that are distributed

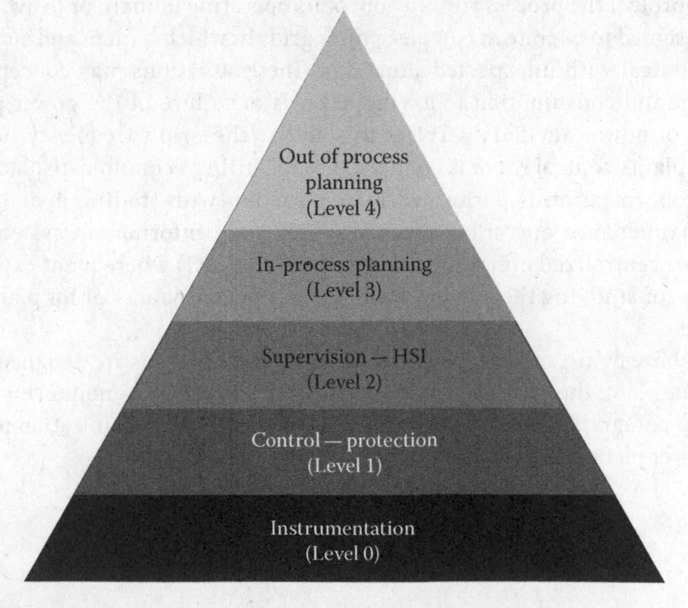

FIGURE 29.2 Functional layers in power plant automation systems architecture.

over the whole installation. These systems are connected to the plant control room through intra-plant communication infrastructure. In the following paragraphs, examples of intra-plant communication in nuclear power generation automation systems are presented because they have numerous and specific requirements comparing to other power generation technologies.

29.2.1 Example of Nuclear Power Plant Automation Systems

In a nuclear power plant, there are, as in fossil fuel power plant, several automation systems (Figure 29.3): turbine control, boiler control, core protection control, switchyard control, and other dedicated control systems (chemical processing, air conditioning, etc.).

The difference with fossil fuel power plants is the mandatory high safety level. For example, the core protection control system must be designed for this purpose following very strong design rules, whereas the safety requirements on the boiler control system authorized the use of commercial off-the-shelf control system if it fulfills the requirements on its class of systems. The other control systems (turbine and switchyard) are often provided by the corresponding system manufacturers and they are not concerned by nuclear-specific safety requirements.

The global automation systems, which are designed to interact with the operators in the control room, are particularly concerned by the safety design rules. These systems involve control functions from level 0 to level 2 as shown in Figure 29.2. In the following paragraph, the nuclear safety design rules of automation system will be detailed.

29.2.2 Safety Requirements and System Classifications (IEC 61226, F1A, F1B, F2)

Safety requirements in nuclear power generation concern many systems. There are strong design rules on the automation systems and particularly on the communication subsystems that are included in the core protection control systems. The required characteristics of the communication subsystems are determinism, redundancy, functional and technical diversity, and fault tolerance. Moreover, the functional architecture of the global control systems must comply with the system classification rules that

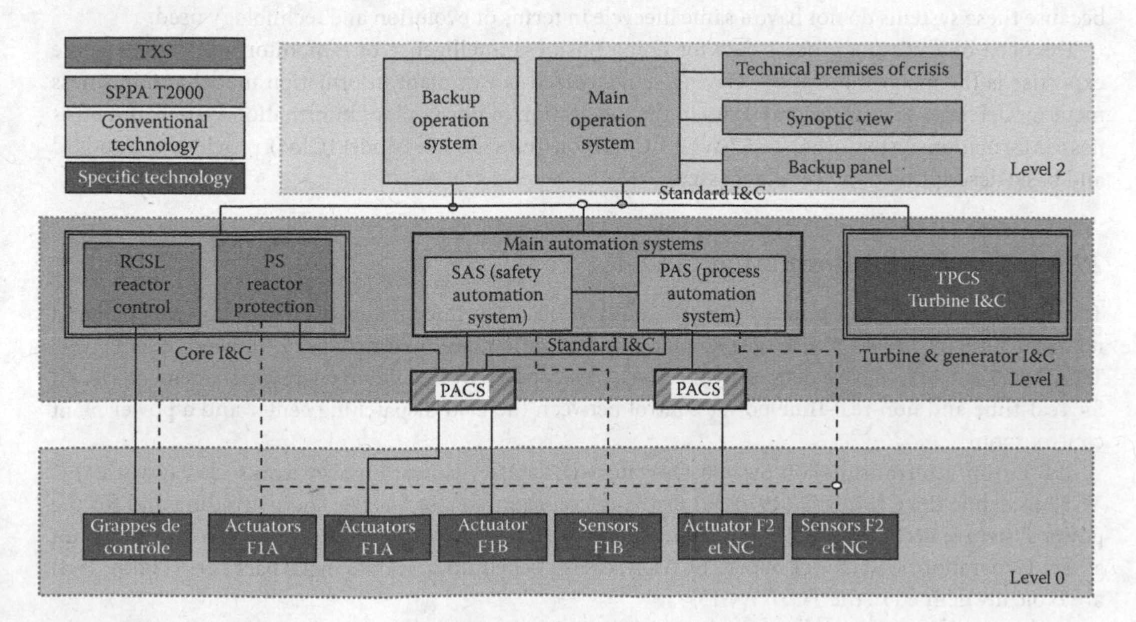

FIGURE 29.3 Example of core protection control system.

are listed in the IEC 61226 recommendations. There are typically three classes of instrumentation and control functions for European pressurized reactor (EPR):

- F1A functions, which are performed by the reactor protection system, RPS (automatic reactor trip, actuation of the engineered safeguard systems and their supporting systems), and by the priority and actuator control system, PACS. This is the strongest safety rule where the behavior must be deterministic.
- F1B functions, which are performed mainly by the safety automation system (SAS) and by the safety information and control system (SICS). In this class, the behavior must be predictive.
- F2 functions, which are performed by the process automation system (PAS); the process information and control system (PICS); and the reactor control, surveillance, and limitation system (RCSL). F2 functions can be either seismic classified (F2E functions) or non-seismic classified (F2N functions).

The priority rules are defined as follows:

- Higher class functions have priority over low-class functions; so, F1A has priority over F1B, which has priority over F2 functions. Higher class functions must not be disturbed by lower class functions.
- The order of priority within each control system is defined as follows: system and component protection has priority over automated functions, which has priority over manual functions. Automatic functions can be switched off if the process conditions allow it.

29.3 Power Plant Information Systems and Extra-Plant Communications

Power plant information system plays an important role in operational performance. A well-designed power plant information system helps to achieve operational efficiency by providing secure access to the right information at the right time to the right person. It concerns in particular the integration of power plant information system with enterprise business information system. It represents a challenging task because these systems do not have a same lifecycle in terms of evolution and technology used.

The need of accessing process data for doing business intelligence or e-monitoring from a remote expertise is the major motivation. Having standardized power plant information model and business meta-model helps to reduce complexity in the integration of power plant information system and business information system. IEC TC57-WG13 Common Information Model (CIM) provides a standard and basic description of power plant assets.

29.3.1 Common Information Model

The most important extra-power plant communication is the information exchange for real-time and non-real-time information, used in the planning, operation, and maintenance of power systems. IEC TC57 WG3 and WG13 have defined, respectively, telecontrol protocols and energy management system for real-time and non-real-time power control between the grid dispatching center and a power plant control room.

The European Transmission System Operators (ETSO) organization also works closely with IEC TC57 to define the CIM, IEC-61970-301 as the reference model for electric energy trading and for the power system operation. The CIM provides, for example, a standard and basic description of power plant assets. Generation, load model, outage, SCADA, energy scheduling, and financial packages (Figure 29.4) are typically defined in the TC57-WG13.

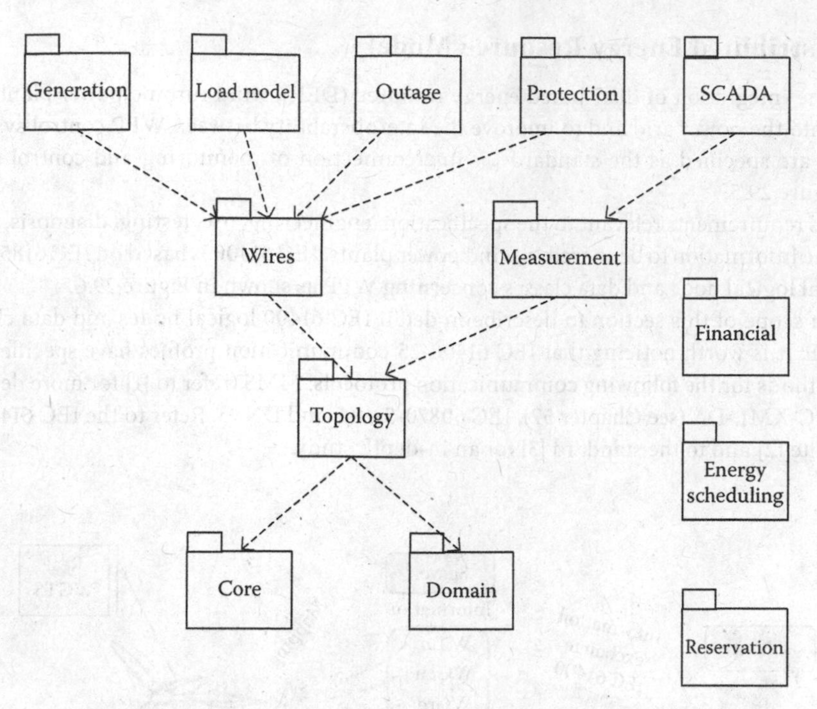

FIGURE 29.4 TC57-WG13 packages.

Concerning the Telecontrol protocols, TASE.2 (Telecontrol Application Service Element)/ICCP (Inter-Control Center Protocol) which is defined in IEC 61870-6 series is used. It is based on Manufacturing Message Specification (MMS), an ISO 9506 standard and it uses the following MMS services:

- MMS variables and variable lists functions:
  - Read
  - Write
  - InformationReport
  - GetVariableAccessAttributes
  - DefineNamedVariableList
  - DeleteNamedVariableList
  - GetNamedVariableListAtributes
- MMS Program Invocations functions:
  - Start
  - Stop
  - Resume
  - Reset
  - Kill
  - GetProgramInvocationAttributes
- Operator Stations functions:
  - Output
- Events functions:
  - EventNotification
  - DefineEventEnrollment
  - DeleteEventEnrollment
  - GetEventEnrollmentAttributes

29.3.2 Distributed Energy Resource Model

To control the integration of distributed energy resource (DER), wind turbine power plant (WPP) in particular, into the power grid and to improve the interoperability between WPP control systems, IEC 61400 series are specified as the standard for interconnection of monitoring and control systems as shown in Figure 29.5.

It provides requirements relevant to the specification, engineering, use, testing, diagnosis, and maintenance of the information to be shared in wind power plants. IEC 61400 is based on IEC 61850 by defining additional logical node and data classes concerning WPP as shown in Figure 29.6.

It is out of scope of this section to describe in detail IEC 61400 logical nodes and data classes concerning WPP. It is worth noticing that IEC 61400-25 communication profiles have specified different mapping methods for the following communication protocols: MMS (refer to [1] for more details), Web Services, OPC-XML-DA (see Chapter 57), IEC 60870-5-104, and DNP3. Refer to the IEC 61400-25 user group Web site [2] and to the standard [3] for an in-depth study.

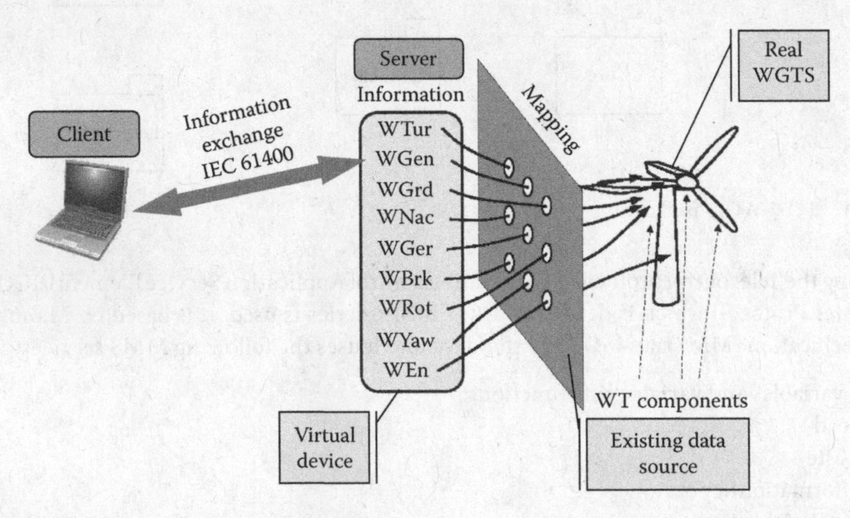

FIGURE 29.5 Illustration of IEC 61400.

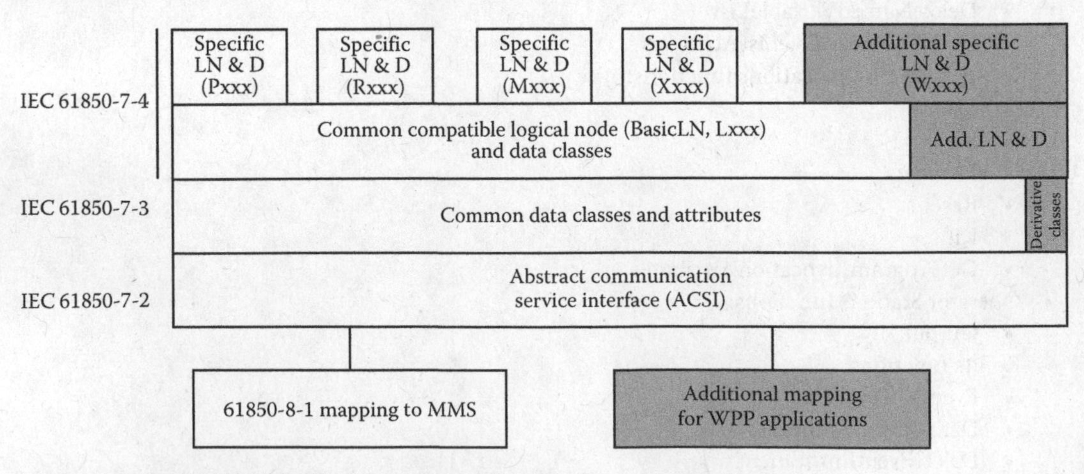

FIGURE 29.6 IEC 61400 model as an extension of IEC 61850.

29.4 Conclusions

In this chapter, an overview of specific requirements and communication protocols that are used in power generation automation and information systems are presented. Safety characteristics are among the strongest requirements that need to be taken into account in the design of communication system architecture. With the development of Internet-based communication protocols in industrial systems, cyber security constitutes an important issue that needs to be considered at present and in the future in the industrial information systems.

References

1. ISO 9506-2 Second edition, Industrial automation systems—Manufacturing message specification, 2003-07-01.
2. IEC 61400-25 User Group Web site, http://www.use61400-25.com, Technical paper section, May 2009.
3. IEC 61400-25-1 Ed. 1.0, Wind turbines—Part 25-1: Communications for monitoring and control of wind power plants—Overall description of principles and models, 2006-12-14.

30

Communications in Medical Applications

Paulo Bartolomeu
University of Aveiro

José Alberto Fonseca
Universidade of Aveiro

Nelson Rocha
University of Aveiro

Filipe Basto
Hospital de Sao João

30.1 Introduction 30-1
30.2 Requirements... 30-2
30.3 Localization 30-4
30.4 Clinical Monitoring.. 30-5
Controller Area Network • Profibus DP • IEEE 802.15.4 • Bluetooth • IEEE 802.11 • Nonindustrial Technologies
30.5 Automation 30-9
Smart Homes • Healthcare Robotics
30.6 Issues and Challenges 30-12
Security and Privacy • Safety and Reliability • Standardization • Timeliness
References.. 30-14

30.1 Introduction

Medicine, as a science, involves a complex and a sophisticated network of real-time communication systems, where safety and security issues are primary concerns. Aggregating information from multiple sources is essential to maximize resource usage, formulate accurate diagnosis, provide adequate treatment, define management plans, and grant critical clinical monitoring. The availability of online electronic medical and health records promotes the coordination of care and the integration of clinical information (e.g., laboratory and imaging results). To harness the aforementioned benefits, subsystems must be interoperable, flexible (e.g., allow the customization of scheduling, tracking, and alert modules), and able to cope with the standards of care—security, safety, timeliness, efficiency, effectiveness, equity, and patient-centered care.

In the last decade, healthcare services have evolved as the result of two main factors: social context and technical innovations. Regarding the first, there is an increasing pressure for cost containment in health organizations and social care institutions that poses demanding challenges in providing cost-effective healthcare services. Besides, the nature of the disease burden is changing as a result of the demographic boom in elderly population [1], which shifts the nature of healthcare delivery from acute, episodic care treatments to long-term, chronic condition interventions. Furthermore, as familiar income and level of education increase, higher demands are put on healthcare services for preventing diseases and on medical treatments allowing lifestyle maintenance and independence.

The second factor fueling the improvement of healthcare services is the broad adoption of microprocessors in medical devices, which enables local data processing resulting in faster response times and enhanced treatment compliance. Examples of microcontroller-based devices are the electronic sphygmomanometer employed in blood pressure monitoring, "smart" infusion pumps applied in the administration of medications intravenously, "smart" camera pills used in digestive tract disease diagnosis, orthopedic implants utilized in post-surgery observation, and remote patient monitoring devices used to collect real-time clinical data for medical analysis [2]. Besides integrating intelligence, medical devices also embed standard-based communication interfaces, thus enabling them to participate in clinical networks sharing information with medical subsystems and healthcare providers.

Healthcare systems encompass a broad spectrum of technologies benefiting from the support of data communications. Web-based imaging systems facilitate data collection and "in time," remote, expert judgment. Telemetry and wireless systems monitor critical vital signs in intensive care units, providing real-time information that can trigger immediate life-saving medical response. Telemedicine and teleconference services provide global communication tools for timely advice, information exchange, monitoring, and decision. Robotic or computer-assisted surgery allows significant advances in remote, minimally invasive, and unmanned surgery, increasing patients' access and surgery precision, while decreasing comorbidities.

Wireless systems have enabled the monitoring of inpatient localization in healthcare institutions, thus improving response times in accidents (e.g., falls), security (e.g., parents remotely see their babies in neonatal intensive care units), and comfort (voice over IP communications between inpatients and their families).

Although there is a broad range of healthcare applications employing data communications (telemonitoring, PACS, etc.), due to length restrictions, this chapter solely focuses on communication technologies addressing healthcare localization, monitoring, and automation applications. Also, this chapter does not provide detailed specifications for standard technologies as they are available within the IEEE and industry standards, and in the corresponding chapters of the book.

The remaining chapter is structured as follows: Section 30.2 introduces the main communication requirements of medical applications by focusing on localization, clinical monitoring, and automation (and control) applications. Section 30.3 provides an overview of localization techniques and presents commercial examples of localization systems by outlining their architecture and operation. Section 30.4 covers clinical monitoring applications and reviews several monitoring systems employing different communication technologies. Section 30.5 summarizes the communication approaches applied in medical automation applications, particularly smart homes and healthcare robotics. Finally, Section 30.6 raises several topics related to the issues and challenges of developing medical communications.

30.2 Requirements

Communications have a wide spectrum of applications in healthcare environments ranging from appliance automation in smart homes to the clinical monitoring of patients in critical condition. As such, requirements are significantly heterogeneous in what concerns the main architectural and performance parameters of communication technologies: range, topology, autonomy, latency, jitter, and throughput. Table 30.1 summarizes the requirements for localization, monitoring, and automation applications in medical environments.

The operation range is classified as body area network (BAN), personal area network (PAN), or local area network (LAN) in accordance with the typical coverage of standard communication technologies. Topology requirements are diversified as the application domain demands the support of multiple network configurations ranging from simple point–point links (e.g., bedside monitoring) to multihop connections (e.g., telemetry). The maximum number of required nodes in home monitoring and healthcare robotic applications is low. Conversely, dense localization and health monitoring scenarios (e.g., found in healthcare institutions) require the deployment of a significant number of devices. The autonomy parameter is defined as the lifetime of a node with regard to battery capacity and power consumption. In effect, this is one of the most important requirements for mobile healthcare devices and has a strong impact on its usefulness. Latency and jitter define the timeliness of communication technology, which has a considerable effect on the reliability and accuracy of the overlying medical applications. For example, the occurrence of a high delay in communicating an alarm condition may result in irreversible damage to a patient's health. Moreover, significant jitter may affect the accuracy of a medical diagnosis due to the utilization of tainted data.

TABLE 30.1 Communication Requirements of Medical Applications

	Monitoring			Automation	
Parameters	Localization	Healthcare Institutions	Home	Smart Homes	Healthcare Robotics
Range	LAN	LAN PAN BAN	LAN PAN BAN	LAN	LAN
Topology	Star Tree Mesh	Star Tree Mesh	Star Tree Mesh	Bus Tree Mesh	Star Tree Mesh
Number of nodes	1–2000	1–2000	1–10	10–100	1–10
Autonomy	Days	Days/weeks	Days/weeks	Years	Days/weeks
Maximum latency (ms)	10	10	200	500	5
Maximum jitter	100 μs	100 μs	1 ms	100 ms	100 μs

TABLE 30.2 Information Rate of Physiological Signals

Biomedical Measurement	Number of Sensors	Sample Rate (Hz)	Resolution (B/Sample)	Information Rate (B/S)
Arterial pressure	1	300	12	3,600
Body temperature	1+	5	16	80
Electrocardiogram (ECG)	5–9	1,250	12	15,000
Electroencephalogram (EEG)	20	350	12	4,200
Electromyogram (EMG)	2+	50,000	12	600,000
Electrooculogram (EOG)	4	100	12	1,200
Heart rate	2	25	24	600
Heart sound	2–4	10,000	12	120,000
Oximetry	1	150	12	1,800
Respiratory rate	1	50	16	800

Sources: Arnon, S. et al., *IEEE Wirel. Commun.*, 10(1), 56, 2003; Monton, E. et al., *IET Commun.*, 2(2), 215, 2008; Gama, O. et al., Quality of service support in wireless sensor networks for emergency healthcare services, in *30th Annual International Conference of the IEEE (EMBS2008)*, Vancouver, British Columbia, Canada, August 20–25, 2008, pp. 1296–1299.

Regarding throughput, Table 30.2 [3–5] provides information on the typical amount of data produced by physiological signals in monitoring applications. The throughput associated with either localization or automation applications is highly dependent on the particular scenario addressed. For example, in receiver signal strength (RSS)-based localization systems, each device being periodically localized transmits a packet with a small payload. However, the transmission rate is dependent on the update rate required by the localization system. Appliance networking in smart homes is another example of varied throughput requirements given its high dependence on human activity. Nevertheless, these throughput requirements are generally low, as events like the opening of a motorized window blind or the turning on of the TV are usually sparse and produce data packets with small payloads.

Regarding privacy and security, healthcare communications have stringent requirements motivated by the prerequisites of maintaining patient information only available to authorized personnel and guaranteeing data integrity. Furthermore, fault tolerance and reliability are also of critical importance as human lives may be at risk. As such, mechanisms avoiding system failures and providing graceful degradation when unavoidable conditions occur must be validated and implemented. The following sections provide an overview of communication protocols and implementations addressing localization, monitoring, and automation applications in healthcare environments.

30.3 Localization

Healthcare environments encompass human resources, such as physicians, nurses, medical staff, and equipment, that must be correctly managed to efficiently satisfy a given set of objectives. Resource localization is an important factor when evaluating management actions, since it provides information that allows the improvement of medical response times, which is of the utmost significance in life-critical scenarios. In addition, localization can potentially have a strong impact concerning security and safety in healthcare environments. Given that many hospital equipments are expensive and can be handheld, the possibility of theft can be avoided by having a real-time localization system tracking their localization and triggering an alarm when carried beyond a predefined physical area. Furthermore, in scenarios where patients must remain in a restricted area (the psychiatric ward or the neonatal ward of a hospital, for example), localization information is highly relevant as it hinders occurrences in which hospitals may be held accountable (violence, child kidnap, etc.). Regarding safety, in healthcare environments having patients with a high risk of brain injury and stroke (e.g., hospitals with elderly inpatients), localization can be of paramount importance, as it can shorten the medical response time and reduce the chances of irreversible physical impairments.

Currently, four types of localization are available: direct identification, time acquisition, angle detection, and receiver signal strength (RSS). Direct identification simply detects if a tag is present/absent in the neighborhood, i.e., the result of a localization round is either true (present) or false (absent). Radio-frequency identification (RFID) [6] is the most common technology employed in this type of localization. Time acquisition localization, as the name suggests, is based on time measurements and includes techniques such as time of arrival (TOA) [7,10,11], time difference of arrival (TDOA) [8,11,12], and differential time difference of arrival (DTDOA) [9]. Angle detection includes the angle-of-arrival (AOA) localization technique [10–12] employing stations capable of determining the direction of propagation of a packet by using multiple receivers. RSS localization [10–12] is supported on the ability to measure the energy of a received packet, which provides a rough estimate of the traveled distance (if the transmission power is known). Several localization systems using these techniques have been proposed in the literature. However, only a few have become commercially available and can currently be used in real deployments. The following paragraphs provide an overview of such systems.

The Exavera eShepherd™ [13] system is supported on direct identification and was specifically designed targeting healthcare environments allowing the tracking of patients (Vera-T Bracelets), medical staff (Staff Badges), and objects (Asset Tags). Besides the RFID-enabled mobile devices (bracelets, badges, and tags), this system includes Vera-Fi devices (RFID reader plus Wi-Fi Access Point) and relay RFID transceivers (RRT), responsible for relaying data packets. Mobile devices periodically transmit RFID packets, which are received by neighbor Vera-Fi devices and transmitted (using Wi-Fi) to a hospital information system identifying the area where the "tag" is localized. The Exavera eShepherd localization system specifies an accuracy of 3 m (95th percentile, herein assumed as default).

Ubisense [14] is a real-time localization solution with subsecond response and proven robustness in challenging manufacturing environments that employs a combination of the AOA and TDOA techniques. Although originally designed targeting industrial applications, it can also be applied in healthcare environments. The Ubisense system is composed of a network of sensors (Ubisensors)—a cell, mobile tags (Ubitags), and a Ubisense software platform (USP) running in a PC. The Ubisensors are equipped with UWB and RFID transceivers installed in known positions, connected by an Ethernet network. A Ubisensor simultaneously transmits a UWB impulse sequence and an RF message with its identification (ID). These messages will be received by the nearby fixed Ubisensors. Each cell has a single Ubisensor playing the role of the master, i.e., responsible for controlling the medium access using time division multiple access (TDMA). In addition to the received RF identification, Ubisensors determine the UWB impulse angle of arrival (and instant of reception) through the use of an array of antennas. The master Ubisensor calculates the TDOA for each Ubisensor relative to a given Ubitag and sends all data (ID, AOA, and TDOA) to the USP, which computes the Ubitag localization. This solution is particularly suited for scenarios requiring high accuracy (30 cm).

The RSS-based approach is employed in localization systems, such as the Ekahau real-time location system [15] or the AeroScout Visibility System [16], both based on IEEE 802.11 technology. The Ekahau system is supported on a Wi-Fi infrastructure using an RSS fingerprinting localization technique. As such, it uses 802.11b access points (AP) that measure the RSS of the packets sent by Wi-Fi-active RFID tags, forwarding this data to a control platform where the localization is computed. The Ekahau system supports a localization accuracy of 2 m. The Visibility System is a real-time localization system operating with both TDOA and RSS techniques supported by Wi-Fi technology. Although the system includes devices known as exciters to capture RFID responses from tags passing nearby, the localization is mainly supported on wireless location receiver (WLR) devices measuring the TDOA and/or the RSS of the transmissions received from tags, PDAs, laptops, etc. The information collected by WLRs and exciters is then forwarded, using Ethernet, to the Aeroscout Engine where the localization is determined. The manufacturer claims accuracies of 3–5 m when the TDOA localization technique is employed and 3–10 m when RSS base localization is used instead.

30.4 Clinical Monitoring

Clinical monitoring is a highly representative application of the support provided by communications in healthcare environments. The remote collection of health data from patients at home, in healthcare environments, and in some cases, in outdoor scenarios, allows healthcare professionals to follow their patients' condition in real time and with minimal intrusion, which facilitates disease management, diagnosis, prediction, and follow-up. This improves the quality of life of elderly people and chronic patients while promoting preventive lifestyles and early diagnosis for the general population. Besides health motivations, clinical monitoring promotes the reduction of medical errors and the cost burden caused by an aging society with a high incidence of chronic patients.

Both wired and wireless communication technologies have been reported in clinical applications, such as bedside monitoring of inpatients or health monitoring using networks of sensors. Although wired protocols are still dominant in safety-critical monitoring applications, wireless communications are becoming a feasible alternative given their increasing performance, availability, and miniaturization.

The following subsections provide examples and an overview of clinical monitoring applications employing wired and wireless communication technologies.

30.4.1 Controller Area Network

An example of the controller area network (CAN) technology applied to healthcare environments is the support of networking for bedside medical instruments [17]. McKneely et al. propose a technology suite named MediCAN™ defining the hardware and the communication protocol to enable medical instrument networking. The network architecture encompasses Hubs (bedside devices) providing connectivity to the physical (MediCAN) instruments and gateways allowing high-level access to these instruments. An Ethernet communication backbone connects gateways to a MediCAN server that maintains a real time list of currently available MediCAN gateways and clients. The MediCAN technology suite defines a set of network protocols that provide real-time client direct access to gateways and their resources on a LAN, remote device control, and automatic remote data acquisition, among other features. The MediCAN protocol supports data rates of up to 1 Mbps and shares the carrier-sense multiple-access protocol with collision detection and arbitration on message priority (CSMA/CD + AMP) defined in CAN.

30.4.2 Profibus DP

A bedside patient monitoring system employing the Profibus DP fieldbus standard was proposed in [18]. The option for this fieldbus is justified by the adequacy of Profibus characteristics (e.g., determinism, robustness, redundancy, and support for high-speed cyclical data transmission) to the requirements of a bedside communication application. The proposed system includes a bedside monitor, an actuator unit, an environment unit, and a central monitor. The bedside monitor collects the patient's vital signals, enabling the measurement of parameters, such as heart rate (number of heart beats per minute, pulse), end-tidal lung volume (liters per minute), respiration rate (number of breaths per minute), and systolic and diastolic arterial blood pressure values (mmHg). The remote control of bedside devices (e.g., infusion pumps) is enabled by the actuator unit, which can drive digital and analog outputs. Environmental parameters (e.g., temperature) and discrete inputs (e.g., nurse call button) are monitored using the environmental unit. These devices (slaves) are placed nearby a patient's bed and are connected to the central monitor (master) by a Profibus DP network operating at 1.5 Mbps. The central monitor provides functions like real-time reception and a display of bedside data of up to 16 patients, signal interpretation, and alarming with event logging, etc. Redundancy can be provided by interconnecting devices using a two-wire cable on a bus topology.

30.4.3 IEEE 802.15.4

The IEEE 802.15.4 is a low-power, low-cost PAN technology with application in smart home health monitoring and in BANs. Both Dagtas et al. [19] and Junnila et al. [20] propose a smart home wireless health monitoring system based on ZigBee technology. However, their architectures are different: the first employs three tiers for data collection and processing while the second uses a two-tiered approach. The first tier of the Dagtas et al. monitoring system corresponds to the mobile device worn by the patient that collects signals from wired and wireless sensors. This device periodically transmits raw data to a local server for storage and processing (second tier). Here, further processing is carried out and the resulting data are transmitted to the service provider center for storage, expert review, and diagnosis (third tier). The UUTE home network [20] encompasses sensor nodes, a coordinator, a home client, and a server. Sensor nodes communicate directly with the coordinator, responsible for setting up and managing the ZigBee network. Additionally, the coordinator provides information to the home client about sensors that joined/left the network and forwards data from/to the ZigBee network. The home client operates as a processing unit for the data coming from the ZigBee sensor network, stores it locally, provides a user interface, and acts as a gateway to a remote UUTE server where data are stored for redundancy (and future analysis). The UUTE home network employs nonbeaconed ZigBee communications, offers no guarantees of data delivery, and as two-way communication is desired, has no support for power consumption optimization.

The architectures of the BANs proposed by Monton et al. [21] and Lo et al. [22] are very similar. A BAN encompasses a set of sensors (ECG, temperature, SpO_2, etc.) equipped with wireless transceivers that periodically transmit data to a local central device (e.g., PDA) where data are processed and transmitted to a server for long-term storage. Monton et al. define a proprietary BAN protocol compatible with IEEE 802.15.4 and ZigBee that operates using the beaconed mode of IEEE 802.15.4. At the beginning, a sensor node associates with the coordinator (central node) and waits for beacons by enabling the receiver the instant immediately before the beacon transmission. In addition to information on the instant of the next beacon transmission, beacons can transport commands for the sensor nodes (configuration, activation/deactivation, and data transmission). After the activation of the sensor node, data collection begins from the attached sensor(s) until a complete data frame is filled. This frame is sent to the central node when a data transmission command is received. The central node stores sensor data and makes it available to the outside world using the Session Initiation Protocol (SIP) over technologies, such as the IEEE 802.11 (Wi-Fi) and GPRS. The BAN proposed by Lo et al. employs standard IEEE 802.15.4 communications, including context-aware sensors (e.g., accelerometer) that are used in multisensor data fusion for false alarm detection and bandwidth usage minimization.

30.4.4 Bluetooth

Although requiring more power than IEEE 802.15.4-based communications, Bluetooth has found applications in ambulatory medical monitoring [23], mobile care alert systems [24], and in physiological monitoring [25]. The ambulatory monitoring solution presented by Zhang and Liu addresses the replacement of traditional holter systems by real-time ones. Traditional holter systems store physiological data in a local memory that is later downloaded onto a PC for diagnosis. This approach endorses the occurrence of a wide range of errors that can only be detected after downloading the holter data (e.g., lead malfunction). As such, the authors propose a monitoring architecture where several patients can be monitored through the use of Bluetooth-enabled sensors communicating with a PC using Bluetooth asynchronous connectionless (ACL) links.

Lee et al. put forward a mobile care alert system allowing the monitoring of hypertension and arrhythmia patients, triggering early warning notifications when abnormal parameters are detected. This system encompasses a front-end personal mobile device and a back-end care center server. The first comprises a physiological (blood pressure, pulse, and ECG) parameter extraction Bluetooth-enabled device and a mobile phone. The second consists of a GSM/GPRS module (able to transmit/receive short messages) and a PC-based care center host with an Internet connection. The system operates by periodically sampling physiological signals performing its transmission to a Bluetooth-enabled mobile phone (device) using the RFCOMM transport protocol, which emulates RS232 serial ports over the (Bluetooth) L2CAP protocol. Besides uploading the physiological data to a database in the healthcare center server, the mobile phone uses a simple algorithm to identify abnormal conditions, which result in the transmission of short messages to the physicians or the other healthcare providers. Abnormal conditions (identified by professional judgment of the stored information in the healthcare center server) trigger an alarm to the related personnel, for example, local officers can be instantly informed or an ambulance can be immediately dispatched to the location of the patient for rescue.

Stojanovic et al. proposed a physiological monitoring system with an architecture comprising Bluetooth-enabled sensors worn by patients, a Bluetooth access point, and a local server. The major difference from the system proposed by Lee et al. is that the patient-worn integration device is here replaced by an access point. As such, if the patient wears multiple Bluetooth-enabled sensors, there will be multiple RFCOMM links to the access point, which acts as a wireless multiserial server that receives serial data over the Bluetooth RFCOMM links and packs it in TCP/IP frames that are sent to the local server and to other remote locations (e.g., hospital), using the Internet, if required.

As suggested by the aforementioned examples, Bluetooth has been identified as a feasible PAN technology to support medical sensor-based applications. In consequence, the Bluetooth Special Interest Group (SIG) developed a specific Medical Device Profile (MDP) [26] aimed at expanding the use of the technology into the medical, health, and fitness markets. The MDP addresses a wide variety of medical devices (blood pressure meters, weight scales, pulse oximeters, glucose monitors, pulse/heart rate monitors, etc.), supports multiple data transfer modes (real-time streaming, periodic, episodic, and batch transfer), and is able to cope with tight synchronization requirements (1 ms). Data communications use the exchange specifications and the data structures defined by the IEEE 11073 [53] protocol on top of the MDP layer, thus enabling data sharing between applications and improving interoperability between medical data endpoints.

30.4.5 IEEE 802.11

The IEEE 802.11 protocol has been widely adopted in wireless LANs addressing home and business application scenarios. Given its target application domain (power consumption characteristics), it is not commonly used for supporting small form factor monitoring devices requiring a high level of autonomy. Nevertheless, there are reported results of using IEEE 802.11 for physiological data collection [27]. However, due to its pervasiveness, it is frequently adopted as the enabling technology for gateways in multitiered monitoring applications [28,29]. Tejero-Calado et al. provide an example of physiological monitoring using IEEE 802.11 technology in [27]. Here, an IEEE 802.11b compliant module supporting rates up to 11 Mbps collects real-time ECG data from an electronic subsystem (analog conditioning plus digital sampling) and performs its transmission to a central server for storage (and analysis) using a TCP/IP communication transport. The real-time ECG data fed to the server can be viewed on display devices, such as PCs, laptops, PDAs, etc., as long as they can communicate with the server using the TCP/IP transport.

The architecture of monitoring applications typically employs (at least) two tiers: one where data are collected and the other where data are made available for professional analysis. In the first tier, the vital sensors worn by patients are connected (through wires or wirelessly) to a local unit (mobile phone, PDA, etc.) that displays signals and related information, performs basic signal processing, and transmits data using the IEEE 802.11 technology. In the second tier, multiple local units transmit data to a central server where it is stored, processed, and further analyzed looking for abnormalities. Examples of monitoring applications supported on IEEE 802.11 enabled gateways are proposed by Karatzanis et al. and Postolache et al. The former presents an acquisition and processing system, based on a wearable textile-based ECG device connected to a PDA that collects heart rate (HR) and respiratory rate (RespR) signals, which are then transmitted to a home gateway using IEEE 802.11 communications. The home gateway produces the corresponding time series, packs, and sends them to a central repository where a clinical decision support system can detect the onset of early decompensation episodes. The latter introduces a portable biosignal measuring system based on a personal digital assistant (PDA) that collects, stores, and analyzes data from the ECG, and the oxygen saturation (SpO2S) and skin temperature sensors. Besides online spontaneous heart rate (HR), the PDA determines the heart rate variability (R-R variation). The detection of a major event (deviation from a given set of parameters) results in the transmission of the last 10 min of physiological data to the host laptop. This transmission is performed using a TCP/IP transport protocol on top of IEEE 802.11b communications. The physiological data are then further analyzed using advanced algorithms (e.g., the fast Fourier transform) in order to obtain an accurate patient diagnosis.

30.4.6 Nonindustrial Technologies

The free industrial, scientific, and medical (ISM) bands are becoming crowded as a result of the massive adoption of wireless communication technologies (e.g., Wi-Fi and Bluetooth). To cope with the problem of interference in medical environments, the US Federal Communications Commission (FCC) has allocated the 402–405 MHz range of frequencies to the medical implant communication systems (MICS) band and the 608–614, 1395–1400, and 1427–1432 MHz frequency ranges to the wireless medical telemetry service (WMTS) band [30]. These unlicensed bands can be used for both short range (MICS) and moderate range (WMTS) communications without the risk of interference from industrial wireless technologies. Furthermore, the MICS band is restricted to transmissions of less than −16 dBm, which reduces the operational range to a few meters but increases the autonomy of MICS-based communication devices. The WMTS band is usually employed in telemetry applications requiring operational ranges of more than 100 m, as the band allows a maximum transmission power of 1.5 W.

Examples of nonindustrial communication technologies are the wireless woven inductor channel described in [31] and the BAN based on the MICS/WMTS proposed in [32]. The former is employed in BANs and addresses the problem of ultra-low-power near-field communication while the latter envisages multitiered vital signal monitoring systems based on MICS/WMTS communications for healthcare environments.

The Wireless Woven Inductor allows a wearable sensor network to communicate with a gateway device through several layers of clothes, i.e., it enables ultralow-power communication between sensors in close contact with the patient's body and a gateway without the inconvenience of having wires passing through multiple clothes. The architecture of the system comprises three sensor nodes that are placed on the inner wear and connected to each other through digital yarn (having conductivity). The sensor node signals are transmitted to the outer wear through the woven inductor channel, being supplied to a personal sensor controller that stores and processes the received information. This approach presents benefits as it reduces power consumption and enables over-the-air energy transmission, which allows a dramatic increase in sensor network autonomy. The authors claim a sensor power consumption of 300 μW under a 3.3 V supply voltage (at 1 Mbps).

The MICS/WMTS-based BAN described by Yuce and Ho [32] is a three-tiered communication system composed of sensor nodes, central control units (CCUs), a base station (composed of a CCU and a PC), and a server. The lowest tier includes sensor nodes that collect and forward data from inpatients (in bed) to a CCU using MICS technology. Then, the CCU conveys this data to the base station using WMTS technology (second tier). The received data are then transmitted to an external server using the internet (third tier). The MICS band is used for network communications at the sensor level due to its low transmission power (25 μW) that enables high-sensor autonomy. Because the WMTS is allowed to operate with higher power transmissions (10 dBm ≤ transmit power < 1.8 dB), it is often applied in application scenarios where higher ranges are required (e.g., CCU communications). The authors claim communications without significant delays.

30.5 Automation

Automation has found application in a broad range of scenarios going from industrial plants with assembly and transport robots, to smart homes adapted for patients with functional impairments, to healthcare institutions employing medical robots in surgery or in patient rehabilitation. Currently, it is playing an increasingly important role in healthcare systems as it improves efficiency, increases the quality of service, saves time and manpower, and supports the execution of dull (e.g., processing blood samples), dirty, or dangerous (e.g., handling chemical substances) tasks. The following subsections introduce two representative healthcare domains where automation is being effectively applied.

30.5.1 Smart Homes

The world population aged over 65 years is growing fast and it is expected to rise above 1490 million people by the year 2050 [1]. Dishman [33] identifies four requirements that should be addressed to cope with the current trends of elderly demographic increase: promote healthy behaviors, detect diseases at an early stage, improve treatment compliance, and provide support for informal caregiving. These requirements can be met by maintaining elderly (or impaired) people living in their homes [34], within their social network, and being assisted by automation systems that, besides helping them with the activities of daily living (ADLs), also monitor their health condition.

Smart homes can be defined as dwellings incorporating a communication network that connects electrical appliances and services and allows them to be remotely controlled, monitored, and accessed [35]. Smart home communication requirements are significantly distinctive from those of technologies addressing large buildings (e.g., BACnet, LON, and KNX/EIB) mainly in what concerns coverage area, number of supported devices, and per module cost. As such, technologies like X10, Z-Wave, and ZigBee were specifically developed targeting home networking.

X10 [36] is an open and an international standard protocol for home communications supporting two physical media: the house electrical network and the radio frequency. The communication of data employs brief radio frequency bursts (120 kHz signals) representing digital information, which are inserted immediately after the mains 50 Hz sine wave origin. This protocol defines no specific medium access mechanism, and to cope with high noise levels, a bit is always sent together with its complement, resulting in a transmission rate of one data bit per cycle of the electricity network.

Z-Wave is a wireless (low-power) standard designed to address remote control applications in residential environments, created by Zensys and standardized by the Z-Wave Alliance [37]. Z-Wave operates in the 900 MHz ISM band (United States, 908.42 MHz; Europe, 868.42 MHz; Hong Kong, 919.82 MHz; and Australia/New Zealand, 921.42 MHz), has a 30 m range in open space, and supports data rates of 9600 or 40 kbps. The technology supports a mesh topology in which nodes communicate with each other directly, when they are in range, or route packets through other nodes, otherwise.

ZigBee is a communication technology designed for low-rate wireless PANs specifying the network and application layers on top of the IEEE 802.15.4 MAC (and PHY) and encompassing application profiles that guarantee interoperability among products from different manufacturers. Currently, two (publicly available) profiles targeting home area networks (HANs) are available: smart energy and home automation. The first focuses on energy management while the second targets appliance control and management in smart homes. Further information about this technology can be found in the corresponding chapter of the book.

Although X10, Z-Wave, and ZigBee protocols were designed envisaging smart home networking applications, there are other standard technologies that can equally meet the requirements of smart homes. Examples of these technologies are the CAN fieldbus [38], Bluetooth [39], and IEEE 802.15.4 [40]. The CAN protocol was developed by Bosch, and besides being widely adopted in the automotive industry, it can be employed in smart/assistive home networking due to its properties and add-ons related to fault-tolerant operation and real-time communication. Bartolomeu et al. [38] describe a commercial home automation system that enables severely disabled users to operate enhanced appliances using custom interfaces. This system is supported in a CAN fieldbus using a producer–consumer communication model in which enhanced home appliances (e.g., window blinds, doors, lights) are operated using human–machine interfaces (HMIs) or specific control appliances (e.g., switches). The trigger of a command results in the transmission of a broadcast message in the CAN fieldbus. Since each message defines its consumers, only they will accept the command and act accordingly. An extension of this communication model employing a wireless technology has been studied in [41] where an assessment of the timeliness of multihop broadcasts in ZigBee networks was conducted. The conclusion was that producer–consumer communications supported on ZigBee multihop broadcasts are adequate to building automation systems of reduced size, operating without retransmissions (i.e., in free RF channels).

Dengler et al. [39] employed Bluetooth for supporting sensor/actuator communications in smart homes. Their approach consists of installing sensor (e.g., temperature, motion, light) and actuator nodes in specific places around the dwelling so that abnormal events can be detected and properly dealt with. Sensors and actuators (BTnodes) are enabled to communicate with each other over multiple hops by a routing protocol implemented on top of the Bluetooth L2CAP layer. This sensor/actuator network employs a simple proprietary protocol (on top of the routing layer) that comprises three types of messages handled with different priorities: emergency, command, and data. Emergency messages (calls) are broadcasted whenever an anomaly is detected and are periodically retransmitted until a handling confirmation (acknowledgment) is received from an actuator. Command calls are (acknowledged) messages used to initiate remote processes in sensor nodes (e.g., recalibration). Data calls are periodic unacknowledged messages transmitted to maintain a chronicle of the environment status, thus allowing information to be analyzed and/or displayed to users in and outside the smart home. Provided that sensor/actuator information must be available beyond the smart home (e.g., alarm notifications, real-time monitoring), Dengler et al. also propose gateways allowing the bridging of the sensor/actuator network with popular communication technologies, such as Wi-Fi, telephony network, etc.

In [40], Khan et al. present the architecture of a clustered hierarchical home area network (HAN) supported on IEEE 802.15.4 technology operating in the beacon-enabled (beaconed) mode. Here, nodes including sensors (e.g., temperature, light, pressure, motion, smoke, humidity) are deployed in different areas of the house (bedroom, lounge, library, pool, etc.) and grouped in clusters. Sensor nodes belonging to a given cluster can only communicate with a central node denoted by cluster head, thus enforcing a star topology at the intracluster level. Cluster heads, on the other hand, can communicate with each other and with the PAN coordinator. Furthermore, cluster heads can route packets from/to their own cluster, other clusters, or the PAN coordinator. Sensor nodes, cluster heads, and the PAN coordinator communicate with each other and with the outside world by employing a superframe structure containing an active part, used for contention-based (CSMA/CA) or contention-free communications, and an inactive part where the transceiver can be turned off for power saving. This superframe is bounded by beacon frames periodically transmitted by the PAN coordinator, which conveys the sensor network information to other systems by means of a (PC) gateway enabled with TCP/IP communications. Additional information regarding the beaconed mode of the IEEE 802.15.4 standard can be found in [42].

30.5.2 Healthcare Robotics

The initially slow acceptance of robotic technology in medicine has been motivated by psychological, technical, and economic reasons. However, given the current high safety insurance of medical robots, their application has spawned a broad set of medical areas, such as orthopedic and cardiac surgery, laparoscopy, endoscopy, among others [43]. Some current healthcare domains where robots are being employed include use in multiple types of surgery (image-guided, minimal invasive, etc.), rehabilitation robots, rehab-manipulators, and mobile systems (for hospitals, smart homes, etc.). In the following paragraphs, healthcare robotic systems are introduced with focus on the employed communication technologies.

Ozkil et al. [44] propose a robotic automation system for transportation of goods in hospitals, specifying mobile robots for carrying containers, and stationary robotic stations for loading/unloading operations. Three layers of decision are defined: supervisory control, traffic control, and vehicle control. The first is responsible for generating and combining transportation requests, and assigning tasks to mobile robots. The second plans the movements of containers in order to cope with required schedules and to avoid conflicts resulting from different user requests. The third generates movement set points that are transmitted to mobile robots in order to meet the planned movements of containers. Besides robots, the systems' architecture encompasses a server providing fleet supervision (storage of order and container information) and a Wi-Fi AP enabling TCP/IP data communication with the mobile robots (transmission of movement setpoints). In addition to Wi-Fi communications, mobile robots also encompass a CAN network allowing high-level set points to be conveyed to the low-level controller, responsible for sensing bumps and emergency events (activated by a switch), while driving the servo motors and brakes.

Regarding cooperative teams of robots, Choi et al. [45] introduced a robotic (BioRobot) platform for clinical tests in small to medium-sized laboratories using mobile agents (MAs) with the ability to transport blood samples, reagents, microplates, and medical instruments. The system architecture includes MAs for transport, SCARA robotic manipulators for handling samples and reagents, an elevator module to supply microplates to the MA, an incubator module, and a photometry scanner. These components are orchestrated to perform a set of tests in an efficient way by managing the available resources according to laboratory requirements. The analysis workflow is initiated by loading samples and reagents in individual tubes, and having each microplate identified by an RFID tag (A clinical module includes a microplate, an RFID tag and a reader). Then, the operator specifies the required test sequences and the schedule. From this moment on, the BioRobot platform operates automatically by having MAs collecting and transporting samples between system modules (SCARA robot, incubator, spectrophotometer, etc.), until the scheduled test sequences are completed. During this process, samples are tracked by the RFID tags embedded in clinical modules. The BioRobot platform employs Bluetooth technology for supporting communications among MAs and a central unit (PC), which enables the provision of updated MA status information (position, samples being carried, etc.) and the transmission of supervisory commands to MAs.

The use of robots in domestic settings is also becoming popular. For example, Dengler et al. [39] present a family of robots (Robertino) allowing the support and the maintenance of smart home communication networks, providing environment and inhabitant monitoring and locating frequently lost objects. This system is also supported on Bluetooth communication technology. The following section presents some issues regarding the application of communications in healthcare environments, as well as the corresponding challenges.

30.6 Issues and Challenges

Over the years, standard industrial communications have been integrated in healthcare equipment for networking support. In consequence, medical and industrial communications share a significant range of features and problems in areas like security, privacy, safety, reliability, and timeliness. Current endeavors in the standardization of different medical communications are also expected to improve the interoperability of communication devices from different manufacturers.

30.6.1 Security and Privacy

Nowadays, security and privacy are two critical concerns, not only for the direct implications of their violation but also because their support, in many healthcare applications, is an open research area. Wireless communications are becoming very popular among home networking applications given their low-cost installation, aesthetics, flexibility, and reconfiguration capabilities [46]. However, they are more exposed to security and privacy breaches than their wired counterparts.

Bergstrom et al. [47] state that, to ensure security, a network must provide confidentiality, authentication, authorization, and integrity. Confidentiality is concerned with the possibility of third parties being able to read the transmitted messages in the network while authentication relates with the concern that somebody may masquerade as a legitimate operator of the system. The system's integrity can be corrupted when an attacker modifies system data or commands. This can occur, when, for example, a malicious user is able to tamper with data by gaining access to the system using false or pirated authorizations. Another important aspect is privacy, not only from the trust perspective [48] but also from an ethical point of view [49], i.e., patients have human rights and civil liberties that must be respected. An example illustrating the multiple dimensions of privacy is the monitoring of elderly people where privacy can be interpreted in a perspective of maintaining personal data secure (inaccessible to third parties) or respecting the right to intimacy of elderly people. These perspectives should be taken into account in the development of legislation concerning the protection of privacy.

30.6.2 Safety and Reliability

The support of safety in healthcare environments is a domain often regarded as solely relying on security and the protection of privacy. However, it is dependent on a broader spectrum of scenarios: system misuse, software bugs, hardware malfunctions, interference on communications, etc. If a user has the ability to perform a given set of actions (commands), among them commands that require human intervention for completion or continuous operation, he/she may utilize the system improperly and cause damage not only to the equipment but also to himself or someone nearby (e.g., patient). Another scenario arises from software bugs or hardware malfunctions. In this case, the system has no predictable behavior and can damage the medical equipment and injure its users. Since a lot of healthcare equipment use buses for communicating data, they may work on unpredictable states due to network malfunctions caused by electromagnetic interference (EMI) from nearby equipment [50]. In order to improve the reliability of wireless medical device networks (WMDN), Gehlot and Sloane [51] propose a WMDN verification and validation toolkit (V2T) that allows warranting safe and reliable WMDN operation. However, due to the broad range of component manufacturers, device models, and user configurations, the support of universally safe and reliable WMDN systems is still in an early stage of development.

30.6.3 Standardization

Standardization promotes the avoidance of duplicate research efforts, the improvement of technical support, an increased volume production leading to a reduced per item cost, a broadening of the application market, etc. To harness these benefits in healthcare environments, specific protocols addressing interoperable bedside devices (ISO/IEEE 11073 [53]) and BANs (IEEE 802.15.6 [54]) have been proposed.

The ISO/IEEE 11073 family of standards specifies the nomenclature, the abstract data model, the service model, and the transport for interoperable bedside devices and aims to "provide real-time plug-and-play interoperability for patient-connected medical devices and facilitate the efficient exchange of vital signs and medical device data, acquired at the point-of-care, in all health care environments" [56]. Currently, only two transport systems have been approved: cable connected and infrared wireless. However, an approved draft version of a set of guidelines for the use of RF wireless technology as transport in ISO/IEEE 11073 systems is available since June 2008.

In the latter part of 2007, the IEEE 802.15 Task Group 6 (BAN) was formed to develop a communication standard optimized for low-power devices operating in (or around) the human body (but not limited to humans), addressing medical healthcare applications (e.g., physiological monitoring), consumer electronics including entertainment (e.g., wireless headphone), and other applications. Besides the unique criteria defined by Task Group 6 [58], issues regarding candidate wireless technologies have been introduced [57]. However, a draft ready for the sponsor ballot is only scheduled for the end of 2009, which will significantly delay the standard adoption.

Although not specifically designed envisaging healthcare environments, the IEEE 1451 [52] has the potential to become a contender in medical monitoring applications. This standard was initially sponsored by the IEEE Instrumentation and Measurement Society and targets industrial wired and wireless distributed monitoring and control applications. The IEEE 1451 family of standards defines a set of open, common, network-independent communication interfaces for connecting transducers (sensors or actuators) to microprocessors, instrumentation systems, and control/field networks [55]. Smart transducer interfaces for sensors (and actuators) employing wireless communication protocols (Wi-Fi, ZigBee, Bluetooth, and LoWPAN) are standardized since 2007 and are already commercially available [59].

Despite the existence of standards specifically addressing communications among medical devices in healthcare environments, commercial modules supporting those standards are still scarce, thus hindering their application in real deployments. Also, standards should be designed to reduce feature redundancy and incompatibility, which would foster interoperability.

30.6.4 Timeliness

The occurrence of high delays and jitter in medical communication systems has a direct impact on the quality of service provided by the associated equipment. For example, a varying (high) latency in real-time patient vital sign communication can interfere with computerized and human interpretation, increasing the probability of misdiagnosis and risking therapeutic interventions. Moreover, delays can originate spurious alarms, such as "lead-loss" or "offline patient," which reduce system trust and increase complaints (and investigations) for "no-problem-found" events.

Because security and privacy are main requirements in healthcare communications, the use of standard encryption mechanisms is highly desirable as it allows securing data and guaranteeing compatibility and continuous improvement of the technology. However, since most medical devices embed small power microprocessors, the delay associated with the encryption/decryption procedure can be significant.

Many of the aforementioned medical communications employed in localization, monitoring, and automation applications have not been fully assessed, i.e., standard protocols are employed for data communication without any experimental or theoretical evaluation of latency and jitter. This approach is inadequate and highly censurable for medical applications with safety-critical real-time communication requirements.

References

1. Division of the Department of Economic and Social Affairs of the United Nations Secretariat, Growth of the Europe population, in *World Population Prospects: The 2006 Revision and World Urbanization*, New York: United Nations, 2006.
2. Cohen, T., Medical and information technologies converge, *IEEE Engineering in Medicine and Biology Magazine*, 23(3), 59–65, May–June 2004.
3. Arnon, S., Bhastekar, D., Kedar, D., and Tauber, A., A comparative study of wireless communication network configurations for medical applications, *IEEE Wireless Communications*, 10(1), 56–61, February 2003.
4. Monton, E., Hernandez, J.F., Blasco, J.M., Herve, T., Micallef, J., Grech, I., Brincat, A., and Traver, V., Body area network for wireless patient monitoring, *IET Communications*, 2(2), 215–222, February 2008.
5. Gama, O., Carvalho, P., Afonso, J.A., and Mendes, P.M., Quality of service support in wireless sensor networks for emergency healthcare services, in *30th Annual International Conference of the IEEE Engineering in Medicine and Biology Society (EMBS2008)*, Vancouver, British Columbia, Canada, August 20–25, 2008, pp. 1296–1299.
6. Shrestha, S., Balachandran, M., Agarwal, M., Phoha, V.V., and Varahramyan, K., A chipless RFID sensor system for cyber centric monitoring applications, *IEEE Transactions on Microwave Theory and Techniques*, 57(5), 1303–1309, May 2009.
7. Yu, K., Guo, Y.J., and Hedley, M., TOA-based distributed localisation with unknown internal delays and clock frequency offsets in wireless sensor networks, *IET Signal Processing*, 3(2), 106–118, March 2009.
8. Lee, H.B., A novel procedure for assessing the accuracy of hyperbolic multilateration systems, *IEEE Transactions on Aerospace and Electronic Systems*, AES-11(1), 2–15, January 1975.
9. Winkler, F., Fischer, E., Grass, E., and Langendorfer, P., An indoor localization system based on DTDOA for different wireless LAN systems, in *Proceedings of the Third Workshop on Positioning, Navigation and Communication (WPNC'06)*, Hannover, Germany.
10. Patwari, N., Ash, J.N., Kyperountas, S., Hero, A.O., III, Moses, R.L., and Correal, N.S., Locating the nodes: Cooperative localization in wireless sensor networks, *IEEE Signal Processing Magazine*, 22(4), 54–69, July 2005.
11. Gezici, S., Zhi, T., Giannakis, G.B., Kobayashi, H., Molisch, A.F., Poor, H.V., and Sahinoglu, Z., Localization via ultra-wideband radios: A look at positioning aspects for future sensor networks, *IEEE Signal Processing Magazine*, 22(4), 70–84, July 2005.
12. Mao, G., Fidan, B., and Anderson, B.D.O., Wireless sensor network localization techniques, *Computer Networks*, 51(10), 2529–2553, July 2007.
13. Exavera Technologies Inc., Portsmouth, NH, [Online], Available: http://www.exavera.com
14. Ubisense, Cambridge, U.K., [Online], Available: www.ubisense.net
15. Ekahau Inc., Reston, VA, [Online], Available: http://www.ekahau.com
16. AeroScout Inc., Redwood City, CA, [Online], Available: http://www.aeroscout.com/
17. McKneely, P.K., Chapman, F., and Gurkan, D., Plug-and-play and network-capable medical instrumentation and database with a complete healthcare technology suite: MediCAN, in *Joint Workshop on High Confidence Medical Devices, Software, Device Plug-and-Play Interoperability (HCMDSS-MDPnP)*, Boston, MA, June 25–27, 2007, pp. 122–130.
18. Varady, P., Benyo, Z., and Benyo, B., An open architecture patient monitoring system using standard technologies, *IEEE Transactions on Information Technology in Biomedicine*, 6(1), 95–98, March 2002.
19. Dagtas, S., Pekhteryev, G., and Sahinoglu, Z., Multi-stage real time health monitoring via ZigBee in smart homes, in *21st International Conference on Advanced Information Networking and Applications Workshops, 2007 (AINAW '07)*, Niagara Falls, Canada, May 21–23, 2007, Volume 2, pp. 782–786.
20. Junnila, S., Defee, I., Zakrzewski, M., Vainio, A.-M., and Vanhala, J., UUTE home network for wireless health monitoring, in *International Conference on Biocomputation, Bioinformatics, and Biomedical Technologies, 2008 (BIOTECHNO '08)*, Bucharest, Romania, June 29–July 5, 2008, pp. 125–130.

21. Monton, E., Hernandez, J.F., Blasco, J.M., Herve, T., Micallef, J., Grech, I., Brincat, A., and Traver, V., Body area network for wireless patient monitoring, *IET Communications*, 2(2), 215–222, February 2008.
22. B. Lo, Thiemjarus, S., King, R., and Yang, G., Body sensor network—A wireless sensor platform for pervasive healthcare monitoring, in *Proceedings of the 3rd International Conference on Pervasive Computing (PERVASIVE 2005)*, Munich, Germany, May 2005, pp. 77–80.
23. Zhang, Z. and Liu, P., Application of Bluetooth technology in ambulatory wireless medical monitoring, in *Conference on 4th International Microwave and Millimeter Wave Technology Proceedings, 2004 (ICMMT)*, Beijing, China, August 18–21, 2004, pp. 974–977.
24. Lee, R.-G, Chen, K.-C, Hsiao, C.-C, and Tseng, C.-L, A mobile care system with alert mechanism, *IEEE Transactions on Information Technology in Biomedicine*, 11(5), 507–517, September 2007.
25. Stojanovic, R., Tafa, Z., and Lekic, N., An approach to monitoring of physiological signals using Bluetooth, in *International Conference on Applied Electronics, 2006 (AE 2006)*, Pilsen, Czech Republic, September 6–7, 2006, pp. 185–188.
26. Bluetooth SIG, Medical Device Profile (MDP), Kirkland Washington, [Online], Available: http://www.bluetooth.org
27. Tejero-Calado, J.C., Lopez-Casado, C., Bernal-Martin, A., Lopez-Gomez, M.A., Romero-Romero, M.A., Quesada, G., Lorca, J., and Rivas, R., IEEE 802.11 ECG monitoring system, in *27th Annual International Conference of the IEEE-EMBS 2005*, Shanghai, China, January 17–18, 2006, pp. 7139–7142.
28. Postolache, O., Postolache, G., and Girao, P.S., Non-invasive mobile homeostasis instrument, in *IEEE International Workshop on Medical Measurement and Applications, 2006 (MeMea 2006)*, Benevento, Italy, April 20–21, 2006, pp. 94–97.
29. Chiarugi, F., Karatzanis, I., Zacharioudakis, G., Meriggi, P., Rizzo, F., Stratakis, M., Louloudakis, S., Biniaris, C. et al., Measurement of heart rate and respiratory rate using a textile-based wearable device in heart failure patients, *Computers in Cardiology, 2008*, Bologna, Italy, September 14–17, 2008, pp. 901–904.
30. Fedral Communication Comission, FCC rules and Regulations, Table of Frequency Allocations, Part 2.106, Washington, September 2008.
31. Lee, S., Jerald, Y., and Yoo, H.-J., A healthcare monitoring system with wireless woven inductor channels for body sensor network, in *5th International Summer School and Symposium on Medical Devices and Biosensors, 2008 (ISSS-MDBS 2008)*, Hong Kong, China, June 1–3, 2008, pp. 62–65.
32. Yuce, M.R. and Ho, C.K., Implementation of body area networks based on MICS/WMTS medical bands for healthcare systems, in *30th Annual International Conference of the IEEE Engineering in Medicine and Biology Society, 2008 (EMBS 2008)*, Vancouver, British Columbia, Canada, August 20–25, 2008, pp. 3417–3421.
33. Dishman, E., Inventing wellness systems for aging in place, *Computer Journal*, 37(5), 34–41, May 2004.
34. Harmo, P., Taipalus, T., Knuuttila, J., Vallet, J., and Halme, A., Needs and solutions—Home automation and service robots for the elderly and disabled, in *Proceedings of the IEEE/RSJ International Conference on Intelligent Robots and Systems (IROS 2005)*, Edmonton, Canada, August 2005.
35. Jiang, L., Liu, D., and Yang, B., Smart home research, in *Proceedings of the International Conference on Machine Learning and Cybernetics*, Shanghai, China, 2004, Volume 2, pp. 659–663.
36. Eurox10, Porto, Portugal, [Online], Available: http://www.eurox10.com/
37. Z-Wave Alliance, Milpitas, CA, [Online], Available: http://www.z-wavealliance.org
38. Bartolomeu, P., Fonseca, J., Santos, V., Mota, A., Silva, V., and Sizenando, M., Automating home appliances for elderly and impaired people: The B-Live approach, in *Software Development for Enhancing Accessibility and Fighting Info-exclusion (DSAI'2007)*, Vila Real, Portugal, November 2007.
39. Dengler, S., Awad, A., and Dressier, F., Sensor/actuator networks in smart homes for supporting elderly and handicapped people, in *21st International Conference on Advanced Information Networking and Applications Workshops, 2007 (AINAW '07)*, Niagara Falls, Canada, May 21–23, 2007, Volume 2, pp. 863–868.

40. Khan, S.A., Aziz, H., Maqsood, S., and Faisal, S., Clustered home area network: A beacon enabled IEEE 802.15.4 approach, in *Fourth International Conference on Emerging Technologies, 2008 (ICET 2008)*, Rawalpindi, Pakistan, October 18–19, 2008, pp. 193–198.
41. Bartolomeu, P., Fonseca, J., and Vasques, F., On the timeliness of multi-hop non-beaconed ZigBee broadcast communications, in *IEEE International Conference on Emerging Technologies and Factory Automation, 2008 (ETFA 2008)*, Hamburg, Germany, September 15–18, 2008, pp. 679–686.
42. Institute of Electrical and Electronics Engineers (IEEE), Wireless Medium Access Control (MAC) and Physical Layer (PHY) Specifications for Low-Rate Wireless Personal Area Networks (WPANs), IEEE Std 802.15.4-2006, Los Alamitos, CA, pp. 1–305, 2006.
43. Gen, L.P., Ellata, A.Y., Zhi, F.L., and Ping, L.X., Review of application of robots in health care, *Pakistan Journal of Information Technology*, 1(3), 269–271, 2002.
44. Ozkil, A.G., Dawids, S., Zhun, F., and Srensen, T., Design of a robotic automation system for transportation of goods in hospitals, in *International Symposium on Computational Intelligence in Robotics and Automation, 2007 (CIRA 2007)*, Jacksonville, FL, June 20–23, 2007, pp. 392–397.
45. Choi, B.J., Jin, S.M., Shin, S.H., Koo, J.C., Ryew, S.M., Kim, M.C., Kim, J.H., Son, W.H. et al., Development of flexible laboratory automation platform using mobile agents in the clinical laboratory, in *IEEE International Conference on Automation Science and Engineering, 2008 (CASE 2008)*, Washington DC, August 23–26, 2008, pp. 918–923.
46. Reinisch, C., Kastner, W., Neugschwandtner, G., and Granzer, W., Wireless technologies in home and building automation, in *Proceedings of the International Conference on Industrial Informatics (INDIN'2007)*, Vienna, Austria, July 2007.
47. Bergstrom, P., Driscoll, K., and Kimball, J., Making home automation communications secure, *IEEE Computer Magazine*, 34, 50–56, 2001.
48. Schaefer, R., Ziegler, M., and Mueller, W., Securing personal data in smart home environments, in *Workshop on Privacy-Enhanced Personalization (PEP'06)*, Montreal, Quebec, Canada, 2006.
49. Welsh, S., Hassiotis, A., O'Mahoney, G., and Deahl, M., Big brother is watching you—The ethical implications of electronic surveillance measures in the elderly with dementia and in adults with learning difficulties, *Aging and Mental Health*, 7, 372–375, 2003.
50. Krishnamoorthy, S., Reed, J.H., Anderson, C.R., Max Robert, P., and Srikanteswara, S., Characterization of the 2.4 GHz ISM band electromagnetic interference in a hospital environment, in *Proceedings of the 25th Annual International Conference of the IEEE (EMBS2003)*, Cancun, Mexico, September 17–21, 2003, Volume 4, pp. 3245–3248.
51. Gehlot, V. and Sloane, E.B., Ensuring patient safety in wireless medical device networks, *Computer*, 39(4), 54–60, April 2006.
52. National Institute of Standards and Technology, IEEE Standard 1451, Gaithersburg, MD, [Online], Available: http://ieee1451.nist.gov
53. ISO/IEEE 11073 standard, Geneva, Switzerland, [Online], Available: http://www.iso.org
54. Institute of Electrical and Electronics Engineers, IEEE 802.15 WPAN Task Group 6 (TG6) Body Area Networks, CA, [Online], Available: http://www.ieee802.org/15/pub/TG6.html
55. Song, E.Y. and Lee, K., Understanding IEEE 1451-networked smart transducer interface standard - What is a smart transducer? *Instrumentation & Measurement Magazine, IEEE*, 11(2), 11–17, April 2008.
56. Yao, J. and Warren, S., Applying the ISO/IEEE 11073 standards to wearable home health monitoring systems, *Journal of Clinical Monitoring and Computing*, 19, 427–436, 2005.
57. Li, H.-B, Takizawa, K., and Kohno, R., Trends and standardization of body area network (BAN) for medical healthcare, in *European Conference on Wireless Technology, 2008 (EuWiT 2008)*, Amsterdam, the Netherlands, October 27–28, 2008, pp. 1–4.
58. Astrin, A.W., Li, H.-B., and Kohno, R., SG BAN project draft 5C, Institute of Electrical and Electronics Engineers, IEEE 802 15-06-0488/r7, CA, July 2007
59. Esensors, Inc., Amherst, NY, [Online], Available: http://www.eesensors.com

III

Technologies

31 Controller Area Network *Joaquim Ferreira and José Alberto Fonseca*........................**31**-1
Introduction • CAN Technology Basics • CAN-Based Upper Layer Protocols • CAN Limitations • List of Acronyms • References

32 Profibus *Max Felser and Ron Mitchell*........................**32**-1
Introduction • Physical Transmissions • Fieldbus Data Link • DP System • Cyclic Data Exchange: MS0—Relation • Acyclic Data Exchange: MS1/MS2 Relations • Application Profiles • References

33 INTERBUS *Juergen Jasperneite and Orazio Mirabella*........................**33**-1
INTERBUS Overview • INTERBUS Protocol • Diagnostics • Performance Evaluation • Summary • References

34 WorldFip *Francisco Vasques and Orazio Mirabella*........................**34**-1
Introduction • Physical Layer • Data Link Layer • Application Layer • Timing Properties of WorldFIP Networks • References

35 Foundation Fieldbus *Carlos Eduardo Pereira, Augusto Pereira, and Ian Verhappen*........................**35**-1
Introduction • Foundation Fieldbus Overview • Topology • Drivers (DD, EDDL, and FDT/DTM) • Cables • Segment Design • FFPS—Fieldbus Power Supplies • Installation of Segment in Safe Areas • Installation of Segments in Classified Areas • Project Documentation • Installations and Commissioning • Maintenance • References

36 Modbus *Mário de Sousa and Paulo Portugal*........................**36**-1
Introduction • Modbus Interaction and Data Models • Modbus Protocol Architecture • Modbus Application Layer • Modbus Serial • Modbus TCP • Example • Acronyms • References

37 Industrial Ethernet *Gaëlle Marsal and Denis Trognon*........................**37**-1
Introduction • Industrial Ethernet • Standardized Solutions of IEC 61158 and IEC 61784 • Features of Major Industrial Ethernet Solutions • Synthesis • Abbreviations • References

38 EtherCAT *Gianluca Cena, Adriano Valenzano, and Claudio Zunino*........................**38**-1
Introduction • Physical Layer • Communication Protocol • Addressing • SyncManager • Distributed Clock • Application Layer • References

39 Ethernet POWERLINK *Paulo Pedreiras, Stefan Schoenegger, Lucia Seno, and Stefano Vitturi*........................**39**-1
Introduction • EPL Protocol • Frame Mapping • Network Configurations • Redundancy Aspects • Security Aspects • Performance Analysis • References

40 PROFINET *Max Felser, Paolo Ferrari, and Alessandra Flammini*........................**40**-1
Introduction • PROFINET IO Basics • IRT Communication in PROFINET IO • Engineering and Commissioning • Integration of Fieldbus Systems and Web Applications • Acronyms • Bibliography

41 LonWorks *Uwe Ryssel, Henrik Dibowski, Heinz Frank, and Klaus Kabitzsch* **41**-1
Introduction • System Components • LonTalk Protocol • The Application Layer Programming Model • Function Block-Based Design and System Integration • Network Design Tools • Automatic Design Approaches • References

42 KNX *Wolfgang Kastner, Fritz Praus, Georg Neugschwandtner, and Wolfgang Granzer* **42**-1
Introduction and Overview • Medium-Independent Layers • Medium-Dependent Layers • Runtime Interworking • Devices • Configuration • Conclusion and Outlook • Abbreviations • References

43 Protocols of the Time-Triggered Architecture: TTP, TTEthernet, TTP/A *Wilfried Elmenreich and Christian El-Salloum* **43**-1
Introduction • The Time-Triggered Paradigm • Time-Triggered Communication • Time-Triggered Protocol (TTP) • Time-Triggered Ethernet • TTP/A • Acknowledgments • References

44 FlexRay *Martin Horauer and Peter Rössler* **44**-1
Introduction • Protocol • System Architecture • System Design Considerations • References

45 LIN-Bus *Andreas Grzemba, Donal Heffernan, and Thomas Lindner* **45**-1
LIN Background • LIN History and Versions • Communication Concept • Physical Layer • LIN Message Frames • Network and Status Management • Transport Layer Protocol • Configuration • Relationship between SAE J2602 and LIN 2.0 • Conclusion • References

46 Profisafe *Ron Mitchell, Max Felser, and Paulo Portugal* **46**-1
Introduction • Profisafe Communication • Deployment • Acronyms • References

47 SafetyLon *Thomas Novak, Thomas Tamandl, and Peter Preininger* **47**-1
Introduction • The General SafetyLon Concept • The Safety-Related Lifecycle • The Hardware • The Safety-Related Firmware • The SafetyLon Tools • Acronyms • References

48 Wireless Local Area Networks *Henning Trsek, Juergen Jasperneite, Lucia Lo Bello, and Milos Manic* **48**-1
Introduction • The 802.11 Family • Physical Layer • Medium Access Control • Limitations of DCF and HCF for QoS Support in Industrial Environments • Security Mechanisms • Fast Handover • Future Enhancements • References

49 Bluetooth *Stefan Mahlknecht, Milos Manic, and Sajjad Ahmad Madani* **49**-1
Introduction • Bluetooth Core Architecture Blocks • Bluetooth Protocol Stack • Bluetooth Profiles • Competitive Technologies • Future of the Bluetooth Technology: Challenges • References

50 ZigBee *Stefan Mahlknecht, Tuan Dang, Milos Manic, and Sajjad Ahmad Madani* **50**-1
Introduction • ZigBee and Mesh Networks • ZigBee in the Context of Other Wireless Networks • ZigBee Stack • IEEE 802.15.4 • Development and Industrial Applications • Conclusion • References

51 6LoWPAN: IP for Wireless Sensor Networks and Smart Cooperating Objects *Guido Moritz and Frank Golatowski* **51**-1
Introduction • Why IP in WSN and For Smart Cooperating Objects? • Introduction in 802.15.4 • 802.15.4 and 6LoWPAN • 6LoWPAN • Summary • References

52 WiMAX in Industry *Milos Manic, Sergiu-Dan Stan, and Strahinja Stankovic* **52**-1
Introduction • The WiMAX Broadband Technology • WiMAX Architecture • The WiMAX Forum and Working Groups • Integration with Other Networks • Conclusion • References

53 **WirelessHART, ISA100.11a, and OCARI** *Tuan Dang and Emiliano Sisinni* **53**-1
Introduction • WirelessHART • ISA 100.11a • OCARI • Coexistence of the Three Protocols • Example of Platform Providers • Conclusions • References

54 **Wireless Communication Standards** *Tuan Dang* **54**-1
Introduction • A Wireless Standards Taxonomy • Regulations and EMC • Conclusion • References

55 **Communication Aspects of IEC 61499 Architecture** *Valeriy Vyatkin, Mário de Sousa, and Alois Zoitl* **55**-1
Introduction • Illustrative Example • Logic Encapsulated in Basic FB • Extension • Distribution • Communication FBs • Communication Using Services of Internet Protocol Suite • Adding Distribution and Communication to the Sample System • Internals of Communication FBs: Modbus • Communication via the CIP • Impact of Communication Semantics on Application Behavior • Failures in Distributed Applications • Conclusion • References

56 **Industrial Internet** *Martin Wollschlaeger and Thilo Sauter* **56**-1
Introduction • Application of Internet Technologies in Industry • Technologies • Application Examples • Conclusions and Outlook • Acronyms • References

57 **OPC UA** *Tuan Dang and Renaud Aubin* **57**-1
Introduction • Overview of OPC UA System Architecture • Overview of UA AddressSpace • Overview of UA Services • Implementations and Products • Conclusion • References

58 **DNP3 and IEC 60870-5** *Andrew C. West* **58**-1
Requirements for SCADA Data Collection in Electric Power and Other Industries • Features Common to IEC 60870-5 and DNP3: Data Typing, Report by Exception, Error Recovery • Differentiation between IEC 60870-5 and DNP3 Operating Philosophy, Message Formatting, Efficiency, TCP/IP Transport • References

59 **IEC 61850 for Distributed Energy Resources** *Sidonia Mesentean, Heinz Frank, and Karlheinz Schwarz* **59**-1
Introduction • Basic Concept • Modeling the Automation Functions • Communication Services • Modeling with System Configuration Language • Different Types of DER • References

31 Controller Area Network

31.1 Introduction 31-1
31.2 CAN Technology Basics 31-2
Physical Layer • Data Link Layer • Detecting and Signaling Errors • Network Topology
31.3 CAN-Based Upper Layer Protocols 31-6
CANopen • DeviceNet • TTCAN • FTT-CAN
31.4 CAN Limitations 31-12
Consequences of Faults at the Node Level
List of Acronyms 31-14
References 31-14

Joaquim Ferreira
University of Averio

José Alberto Fonseca
Universidade of Aveiro

31.1 Introduction

Controller area network (CAN) is a popular and a very well-known bus system, both in academia and in industry. CAN protocol was introduced in the mid-1980s by Robert Bosch GmbH [7], and it was internationally standardized in 1993 as ISO 11898-1 [24]. It was initially designed for distributed automotive control systems as a single digital bus to replace traditional point-to-point cables that were growing in complexity, weight, and cost with the introduction of new electrical and electronic systems. Nowadays CAN is still used extensively in automotive applications, with an excess of 400 million CAN-enabled microcontrollers manufactured each year [14].

The widespread and successful use of CAN in the automotive industry, the low cost associated with high-volume production of controllers, and CAN's inherent technical merit have driven to CAN adoption in other application domains such as industrial communications, medical equipment, machine tool, robotics, and in distributed embedded systems in general.

CAN provides two layers of the stack of the open systems interconnection (OSI) reference model: the physical layer and the data link layer. Optionally, it could also provide an additional application layer, not included on the CAN standard. Notice that CAN physical layer was not defined in Bosch original specification, only the data link layer was defined. However, the CAN ISO specification filled this gap and the physical layer was then fully specified. CAN is a message-oriented transmission protocol, i.e., it defines message contents rather than nodes and node addresses. Every message has an associated message identifier, which is unique within the whole network, defining both the content and the priority of the message. Transmission rates are defined up to 1 Mbps.

The large installed base of CAN nodes with low failure rates over almost two decades led to the use of CAN in some critical applications such as anti-locking brake systems (ABS) and electronic stability program (ESP) in cars. In parallel with the wide dissemination of CAN in industry, the academia also devoted a large effort to CAN analysis and research, making CAN one of the most studied fieldbuses. That is why a large number of books or book chapters describing CAN were published. The first CAN book, written in French by D. Paret, was published in 1997 and presents the CAN basics [32].

More implementation oriented approaches, including CAN node implementation and application examples, can be found in Lorenz [28] and in Etschberger [16], while more compact descriptions of CAN can be found in [11] and in some chapters of [31].

Despite its success story, CAN application designers would be happier if CAN could be made faster, cover longer distances, be more deterministic, and be more dependable [34]. Over the years, several protocols based on CAN were presented, taking advantage of some CAN properties and trying to improve some known CAN drawbacks. This chapter, besides presenting an overview of CAN, describes also some other relevant higher level protocols based on CAN, such as CANopen [13], DeviceNet [6], FTT-CAN [1], and TTCAN [25].

31.2 CAN Technology Basics

The original specification [7] contains few more than a MAC protocol based on decentralized medium access (CSMA type) that uses a particular physical characteristic of the medium to support a nondestructive bit-wise arbitration. The communication is message-oriented since each message receives an identifier which must be unique in the system and which establishes the message priority for the arbitration process. Any transmitted CAN message is broadcast to every node in the network.

Whenever two or more nodes start transmitting at the same time, only the one transmitting the message with the highest priority will proceed with the transmission. All others quit and try again after the current transmission ceases. This deterministic mechanism allows one to compute the worst-case response time of all CAN messages [14,42]. So, from the real-time point of view, CAN is a serial data bus that supports priority-based message arbitration and non-preemptive message transmission. In 1994, Tindell and Burns [42] adapted previous research on fixed priority preemptive scheduling for single-processor systems to the scheduling of messages on CAN, providing a method for calculating the worst-case response times of all CAN messages. Prior to Tindell's results, the bus utilization in automotive applications was typically around 30% or 40%, but still requiring extensive testing to obtain confidence that CAN messages would meet their deadlines [14], since then there were no analysis tools that could guarantee real-time behavior. With the advent of new design techniques, based on schedulability analysis, CAN bus utilization could be increased to around 80% [15]. Tindell's results were transferred to industry in the form of commercial CAN schedulability analysis tools, e.g., Volcano Network Architect, that have been used by a large number of major automotive manufacturers. Unfortunately, the initial CAN schedulability analysis [42] was flawed [14]. It may provide guarantees for messages that, in the worst-case scenario, will in fact miss their deadlines. Davis et al. [14] correct previous flawed analysis and discuss the impact on commercial CAN systems designed and developed using flawed schedulability analysis, while making recommendations for the revision of CAN schedulability analysis tools.

31.2.1 Physical Layer

The CAN ISO standard incorporates the original Bosch specifications [7] as well as part of the physical layer, the physical signaling, the bit timing, and synchronization. There are a small number of other CAN physical layer specifications including

- *ISO 11898-2 High Speed*—ISO 11898-2 [24] is the most used physical layer standard for CAN networks. In this standard, the data rate is defined up to 1 Mbps with a theoretically possible bus length of 40 m at 1 Mbps.
- *ISO 11898-3 Fault Tolerant*—This standard defines data rates up to 125 kbps with the maximum bus length depending on the data rate used and the bus load. Up to 32 nodes per network are specified. The fault-tolerant transceivers support the complete error management including the detection of bus errors and automatic switching to asymmetrical signal transmission.

TABLE 31.1 Practical CAN Bus Length for ISO 11898 Compliant Transceivers and Standard Bus Line Cables

Bit Rate	Bus Length (m)
1 Mbps	30
800 kbps	50
500 kbps	100
250 kbps	250
125 kbps	500
62,5 kbps	1000
20 kbps	2500
10 kbps	5000

Source: CiA, CAN physical layer, CiA (CAN in Automation), Nuremberg, Germany, 2001.

- *SAE J2411 Single Wire*—An unshielded single wire is defined as the bus medium and the communication takes place with a nominal data rate of 33.3 kbps. The standard defines up to 32 nodes per network. The main application area of this standard is in comfort electronics networks in vehicles.
- *ISO 11992 Point-to-Point*—This standard defines a point-to-point connection for use mainly in vehicles with trailers. The nominal data rate is 125 kbps with a maximum bus line length of 40 m.

The most popular CAN physical layer protocol, available in most of the CAN transceivers, is the one defined in the ISO 11898-2 standard. The maximum achievable bus line length in a CAN network, represented in Table 31.1, depends on

- The loop delays of the connected bus nodes and the delay of the bus lines
- The differences of the relative oscillator tolerance between nodes
- The signal amplitude drop due to the series resistance of the bus cable and the input resistance of bus nodes

The CAN physical layer provides a two-state medium, where the bus can be either *dominant* or *recessive*. Whenever two nodes simultaneously transmit bits of opposite value, then all nodes should read *dominant*. Usually the *dominant* state is associated with the binary value 0 and *recessive* with the binary value 1.

The physical layer has a number of built-in fault-tolerant features. CAN provides resilience against a variety of physical faults such as one open wire, the short-circuit of the two signal wires, or even one of the signal wires shorted to ground or power. Notice that not all CAN controllers are able to implement these features, by switching from differential signaling to single-line signaling, at higher bus speeds. The CAN differential electrical signaling mode is very resistant to electromagnetic interference (EMI) since interference will tend to affect each side of a differential signal almost equally. However, the differential electrical signaling does not fully prevent EMI to affect the signal on the bus in such a way that one or more nodes on the bus will simultaneously read a different bit value from that which was transmitted. A node detecting the error (possibly the transmitter) will invalidate the message by transmitting an error frame. The number and the nature of EMI-induced transmission faults and the ability of the physical layer to prevent them depends on several factors such as the cable type and length, the number of nodes, the transceiver type, and the EMI shielding.

31.2.2 Data Link Layer

The data link layer of CAN includes the services and protocols required to assure a correct transfer of information from one node to another.

There are four types of message on CAN: data frames, error frames, remote frames, and overload frames. The latter two types of messages are rarely used in real application and, thus, will not be described further.

The CAN protocol supports two message frame formats, the only essential difference being in the length of the identifier. The CAN *base frame* format supports a length of 11 bits for the identifier (formerly known as CAN 2.0 A), and the CAN *extended frame* format supports a length of 29 bits for the identifier (formerly known as CAN 2.0 B).

Data on the bus are sent in data frames which consist of up to 8 bytes of data plus a header and a footer. The frame is structured as a number of fields, as depicted in Figure 31.1.

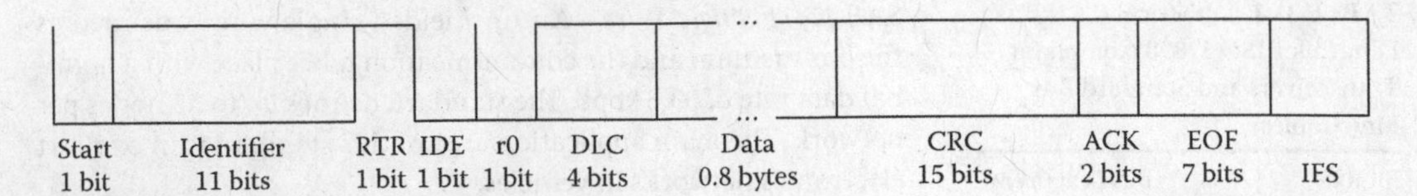

FIGURE 31.1 CAN base frame format.

A CAN base frame message begins with the start bit called *start of frame* (SOF). This bit is followed by the *arbitration field*, which consist of the identifier and the *remote transmission request* (RTR) bit used to distinguish between the data frame and the data request frame called remote frame. The following *control field* contains the *identifier extension* (IDE) bit to distinguish between the CAN base frame and the CAN extended frame, as well as the data length code (DLC) used to indicate the number of following data bytes in the *data field*. If the message is used as a remote frame, the DLC contains the number of requested data bytes. The data field that follows is able to hold up to 8 bytes. The integrity of the frame is guaranteed by the following *cyclic redundant check* (CRC) sum. The *acknowledge* (ACK) field comprises the ACK slot and the ACK delimiter. The bit in the ACK slot is sent as a recessive bit and is overwritten as a dominant bit by those receivers which have at this time received the data correctly. Correct messages are acknowledged by the receivers regardless of the result of the acceptance test. The end of the message is indicated by *end of frame* (EOF). The *intermission frame space* (IFS) is the minimum time in equivalent number of bits separating consecutive messages. Unless another station starts transmitting, the bus remains idle after this.

Associated with every CAN message there is a unique message identifier that defines its content and also the priority of the message. Bus access conflicts are resolved by a nondestructive bitwise arbitration scheme where the identifiers of the involved messages are observed bit-by-bit by all nodes, in accordance with the wired-AND mechanism, by which the dominant state overwrites the recessive state. All those nodes with recessive transmission and dominant observation lose the competition for bus access. The nodes that lost the arbitration automatically become receivers of the message with the highest priority and do not reattempt transmission until the bus is available again. In this way, transmission requests are handled in order of their importance for the system as a whole.

31.2.3 Detecting and Signaling Errors

A CAN node does not acknowledge message reception, instead it signals errors immediately as they occur. For error detection the CAN protocol implements three mechanisms at the message level:

- *Cyclic redundancy check*—This mechanism accounts for message corruption. The transmitter node computes the CRC and transmits it within the message. The receiver decodes the message and recomputes the CRC and if they do not match, there has been a CRC error.
- *Frame check*—This mechanism detects message format violations, i.e., it checks each field against the fixed format and the frame size.
- *Acknowledge errors*—Since the receivers of a message must issue an acknowledgement bit in the ACK field, if the transmitter does not receive an acknowledgment an ACK error is indicated, thus allowing a node to detect isolation from the network.

Besides error detection at the message level, the CAN protocol also implements mechanisms for error detection at the bit level:

- *Transmission monitoring*—Each node transmitting a message also monitors the bus level to be able to detect differences between the bit sent and the bit received. This mechanism allows distinguishing global errors from errors local to the transmitter only.

- *Bit stuffing*—CAN uses a non-return-to-zero codification to prevent nodes from loosing synchronization by receiving long sequences of recessive or dominant bits. A supplementary bit is inserted by the transmitter into the bitstream after five consecutive equal bits. The stuff bits are removed by the receivers that also detect violations of the bit stuffing rule.

If at least one station detects any error, it will start transmitting an error frame in the next bit aborting the current message transmission. This prevents other stations from accepting the message and thus ensures the consistency of data throughout the network. After transmission of an erroneous message that has been aborted, the sender automatically reattempts transmission (automatic retransmission). During the new arbitration process, all nodes compete for bus access and the one with the higher priority message will win arbitration. Thus, the message affected by the error could be delayed. There is, however, a special case where consistency throughout the network is compromised, as it will be discussed later.

To prevent a faulty CAN controller to abort all transmissions, including the correct ones, the CAN protocol provides a mechanism to distinguish sporadic errors from permanent errors and local failures at the station. This is done by statistical assessment of station error situations with the aim of recognizing a station's own defects and possibly entering an operation mode in which the rest of the CAN network is not negatively affected. This may go as far as the station switching itself off (bus-off state).

In CAN each node that detects an error sends an error flag composed of six consecutive dominant bits, i.e., a violation of the bit stuffing rule, enabling all nodes on the bus to be aware of a transmission error. The frame affected by the error automatically reenters into the next arbitration phase. The error recovery time (the time from detecting an error until the possible start of a new frame) varies from 17 to 31 bit times [30].

To prevent an erroneous node from disrupting the functioning of the whole system, e.g., by repetitively sending error frames, the CAN protocol includes fault confinement mechanisms that are able to detect permanent hardware malfunctioning and to remove defective nodes from the network. To do this a CAN controller has two error counters; the transmit error counter (TEC) and the receive error (REC) counter which are incremented/decremented according to a set of rules [7,24]. Each time a frame is correctly received or transmitted by a node, the value of the corresponding counter is decreased. Conversely, each time a transmission error is detected the value of the corresponding counter is increased.

Depending on the value of both counters, the station will be in one of the three states defined by the protocol: error active, error passive, and bus-off. In the error-active state (REC < 128 and TEC < 128), the node can send and receive frames without restrictions. In the error-passive state ([REC > 127 or TEC > 127] and [TEC ≤ 255]), the node can transmit but it must wait 8 supplementary bits after the end of the last transmitted frame and it is not allowed to send active error frames upon the detection of a transmission error, it will send passive error frames instead. Furthermore, an error-passive node can only signal errors while transmitting.

After behaving well again for a certain time, a node is allowed to reassume the error-active status. When the TEC is greater than 255, the node CAN controller goes to the bus-off state. In this state, the node can neither send nor receive frames and can only leave this state after a hardware or software reset and after having successfully monitored 128 occurrences of 11 consecutive recessive bits (a sequence of 11 consecutive recessive bits corresponding to the ACK, EOF, and the intermission field of a correct data frame).

When in the error-passive state, the node signals the errors in a way that cannot force the transmitter to retransmit the incorrectly received frame. This behavior is a possible source of inconsistency that must be controlled. As an example of the consequences this can have, consider the case of an error-passive node being the only one to detect an error in a received frame. The transmitter will not be forced to retransmit and the error-passive node will be the only one not to receive the message. Several authors proposed avoiding the error-passive [23,40] state to eliminate this problem. This is easily achieved [40] using a signal available in most CAN circuits, the error warning notification signal. This signal is generated when any error counter reaches the value 96. This is a good point to switch off the node before it goes into the error-passive state, assuring that every node is either helping to achieve data consistency or disconnected.

31.2.4 Network Topology

CAN network topology is bus based. Replicated busses are not referred in CAN standard; however, it is possible to implement them [39,41].

Over the years, some star topologies for CAN have been proposed [5,26,38]. The solution presented in [26] is based on a passive star network topology and relies on the use of a central element, the star coupler, to concentrate all the incoming signals. The result of this coupling is then broadcast to the nodes. The solution presented by Rucks [38] is based on an active star coupler capable of receiving the incoming signals from the nodes bit by bit, implementing a logical AND, and retransmitting the result to all nodes. None of these solutions is capable of disconnecting a defecting branch and so, from the dependability point of view, they only tolerate spatial proximity faults.

Barranco et al. [5] proposed a solution based in an active hub that is able to isolate defective nodes from the network. This active star is compliant with CAN standard and allows detecting a variety of faults (stuck-at node fault, shorted medium fault, medium partition fault, and bit-flipping fault) that will cause the faulty node to be isolated. The replication of the active star has also been considered [4]. Recently, Silva et al. [41] proposed a set of components, an architecture and protocols to enable the dynamic management of the topology of CAN networks made of several replicated buses, both to increase the total bandwidth and to reconfigure the network upon bus permanent error. In many operational scenarios, the proposed solution could be plugged into existing systems to improve its resilience to bus permanent error, without changing the code running in the nodes.

31.3 CAN-Based Upper Layer Protocols

Several CAN-based higher layer protocols have emerged over the years. Some of them are widely used in industry, e.g., CANopen [13] and DeviceNet [3], while others such as TCAN [8], FlexCAN [35], and FTT-CAN [2] have emerged in academia to overcome some well-known CAN limitations, namely, large and variable jitter, lack of clock synchronization, flexibility limitations, data consistency issues, limited error containment, and limited support for fault tolerance [34]. In 2001, an extension of CAN, the time-triggered CAN (TTCAN) [25] was presented to support explicit time-triggered operation on top of CAN. Although TTCAN is not strictly an upper layer protocol based on CAN, it is considered in this section since it offers some services that can be found in some CAN-based upper layer protocols.

Due to space limitations, this section presents only an overview of a subset of these CAN-based protocols, namely, CANopen, DeviceNet, FTT-CAN, and TTCAN.

31.3.1 CANopen

CANopen, internationally standardized as CENLEC EN 50325-4, is a CAN-based higher layer protocol and its specifications cover application layer and communication profile, a framework for programmable devices, recommendations for cables and connectors, and SI units and prefix representations. In terms of configuration capabilities, CANopen is quite flexible and CANopen networks are used in a very broad range of application fields such as machine control, building automation, medical devices, maritime electronics, etc. [13]. CANopen was initially proposed in an EU-Esprit research project and, in 1995, its specification was handed over to the CAN in automation (CiA), an international group of users and manufacturers.

As depicted in Figure 31.2, the CANopen device model can be divided into three parts: the communication interface and protocol software, the object dictionary, and the process interface and application program. The communication interface and protocol software provide services to transmit and to receive communication objects over the CAN bus. The object dictionary describes all data types, communication objects, and application objects used in the device, acting as the

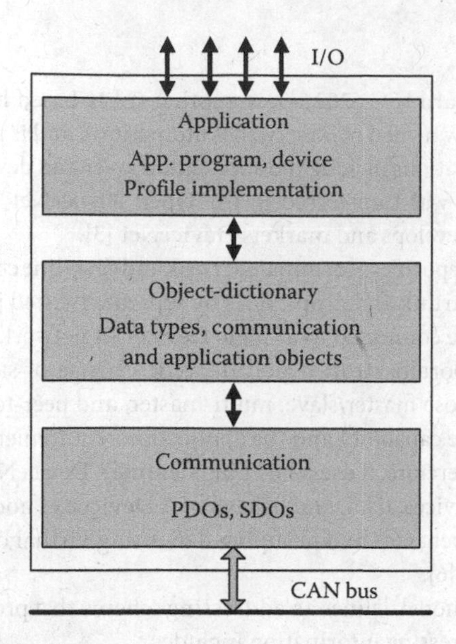

FIGURE 31.2 CANopen device model.

interface to the application software. The application program provides the internal control functionality as well as the interface to the process hardware interfaces.

A device profile is the complete description of the device application with respect to the data items in the object dictionary. The object directory is the central element of every CANopen device describing the device's functionality. The object directory contains all the parameters of a device that can be accessed via the network, for example, device identifiers, manufacturer name, communications parameters, and device monitoring. The device-specific area contains the connection to the process, i.e., the I/O functionality and drivers' parameters. The behavior in the event of an error can also be configured in the object directory. Accordingly, the behavior of a device can be adapted to the respective utilization requirements using the object directory [13].

CANopen uses standardized profiles, and off-the-shelf devices, tools, and protocol stacks are commercially available. In an effort to promote reuse of application software and to potentiate communication compatibility, interoperability, and interchangeability of devices, CANopen provides predefined application objects. Also, manufacturer-specific functionality in devices can be added to the generic functionality described in the profiles. CANopen provides standardized communication objects for real-time data (process data objects, PDO), configuration data (service data objects, SDO), and special functions (time stamp, sync message, and emergency message) as well as network management data (boot-up message, NMT message, and error control).

CANopen differentiates between two data transfer mechanisms: fast exchange of short process data, using process data objects (PDOs); and access to the entries of the object directory, that is done via service data object (SDO). PDOs can be asynchronous, synchronous, or on-demand and the transfer is done without protocol overhead. The synchronous transmission of messages is supported by predefined communication objects (sync message and time stamp message). Synchronous messages are transmitted with respect to a predefined synchronization message, asynchronous messages may be transmitted at any time. SDOs are confirmed data transfers that establish point-to-point communication between two devices, implementing services of handshaking, fragmentation, and reassembly. SDOs are used for parameter passing, configuration, etc.

CANopen supports three types of communication models: master/slave, client/server, and producer/consumer. CANopen networks also provide redundant transfer of safety-oriented information in defined time windows.

31.3.2 DeviceNet

DeviceNet, international standard IEC 62026-3, is another CAN-based higher layer protocol, initially proposed by Allen-Bradley (now owned by Rockwell Automation), and is mostly dedicated to industrial automation. DeviceNet specifications include the application layer and device profiles. The management of the DeviceNet specification was transferred to the Open DeviceNet Vendor Association (ODVA), a nonprofit organization that develops and markets DeviceNet [3].

The DeviceNet physical layer specifies a terminated trunk and drop line configuration. Communication and power are provided in the trunk and drop cables by separate twisted pair buses.

Up to 64 logical nodes can be connected to a single DeviceNet network, using both sealed and open-style connections. There is support for strobed, polled, cyclic, change-of-state, and application-triggered data transfer. The user can choose master/slave, multi-master, and peer-to-peer, or a combination configuration, depending on device capability and the application requirements. DeviceNet is based on the Part A of the CAN standard; therefore, it uses the 11 bit identifier. DeviceNet uses abstract object models to describe communication services, data, and behavior. A DeviceNet node is built from a collection of objects describing the system behavior and grouping data using virtual objects, as any object oriented programming language would [6].

DeviceNet communication model defines an addressing scheme that provides access to objects within a device. The object model addressing information includes

- *Device address*—Referred to as the media access control identifier (MAC ID) is an integer identification value assigned to each node on the network. A test, executed at power-up, guarantees the uniqueness of the value on the network.
- *Class identifier*—Refers to a set of objects that represent the same type of system component. The class identifier is an integer identification value assigned to each object accessible from the network.
- *Instance identifier*—Refers to the actual representation of an object of a class. The instance identifier is an integer identification assigned to an object instance that identifies it among all instances of the same class within a particular device, with a unique identifier value.
- *Attribute identifier*—Attributes are parameters associated with an object. Attributes typically provide some type of status information, represent a characteristic with a configurable value, or control the behavior of an object. The attribute identifier is an integer identification value assigned to an object attribute.

The DeviceNet application layer specifies how CAN identifiers are assigned and how the data field is used to specify services, move data, and determine its meaning. DeviceNet adopts the producer–consumer model, so the source device broadcast the data to the network with the proper identifier. All devices who need data listen for messages and when devices recognize the appropriate identifier, they consume the data.

There are two types of messages usually required by most automation devices: I/O messages and explicit messages. I/O messages are for time critical control-oriented data and provide dedicated communication paths between a producing application and one or more consuming applications and typically use the higher priority identifiers [6]. Explicit messages, on the other hand, provide multipurpose point-to-point communication paths between two devices. They provide the typical request/response-oriented network communications used to perform node configuration and diagnosis and typically use lower priority identifiers. Fragmentation services are also provided for messages that are longer than 8 bytes.

31.3.3 TTCAN

Time-triggered communication on CAN (TTCAN) is an extension of CAN, introducing time-triggered operation based on a high-precision network-wide time base. There are two possible levels in

TTCAN: level-1 and level-2. Level-1 only provides time-triggered operation using local time (Cycle_Time). Level-2 requires a hardware implementation and provides increased synchronization quality, global time, and external clock synchronization. As native CAN, TTCAN is limited to a maximum data rate of 1 Mbps, with typical data efficiency below 50%. TTCAN network topology is bus based. Replicated busses are not referred in TTCAN standard; however, it is possible to implement them [29].

TTCAN adopts a Time-Division Multiple Access (TDMA) bus access scheme. The TDMA bandwidth allocation scheme divides the timeline into time slots or time windows. Network nodes are assigned different slots to access the bus. The sequence of time slots allocated to nodes repeats according to a basic cycle. Several basic cycles are grouped together in a matrix cycle. All basic cycles have the same length but can differ in their structure. When a matrix cycle finishes, the transmission scheme starts over by repeating the matrix cycle (Figure 31.3). The matrix cycle defines a message transmission schedule. However, a TTCAN node does not need to know the whole system matrix, it only needs information of the messages it will send and receive.

Since TTCAN is built on top of native CAN, some time windows may be reserved for several event messages. In such windows, it is possible that more than one transmitter may compete for the bus access right. During these slots, the arbitration mechanism of the CAN protocol is used to prioritize the competing messages and to grant access to the higher priority one. In this sense, the medium access mechanism in TTCAN can be described as TDMA with Carrier Sense Multiple Access with Bitwise Arbitration (CSMA-BA) in some predefined time slots. This feature makes the TTCAN protocol as flexile as CAN during the arbitration windows, without compromising the overall system timeliness, i.e., the event-triggered messages do not interfere with the time-triggered ones.

The TDMA cycle starts with the transmission of a reference message from a time master. The reference messages are regular CAN messages with a special and known a priori identifier and are used to synchronize and calibrate the time bases of all nodes according to the time master's time base, providing a global time for the network.

TTCAN level-1 guarantees the time-triggered operation of CAN based on the reference message of a time master. Fault-tolerance of that functionality is established by redundant time masters, the so called potential time masters. Level-2 establishes a globally synchronized time base and a continuous drift correction among the CAN controllers.

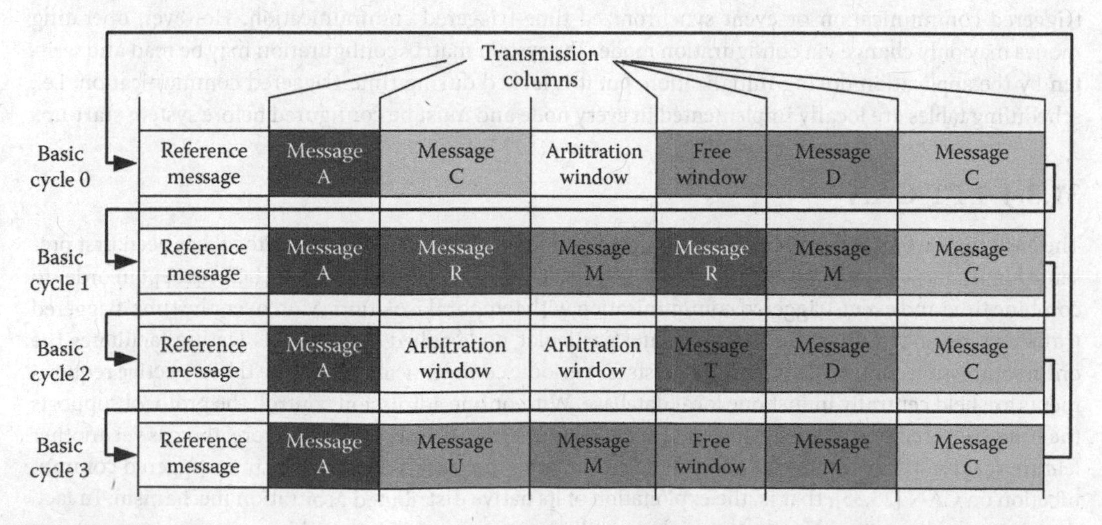

FIGURE 31.3 TTCAN system matrix, where several basic cycles build the matrix cycle. (Adapted from Führer, T. et al., Time triggered communication on CAN, In *Proceedings of the Seventh International CAN Conference*, Amsterdam, the Netherlands, October 2000, CAN in Automation GmbH.)

There are three types of time windows (Figure 31.3):

- *Exclusive time windows*—for periodic messages that are transmitted without competition for the CAN bus. No other message can be scheduled in the same window. The automatic retransmission, upon error, is not allowed.
- *Arbitrating time windows*—for event-triggered messages, where several event-triggered messages may share the same time window and bus conflicts are resolved by the native CAN arbitration. Two or more consecutive arbitrating time windows can be merged. Message transmission can only be started if there is sufficient time remaining for the message to fit in. Automatic retransmission of CAN messages is disabled (except for merged arbitrating windows).
- *Free time windows*—reserved for future extensions of the network. A transmission schedule could reserve time windows for future use, either for new nodes or to assign existing nodes more bandwidth. Notice that a free time window cannot be assigned to a message unless it is previously converted to either an exclusive or an arbitrating time window.

Clock synchronization in a TTCAN network is provided by a time master. The time master establishes its own local time as global time by transmitting the reference message. To compensate for slightly different clock drifts in the TTCAN nodes and to provide a consistent view of the global time, the nodes perform a drift compensation operation. A unique time master would be a single point of failure, thus TTCAN provides a mechanism for time masters' redundancy and replacement whenever the current time master fails to send a reference message. In this case, the CAN bus remains idle and any of the potential time masters will try to transmit a reference message after a certain amount of time. In case two potential time masters try to send a reference message at the same time, the native CAN bit arbitration mechanism ensures that only the one with the highest priority wins. When a failed time master reconnects to the system with active time-triggered communication, it waits until it is synchronized to the network before it may try to become time master again.

The strategy adopted, concerning fault confinement, is very similar to the one followed by native CAN, i.e., error passive and bus-off. Since CAN failures are already considered in ISO 11898-1 standard (data link layer), TTCAN considers mainly scheduling errors (e.g., absence of a message) and each TTCAN controller only provides error detection. Active fault confinement is left to a higher layer or to the application.

TTCAN may work under several operational modes: configuration mode, CAN communication, time-triggered communication or event synchronized time-triggered communication. However, operating modes may only change via configuration mode. The system matrix configuration may be read and written by the application during initialization, but it is locked during time-triggered communication, i.e., scheduling tables are locally implemented in every node and must be configured before system start-up.

31.3.4 FTT-CAN

The basis for the flexible time-triggered communication on CAN (FTT-CAN) protocol has been first presented in [1]. Basically, the protocol makes use of the dual-phase elementary cycle (EC) concept in order to combine time and event-triggered communication with temporal isolation. Moreover, the time-triggered traffic is scheduled online and centrally in a particular node called master. This feature facilitates the online admission control of dynamic requests for periodic communication because the respective requirements are held centrally in just one local database. With online admission control, the protocol supports the time-triggered traffic in a flexible way, under guaranteed timeliness. Furthermore, there is yet another feature that clearly distinguishes this protocol from other proposals concerning time-triggered communication on CAN [25,33], that is, the exploitation of its native distributed arbitration mechanism. In fact, the protocol relies on a relaxed master–slave medium access control in which the same master message triggers the transmission of messages in several slaves simultaneously (master/multislave). The eventual collisions between slave's messages are handled by the native distributed arbitration of CAN.

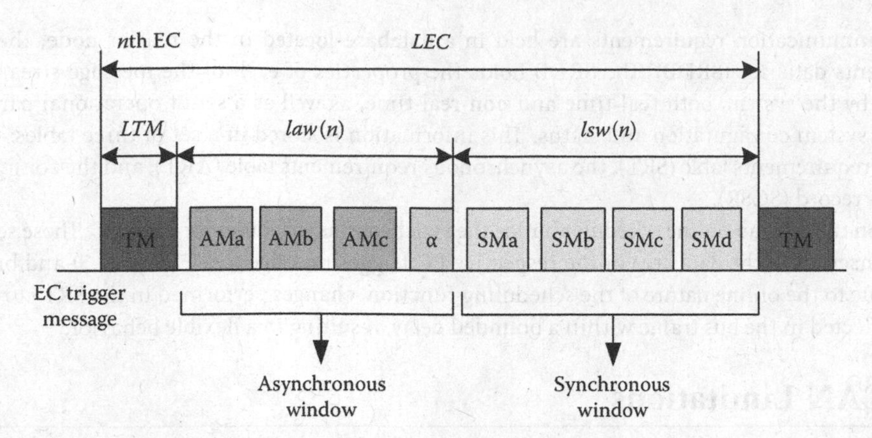

FIGURE 31.4 The elementary cycle (EC) in FTT-CAN.

The protocol also takes advantage of the native arbitration to handle event-triggered traffic in the same way as the original CAN protocol does. Particularly, there is no need for the master to poll the slaves for pending event-triggered requests. Slaves with pending requests may try to transmit immediately, as in normal CAN, but just within the respective phase of each EC. This scheme, similar to the arbitration windows in TTCAN, allows a very efficient combination of time and event-triggered traffic, particularly resulting in low communication overhead and shorter response times.

In FTT-CAN the bus time is slotted in consecutive ECs with fixed duration. All nodes are synchronized at the start of each EC by the reception of a particular message known as EC trigger message, which is sent by the master node.

Within each EC the protocol defines two consecutive windows, asynchronous and synchronous, that correspond to two separate phases (see Figure 31.4). The former one is used to convey event-triggered traffic, herein called asynchronous because the respective transmission requests can be issued at any instant. The latter one is used to convey time-triggered traffic, herein called synchronous because its transmission occurs synchronously with the ECs. The synchronous window of the *n*th EC has a duration that is set according to the traffic that is scheduled for it. The schedule for each EC is conveyed by the respective EC trigger message (see Figure 31.5). Since this window is placed at the end of the EC, its starting instant is variable and it is also encoded in the respective EC trigger message. The protocol allows establishing a maximum duration for the synchronous windows and correspondingly a maximum bandwidth for that type of traffic. Consequently, a minimum bandwidth can be guaranteed for the asynchronous traffic.

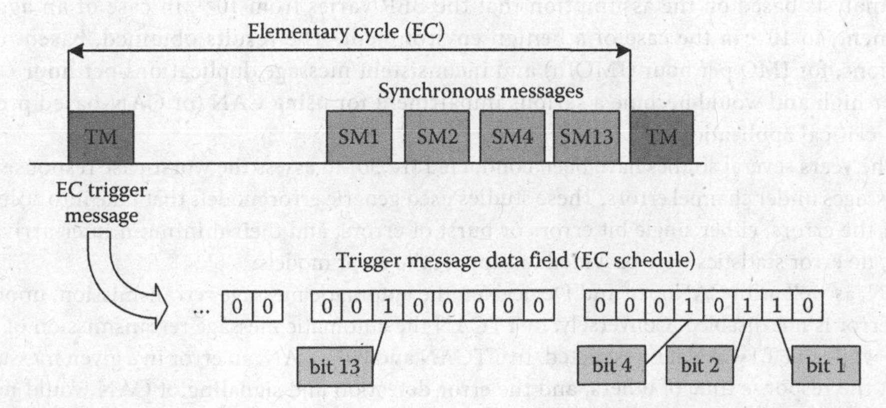

FIGURE 31.5 Master/multislave access control. Slaves produce synchronous messages according to an EC schedule conveyed by the trigger message.

The communication requirements are held in a database located in the master node, the system requirements database (SRDB). The SRDB holds the properties of each of the message streams to be conveyed by the system, both real-time and non-real-time, as well as a set of operational parameters related to system configuration and status. This information is stored in a set of three tables: the synchronous requirements table (SRT), the asynchronous requirements table (ART), and the configuration and status record (SCSR).

Based on the SRT, an online scheduler builds the synchronous schedules for each EC. These schedules are then inserted in the data area of the respective EC trigger message (see Figure 31.5) and broadcast with it. Due to the online nature of the scheduling function, changes performed in the SRT at run-time will be reflected in the bus traffic within a bounded delay, resulting in a flexible behavior.

31.4 CAN Limitations

Judging by the large installed base of CAN controllers in significative application domains, system designers are satisfied with CAN features; however, they would be much happier if CAN could be made faster, cover longer distances, be more deterministic and more dependable [34]. In fact, because of its limited dependability features, there is an ongoing debate on whether the CAN protocol, with proper enhancements, can support safety-critical applications [21,27]. CAN limitations and recent development and research carried out to overcome them are discussed in detail in [34]. These include large and variable jitter, lack of clock synchronization, flexibility limitations, data consistency issues, limited error containment, and limited support for fault tolerance. Over the years, several proposals contributed to overcome some of CAN limitations [4,8,17,18,37,40].

A specially important limitation is related with inconsistent communication scenarios [36,40], that is one of the strongest impairments to achieve high dependability over CAN. These scenarios are due to the protocol specification and may reveal themselves both as inconsistent message omissions (IMO), i.e., some nodes receive a given message while others do not, and as inconsistent message duplicates (IMD), i.e., some nodes receive the same message several times while others receive only once. Inconsistent communication scenarios make distributed consensus in its different forms, e.g., membership, clock synchronization, consistent commitment of configuration changes, or simply the consistent perception of asynchronous events, more difficult to attain.

In principle, and according to the error detection and signaling capabilities of CAN, any frame which suffers an error would be consistently rejected by all the nodes of the network. However, some failure scenarios have been identified [40] that can lead to undesirable symptoms such as inconsistent omission failures and duplicate message reception. The probability of those error scenarios depends on an important factor, the CAN bit error rate (BER). Rufino et al., presented [40] an analysis based on the assumption that the BER varies from 10^{-4}, in case of an aggressive environment, to 10^{-6} in the case of a benign environment. The results obtained, based on these assumptions, for IMO per hour (IMO/h) and inconsistent message duplications per hour (IMD/h) are rather high and would become a serious impairment for using CAN (or CAN-based protocols) in safety-critical applications.

Over the years several studies have been conducted [10,30] to assess the worst-case response time of CAN messages under channel errors. These studies used generic error models that take into account the nature of the errors, either single bit errors or burst of errors, and their minimum inter-arrival time. However, no error statistics were provided to support the error models.

In CAN, as well as in CANopen and DeviceNet, the automatic message retransmission, upon transmission error is not disabled. Conversely, in TTCAN the automatic message retransmission of CAN is disabled, while in FTT-CAN it is restricted. In TTCAN and FTT-CAN, an error in a given message does not affect the response time of others, and the error detection and signaling of CAN would normally ensure that the error would be consistently detected by all network nodes except in the cases where inconsistent message delivery may occur.

TABLE 31.2 Estimated Rates of IMO per Hour in CAN, TTCAN, and FTT-CAN

	CAN			FTT-CAN	
	$\Delta t = 5$ ms		TTCAN	$\rho = 0.5$	
Bit Error Rate	IMD/h	IMO/h	IMO/h	IMD/h	IMO/h
2.6×10^{-7}	7.59	1.05×10^{-9}	7.59	3.79	3.79
3.1×10^{-9}	8.93×10^{-2}	1.24×10^{-11}	8.93×10^{-2}	4.46×10^{-2}	4.46×10^{-2}
3.0×10^{-11}	8.75×10^{-4}	1.22×10^{-13}	8.75×10^{-4}	4.37×10^{-4}	4.37×10^{-4}

Inconsistent message reception scenarios are much more frequent in the case of TTCAN than in native CAN and their frequency depends on the BER of the CAN bus. The assumptions concerning the BER made by Rufino et al., although realistic in other networks, seemed somewhat pessimistic considering the specific case of CAN and specially the characteristics of the CAN physical layer. This led to the design of an experimental setup to measure real values of CAN BER [20]. Experimental results from a set of experiments showed that BER was 2.6×10^{-7} in an aggressive environment (CAN network placed near a high-frequency arc-welding machine), 3.1×10^{-9} in a normal environment (CAN network at an industrial production line), and 3.0×10^{-11} at a benign environment (CAN network at the university laboratory). Please refer to [17] for a detailed analysis and discussion of these issues.

Table 31.2 presents the probability of inconsistencies both for CAN (including CANopen and DeviceNet), using the same assumptions as in [40].

Observing Table 31.2, one may conclude that in native CAN the occurrence of IMO has a lower probability than previously assumed, at least in the environments considered in the experiments. In fact, it is below the 10^{-9} threshold usually accepted for safety-critical applications. However, the probability of IMD (messages are eventually delivered but they could be out of order) is still high enough to be taken into account.

Concerning both TTCAN and FTT-CAN, one cannot neglect IMO because they are rather frequent. However, it should be stressed that this does not mean that native CAN is more dependable than TTCAN or FTT-CAN since native CAN does not bound automatic retransmissions that may origin deadline losses. That is, the probability of IMO is lower in native CAN, but the probability of a message miss its deadline is higher.

31.4.1 Consequences of Faults at the Node Level

CAN protocol does not restrict the *failure semantics* of the nodes, so that they may fail in arbitrary ways. Some of these failures are automatically handled by CAN's native implementation of error detection capabilities and automatic fail-silence enforcement, described previously, leading the erroneous node to a state, the bus-off state, where it is unable to interfere with other nodes. However, in some specific situations, these mechanisms do not fully contain the errors within the nodes.

Specifically, a CAN node only reaches the bus-off state (fail silence) after a relatively long period of time (when the TEC reaches 255). Moreover, a CAN node running an erroneous application can also compromise most of the legitimate traffic scheduled according to a higher layer protocol implemented in software in a standard CAN controller, simply by accessing the network at arbitrary points in time (*babbling idiot* failure mode). Notice that a faulty application running in a node with a CAN controller may transmit high-priority messages at any time without causing any network errors, and consequently the CAN controller will never reach the bus-off state. An uncontrolled application transmitting at arbitrary points in time via a non-faulty CAN controller is a much severe situation than a faulty CAN controller also transmitting at arbitrary points in time because, in the first case, a non-faulty CAN controller has no means to detect an erroneous application transmitting legitimate traffic. In the second case, the CAN controller would enter bus-off state after a while and the error would be eventually confined.

To overcome the *babbling idiot* failure mode, special components located between the node and the bus, the bus guardians, have been proposed for CAN [9,19]. Bus guardians are usually adopted to enforce fail-silence behavior in nodes of a distributed system because they are simpler than using node replication and some sort of voting mechanism. Apart from the cost issue, simplicity is also important because a simpler component is often more dependable than a complex one.

List of Acronyms

CAN Controller area network
TTCAN Time-triggered CAN
FTT-CAN Flexible-time-triggered CAN
BER Bit error rate
CSMA Carrier sense multiple access
OSI Open systems interconnection
IMD Inconsistent message duplicates
IMO Inconsistent message omissions
EMI Electromagnetic interference
SOF Start of frame
RTR Remote transmission request
IDE Identifier extension
DLC Data length code
CRC Cyclic redundant check
TEC Transmit error counter
REC Receive error counter
EC Elementary cycle

References

1. L. Almeida, J. A. Fonseca, and P. Fonseca. Flexible time-triggered communication on a controller area network. In *Proceedings of the Work-in-Progress Session of RTSS'98 (19th IEEE Real-Time Systems Symposium)*, Madrid, Spain, 1998.
2. L. Almeida, P. Pedreiras, and J. A. Fonseca. The FTT-CAN protocol: Why and how. *IEEE Transactions on Industrial Electronics*, 49(6):1189–1201, 2002.
3. Open DeviceNet Vendor Association. DeviceNet technical overview, http://www.odva.org/Portals/0/Library/Publications_Numbered/PUB00026R1.pdf, accessed February 12, 2009.
4. M. Barranco, L. Almeida, and J. Proenza. Experimental assessment of ReCANcentrate, a replicated star topology for CAN. *SAE 2006 Transactions Journal of Passenger Cars: Electronic and Electrical Systems*, 115(7), 437–446, 2006.
5. M. Barranco, G. Rodríguez-Navas, J. Proenza, and L. Almeida. CANcentrate: An active star topology for CAN networks. In *Proceedings of the Fifth IEEE International Workshop on Factory Communication Systems (WFCS 2004)*, Vienna, Austria, 2004.
6. S. Biegacki and D. VanGompel. The application of deviceNet in process control. *ISA Transactions*, 35:169–176, 1996.
7. R. Bosch. CAN specification version 2.0. BOSCH, Stuttgart, Germany, 1991.
8. I. Broster. Flexibility in dependable communication. PhD thesis, Department of Computer Science, University of York, York, U.K., August 2003.
9. I. Broster and A. Burns. An analysable bus-guardian for event-triggered communication. In *Proceedings of the 24th Real-time Systems Symposium*, Cancun, Mexico, December 2003, pp. 410–419, IEEE.
10. I. Broster, A. Burns, and G. Rodríguez-Navas. Probabilistic analysis of CAN with faults. In *Proceedings of the 23rd Real-time Systems Symposium*, Austin, TX, 2002, pp. 269–278.

11. G. Cena and A. Valenzano. Operating principles and features of CAN networks. In *The Industrial Information Technology Handbook*, R. Zurawski (ed.). CRC Press, Boca Raton, FL, 2005, pp. 1–16.
12. CiA. CAN physical layer. CiA (CAN in Automation), Nuremberg, Germany, 2001.
13. CiA. CANopen application layer and communication profile, version 4.02. CiA (CAN in Automation), Nuremberg, Germany, 2002. EN 50325-4 Standard.
14. R. I. Davis, A. Burns, R. J. Bril, and J. J. Lukkien. Controller area network (CAN) schedulability analysis: Refuted, revisited and revised. *Real-Time Systems*, 35(3):239–272, 2007.
15. R. DeMeis. Cars sag under weighty wiring. *Electronic Times*, October 24, 2005.
16. K. Etschberger. *Controller Area Network—Basics, Protocols, Chips and Applications*. IXXAT Automation, Weingarten, Germany, 2001.
17. J. Ferreira. Fault-tolerance in flexible real-time communication systems. PhD thesis, University of Aveiro, Aveiro, Portugal, May 2005.
18. J. Ferreira, L. Almeida, J. Fonseca, P. Pedreiras, E. Martins, G. Rodriguez-Navas, J. Rigo, and J. Proenza. Combining operational flexibility and dependability in FTT-CAN. *IEEE Transactions on Industrial Informatics*, 2(2):95–102, 2006.
19. J. Ferreira, L. Almeida, E. Martins, P. Pedreiras, and J. Fonseca. Components to enforce fail-silent behavior in dynamic master-slave systems. In *Proceedings of SICICA'2003 Fifth IFAC International Symposium on Intelligent Components and Instruments for Control Applications*, Aveiro, Portugal, July 2003.
20. J. Ferreira, A. Oliveira, P. Fonseca, and J. A. Fonseca. An experiment to assess bit error rate in CAN. In *RTN 2004—Third International Workshop on Real-Time Networks Satellite Held in Conjunction with the 16th Euromicro International Conference on Real-Time Systems*, Catania, Italy, June 2004.
21. J. Ferreira, P. Pedreiras, L. Almeida, and J. Fonseca. The FTT-CAN protocol: Improving flexibility in safety-critical systems. *IEEE Micro (Special Issue on Critical Embedded Automotive Networks)*, 22(4):46–55, 2002.
22. T. Führer, B. Müller, W. Dieterle, F. Hartwich, R. Hugel, and M. Walther. Time triggered communication on CAN. In *Proceedings of the Seventh International CAN Conference*, Amsterdam, the Netherlands, October 2000, CAN in Automation GmbH.
23. H. Hilmer, H.-D. Kochs, and E. Dittmar. A fault-tolerant communication architecture for real-time control systems. In *Proceedings of the IEEE International Workshop on Factory Communication Systems*, Barcelona, Spain, October 1997.
24. International Standards Organisation. ISO 11898. Road vehicles—Interchange of digital information—Controller area network (CAN) for high speed communication. ISO, Geneva, Switzerland, 1993.
25. ISO. Road vehicles—Controller area network (CAN)—Part 4: Time triggered communication. ISO, Geneva, Switzerland, 2001.
26. IXXAT. FO-star-coupler, IXXAT, http://www.ixxat.de/fo-star-coupler_en, 7460, 5873.html, accessed February 21, 2005.
27. H. Kopetz. A comparison of CAN and TTP. Technical Report 1998, Technishe Universitat Wien, Vienna, Austria, 1998.
28. W. Lawrenz. *CAN System Engineering: From Theory to Practical Applications*. Springer, New York, 1997.
29. B. Müller, T. Führer, F. Hartwich, R. Hugel, and H. Weiler. Fault tolerant ttcan networks. In *Proceedings of Eighth International CAN Conference*, Las Vegas, NV, October 2002, CAN in Automation GmbH.
30. N. Navet, Y.-Q. Song, and F. Simonot. Worst-case deadline failure probability in real-time applications distributed over controller area network. *Journal of Systems Architecture*, 46(7):607–617, 2000.
31. N. Navet and F. Simonot-Lion (eds.). *Automotive Embedded Systems Handbook*. CRC Press, Boca Raton, FL, 2009.
32. D. Paret. *Le Bus CAN*. Dunod, Paris, France, 1997.
33. M. Peraldi and J. Decotignie. Combining real-time features of local area networks FIP and CAN. In *Proceedings of the ICC'95 (Second International CAN Conference)*, London, England, 1995.

34. J. Pimentel, J. Proenza, L. Almeida, G. Rodríguez-Navas, M. Barranco, and J. Ferreira. Dependable automotive CAN networks. In *Automotive Embedded Systems Handbook*, N. Navet and F. Simonot-Lion (eds.). CRC Press, Boca Raton, FL, 2009.
35. J. R. Pimentel and J. A. Fonseca. FlexCAN: A flexible architecture for highly dependable embedded applications. In *RTN 2004—Third International Workshop on Real-Time Networks sattelite Held in Conjunction with the 16th Euromicro International Conference on Real-Time Systems*, Catania, Italy, June 2004.
36. G. Rodríguez-Navas and J. Proenza. Analyzing atomic broadcast in TTCAN networks. In *Proceedings of the Fifth IFAC International Conference on Fieldbus Systems and Their Applications (FET 2003)*, Aveiro, Portugal, 2003, pp. 153–156.
37. G. Rodríguez-Navas, J. Proenza, and H. Hansson. An UPPAAL model for formal verification of master/slave clock synchronization over the controller area network. In *Proceedings of the Sixth IEEE International Workshop on Factory Communication Systems*, Turin, Italy, 2000.
38. M. Rucks. Optical layer for CAN. In *Proceedings of the First International CAN Conference*, Vol. 2, Mainz, Germany, 1994, pp. 11–18, CiA.
39. J. Rufino, P. Veríssimo, and G. Arroz. A Columbus' egg idea for CAN media redundancy. In *Digest of Papers, the 29th International Symposium on Fault-Tolerant Computing Systems*, Madison, WI, June 1999, pp. 286–293, IEEE.
40. J. Rufino, P. Veríssimo, G. Arroz, C. Almeida, and L. Rodigues. Fault-tolerant broacast in CAN. In *Digest of Papers, 28th International Symposium on Fault Tolerant Computer Systems*, Munich, Germany, pp. 150–159, 1998.
41. V. Silva, J. Ferreira, and J. Fonseca. Dynamic topology management in CAN. In *Proceedings of the 11th IEEE International Conference on Emerging Technologies and Factory Automation*, Prague, Czech Republic, 2006, IEEE.
42. K. Tindell and A. Burns. Guaranteeing message latencies on controller area network (CAN). In *Proceedings of the First International CAN Conference*, Mainz, Germany, 1994, pp. 1–11.

32 Profibus

Max Felser
Bern University of Applied Sciences

Ron Mitchell
RC Systems

32.1 Introduction 32-1
32.2 Physical Transmissions 32-2
Asynchronous (RS-485) • Manchester Bus Powered • Fiber Optics
32.3 Fieldbus Data Link 32-5
Services • Framing • Media Access
32.4 DP System 32-7
DP-Master Class 1: Controllers • DP-Master Class 2: Engineering Stations • DP-Slaves: Field-Devices • Application Relations
32.5 Cyclic Data Exchange: MS0—Relation 32-9
Device Model • Initialization and Supervision of the Relation • Status of the Controller and Fail-Safe Functionality • Diagnostics of the Field-Device • Distributed Database • Synchronization of the Applications
32.6 Acyclic Data Exchange: MS1/MS2 Relations 32-10
Variables • Device Model including Identification and Maintenance • Alarm Handling • Procedure Calls
32.7 Application Profiles 32-12
References 32-12

32.1 Introduction

The Profibus (PROcess FIeld BUS) was defined by a consortium of different manufacturers and research institutions in Germany more than 20 years ago and defined from the beginning as an open national standard DIN 19245. It is a simple fieldbus based on an extended token passing and object-oriented services to provide distributed control based on a simplified and adapted version of ISO 69506 manufacturing message specification (MMS) called the fieldbus message specification (FMS). But the industry did not want a distributed system and adopted a simple remote I/O system with centralized control. So the Profibus-FMS was extended with an application protocol for decentralized periphery (DP) we call here Profibus DP. To cope with the requirements of process automation, additional definitions for transmission media and function blocks are added and collected under the name Profibus PA. All these definitions were brought to the European standard EN50170-2. Finally, the Versions PA and DP are today part of the international standard IEC 61158 [IEC58] and IEC 61784 [IEC84-1] under the designation "Family 3" with subdivisions 3/1 and 3/2. The functionality of Profibus DP can further be split in three levels: DPV0 for simple remote I/O, DP-V1 for additional acyclic data exchange, and DP-V2 for high-performance functions like isochronous cycles and broadcast data exchange. Table 32.1 shows their assignment.

The "PROFIBUS Nutzerorganisation" (PNO), representing the interested enterprises and institutes in this technology, was founded as early as 1989 in Germany. Profibus is today supported by Profibus & Profinet International (PI), an umbrella organization formed out of more than 30 Regional Profibus Associations (RPA) on all continents like the PNO for Germany or the PTO representing the North American market.

TABLE 32.1 Properties of the Communication Family CPF 3 for Profibus

Layer	Profile CP3/1	Profile CP3/1	Profile CP3/2
Application	Manufacturing automation	Motion control	Process automation (Profibus PA)
Data link	IEC 61158 subset DP-V0 asynchronous transmission	IEC 61158 subset DP-V2 asynchronous transmission	IEC 61158 subset DP-V1 synchronous transmission
Physical	RS485 Plastic fiber Glass fiber PCF fiber	RS485 Plastic fiber Glass fiber PCF fiber	MBP

Support for the Profibus technology is provided by more than 30 competence centers (PICC) and certification trainings are offered by training centers (PITC). The conformance of products to the standard and the interoperability of the different products is verified and tested by the test laboratories (PITL) available in all major markets.

32.2 Physical Transmissions

Profibus provides different versions of layer 1 as transmission technology listed in Table 32.2.

32.2.1 Asynchronous (RS-485)

RS485 transmission technology is simple and cost-effective and primarily used for tasks that require high transmission rates. Shielded, twisted pair copper cable with one conductor pair is used. Details to RS 485 installation are described in [P2112,P8022,P8032]. The bus structure allows addition or

TABLE 32.2 Transmission Technologies (Physical Layer) at Profibus

	MBP	RS485	RS485-IS	Fiber Optic
Data transmission	Digital, bit-synchronous, Manchester encoding	Digital, differential signals according to RS485, NRZ	Digital, differential signals according to RS485, NRZ	Optical, digital, NRZ
Transmission rate	31.25 kb/s	9.6–12,000 kb/s	9.6–1,500 kb/s	9.6–12,000 kb/s
Data security	Preamble, error-protected, start/end delimiter	HD = 4, Parity bit, start/end delimiter	HD = 4, Parity bit, start/end delimiter	HD = 4, Parity bit, start/end delimiter
Cable	Shielded, twisted pair copper	Shielded, twisted pair copper, cable type A	Shielded, twisted 4-wire, cable type A	Multimode glass fiber, singlemode glass fiber, PCF, plastic
Remote feeding	Optional available over signal wire	Available over additional wire	Available over additional wire	Available over hybrid line
Protection type	Intrinsic safety (EEx ia/ib)	None	Intrinsic safety (EEx ib)	None
Topology	Line and tree topology with termination; also in combination	Line topology with termination	Line topology with termination	Star and ring topology typical; line topology possible
Number of stations	Up to 32 stations per segment; total of max. 126 per network	Up to 32 stations per segment without repeater; up to 126 stations with repeater	Up to 32 stations per segment; up to 126 stations with repeater	Up to 126 stations per network
Number of repeaters	Max. 4 repeater	Max. 9 repeater with signal refreshing	Max. 9 repeater with signal refreshing	Unlimited with signal refreshing (time delay of signal)

TABLE 32.3 Transmission Range and Range for Cable Type A

Transmission Rate (kb/s)	Range per Segment (m)
9.6, 19.2, 45.45, 93.75	1,200
187.5	1,000
500	400
1,500	200
3,000, 6,000, 12,000	100

The values refer to cable type A with the following properties:

Impedance	135 to 165 Ω
Capacity	≤30 pf/m
Loop resistance	≤110 Ω /km
Wire diameter	>0.64 mm
Core cross-section	>0.34 mm²

removal of stations or the step-by-step commissioning of the system without influencing other stations. Subsequent expansions (within defined limits) have no effect on stations already in operation.

Various transmission rates can be selected between 9.6 kb/s and 12 Mb/s. One uniform speed is selected for all devices on the bus when commissioning the system. The maximum permissible line length depends on the transmission rate and is listed in Table 32.3.

All devices are connected in a bus structure (line). Up to 32 stations (masters or slaves) can be connected in a single segment. The beginning and end of each segment is fitted with an active bus terminator. Both bus terminators have a permanent power supply to ensure error-free operation. The bus terminator is usually switched in the devices or in the bus terminator connectors. If more than 32 stations are implemented or there is a need to expand the network area, repeaters must be used to link the individual bus segments.

Different cable types (type designation A–D) for different applications are available on the market for connecting devices either to each other or to network elements (e.g., segment couplers, links, and repeaters).

When connecting the stations, always ensure that the data lines are not reversed. Always use a shielded data line (type A is shielded) to ensure high interference immunity of the system against electromagnetic emissions. The shield should be grounded on both sides where possible and large-area shield clamps used for grounding to ensure good conductivity. Furthermore, always ensure that the data line is laid separately and, where possible, away from all power cables. Never use spur lines for transmission rates ≥1.5 Mb/s. Commercially available connectors support direct connection of the incoming and outgoing data cable in the connector. This dispenses with the need for spur lines and the bus connector can be connected and disconnected at the bus at any time without interrupting data communications. The type of connector suitable for RS485 transmission technology depends on the degree of protection. A 9-pin D-Sub connector is primarily used for protection rating IP 20. The preferred solution for protection rating IP 65/67 is the M12 circular connector in accordance with IEC 947-5-2.

There exists a configuration of intrinsically safe RS485 solutions with simple device interchangeability. The specification of the interface stipulates the level for current and voltage that must be adhered to by all stations in order to ensure safe functioning during interconnection. An electric circuit permits maximum currents at a specified voltage. When interconnecting active sources, the sum of the currents of all stations must not exceed the maximum permissible current. An innovation of the RS485-IS concept is that, in contrast to the FISCO model, which only has one intrinsically safe source, all stations now represent active sources.

32.2.2 Manchester Bus Powered

Manchester Bus Powered (MBP) stands for transmission technology with the following attributes:

- Manchester Coding (M)
- Bus Powering (BP)

MBP is synchronous transmission with a defined transmission rate of 31.25 kb/s and Manchester coding. This transmission technology is frequently used in process automation as it satisfies the key demands of the chemical and petrochemical industries for intrinsic safety and bus powering using two-wire technology. This means that Profibus can also be used in potentially explosive areas with the attribute intrinsically safe.

The intrinsically safe transmission technology of MBP is usually limited to a specific segment (field devices in hazardous areas) of a plant, which are then linked to the RS485 segment (control system and engineering devices in the control room) over segment coupler or links. Segment couplers are signal converters that modulate the RS485 signals to the MBP signal level. These are transparent from the bus protocol standpoint. In contrast, links have their own intrinsic intelligence. They map all the field devices connected to the MBP segment as a single slave in the RS485 segment. There is no limit to the transmission rate in the RS485 segment, so that fast networks can also be implemented using field devices with MBP connection, e.g., for control tasks.

Tree or line structures (and any combination of the two) are network topologies supported by Profibus with MBP transmission. In a line structure, stations are connected to the trunk cable using tree adapters. The tree topology is comparable to the classic field installation method. The multi-core master cable is replaced by the two-wire bus master cable; the field distributor retains its function of connecting the field devices and detecting the bus terminator impedance. When using a tree topology, all field devices connected to the fieldbus segment are wired in parallel in the field distributor. In all cases, the maximum permissible spur line lengths must be taken into account when calculating the over-all line length. In intrinsically safe applications, a spur line has a maximum permissible length of 30 m.

A shielded two-wire cable is used as the transmission medium. The bus trunk cable has a passive line terminator at each end, which comprises an RC element connected in series with $R = 100\ \Omega$ and $C = 2\ \mu\text{F}$. The bus terminator is already integrated in the segment coupler or link. When using MBP technology, incorrect connection of a field device (i.e., polarity reversal) has no effect on the functionality of the bus as these devices are usually fitted with an automatic polarity detection function.

The number of stations that can be connected to a segment is limited to 32. However, this number may be further determined by the protection type selected and bus powering (if any).

In intrinsically safe networks, both the maximum feed voltage and the maximum feed current are defined within strict limits. But the output of the supply unit is limited even for nonintrinsically safe networks. The FISCO model considerably simplifies the planning, installation, and expansion of Profibus networks in potentially explosive areas.

32.2.3 Fiber Optics

Some fieldbus application conditions place restrictions on wire-bound transmission technology, such as those in environments with very high electromagnetic interference or when particularly large distances need to be covered. Fiber-optic transmission over fiber-optic conductors is suitable in such cases. The transmission characteristics support not only star and ring topology structures but also line structures. In the simplest case, a fiber-optic network is implemented using electrical/optical transformers that are connected to the device and the fiber optics over a RS485 interface. This allows you to switch between RS485 and fiber-optic transmission within a plant, depending on the circumstances.

32.3 Fieldbus Data Link

The Profibus Communication Profiles Family (CPF3) uses a uniform medium access protocol. This protocol is implemented by layer 2 of the OSI reference model and is called fieldbus data link (FDL). This also includes the handling of the transmission protocols and telegrams and the detection and correction of transmission errors.

32.3.1 Services

Profibus layer 2 operates in a connectionless mode. In addition to logical peer-to-peer data transmission, it provides multi-peer communication (broadcast and multicast).

Table 32.4 lists all available FDL services. The services are called up by the higher order layers via the service access points (SAPs). In DP, a precisely defined function is assigned to each service access point. Several service access points can be used simultaneously for all active and passive stations. A distinction is made between source access points (SSAPs) and destination service access points (DSAPs).

32.3.2 Framing

Another important task of layer 2 is to ensure the detection of transmission errors. All telegrams have a Hamming Distance of HD = 4. This is achieved through compliance with the international standard IEC 870-5-1, through special telegram start and end delimiters, slip-free synchronization, a parity bit, and a check byte. An example of Profibus frames is shown in Figure 32.1. If a receiver detects transmission errors the frame is deleted. The sender repeats the transmission automatically. The details of the mechanism are also explained in [F06].

Every station is assigned one unique address in the range of 0–126. The address 0 is typically reserved for engineering tools and the address 126 is the default address for devices. The value 127 addresses all stations on the network.

TABLE 32.4 List of Available FDL Services

Name	Service
SDN	Send data with no acknowledge
SDA	Send data with acknowledge
SRD	Send and request data
MSRD	Send and request data with multicast reply
CS	Clock synchronization

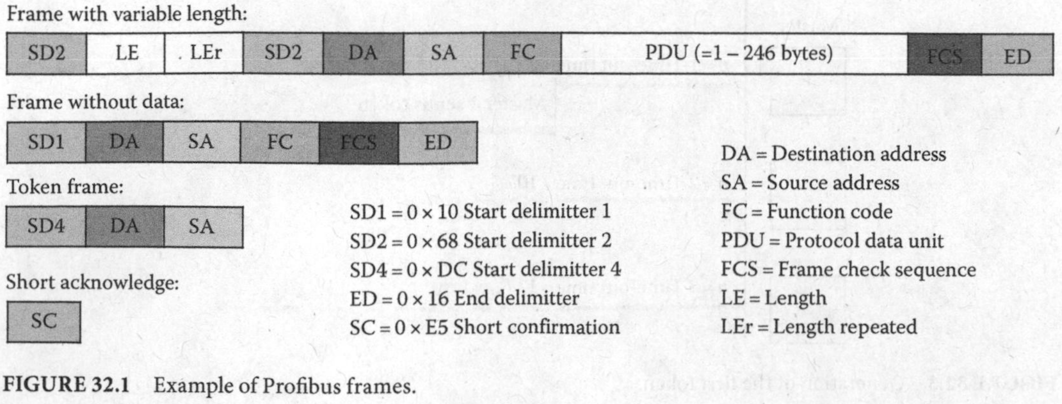

Frame with variable length:

SD2	LE	LEr	SD2	DA	SA	FC	PDU (=1 – 246 bytes)	FCS	ED

Frame without data:

SD1	DA	SA	FC	FCS	ED

Token frame:

SD4	DA	SA

Short acknowledge:

SC

SD1 = 0 × 10 Start delimitter 1
SD2 = 0 × 68 Start delimitter 2
SD4 = 0 × DC Start delimitter 4
ED = 0 × 16 End delimitter
SC = 0 × E5 Short confirmation

DA = Destination address
SA = Source address
FC = Function code
PDU = Protocol data unit
FCS = Frame check sequence
LE = Length
LEr = Length repeated

FIGURE 32.1 Example of Profibus frames.

32.3.3 Media Access

The Profibus medium access control (MAC) includes the token-passing procedure, which is used by complex bus stations (masters) to communicate with each other and the master–slave procedure used by complex bus stations to communicate with the simple peripherals (slaves).

The token-passing procedure ensures that the bus access right (the token) is assigned to each master within a precisely defined timeframe. The token message, a special telegram for passing the token from one master to the next master must be passed around the logical token ring once to all masters within a (configurable) maximum token rotation time. In Profibus, the token-passing procedure is only used for communication between complex stations (masters).

The master–slave procedure permits the master (the active station) which currently owns the token to access the assigned slaves (the passive stations). This enables the master to send messages to, or retrieve them from the slaves. This method of access allows implementation of the following system configurations:

- Pure master–slave system with just one master polling the slaves
- Pure master–master system with token passing and no slaves
- A hybrid system combining multiple masters passing the token and polled slaves (Figure 32.2)

On system startup, every master listens to the bus and watches the transmissions. If the bus is idle a timer is started; the timer is set as a multiple of the unique station address n as shown in formula (32.1). The timer of the master with the lowest address expires first and this master has the first token as shown

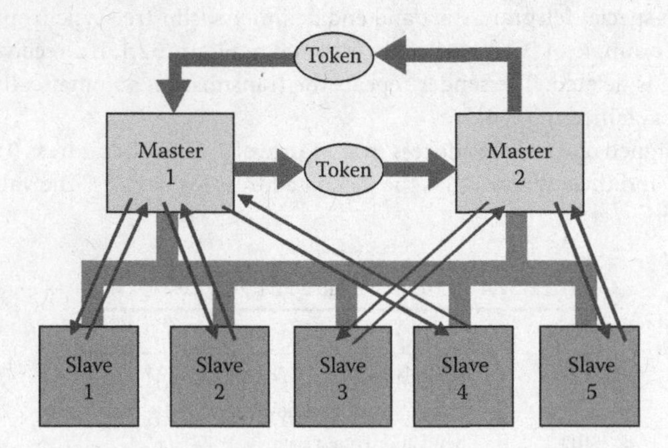

FIGURE 32.2 Hybrid system with two masters and five slaves.

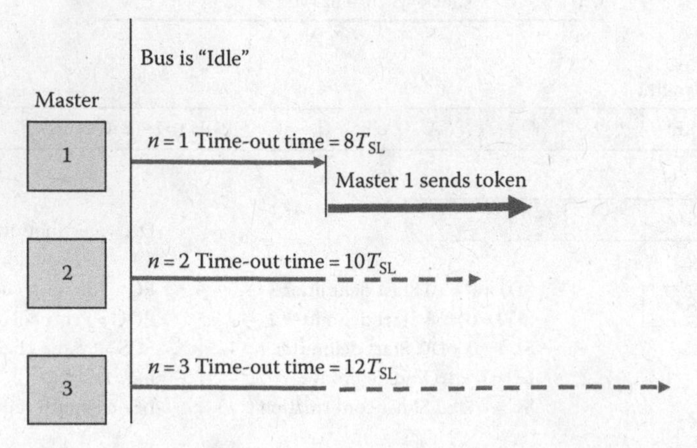

FIGURE 32.3 Generation of the first token.

in Figure 32.3. He signals this to the other stations by sending the token to himself. The active master starts to poll all addresses to detect additional stations on the bus

$$t_{\text{TIMEOUT}} = 6t_{\text{SLOT}} + n2t_{\text{SLOT}} \tag{32.1}$$

where t_{SLOT} = maximum time to answer to a poll request.

If an additional master detects traffic on the bus he has to listen to all token frames and collect the list of all addressed masters. As soon as he has a complete and stable list, he claims the token for himself. When the active master polls, him he replies that he is ready to get the token and that it will be included in the token ring.

The token is passed from one master to the other in an increasing order of the station address number. For every bus is defined a target token rotation time (TTRT). When the active station sends the token, it starts a timer reducing the TTRT. When the token gets back, the rest of the time is the maximum token hold time for this node. During this time, the active station can communicate with all slave stations in a master–slave communication relationship and to all master stations in a master–master communication relationship.

In a Profibus, DP system the configuration tool ensures that the TTRT is large enough that the master can poll assigned slaves in one token hold time. More detailed descriptions and formulas may be found at [CMTV02,V04, and KHU08].

32.4 DP System

DP supports the implementation of both mono-master and multi-master systems. This affords a high degree of flexibility during system configuration. The specifications for system configuration define the following:

- Number of stations
- Assignment of station addresses to the I/O addresses
- Data integrity of I/O data
- The format of diagnostic messages
- The bus parameters used

[P02] explains all these parameters in detail and [WG03] gives an example of a commercial implementation.

Each DP system is made up of different device types, whereby a distinction is made between three types of devices:

- DP master class 1 (DPM1)—controllers
- DP master class 2 (DPM2)—engineering stations
- DP slaves (DPS)—field devices

32.4.1 DP-Master Class 1: Controllers

This is a central controller that cyclically exchanges information with the distributed stations (slaves) at a specified message cycle. Typical DPM1 devices are programmable logic controllers (PLCs) or PCs. A DPM1 has active bus access with which it can read measurement data (inputs) of the field devices and write the setpoint values (outputs) of the actuators at fixed times. This continuously repeating cycle is the basis of the automation function.

In the case of mono-master systems, only one master is active on the bus during the operation of the bus system. This single-master is the central control component. The slaves are decentrally coupled to the master over the transmission media. This system configuration enables the shortest bus cycle times.

In multi-master operation, several masters are connected to one bus. They represent either independent subsystems, comprising one DPM1 and its assigned slaves, or additional configuration and diagnostic devices. The input and output images of the slaves can be read by all DP masters, while only one DP master (the DPM1 assigned during configuration) can write-access the outputs.

32.4.2 DP-Master Class 2: Engineering Stations

Devices of this type are engineering, configuration, or operating devices. They are implemented during commissioning and for maintenance and diagnostics in order to configure connected devices, evaluate measured values and parameters, and request the device status. A DPM2 does not have to be permanently connected to the bus system. The DPM2 also has active bus access.

32.4.3 DP-Slaves: Field-Devices

A slave is a peripheral (I/O devices, drives, HMIs, valves, transducers, analysis devices), which reads in process information and/or uses output information to intervene in the process. There are also devices that solely process input information or output information. As far as communication is concerned, slaves are passive devices, they only respond to direct queries. This behavior is simple and cost-effective to implement (in the case of DP-V0 it is already completely included in the hardware).

32.4.4 Application Relations

The different types of devices may enter different communication relations to each other (Figure 32.4):

MM: Exchange of parameters, configuration data, and diagnostics between a DPM1 and a DPM2
MS0: Cyclic data exchange of a DPM1 controller with a field-device (DP slave)
MS1: Acyclic data exchange between a DPM1 controller and a field-device (DP slave)
MS2: Acyclic data exchange between a DPM2 engineering station and a field-device (DP slave)

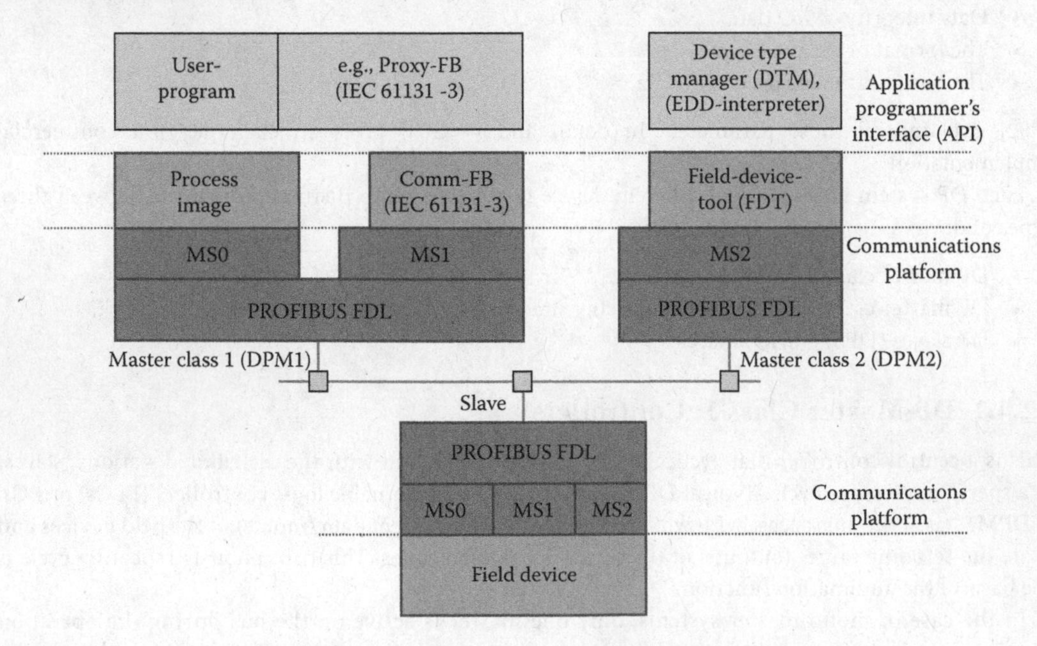

FIGURE 32.4 Device types and relations.

32.5 Cyclic Data Exchange: MS0—Relation

The central control (DPM1) performs the following functions:

- Reads input information from the slaves cyclically
- Writes output information to the slaves cyclically

The bus cycle time should be shorter than the program cycle time of the central automation system, which is approximately 10 ms for many applications.

32.5.1 Device Model

Every field-device is composed of modules, which are located in slots. Every module can have one or several channels. These modules may be real physical modules or only virtual organization units. The configuration and number and version of modules may be fixed or variable. The possible modules, order, and parameters are described in the Generic Station Description (GSD) file, which is provided by the manufacturer of the device and used by the configuration tool of the controller to support the engineer in the planning of the Profibus system.

32.5.2 Initialization and Supervision of the Relation

Data communication between the DPM1 and its assigned slaves is automatically handled by the DPM1 in a defined, recurring sequence. The user defines the assignment of the slave(s) to the DPM1 when configuring the bus system. The user also defines which slaves are to be included or excluded in the cyclic user data communication.

Data communication between the DPM1 and the slaves is divided into three phases: parameterization, configuration, and data transfer. Before the master includes a DP slave in the data transfer phase, a check is run during the parameterization and configuration phase to ensure that the configured setpoint configuration matches the actual device configuration. During this check, the device type, format, and length information and the number of inputs and outputs must also correspond. This provides the user with reliable protection against parameterization errors. In addition to user data transfer, which is automatically executed by the DPM1, the user can also request that new parameterization data are sent to the slaves.

For safety reasons, it is necessary to ensure that DP has effective protective functions against incorrect parameterization or failure of transmission equipment. For this purpose, the DP master and the slaves are fitted with monitoring mechanisms in the form of time monitors. The monitoring interval is defined during configuration.

The DPM1 uses a Data_Control_Timer to monitor the data communication of the slaves. A separate timer is used for each slave. The time monitor is tripped if no correct user data transfer is executed within the monitoring interval. In this case, the user is notified. If the automatic error handling (Auto_Clear = True) is enabled, the DPM1 exits the operate state, switches the outputs of the assigned slaves to the fail-safe state, and shifts to the clear mode.

The slave uses the watchdog control to detect errors of the master or of the transmission. If no data communication with the master occurs within the watchdog control interval, the slave automatically switches its outputs to the fail-safe state.

In addition, access protection is required for the inputs and outputs of the slaves operating in multimaster systems. This ensures that only the authorized master has direct access. For all other masters, the slaves provide an image of their inputs and that can be read without access rights.

32.5.3 Status of the Controller and Fail-Safe Functionality

In order to ensure a high degree of device interchangeability among devices of the same type, the system behavior of the DP has also been standardized. This behavior is determined primarily by the operating state of the DPM1. This can be controlled either locally or over the bus from the configuration device. There are three main states:

Stop No data communication between the DPM1 and the slaves.

Clear The DPM1 reads the input information of the slaves and keeps the outputs of the slaves in a fail-safe state ("0" output).

Operate The DPM1 is in the data transfer phase. In cyclic data communication, inputs are read from the slaves and output information written to the slaves.

The DPM1 cyclically sends its status to all its assigned slaves at configurable intervals using a multicast command. The reaction of the system to a fault during the data transfer phase of the DPM1, e.g., the failure of a slave, is determined by the "auto clear" configuration parameter.

If this parameter is set to True, the DPM1 switches the outputs of all assigned slaves to a fail-safe state the moment a slave is no longer ready for user data transmission. The DPM1 subsequently switches to the clear state. If this parameter is set to False, the DPM1 remains in the operate state even in the event of a fault and the user can control the reaction of the system.

32.5.4 Diagnostics of the Field-Device

The comprehensive diagnostic functions of DP enable the fast location of faults. The diagnostic messages are transmitted over the bus and collected at the master. These messages are divided into three levels:

- *Device-specific diagnostics*: Messages on the general readiness for service of a station, such as "Overheating," "Undervoltage," or "Interface unclear."
- *Module-related diagnostics*: These messages indicate whether a diagnosis is pending within a specific I/O subdomain of a station (e.g., 8 bit output module).
- *Channel-related diagnostics*: These messages indicate the cause of a fault related to an individual input/output bit (channel), such as "Short-circuit at output."

32.5.5 Distributed Database

This function enables direct and thus time-saving communication between slaves using broadcast communication without the detour over a master. In this case, the slaves act as a "publisher," i.e., the slave response does not go through the coordinating master, but directly to other slaves embedded in the sequence, the so-called "subscribers." This enables slaves to directly read data from other slaves and use them as their own input. This opens up the possibility of efficient applications; it also reduces response times on the bus by up to 90%.

32.5.6 Synchronization of the Applications

This function enables clock synchronous control in masters and slaves, irrespective of the bus load. The function enables highly precise positioning processes with clock deviations of less than a microsecond. All participating device cycles are synchronized to the bus master cycle through a "global control" broadcast message. A special sign of life (consecutive number) allows monitoring of the synchronization.

32.6 Acyclic Data Exchange: MS1/MS2 Relations

The key feature of the MS1/MS2 relation is the extended function for acyclic data communication. This forms the requirement for parameterization and calibration of the field devices over the bus during runtime and for the introduction of confirmed alarm messages. Transmission of acyclic data

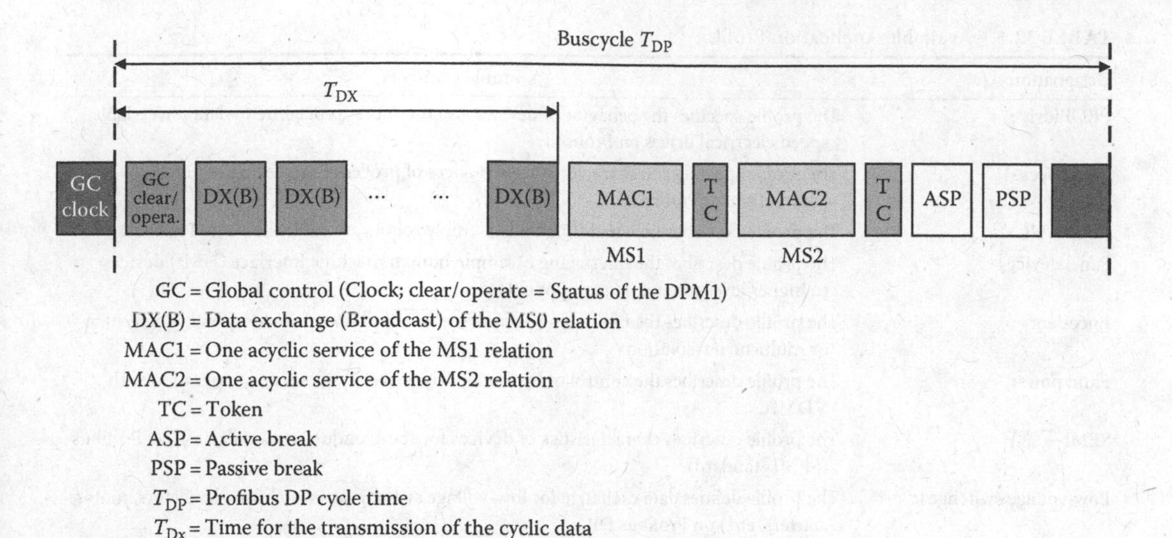

FIGURE 32.5 Timing of an MS1/MS2 connection with isochronous cycle.

is executed parallel to cyclic data communication, but with lower priority. Figure 32.3 shows some sample communication sequences. The master class 1 has the token and is able to send messages to or retrieve them from slave 1, then slave 2, etc., in a fixed sequence until it reaches the last slave of the current list (MS0 relation); it then passes on the token to the master class 2. This master can then use the remaining available time of the programmed cycle to set up an acyclic connection to any slave to exchange records (MS2 relation); at the end of the current cycle time, it returns the token to the DPM1. Similarly, as well as the DPM2, the DPM1 can also execute acyclic data exchange with slaves (MS1 relation) (Figure 32.5).

32.6.1 Variables

The field-device is composed of modules located in slots. Every slot has now a set of variables addressed with an index in the range of 0–255. With the acyclic read and write services, it is possible to read and modify these parameters of the field-device. The format and meaning of these parameters is defined in the device profile.

32.6.2 Device Model including Identification and Maintenance

For the uniform identification and maintenance procedures, every device and module of a Profibus system must be identified. The set of Identification and Maintenance (I&M) variables inside every Profibus DP field-device permits this functionality. These variables are read and also written with the acyclic data exchange services.

32.6.3 Alarm Handling

The alarm-handling function allows a field-device to signal-detected alarm conditions in a controlled procedure to the DPM1 controller. The controller has to acknowledge the signaled alarms.

Possible alarm conditions are pulled or plugged modules, diagnostic information, changed parameters in the field-device, or user defined changes of process values.

TABLE 32.5 Available Application Profiles

Designation	Profile Contents
PROFIdrive	The profile specifies the behavior of devices and the access procedure to data for variable-speed electrical drives on Profibus.
PA devices	The profile specifies the characteristics of devices of process engineering in process automation on Profibus.
Robots/NC	The profile describes how handling and assembly robots are controlled over Profibus.
Panel devices	The profile describes the interfacing of simple human machine interface (HMI) devices to higher level automation components.
Encoders	The profile describes the interfacing of rotary, angle, and linear encoders with single-turn or multi-turn resolution.
Fluid power	The profile describes the control of hydraulic drives over Profibus. In cooperation with VDMA.
SEMI	The profile describes characteristics of devices for semiconductor manufacture on Profibus (SEMI standard).
Low-voltage switchgear	The profile defines data exchange for low-voltage switchgear (switch-disconnectors, motor starters, etc.) on Profibus DP.
Dosage/weighing	The profile describes the implementation of weighing and dosage systems on Profibus DP.
Ident systems	The profile describes the communications between devices for identification purposes (bar codes, transponders).
Liquid pumps	The profile defines the implementation of liquid pumps on Profibus DP. In cooperation with VDMA.
Remote I/O for PA devices	Due to their special place in bus operations, a different device model and data types are applied to the remote I/Os compared to the Profibus PA devices.

32.6.4 Procedure Calls

The procedure call function allows the loading of any size of data area into a field device with a single command. This enables, for example, programs to be updated or devices replaced without the need for manual loading processes.

32.7 Application Profiles

Profibus stands out from other fieldbus systems primarily due to its extraordinary breadth of application options. Table 32.5 shows all current specific Profibus application profiles. Major application profiles are Profisafe as defined in [IEC84-3] and Profibus PA as explained in [M03] and [DB07].

References

[CMTV02] Cavalieri, S., Monforte, S., Tovar, E., and Vasques, F., Evaluating worst case response time in mono and multi-master profibus DP, *Fourth IEEE International Workshop on Factory Communication Systems*, Vasteras, Sweden, 2002, pp. 233–240.

[DB07] Dietrich, Ch, and Bangemann, Th., *Profibus PA, Instrumentation Technology for the Process Industry*, Oldenbourg Industrieverlag, Munich, Germany, 2007, 344 pp., ISBN 978-3-8356-3125-0.

[F06] Felser, M., Quality of profibus installations, *Sixth IEEE International Workshop on Factory Communication System* (*WFCS 2006*), Conference Center Torino Incontra, Torino, Italy, June 27–30, 2006.

[IEC58] IEC 61158, Digital data communications for measurement and control—Fieldbus for use in industrial control systems, IEC, available at www.iec.ch

[IEC84-1] IEC 61784-1, IEC 61784-1: Industrial communication networks—Profiles—Part 1: Fieldbus profiles, IEC, available at www.iec.ch

[IEC84-3] IEC 61784-3-3, Industrial communications networks—Profiles—Part 3-3: Functional safety fieldbuses—Additional specifications for CPF 3, IEC, available at www.iec.ch

[KHU08] Kaghazchi, H., Hongxin, Li, and Ulrich, M., Influence of token rotation time in multi master PROFIBUS networks, *IEEE International Workshop on Factory Communication Systems, 2008* (*WFCS 2008*), Dresden, Germany, May 21–23, 2008, pp. 189–197.

[M03] Mitchell, R., *PROFIBUS: A Pocket Guide (Paperback)*, The Instrumentation, Systems, and Automation Society (ISA), Research Triangle Park, NC, October 2003, 200 pp., ISBN: 978-1556178627.

[P02] Popp, M., The new rapid way to PROFIBUS DP, describing PROFIBUS from DP-V0 to DP-V2, Profibus Organization, November 2002, app. 260 pp., Order No:4.072.

[P2112] Profibus User Organization, Installation guideline for PROFIBUS DP/FMS V1.0, Profibus User Organization, Order at http://www.profibus.com, Order No: 2.112.

[P8022] Profibus User Organization, Handbook PROFIBUS installation guideline for cabling and assembly V1.0.6, Profibus User Organization, Free download at http://www.profibus.com, Order No: 8.022.

[P8032] Profibus User Organization, Handbook PROFIBUS installation guideline for commissioning V1.0.1, Profibus User Organization, Free download at http://www.profibus.com, Order No: 8.032.

[V04] Vitturi, S., Stochastic model of the Profibus DP cycle time, *IEE Proceedings Science, Measurement and Technology*, 151, 335–342, September 4, 2004, ISSN: 1350-2344.

[WG03] Weigmann, J. and Kilian, G., *Decentralization with PROFIBUS DP/DPV1: Architecture and Fundamentals, Configuration and Use with SIMATIC S7*, 2003 edition, Wiley-VCH, Weinheim, Germany, March 1, 2004, 251 pp., ISBN: 978-3895782183.

33

INTERBUS

Juergen Jasperneite
Ostwestfalen-Lippe University of Applied Sciences

Orazio Mirabella
University of Catania

33.1 INTERBUS Overview 33-1
33.2 INTERBUS Protocol 33-2
33.3 Diagnostics 33-7
33.4 Performance Evaluation 33-8
33.5 Summary 33-9
References 33-9

33.1 INTERBUS Overview

INTERBUS was developed by the German company Phoenix Contact and presented in 1987. INTERBUS has been designed as a fast sensor/actuator bus for transmitting process data in industrial environments and is standardized in the IEC-61158 [1]. It presents several peculiarities that are mainly related to the topology and to the protocol that provides fast, cyclic, and time-equidistant process data transmission, diagnostics to minimize downtime, and easy operation and installation [2,3].

The first important feature refers to the topology. INTERBUS is a ring system, that is, all devices are actively integrated in a closed transmission path (see Figure 33.1). This is a large difference with respect to the majority of existing fieldbus systems, which (according to their name) are based on a bus. In a ring, each device amplifies the incoming signal and forwards it, thus providing a better signal quality which, in turn, reduces the noise and enables higher transmission speeds over longer distances.

Unlike other ring systems, the data forward and return lines in the INTERBUS system lead to all devices via a single cable. This means that the general physical appearance of the system is an "open" tree structure. A main line exits the bus master and can be used to form seamless subnetworks up to 16 levels deep. This removes one big limit of ring-based topologies that require the communication line to round back to the starting point in order to close the ring. With an open tree structure, the bus system can be quickly adapted to changing applications and it is easy to add new branches for expanding the system. Moreover, as communication in the ring is unidirectional, it is well suited for fiber optic technology.

The INTERBUS master/slave system enables the connection of up to 512 devices, across 16 network levels. The ring is automatically closed by the last device. Countless topologies can be created. Branch terminals create branches, which enable the connection and disconnection of devices.

The second important peculiarity is inherent to the way nodes are addressed and data are transferred through the ring. Unlike in other systems where data are assigned by entering a bus address for each individual device and it is transferred through dedicated frames, in the INTERBUS system data are automatically assigned to devices using their physical location in the system and all information for all devices are transferred through one total frame called "summation frame." This feature is provided by the physical implementation of the system that can be seen as a long, distributed shift register (SR) with serial/parallel in/out capabilities composed by several small SRs, each one belonging to a node. Data traveling on the ring are simply shifted through the SR cells and are distributed through the various devices. With a suitable data organization, at a specific time instant, each node will find its relevant data

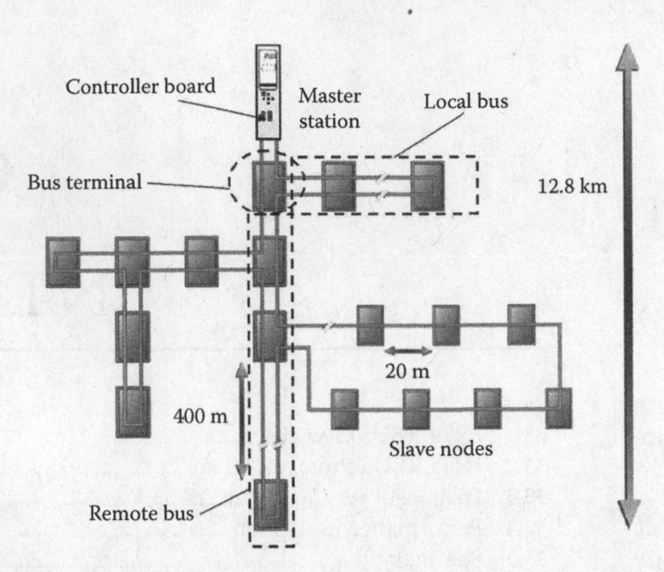

FIGURE 33.1 Topology flexibility allows a cost-efficient cabling in machines and plants.

inside its local SR. This plug and play function is a great advantage with regard to the installation effort and service-friendliness of the system. The problems and errors, which may occur when manually setting device addresses during installation and servicing, are often underestimated.

In order to provide these features, various basic elements must be used in the system as shown in Figure 33.1:

Bus master: The bus master controls all bus operations. It transfers output data to the corresponding nodes, receives input data, and monitors data transfer. In addition, diagnostic messages are displayed and error messages are transmitted to the host system. The controller has the responsibility of preparing the single total frame that provides data distribution among the various nodes.

Remote bus: The controller board is connected to the bus devices via the remote bus. A branch from this connection is referred to as a remote bus branch. Data can be physically transmitted via copper cables (RS-485 standard), fiber optics, infrared data links, slip rings, or other media (e.g., wireless). Special bus terminal modules and certain input/output (I/O) modules or devices such as robots, drives, or operating devices can be used as remote bus devices. Each one has a local voltage supply and an electrically isolated outgoing segment. In addition to the data transmission lines, the installation remote bus can also carry the voltage supply for the connected I/O modules and sensors.

Bus terminal module: The bus terminal modules, or devices with embedded bus terminal module functions, are connected to the remote bus. The distributed local buses branch out of the bus terminal module with I/O modules, which establish the interface between INTERBUS and the sensors and actuators. The bus terminal module divides the system into individual segments, thus enabling to switch individual branches on/off during operation. The bus terminal module amplifies the data signal (repeater function) and electrically isolates the bus segments.

Local bus: The local bus branches from the remote bus via a bus coupler and connects the local bus devices. Branches are not allowed at this level. The communication power is supplied by the bus terminal module, while the switching voltage for the outputs is applied separately at the output modules. Local bus devices are typically I/O modules.

33.2 INTERBUS Protocol

INTERBUS operations are based on two cycle types: The identification cycle for system configuration and error management, and a data transfer cycle for the transmission of user data. Both cycle types are based on a summation frame whose structure is shown in Figure 33.2.

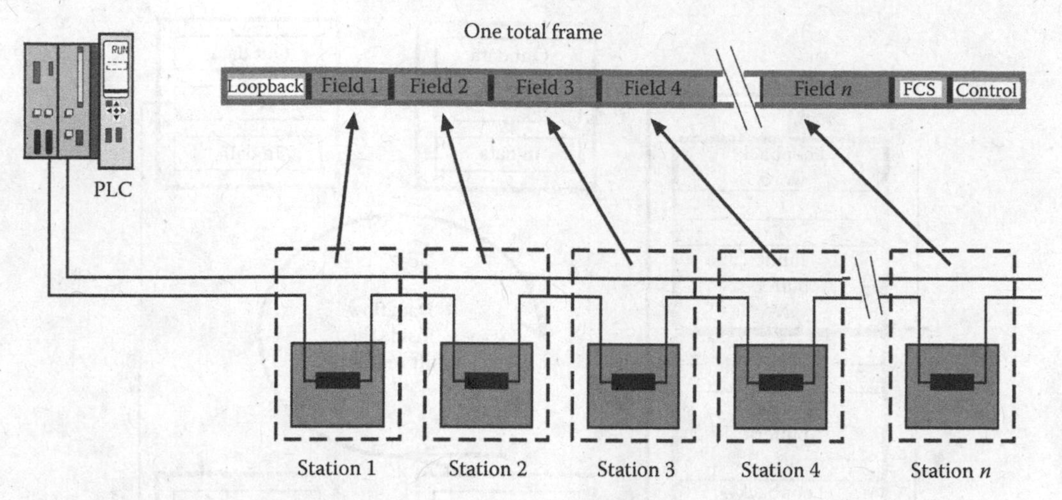

FIGURE 33.2 The layer 2 summation frame structure of INTERBUS.

The layer 2 summation frame consists of a special 16 bit loopback word (i.e., the preamble), the user data of all devices, and a terminating 32 bit frame check sequence (FCS) to control the data integrity. The FCS consists of a 16 bit cyclical redundancy check (CRC) value and further 16 bits for controlling the data exchange between the input and output registers. According to the physical structure, data can be sent and received by the ring structure of the INTERBUS system in a single round (full duplex mode), and this results in very high protocol efficiency.

The logical method of operation of an INTERBUS slave can be configured between its incoming and outgoing interfaces by the register set shown in Figure 33.3. Each INTERBUS slave is part of a large, distributed SR ring, whose start and end point is the INTERBUS master. The register set inside the slave can be used for both data transfer and management. Data transfer is performed through the input and output data registers.

During the data transfer phase, the input data register will contain input data, that is, data that are to be transmitted to the master, and the output data registers will contain data received through the ring.

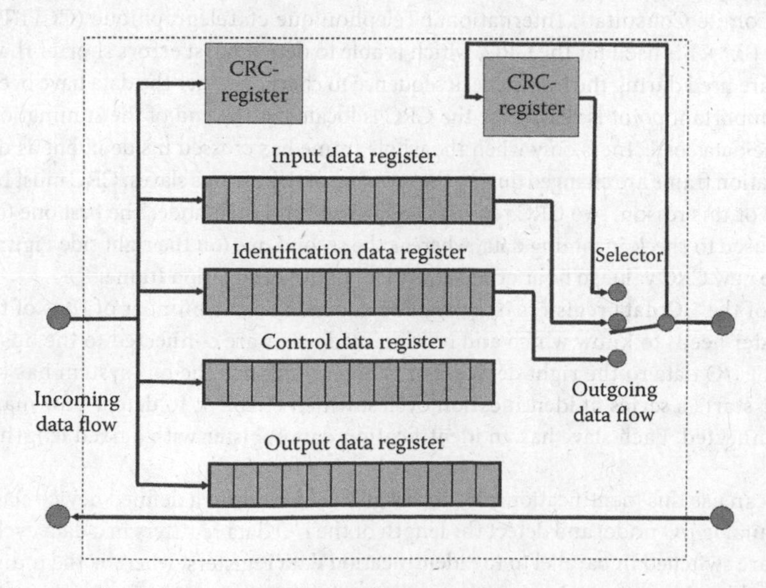

FIGURE 33.3 Basic model of an INTERBUS slave node.

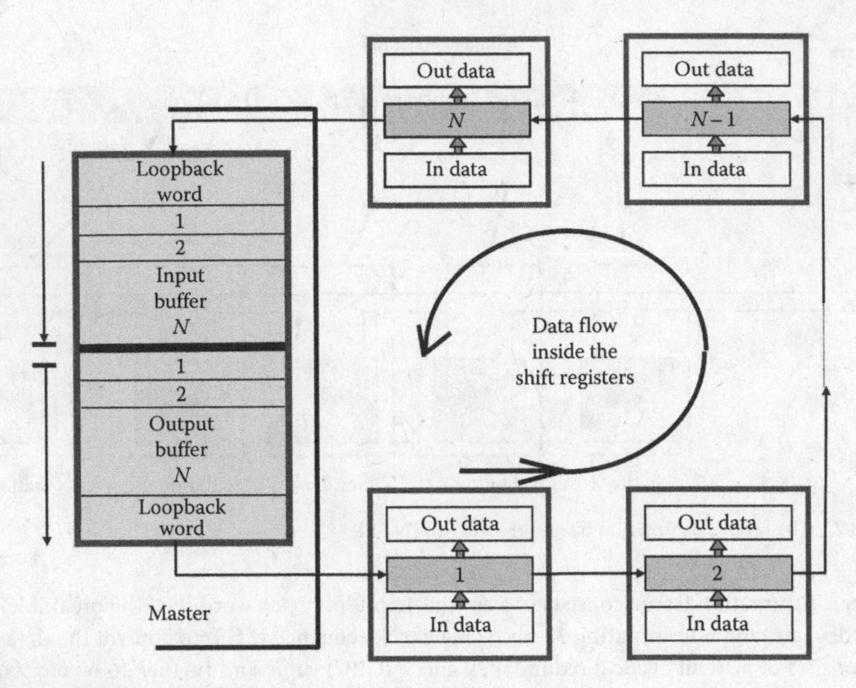

FIGURE 33.4 Logical structure of the ring in INTERBUS.

According to the logical structure shown in Figure 33.4, the summation frame will shift inside all the SRs and this way data sent by the master will cross all the nodes. There is a precise time instant when all output data registers inside each slave will contain the data addressed to that slave. At that moment, data contained in the output data register will be stored into the node, whereas data addressed to the master will be loaded into the input data register and will begin to travel along the ring by substituting the original data into the summation frame. This operating mode is very efficient allowing the ring, into a single round, to transfer data from and to the master.

An important role is played by the CRC register, which is switched to the input data register in parallel. The Comité Consultatif International Téléphonique et Télégraphique (CCITT) polynomial $g(x) = x^{16} + x^{12} + x^5 + 1$ is used for the CRC, which is able to detect burst errors shorter than 16 bits. The CRC registers are used during the frame check sequence to check whether the data have been transmitted correctly. An important point is that, since the CRC is located at the end of the summation frame, each slave checks the data correctness only when the whole frame has crossed inside it, but as data contained into the summation frame are changed during the crossing of the various slaves, CRC must be recalculated several times. For this reason, two CRC registers are present into each node. The first one (on the left side Figure 33.3) is used to check incoming data, whereas the second one (on the right side Figure 33.3) is used to calculate the new CRC value to be attached at the end of the summation frame.

The length of the I/O data registers in each node depends on the number of I/Os of the individual node. The master needs to know which and how many devices are connected to the bus so that it can assign the right I/O data to the right device. For this reason, once the bus system has been switched on, the master starts a series of identification cycles, which enable it to detect how many and which devices are connected. Each slave has an identification data register with a fixed length 16 bit identification code.

The master can use this identification code to assign a slave node to a defined device class (e.g., digital I/O node and analog I/O node) and detect the length of the I/O data registers in a data cycle. The control data registers are switched in parallel to the identification data registers, whereby the individual devices can be managed by the master. Commands are transmitted, for example, for local resets or outgoing interface shutdown. The identification cycle is also used to find the cause of a transmission error and to check the integrity of the SR ring.

The individual registers are switched in the different phases of the INTERBUS protocol via the selector in the ring.

On layer 1, INTERBUS uses two telegram formats with start and stop bits:

- The 5 bit status telegram
- The 13 bit data telegram

The status telegram is used to synchronize the activities on the medium during pauses in data transmission. The slave nodes use the status telegram to reset their internal watchdogs, which are used to control a failsafe state. The data telegram is used to carry a byte of the layer 2 payload. The remaining bits of both telegrams are used to distinguish between data and identification cycles, as well as the data transfer and FCS phases within a cycle. This information is used by the selector shown in Figure 33.4 to switch the relevant register in the ring. INTERBUS uses a physical transmission speed of 500 kbps or 2 Mbps.

The cycle time, that is, the time required for I/O data to be exchanged once with all the connected nodes, depends on the amount of user data in an INTERBUS system, that is, on the number of nodes and on the dimension of the SR inside each node. Depending on the configuration, INTERBUS can achieve cycle times of just a few milliseconds. The cycle time increases linearly with the number of I/O points, as it depends on the amount of information to be transmitted. For more detailed information, refer to the chapter performance evaluation.

The architecture of the INTERBUS protocol is based on the open system interconnection (OSI) reference model according to International Standards Organization (ISO) 7498. As is typical for fieldbus systems, for reasons of efficiency, ISO layers 3–6 are not explicitly used (see Figure 33.5) but are combined in the lower layer interface (LLI) in layer 7. The protocol architecture of INTERBUS provides both periodic and asynchronous communications.

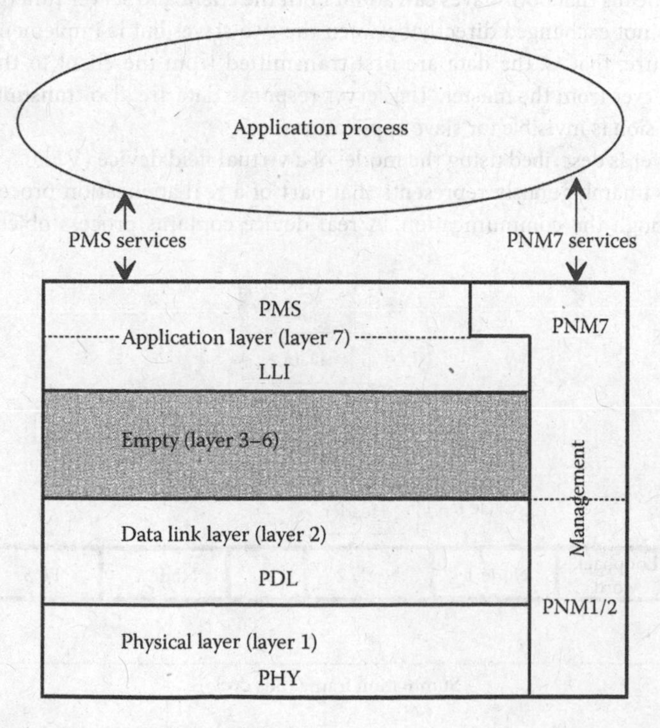

FIGURE 33.5 Protocol architecture of an INTERBUS node.

Periodic communication is used for the exchange of cyclic process data that support the services of the peripheral message specification (PMS) through a process data channel. PMS is a subset of the manufacturing message specification (MMS). This enables direct access to the cyclically transmitted process data and is able to transmit process-relevant data quickly and efficiently. From the application point of view, it acts as a memory interface.

Instead, asynchronous communication is used to transfer configuration files provided by the network management channel (peripheral network management, PNM) through a parameter channel. The data transmitted in the parameter channel have low dynamics and occur relatively infrequently (e.g., updating text in a display) and are used for manufacturer-independent configuration, maintenance, and startup of the INTERBUS system.

Network management is used, for example, to start/stop INTERBUS cycles, to execute a system reset, and fault management. Furthermore, logical connections between devices can be established and aborted via the parameter channel.

INTERBUS must then transmit both parameter data and time-critical process data simultaneously, avoiding any interference between the two data flows. To this aim, the data format of the summation frame has been extended by including a specific time slot inside the time assigned to each node. Whereas process data are fully exchanged at every cycle, parameter data telegrams, which are up to 246 bytes long and are not featured by real-time constraints, will be exchanged in several consecutive bus cycles by inserting a different part of the data in the time slot provided for the addressed devices. The peripherals communication protocol (PCP) performs this task: It inserts a part of the telegram in each summation frame and recombines it at its destination (see Figure 33.6). The parameter channels are activated, if necessary, and do not affect the transfer of I/O data. The longer transmission time for parameter data that is segmented into several bus cycles is sufficient for the low time requirements that are placed on the transmission of parameter information.

INTERBUS uses a master/slave procedure for data transmission. The parameter channel follows the client/server paradigm. It is possible to transmit parameter data between two slaves (peer-to-peer communication). This means that both slaves can adopt both the client and server function. With this function, layer 2 data is not exchanged directly between the two slaves but is implemented by the physical master/slave structure, that is, the data are first transmitted from the client to the master and then forwarded to the server from the master. The server response data are also transmitted via the master. However, this diversion is invisible for slave applications.

The task of a server is described using the model of a virtual field device (VFD).

The VFD model unambiguously represents that part of a real application process, which is visible and accessible through the communication. A real device contains process objects. Process objects

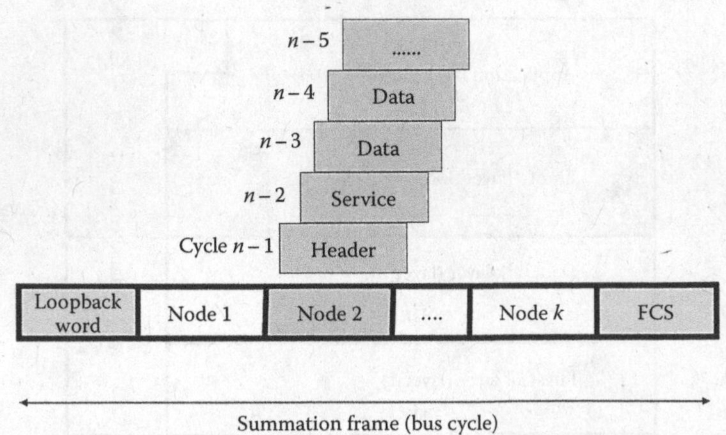

FIGURE 33.6 Transmission of acyclic parameter data with a segmentation and recombination mechanism.

include the entire data of an application process (e.g., measured values, programs, or events). The process objects are entered in the object dictionary (OD) as communication objects. The OD is a standardized public list, in which communication objects are entered with their properties. To ensure that data are exchanged smoothly in the network, additional items must be standardized, in addition to the OD, which can be accessed by each device. This includes device features such as the manufacturer name or defined device functions that are manufacturer independent. These settings are used to achieve a closed and manufacturer-independent representation of a real device from the point of view of the communication system. This kind of modeling is known as a VFD.

33.3 Diagnostics

The system diagnostics plays an important role in real-world applications. In increasingly complex systems, errors must be located quickly using system diagnostics and must be clearly indicated to the user. For this reason, a good error diagnostics, in addition to detecting errors, must include reliable error localization. For usual message-oriented fieldbus systems, which are based on a bus structure, only one message is ever transmitted to a device at any time. An error, which affects the system via a specific device, can even destroy messages, which are not addressed to the faulty device itself, but may be directed toward other remote devices. So it is therefore virtually impossible to determine the exact error location.

INTERBUS uses the CRC algorithm in each device to monitor the transmission paths between two devices and, in the event of CRC errors, can determine in which segment the error occurred.

In order to maintain data communication in the system, the master must be able to cope at least with the following errors:

- Cable break
- Failure of a device
- Short circuit on the line
- Diagnostics of temporary interference (e.g., caused by electromagnetic interference)

In all fieldbus systems, in the event of a line interruption, the devices after the interrupt are no longer reachable. The error localization capability depends on the transmission system used. In linear bus systems, messages are broadcast to all devices. However, in case of a line break, these messages are lost because the devices are no longer able to respond. Bus operation cannot be maintained because, due to the interruption, the condition of the physical bus termination using a termination resistor is no longer met. This can lead to reflections within the bus configuration. The resulting interference level will make correct operations not possible. After a certain period, the master will detect the data loss but will not be able to precisely determine the error location. The wiring diagrams must be consulted, so that the service or maintenance personnel can determine the probable error location (see Figure 33.7).

Unlike bus systems, the individual devices in the INTERBUS system are networked, so that each one behaves as a separate bus segment. Following a fatal error, the outgoing interfaces of all devices are fed back internally via a bypass switch. In the event of a line interrupt between the devices, the master activates each separate device in turn. To do this, the master opens the outgoing interface, starting from the first device up until the error location, thus clearly identifying the inaccessible device. The bus master can then clearly assign the error location. If a device fails, the fieldbus behaves in the same way as for a line interrupt. However, the functional capability of the remaining stations differs in bus and ring systems.

Short circuits on the line are a major challenge in a bus system. In the event of a direct or indirect (e.g., via ground) short circuit on the line, the transmission path is blocked for the entire section.

In bus systems, one transmission line is used for all devices, which means that the master cannot reach segment parts either. Signal reflections due to impedance mismatching will make communication impossible. This considerably reduces further error localization. Linear bus systems also support

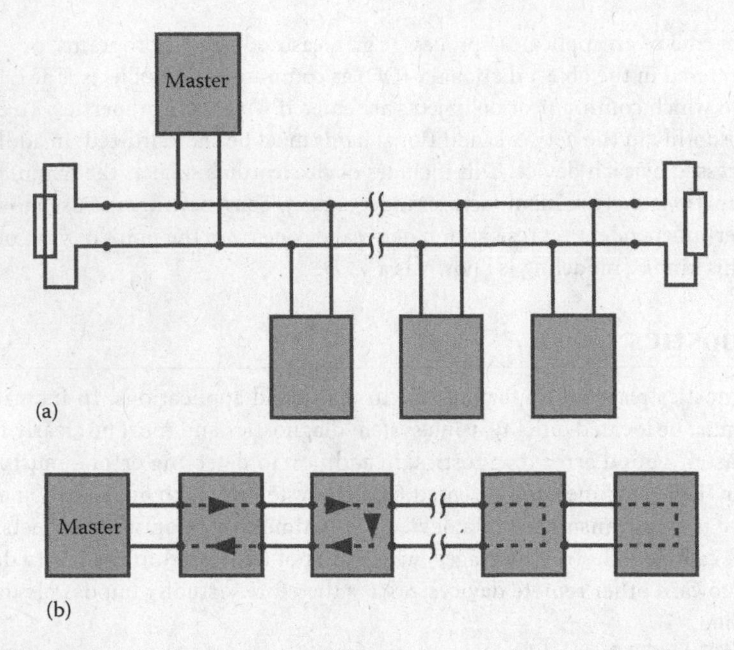

FIGURE 33.7 The behavior of (a) bus systems and (b) ring systems in the event of a cable break.

a division into different segments. Repeaters, which are placed at specific points, can then perform diagnostic functions. However, a repeater cannot monitor the entire system; it can only cover a defined number of devices per segment. Furthermore, the use of repeaters incurs additional costs and increased configuration effort.

In the INTERBUS system, the user is supported by the physical separation of the system into different bus segments. As described for the line interrupt, the devices are activated by the master in turn and the ring is closed prior to the short circuit, which means that subsystems can be started up again. The error location can be reported in clear text at the bus master.

In summary, the INTERBUS diagnostic features are essentially based on the physical segmentation of the network into numerous point-to-point connections. This feature makes INTERBUS particularly suitable for use with fiber optics, which are used increasingly for data transmission in applications with large drives, welding robots, etc. In linear systems, the use of fiber optics—as in bus segmentation—requires expensive repeaters, which simulate a ring structure.

33.4 Performance Evaluation

This section considers the performance of INTERBUS with respect to the relevant parameters, such as the number of I/O nodes and the amount of cyclic I/O data.

The following equation applies to the INTERBUS layer 2 cycle time T_{IB}:

$$T_{IB} = 13 \cdot (6 + n) \times T_{Bit} + T_{SW}$$

where

n is the total length of the SR [bytes]
T_{Bit} the time needed for transmitting a bit
T_{SW} the software runtime of the master

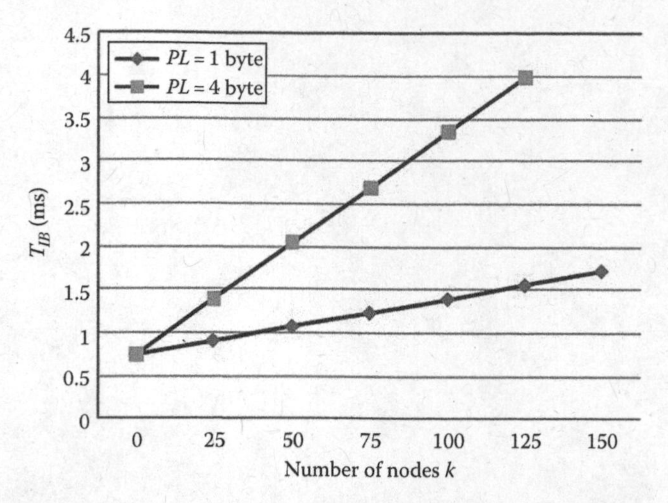

FIGURE 33.8 INTERBUS cycle times as a function of the number of nodes and the used payload size.

The frame overhead of 6 bytes is caused by a 2 byte long loopback word and a 4 byte long FCS. Each octet at layer 2 is transmitted within a 13 bit data telegram at layer 1. Depending on the total amount of input and output data, the total payload size can be calculated by

$$n = \max\left(\sum_{i}^{i=k} PL_i^{Input}, \sum_{i}^{i=k} PL_i^{Output}\right)$$

where

- n is the total payload size: sum of all user data [bytes] of all devices k with $n \le 512$ bytes and $k \le 512$
- T_{Bit} is the bit time: 2 μs at 0.5 Mbps or 0.5 μs at 2 Mbps
- T_{SW} is the software runtime of the master (depending on implementation)
- PL_i is the layer 2 payload size of the ith device [bytes] where $1 \le i \le k$ and $k \le 512$

In Figure 33.8, the INTERBUS cycle time at the MAC level with a bit rate of 2 Mbps and a software runtime of 0.7 ms is illustrated as a function of the number of nodes k and their corresponding payload size PL.

33.5 Summary

INTERBUS is a serial bus system developed for the time-critical data transfer at the field level of industrial automation systems. The physical topology of INTERBUS is based on a ring structure, that is, all devices are actively integrated in a closed transmission path. The data forward and return lines in the INTERBUS system are led to all devices via a single cable. This means that the general physical appearance of the system is an "open" tree structure. Due to a total frame approach, the cyclic data transfer of INTERBUS is very efficient in comparison to fieldbus systems based on an individual frame approach. As a result, small cycle times can be achieved with small bit rates. Acyclic messages are transmitted using preconfigured time slots within the total frame in conjunction with a segmentation and recombination mechanism at layer 2.

References

1. IEC, Digital data communication for measurement and control—Fieldbus for use in industrial control systems, 2001, IEC 61158 Ed.3, Type 8 (INTERBUS), IEC, Geneva.
2. INTERBUS Club, www.interbusclub.com, visited May 2009.
3. Baginski, A. and Müller, M., *INTERBUS Basics and Practice*, Hüthig-Verlag, Heidelberg, Germany, 2002 (German).

34

WorldFip

34.1 Introduction 34-1
34.2 Physical Layer 34-2
34.3 Data Link Layer 34-2
Transmission of Cyclic Traffic • Bus Arbitrator Table • Transmission of Asynchronous Traffic
34.4 Application Layer 34-6
34.5 Timing Properties of WorldFIP Networks 34-7
Concept of Producer/Distributor/Consumer • Buffer Transfer Timings • Bus Arbitrator Table • WorldFIP Aperiodic Buffer Transfers • Setting the WorldFIP BAT: Rate Monotonic Approach
References 34-18

Francisco Vasques
University of Porto

Orazio Mirabella
University of Catania

34.1 Introduction

WorldFIP is a fieldbus that has been mainly designed for time-critical applications featured by periodic processes. Its name comes from the *fieldbus instrumentation protocol* that was initially standardized as French national Standard and subsequently included into the European Fieldbus Standard CENELEC EN50170. The name has been changed into WorldFIP to make it more attractive for the international market.

The basic idea behind WorldFIP is to consider all variables relevant to the application process as a distributed database that is periodically refreshed in order to maintain the values of the updated variables. This provides a deterministic framework for the exchange of process variables produced by sensors, processed by PLCs or process computers, and consumed by actuators. The new concept of producer/consumer, as opposed to the traditional client/server concept, was introduced first time in WorldFip.

As the fieldbus must carry not only periodic process variables (which are usually a few bytes long) but also management data (e.g., configuration files), which can be very long, different kinds of traffic must be considered:

- Periodic traffic for the updating of process variables. In WorldFIP-based applications, it is assumed that about 99% of traffic belongs to this category.
- Aperiodic traffic for the exchange of process variables.
- Aperiodic traffic for the exchange of messages. This last kind of traffic is generated by management files, which according to their dimensions must be handled in a different way with respect to process variables.

The layered architecture adopted in WorldFIP is based on only three layers (as shown in Figure 34.1), which is usual for real-time networks: physical layer (PHL), data link layer (DLL = medium access control [MAC] + logical link control [LLC]), and application layer (AL).

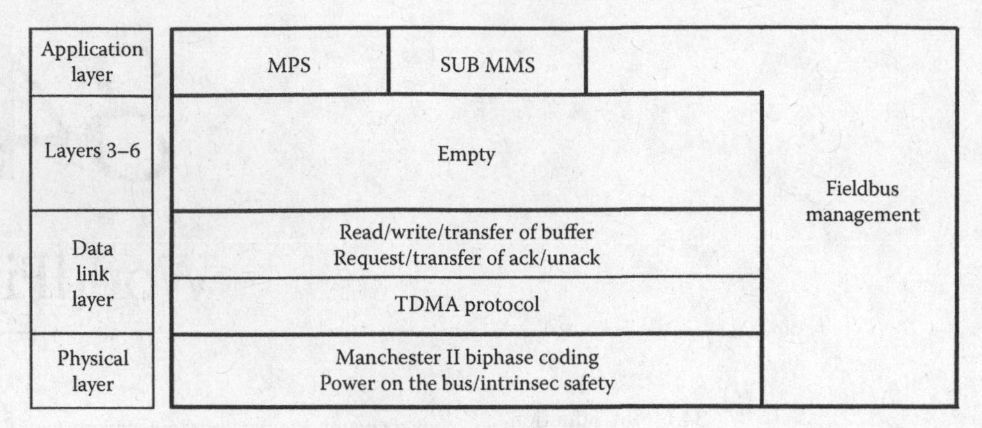

FIGURE 34.1 Layered architecture of WorldFip.

34.2 Physical Layer

PHL uses Manchester type encoding. Three versions have been defined: a first version, called S1, operates at 50 kbps using twisted pairs; the second version, S2, operates at 1 Mbps; and the third version, S3, operates at 2.5 Mbps. Both S2 and S3 versions use coax or fiber optics. Two interesting features of PHL concern the use of intrinsic safety barriers for operations in hazardous environments and the possibility to have power on the bus that can be used by field devices without the need of a dedicated power line.

With reference to the topology adopted in WorldFIP, a minimum system is set up by a single trunk with several stations, of which one has the role of bus arbitrator (BA) and manages the access to the medium. Three kinds of devices can be connected to the bus:

- *Junction boxes*: These devices are multiport repeaters that connect a cluster of field devices to the bus.
- *Field concentrators*: These devices are gateways that connect field devices that use the WorldFIP protocol or different protocols. Moreover, they can be used to interconnect several independent WorldFip trunks.
- *Repeaters*: These devices are used to extend the bus length by connecting two or more trunks.

34.3 Data Link Layer

DLL is split into a MAC and LLC sublayers. MAC is based on a time division multiple access protocol (TDMA), where a special station called the BA makes the scheduling of all transmissions through the distribution of a specific authorization. Any of the several WorldFIP stations active on the bus can operate as the BA (this allows for redundancy in case of a fault in the current BA) but at each time instant only one BA can be in the bus.

The presence of a centralized control for the access to the physical medium is an important feature of the MAC protocol. The BA knows the transmission requirements of all the devices present in the system and can distribute the bandwidth available in such a way so as to satisfy their time constraints. This approach is particularly efficient in cases when most of the traffic is periodic since the BA can authorize the transmission at the correct time. On the other hand, this approach is ineffective for asynchronous traffic that cannot be known a priori and cannot be scheduled. For asynchronous traffic, the protocol must use an approach that allows the BA to became aware of additional requests and update its schedule.

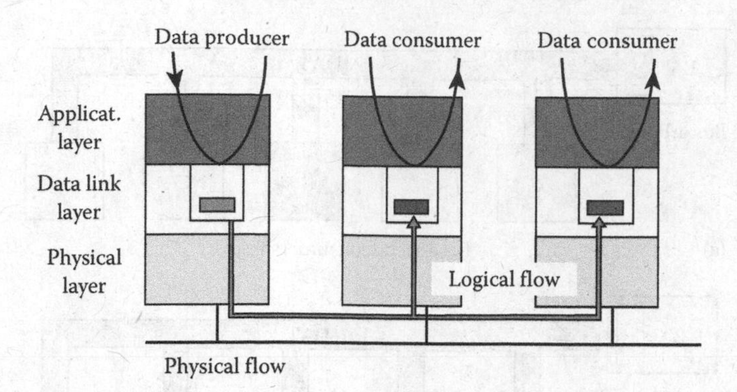

FIGURE 34.2 The basic producer–consumer data transfer.

LLC sublayer exploits the communication channel made available by MAC and provides two main kinds of services:

- Services to read/write/transfer process data that are contained into suitable reception/transmission buffers
- Services to request the confirmed/unconfirmed transmission of message files

Figure 34.2 shows the basic mechanism adopted in WorldFip for data transfer, which is performed in several steps:

- The user (producer) passes to the AL the information to be transferred to remote consumers.
- The AL entity stores the information inside a buffer that can be read by the DLL entity.
- The DLL entity through the physical medium, transfers such information into the remote buffers of the various consumers.
- The information is read by the remote ALs and delivered to the consumer processes.
- This sequence is repeated cyclically and is well suited for periodic data transfer.

34.3.1 Transmission of Cyclic Traffic

As stated earlier, the task of the DLL is to read a buffer and to transfer its data through the bus. Since the bus is a shared medium, in order to avoid collisions, it is necessary to organize suitably the various transmissions. This is the task of the BA, a station that carries out a key role in the system by providing the authorization for each buffer transfer.

The mechanism adopted for the transmission of cyclic traffic is shown in Figure 34.3 and is executed through the following steps:

- The BA broadcasts, on the bus, an identifier (ID_DAT) that implicitly specifies all stations involved in the current transaction. One station will be the producer of the information, whereas the other ones will be the consumers.
- The identifier is received by both the producer and consumers.
- The producer sends its message (RP_DAT) on the bus.
- The consumers that are involved into the transaction accept the data.

This sequence illustrates clearly the role of BA which, through a single message, identifies the role of all stations participating to the next phases. All devices receive the identifier but only one will identify itself as a producer, whereas all the others will be consumers. The architecture of the BA is shown in Figure 34.4. As we can see, the BA is more than simply a scheduler as it must also cope with asynchronous requests from nonscheduled traffic. For this reason, the BA must monitor the bus and check for some additional information that the producer will incorporate in its frame with reference to some additional bandwidth request.

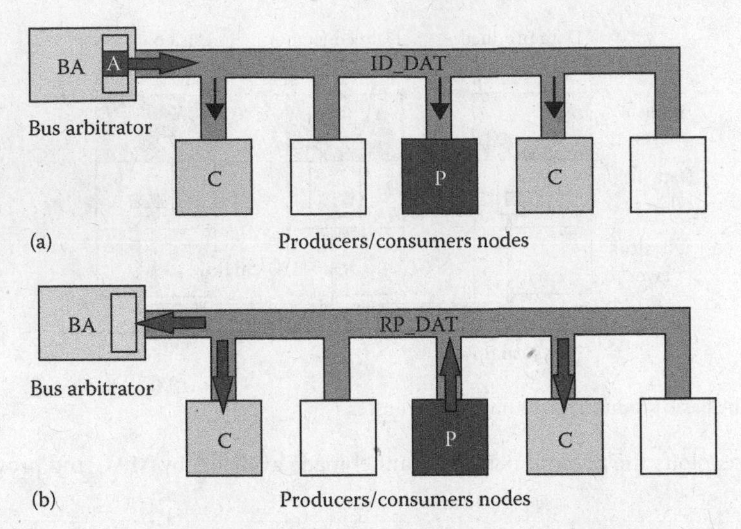

FIGURE 34.3 The sequence of messages required for a transaction. Transmission of the authorization from (a) the bus arbitrator and (b) transmission of a data reply from a producer.

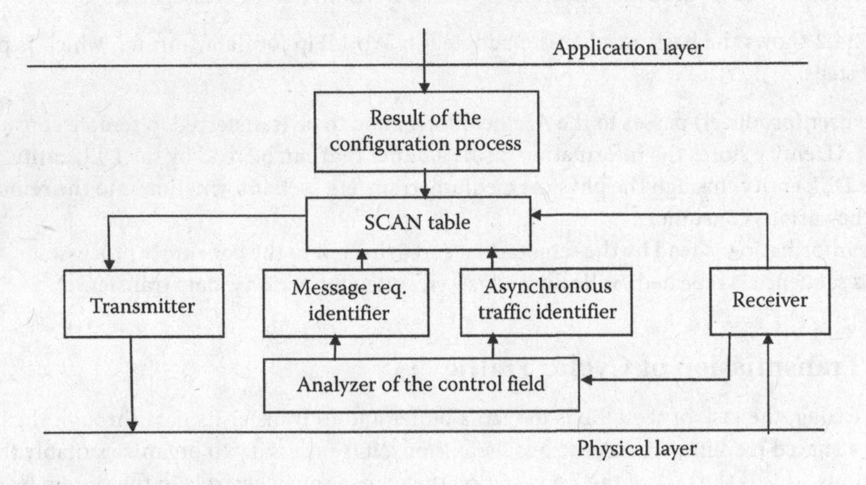

FIGURE 34.4 Architecture of the bus arbitrator.

Through the monitoring of the current transmission, the BA can check some control bits that indicate an additional request from the producer. If any, the BA will proceed through an analysis of the scan table that contains the time-ordered list of all the identifiers to be circulated on the bus. Through this analysis, the BA will identify free portions of bandwidth to be assigned to the asynchronous traffic.

34.3.2 Bus Arbitrator Table

The BA is the network conductor. After the system has been configured, the BA is in possession of the variables list to be analyzed and their time constraints. If the configuration is validated and it satisfies the time constraints imposed by the application process, the BA can generate the Scan table and use it again and again in order to authorize cyclic traffic. Scan of the variables is deterministic, that is, WorldFIP can guarantee that a variable featured by a certain periodicity will be analyzed at the right time instants. As an example, Table 34.1 shows a list of six periodic variables that need to be scheduled.

TABLE 34.1 List of Variables to Be Scheduled

Variable Name	Refresh Period (ms)	Data Type	Transaction Time (μs)
A	5	INT_8	170
B	10	INT_8	170
C	15	INT_16	178
D	15	SFPOINT	194
E	20	INT_16	178
F	30	UNS_32	194

The BA must first analyze the single variables and achieve their refresh period (which will be expressed in milliseconds) and the data type (e.g., 8 bit integer number and short floating point number) in order to compute the time required by each variable. This time is made up of three components: the time required by the BA to transmit the ID frame that polls a producer, the turnaround time (i.e., the time between the transmission of an ID frame by the BA and the reply by an information producer), and the time required by the producer to send the data frame. The values reported in Table 34.1 are based on a transmission bit rate of 1 Mbps and a turnaround time of 20 ms (the range allowed by WorldFIP with a bit rate 1 Mbps is from 10 to 70 ms).

From the values reported in Table 34.1, we can infer that transmission of single variables is not very efficient in WorldFip. In fact, we have a heavy transmission overhead since a single variable requires two frames and some turnaround time. Of course, the situation becomes better when the dimension of the data increases as the rate (data field)/(transmission overhead) is improved.

Figure 34.5 shows a possible schedule of the variables listed in Table 34.1 to be stored into the bus arbitrator table (BAT). Time on *X*-axis is divided into elementary cycles that correspond to the time required to transmit variables with the shortest refresh period. This organization of the data scan simplifies the construction of the scheduling table as we divide the time in fixed slots and then try to allocate all variables into the various time slots without violating their time constraints. Time on *Y*-axis is the sum of the time required to transmit all the variables that are scheduled inside a time slot.

In Figure 34.5, we can see that in the first elementary cycle, all the variables are transmitted. In the subsequent cycles, the variables are transmitted according to a distribution based on their time constraints. We observe that no elementary cycle overcomes the value of 5 ms, so that the schedule does not violate any time requirement.

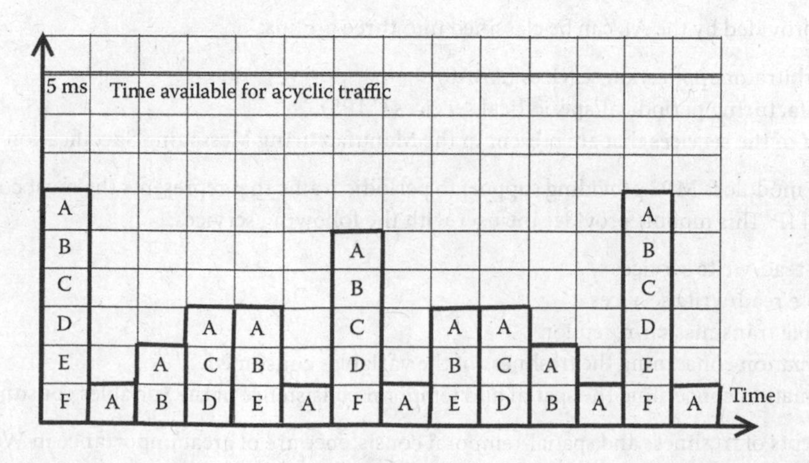

FIGURE 34.5 Possible schedule of the variables listed into Table 34.1.

Moreover, some spare time is available in each microcycle for the transmission of asynchronous traffic. This will allow the BA to consider additional requests for asynchronous traffic and allocate them into a suitable time slot. If no request is received, the BA will transmit padding identifier (i.e., identifiers that do not refer to any station) until the end of the elementary cycle, in order to indicate to the other stations that it is still active.

Looking at Figure 34.5, we can observe that the schedule is repeated after a number of microcycles (12 in the example considered). This interval is called macrocycle.

34.3.3 Transmission of Asynchronous Traffic

Not all the information required by a distributed application can be distributed by the BA through the cyclic scan table. In fact, asynchronous traffic can be produced by both sporadic variables (e.g., alarm signals) and messages (e.g., configuration files). In these cases, as this traffic cannot be foreseen a priori it cannot be scheduled. The solution adopted in WorldFip is to link this kind of traffic to cyclic traffic, which instead can be scheduled. Both asynchronous variables and messages are transmitted by using similar approaches, but with two important differences:

- Variables are shorter than messages, so it is easier for the BA to find a free space in the scan table to allocate them.
- Variables never require to be confirmed, whereas messages can be confirmed.

The transmission of asynchronous traffic is performed in three steps:

1. In the first step, the producer communicates to the BA its need for additional bandwidth, setting a field inside an RP_DAT frame.
2. The BA asks the producer the detailed list (amount) of the asynchronous traffic.
3. The BA sends the transmission authorization to the producer, according to the space available inside its scheduling table.

34.4 Application Layer

In the AL, a set of abstract objects called application objects are visible. The totality of these objects makes up a virtual database that represents the distributed database. These objects can be manipulated through some specialized services that are suitable for periodic/aperiodic data transfer in industrial applications.

Services provided by the AL can be classified into three groups:

- Bus arbitrator application services (ABAS)
- Manufacturing periodical/aperiodical services (MPS)
- Subset of the services that are present in the Manufacturing Messaging Specification (subMMS)

The main module is MPS providing support to periodic traffic that represents the most common traffic in WorldFIP. This module provides the user with the following services:

- Local read/write services
- Remote read/write services
- Variable transmission/reception
- Information concerning the freshness of the variables consumed
- Information concerning the spatial and temporal consistence of the variables consumed

The concepts of freshness and spatial/temporal consistence are of great importance in WorldFIP and are associated to the ability of the protocol to provide a timely communication.

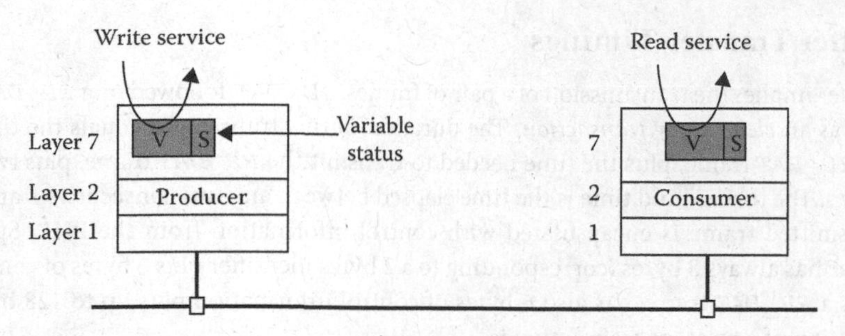

FIGURE 34.6 Transfer of a variable from the producer to the consumer.

Information timeliness is one main feature of WorldFip protocol, and to this aim when a user receives a variable he or she can also receive information regarding its freshness. This is obtained through a Boolean flag that can be bound to each variable to be consumed by the application processes.

Figure 34.6 shows how the timeliness information is processed and transferred from the producer process to the consumer. A producer process (in the left side of Figure 34.6) uses a write service to store the variable value inside a suitable buffer, and a refreshment status check process evaluates the timeliness of this value with reference to a *theoretical refresh period* (TRP). Each time a new value is stored in the buffer, the status flag is set "true" and a timer is started with a duration equal to the TRP. If the timer expires before the variable is refreshed with a new sample, the status flag is reset to "false." In this way, when the consumer process receives the variable, it can know if its value has been correctly refreshed or not.

34.5 Timing Properties of WorldFIP Networks

As stated earlier, timeliness is one fundamental property that strongly features the WorldFIP protocol and differentiates it from other fieldbuses. In this section, we will analyze and evaluate some important timing properties of WorldFIP with reference to the construction of the scan table and the transmission of asynchronous traffic.

34.5.1 Concept of Producer/Distributor/Consumer

As stated earlier, in WorldFIP, the exchange of identified variables services are based on a producer/distributor/consumer (PDC) model, which relates producers and consumers within the distributed system. In this model, for each process variable there is one, and only one producer, and several consumers. In order to manage transactions associated with a single variable, a unique identifier is associated with each variable. The WorldFIP DLL is made up of a set of produced and consumed buffers, which can be locally accessed (through AL services) or remotely accessed (through network services).

The AL provides two basic services to access the DLL buffers: `L_PUT.req`, to write a value in a local produced buffer, and `L_GET.req`, to obtain a value from the local consumed buffer. None of these services generate activity on the bus. Produced and consumed buffers can be also remotely accessed through a network transfer service (also known as *buffer transfer*). The BA broadcasts a question frame `ID_DAT`, which includes the identifier of a specific variable. The DLL of the station that has the corresponding produced buffer responds using a response frame `RP_DAT`. The DLL of the station that contains the produced buffer then sends an indication to the AL (`L_SENT.ind`). The DLL of the station(s) that has the consumed buffers accepts the value contained in the `RP_DAT`, overwriting the previous value and notifying the local AL with a `L_RECEIVED.ind`.

34.5.2 Buffer Transfer Timings

A buffer transfer implies the transmission of a pair of frames: ID_DAT, followed by a RP_DAT. We denote this sequence as an *elementary transaction*. The duration of this transaction equals the time needed to transmit the ID_DAT frame, plus the time needed to transmit the RP_DAT frame, plus twice the turnaround time (t_r). The turnaround time is the time elapsed between any two consecutive frames.

Every transmitted frame is encapsulated with control information from the PHL. Specifically, an ID_DAT frame has always 8 bytes (corresponding to a 2 bytes identifier plus 6 bytes of control information), whereas a RP_DAT frame has also 6 bytes of control information plus up to 128 bytes of useful data. The duration of a message transaction is

$$C = \frac{len(\text{id\_dat}) + len(\text{rp\_dat})}{bps} + 2 \times t_r \tag{34.1}$$

where

bps stands for the network data rate
len(<frame>) is the length, in bits, of frame <frame>

For instance, assume that all variables have a data field with 4 bytes (resulting in ID_DAT: 8 bytes and RP_DAT: 10 bytes). If $t_r = 20\,\mu s$ and the network data rate is 2.5 Mbps then, the duration of an elementary transaction is $(64 + 80)/2.5 + 2 \times 20 = 97.6\,\mu s$ (Equation 34.1).

34.5.3 Bus Arbitrator Table

In WorldFIP networks, the *bus arbitrator table* regulates the scheduling of all buffer transfers. The BAT imposes the timings of the periodic buffer transfers and also regulates the aperiodic buffer transfers. In Section 34.3.2, we have discussed the importance of the BAT as an indispensable tool for the scheduling of periodic transmissions. In this section, we will deal with this in depth and will discuss the problem of how to dynamically organize the BAT when new requests for asynchronous transmission are received by the BA. Assume here a WorldFIP system where six variables are to be periodically scanned, with scan rates as shown in Table 34.2. The WorldFIP BAT must be set in order to cope with these timing requirements. Two important parameters are associated with a WorldFIP BAT: the microcycle (elementary cycle) and the macrocycle. The microcycle imposes the maximum rate at which the BA performs a set of scans. Usually, the microcycle is set equal to highest common factor (HCF) of the required scan periodicities. Using this rule, and for the example shown in Table 34.2, the value for the microcycle is set to 1 ms. A possible schedule for all the periodic scans can be as illustrated in Figure 34.7, where we consider $C = 97.6\,\mu s$ for each elementary transaction.

It is easy to depict that, for example, the sequence of microcycles repeats each 12 microcycles. This sequence of microcycles is said to be a macrocycle and its length is given by the lowest common multiple (LCM) of the scan periodicities.

The HCF/LCM approach for building a WorldFIP BAT has the following properties:

1. The scanning periods of the variables are multiples of the microcycle.
2. The variables are not scanned at exactly regular intervals. For the given example, only variables *A* and *B* are scanned exactly in the same "slot" within the microcycle. All other variables suffer from a slight communication jitter. For instance, concerning variable *F*, the interval between microcycles 1 and 7 is 5.80 ms, whereas between microcycles 7 and 13 it is 6.19 ms.

TABLE 34.2 Example Set of Periodic Buffer Transfers

Identifier	A	B	C	D	E	F
Periodicity (ms)	1	2	3	4	4	6

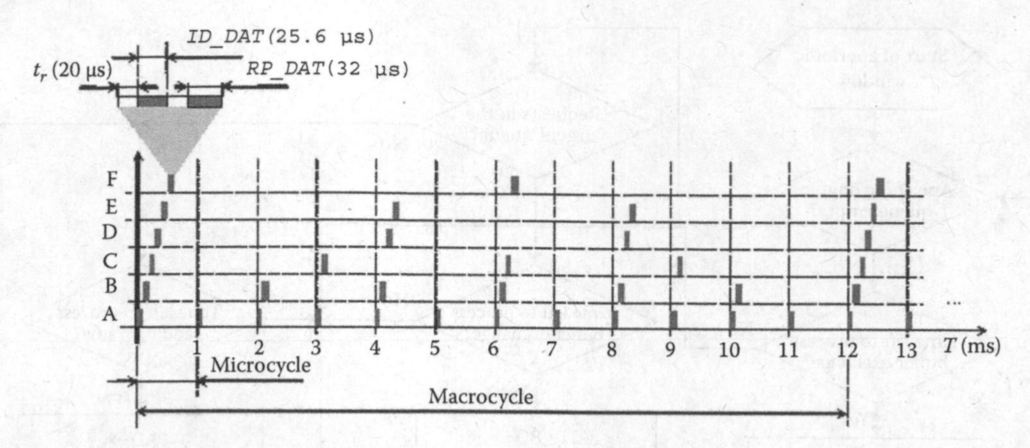

FIGURE 34.7 Schedule for the periodic transfers of Table 34.2.

3. The length of the macrocycle can induce a memory size problem, since the table parameters must be stored in the BA. For instance, if the scanning periodicities of variables *E* and *F* were, respectively, 5 and 7 ms, the length of the macrocycle would be 420 microcycles instead of just 12 microcycles.

Both the communication jitter and memory size problems have been addressed in the literature. In [1], the authors discuss different methodologies to reduce the BAT size, without penalising the communication jitter. The idea is very simple, and it basically consists on reducing some of the scan periodicities in order to have a harmonic pattern. The problem of table size has also been addressed in other works [2,3], however, in a different perspective. In the referred work, the authors discuss an online scheduler (instead of storing the schedule in the BA's memory), which is not directly applicable to the WorldFIP case.

It is also worth mentioning that Figure 34.7 represents a macrocycle composed of synchronous microcycles, that is, for the specific example, each microcycle starts exactly 1 ms after the previous one. Within a microcycle, the spare time can be used by the BA to process aperiodic requests for buffer transfers, message transfers, and padding identifiers. A WorldFIP BA can also manage asynchronous microcycles, not transmit padding identifiers. In such case, a new microcycle starts as soon as the periodic traffic is performed and there are no pending aperiodic buffer transfers or message transfers. Initial periodicities are not respected since identifiers may be more frequently scanned.

34.5.4 WorldFIP Aperiodic Buffer Transfers

The BA handles aperiodic buffer transfers only after processing the periodic traffic in a microcycle. The portion of the microcycle reserved for the periodic buffer exchanges is denoted as the *periodic window* of the microcycle. The time left after the periodic window until the end of the microcycle is denoted as the *aperiodic window* of the microcycle. The aperiodic buffer transfers take place in three stages:

1. When processing the BAT schedule, the BA broadcasts an *ID_DAT* frame concerning a periodic variable, say identifier *X*. The producer of variable *X* responds with a *RP_DAT* and sets an aperiodic request bit in the control field of its response frame. The BA stores variable *X* in a queue of requests for variable transfers. Two priority levels can be set when the request for aperiodic transfer is made: urgent or normal. The BA has two queues, one for each priority level.
2. In the aperiodic window, the BA uses an identification request frame (*ID_RQ*) to ask the producer of the identifier *X* to transmit its list of pending aperiodic requests. The producer of *X* responds with a *RP_RQ* frame (list of identifiers). This list of identifiers is placed in another BA's queue, the *ongoing aperiodic queue.*

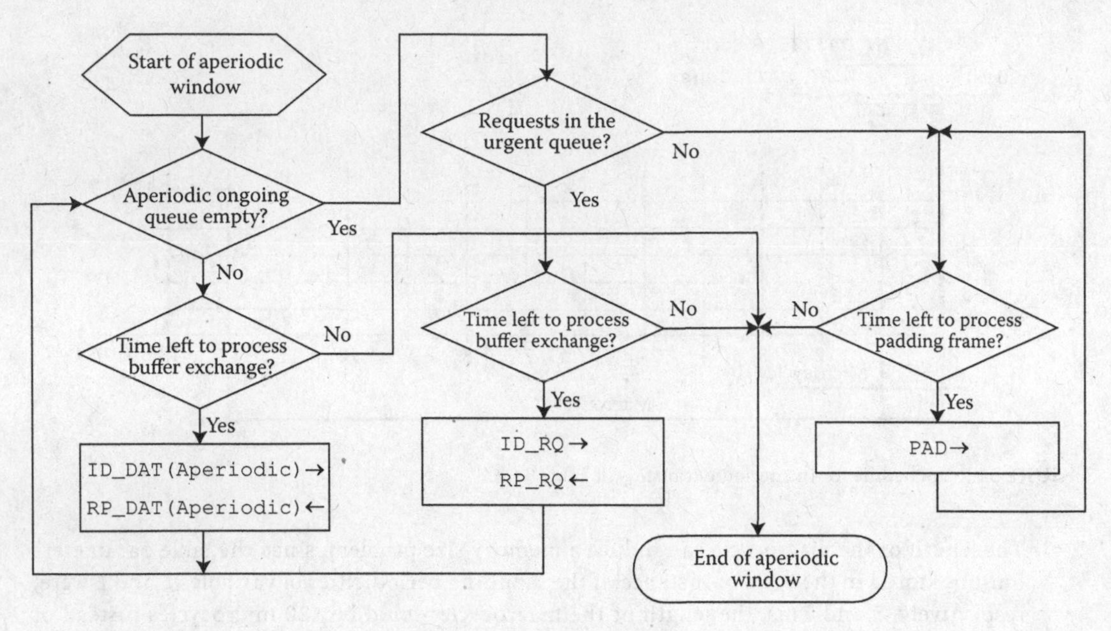

FIGURE 34.8 Sequence of transactions during the aperiodic window.

3. Finally, the BA processes requests for aperiodic transfers that are stored in its ongoing aperiodic queue. For each transfer, the BA uses the same mechanism as the used for the periodic buffer transfers (*ID_DAT* followed by *RP_DAT*).

It is important to note that a station can only request aperiodic transfers using responses to periodic variables that it produces and that are configured in the BAT. For the sake of simplicity, in the following sections we consider that all aperiodic requests concern the use of the AL service *L_FREE_UPDATE.req(ID_*,Urgent)*, thus, only the urgent queues (both at the requesting station and at the BA) are considered. Finally, it is important to stress that the urgent queue in the BA is only processed if, and only if, the BA's ongoing aperiodic queue is empty, as detailed in Figure 34.8. Also, it is important to note that at the end of a microcycle, a transaction is processed if, and only if, there is still time to complete it.

34.5.5 Setting the WorldFIP BAT: Rate Monotonic Approach

34.5.5.1 Model for the Periodic Buffer Transfers

Assume a system with np periodic variables (Vp_i, $i = 1, \ldots, np$). Each periodic variable is characterized as

$$Vp_i = (Tp_i, Cp_i) \tag{34.2}$$

where

Tp_i corresponds to the periodicity of Vp_i (assume a multiple of 1 ms)

Cp_i is the length of the transaction corresponding the buffer transfer of Vp_i (as given by Equation 34.1)

34.5.5.2 Building the WorldFIP BAT (RM Approach)

For the periodic traffic, end-to-end communication deadlines can be guaranteed by an a priori test since the BAT implements a static schedule of the periodic variables. In this section, we provide tools and analysis for setting the WorldFIP BAT according to a rate monotonic (RM) approach [4].

TABLE 34.3 BAT (Using RM) for Example of Table 34.1

	Microcycle											
	1	2	3	4	5	6	7	8	9	10	11	12
bat [*A, cycle*]	1	1	1	1	1	1	1	1	1	1	1	1
bat [*B, cycle*]	1	0	1	0	1	0	1	0	1	0	1	0
bat [*C, cycle*]	1	0	0	1	0	0	1	0	0	1	0	0
bat [*D, cycle*]	1	0	0	0	1	0	0	0	1	0	0	0
bat [*E, cycle*]	1	0	0	0	1	0	0	0	1	0	0	0
bat [*F, cycle*]	1	0	0	0	0	0	1	0	0	0	0	0

Considering the WorldFIP characteristics, the BAT can be built as follows:

1. From the variable with the shortest period till the variable with the longest period
   a. If the load in each cycle plus the variable's length (buffer transfer length) is still shorter than the value of the microcycle, then schedule a scan for that variable in each of the microcycles (within the macrocycle) multiple of the period of the variable.
   b. If the load in some of the microcycles does not allow scheduling a scan for that variable, schedule it in a subsequent microcycle up to the microcycle in which a new scan for that variable should be made. If this is not possible, the variable set is not scheduled.

For the example of Table 34.2 ($Cp_i = 0.0976$ ms, $\forall_i$) and considering the RM algorithm, the BAT becomes as represented in Table 34.3, where *bat*[*i*, *j*] is a table of booleans with *i* ranging from 1 up to *np*, and *j* ranging from 1 up to *N* (number of microcycles in a macrocycle).

Below, we describe an algorithm for building the BAT using the RM algorithm. In the algorithm, the vector *load*[·] is used to store the load in each microcycle as the traffic is scheduled. It also assumes that the array *Vp*[,] is ordered from the variable with the shortest period (*Vp*[1,]) to the variable with the longest period (*Vp*[*np*,]). The proposed RM-based algorithm already gives an indication whether all traffic is schedulable or not (line 22).

```
------------------------------------------------------------------
- Rate Monotonic for Building the BAT
------------------------------------------------------------------
function rm_bat;
input: np    /* number of periodic variables */
 Vp[i,j]/* array containing the periodicity
     /* and length of the variables */
 μCy      /* value of the microcycle */
 N     /* number of microcycles */
output:
 bat[i,cycle]   /* i ranging from 1 to np */
       /* cycle ranging from 1 to N */
begin
1: for i = 1 to np do
2: cycle = 1;
3: repeat
4:    if load[cycle] + Vp[i,2] <= μCy then
5:      bat[i,cycle] = 1;
6:      load[cycle] = load[cycle] + 1;
7:      cycle = cycle + (Vp[i,1] div μCy)
```

```
8:    else;
9:       cycle1 = cycle1 + 1;
10:      ctrl = FALSE;
11:      repeat
12:        if load[cycle1] + Vp[i,2] <=
                      μCy then
13:           ctrl = TRUE
14:        end if;
15:      until (ctrl = TRUE) or (cycle1 >=
           (cycle + (Vp[i,1] div μCy)));
16:      if cycle1 >= (cycle + (Vp[i,1]
                   div μCy) ) then
17:        bat[i,cycle1] = 1;
18:      load[cycle1] = load[cycle1] + 1;
19:      cycle = cycle + (Vp[i,1] div μCy)
20:      else
21:        /* MARK Vpi NOT SCHEDULABLE with
                  RM Algorithm */
22:        cycle = cycle + (Vp[i,1] div μCy)
23:      end if
24:    end if
25:until cycle > N
26:end for
return bat;
```

Note that by using the RM algorithm some of the variables with longer periods can be scheduled in subsequent microcycles, thus inducing an increased communication jitter. For example, if the network data rate is 1 Mbps instead of 2.5 Mbps ($Cp_i = (64 + 80)/1 + 2 \times 20 = 184\,\mu s$), a microcycle is only able to schedule up to five periodic buffer transfers (Table 34.4, Figure 34.9).

34.5.5.3 Setting the WorldFIP BAT: Earliest Deadline Approach

Assume now the set example shown in Table 34.5, in which all $Cp = 0.30$ ms.

With the RM approach, the first request for Vp_F would miss the deadline. In this case, the macrocycle is six microcycles (Table 34.6). In fact, there is no empty slot for a variable in the first three microcycles. When compared to the RM scheduling, one of the known advantages of the earliest deadline first (EDF) scheduling is that it allows a better utilization [4].

TABLE 34.4 BAT (RM) for Modified Example of Table 34.1

	Microcycle											
	1	2	3	4	5	6	7	8	9	10	11	12
bat [*A*, cycle]	1	1	1	1	1	1	1	1	1	1	1	1
bat [*B*, cycle]	1	0	1	0	1	0	1	0	1	0	1	0
bat [*C*, cycle]	1	0	0	1	0	0	1	0	0	1	0	0
bat [*D*, cycle]	1	0	0	0	1	0	0	0	1	0	0	0
bat [*E*, cycle]	1	0	0	0	1	0	0	0	1	0	0	0
bat [*F*, cycle]	0	1	0	0	0	0	1	0	0	0	0	0

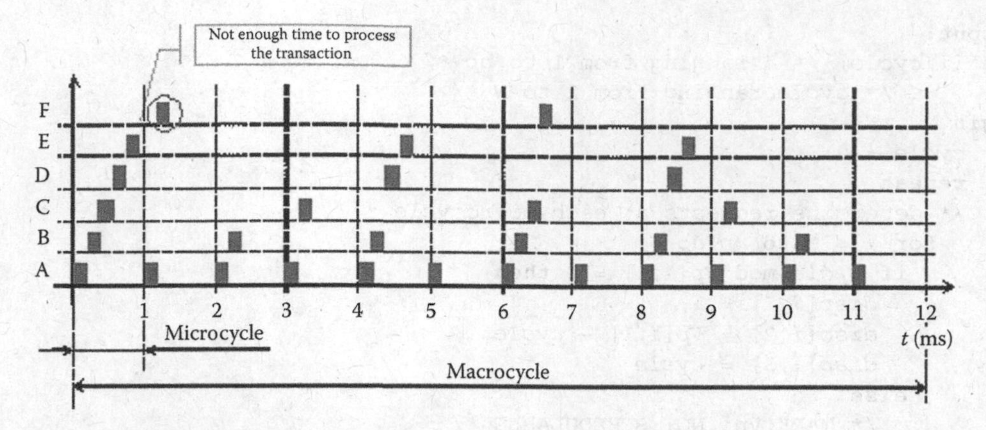

FIGURE 34.9 Schedule (RM) for the example of Table 34.4.

TABLE 34.5 Set of Periodic Buffer Transfers (Cp = 0.30 ms)

Identifier	A	B	C	D	E	F
Periodicity (ms)	1	2	2	3	3	3

TABLE 34.6 BAT (Using RM) for Example of Table 34.4

	Microcycle					
	1	2	3	4	5	6
bat [*A, cycle*]	1	1	1	1	1	1
bat [*B, cycle*]	1	0	1	0	1	0
bat [*C, cycle*]	1	0	1	0	1	0
bat [*D, cycle*]	0	1	0	1	0	0
bat [*E, cycle*]	0	1	0	1	0	0
bat [*F, cycle*]	X	0	0	0	0	1

34.5.5.4 Building the WorldFIP BAT (EDF Approach)

With the EDF Approach, variables are scheduled according to the earliest deadline (in terms of microcycle). If several variables have the same deadline, priority is given to the one with the earliest request.

Below, we describe of an algorithm for building the BAT using the EDF approach. The array *disp*[,] is used to store in *disp*[*i*, 1] if there is a pending request for variable *i*, in *disp*[*i*, 2] the deadline (multiple of the microcycle) and in *disp*[*i*, 3] the microcycle at which the request is made. Note that for algorithmic convenience, the first requests (for all the variables) appear in *cycle* = 0, and from those, some will be scheduled in the first microcycle (*cycle* + 1).

```
--------------------------------------------------------------------------
- Earliest Deadline First for Building the BAT
--------------------------------------------------------------------------
function edf_bat;
input: np    /* number of periodic variables */
 Vp[i,j]/* array containing the periodicity
     /* and length of the variables */
 μCy /* value of the microcycle */
 N     /* number of microcycles */
```

```
output:
 bat[i,cycle] /* i ranging from 1 to np */
        /* cycle ranging from 1 to N */
begin
1:  cycle = 0;
2:  repeat
3:  /* determine requests at each microcycle */
4:    for i = 1 to np do
5:      if cycle mod Vp[i,1] = 0 then
6:        disp[i,1] = 1;
7:        disp[i,2] = Vp[i,1] + cycle;
8:        disp[i,3] = cycle
9:      else
10:       /* MARK Vpi NOT SCHEDULABLE */
11:     end if
12:  end for;
13:
14:/* schedule vars in current microcycle */
15:  load_full = FALSE;
16:  repeat
17:    no_rq = TRUE;
18:    earliest = MAXINT;
19:    var_chosen = 0; /* no var. chosen */
20:
21:  /* chose the earliest deadline from
               pending requests */
22:    for i = 1 to np do
23:      if disp[i,1] = 1 then
24:        no_rq = FALSE;
25:        if disp[i,2] <= earliest then
26:          if var_chosen = 0 then
27:            earliest = disp[i,2];
28:            var_chosen = i
29:          else
30:          /* decide earliest req. */
31:            if disp[i,3] <
            disp[var_chosen,3] then
32:              earliest = disp[i,2];
33:             var_chosen = i
34:            end if
35:          end if
36:        end if
37:      end if
38:    end for;
39:
40:     /* Verify Load in Current Microcycle */
41:     if load[cycle + 1] + Vp[i,2] <= μCy then
42:        bat[var_chosen, cycle + 1] = 1
43:        load[cycle + 1] = load[cycle + 1]
                 + Vp[var_chosen,2];
44:        disp[var_chosen,1] = 0;
45:        disp[var_chosen,2] = 0;
46:        disp[var_chosen,3] = 0;
```

```
47:     else
48:         load_full = TRUE
49:     end if;
50:   until (load_full = TRUE) or (no_rq =TRUE)
51:   cycle = cycle + 1;
52:until cycle = N;
return bat;
```

Using the EDF approach, the BAT concerning the variables set represented in Table 34.5 becomes as shown in Table 34.7. This set is now schedulable, which was not possible with the RM approach (Table 34.6). However, the communication jitter is increased, since some of the buffer transfers with more stringent deadlines are occasionally "delayed" by buffer transfers with less stringent deadlines. For instance, variable Vp_C is scheduled for the fourth microcycle, whereas with the RM approach would be scheduled for the third microcycle.

34.5.5.5 Response Time Analysis for the Aperiodic Traffic

We define the worst-case response time for an aperiodic transfer as the time interval between placing, at time instant t_0, the *L_FREE_UPDATE.req(ID_ Va_i, urgent)* in the local urgent queue and the completion of the buffer transfer concerning the aperiodic variable Va_i in a BA's aperiodic window. The response time associated to an aperiodic buffer transfer includes the following three components:

1. The time elapsed between t_0 and the time instant when the requesting station is able to indicate the BA (via RP_DAT, with the request bit set) that there is an aperiodic transfer request pending. We define this time interval as the dead interval of a producer station.
2. The time that the request indication stays in the BA's urgent queue till the related ID_RQ/RP_RQ pair of frames is processed in an aperiodic window.
3. And the time the buffer exchange request for variable Va_i stays in the BA's ongoing aperiodic queue till the related ID_DAT_Ap/RP_DAT pair of frames is processed in an aperiodic window.

34.5.5.6 Upper Bound for the Dead Interval

The upper bound for the dead interval in a station k is related to the smallest scanning period of a produced variable in that station. It is important to note that a periodic variable (Vp_i) is not polled at regular

TABLE 34.7 BAT (using EDF) for example of Table 34.5

	Microcycle					
	1	2	3	4	5	6
bat [*A*, *cycle*]	1	1	1	1	1	1
bat [*B*, *cycle*]	1	0	1	0	1	0
bat [*C*, *cycle*]	1	0	0	1	1	0
bat [*D*, *cycle*]	0	1	0	1	0	0
bat [*E*, *cycle*]	0	1	0	0	0	1
bat [*F*, *cycle*]	0	0	1	0	0	1

intervals since there is a communication jitter inherent to the BAT setting. Therefore, the upper bound for the dead interval in a station k is

$$\sigma^k = Tp_j + J_{Vp_j} + Cp_j, \quad \text{with } Vp_j : Tp_j = \min_{Vp_i \text{ produced in } k} \{Tp_i\} \tag{34.3}$$

where J_{Vp_j} is the maximum communication jitter of Vp_j.

The following assumption is inherent to Equation 34.3: a local aperiodic request is only processed (setting the request bit in the *RP _ DAT* frame) if it arrives before the start of the related *ID _ DAT*. Hence, the term Cp_j is included in Equation 34.3.

34.5.5.7 Aperiodic Busy Interval

The worst-case response time for an aperiodic variable transfer occurs if, when the request is placed in the BA's urgent queue (σ^k after t_0), the queue already has requests for all the other aperiodic variables in the network. We consider that

1. For each aperiodic variable, a request for identification must be made, and thus the network load is maximized
2. And that those requests will begin to contend for the medium access at the start of the macro-cycle: *critical instant.*

We also consider that all aperiodic traffic has a minimum inter-arrival time between requests, which is greater than its worst-case response time. Therefore, no other aperiodic request appears before the completion of a previous one. Hence, the maximum number of aperiodic requests pending in the BA is *na*, *na* being the number of aperiodic requests (different station can require an aperiodic buffer transfer of a same variable) that can be made in the network.

We define the time span between the critical instant and the end of the processing of aperiodic requests that are pending at the critical instant as an *aperiodic busy interval* (ABI) since all aperiodic windows within the microcycles are used to process aperiodic traffic.

It is also clear that to process all those *na* requests, the aperiodic windows will perform alternately sequences of (*ID_RQ/RP_RQ*) and (*ID_DAT/RP_DAT*), as the BA gives priority to the ongoing aperiodic queue (see Figure 34.2). If all the aperiodic variables have a similar length, Ca^* may be defined as the maximum length of all the (*ID_RQ/RP_RQ*) and (*ID_DAT/RP_DAT*) transactions concerning the aperiodic traffic. Therefore, the number of transactions to be processed during the ABI is $2 \times na$, corresponding to the set of *ID_RQ/RP_RQ* transactions and the set of *ID_DAT/RP_DAT* transactions.

With these assumptions, the analysis for the worst-case response time for the aperiodic traffic is as follows.

34.5.5.8 Worst-Case Response Time

The worst-case response time of aperiodic transfers is the function of the network periodic load during the ABI since it bounds the aperiodic windows length. The length of the aperiodic window in the lth cycle ($l = 1, \ldots, N, N+1, \ldots$) may be evaluated as follows:

$$aw(l) = \mu Cy - \sum_{i=1}^{np} (bat[i,l] \times Cp_i) \tag{34.4}$$

Therefore, the number of aperiodic transactions that fit in the lth aperiodic window is

$$nap(l) = \left\lfloor \frac{aw(l)}{Ca^*} \right\rfloor \tag{34.5}$$

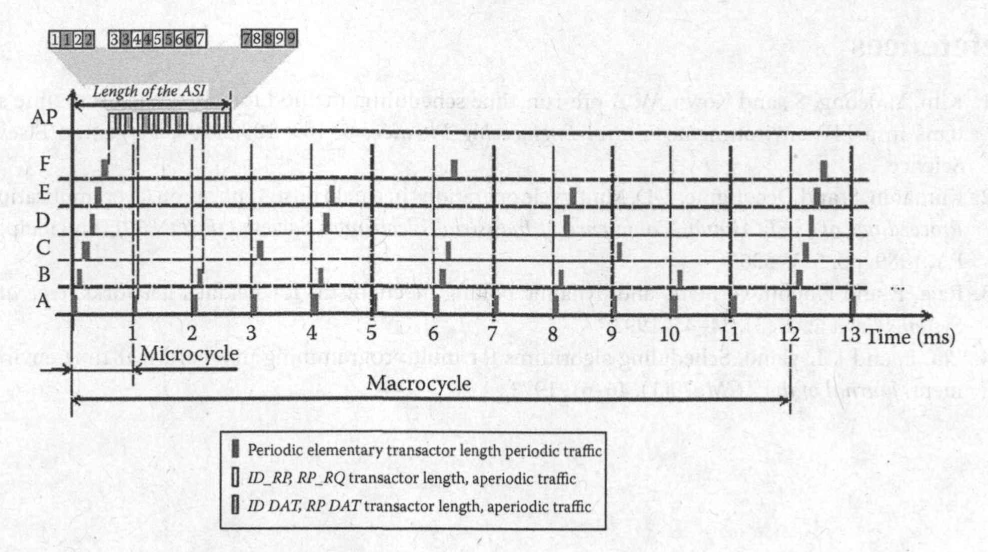

FIGURE 34.10 Example of aperiodic traffic schedule.

The number of microcycles (N') in an ABI is then

$$N' = \min\{\Psi\}, \quad \text{with } \Psi = \sum_{l=1}^{N'} nap(l) \wedge \Psi \geq 2 \times na \tag{34.6}$$

that is, the minimum number of microcycles within which the number of available "slots" (each "slot" with the length of Ca^*) is at least $2 \times na$.

Knowing N', the length of the aperiodic busy interval (len_abi) may be evaluated as follows:

$$len\_abi = (N'-1) \times \mu Cy + \sum_{i=1}^{np} (bat[i,N] \times Cp_i) + \left(2 \times na - \sum_{l=1}^{N'-1} nap(l)\right) \times Ca^* \tag{34.7}$$

where

$\Sigma_{i=1,\ldots,np}(bat[i,N] \times Cp_i)$ gives the length of the periodic window in microcycle N'

$(2 \times na - \Sigma_{l=1,\ldots,N'-1} nap(l)) \times Ca^*$ gives the length of the aperiodic window, with respect to the aperiodic busy interval, also in microcycle N'

Therefore, the worst-case response time for an aperiodic transfer requested at station k is

$$Ra^k = \sigma^k + len\_abi \tag{34.8}$$

and the maximum admissible interval between consecutive aperiodic requests of an aperiodic variable in a station k is

$$MIT(Va^k) \geq Ra^k = \sigma^k + len\_abi \tag{34.9}$$

34.5.5.9 Example of Aperiodic Traffic Scheduling

Assume that a system configured for six periodic variables (Table 34.1) must also support nine aperiodic buffer exchanges. Assume also that the length of each Vp_i, $\forall_i$, is $Cp_i = Cp = 0.0976$ ms, and that for sporadic traffic $Ca^* = 0.1$ ms. If the BAT is implemented as shown in Table 34.2, transactions concerning the nine aperiodic variables are scheduled as shown in Figure 34.10.

References

1. Kim, Y., Jeong, S., and Kown, W. A pre-run-time scheduling method for distributed real-time systems in a FIP environment. *Control Engineering Practice*, 6, 103–109, 1998, Pergamon/Elsevier Science.
2. Kumaran, S. and Decotignie, J.-D. Multicycle operations in a field bus: Application layer implications. *Proceedings of IEEE Annual Conference of Industrial Electronics Society (IECON'89)*, Philadelphia, PA, 1989, pp. 531–536.
3. Raja, P. and Noubir, G. Static and dynamic polling mechanisms for fieldbus networks. *Operating Systems Review*, 27(3), 34–45, 1993.
4. Liu, L. and J. Layland. Scheduling algorithms for multiprogramming in a hard real-time environment. *Journal of the ACM*, 20(1), 46–61, 1973.

35 Foundation Fieldbus

35.1 Introduction 35-1
35.2 Foundation Fieldbus Overview 35-1
35.3 Topology 35-2
35.4 Drivers (DD, EDDL, and FDT/DTM) 35-3
35.5 Cables 35-3
35.6 Segment Design 35-4
35.7 FFPS—Fieldbus Power Supplies 35-5
35.8 Installation of Segment in Safe Areas 35-6
35.9 Installation of Segments in Classified Areas 35-7
FISCO—Fieldbus Intrinsically Safe COncept • High-Energy Trunk–Fieldbus Barrier Solution
35.10 Project Documentation 35-8
35.11 Installations and Commissioning 35-8
35.12 Maintenance 35-10
References 35-10

Carlos Eduardo Pereira
Federal University of Rio Grande do Sul

Augusto Pereira
Pepperl-Fuchs

Ian Verhappen
Industrial Automation Networks Inc.

35.1 Introduction

This chapter describes the foundation fieldbus (FF) protocol, proposed by the Fieldbus Foundation organization and standardized as IEC norm [IEC00a,IEC00b,VP02,PF03,PN09]. This protocol, which has a strong influence from the World FIP protocol and is mainly used in process automation applications [BE01], has as its main characteristic a full distribution of control functions, which can run on processing units embedded on sensors and actuators. In the open, nonproprietary Foundation architecture, each field device has its own "intelligence" and communicates via an all-digital, serial, two-way communications system. For that purpose, standardized function blocks, such as PID control, analog and digital inputs and outputs, etc., are available. This section gives an overview on the physical layer (Manchester-encoded bus-powered communication), the communication stack capabilities, and on the "user layer" (on which function blocks are interconnected and control strategies are defined).

35.2 Foundation Fieldbus Overview

Every FF field device is capable of executing those functionalities encapsulated in function blocks that are the basic elements for programming FF applications and executing control and automation algorithms, such as PID control, arithmetic operations, analog and digital input/output, etc. Another major characteristic of the FF protocol is its advanced diagnostic capability, which defines not only the health of the device but now also in a standard format that can be easily integrated into asset-management systems.

In order to guarantee that messages exchanged between function blocks residing on different devices are successfully transferred to their destinations within precise and bounded temporal intervals, FF relies on schedule timetables. These can be generated online or off-line and are used by the link active scheduler (LAS). The LAS manages all message transfers between different field devices on a bus and it is also in charge of starting the execution of function blocks in each device. Due to its importance, in case of failure of the device running the LAS, another previously assigned device will automatically assume responsibility for message scheduling. The LAS is an important concept toward achieving a deterministic real-time behavior and overcoming common deficiencies of fieldbus technologies, in relation to timing constraints, high-processing load, and communication schedulability [FA00].

Fieldbus network traffic is periodic and a single interaction of a schedule in a link is called a macrocycle. During the scheduled communications portion of the macrocycle, all the data related to process control strategies are transmitted. One part of the macrocycle is used for periodic data (cyclical communications) publishing (external links and control data) and another part is used for acyclic background activities (supervision and network management messages). Periodic data are considered high priority and their timing requirements must be respected during the execution of the control application. Asynchronous requests (both initiated by devices or by external users) have lower priority and are only transmitted between the transmissions of periodic data if and only if they do not delay the periodic communications. Usually, the complete supervision information may require several macrocycles to be communicated from the field devices to the supervision workstations. This kind of information is nonperiodic.

The LAS selects between periodic data and acyclic messages using two types of tokens. Periodic data are requested using message frames named compel data (CD), which are sent to the target publisher nodes (scheduled token to periodic data) according to the schedule timetable. Once the publisher node receives the CD, it places its data on the bus using a data transfer (DT) datagram and any devices on the network can then read the message (publish–subscriber pattern).

In portions of the communications cycle occupied by neither the transmission of CD nor DT, the LAS sends an unscheduled token called pass token (PT) to one of those devices included in a live list (a list with all devices that are currently connected to the local link). Each unscheduled token has a bounded interval, during which the receiving node can use the bus to transmit a pending message until no further nonperiodic messages exist or the interval expires. This bounded interval precludes that the nonperiodic message exceeds the gap before the next schedule transmission. When no waiting nonperiodic message or unscheduled interval expires, the receiving device sends the token back to the LAS by means of a return token (RT).

An FF application may include several process control strategies. One stand-alone function block or many function blocks linked to other function blocks are used to implement each process control strategy designed by the users. These function blocks consist of a set of data and algorithms that are intended to provide a given functionality to the application. They may represent a control or I/O function block. Examples are PID controllers, input and output blocks, transducer blocks, etc. Currently, there are commercially available devices that can contain up to 19 function blocks. Each specific device runs the function blocks at a specific speed, depending on its internal computing resources (processor, memories, etc.). Consequently, the functionality provided by a given function block may have different execution times depending on each device in which it is executed. This aspect should be taken into account by the scheduler by reading this information from the DD file.

35.3 Topology

The FF specification defines two levels of networks: H1, network for control with 31.25 kbps; and HSE, data highway with 100 Mbps. H1 is a Manchester bus–powered and is usually applied at the field level and can also used in ex-applications with safety barriers. The original specification includes a higher speed serial communications protocol that was called H2 but it has never been developed. The high-speed Ethernet (HSE) bus is adopted at higher levels, allowing an easier connection to Ethernet-based control networks.

Four device classes are specified for HSE networks: (1) host devices (HD) are PCs or controller devices with an Ethernet port, (2) a link device (LD) allows an connection from an Ethernet network to multiple H1 segments, (3) foreign I/O gateways are components for integrating third-party fieldbuses, and (4) Ethernet devices (ED) that allow a direct integration to Ethernet networks.

35.4 Drivers (DD, EDDL, and FDT/DTM)

The majority of today's "smart" analog instruments use EDDL (electronic device description language), often shortened to DD (device description) in conversation. EDDL is the base technology underlying the HART, Profibus PA, and FF protocols by defining a significant part of their user layer. DD files for all approved FF devices are accessible from the Fieldbus Foundation Web site. Each device has a unique set of three or four files in the zip file associated with the device. All of these files except the CFF, or capabilities file, which is in plain ASCII are binary files.

The original EDDL specification as defined by IEC 61804 Part 2 contains little support for graphics; thus, the interface and degree of integration between field devices and hosts was not only limited, but the level of integration and information that could be presented to a user varied between each installation depending on the host being used.

One technology that was developed to attempt to overcome these limitations was FDT/DTM (field device tool/device-type manager). The FDT/DTM technology uses EDDL as the basis for its definitions, so that device manufacturers can define the "look and feel" of how the user accesses the device information for other than process variable (PV) and related signals, such as those used to configure and calibrate the device.

35.5 Cables

The Fieldbus Foundation uses a pair of wires to supply the DC voltage and the communication signal of the protocol. This cable is normally shielded to minimize noise on the segment. As shown in Table 35.1, the fieldbus standard defines four possible cable types. The most commonly used cable is type "A." When type "A" cable is used, networks, or segments as they are commonly called, with total combined length (sum of the trunk plus all spurs) of up to 1900 m are possible without the use of repeaters. The specification does, however, allow for up to four repeaters to be used on a single segment.

If the designer chooses another cable type than type "A," the available segment length will be reduced. For example, if a type "B" cable (twisted pair with overall shield) were to be used, the total cable budget available is reduced to 1200 m. Experience has shown that Ohm's law is often the determining factor in the limitation on overall length, so an effective way to be able to increase the length of a network to these maximums is to use a larger diameter cable with less resistance.

The IEC 61158-2 specifies in Section 11.7.2 that "The cable used for testing Fieldbus devices with a 31.25 kbps voltage-mode MAU for conformance to the requirements of this part of IEC 61158 shall be a single twisted-pair cable with overall shield meeting the following minimum requirements at 25 deg C… subsection d) maximum d.c. resistance (per conductor) = 24 Ohm/km." Then later in the

TABLE 35.1 Foundation Fieldbus Cables

Type	Description	Size	Maximum Length
A	Shielded, twisted-pair	#18 AWG (0.8 mm^2)	1900 m (6232 ft.)
B	Multiple, twisted-pair with shield	#22 AWG (0.32 mm^2)	1200 m (3963 ft.)
C	Multiple, twisted-pair without shield	#26 AWG (0.13 mm^2)	400 m (1312 ft.)
D	Multi-core, w/o twisted pairs and having overall shield	#16 AWG (0.1.25 mm^2)	200 m (656 ft.)

specification, B.1 Cable description and specifications, "The preferred Fieldbus cable is specified in 11.7.2 for conformance testing, and it is referred to as type 'A' Fieldbus cable." Therefore, it is possible to, as we just described, use larger diameter cable and remain in compliance with the standards.

To reduce the possible confusion in this area, the Fieldbus Foundation has recently released specification FF-844 "H1 Cable Test Specification." When manufacturers submit their cable for testing against this standard, they can obtain an FF "check mark" for compliance and this will provide the end user a high level of confidence that when properly installed, this cable will result in reliable fieldbus communications. The FF-841 standard supports the use of single and multi-pair cables for fieldbus H1 communications.

35.6 Segment Design

One of the unique features of the FF technology is that it allows "control in the field." However, to do so, just like was the case with pneumatic control, all the elements of the control loop must reside on the same segment or cable. Because the analog input (AI) and analog output (AO) devices must be on the same cable for "control in the field," even if you are not planning to implement this feature from the beginning of the project, the segments should be designed so that you can do so in the future without needing to physically rearrange your network.

When starting the segment design component of a FF project, you should have on hand the following minimum information:

- Plant layout—for the approximate position of the instruments and the routing of the cable trays in 3D, to determine approximate cable lengths.
- Process and instrumentation diagram (P&ID)—for the assignment of appropriate field devices to the same segment.
- Area classification drawing—because it has an impact on the type of power conditioners that can be selected and used.

Other decisions that you will need to make prior to starting the design include

- Maximum number of points per segment
- Protection method to be used. Will it be intrinsic safety (IS) or its equivalent, explosion proof or other means
- Number of spares for future expansion
- Basic network design (tree/chicken foot) or some other arrangement

Figure 35.1 shows the tree or chicken foot arrangement, which is the most widely used design basis for fieldbus networks.

The configuration has a "trunk" defined as the wire pair between the two terminators, with spurs to each of the individual field devices.

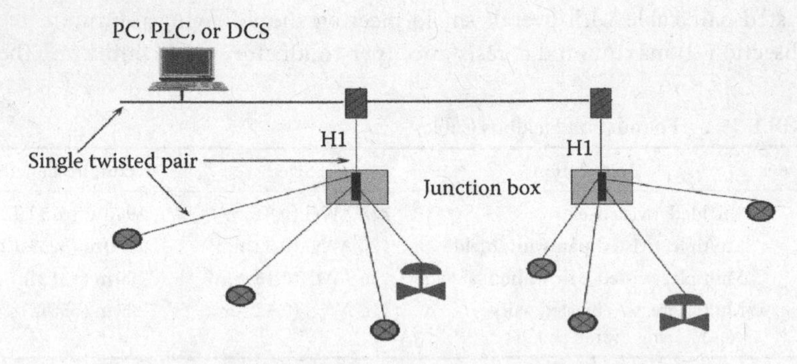

FIGURE 35.1 Foundation fieldbus network topology.

Many designs also endeavor to take into account the criticality of the individual field devices, so that in the unlikely event of a communications failure, the impact to the control system is minimized. It is recommended that you follow similar rules as are being used to manage this risk to determine the assignment of devices to I/O cards today. By using these guidelines, the impact on a single component failure on a design will be similar or less than for an analog control loop, especially if you have "control in the field." With control in the field, as long as the field devices have power, it is possible for them to continue to control at the last set point, though there will be "loss of view" by the DCS and process operator. Fieldbus is also designed to "fail gracefully" with the additional options on loss of contact with the host of holding last position, or moving to a mechanical fail position.

The following formula is used to determine the voltage available at any device or node (V_d) in the fieldbus network.

$$V_d = V_p - \left[\sum I_d + I_{HH} + \left(I_{SC} - I_{Dmin}\right)\right] \times R$$

where

V_p is the voltage at power supply
I_d is the current required to power device
I_{HH} is the current required to power handheld unit(s), typically 10 mA
I_{SC} is the current load on segment when short circuit is activated
I_{Dmin} is the current demand of device on network with minimum current requirement
R is the resistance of cable

35.7 FFPS—Fieldbus Power Supplies

FF H1 networks require fieldbus power supplies (FBPS)/power conditioners because the fieldbus signal itself is normally between 0.75 and 1 V (approximately 9 mA) peak to peak. This Manchester-encoded signal changes every 16 or 32 ms, so a traditional 24 V bulk power supply would be working full time to absorb this signal to maintain a steady DC voltage. FBPS/conditioners typically source their power from a 24 V DC bulk power supply, as shown in Figure 35.2.

FBPS or power conditioners must comply with Fieldbus Foundation standard FF-831 "fieldbus power supply test specification." After a FBPS has passed the tests outlined in this specification, they will have the fieldbus "check mark" confirming their suitability for use on H1 networks.

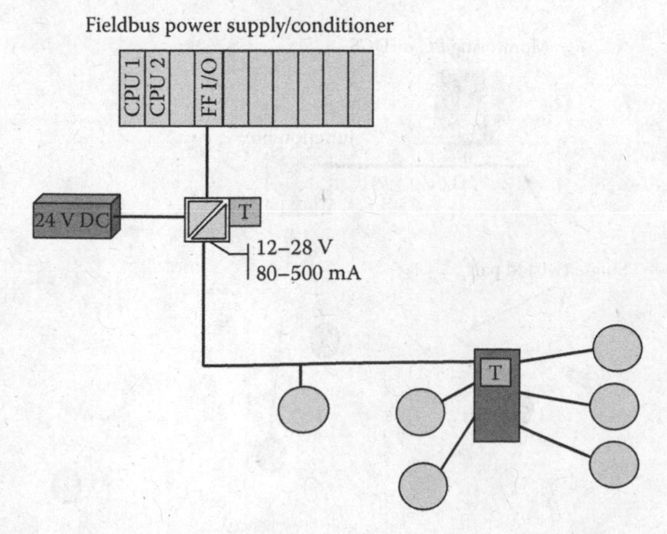

FIGURE 35.2 Foundation fieldbus power supply/conditioner.

The majority of today's FBPS have galvanic isolation between the fieldbus communication module to the host and the field segment where the instruments are installed as this prevents disturbances and noises from the field or host system and the bulk power supply, propagating from one system to another; thus, minimizing the risk of damage to the host I/O card.

FBPS are designed for a range of conditions, and as a result, the power conditioner may have voltage output between 12 and 28 V and 80–500 mA. The predominant factor affecting the available power range and associated voltage/current available is the area classification into which it is sourced if you wish to have live working on the full system, partial system (spurs only), or use gas testing and similar restrictions at all times when doing maintenance.

35.8 Installation of Segment in Safe Areas

The fieldbus specifications state that a maximum of 32 devices can be installed on a single H1 network; however, doing so would result in a very long macrocycle and higher risk than most end users would be willing to accept should something happen to communications on that port. Consequently, and in part because many host systems reach a parameter/port limit at this level, most designs use a maximum of 16 loop powered devices per segment.

Many projects therefore start their design basis with the maximum number of devices per network of 12 devices per segment. Many designs have a corollary design rule that of these 12 devices, no more than four control loops (valves) should be used.

Because fieldbus is wired in parallel, it is strongly recommended that the spurs have short-circuit protection for all spur connections. This is so that a short circuit on a spur or in an instrument will not cause a shutdown of the entire segment. Fortunately, almost all field device couplers have this capability.

Many projects also develop a number of "standard" fieldbus enclosure configurations with a fixed number of spurs and home run cables per junction box. This standardization makes the project design easier, simplifies the installation, and optimizes the number of items that need to be carried in stock. One other benefit is that many fieldbus device coupler manufacturers now also offer a standard range of "preconfigured" junction boxes, so that all the project needs to do is mount the enclosure, which has been specified and designed for the environment where it is to be installed and run the cables to the appropriate terminations. The drawing here (Figure 35.3) shows the minimum size junction box specified, which is a four-spur configuration.

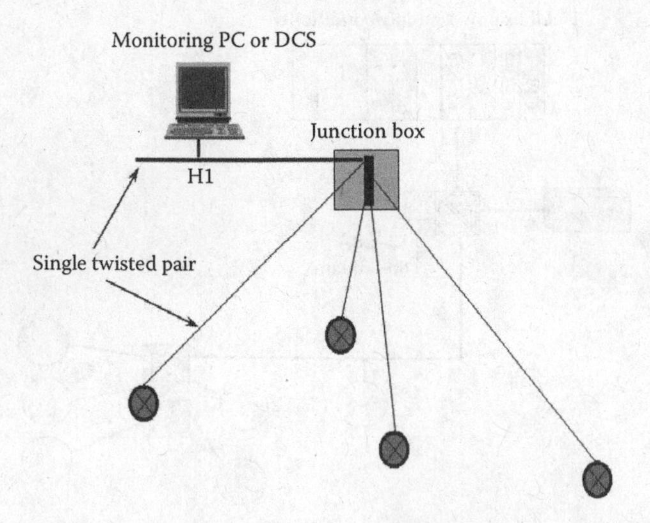

FIGURE 35.3 Fieldbus foundation installation in the field.

35.9 Installation of Segments in Classified Areas

Because FF requires that the device be connected to a network to perform maintenance, the majority of fieldbus installations with equipment in classified areas are using the IS technology. IS technology allows "live working," meaning all field work can be done "hot," with the network energized and communications with the host at all times. "Live working" means you can work without continuous gas testing and is therefore the safest way to protect your plant. An alternative to IS, or its equivalents as described below, is to use explosion-proof technology.

One advantage of fieldbus technology and digital communications, however, is that a significant portion of the work that in the past had to be done at the device for analog systems (i.e., range change) can now be completed over the network without actually having to go to the physical device in the field.

IS technology is based in the reduction of the energy to the field to levels below the ignition curves shown in Figure 35.4, plus a safety margin. For analog circuits, this is typically done using IS barriers; however, in the case of FF, this circuitry is either incorporated in the fieldbus power conditioner or in the fieldbus barrier mounted in the field junction box.

A typical FFPS delivers approximately 24–28 V DC and currents of up to 360 mA power to the field segment. This is well above all the explosion limit curves shown on Figure 35.4. FF offers two additional technologies other than an IS power supply that is limited to approximately 80–110 mA at around 12 V to address this problem. One solution is with IS-equivalent power supplies and the other is to reduce the available energy level in the field using the above-mentioned fieldbus barrier technology.

35.9.1 FISCO—Fieldbus Intrinsically Safe COncept

Unlike the IS entity concept, which is derived from theoretical calculations, the FISCO is based on actual field trials and experiments by the Physikalisch-Technische Bundesaanstalt (PTB) research center in Germany. Because this IEC-60079-15 standard is based on experimentation, all installations must operate within the maximum limits within which the experiments were conducted. These limits are that the total cable length of the system is limited to a maximum of 1000 m in IIC/Groups A, B gases and 1900 m in IIB/Groups C, D (limited by FF-831 and therefore identical to any fieldbus system). The maximum spur length for any FISCO installation is 60 m per spur.

Because the FISCO power supply must operate at voltages below the ignition curve, FISCO systems are not able to connect to trunk and cable combinations that are as long as non-IS systems. Despite this, FISCO is able to support trunk lengths greater than 500 m with full system current loads. It should be

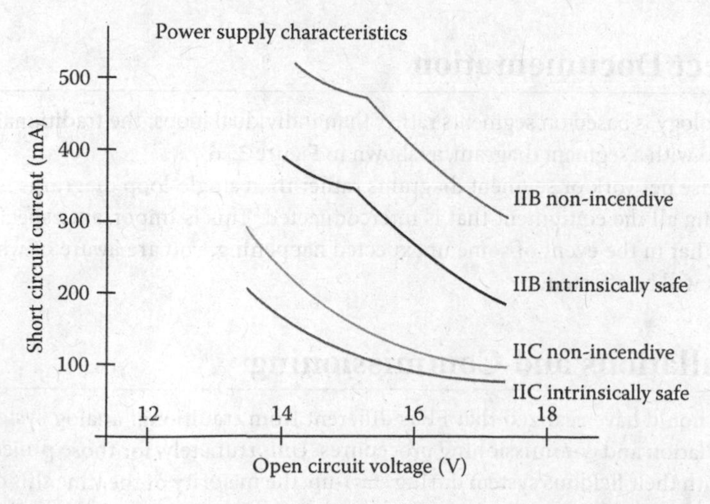

FIGURE 35.4 Power supply characteristics.

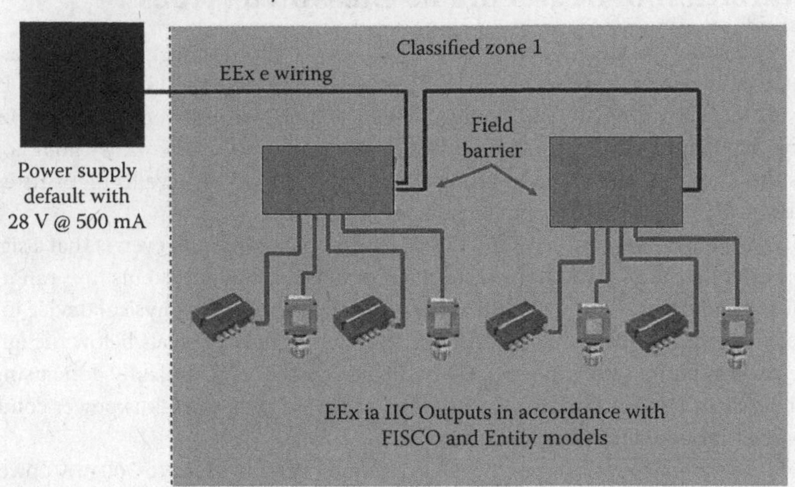

FIGURE 35.5 Intrinsically safe barriers.

noted, however, that because FISCO power conditioners are also repeaters, if desired it is possible to install up to four FISCO power conditioners in a single fieldbus network. Because FISCO uses IS-equivalent circuitry throughout, the resulting segment is similar to that shown in Section 35.6 with simple field device couplers in the field junction boxes.

35.9.2 High-Energy Trunk–Fieldbus Barrier Solution

For those cases in which longer trunk length is required without the need for live working of the trunk cable, the high-energy trunk–fieldbus barrier solution is often used. As mentioned earlier, in this configuration, the individual spurs from the fieldbus barrier are IS or IS equivalent as the energy-limiting circuitry is contained in the fieldbus barrier devices. As shown in Figure 35.5, this technology has a high-voltage "conventional" FF power conditioner at one end of the trunk, and then a number of spurs from each fieldbus barrier. Each of these spurs can be the full 120 m specified by the Foundation documents.

35.10 Project Documentation

Because FF technology is based on segments rather than individual loops; the traditional "loop diagram" should be replaced with a segment diagram, as shown in Figure 35.6.

The reason to use network or segment diagrams rather than single-loop diagrams is so that you can see on one drawing all the equipment that is interconnected. This is important especially when doing maintenance, so that in the event of some unexpected happening, you are aware of what other devices and control loops will be affected.

35.11 Installations and Commissioning

By now, readers should have realized that FF is different from traditional analog systems. The same is true for the installation and commissioning procedures. Unfortunately, for those projects that do experience troubles with their fieldbus system during start-up, the majority of the time this occurs is because of the fieldbus network. Fortunately, these errors are easily fixed through the use of good practices

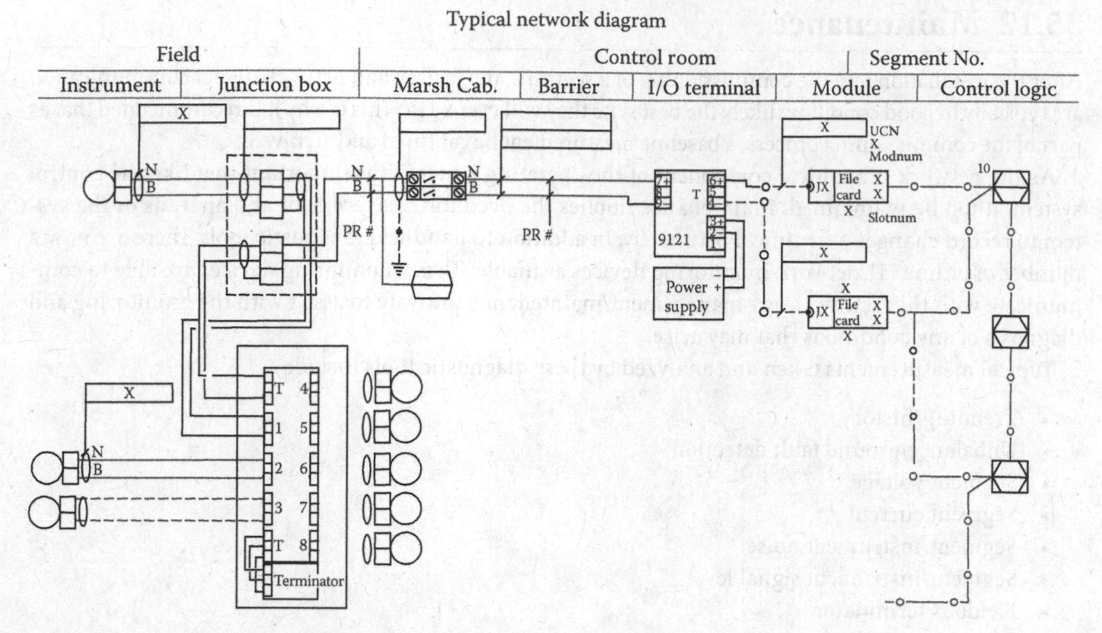

FIGURE 35.6 Typical project diagram.

during construction with simple procedures such as insuring that all connections are properly tightened and checking that wires are properly terminated, without short circuits and so on. However, below are some recommendations to help you get your system operational faster and start making profits.

1. Electrical commissioning
   a. Inspect the FF cable coil, check if the ends are protected against water leaks or rain. Measure the isolation and confirm if it matches the value of the commissioning report.
   b. Before connecting the home run cable to any electronics, confirm isolation and the capacitance measurements and record the values in the commissioning report.
   c. Repeat this procedure for each section of the trunk and for all its spurs and record the readings in the electrical commissioning report.
2. Electronic commissioning
   a. Connect the trunk to the coupler installed in the cabinet of the control system and in all the field junction boxes. Confirm the continuity of the conductor wires and the cable shield.
   b. Check the power for each input and output of the trunk of the field junction boxes; record the values in the commissioning report.
   c. Connect the instruments to the field junction boxes, taking care to isolate the shield at the instrument.
   d. For each spur, check the power at the field junction box terminals and at the instrument terminals; record these values in the report.
3. Configuration download
   a. Contact the people in charge of the configuration of the control system and ask for the download of the configuration to all the instruments in the segment.
4. Digital communication certification
   a. After the download of the configuration to all the instruments of the segment, confirm digital communications using a fieldbus diagnostic tool. Record results for each segment.
   b. Create a digital and paper file of the diagnostic tool reports for each segment and complete the delivery of the segment certification report to the plant commissioning team.

35.12 Maintenance

After the installation and the commissioning of a segment, the facility and network digital communication are typically in good condition; likely the best state they will ever be in. This is why it is recommended that as part of the commissioning process, a baseline measurement be captured and archived.

As the network is a critical component of the operating system, it is important that like the control system, it too be maintained. Maintenance implies the need for measurement and analysis of the system to record changes over time. Fortunately, in addition to handheld diagnostic tools, there are now a number of online H1 network-monitoring devices available. These monitoring devices are able to communicate with the facility's asset-management/maintenance software to assist with the monitoring and diagnosis of any conditions that may arise.

Typical measurements taken and analyzed by these diagnostic tools include

- Trending/history
- Unbalance/ground fault detection
- Segment voltage
- Segment current
- Segment/instrument noise
- Segment/instrument signal level
- Fieldbus terminator
- Segment and device communication error statistics
- Segment FF wave form

Thus, for the reasons stated above that your H1 network is part of your control system, it is strongly recommend that fieldbus projects include diagnostic tools in order to monitor the digital communication during the life cycle of the plant.

References

[BE01] Berge, J. *Process Fieldbuses: Engineering, Operation and Maintenance*. ISA Press, New York, 2001, ISBN 1-55617-760-7.

[FA00] Fayad, C. A. Process control performance on a foundation fieldbus system. *Proceedings, ISA Expo 2000*, Houston, TX, November 2000.

[HP07] Husemann, R. and Pereira, C. E. A multi-protocol real-time monitoring and validation system for distributed fieldbus-based automation applications. *Control Engineering Practice*, 15, 955–968, 2007.

[IEC00a] International Electrotechnical Commission (IEC) 61158.3. Fieldbus Specification, Data Link Layer Service Definition, Geneva, Switzerland, 2000.

[IEC00b] International Electrotechnical Commission (IEC) 61158.4. Fieldbus Specification, Data Link Layer Protocol Definition, Geneva, Switzerland, 2000.

[PF03] Pereira, C. E. and Franco, L. R. Real-time characteristics of the foundation fieldbus protocol. In: N. P. Mahalik (Ed.), *Fieldbus Technology: Industrial Network Standards for Real-Time Distributed Control*. Springer Verlag, Berlin, Germany, 2003, pp. 57–94.

[PN09] Pereira, C. E. and Neuman, P. Industrial communication protocols. In: S. Y. Nof (Ed.), *Springer Handbook of Automation*. Springer, Berlin, Germany, 2009, Chapter 56, pp. 981–999.

[VP02] Verhappen, I. and Pereira, A. *Foundation Fieldbus: A Pocket Guide, ISA—The Instrument, Systems and Automation Society*, ISA Press, Research Triangle Park, NC, 2002.

36

Modbus

36.1 Introduction 36-1
36.2 Modbus Interaction and Data Models 36-1
36.3 Modbus Protocol Architecture 36-3
36.4 Modbus Application Layer 36-4
Data Access Functions • Diagnostic Functions • Device Classes • Error Handling
36.5 Modbus Serial 36-8
Frames • RTU Mode • ASCII Mode • Error Detection • Physical Layer
36.6 Modbus TCP 36-12
Frames
36.7 Example 36-13
Acronyms 36-15
References 36-16

Mário de Sousa
University of Porto

Paulo Portugal
University of Porto

36.1 Introduction

The Modbus protocol was created in 1979 by Modicon* as a means of sharing data between their PLCs (programmable logic controllers). Modicon took the approach of openly publishing its specification and allowing its use by anyone without asking for royalties. These factors, along with the simplicity of the protocol itself, allowed it to become the first widely accepted *de facto* standard for industrial communication.

Although initially a proprietary protocol controlled only by Modicon, it is since 2004 controlled by a community of users and suppliers of automation equipment, known as Modbus-IDA. This nonprofit organization oversees the evolution of the protocol and seeks to drive its adoption by continuing to openly distribute the protocol specifications [1–4] and providing an infrastructure for device compatibility certification.

Currently, Modbus is used in equipment ranging from the smallest network-enabled sensors and actuators to complex automation controllers and SCADA (supervisory control and data acquisition) systems. Additionally, many implementations are freely available in source code form in a great variety of programming languages. Readers interested in obtaining the Modbus standards or one of the many available implementations of the protocol are suggested to access the Modbus-IDA Web site at http://www.mobus.org.

36.2 Modbus Interaction and Data Models

The Modbus communication protocol provides a means whereby one device may read and write data to memory areas located on another remote device. The typical example of a remote device is an RTU (remote terminal unit) providing a physical interface (inputs and outputs) to an industrial process, with

* At the time, Modicon was a PLC Manufacturer. It is now a brand of Schneider Electric.

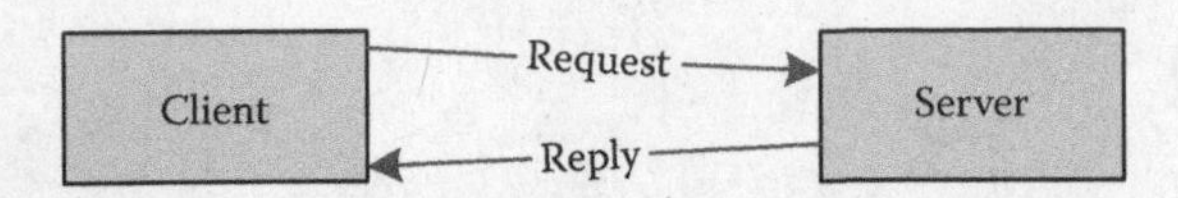

FIGURE 36.1 Client–server interaction model.

little to no processing capability, whereas the device taking the initiative to read and write data is usually a PLC. These two devices interact over the Modbus protocol using the client–server interaction model (Figure 36.1). The client (PLC) takes the initiative of asking the server (RTU) to write and read data that is present in the RTU. The server merely responds to these requests, never taking the initiative to start a data exchange.

A client may make requests to several distinct Modbus servers, usually one at a time. Likewise, the server may respond to requests coming from several clients. Whether or not the server is capable of processing simultaneous requests coming from several clients depends on the server, although typically these are processed one after the other.

The organization of the memory areas on the server to which the client may write to or read from is defined by the Modbus protocol itself. There are four distinct memory areas (Table 36.1); two of them are organized as 16 bit registers, whereas the other two are composed of arrays of single bits. Likewise, two of the memory areas provide read and write access permissions, while the other two may only be read from.

Typically, an RTU acting as a Modbus server will allow its physical digital inputs to be read through the "discrete inputs" memory area, the digital outputs to be read and written to through the "coils" memory area, the status of any analog inputs will be read using the "input registers" memory area, and the values of any analog outputs will be read and changed through the "holding registers" memory area. Nevertheless, it is completely up to the server manufacturer to decide the semantics attributed to the data located in these memory areas. In other words, the Modbus protocol does not define what will happen on the server when a specific memory address is written with some determined value. How the memory areas are organized, and where the data are actually stored, will depend on the type of device acting as a Modbus server.

An example would be a variable speed drive that controls the rotation speed of an electrical motor. When acting as a Modbus server, it may map the holding registers memory area to the area where the drive stores its configuration parameters (maximum and minimum motor voltage, motor speed, etc.), and the input registers memory area to the speed drive internal status registers (current speed, current output voltage, etc.).

Another example would be a graphical HMI (human–machine interface) terminal, which graphically displays a representation of the current state of an industrial process and permits an operator to modify operating parameters using specific buttons and dials. The HMI terminal, acting as a Modbus client, obtains the data values corresponding to the industrial process' state from the controlling PLC, which acts as the Modbus server. This PLC might use the holding register memory area to store the current state of a continuous variable of the process under control (e.g., volume of liquid in a tank), and the coil's memory area to store the flags indicating which on-screen buttons the operator is currently pressing.

TABLE 36.1 Memory Areas of the Modbus Data Model

Memory Area Name	Address—Range	Register Size (Bits)	Access Permissions
Input registers	1–65,536	16	Read
Holding registers	1–65,536	16	Read/write
Discrete inputs	1–65,536	1	Read
Coils	1–65,536	1	Read/write

The devices acting as Modbus servers may even decide to map several Modbus memory areas to the same internal physical memory. In these cases, the Modbus memory areas may overlap, and writing to a holding register, for example, may also change the state of the corresponding input register residing on the same physical memory.

Note, however, that although the Modbus Data Model specifies that elements in each memory area are addressed using values ranging from 1 to 65,536, unfortunately these address values are in actual fact encoded on the frames sent over the "wire" between the client and server as ranging from 0 to 65,535. This usually leads to many errors of "off by one," as the manuals of some device manufacturers specify how the Modbus memory areas are mapped onto to the device's internal architecture assuming the 1–65,536 addressing range, whereas other device manufacturers assume the 0–65,535 addressing range. When reading a manual one can never be sure which addressing range is being used, unless this is explicitly stated, which is rare. However, if the manual makes any reference to an address value of "0," then usually one may safely assume that the 0–65,535 addressing range is being considered.

36.3 Modbus Protocol Architecture

The Modbus protocol is organized as a two-layer protocol (Figure 36.2).

The upper layer, called the "Modbus application layer," defines the functions or services that a Modbus client may request of a Modbus server, and how these requests are encoded onto a message or APDU (application layer protocol data unit). It also defines how the servers have to reply to each function, and the actions they should take on behalf of the client. This layer is further explained in Section 36.4.

The lower layer defines how the upper layer APDUs are encapsulated and encoded onto frames to be sent over the wire by the underlying physical layer, and how the server devices are addressed. There are three distinct versions of this lower layer. The ASCII (American Standard Code for Information Interchange) version encodes the upper layer APDU as ASCII characters, whereas the RTU and TCP versions use direct byte representation. The RTU and ASCII versions are expected to be sent over EIA/TIA-232 or EIA/TIA-485 (commonly known as RS232 and RS485). The TCP version, as the name implies, is sent over TCP connections established over the IP protocol. Although it is common that the IP protocol frames are then sent over an Ethernet LAN (local area network), there is no reason that they may not use any other underlying network, including the global Internet.

As the RTU and ASCII versions are very similar, they are both described in the same Section 36.5. Section 36.6 will explain the TCP version.

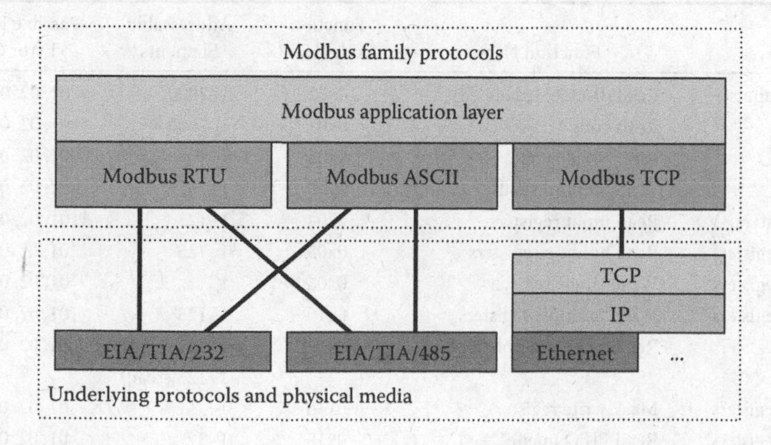

FIGURE 36.2 Organization of Modbus protocols.

36.4 Modbus Application Layer

The Modbus protocol defines three distinct APDUs (Figure 36.3) used in the Modbus application layer. All three APDUs start with a single-byte value indicating the Modbus function being requested or to which a reply is being made. Following the "function" byte value come all data and parameters of that specific function. Data and parameters have a variable number of bytes, depending on the function in question, and the number of memory registers that are being accessed.

All APDUs are, however, limited in size to a maximum of 253 bytes, due to limitations imposed by the underlying EIA/TIA-485 layer. The Modbus protocol also specifies that all 16 bit addresses and data items are encoded using big-endian representation. This means that when a numerical quantity larger than a single byte is transmitted, the most significant byte is sent first. Nevertheless, some device manufacturers allow the user to specify whether the device should use big or little-endian encoding.

The request APDU is sent by the client to the server. Upon successful completion of the desired function, the server replies with a response APDU. If the server encounters an error, the server notifies the client with an exception response APDU.

36.4.1 Data Access Functions

The Modbus protocol defines a large list of functions. The most often used functions are those associated with accessing the memory areas (Table 36.2).

All functions listed in Table 36.2, with the exception of functions 0x14, 0x15, 0x16, and 0x18, simply request that some data be read or written to a specific memory area. Function codes 0x05 and 0x06 are used to write to a single element (coil or holding register, respectively). The remaining functions allow the client to read from or write to multiple contiguous elements of the same memory area.

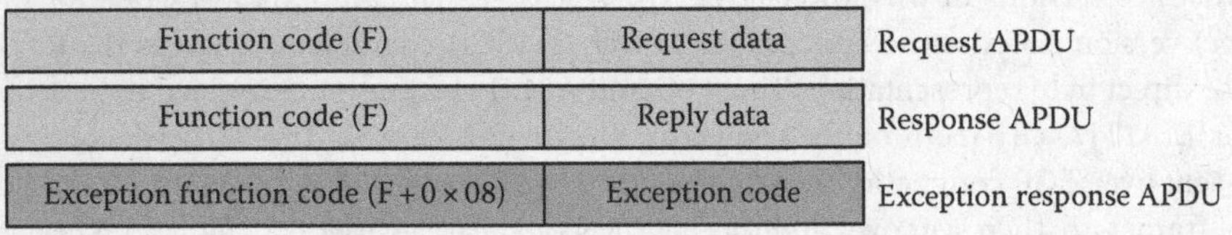

Function code (F)	Request data	Request APDU
Function code (F)	Reply data	Response APDU
Exception function code (F + 0 × 08)	Exception code	Exception response APDU

FIGURE 36.3 General format of APDU frames.

TABLE 36.2 Functions Used for Data Access

Memory Area	Function Name	Function Code (Hex)	Addressable Elements	Possible Response Error Codes
Discrete Inputs	Read discrete inputs	0x02	1–2000	01, 02, 03, 04
Coils	Read coils	0x01	1–2000	01, 02, 03, 04
Coils	Write single coil	0x05	1	01, 02, 03, 04
Coils	Write multiple coils	0x0F	1–1976	01, 02, 03, 04
Input registers	Read input registers	0x04	1–125	01, 02, 03, 04
Holding registers	Read holding registers	0x03	1–125	01, 02, 03, 04
Holding registers	Write single register	0x06	1	01, 02, 03, 04
Holding registers	Write multiple registers	0x10	1–123	01, 02, 03, 04
Holding registers	Read/write multiple registers	0x17	1–121 (write) 1–125 (read)	01, 02, 03, 04
Holding registers	Mask write register	0x16	1	01, 02, 03, 04
Holding registers	Read FIFO queue	0x18	1–32	01, 02, 03, 04
Files	Read file record	0x14		01, 02, 03, 04, 08
Files	Write file record	0x15		01, 02, 03, 04, 08

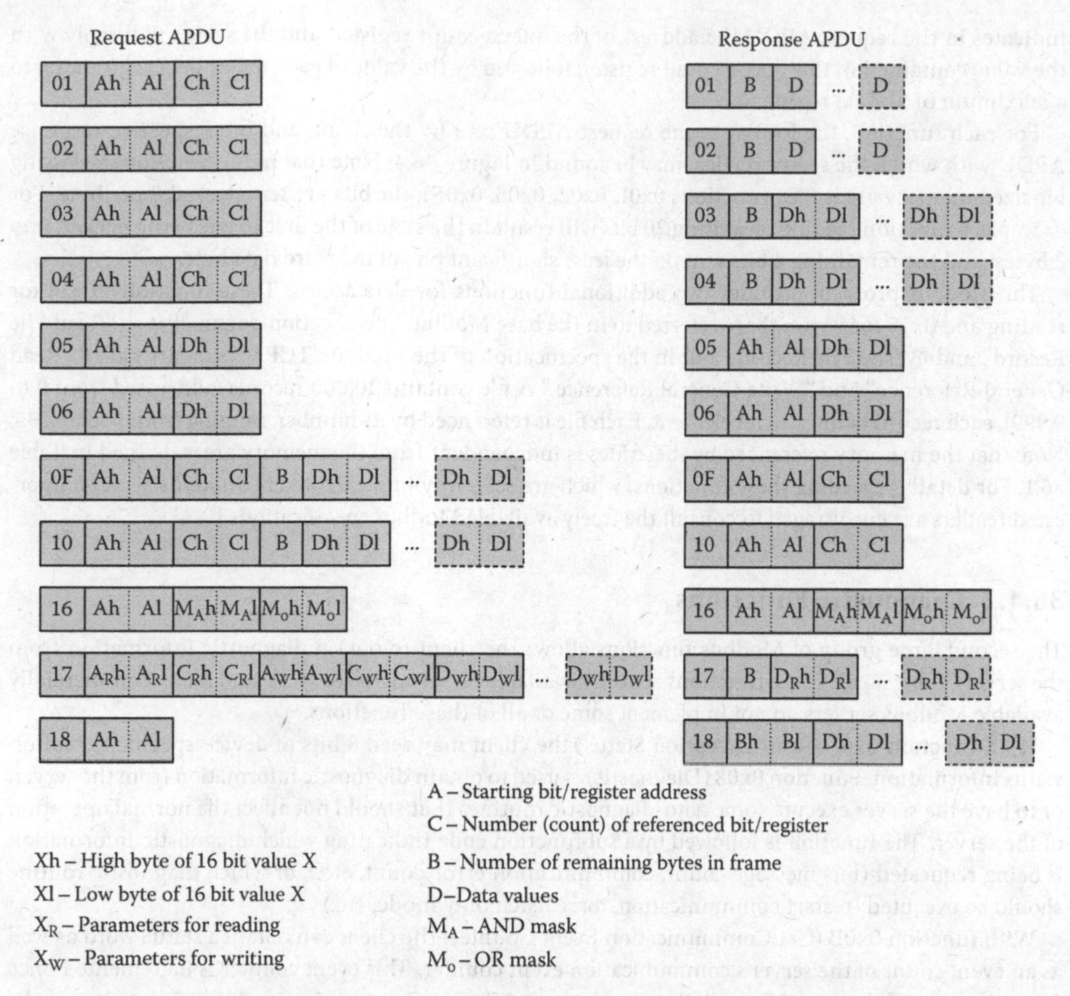

FIGURE 36.4 Format of data access functions APDUs.

Note that the maximum number of addressable elements is limited by the maximum size of the APDU. For example, for function code 0x0F, the request APDU starts off with 6 bytes containing the function code and respective parameters (see Figure 36.4), leaving a maximum of 247 (equal to 253 − 6) available bytes in which to send the data. The data, consisting of 1 bit elements, are packed 8 per byte, resulting in a maximum of 1976 (247 × 8) addressable elements.

Function 0x16 (Mask Write Register) changes the value stored in a single holding register. However, unlike the other functions, the new value is not obtained directly from the data sent in the APDU, but rather is obtained by applying a logical operation using the current value of the holding register with two masks, an AND mask and an OR mask, sent with the APDU.

$$\text{New\_value} = (\text{Current\_value AND And\_mask}) \text{ OR } (\text{Or\_mask AND (NOT And\_mask)})$$

Using this function the client can switch off certain bits and set other bits of the holding register, while simultaneously leaving the remaining bits unchanged, in a single atomic operation.

Function 0x18 (Read FIFO Queue) allows the client to request reading a FIFO queue from the server. The FIFO queue must be stored within the holding registers memory area in the server and consists of a first register containing the number of elements in the queue (queue count register), followed by the values of each element in the queue in the following registers (queue data registers). The client merely indicates in the request APDU the address of the queue count register, and the server will reply with the value contained in this queue count register, followed by the value of each queue data register, up to a maximum of 31 data registers.

For each function, the format of the request APDU sent by the client, and the respective response APDU with which the server replies, may be found in Figure 36.4. Note that in the functions accessing bit-sized memory areas (i.e., functions 0x01, 0x02, 0x05, 0x0F), the bits are sent packed 8 per byte. For example, a function reading or writing 20 bits will result in the state of the first 16 bits being packed into 2 bytes, and the remaining 4 bits sent on the least significant bits of the third data byte.

The Modbus protocol includes two additional functions for data access. These functions (0x14 for reading and 0x15 for writing) are referred to in the base Modbus specification documents as "Read File Record" and "Write File Record," but in the specification of the Modbus/TCP version are called "Read General Reference" and "Write General Reference." A file contains 10,000 records (addressed from 0 to 9,999), each record being a 16 bit element. Each file is referenced by its number, ranging from 0 to 65,535. Note that the memory referenced by these files is independent from the memory areas defined in Table 36.1. For details regarding these functions, which are seldomly implemented by Modbus servers, interested readers are encouraged to consult the freely available Modbus specifications [1–4].

36.4.2 Diagnostic Functions

The second large group of Modbus functions allows the client to obtain diagnostic information from the server (Table 36.3). These functions are only available on serial line devices, and many commercially available Modbus servers do not implement some or all of these functions.

With function 0x07 (Read Exception Status) the client may read 8 bits of device-specific exception status information. Function 0x08 (Diagnostic) is used to obtain diagnostic information from the server, or to have the server execute some auto-diagnostic routines that should not affect the normal operation of the server. The function is followed by a subfunction code indicating which diagnostic information is being requested (bus message count, communication error count, etc.), or which diagnostic routine should be executed (restart communication, force listen only mode, etc.).

With function 0x0B (Get Communication Event Counter) the client can obtain a status word as well as an event count of the server's communication event counter. This event counter is incremented once for each successful message completion. Function 0x0C (Get Communication Event Log) returns the same data as function 0x0B, plus the message count (number of messages processed since last restart) and a field of 64 event bytes. These event bytes contain the status result of the last 64 messages that were either sent or received by the server.

With function 0x11 (Report Slave ID), a client may obtain the run status of the server device (run/stop), a device-specific identification byte, and some additional device-specific 249 bytes of data.

The Modbus specification includes one additional function, 0x2B, with two subfunctions. One subfunction (0x0E) allows the client to obtain device identification (*Mandatory*: vendor name, product code, major and minor revision; *Optional*: vendor URL, product name, model name, user application name) all in ASCII string format. The second subfunction may be used to tunnel other communication protocols inside a Modbus connection.

TABLE 36.3 Modbus Function Codes for Diagnostic Purposes

Function Name	Function Code (Hex)	Possible Response Error Codes
Read exception status	0x07	01, 04
Diagnostic	0x08	01, 03, 04
Get communication event counter	0x0B	01, 04
Get communication event log	0x0C	01, 04
Report slave ID	0x11	01, 04

For more details regarding these functions, as well as their corresponding frame formats, the interested reader is once again encouraged to consult the Modbus specification [1].

36.4.3 Device Classes

Modbus server and client devices are classified into classes [3], depending on which of the above functions they implement.

Class 0 includes the minimum useful set of functions:

- Read holding registers (0x03)
- Write multiple registers (0x10)

Class 1 defines a more complete set of the most commonly used functions:

- Read coils (0x01)
- Read discrete inputs (0x02)
- Read input registers (0x04)
- Write single coil (0x05)
- Write single register (0x06)
- Read exception status (0x07)

Class 2 devices support all data transfer functions:

- Write multiple coils (0x0F)
- Read file record (0x14)
- Write file record (0x15)
- Mask write register (0x16)
- Read/write multiple registers (0x17)
- Read FIFO queue (0x18)

36.4.4 Error Handling

Error checking starts as soon as the server receives a request APDU. At this time it will start off by verifying the validity of the function code, address values, and data values (in this order) and, upon the first error encountered, will reply with an exception response APDU with error codes 1, 2, or 3, respectively (Table 36.4). After passing these initial tests, the actions corresponding to the function indicated in the request APDU are executed. If an error occurs during this processing, additional error codes may be generated and sent to the client using the exception response APDUs.

TABLE 36.4 Modbus Exception Codes

Code (Hex)	Name	Comments
0x01	Illegal function	
0x02	Illegal data address	
0x03	Illegal data value	
0x04	Slave device failure	
0x05	Acknowledge	Not commonly used. Indicates that the slave will reply later to the request.
0x06	Slave device busy	
0x08	Memory parity error	Used only in function codes 0x14 and 0x15.
0x0A	Gateway path unavailable	Only used by gateways.
0x0B	Gateway target device failed to respond	Only used by gateways.

36.5 Modbus Serial

A Modbus serial network comprises multiple devices connected to a physical medium (e.g., a serial bus). Due to the shared nature of the physical medium, a mechanism is required to regulate medium access. For this reason, Modbus serial follows the master–slave interaction model (Figure 36.5), with the Modbus client becoming the master, and the Modbus servers taking the role of slaves. The master is responsible for initiating the communication by sending requests to the slaves, one request at a time. These, in turn, reply to the master with the requested data. The request/reply exchange can be performed in one of two ways:

- *Unicast mode*: The master sends a request to a specific slave. The slave processes this request and replies to the master.
- *Broadcast mode*: The master sends a request to all slaves. The slaves process this request, but do not reply to the master.

The master only starts a request/reply exchange once the previous exchange has finished.

Each Modbus serial network may only have one master. The number of slaves is limited to 247. A device can be both master and slave, but not at the same time. Therefore, a device can change its role whenever appropriate. In practice, roles are usually static since the Modbus protocol itself does not specify how the changing of roles may be managed.

36.5.1 Frames

The request/reply exchange is performed using request and reply APDUs, which are transmitted within frames that always have the same structure (Figure 36.6):

- *Address*: This field contains the slave address. In a request frame, it identifies the destination device, while in a reply frame it identifies the sender. Each slave has a unique address in the range 1–247. Address 0 is used (by the master) for broadcasting messages. The range 248–255 is reserved.
- *Modbus APDU*: This field is the application layer PDU (Section 36.4).
- *CRC or LRC*: This field is used for error detection purposes. The content depends on the transmission mode (RTU or ASCII) being used. For details, see Section 36.5.4.

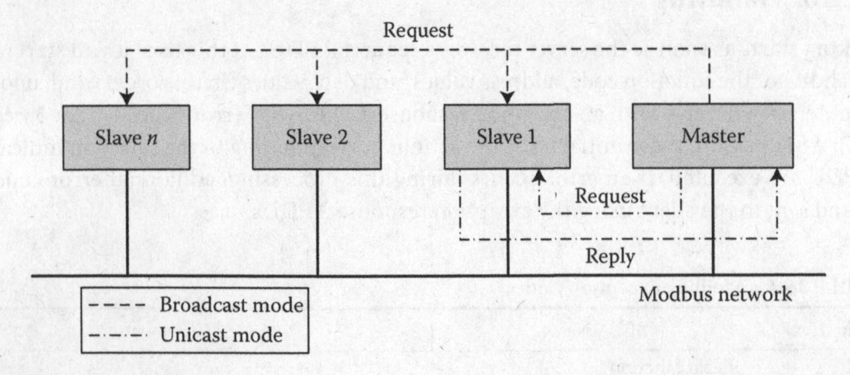

FIGURE 36.5 Request/reply exchange.

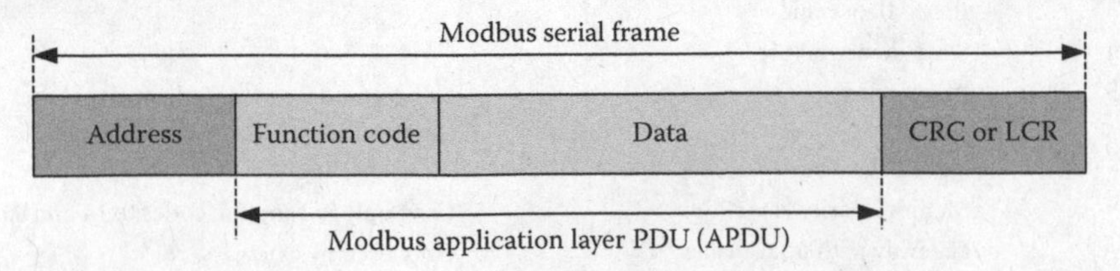

FIGURE 36.6 Modbus serial frame.

In Modbus serial, frames exchanged between master and slave devices can be transmitted in one of two modes: RTU or ASCII. Each mode defines different ways to code the frame contents (i.e., bytes) for transmission. What most distinguishes both modes are the smaller frames produced by the RTU mode. Consequently, for the same baud rate, the RTU mode achieves a higher throughput. There are, however, circumstances that may lead one to choose the ASCII mode. These will be discussed below.

All devices must implement the RTU mode, with the ASCII mode being optional. A device may support both modes, but never simultaneously. Within a network all devices must use the same transmission mode; the communication cannot occur otherwise since both modes are incompatible.

36.5.2 RTU Mode

The RTU mode uses an asynchronous approach for data transmission. Each byte, within a frame, is transmitted using an 11 bit character (Figure 36.7):

- 1 start bit (ST), used for the initial synchronization
- 8 bits, the data coded in binary with the least significant bit sent first
- 1 parity bit (PT), used for error detection
- 1 stop bit (SP), to ensure a minimum idle time between consecutive character transmissions

All devices must support the even parity checking method, but the user may choose any other method (even, odd, or no parity) on devices that support them. However, all devices within the same network must all be configured to use the same method. If the no parity method is used, then the number of stop bits must be 2. This guarantees that 1 byte will always be coded with 11 bits.

The absence of frame delimiters (header or trailer) may lead to synchronization errors during frame reception, i.e., the receiver may have difficulty in identifying the beginning or the end of each frame. To overcome this, it becomes necessary that the time interval between consecutive characters (within the same frame) does not exceed a certain value; otherwise the next character could be easily interpreted by the receiver as the beginning of a new frame. In the same manner, if consecutive frames are separated by a very small time interval, they could be interpreted as a single frame. In both cases, received data will be misinterpreted which will lead to communication errors. To avoid this issue, the transmission of a frame must meet the following timing requirements (Figure 36.8):

- Each frame must be transmitted, one character after the other with a minimum idle interval between them of 1.5 character times.* If this interval is exceeded, the receiver considers that the frame is incomplete and should discard it.
- Consecutive frames must be separated by an idle interval of at least 3.5 character times.

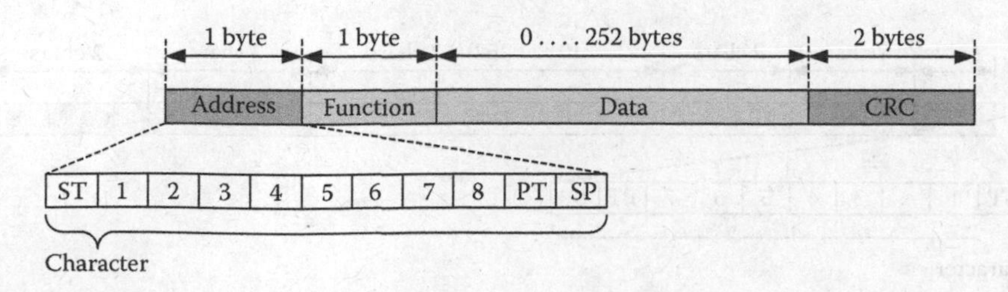

FIGURE 36.7 RTU frame transmission.

* Time necessary for the transmission of 1.5×11 bits.

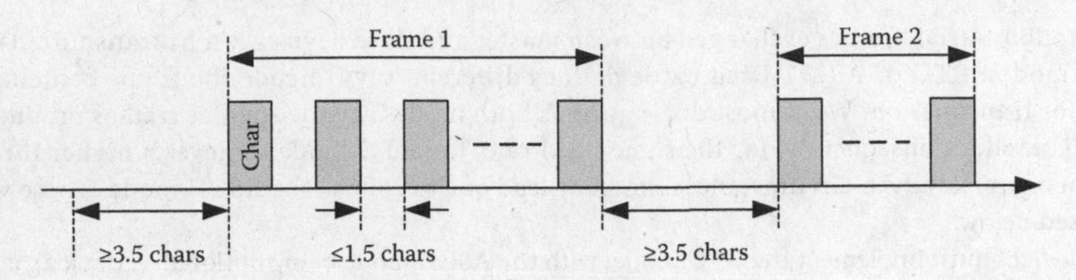

FIGURE 36.8 RTU timing requirements.

From this description, it is easy to understand that some care must be taken when implementing an RTU driver, particularly with regard to the accuracy of the timers used. Although efficient, the RTU mode requires a tight control of timing. For more details on implementation aspects, readers are suggested to consult [4].

36.5.3 ASCII Mode

The ASCII mode also employs an asynchronous transmission. In this case, however, each byte is transmitted as two ASCII characters (Figure 36.9).

The coding process is as follows: each 8 bit byte is divided into two nibbles of 4 bits each. One nibble contains the 4 most significant bits in the byte, and the other the 4 least significant bits. Each nibble is then coded using its hexadecimal representation on the ASCII alphabet. For example, a byte with value 0xA1 (or 161 in decimal) is transmitted as the characters "A" followed by "1." The character "A" is sent as the byte with value 0x41 which is the ASCII representation of the "A" character. Likewise, the character "1" is transmitted as its ASCII value of 0x31.

Each ASCII character is transmitted using 10 bits as follows:

- 1 start bit (ST)
- 7 data bits,* the ASCII character ("0"-"9," "A"-"F") with the least significant bit sent first
- 1 Parity bit (PT)
- 1 Stop bit (SP)

The requirements for parity checking are the same as those of the RTU mode.

Contrary to the RTU mode, a frame transmitted in the ASCII mode has a header and a trailer. The header identifies the beginning of a frame and is represented by the ":" ASCII character (0x3A). The trailer identifies the end of the frame, and comprises two ASCII characters: CR (Carriage Return—0x0D) and LF (Line Feed—0x0A), transmitted in this order.

The use of headers and trailers helps the reception process, since the receiver has only to wait for the ":" character to detect the beginning of a frame. Conversely, the end of a frame is detected when the

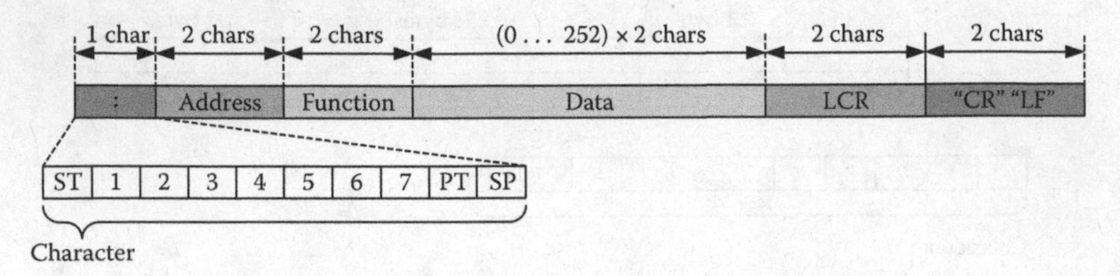

FIGURE 36.9 ASCII frame transmission.

* Each ASCII character is coded using 7 bits.

sequence "CR" "LF" occurs. Since this mode does not specify a maximum time interval between consecutive characters, this implies that this interval could theoretically assume any value. However, this could lead to a situation in which the receiver would block, e.g., when the sender crashes during a frame transmission, whereupon the receiver would wait indefinitely for the remaining characters. In order to prevent these situations, the receiver should implement a timeout mechanism. When a timeout occurs, the received data are discarded and the receiver waits for the next frame. In [4], it is suggested to use 1 s for the timeout value. However, it is possible to use other values depending on the network topology. Further details can be found in [4].

From the discussion, it is easy to understand that the ASCII mode is simpler to implement and does not pose any particular timing requirements. However, the price to pay is the doubling of the frame size when compared to the RTU mode.

36.5.4 Error Detection

Error detection mechanisms are located at two levels, namely, the character level and frame level.

At the character level, a parity checking method is used. The sender computes the parity of each character using the parity method chosen (even, odd, or no parity) and sets the parity bit accordingly. The receiver, using the same method of the sender, computes the expected parity bit, and compares it with the received parity bit. If they are different, an error has occurred and the entire frame is discarded. It is easy to see that the error detection capability of this method is quite limited. For example, when two bits within a character are flipped (due to a transmission error), the error will not be detected.

Error detection at the frame level uses a CRC (cyclic redundancy check) in the RTU mode or LRC (longitudinal redundancy check) in the ASCII mode. Although these methods are conceptually different, the methodology for their use is the same. The sender computes the CRC or LRC value using a predefined algorithm which is applied to the frame contents: Address + APDU fields. This value is then appended to the frame. The receiver computes the CRC or the LRC using the same algorithm, and compares the computed results with the received ones. If they are different, it concludes that the frame has errors and should therefore be discarded. The CRC value is computed through a polynomial division of the frame using $X^{16} + X^{15} + X^2 + 1$ as the generator polynomial. It produces a 16 bit value, which is transmitted with the lower byte first. The LRC value is computed using a modulo-2 sum (XOR-sum) of all bytes of the frame and produces an 8 bit value. A comparison between both methods shows that the CRC has a higher error detection capability. The CRC and LRC values are transmitted using the same methodology used for the remaining characters. Details about CRC or LRC computation can be found in [4].

36.5.5 Physical Layer

Modbus serial supports two physical layers: RS485 [5] and RS232 [6]. The RS485 consists of a multi-drop network able to support multiple devices (typically 32 devices per network segment). Several segments can be connected by using repeaters, thus increasing the maximum number of devices in the network. It supports 1 master (at a time) and up to 247 slaves, and it can be used for larger distances (up to 1000 m) with high baud rates. The RS232 is a point-to-point solution with just 1 master and 1 slave. It is used for short distances (<20 m) with low to medium baud rates.

All devices must support an RS485 interface, while the RS232 is optional. Both transmission modes (RTU and ASCII) can be used on either interface type. Moreover, all devices within a network must have the same configuration, comprising the following parameters: transmission mode (RTU or ASCII), parity type (even, odd, or no parity), and baud rate (bits/s). All devices must support 9.6 and 19.2 kbps rates, with the latter being the default value. Further details about the characteristics of each interface and their application to Modbus networks can be found in [4].

36.6 Modbus TCP

A Modbus TCP network consists of multiple devices connected through a TCP/IP network, interacting following the client–server model. A client sends a request to a server, which in turn responds to the client with the requested data. This transaction (request/response exchange) is performed by sending Modbus TCP frames through a TCP connection previously established between the client and the server. Connection establishment and management are handled by the TCP/IP protocol and occur independently of the Modbus protocol. A Modbus server listens on port 502 for requests from clients that wish to establish a new connection with the server. This port is presently reserved (and registered) for Modbus applications [2].

Some Modbus server devices may support multiple connections from distinct clients simultaneously. Similarly, clients may establish multiple connections to distinct servers. Additionally, the Modbus TCP protocol allows a client to send multiple Modbus requests to the same server over the same connection, even before the arrival of the replies to previous requests. A device may be both a client and server at the same time.

36.6.1 Frames

Modbus TCP frames (request and response) consist of a MBAP header (Modbus Application Protocol header) plus the application layer APDU (Figure 36.10).

The MBAP header comprises several fields:

- *Transaction identifier*: Since a client can issue several concurrent transactions over the same TCP connection, the responses are not guaranteed to arrive in the same order that the requests were sent. It is therefore necessary to have an identifier for each transaction in order to match the request and response frames. The client initializes this field when it performs a request. The server echoes this value in the response frame.
- *Protocol identifier*: This is used to identify the protocol; it currently always has the value 0.
- *Length*: This field indicates the size (in bytes) of the unit identifier field plus the APDU. Data transfer in a TCP connection is performed as a stream of bytes, which could lead to a situation where several frames are waiting to be read in the reception buffer. To identify frame boundaries in these situations, the frame length must be known.
- *Unit identifier*: This field is used to identify the destination device (a slave). It is used mainly by gateways between Modbus/TCP and Modbus serial networks, where the gateway, upon receiving a Modbus/TCP frame, needs to know the identification of the slave on the Modbus serial network that should receive that frame. Modbus/TCP devices that are not gateways usually ignore this field.

Unlike Modbus serial, Modbus TCP frames do not have an error detection field. This was considered unnecessary as the TCP/IP stack already includes several error detection mechanisms. Further details about the implementation of Modbus TCP can be found in [2].

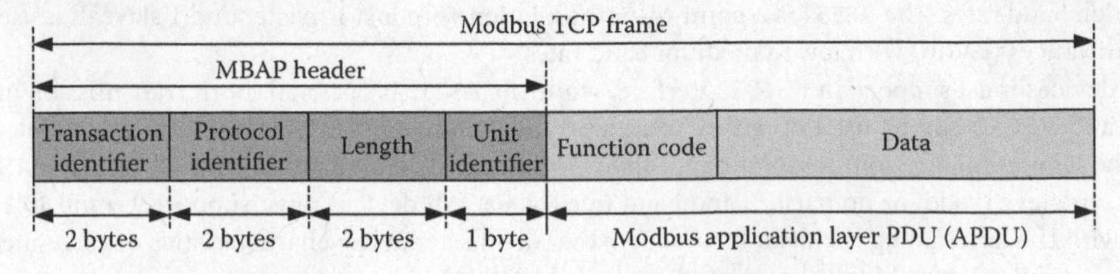

FIGURE 36.10 Modbus TCP frame.

36.7 Example

A typical example of a small-scale project using a Modbus network is the automation of a rope-making machine. This machine consists of two physically separated parts: a braiding component that winds the twines into a rope and a traction component that pulls out the built rope at the correct speed. Each component is driven by an electrical motor connected to a variable speed drive. The braiding component additionally contains several digital and analog sensors to detect possible faults (e.g., twine breaking) and digital and analog actuators (e.g., status signaling lights). The automation logic is handled by a single small PLC, and an HMI graphical terminal is used to interact with the operator (Figure 36.11). A SCADA graphical interface on the plant manager's computer also connects to the PLC over the Modbus TCP network in order to obtain manufacturing data (e.g., up-time, meters of rope produced).

In this project, the PLC communicates with the two variable speed drives over a Modbus serial network, with the PLC acting as the master and the drives acting as slaves. The digital and analog I/Os are handled by an RTU, which, along with the graphical terminal and the SCADA, connects to the PLC over a Modbus TCP network (TCP/IP over Ethernet). The PLC acts as a client to the Modbus server in the RTU and as a server to the Modbus clients in the graphical terminal and SCADA. Note that on the Modbus TCP network: (1) the PLC acts as both a client and a slave simultaneously, (2) three clients coexist on the same network, and (3) the server on the PLC is accessed simultaneously by two clients (the graphical terminal and the SCADA).

In order to show in more detail how the data are actually accessed remotely over the Modbus protocol the connection between the PLC and the RTU will be explained further. A typical example of a Modbus slave device is the RTU of the STB series of products by Schneider Electric. These RTUs are modular, consisting of a network adapter, followed by one or more physical interface modules. The following example (Figure 36.12) considers an RTU with a Modbus/TCP network interface (NIP 2212), followed by the following modules: power supply (PDT 3100), four digital inputs (DDI 3420), four digital outputs (DDO 3410), two analog inputs (AVI 1270), and two analog outputs (AVO 1250).

Although the Modbus protocol defines bit-sized memory areas, this particular RTU does not make use of them; in fact, it only uses the holding registers memory area. It maps the four digital inputs onto the four least significant bits of one 16 bit holding register, the four digital outputs onto the four least significant bits of another register, and each of the analog inputs/outputs onto their own register. Additionally, it uses some registers to store status information.

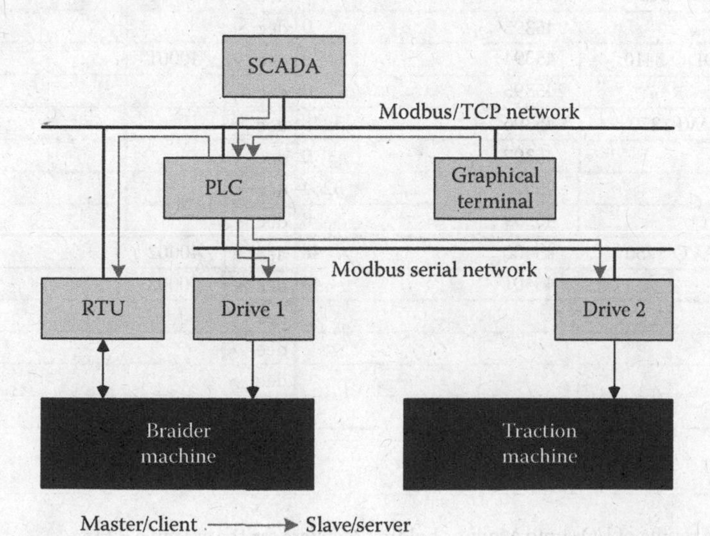

FIGURE 36.11 Example machine and automation equipment.

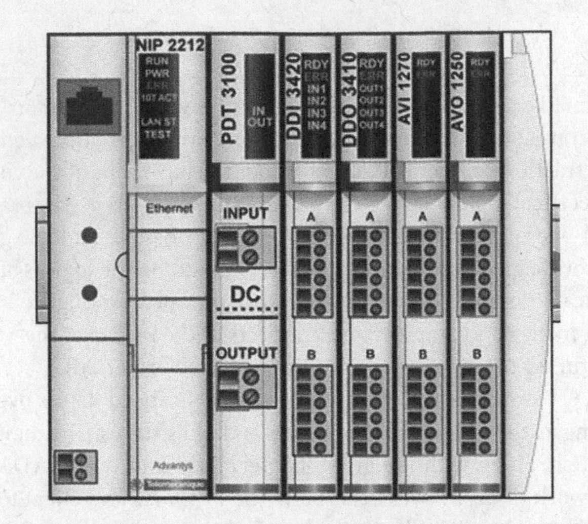

FIGURE 36.12 Physical aspect of RTU with configuration as described in the example.

Figure 36.13 shows a table with the exact mapping of I/Os and status data to the holding register memory area, for an RTU configured as in Figure 36.11. All addresses in this table start with the digit 4, which is the method used by this RTU's supplier to indicate that these refer to registers in the holding register memory area. The remaining digits represent the real register address.

The holding registers used are grouped into two contiguous areas. One area, starting at address 0001, will be used as an output memory area, i.e., the Modbus client may write new data to these registers, which will then be used by the RTU to drive the physical outputs. The data stored in the second contiguous area, starting at address 5392, will be updated by the RTU in such a way as to reflect the actual values in the physical inputs and outputs. Although the Modbus protocol allows a Modbus client to write new data values to

I/O data values

Node Number	Module Name	Input Address	Input Value	Format	Output Address	Output Value	Format
1	STB DDI 3420	45392	0	dec			dec
		45393	0	dec			dec
2	STB DDO 3410	45394	0	dec	40001	0	dec
		45395	0	dec			dec
3	STB AVI 1270	45396	3248	dec			dec
		45397	0	dec			dec
		45398	3264	dec			dec
		45399	0	dec			dec
4	STB AVO 1250	45400	48	dec	40002	0	dec
		45401	48	dec	40003	0	dec
				dec			dec
				dec			dec
				dec			dec
				dec			dec
				dec			dec

FIGURE 36.13 Mapping of I/Os onto Modbus holding registers for the example RTU.

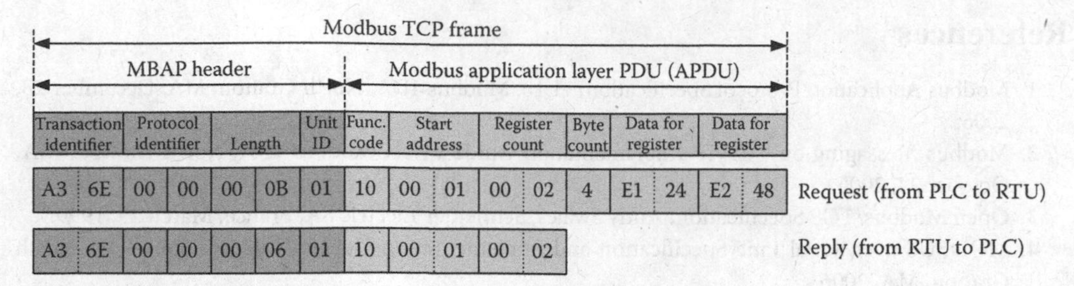

FIGURE 36.14 Example Modbus transaction between the PLC and the RTU.

holding registers, this particular RTU will ignore any attempt to write to registers in this second contiguous area. Registers in this second contiguous area should therefore be considered as being "read only."

As may be seen from this table, the four digital inputs of the first DDI 3420 module are mapped onto the holding register at address 5392. The least significant 4 bits of the following register at address 5393 will contain status information (short-circuit, power failure) related to these same inputs.

A Modbus master changes the outputs on the next module, the DDO 3410 with four digital outputs, by writing the desired state to the holding register at address 0001. This DDO 3410 module uses another two registers in the input area: the first (at address 5394) echoes the value stored in the register 0001, while the following (at address 5395) stores the actual status of the physical outputs.

The AVI 1270 module uses the registers 5396 and 5398 to store the input voltage of each analog input, while the other two registers, at addresses 5397 and 5399, store status information (over and under-voltage errors and warnings). The AVO 1250 uses the registers at addresses 0002 and 0003 to receive the voltage values that should be applied to the analog outputs, and the registers at 5400 and 5401 store status information (over and under-voltage errors).

An example of a possible exchange of Modbus message frames between the PLC and this particular RTU (functioning as Modbus server) is described in Figure 36.14. In this exchange, the PLC uses the Modbus function "Write Multiple Registers" to change the analog output values stored in registers 0002 and 0003 to the values 0xE124 and 0xE248, respectively. Note that the starting address of 0002 is sent on the wire as 0001 since this particular RTU follows the Modbus specification to the letter, and all documentation assumes that the memory areas start at address 1, which is then sent on the wire as 0.

Acronyms

APDU	Application layer protocol data unit
ASCII	American Standard Code for Information Interchange
CRC	Cyclic redundancy check
EIA	Electronics Industries Association
FIFO	First in first out
HMI	Human–machine interface
IP	Internet protocol
LAN	Local area network
LRC	Longitudinal redundancy check
MBAP	Modbus application protocol
PLC	Programmable logic controller
RTU	Remote terminal unit
SCADA	Supervisory control and data acquisition
TCP	Transmission control protocol
TIA	Telecommunications Industry Association

References

1. Modbus Application Protocol Specification, v1.1b, Modbus-IDA, North Grafton, MA, December 28, 2006.
2. Modbus Messaging on TCP/IP Implementation Guide v1.0b, Modbus-IDA, North Grafton, MA, October 24, 2006.
3. Open Modbus/TCP Specification, Andy Swales, Schneider Electric SA, France, March 29, 1999.
4. MODBUS over Serial Line Specification and Implementation Guide V1.02, Modbus-IDA, North Grafton, MA, 2006.
5. Electrical Characteristics of Generators and Receivers for Use in Balanced Digital Multipoint Systems, ANSI/TIA/EIA-485-A-1998.
6. Interface Between Data Terminal Equipment and Data Circuit-Terminating Equipment Employing Serial Binary Data Interchange, ANSI/TIA/EIA-232-F-1997.

37

Industrial Ethernet

37.1 Introduction 37-1
37.2 Industrial Ethernet 37-2
What Does Ethernet Mean? • What Does Industrial Mean? • Classification of Industrial Ethernet Solutions
37.3 Standardized Solutions of IEC 61158 and IEC 61784 37-5
EtherNet/IP • Foundation Fieldbus High-Speed Ethernet • SERCOS III • "Exotic Solutions": EPA, Tcnet, Vnet/IP, PNET on IP
37.4 Features of Major Industrial Ethernet Solutions 37-9
37.5 Synthesis 37-9
Abbreviations 37-9
References 37-10

Gaëlle Marsal
EDF Research and Development

Denis Trognon
EDF Research and Development

37.1 Introduction

The distribution of automation functions in instrumentation and control (I&C) architectures arose in the late 1970s with proprietary network solutions designed to fit an application's performance. To ensure interoperability between different products from different vendors, the IEC 61158 standard has been created. However, due to manufacturers' pressure, this lead only to a collection of several different specifications corresponding to each vendor's commercial offer.

Meanwhile, Ethernet (IEEE 802.3 [12]) became the standard solution in desktop communication systems that implements the first two layers of the OSI Model (see Chapter 1) while TCP/IP protocols (RFC 791 [15] and 793 [16], cf. Chapter 61 of this book) did so for layers 3 and 4.

Until the late 1990s, Ethernet could not be used at the plant floor because of its nondeterministic medium access control method, called CSMA/CD (cf. Chapter 2). Even though the apparition of full-duplex links and switching technology turned Ethernet into a candidate for plant floor communication, it still could not guarantee a deterministic response time, mainly because frames could still be delayed or lost inside switches. Such a behavior would not meet the requirements of real-time control systems. Hence, to overcome this problem, vendors developed several different "Industrial Ethernet" solutions.

Figure 37.1 illustrates the usual network segmentation in an industrial information system. The two highest levels—corporate LAN and Industrial LAN level 3—do not support time-constrained application. Standard Ethernet is generally deployed at those levels. On the contrary, Industrial Ethernet solutions are designed to address the Industrial LAN requirements of the automation network and fieldbuses. Indeed, real-time and critical applications are performed at those automation levels.

This chapter proposes not to focus on one specific solution but to give an overview of several available commercial solutions.

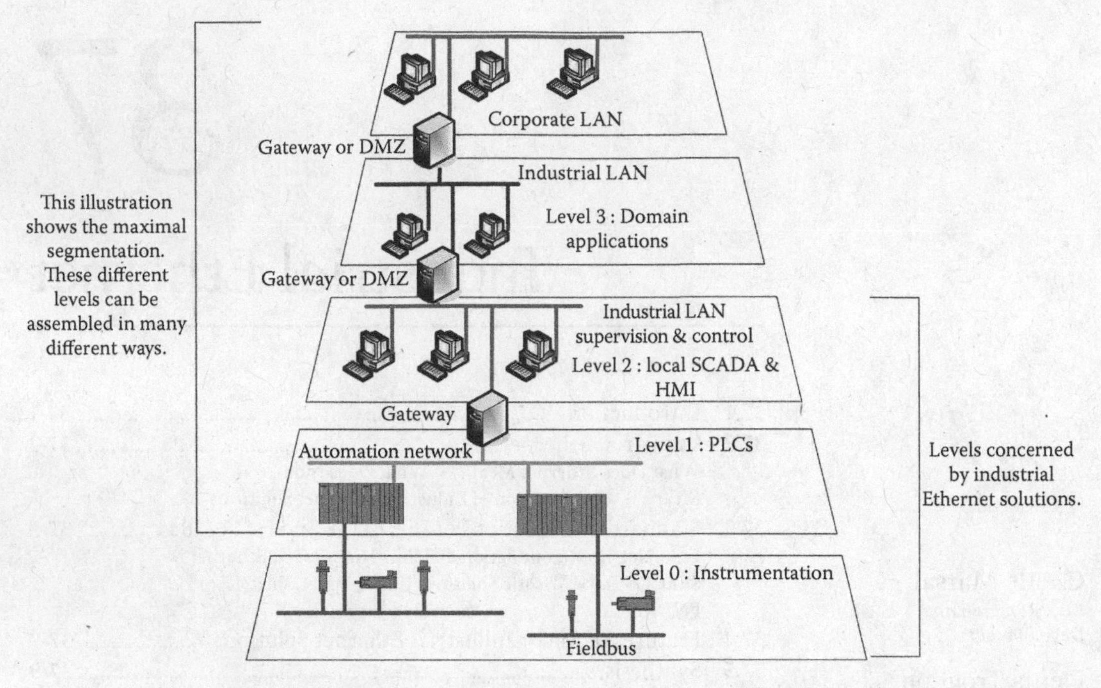

FIGURE 37.1 Theoretical network segmentation from Corporate LAN (top) to Fieldbus (bottom).

First, a general definition of Industrial Ethernet is built, as well as a technological classification of the solutions into four categories. Then a focus is given on standardized solutions of the IEC 61158.

Finally, a synthesis table is proposed to give the main features of each solution. Indeed, there is no solution adapted for all problems, and industries have to be aware of both functional and technical requirements to choose the best fitted Industrial Ethernet solution.

37.2 Industrial Ethernet

In the IT domain, Ethernet, TCP (or UDP), and IP are now the de facto standards. Most higher layer protocols, open or proprietary, rely on those protocols for access control, network, and transport services. Thus, higher layer communications protocols can easily coexist on the same network. Moreover, many standard protocols over TCP/IP, for instance, FTP for file transfer or HTTP for transporting web pages, commonly provide the necessary network services needed by applications. This is far from being the case in the industrial networking solutions, as it is shown in the two following subsections. Section 37.2.3 then suggests a classification of "Industrial Ethernet" solutions.

37.2.1 What Does Ethernet Mean?

Ethernet is defined in the IEEE 802.3 standard. It provides most of the functions and services corresponding to layers 1 and 2 of the OSI model (cf. Chapter 1) and is well known for its famous and convenient medium access control method: CSMA/CD (cf. Section 2.1).

However, when associated with the adjective "industrial," Ethernet can refer to three different exclusive meanings. Such designation can either only use Ethernet standard products, including network interface cards (NICs), switches, and cables; be used in a network with Ethernet standard hardware but with specific NICs; or only use Ethernet cables, everything else being specific.

As a consequence, different Industrial Ethernet types are not interoperable and may or may not coexist on the same network, depending on the solutions.

37.2.2 What Does Industrial Mean?

"Industrial" refers to the compatibility with harsh industrial environment. Users are often more concerned with and watchful to this feature.

Let us detail some properties that an industrial network should satisfy.

Connectors and hardware compatible with industrial environment
Communication hardware—electronic components, shielding, hardware protections and connectors—must be able to operate in an industrial environment. IP6x certification, M12 connectors, EMI and IECEx certification are examples of requirements.

Service guaranties (cf. Chapter 17)
As Ethernet uses the CSMA/CD mechanism to manage the media access; it cannot guarantee either the arrival of a frame or its transfer time. To be used at the plant floor, Industrial Ethernet solutions should give some guaranties detailed hereafter.

Transmission time guaranties (cf. Chapter 19)
Control systems are real-time systems, so information must be transmitted in a bounded time. However, Ethernet, even switched and using full-duplex links, cannot guarantee bounded transfer time and jitter. For multimedia applications, this problem has been solved thanks to prioritization of flows and virtual LAN (VLAN), now standardized in the IEEE 802.1Q [14] and IEEE 802.1D [13]. This solution enables to reduce jitter and guarantee transfer time as low as about 10 ms (for high-priority traffic). However, these are still statistical assessments.

For traffic needing lower or deterministic transfer time, Ethernet has to be modified. In this case, two solutions can be found:

- The protocol itself is modified
- Or network components implement specific communication management functions

In both solutions, this modification is often based on the principle of time division multiple access (TDMA) (cf. TDMA Chapters 17 and 43).

37.2.2.1 Redundancy

Standard Ethernet does not support physical link redundancy feature.

The Spanning Tree Protocol has been developed and standardized in the IEEE802.1D. This protocol makes it possible for several physical paths to exist between two machines. However, the time needed for a network reconfiguration with this protocol is really slow (over 10 s), even for enterprise information applications. Consequently, the Rapid Spanning Tree has been developed. With this protocol, a network reconfiguration requires about 1 s.

However, in case of failure in the communication path, this can still be too slow for control system networks. As a consequence, automation vendors have developed proprietary solutions of "virtual rings," often based on specific network devices. One such solution has been standardized under the name "Media Redundancy Protocol" (MRP) in the IEC 62439 [11].

37.2.2.2 Definition of Application Domain Profiles

Most Industrial Ethernet solutions include application profiles. These solutions make possible an engineering software suite that takes into account device specificities.

These profiles are carried by either text files such as EDDL or software pieces such as DTM. They define the behavior and use the operating mode of equipments in order to have standard interfaces and services for communication.

Safety (cf. Chapter 21)

The IEC 61784-3 standard specifies dedicated profiles for safety. Presently, only two Industrial Ethernet solutions have a profile in this standard, EtherNet/IP (CIP Safety) and ProfiNet IO (ProfiSafe). Other Industrial Ethernet solutions propose a safety profile but not a standardized one, such as "safety over EtherCAT."

Cyber Security (cf. Chapter 22)

Vendors often complete their automation offer with solutions to secure the network. In particular, for the connection between industrial and corporate domains, vendors propose firewalls with a configuration adapted to automation technicians' skills.

37.2.3 Classification of Industrial Ethernet Solutions

In the previous composite definition of Industrial Ethernet, it appears that there are different Industrial Ethernet solutions. Before listing and comparing the different standardized solutions, four categories are proposed to classify Industrial Ethernet solutions according to the technology used.

1. "Full Ethernet"
   These networks use the Ethernet standard IEEE802.3. To avoid uncertainty due to collisions, all these networks are switched and links are used in full duplex. They can also be based on standard prioritization of data over Ethernet and the establishment of VLAN (IEEE 802.1p and IEEE 802.1Q). However, without any management of data flows by the emitting stations (emission rate, minimum and maximum size of frames, etc.) they can only provide statistical guaranties of service.
   Advantage: The use of COTS network devices can coexist with other Industrial Ethernet solutions and with standard Ethernet stations.
   Drawback: No guarantee of deterministic services
   Examples: Modbus/TCP, Ethernet/IP, ProfiNet IO RT, FF HSE (Fieldbus Foundation High-Speed Ethernet)
2. "Ethernet compatible" but using specific devices
   To guarantee hard real-time performances, these kinds of solutions usually implement proprietary management of traffic inside switches based on time slicing. Hence, it is mandatory to use only specific network devices.
   Advantage: Can coexist with standard Ethernet stations; can provide deterministic guaranties
   Drawback: Uses specific network devices
   Example: ProfiNet IO IRT
3. Implemented on common Ethernet devices
   These networks can be used with standard Ethernet hardware. They use standard layers defined in 802.3. Upper layers are modified in order to manage the network traffic, so that guaranteed services can be provided. Thus, it is not possible to connect a standard Ethernet station directly to the network since it would not implement the layers managing the communications on the bus. A gateway is generally offered to provide connectivity to standard Ethernet LANs.
   Advantage: Uses COTS network devices; can provide deterministic guaranties
   Drawback: Uses specific NICs or requires specific communication management layers and cannot coexist with other Industrial Ethernet or standard Ethernet stations
   Example: Ethernet Powerlink
4. New fieldbuses using Ethernet links
   These networks define a MAC layer different from Ethernet to provide real-time capabilities. They require the use of specific devices. These are the latest generation fieldbuses, often providing better performances than those of other networks. They usually provide, as did classical fieldbuses, an acyclic traffic service that enables encapsulating other flows, such as Internet flow (http, etc.), by means of gateways.

Advantage: Provides deterministic guaranties and hard real-time performances
Drawback: Uses specific network devices and specific stations; cannot coexist with other Industrial Ethernet and standard Ethernet stations
Examples: EtherCAT, SERCOS III

37.3 Standardized Solutions of IEC 61158 and IEC 61784

Industrial Ethernet profiles have been included in the IEC 61158 standard (Industrial communication networks—Fieldbus specifications) and IEC 61784 (Industrial communication networks—Profiles). Indeed, except EtherCAT and Ethernet Powerlink, Industrial Ethernet profiles are added to existing profile families (or CPF for Communication Profile Families). In this standard, a profile (or CP for Communication Profile) is defined as a set of all-layer specifications while a family is a set of profile with similar application layers.

Under the promise of better interoperability, 16 different Industrial Ethernet profiles are standardized grouped in 11 families in IEC 61158 [1–6] and IEC 61784 [7–9], which are not more interoperable than their ancestors, fieldbuses.

The Table 37.1, extracted from IEC 61158-1 [1], synthesizes all the industrial communication profiles standardized in IEC 61158 and 61784. The Industrial Ethernet solutions are outlined in red.

Among these Industrial Ethernet, ProfiNet* is detailed in Chapter 40, EtherCAT in Chapter 38, Ethernet Powerlink in Chapter 39, and Modbus in Chapter 36.

Before synthesizing the main features of standardized solutions, the ones not detailed in other chapters of this section are briefly described.

37.3.1 EtherNet/IP

EtherNet/IP [18] is a communication standard supported by Open DeviceNet Vendor Association (ODVA) and ControlNet International Association. As DeviceNet and ControlNet, EtherNet/IP uses a messaging Common Industrial Protocol (CIP). This protocol is supported by Rockwell and Schneider Electric. While ControlNet sets (in IEC 61158 part 6) to support its own messaging and DeviceNet uses CAN as a carrier, EtherNet/IP stack uses Ethernet/TCP/IP as a means of transporting CIP messaging. CIP is the higher layer protocol and is based on object-oriented approach. It defines objects for the messaging application and objects to describe the devices.

The use of CIP messaging provides, in addition to the CIP team model, a set of features including

- CIPsync, which allows synchronizing clocks of network equipments to the milliseconds and is based on the standard IEC 61588 protocol that defines the Precision Time Protocol (PTP) [10]
- CIPSafety, which is an upper layer for functional security for CIP applications, standardized in the IEC 61784-3 [9]

Ethernet/IP provides

- Services for cyclic exchanges of variables, called implicit transfers, based on multi-cast traffic in UDP/IP
- Services to exchange event-based messages, called explicit transfers, based on TCP/IP traffic

The event-based messages over TCP/IP are used both for transmission of system information (configuration and diagnostic operation) and for information in applications that do not require fast response times.

* The classes listed in the standard (CC-A, CC-B, and CC-C) do not correspond to the real-time profiles ProfiNet RT, or IRT, but are conformance classes reporting to the scope of functions needed.

TABLE 37.1 "Industrial Ethernet" Outlined in Thick Black Lines

Communication Profiles (CP) in IEC 61784			IEC 61158 Types Corresponding to CP	
Family CPFs	Technology Name	61,784 Parts	CP Number	Type Number
1	FOUNDATION Fieldbus™	1, 5-1	CP 1/1 FF H1	1,9
		1, 5-1	CP 1/2 FF HSE	5
		1, 5-1	CP 1/3 FF H2	1,9
2	CIP™	1, 5-2	CP 2/1 ControlNet™	2
		1, 2, 5-2	CP 2/2 EtherNet™	2
		1, 5-2	CP 2/3 DeviceNet™	2
3	PROFIBUS, PROFINET	1, 5-3	CP 3/1 PROFIBUS DP	3
		1, 5-3	CP 3/2 PROFIBUS PA	3
		1, 5-3	CP 3/3 PROFINET CBA	10
		2, 5-3	CP 3/4 PROFINET IO CC-A	10
		2, 5-3	CP 3/5 PROFINET IO CC-B	10
		2, 5-3	CP 3/6 PROFINET IO CC-C	10
4	P-NET®	1, 5-4	CP 4/1 P-NET RS-485	4
		1, 5-4	CP 4/2 P-NET RS-232	4
		2, 5-4	CP 4/3 P-NET on IP	4
5	WorldFIP®	1, 5-5	CP 5/1 WorldFIP	7
		1, 5-5	CP 5/2 WorldFIP with subMMS	7
		1, 5-5	CP 5/3 WorldFIP minimal for TCP/IP	7
6	INTERBUS®	1, 5-6	CP 6/1 INTERBUS	8
		1, 5-6	CP 6/2 INTERBUS TCP/IP	8
		1, 5-6	CP 6/3 INTERBUS minimal subset of CP 6/1	8
		2, 5-6	CP 6/4	8
		2, 5-6	CP 6/5	8
		2, 5-6	CP 6/6	8
7	—	—	This CPF and the associated Type 6 are deleted for lack of market relevance	6
8	CC-Link	1, 5-8	CP 8/1 CC-Link/V1	18
		1, 5-8	CP 8/2 CC-Link/V2	18
		1, 5-8	CP 8/3 CC-Link/LT	18
9	HART	1, 5-9	CP 9/1 HART	20
10	Vnet/IP	2, 5-10	CP 10/1 Vnet/IP	17
11	TCnet	2, 5-11	CP 11/1 TCnet	11
		2, 5-11	CP 11/2 TCnet-Loop	11
12	EtherCAT	2, 5-12	CP 12/1	12
		2, 5-12	CP 12/2	12
13	ETHERNET Powerlink	2, 5-13	CP 13/1 EPL	13
14	EPA	2, 5-14	CP 14/1	14
		2, 5-14	CP 14/2	14

TABLE 37.1 (continued) "Industrial Ethernet" Outlined in Thick Black Lines

Communication Profiles (CP) in IEC 61784			IEC 61158 Types Corresponding to CP	
Family CPFs	Technology Name	61784 Parts	CP Number	Type Number
15	MODBUS®-RTPS	2, 5-12	CP 15/1 MODBUS TCP	15
		2, 5-12	CP 15/2 RTPS	15
16	SERCOS	1, 5-16	CP 16/1 SERCOS I	16
		1, 5-16	CP 16/2 SERCOS II	16
		2, 5-16	CP 16/3 SERCOS III	19

To avoid saturation phenomena of the network traffic due to cyclic multi-cast over UDP, Ethernet/IP requires that the network components implementing the IGMP standard (RFC 3376 [17]) for filtering multi-cast and 802.1Q, support for VLANs.

37.3.2 Foundation Fieldbus High-Speed Ethernet

FF HSE is developed by the Fieldbus Foundation for process control [20].

FF HSE is an application protocol (layer 7 of the OSI model) that uses UDP/IP rather than TCP/IP for communications in "real time," and TCP/IP for other exchanges.

FF HSE is an adaptation of the Foundation Fieldbus for Ethernet. It keeps the same data model oriented for process applications.

FF HSE includes a real protocol for managing the "SNCC-oriented" physical redundancy. It allows the physical redundancy of machines and ports, in contrast to the redundancy offered by the logical ring topologies.

FF HSE implements scheduling communications to ensure that two components do not publish their data at the same time. This mechanism is implemented in the application layer to enable the use of standard Ethernet and TCP/IP protocols. With FF HSE, it is possible to achieve a cycle time of 100 ms.

37.3.3 SERCOS III

SERCOS III is supported by SERCOS trade associations [19].

SERCOS III translates the application layers of SERCOS fieldbus on Ethernet and implements mechanisms similar as EtherCAT. It focuses on hard real-time applications, in particular, motion control.

To obtain deterministic cycle time guaranties, this solution only uses Ethernet cables (category 4, Section 37.2.3), and is based on a daisy chain topology where only one frame circulates from devices to devices. This frame contains data from and for all devices. There is one master by-chain, which sends the frame all the required data to be read by slaves in a given place. Then, the devices receive this frame one after the other, read the data they need, and write data if necessary.

Recently, SERCOS International (parent organization of SERCOS trade associations) joined the Open Source Automation Development Lab (OSALD) [24]. This promises an open source version of SERCOS III.

37.3.4 "Exotic Solutions": EPA, Tcnet, Vnet/IP, PNET on IP

These last four solutions are less known either because they are not supported by one major automation vendor (EPA and PNET on IP) or because vendors mainly use them as internal communication protocols for their own products (Tcnet and Vnet/IP).

Ethernet for plant automation (EPA) is supported by the Chinese industry. It is built on the standard Ethernet layers (802.3 and 802.2 or Ethernet II) and adds a top layer called EPA communication scheduling management entity (ECSME). This layer manages network traffic on a macro cycle consisting of a periodic transfer phase and an aperiodic transfer phase. The regularity of traffic is based on synchronization of equipment by IEC 61588 (PTP).

TABLE 37.2 Features of Major Industrial Ethernet Solutions

		Type (cf. Section 37.2.3)	Real-Time Capability	Standards Compliance	Determinism	QoS	Transmission Time	Application Domain
EtherCAT		4	Hard real-time	61158 Ethernet cables	Deterministic with short cycle time	Ensured thanks to the transfer of all data in a unique frame	Typically around 100 µs	Motion control
EtherNet/IP		1	Soft real-time	61158 IEEE802.3 TCP/IP (RFC791 and 793)	Nondeterministic	Use of 802.1Q and 802.1D standards	No guarantee offered by the protocol if not using CIPSync	Process, batch, manufacturing
FF HSE		1	Soft real-time	61158 IEEE802.3 TCP/IP (RFC791 and 793)	Nondeterministic	Possible use of 802.1Q and 802.1D standards	No guarantee offered by the protocol ~100 ms	Process
Modbus/TCP		1	Soft real-time	61158 IEEE802.3 TCP/IP (RFC791 and 793)	Nondeterministic	No QoS	No guarantee offered by the protocol	Manufacturing
EPL		3	Soft real-time or hard real-time	61158 IEEE802.3 TCP/IP (RFC791 and 793)	Deterministic with short cycle time	QoS ensure by use of TDMA mechanism	Guaranteed (values depend on devices: hub or switch)	Motion control Electricity production (fossil fuel, hydraulic)
ProfiNet	RT	1	Soft real-time	61158 IEEE802.3	Nondeterministic	Prioritization of flows, limitation of broadcast, and use of VLANs	No guarantee offered by the protocol	ProfiNet
	IRT	2	Hard real-time	61158 IEEE802.3	Deterministic for IRT flows	Flow prioritization inside the switch	Guaranteed for IRT flows thanks to time slot managed by switches	
SERCOS III		4	Hard real-time	61158 Ethernet links	Deterministic with short cycle time	Ensured thanks to the transfer of all data in a unique frame	Typically around 100 µs	Motion control

Tcnet [21] is supported and used by Toshiba, e.g., in the transmission module of the Toshiba 3000. It is a master/slave protocol, with a prioritization of messages. Each slave sends the messages to the master in an available time period (time slot). Messages with lower priority use TCP/IP and those of medium and high priorities use EPL.

Vnet/IP [22] is supported by Yokogawa. This would work with HTTP protocol over TCP/IP. There are real-time extensions and safety extensions, Reliable Datagram Protocol (RTP). No more information is freely available about this protocol and its extensions.

PNET on IP [23], originally developed by a Danish company (Process Data Aps), is supported by the International P-NET User Organization (IPUO). This would be the translation of the P-NET protocol on Ethernet UDP/IP. The structure of the messages remains the same as for P-NET over RS-485.

37.4 Features of Major Industrial Ethernet Solutions

The different Industrial Ethernet solutions are so different because they address different problems. For instance, there is no quick and easy solution that can ensure hard real-time communication (100 μs) based on nondeterministic protocols and COTS devices. Here are seven features, not necessarily independent, that could help choosing an Industrial Ethernet solution compliant to different application requirements:

The category of the solution, in the sense of Section 37.2.3
The real-time capability
The compliance to communication standards
The determinism of medium access and transport layer
The QoS mechanisms available
The common transmission time
The designated application domain

The features of six major Industrial Ethernet solutions are synthesized and compared in the following table (Table 37.2).

37.5 Synthesis

Under the promise of interoperability, "Industrial Ethernet" refers to many different solutions that are mostly adapted from fieldbuses protocols to be used over Ethernet links. Even for nonspecific solutions (e.g., non-hard real-time), they are not interoperable but can possibly coexist on the same network.

On the one hand, the choice of an Industrial Ethernet is then not trivial and can be linked to an automation supplier, as was the case when using fieldbuses.

On the other hand, solutions adapted for many different types of systems now exist. However, it is compulsory to well define the requirements of the system and to traduce them into network features before, to be able to choose the best solution.

Abbreviations

CIP	Common Industrial Protocol
COTS	Commercial Off the Shelf
CSMA/CD	Carrier Sense Multiple Access/Collision Detection (IEEE 802.3)
DTM	Device Type Manager
EDDL	Electronic Device Description Language
EPL	Ethernet PowerLink
FF	Fieldbus Foundation
FF HSE	Fieldbus Foundation High-Speed Ethernet
I&C	Instrumentation and Control

IP	Internet Protocol (RFC 791)
IRT	Isochronous Real Time (specific to ProfiNet solution)
IT	Information Technology
LAN	Local Area Network
MRP	Media Redundancy Protocol
NIC	Network Interface Card
OSADL	Open Source Automation Development Lab
OSI	Open System Interconnection
QoS	Quality of Service
RSTP	Rapid Spanning Tree Protocol
RT	Real Time
STP	Spanning Tree Protocol
TCP	Transport Control protocol (RFC 793)
TDMA	Time Division Multiple Access
UDP	User Datagram Protocol
VLAN	Virtual Local Area Network

References

1. IEC/TR 61158-1 Edition 2.0 2007-11 Industrial communication networks—Fieldbus specifications—Part 1: Overview and guidance for the IEC 61158 and IEC 61784 series, International Electrotechnical Commission, Geneva, Switzerland.
2. IEC 61158-2 Edition 4.0 2007-12 Industrial communication networks—Fieldbus specifications—Part 2: Physical layer specification and service definition, International Electrotechnical Commission, Geneva, Switzerland.
3. IEC 61158-3 Edition 2004-12 Digital data communication for measurement and control—Fieldbus for use in industrial control systems—Part 4: Data link service definition, International Electrotechnical Commission, Geneva, Switzerland.
4. IEC 61158-4 Edition 2004-12 Digital data communication for measurement and control—Fieldbus for use in industrial control systems—Part 4: Data link protocol specification, International Electrotechnical Commission, Geneva, Switzerland.
5. IEC 61158-5 Edition 2004-12 Digital data communication for measurement and control—Fieldbus for use in industrial control systems—Part 5: Application layer service definition, International Electrotechnical Commission, Geneva, Switzerland.
6. IEC 61158-6 Edition 2004-12 Digital data communication for measurement and control—Fieldbus for use in industrial control systems—Part 6: Application layer protocol specification, International Electrotechnical Commission, Geneva, Switzerland.
7. IEC 61784-1 Edition 2.0 2007-12 Industrial communication networks—Profiles—Part 1: Fieldbus profiles, International Electrotechnical Commission, Geneva, Switzerland.
8. IEC 61784-2 Edition 1.0 2007-12 Industrial communication networks—Profiles—Part 2: Additional fieldbus profiles for real-time networks based on ISO/IEC 8802-3, International Electrotechnical Commission, Geneva, Switzerland.
9. IEC 61784-3 Edition 1.0 2007-12 Industrial communication networks—Profiles—Part 3: Functional safety fieldbuses—General rules and profile definitions, International Electrotechnical Commission, Geneva, Switzerland.
10. IEC 61588 First edition 2004-09 (IEEE 1588™) Precision clock synchronization protocol for networked measurement and control systems, International Electrotechnical Commission, Geneva, Switzerland.
11. IEC 62439 Edition 1.0 2008-05 High availability automation networks, International Electrotechnical Commission, Geneva, Switzerland.

12. IEEE 802.3 edition 2005 IEEE Standard for Local and metropolitan area networks—Specific requirements Part 3: Carrier sense multiple access with collision detection (CSMA/CD) access method and physical layer specifications, Institute of Electrical and Electronics Engineers, New York.
13. IEEE 802.1D Edition 2004 IEEE Standard for Local and metropolitan area networks—Media Access Control (MAC) Bridges, Institute of Electrical and Electronics Engineers, New York.
14. IEEE 802.1Q Edition 2005 IEEE Standard for Local and metropolitan area networks—Virtual Bridged Local Area Networks, Institute of Electrical and Electronics Engineers, New York.
15. RFC 791 Internet Protocol, 1981, Internet Engineering Task Force (IETF), Reston, VA.
16. RFC 793 Transmission Control Protocol, 1981, Internet Engineering Task Force (IETF), Reston, VA.
17. RFC 3376 Internet Group Management Protocol, Version 3, 2002, Internet Engineering Task Force (IETF), Reston, VA.
18. ODVA, The organization that supports network technologies built on the Common Industrial Protocol (CIP™)—DeviceNet™, EtherNet/IP™, CompoNet™, and ControlNet™, December 2009, available at http://www.odva.org/
19. SERCOS, International E.V. and SERCOS North America, group of associations dedicated to developing, promoting and expanding the use of the SERCOS digital interface, December 2009, available at http://www.sercos.com/
20. Fieldbus Foundation, A global not-for-profit corporation consisting of process end users and automation companies in order to develop an automation infrastructure on the Foundation technology, December 2009, available at http://www.fieldbus.org/
21. TOSHIBA, A corporation dealing with their solution for time-critical information and control network, December 2009, available at http://www3.toshiba.co.jp/sic/english/seigyo/tcnet/
22. Yokogawa Electric Corporation, A corporation dealing with their control network technology, December 2009, available at http://www.yokogawa.com/dcs/products/vnetip/dcs-vnetip-02-en.htm
23. International P-NET User Organization, December 2009, available at http://www.p-net.org/
24. Open Source Automation Development Lab, The Open Source Automation Development Lab are promoting and supporting the usage of open-source software in the context of machine and plant control systems, December 2009, available at http://www.osadl.org/

38

EtherCAT

Gianluca Cena
Istituto di Elettronica e di Ingegneria dell'Informazione e delle Telecomunicazioni

Adriano Valenzano
Istituto di Elettronica e di Ingegneria dell'Informazione e delle Telecomunicazioni

Claudio Zunino
Istituto di Elettronica e di Ingegneria dell'Informazione e delle Telecomunicazioni

38.1 Introduction 38-1
38.2 Physical Layer 38-1
38.3 Communication Protocol 38-2
Commands
38.4 Addressing 38-5
Physical Addressing • Logical Addressing: FMMU
38.5 SyncManager 38-7
38.6 Distributed Clock 38-8
38.7 Application Layer 38-9
Mailbox Services
References 38-10

38.1 Introduction

Ethernet for control automation technology (EtherCAT) [ETG04,ETG05] is a high-performance Ethernet-based fieldbus system. Its main development goal was to apply Ethernet to automation applications, which require short-cycle times and low-communication jitters. EtherCAT is nowadays a popular solution for connecting control applications to field devices in industrial environments, including motion-control applications, and communication equipment and devices can be easily found off-the-shelf [ESC20].

The EtherCAT protocol is an open standard currently managed by the EtherCAT technology group (ETG). Its specification has been recently integrated into the international fieldbus standards IEC 61158 [IE158] and IEC 61784 [IE784]. EtherCAT is based on a master/slave approach (where only one master is allowed in the network) and relies on a ring topology at the physical level. It can interoperate with both common TCP/IP-based networks and other Ethernet-based solutions, such as EtherNet/IP or PROFInet.

38.2 Physical Layer

The EtherCAT master processes data via standard hardware (full-duplex Ethernet network interface controllers) and dedicated software (e.g., Beckhoff TwinCAT), but it supports also open source solutions based on Linux-like operating systems. On the contrary, purposely designed hardware has to be used on slaves. The master node completely controls traffic over the network by initiating all transmissions. Every slave, when receiving a datagram, processes it (in hardware) and then forwards it to the next slave in the physical ring. Unlike Ethernet switches and bridges, frames are not managed according to a store-and-forward scheme (i.e., received, interpreted, and then copied as process data at every connection). Instead, every frame is processed on-the-fly at the data-link layer.

To further improve communication efficiency, a fieldbus memory management unit (FMMU) is provided in each slave device that reads/writes portions of data included in the frame while it is being forwarded to the next device.

EtherCAT supports two different types of physical layer, namely Ethernet and EBUS. The former is used for connecting to an external Ethernet network according to IEEE 802.3; it is typically used for the connection between the master and the slave network segment. Indeed, an EtherCAT network can be seen as a single, large, Ethernet device that receives and sends Ethernet frames. However, this "device" is not made up of a single Ethernet controller but includes a (possibly large) number of EtherCAT slaves.

Conversely, EBUS can be used as a backplane bus (i.e., it is not qualified for wire connections).

In particular, EBUS is a physical layer designed to reduce pass-through delays inside the nodes; typically, frames are only delayed by 60÷500 ns [PRY08,BEC06], whereas a greater delay (about 1 μs) is introduced by Ethernet interfaces between segments. It uses the Manchester (biphase L) encoding and encapsulates frames between start-of-frame (SOF) and end-of-frame (EOF) identifiers. The beginning of a frame is defined by a Manchester violation with positive level followed by a "1" bit, which is also the first bit of the Ethernet preamble. Then, the whole Ethernet frame is transmitted (including Ethernet SOF at the end of the preamble, up to the CRC). The frame finishes with a Manchester violation with negative level followed by a "0" bit, which is also the first bit of the IDLE phase. It uses low-voltage differential signaling (LVDS), and the data rate is 100 Mbps to accomplish the fast Ethernet specification. The EBUS protocol simply encapsulates Ethernet frames; thus, EBUS can carry any Ethernet frame.

The maximum number of addressable devices is 2^{16} for each segment, while the distance between any two adjacent nodes (i.e., the cable length) can be up to 10 m for EBUS and (as usual) up to 100 m for Ethernet connections. The whole network size is practically unlimited (as, in theory, up to 2^{16} devices can be connected using 100 m Ethernet cables).

The topology of a communication system is one of the crucial factors for the successful application in automation environments. The topology has significant influence on cabling efforts, diagnostic features, redundancy options, and hot-plug-and-play features. Since the star topology commonly used for Ethernet leads to increased cabling efforts and infrastructure costs, a line or tree topology is preferable. Typically, the slave node arrangement represents an open ring. At one of the open ends, the master device sends frames, either directly or via Ethernet switches, and receives them at the other end after they have been processed. All frames are relayed from each slave to the next one.

The last slave in the network returns the frame back to the master. Thanks to the full-duplex capabilities of Ethernet, which uses two wire-pairs to carry out communications in both directions at the same time, the resulting topology visually resembles a physical line. Branches, which in principle are possible anywhere, can be used to enhance the basic line structure by making it possible to set up tree network topologies.

38.3 Communication Protocol

The EtherCAT communication protocol aims at maximizing the utilization of the Ethernet bandwidth, thanks to its very high communication efficiency. As mentioned above, the medium access mechanism (MAC) is based on the master/slave principle, where the master node (typically, the control system) sends Ethernet frames to the slave nodes over the physical ring, and each of the slave extracts data from (and inserts data into) the payload of these frames.

The frames sent over the network are standard Ethernet frames, whose data field encapsulates the EtherCAT frame. This one, in turn, is made up of a header and one or more EtherCAT datagrams or protocol data unit (PDUs), as to obtain a more efficient use of the large Ethernet data field available. EtherCAT PDUs are packed together without inter-PDU gaps. The frame is completed with the last EtherCAT PDU, unless the frame size is less than 64 octets, in which case the frame is padded to 64 octets in length. Each

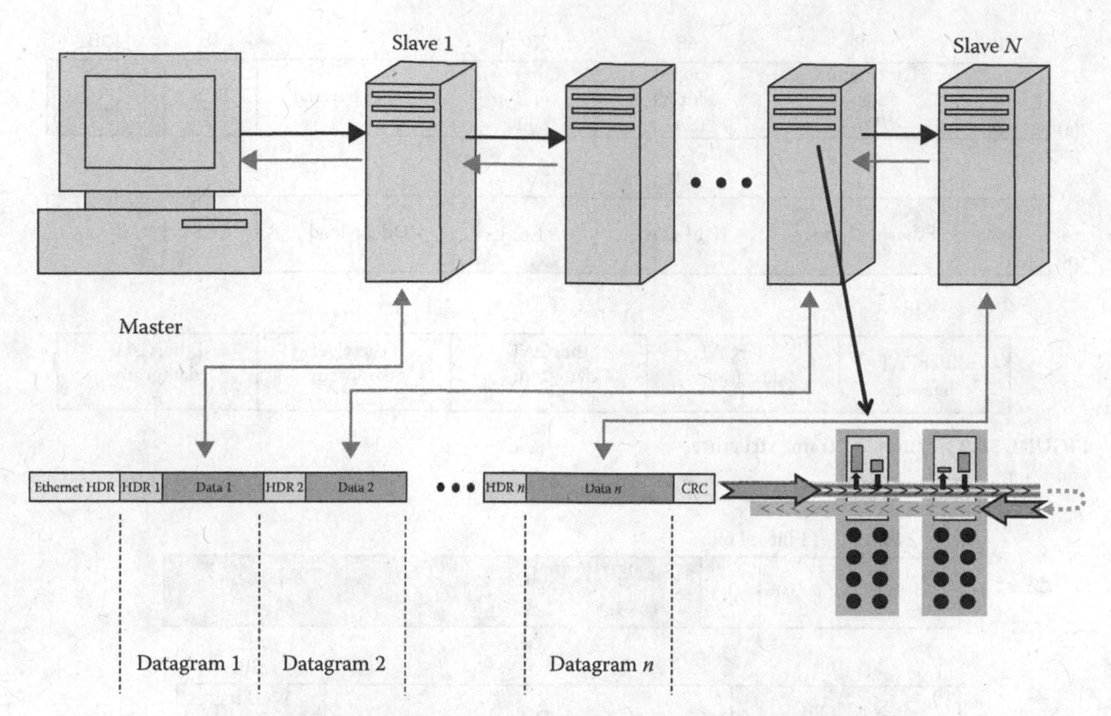

FIGURE 38.1 EtherCAT typical topology, with the "on-the-fly" frame processing.

EtherCAT PDU (that corresponds to a separate command) contains the header, data, and working counter (WC) fields. The standard Ethernet CRC is used to check the integrity of messages.

From the point of view of the master, the whole EtherCAT segment is seen as a single Ethernet device, which receives and sends standard Ethernet frames with the EtherType field set to 0x88A4 to distinguish it from other Ethernet frames. In this way, EtherCAT can coexist with other Ethernet protocols. The last EtherCAT slave device in the segment sends the fully processed frame back and the same occurs for each device. This procedure is based on the full-duplex mode of Ethernet: both communication directions are operated independently. In order to achieve the maximum performance, Ethernet frames have to be processed on-the-fly, as depicted in Figure 38.1. Slave nodes implemented in this way recognize relevant commands and execute them accordingly while the frames are passing through.

As said before, several EtherCAT datagrams can be embedded within the same Ethernet frame, each one addressing different devices and/or memory areas. As shown in Figure 38.2, the EtherCAT datagrams are transported either

1. Directly in the data area of the Ethernet frame or
2. Within the data section of a UDP datagram carried via IP

The first variant is limited to one Ethernet subnet since associated frames are not relayed by routers. For machine control applications, this usually is not a limitation. Multiple EtherCAT segments can be connected to one or several switches. The Ethernet MAC address of the first node in the segment is used for addressing the EtherCAT segment.

The second variant (via UDP/IP) implies a slightly larger overhead (because of the IP and UDP headers), but for less time-critical applications such as process control it enables the use of IP routing. On the master side, any standard UDP/IP implementation can be used in this case.

38.3.1 Commands

Each EtherCAT PDU (see Figure 38.3) consists of an EtherCAT header, the data area, and a subsequent counter area (WC).

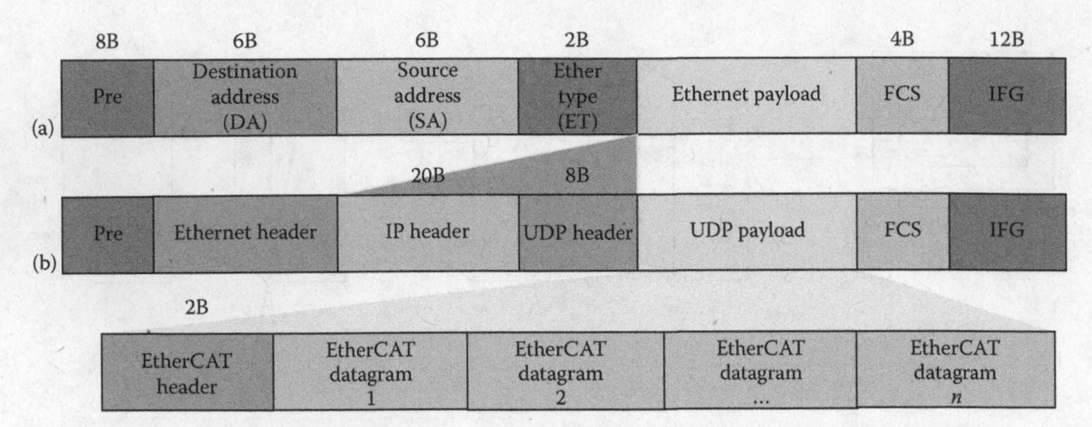

FIGURE 38.2 EtherCAT frame structure.

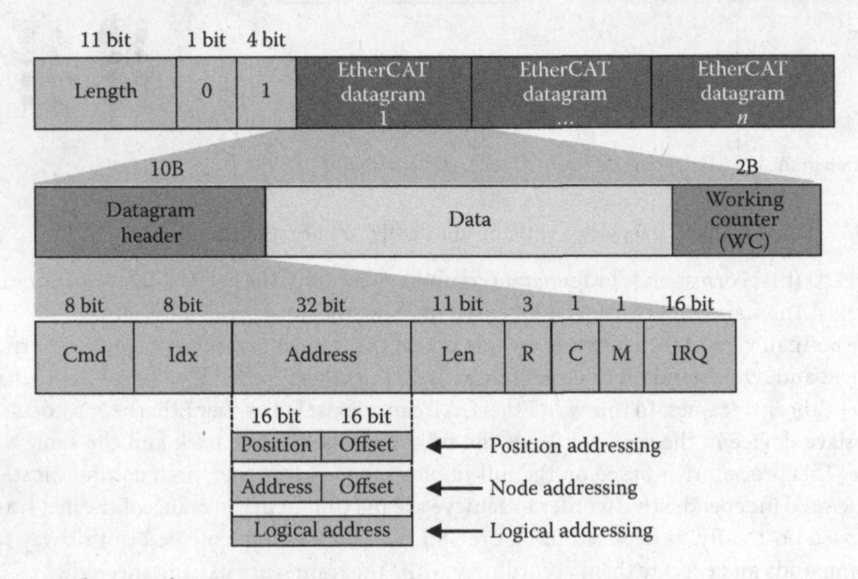

FIGURE 38.3 EtherCAT datagram structure.

In particular, each datagram contains the fields listed in Table 38.1.

The parameter "command" encodes the service command. Different types of commands can be used to optimize read and write operations on slave devices, which can be classified depending on the access type:

- *Broadcast read/write*: All slaves carry out a logical "OR" operation of data of the memory area and data of the incoming EtherCAT datagram, insert this information into the outgoing EtherCAT datagram, and write it back into the local memory area.
- *Read, write, or read/write actions*: Exchange of incoming data and local data. For read and write operations, the read operation is performed before the write operation.
- *Read multiple write actions* (*RMW*): Addressed slave will read while others will write.

The IDX field is the identifier of the datagram; it is left unchanged by the slaves. The address field can contain different information according to the addressing mode chosen (a more detailed description is provided in Section 38.4).

The parameter LEN shall contain the size in byte of the data to be read or written.

TABLE 38.1 EtherCAT Datagram Format

CMD	Unsigned8	EtherCAT command type
IDX	Unsigned8	Index for identification of duplicates(lost) datagrams
Address	DWORD	Address (autoincrement, configured station address, or logical address, see Section 38.4)
LEN	Unsigned11	Length of the DATA field
R	3 bit	Reserved 0x00
C	1 bit	Circulating frame 0x00: Frame is not circulating 0x01: Frame has circulated once
M	1 bit	0x00: Last EtherCAT PDU in EtherCAT frame 0x01: EtherCAT PDU in EtherCAT frame follows
IRQ	WORD	External EtherCAT event request registers of all slaves combined with a logical OR
DATA	OctetString LEN	Data
WC	WORD	Working counter

The field C is used in the case of a network with a link failure between two slaves; a frame currently traveling through the part of the ring on the isolated side of the network might start circulating. Indeed, each slave has a mechanism called loop-back function that forwards Ethernet frames to the next internal port if either there is no link at a port or the port is not available. This feature is needed to create the ring on the last slave. Thus, to prevent an endless frame from circulating, a slave with no active link through the master has to do the following:

- If the circulating bit of the EtherCAT datagram is 0, set the circulating bit to 1.
- If the circulating bit is 1, do not process the frame and destroy it.

M specifies whether or not the EtherCAT PDU is the only one in the frame.

An important mechanism is based on the IRQ field. EtherCAT event requests are used to inform the EtherCAT master of slave events. The EtherCAT event request register is combined using a logical AND operation with a predefined EtherCAT event mask register. This one is used for selecting the interrupts which are relevant for the EtherCAT master. The resulting bits are then combined with the EtherCAT IRQ field present on the incoming datagram with a logical OR operation and written into the EtherCAT IRQ field of the outgoing datagram. It is worth noting that the master is not able to distinguish which slave(s) was/were originating an interrupt.

Finally, process data that can be read or written are placed in the DATA field. In the case a slave is writing information in the data field, it must recompute and modify the CRC field at the end of the packet to make it correct for next slaves.

The last field is the WC. This field is used to count the number of devices that were successfully addressed by an EtherCAT datagram. Each addressed slave increments the WC in hardware. Because each datagram has an expected WC value computed by the master, the valid processing of datagrams can be checked by comparing the received WC with the expected value. The WC is increased if at least one byte/bit of the data field was successfully read and/or written. Complex operations like RMW commands are treated like either a read or a write command, depending on the address matching mechanism described above.

38.4 Addressing

Different addressing modes are supported for accessing slaves, as shown in Figure 38.3. The header within the EtherCAT PDU contains a 32 bit address, which is used for both physical node addressing and logical addressing.

38.4.1 Physical Addressing

In the physical addressing scheme, the 32 bit address field within each EtherCAT PDU is split into a 16 bit slave device address and a 16 bit physical address within the slave device, thus leading to 2^{16} slave device addresses, each one with an associated 16 bit local address space. With device addressing, each EtherCAT PDU uniquely addresses one single slave device.

This mode is mostly suitable for transferring parameterization information for devices. There are two different device addressing mechanisms, namely, position addressing and node addressing. There is also a third mode for broadcast messages.

Now let us consider the node addressing mode (also defined as configured station address); because EtherCAT slaves can have up to two configured station addresses (if this functionality is enabled), the address may be either assigned by the master (configured station address) or read from the internal EEPROM that can be changed by the slave application (configured station alias address).

The following addressing alternatives are available:

- *Position address/autoincrement address*: Position addressing is used to address each slave device via its physical position within the segment. Each slave device increments the 16 bit address field as the datagram transits through the slave device; the device which receives a frame with an address field with value 0 is the one being addressed. This means that the field contains the negative position of the slave in the loop (the master position is marked "0"). Due to the mechanism employed to update this address while the frame is traversing the node, the slave device is said to have an autoincrement address. Position addressing should be used during the start-up phase of the EtherCAT system only to scan the fieldbus; it can be adopted occasionally to detect newly attached slaves because it can cause problems if loops are closed temporarily due to link problems.
- *Node address/configured station address and configured station alias*: In this case, the slaves are addressed via configured node addresses assigned by the master during the data-link start-up phase. This ensures that, even if the segment topology is changed or devices are added or removed, the slave devices can still be addressed via the same configured addresses. Node addressing is typically used for register access to individual and already identified devices.
- *Broadcast*: Each EtherCAT slave is addressed. Broadcast addressing is used for initialization of all slaves and for checking the status of all slaves if they are expected to be identical.

Each slave device has a 16 bit local address which is addressed via the offset field of the EtherCAT datagram.

38.4.2 Logical Addressing: FMMU

Using the datagram structure described above, several EtherCAT devices can be separately addressed via a single Ethernet frame, each one by means of one EtherCAT datagram, which leads to a significant improvement of the system bandwidth. However, for small-size input terminals, for example, with 2 bit input devices that map precisely 2 bit of user data, the overhead of a single EtherCAT command might be excessive.

The FMMU lessens this problem, and the available utilization ratio is more than 90%—even for devices with only 2 bits of user data, as mentioned before. Similar to the memory management units (MMUs) in modern processors, the FMMU converts a logical address into a physical one via an internal table. The FMMU is integrated in the EtherCAT slave architecture and enables individual address mapping for each device. In contrast to processor-internal MMUs that typically map complete memory pages, the FMMU also supports bit-wise mapping. This enables, for example, the 2 bits of an input terminal to be inserted individually anywhere within a logical address space. If an EtherCAT command is sent to read or write a certain memory area, instead of addressing a particular EtherCAT device, the 2 bit input terminal inserts its data at the right place within the data area. In the same way, other terminals

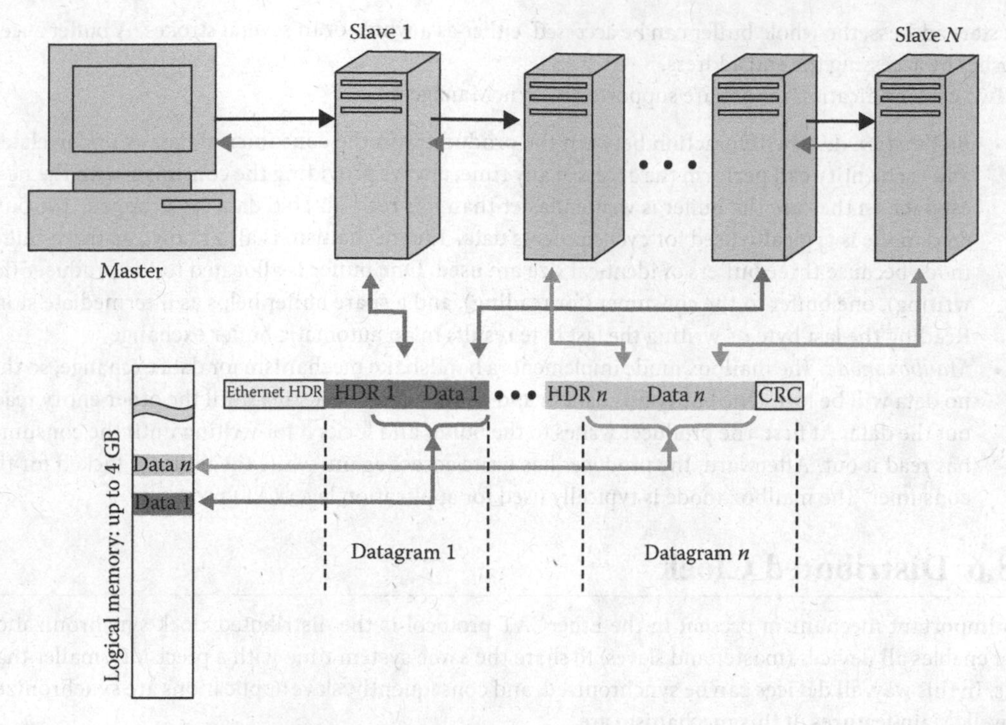

FIGURE 38.4 FMMU mapping example.

that are addressed in the FMMU also insert their data, so FMMUs allow to use logical addressing for data segments that span several slave devices: one command addresses data within several arbitrarily distributed slaves, as shown in Figure 38.4. The access type supported by an FMMU is configurable to be read, write, or read/write. Configuration of the FMMU entities is performed by the master device and transferred to the slave devices during the data-link start-up phase. For each FMMU entity, the following items are configured: a logical, bit-oriented start address, a physical memory start address, a length, and a type that specifies the direction of the mapping (input or output). When an EtherCAT datagram with logical addressing is received, the slave device checks whether one of its FMMU entities exhibits an address match. If appropriate, it inserts data at the associated position of the data field into the datagram (read operation) or extracts data from the associated position (write operation).

38.5 SyncManager

EtherCAT provides a mechanism to synchronize slave memory access. Because the memory of a slave can be used for exchanging data between the master and the local application without any restrictions, some problems can arise. First of all, data consistency is not guaranteed, and a mechanism like semaphores has to be implemented for exchanging data in a coordinated way. Moreover, both master and slave applications have to access the memory in order to know when the memory is no longer used on the other side.

Thus, SyncManager enables consistent and secure data exchanges between the EtherCAT master and the local application, generating interrupts to inform both sides of changes. It is organized in channels, and each channel defines a consistent area of the memory.

The SyncManager system is configured by the EtherCAT master. The communication direction is configurable, as well as the communication mode (buffered mode and mailbox mode). The SyncManager system uses a buffer located in the memory area for exchanging data and controls the access to this buffer. All accesses to the buffer begin from the start address, otherwise the access is denied. After the access to the start address, the whole buffer can be accessed, either as a whole or in several strokes. A buffer access finishes by accessing the end address.

Two communication modes are supported by SyncManagers:

- *Buffered mode*. The interaction between the producer and the consumer of data is uncorrelated, and each entity can perform the access at any time, always providing the consumer with the newest data. In the case the buffer is written faster than it is read out, old data are dropped. The buffered mode is typically used for cyclic process data. The mechanism is also known as three-buffer mode because three buffers of identical size are used. One buffer is allocated to the producer (for writing), one buffer to the consumer (for reading), and a spare buffer helps as intermediate store. Reading the last byte or writing the last byte results in an automatic buffer exchange.
- *Mailbox mode*. The mailbox mode implements a handshake mechanism for data exchange, so that no data will be lost. One entity fills data in and cannot access the area until the other entity reads out the data. At first, the producer writes to the buffer and locks it for writing until the consumer has read it out. Afterward, the producer has write access again, while the buffer is locked for the consumer. The mailbox mode is typically used for application layer (AL) protocols.

38.6 Distributed Clock

An important mechanism present in the EtherCAT protocol is the distributed clock synchronization that enables all devices (master and slaves) to share the same system time with a precision smaller than 1 μs. In this way, all devices can be synchronized, and consequently, slave applications are synchronized as well. Main features of this mechanism are

- Generation of synchronous output signals
- Precise time stamping of input events
- Synchronous digital output updates
- Synchronous digital input sampling

Typically, the clock reference (system time) is the first slave with distributed clock capability after the master within one segment. This system time is used as the reference clock to synchronize the DC slave clocks of other devices and of the master itself. The propagation delays, local clock source drifts, and local clock offsets are taken into account for the clock synchronization. All settings are done during the clock synchronization process made of three steps:

1. *Propagation delay measurement*: The master sends a synchronization datagram at certain intervals (as required in order to avoid the slave clock diverging beyond application-specific limits), and each slave stores the time of its local clock twice—once when the message is received and once when it is sent back (remember EtherCAT has a ring topology). The master reads all time stamps and computes the propagation delays between all slaves. It is worth noting that clocks are not synchronized for the delay measurement, and only local clock values are used. So the propagation delay calculation is based on receive time differences.
2. *Offset compensation to reference clock (system time)*: Because the local time of each device is a free running clock that has not the same time as the reference clock, a compensation mechanism is necessary. To achieve the same absolute time, all device difference between the reference clock and every slave device's clock is computed by the master. Then the offset time is written to a specific slave register "system time offset." Each slave computes its local copy of the system time using its local time and the local offset value. Moreover, small offset errors are eliminated by the drift compensation.
3. *Drift compensation to reference clock*: After the delay time between the reference clock and the slave clocks has been measured, and the offset between both clocks has been compensated, the natural drift of every local clock has to be compensated by a time-control loop. This mechanism readjusts the local clock by regularly measuring the differences.

38.7 Application Layer

The EtherCAT state machine is responsible for the coordination of master and slave applications at start-up and during operation.

State changes are typically initiated by requests of the master. They are acknowledged by the local application after the associated operations have been executed. Unsolicited state changes of the local application are also possible. Simple devices without a microcontroller can be configured to use EtherCAT state machine emulation. These devices simply accept and acknowledge any state change automatically.

The state machine is controlled and monitored using some registers present on the slave. The master requests state changes by writing to the AL control register. The slave indicates its state writing in the AL status register and puts possible error codes into the AL status code register.

As shown in Figure 38.5, there are four basic states an EtherCAT slave shall support, plus one optional state:

- Init
- Pre-operational
- Safe-operational
- Operational
- Bootstrap (optional)

All state changes are possible except for the "Init" state and for the "Pre-operational": for the former the only possible transition is the one to the "Pre-operational" while for the latter there is no direct state change to "Operational."

State changes are normally requested by the master. The master requests a write to the AL control register, the slave responds through a local AL status register write operation after a successful or a failed state change. If the requested state change failed, the slave also responds by setting an error flag.

The Bootstrap state is optional, and there is only a transition from or to the Init state. The only purpose of this state is to download the device's firmware. In the Bootstrap state, the mailbox is active but restricted to the file access over EtherCAT service (FoE) protocol.

38.7.1 Mailbox Services

Another important characteristic of EtherCAT is the possibility to feature multiprotocol capability, consolidating high- or same-level protocols in a standardized mailbox. In particular, the following capabilities are supplied:

- *Ethernet over EtherCAT (EoE)*: tunnels standard Ethernet frames over EtherCAT
- *CANopen over EtherCAT (CoE)*: access to CANopen devices and their objects, with also an optional event-driven PDO (process data object) message mechanism

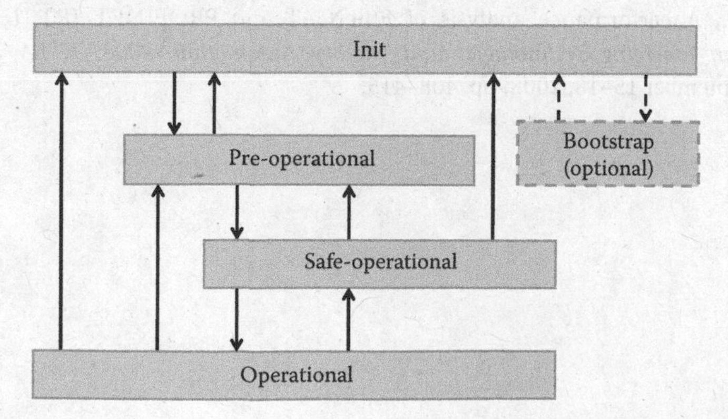

FIGURE 38.5 EtherCAT state machine.

- *File access over EtherCAT (FoE)*: download and upload of firmware and other files
- *Servo drive over EtherCAT (SoE)*: access to servo profile identifier (IDN)

The EoE and FoE protocols provide options for integrating a Web server using standard IP-based protocols such as TCP/IP, UDP/IP, and all higher protocols based on these (HTTP, FTP, SNMP, etc.). For example, EtherCAT can use a mechanism in which the Ethernet datagrams are tunneled and reassembled in a associated device, before being relayed as complete Ethernet datagrams. This procedure does not restrict the achievable cycle time since the fragments can be optimized according to the available bandwidth.

Another important aspect for a fieldbus system is to support communication protocols used for compatibility and efficient data exchange between a controller and a drive; for these purposes, EtherCAT uses well-known and established technologies. Indeed, the "CANopen over EtherCAT" (CoE) protocol enables the complete CANopen profile family to be used via EtherCAT. The SDO (service data object) protocol is used directly, so that existing CANopen stacks can be left practically unchanged. The process data are organized in PDOs, which are transferred using the efficient support of EtherCAT. Moreover, optional extensions are defined that lift the 8 byte limit and enable complete readability of the object list.

Another protocol supported is the "Servo profile over EtherCAT" (SoE). This capability enables the proven servo profile, which is specialized for demanding drive technology, to be used. The servo service channel, and therefore access to all parameters and functions residing in the drive, is mapped to the EtherCAT mailbox.

References

[BEC06] Beckoff Automation GmbH, EtherCAT per Motion Control, presented at MOTION CONTROL for, ilB2B.it, Bologna, Italy, February 23, 2006, [Online], available: http://www.ilb2b.it/mc4_2006/atti/BECKHOFF.pdf

[ESC20] Hardware Data Sheet ESC20 Slave Controller, [Online], available: Beckhoff website http://www.beckhoff.com/

[ETG04] ETG.1000.4: Data link protocol specification, available on EtherCAT Technology Group: http://www.ethercat.org/

[ETG05] ETG.1000.5: Application layer service definition, available on EtherCAT Technology Group: http://www.ethercat.org/

[IE158] International Electrotechnical Commission, Industrial communication networks—Fieldbus specifications—Part 3-12: Data-link layer service definition—Part 4-12: Datalink layer protocol specification—Type 12 elements, IEC 61158-3/4-12 (Ed.1.0).

[IE784] IEC 61784-2, Industrial communication networks—Profiles—Part 2: Additional fieldbus profiles for real-time networks based on ISO/IEC 8802-3.

[PRY08] Prytz, G., A performance analysis of EtherCAT and PROFINET IRT, *IEEE International Conference on Emerging Technologies and Factory Automation, 2008 (ETFA 2008)*, Hamburg, Germany, September 15–18, 2008, pp. 408–415.

39

Ethernet POWERLINK

Paulo Pedreiras
University of Aveiro

Stefan Schoenegger
B&R Industrial Automation

Lucia Seno
Italian National Research Council

Stefano Vitturi
Italian National Research Council

39.1 Introduction 39-1
39.2 EPL Protocol 39-1
39.3 Frame Mapping 39-3
39.4 Network Configurations 39-4
39.5 Redundancy Aspects 39-5
Medium Redundancy • MN Redundancy
39.6 Security Aspects 39-7
POWERLINK Safety
39.7 Performance Analysis 39-8
Jitter • Turn-Around Time • Cycle Time • Acyclic Traffic
References 39-10

39.1 Introduction

Ethernet POWERLINK (EPL) is a real-time Ethernet (RTE) network originally developed by B&R GmbH [1] and currently managed by the Ethernet POWERLINK Standardization Group (EPSG) [2]. The first version of EPL was developed by B&R in 2001 and subsequently published as an open standard, namely version 2.0 [11], by the EPSG in 2003. Such a version has been then encompassed in the IEC 61784 International Standard [7] where it is referred as Communication Profile (CP) #1 of the Communication Profile Family (CPF) #13. A more recent version of the EPL specification [6] is currently downloadable from the EPSG Web site. Starting from the tight timing requirements of industrial communication systems [15], the EPL protocol has been developed in order to provide very fast communication cycles (down to 100 μs) with low jitter (below 1 μs) and, at the same time, maintaining compatibility with legacy Ethernet. An application layer protocol has been introduced on top of the EPL communication stack. In particular, the well-established CANOpen standard [10] has been chosen, ensuring in such a way compatibility with a wide number of already deployed communication systems. In this context, EPL is often referred as "CANOpen over Ethernet."

39.2 EPL Protocol

The communication architecture of EPL is shown in Figure 39.1.

As can be seen, EPL makes use of the native Ethernet stack and, in particular, it is defined as a data-link layer (EPL DLL) protocol, placed on top of the Ethernet medium access control (MAC) [8] that, as it is well known, relies on a CSMA/CD (carrier sense multiple access with collision detection) technique. The physical layer of the EPL is the same as that of the Ethernet. Specifically, the 100BASE-X half-duplex transmission mode has been chosen. Moreover, both RJ-45 and M12 connectors can be used at the user's discretion. The EPL specification also recommends the use of standard patch cables (twisted pair, screened twisted pair, AWG26). Finally, concerning installation guidelines, EPL follows the recommendations of IAONA [5].

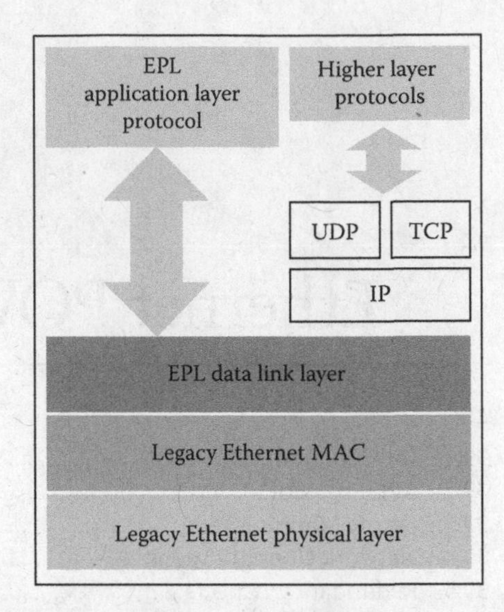

FIGURE 39.1 Ethernet POWERLINK communication architecture.

EPL defines two types of stations, namely the managing node (MN) and controlled nodes (CNs). The MN represents, usually, the "controller" of an automation system (i.e., the device that implements the automation tasks), whereas the CNs are typically field devices such as, for example, sensors/actuators. Each network contains exactly one MN and up to 240 CNs connected in different topologies, as will be illustrated in the next section. Several EPL compliant devices are off-the-shelf available from different vendors which encompass most of the components used in industrial automation systems such as, for example, programmable logic controllers (PLCs), computer numerical control (CNC) machines, sensors, actuators, etc. Moreover, since April 2008, the open source software codes of both POWERLINK MN and CN are fully downloadable [4], allowing for the free implementation of POWERLINK nodes.

Thanks to its performance figures, EPL may be employed in several fields and, in particular, for applications with very tight timing constraints such as, for example, coordinated motion control systems [9].

The DLL of EPL relies on a time division multiple access (TDMA) technique managed by the MN, which allows stations to orderly access the physical medium and hence avoiding collisions even if nonswitched Ethernet configurations are employed. In practice, the network operation is based on a periodic cycle (EPL cycle) with constant duration. As can be seen in Figure 39.2, the EPL cycle is split into four sections.

The first section, designated start period, is initiated by the MN, which broadcasts the SoC (start of cycle) frame to synchronize all the CNs. Subsequently, the isochronous period is entered,

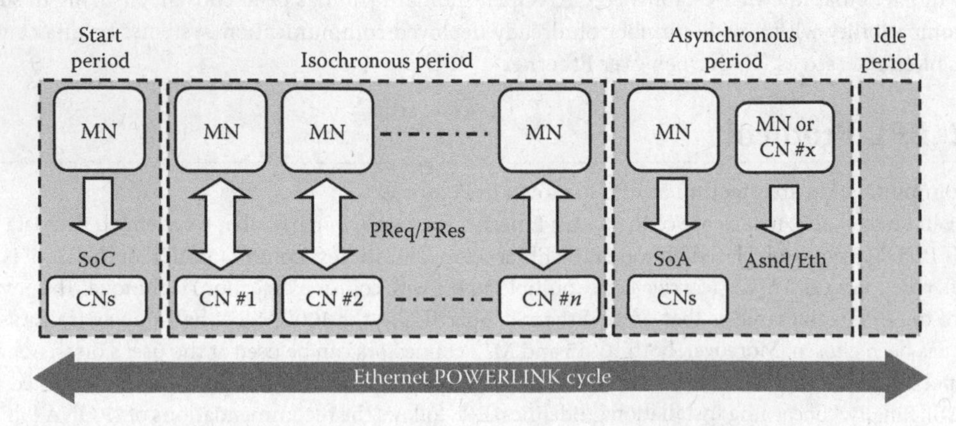

FIGURE 39.2 Ethernet POWERLINK cycle.

TABLE 39.1 EPL Parameters

Parameter Description	Value
Transmission speed	10–100 Mbps
Maximum number of CNs	240
Duration of the start period	45 μs
Quiet time of MN	1 μs
Elaboration time of each CN	8 μs

where MN and CNs exchange cyclic data. This is accomplished by means of two different frames, namely PReq (Poll Request) and PRes (Poll Response). PReq is issued by the MN toward a specific CN and carries the output data. The addressed CN, after an elaboration time, sends as a response the PRes frame containing the input data. Such a latter frame is not sent exclusively to the MN but, instead, it is broadcasted on the network so that all CNs receive it. After receiving the PRes frame, the MN waits for a quiet time, and then it moves on the next CN. It is worth observing that not all the CNs are necessarily polled at each EPL cycle. Indeed, the standard supports two communication classes, referred respectively as *continuous* and *multiplexed*, which specify the way in which CNs are addressed. In the continuous class, a CN is polled at every cycle, whereas in the multiplexed class a CN is polled every n cycles (with $n > 1$). Both communication classes may be handled contemporaneously in an EPL cycle, so that it is possible to have a polling scheme in which some CNs that are queried continuously (i.e., every EPL cycle), whereas some others are polled with a period which is an integer multiple of the EPL cycle.

At the end of the isochronous period, the MN broadcasts the SoA (start of acyclic) frame informing all CNs that the asynchronous period is started. In this period, only one asynchronous message may be sent either by the MN or by one of the CNs. In practice, during the isochronous period, the MN may collect requests for asynchronous transmissions (from itself and/or the CNs) and then, according to a priority scheme, it selects the node which is granted to transmit. The selected node, which is informed by the MN via the SoA frame, may transmit a frame belonging to one of two different categories, namely EPL AsyncSend frame and legacy Ethernet message. The AsyncSend frame type is defined by the EPL protocol; it can be used for acyclic process data (typically alarms) and it may be delivered either in unicast or in multicast mode. Concerning legacy Ethernet messages, they are used for general purpose communication, and they may be of whatever type although, to this regard, the EPL standard recommends the use of the UDP/IP protocol suite (rather than the most popular TCP/IP) since UDP introduces a lower overhead.

Finally, the idle period is entered at the end of the asynchronous period. The purpose of this time interval is to guarantee that there are no message overruns, i.e., no messages cross the EPL cycle boundaries. Thus, in this period, there is no activity on the network since all stations simply wait for the beginning of a new cycle. Table 39.1 summarizes the most relevant parameters of the EPL protocol as derived from the standardization documents.

The correct execution of the EPL cycle requires that all stations connected to the network are compliant with the EPL protocol. This mode of operation is referred in the standard document as "EPL mode." Nevertheless, a specific mode of operation (namely "basic Ethernet mode") allows for the coexistence on the same network of EPL stations as well as legacy Ethernet ones. For these configurations, however, as mentioned in the standard document, the strict determinism of the EPL cycle is no longer ensured.

39.3 Frame Mapping

The EPL frames are encapsulated in standard Ethernet protocol data units as shown in Figure 39.3.

As can be seen, the Ethernet type field is set to the hexadecimal value 88ABh which uniquely identifies EPL. The first octet of the EPL frame ("message type," i.e., the first octet in the Ethernet data field) specifies the type of frame. The codes used in order to specify the different frames are reported in Table 39.2.

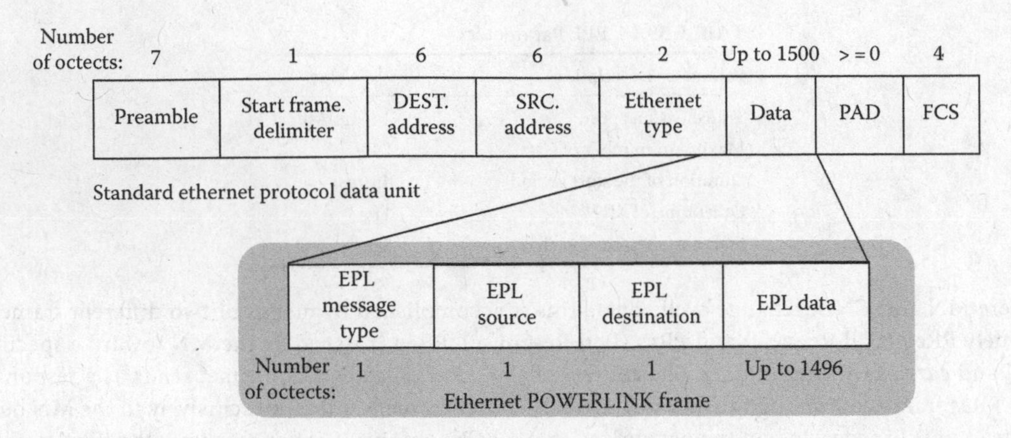

FIGURE 39.3 Ethernet POWERLINK frame.

TABLE 39.2 Codes of the Ethernet POWERLINK Frames

Ethernet POWERLINK Frame	Hexadecimal Code
Start of cyclic (SoC)	01_h
Poll request (PReq)	03_h
Poll response (PRes)	04_h
Start of asynchronous (SoA)	05_h
Asynchronous send (ASnd)	06_h

The second and third octets contain, respectively, the source and destination nodes of the message in the EPL network. The subsequent octets accommodate the data to be exchanged.

39.4 Network Configurations

The connections between the EPL nodes can be realized by means of either traditional Ethernet hubs or switches. Actually, the EPL standard recommends the use of hubs since they ensure both limited jitter (about 500 ns) and low path delay (about 40 ns). Nonetheless, switches can also be employed as connection devices, provided that additional jitter and latency are taken into account during the EPL system configuration.

The availability of standard Ethernet components allows for the implementation of several different EPL configurations. Moreover, particular CNs may be equipped with an integrated hub in order to simplify the implementation of complex network topologies with various hub levels in which traditional star configurations are combined with linear chains of CNs as shown, for example, in Figure 39.4.

Since the EPL protocol is purposely developed to avoid frame collisions, the Ethernet standard constraint of a maximum round trip time of 5.12 μs (at 100 Mbps) has not to be necessarily fulfilled. This reveals particularly helpful for linear configurations, often deployed in low-level industrial applications that, due to their structures, may have relevant round trip times.

On the other hand, the tight timing of the EPL protocol introduces further important constraints. In particular, in order to detect possible transmission errors and/or device failures, the correct operation of the protocol requires that, after issuing a PReq frame, the MN has to receive the PRes from the addressed CN within a specified time-out. The default time-out value is 25 μs. However, since it clearly

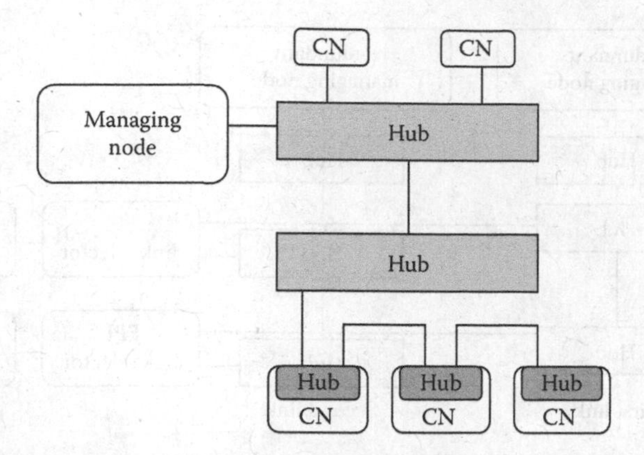

FIGURE 39.4 Example of Ethernet POWERLINK configuration.

depends on the network configuration as well as on the employed components, the EPL standard specifies that it can be either globally overwritten or set appropriately for each single CN.

39.5 Redundancy Aspects

The redundancy features of EPL are explicitly defined by the POWERLINK high-availability services [12] which represent an extension to the basic specification [11] and which maintain full compatibility with it. The high-availability services of POWERLINK have the objective of providing full network functionality even in case of any upcoming single failure on any of the POWERLINK components (MN, CNs, and physical infrastructure). The main focus of this concept is to provide a very fast switchover time in case of failure. In particular, POWERLINK high availability includes a *hot standby* functionality. The set of services provided by POWERLINK high availability rely on two specific aspects, namely medium redundancy and MN redundancy.

39.5.1 Medium Redundancy

This feature is provided by introducing a second physical cabling in the network. All devices are connected to both cables, and all information can be transmitted on both links by each node (MN as well as CNs) implementing the high-availability service set. The receive direction is bundled in the so-called link selector functionality that is part of the slave device. This functionality receives both incoming data streams and evaluates, based on predefined algorithms, which signal and information to choose. The decision algorithm can take a single attribute, such as the incoming time, or can use several attributes, such as CRC or link quality. Each link selector is requested to include its link status into the PRes frame. This information can be used by the MN to give feedback to the application regarding the quality of the network. The functionality of the link selector requires that the maximum variation of the path delay on the network, taken between any two specific nodes in both networks, shall not exceed 5.12 μs (transmission time for 64 byte on a Fast Ethernet network).

Figure 39.5 shows an example of medium redundancy. As can be seen, two links are used for communicating with the CNs. If a problem occurs on the first link, then the link selectors of the CNs that do not receive the signal correctly on the first link switch immediately to the second one.

39.5.2 MN Redundancy

MN redundancy introduces an extension to the standard MN functionality to provide high-availability services at the application layer. In order to realize a high-availability system, two or more redundant

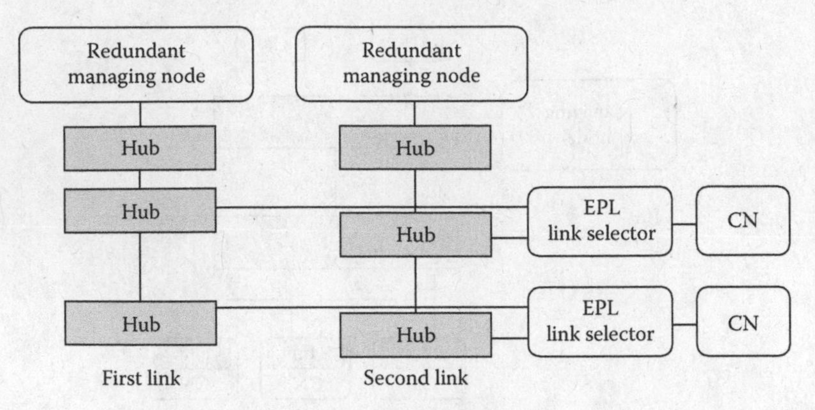

FIGURE 39.5 Example of Ethernet POWERLINK medium redundancy.

managing nodes (RMNs) need to be available in the network. In this direction, two new entities have been defined. The active managing node (AMN) is one (and the unique) of the RMNs that currently hosts and executes the functionality of the MN. The stand-by managing node (SMN) is one of the RMNs that is in stand-by state regarding MN functionality. From the network point of view, a SMN behaves like a CN, with the difference that the SMN is constantly monitoring the network and hosting also the current status of all participants. In case of any failure of the AMN, one of the RMNs will immediately takeover the functionality of the AMN. The takeover process will not last more than one single EPL cycle. The takeover process can introduce jitter, upper bounded by the maximum path delay of the network. All devices connected to the network shall be capable of tolerating this single jitter of the network. Since more than one SMN could be available in the network, an election mechanism has to executed before the beginning of network operation. Such a technique is based on the (unique) priorities of the different SMNs. The priorities could be either assigned with an engineering tool (and configured within the object dictionary) or derived from the unique node address of the SMNs.

A further way of providing basic redundancy service is to use a ring structure for the cabling architecture. A closed ring always offers the possibility of having two independent transmission paths from every node to every other node in the network. For some branches, this is a cost-effective approach, as the critical section is the physical connection between the cable and the socket, influenced by such factors of the rough industrial environment as, for example, temperature, humidity, and vibration. This feature should not be confused with the full redundant network described above. The ring redundancy for EPL networks is actually an optional product feature of the MN.

As can be noticed in Figure 39.6, EPL products (MN as well as CNs) that support ring redundancy need to be equipped with two Ethernet ports. In case of a noninterrupted transmission path, the closed ring passes the sent frame through the network back to the MN. The MN will recognize this frame as "recently sent" and simply will filter it. If such a frame is not received within a time-out, then the MN will conclude that a problem has occurred (typically a cable break). In this case, the MN will send the

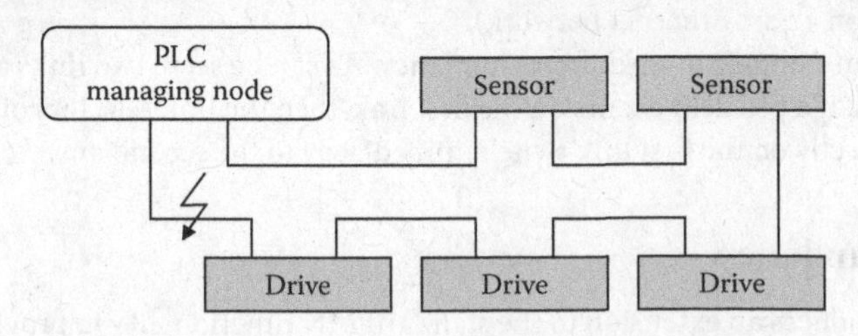

FIGURE 39.6 Example of Ethernet POWERLINK ring redundancy.

data of the next EPL cycle on both connections and make use of the alternative transmission path. If the ring is reestablished, the MN will detect the obvious collision and logically unlock the ring topology again directly.

39.6 Security Aspects

The functional principle of POWERLINK, when operating in EPL mode, clearly separates the real-time domain from any non-real-time (office) one. The access to the POWERLINK real-time domain is granted by dedicated gateway functionality of the network. These gateways typically provide a basic NAT (network address translation) mechanism and therefore hides the POWERLINK IP address from being accessed without having the exact knowledge about the gateway configuration. This principle provides basic security measures already by design, though most gateway products could include further firewall functionality on top.

39.6.1 POWERLINK Safety

POWERLINK safety is a protocol for a so-called one channel safe data transmission up to safety integrity level (SIL) 3. The protocol has been specified in 2007 by the Safety Working Group of EPSG, and the first products compliant with the specification have been certified by the TÜV Rheinland Group in 2008.

POWERLINK safety works according to the "Black Channel" principle, meaning that the safety measures are completely independent of the underlying protocol. As a restriction, however, the underlying protocol has to provide security measures for the network. Obviously, POWERLINK safety has been developed in order to optimize its integration with the EPL protocol (in this direction, both the domain separation and the high cyclic update rate revealed particularly helpful). Nonetheless, it may be employed effectively on top of any other protocol as well. POWERLINK safety allows to implement safety network management (SNMT) techniques, a safe configuration manager (SCM) for network participant and the safe exchange of process information via safe process data objects (SPDO). With POWERLINK safety, it is possible to exchange devices and they will be automatically booted and configured by the system without any user intervention. The SCM is responsible for parametrization and configuration of all safety nodes in the network. Its main functionality is to verify during boot-up of the network if the current parameter set of each node matches the expected parameters as configured by the users. In the case there is a mismatch, the SCM downloads the expected parameter set to the device and reboots it. All safety parameters are stored on a database on the SCM. For any given safe configuration of the system, the SCM is able to identify wrongly plugged modules by comparing the vendor ID and the product code of any booting device with the centrally located database. A mismatch would prevent the device from booting, and the protocol may then indicate the failure to the application layer.

POWERLINK safety devices can be used in a mixed mode with both safe and nonsafe devices connected to the same network infrastructure. This is often referred to as integrated safety technology.

The communication model of the POWERLINK safety protocol is based on the well-known producer/consumer technique. Every safe node can be producer and/or consumer of any safety-related information. The unique identifier of safe nodes as well as safe data is provided by the safety address (SADR). POWERLINK safety systems can be freely set up in every possible topology provided by the transport media. Each POWERLINK safety network requires one single SCM. The number of safety nodes for one SCM is limited to a maximum of 1023. This is then referred to as one safety domain (SD). For a huge safety network, a maximum of 1023 SDs can be connected via safety domain gateways (SDGs). From the point of view of one SD, the SDG appears simply as a safety node.

Data transmission of safety-related data is provided by the dedicated safety frame format. The maximum data that can be transferred with one frame is 254 byte. Each user data are duplicated and transmitted within two subframes. The CRC type used (either CRC8 or CRC16) depends on the size of the user data. All systematic failures will be recognized by the POWERLINK safety protocol including delay, repetition, wrong sequence, insertion, loss, and masquerade [11].

39.7 Performance Analysis

While the performance of real-time networks can be assessed according to a multitude of performance criteria, in the context of EPL the most common metrics are jitter, turn-around time (TAT), and cycle time. Other parameters, such as the latency of the asynchronous messages, are also commonly used to assess the protocol performance. This section addresses the performance analysis of the EPL protocol, illustrating how these parameters can be estimated and how key network design options affect them.

39.7.1 Jitter

Jitter, defined as the maximum time deviation that may occur on cyclic events, is defined in the EPL standard as being in the submicrosecond range. The jitter in EPL networks is determined by the MN, which is responsible for correct sending of the SoC message, as well as by the network components. While the MN jitter is intrinsic to the hardware/software implementation characteristics, the jitter due to the network components depends on the nature of the components themselves and of the network topology. Although the jitter induced by a single hub (or by a limited number of such devices) may seem negligible, as stated above EPL permits the violation of the IEEE 802.3 topology guidelines, opening the way for the use of tree or linear topologies with several cascaded hubs. Despite their convenience, such topologies have a negative impact on the global network jitter, which depends on the highest number of hubs that exist between the MN and any CN. The jitter upper bound can be computed as

$$J_{NET} = J_{mn} + N_{hub} \times J_{hub} \tag{39.1}$$

where

J_{mn} is the MN jitter
J_{hub} is the hub-induced jitter (homogeneous hubs assumed)
N_{hub} is the number of hubs between the MN and any CN

For illustrative purposes, consider a MN and a CN interconnected by a single hub. In this configuration, considering, for example, the POWERLINK interface described in [3], J_{mn} = 50 ns, whereas a typical value for the hub-induced jitter is J_{hub} = 70 ns. Thus, the jitter upper bound becomes, trivially, J_{mn} = 120 ns. For a two-level tree topology, the jitter upper bound becomes J_{mn} = 190 ns, with an increase of 58.3%. In these conditions, a linear topology has to be limited to 13 devices in order to guarantee a jitter value below 1 μs.

39.7.2 Turn-Around Time

Another important performance criterion is the TAT, defined as the time elapsed between the transmission of a request and the reception of the corresponding response. The TAT has a noticeable impact on the bandwidth utilization efficiency since this corresponds to network idle time.

As for jitter, the TAT depends on constructive aspects of the network components and devices as well as on the network topology. In software-based implementations, the EPL stack is entirely executed in a microprocessor, requiring several processing steps to decode the requests and issue the corresponding replies. The associated latency depends on the delays introduced by several factors such as code execution times, DMA techniques, interrupt handling procedures and bus protocols that may lead to response times typically in the order of some microseconds. The EPL protocol may also have a total hardware support, typically via FPGA technology. In this case, the EPL packet handler, the Ethernet controller, and the hub (if included) are completely embedded in hardware. The processing of incoming requests, in this case, is entirely carried out in hardware, and the interconnection between the Ethernet controller and the EPL packet handler can use dedicated buses, leading to extremely shorter response times, below 1 μs.

Similarly to the jitter, the use of tree or linear topologies with several cascaded hubs also has a negative impact on the TAT since each hub situated between the MN and a CN adds its own delay. Moreover, it has to be noticed that the hub-induced delay affects both the MN requests and the CN replies, doubling in such a way its impact in each query of the CN.

In general, the TAT in a transaction involving node i can be upper bounded as

$$\mathrm{TAT}^i = D^i_{\mathrm{CN}} + 2 \times D_{\mathrm{hub}} \times N^i_{\mathrm{hub}} + t_{\mathrm{prop}} + C_{\mathrm{Req}} + C_{\mathrm{Res}} \quad (39.2)$$

where

D^i_{CN} is the nominal response time of the ith CN
D_{hub} is the nominal hub delay (homogeneous hubs assumed for simplification)
N^i_{hub} is the number of hubs involved in a transaction between the MN and the ith CN
t_{prop} is the wire-propagation time
C_{Req} is the transmission time of the request message
C_{Res} is the transmission time of the corresponding response message

As an example, a transaction in which the MN polls a CN having a short amount of data to transmit (up to 36 data bytes) may be considered. In this case, $C_{\mathrm{Req}} = C_{\mathrm{Res}} = 6.72\,\mu\mathrm{s}$. Assuming also a fast CN (i.e., with hardware support) yielding a 1 μs response time, a single hub with 400 ns delay, and that the segment (wire) length is of 100 m (leading to $t_{\mathrm{prop}} = 500\,\mathrm{ns}$), the TAT, given by Equation 39.2, is (in seconds):

$$\mathrm{TAT} = 1\times10^{-6} + 2\times(400\times10^{-9} + 500\times10^{-9}) + 2\times6.72\times10^{-6} = 16.24\times10^{-6}$$

If a two-hub topology is considered, the above computed value of TAT is increased by 800 ns, corresponding to a relative degradation of about 4.9% on the TAT. For a linear topology with 10 devices, the total increase would be around 22.2%. These performance figures show how the network topology design may have a substantial impact on the TAT.

The use of switch devices may significantly worsen the performance. Indeed, cut-through switches induce a latency of at least 12 μs since message forwarding only starts after the reception of the destination address Ethernet field, thus multiplying the latency for a factor of at least 3. For store-and-forward switches, the latency time depends on the message size and can be very relevant (up to 120 μs per switch, if maximum size Ethernet messages are employed).

39.7.3 Cycle Time

The cycle time is a factor of paramount importance since EPL applications (for example, those concerned with motion control issues) may require very short cycle times (well below 1 ms, as discussed in [14]). The cycle time of EPL may be trivially calculated as the sum of the periods shown in Figure 39.2:

$$T_C = T_{st} + T_{is} + T_{ac} + T_{id} \quad (39.3)$$

where the addends in Equation 39.3 represent the durations of the start, isochronous, asynchronous, and idle periods, respectively. However, the EPL cycle time is mostly determined by the duration of the isochronous period which, in its turn, is a direct function of the number of queries to the configured CNs.

Indeed, the size of each communication slot depends on the amount of data transmitted as well as on the network topology and latency induced by the network components, which therefore also influence the cycle time. Moreover, the adoption of the multiplexed communication class (either alone or in conjunction with the continuous one) allows to reduce the maximum number of slots in each cycle and consequently the cycle time as well.

Finally, it is also important to remember that the CNs have also to perform some protocol-related tasks like, for example, SoC processing or the maintenance of the EPL cycle state machine. Thus, the time required to handle the different node activities also has to be considered. In this context, the EPL protocol provides a (limited) level of control since it allows the MN to control the poll order permitting, for example, to avoid polling the CNs with highest SoC latency immediately after the sending of the SoC frame.

39.7.4 Acyclic Traffic

According to the EPL specification, the asynchronous period aims at the transfer of non time-critical data, typically related with the network management and supervision (ASnd messages) as well as with non-EPL traffic (e.g., TCP or UDP/IP messages). Despite not being addressed in the EPL specification, the use of asynchronous services for carrying infrequent real-time data, e.g., alarms, could also bring efficiency gains provided that suitable latency values can be guaranteed.

The latency associated with the asynchronous traffic is a relevant performance figure since it is important to have acceptable response time in the network management services as well as on the user-level applications supported on IP traffic.

The latency experienced by asynchronous requests depends on the cycle duration, on the period of the real-time data associated with the CN, and on the particular scheduling algorithm used by the MN. The cycle duration defines the number of requests that can be handled by time unit. The period of the real-time data associated with the CN constraints the signaling latency since CNs notify the MN about the existence of requests via PRes messages. Thus, in the worst-case situation, the CN may have to wait one entire poll period prior to be able to inform the MN of the existence of requests. Finally, the latency also depends on the scheduling decisions at the MN. The user can influence these decisions by assigning higher priorities to the more time-sensitive asynchronous messages.

As an example, an EPL network in which N different asynchronous requests occur at the same time at one or more CNs can be considered. The CNs can only notify that event to the MN in the following cycle, and the MN shall schedule at most one request in each following cycle, thus requiring a total of $N + 2$ cycles to handle all the requests. It has to be noticed that the number of cycles necessary to process the pending requests does not depend on the cycle duration; thus decreasing the cycle time (if possible) leads to shorter response times.

On the other hand, the latency of acyclic messages could be reduced, allowing multiple ASnd transmissions in the same cycle. This is actually addressed by the latest version of the EPL specification [6], which makes such an opportunity possible, provided that there is enough remaining time in the cycle. Another possible way of improving the response time for asynchronous time-sensitive messages would consist in reserving part of the asynchronous phase exclusively for alarms, as discussed in [13].

References

1. Bernecker & Rainer Industrie-Elektronik GmbH, Eggelsberg, Austria [Online]: http://www.br-automation.com
2. Ethernet POWERLINK Standardization Group. [Online]: http://www.ethernet-powerlink.org
3. IXXAT Automation GmbH: POWERLINK interface board for PCI bus systems PL-IB 200/PCI, Technical description. [Online]: http://www.ixxat.com/powerlink_pci_interface_en.html
4. OpenPOWERLINK: Open source solution for POWERLINK MN and CN. [Online]: www.sourceforge.net/projects/openPOWERLINK

5. The IAONA Industrial Ethernet Planning and Installation Guide, IAONA Guide. [Online]: http://www.iaona-eu.com/home/downloads.php
6. Ethernet POWERLINK Communication Profile Specification V.1.1.0, EPSG Draft Std 301. [Online]: http://www.ethernet-powerlink.org, 2008.
7. IEC61784 International Standard: Digital data communications for measurement and control. Part 1: Profile sets for continuous and discrete manufacturing relative to fieldbus use in industrial control systems. Part 2: Additional profiles for ISO/IEC8802-3 based communication networks in real-time applications, International Electrotechnical Commission Std., November 2007.
8. IEEE 802.3 Standard: Carrier sense multiple access with collision detection (CSMA/CD) access method and physical layer specifications, IEEE Std., Institute of Electrical and Electronics Engineers, Inc., October 2000.
9. F. Benzi, G. Buja, and M. Felser. Communication architectures for electrical drives. *IEEE Transactions on Industrial Informatics*, 1(1):47–53, February 2005.
10. CAN In Automation, International Users and Manufacturers Group e.V. Std. CANopen Application Layer and Communication Profile, CiA/DS301, Version 4.01, June 2000.
11. Ethernet POWERLINK Standardization Group. Ethernet POWERLINK Communication Profile Specification V.0.1.0, EPSG Draft Std. [Online]: http://www.ethernet-powerlink.org, 2003.
12. Ethernet POWERLINK Standardization Group. High Availability Service Specification Part A V1.0.0, EPSG Draft Standard 302. [Online]: http://www.iaona-eu.com/home/downloads.php, 2007.
13. S. Vitturi and L. Seno. A simulation study of Ethernet powerlink networks. *IEEE Conference on Emerging Technologies and Factory Automation*, pp. 740–743, Patras, Greece, September 2007.
14. A. Pfeiffer. Ethernet powerlink: Real-time industrial ethernet is real. [Online]: http://www.ict.tuwien.ac.at/wfcs2004, *Workshop on Factory Communication Systems, WFCS 2004 Industry Day*, Vienna, Austria, September 22–24, 2004.
15. R. Zurawsky. Industrial communication systems. *The Industrial Information Technology Handbook*, CRC Press, Boca Raton, FL, pp. 37.1–47.16, 2005.

40 PROFINET

40.1 Introduction 40-1
Device Classes in PROFINET IO • Performance • Conformance Classes • Prerequisites
40.2 PROFINET IO Basics 40-5
Device Model • Address Resolution • Cyclic Data Traffic • Acyclic Data Traffic • Diagnostics
40.3 IRT Communication in PROFINET IO 40-8
Flexible Communication Based on RT_CLASS_2 (Orange Interval) • Communication Based on RT_CLASS_3 (Red Interval) • Cycle Duration and Constrains
40.4 Engineering and Commissioning 40-11
GSD File • Device Addressing • System Power-Up • Neighborhood and Topology Detection • Redundancy
40.5 Integration of Fieldbus Systems and Web Applications 40-14
Integration via Proxy • Web Integration
Acronyms 40-15
Bibliography 40-15

Max Felser
Bern University of Applied Sciences

Paolo Ferrari
University of Brescia

Alessandra Flammini
University of Brescia

40.1 Introduction

PROFINET is the real-time (RT) Ethernet solution developed by PROFIBUS International (PI), the world association of PROFIBUS and PROFINET manufacturers and users. The PROFINET protocol is an international standard since it has been incorporated in the current edition of the IEC 61158 and IEC 61784 standards. In details, the IEC61784 defines the subsets of the services specified in IEC 61158 that are to be applied for PROFINET under the designation "Family 3" with subdivisions 3/3, 3/4, 3/5, and 3/6.

From the abstract point of view, the PROFINET concept is a modular concept that allows the user to choose the functionality he requires. Generally, the functionality differs mainly in terms of the type of data exchange; for instance, some applications have very stringent requirements for data transmission speed, while other applications need data refresh cycles of tens of milliseconds.

The development of PROFINET started in 2000. The first released specifications described the so-called **PROFINET CBA** (component-based automation) protocol. PROFINET CBA is suitable for component-based machine-to-machine communication via TCP/IP and object-oriented programming. It enables a simple modular design of plants and production lines based on distributed intelligence using graphics-based configuration of communication between intelligent modules. The basic idea behind CBA is that whole automation systems can often be grouped into autonomously operating units. The structure and functionality can be repeated in identical, or slightly modified, form in multiple plants. These so-called PROFINET CBA components are generally controlled by an easily identified set of input signals. Within the component, a control program written by the user executes the required functionality of the component and sends the corresponding output signals to another controller. Thanks to its component-based architecture, the communication in PROFINET CBA is configured rather than programmed. On the other hand, the communication with PROFINET CBA offers bus cycle times of approximately 50–100 ms (when the sole TCP/IP services are used), roughly compatible with RT requirements in modular plant manufacturing. Faster bus cycles (on the order of milliseconds) are possible if an additional data-exchange modality that also uses Ethernet frames is activated.

A more recent version of the PROFINET specifications introduced **PROFINET IO** (input output) in order to deal with the more-demanding distributed I/O. PROFINET IO features RT communication (best effort paradigm) and isochronous real-time (IRT) communication (reserved band and time paradigm) with the distributed I/O. The designations RT and IRT are used solely to describe the RT properties of communication within PROFINET IO; the user can scale the performance following the application requirements. PROFINET IO describes the overall data exchange between controllers and devices as well as the parameterization and diagnostic options. The bus cycle times for the data exchange are much faster than in PROFINET CBA, e.g., cycle time in the order of few hundred microseconds is possible.

In the following of this chapter, PROFINET IO will be discussed in details, since it is the more innovating part of the PROFINET concept. Currently, PRONIFET CBA is diffused and used mainly in the North American market, while it is not well established in Europe and other continents. On the contrary, PROFINET CBA is the optimal solution for plant supervising and control. PROFINET CBA and PROFINET IO can be operated separately and in combination such that a PROFINET IO plant appears in the plant view as a PROFINET CBA unit.

40.1.1 Device Classes in PROFINET IO

PROFINET follows the provider/consumer model for data exchange. The provider (i.e., the source of data as defined by IEC Standard), usually the field device at the process level, provides process data to a consumer (normally a PLC with a processing program). In principle, a PROFINET IO field device can contain any arrangement of provider/consumer functions. Figure 40.1 presents the device classes (IO controller, IO supervisor, IO device) and the communication services. The following devices classes are defined to facilitate structuring of PROFINET IO field devices:

IO Controller: This is typically the programmable logic controller (PLC) on which the automation program runs (corresponds to the functionality of a class 1 master in PROFIBUS).

IO Supervisor (e.g., engineering station): This can be a programming device (PG), personal computer (PC), or human machine interface (HMI) device for commissioning or diagnostic purposes.

IO Device: An IO device is a distributed I/O field device that is connected via PROFINET IO (corresponds to the function of a slave in PROFIBUS).

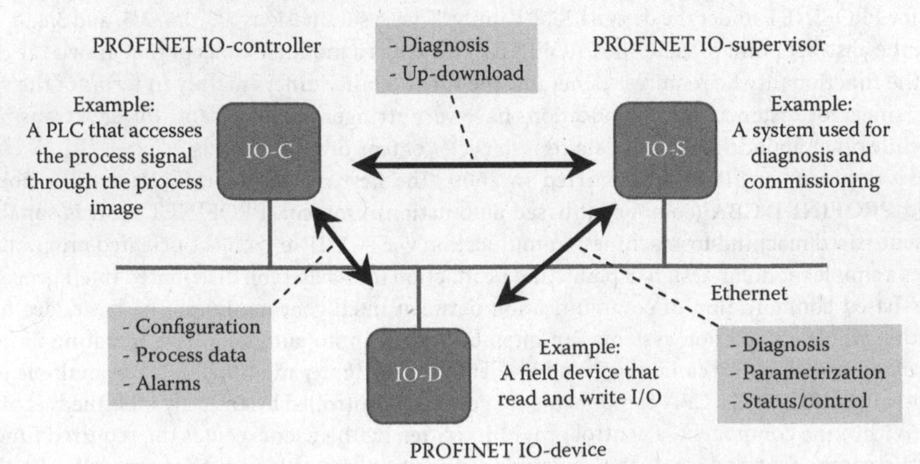

FIGURE 40.1 Device classes and their relations.

A plant unit contains at least one IO controller and one or more IO devices. An I/O device can exchange data with multiple I/O controllers. IO supervisors are usually integrated only temporarily for commissioning or troubleshooting purposes.

40.1.2 Performance

RT communication constitutes the basis for data exchange in PROFINET IO. RT data are handled with higher priority compared to TCP (UDP)/IP data. This method of data exchange allows bus cycle times in the range of a few tens of milliseconds to be achieved.

Standard Ethernet communication via TCP (UDP)/IP communication is sufficient for data communication in some cases. In industrial automation, however, requirements regarding time behavior and isochronous operation exist that cannot be fully satisfied using the UDP/IP channel.

For these reasons, PROFINET IO adopts a scalable RT approach. A PROFINET IO system with RT can be realized with standard network components, such as switches and standard Ethernet controllers. RT communication takes place without TCP/IP information, i.e., the transmission of RT data is based on cyclical data exchange using a provider/consumer model and best effort paradigm. The communication mechanisms of layer 2 (according to the ISO/OSI model) are sufficient for this. For optimal processing of RT frames within an IO device, the VLAN tag according to IEEE 802.1Q (prioritization of data frames) has been supplemented with a special Ethertype that enables fast channelization of these PROFINET frames in the higher level software of the field device. Ethertypes are allocated by IEEE and are, therefore, an unambiguous criterion for differentiation among Ethernet protocols. Ethertype 0x8892 is specified in IEEE and is used for fast data exchange in PROFINET IO.

Isochronous data exchange with PROFINET is defined in the IRT concept. Data exchange cycles are normally in the range of a few hundred microseconds to 1 ms. IRT communication differs from RT communication mainly in its isochronous behavior, meaning that the bus cycles are started with maximum precision. For example, the start of a bus cycle can deviate by a maximum of 1 μs. IRT is required in motion control applications (positioning operations).

To enable enhanced scaling of communication options and, thus, also of determinism in PROFINET IO, RT classes have been defined for data exchange. From the user perspective, these classes involve unsynchronized and synchronized communication. The details are managed by the field devices themselves. RT frames are automatically prioritized in PROFINET compared to UDP/IP frames. This is necessary in order to prioritize the transmission of data in switches to prevent RT frames from being delayed by UDP/IP frames. PROFINET IO differentiates the following classes for RT communication. They differ not in terms of performance but in determinism.

RT_CLASS_1: Unsynchronized RT communication within a subnet. No special addressing information is required for this communication. The destination node is identified using the "destination address" only. Unsynchronized RT communication within a subnet is the usual data transmission method in PROFINET IO. If the RT data traffic has been restricted to one subnet (same network ID), this variant is the simplest. This communication path is standardized in parallel to UDP/IP communication and implemented in each IO field device. A deliberate decision was made here to eliminate the management information of UDP/IP and RPC. The RT frames received are already identified upon receipt using the Ethertype (0x8892) and forwarded to the RT channel for immediate processing. Any (industrial-grade) standard switch can be used in this RT class.

RT_CLASS_2: Frames can be transmitted via synchronized or unsynchronized communication. Unsynchronized communication in this case can be viewed exactly the same as RT_CLASS_1 communication. In synchronized communication, the start of a bus cycle is defined for all nodes. This specifies exactly the allowable time base for field device transmission. For all field devices participating in RT_CLASS_2 communication, this is always the start of the bus cycle. The switches used for this communication class must support this synchronization mechanism. This type of data transmission, which has been designed for performance, brings with it specific hardware requirements (Ethernet controller/switch with support of isochronous operation).

RT_CLASS_3: Synchronized communication within a subnet. During synchronized RT_CLASS_3 communication, process data are transmitted with maximum precision in an exact order specified during system engineering (maximum allowable deviation from start of bus cycle of 1 μs). With the aid of topology-optimized data transmission, this is also referred to as IRT functionality. In RT_CLASS_3 communication, there are no wait times. In order to take advantage of the data transmission designed for maximum performance, special hardware requirements apply (Ethernet controller with support of isochronous operation).

RT_CLASS_UDP: The unsynchronized cross-subnet communication between different subnets requires addressing information via the destination network (IP address). This variant is also referred to as RT_CLASS_UDP. Standard switches can be used in this RT class.

For RT frames, data cycles of 5 ms at 100 Mbps in full-duplex mode with VLAN tag are sufficient. This RT communication can be realized with all available standard network components.

40.1.3 Conformance Classes

Since the complete scope of functions implemented in PROFINET IO is not required in every automation system, the conformance classes/application classes are introduced. The objective is to simplify the application areas of PROFINET IO. The resulting application classes enable plant operators to easily select field devices and bus components with explicitly defined minimum properties. The minimum requirements for three conformance classes (CC-A, CC-B, CC-C) have been defined from the perspective of the plant operator. In addition to the three application classes, additional specifications have been made for device types, type of communication, transmission medium used, and redundancy behavior. This specification ensures the interoperability in an automation system with regard to the scope of functions and performance parameters. It is assured that all field devices within the selected CC meet the same minimum requirements. The application areas currently are

CC-A: Use of the infrastructure of an existing Ethernet network including integration of basic PROFINET functionality. All IT services can be used without restrictions. Examples of typical applications are in building automation and process automation. Wireless communication is only possible in this class.

CC-B: In addition to the functions of CC-A, the scope of functions of CC-B supports easy and user-friendly device replacement without the need for an engineering tool. To increase the data security, a media redundancy protocol for TCP (UDP)/IP data is integrated. Examples of typical applications are in automation systems with a higher-level machine controller that place relatively low demands for a deterministic data cycle.

CC-C: In addition to the functions of CC-B, the scope of functions of CC-C supports high-precision and deterministic data transmission, including for isochronous applications. The integrated media redundancy enables smooth switchover of the I/O data traffic if a fault occurs. An example of a typical application is the field of motion control.

40.1.4 Prerequisites

PROFINET IO field devices are addressed using MAC addresses and IP addresses. Each subnet is represented by a different network_ID (subnet mask). PROFINET IO field devices are always connected as network components via switches. This takes the form of a star topology with separate multiport switches or a line topology with switches integrated in the field device (two ports occupied). Within a network, a PROFINET IO field device is addressed by its device MAC address.

PROFINET transmits some message frames (e.g., for synchronization, neighborhood detection) with the MAC address for the respective port and not the device MAC address. For this reason, each switch port in a field device requires a separate port MAC address. Therefore, a two-port field device has three MAC addresses in the as-delivered condition. However, these port MAC addresses are not visible to users. Because the field devices are connected via switches, PROFINET always sees only point-to-point connections (same as Ethernet). That is, if the connection between two field devices in a line is interrupted, the field devices located after the interruption are no longer accessible. If increased availability is required, provision must be made for redundant communication paths when planning the system, and field devices/switches that support the redundancy concept of PROFINET must be used. Any switch that supports "autonegotiation" (negotiating of transmission parameters) and "autocrossover" (crossing of send and receive lines in the switch) is suitable for PROFINET IO RT communication. On the other hand, IRT communication requires switch with embedded synchronization capabilities and PROFINET frame managing. Any Ethernet cable is suitable for PROFINET applications from the electrical point of view; however, further requirements due to the factory/industry environment could apply.

40.2 PROFINET IO Basics

PROFINET IO protocol provides definitions for the following services: device model; address resolution for field devices; cyclic transmission of I/O data (RT and IRT); acyclic transmission of alarms to be acknowledged; cyclic transmission of data (parameters, detailed diagnostics, I&M data, information functions, etc.) on an as needed basis; redundancy mode for RT frames.

The combination of these communication services in the higher level controller makes it possible to implement convenient system diagnostics, topology detection, and device replacement.

40.2.1 Device Model

The overview of the device modeling, given in Figure 40.2, is advantageous to facilitate understanding of process data addressing in a PROFINET IO field device. For field devices, a distinction is made between *compact field devices* (the degree of expansion is defined in the as-delivered condition and cannot be

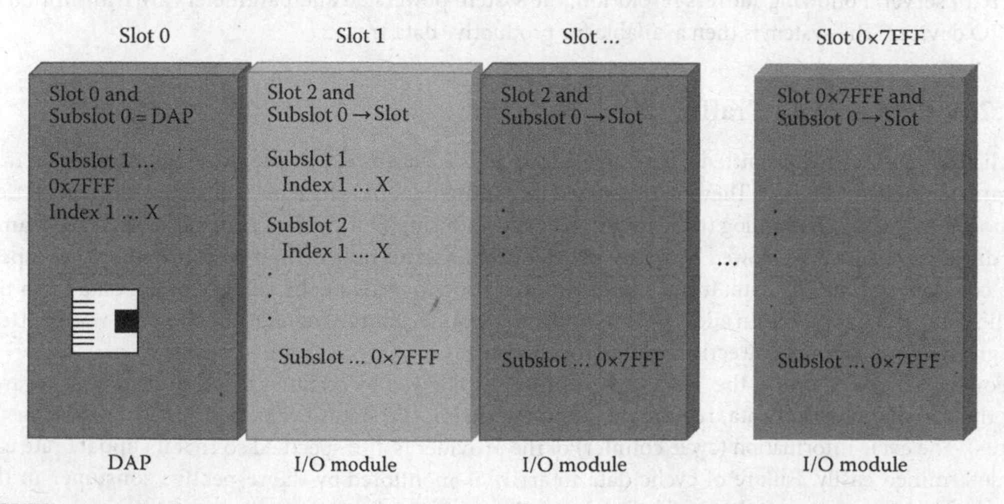

FIGURE 40.2 I/O data are addressed in PROFINET on the basis of slots and subslots.

changed to meet future requirements) and *modular field devices* (for different applications, the degree of expansion can be customized to the use case when configuring the system).

All field devices are described in terms of their available technical and functional properties in a general station description (GSD) file to be created by the field device developer. It contains, among other things, a representation of the device model that is reproduced by the device access point (DAP) and the defined modules for a particular device family. A DAP is, so to speak, the bus interface (access point for communication) to the Ethernet interface and the processing program. It is defined along with its properties and available options in the GSD file. A variety of I/O modules can be assigned to it in order to manage the actual process data traffic.

The slot designates the physical slot of an I/O module in a modular I/O field device in which a module described in the GSD file is placed. Within a slot, the subslots form the actual interface to the process (inputs/outputs). The granularity of a subslot (bitwise, bytewise, or wordwise division of I/O data) is determined by the manufacturer. The data content of a subslot is always accompanied by status information, from which the validity of the data can be derived.

The index specifies the data within a slot/subslot that can be read or written acyclically via read/write services. For example, parameters can be written to a module or manufacturer-specific module data can be read out on the basis of an index. Cyclic I/O data are addressed by specifying the slot/subslot combination. These can be freely defined by the manufacturer.

For acyclic data traffic via read/write services, an application can specify the exact data to be addressed using slot and subslot. For demand-oriented data exchange, the third addressing level, i.e., the index is added. The index defines the function that is to be initiated via the slot/subslot combination (e.g., reading of input data of a subslot, reading of I&M functions, and reading of actual/desired configuration).

40.2.2 Address Resolution

For PROFINET IO field devices, address resolution is based on the symbolic name of the device, to which a unique MAC address is assigned.

After the system is configured, the engineering tool loads all information required for data exchange to the IO controller, including the IP addresses of the connected IO devices. Based on the name (and the associated MAC address), an IO controller can recognize the configured field devices and assign them the specified IP addresses using the discovery and configuration protocol (DCP) integrated in PROFINET IO. Alternatively, addressing can be performed via a dynamic host configuration protocol (DHCP) server. Following address resolution, the system powers up and parameters are transmitted to the IO devices. The system is then available for productive data traffic.

40.2.3 Cyclic Data Traffic

Cyclic I/O data are transmitted unacknowledged as RT data between provider and consumer in a parameterizable resolution. They are organized into individual I/O elements (subslots). The connection is monitored using a watchdog (time monitoring mechanism). During data transmission in the frame, the data of a subslot are followed by a provider status. This status information is evaluated by the respective consumer of the I/O data. It can use this information to evaluate the validity of the data from the cyclic data exchange alone. In addition, the consumer statuses for the counter direction are transmitted. Diagnostics are no longer directly required for this purpose.

For each message frame, the "data unit" (trailer) is followed by accompanying information regarding the global validity of data, redundancy, and the diagnostic status evaluation (data status, transfer status). The cycle information (cycle counter) of the provider is also specified so that its update rate can be determined easily. Failure of cyclic data to arrive is monitored by the respective consumer in the communication relation. If the configured data fail to arrive within the monitoring time, the consumer sends an error message to the application.

For data exchange with multiple parameters, multicast communication relation (MCR) has been defined. This allows direct data traffic from a provider to multiple nodes (up to all nodes) as direct data exchange. MCRs within a segment are exchanged as RT frames. Cross-segment MCR data follow the data exchange of the RT class.

40.2.4 Acyclic Data Traffic

Acyclic data exchange can be used to parameterize and configure IO devices or to read out status information. This is accomplished with read/write frames via standard IT services using UDP/IP.

In addition to the data records available for use by device manufacturers, the following system data records are also specially defined: diagnostic information can be read out by the user from any device at any time; error log entries (alarms and error messages), which can be used to determine detailed timing information about events within an IO device; identification information as specified in PI guideline "I&M functions"; information functions regarding real and logical module structuring; readback of I/O data.

An index is used to distinguish which service is to be executed with the read/write services.

In PROFINET IO, the transmission of events is modeled as part of the alarm concept. These include both system-defined events (such as removal and insertion of modules) and user-defined events detected in the control systems used (e.g., defective load voltage) or occurring in the process being controlled (e.g., temperature too high). When an event occurs, sufficient communication memory must be available for data transmission to ensure against data loss and to allow the alarm message to be passed quickly from the IO device. The application in the data source is responsible for this. Alarms are included in acyclic RT data.

The ability to read out basic information from a field device is very helpful in many cases. For example, this allows inferences to be drawn in response to incorrect behavior or regarding unsupported functionality in a field device. Therefore, IO devices must supply at least the following data: order ID, MAC address, hardware revision, software revision, device type, vendor ID, all I&M0 data. These data are necessary for addressing the field device as well as for reading out the I&M functions. (*Note*: Each IO device must support at least the IM0 function.)

40.2.5 Diagnostics

PROFINET IO transmits high-priority events mainly as alarms. These include both system-defined events (such as removal and insertion of modules) and user-defined events (e.g., defective load voltage) detected in the control systems used or occurring in the process (e.g., boiler pressure too high). Diagnostic and status messages represent another means of forwarding information regarding incorrect behavior in a system. These are not transmitted actively to the higher level controller. In order to assign them explicitly, PROFINET distinguishes between process and diagnostic alarms.

Process alarms must be used if the message originates from the connected process, e.g., a limit temperature was exceeded. In this case, the IO device may still be operable. The data are not saved locally in the submodule.

Diagnostic alarms must be used if the error or event occurs within an IO device (or in conjunction with the connected components, such as a wire break). Diagnostic and process alarms can be prioritized differently by the user. In contrast to process alarms, diagnostic alarms are identified as incoming or outgoing.

For system power-up, the IO controller transfers the connect frame containing the "CMInactivityTimeout-Factor," which is used to monitor the system power-up. This monitoring time ends after the first valid data exchange between IO controller and IO device and is then replaced by the watchdog function. In PROFINET IO communication, the cyclic data traffic between provider and consumer is monitored by the watchdog function, which is integrated by default. Cyclic data including status information are transmitted between the IO controller and IO device. A consumer detects failure of the communication connection based on expiration of the watchdog and the application in the consumer is thereby informed. The response to this must be defined on a user-specific basis. The network diagnostics is part of the diagnostics management and contributes significantly to the reliability of the network operation.

For maintenance purposes and for monitoring the network components, simple network management protocol (SNMP) has been established as the international standard. SNMP allows both read access and write access (for administration) of network components in order to read out statistical data pertaining to the network as well as port-specific data and information regarding the neighborhood detection. In PROFINET, only read access to device parameters has been initially specified. Like the DHCP of IP management, SNMP will also be optional (mandatory for conformance classes CC-B and CC-C). When SNMP is implemented in components, only the standard information usual for SNMP (MIB 2) is accessed. It should be noted that SNMP will not open another diagnostic path but rather enable integration into network management systems that generally do not process PROFINET-specific information. The SNMP software can be integrated in the PROFINET stack at the user level and used without restrictions. When standard switches are used, the switch directly forwards the diagnostic information from the connected PROFINET devices to the controller. However, a switch can also be configured as a PROFINET IO device and relay the detected network errors of a lower level Ethernet line directly to the controller.

40.3 IRT Communication in PROFINET IO

PROFINET IO provides scalable RT classes for cyclic transmission of process data. In addition to the requirements for RT capability, there are also processes that require isochronous I/O data transfer (reserved band and time slot paradigm). For this reason, synchronized PROFINET communication, also called IRT communication or isochronous communication, was introduced. For isochronous data exchange, PROFINET offers a scalable concept that, on the one hand, provides a very flexible method of communication that uses the synchronized RT_CLASS_2 communication. On the other hand, PROFINET offers communication designed for maximum performance, which requires precise planning of communication paths in advance. The available bandwidth is utilized optimally in this case because waiting times can never occur during data transmission. This modality uses a synchronized RT_CLASS_3 communication. The communication is divided into a reserved interval and an open interval. Only the time-critical I/O data are transferred in the reserved interval, while all other data are sent in the open phase. No additional lower level protocol is required for this. All field devices participating in IRT communication are synchronized by the same clock master, which is generally integrated in the IO controller. In order to achieve better synchronization, the *line delay* between the neighboring nodes and the current synchronization is also determined. IRT communication is based on the following conditions:

1. The communication takes place exclusively within one subnet, i.e., the existing addressing mechanisms have been reduced (also for unsynchronized communication).
2. The bus cycle is divided into a reserved IRT phase and an open phase. These are defined as follows:
   In the "reserved interval" (IRT phase), only IRT jobs can be processed.
   In the "open interval," job processing is managed according to the rules in IEEE 802 (based on priorities).
3. All field devices within an IRT domain must support isochronous operation, even if the application is not operating synchronously.

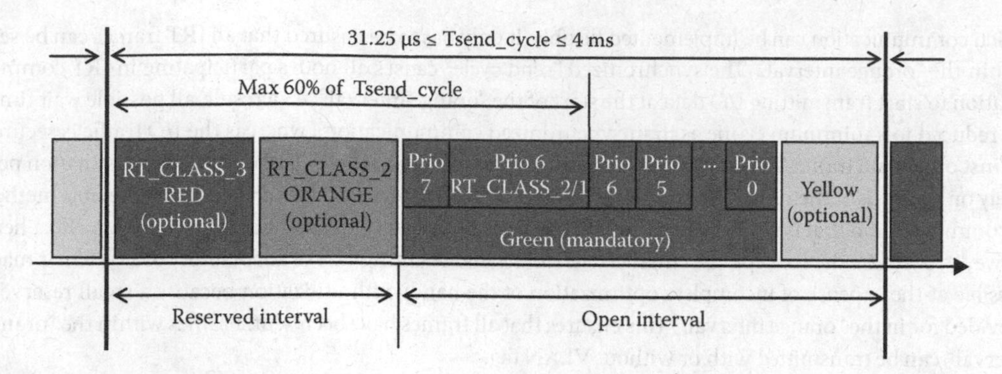

FIGURE 40.3 Bus cycle of PROFINET IO with IRT communication.

The division of the cycle is shown in Figure 40.3.

The following intervals are defined along with their properties in PROFINET:

Red interval: Only RT_CLASS_3 frames may be forwarded through switches in this interval. The forwarding rules defined in IEEE 802.1D do not apply here. Instead, the forwarding rules defined in IEC 61158 are used. The start time of the red interval is constantly synchronized. The chronological sequence of all RT_CLASS_3 frames is defined during engineering. If UDP/IP frames arrive or are generated (because the application is not IRT-capable) during a "red interval," they are temporarily saved in an IRT-capable switch and are sent only after completion of the "reserved interval." The frame IDs used to identify the different frames is specified during plant configuration in the engineering tool. The receipt of the cyclic data is timed exactly such that the synchronous application can be started directly without delays.

Orange interval: Only RT_CLASS_2 frames may be forwarded through switches in this interval. The forwarding rules defined in IEEE 802.1D are used here. The "orange interval" starts (if present) immediately at the start of a "send clock" or after the "red interval." RT_CLASS_2 frames require no prior planning. As a result, the available band width is not optimally used. Receipt of the cyclic data is not timed exactly. A safety reserve must therefore be included.

Green interval: For forwarding of data frames in switches, the rules defined in IEEE 802.1D apply. Prioritization can occur based on IEEE 802.1Q (VLAN tag). If IRT frames arrive during the "green interval," they are destroyed and an alarm message is generated. The important thing is that no jobs are still active at the end so that the reserved interval can start unhindered. A "green interval" does not have to exist within a phase.

Yellow interval: The transition from the "green interval" to the "red interval" is preceded by a "yellow interval" in which an IRT-suitable switch accepts only jobs that can be completely transported before the start of the next "red interval."

For forwarding data frames in switches, the rules defined in IEEE 802.1D may be disabled to ensure the start of the next reserved phase. Prioritization can occur based on IEEE 802.1Q (VLAN tag). If the forwarding of these frames before the start of the next reserved interval is not assured, these frames are stored temporarily and sent in the next "green interval."

40.3.1 Flexible Communication Based on RT_CLASS_2 (Orange Interval)

For communication during the "orange interval" in "switched Ethernet networks," the configuration of end nodes is sufficient. During the power-up phase, all network components (switches) in between set up address tables can be used to forward the received frames to their appropriate destinations. The communication is trained in a quasi manner using "source and destination MAC addresses." The rules defined in IEEE 802.1D are used here. In the "orange interval," data must always be exchanged in conjunction with bus synchronization (synchronized RT_CLASS_2). Frames are transmitted within one Send clock in the "orange interval," in which communication can be implemented flexibly. It only has to be ensured that all IRT frames can be sent within the "orange interval." The synchronized "send cycle" causes all nodes participating in IRT communication to start transmitting I/O data at the start of the "orange interval." As a result, all possible wait times are reduced to a minimum (same as in unsynchronized communication), whereas the I/O traffic is secured against other data traffic. The data frames are transmitted to the end node via the respective destination port solely on the basis of their MAC address (and corresponding Frame_ID). This enables a very flexible method of communication that is not subject to any special rules. Changes in the plant topology have no effect here. However, based on the concept, the enhanced flexibility and resulting ease of adaptability of a system is made possible at the expense of incomplete optimization of the bandwidth utilization because a small reserve is provided for in the "orange interval." This ensures that all frames have been sent. Frames within the "orange interval" can be transmitted with or without VLAN tag.

40.3.2 Communication Based on RT_CLASS_3 (Red Interval)

In this case, the communication in the "red interval" is based on a schedule configured in advance (during engineering phase), i.e., in addition to the information for the end nodes, the network components located in between require information defining the forwarding of frames. Frames are forwarded based exclusively on the planning algorithm defined in IEC 61158. As a result of planning, an Ethernet controller (or more precisely, the integrated switch in an Ethernet controller) knows exactly which frame arrives at which port and when it must be forwarded to where. This enables a very high utilization of the bandwidth available. If a system requires RT_CLASS_3 communication, the bus cycle must be divided into a "red interval" and a UDP/IP part (green interval) during engineering. Here, the timing and length of each frame to be sent is specified on a port-by-port basis. The plant topology, the respective frame length, and the cable lengths between the individual nodes are critical factors in the timing for the purpose of its optimal utilization. If the system is changed, the planning algorithm must therefore be repeated. Data transmission in the IRT portion is always scheduled. The "schedule" is geared only to the sequence of arriving frames, which is determined by their Frame_ID and the frame length. The time-controlled processing of jobs within the "red interval" aids in eliminating the final sources of inaccuracies. Since RT_CLASS_3 communication is oriented only on the basis of timing, the throughput times of data frames through a switch are significantly shorter. Likewise, the performance in branched networks can be increased by optimized use planning of the same communication path. The topology information is sent to the respective IO controller during system power-up. RT_CLASS_3 frames are always sent without VLAN tag since the chronological position is always known.

40.3.3 Cycle Duration and Constrains

The maximum frame length in Ethernet/PROFINET yields a minimum duration of the "green interval" of 125 μs (4 × 31.25 μs). However, certain rules must be defined in order to carry out efficient data communication involving different RT classes, ensuring that the timing and the isochronous operation are adhered to in every configuration. The general rule is that at least two TCP/IP frames with maximum length can be sent per millisecond. This corresponds to a transmission time of approximately 250 μs. For bus cycles <= 500 μs, the rule is reduced to one TCP/IP frame. Furthermore, the transmission time of cyclic data should not exceed 60% of the bus cycle in order to allow sufficient time for TCP/IP communication (see Figure 40.5).

Only field devices that support synchronization measures can participate in synchronized communication (bus synchronization). Otherwise, the timing of the different phases cannot be adhered to. Each device has its own update time. The advantage of the device-granular update specification is that the bandwidth can be shared by fast nodes and slow nodes. As a result, the update rate is no longer determined by the total number of nodes but instead can be adapted according to the application. Under certain conditions, it is possible to have a mixture of isochronous and nonisochronous applications in field devices in one automation system. However, between two devices that run isochronously, only switches with support to isochronous operation can be used.

40.4 Engineering and Commissioning

The PROFINET support to the system engineering is very powerful in order to reduce the commissioning phase of the network. For instance, the PROFINET functionalities of a certified device are described in a portable file (the GSD file), and network management (addressing, redundancy, etc.) are included in the protocol stack.

40.4.1 GSD File

The functionality of a PROFINET IO device is always described in a GSD file. This file contains all data that are relevant for engineering as well as for data exchange with the IO device. PROFINET IO devices can be described using XML-based GSD. The description language of the GSD file, i.e., Generic Station Description Markup Language (GSDML), is based on international standards. As the name suggests, the GSD file is a language-independent eXtensible Markup Language (XML) file. Many XML parsers are currently available on the market for interpreting XML files. Every manufacturer of a PROFINET IO device must supply an associated GSD file according to the GSDML specification. This file is tested as part of certification testing. To describe PROFINET IO devices, PI provides an XML schema to each manufacturer. This allows a GSD file to be created and tested easily. The need for numerous subsequent input checks is therefore omitted. In addition, the device model of PROFINET IO exhibits a further hierarchy level for data addressing when compared to PROFIBUS. Thus, e.g., addressing within a field device (in PROFIBUS: slot and index) has been expanded to include the identifier of a subslot. In PROFINET IO, addressing within a field device can be performed with finer granularity (slot and subslot). In addition, this type of addressing could not be described with the GSD file for PROFIBUS. To enable system engineering, the GSD files of the field devices to be configured are required. The field device manufacturer is responsible for supplying these. During system engineering, the configuring engineer joins together the modules/submodules defined in the GSD file to map them to the real system and to assign them to slots/subslots. The configuring engineer configures the real system, so to speak, symbolically in the engineering tool.

40.4.2 Device Addressing

A logical name is assigned to every field device. It should reference the function or the installation location of the device in the plant and ultimately lead to assignment of an IP address during address resolution. The name can always be assigned with the DCP integrated by default in every PROFINET IO field device. PROFINET provides also the option for address setting via DHCP or other manufacturer-specific mechanisms. The addressing options supported by an IO field device are defined in the GSD file for the respective device. An IO controller has all the information needed for addressing the IO devices and for data exchange after all the necessary information have been downloaded by the manufacturer-specific engineering tool.

Before it can perform data exchange with an IO device, the IO controller must assign the IO device an IP address based on the device name. This must take place prior to system power-up ("Power on" or a "Reset"). The IP address is assigned within the same subnet using the DCP protocol integrated by default in every PROFINET IO device. An IO controller always initiates system power-up after a startup/restart based on the configuration data without any intervention by the user.

40.4.3 System Power-Up

Following a "power on," the following steps are performed in the field device: initializing the physical interfaces in an IO device in order to accommodate the data traffic; negotiating the transmission parameters; determining the degree of expansion in the field device and communicating the

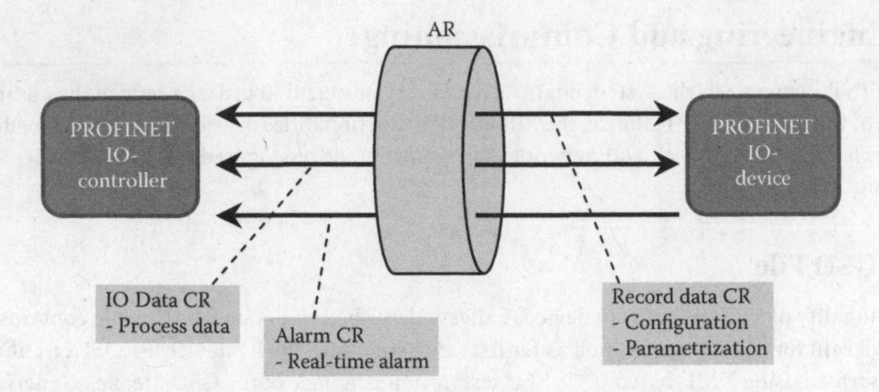

FIGURE 40.4 Data communication in PROFINET IO is encapsulated in "application and communication relations."

information to the context management; starting the exchange of the neighborhood information; address resolution on the side of the IO controller; establishment of communication between IO controller and IO device; parameterizing the submodules in the device (write records); retentive saving of port information to the physical device (PDev); completing and checking the parameterization, and starting the data exchange.

To establish communication between the higher level controller and an IO device, the communication paths must be established. These are set up by the IO controller during system startup based on the configuration data in the engineering system. This specifies the data exchange explicitly. As shown in Figure 40.4, every data exchange has an embedded "application relation" (AR). This establishes a precisely specified application (connection), i.e., the AR, between the higher level controller (IO controller or IO supervisor) and the IO device. Within the AR, "communication relations" (CR) specify the data explicitly. An IO device can have multiple ARs established from various IO controllers.

The IO controller initiates setup of an AR during system power-up. As a result, all data for the device modeling, including the general communication parameters, are downloaded to the IO device. At the same time, the communication channels for cyclic/acyclic data exchange (IO data CR, record data CR), alarms (alarm CR), and multicast communication relations (MCR) are set up. CR for data exchange must be set up within an AR. These specify the explicit communication channel between a consumer and a provider.

40.4.4 Neighborhood and Topology Detection

PROFINET specification describes a method to replace a device without using an engineering tool. This task is done by means of neighborhood detection with the link layer discovery protocol (LLDP) according to IEEE 802.1 AB. This requires the ability to determine the data of neighboring devices on a port-by-port basis using LLDP services and to provide these data to the higher level controller. Together, these conditions enable modeling of a plant topology and convenient plant diagnostics as well as device replacement without additional tools. PROFINET IO field devices exchange existing addressing information with connected neighbor devices over each switch port. The neighbor devices are thereby unambiguously identified and their physical location is determined. The LLDP protocol is implemented in software and therefore requires no special hardware support. LLDP is independent of the network structure (line, star, etc.). Automation systems can be configured with a line, star, or tree structure. For this reason, it is important to know which field devices are connected to which switch port and the identity of the respective port neighbor. The higher level controller can then reproduce the plant topology accordingly. In addition, if a field device fails, it is possible to check whether the replacement device has been reconnected in the proper position.

40.4.5 Redundancy

The goal of redundancy in automation systems is to increase the system availability significantly, thus a redundancy manager (RM) and several clients should be used to assure the correct operation in this cases.

The media redundancy protocol (MRP), according to IEC 61158 and IEC 61784, describes PROFINET redundancy (for TCP/IP and RT frames) with a typical reconfiguring time of communication paths after a fault of less than 200 ms. The MRP operates with a physical ring structure where the RM is the coordinator. The task of an RM is to check the functional capability of the configured ring structure. This is done by sending out cyclic test frames: as long as the ring structure is intact, it receives all the test frames again after one circulation through the ring. The architecture allows an RM to prevent standard frames from circulating and to convert a physical ring structure into a "logical" line structure. The RM must communicate changes in the ring to all clients involved (switches as so-called passers) through special "change in topology" frames.

In principle, even in PROFINET IO redundant installations, the same previously described data transmission mechanisms are used. The only difference is the communication path used to transmit the frames (UDP/IP and RT frames). These frames are only transmitted via the "healthy" channel (single channel) that looks like a line structure. A redundancy client (RC) is a switch that acts only as a "passer" of frames and generally does not assume an active role. It must have two switch ports in order to connect to other clients or the RM in a single ring. In PROFINET IO, only "managed switches" that support MRP are used for implementing media redundancy. Redundant switches can be configured, e.g., via SNMP or Web services.

The media redundancy for real-time (MRRT) protocol defined in IEC 61158 describes the handling of RT frames of RT_CLASS_1 and RT_CLASS_2 for redundancy operation. Operation of MRP is always a prerequisite for operation of MRRT. IEC 61784 describes the procedure for using the MRRT protocol. With RT communication, the MRRT protocol enables a virtually smooth switchover of communication paths if a fault occurs. This is accomplished by redundant transmission of RT frames (i.e., via two channels) if the destination port is designed as a redundant port (see Figure 40.5). On the receiving side, two RT frames always arrive, provided the redundant transmission is error-free. Only the first frame to arrive is forwarded to the application. Also in this case, the RM must check the functional capability of the system by means of test frames.

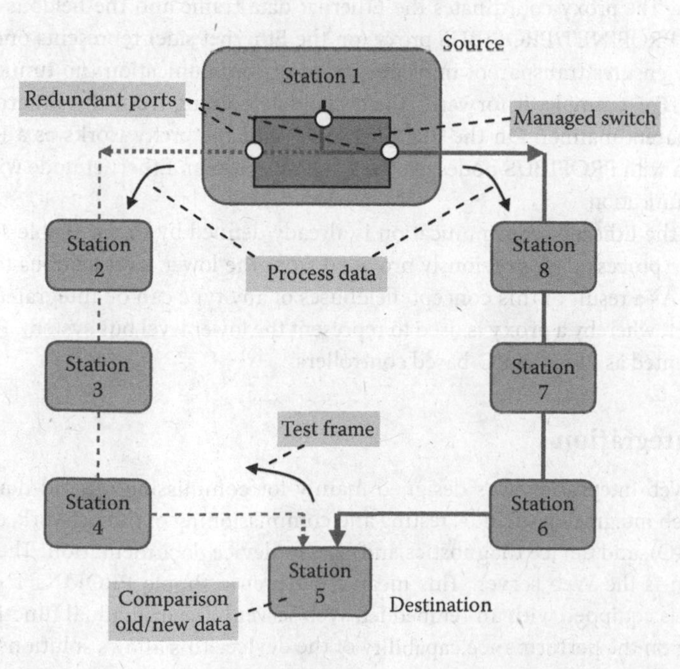

FIGURE 40.5 Media redundancy with MRP increases the plant availability.

IEC 61158 describes the redundancy concept for RT_CLASS_3 frames as "media redundancy for planned duplication." IEC 61784 describes the use of redundancy class 3 for RT_CLASS_3 communication with smooth switchover of communication paths if a fault occurs. During system power-up, the IO controller loads the data of the communication paths for both communication channels (directions) in a communication ring to the individual nodes. Thus, it is unimportant which node fails because the loaded "schedule" for both paths is available in the field devices. Loading of the "schedule" alone is sufficient to exclude frames from circulating in this variant, because the destination ports are explicitly defined.

40.5 Integration of Fieldbus Systems and Web Applications

PROFINET specifies a model for integrating existing PROFIBUS and other fieldbus systems such as INTERBUS and DeviceNet. This means that any combination of fieldbus and Ethernet-based subsystems can be configured in a simple way. The following constrains are taken into consideration: the plant operator would like the ability to easily integrate his existing installations into a newly installed PROFINET system; the plant and machine manufacturer would like the ability to use the well-proven devices it is familiar with for PROFINET automation projects, as well, without the need for any modifications; the device manufacturer would like the ability to integrate its existing field devices into PROFINET systems without expending any effort for modifications. Fieldbus solutions can be easily and seamlessly integrated into a PROFINET system using proxies and gateways. Anyway, the more powerful way is using a proxy.

40.5.1 Integration via Proxy

In simplest terms, a proxy is a representative for a lower-level fieldbus system. In PROFINET, a proxy represents a fieldbus (e.g., PROFIBUS, Interbus, Foundation Fieldbus, DeviceNet). As a result, the advantages of fieldbuses, such as high dynamic response, pinpoint diagnostics, and automatic system configuration without settings on devices, can be utilized in the PROFINET world. Moreover with the proxy, devices and software tools are also supported in the accustomed manner and integrated into the handling of the PROFINET system. The proxy coordinates the Ethernet data traffic and the fieldbus-specific data traffic. For instance, a PROFINET/PROFIBUS proxy (on the Ethernet side) represents one or more fieldbus devices. This proxy ensures transparent implementation of communication (no tunneling of protocols) between networks. For example, it forwards the cyclic data coming from the Ethernet to the fieldbus devices in a transparent manner. On the PROFIBUS DP side, the proxy works as a PROFIBUS master that exchanges data with PROFIBUS nodes. At the same time, it is an Ethernet node with Ethernet-based PROFINET communication.

In this process, the Ethernet communication is already defined by the available software. Only the initialization of the process data previously provided from the lower level fieldbus to PROFINET still has to be ensured. As a result of this concept, fieldbuses of any type can be integrated into PROFINET with minimal effort, whereby a proxy is used to represent the lower level bus system. For instance, proxies can be implemented as PLCs or PC-based controllers.

40.5.2 Web Integration

The PROFINET Web integration was designed mainly for commissioning and diagnostics. Possible applications for Web integration include: testing and commissioning of the network, overview of device data (PROFINET IO), and device diagnostics and system/device documentation. The basic component of Web integration is the Web server. This means that even a simple PROFINET devices with Web integration option is equipped with an "embedded Web server." The individual functions can be implemented depending on the performance capability of the device. This allows solutions to be customized to each use case.

Acronyms

AR	Application relation
CR	Communication relations
DCP	Discovery and configuration protocol
DHCP	Dynamic host configuration protocol
IRT	Isochronous real-time communication
LLDP	Link layer discovery protocol
MRP	Media redundancy protocol
MRRT	Media redundancy for real-time
PI	PROFIBUS International
RT	Real-time communication

Bibliography

1. International Electrotechnical Commission, IEC 61158. Digital data communications for measurement and control—Fieldbus for use in industrial control systems, International Electrotechnical Commission, Geneva, Switzerland, 2007.
2. International Electrotechnical Commission, IEC 61784-1. Industrial Communication Networks—Profiles—Part 1: Fieldbus profiles, International Electrotechnical Commission, Geneva, Switzerland, 2007.
3. International Electrotechnical Commission, IEC 61784-2. Industrial Communication Networks—Profiles—Part 2: Additional fieldbus profiles for real-time networks based on ISO/IEC 8802-3, International Electrotechnical Commission, Geneva, Switzerland, 2007.
4. Profibus International, *GSDML Specification for Profinet IO*, Version 2.20, Order No. 2.352, PROFIBUS Nutzerorganisation e.V., Karlsruhe, Germany, 2008.
5. Profibus International, *Profile Guidelines Part 1: Identification & Maintenance Functions*, Version 1.1.1, Order No. 3.502, PROFIBUS Nutzerorganisation e.V., Karlsruhe, Germany, 2005.

41

LonWorks

Uwe Ryssel
Dresden University of Technology

Henrik Dibowski
Dresden University of Technology

Heinz Frank
Reinhold-Würth-University

Klaus Kabitzsch
Dresden University of Technology

41.1 Introduction **41**-1
41.2 System Components **41**-1
41.3 LonTalk Protocol **41**-4
Physical Layer • Link Layer • Network Layer • Transport and Session Layer • Application and Presentation Layer
41.4 The Application Layer Programming Model **41**-7
41.5 Function Block-Based Design and System Integration **41**-8
41.6 Network Design Tools **41**-11
41.7 Automatic Design Approaches **41**-12
References **41**-13

41.1 Introduction

LonWorks or local operating network (LON) is an event-triggered control network system originally designed by Echelon Corporation in 1990. It consists of a communication protocol called LonTalk, a special controller (the Neuron Chip), transceivers for bus access, and a set of development and management tools.

The main application of LonWorks is building and home automation, where it is used to control and monitor, among others, heating, ventilating, and air-conditioning (HVAC), lighting, security, and elevators. But also other application areas exist, such as industrial automation, transportation automation, and street lighting.

The LonTalk communication protocol, which implements the OSI reference model, was standardized as ANSI/EIA-709 and ANSI/CEA-709 in 1999. Both standards also include the channel specifications of free-topology twisted-pair, power line and fiber-optics, which are used for LonWorks networks. The LonTalk protocol ANSI/CEA-709 was revised in 2002 [1] and was released as European Standard EN-14908 [2] in 2005. The corresponding international standard ISO/IEC-14908 is currently in draft state [3].

Guidelines for the implementation of the applications on the Neuron Chips are defined by the LonMark, a global membership organization that promotes and advances the LonWorks platform.

41.2 System Components

LonWorks networks normally consist of network nodes, based on the Neuron Chip. In the first years, these chips were manufactured by Motorola [4] and Toshiba [5]. Today's models are built by Cypress [6] and Echelon itself. Echelon offers so-called smart transceivers, which integrate a Neuron Chip and a transceiver into one chip [7].

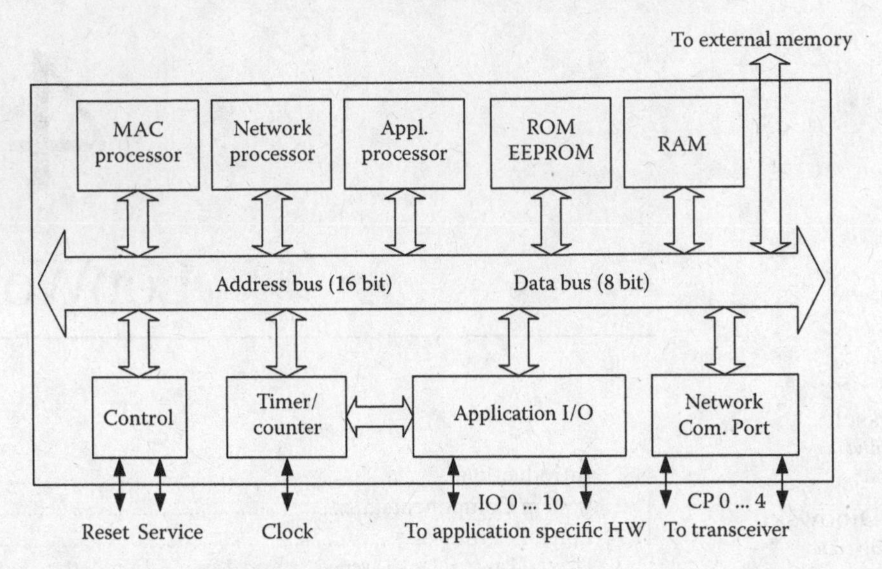

FIGURE 41.1 Structure of a Neuron Chip.

As shown in Figure 41.1, the Neuron Chip implements a three-processor architecture with shared memory, which includes the firmware ROM, an EEPROM for the user application program, and RAM. Each of the three processors handles a certain task:

The *media access control* (MAC) *processor* sends and receives messages to and from the communication medium with support of the transceiver, which handles the line coding for a specific medium. Since many different transceiver types exist, a wide range of communication media can be used, such as twisted pair, power line, fiber optics, and radio frequency. The *network processor* handles the upper layers of the LonTalk protocol, which are described in the next section. And the *application processor* runs the user application implementing the specific task of the node.

The communication among the three processors is realized in shared RAM by communication buffers and flags. There are input and output buffer queues storing messages to be forwarded to the other processors. Special flag bytes indicate that the other processor should do a certain action, such as creating, sending, or receiving a message.

Peripheral hardware, such as sensors and actuators, are connected to a set of I/O pins, which can be configured in a range of single bit input and output for switches or LEDs via nibble and byte input and output up to higher I/O protocols, such as serial and parallel interfaces, I^2C, and Magcard.

There are also alternatives to the Neuron Chip. For example, LOYTEC offers a more powerful ARM7-based embedded controller (LC3020) that implements the LonTalk protocol according to CEA-709 as well. It also supports LonWorks/IP (standardized as CEA-852 [8]), where LonTalk messages are tunneled over LAN or Internet in IP packets. Such controllers are used for more complex tasks, like panels with a graphical user interface or IP routers.

To identify the LonWorks nodes in the network, each Neuron Chip has a fixed 48 bit identifier called *Neuron ID* assigned by the manufacturer. A LonWorks node can actively transmit its Neuron ID in a *service pin message*, when a special service pin is pressed on the node. This simplifies the network integration, when new nodes have to be added to the network, because it is not required to enter the Neuron ID manually in the management tool. Instead, the management tool simply has to wait for the service pin message.

During configuration, a logical address is assigned to each node. It consists of the *domain*, *subnet*, and *node ID*. The *domain* represents a whole LonWorks network, which is identified by the domain ID with up to 6 bytes length. Direct message passing among domains is not possible, but it can be done via gateways.

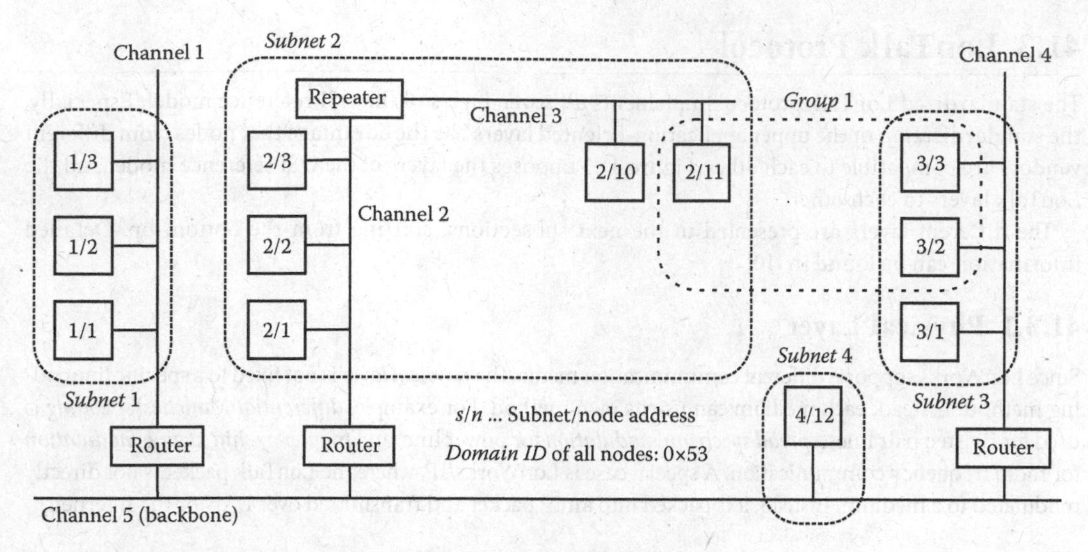

FIGURE 41.2 Physical and logical segmentation in LonWorks systems.

The physical and logical view of a LonWorks network example is shown in Figure 41.2. The network physically consists of several channels connected by routers and repeaters. Logically, it is separated into *subnets*. One domain can consist of up to 255 subnets with a maximum of 127 nodes each, which results in a maximum of 32,385 nodes per domain. Additionally, nodes can be grouped. Per domain 256 groups can be defined. The example contains one group, which spans three nodes in different subnets and channels.

As mentioned before, different transceivers support different communication media. A common medium is twisted pair with a data rate of 78 kbps, where the signal is coded as voltage difference between the two conductors. A compatible variant of twisted pair is Link Power, which also provides the power supply (42.4 V DC) for the nodes over the twisted pair medium. For backbones, twisted pair can also be run with 1250 kbps, but only with a reduced maximum distance of 130 m among the nodes.

The topology of twisted pair networks does not have to be a strict line or star. Networks can also be built as free topology containing arbitrary branches and loops. However, the maximum distance decreases when using free topology. So, for example, a line bus with short branches allows a maximum distance of 2700 m among the nodes, whereas in a free topology, it is limited to 500 m.

Other supported media are power line with up to 10 kbps, fiber optics (1.25 Mbps), radio frequency (up to 19.5 kbps), and the above-mentioned LonWorks/IP (100 Mbps with current IP routers).

To increase the maximum distance of LonWorks networks, repeaters can be used. They simply refresh all incoming signals and send them to the other network segment, which has to be the same medium and has to use the same message coding.

Since LonTalk implements the OSI reference model, LonWorks networks can be connected at higher layers, too. Bridges, working on the data link layer, can be used to connect networks based on different media and coding. They also ensure that only valid messages are forwarded to the other side.

More common than pure LonTalk bridges are routers, which work on the network layer. Routers have the advantage that they do not only increase the range of the signals or connect different media, but they also separate the network traffic of different subnets because routers only forward those messages, whose destination is on the other side of the router. Therefore, routing tables are used. They describe which subnets or groups are connected on which port of the router. There are two methods how routing tables can be obtained: In *configured router* mode, the tables have to be defined by the system integrator and in *learning router* mode, the network traffic is used to determine which subnets are connected to which port. LonWorks routers can also operate in bridge and repeater mode.

41.3 LonTalk Protocol

The standardized LonTalk protocol implements all seven layers of the OSI reference model. Especially, the standardization of the upper application-oriented layers has the advantage that nodes from different vendors are compatible to each other. Figure 41.3 opposes the layers of the OSI reference model and the LonTalk layers to each other.

The different layers are presented in the next subsections, starting from the bottom up. Detailed information can be found in [10].

41.3.1 Physical Layer

Since LonWorks supports different communication media, the *physical layer* is not fixed to a specific line coding method. Instead, each medium can use its own method. For example, *differential Manchester coding* is used for twisted pair lines, *spread spectrum modulation* for power line, and *frequency shift keying modulation* for radio frequency communication. A special case is LonWorks/IP, where the LonTalk packet is not directly modulated to a medium. Instead, it is packed into an IP packet and transmitted over LAN or the Internet.

41.3.2 Link Layer

The next layer is the *link layer* containing the MAC sublayer. This sublayer handles the medium access, which is realized by the so-called predictive p-persistent carrier sense multiple access (CSMA) method

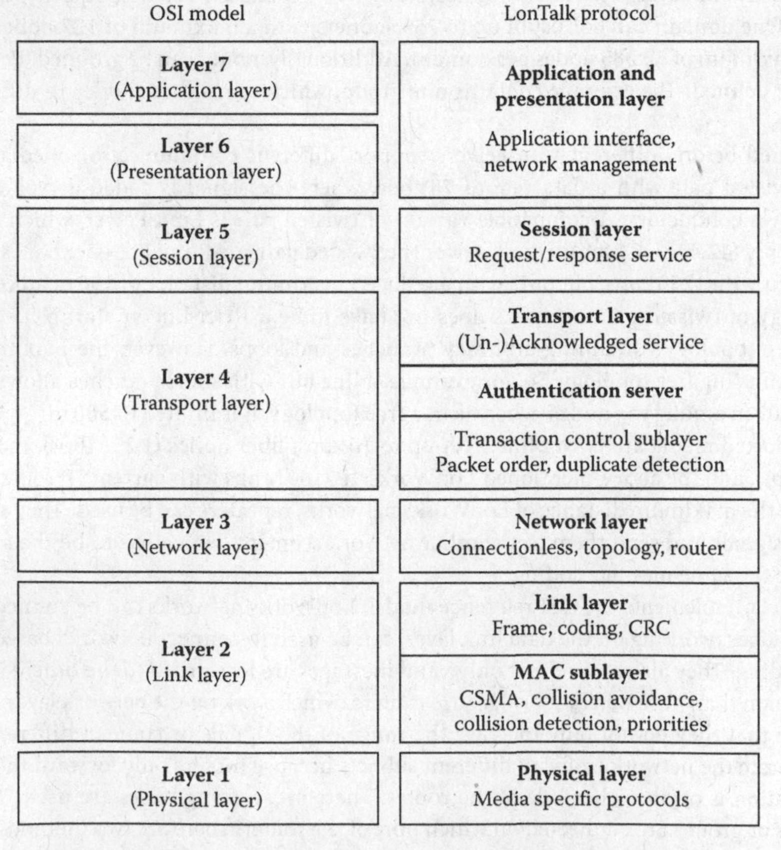

FIGURE 41.3 The OSI reference model and their corresponding LonTalk layers. (From Loy, D. et al. (eds.), *Open Control Networks—LonWorks/EIA-709 Technology*, Kluwer Academic Publishers, Boston, MA, p. 76, figure 4-2, 2001. With kind permission of Springer Science and Business Media.)

in LonWorks. In p-persistent CSMA, the bus is checked for activity before the frame is transmitted. As it becomes idle, the sender can start to transmit its frames. To avoid collisions, there are n time slots, from which each sender chooses one with a probability of $p = 1/n$. In LonTalk, the default value of n is 16. Each sender will start to transmit its frame in its chosen time slot as long as the bus is idle. Thus, the sender who takes the earliest time slot will transmit its frame and all others have to wait for the next idle phase. Collisions can only occur when two or more senders have chosen the same time slot. To lower the probability of collisions, the number of time slots n is *predictively* increased in intervals of 16 according to the estimated channel backlog, which is the estimated number of return messages.

Optionally, the first time slots can be reserved for priority messages. The number of priority time slots is specified in each node and should be the same for all nodes on a channel to prevent collisions with non-priority slots. Each node, which wants to send priority messages, gets its dedicated priority slot, whose number is configured in the node. So when n_p priority slots are reserved, n_p nodes can send priority messages. The remaining time slots are used for non-priority frames as mentioned above. So, priority messages always suppress the other messages, which can theoretically result in starvation of the non-priority frames. But this would require permanently high network traffic, which is atypical for a well-configured network. To prioritize the complete message transaction, priority messages are sent with a set priority bit in the frame. This signals routers to forward messages and receivers to send the responses prioritized, too.

Optionally, the Neuron Chip supports the processing of *collision detections*, which results in canceling and retransmitting of the frame. But the transceiver has to detect the collision and signal it to the Neuron Chip, which is not supported by all transceivers.

The link layer has to ensure that frames are transmitted correctly. Therefore, a 16 bit cyclic redundancy check (CRC) sum is appended. Additionally, the frame contains the preamble, the priority bit, the change of the backlog, and the alternate path bit. A set alternate path bit is an indicator for the transceiver to send the frame via an alternate channel. This increases the reliability on unreliable media, such as power line and radio frequency. By default, the bit is cleared, but based on the node configuration, the bit can be set automatically for the last two transmission attempts. For example, using a power line transceiver, a set alternate path bit implicates that the modulation method will be changed and the data transfer rate will be decreased to increase the signal quality.

41.3.3 Network Layer

The *network layer* offers connectionless, unacknowledged transmissions of packets within a domain. Following services are supported:

- *Unicast*
  The destination of the packet is one single node. It can be addressed by the unique Neuron ID or by the logical address consisting of domain, subnet, and node ID.
- *Multicast*
  The destinations of the packet are a group of nodes defined by its group ID.
- *Broadcast*
  The destinations of the packet can be all nodes belonging to a specific domain or all nodes belonging to a specific subnet.

41.3.4 Transport and Session Layer

The network layer does not handle the repeated transmission of lost packets or the detection of packet duplicates. This is done by the *transport layer* and the *session layer.* The transport layer offers the following services:

- *Unacknowledged service*
  The packet is sent exactly one time. Since there are no acknowledgments, a loss of the message cannot be recognized. This service is usually used for broadcasts but can be used for unicast and multicast packets as well.

- *Acknowledged service*
  Each destination node returns an acknowledge message to the source to confirm the transmission. If it is missing, the transmission will be repeated after a certain time period. Missing acknowledges from individual group members will be inquired by resending the message accompanied with special reminder messages. Duplicated messages are recognized by a transaction ID contained in each packet. This service can be used for unicast and multicast packets.
- *Unacknowledged repeated service*
  If acknowledge messages should be avoided, especially group messages can create many of them, and additionally the probability of unrecognized message losses should be lowered, the unacknowledged repeated transmission can be used. This service will send the message consecutively several times. The number of repetitions is configured by a retry count property. As in the acknowledged service, transaction IDs avoid the duplicate processing of messages. Also, this service can be used for unicast and multicast packets.

The session layer offers a *request/response service*. This service is used to execute actions, which return data to the sender, such as most of the network management messages. This data, contained in the response packet, consists of a success or fail code and the requested data. Like in acknowledged service, requests will be repeated if the response was not received for a certain time. This request/response service is used alternatively to the services of the transport layer.

To avoid unauthorized execution of commands and requests, the *authentication service* can be used in addition to the acknowledged and request/response service. The authentication service uses a *challenge–response authentication method* to test the authorization of the sender. Therefore, each LonWorks node in the network has the same shared secret key, set in the configuration phase during the assignment of the node's logical address (domain, subnet, and node ID). After receiving a message, the receiver transmits a random number (the challenge) to the initial sender, who encrypts it with his secret key and sends it back (the response). The receiver compares the returned value with the value he has calculated locally. If they match, the authorization is confirmed and the initially received command or request is executed.

In LonWorks, a 48 bit secret key is used to encrypt a 64 bit random number resulting in a 64 bit encrypted value. Since the encryption algorithm is not published, its quality cannot be evaluated. Moreover, a 48 bit key is not strong enough for brute force attacks on high-bandwidth channels. Another weakness is the distribution of the shared secret key, which has to be sent in unencrypted network management messages over the unsecure bus.

Authentication can be used for all network management transactions and can be activated for each network variable connection, too.

41.3.5 Application and Presentation Layer

The *application and presentation layers* form a unit in LonWorks. They offer services to support the application on the, one hand and to execute network management and diagnostic functions, on the other hand. These services and functions are

- *Network variable propagation*
  Network variables are basic communication objects, which define the logical datapoints of the application. *Input network variables* receive values from the network and *output network variables* send values to the network. The connections between output and input variables are realized by *bindings*, which are configured by setting special table entries in the source node via a system integration tool. Whenever the value of an output network variable is changed, these table entries are used for sending a network variable propagation message to those nodes, which contain the bound input variables. According to these messages, the bound nodes update the values of their input variables.

- *Network management and diagnostic messages*
  These messages are used for the node configuration, including the setting of the logical addresses (domain, subnet, node ID), writing to EEPROM (for application loading), and setting of bindings. Router configuration messages belong to this group, too.
- *Generic message passing*
  Applications can create and send any messages, including the above-specified ones. With the help of application-specific message codes, own application-specific protocols can be implemented.

41.4 The Application Layer Programming Model

User applications for Neuron Chips are programmed in *Neuron C*, an extended ANSI C dialect. Since applications are event based, Neuron C contains a construct called *task*, which is called whenever an event has occurred or a condition is fulfilled. The event or condition is defined by a when clause. Events are, for example, changes of network variable inputs (`nv_update_occurs`), the expiration of timers (`timer_expires`), the changing of I/O pins (`io_changes`), or the reset event, which is called after a reset. The following listing shows a simple application in Neuron C, which controls the state of a LED using a network variable.

```
network input SNVT_switch nviSwitch;
IO_0 output bit ledOutput; // specifies LED output
#define LED_ON  1
#define LED_OFF 0
when (reset)
{
  io_out(ledOutput, LED_OFF);
}

when (nv_update_occurs(nviSwitch))
{
  // set LED according to NV value
  if (nviSwitch.state == 1)
   io_out(ledOutput, LED_ON);
  else
   io_out(ledOutput, LED_OFF);
}
```

The first two lines declare an input network variable named nviSwitch and an output bit object using I/O pin 0. After that, two tasks are defined by when clauses. The first one is called after the reset of the node and sets the output pin to a defined state (LED off). The second task is called when the above-declared network variable is changed externally. Since `nviSwitch` has the structured type `SNVT_switch`, the concrete switch state is taken by the appropriate structure member. According to the state value, the LED is switched on or off.

As shown in the example, special directives declare network variables and I/O objects. Further directives are used for configuration parameters, function objects, and other node-specific properties.

The interface between the Neuron C program and the application layer of the Neuron Chip is offered as a set of library functions. Important functions are implemented directly in the firmware. For example, there are functions to process application-specific messages, for the network management of the node itself, and for accessing the I/O objects. Other little-used functions, like floating point arithmetic functions, are linked to the application image by the Neuron C linker, if required.

By running the Neuron C compiler, two kinds of files are created: First, *application images* contain the compiled code, which is loaded to the node later, and second, the associated *external interface files* (XIF) define the logical interface of the node that means the function blocks and their network variables and configuration properties. These interface files are used later by system integration tools for configuration and network variable binding.

Since the Neuron Chip does not support interrupts, the events and conditions of the tasks have to be checked periodically by a scheduler. The sequence of checks depends on the priority of the tasks. In each scheduler loop, all priority tasks are checked first. If none of these has to be processed, one of the non-priority tasks will be checked and processed if necessary. The other remaining non-priority tasks are checked at the end of the next loops, one in each loop. The scheduler jumps to the head of the non-priority task list again, when the tail is reached.

41.5 Function Block-Based Design and System Integration

The function block-based design and the system integration shall be explained with the example of a single-room sunblind control, which should realize an antiglare functionality. Such a room control typically consists of the following functions:

- *Antiglare control*
  The antiglare function controls the position of the sunblind and the angle of the sunblind's lamellas to prevent the glaring of persons, which are present in the room. A measurement of the current glare is the luminance level in the room. As soon as the luminance exceeds a certain threshold, the sunblinds will move down. If there is nobody in the room, the antiglare function will be switched off.
- *Sunblind actuator*
  The sunblind actuator drives the motor of the sunblind. It receives the actuating values from the antiglare control function.
- *Luminance sensor*
  The luminance sensor measures the current luminance level in the room in lux.
- *Occupancy sensor*
  The occupancy sensor detects whether a person is within the room or not. Based on this information, it activates or deactivates the antiglare controller.
- *Manual control*
  The position of the sunblind can be set manually using a sunblind switch, which overrides the antiglare control. Additionally, the building management system can override the position, for example, in the night or in case of fire.

As defined above, network variables are basic communication objects to exchange values over the network. To separate specific automation functions and their related network variables within a node, the application of a node is subdivided into *functional blocks* [11]. Figure 41.4 shows the functional block of the sunblind controller, which realizes the antiglare functionality and, bound to this, the other needed functional blocks. These are the sunblind switch for the manual control, the light sensor for measuring the luminance, the occupancy sensor for the presence detection, and the sunblind actuator, which drives the sunblind. Functional blocks define an interface, which consists of input (`nvi...`) and output (`nvo...`) network variables and configuration properties. Configuration properties can be implemented as *configuration network variables* or as properties in *configuration files*. Configuration network variables (`nci...`) can be altered by other LonWorks devices using standard network variable propagation, but their size and count is limited. Properties in configuration files have no such limitations, but they can only be set by network tools.

Bindings are not restricted to one-to-one interconnections. They are allowed to be of kind one-to-*n* and *n*-to-one, too. One-to-*n* bindings are used if more than one input variable needs the value of one output variable. *N*-to-one bindings, however, can be problematical since the input variable receives values from different outputs, which can overwrite each other concurrently.

To ensure the interoperability among the LonWorks nodes, especially if they are from different manufacturers, their functions and their communication interfaces have to be standardized. Therefore, the

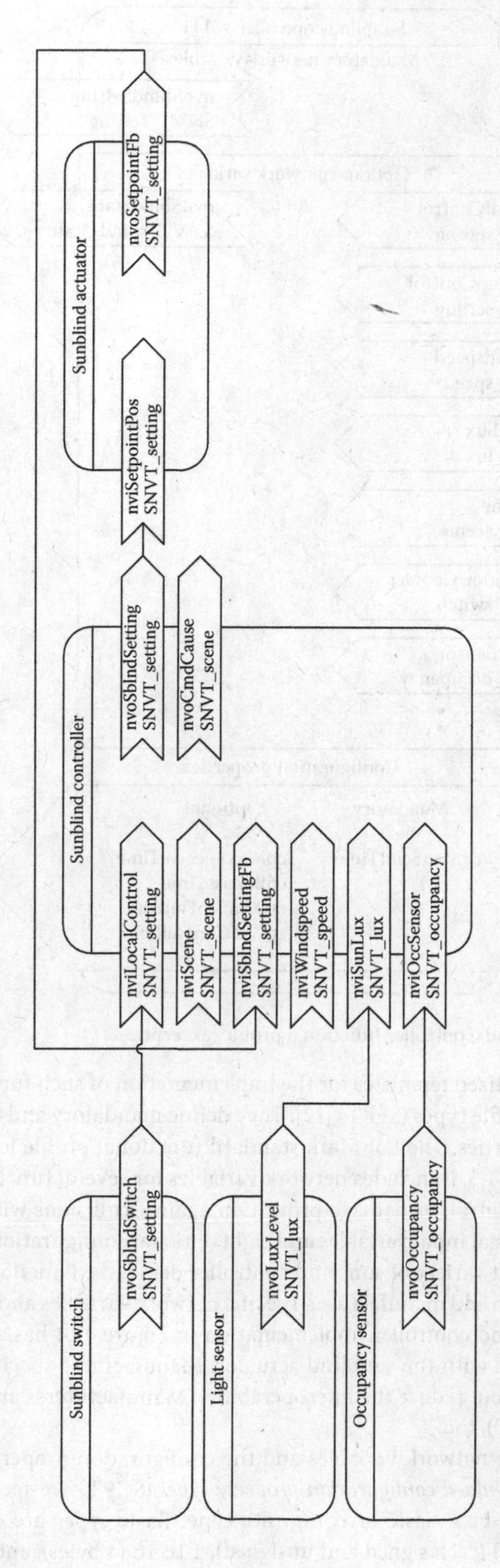

FIGURE 41.4 Functional blocks implementing antiglare control.

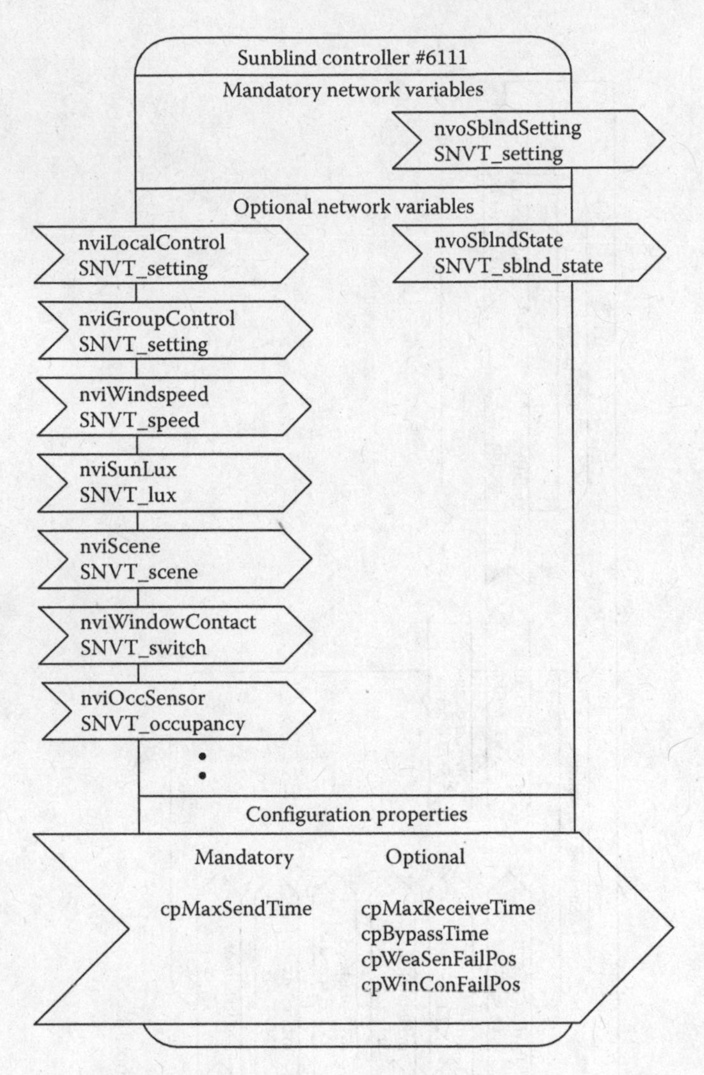

FIGURE 41.5 LonMark Sunblind Controller functional profile (excerpt).

LonMark has defined standardized templates for the implementation of such functional blocks, the so-called standard functional profile types (SFPT) [12]. They define mandatory and optional network variables and configuration properties. The LonMark standard functional profile for sunblind controllers is shown as excerpt in Figure 41.5. It includes network variables for several functions, such as antiglare control, weather protection control, or damage protection against collisions with open windows. The entire profile contains 27 optional input variables and eight optional configuration properties to provide a template, which meets a great variety of sunblind controller demands. Functional blocks implementing a functional profile can also add manufacturer-specific network variables and configuration parameters. For instance, the sunblind controller implementation in Figure 41.4 has an additional feedback input variable, which is bound with the sunblind actuator. Manufacturer-specific extensions increase the flexibility of the standard, but reduce the interoperability. Manufacturers can define their own user functional profile types (UFPT), too.

For the compatibility of the network variables and the configuration properties, *standard network variable types* (SNVT) and *standard configuration property types* (SCPT) are specified by the LonMark [12]. Each of these types can be a basic or composite type. Basic types are char, `short` (1 byte), `long` (2 bytes), `quad` (4 bytes) (each signed and unsigned), `float` (4 bytes), and enumeration (`enum`).

TABLE 41.1 Definition of Standard Network Variable Types Used in the Example

SNVT	Purpose	Data Type	Range
SNVT_lux	Illumination level	unsigned long	0…65,535*lx* (raw: 0…65,535)
SNVT_occupancy	Occupancy state	enum occup_t	OC_OCCUPIED, OC_UNOCCUPIED,…
SNVT_scene	Calling scenes	*struct of*:	
		enum scene_t *function*	SC_RECALL, SC_LEARN,…
		unsigned short *scene_number*	1…255
SNVT_setting	Setting value (e.g., for sunblind)	*struct of*:	
		enum setting_t *function*	SET_DOWN, SET_UP, SET_STOP,…
		unsigned short *setting*	0.0…100.0% (raw: 0…200)
		signed long *rotation*	−359.98…360.00° (raw: −17,999…18,000)
SNVT_speed	Linear velocity	unsigned long	0…6,553.5 *m/s* (raw: 0…65,535)

Composite types are structures (`struct`) and unions (`union`) of basic types. Each SNVT field has a scale and an offset to represent values with a resolution lower than one. `SNVT _ temp` has, for example, a scale of 0.1 and an offset of −2740 to represent a range of −274.0°C to 6,279.5°C by raw values 0–65,535 (2 bytes). Other SNVTs, which are used in the example, are described in Table 41.1.

SCPT types often reference SNVT types. For example, `SCPTtempOffset`, which is used as calibration offset of temperature probes, references `SNVT _ temp _ diff _ p`, which describes a temperature difference.

Manufacturers can use their own *user-defined network variable* and *configuration types* (UNVT/UCPT), which can be new types or referenced SNVT/SCPTs. These user types are always manufacturer specific, so they are only valid for one manufacturer. They additionally have a scope, which defines the range of application of a type. The range can be further limited to a device class, usage, channel type, or an application itself. System integration tools ensure that only network variables of the same type (SNVT or UNVT) can be bound.

41.6 Network Design Tools

Traditionally, each assembly section, such as lighting, sunblind, and HVAC, has its own designers and system integrators. Since LonWorks covers all these assembly sections, the bus can be shared. If all nodes are installed on one shared bus and the system integrators do not coordinate each other, which often was the case, the completed system will not work: Logical addresses are used more than once or bindings will be overwritten. A solution to this problem is to use a central database, where all assembly sections are working on. On this account, Echelon released LonWorks network services (LNS) in 1996, which is a de facto standard in the LonWorks world now. The LNS Server offers a database, where networks can be stored, and an API, which provides network management methods for adding devices to a network, binding network variables, and setting node and application-specific configuration properties.

With the first release of the LNS database, Echelon created the prerequisites for the development of a wide range of LNS-based system integration tools for LonWorks networks. Today, tools from several manufacturers are available on the market, which use the LNS database as backend [13]. Examples are the LonMaker from Echelon [14], NL 220 and NL Facilities from Newron System [15], Alex TE from Spega [16], and the Network Integrator from Circon Systems [17]. They all incorporate mostly all LNS functions and enable the design, parameterization, commissioning, and test of open multi-vendor LonWorks systems.

Besides, also LonWorks system integration tools exist that are specialized to devices from a specific manufacturer. This limitation to a small number of devices makes an easier, semiautomatic development possible but fails for multi-vendor LonWorks systems. Because of this restriction, only the before-mentioned multi-vendor system integration tools are regarded here.

LNS-based system integration tools support two different network installation scenarios. In the *engineered mode*, the entire network is designed without commissioning any devices until the design is complete, and in the *ad-hoc mode*, all network configuration information is immediately loaded into the devices. Thus, the engineered mode is the only way to design a network without having access to the physical devices, whereas the ad-hoc mode is only possible for onsite network installations.

With the LNS database as backend, the different system integration tools offer specific user interfaces. LonMaker follows a graphical block-oriented design that allows a drag-and-drop–based placement and connection of functional blocks and devices. In contrast, NL220, Alex TE, and Network Integrator are tree-view oriented and do not support graphical views of the logical network.

Despite different user interface representations, all mentioned tools widely support the same design tasks. Typically, the proceeding starts with the definition of a domain and one or more subnets and channels. Devices can be added to a channel in the next step. This involves the creation and instantiation of LNS device templates for each device type based on their XIF files or per download of the devices' self-description (in ad-hoc mode only). The device templates contain essential information for the LNS database and system integration tools. They describe various device hardware criteria like transceiver or processor type, communication settings, functional blocks, their interfaces, network variables, and configuration properties.

Once a device has been added, its functional blocks can be used and connected via bindings with other blocks. Functional blocks may need to be parameterized, which is possible either by directly setting the configuration properties or via comfortable LNS plugins, delivered by the device manufacturers. Furthermore, various other settings can be made, for example, adding routers and gateways or setting up authentication.

Among all LNS-based system integration tools, NL Facilities deserves an exceptional position since it mechanizes many repeated actions. The key elements are semiautomatic approaches based on solution libraries for reuse. NL Facilities provides a graphical representation of the floor plan, where the user just has to place devices into rooms and define interrelationships among devices. It fully hides the technology from the user and thus makes it appropriate also for nonprofessionals. However, this procedure works fine only for networks with many identical, predefined devices, whereas a high variability restricts its applicability.

41.7 Automatic Design Approaches

More advanced than NL Facilities are the automated design approaches from the AUTEG project [18]. The automatic design tools developed there are able to cope with the variety of all market available devices from all manufacturers by using semantic device descriptions [19]. Industry spanning, multi-vendor LonWorks-based building automation systems are generated automatically [20,21]. The design process starts with a formal requirement specification, where functional and nonfunctional requirements for the building automation system are gathered. All subsequent design tasks are done automatically, including the generation of functional schematics for all rooms [22], the selection of appropriate devices fulfilling the requirements, the evaluation of interoperability, the definition of bindings, parameterization, and layout of the physical network. Evolutionary algorithms are used to obtain optimized LonWorks systems according to multi-objectives such as costs, correctness, completeness, and interoperability [23].

Also, quality-oriented aspects are incorporated here. By forecasting the resulting network load already at the early design phase, bottlenecks and network overloads can be identified before the system is built [24,25]. Modifications of the physical network and the parameterization are done to avoid these problems.

References

1. ANSI/CEA 709.1-B:2002. *Part 1: Control Network Protocol Specification*. CEA, 2002.
2. EN 14908-1:2005. *Open Data Communication in Building Automation, Controls and Building Management—Control Network Protocol—Part 1: Protocol Stack*. CEN, 2005.
3. ISO/IEC DIS 14908-1:2008. *Open Data Communication in Building Automation, Controls and Building Management—Control Network Protocol—Part 1: Protocol Stack*. ISO, 2008.
4. Motorola, Inc. *LonWorks—Technology Device Data, Rev. 5—Volume 1: Device Data*, 1998.
5. Toshiba Corporation. TAEC Neuron Chip Databook—TMPN3120/3150, 2001. Available at http://www.toshiba.com/taec/
6. Cypress Semiconductor Corporation. *Neuron Chip Technical Reference Manual*, 2002. Available at http://www.cypress.com
7. Echelon Corporation. *FT 3120/3150 Smart Transceiver Data Book*, 2006. Available at http://www.echelon.com
8. ANSI/CEA-852:2004. *Tunneling Device Area Network Protocols over Internet Protocol Channels*. CEA, 2004.
9. D. Loy, D. Dietrich, and H.-J. Schweinzer, editors. *Open Control Networks—LonWorks/EIA-709 Technology*, p. 76, figure 4–2. Kluwer Academic Publishers, Boston, MA, 2001.
10. D. Loy, D. Dietrich, and H.-J. Schweinzer, editors. *Open Control Networks—LonWorks/EIA-709 Technology*. Kluwer Academic Publishers, Boston, MA, 2001.
11. LonMark International. *LonMark Application-Layer Interoperability Guidelines Version 3.4*, September 2005. Available at http://www.lonmark.com/technical_resources/guidelines/docs/LmApp34.pdf
12. LonMark International. *LonMark Resource Files Version 13*, 2006. Available at http://types.lonmark.org
13. Echelon Corporation. *LNS Powered Tools Overview Website*. Available at http://www.echelon.com/products/development/lns/pwrtools.htm
14. Echelon Corporation. *LonMaker Integration Tool Turbo Edition Website*. Available at http://www.echelon.com/Products/networktools/lonmaker/default.htm
15. Newron System. *Installing and Commissioning Tools Website*. Available at http://www.newron-system.com/index.php?Page=installation
16. Spelsberg Gebäudeautomation GmbH + Co. KG. *Alex TE Website*. Available at http://www.spega.com/html/products/p_k_software.html
17. Circon Systems Corporation. *Network Integrator Website*. Available at http://www.circon.com/products/network-integrator_LNS.asp
18. Dresden University of Technology. *Automated Design for Building Automation (AUTEG) Website*. Available at http://www.ga-entwurf.de
19. H. Dibowski and K. Kabitzsch. Semantic device descriptions based on standard semantic web technologies. In *Proceedings of the Seventh IEEE International Workshop on Factory Communication Systems (WFCS'08)*, pp. 395–404, Dresden, Germany, May 2008.
20. S. Runde, H. Dibowski, A. Fay, and K. Kabitzsch. Integrated automated design approach for building automation systems. In *Proceedings of the 13th IEEE International Conference on Emerging Technologies and Factory Automation (ETFA'08)*, pp. 1488–1495, Hamburg, Germany, September 2008.
21. H. Dibowski and K. Kabitzsch. Automated design of LON based building automation systems. *LonMark Magazine International Edition*, 5(1):28–31, 2009.
22. U. Ryssel, H. Dibowski, and K. Kabitzsch. Generation of function block based designs using semantic web technologies. In *Proceedings of the 14th IEEE Conference on Emerging Technologies & Factory Automation (ETFA'09)*, Palma de Mallorca, Spain, September 2009.
23. A.C. Oezluek, H. Dibowski, and K. Kabitzsch. Automated design of room automation systems by using an evolutionary optimization method. In *Proceedings of the 14th IEEE Conference on Emerging Technologies & Factory Automation (ETFA'09)*, Palma de Mallorca, Spain, September 2009.

24. J. Ploennigs, P. Buchholz, M. Neugebauer, and K. Kabitzsch. Automated modeling and analysis of CSMA-type access schemes for building automation networks. In *IEEE Transactions on Industrial Informatics*, 2(1):103–111, May 2006.
25. J. Ploennigs, M. Neugebauer, and K. Kabitzsch. Diagnosis and consulting for control network performance engineering of CSMA-based networks. In *IEEE Transactions on Industrial Informatics*, 4(2):71, May 2008.

42

KNX

Wolfgang Kastner
Vienna University of Technology

Fritz Praus
Vienna University of Technology

Georg Neugschwandtner
Vienna University of Technology

Wolfgang Granzer
Vienna University of Technology

42.1 Introduction and Overview 42-1
42.2 Medium-Independent Layers 42-2
42.3 Medium-Dependent Layers 42-5
42.4 Runtime Interworking 42-6
42.5 Devices 42-11
42.6 Configuration 42-12
42.7 Conclusion and Outlook 42-12
Abbreviations 42-13
References 42-13

42.1 Introduction and Overview

The KNX standard describes an extensive open system concept for distributed home and building automation and control. KNX covers the full scope of related applications, including lighting, shading, shutters and blinds, heating, cooling, ventilation, and air conditioning (HVAC), and remote meter reading. All relevant communication media are supported. The KNX specification results from a formal merger of three technologies dedicated to this area: European Installation Bus (EIB), Batibus, and European Home System (EHS).

Significant parts of the KNX specification are published as European standards, mainly the EN 50090 family but also EN 13321-1:2006 and EN 13321-2:2006. KNX technology is also covered by ISO/IEC 14543-3-x (2006/2007), which has been published as GB/Z 20965-2007 by the Standardization Administration of China as well.

The KNX Association is a for-profit organization governed by Belgian law. Its members are mostly manufacturers and developers of home and building electronic devices, building services equipment, and related software. They can freely make use of patents held by fellow KNX members in certified KNX products if this is necessary to make them KNX compatible. The main task of the KNX Association is to maintain the system master specification frequently referred to as the "KNX Handbook." Moreover, the association is responsible for the development and maintenance of a manufacturer-independent PC-based commissioning tool, trademark licensing (bound to conformance tests), and the certification of training centers and testing laboratories. Also, the association maintains partnership programs for national associations, user clubs, and scientific partners.

KNX is more than just a simple network protocol specification. Besides merely defining how data is transferred, the standard includes rules and definitions of how a KNX system is managed and how devices implemented by different vendors have to behave to achieve interworking.

Certainly, the *user applications* that implement the desired automation and control functionality are central components within a KNX system. However, their internal structure and implementation is up to the device manufacturer and thus not defined by the KNX standard. The *communication system*, on the other hand, which specifies the services that are used by the user applications to communicate with each other, is part of the KNX specification. The KNX protocol stack is based on the ISO/OSI reference model. Since different communication media are supported, the protocol stack is divided into a medium-dependent and a medium-independent part. While the former consists of the physical layer and the lower level of the data link layer (DL), the latter includes the upper level of the DL that is common for all media, a lean network layer (NL) and transport layer (TL) as well as the application layer (AL). The session and presentation layers are left empty. In addition to data exchange over the network, the KNX protocol specification also covers local point-to-point connections to KNX devices, for example, via EIA-232 or USB.

The *interworking and application model* specifies how data are represented in KNX and how they are accessible via the network. *Datapoints*, associated with *functional blocks* (FBs), are a central concept in this model. A KNX datapoint may be related to a sensor value/actuator state (e.g., input and output) or it may be a parameter that controls the behavior of the user application. The necessary association between datapoints is established via bindings. KNX also specifies the *application interface* that is presented to user applications for interacting with remote datapoints. The interworking and application model as well as the communication system need schemes for *configuration and maintenance*. Therefore, KNX also specifies related resources and management procedures (together with the functionality of the server entities within the devices that handle them). This includes, for example, services for setting up the user applications, for configuring the necessary bindings, and for initializing the communication system of a device (e.g., address assignment). To achieve this, different configuration modes are defined.

Finally, KNX also specifies various device models with different features, mainly regarding supported configuration procedures. *Profiles* define which parts of the KNX specification have to be implemented and ensure interworking of devices with the same profile that are provided by different manufacturers. The conformance tests defined in the KNX specification, which are the basis for KNX compliance certification, are also based on these profiles.

42.2 Medium-Independent Layers

The medium-independent part of the KNX network protocol includes the upper level of the DL, the NL, TL, and the AL. The DL defines services to send and receive frames over the network. Acknowledgment is available as an option for some network media. The two most important services defined are `L_Data` for peer-to-peer data frame transfer and `L_Poll_Data` for a master collecting data from slaves in a so-called polling group. Furthermore, the DL defines a generic addressing scheme that is common to all available network media. This addressing scheme distinguishes between four different kinds of addresses. First, each node has a so-called individual address. The assignment and structure of the individual address are based on the location of the device within the three-level topology that is used in KNX. In this topology, the network segments at the lowest level are called *lines*. Depending on the network medium, different devices can be used to overcome physical range restrictions within a line. On the twisted-pair medium, *bridges* (or *repeaters*, a traditional KNX term that does not relate to their function with respect to the OSI layers) can be used to extend the physical length of a line. A line can contain up to 256 devices. Up to 16 lines can be interconnected by *main lines* to form an *area*. Routers linking the lines to the main line are referred to as *line couplers* (LC). Again, up to 15 main lines can be interconnected by a common *backbone line* using routers called *backbone couplers* (BC). The entire address space is referred to as a *domain*. The resulting topology is shown in Figure 42.1 (bridges are labeled "B").

To address multiple nodes at the same time, they can be arranged into so-called communication groups. Each group has a dedicated *group address* that is two octets long. Group addresses are globally defined for the whole domain, that is, they are independent of the physical location of the group members. Devices may be part of more than one group.

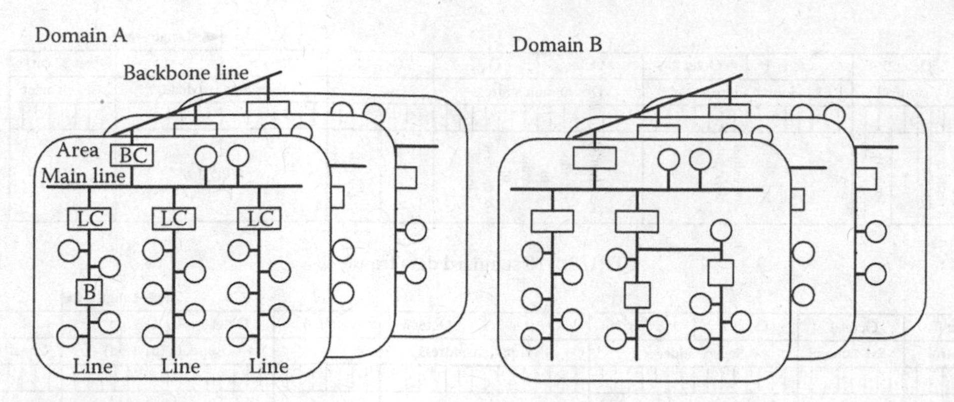

FIGURE 42.1 Topology.

In addition to individual and group addresses, KNX also defines the use of *polling addresses* allowing a device to request status data from up to 15 other nodes with minimal protocol overhead. This communication scheme is useful for high-frequency node liveness checking and rests upon the service `L_Poll_Data`. It is limited to a single physical segment. Finally, to avoid interference between different KNX installations on open media, KNX supports the use of *domain addresses*.

The *network layer* (NL) uses the services provided by the DL and offers four different NL services: a unicast service (`N_Data_Individual`), a multicast service (`N_Data_Group`), a domain-wide broadcast service (broadcast within a single domain; `N_Data_Broadcast`), and a system broadcast service (broadcast across domain borders; `N_Data_SystemBroadcast`). For all four services, the individual address of the sender is used as the source address. The destination address type depends on the used service: for unicast communication, the individual address of the receiver is used, while for multicast communication, the group address of the destination group is used. The broadcast services are implemented by using a DL group address of "0." Additionally, the NL introduces a *hop count*, which is decremented and examined by routers and repeaters to perform filtering based on the amount of elapsed hops of a packet.

The TL uses the NL services and enriches them by providing a connection-oriented unicast service (`T_Data_Connected`). Using this service, a device is able to establish a reliable unicast connection to another device. The state machine used implements an acknowledgment mechanism where data packets are retransmitted in case of negative or absent acknowledgments. The other NL services are transparently passed through (`T_Data_Individual`, `T_Data_Group`, `T_Data_Broadcast`, and `T_Data_SystemBroadcast`).

The AL on top of the protocol stack supports a multitude of AL services. Generally, these services can be divided into two different service classes. The first one is dedicated to exchanging process data (*process data communication*). Beside other rarely used opportunities to exchange process data in KNX (e.g., using the polling mechanism), the most important ones are the *group communication* services (`A_GroupValue_Read` and `A_GroupValue_Write`). Since they are based on `T_Data_Group`, they are multicast services. The second class of services is used for configuration and maintenance tasks (*management communication*). It includes services for uploading user applications, assigning individual and group addresses and accessing diagnostic information.

KNX defines different frame formats depending on the communication medium and service. The most commonly encountered format by far is the twisted-pair *standard data frame* `L_Data`, shown at the top of Figure 42.2.

The frame starts with the control octet. This octet indicates the frame format (Frame Type) and contains the priority field. It also holds a repeat flag that is set for retransmitted frames. The following

Octet 0	Octet 1	Octet 2	Octet 3	Octet 4	Octet 5	Octet 6	...	Octet 22 (max)
Control	Source address		Destination address			User data (payload)		Check octet
Frame type, Repeated, Priority	Sender: Individual address		Receiver(s): Individual or group or broadcast address		Address type, Hop count, Length	TPDU		FCS

TP1/PL110 standard data frame

Octet 0	Octet 1	Octet 2	Octet 3	Octet 4	Octet 5	Octet 6	Octet 7	...	
Control	Ext. control	Source address		Destination address			User data (payload)		Check octet
Frame type, Repeated, Priority	Address type, Hop count, Extended frame format	Sender: Individual address		Receiver(s): Individul or group or broadcast address		Length	TPDU		FCS

TP1/PL110 extended data frame

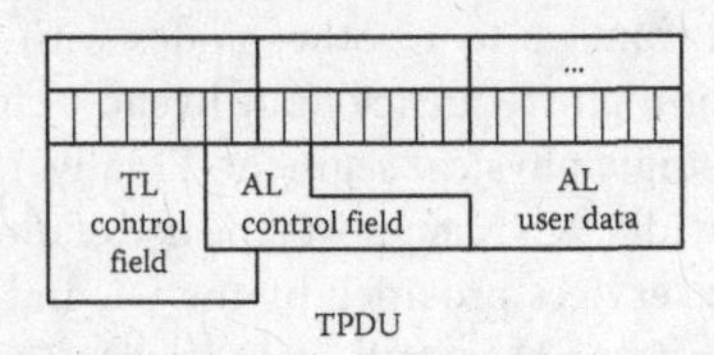

FIGURE 42.2 Frame formats.

two octets specify the source address. Regardless of the used communication service, this is always the individual address of the sending device. The succeeding two octets identify the destination of the message. The address type (AT) bit indicates whether the destination is an individual address or a group address. Together with the hop count used by the NL and the frame length, the AT bit makes up the last octet before the NL payload. The value of the Length field is the number of NL payload octets minus one. The frame ends with a check octet (FCS, frame checking sequence), which holds the vertical odd parity of all preceding octets (each Bit is calculated as the logical XOR over all Bits of the same significance, then inverted).

The format of the TL protocol data unit (TPDU), that is, the NL payload, is shown at the bottom of Figure 42.2. On KNX, it is the lowest layer PDU whose format is independent of the communication medium. The width and alignment of the AL protocol control information and user data differ depending on the TL and AL service type as well as the size of the AL payload.

The standard data frame supports TPDUs with a maximum length of 16 octets. This corresponds to a maximum AL payload of 14 octets. For the transmission of longer TPDUs, the *extended data frame* format has been defined. It is shown in the middle of Figure 42.2. Extended frames have an additional control field (Extended Control) as their second octet. Besides the AT Bit and the hop count, this octet contains a 4 bit field that holds additional format information (Extended Frame Format). The length field is expanded from 4 to 8 bit.

Standard and extended data frames are usually followed by a layer 2 acknowledgment frame sent by the receiver, which is 1 octet long. If group addressing is used on the twisted-pair medium, all receivers send this frame at the same time; negative acknowledgments will cancel out positive ones. The `L_Poll_Data` frame format (which is rarely used) is roughly similar to `L_Data`, with the master transmitting a request consisting of the control and address part followed by a check octet and the slaves responding with one octet of user data each.

On the powerline medium, the formats of standard and extended data frames are identical to those defined for twisted-pair segments. However, a preamble and receiver training sequence are added in front, and the domain address follows as a postamble. Acknowledgment frames can only be sent by a single group member. The total length of extended frames (which are practically never used on this medium) is limited to 73 octets.

The KNX radio frequency (RF) frame format has supported TPDUs longer than 16 octets from its inception. Therefore, the difference between standard and extended data frames does not exist on this medium, and a slightly different format is used. The cEMI frame format that is used for tunneling KNX data over IP networks is similar to the extended data frame format as described.

In addition to network protocol messages, the KNX standard defines the use of so-called external messages which are specified in the *external message interface* (EMI). EMI specifies a standard protocol for accessing services of the KNX network stack and management server entities. Its messages can be sent via various lower layer protocols, which are also part of the KNX specification. EMI is mainly used for local point-to-point connections where external user applications communicate with a connected bus attachment unit (BAU). A typical example would be a management client like a workstation with a USB or EIA-232 connection to a device that provides a connection to the KNX bus medium. Various EMI versions that differ in the available service types and in the used encodings exist: EMI1, EMI2, and cEMI (common EMI). In contrast to other EMI versions, cEMI is also used for KNXnet/IP and thus the only version to be used outside the context of local point-to-point connections.

42.3 Medium-Dependent Layers

The current release 2.0 of the KNX Handbook specifies three different physical layers for communication. Communication over IP networks as a first class medium is included as a draft specification. The resulting four options are described in the following. All media can be deployed and combined using routers according to an integrator's need.

Twisted Pair 1 (TP1) is the oldest and still most popular medium for KNX. It was taken from EIB. A single-wire pair (e.g., ½ YCYM 2 × 2 × 0.8) is used for both data and power transmission. A TP1 segment can accommodate up to 256 devices in free topology. It may extend up to an accumulated cable length of 1000 m and must contain at least one bus power supply unit to provide the link power of about 30 V DC. The contention protocol being used is carrier-sense multiple access (CSMA) with collision avoidance by bit-wise arbitration. The dominant logical 0 is encoded as a negative voltage imprinted on the DC supply voltage, whereas a logical 1 corresponds to the idle state of the medium. Data are transferred character oriented via half-duplex bidirectional communication at a transmission rate of 9600 bit/s. Characters contain eight data bits, even parity, and one stop bit (8e1) and are separated by a minimum idle period of 2 bit times. Data frames of variable length may be transmitted after a line idle time of 50 bit times. An immediate acknowledgment frame may follow after a gap of 15 bit times. TP1 is the only medium to support polling groups. Domain addresses are not used on TP1. The Batibus twisted-pair medium was also part of the KNX standard under the name of TP0, but was removed from release 2.0.

The design of the KNX RF physical layer and DL is compatible with Wireless M-Bus, a European metering protocol standard, allowing tight integration. KNX RF operates at a center frequency of 868.3 MHz, which is located within the band reserved for SRDs (Short Range Devices). In the 868.0–868.6 MHz subband, a duty cycle limitation to less than 1% applies. Data are transmitted at a data rate of 16.4 kbps using frequency shift keying (FSK) modulation and Manchester encoding. The KNX RF frame format is based on FT3 as specified in IEC 60870-5. It provides the capabilities of the extended frame format as described in the previous section and contains additional medium-related information such as signal quality or battery state. Retransmitters can be used to extend the range of the network. To allow detection and elimination of duplicate frames that can occur in particular when such retransmitters are employed, the TP1/PL110 repeat flag is replaced by a 3 bit link layer frame number (LFN). KNX RF allows the use of unidirectional, transmit-only devices. Such devices are cheaper to produce and have a greatly extended battery lifetime due to reduced power consumption, but accommodating them requires an adapted addressing scheme. Therefore, an extended address format composed of the unique serial number of the device concatenated with either the KNX individual or group address (correspondingly referred to as an extended individual or an extended group address) was introduced. Transmit-only devices also use preassigned group addresses. To allow an adequate battery lifetime for battery driven bidirectional devices as well, a synchronous operation mode was defined. In this mode, devices need only have their receivers enabled during certain time windows which are known by all nodes. Also, the specification includes a draft for extending KNX RF to multiple channels, different signalling speeds, additional frequency bands, and providing additional features for power-saving asynchronous receiver operation.

KNX Powerline 110 (PL110) uses the 230 V/50 Hz electrical power supply network for data and power transmission (in compliance with EN 50065-1). Half-duplex bidirectional communication is supported. KNX data are modulated using spread frequency shift keying (SFSK) with a center frequency of 110 kHz. The signal is injected between phase and neutral and is superimposed to the sinusoidal oscillation of the mains. No restrictions are made on the physical topology of the installation. Repeaters can be installed in three-phase networks if passive phase coupling is not sufficient. Owing to the relatively poor transmission properties of the mains network, the data transmission rate is limited to 1200 bit/s. Each data link octet is coded into a 12 bit character (8 bit data, 4 bit error correction). Medium access control is based on a slotted technique to reduce the probability of collisions: After the minimum silence period between two frames has elapsed, two time slots are reserved for pending high-priority transmissions, followed by seven more from which nodes with pending standard-priority transmissions choose one at random as their starting time. PL110 was taken over from EIB. The EHS powerline medium was also part of the KNX standard under the name of PL132, but was removed from release 2.0.

Regarding the transportation of KNX telegrams on top of IP networks, KNXnet/IP (formerly known as EIBnet/IP) currently focuses on scenarios for enhancing central and/or remote management. KNXnet/IP specifies several service protocols. The KNXnet/IP *Core Services* define the packet structure and methods required for discovery and self-description of a KNXnet/IP server and for setting up and maintaining a communication channel between the client and the server.

KNXnet/IP *tunneling* describes the point-to-point exchange of KNX data over the IP network. Its main purpose is to replace USB or EIA-232 connections between KNX network interfaces and PC workstations or servers (e.g., for configuration and diagnostics of devices on the KNX network or visualization) by tunneling `L_Data` frames. Acknowledgments, sequence counters, and a heartbeat mechanism are used to ensure the robustness expected in comparison with such a traditional connection. KNXnet/IP *routing* is a point-to-multipoint protocol for routing messages between KNX lines over a high-speed IP backbone. KNXnet/IP routers send UDP/IP multicast messages to other KNXnet/IP routers on the same IP network, which in turn filter the messages according to their destination address or group address and eventually pass them to the "native" KNX segment. KNXnet/IP routers can replace traditional KNX LCs and BCs. KNXnet/IP *device management* allows configuration and diagnosis of KNXnet/IP tunneling interfaces or routing devices via the IP network. Within all these three service protocols, the actual KNX data is carried within cEMI messages.

While KNXnet/IP is designed for devices with one IP and one "traditional" KNX network interface, the upcoming KNX IP is also intended for end devices that are solely connected to the IP medium. A typical bandwidth of 10–100 Mbps for IP over Ethernet allows the integration of devices requiring higher data rates (e.g., telecommunication and multimedia), which is a significant improvement compared to the 9.6 kbps of TP1. Largely based on KNXnet/IP routing, KNX IP will also include mechanisms to counteract possible problems arising from this throughput difference. Both configuration and runtime interworking remain unchanged from other KNX media.

42.4 Runtime Interworking

In KNX, *runtime interworking* refers to the definitions that enable devices to communicate for the purpose of exchanging process control data. The communication endpoints of a KNX device application relevant for this purpose (such as the illuminance level measured by a light sensor or the control input of a load switch) are called *group objects* (GOs). The KNX AL provides services to propagate changes to these GOs using the at-most-once multicast service `Data_Group` provided by the KNX TL. For this purpose, it maintains an *association table* between AL service access points (SAPs), that is, GO identifiers, and TL SAPs (which correspond one-to-one to DL group addresses). Thus, a group address describes a group of GOs in a communication relationship with each other. Every node is aware which groups it belongs to and treats messages accordingly.

Usually, data sources will actively publish new data via the AL `GroupValue_Write` service. If a user application changes, the value of a GO and `A_GroupValue_Write.req` is invoked, the network stack determines the group address this information is to be sent to. The user application is not aware of the group addresses associated with a GO (implicit addressing). In receivers that recognize this group address as one of their own, the AL generates an `A_GroupValue_Write.ind` for the AL SAPs bound to this group address, providing the user data from the `GroupValue_Write` PDU (protocol data unit). The application environment (KNX *application interface layer*) updates the GOs accordingly and notifies the user application. This mechanism allows passing a piece of information to an arbitrary number of recipients by way of a single message. For a single receiver only, it is shown in Figure 42.3. `GroupValue_Write` is an unconfirmed service. Message delivery is not guaranteed. If exactly-once semantics are required, confirmations have to be obtained by the user application.

In this mode of communication, senders do not need to know which nodes will actually be receivers of their messages. The knowledge concerning which nodes participate in any particular communication relationship is distributed over all nodes in the system. In KNX, it has traditionally even been impossible (or at least infeasible) for any node to determine which other nodes will accept and process messages to any specific group address. This pattern is sometimes referred to as the producer/consumer

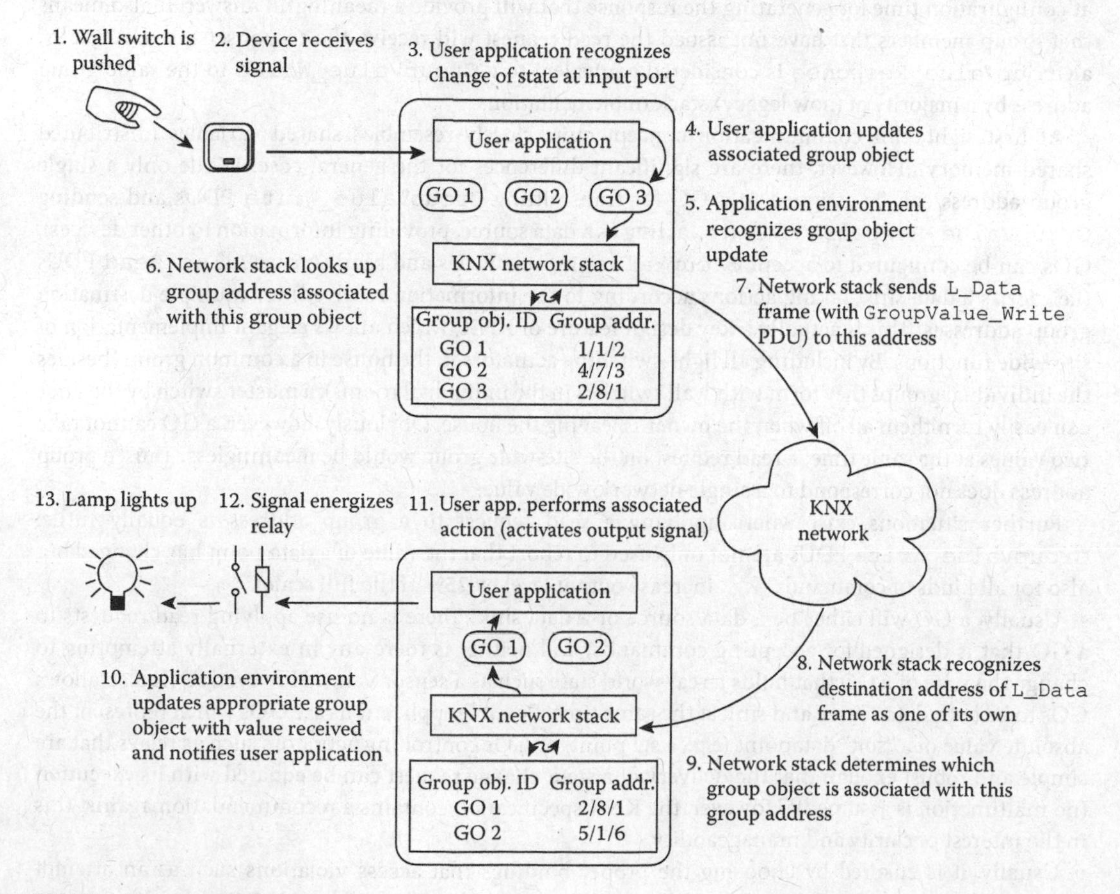

FIGURE 42.3 `GroupValue_Write`—end-to-end overview.

principle. Note that there is no implicit synchronization, however. No central instance exists at runtime that would manage subscriptions. The mechanism is entirely peer-to-peer.

To the network stack, the group address is an entirely opaque identifier that bears no relation to either the content of the message or the receivers' individual addresses. Still, since KNX groups are highly static, group addresses can be considered identifiers for a particular class of information traveling through the network. Receivers pick up the messages if this class of information is relevant to them. Especially in the context of Controller Area Network technology, this mechanism has been called subject-based addressing.

In practice, integrators actually assign group addresses based on the information provided or the function to be effected by the associated GOs. For this purpose, group addresses are divided into either two fields (5 and 11 bit wide) or three fields (5, 3, and 8 bit). This division is used for hierarchical grouping (e.g., site-wide functions/panic alarm). No binding standard interpretation of the group-address space is specified by the KNX standard, however.

Since no other transport mechanism than multicast is specified, datapoints (i.e., GOs) cannot be addressed individually unless appropriate provisions are made at configuration time. It also means that protocol security mechanisms are more difficult to implement.

While `GroupValue_Write` is the mainstay of KNX communication, a confirmed AL service for querying the value of a GO is provided as well. It involves the PDUs `GroupValue_Read`, representing the query (request), and `GroupValue_Response`, carrying the answer (response). Both `GroupValue_Read` and `GroupValue_Response` again use multicast communication via group addresses exclusively. This means that multiple responses—with different values—may be prompted by a single `GroupValue_Read` PDU. It is the system integrator's responsibility to preselect a node at configuration time for generating the response that will provide a meaningful answer. It also means that group members that have not issued the read request will receive the response(s), too. Actually, a `GroupValue_Response` is considered equivalent to a `GroupValue_Write` to the same group address by a majority of (now legacy) stack implementations.

At first sight, this communication concept quite closely resembles shared variables (distributed shared memory). However, there are significant differences for the general case. While only a single group address can be assigned to a GO for transmitting `GroupValue_Write` PDUs and sending `GroupValue_Response` PDUs (i.e., acting as a data source, providing information to other devices), GOs can be configured to accept `GroupValue_Write` PDUs and send `GroupValue_Read` PDUs (i.e., act as a data sink, taking actions according to the information received) for multiple destination group addresses. This is actually a key design feature of KNX, which allows elegant implementation of site-wide functions. By including all light-switching actuators in the house in a common group (besides the individual groups they form with wall switches in the individual rooms), a master switch by the door can easily turn them all off when the owner is leaving the house. Obviously, however, a GO cannot take two values at the same time; a read request on the site-wide group would be meaningless. Thus, a group address does not correspond to a single network-wide value.

Further situations exist where applying a read request to a group address is equally futile: `GroupValue_Write` PDUs are not only used to report that the value of a data point has changed but also for all kinds of commands (e.g., increase output level by 25% of the full scale).

Usually, a GO will either be a data source or a data sink. There is no use applying read requests to a GO that is designed for accepting commands, and neither is there any in externally attempting to change the state of a GO that holds a real-world state such as a sensor value. However, KNX also allows GOs to be both data source and sink at the same time. Possible applications are GOs which represent the absolute value of a "soft" datapoint (e.g., a set point) or GOs controlling actuators such as relays that are simple and robust enough that the delivery of a state-change request can be equated with its execution (no malfunction is assumed). However, the KNX specification contains a recommendation against this in the interest of clarity and manageability.

Usually, it is ensured by choosing the proper bindings that access violations such as an attempt to change a sensor value do not happen. In addition, it is determined for each GO at binding time if

`GroupValue_Write` PDUs should be sent; if `GroupValue_Write` PDUs should lead to the value of the GO being updated; if `GroupValue_Read` PDUs should be answered; and, except for legacy stack implementations, if `GroupValue_Response` PDUs should be treated as `GroupValue_Write` PDUs with the same destination address. Device applications come with sensible defaults and usually provide separate GOs for command input and status output, but the final responsibility for setting up the proper communication relationships remains with the installer.

Enabling read-back of a GO that is actually a data sink would, in theory, also allow checking if messages requiring exactly-once delivery semantics were delivered successfully by the at-most-once transport service. However, one will rather bind to a status feedback GO instead in such a case as this allows checking successful execution instead of delivery only. Usually, there are reasons besides a failed transmission that may cause a destination to ignore a message (e.g., wind alarm may block sunblind operation even if the command to move was received properly). If such exceptions are to be caught properly, end-to-end status feedback is mandatory anyway.

AL PDUs can be sent with different priorities (as provided by the DL). The common configuration tool (ETS) allows choosing the priority level on a per GO basis among the lower three; the KNX specification suggests only using the lower two.

While the AL `GroupValue` services deal with transferring new values from one GO to another, user applications still need to agree on how to interpret these values. In the KNX runtime interworking model, *applications* are split in FBs to be implemented on individual KNX networked devices. *Application models* describing the interplay of FBs are typically specified for a single application domain or a domain substructure (e.g., hot water heating) but can (and do) include points of contact with other applications from the same or other domains. Note that the term "profile" refers to something different in KNX.

FBs correlate device behavior with activity regarding their network communication endpoints. In KNX, communication endpoints of FBs are termed *datapoints*, irrespective of whether they are message sources or sinks for the exchange of process control data between devices during normal operation (*input* and *output* datapoints) or *parameters* that are only changed in the configuration phase. Traditionally, parameter datapoints are accessed following a manufacturer proprietary memory mapped model, but recent FB definitions include provisions for using a more open object-oriented model (*interface objects* and *properties*). Input and output datapoints may also be reflected as interface object properties (and accessed via unicast TL services) or implemented using the TP1 master/slave polling mechanism. In practice, however, the only communication mechanism that is relevant for input/output datapoints is the group-object model as described in detail above.

Standard *data point types* (DPTs) define the data type (e.g., integer) and bit-level encoding (e.g., two's complement) of both input/output and parameter datapoints. The unit of measurement to be associated with a datapoint value and range restrictions are also reflected in the DPT. A wide variety of DPTs exists, covering, among others, Boolean values, signed and unsigned integers of multiple widths, time, date, and floating-point values. DPTs may also consist of multiple fields ("structured" DPTs). Thus, depending on the context and AL service used to transport it, a single DPT can represent the present value of a datapoint together with associated status information or a command with multiple parameters.

Traditionally, the meta information contained in FB definitions or DPTs is no longer visible at the level of network messages exchanged during normal operation. Just as for group addresses, a project specific (external) database containing the semantic information is required for interpreting monitored communication. Such a database also must contain device configuration information, since control information in KNX is usually reduced to trigger information. Instead of "change brightness level to 100%, taking 10 s for the transition from the current level," a typical message will say "level 100%" or maybe even only "on," with the precise interpretation governed by the device configuration.

In EIB (which can be regarded as the main predecessor system to KNX), hardly any behavioral aspects were standardized at all, with control of actuators for light dimmers and motorized blinds being notable exceptions. Devices interacted chiefly by exchanging Boolean values. This concept actually was very close to powering traditional electric circuits on and off, with the actual outcome depending on the devices connected. Consequently, there was only very basic type compatibility checking even at setup time, limited to value encodings. There were no provisions to prohibit binding the output of a motion detector to the slat angle adjustment input of a sunblind actuator. Still today with KNX, integrators can bind GOs individually and with minimal restrictions (*free binding*)—taking the responsibility for the devices to work together as intended.

Only recently, the key function blocks for the electrical trade (lighting and shutters/blinds sensors and actuators) have been thoroughly revised. Parameters were standardized; still, process data exchange remains not only backward compatible but essentially unchanged. Besides lighting and blinds, typical EIB applications included simple room temperature control and basic energy management. As the focus of KNX has expanded beyond the electrical trade, the amount of interworking specifications has increased considerably.

Today, function blocks for the HVAC domain outnumber all others. Their definitions are so far the only ones to include maximum and typical message generation rates, allowing the network load generated by devices implementing them to be estimated a priori and serve as a guideline for network planning.

Mappings for the CHAIN (CECED Home Appliances Interoperating Network) white goods profiles specified in EN 50523 have also been developed but are not contained in the KNX Handbook. Most recently, metering has been added. The specification includes parts of the M-Bus RF protocol stack as well as function blocks for representing the metering data obtained this way within the KNX system, effectively defining a gateway. A similar approach was taken for integrating OpenTherm, a protocol standard for connecting home heating appliances and room thermostats: Essential control information was mapped to KNX FBs, while a tunneling mechanism allows OpenTherm communication as usual.

What was described so far is the traditional mode of runtime interworking, also known as "S-Mode." Another mode has been designed and is used for HVAC applications (although it is not necessarily limited to them). It is known as "LTE" (logical tag extended). While S-Mode and LTE devices can coexist on the same KNX network, their message formats are different. Therefore, LTE devices additionally implement S-Mode communication for a subset of their functions. Instead of free addressing and binding, LTE uses a "tagged" assignment scheme. In it, zoning information is statically mapped into the group-address space, defining a fixed correlation between certain addresses and locations (as well as domain substructures). A hierarchical structure allows addressing supersets of destinations at once. The group-address range was quadrupled by making use of the Extended Frame Format field. This means that the TP1/PL110 standard frame format cannot be used (see Figure 42.2).

Although LTE input/output datapoints are connected via multicast communication, they make use of the interface object/property concept (which is otherwise only used for parameter datapoints) for sending semantic information along. In every message, the object number identifying the function block and the property number identifying the datapoint within the block are included. As an example of LTE runtime interworking, consider a room temperature sensor sending the current measured value to a fancoil controller.

Since the sensor includes the object number of the Room temperature sensor (RTS) FB and the property number of the TempRoom datapoint with its transmission, the controller knows what to do with the value without further configuration. To establish the binding between a particular sensor and contoller, the installer sets the same location tag on both devices. This tag uniquely defines the group address that the sensor will use as the destination address and that the controller will listen for.

LTE also distinguishes between reporting the state of a value and ordering it changed. To accommodate these changes and additions, the AL services `GroupPropValue_Read/-Response`, `GroupPropValue_Write`, and `GroupPropValue_InfoReport` were introduced. All of them use the `Data_Tag_Group` TL service, which was defined for this purpose.

Finally, the recent addition of a file system access and bulk data transfer protocol (e.g., for trend logs) that is closely modeled on FTP deserves mention.

42.5 Devices

Various building blocks exist for developing KNX compatible devices. Different transceivers are required depending on the transmission medium. For TP1, four different transceiver solutions allow optimization of the KNX device design in accordance with the area of application. A transceiver circuit constructed from discrete components offers low-cost bus access, but requires high certification effort. The use of the FZE 1065 transceiver IC with transformer coupling and bit interface allows a highly resistant solution, while the FZE 1066 transceiver with direct bus coupling enables a resistant and miniaturized solution. Both transceiver ICs offer a bit interface and allow implementing a certified physical layer with minimum effort. Finally, the TP-UART IC is a transceiver with direct bus coupling and a UART host interface. It allows certified and miniaturized bus access with relaxed timing requirements on the microcontroller, as it handles most of the KNX TP1 DL. For the PL110 medium, only a single type of transceiver ASIC is available. The use of the RF and IP media does not require hardware that is specifically designed for KNX.

To further ease system development and certification, KNX defines standardized communication modules (i.e., bus attachment units, BAUs) such as the bus coupling unit (BCU), the bus interface module (BIM), or the powerline interface module (PIM), which provide an implementation of the complete network stack and application environment. They can host simple user applications, supporting the use of GOs in a way similar to local variables. BIMs/PIMs are soldered or detachable communication modules that are directly attached to the main device circuit board whereas BCUs are equipped with housing and shielding against electromagnetic interference.

Most KNX devices come either in rail-mount housings for mounting in distribution boxes or as flush-mount devices. Flush-mount end devices based on BCUs are typically split into two parts, the communication module (i.e., BCU) and the application module. Both are connected via the standardized physical external interface (PEI), which can be configured in a number of ways. Simple application modules such as wall switches may use it for parallel digital I/O or as inputs to the BCU A/D converter. More complex user applications requiring the use of a separate microprocessor may use the PEI for high-level access to the network stack via EMI wrapped in an asynchronous serial protocol (BCU 1 devices use a protocol with proprietary hardware handshaking, while the protocol supported by BCU 2 devices is based on FT1.2 as specified in IEC 60870-5).

Different generations of TP1 communication modules exist. The traditional BCU 1, BIM M111, and BIM M115 modules feature a Motorola 68HC05B6 or compatible type microcontroller running at 2 MHz with 176 bytes RAM, 256 bytes EEPROM, and 5936 bytes ROM. BCU 2 and BIM M113 use a 68HC05BE12 microcontroller running at 2.4576 MHz. They have 384 bytes RAM, 991 bytes EEPROM, and 11904 bytes ROM. Most of the RAM and ROM are occupied by the system software. To overcome the severe performance limitations of these out-dated platforms, a new generation of BIMs has been developed. They are based on the NEC 78K0/Kx2 microcontroller and offer from 8 up to 48 kB flash memory. Additionally, a far more comfortable software development toolchain (including a C compiler and debugger) is available for this new platform. For PL110, 68HC05B16 microcontroller based BAUs (e.g., PIMs) exist that are quite similar to the 68HC05B6 series but feature larger ROM and RAM and a clock frequency of 4 MHz.

Apart from these standardized BAUs, various other KNX certified interface modules exist. For example, so-called serial interface modules (SIMs) are available, which can be directly soldered to printed circuit boards in a way similar to BIMs. They contain the complete communication stack and the application can interface with the KNX network using a simple, proprietary serial protocol. Complete software stacks for several microcontroller families (e.g., Atmel ATmega and ARM7, Texas Instruments MSP430, NEC 78K0) can also be obtained from various vendors.

42.6 Configuration

In KNX, considerable attention is given to uniform, manufacturer-independent configuration, parameterization, and binding. This is referred to as *configuration interworking*. Central to the KNX configuration interworking concept is a single, universal PC software package for planning and commissioning KNX installations. It is maintained and sold by the KNX Association, based on Microsoft Windows and called ETS (Engineering Tool Software). ETS assists with defining the project in a structured way and is used to configure the behavior and communication relationships of KNX devices. This includes loading application programs and setting application parameters and can be done over the network or the PEI, if available. ETS also provides bus monitoring functions for troubleshooting.

Manufacturers are required to supply the necessary device and user-application descriptions along with their hardware. Tools for their creation are provided by the KNX Association as well. ETS can also be extended by way of plug-ins, should the configuration of a device require it. Only a minor number of devices need further setup tools in addition to ETS. For interacting with such external tools, ETS supports the export of project files.

For accessing the configurable resources within KNX devices, both a memory-mapped approach and a more transparent object/property-oriented model are defined. The resources themselves (e.g., memory locations, property identifiers, and their purpose) are also standardized as far as the communication system is concerned. The *management procedures* applicable for a particular device and its resource layout are specified by the device profile it conforms to. Access to the user application still largely requires ETS and the device manufacturers' descriptions.

Increasingly, KNX devices that do not require ETS for commissioning are available. This is referred to as E-Mode (for "Easy"), while configuration via ETS is called S-Mode (for "System"). A variety of E-Modes exists. Their aim is to provide installers with easier and less error-prone (albeit less powerful and flexible) ways of creating communication bindings. This can be accomplished by locally bringing the binding partners (and only them) into a special configuration mode (push button mode), setting tag numbers on devices that are to cooperate to identical values (LTE mode), or a hand-held, dedicated configuration tool that enumerates possible binding partners (controller mode). While in S-Mode, GOs are bound individually, E-Modes usually perform binding for a group of logically corresponding GOs (referred to as channels) together.

Although the ability to be configured with ETS is a certification requirement for all KNX devices, this requirement is temporarily suspended since ETS does not yet support E-Mode devices. A further configuration mode (A-Mode) was also part of the KNX standard, but was removed from release 2.0.

42.7 Conclusion and Outlook

When the convergence process between EIB, Batibus, and EHS started in 1997, EIB was assigned a leading role: It was to be enhanced by features that were present in EHS and Batibus but not yet in EIB. In 2002, a new standard emerged as a result, attempting to bridge three different domains by the combination of three technologies: electrical installation (lighting/shading) via EIB, HVAC via Batibus, and household appliances (white goods) via EHS. This standard was later labeled KNX. EIB devices were automatically KNX compliant; this was not the case for Batibus and EHS devices, however. Today, Batibus and EHS heritage has largely been eliminated from KNX since no KNX products based on these parts of the standard were made available commercially. EHS 1.3a has become a KNX-associated standard, but the products on the market that are based on it are not KNX compatible.

Nevertheless, the KNX standard has been steadily improved and augmented. The most visible improvements over EIB are its RF medium, the IP tunneling protocol, new configuration modes that do not require ETS, and the addition of metering as well as full-fledged HVAC control for home as well as light commercial environments. However, the functions added for the HVAC trade are not fully and seamlessly integrated with those that have grown from the electrical trade. As a consequence from these developments, the KNX specification has become complex to read, although considerable efforts toward consolidation were taken for the current release of the KNX master specification (*KNX Handbook* 2.0).

Nevertheless, there is ample room for further activities: Open and stable formats for the description of product configuration data (user application resources) and project specific data are still missing, although an XML-based proposal has been in its draft stage for years and the upcoming ETS version 4 stores project and product data in XML. Efforts should also be directed into a Web services mapping. While protocol security is slowly being addressed by manufacturers, safety issues have not been taken into account so far.

Abbreviations

AL	Application layer
APDU	Application layer protocol data unit
AT	Address type
BAU	Bus attachment unit
BCU	Bus coupling unit
BIM	Bus interface module
cEMI	Common external message interface
DL	Data link layer
DPT	Data point type
EIB	European Installation Bus
EMI	External message interface
ETS	Engineering tool software
FB	Functional block
FCS	Frame checking sequence
GO	Group object
IP	Internet protocol
LTE	Logical tag extended
NL	Network layer
PDU	Protocol data unit
PIM	Powerline interface module
PL110	Powerline 110
RF	Radio frequency
SAP	Service access point
TL	Transport layer
TP1	Twisted Pair 1
TPDU	Transport layer protocol data unit

References

1. *KNX Specifications, Version 2.0*, KNX Association, Diegem, Belgium, 2009.
2. EN 50090, Home and building electronic systems (HBES), CENELEC, Brussels, Belgium, 1994–2009.
3. EN 13321-1, Open data communication in building automation, controls and building management—Home and building electronic system—Part 1: Product and system requirements, CEN, Brussels, Belgium, 2006.
4. EN 13321-2, Open data communication in building automation, controls and building management—Home and building electronic systems—Part 2: KNXnet/IP communication, CEN, Brussels, Belgium, 2006.

5. ISO/IEC 14543-3, Information technology—Home electronic system (HES) architecture—Part 3, ISO/IEC, Geneva, Switzerland, 2006–2007.
6. *Handbook for Home and Building Control, Basic Principles*, 5th edn., ZVEI/ZVEH (eds.), KNX Association, Diegem, Belgium, 2006.
7. W. Kastner and G. Neugschwandtner, EIB: European Installation Bus, in *Industrial Communication Technology Handbook*, CRC Press, Boca Raton, FL, 2005.
8. ERC recommendation 70-03 relating to the use of short range devices (SRD), CEPT, European Communications Office, Copenhagen, Denmark, 1997–2010.
9. EN 13757, Communication systems for meters and remote reading of meters, CEN, Brussels, Belgium, 2002–2008.
10. EN 50065-1, Specification for signalling on low-voltage electrical installations in the frequency range 3 kHz to 148.5 kHz. General requirements, frequency bands and electromagnetic disturbances, CENELEC, Brussels, Belgium, 2001.
11. EN 50523, Household appliances interworking, CENELEC, Brussels, Belgium, 2009.
12. IEC 60870-5-1, Telecontrol equipment and systems. Part 5: Transmission protocols—Section 1: Transmission frame formats, IEC, Geneva, Switzerland, 1990.
13. IEC 60870-5-2, Telecontrol equipment and systems. Part 5: Transmission protocols—Section 2: Link transmission procedures, IEC, Geneva, Switzerland, 1992.

43

Protocols of the Time-Triggered Architecture: TTP, TTEthernet, TTP/A

Wilfried Elmenreich
University of Klagenfurt

Christian El-Salloum
Vienna University of Technology

43.1 Introduction 43-1
43.2 The Time-Triggered Paradigm 43-2
Sparse Time • Flow Control and Temporal Firewall
43.3 Time-Triggered Communication 43-3
43.4 Time-Triggered Protocol (TTP) 43-4
Fault Hypothesis and Fault Handling • Fault Tolerance • Membership
43.5 Time-Triggered Ethernet 43-5
Principles of Operation • Time Format • Periods • Fault-Tolerant TTEthernet Configuration • Clock Synchronization
43.6 TTP/A 43-7
Interface File System • The Three Interfaces of a Smart Transducer • Principles of Operation
Acknowledgments 43-11
References 43-11

43.1 Introduction

Time-triggered systems derive control by the progression of time and thus use the concept of time in the problem statement as well as in the provided solution. This approach supports a specification of interfaces including the temporal domain (e.g., send and receive instants of messages) and the implementation of "temporal firewalls," which prevent error propagation via control signals. Time-triggered systems support membership identification, interoperability, and replica determinism.

The concepts of time-triggered systems have been composed in the *time-triggered architecture* [KB03], establishing a framework for the implementation of dependable distributed real-time embedded applications. Besides protocols that employ a time-triggered scheduling like Flexray [Fle05] and TT-CAN [HMFH00], there exist three protocols that are especially designed according to the time-triggered architecture: time-triggered protocol (TTP), time-triggered Ethernet (TTEthernet), and TTP for SAE Class A [SAE95a] applications (TTP/A).

The TTP (also known as TTP/C protocol since it suites SAE class C application requirements [SAE95b]) provides a highly dependable real-time communication service with a fault-tolerant clock synchronization and membership service. TTP is suitable for X-by-wire systems in the automotive and avionics domain.

TTEthernet is a uniform communication architecture covering a whole spectrum of distributed applications reaching from simple non-real-time applications to multimedia systems up to the most demanding safety-critical hard real-time products.

The time-triggered fieldbus TTP/A is intended for the integration of smart transducers in all types of distributed real-time control systems. Although the first targets are automotive applications, TTP/A has been designed to meet the requirements of process control systems as well. TTP/A supports low-cost implementations on a wide set of available component-off-the-shelf microcontrollers.

In this chapter, we introduce these protocols by first giving a short overview on the common underlying concepts of the time-triggered architecture. A reader that is familiar with the time-triggered paradigm may directly go to the sections on TTP, TTEthernet and TTP/A, where we introduce the application domains, requirements, and principles of operation of the specific protocols.

43.2 The Time-Triggered Paradigm

The time-triggered paradigm encompasses a set of concepts and principles that support the design of highly dependable hard real-time systems.

43.2.1 Sparse Time

When global physical time is used to deduce causality of distributed events, it is necessary to synchronize the local clocks precisely. Clock synchronization is concerned with bringing the time of clocks in a distributed network into close relation with respect to each other. Measures of the quality of clock synchronization are *precision* and *accuracy*. Precision is defined as the maximum offset between any two clocks in the network during an interval of interest. Accuracy is defined as the maximum offset between any clock and an absolute reference time.

Due to the impossibility to perfectly synchronize clocks and the digitalization error, it is impossible to guarantee that two observations of the same event will yield the same timestamp. A solution to this problem is provided by introducing the concept of a *sparse time base* [Kop92]. In this model, the timeline is partitioned into an infinite sequence of alternating intervals of *activity* and *silence*. Figure 43.1 depicts the intervals of silence (s) and activity (a). The duration of the silence intervals depends on the precision of the clock synchronization.

The architecture must ensure that significant events, such as the sending of a message or the observation of an event, occur only during an interval of activity. Events occurring during the same segment of activity are considered to have happened at the same time. Events that are separated by at least one segment of silence can be consistently assigned to different timestamps for all clocks in the system.

43.2.2 Flow Control and Temporal Firewall

In order to transfer data between two components, they must agree on the flow-control mechanism to use and the direction of the transfer. Commonly, a communication between two subsystems is either controlled by the sender's request (*push* style) or by the receiver's request (*pull* style) [AFI+00].

Figure 43.2 shows the push method. The producer is allowed to generate and send its message at any time, thus flow control is managed by the producer. This method is very comfortable for the push producer, but the push consumer has to be watchful for incoming data messages at any time, which may

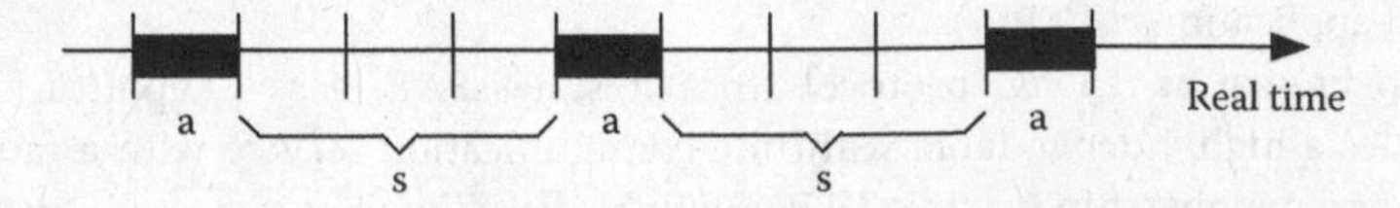

FIGURE 43.1 Sparse time base.

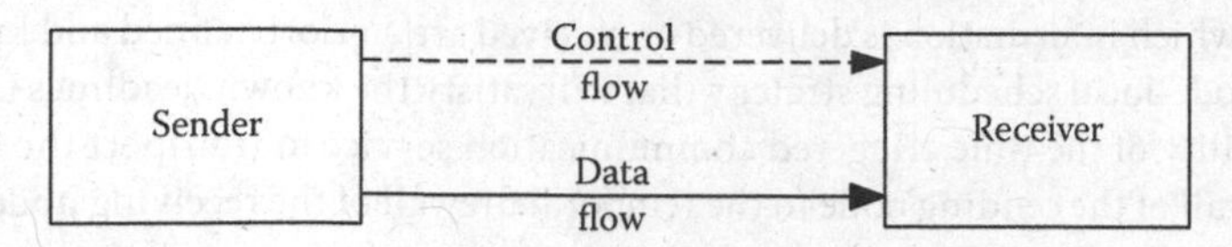

FIGURE 43.2 Push communication model (implicit flow control).

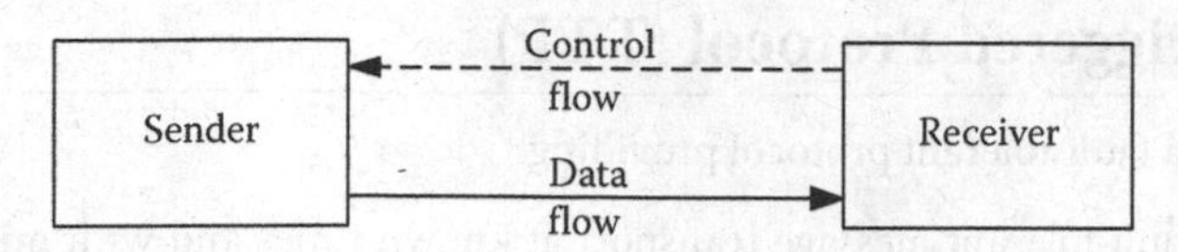

FIGURE 43.3 Pull communication model (explicit flow control).

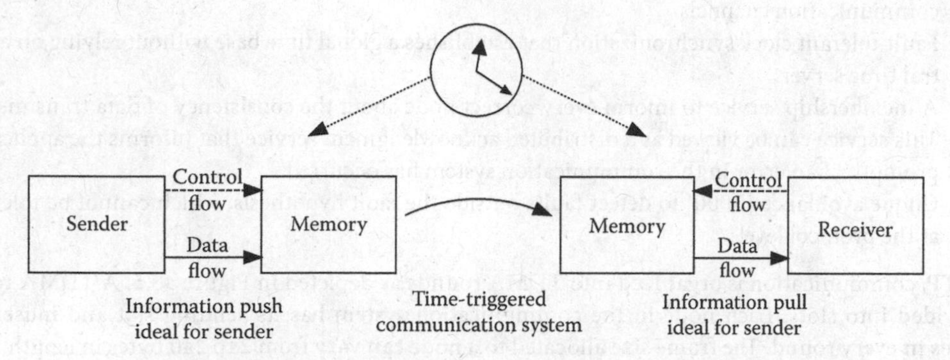

FIGURE 43.4 Temporal firewall.

result in high resource costs and difficult scheduling [EHK01]. Popular "push" mechanisms are messages, interrupts, or writing to a file [DeL99]. The push style communication is the basic mechanism of event-triggered systems.

In Figure 43.3, the flow control is on the consumer. Whenever the consumer asks to access the message information, the producer has to respond on the request. This facilitates the task for the pull consumer, but the pull producer has now to be watchful for incoming data requests [EHK01]. Popular "pull" mechanisms are reading a file, polling, state messages, or shared variables [DeL99]. The pull-style communication is the basic mechanism of client-server systems. The time-triggered paradigm implements a hybrid approach, which is explained in Figure 43.4.

A temporal firewall [KN97] enables different control flow directions between sender and receiver (Figure 43.4).

This model uses a combination of "push" and "pull" communication. Each component possesses a memory object that acts as a data source and sink for communication activities. Components that want to submit data are able to write the data into this memory using a *producer's push* interface. The transmission of data between the sender and the receiver is handled autonomously by the communication subsystem according to the time-triggered paradigm. After transmission, the consumer component accesses the data using a *consumer's pull* interface.

The values in the memory are state messages that keep their content until they are updated and overwritten. Since in the time-triggered architecture all nodes have knowledge about transmission schedules and access to a global time base, the instant when the protocol updates a value in the memory element is known to all nodes.

43.3 Time-Triggered Communication

The very basic principle of time-triggered communication is that the events of message transmission depend only on the progression of time and not on the availability of new information. Therefore, time-triggered communication requires a sufficient synchronization among all nodes' clocks.

The instants at which information is delivered or received are a priori defined and known by all nodes of a cluster. Any node-local scheduling strategy that will satisfy the known deadlines is "fit for purpose." It is the responsibility of the time-triggered communication service to transport the information from the temporal firewall of the sending node to the temporal firewall of the receiving node within the interval delimited by these a priori known fetch and delivery instants.

43.4 Time-Triggered Protocol (TTP)

The TTP [TTA03] is a fault-tolerant protocol providing

- Autonomous fault-tolerant message transport at known times and with minimal jitter between the nodes of a cluster by employing a time division multiple access (TDMA) strategy on replicated communication channels
- Fault-tolerant clock synchronization that establishes a global time base without relying on a central time server
- A membership service to inform every correct node about the consistency of data transmission. This service can be viewed as a distributed acknowledgment service that informs the application promptly if an error in the communication system has occurred
- Clique avoidance [BP00] to detect faults outside the fault hypothesis, which cannot be tolerated at the protocol level

In TTP, communication is organized into TDMA rounds as depicted in Figure 43.5. A TDMA round is divided into slots. Each node in the communication system has its sending slot and must send frames in every round. The frame size allocated to a node can vary from 2 to 240 bytes in length, each frame usually carrying several messages. The cluster cycle is a recurring sequence of TDMA rounds; in different rounds different messages can be transmitted in the frames, but in each cluster cycle the complete set of state messages is repeated. The data are protected by a 24 bit cyclic redundancy check (CRC). The schedule is stored in the message descriptor list (MEDL) within the communication controller of each node.

The clock synchronization is necessary to provide all nodes with an equivalent time concept. In doing so, it makes use of the common knowledge of the send schedule. Each node measures the difference between the a priori known expected and the observed arrival time of a correct message to learn about the difference between the sender's clock and the receiver's clock. A fault-tolerant average algorithm needs this information to periodically calculate a correction term for the local clock, so that the clock is kept in synchrony with all other clocks of the cluster. The membership service uses a distributed agreement algorithm to determine whether, in case of a failure, the outgoing link of the sender or the incoming link of the receiver has failed.

The core algorithms of TTP have been formally verified to prove their correctness. Particular controller hardware has been tested with multimillion fault injection and heavy ion radiation experiments and experiments with electromagnetic interferences have been carried out successfully.

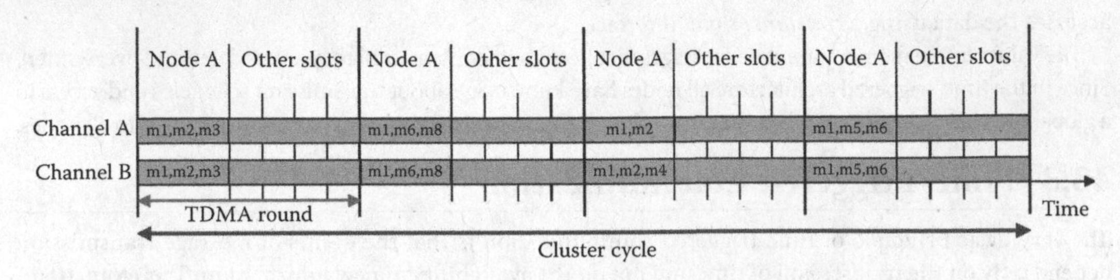

FIGURE 43.5 Frames, messages, slots, TDMA round, and cluster cycle in TTP communication.

43.4.1 Fault Hypothesis and Fault Handling

Provided that the components of a properly configured TTP-based system are in different fault containment regions, each can fail in an arbitrary way. Under this assumption, the probability of two concurrent independent component failures is small enough to be considered a rare event that can be handled by an appropriate never-give-up (NGU) strategy [TTA03, p. 27].

As for hardware faults, TTP is designed to isolate and tolerate single-node faults. By introducing a bus guardian, it is guaranteed that a faulty node cannot prevent correct nodes from exchanging data. The bus guardian ensures that a node can only send once in a TDMA round, thereby eliminating the problem of babbling idiots that monopolize the communication medium.

43.4.2 Fault Tolerance

The mechanisms described above ensure fault tolerance at the communication subsystem level in TTP. These mechanisms of the communication subsystem guarantee that faulty nodes cannot prevent correct nodes from communicating and serve as a communication platform for the application. At the application level, fault tolerance needs to be implemented by a fault-tolerance layer and an appropriate application design. Fault tolerance can be realized by replicating a software subsystem on two *fail-silent* nodes. Tolerance of a single *arbitrary node failure* can be ensured by triple modular redundancy (TMR) voting. Both mechanisms will tolerate single-component faults with the respective failure semantics and are thus fit to handle both transient and permanent hardware faults.

43.4.3 Membership

A major objective in the design of TTP is that the protocol should transmit data consistently to all correct nodes of the distributed system and that, in case of a failure, the communication system should decide on its own which node is faulty. These properties are achieved by the membership protocol and an acknowledgment mechanism.

Each node of a TTP-based cluster maintains a membership list with all nodes that are considered to be correct. This information is updated locally in accordance with successful (or unsuccessful) data transmissions and thus reflects the local view of the receiving node on all other nodes. With each transmission, each receiver sees and checks the sender's membership that is included in the sender's transmission or hidden in the CRC calculation.

An inconsistent view on the membership can only be caused by faults exceeding the fault hypothesis (e.g., multiple concurrent faults due to heavy electromagnetic interference). In this case, a clique avoidance mechanism establishes consistency by restarting the nodes which have inconsistent view with the majority of nodes.

43.5 Time-Triggered Ethernet

TTEthernet is a uniform communication architecture covering the whole spectrum of real-time applications [KAGS05]. It meets the requirements of simple non-real-time applications, multimedia systems, and safety-critical hard real-time systems. TTEthernet is fully backward-compatible with the Ethernet standard and combines the properties of standard Ethernet and TTP.

An important feature of TTEthernet is that it enables the integration of noncritical best-effort applications and highly demanding safety-critical control applications into a single network. Due to its backward compatibility to the Ethernet standard, existing Ethernet-based legacy applications can be integrated into a TTEthernet network without any modification. It is guaranteed by design, that these legacy applications cannot disrupt the temporally predictable communication of hard real-time applications in the TTEthernet network.

43.5.1 Principles of Operation

In order to allow deterministic real-time communication to coexist with competing senders communicating in a best-effort manner, TTEthernet distinguishes between two fundamentally different categories of traffic: *standard event-triggered* (ET) *Ethernet traffic* and *time-triggered* (*TT*) *Ethernet traffic*. In the case of TTEthernet traffic, the senders are cooperating according to a consistent message schedule in a way that no conflicts will occur between TTEthernet messages in a fault-free TTEthernet system. TTEthernet traffic provides dependable hard real-time communication traffic which coexists with standard Ethernet traffic but is guaranteed not to be disrupted by standard Ethernet messages.

Since standard Ethernet traffic typically originates in an uncontrollable open world, no temporal guarantees for the standard Ethernet traffic can be given. Whenever a standard Ethernet message (which is typically uncoordinated) conflicts with a TTEthernet message, the transmission of the standard Ethernet message will be preempted in order to be able to transmit the TTEthernet message with an a priori established constant delay (i.e., this delay is the same in the case where a standard Ethernet message has to be preempted in order to resolve a conflict as it is in the case where no conflict has occurred). When the transmission of the TTEthernet message has been completed, the preempted standard Ethernet message will be automatically retransmitted.

The preemption of standard Ethernet messages is performed by a dedicated *TTEthernet switch* [SGAK06]. The switch handles the standard Ethernet traffic according to the *store-and-forward* paradigm and a best-effort delay according to the standard Ethernet specification. The time-triggered traffic is handled according to the *cut-through* paradigm with an a priori known constant delay and minimal jitter.

43.5.2 Time Format

A digital time format can be characterized by three parameters: the *granularity*, the *horizon*, and the *epoch*. The granularity determines the minimum interval between two adjacent ticks of a clock, that is, the smallest interval that can be measured with this time format. The reasonable granularity can be derived from the achieved precision of the clock synchronization [KO87]. The horizon determines the instant when the time will wrap around. The epoch determines the instant when the measuring of time starts.

The time format of TTEthernet is a binary time format, which is based on the physical second represented by 64 bits. The fractions of a second are represented as 24 negative powers of 2, which results in a granularity of about 60 ns, and the full seconds are represented as 40 positive powers of 2 which results in a horizon about 30,000 years. The time format has been standardized by the OMG in the small transducer interface standard [OMG02]. The representation of the binary time format can be translated to the wall clock time by a standard Gregorian calendar function.

43.5.3 Periods

In TTEthernet, the period durations of time-triggered messages are restricted to the positive and negative powers of 2 of the second (1 s, 2 s, 4 s … or 1/2 s, 1/4 s, 1/8 s …). Due to this restriction, each period duration can be represented by one of the 64 bits of the binary time format. The bit representing the period duration of a time-triggered message is called the *period bit*. The TTEthernet implementation (version 1.9) supports 16 different durations and thus requires 4 bits to encode the period bit (see Figure 43.6). The offset of the send instant of a message to the start of the period is called the *phase* of the message. The phase is represented by a pattern of 12 bits right to the period bit. Thus, in TTEthernet 2 bytes are required to store the period duration and the phase of a message. These 2 bytes are called the *period identifier*.

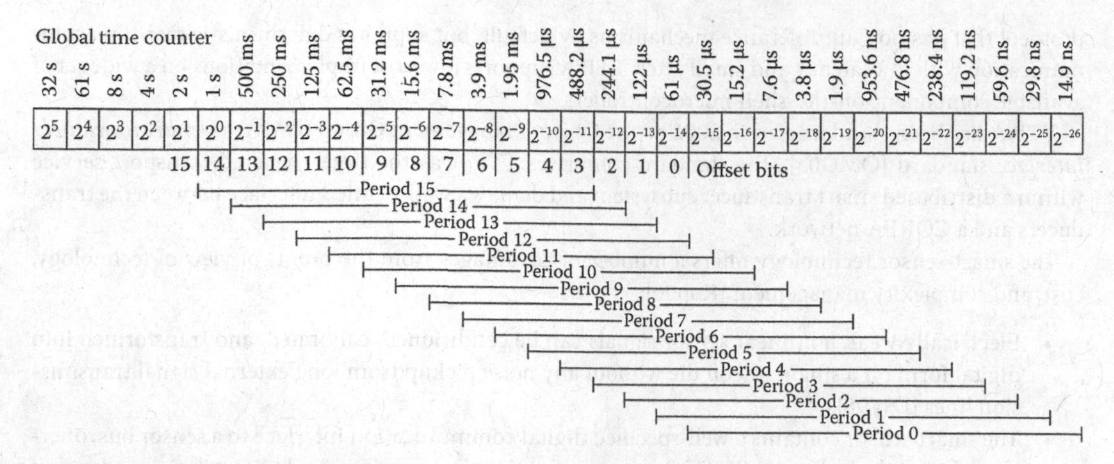

FIGURE 43.6 Period and phase (offset).

43.5.4 Fault-Tolerant TTEthernet Configuration

In the safety-critical configuration of TTEthernet, the network provides its services in the presence of node or communication channel failures [AKG+06]. To tolerate a fault in a communication channel, a *safety-critical TTEthernet controller*, which transmits and receives redundant messages using two distinct communication channels, is required. A message that should be redundantly sent on two communication channels is called *protected TT message*. In a safety-critical configuration, a TTEthernet switch is accompanied by a guardian controlling the input and output ports of the switch. The guardian has knowledge about the time-triggered schedule of all *protected TT messages* and dynamically disables all input ports that could possibly interfere with the transmission of a protected TT message.

43.5.5 Clock Synchronization

The clock synchronization mechanisms in TTEthernet can be adapted to the specific requirements of the actual applications. The initial establishment of the global time after start-up and the maintenance of this time during operation is performed by a subset of TTEthernet controllers called *time-keeping controllers* [KAGS05]. The most basic form that is supported is the *central master algorithm*. In this case, a single time-keeping controller—the master—will periodically send out a resynchronization message containing its local time. Being aware of the transmission latency of the clock synchronization message, the individual nodes can adjust their clock according to the time value in the synchronization message.

To tolerate a *fail-stop failure* of the time-master, a shadow master can be employed that takes over when the current master stops sending synchronization messages.

In the most dependable configuration, multiple time-keeping controllers can execute a *Byzantine resilient* clock synchronization algorithm that is able to tolerate an arbitrary failure of any controller.

43.6 TTP/A

The time-triggered fieldbus TTP/A is intended for the integration of smart transducers in all types of distributed real-time control systems. Although the first targets are automotive applications, TTP/A has been designed to meet the requirements of process control systems as well. TTP/A provides a well-defined smart transducer interface to the sensors and actuators of a network. TTP/A is a master/slave protocol that has no fault-tolerance mechanisms by default, but supports deterministic real-time communication with low latency and small jitter. TTP/A supports low-cost implementations on a wide set of available component-off-the-shelf microcontrollers.

TTP/A has been standardized in 2002 by the Object Management Group (OMG) as *smart transducer interface* standard [OMG03]. The standard comprises TTP/A as the time-triggered transport service within a distributed smart transducer subsystem and defines a well-defined interface between the transducers and a CORBA network.

The smart-sensor technology offers a number of advantages from the points of view of technology, cost, and complexity management [Kop00]:

- Electrically weak nonlinear sensor signals can be conditioned, calibrated, and transformed into digital form on a single silicon die without any noise pickup from long external signal transmission lines [DW98].
- The smart sensor contains a well-specified digital communication interface to a sensor bus, offering "plug-and-play" capability if the sensor contains a reference to its documentation in form of an electronic data sheet as it is proposed in the IEEE 1451.2 Standard [Ecc98].
- It is possible to monitor the local operation of the sensing element via the network and thus simplify the diagnosis at the system level.

The internal complexity of the smart-sensor hardware and software and internal failure modes can be hidden from the user by well-designed fully specified smart-sensor interfaces that provide just those services that the user is interested in.

43.6.1 Interface File System

The information transfer between a smart transducer and its client is achieved by sharing information that is contained in an internal interface file system (IFS), which is encapsulated in each smart transducer.

The IFS provides a unique address scheme for transducer data, configuration data, self-describing information, and internal state reports of a smart transducer [KHE01]. It establishes a stable intermediate structure that is a solid base for smart transducer services. The implementation of the IFS is economically feasible to assign local intelligence even to low-cost I/O devices like 8 bit microcontrollers with 4 kbytes Flash ROM and less than 64 bytes of RAM memory.

The IFS is the source and sink for all communication activities and acts as a temporal firewall [KN97] that decouples the local transducer application from the communication activities. A time-triggered sensor bus will perform a periodical time-triggered communication to copy data from the IFS to the fieldbus and write received data into the IFS. The RODL—a predefined communication schedule—defines time, origin, and destination of each protocol communication. The instants of updates are specified a priori and known by the communicating nodes. Thus, the IFS acts as a *temporally specified interface* that decouples the local transducer application from the communication task.

Each transducer can contain up to 64 files in its IFS. An IFS file is an indexed sequential array of up to 256 records. A record has a fixed length of 4 bytes (32 bits). An IFS record is the smallest addressable unit within a smart transducer system. Every record of an IFS file has a unique hierarchical address (which also serves as the global name of the record) consisting of the concatenation of the cluster name, the logical node name, the file name, and the record name.

The local applications in the smart transducer nodes are able to execute a clusterwide application by communicating directly with each other. Figure 43.7 depicts the logical network view for such a clusterwide application.

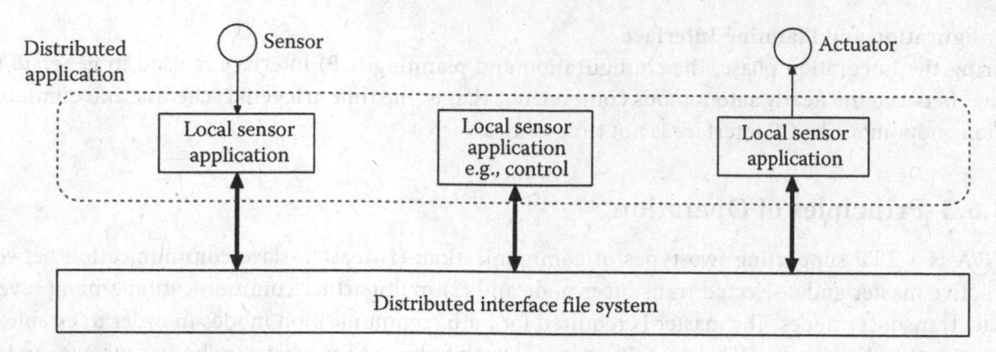

FIGURE 43.7 Logical network structure.

43.6.2 The Three Interfaces of a Smart Transducer

In order to support complexity management and composability, it is useful to specify distinct interfaces for functionally different services [Kop00]. As depicted in Figure 43.8, a smart transducer node can be accessed via three different interfaces.

Real-Time Service Interface

The real-time service (RS) interface provides the timely real-time services to the component during the operation of the system.

Diagnostic and Maintenance Interface

The diagnostic and maintenance (DM) interface opens a communication channel to the internals of a component. It is used to set parameters and to retrieve information about the internals of a component, for example, for the purpose of fault diagnosis. The DM interface is available during system operation without disturbing the real-time service. Usually, the DM interface is not time-critical.

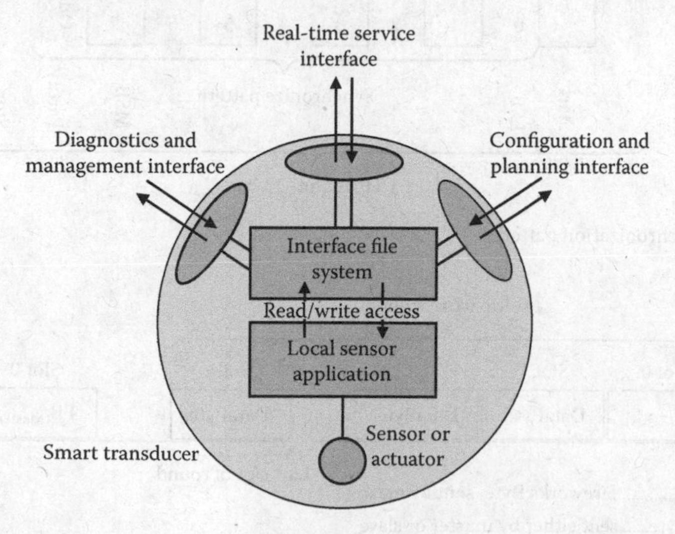

FIGURE 43.8 Three interfaces to a smart transducer node.

Configuration and Planning Interface

During the integration phase, the configuration and planning (CP) interface is used to generate the "glue" between the nearly autonomous components, that is, the time-triggered schedule and communication endpoints. The CP interface is not time-critical.

43.6.3 Principles of Operation

TTP/A is a TTP supporting two types of communication: (1) master–slave communication between an active master and a selected transducer node and (2) multipartner communication among several smart transducer nodes. The master is required for both communication modes in order to establish a common time base among the nodes. In case of a crash failure of the master, a shadow master can take control. Every node in this cluster has a unique *alias*, an 8 bit (1 byte) integer, which can be assigned to the node a priori or set via the configuration interface.

The bus allocation is done by a TDMA scheme. Communication is organized into rounds consisting of several TDMA slots. A slot is the unit for transmission of one byte of data. Data bytes are transmitted in a standard UART format. Each communication round is started by the master with a so-called *fireworks* byte. The fireworks byte defines the type (e.g., multipartner or master–slave) of the round and is a reference signal for clock synchronization. Followed by the fireworks byte are a number of data bytes. The communication pattern for each round is predefined via the RODL, which is distributively stored in all nodes.

There are eight different firework bytes encoded by a redundant bit code supporting error detection. One fireworks byte is a regular bit pattern, which is also used by slave nodes with an imprecise on-chip oscillator for startup synchronization (see Figure 43.9). Between 2 data bytes, there is an extra Inter-Byte Gap (IBG) in order to account for a switching time between different senders. The IRG defines a time of silence between two consecutive rounds.

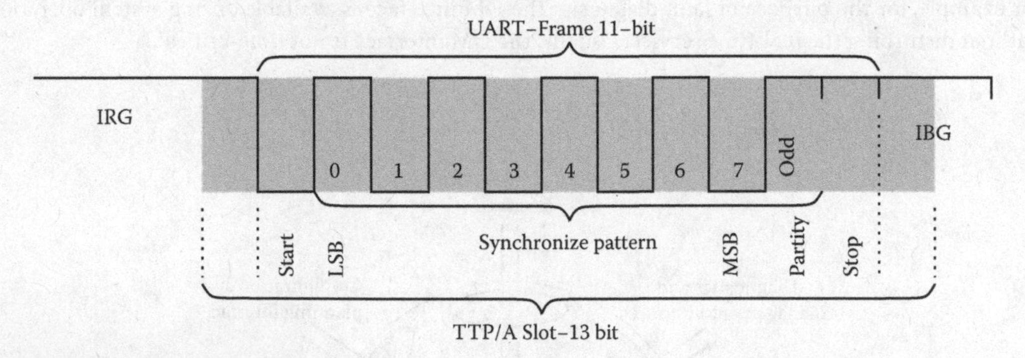

FIGURE 43.9 Synchronization pattern.

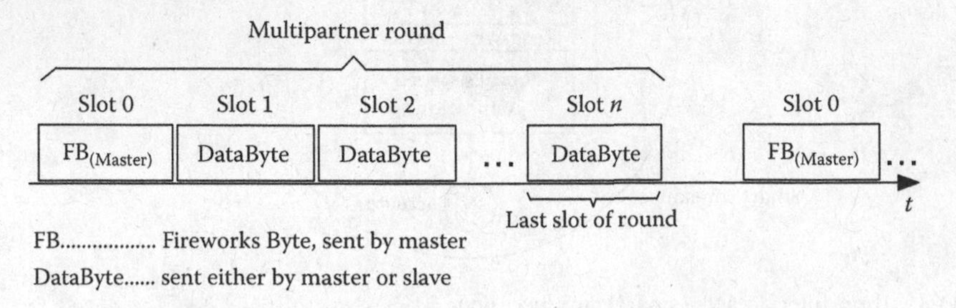

FIGURE 43.10 Example for a TTP/A multipartner round.

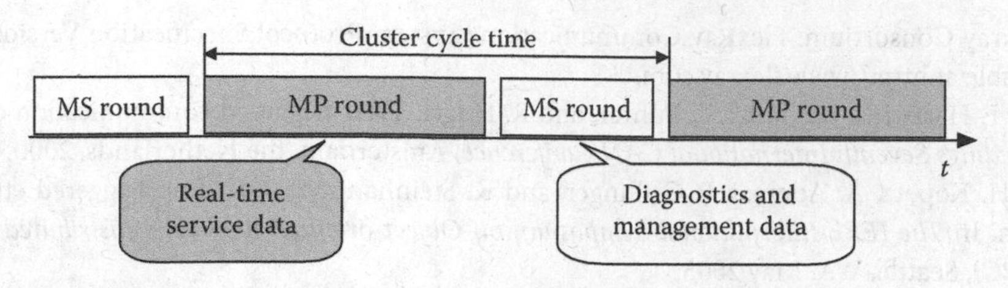

FIGURE 43.11 Recommended TTP/A schedule.

Generally, there are two types of rounds:

Multipartner round

This round consists of a configuration dependent number of slots and an assigned sender node for each slot. The configuration of a round is defined in a data structure called "RODL" (ROund Descriptor List). The RODL defines which node transmits in a certain slot, the operation in each individual slot, and the receiving nodes of a slot. RODLs must be configured in the slave nodes prior to the execution of the corresponding multipartner round. An example for a multipartner round is depicted in Figure 43.10.

Master/slave round

A master/slave round is a special round with a fixed layout that establishes a connection between the master and a particular slave for accessing data of the node's IFS, for example, the RODL information. In a master/slave round, the master addresses a data record using a hierarchical IFS address and specifies an action like reading of, writing on, or executing that record.

The master/slave rounds are used to access the DM and the CP interface to the transducer nodes. Communication via the RS interface is established by periodical multipartner rounds. Master/slave rounds are scheduled periodically between multipartner rounds as depicted in Figure 43.11 in order to enable maintenance and monitoring activities during system operation without a probe effect.

Acknowledgments

This survey was supported by the Austrian FWF project TTCAR under contract No. P18060-N04.

References

[AFI+00] Alcatel Corp., Fujitsu Ltd., IBM, NEC Corp., NTT Corp., and IONA Tech. Management of event domains. OMG TC Document telecom/2000-01-01, January 2000. Available at http://www.omg.org.

[AKG+06] A. Ademaj, H. Kopetz, P. Grillinger, K. Steinhammer, and A. Hanzlik. Fault-tolerant time-triggered ethernet configuration with star topology. In *Dependability and Fault Tolerance Workshop*, Frankfurt, Germany, March 2006.

[BP00] G. Bauer and M. Paulitsch. An investigation of membership and clique avoidance in ttp/c. In *the IEEE Symposium on Reliable Distributed Systems*, Nürnberg, Germany, October 16–18, 2000.

[DeL99] R. DeLine. Resolving packaging mismatch, PhD thesis, Computer Science Department, Carnegie Mellon University, Pittsburgh, PA, June 1999.

[DW98] P. Dierauer and B. Woolever. Understanding smart devices. *Industrial Computing*, 47–50, October 1998.

[Ecc98] L. H. Eccles. A brief description of IEEE P1451.2. In *Sensors Expo*, San Jose, CA, May 1998.

[EHK01] W. Elmenreich, W. Haidinger, and H. Kopetz. Interface design for smart transducers. In *IEEE Instrumentation and Measurement Technology Conference*, Budapest, Hungary, May 2001, Vol. 3, pp. 1642–1647.

[Fle05] Flexray Consortium. FlexRay Communications System Protocol Specification Version 2.1, 2005. Available at http://www.flexray.com.

[HMFH00] F. Hartwich, B. Müller, T. Führer, and R. Hugel. Time triggered communication on CAN. In *Proceedings Seventh International CAN Conference*, Amsterdam, the Netherlands, 2000.

[KAGS05] H. Kopetz, A. Ademaj, P. Grillinger, and K. Steinhammer. The time-triggered ethernet (tte) design. In *The IEEE International Symposium on Object-oriented Real-time distributed Computing (ISORC)*, Seattle, WA, May 2005.

[KB03] H. Kopetz and G. Bauer. The time-triggered architecture. *Proceedings of the IEEE*, 91(1): 112–126, January 2003.

[KHE01] H. Kopetz, M. Holzmann, and W. Elmenreich. A universal smart transducer interface: TTP/A. *International Journal of Computer System Science & Engineering*, 16(2): 71–77, March 2001.

[KN97] H. Kopetz and R. Nossal. Temporal firewalls in large distributed real-time systems. *Proceedings of the Sixth IEEE Workshop on Future Trends of Distributed Computing Systems (FTDCS'97)*, Tunis, Tunisia, 1997, pp. 310–315.

[KO87] H. Kopetz and W. Ochsenreiter. Clock synchronization in distributed real-time systems. *IEEE Transactions on Computers*, C-36(8): 933–940, 1987.

[Kop92] H. Kopetz. Sparse time versus dense time in distributed real-time systems. In *Proceedings of the 12th International Conference on Distributed Computing Systems*, Yokohama, Japan, June 1992.

[Kop00] H. Kopetz. Software engineering for real-time: A roadmap. In *Proceedings of the IEEE Software Engineering Conference*, Limmerick, Ireland, 2000.

[OMG02] OMG. Smart Transducers Interface. Specification ptc/2002-05-01, Object Management Group, May 2002. Available at http://www.omg.org/.

[OMG03] Object Management Group (OMG). Smart Transducers Interface V1.0, January 2003. Specification available at http://doc.omg.org/formal/2003-01-01 as document ptc/2002-10-02.

[SAE95a] Class A application/definition (sae j2057/1 jun91). In *SAE Handbook*, Vol. 2, pp. 23.478–23.484. Society of Automotive Engineers, Inc., Warrendale, PA, 1995. Report of the SAE Vehicle Network for Multiplex and Data Communication Standards Committee approved June 1991.

[SAE95b] SAE. Class C application requirements, survey of known protocols, J20056. In *SAE Handbook*, pp. 23.437–23.461. SAE Press, Warrendale, PA, 1995.

[SGAK06] Klaus Steinhammer, Petr Grillinger, Astrit Ademaj, and Hermann Kopetz. A time-triggered ethernet (tte) switch. *Design, Automation and Test in Europe*, Munich, Germany, March 6–10, 2006.

[TTA03] TTAGroup. Specification of the TTP/C Protocol V1.1, 2003. Available at http://www.ttagroup.org.

44 FlexRay

44.1 Introduction 44-1
44.2 Protocol 44-1
Communication Cycles • Framing • Startup of a Cluster • Physical Layer
44.3 System Architecture 44-4
Topologies • Node Architecture • Star Couplers
44.4 System Design Considerations 44-5
Configuration • AUTOSAR
References 44-7

Martin Horauer
University of Applied Sciences Technikum Wien

Peter Rössler
University of Applied Sciences Technikum Wien

44.1 Introduction

Today, electronic control units (ECUs), and especially their interoperation, are the driving force behind most automotive innovations. They allow the establishment of extended and improved functionality with regard to safety, reliability, environmental efficiency, and comfort (e.g., combining multiple sensor information to a comprehensive picture of the car's surrounding). In recent years, more than 50 ECUs interconnected with different communication subsystems have become common [H05]. This in turn led to higher demands on communication resulting in a strong increase on the bus traffic. In fact, the communication subsystem became a bottleneck with regard to bandwidth and reliability.

In 2000, some leading automotive original equipment manufacturers (OEMs) founded an industrial consortium to develop a new bus protocol termed FlexRay* in order to serve the needs of future automotive applications. Relying on the time-triggered paradigm, FlexRay addresses reliability and fault-tolerance aspects, namely by implementing a time division multiple access (TDMA) channel access method and features like bus-guardians or support for redundant physical media. Additional provisions also support event-triggered communication for an effective coupling of sensor/actuator systems with lower requirements on reliability.

44.2 Protocol

The FlexRay consortium provides specifications and applications for the electrical physical layer and a protocol specification [FE05,FP05] as well as conformance test specifications that essentially cover the physical and data link layer when compared to the ISO/OSI reference model. Network, transport, session, and presentation layer functionality may be implemented according to the AUTomotive Open System ARchitecture (AUTOSAR)† standard.

* http://www.flexray.com
† http://www.autosar.org

44.2.1 Communication Cycles

The communication in FlexRay is arranged in periodic cycles where every cycle consists of a static segment, an optional dynamic segment, a symbol window, and a network idle time phase, cf. Figure 44.1. Communication cycles are numbered sequentially from 0 to cycle 63; after cycle 63, the cycle count starts anew at 0. Every node transmits data in the static or the (optional) dynamic segment. While the static segment provides a communication mechanism for reliable data transmissions, the dynamic segment is intended to be used for nonsafety critical sporadic messages. The symbol window is used to transmit media test symbols in order to check the integrity of the optional bus-guardians, cf. Section 44.3. The network idle time is used for the offset correction of the local clock of every node. Therefore, FlexRay uses a distributed clock synchronization scheme by employing a fault-tolerant midpoint algorithm; see [WL88] for further details. Selected nodes send synchronization messages (having the sync bit set, see below); these messages get timestamped at every receiver and form the input for the clock synchronization algorithm along with the node's local clock.

In the static segment, communication follows a TDMA scheme. Herein, frames are transmitted within one or multiple preassigned time slots all having an equal length. In particular, transmission within an assigned slot starts after the so-called action-point offset (see Figure 44.1) and ceases some time before the end of the slot is reached. These times as well as the assignment of slots to nodes is configured at design time, that is, before a FlexRay cluster is put into operation. The number of slots n within the static segment can be configured between 2 and 1023.

Using a *minislotting* access scheme, the dynamic segment is divided into multiple slots of equal length numbered in increasing sequence. A node may send a message whenever its configured frame identifier matches the slot number. In this case, the actual slot gets elongated by a multiple number of minislots up to the end of the dynamic segment. However, when no message is transmitted, a slot takes up only the length of one minislot. In both situations, the slot counter is incremented after one slot. Nodes may not be able to send frames with higher identifiers in the dynamic segment when other frames with lower identifiers have been sent before.

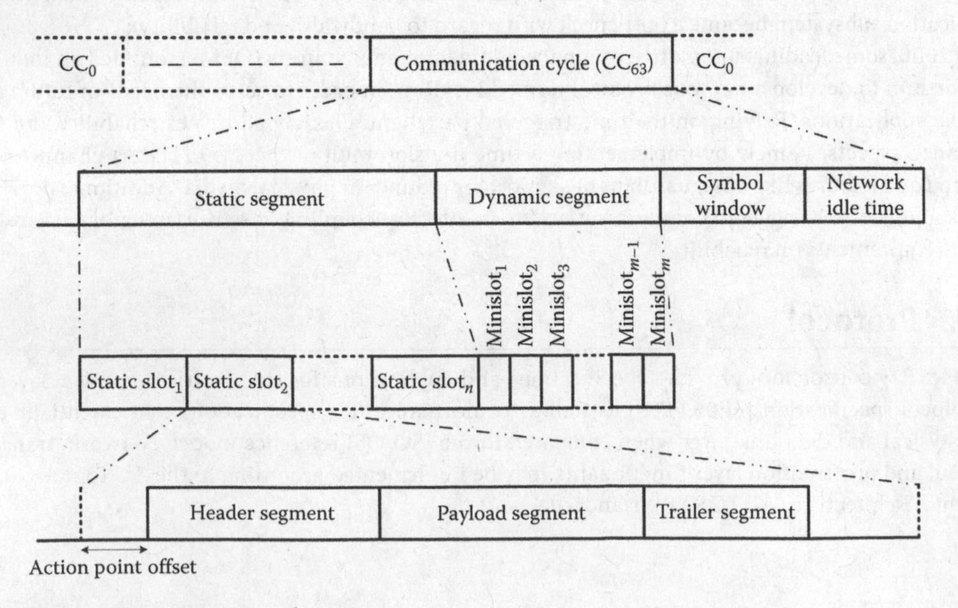

FIGURE 44.1 Communication cycle.

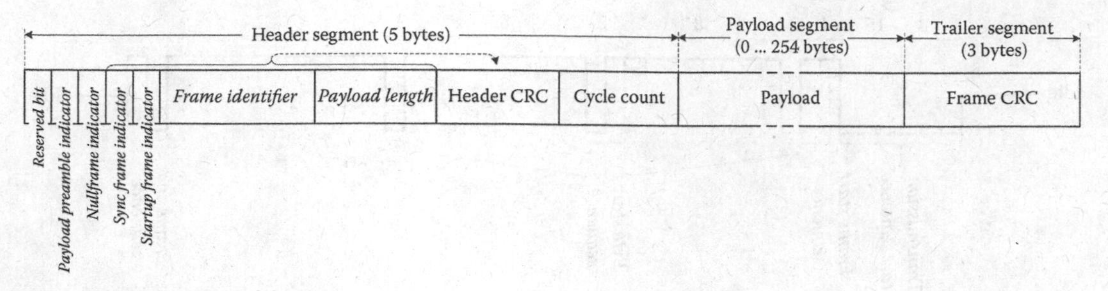

FIGURE 44.2 Frame format.

44.2.2 Framing

A frame sent in either the static or dynamic segment consists of a header, the payload, and a trailer. The 5-byte long header is made up of five control bits, an 11 bit long frame identifier, a payload length field, a header cyclic redundancy checksum (CRC), and a cycle count field; cf. Figure 44.2. The header CRC secures the *sync frame indicator* bit, the *startup frame indicator* control bit, the frame identifier, and the *payload length* field. The payload holds an even number of bytes (between 0 and 254) that carry the application data. When the payload preamble indicator control bit in the header is set, a part of the payload in the static segment is reserved for network management information. Special cases are nullframes that are identified by the *nullframe indicator* control bit set in the header; these frames have all the payload bits set to zero. Null frames are sent autonomously by the network controller whenever no new data are at hand from the node's host. The trailer contains a CRC over the entire frame with a Hamming distance of 6 for frame lengths up to 248 bytes payload and a Hamming distance of 4 for frame lengths between 249 and 254 bytes payload. With a Hamming distance of 6, up to 5 arbitrary bit flips can be detected.

44.2.3 Startup of a Cluster

When the cluster is not running (e.g., after power-up) the host of a dedicated, so-called coldstart node may try to start the cluster. It sends a *collision avoidance symbol* and waits for a response within a configured time. When no response is received, the node stops its start-up attempts and waits for an attempt from another node to start the cluster. When the latter occurs, the leading coldstart node will reply and synchronize its timebase. Both nodes will now make the transition to an active state and other noncoldstart nodes may join the cluster, see [FE05] or [R08] for details.

44.2.4 Physical Layer

FlexRay employs a variant of the inverted non-return-to-zero (NRZ-I) coding for bit transmission. A standard NRZ-I code represents a "One" with a change in the physical level whereas a "Zero" is with no change. The modification of the employed coding consists of a *byte-start sequence* (the bit combination "10") that is prepended to every byte. The falling edge between these two bits is used for bit-resynchronization. Furthermore, every frame transmission is prepended by a *transmission start* and a *frame start sequence*. The transmission start sequence consists of 3–15 bits and is used to activate star couplers. Since activation takes some time, a few bits of the transmission start sequence are usually consumed by them. Furthermore, the end of a frame is followed by a *frame end sequence* consisting of the bits "01," and an idle phase where the transmitter is turned off. Figure 44.3 illustrates the coding of an entire FlexRay frame.

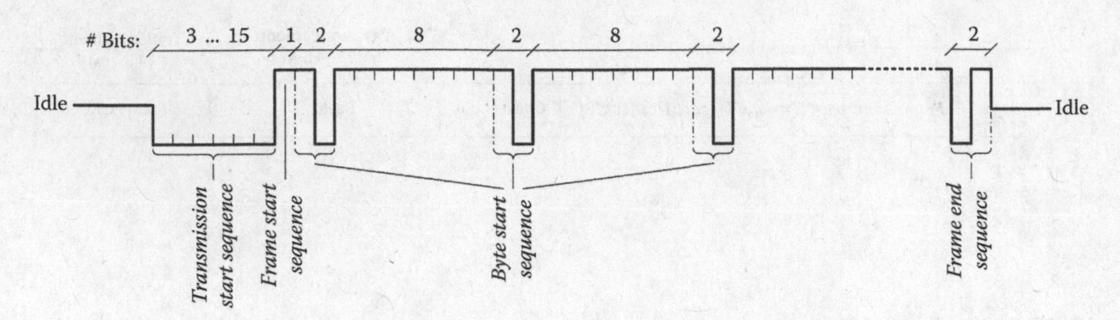

FIGURE 44.3 Coded frame (static segment).

In the dynamic segment, an additional *dynamic trailing sequence* is inserted after the frame end sequence. The dynamic trailing sequence is a low period of variable length followed by a high bit; it extends a frame until the next action point.

Termination needs to be applied at the two nodes that are set apart with the longest wiring distance. It is calculated as equal to the nominal cable impedance between the two points. One concept is split termination, where one half of the cable impedance to ground is connected between the two bus lines. Furthermore, a common mode choke should be applied between bus driver and termination circuitry.

The maximum transfer rate is 10 Mbps/channel allowing for a maximum cable length of up to 24 m between two active components. Hence, when two cascaded star couplers are in use up to 72 m can be spanned at 10 Mbps. Lowering the data transfer rates allows elongating the cabling; in particular, reducing the data rate, for example, to 5 Mbps, allows doubling the cabling distance.

The standard specifies an electrical physical layer using either shielded or unshielded twisted pair cables. An optical medium can be used as well; however, the latter is not yet specified in the standard.

44.3 System Architecture

To address the diverse requirements of the automotive industry, the FlexRay specification allows for a lot of different topologies and options in the node design.

44.3.1 Topologies

A communication system of multiple nodes connected via at least one communication channel directly (bus topology) or by star couplers (star topology) is called a *cluster*. FlexRay supports point-to-point, linear bus, star topologies, or hybrids thereof cascaded with up to two star couplers, cf. Figure 44.4. Furthermore, communication can take place in a redundant fashion via two separate channels or in a nonredundant fashion.

A maximum of 22 active nodes may be attached to one segment (a linear bus or a passive star); there is, however, no guarantee that such a system is operational. It is up to the system designer to verify the electrical interplay between bus drivers, cables, and termination circuitry.

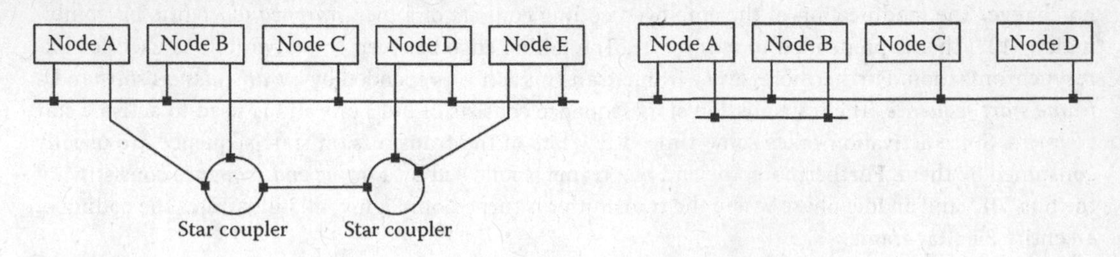

FIGURE 44.4 Topology examples.

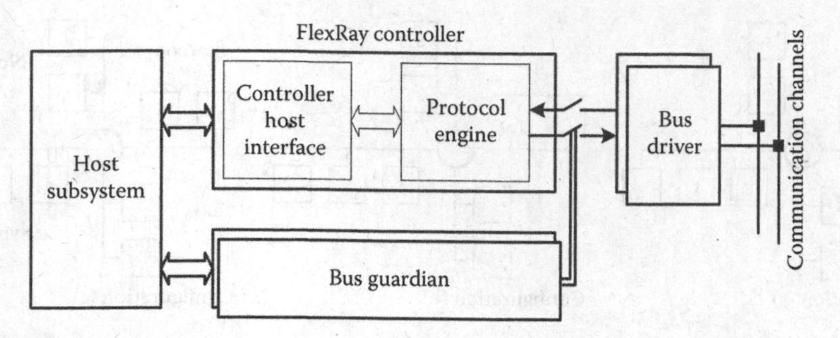

FIGURE 44.5 Node architecture.

44.3.2 Node Architecture

A typical node architecture consists of a host CPU subsystem, a network controller, and bus drivers for the separate channels; see Figure 44.5. Interfacing between the host and the network subsystem is facilitated via a controller host interface employing some kind of dual-ported memory. This choice of an interface allows separating control from data flow, creating some kind of temporal firewalls restricting error propagation to defined areas. Optional bus-guardians may be installed either at the nodes or at central, active star couplers. The bus-guardians oversee and inhibit bus access whenever a node is not allowed to transmit in order to avoid so-called babbling idiot failures; situations where a node playing havoc disrupts the entire communication.

44.3.3 Star Couplers

Both passive and active star couplers are supported by the FlexRay protocol. Whereas passive star couplers are simply circuits that relay the signal received on one branch to all other branches; active star couplers provide logic and circuitry for power-down and wakeup, error detection and isolation, as well as signal reshaping. Typically, an active star coupler listens on all branches for traffic and when a signal is detected it is relayed—in an error free case—to all other branches. Whenever an error occurs or before a cluster is synchronized, collisions of data frames or symbols may occur. In this case, an active star-coupler relays the superimposed signal to the other branches.

44.4 System Design Considerations

Guaranteed bandwidth and reliable operation are some of the benefits of time-triggered communication systems. On the other hand, such systems may become less flexible (especially when frequent system changes occur) and require higher efforts at design time.

44.4.1 Configuration

Next to topology considerations, one has to configure which nodes act as coldstarters and participate at the clock synchronization (sending of sync frames). At least two or more nodes should be assigned therefore; in fact, more of these nodes enhance the availability of the cluster but increase, however, the likeliness for the establishment of cliques (cliques are groups of nodes that are able to communicate with each other, however, not with other nodes), cf. [MHS08].

Furthermore, the physical distribution of these nodes as well as the distribution of their slots impairs the overall robustness of the cluster. In fact, having these nodes and slots evenly dispersed over the cluster and schedule, respectively, will likely improve the overall availability. Furthermore, configuring more nodes as the minimum to transmit the specified wakeup patterns and perform coldstart attempts will improve the startup mechanism. When one of these so configured nodes, however, is faulty, a higher

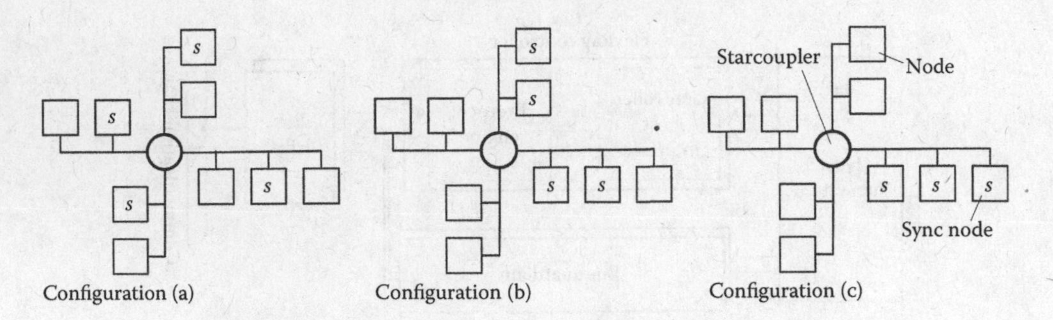

FIGURE 44.6 Sync node distribution examples.

TABLE 44.1 Consequences of Faults in the Configurations Shown in Figure 44.6

Configuration (a)	(i) The affected branch loses communication (ii) All nodes lose communication
Configuration (b)	(i) The affected branch loses communication (ii) Two cliques are formed; the branches without sync nodes lose communication
Configuration (c)	(i) In the case when the branch with the sync nodes is affected, all nodes will lose communication; when another branch is affected, the respective branch will lose communication (ii) The branch with the sync nodes stays synchronized; all other branches lose communication

number of attempts will elongate the startup time. Figure 44.6 along with Table 44.1 depicts some examples that illustrate the effect of sync node distribution on the system behavior when either a fault affects one branch (i) or the star-coupler (ii).

All these issues have to be considered when determining the large number of configuration parameters and constants of the FlexRay protocol for an actual cluster implementation. Here, proper planning and tool support are of utmost importance.

44.4.2 AUTOSAR

In practice, when designing a cluster that shall be built with hardware and software from different vendors, the AUTOSAR standard must be considered. The goals of the latter are to provide for system-wide

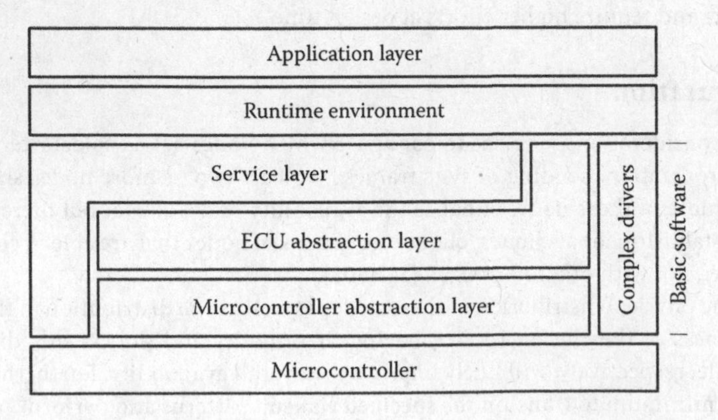

FIGURE 44.7 AUTOSAR layer model.

optimizations with regard to modularity, scalability, interchangeability, and reuse. To that end it provides standard interfaces and abstractions between the application software and the basic system software as well as an implementation of the latter.

In a functional model, software components communicate with each other via a virtual functional bus using standardized interfaces. In the deployment phase, the functional model is mapped to a structural model; hereby, the components are mapped to the application layer and the virtual function bus to the run-time environment at discrete ECUs. The basic software is composed of four layers and one driver module as depicted in Figure 44.7 and implements abstractions, drivers, and services for both FlexRay and controller area network (CAN).

Herein, the *microcontroller* and *ECU abstraction layers* provide abstractions from the specific underlying hardware, whereas the *service layer* adds things like, transport protocols, diagnostic services, and network management services. Finally, the *complex drivers*' module allows circumventing these modules by implementing specific drivers whenever time critical implementations are required.

References

[FE05] FlexRay Communications Systems—Electrical Physical Layer Specification Version 2.1., available at http://www.flexray.com, FlexRay Consortium, 2005.

[FP05] FlexRay Communications Systems—Protocol Specification Version 2.1, available at http://www.flexray.com, FlexRay Consortium, 2005.

[H05] Hansen, P., New S-Class Mercedes: Pioneering electronics, *The Hansen Report on Automotive Electronics*, 18(8), 1–2, October 2005.

[MHS08] Milbredt, P., Horauer, M., and Steininger A., An investigation of the clique problem in FlexRay, *Proceedings of the Third IEEE Symposium on Industrial Embedded Systems* (*SIES 2008*), Montpellier, France, June 11–13, 2008, pp. 224–231.

[R08] Rauch, M., FlexRay, Hanser, Munich, Germany, 2008.

[WL88] Welch, J.L. and Lynch, N.A., A new fault-tolerant algorithm for clock synchronization, *Information and Computation*, 77(1), 1–36, 1988.

45

LIN-Bus

45.1 LIN Background 45-1
45.2 LIN History and Versions 45-2
45.3 Communication Concept 45-3
45.4 Physical Layer 45-3
Signal Specification • Topology
45.5 LIN Message Frames 45-6
Break Field • Sync Byte Field • Identifier • Data Field • Checksum • Frame Length • Time-Triggered Data Transmission • Frame Types • Diagnostic Frame
45.6 Network and Status Management 45-10
45.7 Transport Layer Protocol 45-11
45.8 Configuration 45-12
45.9 Relationship between SAE J2602 and LIN2.0 45-12
45.10 Conclusion 45-12
References 45-13

Andreas Grzemba
University of Applied Sciences Deggendorf

Donal Heffernan
University of Limerick

Thomas Lindner
BMW Group

45.1 LIN Background

Modern cars contain a large number of electronic control units (ECU). For example, the BMW AG 7-series (model 2008) has over 70 ECUs and about 100 different electric motors. The car's comfort domain has been much improved in recent years. Comfort functions such as electric window lifters, electric outside mirrors, central locking, air conditioning, electric seat adjustment, and an electrical sliding roof are common features in today's cars. In the past, when the number of features was low, the various electrical devices were conventionally wired. However, as the number of features grew, the wiring harness became unmanageable, and the diagnostics for such complex systems became complex. For these reasons, the car manufacturers introduced multiplexed wired communication systems. However, an uncontrolled growth evolved as the various manufacturers developed proprietary protocols for in-vehicle control networks (e.g., BMW AG's K/I-BUS, Toyota's BEAN, GM's Sinebus, and E&C).

The proprietary protocols proved expensive and often had dependence on single suppliers. In an attempt to devise a standardized solution, the car makers BMW Group, Daimler-Benz (now Daimler), Audi, and VW, the semiconductor company Motorola (now Freescale), and the network specialist Volcano Communications Technologies (now Mentor Graphics) founded the LIN Consortium to develop a low-cost communications standard for the so-called Society of Automotive Engineers (SAE) communication class A (systems up to 20 kbps) systems. The Local Interconnect Network (LIN) is a low-cost solution, which is a universal asynchronous receiver transmitter (UART) based, single-master, multiple-slave architecture, developed for automotive sensor and actuator applications. The LIN master node connects the LIN network with higher level networks, usually the controller area network (CAN). The LIN is positioned below CAN and it does not replace CAN.

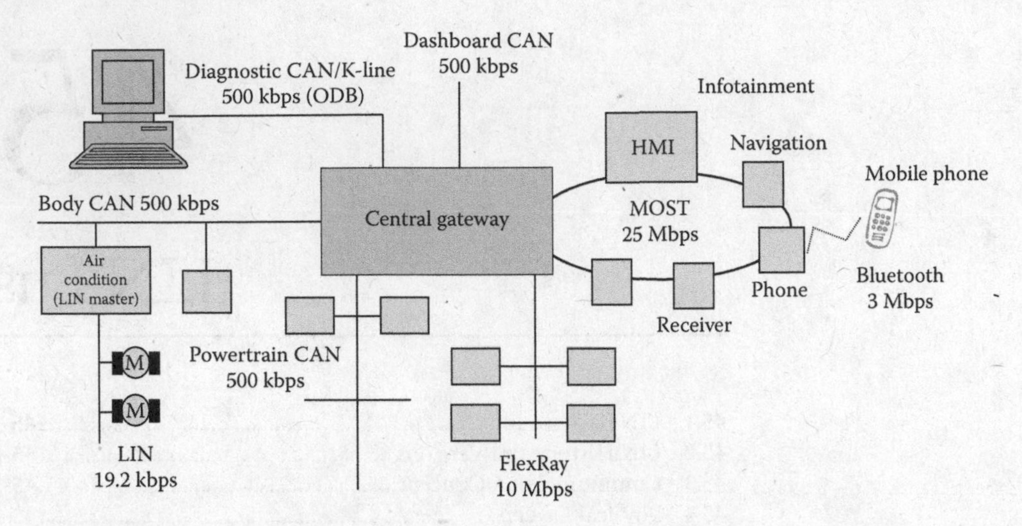

FIGURE 45.1 Automotive communication architecture with a central gateway.

Figure 45.1 shows an example of an automotive communication architecture with a central gateway. Here, LIN controls the damper motors of the air-conditioning system. The LIN specification denotes a LIN subbus as a LIN cluster.

45.2 LIN History and Versions

In the summer of 1999, the first version [LIN1.0] of the LIN specification package was published. It contained specification information to describe a full network including the parts protocol, the physical layer, and the application program interface. It also specified the configuration language description (CLD) with which a LIN cluster can be described in a LIN description file (LDF). This specification package enforced a development workflow to allow the efficient development of a LIN cluster, where all the communication parameters of a LIN cluster are defined in the LDF, which is also machine readable.

The LIN Consortium was constituted in March 2000 at the SAE congress in Detroit. Version 1.1 of the LIN specification was released in April 2000, followed by version 1.2 in November. Both versions contained only editorial modifications. The next important step was the release of the version 1.3 [LIN1.3] in November 2002, which made LIN more versatile. Version 2.0 [LIN2.0] realized a comprehensive expansion of the LIN standard, although it retained backward compatibility with previous versions. One of the main innovations of LIN2.0 was the introduction of the so-called off-the-shelf slave nodes. LIN slave nodes can be simply integrated into a new cluster without any modifications to the slave nodes. To allow this concept to be supported by appropriate network defining tools, LIN2.0 specifies the slave node configuration, which allows the assignment of a node address (NAD) and frame identifiers. It also defines the syntax of the node capability file (NCF), which contains all parameters of the off-the-shelf slave nodes, using the node capability language description (NCLD). These features allow plug-and-play functionality to be used in a LIN cluster. Figure 45.2 illustrates the interaction.

The latest LIN version 2.1 [LIN2.1] was released in November 2006. Mainly, the document has a better structure and mostly contains clarifications on the functionality. But there are also some specific modifications, including diagnostic and configuration classes for the slave nodes.

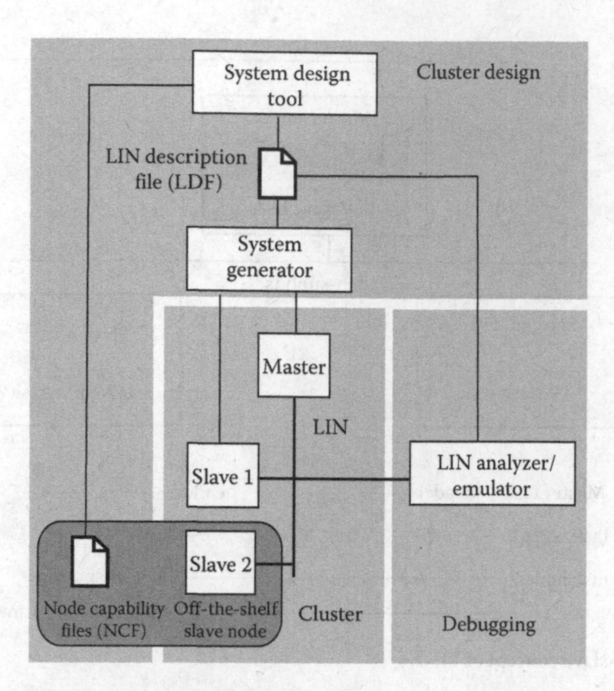

FIGURE 45.2 Workflow with off-the-shelf slave nodes.

45.3 Communication Concept

LIN uses the master–slave principle for the medium access control feature. This principle allows master and slave nodes to be implemented with minimal hardware resources. The master node is usually implemented in a control unit or in a gateway that possesses the necessary resources.

The specification also defines a master task and a slave task. The master task initiates every communication. A message will always consist of a header generated by the master task and a response from the slave task. As the master node should also be able to send messages to the slave nodes, along with the master task a slave task can also be integrated into the master node (see Figure 45.3) [Grz05]. This provides the possibility of integrating several slave tasks onto a hardware platform.

The LIN communication scheme is based on the producer–consumer model. Therefore, the specification denotes the transmitter as the "publisher" of the message and the receiver(s) as "subscriber(s)." LIN, like CAN, uses identifiers to uniquely identify a message. As well as a point-to-point communication scheme, multicast and broadcast communication schemes are also supported. Figure 45.4 shows the three possible communication relationships:

a. A slave module responds to a master request
b. The master sends a message to one or more slave nodes
c. The master initiates the communication between two slave nodes

45.4 Physical Layer

It was a goal of the LIN design to achieve a simplistic wiring topology. The simple single-wire bus connects to each node in the cluster and switches from ground to battery-level voltage.

The physical layer is based on the ISO 9141 standard [ISO9141]. It consists of the bidirectional single-wire bus line (LIN) that is connected via a pull-up resistor (*R*) and a diode to the internal supply voltage (V_{SUP}) via a transistor (component of the transceiver) to the ground (GND) (see Figure 45.5a). The diode

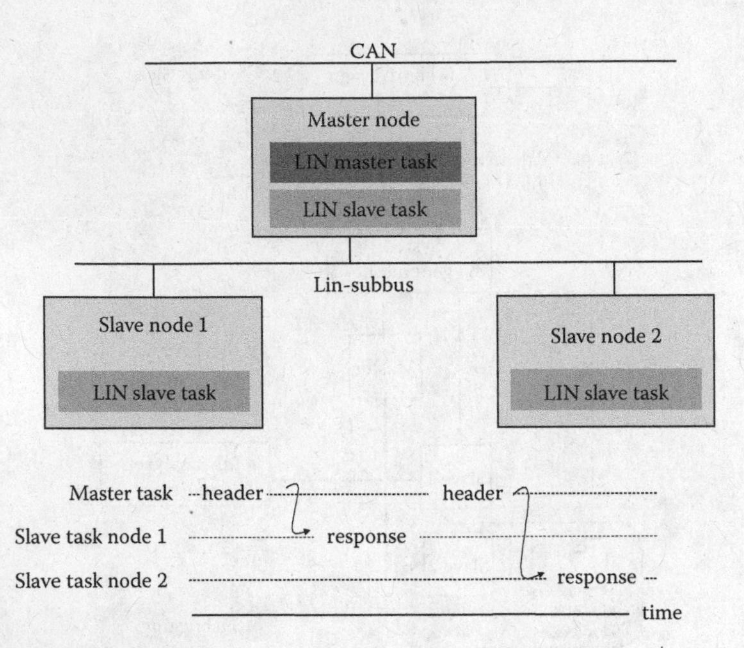

FIGURE 45.3 Master–slave concept of LIN.

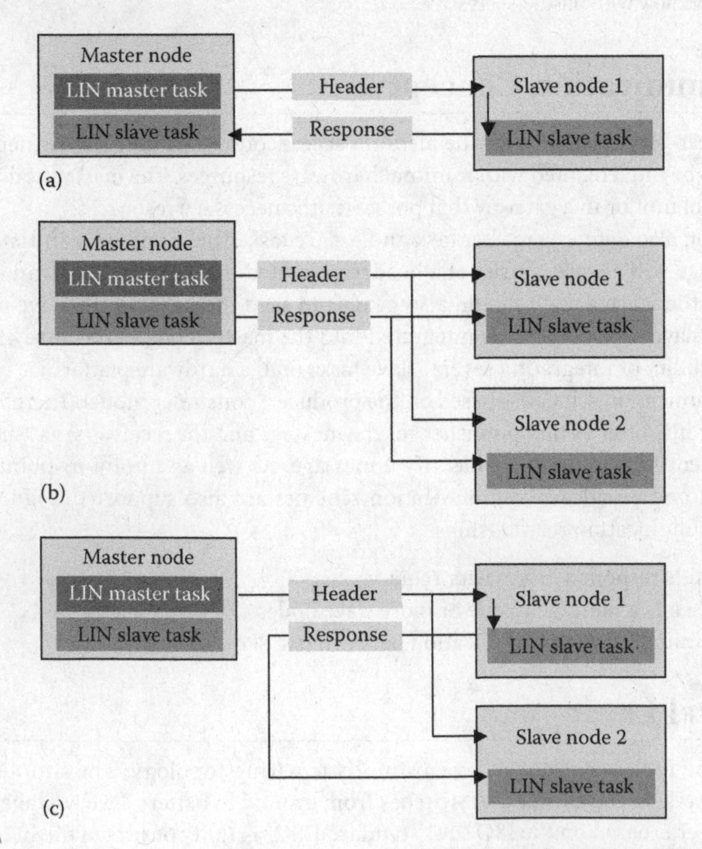

FIGURE 45.4 Communication relationships within a LIN cluster: (a) a slave responds to a master's request, (b) the master sends a message to one or more slaves, and (c) the master initiates the communication between the two slaves.

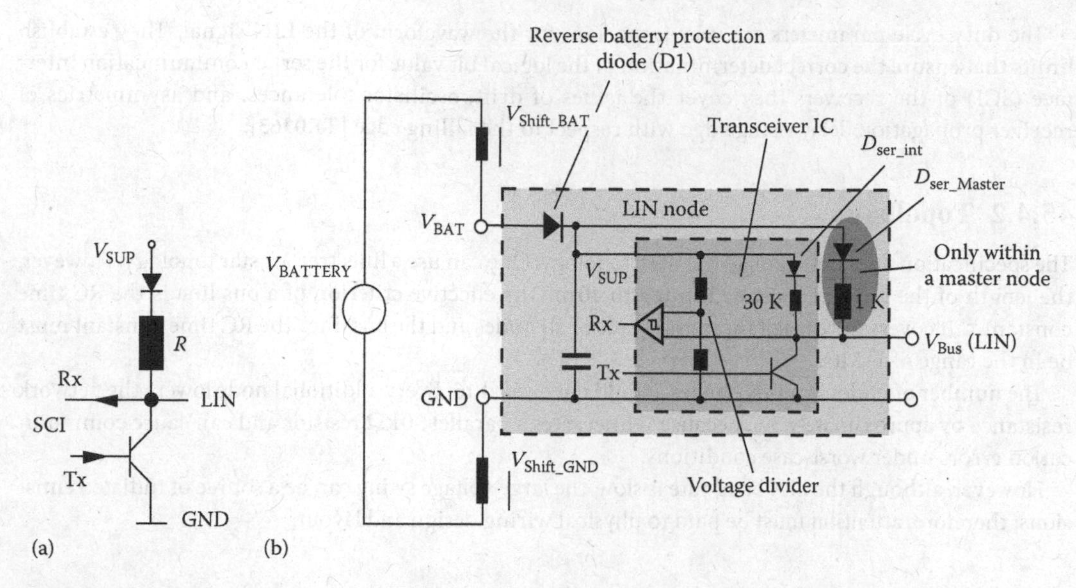

FIGURE 45.5 (a) Basic principle of the physical layer according to ISO 9141. (b) Basic structure of a LIN master node.

is mandatory to prevent uncontrolled powering of the node from the bus line, in the case of a battery loss, that is, if V_{BAT} is disconnected in Figure 45.5b. The pull-up resistor is specified at 30 kΩ on a slave node and at 1 kΩ on a master node. The baud rate is specified in the range of 1–20 kbps but typical baud rates are 9.6 or 19.2 kbps, or 10.4 kbps according to SAE J2602, as will be discussed later.

45.4.1 Signal Specification

The V_{SUP} and GND form the reference potential of the voltage divider to determine the logical value of a bit. A voltage level below 40% V_{SUP} represents logical 0 and a level above 60% V_{SUP} represents a logical 1. The transmitter has to ensure that the receiver receives valid voltage levels. The specification does not define transmitter voltage levels.

Because any node determines the logical bit values in reference to its own V_{SUB} and GND, a drift has an impact on this process (see Figure 45.6). The specification margins the drift of V_{BAT} (V_{Shift_BAT}) and GND (V_{Shift_GND}) at about 10%. Note that V_{SUP}, seen in Figure 45.5b, can be computed as

$$V_{SUP} = V_{BAT} - V_{D1} \tag{45.1}$$

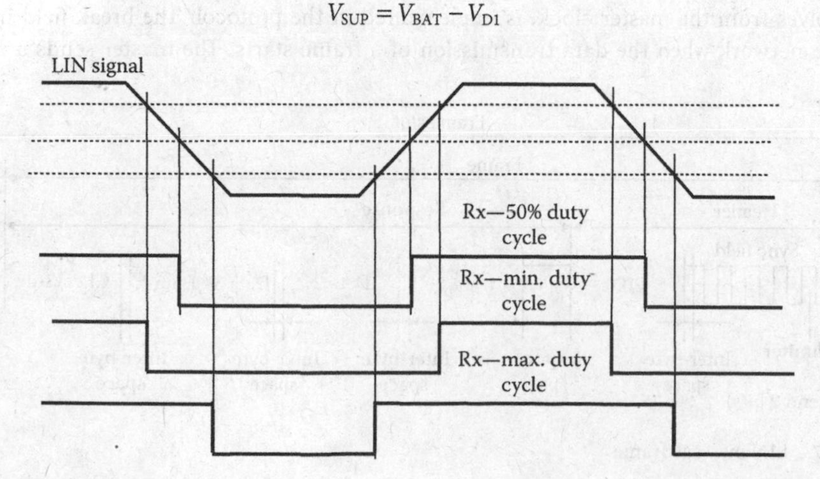

FIGURE 45.6 Influence of the drift.

The duty cycle parameters are very important for the waveform of the LIN signal. They establish limits that ensure the correct determination of the logical bit value for the serial communication interface (SCI) of the receiver. They cover the issues of drift, oscillator tolerances, and asymmetries of receiver propagation delay rising edge with respect to the falling edge [TR0363].

45.4.2 Topology

The specification does not regulate the wire topology. One can use a line, tree, or star topology. However, the length of the bus line must be limited to 40 m. The effective criterion of a bus line is the RC time constant τ. It covers the capacitances and loads of all nodes and the bus line. The RC time constant must be in the range of 1–5 μs.

The number of nodes in a LIN cluster should not exceed 16. Every additional node lowers the network resistance by approximately 3%, because it integrates a parallel 30 kΩ resistor and can cause communication errors under worst-case conditions.

However, although the LIN baud rate is slow, the large voltage swing can be a source of radiated emissions; therefore, attention must be paid to physical wiring design and layout.

45.5 LIN Message Frames

The LIN message frame, as represented in Figure 45.7, is composed of a header and a message field, the so-called response. The header field consists of the break field (Synch_Break), the synch byte field (Synch_Field), and the identifier field, which is designated as the protected identifier (PID). The message field (response) consists of the data field with one to eight data bytes and the checksum. All fields except the break field are byte oriented and can be generated with the integrated serial interface of a microcontroller.

Message activity is coordinated by the periodic execution of the LIN message schedule. For each frame slot, the master task transmits a frame header with a frame identifier (ID). All slaves evaluate the frame header, and a slave immediately transmits the frame response that is associated with the ID. The LIN frame is the combined frame header and response.

45.5.1 Break Field

A synchronization mechanism, which permits slaves without stabilized clock frequencies to synchronize themselves from the master clock, is implemented in the protocol. The break field informs the slaves in the network when the data transmission of a frame starts. The master sends a zero, 13 bit

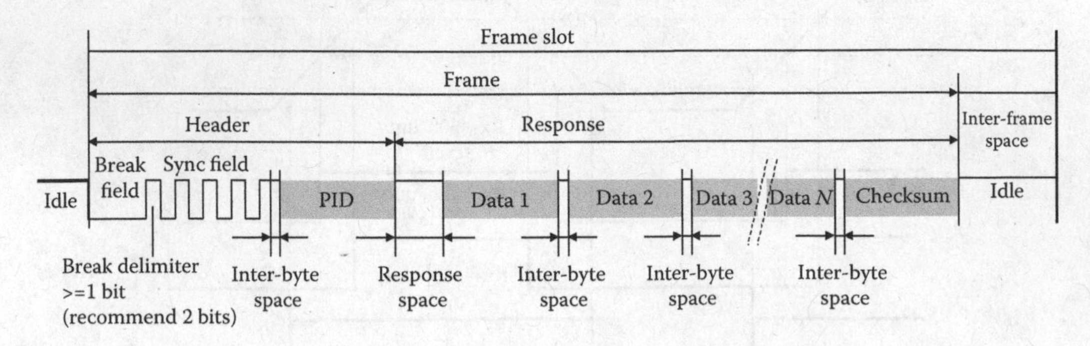

FIGURE 45.7 LIN message frame.

times, followed by a so-called delimiter that must be at least 1 bit time long (2 bits are recommended). The slaves recognize this break after 11 bit times. This recognition can be carried out with the break detection feature of a standard SCI. The master generates a break signal that is about two bit times longer to ensure a safe synchronization to the master clock, for slaves without a crystal.

45.5.2 Sync Byte Field

The sync byte field follows the break field and supports the baud rate estimation. The master sends a 0x55 data byte on which the slaves synchronize themselves. The slaves measure the time period between the falling edge of the start bit and those of the seventh data bit. The value is divided by 8, for example, by simply shifting the measured value three bits to the right. Only the falling edges are allowed to be used for the measurement of the baud rate because they are considerably steeper than rising ones and therefore provide more exact results. The different edge steepness is due to the characteristics of the physical layer, as the bus level is actively drawn to "zero" level by a transistor, while the "one" level is raised by a pull-up resistor.

When the baud rate is adjusted on a standard SCI device, it must be ensured that it does not deviate by more than the allowed 2% from the master clock.

45.5.3 Identifier

LIN uses a message ID in the third and final byte of the header. This byte consists of 6 bits for the ID and two parity bits, and thus it is designated as the PID. The 64 possible IDs are shown in Table 45.1.

Up until version 1.3 of the LIN specification, bits four and five of the ID had a special meaning. When the IDs 0–59 were used for the transmission of signals, bit four and five defined the number of data bytes, as seen in Table 45.2. This table does not exist for LIN version 2.0 and later.

The implicit coding of the data length in the identifier is no longer obligatory since LIN version 1.2. Now a field length of one to eight data bytes is possible for every ID. The field length must be indicated at the message definition in the section "Frames" of the LDF.

TABLE 45.1 Partitioning of the Range of Values of the "Protected Identifiers"

Identifier	Meaning
0x0–0x3B (0–59)	Transmission of signals
0x3C (60)	Master request frame (MRF) for the configuration and diagnostics
0x3D (61)	Slave response frame (SRF) to ID 0x3C
0x3E/0x3F 62/63	Reserved for future expansions of the protocol

TABLE 45.2 Meaning of Bit 4 and 5 of the Identifiers (Mandatory until Version 1.2; Optional from Version 1.3; Invalid from Version 2.0)

ID5	ID4	Number of Data Bytes
0	0	2
0	1	2
1	0	4
1	1	8

The two parity bits for the ID are calculated using the following equations:

$$P0 = ID0 \oplus ID1 \oplus ID2 \oplus ID4 \quad (45.2)$$

$$P1 = \overline{ID1 \oplus ID3 \oplus ID4 \oplus ID5} \quad (45.3)$$

45.5.4 Data Field

The data field is the first part of the response. The response can also be generated in the master by the slave task and can consist of one to eight data bytes. As described above, the ID determines the number of bytes unless another length is indicated. The signals are transmitted in the data bytes, starting with the least significant bit (LSB). A data byte consists of a start bit, eight data bits, and a stop bit. A parity bit is not transferred.

45.5.5 Checksum

There is a distinction in LIN between a so-called classic checksum and an enhanced checksum. The LIN protocol version 1.3 uses the classic checksum. LIN version 2.x uses the enhanced checksum and includes the PID. The enhanced checksum is only used for the IDs 0x0–0x3B (0–59).

45.5.6 Frame Length

The minimum and the maximum header and frame length are specified in the LIN specification. The maximum values are 140% of the minimal times. An under run of the minimum times is not possible because they exactly correspond to at least the necessary bit times. Table 45.3 summarizes these parameters.

45.5.7 Time-Triggered Data Transmission

Modern automotive networking solutions favor the time-triggered model for message scheduling so that the overall timing of messages in the system can be deterministic. In a time-triggered network, messages are transmitted against a predefined message schedule table and there is a notion of a global clock to ensure proper time synchronization. The FlexRay network is an example of such an automotive time-triggered network. Since LIN is a subnetwork that can be used in conjunction with larger networks in a system, LIN supports time-triggered operations.

LIN controls the time-triggered transmission by using schedule tables, which are implemented in the master. Thus, the system can be deterministic, that is, the time when a message is transmitted can be exactly predicted and guarantees the periodicity of signals. The schedule tables define a time slot for every scheduled frame, as are described in the LDF by the system designer.

TABLE 45.3 Nominal and Maximum Transmission Time of Headers and Frames

Name	
$T_{Header\_Nominal}$	$34\ T_{Bit}$
$T_{Response\_Nominal}$	$10^*(N_{Data} + 1)\ T_{Bit}$
$T_{Frame\_Nominal}$	$T_{Header\_Nominal} + T_{Response\_Nominal}$
$T_{Header\_Maximum}$	$1.4^*T_{Header\_Nominal}$
$T_{Response\_Maximum}$	$1.4^*T_{Response\_Nominal}$
$T_{Frame\_Maximum}$	$T_{Header\_Maximum} + T_{Response\_Maximum}$

45.5.8 Frame Types

There are three different ways of transmitting frames on the LIN bus: unconditional, event-triggered, and sporadic frames. Diagnostic frames are also supported.

45.5.8.1 Unconditional Frame

Unconditional frame transfer signals may use the IDs 0x0–0x3B (0–59). Every frame can have only one transmitter (publisher) but one or several receivers (subscribers). The setup of the frame is described in the LIN description file. Unconditional frames can be transferred, as shown in the relationships of Figure 45.4.

45.5.8.2 Event-Triggered Frame

The unconditional frame can be considered as the normal LIN frame. LIN2.0 also defines a so-called event-triggered frame. Event-triggered frames improve the response time of the master–slave system by mapping unconditional frames with rare-event data of multiple slaves into one event-triggered frame. This reduces the cycle time of the time schedule table because the master does not have to poll the slave nodes individually. Several frame IDs can be defined as "event triggered" within the network. If the master requests such an "event-triggered" frame, all corresponding slaves will begin to transmit their data. However, corresponding slave nodes only provide their frame response when the values in their data fields have changed. But, multiple slaves can provide a frame response to a single event-triggered frame. To enable the master to recognize which slave node has responded, the first byte of the associated unconditional frames have to contain their frame ID.

As more than one slave can answer, by updating data in the frame, it is possible that a collision can occur on the bus line. The master recognizes such a collision situation and resolves it by switching automatically to the so-called collision resolving schedule table. The table will, at a minimum, contain the associated unconditional frames and it will be a member of schedule table set.

45.5.8.3 Sporadic Frame

The sporadic frames were introduced in the LIN version 2.0 for better utilization of bandwidth. This method provides some dynamic behavior to the otherwise static LIN protocol. They are in principle placeholders in the time schedule table in which the master can send several different messages that carry signals that seldom change, for example, commands. The master sends the message only if a signal carried by this message is changed. Hence the time slot can remain empty. The assignment of the unconditional frames to the sporadic frame takes place in the LIN description file. Figure 45.8 shows a simple example. In slot three, the master can transfer either the unconditional frame "Master Command" or "Outdoor Temperature" or it can leave the slot empty.

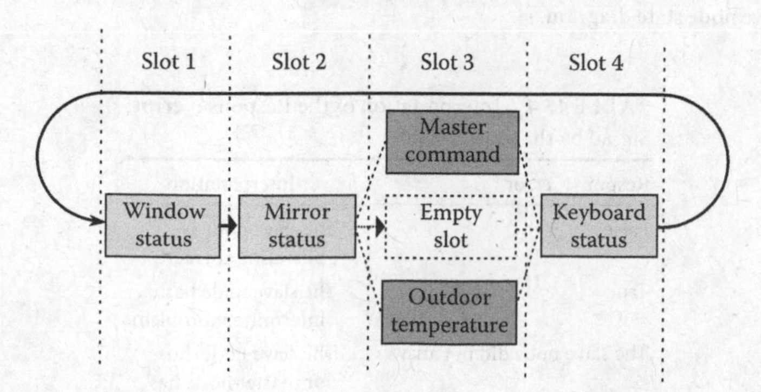

FIGURE 45.8 Principle of sporadic frames.

45.5.9 Diagnostic Frame

Diagnostic frames always carry transport layer data. There are two types, the master request frame (MRF) and the slave response frame (SRF). The transmitter is always the master. The MRF has the ID 0x3C (60) and always has eight data bytes. It is the "go-to-sleep" command if the first byte carries 0x00, otherwise it carries a transport layer protocol. The SRF transfers the slave response of the transport layer protocol to the master. It has the ID 0x3D (61) and also has eight data bytes.

45.6 Network and Status Management

The network and the status management features are very simple. The network management feature supports only the "wake-up" and "go-to-sleep" signals. The corresponding slave state diagram is shown in Figure 45.9. The slave node enters the Initializing state after power-on, reset or wake-up. In the operational mode state that the slave node can transmit and receive frames. In the bus sleep mode state, the node is normally in power save mode. The bus line is set to the recessive level (logical 1). Any node can send a wake-up request for the cluster. Each slave node can detect the wake-up request and listen for bus commands for a short period. The master node can also wake-up and send frame headers to establish the cause of the wake-up.

The slave nodes enter the bus sleep node if they receive the go-to-sleep command from the master or if the bus is inactive for a minimum of 4 s. Bus inactivity means no transitions between logical 0 and 1 bit values and so covers the IDLE state and a short circuit condition with V_{SUP} or GND.

For status management each slave node monitors its communication. The slave provides two status bits for status management within its own node: error_in_response and successful_transfer. If it detects an error (bit error, framing error, or checksum error), it sets the response_error signal. This is carried as a one bit scalar signal in an unconditional frame to the master. The position in the frame is defined in the LDF. Based on this single bit, the master node can interpret the error states as listed in Table 45.4.

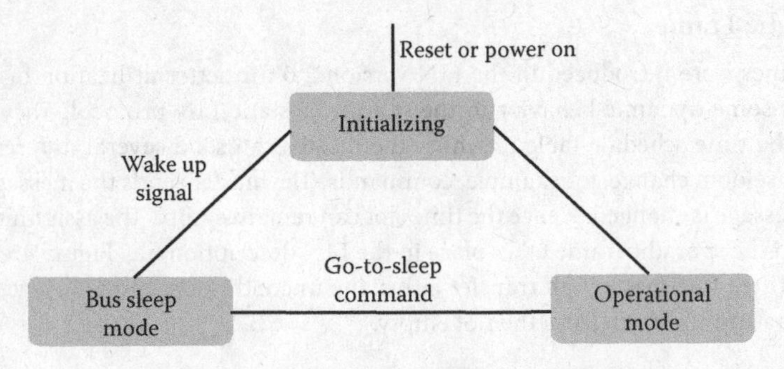

FIGURE 45.9 Slave node state diagram.

TABLE 45.4 Interpretation of the Response_error Signal by the Master Node

Response_error	Interpretation
False	The slave node is operating correctly
True	The slave node has intermittent problems
The slave node did not answer	The slave node, bus, or master node has serious problems

The successful_transfer status is set when a frame has been successfully transferred by the slave, that is, a frame has either been received or transmitted.

45.7 Transport Layer Protocol

The LIN transport layer provides a data transfer between master and slave node for diagnostic, slave node configuration, and identification purposes. It achieves segmentation/desegmentation and flow control. The diagnostics message transfer is compatible with CAN-based diagnostic protocols. Typically diagnostics are performed from a tester where a LIN master can pass the received tester requests to the relevant LIN slaves. A LIN slave receives requests, routes them to the appropriate service, and constructs the response.

LIN version 2.0 defines a transport layer protocol that was deduced from the ISO 15765-2 protocol [ISO15765-2]. It tunnels Keyword Protocol 2000 (KWP-2000) and Unified Diagnostic Services (UDS), as defined in ISO 15765-2 and ISO 14229-1 [ISO14229-1], and provides the configuration and diagnostic service for LIN slaves. Therefore, the full transport layer specification does not have to be implemented in all slave nodes; LIN version 2.1 specifies three diagnostic classes as:

Class 1 supports single frame transport protocol only, with fault indication by signals.

Class 2 provides node identification support, implementing the multi-frame transport protocol, but supporting only the UDS/KWP-2000 "read by identifier" service.

Class 3 slave nodes are the most complex nodes. They execute additional tasks beyond the basic sensor and support the services: diagnostic session control, input/output control, read/clear diagnostic trouble code information, as well as optional flash programming features.

The transport layer protocol uses two diagnostic frames: the MRF and the SRF. The MRF (ID = 0x3C) is the diagnostic request, where the master transmits both the frame header and the frame response to a slave node. The SRF (ID = 0x3D) represents the diagnostic response, where the master transmits the header, and a slave transmits the response to the master that contains the answer of the preceding MRF. The protocol uses single frames (SF), but for multi-frame support, first frames (FF) and consecutive frames (CF) are mapped into the eight data bytes of a LIN frame. A MRF is always transmitted to all slave nodes (broadcast traffic). Therefore, the slave node address, the so-called node address for diagnostics (NAD) is stored in the first data byte. The second data byte stores the protocol control information (PCI) of the transport layer and the third data byte stores the service identifier (SID) defined in the standards ISO 15765-3 (KWP-2000) [ISO15765-3] and ISO 14229-1 (UDS) [ISO14229-1].

A simple example for use of a single frame transport protocol is the node configuration service "assign NAD" (see Figure 45.10). With this service, the slave nodes in a cluster can assign a unique NAD.

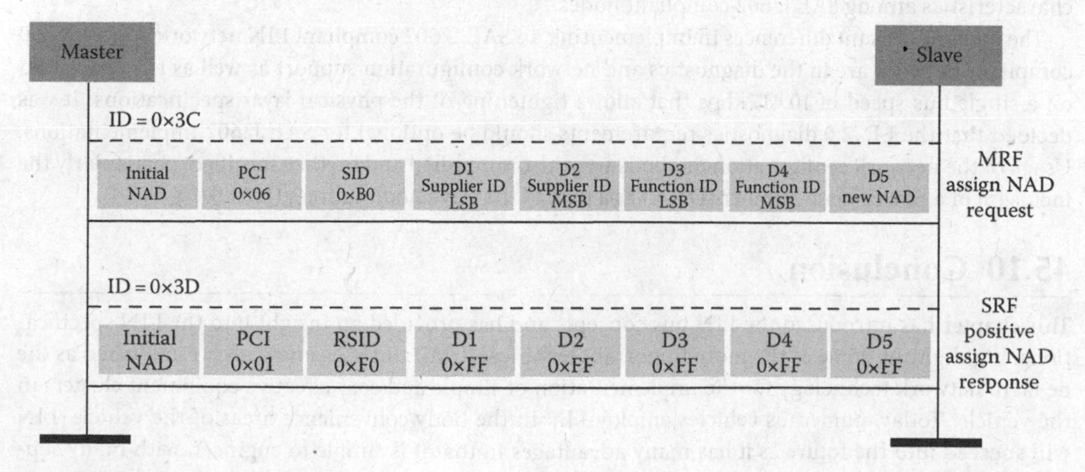

FIGURE 45.10 The "assign NAD" service as an example for a single frame.

The identification of the slaves is provided by the LIN Product Identification, where a 32 bit value identifies the supplier ID, the function ID, and the variant. The data field of the "assign NAD" request (it is a MRF) carries the following information: first byte is the initial NAD (old NAD); high nibble of second byte is the ID of SF (0x0); low nibble of second byte is the data length plus one (for the SID or RSID); third byte, the SID (0xB0—assign NAD); byte four and five is the supplier ID; byte six and seven is the function ID and byte eight is the new NAD. If the NAD, the supplier ID, and the function ID match, the slave answers to the header of the SRF with a positive "assign NAD" response. The third byte carries the response service identifier (RSID). The SID + 0x040 represents a positive acknowledgment (here, 0xF0 for positive assign NAD response).

45.8 Configuration

As stated earlier, a LIN cluster can be configured using automated tools that use the CLD to describe the cluster, in an LDF. A first step is to create a cluster message strategy to completely describe the required communication between all units in the cluster. A list of all frames is generated that includes frame IDs, publishers and subscribers, data content, etc. A schedule table can also be described for the cluster. All of these parameters are described by the LDF.

LIN2.x also defines the "off-the-shelf" approach for a dynamic configuration, which affords a simple plug-and-play feature for the slave nodes. This approach is based on the node configuration and identification services. The main services for the node configuration are the above-mentioned "assign NAD" as well as the "assign frame ID" (LIN2.0) and "assign frame ID range" (LIN2.1) services. With the latter service, one can assign and change the protected identifier of a frame.

If more identical slave nodes are used in a cluster (e.g., damper motors in an air-conditioning system), the address conflict can be resolved via a slave node position detection (SNPD) procedure only. One possible algorithm is the Bus Shut Method (also called "cool-LIN") that is described in the patent WO 03/094001 A1 [WO03]. Another algorithm is the Extra Wire Daisy Chain method. (In this area, there are more patents, e.g. [US59].)

45.9 Relationship between SAE J2602 and LIN2.0

In March of 2004, the SAE approved the SAE J2602 recommended practice for the use of LIN, based upon the released LIN2.0 specification.

The SAE J2602 recommended practice narrows down some of the choices allowed in the LIN2.0 specification to ensure a greater degree of interoperability and to provide a minimum level of performance characteristics among SAE J2602 compliant nodes.

The most significant differences in implementing an SAE J2602 compliant LIN network and a LIN2.0 compliant network are in the diagnostics and network configuration support as well as the restriction on a single bus speed of 10.417 kbps that allows tightening of the physical layer specifications. It was decided that the LIN2.0 diagnostics requirements should be optional for SAE J2602 implementations. Usage of the network configuration methods were also simplified and further specified, particularly the inclusion of a device node number (DNN) as a part of the slave node address (NAD).

45.10 Conclusion

This chapter has introduced the LIN bus concepts and has provided an insight into the LIN specifications, highlighting some of the more important technical details. LIN has now been established as the de facto network technology for the implementation of simple and cost-effective equipment clusters in the vehicle. Today, numerous vehicles employ LIN in the body/convenience areas of the vehicle. LIN will succeed into the future as it has many advantages in that it is simple to engineer, with many supporting design tools. Hardware components are readily available, the implementations are very cost effective, and there is no protocol license required. The main versions are LIN1.3 and LIN2.0, and in future LIN2.1 as well. If nodes are used with different LIN versions within a cluster, the master's version level must be at the same version level or at a higher version level than any slave node. The master has to respect the diverse properties of the slave nodes.

References

[Grz05] A. Grzemba, C v.d. Wense: LIN-Bus—Systeme, Protokolle, Tests von LIN-Systemen, Tools, Hardware, Applikationen; Franzis-Verlag, München, Germany; 2005; ISBN 3-7723-4009-1.

[ISO14229-1] Road vehicles—Diagnostic systems—Part 1: Diagnostic services; International Standard ISO 14229-1.6, Issue 6, 2001-02-22.

[ISO15765-2] Road vehicles—Diagnostics on Controller Area Network (CAN)—Part 2: Network layer services; International Standard ISO 15765-2.4, Issue 4, 2002-06-21.

[ISO15765-3] Road vehicles—Diagnostics on controller area network (CAN)—Part 3: Implementation of diagnostic services; International Standard ISO 15765-3.5, Issue 5, 2002-12-12.

[ISO9141] Road vehicles—Diagnostic systems—Requirement for interchange of digital Information; International Standard ISO9141, 1st edn., 1989.

[LIN1.0] LIN Specification Package Revision 1.0, 1999, www.lin-subbus.org

[LIN1.3] LIN Specification Package Revision 1.3. December 12, 2002. www.lin-subbus.org

[LIN2.0] LIN Specification Package Revision 2.0. September 23, 2003. www.lin-subbus.org

[LIN2.1] LIN Specification Package Revision 2.1. November 24, 2006. www.lin-subbus.org

[TR0363] TR0363—The LIN Physical Layer; Rev. 2.0;July 7, 2008; Technical Report; nxp.

[US59] Automatic node configuration with identical nodes; U.S. patent 5,914,957.

[WO03] Method for Addressing the users of a bus system by means of identification flows; patent WO 03/094001 A1; 2003-11-13; or U.S. patent 7,0918,76.

46
Profisafe

Ron Mitchell
RC Systems

Max Felser
Bern University of Applied Sciences

Paulo Portugal
University of Porto

46.1 Introduction **46**-1
Standardization Framework • Black Channel Principle
46.2 Profisafe Communication **46**-3
Error-Detection Requirements • Error Types and Safeguards • Cyclic/Acyclic Communication • Cyclic Communication PDU • Virtual Consecutive Number • Time-Out with Receipt • Code Name for Sender/Receiver • Data Consistency Check • Detected Safety Data Failures
46.3 Deployment **46**-11
Power Supplies and Electrical Safety • Increased Immunity • Installation Guidelines • Wireless Transmission and Security • Response Time
Acronyms...**46**-14
References...**46**-14

46.1 Introduction

Profisafe is a comprehensive and integrated solution with the aim to support safe communication in fieldbus networks (Figure 46.1). Profisafe comprises the following principles [1]:

- Integration of safety-related applications into standard solutions without any impact for the latter ones.
- Standard and safety data coexist in the same network—Profibus (see Chapter 32) or Profinet (see Chapter 40)—without modification of the standard protocols.
- Safety data is transmitted between safety equipment—controllers (F-Host) and devices (F-Device)—using the highest integrity level (SIL 3 or PL "e") required by current applications in factory and process automation.
- Configuration, parameterization, diagnosis, and maintenance of safety devices is performed using engineering tools that are similar to those employed in standard applications, thus facilitating the development and integration process.

46.1.1 Standardization Framework

The development of safety-related systems for factory and process automation is a complex task, which involves the employment of multiple and interrelated standards. Figure 46.2 presents a selection of the most relevant ones.

The development of Profisafe was performed based on two standards: IEC 61508 [2] and EN 50159-1, now IEC 62280-1 [3]. IEC 61508 outlines the requirements when electrical/electronic/programmable electronic systems are used to perform safety functions (further details can be found in Chapter 2),

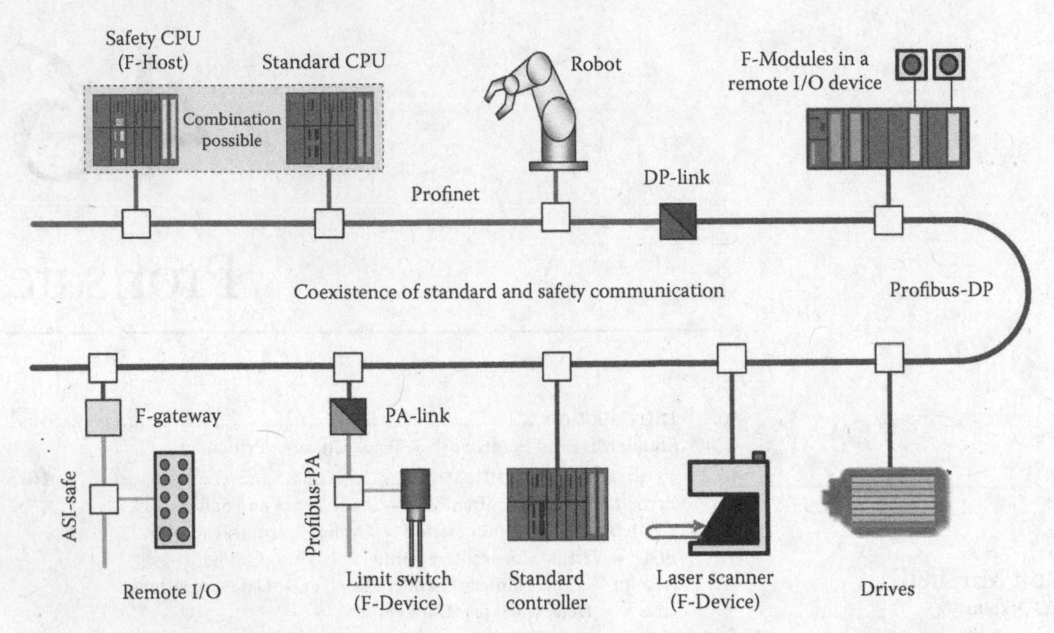

FIGURE 46.1 The Profisafe approach. (Adapted from PROFIsafe—Safety technology for PROFIBUS and PROFINET—System description, PROFIBUS Nutzerorganisation e.V. PNO, 2007.)

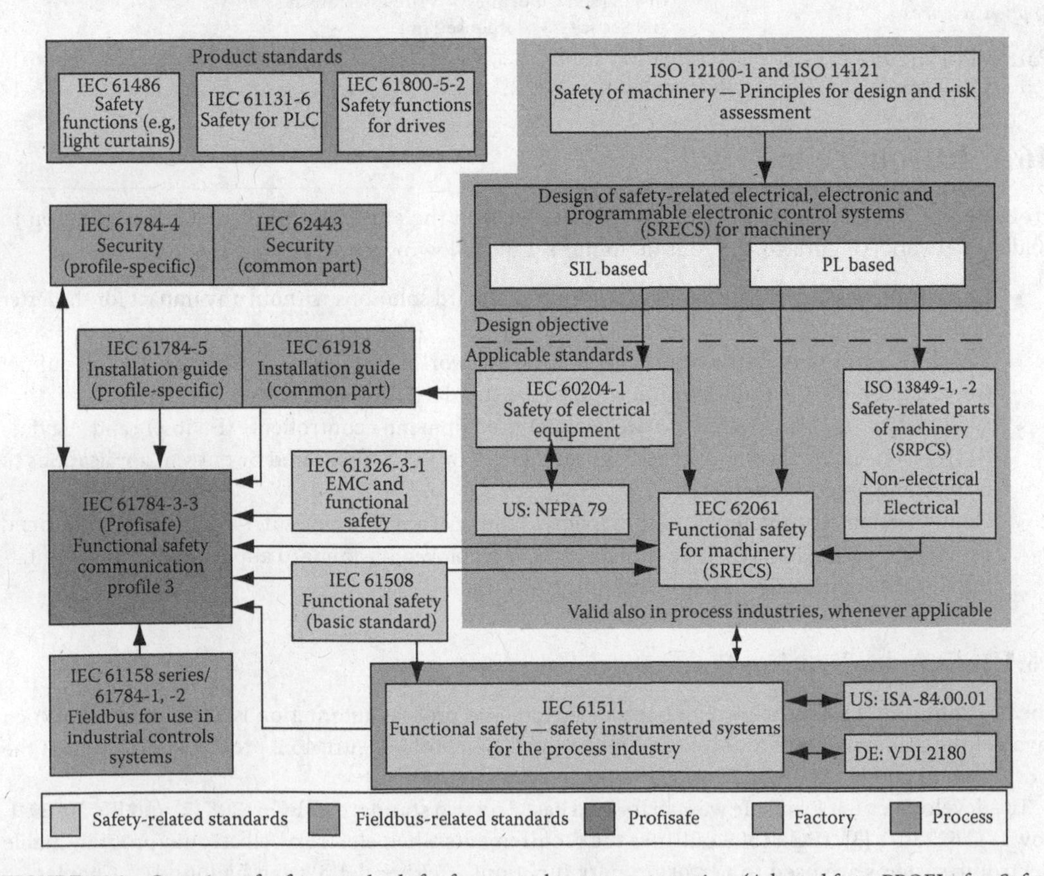

FIGURE 46.2 Overview of safety standards for factory and process automation. (Adapted from PROFIsafe—Safety technology for PROFIBUS and PROFINET—System description, PROFIBUS Nutzerorganisation e.V. PNO, 2007.)

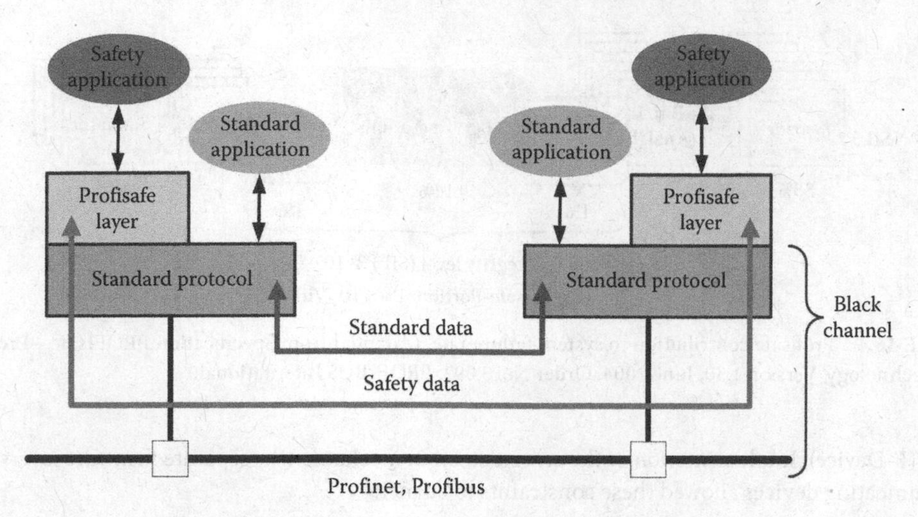

FIGURE 46.3 Black channel approach. (Adapted from PROFIsafe—Safety technology for PROFIBUS and PROFINET—System description, PROFIBUS Nutzerorganisation e.V. PNO, 2007.)

while EN 50159-1 defines the requirements needed to support safe communication between safety-related equipment. Many of the solutions employed by Profisafe are based on EN 50159-1 proposals, the "black channel" principle being the most relevant example. Nowadays, Profisafe is part of a series of standards related to fieldbus systems—IEC 61158 [4] and its companion IEC 61784 [5]. Within this series, the IEC 61784-3 [6] defines a set of profiles to support safe communication in fieldbus networks. Profisafe is standardized as IEC 61784-3-3 [7], resulting from additional specifications proposed within IEC 61784-3 to encompass Profibus and Profinet networks. According to its developers [1], it is certified for use in safety-related applications up to SIL 3 (Safety Integrity Level, IEC 61508 or IEC 62061) or PL "e" (Performance Level, ISO 13849-1). This certification is also extended to wireless networks such as IEEE 802.11 and Bluetooth, but with additional security requirements (see Section 46.3.4).

46.1.2 Black Channel Principle

The "black channel" principle is based on the following requirement: the transmission of safety data is performed independently of the characteristics of the transmission system, e.g., medium, topology, communication stack, network devices, etc., and without trusting the internal safety mechanisms provided by it. In order to achieve this goal, its implementation is undertaken as an additional layer—*the safety layer*—on top of the application layer. The safety layer considers that safety data are subject to various threats (i.e., errors), and for each one defines a set of defense measures in order to protect this data [3] (see Section 46.2.2). In the Profisafe case, the safety layer is implemented as a safety profile (Figure 46.3). Therefore, safety data are encapsulated in standard Profinet or Profibus frames, jointly with standard data, and transmitted in accordance with the usual rules defined by the communication protocol.

46.2 Profisafe Communication

From the beginning, the Profisafe solution was subject to certain design constraints: compatibility with existing Profibus/Profinet physical layers, interoperability with existing devices, and compatibility with existing modular devices. Any solution must attempt to guarantee the error-free delivery of data, or the detection of data with errors, between communicating devices, such as a PLC (F-Host) and an I/O

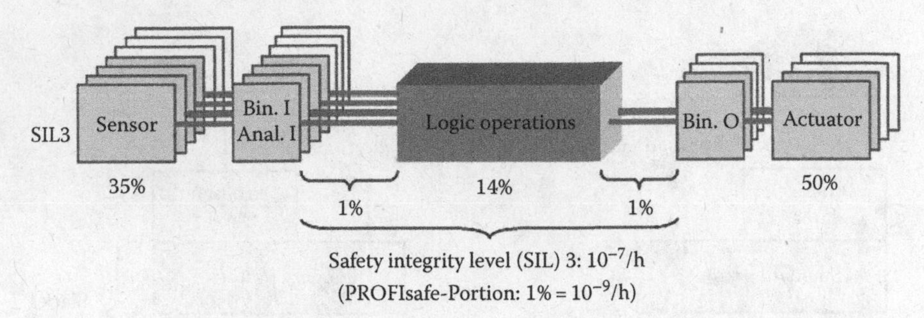

FIGURE 46.4 Profisafe contribution to system failure rate. (Adapted from Specification PROFIsafe—Profile for safety technology, Version 1.30, June 2004, Order No: 3.092, PROFIBUS International.)

device (F-Device). Implementation of the error-detection mechanism as separate firmware layers in the communicating devices allowed these constraints to be met.

46.2.1 Error-Detection Requirements

The requirement for being able to designate a system as capable of operating at SIL 3 is that the system must exhibit a failure rate less than 10^{-7}/h. This failure rate is for the entire system. However, a system is made up of several components and it must be determined what each component is allowed to contribute to the overall system failure rate. There were no precedents for safety systems involving digital communication systems. The IEC 61508 specifies the contributions to the overall system failure rate as 35% for sensors, 15% for logic controllers, and 50% for actuators. Therefore, the Working Group developing the Profisafe specifications chose a maximum of a 1% contribution allowed for Profisafe. This percentage contribution was deducted from the contribution of the logic controller and is now specified in IEC 61784-3. As one can see in Figure 46.4, this results in a required failure rate for the Profisafe contribution of only 10^{-9}/h ($10^{-2} \times 10^{-7}$).

46.2.2 Error Types and Safeguards

Various errors can occur when messages are transmitted, whether due to hardware failures, extraordinary electromagnetic interference, or other influences. A message can be lost, occur repeatedly, be inserted from somewhere else, appear delayed or in an incorrect sequence, and/or show corrupted data. In the case of safety communication, there may also be incorrect addressing—a standard message erroneously appears at an F-Device and pretends to be a safety message. Different transmission rates may additionally cause storage effects to occur. Of the numerous error types and safeguards known from literature, Profisafe concentrates on those shown in Figure 46.5.

Although it can be detected by use of the consecutive number technique, an additional error type was identified for Profinet communication—that of the failure of revolving memory in switches. Since there is little control of what Ethernet switches a customer actually uses, some COTS (commercial, off-the-shelf) switches may be unreliable and lead to revolving memory failures as shown in Figure 46.6. Accidental "jumping" of the send pointer, for example, can cause emptying/sending of the entire queue of messages.

46.2.3 Cyclic/Acyclic Communication

Cyclic data exchange between a controller and its field devices utilizes a one-to-one communication relationship as shown in Figure 46.7. This figure illustrates that a controller (F-Host) can operate any mix of standard and safety devices (F-Devices) connected to the same network. Safety tasks and

Remedy: / Error type:	(Virtual) consecutive number	Time out with receipt	Code name for sender and receiver	Data consistency check
Repetition	✓			
Deletion	✓	✓		
Insertion	✓	✓	✓	
Resequencing	✓			
Data corruption				✓
Delay		✓		
Masquerade (standard message mimics failsafe)		✓	✓	✓
Revolving memory failures within switches	✓			

FIGURE 46.5 Error types and safeguards. (Adapted from Specification PROFIsafe—Profile for safety technology, Version 1.30, June 2004, Order No: 3.092, PROFIBUS International.)

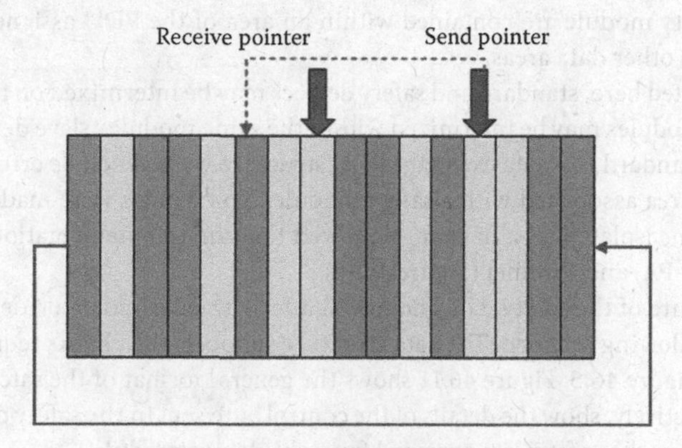

FIGURE 46.6 Possible queuing problems in switches. (Adapted from Specification PROFIsafe—Profile for safety technology, Version 1.30, June 2004, Order No: 3.092, PROFIBUS International.)

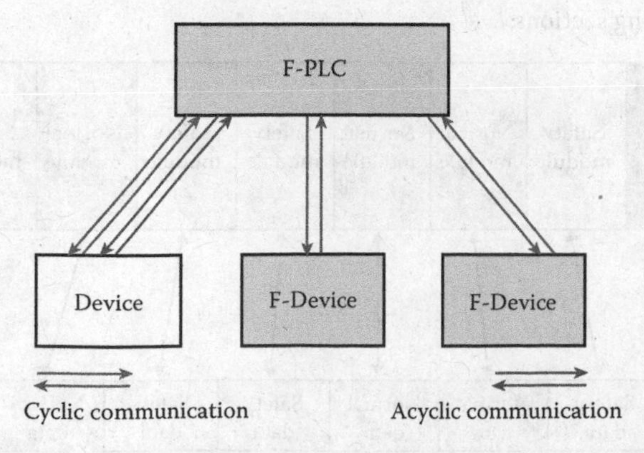

FIGURE 46.7 Profisafe cyclic/acyclic communication. (Adapted from Stripf, W. and Barthel, H., Comprehensive safety with PROFIBUS DP and PROFINET IO—Training for "PROFIsafe certified designers.")

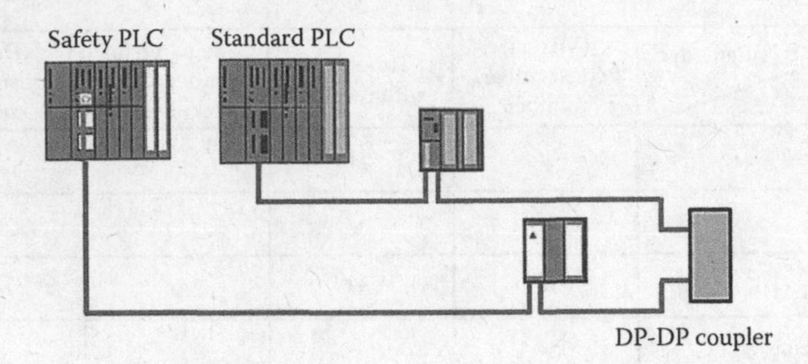

FIGURE 46.8 Split assignment of safety and standard tasks.

standard tasks may also be assigned to different controllers and networks as shown in Figure 46.8. Any so-called acyclic communications between devices and controllers or supervisors such as programming devices are intended for configuration, parameterization, diagnosis, and maintenance purposes.

46.2.4 Cyclic Communication PDU

The PDU (protocol data unit) structure of the Profibus I/O data exchange message is exactly the same whether or not there is any safety data in the message. This structure has separate data areas assigned to each module in a modular slave device, facilitating the intermixing of safety and standard modules as shown in Figure 46.9, where the safety modules map onto the Profisafe data areas in the PDU. Data associated with a safety module are contained within an area of the PDU assigned only to that safety module and affects no other data areas.

As has been illustrated here, standard and safety devices may be intermixed on the same network and standard and safety modules may be intermixed within the same modular slave device. The only change required within the standard I/O data exchange PDU structure was to include error-detection information within the data area associated with a safety module. No changes were made to the data areas of standard modules. This isolated area of change allowed the same implementations of Profisafe across Profibus-DP, Profibus-PA, and Profinet (Figure 46.10).

The detailed structure of the safety data and how it affects the desired error-detection solutions will be described in the following sections. The data structure supports the checks required to detect all the error types listed in Figure 46.5. Figure 46.11 shows the general format of the safety data while Figures 46.12 and 46.13, respectively, show the details of the control byte sent in the safety data from the controller and the status byte in the safety data returned from the device/module.

The mechanisms used for error detection, i.e., consecutive numbering, timeout with receipt, code name for sender/receiver, and data consistency check (cyclic redundancy check, CRC), will be briefly discussed in the following sections.

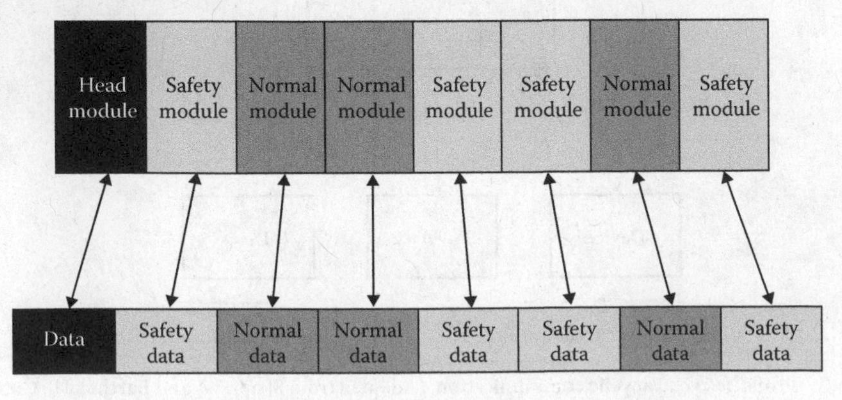

FIGURE 46.9 Intermixed standard and safety I/O modules.

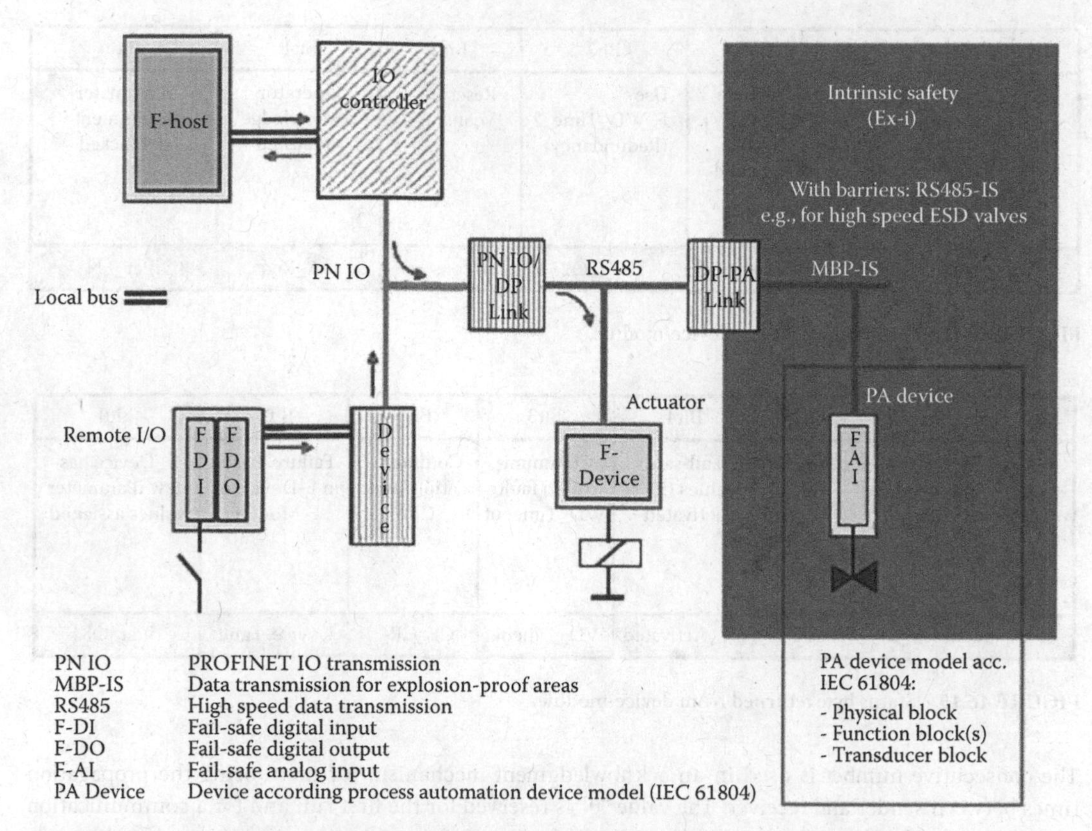

FIGURE 46.10 Profisafe is protocol-independent. (Adapted from Stripf, W. and Barthel, H., Comprehensive safety with PROFIBUS DP and PROFINET IO—Training for "PROFIsafe certified designers.")

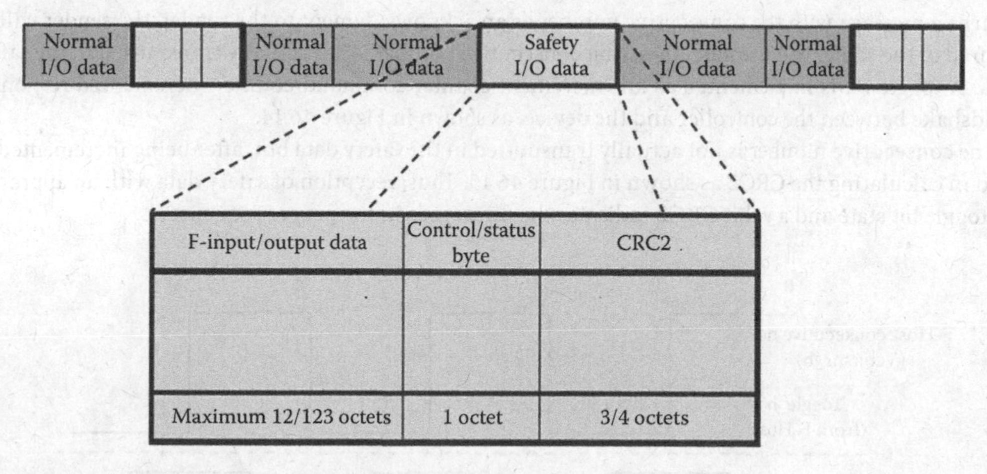

FIGURE 46.11 General safety data format.

46.2.5 Virtual Consecutive Number

Only the V2-mode (virtual) consecutive number will be described here. For a description of the V1-mode, see [9]. The consecutive numbering uses a range that is big enough to detect any error caused by message storing network elements, e.g., switches for Profinet, links or repeaters for Profibus. The receiver uses the consecutive number to monitor whether the sender and the communication channel are still alive.

Bit7	Bit6	Bit5	Bit4	Bit3	Bit2	Bit1	Bit0
res	res	Toggle bit	Fail-safe values (FV) to be activated	Use F_WD_Time_2 (Redundancy)	Reset Vconsnr_d	Operator acknowledge requested	iParameter assignment deblocked
–	–	Toggle_h	Toggle_FV	Use_TO2	R_cons_nr	OA_Req	iPar_EN

FIGURE 46.12 Control byte sent to device/module.

Bit7	Bit6	Bit5	Bit4	Bit3	Bit2	Bit1	Bit0
res	Vconsnr_d has been reset	Toggle Bit	Fail-safe values (FV) activated	Communi-cation fault: WD_Timeout	Communi-cation fault: CRC	Failure exists in F-Device or F-Module	F-Device has new iParameter values assigned
–	cons_nr_R	Toggle_d	FV_Activated	WD_Timeout	CE_CRC	Device_Fault	iPar_OK

FIGURE 46.13 Status byte returned from device/module.

The consecutive number is used in an acknowledgment mechanism for monitoring the propagation times between sender and receiver. The value "0" is reserved for the first run and for a communication error reaction. The V2-mode uses 24 bit counters for consecutive numbering, counting in a cyclic mode from 1…0×FFFFFF and wrapping over back to 1 at the end. Using the consecutive number, a receiver can see whether or not it received the messages completely and within the correct sequence. When it returns a message with the consecutive number as an acknowledgment to the sender, the sender will be assured of the same. In V2-mode, the 24 bit consecutive number is not actually transmitted in the safety data. A toggle bit is implemented as an "increment counter command/counter incremented response" handshake between the controller and the device, as shown in Figure 46.14.

The consecutive number is not actually transmitted in the safety data but, after being incremented, is used in calculating the CRC2 as shown in Figure 46.15. Thus, reception of safety data with an appropriate toggle bit state and a valid CRC2 indicates the message is in the proper sequence.

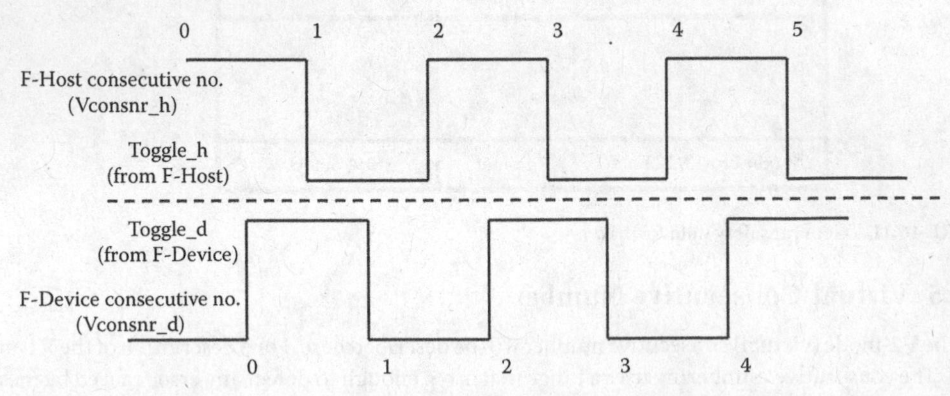

FIGURE 46.14 Toggle bit functionality. (Adapted from Stripf, W. and Barthel, H., Comprehensive safety with PROFIBUS DP and PROFINET IO—Training for "PROFIsafe certified designers.")

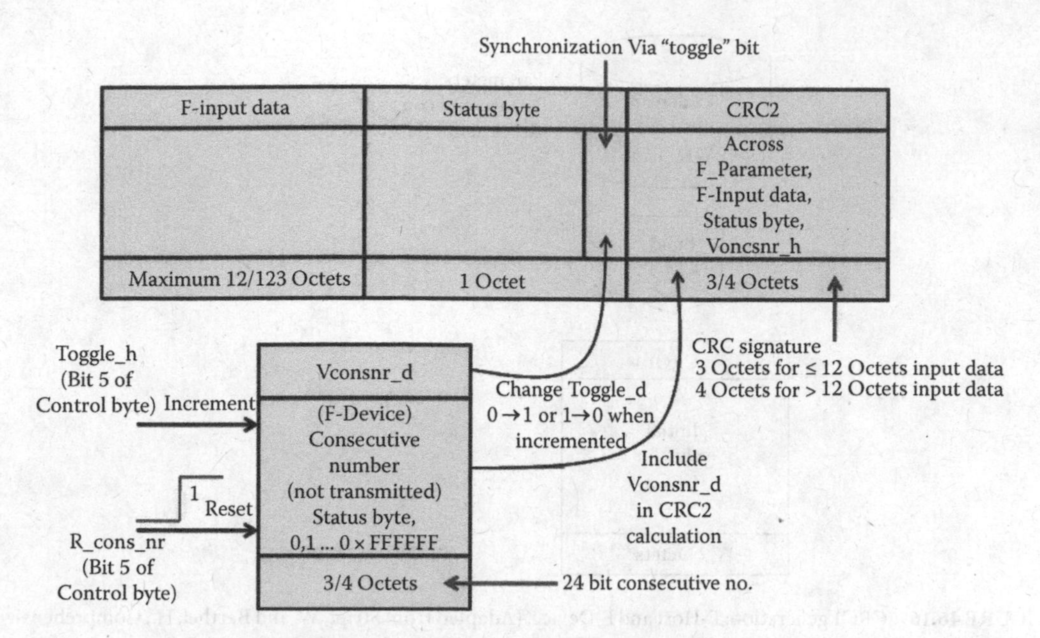

FIGURE 46.15 F-Device (virtual) consecutive number. (Adapted from Stripf, W. and Barthel, H., Comprehensive safety with PROFIBUS DP and PROFINET IO—Training for "PROFIsafe certified designers.")

46.2.6 Time-Out with Receipt

In safety technology, it not only matters that a message transfers the correct process signals or values. Additionally, the updated actual values must arrive within a fault tolerance time. Recognition of a timeout enables the respective F-Device to automatically initiate any necessary safety reactions, for example, stopping a motor. For monitoring the time, the F-Devices use a watchdog timer that is restarted whenever new Profisafe messages with incremented consecutive numbers are received.

46.2.7 Code Name for Sender/Receiver

The 1:1 relationship between the F-Host and a F-Device facilitates the detection of misdirected message frames, e.g., an inserted or masquerading message. The sender and receiver must have an identification scheme (code name) that is unique in the network that can be used for verifying the authenticity of a Profisafe message. Profisafe uses an "F-Address" as a code name. CRC1 is constructed from this code name, the assigned SIL, and the watchdog time as shown in Figure 46.16.

46.2.8 Data Consistency Check

Once the F-Parameters (code name, SIL, watchdog time) have been transferred to the F-Device, these identical parameters are utilized in identical calculations in both the F-Host and in the F-Device/F-Module for producing a 2-octet CRC1. This CRC1 signature provides the initial value for CRC2. Subsequently, this CRC1, the F I/O data, the status or control byte, and the corresponding consecutive number (host or device) are used for producing another CRC2 signature within the F-Host as shown in Figure 46.17. In the F-Device, the identical CRC signature is generated and the signatures are compared.

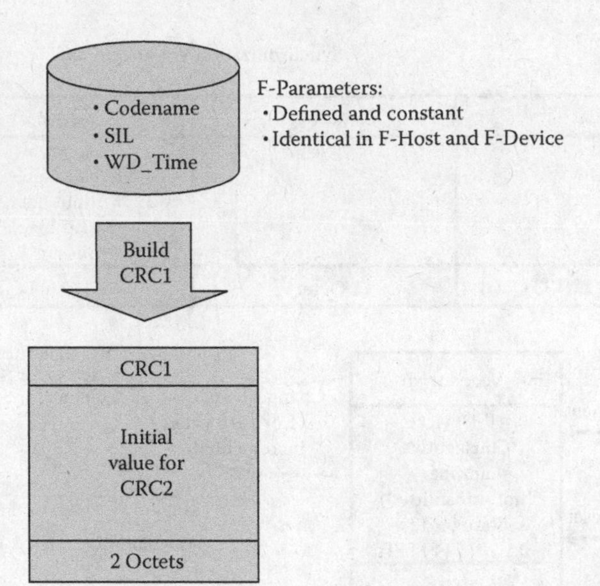

FIGURE 46.16 CRC1 generation: F-Host and F-Device. (Adapted from Stripf, W. and Barthel, H., Comprehensive safety with PROFIBUS DP and PROFINET IO—Training for "PROFIsafe certified designers.")

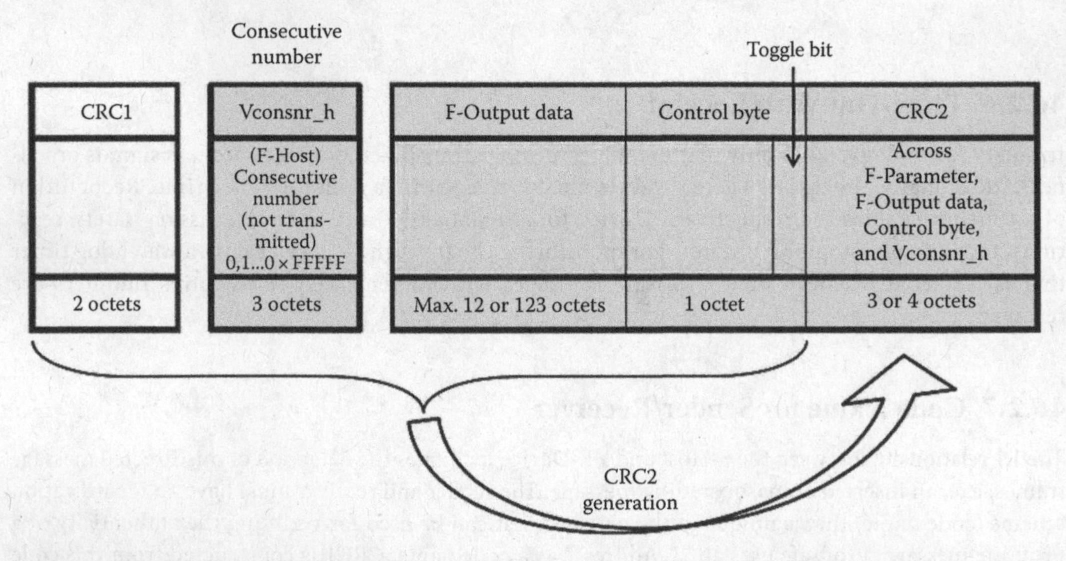

FIGURE 46.17 CRC2 generation: F-Host and F-Device. (Adapted from Stripf, W. and Barthel, H., Comprehensive safety with PROFIBUS DP and PROFINET IO—Training for "PROFIsafe certified designers.")

46.2.9 Detected Safety Data Failures

The ultimate purpose of Profisafe is to prevent a system failure caused by an undetected communication error. In order to determine when the system may be in a state in which there is an increasing probability of a communication error going undetected, an SIL monitor is implemented in the F-Host. Every corrupted message (CRC2 or virtual consecutive number fault) detected by the F-Host or F-Device will be counted during a configurable SIL monitor time period (*T*). The SIL monitor time period depends upon the SIL value of the system and the CRC2 length, which depends upon the amount of safety data being transferred. If 12 or fewer bytes of safety data are being transferred, a 3-octet (24 bit) CRC2 is used, while a 4-octet (32 bit) CRC2 is used for 13–123 bytes. The monitor will cause the fail-safe values to be set whenever more than one such fault is detected, i.e., one detected corrupted message can be tolerated. A second corrupted message detected within time *T* indicates a serious cause, e.g., excessive EMI, hardware failures, etc., that requires immediate attention.

46.3 Deployment

Profisafe would not be complete if there were only a specification of the safety communication protocol. For F-Devices questions would be raised about voltage protection, installation rules, and security.

46.3.1 Power Supplies and Electrical Safety

The fieldbus standards IEC 61158 and IEC 61784-1, IEC 61784-2 require all devices within the network to comply with the legal requirements of that country where they are deployed. These measures are called PELV (protected extra-low voltage) and limit the permitted voltages in case of one failure to ranges that are not dangerous for humans. Due to this normally legal requirement, it is possible to limit the protection effort within an F-Device or an F-Host. It is also possible to use the same 24 V power supplies for standard and F-Devices/F-Hosts. In both cases, the power supplies shall provide PELV due to legal requirements.

46.3.2 Increased Immunity

For each safety application, the corresponding SRS (safety requirements specification) shall define electromagnetic immunity limits (see IEC 61000-1-1 [10]) which are required to achieve electromagnetic compatibility. These limits should be derived taking into account both the electromagnetic phenomena (see IEC 61000-2-5 [11]) and the required safety integrity levels. For general industrial applications, the IEC 61326-3-1 [12] defines immunity requirements for equipment performing or intended to perform safety-related functions.

46.3.3 Installation Guidelines

It is the goal of Profisafe to integrate safety communication into the standard Profibus and Profinet networks with minimal impact on the existing installation guidelines. All F-Devices shall be certified according to IEC 61508 and in the case of process automation according to IEC 61511 [13]. They shall be tested and approved for Profisafe conformity by PI (Profibus International) Test Laboratories. All other standard devices within a Profisafe network shall prove conformity to Profibus or Profinet via a PI certificate or equivalent evidence. For Profibus, no spurs or branch lines are permitted. For Profinet less than 100 switches in a row, only one F-Host per sub module and no single-port routers are permitted to separate Profisafe islands (characterized by unique F-Addresses).

Profibus and Profinet both specify the use of shielded cables and double-sided connection of the shield with the connector housing for best electromagnetic immunity. As a consequence, equipotential bonding is usually required. If this is not possible, fiber optics may be used. Even with shielded cables, unacceptable signal noise may be introduced onto the data lines of a device if, for example, the intermediate DC link of a frequency inverter is not filtering well enough. Other sources of unacceptable signal quality may be due to missing terminating resistors and the like. This is not a safety issue but an availability issue. Sufficient availability of the control functions is a precondition of safety. Safety functions on equipment with insufficient availability may cause nuisance trips and as a consequence may tempt production managers to eventually remove these safety functions ("Bhopal effect").

Profisafe has been the enabling technology for many safety devices, especially for drives with integrated safety. Nowadays drives can provide safe states without deenergizing the motor. For example, the

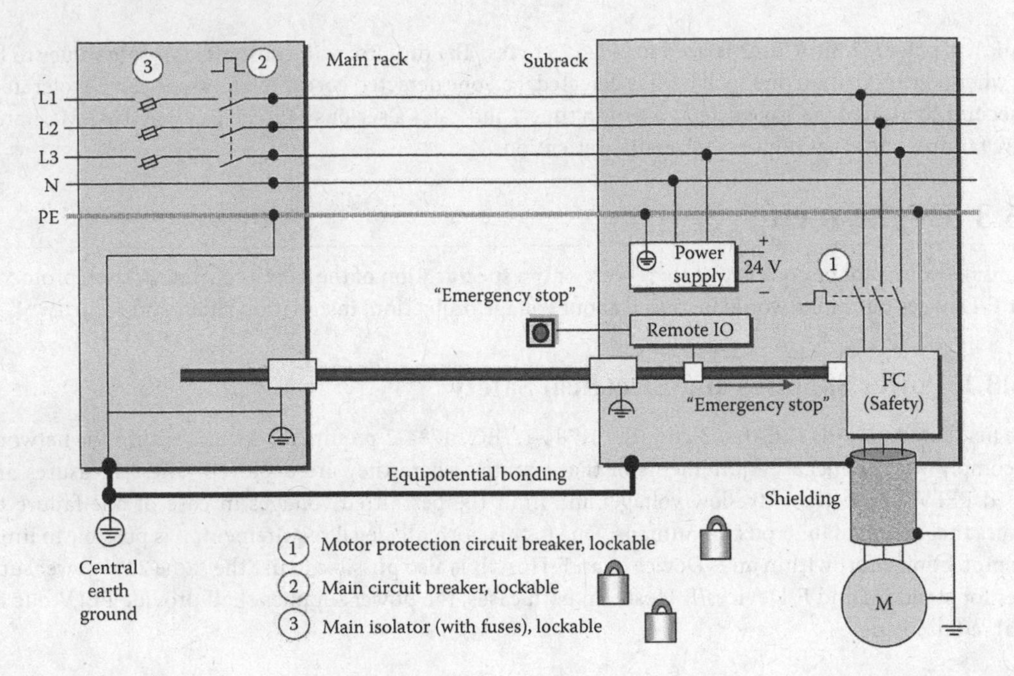

FIGURE 46.18 Emergency switching off concept.

new safety feature "SOS" (Safe Operating Stop) holds the motor under closed loop control in a certain position. This new possibility requires a paradigm shift for the user. In earlier times, pushing an emergency stop button caused the power lines to be physically disconnected from the motor and, therefore, there was no electrical danger for a person exchanging the motor.

The new IEC 60204-1 [14] provides concepts on how to protect against electrical shock (emergency switch-off) with lockable motor protection circuit breakers, main circuit breakers, and main isolators with fuses. Figure 46.18 demonstrates these concepts. It also shows the recommended 5-wire power line connections (TN-S) with separated N and PE lines and the shielded cables between drives and motors. The IEC 60204-1 is a valuable source for many other safety issues complementing the Profisafe technology.

46.3.4 Wireless Transmission and Security

More and more applications such as AGV (automated guided vehicles), rotating machines, gantry robots, and teach panels use wireless transmission in Profibus and Profinet networks. Profisafe, with its error-detection mechanism for bit error probabilities up to 10^{-2}, is approved for both "Black Channels." However, the security issues below must be considered in addition. With Profinet being based on Industrial Ethernet as an open network and in the context of wireless transmission, the issue of security has been raised.

PI published [15] the concept of building so-called security zones which can be considered to be closed networks (Figure 46.19). The only possibility to cross open networks such as Industrial Ethernet Backbones from one security zone to another is via security gates. The security gates use generally accepted mechanisms such as VPN (virtual private network) and Firewalls to protect themselves from intrusion. Profisafe networks always shall be located inside security zones and protected by security gates if connections to open networks cannot be avoided. For wireless transmission, the IEEE 802.11i standard provides sufficient security measures for Profisafe networks. Only the infrastructure mode is permitted; the ad hoc mode shall not be used.

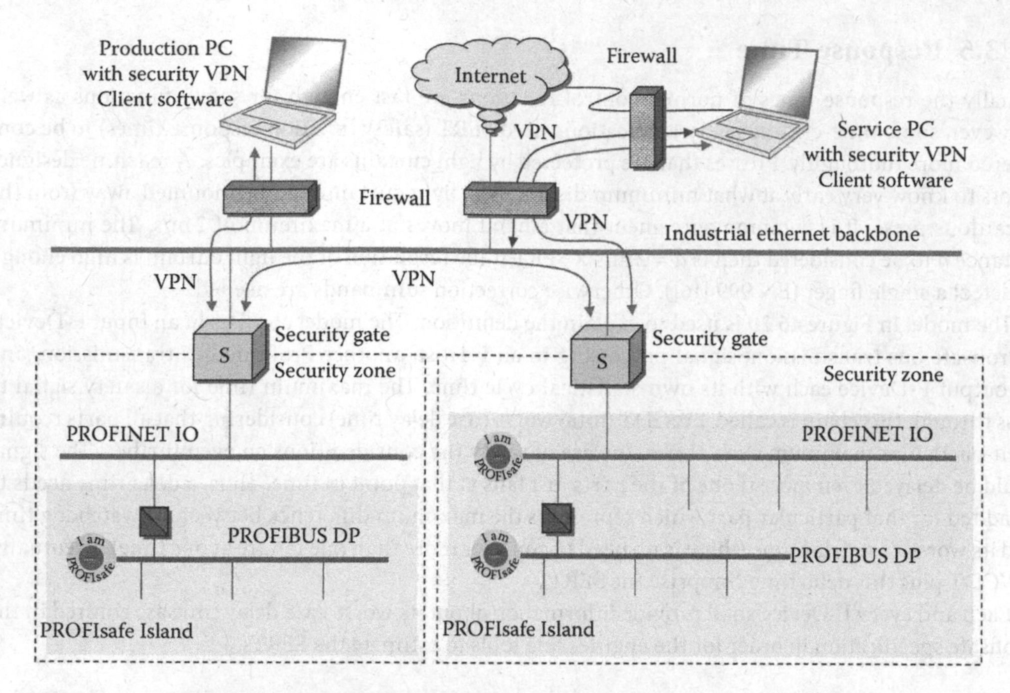

FIGURE 46.19 Security concept for "closed" and "open" networks.

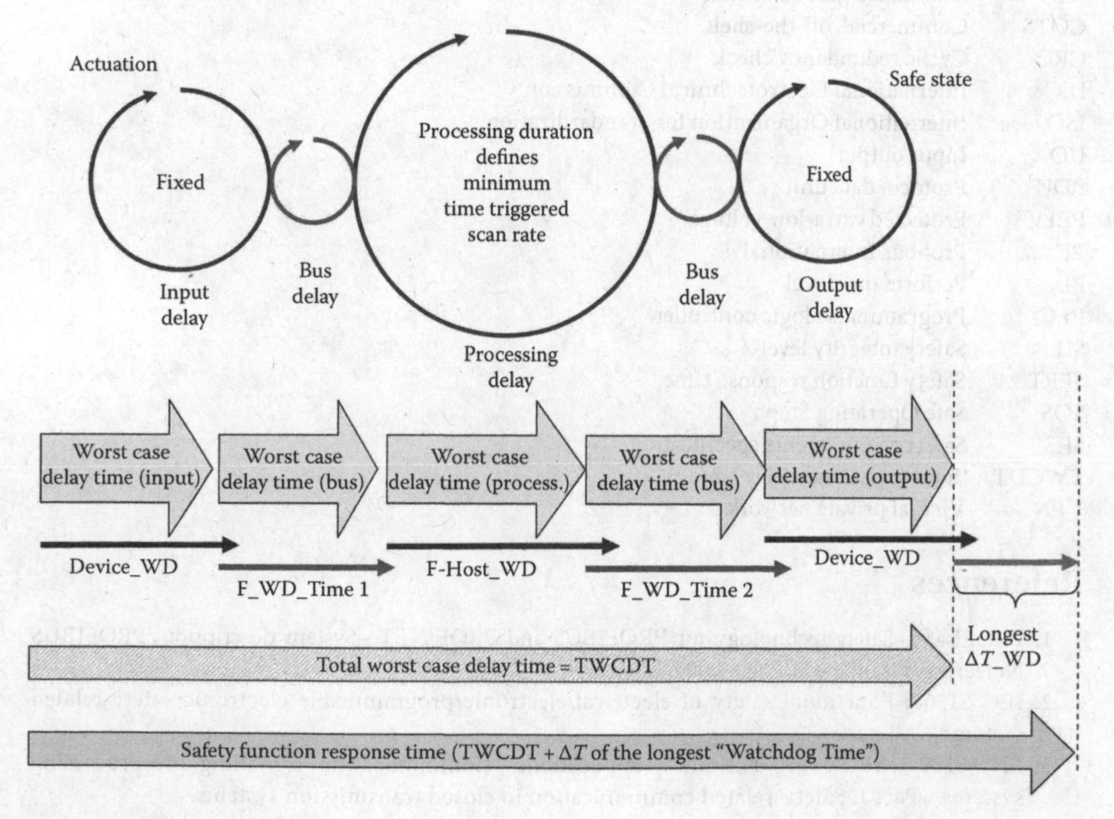

FIGURE 46.20 Safety function response time (SFRT).

46.3.5 Response Time

Usually the response times of normal control functions are fast enough for safety functions as well. However, some time-critical safety applications need SFRT (safety function response times) to be considered more thoroughly. Presses that are protected by light curtains are examples. A machine designer wants to know very early at what minimum distance the light curtain shall be mounted away from the hazardous press. It is common agreement that a hand moves at a maximum of 2 m/s. The minimum distance d to be considered then is $d = 2\,\text{m/s} \times \text{SFRT}$ if the resolution of the light curtain is high enough to detect a single finger (EN 999 [16]). Otherwise correction summands are needed.

The model in Figure 46.20 is used to explain the definition. The model consists of an input F-Device, a Profisafe bus transmission, signal processing in an F-Host, another Profisafe bus transmission, and an output F-Device each with its own statistical cycle time. The maximum time for a safety signal to pass through this chain is called TWCDT (total worst-case delay time) considering that all parts require their particular maximum cycle times. In case of safety the considerations go even further: The signal could be delayed even more if one of the parts just fails at that point in time. Thus, a delta time needs to be added for that particular part which represents the maximum difference between its watchdog time and its worst-case delay time (there is no need to consider more than one failure at one time). Eventually, TWCDT plus this delta time comprise the SFRT.

Each and every F-Device shall provide information about its worst-case delay time as required in the Profisafe specification in order for the engineering tools to estimate the SFRTs.

Acronyms

AGV	Automated guided vehicle
COTS	Commercial, off-the-shelf
CRC	Cyclic redundancy check
IEC	International Electrotechnical Commission
ISO	International Organization for Standardization
I/O	Input/output
PDU	Protocol data unit
PELV	Protected extra-low voltage
PI	Profibus International
PL	Performance level
PLC	Programmable logic controller
SIL	Safety integrity level
SFRT	Safety function response time
SOS	Safe Operating Stop
SRS	Safety requirements specification
TWCDT	Total worst-case delay time
VPN	Virtual private network

References

1. PROFIsafe—Safety technology for PROFIBUS and PROFINET—System description, PROFIBUS Nutzerorganisation e.V. PNO, 2007.
2. IEC 61508, Functional safety of electrical/electronic/programmable electronic safety-related systems—All parts.
3. EN 50159-1, IEC 62280-1, Railway applications—Communication, signalling and processing systems—Part 1: Safety-related communication in closed transmission systems.
4. IEC 61158, Industrial communication networks—Fieldbus specifications—All parts.

5. IEC 61784, Industrial communication networks—Profiles—All parts.
6. IEC 61784-3, Industrial communication networks—Profiles—Part 3: Functional safety fieldbuses—General rules and profile definitions.
7. IEC 61784-3-3, Industrial communication networks—Profiles—Part 3-3: Functional safety fieldbuses—Additional specifications for CPF 3.
8. W. Stripf and H. Barthel, Comprehensive safety with PROFIBUS DP and PROFINET IO—Training for "PROFIsafe certified designers".
9. Specification PROFIsafe—Profile for safety technology, Version 1.30, June 2004, Order No: 3.092, PROFIBUS International.
10. IEC 61000-1-1, Electromagnetic compatibility (EMC)—Part 1: General—Section 1: Application and interpretation of fundamental definitions and terms.
11. IEC 61000-2-5, Electromagnetic compatibility (EMC)—Part 2: Environment—Section 5: Classification of electromagnetic environments, Basic EMC publication.
12. IEC 61326-3-1, Electrical equipment for measurement, control and laboratory use—EMC requirements—Part 3-1: Immunity requirements for safety-related systems and for equipment intended to perform safety-related functions (functional safety)—General industrial applications.
13. IEC 61511, Functional safety—Safety instrumented systems for the process industry sector—All parts.
14. IEC 60204-1, Safety of machinery—Electrical equipment of machines—Part 1: General requirements.
15. PROFINET Security Guideline, Version 1.0, March 2005.
16. EN 999, Safety of machinery—The positioning of protective equipment in respect of approach speed of parts of the human body.

47 SafetyLon

Thomas Novak
SWARCO Futurit Verkehrssignalssysteme GmbH

Thomas Tamandl
SWARCO Futurit Verkehrssignalssysteme GmbH

Peter Preininger
LOYTEC Electronics GmbH

47.1 Introduction 47-1
47.2 The General SafetyLon Concept 47-1
47.3 The Safety-Related Lifecycle 47-2
47.4 The Hardware 47-4
47.5 The Safety-Related Firmware 47-7
47.6 The SafetyLon Tools 47-10
Acronyms 47-13
References 47-13

47.1 Introduction

SafetyLon is a safety-related automation technology based on LonWorks used to realize a safety-related communication system. The idea is to integrate safety measures into the existing LonWorks to ensure a high level of integrity.

The type and implementation of the safety measures is specified by the requirements of the international standard IEC 61508. It is necessary that a malfunction of the system does not lead to serious consequences, such as injury or even death of people, with a very high probability. Requirements and the corresponding measures have a great impact on the various entities of the communication system: the node hardware, the communication protocol, the node firmware, and the development, installation, commission, and maintenance process.

In short, in a safety-related communication system integrity of data (management or process data) exchanged among nodes, or between a node and a management unit, must always be ensured with a high probability. For that reason, the nodes processing the data must meet distinct safety requirements. The hardware is therefore enhanced with additional integrated circuits, and online software self tests are executed. A protocol used to exchange data is integrated into the existing protocol. The tools applied to exchange management data that, in turn, are the base for process data exchange are also enhanced with a safety-related process.

The result is a safety-related automation technology that meets the requirements of IEC 61508 safety integrity level 3 (SIL 3) [1]. It can be integrated into an existing LonWorks and allows safe and non-safe communication within a single communication network.

47.2 The General SafetyLon Concept

A common way to realize a safety-related automation technology is to use a standard technology—in case of SafetyLon LonWorks—and enhance it with safety features. Such a technology can be realized by two different approaches: integration of safety measures directly into the existing technology like CANopen Safety; or treating the non-safe technology as a black-box and base the safety-related measures logically totally isolated from the remainder on the standard technology.

The second strategy is pursued not only in ProfiSafe but also in SafetyLon. It uses standard LonWorks hardware to access the network. Hence, safe and non-safe nodes can be connected to the same network. Only safe nodes can communicate by means of the SafetyLon protocol. But all devices are able to exchange messages that are non-safety related.

Following the second approach, the communication channel is split in the black and yellow channel. The black channel comprises all standard LonWorks technology including the network chip (Neuron Chip or LC3020) and the network (cf. dark gray part in Figure 47.2). These components allow the communication of two or more network nodes in a non-safe manner. In the black channel, it is not guaranteed that the received message is correct. A message can be delayed, corrupted, inserted, or lost.

In contrast to the black channel, the yellow channel is associated with the safety related measures. It uses the features of the black channel to transport messages, but additional effort is made to ensure the integrity of the safety-related payload. In addition, special hardware offers the required hardware integrity necessary to ensure a high level of integrity of data processing.

The tools for the non-safe installation, commissioning, and maintenance of LonWorks are also used in SafetyLon. Whereas the firmware on node side is realized to be safety related, the tools are not made safe. However, safety-related processes are supported by the tools being an extension to the standard tools and the user is taken as authority to ensure a high level of integrity.

Applying the black-channel concept (i.e., setting up SafetyLon on the basis of LonWorks) allows safe and non-safe nodes to be in the same system. In detail, every safe node can handle non-safe and safe messages. Hence, even safe and non-safe network variables can coexist on the same node which leads to a tight integration of non-safe and safe applications. In addition, SafetyLon is compliant to the European standard EN 14908 [2] since it does not change the standard LonWorks in any way.

In the following, the three parts to be enhanced due to safety reasons are outlined: the node hardware, the node firmware, and the tools. Development and use of all the devices is embedded into the safety-related SafetyLon lifecycle presented first and in accordance with IEC 61508.

47.3 The Safety-Related Lifecycle

SafetyLon is not only developed according to the requirements of IEC 61508 but it also specifies a SafetyLon lifecycle in which it supports the whole application during the different phases of the safety-related lifecycle [3]. The SafetyLon lifecycle can be considered as an implementation of the IEC 61508 lifecycle with regard to LonWorks.

The safety-related communication (e.g., how to address communication faults with adequate measures) is just one part. In addition to this, processes related to developing a user-application, or configuring and maintaining the network have to be specified.

The processes presented in the following must be documented in a coordinated way within the functional safety management that is based on the safety plan covering all processes and activities to improve safety. The safety plan includes topics such as organization and responsibilities, documentation requirements, or verification and validation activities [4].

On developing a SafetyLon user-application, various characteristics of SafetyLon must be taken into consideration:

- The functionality of the safety-related firmware (see Figure 47.6 and for details Section 47.5), especially of the application layer interface
- Safe and non-safe network variables on the same node (design requirements as mentioned in Section 47.2)
- Configuration of the safety-related network via a single management unit (see Section 47.6)

The coding and compilation of a SafetyLon user-application is performed by using regular non-safety-related tools. However, the source code of the user-application written in ANSI C must adhere to detailed programming guidelines as mentioned in [5]. The binary received after the compilation process is linked to the safety-related firmware and the output is another binary. Such a binary is flashed into a node and must reside unchanged in the node. A change of the binary in the field is explicitly not allowed and requires safe tools for development.

After making the node ready for operation, they are installed in the field and standard tools are used to setup a network. The next step is to configure the nodes, so that a safety-related application can be performed. The following parameters are required to configure SafetyLon.

- Safe address (node): Each node is assigned a safe address to prevent a node from masquerade. The safe address identifies a SafetyLon uniquely so that parameters are sent to the intended node.
- Binding* parameters: This allows exchange of a safety-related data.
- SafetyLon user-application identifier: Each user-application gets a unique ID. According to the ID it is possible to get detailed information on the functionality and the set of safety-related data points to be configured.

The parameters are transferred to the corresponding node in a three-step process:

1. The management unit sends a request to a node. The node returns its safe address and a transaction ID. The management unit verifies the safe address.
2. The management unit uses the safe address and the transaction to send the configuration parameter to the node. The node stores the configuration parameter temporarily, reads the data back and returns it to the management unit.
3. The management unit checks if the sent and received data are equal. If so, it sends a request to the node to store the parameter permanently and use it for communication purpose.

Following a successful configuration (idle mode as shown in Figure 47.1), the nodes have to be commissioned in order to get them ready for performing safety-related applications [3]. First, the configuration parameters are validated. This refers to checking if the nodes used in the safety-related application perform as expected (test mode). In case of a successful validation, the user has to confirm it (pre-run and wait mode). Next, the nodes are waiting to be activated. Such a mode is implemented to allow a coordinated start-up of a network. In run mode, the node performs its functionality according to the safety-related firmware and SafetyLon user-application. In that mode, the safety-related data transfer among the nodes is handled.

The mode is left in two cases: first, if a safety-critical failure has occurred. Then the safe mode is entered. In case of a recoverable failure, the node leaves the safe mode, otherwise the node remains in that state until an operator checks the failure cause.

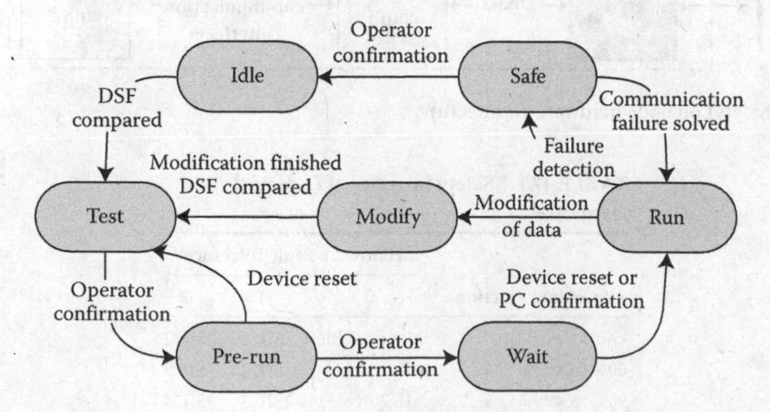

FIGURE 47.1 Safety-related overall state diagram.

* Binding means that a logic connection between two or among many network variables is established.

Maintenance (e.g., removing a safe address of a network variable) and decommissioning of the node is performed when the node is in modify mode. The node is set to that mode explicitly by the management unit. The same three-step process as described before on outlining the configuration process is used.

Such detailed processes for configuration, commissioning, maintenance, and decommissioning are required and several steps have to be authorized by a user because the tool running at the management unit is a non-safety-related tool. The advantage is that a standard PC with COTS software can be used.

47.4 The Hardware

Node hardware designed for a safety-related automation technology has to follow special design rules. It must be designed in a way to satisfy the requirements caused by a SIL defined in IEC 61508 [1]. As a design goal of SafetyLon is to reach a SIL3 compliant system, a two-channel architecture (1oo2) was chosen for the node hardware additional to the EN 14908 chip (Neuron chip or LC3020) as illustrated in Figure 47.2 [6].

The term 1oo2 (speak: one out of two) indicates that two independent channels perform the same actions and finally compare the result. In case of SafetyLon, the 1oo2 is realized by two hardware channels. The result of both channels must be equal that an action such as setting an output can take place. In case of a mismatch of both channels, the system has to take predefined actions such as discarding a message or entering a safe state.

IEC 61508 [7] defines different ways of reaching a SIL 3 compliant system as shown in Table 47.1. There is a trade-off possible between the safe failure fraction (i.e., number of failures not resulting in a dangerous situation) and the hardware architecture.

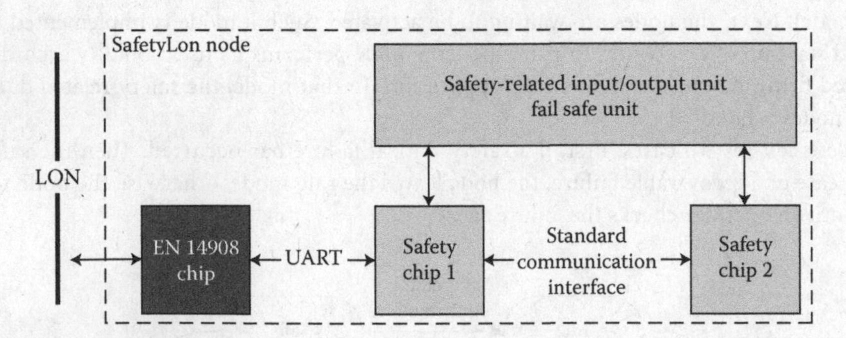

FIGURE 47.2 SafetyLon node hardware architecture.

TABLE 47.1 Safety Integrity of Deployed Hardware

	Hardware Fault Tolerance[a]		
Safe Failure Fraction	0	1	2
<60%	Not possible	SIL 1	SIL 2
60% to <90%	SIL 1	SIL 2	SIL 3
90% to <99%	SIL 2	SIL 3	SIL 4
≥99%	SIL 3	SIL 4	SIL 4

[a] A hardware fault tolerance of N denotes that $N + 1$ faults cause a loss of the safety status of the system.

In detail, a hardware fault tolerance of 1 results in a required safe failure fraction of 90%–99%. The safe failure fraction (SFF) defines the percentage of failures that do not result in a dangerous situation. It is defined as follows:

$$SFF = \frac{\lambda_{SD} + \lambda_{SU} + \lambda_{DD}}{\lambda_{SD} + \lambda_{SU} + \lambda_{DD} + \lambda_{DU}} \quad (47.1)$$

where

λ_{SD} is the safe detected failure
λ_{SU} is the safe undetected failure
λ_{DD} is the dangerous detected failure
λ_{DU} is the dangerous undetected failure

Hence, there must be a trade-off between the quality of the hardware to reduce the inherent risk and the effort performed to detect dangerous failures. As defined in the formula, λ_{DU} reduces the SFF, whereas dangerous failures that can lead into a hazardous state can be used to increase the SFF if they are detected. Hence, it is reasonable to perform hardware self tests (i.e., detecting dangerous faults that therefore result in non-dangerous failures) in order to increase the SFF. The impacts of the above described facts result in extensive testing of all components such as input and output, or volatile and nonvolatile memory [8]. In the following, an example is given on how to design and test the safe input and output logic.

In a standard, non-safe environment a digital input is simply connected to the evaluating device. Connecting a digital IO directly to a microcontroller is often not suitable for safety-related devices. First, the hardware inside the device must be protected against hazardous effects from outside, therefore mostly a galvanic isolation, an optocoupler, is used. Second, additional components are added to setup a testable input. The input is tested only if it is active, which means that a defined voltage is applied to the input terminal. It is not necessary to test an input that is in an inactive state because it is per definition the safe state.

Figure 47.3 depicts the principles of a testable input; several elements were omitted in the picture in order to focus on the functionality. In addition to the first optocoupler, a second one is added. For testing purpose, this device is able to switch off the signal transmission to the microcontrollers for a very short period of time. This interruption of the input signal must be detected by the microcontroller.

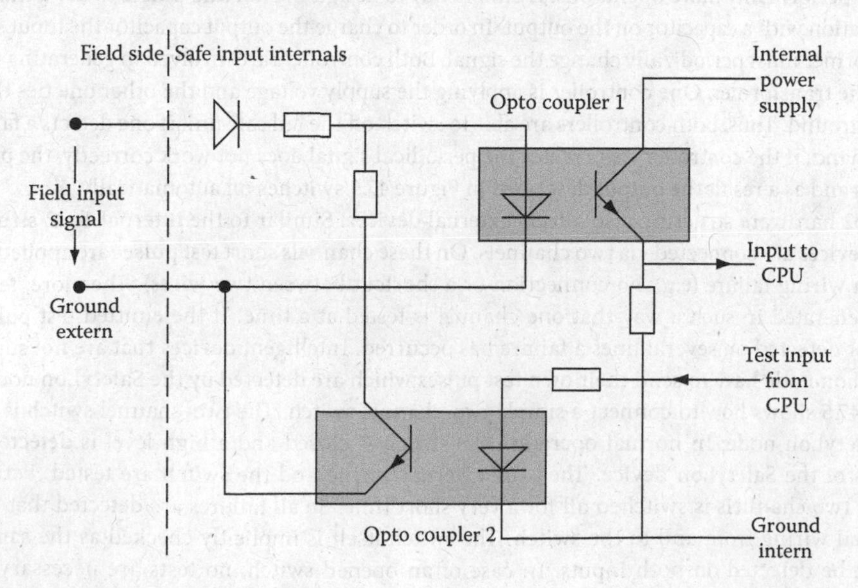

FIGURE 47.3 Schematic of safety-related input.

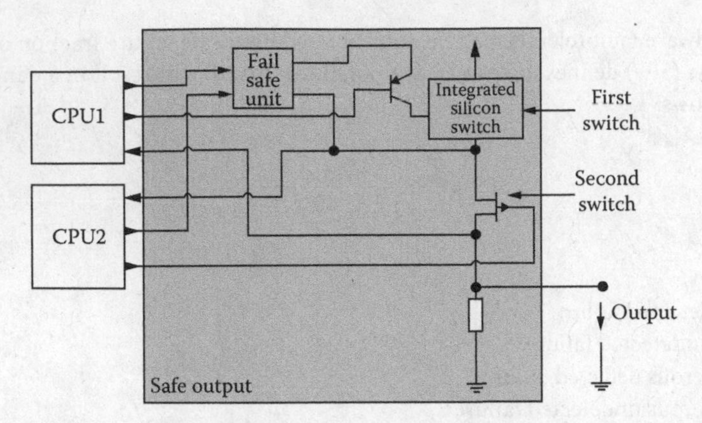

FIGURE 47.4 Schematic of safety-related output.

By performing this procedure, the microcontroller is able to check the switching capability of the input optocoupler and the functional connectivity of all components.

According to the two-channel architecture, the output stage is split into two switching elements as shown in Figure 47.4. Only if both stages switch on, the output is activated. It is preferable that both switching elements use different technologies in order to avoid common cause failures. Hence, the decision to switch on must be done from both channels together. Only then the output is switched on. For testing purpose the state of the two stages is read back and additional in one state the output is switched off for a short period of time in order to test the switching capability. Obviously, the off-time must be shorter than the reaction time of the switched element.

Even if there are two stages that are able to switch off, there is the possibility that both channels are not able to perform any actions but are stuck at on state (e.g., due to a common cause of power supply failure). To overcome that static issue, a frequently triggered control element is added. The fail safe unit only enables its output if it is permanently triggered from the controlling device. The output of the fail safe unit supplies the two switching elements of the output stage. Hence, if the fail safe unit is not triggered anymore, there is no supply control voltage for the switching elements and the output is switched off. So it is guaranteed that the output enters the safe state if both control channels are stuck at one state and are not able to perform any more operations. A simple way to design the fail safe unit is to use a transformer in combination with a capacitor on the output. In order to charge the output capacitor the input signal for the transformer must periodically change the signal. Both controllers are involved in generating the input signal of the transformer. One controller is applying the supply voltage and the other one ties the transformer to ground. Thus, both controllers are able to switch off the fail safe unit if one detects a failure. On the other hand, if the controller that creates the periodical signal does not work correctly, the periodical signal fails and as a result the output described in Figure 47.5 switches off automatically.

The 1oo2 hardware structure also affects external devices. Similar to the internal 1oo2 structure all external devices are connected via two channels. On these channels short test pulses are applied in order to detect a wiring failure (e.g., no connection or a shortcut between two wires). Therefore, test pulses must be generated in such a way that one channel is tested at a time. If the emitted test pulse is not detected or detected on several lines a failure has occurred. Intelligent devices that are not supplied by the SafetyLon node have to send their own test pulses which are detected by the SafetyLon node.

Figure 47.5 shows how to connect a simple two-channel switch. The two-channel switch is supplied by the SafetyLon node. In normal operation the switch is closed and a high level is detected on the input pins of the SafetyLon device. Then the external wiring and the switch are tested. Periodically, one of the two channels is switched off for a very short time. So all failures are detected that occur on the external wiring from and to the switch. The switch itself is implicitly checked as the same signal level must be detected on both inputs. In case of an opened switch, no tests are necessary because low signals on both inputs are defined as the safe state. In that state, the SafetyLon device has to enter

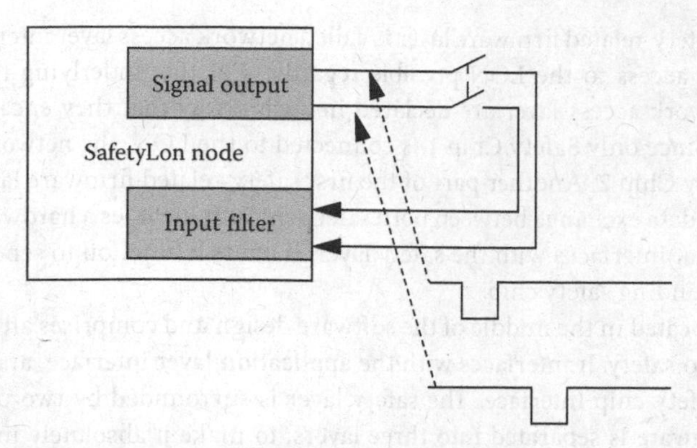

FIGURE 47.5 Switch connected to safety-related output.

a safe state. The activities to perform when entering the safe state depend on the defined application; but it is mandatory that there must be no direct danger to life or indirect to the environment.

47.5 The Safety-Related Firmware

The firmware of a SafetyLon node is organized in a three-layered architecture and located above ISO/OSI layer 7—equal to the approach chosen in PROFIsafe—as shown in Figure 47.6. Since only Safety Chip 1 is connected to the EN 14908 chip (cf. Figure 47.2), the lower layer differs between the chips. Shortstack API and Safety Chip Orion Stack API, respectively, are third-party ISO/OSI layer 7 software located below the first layer of the firmware. They care for the transmission of data from and to the chosen EN 14908 chip. Their software is not part of safety considerations. They are already part of the "black channel" and hence not outlined in the following.

On top of the firmware is the application layer interface. It offers the application programmer functions that are used to realize safe user-application software. Typical functions are sending and receiving functions, receiving value from the safety-related input or setting the safety-related output, or write and read access to the sensor and actuator data stored in a table being part of the safe software. The application layer interface provides a convenient way of programming safety-related user-applications without taking care of safety functions encapsulated in the safety layer.

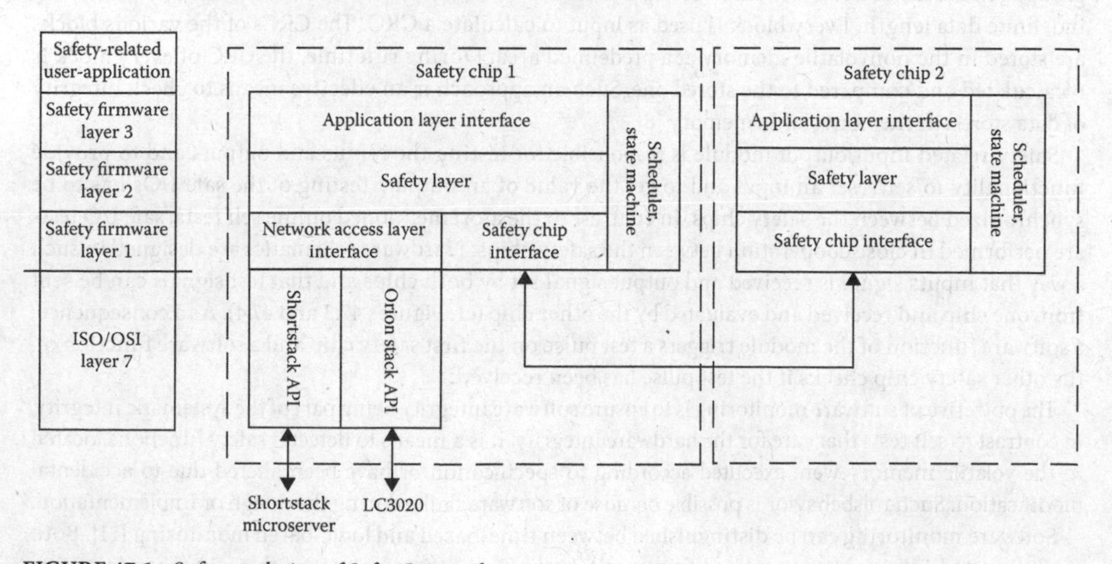

FIGURE 47.6 Software design of SafetyLon node.

A part of the first safety-related firmware layer is called network access layer interface. It is an abstraction layer that makes access to the LON possible regardless of the underlying third-party software. Functions of the network access layer are declared in such a way that they encapsulate functions of the third-party API. Since only Safety Chip 1 is connected to the LON, the network access layer is not implemented on Safety Chip 2. Another part of the first safety-related firmware layer is the safety chip interface that handles data exchange between both safety chips. It includes a hardware dependent driver and a software API that interfaces with the safety layer. It offers a function to send data to and receive data from the corresponding safety chip.

The safety layer is located in the middle of the software design and comprises all software functionality directly referring to safety. It interfaces with the application layer interface, and the network access layer interface and safety chip interface. The safety layer is surrounded by two other layers; in other words, the safety firmware is separated into three layers, to make it absolutely independent from the third-party software. Second, the third layer is specified to hide safety functionality from the application programmer. Such a layer eases programming and avoids misuse of safety functions since it must be assumed that application programmers are not familiar with details of the firmware functionality.

The safety firmware layer 2 consists of multiple parts related to safety, i.e., so-called primary functions: online self test module, safety-related input/output module, software monitoring, or the SafetyLon protocol stack. Other parts are supporting the desired functionality called supporting functions like the Safety Chip interface. Albeit not part of layer 2, the scheduler and state machine, and the safety chip interface are also supporting functions.

The online self test module includes online tests that are executed to guarantee a high integrity of the hardware by revealing faults in the different parts of the hardware. Tests are separated into volatile memory (RAM), nonvolatile read only memory (FLASH), and CPU tests. In [8], implementation examples are presented. In [9] different test algorithms are outlined.

In general, volatile memory test algorithms differ in test effort and diagnostic coverage. A high test effort and a high diagnostic coverage is ensured when using the galloping pattern test, a low one when implementing the marching bit test [10]. The level of the diagnostic coverage depends on the faults revealed by the test. Test with a high-level detect faults according to the DC-fault model [1] others only detect stuck-at faults.

Tests of the nonvolatile read only memory rely on parity bits, checksums or CRCs whereas diagnostic coverage of the first is low and the last is high. In case of a SafetyLon node, the nonvolatile memory is grouped in blocks of 256 bytes since CRC polynomials do not guarantee a defined level of integrity for indefinite data length. Every block is used as input to calculate a CRC. The CRCs of the various blocks are stored in the nonvolatile memory at a predefined area. During run time, the CRC of every block is recalculated and compared to the stored one. Such an approach is an effective means to check integrity of data stored in the nonvolatile memory [8].

Safety-related input/output module is responsible for testing the inputs and outputs and to provide functionality to set/reset an input and to get the value of an output. Testing of the safe I/Os has to be synchronized between the safety chips. In contrast to the aforementioned online self tests, safe I/O tests are performed in close cooperation between the safety chips. Hardware schematics are designed in such a way that inputs signal is received and output signal set by both chips and that test signals can be sent from one chip and received and evaluated by the other chip (cf. Figures 47.3 and 47.4). As a consequence, a software function of the module triggers a test pulse on the first safety chip and a software function on the other safety chip checks if the test pulse has been received.

The objective of software monitoring is to ensure software integrity being part of the systematic integrity, in contrast to self tests that care for the hardware integrity. It is a means to detect if safety functions located in the volatile memory were executed according to specification or have been altered due to accidental modification. Such misbehavior is possible because of software faults during the design or implementation.

Software monitoring can be distinguished between time-based and logic-based monitoring [11]. Both are integrated into the safety-related firmware. The first type uses a timer with an independent time base. Typically, such a timer is called watchdog timer and realized in hardware. So software functions are monitored by measuring the execution time. After completion of a function the watchdog is reset by a software command. If the execution takes too long, the watchdog is triggered and predefined actions are taken. The type of monitoring is integrated to check if the system is blocked or modified in a way, so that execution takes much longer than expected.

Logic-based monitoring is used to check if functions have not been bypassed. Therefore, a counter is implemented that is increased every time the function has been executed. Such a counter is available for every safety function. The counter values are exchanged within fixed periods of time between the safety chips to detect a fault in the firmware. If the counter values are not equal on both safety chips, predefined actions are taken.

The SafetyLon protocol stack incorporates functionality to send and receive sensor/actuator data in a safe way. Additionally, it supports network management activities such as configuring a node and allows time synchronization [12]. The message structure used for the different tasks is shown in Figure 47.7.

Every safe message starts with a specific header called ID. It specifies the message type (data or command message) and the data length *n*. The ID follows a 3 byte address field. It includes the safe source address of a node. Every safe node holds a table with a list of valid sources. The safe address (of network variables that is different from a safe node address mentioned in Section 47.3) guarantees that only safe devices (valid sources) can exchange safety-related messages. The ID and safe address prevent that unsafe messages look like safe messages.

The next field consists of the upper 2 bytes of the timestamp in the first part of the message and the lower 2 bytes in the second part of the message. By checking the timestamp at the destination, a delay, repetition, wrong sequence, and, in conjunction with safe addresses, an insertion of messages is avoided.

In addition, for detecting data corruption during transmission, two CRCs with different generator polynomials are used. In case of a payload smaller than 8 bytes a 1-byte CRC, otherwise a 2-byte CRC is appended. The CRCs and the comparison of the duplicated message parts (i.e., ID, safe address, and data) finally satisfy the requirements for a safe data transmission sufficiently.

In the case of sending a sensor value, each safety chip builds message part 1 and message part 2 and calculates the CRC. Safety Chip 1 receives the complete message from the other chip and compares the whole message. If the CRCs and the two message parts are identical, the message is sent, otherwise discarded. Consequently, faulty messages due to a node internal failure are not sent. That avoids wastage of bandwidth and saves computational resources on receiver side since it need not process the faulty message.

On the receiver side, the message is forwarded from Safety Chip 1 to Safety Chip 2 and processed by both (two-channel structure): first, the CRC is checked in order to verify integrity; second, the timestamp is used to check for insertion, repetition, and wrong sequence of a message; third, the payload field is compared bit by bit to detect other integrity failures not being revealed by the CRC. Results on the checks are exchanged between both safety chips. Only if both agree on a positive result, the payload is released for further processing for example by the user-application software.

Safety functions must be called and executed on a regular base. For that reason, a scheduler and a state machine are included in the firmware. To avoid computational overhead and to ease the integration of safety requirements, no commercial operating system is used. However, a static scheduling mechanism is realized with a fixed cycle time and a static sequence of functions, i.e., a single-task scheduling. Such an approach first of all ensures a deterministic timing behavior. It guarantees that test pulses are sent or the RAM test is executed in fixed time intervals that cannot be ensured for example by

ID	Safe address	Time stamp MSWord	Safety-related data: *n* byte	CRC a	ID	Safe address	Time stamp LSWord	Safety-related data: *n* byte	CRC b

FIGURE 47.7 SafetyLon message structure.

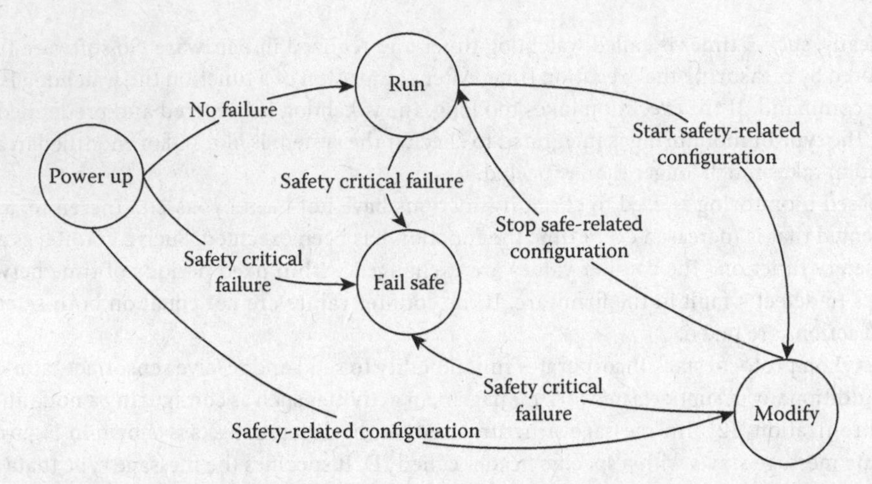

FIGURE 47.8 Node state machine.

the earliest-deadline-first scheduling mechanism [13]. Second, static scheduling eases synchronization between safety chips mandatory for the close cooperation between the safety chips. Both chips start at the same time with the execution of function in the same order.

The state machine controls the behavior of the node. According to inputs received, it decides to which state to switch to. In a SafetyLon node four states are specified as illustrated in Figure 47.8. After a reset the node is in POWER UP state and runs through the start-up procedure. For example, the hardware is tested, the hardware interfaces are initialized and configuration parameters are copied from the flash memory to the RAM due to performance reasons. In case of no error, the node enters the RUN state where the node is operating. If safety-critical failures are detected, the node switches to FAIL SAFE state. In this state, the functionality of the node is limited to a minimum that does not jeopardize safety. The fail safe state is only left when the critical fault was eliminated by an operator. MODIFY state is used to configure the node (e.g., to make a safe binding). In this state, the node provides only such functionality necessary to execute configuration requests and send the responses.

47.6 The SafetyLon Tools

SafetyLon provides two types of tools: a development tool called application builder that gives the user the possibility to configure the user-application (e.g., specifying the safe and non-safe network variable types). Moreover, a SafetyLon management tool makes the safety-related commissioning, configuration of the firmware and user-application, maintenance of the safety-related application, and decommissioning of nodes or removal of bindings possible.

The development tool is a non-safe tool that gets a script file as input (see Figure 47.9) which among LonWorks specific parameters includes

- The type of EN 14908 controller
- The name and type of non-safe network variables and the name and type of safe network variables

Additionally, the application builder interfaces with the LonMark device resource API where the format and size of the network variables is taken from.

The output of the application builder are plain text c- and header-files containing all information required to run the user-application as intended (e.g., a file with the network variable table that holds the SafetyLon message as well as the safety-related payload of the message). The files are

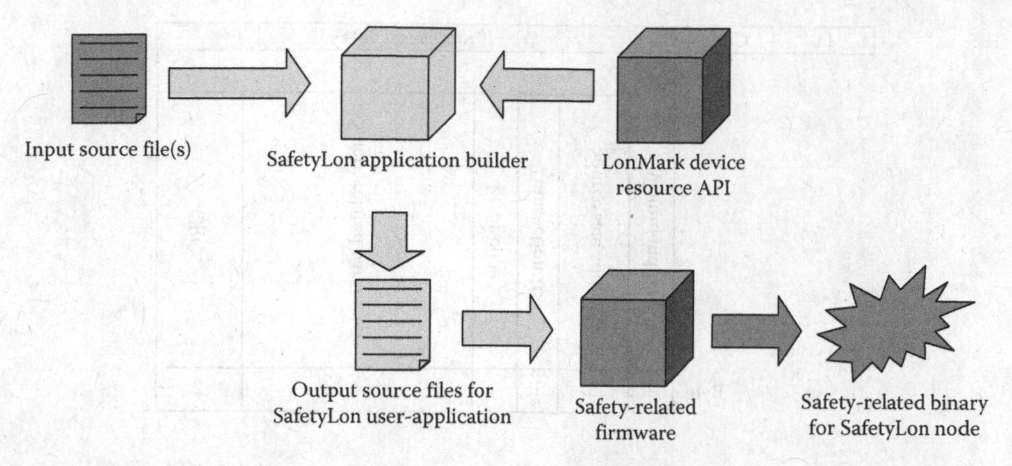

FIGURE 47.9 Application builder.

linked to the user-application and the safety-related firmware. Since a non-safe tool is used, it is not allowed to upload all the user-application remotely as safety integrity cannot be ensured with a high probability.

The SafetyLon management tool is divided into a SafetyLon application (SLA) interfacing with the local network service (LNS) platform [14]. Additionally, a SafetyLon library (SLL) is used to store all SafetyLon specific data and provide SafetyLon functionality as shown in Figure 47.10.

SLA provides services to configure one or several safe nodes, depending on the network system design. The application includes an interface to the SLL and it is instantiating the SLL within the initialization process. With functions used from this library, the application configures the safe nodes. The user-interface shows the operator the resources of every safe node that are changeable. To obtain information from devices, the application uses the interface to the LNS Object Server API. It is required because the SLA is a standard LNS plug-in.

The SLA instantiates the LNS Object Server and then it is passed through to the SLL together with at least LNS network and system object during initialization. As a consequence, both software parts interact with the same instance of LNS Object Server. The event-sink for handling LNS events is implemented in the SLA. Thus, the changes in the LNS Object Server event handling should not influence the functionality of the SLL.

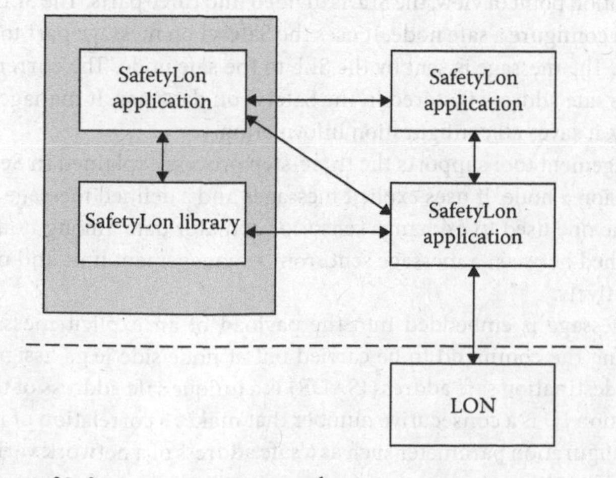

FIGURE 47.10 Structure of SafetyLon management tool.

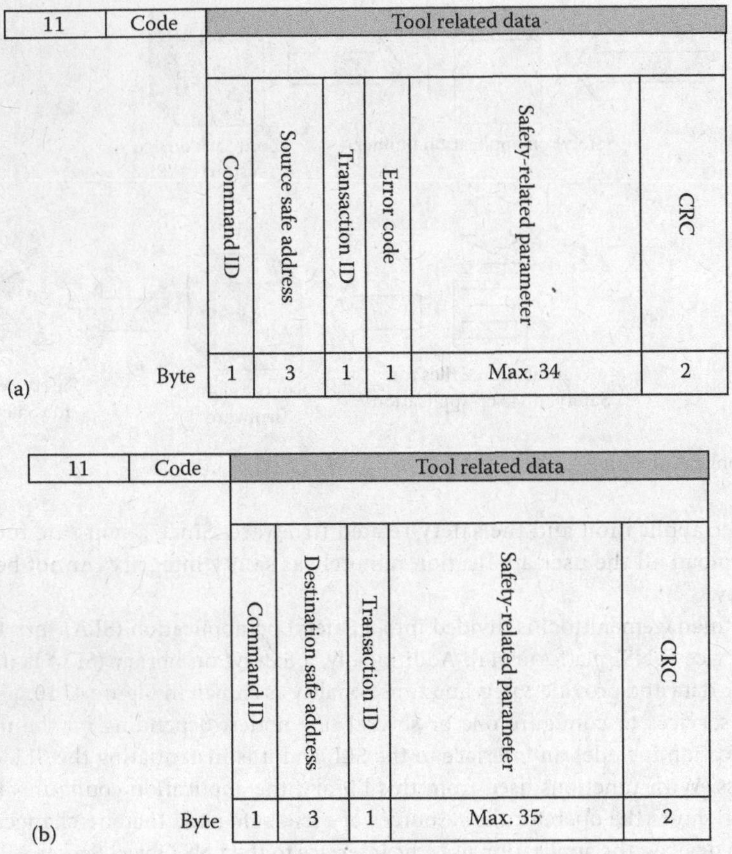

11	Code	Tool related data					
		Command ID	Source safe address	Transaction ID	Error code	Safety-related parameter	CRC
	Byte	1	3	1	1	Max. 34	2

(a)

11	Code	Tool related data				
		Command ID	Destination safe address	Transaction ID	Safety-related parameter	CRC
	Byte	1	3	1	Max. 35	2

(b)

FIGURE 47.11 (a) Request message of SafetyLon management tool. (b) Response message of SafetyLon management tool.

The SLL software architecture consists of two interfaces and one application. An API handles the connection to LNS Object Server, another API is the interface for the SLA to the SLL. The handling of LNS Object Server is described in the Echelon guidelines of the LNS application developer's kit [15].

From the implementation point of view, the SLL is divided into three parts. The SLL is used by the client. It provides functions to configure a safe node. It uses the SafetyLon message part to build and check the management messages. The message is sent by the SLL to the safe node. The current configuration of a safe node including the safe address is stored in the SafetyLon database. It manages and stores the safe addresses. Additionally, it saves all configuration information.

The SafetyLon management tool supports the three-step process explained in Section 47.3 applied to configure and commission a node. It uses explicit messages and a defined message structure (cf. Figure 47.11) different from the one used to exchange sensor or actuator data among nodes. Therefore, it can be explicitly distinguished between a message sent from a management unit and messages sent from a SafetyLon node, respectively.

The management message is embedded into the payload of an explicit message. It starts with a Command ID specifying the command to be carried out at node side (e.g., assign a safe address to a network variable). The destination safe address (SADR) is a unique safe address of the node that shall be addressed. The transaction ID is a consecutive number that makes a correlation of request and response possible. Data are a configuration parameter such as a safe address of a network variable. CRC16, finally, is a 2 byte CRC appended to the message to ensure data integrity. In case of a response message, an additional byte is included: an error byte that is set by the node (e.g., the network variable is not available).

Acronyms

1oo2	One out of two
COTS	Commercial off the shelf
CRC	Cyclic redundancy check
LNS	Local network service
LON	Local operating network
LSWord	Leas significant word
MSWord	Most significant word
SADR	Safe address
SFF	Safe failure fraction
SIL	Safety integrity level
SLA	SafetyLon application
SLL	SafetyLon library
DU	Dangerous undetected
DD	Dangerous detected
SD	Safe detected
SU	Safe undetected

References

1. International Electrotechnical Commission. IEC 61508—Functional safety of electric/electronic/programmable electronic safety-related systems. IEC, Geneva, Switzerland, 1998.
2. European Norm. EN 14908—Open data communication in building automation, controls and building management—control network protocol. CEN, 2005.
3. T. Novak, P. Fischer, M. Holz, M. Kieviet, and T. Tamandl. Safe commissioning and maintenance process for a safe system. In *Proceedings of the Seventh IEEE International Workshop on Factory Communication Systems*, Dresden, Germany, pp. 225–232, 2008.
4. Institute of Electrical and Electronics Engineers Computer Society. IEEE Standard for Software Safety Plans. IEEE Std. 1228, 1994.
5. C. Hatton. *Safer C—Developing Software for High-Integrity and Safety-Critical Systems*. McGraw-Hill Book Company Europe, Berkshire, U.K., 1995.
6. T. Novak and T. Tamandl. Architecture of a safe node for a fieldbus system. In *Proceedings of the Fifth IEEE International Conference on Industrial Informatics*, Vol. 1, Vienna, Austria, pp. 101–106, 2007.
7. International Electrotechnical Commission. IEC 61508—Functional safety of electric/electronic/programmable electronic safety-related systems—Part 2: Requirements for electrical/electronic/programmable electronic safety-related systems. IEC, Geneva, Switzerland, 2000.
8. T. Tamandl and P. Preininger. Online self tests for microcontrollers in safety related systems. In *Proceedings of the Fifth IEEE International Conference on Industrial Informatics*, Vol. 1, Vienna, Austria, pp. 137–142, 2007.
9. P. Wratil and M. Kieviet. *Sicherheitstechnik für Komponenten und Systeme*. Hüthig Verlag, Heidelberg, Germany, p. 203, 2007.
10. H. Hölscher and J. Rader. *Microcomputers in Safety Technique, an Aid to Orientation for Developer and Manufacturer*. TÜV Rheinland, Cologne, Germany, p. 50, 1986, Chapter 7.
11. H. Hölscher, J. Rader. *Microcomputers in Safety Technique, an Aid to Orientation for Developer and Manufacturer*. TÜV Rheinland, Cologne, Germany, p. 86, 1986, Chapter 7.
12. T. Novak and B. Sevcik. Network time synchronization in a safe automation network. In *Proceedings of the Seventh IEEE International Workshop on Factory Communication Systems*, Dresden, Germany, pp. 305–313, 2008.

13. W. Wolf. *Computers as Components, Principles of Embedded Computing System Design*. Morgan Kaufman Publishers, San Francisco, CA, pp. 377, 2001.
14. P. Fischer, M. Holz, and M. Mentzel. Network management for a safe communication in an unsafe environment. In *Proceedings of the Fifth IEEE International Conference on Industrial Informatics*, Vol. 1, Vienna, Austria, pp. 131–136, 2007.
15. Echelon Corporation, LNS programmer's guide, Turbo edition, Echelon, San Jose, CA, 2004.

48

Wireless Local Area Networks

Henning Trsek
Ostwestfalen-Lippe University of Applied Sciences

Juergen Jasperneite
Ostwestfalen-Lippe University of Applied Sciences

Lucia Lo Bello
University of Catania

Milos Manic
University of Idaho Idaho Falls

48.1 Introduction 48-1
48.2 The 802.11 Family 48-2
48.3 Physical Layer 48-2
Frequency Bands • Modulation Techniques
48.4 Medium Access Control 48-3
Distributed Coordination Function • Point Coordination Function Enhanced Distributed Channel Access • HCF Controlled Channel Access • Direct Link Protocol and Block ACK
48.5 Limitations of DCF and HCF for QoS Support in Industrial Environments 48-8
48.6 Security Mechanisms 48-9
48.7 Fast Handover 48-9
Mechanisms on the AP Side • Mechanisms on the Client Side
48.8 Future Enhancements 48-11
References 48-11

48.1 Introduction

Many industrial automation applications are designed to be very flexible in order to manufacture products with increased efficiency. Thus, a flexible communication infrastructure is required, leading to a growing demand for wireless networks. Wireless networks fulfill these requirements because they offer more flexibility, cost reductions, and higher mobility to the automation system as compared to their wired counterparts.

Since communication in future industrial automation systems will most likely be realized by real-time Ethernet protocols like PROFINET [IEC06a,IEC06b], Ethernet/IP [Eth09], etc. (cf. Chapter 37 of this book), the corresponding wireless technology at the field level has to be suitable and carefully chosen with respect to the application requirements, in terms of both dependable and temporal behavior. Therefore, IEEE 802.11 [IEE07] wireless local area networks (WLANs) are an interesting option because they were initially designed to be the wireless extension of Ethernet [IEE05]. Some interesting application scenarios are overhead monorail systems or automated guided vehicles (AGVs) that can be found in the area of logistics. For instance, AGVs are mobile transport systems that can operate autonomously. They carry loads and use either fork lifts or a conveyor system to move the objects to be transported. Basically, they are used to deliver work pieces to specific manufacturing processes and have to be quite flexible by definition. In a typical scenario, the mobile system consists of a local programmable logic controller (PLC) to control the vehicle. The local PLC also communicates with the central controller that is responsible for overall coordination and task assignment. The tasks are then independently executed by the autonomous systems. The requirements are determined by the velocity and the number of vehicles in the entire system.

TABLE 48.1 WLAN Technology Overview

Technology	Release Date	Spectrum (GHz)	Max Data Rate (Mbps)	Outdoor Range (m)
IEEE 802.11a	Oct. 1999	5.0	54	35
IEEE 802.11b	Oct. 1999	2.4	11	42
IEEE 802.11g	June 2003	2.4	54	42

48.2 The 802.11 Family

The first wireless standard of the IEEE 802 family was the 802.11 one, approved by the IEEE in 1997, and thus referred to as the 802.11-1997 [IEE97]. This standard, approved in 1999, is now obsolete. The most popular standards in the 802.11 family are those defined by the 802.11a, 802.11b, and 802.11g protocols as amendments to the original 802.11-1997 standard. They are summarized in Table 48.1.

Several important amendments to the original standard followed, such as the 802.11e [IEE05e] defining a set of quality of service (QoS) enhancements to support delay-sensitive applications, the 802.11i [IEE04i] introducing security enhancements, such as key management and distribution, encryption, and authentication, and 802.11n adding multiple-input multiple-output (MIMO) and other newer features. Other standards in the family (c–f, h, j) are service amendments and extensions or corrections to previous specifications.

The current IEEE 802.11-2007 standard [IEE07] is a single document that merges eight amendments (802.11a, b, d, e.g., h, i, j) with the 1999 version of the 802.11 standard [IEE97]. However, two new amendments to this version, which are 802.11k for radio resource management [IEE08k] and 802.11r for a fast basic service set (BSS) transition [IEE08r], have been ratified in the meantime, and others are already available as first drafts.

48.3 Physical Layer

The physical layer of the original 802.11 [IEE97] standardized three wireless data-exchange techniques. They were the infrared (IR), the frequency-hopping spread spectrum (FHSS), and the direct sequence spread spectrum (DSSS). The physical layer in 802.11 is split into the Physical Layer Convergence Protocol (PLCP) and the Physical Medium Dependent (PMD) sub-layers. The PLCP prepares/parses data units transmitted/received using various 802.11 media access techniques. The PMD performs data transmission/reception and modulation/demodulation directly accessing air under the guidance of the PLCP.

48.3.1 Frequency Bands

The original intention of the internationally designated Industrial, Scientific, and Medical (ISM) radio bands was to generate and to use local, unlicensed Radio Frequency (RF) electromagnetic fields for industrial, scientific, medical, and domestic purposes excluding communications.

The International Telecommunication Union (ITU) extends the original intention to radio communication services operating within these bands that must accept harmful interferences that may be caused by ISM applications [ITU09b]. The definition, as well as the precise frequency ranges, center frequencies, and the availability are defined by the ITU-Recommendations (ITU-R) on radio regulations: ITU-R 5.138, 5.150, 5.280, and 15.13 [ITU09a].

Although typically associated with the microwave oven operating frequency of 2.45 GHz, ISM bands have also been shared with license-free, error-tolerant communications applications, such as wireless LANs (IEEE 802.11b/g at 2.45 GHz) [IEE07], Bluetooth (IEEE 802.15.1 at 2.45 GHz) [IEE05a], ZigBee (IEEE 802.15.4 at 2.45 GHz) [IEE06], and cordless phone communications operating at 900 MHz, 2.45 GHz, 5.8 GHz, and other frequencies.

The a, b, and g encoding protocols of working group 11 of the IEEE 802 wireless LAN standards committee use the ISM bands (802.11a operates at 5 GHz, while 802.11b/g operates in the unlicensed 2.45 GHz band, and 802.11n at 2.45 and 5 GHz). This implies that 802.11b/g/n, operating in an unregulated frequency band can incur interference from microwave ovens, cordless phones, and other appliances using the same 2.4 GHz range. Therefore, coexistence has to be considered in any case, and is of major importance for the successful deployment of WLANs.

48.3.2 Modulation Techniques

Modulation techniques are techniques typically used in telecommunications to transmit a message. A high frequency periodic (sinusoid) waveform is used as a carrier signal, with amplitude, phase, and frequency modulation. The receiving unit performing the inverse operation of modulation is known as the demodulator. The modem (Modulator-Demodulator) is capable of performing both operations.

Wireless communication protocols use various modulation techniques. While Bluetooth and the early 802.11 LANs use the Frequency-Hopping Spread Spectrum signaling method (FHSS), 802.11b and ZigBee 802.15.4 use Direct Sequence Spread Spectrum (DSSS) signaling. The 802.11a uses the Orthogonal Frequency-Division Multiplexing (OFDM) system.

Frequency hopping was the first step in the evolution to DSSS and other data transmission techniques. The idea was to transmit via a predefined frequency hopping pattern known to both the transmitter and the receiver. The 802.11 frequency hopping separates the whole 2.4 GHz ISM band into 1 MHz-spaced channels. The transmitter has to change channels at least 2.5 times per second (every 400 ms or less). This allows dealing with high energy interference in a narrow band, as well as the mutual interference of two FHSS transmitters positioned close to each other.

With Direct Sequence Spread Spectrum (DSSS) signaling, the carrier signals occur over the full bandwidth (spectrum) of a device's transmitting frequency. The data signal at the sending station is combined with a higher data rate bit sequence, or a chipping code that divides the user data according to a spreading ratio. The chipping code is a redundant bit pattern for each bit that is transmitted, which increases the signal's resistance to interference. If one or more bits in the pattern are damaged during transmission, the original data can be recovered due to the redundancy of the transmission.

Complementary Code Keying (CCK) is used in conjunction with DSSS technology. CCK is a set of 64 8-bit code words used to encode data for 5.5 and 11 Mbps data rates in the 2.4 GHz band of 802.11b wireless networking. The code words have unique mathematical properties that allow them to be correctly distinguished from one another by a receiver even in the presence of substantial noise and multipath interference. CCK applies sophisticated mathematical formulas to DSSS codes, permitting the codes to represent a greater volume of information per clock cycle (11 Mbps of data rather than the 2 Mbps in the original standard). CCK does not work with FHSS.

Orthogonal frequency-division multiplexing (OFDM) is a Frequency-Division Multiplexing (FDM) scheme utilized as a digital multicarrier modulation method. OFDM splits the radio signal into multiple smaller sub-signals that are then transmitted simultaneously at different frequencies to the receiver. OFDM, therefore, reduces the amount of cross talk in signal transmissions. OFDM is used in the 802.11a/g, the European alternative for the IEEE 80211 High Performance Radio LAN (HIPERLAN/2), and in 802.16 (WiMAX).

48.4 Medium Access Control

According to the original IEEE 802.11 MAC protocol [IEE97], the architecture of the MAC sub-layer includes a mandatory Distributed Coordination Function (DCF) and an optional Point Coordination Function (PCF). Moreover, due to the limitations of this architecture for transmitting time-critical traffic flows, real-time enhancements were introduced. They are addressed by the standard amendment IEEE 802.11e [IEE05e]. It provides advanced QoS capabilities by adding the hybrid coordination function (HCF), which defines two new access mechanisms. The overall MAC architecture is shown in Figure 48.1.

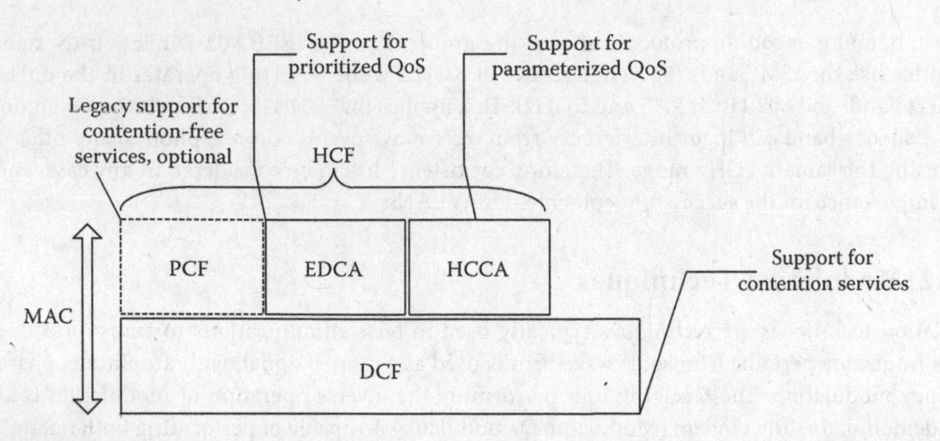

FIGURE 48.1 MAC architecture with HCF.

The DCF is totally distributed and can be used within both ad hoc and infrastructure network configurations. Conversely, the PCF is centralized and can be used only in infrastructure network configurations. In the DCF, each station senses the shared channel and transmits when it finds a free channel according to a carrier-sense mechanism. In the PCF, a coordinator station polls the other nodes, thus enabling them to transmit in a collision-free way. The Enhanced Distributed Channel Access (EDCA) mechanism is based on DCF medium access, except for the ability to assign different priorities to various kinds of traffic flows. HCF-Controlled Channel Access (HCCA) defines a parameterized QoS support. It also relies on a polling procedure with an underlying time-division multiple access (TDMA) principle but is more flexible when compared to the legacy of the PCF.

Currently, the most widely used channel access mechanisms are the DCF and the EDCA, while the PCF and the HCCA have received little attention so far, especially where commercial products are concerned, because the available chipsets still lack support for these mechanisms. In the following, the four different access methods are outlined.

48.4.1 Distributed Coordination Function

The IEEE 802.11 DCF operating mode is based on Carrier-Sense Multiple Access with a Collision Avoidance (CSMA/CA) protocol and on a random backoff time following a busy medium condition. The random backoff time is intended to reduce the collision probability between multiple stations accessing a shared medium at the point where collisions would most likely occur. The highest probability of a collision exists just after the medium becomes idle following a busy medium because multiple stations could have been waiting for the medium to become available again. This is the situation in which the random backoff procedure comes into action to resolve medium contention conflicts.

Before starting transmission, a node listens to the channel for a time called, a Distributed Interframe Space (DIFS), to assess whether the channel is idle or not. If the channel is idle, each node generates a random backoff interval in order to reduce the probability of collisions with other nodes trying to access the sensed idle channel at the same time. Each node decreases its backoff counter as long as the wireless channel is sensed to be idle. If the counter has not reached zero, and the channel becomes busy again, the backoff counter is frozen and reloaded as soon as the channel is sensed to be idle again for a DIFS.

When the backoff interval is over, and if the channel is still idle, the transmission starts. The random backoff interval, expressed as a number of time slots, is generated in the set {0,CW – 1}, where CW denotes the contention window size. The initial value of the contention window is *CWmin.* In the case of an unsuccessful transmission (due to collisions or losses), the CW is doubled up to a maximum value *CWmax.*

After experiencing a maximum number of collisions, a packet is dropped. In the case of a successful transmission, the CW value is reset to *CWmin* before the random backoff interval is selected.

According to the IEEE 802.11 MAC protocol, an explicit acknowledgment (ACK) frame has to be sent by the receiver to notify the transmitter of the successful reception of a data frame. The time interval between the reception of a data frame and the transmission of the relevant ACK is defined as the Short Interframe Space (SIFS). This small gap between transmissions gives priority to the ongoing frame-exchange sequence, by preventing other nodes that have to wait for the medium to be idle for a longer time interval (e.g., at least for the DIFS time), from accessing the channel.

Physical and virtual carrier-sense functions are used to determine the state of the medium. When either function indicates a busy medium, the medium is considered busy, otherwise it is considered idle. The physical carrier-sense mechanism is provided by the 802.11 PHY, while the virtual carrier-sense mechanism is provided by the MAC and is referred to as the Network Allocation Vector (NAV). The NAV maintains a prediction of future traffic on the medium based on the duration information that is either announced in Request to Send/Clear to Send (RTS/CTS) frames prior to the actual data exchange or can be used by the PCF and the HCF.

The RTS/CTS mechanism is very effective in reducing the length of the frames involved in the contention process. In fact, assuming perfect channel-sensing by every station, collisions may only occur when two or more stations start RTS transmission within the same time slot. If both sources employ the RTS/CTS mechanism (the decision to use RTS/CTS depends on the packet length, as will be explained subsequently), collisions would only occur while transmitting the RTS frames and would promptly be detected by the source lacking the CTS responses. The RTS/CTS therefore, significantly lowers the temporal overhead of a collision (i.e., collisions are much shorter in time). This feature is beneficial for time-constrained traffic.

However, the RTS/CTS mechanism has some drawbacks as the additional RTS and CTS frames add a protocol overhead and thus reduce protocol efficiency especially for short data frames. For this reason, each station can be configured to use the RTS/CTS mechanism either always, or never, or only on frames longer than a specified length.

As the typical control traffic exchanged on the factory floor consists of short frames, the use of the RTS/CTS mechanism to transmit this kind of traffic is not advisable. In addition, the RTS/CTS mechanism cannot be used for broadcast and multicast transmissions, as there are multiple recipients for the RTS, and thus potentially, multiple concurrent senders of the CTS in response.

The RTS/CTS mechanism can be successfully exploited to cope with the *hidden node problem*, which arises when there are nodes that are out of range of some other nodes belonging to the same wireless network, thus creating a potential for collisions with transmissions from those nodes. Before sending a packet, the transmitter sends an RTS frame and waits for a CTS frame from the NAV.

The reception of a CTS frame notifies the transmitter that the channel is clear in the receiver area, so the receiver is able to receive the packet. Any other node in the receiver area will hear the CTS even if it cannot hear the RTS and will avoid accessing the channel after hearing the CTS, even if its carrier sense mechanism indicates that the channel is idle. The source is only allowed to transmit the data packet if the CTS frame is correctly received within a duration called the CTS Timeout. More details on the DCF function can be found in [IEE97].

48.4.2 Point Coordination Function

The PCF access method uses a point coordinator (PC), acting as the polling master, which determines which station currently has the right to transmit. The operation of the PCF may require additional coordination that is not specified in the standard [IEE97], to permit efficient operation in cases where multiple point-coordinated networks are operating on the same channel, and to share an overlapping physical space.

The PCF uses a virtual carrier-sense mechanism aided by an access priority mechanism. The PCF distributes information using Beacon management frames to gain control of the medium by setting the NAV in the stations. In addition, all frame transmissions under the PCF may use a Point Interframe Space (PIFS) that is smaller than the DIFS for frames transmitted via the DCF. This means that point-coordinated traffic has priority in accessing the medium over stations in overlapping networks operating in the DCF mode. Such an access priority may be utilized to realize a contention-free (CF) access method. The PC controls the frame transmissions of the stations, so as to eliminate contention for a limited period of time. Although the ability of providing support for collision-free transmissions would be very beneficial to industrial communication, especially when handling time-constrained traffic, the PCF is implemented only in very few hardware devices, as it is not included in the Wi-Fi Alliance's [Wi09] interoperability standard.

48.4.3 Enhanced Distributed Channel Access

The EDCA is an extension of the DCF and defines eight different priority levels. The priorities are mapped into four access categories (AC) in compliance with the IEEE 802.1D standard [IEE04], thereby providing differentiated and distributed channel access. Within a wireless station with QoS support (QSTA) and the Access Point (AP) with QoS support (QAP), every AC is represented as an independent transmission queue. Every single queue contends for the medium separately and has different parameter sets for accessing the channel.

The parameter set consists of the arbitration interframe space number (AIFSN), the two bounding values for the contention window *CWmin*[*AC*] and *CWmax*[*AC*], and the maximum allowed transmission time for one station $TXOP_{limit}$. The AIFSN depends on the AC and is used to derive the arbitration interframe space (AIFS) with Equation 48.1, where *aSlotTime* and *SIFS* are determined by the used PHY layer (e.g., 9 and 16 μs for 802.11g).

$$\text{AIFS[AC]} = \text{AIFSN[AC]} * \text{aSlotTime} + \text{SIFS} \tag{48.1}$$

Similar to the DCF, a station always has to wait an AIFS before it can contend for the medium. The backoff procedure is also similar to the DCF, except for different values for *CWmin* and *CWmax*. The contention for the medium with different priorities is shown in Figure 48.2. As a result, the EDCA parameter sets of the four ACs cause different waiting times, i.e., a decreased waiting time for high priority frames and a longer waiting time for low priority frames. Therefore, the probability of the successful data transmission of high priority traffic is increased.

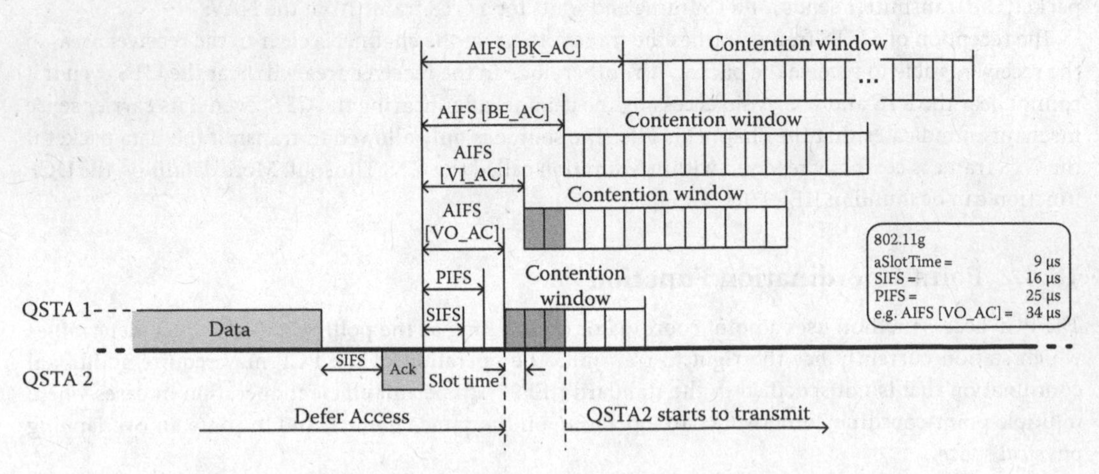

FIGURE 48.2 Channel access with different priorities.

48.4.4 HCF Controlled Channel Access

Contention-free medium access in 802.11e [IEE05e] is realized with HCF Controlled Channel Access (HCCA). HCCA is the replacement of the previously defined PCF and used for controlled channel access. The time between two consecutive beacon frames is called a *superframe*. It is divided into an optional contention-free period (CFP) and a contention period (CP). During the CFP, the hybrid coordinator (HC) controls the access to the channel by polling its associated stations with QoS requirements. However, the HC is also allowed to initiate a controlled access phase (CAP) during the CP after detecting that the channel is idle for a time interval longer than a PIFS (PIFS = SIFS + aSlotTime), and whenever there is a need to transfer time-critical data. An example for a superframe is shown in Figure 48.3. A polled station is granted a transmission opportunity (TXOP) allowing the station to occupy the channel for a time period equal to the TXOP value. This concept elevates HCCA with greater flexibility than its predecessor, although the time for generating CAPs is limited to a maximum duration in order to leave space for stations operating under EDCA.

A very important concept for QoS support in 802.11e is the *admission control*. Whenever a station wants to associate with a certain BSS, it has to specify its requirements during a TSPEC negotiation as was introduced in the integrated services architecture [IET97]. The negotiation is done with a traffic specification element (TSPEC) that may contain parameters related to the time-critical traffic flow, such as the mean data rate or the delay bound: They are exchanged between the QAP and the QSTAs to establish a traffic stream (TS). The admission or the rejection of the new TS depends on the adherence of its requirements. Every station can have up to eight different TSs with different QoS requirements.

After the negotiation, the polling schedule for the stations is calculated by the HC based on the previously defined requirements. The ability of the HC to perform the scheduling depends on several mandatory parameters, such as the *Nominal MSDU Size*, the *Mean Data Rate*, the *delay bound*, etc. Although the applied algorithm for scheduling and admission control has been completely left open to the implementer, i.e., it can be adapted to the needs of specific applications and kinds of traffic flows, a sample scheduler can be found in Annex K of the 802.11e [IEE05e].

48.4.5 Direct Link Protocol and Block ACK

In order to decrease the protocol overhead, the 802.11e [IEE05e] also defines a way to directly communicate with other clients in infrastructure networks and to acknowledge more than one frame with a block acknowledgment. The direct link setup (DLS) enables two stations belonging to the same BSS to communicate directly to each other without directing the frames through the AP. Before the transmission starts, the direct link has to be set up via a request to the AP that is forwarded to the intended other station.

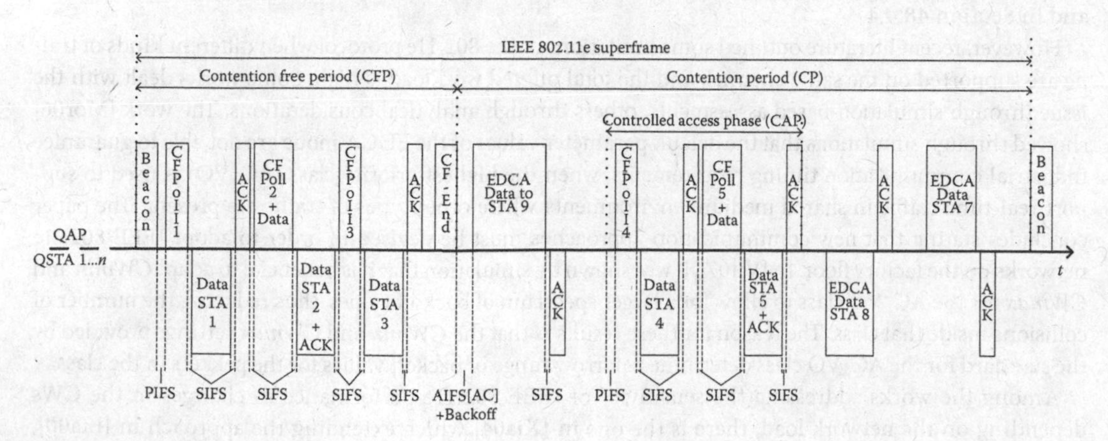

FIGURE 48.3 Example of an 802.11e superframe.

The *block acknowledgment* is also an optional feature that increases the throughput efficiency. With this option enabled, a station is allowed to transmit several frames within one TXOP. All transmitted frames form a single block that is acknowledged by only one block acknowledgment frame in the end, leading to a reduction of the necessary control message exchanges.

48.5 Limitations of DCF and HCF for QoS Support in Industrial Environments

The DCF mechanism suffers from various limitations from the point of view of supporting the traffic flows typically found in industrial environments, for several reasons. First, as collisions may occur, the available bandwidth is lowered and the medium access time is nonpredictable. The lack of predictability represents a major problem when time-constrained traffic flows have to be handled, as there is no way to provide guarantees on deadline meeting, even in statistical terms. Moreover, there is a potential for *capture effect*, as once a station gains access to the channel, it may keep the channel busy for as long as it chooses. A way to prevent a station with heavy non-real-time traffic to monopolize the channel, thus significantly reducing the probability that real-time traffic will suffer from heavy contention and be significantly delayed due to non-real-time traffic, is wireless traffic smoothing. The traffic smoother is at the middleware level, located between the network layer (IP) and the wireless MAC that handles RT and NRT traffic differently, injecting them in the MAC layer at a different rate. An adaptive traffic smoother that implements the harmonic increase multiplicative decrease (HIMD) algorithm to regulate non-real-time traffic bursts before they are sent over the network, thus privileging real-time flows, is proposed in [Jai03]. In [LoB06], a Wireless Traffic Smoother (WTS) for industrial WLANs is presented and evaluated. The WTS, based on a fuzzy controller, is able to provide end-to-end soft real-time communications with low round-trip times. Both approaches, [Jai03] and [LoB06], do not entail any modification in the 802.11 protocol.

Many works have addressed the limitations of the DCF for real-time traffic in industrial and robotic applications. For example, in [San05], the performance of the 802.11b protocol using broadcast and unicast transmissions in different uncontrolled load scenarios was analyzed in the context of real-time communication between mobile robots. The paper concludes that as long as the non-real-time traffic sharing the channel with the real-time traffic is not excessively bursty, broadcast transmissions outperform unicast ones in terms of packet losses. However, when the non-real-time traffic features very bursty patterns, the broadcast reliability significantly degrades.

The lack of mechanisms to provide QoS guarantees is another limitation of the IEEE 802.11 standard. To solve this issue, IEEE (Task Group E) published the IEEE 802.11e standard [IEE05e] as an amendment to the original 802.11 standard intended to enhance the MAC support for applications with QoS requirements. The proposed mechanisms of this standard were discussed in Section 48.4.3 and in Section 48.4.4.

However, recent literature outlined some limitations of the 802.11e protocol when different kinds of traffic are supported on the same channel, and the total offered workload is high. Some works dealt with the issue through simulation-based assessments, others through analytical considerations. The work [Mor06] showed through simulations that the default parameter values of the EDCA mode are not able to guarantee industrial communication timing requirements, when the highest priority class (AC_VO) is used to support real-time traffic in shared medium environments where other types of traffic are present. The paper concludes stating that new communication approaches must be devised in order to adopt IEEE 802.11e networks on the factory floor. In [Vit07], it was shown by simulation that it is beneficial to adapt *CWmin* and *CWmax* for the AC_VO class to allow for a larger spectrum of backoff values, thus reducing the number of collisions inside that class. The reason for these results is that the *CWmin* and *CWmax* settings provided by the standard for the AC_VO class determine a narrow range of backoff values for the packets in the class.

Among the works addressing the sensitivity of IEEE 802.11e performance to changes in the CWs depending on the network load, there is the one in [Xia04] which extending the approach in [Bia00], models the EDCA protocol by means of three-dimensional Markov chains and analyzes the network performance with a varying CW size. Another relevant work is [Kon04], which uses three-dimensional Markov chains to analyze the behavior of the different ACs when both the CWs and the AIFS are varied. Among the approaches proposed in the literature to tune the CW, there are the Adaptive EDCF (AEDCF) proposed in [Nao05], the Adaptive EDCA (AEDCA) proposed in [Rom03], and the adaptive technique presented in [Vit08]. While the AEDCA and AEDCF approaches do not provide for changing the values of *CWmin* and *CWmax*, but simply choose the best one in a suitable calculated range, the work [Vit08] proposes an adaptive technique to increase the channel-access probability of the highest priority AC in a general industrial scenario, using a fuzzy controller to dynamically find the most appropriate CW range, on the basis of the observed network conditions.

In [TJK06], the performance of the 802.11e channel access mechanisms HCCA, using the proposed simple scheduler, and EDCA in industrial automation systems were compared and evaluated. Even though it was found that HCCA is superior to EDCA in scenarios with a large number of stations, improving and adapting the HCCA scheduling algorithm would lead to much better results, which has also been proven in several works, [CLM07], [GMN03].

48.6 Security Mechanisms

Although advantageous to local area networks (LANs), WLANs in industrial environments introduce additional security challenges. Compromising of industrial WLANs can vary from costly downtimes and decreased system performance, to more serious consequences, such as industrial espionage, physical infrastructure damage, and even loss of human lives.

Some of the typical WLAN security issues entail authentication, access control, encryption, jamming, interception, and hijacking. Security mechanisms include various encryption protocols, such are Wired Equivalent Privacy (WEP), WiFi Protected Access (WPA and WPA2), Temporal Key Integrity Protocol (TKIP), Counter Model with CBC-MAC Protocol (CCMP), and WLAN Authentication and Privacy Infrastructure (WAPI).

Although deprecated, WEP (based on Rivest Cipher 4 [RC4] encryption algorithm) is still in wide use. WPA and WPA2 protocols introduced improved security relative to the TKIP-based WEP approach by introducing a larger key length and a new Advanced Encryption Standard (AES) based algorithm known as the Counter Mode with Cipher Block Chaining Message Authentication Code Protocol (CCMP). WEP2 also introduced TKIP. TKIP was designed with a goal of replacing the WEP protocol allowing legacy hardware to remain in use. TKIP introduced multiple master keys, a unique RC4 for each frame generated by a maser key, a new integrity check hashing algorithm, etc. CCMP was designed with the goal of replacing both TKIP and WEP.

WLAN Authentication and Privacy Infrastructure (WAPI) is the Chinese National Standard for the Wireless LAN based on the Authentication Service Unit (ASU).

Following the guidelines for a secure WLAN deployment, common security measures (e.g., signal strength limitation, MAC address filtering, SSID broadcast disabling, and using the most recent encryption protocols) are highly recommended and should not be ignored.

Similarly to WLAN, the ubiquitous wireless personal area networks (WPAN) formed of Bluetooth, Zigbee, and mesh networks suffer from certain security issues. While these issues include those of encryption and authentication, they also relate to concerns relative to the "discovery mode," "pairing," and elements of eavesdropping and location tracking.

48.7 Fast Handover

Especially when it comes to widely distributed industrial systems, a WLAN has to consist of more than one AP to cover the whole area. Hence, special attention has to be paid to the *handover* procedure between APs, because it causes a connection disruption that might not be tolerable for specific industrial real-time applications. Usually, the handover consists of four different phases: the search phase, the

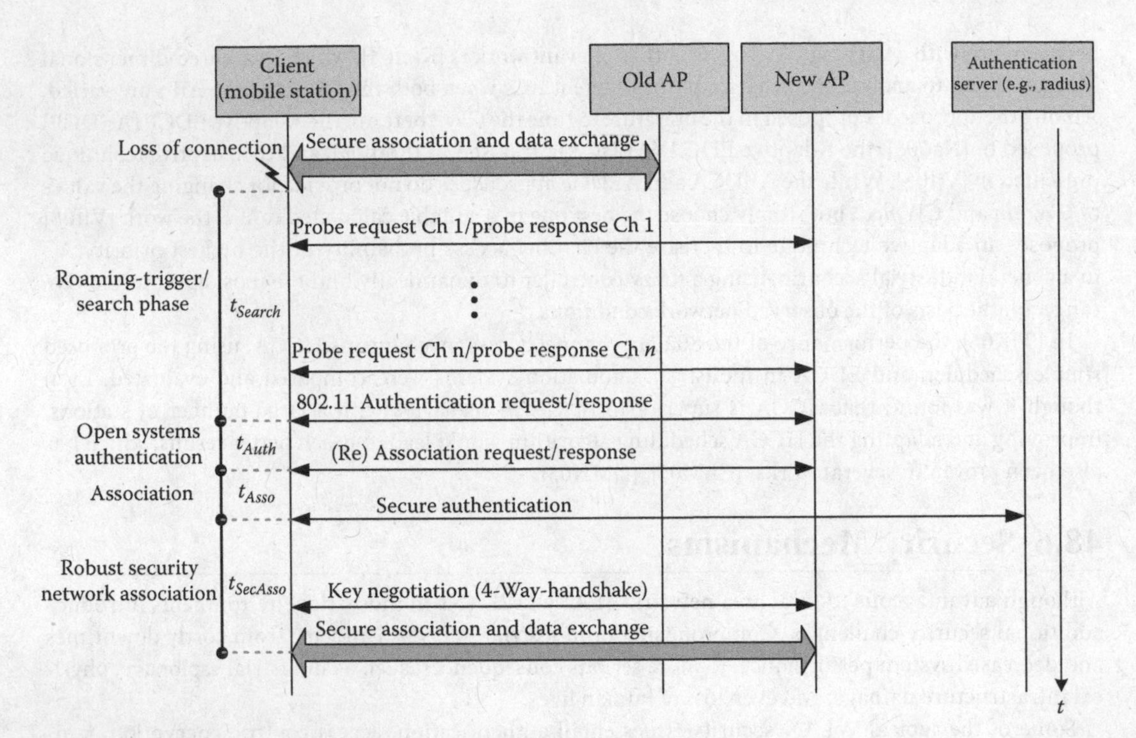

FIGURE 48.4 Handover procedure in 802.11 WLANs.

open authentication phase, the association phase, and the *robust security network association* (RSNA) phase as shown in Figure 48.4. First, the connection loss will be detected and the search phase starts by means of scanning all channels for available APs. Whenever a suitable new AP is found, the client starts the authentication frame exchange followed by the association. Finally, RSNA is established and data can be transmitted again using a secure connection.

In order to minimize the overall handover duration, either the search phase or RSNA can be optimized. RSNA depends mainly on the AP side, whereas the search phase is determined by the client side. In [PCK07], a good survey of recent works in the area of fast handover and currently achievable times is provided.

48.7.1 Mechanisms on the AP Side

The mechanisms on the AP side mainly address a reduction of RSNA. In order to reduce time-consuming authentication against an authentication server, the IEEE 802.11i [IEE04i] standard amendment specifies a *preauthentication* mechanism and allows a caching of the corresponding pair-wise master keys (PMKs). Preauthentication provides an option for the client to perform a full IEEE 802.1X authentication with other APs in range, while it is still associated to its current AP. The PMK caching feature allows both AP and the client to store the results of a first full 802.1X authentication. In other words, whenever the client roams back to an AP it had been previously associated to, only a four-way handshake is necessary to establish the temporal keys for encryption.

A further improvement is achieved by methods specified in the 802.11r standard amendment [IEE08r] for a fast BSS transition. Once a station joins the wireless network, a full 802.1X authentication is done with the result of a generated PMK that is then distributed to all APs within the same mobility domain (MD). Hence, when a station decides to handover to a new AP, PMK can be assumed to be already present. However, the important question of a feasible key distribution has been left open by the amendment. After this the four-way handshake and the resource reservation for QoS are embedded into the mandatory reassociation procedure. To sum up, three enhancements can be identified in the 802.11r amendment.

First of all, an elimination of the 802.1x key exchange as described previously. Second, the four-way handshake for session key establishment is integrated into the mandatory 802.11 open systems authentication. And third, QoS resource requests are also exchanged before the reassociation with the new AP starts.

48.7.2 Mechanisms on the Client Side

The client-side mechanisms mostly optimize the search phase. It is desirable to know the set of possible new APs a priori to get rid of the time-consuming scanning process during the handover. Background scanning is a very promising solution to this problem. It eliminates the channel scanning delay completely because a station actively probes available channels during the normal data exchange. That is, as long as no data has to be transferred, the client continuously probes new channels for potential APs and gets back immediately to the current one. If a handover is necessary, all information about neighboring cells is already present.

Moreover, an additional radio interface can also be used on the client side to both passively monitor the channels and to gather information about neighboring cells. This approach is called the dual radio client and has a second advantage. It also allows a seamless handover using both radio interfaces for data communication. The data transfer is initiated with radio interface A in the first BSS. Whenever the signal quality of radio interface A decreases, the second radio interface B starts to find a new AP and gets associated to it. As soon as the signal quality of radio interface A is below a certain threshold, the data communication will be directed to radio interface B, which is already connected to a new AP, and no service interruption occurs.

48.8 Future Enhancements

Some of the latest additions to 802.11 is the IEEE 802.11n amendment [IEE08n], which aims at increasing the throughput (from 20 Mbps up to approx. 200 Mbps data throughput), as well as extending the range of reception (through reducing signal fading) in comparison to the legacy 802.11a/g standards. One of the most important concepts to obtain these advantages is MIMO. Several MIMO features are supported by the 802.11n amendment, including transmitter beamforming, spatial division multiplexing (SDM), and space time block coding (STBC). Furthermore, using an advanced coding with low density parity check codes (LDPC) and applying a channel bonding to increase the channel bandwidth to 40 MHz is possible. All characteristics mentioned above allow data rates up to 600 Mbps to be specified in 802.11n, which is roughly 10 times more as compared to 802.11a/g.

Moreover, the Task Group v is currently working on a proposal for a wireless network management framework, which is contained in the 802.11v draft [IEE08v]. Wireless network management mainly deals with implementing system-wide functionalities, such as load management among different cells and coordinating radio properties between neighboring APs. The load management algorithms basically comprise load balancing and admission control approaches. The management can be centralized, decentralized, or hybrid depending on the targeted architecture of the network. However, the centralized architecture is most commonly deployed and also addressed by an Internet Engineering Task Force (IETF) working group. The working group specifies an interoperable protocol called control and provisioning of wireless access points (CAPWAP) [IET06]. It aims at reducing the complexity of managing a large numbers of APs and mainly encompasses administrational tasks to ease operation, maintenance, and configuration, and for centralizing client management.

References

[Bia00] G. Bianchi, Performance analysis of the IEEE 802.11 distributed coordination function, *IEEE Journal on Selected Areas in Communications*, 18(3):535–547, March 2000.

[CLM07] C. Cicconetti, L. Lenzini, E. Mingozzi, and G. Stea, Design and performance analysis of the real-time HCCA scheduler for IEEE 802.11e WLANs, *ACM Computer Networks*, 51(9):2311–2325, 2007.

[Eth09] Ethernet/IP. Ethernet industrial protocol (EtherNet/IP). Ethernet/IP, February 2009, www.ethernet-ip.org

[GMN03] A. Grilo, M. Macedo, and M. Nunes, A scheduling algorithm for QoS support in IEEE802.11e networks, *IEEE Wireless Communications*, 10(3):36–43, 2003.

[IEC06a] IEC, IEC 61158-5-10, Digital data communications for measurement and control—Fieldbus for use in industrial control systems—Parts 5–10: Application layer service definition—Type 10 elements, IEC, Geneva, Switzerland, 2006.

[IEC06b] IEC, IEC 61158-6-10, Digital data communications for measurement and control—Fieldbus for use in industrial control systems—Parts 6–10: Application layer service definition—Type 10 elements, IEC, Geneva, Switzerland, 2006.

[IEE97] IEEE Standard for Information Technology—Telecommunications and information exchange between systems—Local and metropolitan area networks—Specific requirements—Part 11: Wireless LAN medium access control (MAC) and physical layer (PHY) specifications, IEEE Std. 802.11-1997, pp. 1–445, November 1997.

[IEE04] IEEE Standard for local and metropolitan area networks media access control (MAC) Bridges. IEEE Std. 802.1D-2004, pp. 1–269, June 2004.

[IEE04i] IEEE Standard for Information Technology—Telecommunications and information exchange between systems—Local and metropolitan area networks—Specific requirements—Part 11: Wireless LAN medium access control (MAC) and physical layer (PHY) specifications amendment 6: Medium access control (MAC) security enhancements. IEEE Std. 802.11i-2004 (Amendment to IEEE Std. 802.11, pp. 1–175, 1999 Edition (Reaff 2003).

[IEE05] IEEE standard for Information Technology—Telecommunications and information exchange between systems—Local and metropolitan area networks—Specific requirements—Part 3: CSMA/CD access method and physical layer specification, IEEE Std. 802.3, June 2005.

[IEE05a] IEEE Standard for Information Technology—Telecommunications and information exchange between systems—Local and metropolitan area networks—Specific requirements—Part 15.1: Wireless medium access control (MAC) and physical layer (PHY) specifications for WPANs, IEEE Std. 802.15.1-2005, June 2005.

[IEE05e] IEEE Standard for Information Technology—Telecommunications and information exchange between systems—Local and metropolitan area networks—Specific requirements Part 11: Wireless LAN medium access control (MAC) and physical layer (PHY) specifications amendment 8: Medium access control (MAC) quality of service enhancements, IEEE Std. 802.11e-2005 (Amendment to IEEE Std. 802.11, 1999 Edition (Reaff 2003), pp. 1–189, 2005.

[IEE06] IEEE Standard for Information Technology—Telecommunications and information exchange between systems—Local and metropolitan area networks—Specific requirements—Part 15.4: Wireless medium access control (MAC) and physical layer (PHY) specifications for low-rate WPANs, IEEE Std. 802.15.4-2006, September 2006.

[IEE07] IEEE Standard for Information Technology—Telecommunications and information exchange between systems—Local and metropolitan area networks—Specific requirements—Part 11: Wireless LAN medium access control (MAC) and physical layer (PHY) specifications, IEEE Std. 802.11-2007, C1–1184, June 2007.

[IEE08k] IEEE Standard for Information Technology—Telecommunications and information exchange between systems—Local and metropolitan area networks—Specific requirements—Part 11: Wireless LAN medium access control (MAC) and physical layer (PHY) specifications amendment 1: Radio resource measurement of wireless LANs, IEEE Std. 802.11k-2008 (Amendment to IEEE Std. 802.11, 2007).

[IEE08n] IEEE Draft Amendment for Standard for Information Technology—Telecommunications and information exchange between systems—Local and metropolitan area networks—Specific requirements—Part 11: Wireless LAN medium access control (MAC) and physical layer (PHY) specifications amendment 4: Enhancements for higher throughput. IEEE Std. 802.11n/D7.0, December 2008.

[IEE08r] IEEE Standard for Information Technology—Telecommunications and information exchange between systems—Local and metropolitan area networks—Specific requirements—Part 11: Wireless LAN medium access control (MAC) and physical layer (PHY) specifications amendment 2: Fast basic service set (BSS) Transition, IEEE Std. 802.11r-2008 (Amendment to IEEE Std. 802.11, 2007), 2007.

[IEE08v] IEEE Draft Amendment Standard for Information technology—Telecommunications and information exchange between systems—Local and metropolitan area networks—Specific requirements—Part 11: Wireless LAN medium access control (MAC) and physical layer (PHY) specifications wireless network management, IEEE Std. 802.11v/D3.0, June 2008.

[IET97] IETF Standard, Specification of guaranteed quality of service, IETF RFC 2212, 1997.

[IET06] IETF Network Working Group, RFC 4564, Objectives for control and provisioning of wireless access points (CAPWAP), IETF, http://tools.ietf.org/html/rfc4564, July 2006.

[ITU09a] International Telecommunication Union (ITU), Terrestrial services, ISM applications, ITU, URL from Jan. 09, http://www.itu.int/ITU-R/terrestrial/faq/index.html#g013

[ITU09b] International Telecommunication Union (ITU), ITU ISM Applications, URL from January 09, http://www.itu.int/net/home/index.aspx

[Jai03] A. Jain, D. Qiao, and K. G. Shin, RT-WLAN: A soft real-time extension to the ORiNOCO linux device driver, in *Proceedings the IEEE International Symposium on Personal, Indoor and Mobile Radio Communications* (*PIMRC 2003*), Beijing, China, September 2003.

[Kon04] Z. Kong, D. Tsang, B. Bensaou, and D. Gao, Performance analysis of IEEE 802.11e contention-based channel access, *IEEE Journal on Selected Areas in Communications*, 22(10):2095–2106, December 2004.

[LoB06] L. Lo Bello, G. Kaczynski, F. Sgro', O. Mirabella, A wireless traffic smoother for soft real-time communications over IEEE 802.11, in *Proceedings of ETFA06, the 11th IEEE International Conference on Emerging Technologies and Factory Automation*, Prague, Czech Republic, pp. 1073–1079, September 2006.

[Mor06] R. Moraes, P. Portugal, and F. Vasques, Simulation analysis of the IEEE 802.11e EDCA protocol for an industrially-relevant real-time communication scenario, in *IEEE Conference on Emerging Technologies and Factory Automation* (*ETFA'06*), Prague, Czech Republic, pp. 202–209, September 2006.

[Nao05] J. Naoum-Sawaya, B. Ghaddar, S. Khawam, H. Safa, H. Artail, and Z. Dawy, Adaptive approach for QoS support in IEEE 802.11e wireless LAN, *IEEE International Conference on Wireless And Mobile Computing, Networking and Communications*, (*WiMob'2005*), Vol. 2, Montreal, Canada, pp. 167–173, August 2005.

[PCK07] S. Pack, J. Choi, T. Kwon, and Y. Choi. Fast handoff support in IEEE 802.11 wireless networks, *IEEE Communications Surveys and Tutorials*, 9, 2–12, 2007.

[Rom03] L. Romdhani, Q. Ni, and T. Turletti, Adaptive EDCF: Enhanced service differentiation for IEEE 802.11 wireless ad-hoc networks, *IEEE Wireless Communications and Networking* (*WCNC 2003*), Vol. 2, New Orleans, LA, pp. 1373–1378, March 2003.

[San05] F. Santos and L. Almeida, On the effectiveness of IEEE 802.11 broadcasts for real-time communication, in *Proceedings of the Fourth International Workshop on Real-Time Networks* (*RTN 05*), *in Conjunction with the 17th Euromicro International Conference on Real-Time Systems*, Palma de Mallorca, Spain, July 2005.

[TJK06] H. Trsek, J. Jasperneite, and S. P. Karanam, A simulation case study of the new IEEE 802.11e HCCA mechanism in industrial wireless networks, in *11th IEEE International Conference on Emerging Technologies and Factory Automation* (*ETFA'06*), Prague, Czech Republic, September 2006.

[Vit07] S. Vittorio and L. Lo Bello, An approach to enhance the QoS support to real-time traffic on IEEE 802.11e networks, in *Sixth International WORKSHOP on Real Time Networks* (*RTN 07*), Pisa, Italy, 2007.

[Vit08] S. Vittorio, E. Toscano, and L. Lo Bello, CWFC: A contention window fuzzy controller for QoS support on IEEE 802.11e EDCA, in *Proceedings of the 13th IEEE International Conference on Emerging Technologies and Factory Automation* (ETFA'08), September 15–18, 2008, Hamburg, Germany, pp. 1193–1196, IEEE 2008.

[Wi09] Wi-Fi Alliance, Wi-Fi Alliance Certification, February 2009, http://wi-fi.org/

[Xia04] Y. Xiao, Performance analysis of IEEE 802.11e EDCF under saturation condition, in *2004 IEEE International Conference on Communications*, Vol. 1, Paris, France, pp. 170–174, June 2004.

49 Bluetooth

Stefan Mahlknecht
Vienna University of Technology

Milos Manic
University of Idaho Idaho Falls

Sajjad Ahmad Madani
COMSATS Institute of Information Technology

49.1 Introduction 49-1
History and Technical Background • Bluetooth Specifications
49.2 Bluetooth Core Architecture Blocks 49-3
Channel Manager • L2CAP Resource Manager • Device Manager • Link Manager • Baseband Resource Manager • Link Controller • Radio Frequency • Bluetooth Networks • Bluetooth Security
49.3 Bluetooth Protocol Stack 49-6
49.4 Bluetooth Profiles 49-7
Four Bluetooth General Profiles • Bluetooth General Profiles
49.5 Competitive Technologies 49-9
49.6 Future of the Bluetooth Technology: Challenges 49-10
References 49-11

49.1 Introduction

Bluetooth is the widely spread short-range wireless technology that has replaced existing cables in many fixed and mobile application scenarios with high measures of security and an enhanced data rate. Its key features are robustness, low power, small size, and low cost. Bluetooth technology is widely accepted now as it can be found in many cell phones and computer peripherals that can connect with any other such device residing in proximity.

Bluetooth specifications were developed by the Bluetooth Special Interest Group (SIG). The Bluetooth SIG consists of companies in the areas of telecommunication, computing, networking, and consumer electronics, which try to define new application profiles and enhancements to the communication protocols as new demands arise.

Interoperability between devices supporting the same, so-called Bluetooth profile, is guaranteed by a certification process of manufacturer devices functionality.

The basic strength of Bluetooth technology is to handle both voice and data transmissions, simultaneously. This helps the user to carry out various personal activities innovatively. For example, the user can be hands-free when listening to voice calls, and at the same time print, share data among peers, synchronize the PDA, the laptop, the camera, and other Bluetooth-enabled devices.

49.1.1 History and Technical Background

In 1994, Ericsson Mobile Communications was the first company to realize the need to cut off the cable between cellular phones and their accessories. The name Bluetooth comes from *Harald Blåtand*, the King of Denmark (and later, the King of Norway), known for trying to unite and Christianize his dominions. A Scandinavian name was chosen because of Ericsson's large contribution to the development of Bluetooth.

In 1998, Ericsson, Nokia, IBM, Toshiba, and Intel founded the Bluetooth Special Interest Group (SIG) to work out a standard and to promote the technology. The first standard was published as Version 1.0a in July 1999. Version 1.0b followed in December of the same year. Only in February 2001, was the standard of Version V1.1 presented, which was considered the first reliable standard on which companies could build products in line with the market, as the previous standards had significant flaws and inaccuracies. Of late, the standard has come to be known as the IEEE 802.15.1 short-range communication.

Bluetooth devices communicate as short-range devices (1–100 m) in the license-free ISM band at 2.4–2.4835 GHz. It uses a radio technology called the frequency-hopping spread spectrum. The initial data rate was specified at 1 Mbps, but the newest enhanced data rate specifications achieve up to 3 Mbps. Bluetooth defines not only a radio but also a protocol stack with profiles. Profiles define the capabilities and the usage scenarios devices have to implement in order to be interoperable with other devices supporting the same profile. With certification required for any Bluetooth device, interoperability shall be guaranteed. However, as it has been shown in practice, consumers have also experienced many problems with certified devices in the past.

49.1.2 Bluetooth Specifications

The specification process is an ongoing process where features are improved while continuing to maintain backward compatibility with previous specifications.

49.1.2.1 Bluetooth 1.0 and 1.0B

These initial versions never got out into the market in significant numbers as the releases had several problems for vendors in developing interoperable devices. A mandatory Bluetooth device hardware address (BD_ADDR) was included while connecting the devices, which proved to be a set back for various Bluetooth planned services.

49.1.2.2 Bluetooth 1.1

Various errors were removed from the previous release. The Bluetooth 1.1 short-range technology was standardized as the IEEE 802.15.1-2002. Additionally, features of support for the nonencrypted channels and the Received Signal Strength Indicator (RSSI) were added.

49.1.2.3 Bluetooth 1.2

Users of Bluetooth 1.1 were confronted with interference problems with Wireless LAN devices and experienced long connection set-up times that had to be reduced. The key features of the release are as follows:

- An adaptive frequency-hopping spread spectrum, to avoid using occupied frequency channels
- A higher effective data throughput, up to 721 kbps
- Extended Synchronous Connections (eSCO) that increase latency for concurrent audio data transfer, improve the voice quality of audio links, and allow retransmissions of corrupted packets
- The previous standard was upgraded as the IEEE 802.15.1-2005

49.1.2.4 Bluetooth 2.0 + EDR

Besides some modifications and fixing of errors in the previous release, this specification introduced the optional enhanced data rate (EDR) for faster data transfer, which was 3 Mbps. The EDR provides the following features:

- Three times faster transmission speed, up to (2.1 Mbps); hence, high quality audio links were possible
- Backward compatibility to the V1.x specifications by allowing simultaneous EDR and non-EDR connections
- Lower power consumption due to a reduced duty-cycle capability

49.1.2.5 Bluetooth 2.1 + EDR

The additional features are the added security features and the improved latency for encrypted channels and power consumption, as well as an enhanced quality of service.

49.1.2.6 Bluetooth 3.0 + HS

Bluetooth 3.0 + HS (high speed) is the latest Bluetooth technology standard operating since April 22, 2009. The interesting fact is that it did not deprecate any existing Bluetooth technology-supported feature. Bluetooth 3.0 has significantly improved its speed, and now, it is no longer a mere dream to be able to wirelessly transfer a bulk of music libraries between a PC and a music player, to download a bulk of images from a camera to a PC, and to send video files from a cameraphone to a computer or a television. The said standard has incorporated the following major improvements:

- An 802.11 protocol adaptation layer—to offer compatibility with existing high-speed wireless networks, and which can now support up to a 24 Mbps data transfer rate
- Unicast connectionless data—a challenge for wireless communications media to support Bluetooth, instead of multicast data
- An enhanced retransmission and a streaming mode—to support real time services (audio and video) over a low-speed and a short-range network

49.2 Bluetooth Core Architecture Blocks

The Bluetooth core system covers the four lowest layers and the associated protocols defined by the Bluetooth specification, as well as one common service layer protocol, the service discovery protocol (SDP), and the overall profile requirements that are specified in the generic access profile (GAP). A complete Bluetooth application requires a number of additional services and higher layer protocols that are defined in the Bluetooth specification [1].

The common implementation of Bluetooth is shown in Figure 49.1. The three lowest layers are known as the Bluetooth controller and are typically found on a single chip, while the uppers are known as the Bluetooth host. In order to achieve interoperability between different Bluetooth device interfaces, the host controller interface (HCI) is specified, which uses a UART or a USB interface as the transport layer. The implementation of the protocols should conform to the following logical structure and not exactly what is shown in the diagram. These logical units work together as the core controllers of Bluetooth technology.

49.2.1 Channel Manager

The channel manager block has the following responsibilities:

- It uses the L2CAP protocol to interact with a remote device and to connect its end points to the appropriate entities.
- It interacts with its local link manager to create new logical links.
- It configures the above links for the provision of quality of service of the transported data.

49.2.2 L2CAP Resource Manager

This block is responsible for the following:

- Managing the ordering of the PDU fragments sent to the baseband.
- Maintaining some scheduling between channels for QoS. This is required as the Bluetooth controller does not have limitless buffering or an HCI, which is a pipe of infinite bandwidth.
- Offering policing services by the L2CAP resource manager to the submitted L2CAP SDUs against their QoS settings.

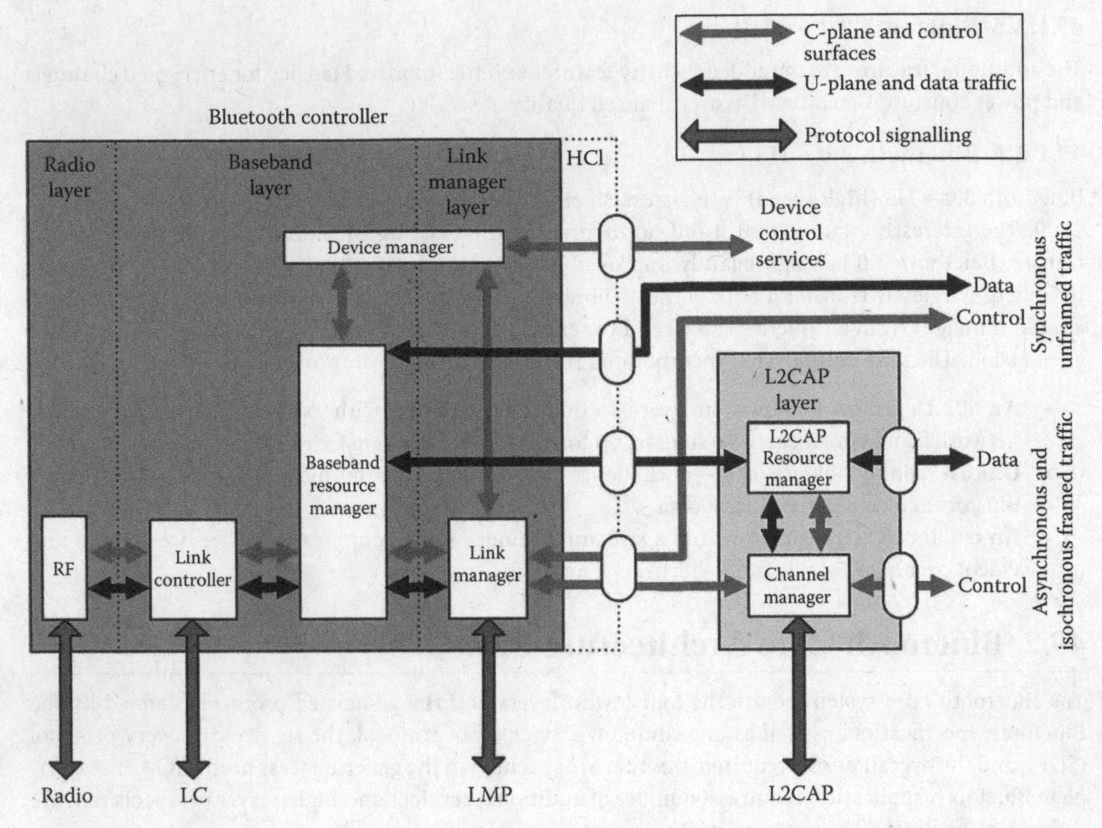

FIGURE 49.1 Bluetooth core system architecture. (From Bluetooth Special Interest Group, URL from January 2009: http://www.bluetooth.com/Bluetooth/SIG/.)

49.2.3 Device Manager

This block controls the general behavior of the Bluetooth-enabled device. The behavior includes

- Indirect data transport operations, for example, inquiring for the presence of another Bluetooth-enabled device, Bluetooth device discovery, connecting the devices, etc.
- Performing functions, such as requesting access to the transport medium from the baseband resource controller
- Controlling local device behavior by using various HCI commands, for example, managing the device, its stored keys, etc.

49.2.4 Link Manager

The link manager block is responsible for managing the logical links and the updates of the parameters related to the physical link between them. This can be achieved by the use of the Link Management Protocol (LMP). It allows the creation and the management of logical links. Additionally, it manages the encryption on the logical transport and guarantees QoS transport at the link.

49.2.5 Baseband Resource Manager

The baseband resource manager ensures access to the radio medium. For access to the channel, it runs a scheduler to grant time to the negotiated entities by a contract. The other major function is to control the contracts among entities. A contract means the guaranteed QoS contents delivery. The Bluetooth behavior for such type of contracts management includes the normal exchange of data between connected devices. It also manages the alignment between slots on the physical channel.

49.2.6 Link Controller

The link controller is responsible for the encoding and the decoding of Bluetooth packets from the data. It also carries out link control protocol signaling that is used to communicate flow control, acknowledgement, and retransmission request signals. The signal characteristics of logical transport is associated with the baseband packet.

49.2.7 Radio Frequency

The radio frequency (RF) block is responsible for transmitting and receiving data packets on the physical channel. The baseband block controls the timing and the frequency carrier of the RF block, which is between the baseband and the RF block.

Bluetooth works as a WLAN in the license-free ISM band at 2.4–2.4835 GHz, which is split up into 79 channels of 1 MHz bandwidths. There are some limitations for a few countries where only 23 channels can be used. Bluetooth switches among 79 frequency channels 1600 times per second escaping possible interferences very quickly. In its basic mode, the modulation is Gaussian frequency-shift keying (GFSK). The raw wireless data rate is 1 and 3 Mbps supported for Version 2.x + EDR.

There are different device classes defined differentiating the maximum RF power output, and hence the range (Table 49.1).

From specification 1.2, adaptive frequency hopping (AFH) was designed to reduce interference between wireless technologies. It uses the available frequency in the spectrum. It is possible to choose the unused frequency by detecting the used frequencies in the spectrum. This adaptive hopping allows for a more efficient transmission within the spectrum, providing users with greater performance, even if using other technologies along with Bluetooth technology.

The devices access the physical channel using assigned time slots. The consecutive time slots can be assigned dynamically to the same device depending on the circumstances. It uses the full-duplex transmission mode under the time-division duplex (TDD) mechanism.

49.2.8 Bluetooth Networks

49.2.8.1 Piconet

When Bluetooth devices establish connection between themselves and form an ad hoc network, this personal area network (PAN) is known as a piconet. Each piconet can contain up to seven slaves and one

TABLE 49.1 Bluetooth Classes

Class	Range	Power Output (dBm)
Class 1 radios	100 m or 300 ft	20 (100 mW)
Class 2 radios	10 m or 33 ft	4 (4 mW)
Class 3 radios	1 m or 3 ft	0 (1 mW)

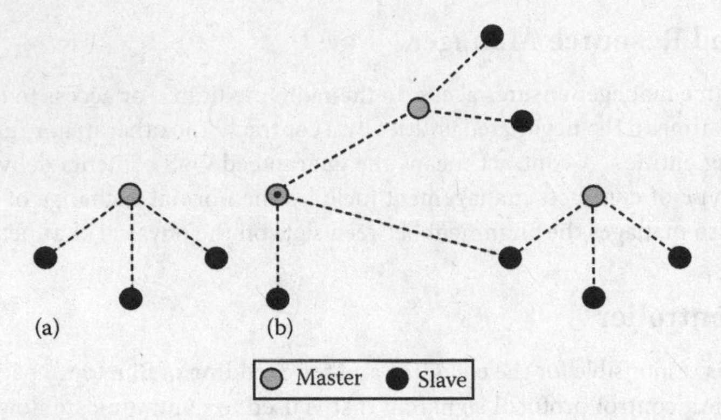

FIGURE 49.2 Bluetooth piconets.

master device. The master device is used as the reference for communication between all the piconets, while any device can belong to multiple piconets. A piconet offers the same frequency-hopping channel to the connected devices.

49.2.8.2 Scatternet

Once a device belongs to multiple piconets, the master devices of the corresponding piconets need to be connected among themselves in order to make this access available, which ultimately forms a scatternet. The frequency distribution is simple and can work without interfering with others, as each piconet is responsible for the distribution to its connected devices (Figure 49.2).

49.2.9 Bluetooth Security

There are three modes of security for Bluetooth access between two devices.

- Security Mode 1: Nonsecure
- Security Mode 2: Service-level enforced security
- Security Mode 3: Link-level enforced security

It is the responsibility of the vendor to implement the security mode. The devices have a two-level security consisting of "trusted devices" and "un-trusted devices." The trusted device can pair with any of the devices and has unrestricted access to the services.

There are different levels of security in services as well, such as

- Services that require authorization and authentication
- Services that require authentication only
- Services that are open to all devices

In general, a personal identification number (PIN) is selected by the user, which must be a 48 bit number. Additionally, a private link key (128 bit random numbers) and a private encryption key (8–128 bit) are used as a security measure for data transfer.

49.3 Bluetooth Protocol Stack

A protocol stack (often referred to as a communications stack) designates a software implementation of a computer networking protocol suite. The modularization within the definition (suite) of protocols enhances design and evaluation. These protocols are commonly described as layers in a stack of protocols, the lowest protocol always being the low-level, physical interaction of the hardware, all the way to

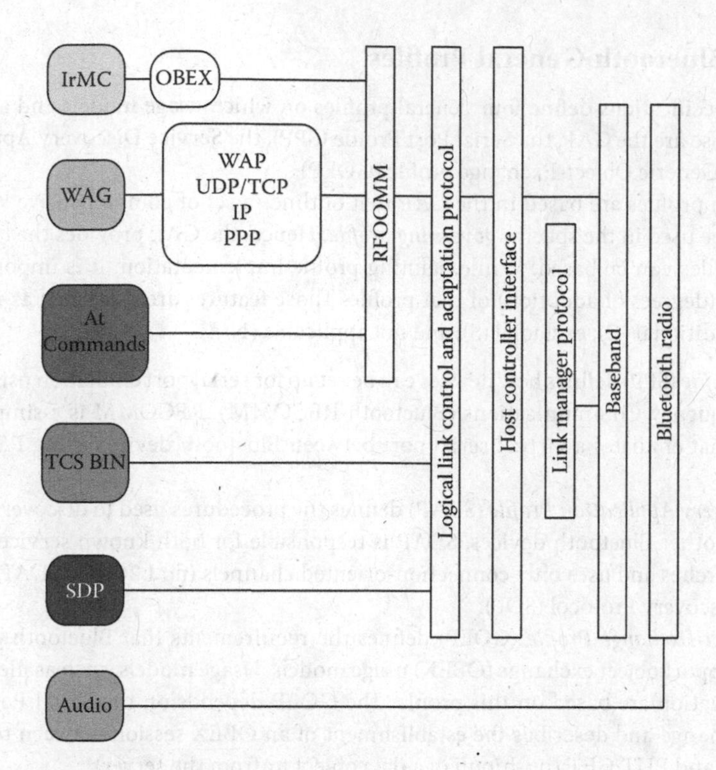

FIGURE 49.3 The Bluetooth protocol stack.

the top, with the user application layer typically dealing with the topmost layers. The underlying generic standard, the *Open Systems Interconnection* (OSI), was developed by the *International Organization for Standardization* (ISO) in 1974. The Bluetooth protocol stack is illustrated in Figure 49.3.

Based on the functionality of the protocols, they have been classified into four basic types of protocols: core, telephony, cable replacement, and adopted protocols [2,3]. This section will only briefly list the classification of the protocols. More details on each of these can be found in Sections 49.2 and 49.4.

Bluetooth Core Protocols entail the Base band, the Link Manager Protocol (LMP), the Logical-Link Control Adaptation Protocol (L2CAP), and the Service Discovery Protocol (SDP).

Telephony Control Protocols entail the Telephony Control Specification Binary (TCS-BIN) and the AT Commands.

Cable Replacement Protocols incorporate Radio Frequency Communication (RCCOMM).

Adopted Protocols comprise the Point-to-Point Protocol (PPP), the User Datagram, the Protocol/Transmission Control Protocol/Internet Protocol (UDP/TCP/IP), the Wireless Application Protocol (WAP), the Object Exchange Protocol (OBEX), and the Infrared Mobile Communications (IrMC).

49.4 Bluetooth Profiles

Profiles enable end-user functionality by defining usage models and behavior characteristics [4]. One of the most important objectives while defining profiles are the expectations on achieving a certain level of interoperability between various manufacturers of Bluetooth devices. The profiles define a minimum set of characteristics that should be inherent in all products.

49.4.1 Four Bluetooth General Profiles

The Bluetooth specifications define four general profiles on which usage models and all other profiles are based [3]. Those are the GAP, the Serial Port Profile (SPP), the Service Discovery Application Profile (SDAP), and the Generic Object-Exchange profile (GOEP).

All foundation profiles are based in the GAP that outlines a set of comprehensive key features that may or may not be used in the specific *governing profile*. Hence, the GAP provides the foundation upon which other profiles can be based. While defining profile implementation, it is important to emphasize the features (degrees of adoption) of that profile. Those features are described as mandatory (M), optional (O), conditional (C), excluded (S), and not applicable (N/A).

Serial Port Profile (SPP) defines how devices can be set up for serial port emulation using Bluetooth Radio Frequency Communications (Bluetooth RFCOMM). RFCOMM is a simple transport protocol that emulates a virtual serial port between Bluetooth devices using RS-232 control signaling.

Service Discovery Application Profile (SDAP) defines the procedures used to discover services registered on other Bluetooth devices. SDAP is responsible for both known service and general service searches and uses only connection-oriented channels (no L2CAP). SDAP requires the Service Discovery Protocol (SDP).

Generic Object-Exchange Profile (GOEP) defines the requirements that Bluetooth devices must meet to support object exchange (OBEX) usage models. Usage models, such as file transfer and synchronization are based on this profile. The GOEP depends on the Serial Port Profile for object exchange and describes the establishment of an OBEX session between the client and the server, and PUT/GET (push/pull of a data object to/from the server).

49.4.2 Bluetooth General Profiles

Bluetooth profiles as defined by *Bluetooth SIG* are general behaviors through which Bluetooth-enabled devices communicate with other devices [1]. By following the guidance provided in Bluetooth specifications, developers can create applications to work with other devices also conforming to the Bluetooth specification. At a minimum, each profile specification contains information on the following topics: dependencies on other profiles, suggested user interface formats, and specific parts of the Bluetooth protocol stack used by the profile [5–7].

There are currently 25 Bluetooth officially adopted profiles. Most of the Bluetooth profiles were introduced with the Bluetooth Core Specification Version 1.1. The Bluetooth SIG governs the development of new Bluetooth profiles.

In order for one Bluetooth device to connect to another, both devices must share at least one of the same Bluetooth profiles. For example, if you want to use a Bluetooth headset with your Bluetooth-enabled cell phone, both devices must use the Headset (HS) profile (defined below). To perform its task, each profile uses particular options and parameters at each layer of the stack. This may include an outline of the required service record, if appropriate. A brief outline of currently defined profiles within the current Bluetooth specification follows [8]:

Advanced Audio Distribution Profile (A2DP) describes how stereo quality sound can be streamed from a media source to a sink.

Audio/Video Control Transport Protocol (AVRCP) is designed to provide a standard interface to control TVs, hi-fi equipment, etc. A single remote control (or another device) should be able to control all the A/V equipment to which a user has access.

Basic Imaging Profile (BIP) defines how an imaging device can be remotely controlled, how an imaging device may print, as well as how an imaging device can transfer images to a storage device.

Basic Printing Profile (BPP) allows devices to send texts, e-mails, vCards, images, or other items to printers based on print jobs.

Common ISDN Access Profile (CIP) defines how ISDN signaling can be transferred via a Bluetooth wireless connection.

Cordless Telephony Profile (CTP) defines how a cordless phone can be implemented over a Bluetooth wireless link.

Dial-Up Network Profile (DUN) provides a standard to access the Internet and other dial-up services via Bluetooth technology.

Fax Profile (FAX) defines how a FAX gateway device can be used by a terminal device.

File Transfer Profile (FTP) defines how folders and files on a server device can be browsed by a client device.

General Audio/Video Distribution Profile (GAVDP) provides the basis for an ADP and a VDP, the basis of the systems designed for distributing video and audio streams using Bluetooth technology.

Generic Object Profile (GOEP) is used to transfer an object from one device to another.

Hands-Free Profile (HFP) describes how a gateway device can be used to place and receive calls for a hand-free device.

Hard Copy Cable Replacement Profile (HCRP) defines how driver-based printing is accomplished over a Bluetooth wireless link.

Headset Profile (HSP) describes how a Bluetooth-enabled headset should communicate with a Bluetooth-enabled device.

Health Device Profile (HDP) is an application profile to connect health-monitoring source devices, such as blood pressure monitors, glucose meters, and pulse oximeters to sink devices, such as computers, laptops etc.

Human Interface Device Profile (HID) defines the protocols, procedures, and features to be used by Bluetooth HID, such as keyboards, pointing devices, gaming devices, and remote monitoring devices.

Intercom Profile (ICP) defines how two Bluetooth-enabled mobile phones in the same network can communicate directly with each other without using the public telephone network.

Object Push Profile (OPP) defines the roles of the push server and the push client.

Personal Area Networking Profile (PAN) describes how two or more Bluetooth-enabled devices can form an ad hoc network, and how the same mechanism can be used to access a remote network through a network access point.

Service Discovery Application Profile (SDAP) describes how an application should use SDP to discover services on a remote device.

Service Port Profile (SPP) defines how to set up virtual serial ports and to connect two Bluetooth-enabled devices.

Synchronization Profile (SYNC) used in conjunction with GOEP enables the synchronization of calendar and address information (personal information manager (PIM) items) between, Bluetooth-enabled devices.

Video Distribution Profile (VDP) defines how a Bluetooth-enabled device streams videos over Bluetooth wireless technology.

49.5 Competitive Technologies

Besides Bluetooth technology, there is a plethora of technologies intended to work in local and PANs to replace wired communication. These technologies are discussed in the following paragraphs.

IEEE 802.11 is the most popular WLAN technology. There are different working groups currently working for IEEE 802.11. IEEE 802.11*a* uses OFDM, operates in a 5 GHz range, and offers data rates up to 54 Mbps. IEEE 802.11*b* uses DSSS, operates in a 2.4 GHz range, and offers data rates up to 11 Mbps.

IEEE 802.11*g* uses OFDM, operates in a 2.4 GHz range, and offers data rates up to 54 Mbps. IEEE 802.11*e* enhances the quality of the service parameters. IEEE 802.11*h* supports dynamic frequency selection; transmits power control, spectrum, and power management. IEEE 802*i* uses an advanced encryption standard to enhance security. IEEE 802*k* focuses on radio resource management. 802.11*n* operates in 5 GHz ranges, offers data rates up to 100 Mbps, and focuses on multimedia applications [9].

The high-performance radio LAN (HIPERLAN) is a European counterpart of IEEE 802.11 defined by the European Telecommunication Standards Institute (ETSI). HIPERLAN/1 operates in a 5.15 and a 17.1 GHz range and offers data rates up to 23.5 Mbps. HIPERLAN/2, intended for short-range communication, operates in a 5 GHz range, offers data rates up to 54 Mbps, and focuses on wireless multimedia services. HIPERACCESS, intended to be the-last-mile, offers data rates up to 25 Mbps [10].

HomeRF is a short-range communication technology intended to be used in small areas like houses or small buildings. It uses FHSS, operates in a 2.4 GHz range, and offers data rates up to 10 Mbps [11].

Infrared (irDA) is a point-to-point, ultra low power, data transmission standard designed to operate over a distance of 1 m (the distance can be increased with a higher power) [12]. It can achieve data rates up to 16 Mbps. Being cheaper and of low power, irDA is widely used.

Ultra-Wideband (UWB) offers a good combination of low power consumption (~1 mW/Mbps) and a high data throughput (up to 480 Mbps). WiMedia UWB is an internationally recognized standard (ECMA-368, ISO/IEC 26970, and ECMA-369, ISO/IEC 26908). Unlike irDA, UWB allows for data rates up to 480 Mbps at ranges of several meters and a data rate of approximately 110 Mbps at a range of up to 10 m [13].

The certified wireless USB is a short-range communication technology that offers data rates up to 480 Mbps for up to a 2 m range, and 110 Mbps for up to a 10 m range. It uses the same radio as the WiMedia Alliance [14].

Radio Frequency Identification (RFID) is used for tracking purposes. There are over 140 different ISO standards for a variety of applications. It operates in low-frequency, high-frequency, and ultrahigh-frequency ranges. Near-field-communication is based on RFID technology that offers a data rate of 212 kbps over an extremely short distance of up to 20 cm [15].

The ZigBee standard is a short-range wireless communication standard for PANs. It offers date rates up to 250 kbits at 2.4 GHz, 40 kpbs at 915 MHz, and 20 kpbs at 868 MHz with a range of 10–100 m. ZigBee uses IEEE 802.15.4 as physical and medium-access control layers. The upper layers, as well as ZigBee security architecture, is defined by the ZigBee standard [16].

WiBree is an evolved Bluetooth standard used for connecting ultralow power consumption devices (button cell batteries). It was initiated by Nokia in 2001 and was ultimately adapted by Bluetooth SIG as the Bluetooth low-energy standard in 2008. The key benefit is that it provides people with new features, like sports and entertainment. It operates in a 2.4 GHz ISM band with a 1 Mbps physical-link data rate. The example devices can be wrist watches, toys etc. [17].

49.6 Future of the Bluetooth Technology: Challenges

Although Bluetooth technology has already established a well-known standard in personal communication systems, when user demands and requirements become ubiquitous and mobile in nature, it needs more robust, secure, and data-intensive technologies to support them. Bluetooth SIG has proposed a future specification that is expected to offer the same set of services and data rate.

It is intended that Bluetooth channels will work in the push information model instead of in the pull information model, as it will enable users to use any shared information points at various places, e.g., while buying some items and appliances.

The topology management of the piconet and the scatternet should be dynamic and seamless. The user should not be bothered to make such configurations and changes.

An alternative MAC and PHY medium is to be used while working in high-performance networks, for example, while transferring bulky data. The system should be compatible and adaptable enough to work with any MAC or PHY in order to achieve high performance in PANs.

Another future enhancement will be improvement in the quality of service, especially regarding audio and video data transfer at high speed with high, quality measures.

It is also intended that Bluetooth will use more adaptable protocols in order to work in dynamic environments to meet the challenges of active and configurable systems. The future Bluetooth Radio 2.0 will offer up to a 480 Mbps data rate.

References

1. Bluetooth Special Interest Group, URL from January 2009: http://www.bluetooth.com/Bluetooth/SIG/.
2. Prabhu, C.S.R., Prathap A.R., *Bluetooth Technology and Its Applications with JAVA and J2ME*, Prentice-Hall of India, New Delhi, 2006.
3. Bakker, D., McMichael, D., Gilster, R., *Bluetooth End to End*, Wiley, New York, 2002.
4. Gratton, D.A., *Bluetooth Profiles*, Prentice Hall PTR, Upper Saddle, NJ, 2003.
5. Huang, A.S., Rudolph, L., *Bluetooth Essentials for Programmers*, Cambridge University Press, New York, 2007.
6. Labiod, H., Afifi, H., De Santis, C., *Wi-Fi, Bluetooth, Zigbee and WiMax*, Springer, Dordrecht, the Netherlands, 2007.
7. Morrow, R., *Bluetooth: Operation and Use*, McGraw-Hill Professional, New York, 2002.
8. Bluetooth Profiles, Bluetooth Special Interest Group, URL from January 2009: http://bluetooth.com/Bluetooth/Technology/Works/Profiles_Overview.htm.
9. IEEE 802.11 Standard, http://standards.ieee.org/getieee802/802.11.html
10. HIPERLAN Specification, http://portal.etsi.org/bran/Summary.asp
11. IEEE HomeRf Tutorial IEEE 802.11-98/54, http://grouper.ieee.org/groups/802/15/pub/1999/Mar99/90547S_WPAN-HomeRF-Tutorial.ppt
12. IrDA Specifications and technical notes, http://www.irda.org/displaycommon.cfm?an=1&subarticlenbr=7
13. ECMA Standard- 368, http://www.ecma-international.org/publications/standards/Ecma-368.htm
14. WiMedia Alliance, http://www.wimedia.org/en/resources/index.asp?id=res
15. RFID in U.S. Libraries, NISO RP-6-2008, http://www.niso.org/publications/rp/RP-6-2008.pdf
16. ZigBee Specification Download, http://www.zigbee.org/ZigBeeSpecificationDownloadRequest/tabid/311/Default.aspx
17. WIBREE FORUM MERGES WITH BLUETOOTH SIG, http://www.bluetooth.com/Bluetooth/Press/SIG/NOKIAS_LOW_POWER_WIBREE_BROUGHT_INTO_BLUETOOTH_SIG.htm

50

ZigBee

Stefan Mahlknecht
Vienna University of Technology

Tuan Dang
EDF Research and Development

Milos Manic
University of Idaho Idaho Falls

Sajjad Ahmad Madani
COMSATS Institute of Information Technology

50.1 Introduction 50-1
50.2 ZigBee and Mesh Networks 50-1
50.3 ZigBee in the Context of Other Wireless Networks 50-2
50.4 ZigBee Stack 50-2
50.5 IEEE 802.15.4 50-3
Physical Layer • MAC Layer • Network Layer • Application Layer
50.6 Development and Industrial Applications 50-8
ZigBee Development Platforms • Industrial Applications
50.7 Conclusion 50-10
References 50-10

50.1 Introduction

ZigBee is a suite of very low-cost and low-power, low data rate, and short-range digital radios based on the IEEE 802.15.4-2006 standard for low-rate wireless personal area networks (LR-WPANs). The ZigBee technology was initially aimed at radio-frequency (RF) applications, with the idea of producing protocol that is simpler and more cost-effective than those of comparative WPANs such as ubiquitous 802.15.1 (Bluetooth).

ZigBee suite of protocols is regulated by the ZigBee Alliance, an association of companies (list available on ZigBee Alliance site) working together on globalization of an open ZigBee standard [ZigBee Alliance]. The standard emerged as response to a growing demand for self-organizing ad hoc digital radio networks. The name "ZigBee" is derived from the erratic zig-zag patterns that bees would typically make between flowers while collecting pollen and sharing critical information among its fellow hive members. The name was never confirmed in scientific studies but was very evocative of the invisible webs of connections existing in a fully wireless environment.

50.2 ZigBee and Mesh Networks

ZigBee has a wide application scope and may be used in building automation such as access control, smoke detection, HVAC, and lighting. Other applications include industrial monitoring, automatic meter reading, medical sensing, and environmental data collection. Security is an important issue in these control and data acquisition applications. Electrical appliances using ZigBee technology can be linked together into a ZigBee network of virtually unlimited distance.

ZigBee is built on top of the IEEE 802.15.4 protocol stack as shown in Figure 50.1. 802.15.4 defines the medium access control and physical layers while the higher layers are defined by the ZigBee specification. The ZigBee specification from the ZigBee alliance is not openly available. ZigBee supports multiple

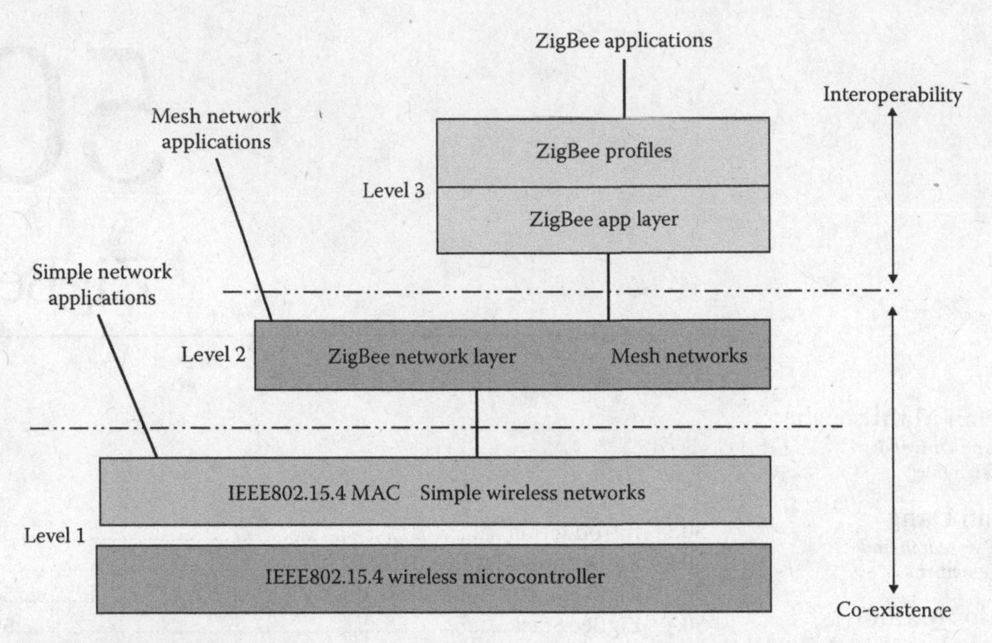

FIGURE 50.1 IEEE 802.15.4 and ZigBee protocol stacks.

network topologies such as star, tree, and mesh. ZigBee devices can act as coordinators that initiate a network, a device with routing capability, or as an end device with no routing capability.

802.15.4 is also the basis for wireless sensor networks (WSN) and mesh networking (see Section 50.5). Wireless mesh networks (WMN) are an effective solution for building automation and control. A WMN would enable maintenance workers using mobile handheld devices such as smartphones or personal digital assistants to gather data from wherever they are in a building.

The mesh access points (MAPs) communicate only with their adjacent routers. In larger mesh network deployments, some of the access points (APs) would act as intermediate routers or relays between APs (multi-hopping) not within wireless range (non-line-of-sight) of one another. When an intermediate router becomes faulty, other routers/APs adapt and pick up the routing; i.e., the mesh network is self healing, ensuring constant availability of information.

50.3 ZigBee in the Context of Other Wireless Networks

Zigbee belongs to lowest range/data rate wireless technologies. Figure 50.2 illustrates some general differences among wireless personal area networks (WPANs), wireless metropolitan area networks (WMANs), wireless local area networks (WLANs), and wireless wide area networks (WWANs).

It is also beneficial to compare ZigBee against other wireless technologies relative to its range, throughput, power consumption, and cost. Table 50.1 provides such brief overview.

50.4 ZigBee Stack

Although based upon standard seven-layer open systems interconnection (OSI) mode, ZigBee stack defines specific layers responsible for performing of specific set of services to the layer above it through a service access point (SAP). While IEEE 802.15.4 standard defines two lower layers—the physical (PHY) and the medium access control (MAC) sub-layer, the ZigBee Alliance builds on this foundation by providing the network (NWK) layer and the framework for the application layer (APL)—application support (APS) sub-layer, the ZigBee device object (ZDO), and the manufacturer-defined application objects (Figure 50.3). The details on these layers will be further elaborated in following sections.

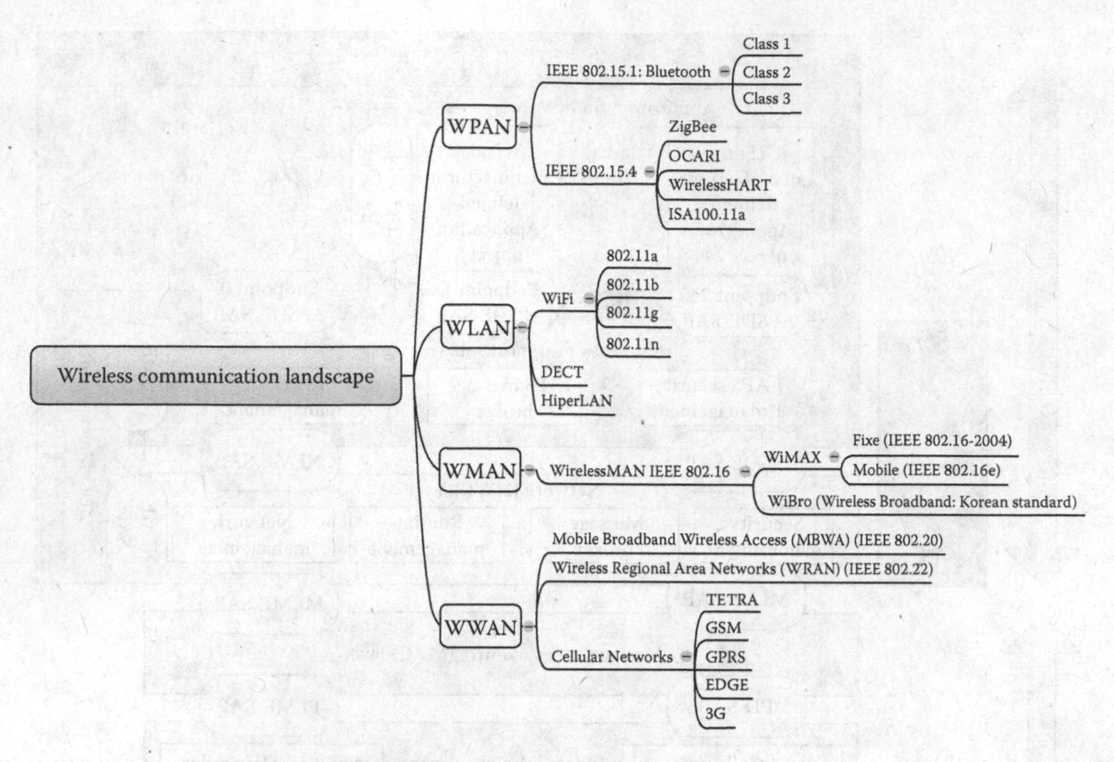

FIGURE 50.2 Wireless technology landscape.

TABLE 50.1 A Comparison of LR-WPAN with Other Wireless Technologies

	WLAN (802.11)	Bluetooth-Based WPAN (802.15.1)	Low-Rate WPAN (802.15.4)
Range	~100 m	~10–100 m	~10 m
Data throughput	~2–11 Mbs	~1 Mbs	~0.25 Mbs
Power consumption	Medium	Low	Ultralow
Size	Larger	Smaller	Smallest
Cost/complexity	>6	1	0.2

50.5 IEEE 802.15.4

The goal of the IEEE 802.15.4 task group is to define the PHY and MAC layers specifications for low-cost, low data rate and hence, low-complexity wireless connectivity of mobile or stationary devices with limited battery capability.

The relationship between IEEE 802.15.4 and ZigBee is that ZigBee makes use of the IEEE 802.15.4 standard for the lower two OSI layers (PHY and MAC layers) and specifies the upper layers as well as being the name for the whole technology from a marketing perspective where ZigBee stands for the full technological solution based upon IEEE 802.15.4.

The key features of IEEE 802.15.4 are

- 2.4 GHz and 868/915 MHz frequency band options
- Support for star and peer-to-peer topologies; therefore, supports a wide variety of routing protocols and topologies

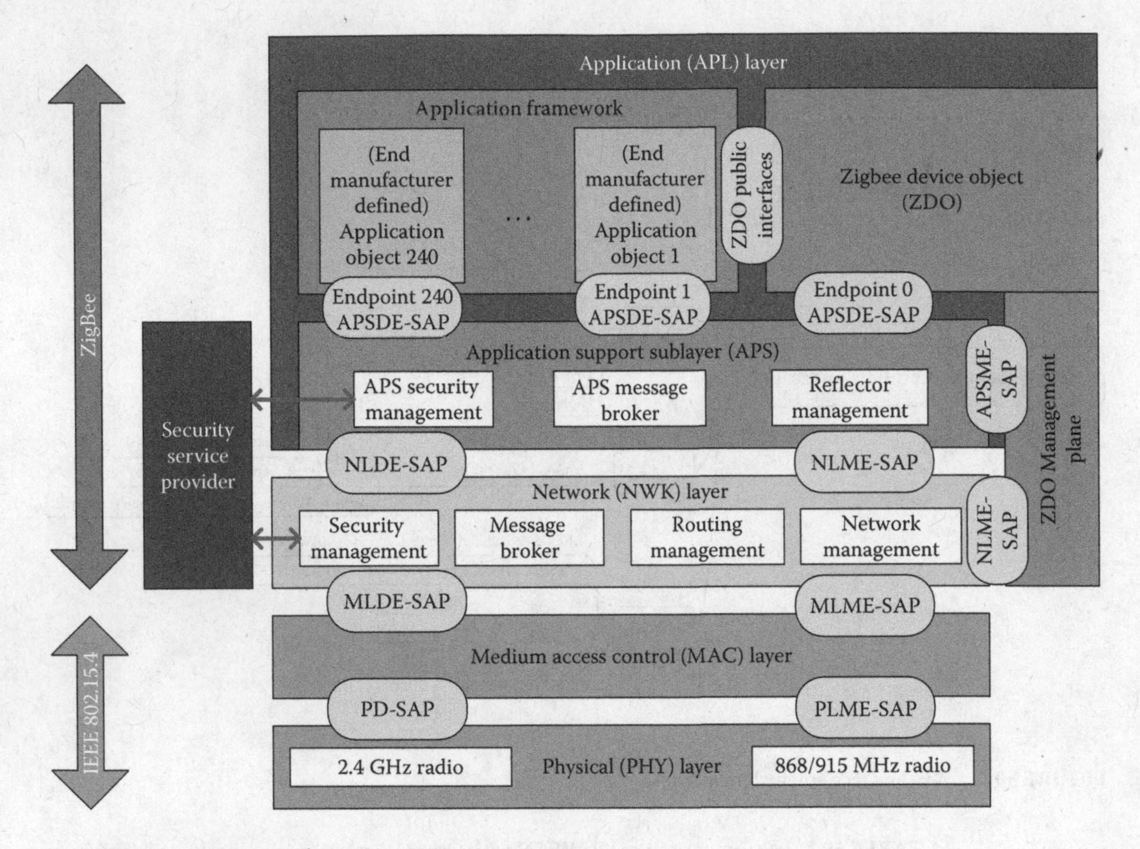

FIGURE 50.3 ZigBee stack.

- Contention-based channel access, CSMA/CA
- Message transmission can be fully acknowledged
- Beacon from 15 ms up to 4 min, allowing very low duty cycles.
- Optional guaranteed time slots (GTSs) can be reserved
- 16 bit address field or extended 64 bit addresses
- Link quality indication byte is attached to each receiver

50.5.1 Physical Layer

There are two groups of frequency bands specified: 2.45 GHz and 868/915 MHz. The popularity of 2.45 GHz is very high compared to the 868/915 MHz solutions that require different hardware and cannot just be switched by software (Figure 50.4).

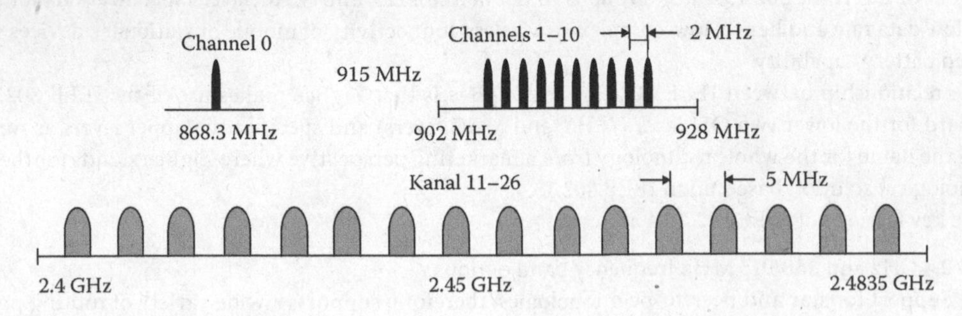

FIGURE 50.4 Frequency spectrum.

TABLE 50.2 Physical Channel Properties

Channel Parameters				Data Parameters[a]		Spreading Parameters[a]		
Frequency Band	Bandwidth	Ch. Nr.	Region	Raw Data Rate	Symbol Rate	Symbols	Chiprate	Modulation
868–868.6 MHz	300 kHz	0	Europe	20 kbps	20 kBaud	Binary	300 kchips/s	BPSK
902–928 MHz	600 kHz	1–10	USA	40 kbps	40 kBaud	Binary	600 kchips/s	BPSK
2400–2483.5 MHz	2 MHz	11–26	Worldwide	250 kbps	62,5 kBaud	16 orthog.	2 Mchips/s	Offset-QPSK

[a] Shows the physical channel properties from the IEEE 802.15.4-2003. The updated IEEE 802.15.4-2006 has only been implemented on a few chip solutions like the Atmel AT86RF231 and the AT86RF212. Among other things, it enhances the data rate in the 868 and 915 MHz band to 100 and 250 kbps, respectively, by using Offset-QPSK modulation with chip rates of 400 and 1000 kchips/s.

The radios in the 2.4 GHz spectrum use direct sequence spread spectrum (DSSS), which is managed by the digital stream into the modulator. Binary phase shift keying (BPSK) is used in the 868 and 915 MHz bands, and orthogonal quadrature phase shift keying (QPSK) that transmits two bits per symbol is used in the 2.4 GHz band (Table 50.2).

Transmission range may vary between 30 m indoor and 100 m outdoor as a rule of thumb, although it is heavily dependent on the particular environment and may vary accordingly. The maximum output power of the radios is generally +3 dBm and can be increased with an external amplifier up to 20 dBm. Solutions with high-sensitivity transceivers (–101 dBm in 2.4 GHz and –110 dBm in 868/915 MHz) and power amplifiers (+10–20 dBm) have proven to reach over 4 km outdoors [Meshnetics].

50.5.2 MAC Layer

From the data link layer after OSI, the IEEE 802.15.4 defines only the lower part of the layer, which is the MAC layer. The implementation of the logical link control (LLC) has not been specified in IEEE 802.15.4.

The basic channel access mode is carrier sense multiple access with collision avoidance (CSMA/CA). There are three notable exceptions to the use of CSMA. Beacons are sent on a fixed timing schedule and do not use CSMA. Message acknowledgments also do not use CSMA. Finally, devices in beacon-oriented networks that have low-latency real-time requirements may also use GTSs, which by definition do not use CSMA.

There are two basic modes of operation that affect the channel access strategy:

- Unslotted mode (non-beacon-enabled)
- Slotted mode (beacon-enabled)

50.5.2.1 Unslotted Mode

In non-beacon-enabled networks, an unslotted CSMA/CA channel access mechanism is used. In this type of network, ZigBee routers typically have their receivers continuously active, requiring a more robust power supply. However, this allows for heterogeneous networks in which some devices receive continuously, while others only transmit when an external stimulus is detected. Before each transmission, the sender checks by CSMA/CA if the channel is occupied and sends its data by a random time offset if the channel becomes free. Optionally, it can indicate to the receiver if an ACK is desired. In this mode, no network maintenance by a PAN coordinator node is required. There are three basic communication scenarios:

- Data are sent from participating nodes to the PAN coordinator. In this case, the PAN coordinator is always in listening mode and is not allowed to enter sleep mode. PAN coordinators are typically not battery powered.

- Data are sent from one participant to another (peer-to-peer-topology). As in the first scenario, the receiver must always be capable of receiving.
- Data are sent from the PAN coordinator to one participant. The participant asks periodically for data at the coordinator. The coordinator sends an ACK and the data if something is available, otherwise it sends a null packet. The coordinator is always in receiving mode when not transmitting.

50.5.2.2 Slotted Mode

In the slotted mode, there is a superframe structure coordinated by the PAN coordinator as the transmitter of periodic beacons. The superframe is started by the beacon (without using CSMA/CA) and followed by 16 equal time slots. The beacons are used to synchronize the attached devices to identify the PAN and to describe the structure of the superframes.

For low-latency applications, the PAN coordinator may optionally dedicate portions of the active superframe to GTSs. The GTSs form the contention-free period (CFP), which always appears at the end of the active superframe. The other slots are right after the beacon and are the so-called contention access slots. Any device wishing to communicate during the contention access period (CAP) between two beacons has to compete with other devices using a slotted CSMA/CA mechanism. All transactions have to be completed by the time of the next network beacon. The first nine slots shall always be CAP slots, which can be used by any device, while the following optional up to seven slots may be allocated by the coordinator for individual devices upon a request. Figure 50.5 shows the concept of the superframe structure.

With the superframe order (SO) parameter, the superframe can be divided in an active and inactive phase. The active phase consists of 16 identical time slots as described above followed by the inactive phase where the PAN coordinator as, of course, also the other participants can enter a low power sleep mode to save batteries. This is an improvement compared to the unslotted mode; however, it comes with a higher synchronization effort. Also here, there are three basic communication scenarios:

- Data are sent from participating nodes to the PAN coordinator. The participant tries to obtain a CAP slot with CSMA/CA if no GTS for it has been defined in the superframe structure. If a GTS has been reserved, it can send data immediately after synchronizing to its slot.
- When a peer-to-peer communication is initiated, participants synchronize themselves before transmitting the data in one of the CAP slots.
- If data are sent from the PAN coordinator to one participant, it shows this information in the following beacon. The participant listens periodically to the beacons and responds in the CAP slots to pick up the data.

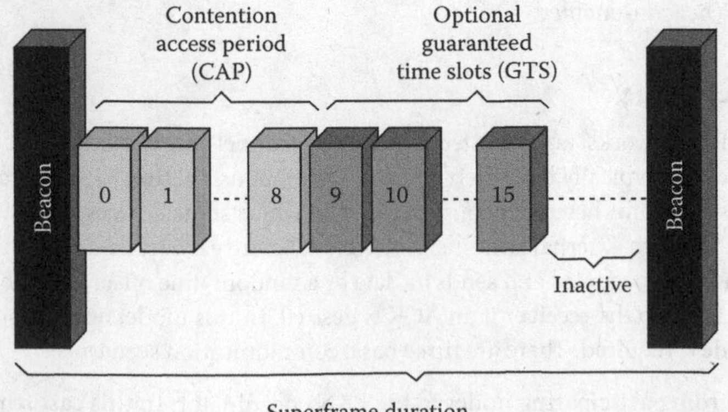

FIGURE 50.5 Superframe.

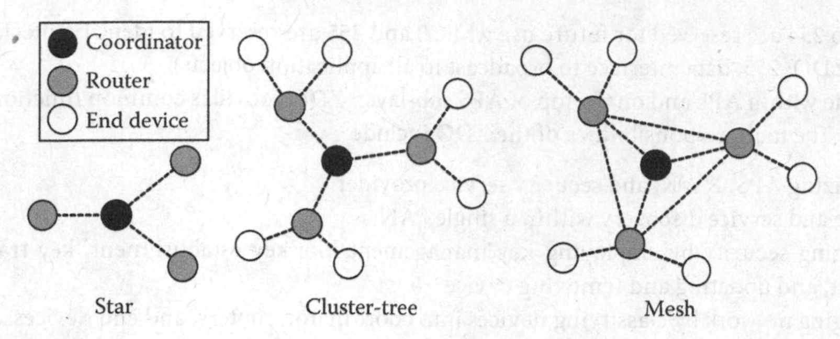

FIGURE 50.6 ZigBee network topologies and classes of device.

50.5.3 Network Layer

Network layer (NWK) enables routing and multi-hop communication between devices in star, mesh, or tree topology. The NWK provides two types of basic services; data services and management services. The data services include a network layer protocol data unit (NPDU), transmitting NPDU to the next hop, and applying NWK security. The management services allow applications to interact with the ZigBee protocol stack. The management services include configuring a new device, establishing a new network, joining and leaving a network, assigning address to devices joining the network, neighborhood discovery, and route discovery.

The devices in a ZigBee network are either reduced function device (RFD) or full function device (FFD). RFD is simple in terms of hardware and software complexity. It is only restricted to start topology, is incapable of network routing, is low power, and is potentially low cost because of reduced functionality. On the other hand, FFD is more resourceful, has routing capability, and is generally powered by mains power supply. RFD only communicates with FFD while FFD can communicate with another FFD and/or RFD. ZigBee devices are logically divided into three different classes; an end device, a router, and a coordinator (Figure 50.6).

The responsibilities of a PAN coordinator comprise initiating new network, managing the network nodes, and storing network information, for instance, security keys. There is exactly one PAN coordinator per ZigBee network and is an FFD. Router is an optional FFD component in ZigBee network and is used to extend network coverage, allocation and de-allocation of addresses, and routing between nodes. End devices implement user applications and can either be FFD or RFD but do not participate in data forwarding.

In start topology, communication is done only through PAN coordinator and follows a master/slave paradigm. In mesh topology, any device can talk to any other device. Such a topology is ad hoc, self-organizing, and self-healing, and employs multi-hop communication. Mesh topology allows multiple paths from a given source to destination and thus increases reliability. If one breaks, the data are sent via another path transparently. Cluster-tree topology is a hybrid (star plus mesh) approach where PAN coordinator designates itself as a cluster-head and forms a tree around itself. Although such a topology increases coverage area, it introduces additional delays.

The routing algorithms that are used in ZigBee networks are hierarchical and have potential for table optimizations. Commonly used algorithms include a routing algorithm similar to Ad-hoc on-demand distance vector (AODV) and cluster-tree algorithm.

50.5.4 Application Layer

The APL of ZigBee protocol stack is composed of application support (APS) sub-layer, application framework, and ZDO. The APS provides an interface to the NWK. The services provided by APS comprise data transmission, fragmentation and reassembly, reliable data transport, security, device binding, and maintenance of APS information base (a database containing information about managed objects). Application objects (manufacturer-defined components that implement the application) reside in application framework. A total of 240 distinct application objects can be defined with IDs ranging from 1 to 240; 241 to 254 are reserved for future use while 0 and 255 are reserved to identify interfaces (0, data interface to ZDO; 255, data interface to broadcast to all application objects).

ZDO reside within APL and on the top of APS sub-layer. ZDO provides common functionality for all applications. The main responsibilities of the ZDO include

- Initializing APS, NWK, and security service provider
- Device and service discovery within a single PAN
- Managing security by employing key management like key establishment, key transport, key request, and updating and removing device
- Managing network by classifying devices into coordinator, routers, and end devices
- Creating and maintaining binding tables
- Node management for FFDs, e.g., retrieval of routing and/or binding table

50.6 Development and Industrial Applications

The ZigBee specifications were developed by the ZigBee Alliance as an effort to establish a standardized base set of solutions for sensor and control systems requiring low data rates and low power consumption. Typical applications that are targeted by ZigBee are

- Home and building automation
- Smart energy

TABLE 50.3 Development Platforms of ZigBee Chipset Manufacturers

	Features				
Manufacturer	RF Transceiver for IEEE 802.15.4 PHY	Microcontroller	ZigBee Stack	ZigBee Pro Stack	Embedded Operating System for Application Layer
Atmel	AT86RF212 for 800/900 MHz band AT86RF231 for 2.4 GHz band	AVR/AVR32 Risc controller with in-system programmable flash	BitCloud	BitCloud	FreeRTOS TinyOS μC/OS-II NORTi RTOS
Freescale	MC13192 for 2.4 GHz band MC13202 for 2.4 GHz band	MC9S08GT Flexis ultralow-power MC9S08QE128 (8 bit) and MCF51QE128 (32 bit)	BeeStack	BeeStack	FreeRTOS μC/OS-II
	MC1321x System in Package (SiP) MC1322x Platform in a Package (PiP)				
TI	CC2420 for 2.4 GHz band CC2520 for 2.4 GHz band	8051 MSP430	Z-Stack	Z-Stack	FreeRTOS TinyOS μC/OS-II
	CC2430 System on Chip (SoC) CC2431 System on Chip (SoC)				
	CC2480 ZigBee Network Processor with Z-Stack + MSP430				
Ember	EM250 System on Chip		EmberZNet Stack	EmberZNet Pro Stack	FreeRTOS TinyOS
Microchip	MRF24J40/MA/MB	PIC18, PIC24, dsPIC, and PIC32 families	Microchip ZigBee 2006 Protocol Stack (AN1232)	Microchip ZigBee PRO (AN1255)	FreeRTOS

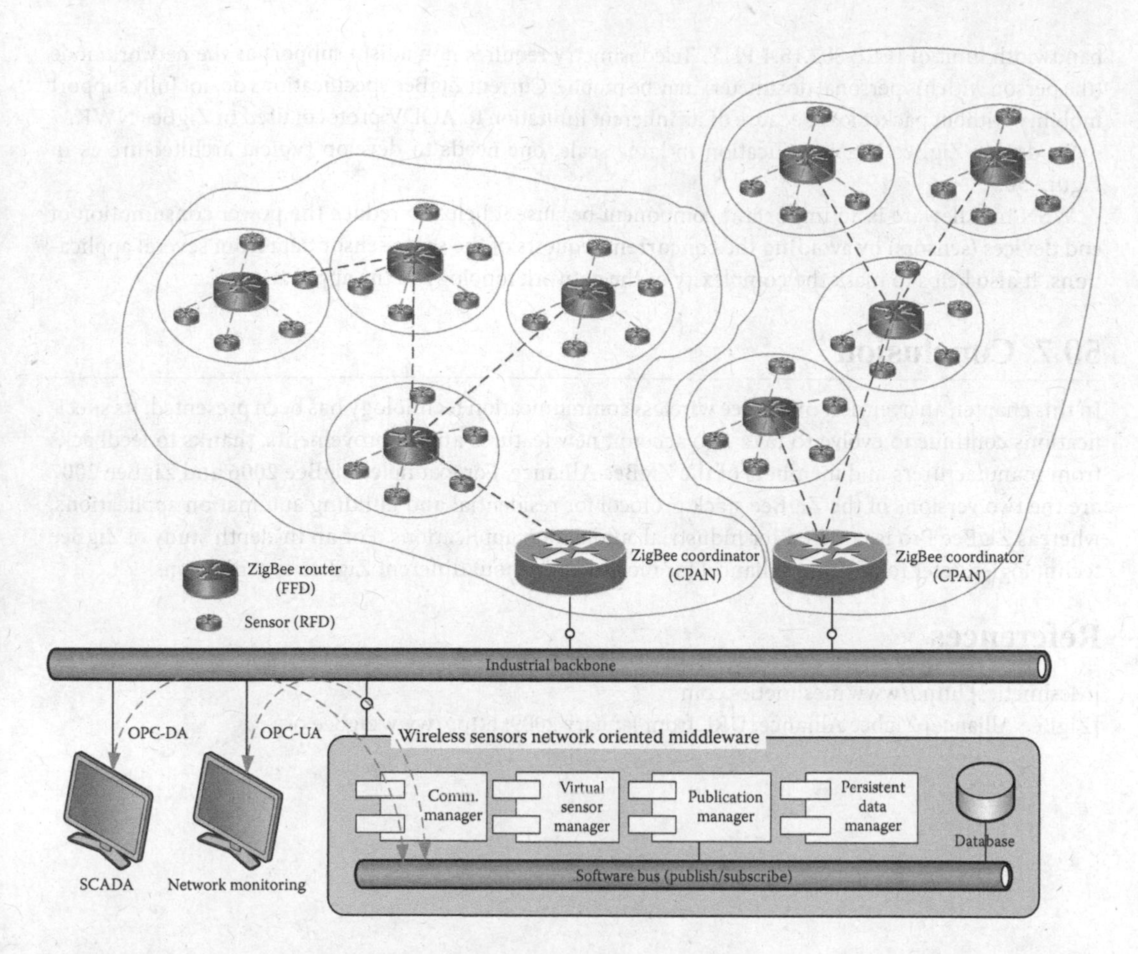

FIGURE 50.7 ZigBee integration with global intranet/Internet.

- Personal health care
- Telecom services
- Industrial asset management

50.6.1 ZigBee Development Platforms

There are several development platforms that are available from chipset manufacturers (Table 50.3).

50.6.2 Industrial Applications

ZigBee is particularly suitable for WSN application. Examples of industrial applications of ZigBee are

- Condition-based maintenance for plant asset management
- Environmental and structural monitoring
- Teledosimetry in nuclear power plants

These applications typically require scalability and low power consumption support. Condition-based maintenance possibly requires vibration analysis that may need the development of specific application profile in order to deal with data fragmentation because accelerometer sampling rate can reach the bandwidth limit of IEEE 802.15.4 PHY. Teledosimetry requires nomadism support as the network node (the person with his personal dosimeter) may be mobile. Current ZigBee specifications do not fully support mobility without packet loss because of its inherent limitation to AODV protocol used in ZigBee NWK.

To deploy ZigBee WSN application in large scale, one needs to develop typical architecture as in Figure 50.7.

WSN middleware is an important component because it helps to reduce the power consumption of end devices (sensors) by avoiding the concurrent requests of the same sensor data from several applications. It also helps to mask the complexity of the network topology to the applications.

50.7 Conclusion

In this chapter, an overview of ZigBee wireless communication technology has been presented. Its specifications continue to evolve to take into account new features and improvements, thanks to feedbacks from manufacturers and members of the ZigBee Alliance. For example, ZigBee 2006 and ZigBee 2007 are the two versions of the ZigBee stack protocol for residential and building automation applications; whereas ZigBee Pro is targeted for industrial automation applications. For an in-depth study of ZigBee technologies, refer to [ZigBee Alliance] for more details about different ZigBee specifications.

References

[Meshnetics] http://www.meshnetics.com
[ZigBee Alliance] Zigbee Alliance, URL from January 2009: http://www.zigbee.org

51

6LoWPAN: IP for Wireless Sensor Networks and Smart Cooperating Objects

51.1 Introduction 51-1
51.2 Why IP in WSN and For Smart Cooperating Objects? 51-1
IP as Open Standard Protocol Instead of Proprietary Protocols • IP Routers Instead of Complex Gateways • Use of Already Existing Protocols, Tools, and Applications • New Architectural Styles like Service-Oriented Architectures • IP Support in Operating Systems • Lessons Learned From Maintaining the Internet
51.3 Introduction in 802.15.4 51-3
51.4 802.15.4 and 6LoWPAN 51-5
51.5 6LoWPAN 51-7
Address Autoconfiguration • Frame Types and Fragmentation • Mesh Frame Type • Broadcast and Multicast Address Mapping • Header Compression • Scopes • Summary of Frame Types and Compression Schemes • Security
51.6 Summary 51-14
References 51-14

Guido Moritz
University of Rostock

Frank Golatowski
University of Rostock

51.1 Introduction

In the last decade, plenty of research has been conducted in the field of wireless sensor networks (WSN). One important part of a WSN node is its communication interface. Starting with proprietary communication systems in early days of WSN, standardized communication interfaces are required today. IEEE 802.15 wireless personal area network (WPAN) task Group 4 has defined a low data rate communication protocol with low energy consumption. IEEE 802.15.4 defines physical (PHY) and medium access control (MAC) layers and is widely used in current WSN. This standard forms a basis for other emerging upper layer standard protocols, like ZigBee, WirelessHART, and SP100. While those protocols are established in home automation, building automation, and factory automation, Internet protocol (IP) connectivity is still desired.

51.2 Why IP in WSN and For Smart Cooperating Objects?

IP is the predominating protocol in network infrastructures for information and communication systems, ranging from local area networks (LANs) to wide area networks (WANs). However, IPv4 is obsolete considering the envisioned trillions of devices, which will appear in the near future. IPv6, the next-generation IP, expands address space of IP to 2^{128} addresses and includes miscellaneous improvements based on experiences with IPv4.

51.2.1 IP as Open Standard Protocol Instead of Proprietary Protocols

On low-power wireless devices like WSN nodes and for smart and cooperating objects (SCO), Zigbee over IEEE 802.15.4 is the predominating protocol today. Alternatively, for machine-to-machine communication (M2M) and for integration purposes in enterprise systems, IP would be a suitable contributor. Until recently, IP was not an option because there were no IP stacks available, which would run on resource-constrained WSN nodes. However, today resource optimized IP stacks are available which run on tiny 8 bit microcontrollers like TI's MSP 430—a commonly used microcontroller in WSN. The footprint of such IP stacks for microcontrollers is around 12 k byte (cf. uIPv6 [Dur08]).

51.2.2 IP Routers Instead of Complex Gateways

WSN lie on the edge of networks or at the field level in process automation. Today, specific gateway or proxy concepts are used to integrate WSN with their proprietary protocols into enterprise systems or higher value services in IP-based networks. Changes in protocol data representation and data encoding require high efforts on the necessary application layer (APL) gateways. In contrast to gateways, routers just forward packets and do not need to have a deep knowledge about the mission of the specific WSN and applied protocols.

51.2.3 Use of Already Existing Protocols, Tools, and Applications

Hooked on IP, other higher level protocols, tools, and applications have been established. Protocols like DNS, DHCP, Zeroconf, etc., can be used for naming, addressing, discovery, and configuration. Tools like SNMP, ping, arp, traceroute, etc., are usable for management purposes. Even on APL, matured protocols, data models, and services, such as HTTP, HTML, XML, SOAP, REST, etc., are well established and fundamental for important IT developments.

51.2.4 New Architectural Styles like Service-Oriented Architectures

New methodologies and paradigms like service-oriented architectures (SOA) would benefit from IP-based communication [Boh09]. Software and service industry uses Web services as glue between different applications for integration and interoperability. WSN are integrated in networking environments by using Web services via complex APL gateways. But this is not a good solution on a long term. In future applications, new concepts and architectures are possible and much research is necessary to find acceptable solutions for service orientation in WSN. Challenges are, e.g., to find acceptable compression for XML data representation, to specify Web service-compliant profiles (similar to the devices profile for Web services—DPWS—for WSN), to realize cost-effective SOA solutions, and to reduce energy consumption for service-oriented solutions for WSN [Mor09,Mor09_2].

51.2.5 IP Support in Operating Systems

Today, IP is supported by all classes of operating systems ranging from desktop operating systems, embedded and real-time operating systems, to operating systems for WSN like Contiki [Dun04]. IP protocol stacks are available as an integral part of the operating system, as open-source software stacks and commercial solutions. The uIPv6 [Dun07,Dur08] stack is an implementation of the 6LoWPAN protocols described in this section. For proprietary protocols, there are only a few open-source solutions available. They are limited to specific application areas and application niches. Future networks of SCO are not for niche market only. They will pave the way for very new and broad application areas. So, IP radically helps to enforce that development.

51.2.6 Lessons Learned From Maintaining the Internet

New protocols require a deep understanding and a profound knowledge of the internals. By decades of iterations, IP and further protocols are matured, wide ranging, and well understood. This results in reduced educational, implementation, installation, and maintenance costs.

51.3 Introduction in 802.15.4

Developments of new protocols for WSN and SCO have just started and are in progress. Based on IEEE 802.15.4 [IEEE154], which defines the PHY and MAC communication layers as depicted in Figure 51.1, various upper layer protocols are defined by large industrial consortia, e.g., WirelessHART, ISA SP100, ZigBee. Open standards and proprietary industrial consortia standards will be in a close competition. Key features of all protocols are

- Long battery life through low-power concepts
- Self-healing networks
- Scalability and support of large networks
- Low-cost node
- Low data rate

IEEE 802.15.4 is a standard for WPANs with low data rates (LR WPAN). While technologies like Wi-Fi (802.11) and Bluetooth have been developed for high data throughput, the focus of IEEE 802.15.4 networks was power-aware and low-cost design, which results in low data rates (up to 250 kbps).

IEEE 802.15.4 standard distinguishes between two device classes: full function devices (FFD) and reduced function devices (RFD). In accordance to IEEE 802.15.4, a device is "any entity (RFD or FFD) containing an implementation of IEEE 802.15.4 MAC and PHY interface to the wireless medium." A coordinator is an FFD with network device functionality that provides synchronization and other services to the network. A PAN coordinator is the principal controller of a personal area network (PAN). An IEEE 802.15.4 network has one PAN coordinator.

The main difference between RFDs and FFDs is the requirements for resources as RFDs are only applied for simple applications. An FFD can operate in any topology and can run as PAN coordinator, coordinator, and device. RFDs are limited to star topologies and end devices. They are communicating directly with a coordinator, while FFDs can both interact with any type of device (RFD and FFD) and operate in star and peer-to-peer networks. In star topology, all devices are communicating with a PAN coordinator. In peer-to-peer topologies—also known as mesh networks—all devices can communicate with each other (cf. Figure 51.2).

For the communication of FFDs and RFDs, 802.15.4 describes a beacon-enabled mode. RFDs can stay in low-power sleep states for predefined time slots. They only wake up to receive the periodical beacons of the coordinator (FFD). Additional to the beacon signal, further information is included if

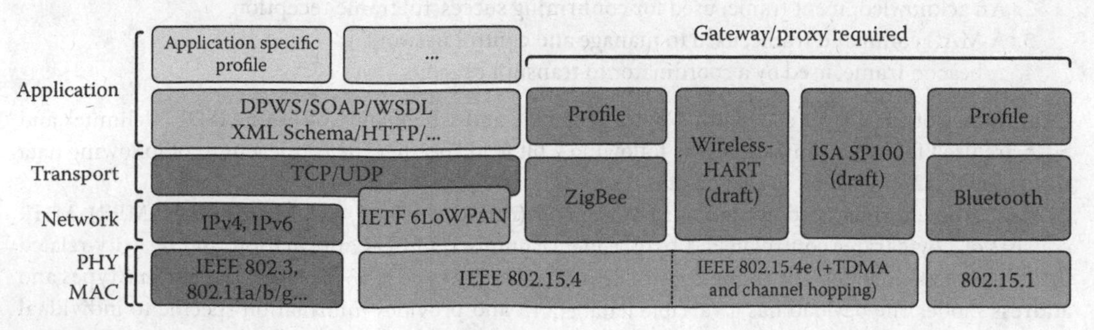

FIGURE 51.1 802.15.4 defined PHY and MAC layers.

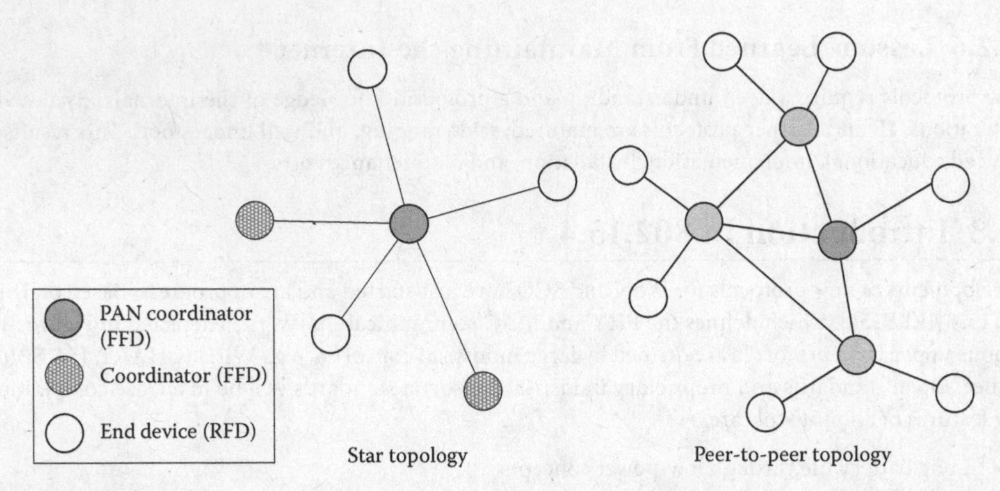

FIGURE 51.2 IEEE 802.15.4 network topologies.

data is pending to be received by the specific RFD and the time slot reserved for the data transmission. Thus, RFDs only need to wake up and activate power-consuming radio for receiving of beacons and data addressed directly to them.

An IEEE 802.15.4 network requires at least one FFD as a network coordinator; all other devices can be RFDs. Because transmission range of LR WPAN is very limited, by means of integrating additional routers, meshed networks can be built. These routers are FFDs. They forward data hop wise and enlarge the dimension of the network.

All devices shall have unique 64 bit IEEE end-system unique identifier (EUI) addresses [EUI64], typically hard coded by manufacturers. This 64 bit address may be used for direct communication in the network. To elide protocol and addressing overhead, alternatively a PAN coordinator may allocate a short 16 bit address to devices that allows theoretically 65,535 end devices to be included in one network. Short addresses are unique for one subset of the network and the same short address can be used in different subset networks. In such a case, the additional 16 bit PAN-ID is used to differentiate between the networks.

The IEEE 802.15.4 standard defines two PHY layers: the 2.4 GHz and 868/915 MHz band PHY layers. The PHY layer depends on local regulation and user preference. The unlicensed 2.4 GHz is valid worldwide and has a higher data rate (240 kbps) compared to 915 MHz band with 40 kbps (United States) and 868 MHz (EU) band with 20 kbps data rate. The radio can operate on 16 channels in the 2.4 GHz band, 10 channels in the 915 MHz band, and 1 channel in the 868 MHz band.

The IEEE 802.15.4 MAC defines four frame structures:

1. A data frame, used for all transfers of data and all later-described 6LoWPAN transmissions (see Figure 51.3)
2. An acknowledgment frame, used for confirming successful frame reception
3. A MAC command frame, used to manage and control network
4. A beacon frame, used by a coordinator to transmit beacons

Data frames of PHY layer start with 4 bytes preamble and 1 byte Start-of-Frame (SOF) delimiter and both are used for synchronization. The following 7 bit field specifies the frame length of following data frame (up to 127), a last bit in that octet is reserved.

MAC frame format is composed of MAC Header (MHR), Payload, and MAC Footer (MFR). MHR consists of 2 byte frame control field, 1 byte sequence number, address information, and security-related information (4–20 byte). The frame control field specifies the type of above-mentioned frame types and address mode. The payload has a variable data length and provides information specific to individual frame types. The frames end with 2 byte checksum.

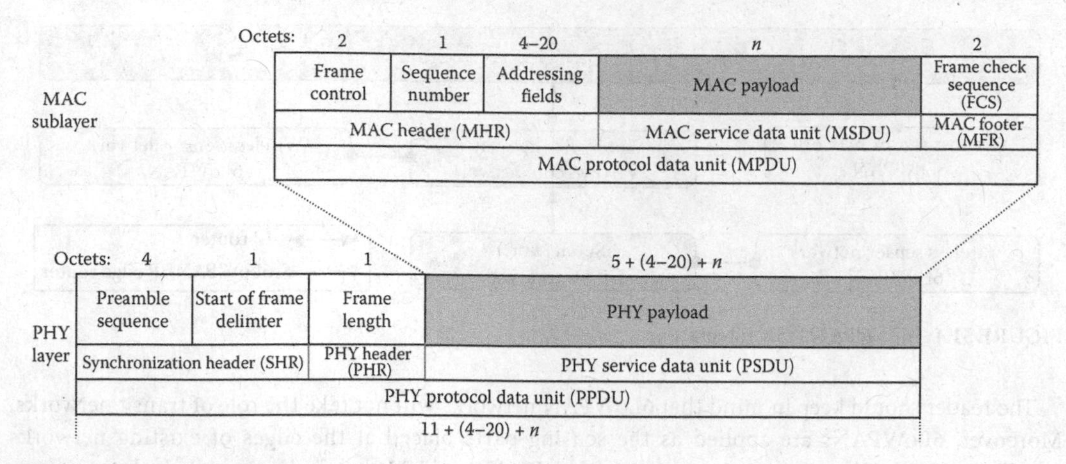

FIGURE 51.3 IEEE 802.15.4 PHY and MAC frame format.

51.4 802.15.4 and 6LoWPAN

The number of wireless networking nodes is expected to rise in future, whereby low power and low cost "are essential enablers toward their deployment in networks" [IETF6LoW]. These networks are characterized by significantly more devices than current LANs, severely limited code and RAM space, and unobtrusive but very different user interfaces for configuration [IETF6LoW]. The assumptions for IEEE 802.15.4 made above like peer-to-peer network topology and limited frame size require further efforts to enable emerging IPv6 [RFC2460] on top of 802.15.4 as MAC and PHY layers. Thus, the IETF founded the 6LoWPAN [IETF6LoW] working group (IPv6 over low-power WPAN), which has produced currently two RFCs:

- IPv6 over low-power wireless personal area networks (6LoWPANs): overview, assumptions, problem statement, and goals (RFC4919)
- Transmission of IPv6 packets over IEEE 802.15.4 networks (RFC4944)

The RFC4919 document describes the requirements that have to be considered for realizing IPv6 in low-power wireless networks and focuses on IEEE 802.15.4 based on application scenarios of 6LoWPAN protocols. The RFC4944 instead describes different dedicated compression formats and required adaptations and enhancements for IPv6 on IEEE 802.15.4.

Additionally, the working group has published four documents "to ensure interoperable implementations of 6LoWPAN networks" and to "define the necessary security and management protocols and constructs for building 6LoWPAN networks, *paying particular attention to protocols already available*" [IETF6LoW]. These documents and further efforts in and around the 6LoWPAN working group focus on applicability of, e.g., neighbor discovery (ND), MIB, and SNMP. Reuse of existing technologies and protocols is a core characteristic of 6LoWPAN and related efforts. The 6LoWPAN working group and miscellaneous draft documents are still active at the time of writing this section. Therefore, this section concentrates on the existing RFCs, explanations of the core concepts, and not on still active and intensively discussed drafts. Routing protocols for 6LoWPAN networks are not focused by the 6LoWPAN working group at all and will be defined by the routing over low-power and lossy networks (ROLL) [IETFROLL] working group. Hence, the 6LoWPAN working group defines only recommendations for the ROLL working group based on their experiences and provides header fields that may be used by future routing protocols. Further deployments of existing and adopted application layer protocols like HTTP, SOAP, etc., are examined in the 6LoWAPP activities of the IETF. 6LoWAPP is not an IETF working group, but the intention is the creation of new working groups or further activities, which will lead to specifications, for 6LoWPAN networks.

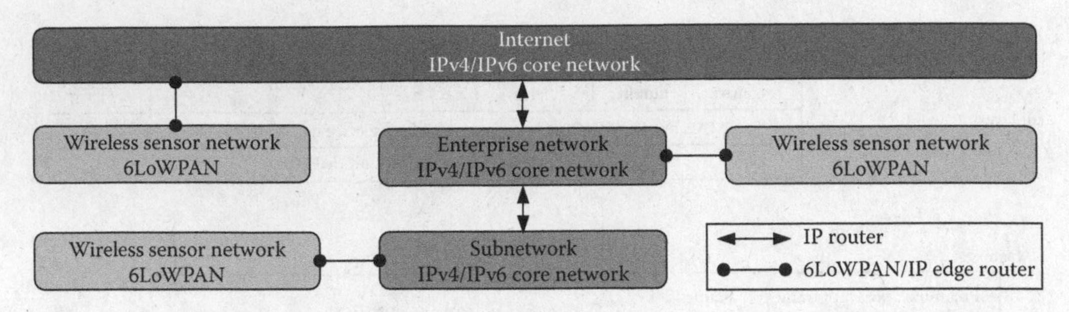

FIGURE 51.4 6LoWPAN WSN integration.

The reader should keep in mind that 6LoWPAN networks will not take the role of transit networks. Moreover, 6LoWPANs are applied as the sensing parts placed at the edges of existing networks (see Figure 51.4). Hence, routers connecting 6LoWPANs and IPv6 networks are called edge routers.

To deploy IPv6 on 802.15.4, 6LoWPAN concentrates on two major points: fragmentation and header compression. For a comprehensive specification, 6LoWPAN also includes rules for stateless address autoconfiguration, address mapping, and security.

Stateless Address Autoconfiguration. Because of the expected high number of nodes within WSNs, stateless address autoconfiguration is required. 6LoWPAN examined how to use the EUI-64 device identifier, assigned to every 802.15.4 device, to generate an identifier for the IPv6 interface.

Addressing Mapping. As 802.15.4 does not support multicast natively, a mapping of IPv6 multicast to the 802.15.4 link layers have to be examined. For an optimized header compression, further mapping even of unicast IPv6 addresses is required.

Fragmentation. The maximum PHY layer packet size of IEE 802.15.4 is 127 bytes. The maximum MAC layer frame size is 102 octets. If link layer security is enabled, further overhead is imposed and 81 octets are left for upper layers. This is far below the minimum packet size (MTU) of 1280 octets required by the IPv6 specification. Thus, 6LoWPAN must provide a fragmentation and reassembly layer between the MAC and the IP layer.

Header Compression. Due to the limited packet size of 802.15.4, protocol headers have to be reduced to provide a suitable payload to header ratio and to avoid protocol overhead. IPv6 basic headers have a size of 40 octets. UDP headers use 8 octets and TCP headers need 20 octets. This leaves 33 octets for UDP and only 21 octets for TCP communication. Thus, a header compression scheme is defined by the 6LoWPAN working group.

Security. The application of WPAN implies requirements for data security, e.g., for personal health records. The working group has also considered the secure exchange of messages and analyzed how the available link layer security of 802.15.4 can be used to ensure secure IP traffic.

Routing. As stated above, 6LoWPAN does not focus on routing of IPv6 on 802.15.4. Hence, the 6LoWPAN does not provide a full specification for mesh routing. Nevertheless, existing routing over IP protocols and routing under IP over MAC layer in a mesh network are examined and specific frame types are defined to be used by the mesh routing under protocols. For more detailed information about *route over* (IP) and *mesh under* (IP) routing, please refer to the ROLL working group [IETFROLL], which is still examining different strategies for routing concepts in the time of writing this section.

A common goal of all efforts in the 6LoWPAN working group is the minimization of required bandwidth and power consumption, protocol overhead, packet overhead, and processing requirements due to the limited resources. Thus, the presented issues could not have been solved one after the other but requires cross layer design. For example, address autoconfiguration has to collaborate with header compression schemes and network management as well. As key concept, 6LoWPAN elides any redundant information/header fields, which already exist in the lower layers or may be derived by their combination.

51.5 6LoWPAN

In general, IEEE 802.15.4 defines four types of frame formats: beacon frames, MAC command frames, acknowledge frames, and data frames. For 6LoWPAN-compliant transmission, only data frames are used. Furthermore, 802.15.4 provides beaconed and non-beaconed data transmission. It is recommended to use the beacon-enabled mode because the beacons are useful for the device discovery with upper layers. Nevertheless, no dedicated mapping of beacons on device discovery protocols is described by 6LoWPAN.

The following subsections are derived from the RFC4944, submitted by the 6LoWPAN working group. The concepts and meanings are explained and additional comments are added. At the end of this section, the different described frame and header types are summarized and compared. Further, header compression and improved multicast mapping will be defined by the working group in their currently active drafts. Because the drafts are not matured in the time of writing this section, they cannot be fully examined here.

51.5.1 Address Autoconfiguration

An IPv6-compliant interface address is 128 bit wide with 64 bit for the interface identifier and a 64 bit prefix. Two different address types are defined by 802.15.4 that can be used to generate an IPv6-compliant interface identifier: the 16 bit short address and the extended IEEE 64 bit address. The 64 bit interface identifier may base on the—for every device unique—EUI-64 identifier [EUI64] but can also be changed by the software network stack. If this 64 bit address is applied, the interface identifier can be formed in accordance to the "IPv6 over Ethernet" (RFC2464) specification [RFC2464]. The shorter 16 bit addresses are used by 802.15.4 nodes inside a specific PAN after an association event with the coordinator. If 16 bit short addresses are applied, a 48 bit address can be generated, which in turn is converted into a 64 bit address following RFC2464 also [RFC2464].

Generating a 64 bit interface identifier out of 16 bit short addresses cannot grant that the derived addresses are globally unique, which has to be announced by setting the universal/local bit—the seventh bit in an IPv6 interface identifier—to zero. When generating an interface identifier out of the 16 bit short address used in one 802.15.4 PAN, the derived address is limited to the lifetime of the connection with the PAN coordinator. Thus, it might be applicable to use 64 bit addresses directly if possible.

The 64 bit prefix for the IPv6 address is obtained by the (edge) router in accordance to RFC4862 "IPv6 Stateless Address Autoconfiguration" [RFC4862]. "Neighbor Discovery (ND) for IP version 6 (IPv6)" [RFC4861] e.g., duplicated address detection and finding of routers, makes extensive use of IP multicast. There is no native support for multicast in IEEE 802.15.4 and thus further efforts are required. Furthermore, RFC4861 assumes all devices of a subnet communicating on one single link. These two points make standard ND not applicable to be applied in 6LoWPAN networks. It should be noted that there is an active document of the 6LoWPAN working group, which describes an alternative ND for 6LoWPAN networks. The routers advertise their presence with periodical multicast messages. This requires almost all nodes to wake up periodically for power consuming receiving and/or forwarding of multicast messages. The implicit unreliable communication in LoWPANs and the absence of multicast messaging capabilities on the link layer make standard ND protocols not practical.

51.5.2 Frame Types and Fragmentation

The 802.15.4 specification defines a maximum packet size of 127 byte for physical packets. The number of available octets is further reduced by the MAC layer header and footer. IPv6 requires a minimum MTU of 1280 octets. To fit the required IPv6 MTU, a separate fragmentation and reassembly layer has to be provided on top over MAC layer and below IP layer (known as layer 2.5). Thus, it is required to define additional frame formats to the standard IPv6 frame. All frame types are prefixed by an 8 bit wide

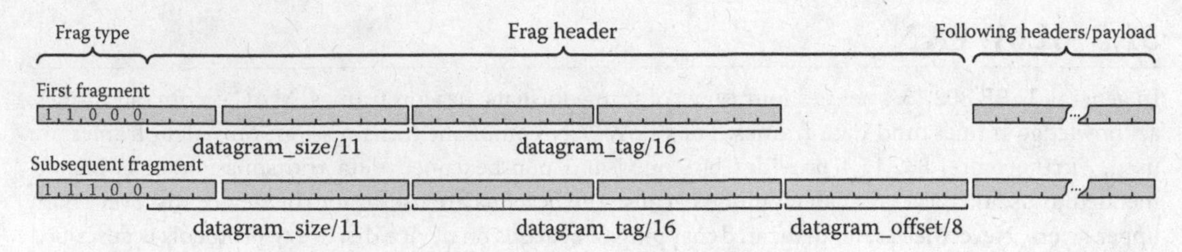

FIGURE 51.5 Fragmentation header structure.

dispatch field, which specifies the frame type and what specific frame type format follows this dispatch field. Four groups are defined and identified by the first two bits of the dispatch field: Not a LoWPAN frame (NALP), mesh frame (MESH), first fragmentation frame (FRAG1), and subsequent fragmentation frame (FRAGN).

If the overall packet size exceeds the limit of the MAC layer, the fragmentation frame as depicted in Figure 51.5 has to be applied. Because the FRAG frame header is located at layer 2.5 between link layer and IP, the FRAG header encloses the IP frame. The IP header is included in the first fragment only and interpreted on the destination node after reassembling the IPv6 packet and not on intermediate nodes. A maximum 2.048 bytes can be transmitted within one IP packet. Packets with a bigger size have to be fragmented on IP layer or above. The MTU of IPv6 requires only 1280 bytes. Different existing IPv6 stacks for computer and server platforms autonomously fragment larger frames independent of the minimum available MTU on the route to the destination to avoid the need of resending packets. This can reduce communication latency and memory requirements because it is assured that the frame size complies with all link layer technologies on the route to the destination. It might be applicable not to exceed the MTU of 1280 bytes per frame in 6LoWPANs also and realize a fragmentation on IP layer or above. The complete fragmentation frame header size differs between 32 bit for the first fragment and 40 bit for subsequent fragments. In accordance to the IPv6 reassembly timeout, a receiver of a fragmented packet must discard received fragments if the packet could not be reassembled within 60 s after the first fragment was received. The 802.15.4 specification defines a beacon-enabled mode, where RFDs only wake up in defined time slots to receive a beacon from their coordinator. This beacon contains information if data are pending to be received by the endpoint. The beacons are transmitted in periodical time slots between 15 ms and 252 s. If beacon-enabled mode is deployed on the 802.15.4 layers, the beacon distance has to fit—particularly in multi-hop scenarios depending on the network topology—to ensure a maximum delay of 60 s between the endpoints.

51.5.3 Mesh Frame Type

For mesh network topologies (named peer-to-peer in IEEE 802.15.4) and multi-hop scenarios, an additional mesh frame header is applied that encloses all other frame types including FRAG type, which in turn encloses the IP frame (cf. Figure 51.6). The additionally required mesh header includes three parts. The first part defines the format of the addresses for originator and final interface, which can be either the 16 bit short addresses or 64 bit EUI-64 addresses format. The second part is the Hops Left field that defines after how many hops the packet can be discarded by the forwarding node to avoid routing loops and can have a maximum value of 256. The third field includes the originator and final address in the format specified by the first field. Depending on the address format, the complete MESH frame header has a size of 5–17 bytes.

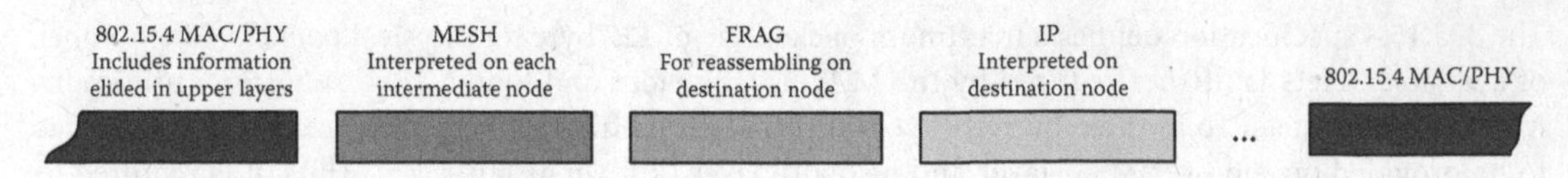

FIGURE 51.6 Frame formats and order.

The mesh frame includes address information in addition to the MAC and PHY layers address information. This is required because in multi-hop network topologies, every forwarding node of a mesh frame has to adapt the source and destination addresses in the link layer. The overall originator address and the overall final destination address are given by the mesh frame address field and stay unchanged. Before transmitting, the forwarding node analyzes the mesh final destination address field to determine the overall destination. If the node itself is not the final destination, the Hops Left field is reduced. If the result is zero, the packet is discarded. Otherwise, the node invokes the link layer routing table to retrieve the next node toward the final destination. This address of the next hop is put into the link layer address field. The source address in the link layer is changed to the address of the current node and the packet is transmitted.

The mesh frame type can be used by routing protocols to be developed in particular by the IETF ROLL working group. Later on (cf. Section 51.5.6), applicability of the mesh header and, thus, routing strategies of IP packets in 6LoWPAN networks are discussed.

51.5.4 Broadcast and Multicast Address Mapping

As IPv6 makes extensive use of multicast messaging, 6LoWPAN specifies an additional compressed multicast mapping as well. There are no multicast concepts defined in IEEE 802.15.4 natively. IPv6 multicasts must be carried by 802.15.4 link layer broadcast messages, if no routing on IP layer is available. The broadcast message includes a specific PAN-ID. Thus, the messages are forwarded by the routers to the PAN in question only. IEEE 802.15.4 defines 0xFFFF as 16 bit short address for broadcast messaging. In addition to this broadcast mapping of multicast messages, for mesh-enabled networks, support of further address transformations are possible. Like all IPv6 addresses, the multicast addresses have a size of 128 bit (16 octets). For 6LoWPAN messaging, these addresses are converted to 16 bit addresses to allow integration into the short address scheme of IEEE 802.15.4 (cf. Figure 51.7). Each of these 16 bit addresses starts with the predefined 3 bit preamble (100, should be common for all multicast addresses in 6LoWPANs), followed by the last 5 bits of the 15th octet of the IPv6 multicast address. The last 8 bits of the mapped address are copied from the 16th octet of the 128 bit address. The full specification of a multicast mesh routing algorithm is out of scope of the 6LoWPAN specification. Nevertheless, such a specification is a vital requirement to avoid flooding low-power WSNs with multicasts mapped on broadcasts (cf. broadcast storm) and will be focused in particular by the ROLL working group.

51.5.5 Header Compression

The transport-specific IPv6 addresses, which are used in 6LoWPAN networks, can be generated by the EUI-64 link layer identifier that is unique for every device. Hence, the additional addressing information in the IP layer are redundant. Thus, 6LoWPAN provides header compression schemes for address information and additional fields like length indications to avoid redundant data in the different layers. All header fields, which are already present in the 802.15.4 fields, are elided in further headers, which is a key design decision in 6LoWPAN. Basically, the header compression is applied without assuming context information for the payload or additional information about the frame (stateless header compression). A full compressed 6LoWPAN header compliant with RFC4944 has a size of only 2 octets in

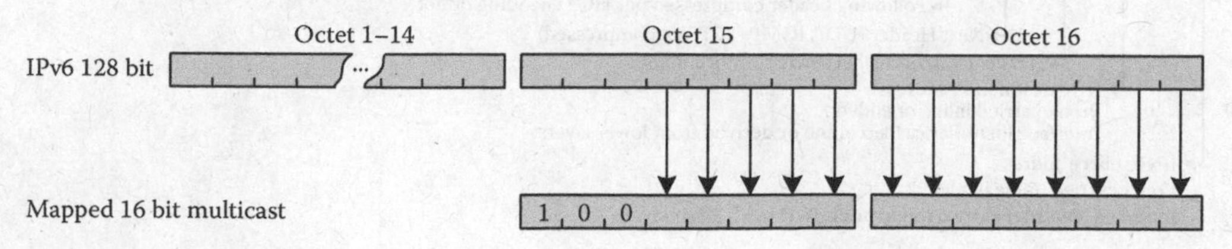

FIGURE 51.7 Derive 16 bit short address for multicast addressing.

opposite to 40 octets for standard IPv6 headers. Because the following described header compression is not fully compliant with the IPv6 specification, compressed headers might be used for messaging inside the 802.15.4 network only. For message exchange with external networks via dedicated edge routers, compliant IPv6 headers might be used. If compressed headers are applied for external messaging also, the edge router is responsible for translating the compressed headers in compliant IPv6 headers. The current 6LoWPAN specification describes header compression for IPv6 and UDP headers also. Compression schemes for TCP and further protocols are not described yet.

51.5.5.1 IPv6 Header Compression

The IPv6 header compression is described by the HC1 encoding. This HC1 encoding scheme is independent from IPv6 and unique for 6LoWPANs. The 128 bit IPv6 transport addresses are separated in 64 bit prefixes derived from the (edge) router of the subnet and 64 bit for the interface identifier. It is possible to define separate compression for the four different address parts of originator and destination: prefix of originator, interface identifier of originator, prefix of destination, and interface identifier of destination. Because 6LoWPANs are estimated to operate within one subnet with one unique prefix, it is not applicable to include the prefixes into the headers for internal communication and can be elided. For communication out of the 6LoWPAN, prefixes can be carried in-line. The interface identifier might be carried in-line also. If elided, in opposite to the prefixes, the interface identifier for the destination address might be either derived from the link layer directly or in meshed networks from the addressing fields in the according MESH header. A compliant IPv6 header additionally includes the IPv6 version identifier, the traffic class field, the flow label field, the hop limit field, and a payload length field. The length field is omitted and derived from the 801.15.4 layers or from fragmentation frame header. The traffic class and the flow label field can be compressed and thereby defined as zero. The IPv6 version identifier is omitted in general. The 8 bits long Hop Limit field is the only field in the IPv6 header that cannot be compressed and must be completely included in-line. The Hop Limit defines the maximum number of IP hops a packet can be forwarded before it is dropped. One IP hop must not be equal with a physical hop because not every intermediate node in a mesh network must be an IP-enabled node. The additional next header field specifies the header type following the IP header, e.g., UDP, TCP, or ICMP. Such next header fields are derived from the evolving IPv6 standard. In opposite to IPv4, IPv6 does not always include all required parts, e.g., fragmentation and routing information in the header. Headers, known as extension headers in IPv6, are added separately depending on the scenario and refer to each other with defined next header fields, comparable to that ones used by the 6LoWPAN compression formats (header chaining). To sum up, the complete RFC4944 HC1 encoded IPv6 header includes at least 2 octets for 8 bit Hop Limit and 8 bit for the LOWPAN_HC1 field (cf. Figure 51.8). The size of the complete header can be between 2 octets and 37 octets plus 4 bit (cf. MTU of IEEE 802.15.4 is 102 byte). The 4 bits are necessary to complete the octet with zero padding and allow routers a more efficient and byte-size-optimized packet handling.

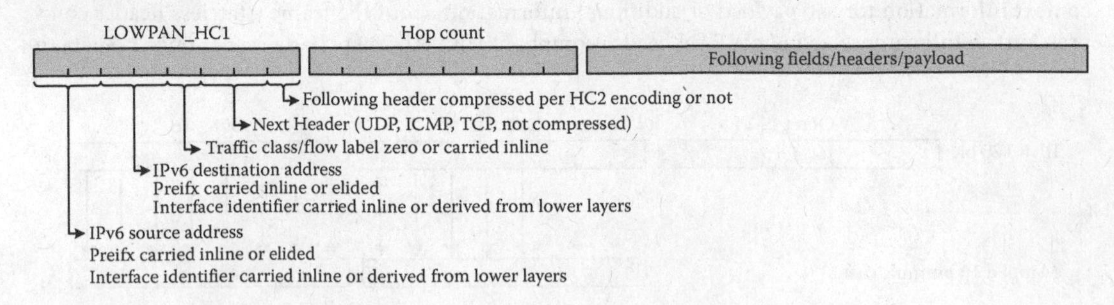

FIGURE 51.8 LOWPAN_HC1 encoding defined in RFC4944.

51.5.5.2 UDP Header Compression

Further, protocol-specific compressions in addition to the above-presented IPv6 header compression are summarized in the RFC4944 HC2 encoding format. In the current version of the 6LoWPAN specification, only the HC_UDP scheme is described for HC2 encoding.

A standard UDP header consists of 2 octets for the source port of the message, 2 octets for the destination port, 2 octets indicating the length of the frame, and 2 octets for the checksum field. This sums up in 8 octets, which can be compressed to 4 octets in 6LoWPAN networks. Like the Hop Limit field in the HC1 header, the checksum field cannot be compressed and must be carried in-line full. According to the IP length field, the UDP length field is omitted by referring to other length fields included in the link layer or the mesh header. The source and destination port can be separately carried full in-line or be compressed to 4 bits. If compressed, the 4 bits determine a port between 61,616 and 61,631. The resulting overall length of the compressed RFC4944 UDP header is between 4 octets and 9 octets.

51.5.6 Scopes

The architecture of wireless sensor mesh networks differs significantly from existing IP-based networks. Networks using wired technologies like Ethernet use one single-hop shared medium and differentiate between unicast messages sent to only one endpoint or broadcast sent to all endpoints. Multicast messages are hybrids carried to multiply endpoints, controlled by the routers in the network. The routers are also responsible for forwarding packages depending on their target address into the correct subnets. Wireless technologies like Wi-Fi achieve a similar behavior by abstraction and link emulation of the used wireless channel. Wi-Fi endpoints can only be connected to one other endpoint (peer-to-peer) or to one access point (AP). This causes the required abstraction and thus the applicability of existing IP-based protocols and technologies. Furthermore, Wi-Fi devices are not subjected by the same energy constraints like WSN.

Emerging meshed sensor networks have more complex topologies with one common wireless communication medium, but in contrast to Ethernet and Wi-Fi, also with limited communication range. Unicast messages cannot be transmitted only to one endpoint without an effect on other endpoints in communication range and might be forwarded by endpoints also and not only by dedicated routers. This has a deep impact on active discussions on scopes and their effect as used in IPv6. IPv6 differs between link-local scope, site-local scope, and global scope. Link-local scope addresses are not forwarded by routers. Global scope addresses are globally unique and thus can be used as target address by every other endpoint. If routing of IP packets is performed under IP layer (route under), the complete 6LoWPAN network might communicate in one single link-local scope and new routing concepts must be developed. Especially for multicast messages, a proper mapping must be found, to avoid flooding of the network. The 6LoWPAN header compression makes use of existing information in the link layers to omit redundant data. Route under concepts require further mappings of other link layers like Ethernet or Wi-Fi to fit the 6LoWPAN requirements or cross-layered solutions. If routing is done on the IP layer (route over), existing concepts might be used but requires every node in the 6LoWPAN, which is not an end node (leaf) in the routing tree, to perform routing. But route over concepts limit the link-local scope to all endpoints in direct communication range. In current developments of IETF ROLL working group, route over concepts are preferred, which would make the mesh header described in this chapter and the broadcast/multicast mapping obsolete.

51.5.7 Summary of Frame Types and Compression Schemes

The miscellaneous frame formats and compression schemes result in various different frame types and combinations depending, e.g., on the scenario, application, and scope of communication. A vital requirement for IPv6 traffic on IEEE 802.15.4 is the adaption to fit into the maximum packet size of 102 byte that is defined for the MAC layer by 802.15.4. Keep in mind that with security on MAC layer enabled, this size is decreased to 81 byte. In this section, different resulting frame header sizes and their combination are summarized. The comparison is presented in Table 51.1 and derived from the currently

TABLE 51.1 Header Comparison Derived from RFC4944

Frame Header	Minimum Size	Maximum Size	Min. Remaining (Assuming 102 byte MAC Frame)	Comment
IPv6-compliant datagram (full standard IPv6 basic header)	40 byte	—	62 byte	Plus optional extension headers
LOWPAN_HC1 compressed (compressed IPv6 header)	3 byte	39 byte	99 byte	Including 4 bit zero padding
Mesh frame (required for multi-hop mesh networking and forwarding)	6 byte	18 byte	96 byte	Including 1 byte for identification of mesh header
LOWPAN_HC1 compressed requiring mesh addressing (compressed IPv6 header + mesh header)	9 byte	56 byte	91 byte	Including 4 bit zero padding and 2 dispatch bytes (for HC1 and MESH)
Fragmentation frame (required if packet size exceeds 802.15.4 MTU and to provide required MTU of IPv6)	5 byte for first fragment	6 byte for subsequent fragments	97 or 96 byte	1 byte for identification of FRAG header
LOWPAN_HC1 compressed requiring fragmentation (compressed IPv6 header + fragmentation header)	8 byte	45 byte	94 byte	Including 4 bit zero padding and 2 dispatch bytes (for HC1 and FRAG) LOWPAN_HC1 compressed header only in first fragment
LOWPAN_HC1 compressed requiring mesh addressing and fragmentation (compressed IPv6 header + mesh header + fragmentation header)	14 byte first fragment 12 byte subsequent fragments	61 byte first fragment 24 byte subsequent fragments	88 or 90 byte	LOWPAN_HC1 compressed header only in first fragment Including 4 bit zero padding
Broadcast/multicast mapping (mapping to keep 16 bit short addressing scheme of 802.15.4)	—	3	Standalone not applicable	Optional, non-mapped addressing can be used with 64 bit destination broadcast/multicast address carried in-line
LOWPAN_HC1 compressed requiring mesh addressing and broadcast/multicast header support (compressed IPv6 + mesh header + broadcast/multicast header)	12 byte	59 byte	90 byte	To support mesh broadcast/multicast Including 4 bit zero padding
UDP (full standard UDP header; additional to underlying layers/headers)	—	8	Standalone not applicable	Excluding IP and other headers
HC_UDP (compressed UDP header; additional to underlying layers/headers)	4	9	Standalone not applicable	Excluding IP and other headers
TCP (full standard TCP header; additional to underlying layers/headers)	20	20 + multiples of 32 bit	Standalone not applicable	Excluding IP and other headers TCP options are optional and multiples of 32 bit
ICMP (full standard ICMP header; additional to underlying layers/headers)	4	—	Standalone not applicable	Excluding IP and other headers Data field is optional

active RFC4944 of the 6LoWPAN working group. It should be mentioned here that active discussions in the time of writing this chapter of 6LoWPAN working group are heading to an update of RFC4944. Hence, the values might differ if the RFC is updated but might provide an overview about approximate header sizes. The third column in the table shows the remaining space for the payload or further protocol headers. The combination of MESH, FRAG, and IP headers with the listed TCP, UDP, and ICMP headers is omitted for clarity.

51.5.8 Security

Specific application domains and scenarios require secure data transmission and data integrity. For example, for health care applications, recorded vital parameters of patients sent via 6LoWPAN over IEEE 802.15.4 should be exchanged with trusted endpoints only. The 6LoWPAN working group analyzed different possibilities to meet the necessary security constraints.

The first security issue that has to be considered is the secure data transmission itself. IEEE 802.15.4 provides AES encryption security on the link layers for secure point-to-point communication. System engineers should always keep in mind that it is not specified in IEEE 802.15.4, how the key for the encryption is deployed. The 6LoWPAN working group proposes to use the built-in AES security for communication of RFDs in terms of IEEE 802.15.4.

6LoWPAN does not exclude the usage of existing security mechanisms and protocols like IPsec and TLS, which allow end-to-end security. Because of the estimated higher energy consumption due to the security on IP layer or above, these security mechanisms are proposed to be used by FFDs only. FFDs are expected to have lower constraints concerning energy consumption. Protocols like IPsec and TLS permit interoperability with external IP-based networks also (cf. Figure 51.9).

The second security issue bases on the method how IPv6-compliant addresses are derived from the 802.15.4 built-in EUI-64 bit device identifier. Hence, this identifier should be unique for every 802.15.4 device, the generated IP addresses are unique as well and might be traced by attackers. There is no protection that the generated IPv6 address is not globally unique. In specific scenarios, persons do not want to be identified through their surrounding wireless body area network based on 6LoWPAN devices. Thus, the 6LoWPAN specifications do not forbid the usage of customizable device identifiers for IPv6 address generating.

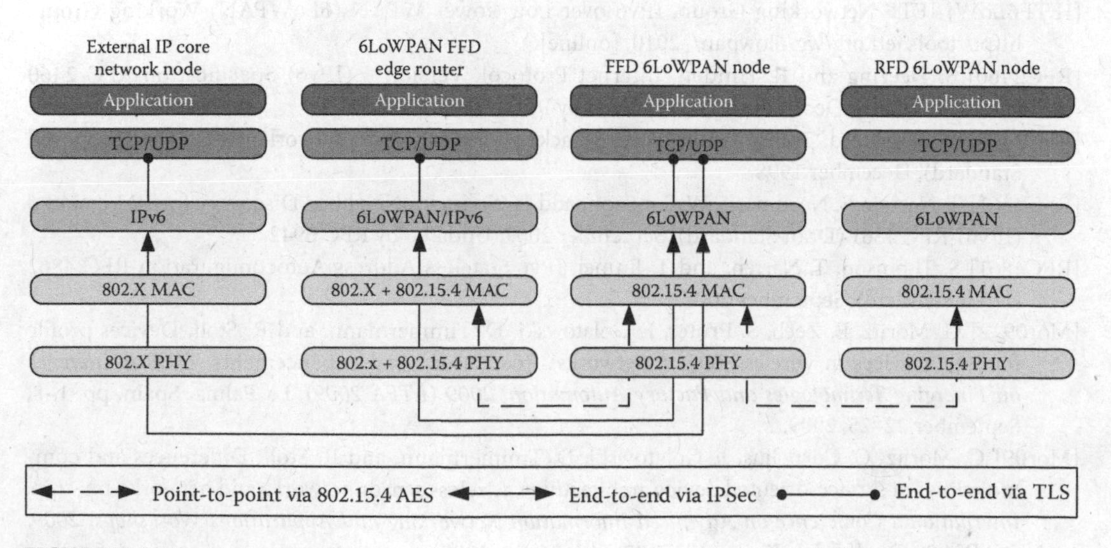

FIGURE 51.9 Security protocols.

51.6 Summary

6LoWPAN realizes the basis of a massively scalable networking and bridges the gap between enterprise networks and massively deployed SCO like wireless sensor and actor networks. 6LoWPAN has shown applicability of IPv6 for low-power, low-rate wireless radio communication based on IEEE 802.15.4 standard. Therefore, 6LoWPAN provides protocols for IP header compression, address autoconfiguration, adoption layer to fit IPv6 specific MTU requirements, etc.

But further efforts are still active to develop new routing concepts tailored for 6LoWPAN networks, focused by the IETF ROLL working group. Additionally, applicability of existing application layer protocols and required adoptions and enhancements to fit the requirements are still missing at the time of writing this section. But 6LoWAPP activities in IETF will bear different solutions based on known and matured protocols instead of developing new ones.

References

[Boh09] H. Bohn and F. Golatowski, Web Services for embedded devices, In R. Zurawski (Ed.), *Embedded Systems Handbook: Embedded Systems Design and Verification*, ISBN: 978-1-4398-0755-2, CRC Press, Boca Raton, FL, S. 19-1–19-31, June 2009.

[Dun04] A. Dunkels, B. Gronvall, and T. Voigt, Contiki—A lightweight and flexible operating system for tiny networked sensors, in *Proceedings of the 29th Annual IEEE International Conference on Local Computer Networks*, Washington, DC, pp. 455–462, 2004, November 16–18, 2004.

[Dun07] A. Dunkels, uIP, Technical report, http://www.sics.se/~adam/uip/index.php, 2007.

[Dur08] M. Durvy, J. Abeillé, P. Wetterwald, C. O'Flynn, B. Leverett, E. Gnoske, M. Vidales et al., Making sensor networks IPv6 ready, in *SenSys'08*, Raleigh, NC, November 5–7, 2008.

[Dur09] M. Durvy, J. Abeillé, P. Wetterwald, C. O'Flynn, B. Leverett, E. Gnoske, M. Vidales, G. Mulligan, N. Tsiftes, N. Finne, and A. Dunkels, Making sensor networks IPv6 ready, in *Proceedings of the 6th ACM Conference on Embedded Network Sensor Systems*, Raleigh, NC, November 05–07, 2008, SenSys'08, ACM, New York, pp. 421–422, 2008.

[EUI64] IEEE, Guidelines for 64-Bit Global Identifier (EUI-64) Registration Authority, Technical Report, 2010.

[IEEE154] IEEE Computer Society, *IEEE Std. 802.15.4–2003*, IEEE Computer Society, New York, 2003.

[IETFROLL] IETF Networking Group, Routing Over Low Power and Lossy Networks (ROLL) Working Group, http://tools.ietf.org/wg/roll/, 2010, [online].

[IETF6LoW] IETF Networking Group, IPv6 over Low Power WPAN (6LoWPAN) Working Group, http://tools.ietf.org/wg/6lowpan/, 2010, [online].

[RFC2460] S. Deering and R. Hinden, Internet Protocol, Version 6 (IPv6) Specification-RFC 2460 (Draft Standard), December 1998, Updated by RFCs 5095, 5722, 5871.

[RFC2464] M. Crawford, Transmission of IPv6 Packets over Ethernet Networks-RFC 2464 (Proposed Standard), December 1998.

[RFC4861] T. Narten, E. Nordmark, W. Simpson, and H. Soliman, Neighbor Discovery for IP Version 6 (IPv6)-RFC 4861 (Draft Standard), September 2007, Updated by RFC 5942.

[RFC4861] S. Thomson, T. Narten, and T. Jinmei, IPv6 Stateless Address Autoconfiguration-RFC 4862 (Draft Standard), September 2007.

[Mor09_2] G. Moritz, E. Zeeb, S. Pruter, F. Golatowski, D. Timmermann, and R. Stoll, Devices profile for web services in wireless sensor networks: Adaptations and enhancements, *IEEE Conference on Emerging Technologies and Factory Automation, 2009 (ETFA 2009)*, La Palma, Spain, pp. 1–8, September 22–25, 2009.

[Mor09] G. Moritz, C. Cornelius, F. Golatowski, D. Timmermann, and R. Stoll, Differences and commonalities of service-oriented device architectures, wireless sensor networks and networks-on-chip, *International Conference on Advanced Information Networking and Applications Workshops, 2009. WAINA'09*, Bradford, U.K., pp. 482–487, May 26–29, 2009.

52

WiMAX in Industry

Milos Manic
University of Idaho
Idaho Falls

Sergiu-Dan Stan
Technical University of Cluj-Napoca

Strahinja Stankovic
Ninet Company Wireless ISP

52.1 Introduction 52-1
52.2 The WiMAX Broadband Technology 52-3
Backhaul/Access Network Applications • Relationship with Other Wireless Technologies • WiMAX vs. Wi-Fi
52.3 WiMAX Architecture 52-4
MAC Layer/Data Link Layer • PHYsical Layer • WiMAX Equipment
52.4 The WiMAX Forum and Working Groups 52-6
52.5 Integration with Other Networks 52-6
WiMAX-DSL Integration • WiMAX-3GPP Integration
52.6 Conclusion 52-8
References 52-8

52.1 Introduction

Most recent statistics presented by International Telecommunication Union (ITU) show that wireless mobile services exploded from 11 million subscribers worldwide in 1990 to over 2 billion in 2005 [1]. Worldwide interoperability for microwave access (or WiMAX) has materialized as reaction to the everlasting need for high-speed broadband access, relieving the Internet providers from needed infrastructure costs, especially apparent in rural and remote areas. WiMAX technology merges together the convenience of wireless technology with compelling performance of broadband access and can be seen as the next logical step in the evolution of wireless and mobile technologies, with a goal of providing Internet connectivity to the widest array of ubiquitous mobile devices ranging from laptop PCs, cellular and smartphones, palm pilots, handsets, input/output devices to consumer electronics such as gaming devices, cameras, musical devices like iPODs, and others.

Following the steps of the Wi-Fi technology, the WiMAX has now become available and affordable technology, becoming a name for the standard itself. WiMAX is based upon the IEEE 802.16 wireless metropolitan area networks (WirelessMAN) technology family, which provides specifications of the media access control (MAC) layer and the PHYsical (PHY) layer [2]. The exact term "WiMAX" is not a standard, it is rather a marketing term (produced by WiMAX forum®) that has become synonymous with IEEE 802.16-based networks, in a similar fashion as the "Wi-Fi" has become a synonym for the IEEE 802.11-based wide area networks (WANs). The WiMAX really refers to the subset of IEEE 802.16 capabilities, briefly addressed in following paragraphs.

The WiMAX technology is a standards-based wireless technology for providing high-speed broadband connectivity to not only homes and businesses but also for mobile wireless networks (Figures 52.1 and 52.2). The WiMAX therefore established itself as the so-called "last-mile" (short distance, broadband to home) connection.

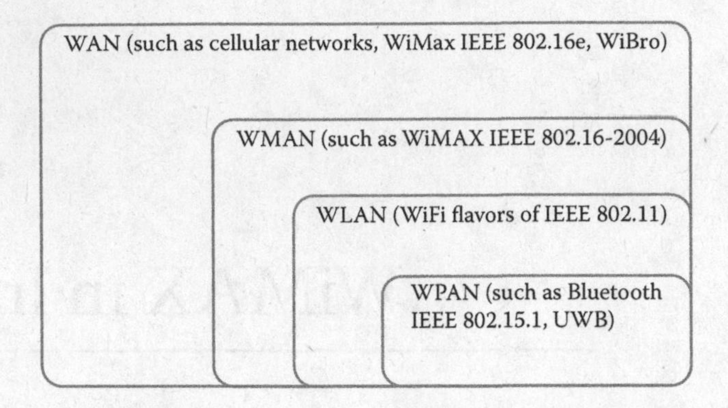

FIGURE 52.1 Comparison of networks types.

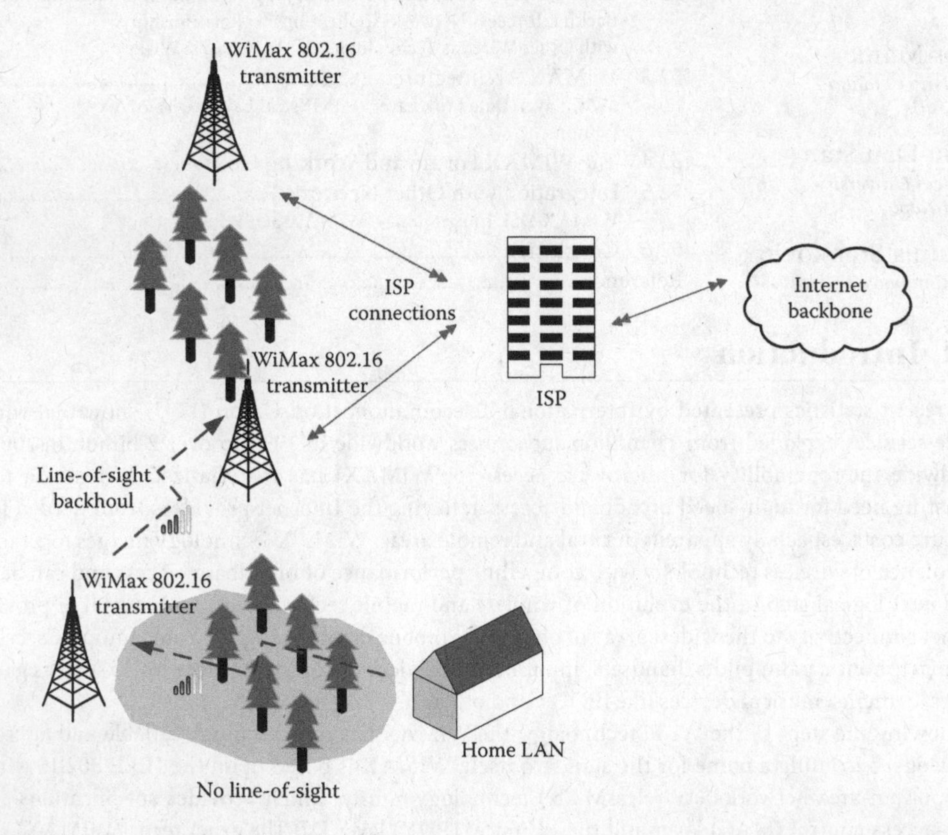

FIGURE 52.2 The example of WiMAX configuration.

Figure 52.1 describes the comparison of several widely used network types: WANs, wireless broadband (WiBro), WirelessMAN, wireless local area networks (WLANs), and wireless personal area networks (WPANs).

WiMAX technology is available in two versions: fixed wireless and mobile. The fixed wireless version (point-to-point), known as 802.16d-2004, was designed as replacement for broadband cable access or DSL. The mobile version (point-to-multipoint), 802.16e-2005 supports roaming in addition to fixed wireless applications. For this reason, the two standards are generally known as *fixed* WiMAX and *mobile* WiMAX.

However, the terms *fixed* WiMAX, *mobile* WiMAX, "802.16," "802.16d," and "802.16e" are frequently used incorrectly. The clarifications follow:

- The IEEE 802 Standards Committee develops LAN standards and MAN standards [3], available for download at http://standards.ieee.org/getieee802/.
- The IEEE 802.16 Working Group on broadband wireless access standards develops standards and recommended practices to support the development and deployment of broadband WirelessMAN [4].
- 802.16-2004 is often called 802.16d, named after the IEEE Standard "Task Group d" that produced it and covers the "Air Interface for Fixed Broadband Wireless Access Systems." Specifically, IEEE Standard 802.16 revised by the Task Group d resulted in 802.16-REVd, later on published as 802.16-2004 (replacing IEEE Standards 802.16-2001, 802.16c-2002, and 802.16a-2003) [5]. It is also frequently referred to as "*fixed* WiMAX" since it has no support for mobility.
- 802.16e-2005 is often called 802.16e, named after the "Task Group e" that developed it. This amendment to IEEE Standard 802.16 is covering "PHY and MAC layers for combined fixed and mobile operation in licensed bands." It introduced support for mobility, among other things and is, therefore, also known as "*mobile* WiMAX" [6,7].

52.2 The WiMAX Broadband Technology

Growing popularity of WiMAX broadband access technology, similarly to cellular phone technologies such as global system for mobile communications or GSM (originally from groupe spécial mobile) and code division multiple access or CDMA, is becoming especially evident in rural areas, as it does not require wired broadband connections.

It is to be expected that WiMAX will, in addition to being used as a wireless backhaul technology for 2G, 3G, and 4G networks, be also instrumental in further deployment and growth of Wi-Fi hotspots. While current operation of most Wi-Fi hotspots is based on wired broadband connections, in future, WiMAX could serve as a faster and cheaper alternative to wired backhaul for these hotspots. Using the point-to-multipoint transmission capabilities of WiMAX to serve as backhaul links to hotspots could substantially improve the business case for Wi-Fi hotspots and provide further momentum for hotspot deployment [8].

52.2.1 Backhaul/Access Network Applications

Backhaul, in telecommunications, typically refers to intermediate links between the backbone and smaller subnetworks of the overall network (local access points [APs] and the backbone network) [2]. The potential to serve as the backhaul connection to the rapidly increasing market of Wi-Fi hotspots paved the way for the WiMAX in the urban but also in non-urban areas. An exploding number of Wi-Fi hotspots is being deployed in public areas such as convention centers, hotels, airports, and coffee shops in the United States and other developed markets [8]. In addition, WiMAX technology can serve as overlay to cable, DSL, data and telecommunication services. Often advertised as the "combination of the mobility you love about your cell phone with the speed you want from broadband," WiMAX providers are continually growing the coverage of "portable Internet" and "Broadband On The Go" [9–11].

Deploying WiMAX in rural areas with limited or no Internet backbone will be challenging as additional methods and hardware will be required to procure sufficient bandwidth from the nearest sources—the difficulty being in proportion to the distance between the end user and the nearest sufficient Internet backbone.

A WiMAX tower station can connect directly to the Internet either via a broadband wired connection or to another WiMAX tower using a microwave, line-of-sight link (LOS). The advantageous characteristic of WiMAX is exactly the backhaul, i.e., the ability to bridge the rural areas via LOS connection to another tower.

TABLE 52.1 The Mobile Standards Comparison

	3G	Wi-Fi: 802.11	WiMAX: 802.16	Mobile-Fi
Max speed	2 Mbps	54 Mbps	100 Mbps	16 Mbps
Area coverage	Several miles	300 ft	50 miles	Several miles
Licensing model	Licensed	Unlicensed	Either	Licensed
Advantages	Range, mobility	Speed, price	Speed, range	Speed, mobility
Disadvantages	Slow, expensive	Short range	Interference issues	High price

TABLE 52.2 WiMAX vs. WLAN and PAN

	PAN	WLAN	WiMAX
IEEE standard	802.15	802.11 a/b/g	802.16 a/d/e
Radius	<30 ft	<500 ft	~5 miles
Data rate	<1 Mbps	~11–54 Mbps	~100 Mbps

Today's market of mobile devices, personal computers, Pocket PCs, and netbooks is increasingly coming as "WiMAX ready," similarly to the devices advertised in the past as "Wi-Fi ready." On April 18, 2005, Intel debuted its first WiMax chip [12]. It featured the plug-in radio unit that could have been replaced to fit to different frequencies while at the same time providing a connection for external antenna.

52.2.2 Relationship with Other Wireless Technologies

The most constructive approach is that Wi-Fi and WiMAX are strongest when working together, however. The mobile standards comparison is presented in Table 52.1.

The IEEE 802.16 radius and data rates are also superior to PAN and WLAN [2] (Table 52.2).

52.2.3 WiMAX vs. Wi-Fi

WiMAX and Wi-Fi frequently get confused due to the wireless Internet access capabilities that both of these technologies provide. For this reason, it is important to recognize their distinctive features:

- Both standards define P2P and ad hoc networks, where an end user communicates to users or servers on another LAN using its AP (Wi-Fi) or base station (WiMAX).
- While WiMAX uses licensed and unlicensed spectrum, Wi-Fi uses unlicensed spectrum. While WiMAX delivers point-to-point (PTP) Internet to user (or another tower), predominantly for WMANs, with a transmission range of several miles, Wi-Fi aims at WLANs, with a transmission range of up to 100 m (such as hotspots).
- While the Wi-Fi standard is based on the contention access MAC called carrier sense multiple access/collision avoidance (CSMA/CA) protocol, WiMAX MAC uses a connection-oriented MAC algorithm where subscriber initially gets a stable time slot assigned by the base station (BS).

52.3 WiMAX Architecture

As we noted at the beginning, WiMAX is not a standard, rather a term grown to become a synonym for the IEEE 802.16 standard, similarly to how Wi-Fi became the term interchangeably used for interoperable implementations of the IEEE 802.11 wireless LAN standard.

The IEEE 802.16 WiMAX architectural model is similar to the one found with cellular telephone networks [2]. Each IEEE 802.16 coverage area consists of one BS supporting several subscriber stations (SSs) that provide access for the end user into the broadband network. The described architecture represents a single cell of network coverage. These cells are further connected through a core network (CN). Transmissions can be PTP, point-to-multipoint (PMP0), or point-to-consecutive-point (PTCP), where SSs can relay data (act as routers), supporting the nodes that do not have the LOS connectivity with BSs [2]. It is important to note that BSs are equipped with wide-beam antennas, where SSs are equipped with highly directional antennas pointing toward the BSs. This is why BSs can provide superior data rates relative to the IEEE 802.11 WLANs, typically equipped with omnidirectional antennas.

The SS-to-BS link is known as uplink, while BS-to-SS is typically referred to as the downlink.

52.3.1 MAC Layer/Data Link Layer

While the IEEE 802.11 Wi-Fi standard is based on the connectionless and contention-based access (MAC) where the subscribers are connecting to a wireless AP (AP) on a random interrupt basis, 802.16 MAC uses a scheduling algorithm for which the subscriber station needs to compete with other subscribers only once (for initial entry into the network). This entails that with Wi-Fi connection, subscribers further away from the AP tend to be repeatedly interrupted by closer subscribers, greatly reducing their throughput. This makes high data rate-dependent services such as Voice over Internet Protocol (VoIP) or Internet Protocol Television (IPTV) difficult to maintain for larger number of simultaneous users.

Unlike with Wi-Fi technology, the 802.16 MAC is based on a scheduling algorithm where subscriber initially gets a time slot assigned by the BS. In addition to being stable under overload and oversubscription, the 802.16 scheduling algorithm unlike 802.11, can also be more bandwidth efficient. The scheduling algorithm can balance the time-slot assignments among the subscribers, hence controlling the Quality of Service (QoS) parameters.

52.3.2 PHYsical Layer

The IEEE 802.16 standard (IEEE, 2004, 2005b) reference model comprises the data, control, and management planes. The data plane protocol stack includes the PHY layer and the MAC layer. Multiple PHY layers are supported, operating in the 2–66 GHz frequency spectrum, and support single- and multicarrier air interfaces, depending on the particular operational environment [13].

The IEEE 802.16 PHY layer provides the following options [2,14]:

- Wireless MAN-SC (Metropolitan Area Network-Single Carrier) is a single-carrier option and is intended for use in LOS environments.
- Wireless MAN-OFDM (Orthogonal Frequency-Division Multiple Access) with 256 subcarriers, which is mandatory in the unlicensed bands. This mode features Time Division Multiplexing (TDM) in the downlink direction and Time Division Multiple Access (TDMA) in the uplink direction.
- Wireless MAN-OFDMA with 2048 carriers. This mode provides OFDMA access by separating the users using FDD in the uplink direction in addition to TDMA.

While the original version of the IEEE 802.16 WiMAX standard, specified a PHY layer operating in the 10–66 GHz range, 802.16a updated in 2004 (802.16-2004), added specifications for the 2–11 GHz range. In 2005, IEEE 802.16-2004 was updated by 802.16e-2005, and introduced the scalable orthogonal frequency-division multiple access (SOFDMA) as opposed to the orthogonal frequency-division multiplexing (OFDM) with 256 subcarriers (of which 200 are used) in 802.16d. The following revisions of IEEE 802.16 such as 802.16e introduced the multiple antenna support through MIMO, which improved the coverage, power consumption, frequency reuse, and bandwidth efficiency, with the added full mobility support. Finally, the WiMAX certification allowed vendors with 802.16d products to sell their equipment as WiMAX certified.

52.3.3 WiMAX Equipment

WiMAX Forum Certified™ equipment is designed and configurable for a range of broadband wireless access deployment scenarios [15]. These scenarios include longer-range LOS deployments in low-density outdoor environments and shorter-range non-LOS deployments in cluttered urban environments. Services can be fixed, portable or mobile, or some combination thereof. Over this range of conditions, the common goal is the capability to reliably deliver broadband connectivity to business and home users. Most radio spectrum that is applicable to WiMAX networks is limited in application and usage. The ITU-R and National Administrations use various well-known band designations to determine which applications are permitted in different frequency bands. Allocations to WiMAX service providers can be typically characterized as either fixed or land mobile. Most National Administrators typically restrict fixed spectrum allocations to fixed and, sometimes, nomadic applications. Conversely, some administrations allow land mobile spectrum allocations to be used for a broader set of applications, which may cover all WiMAX usage scenarios (fixed, nomadic, portable, simple mobility, and full mobility). Land mobile spectrum allocations are also generally linked to IMT-2000 systems. The WiMAX Forum is lobbying for the relaxation of application (usage scenario) limitations. A few administrations are already technology and application neutral. These administrations allow any WiMAX usage scenario to be deployed in certain bands. Over time, it is expected that bands that are initially limited to specific usage scenarios will be relaxed and WiMAX service providers will then offer multiple usage scenarios if the business cases look attractive [15].

52.4 The WiMAX Forum and Working Groups

The WiMAX Forum is an industry-led, nonprofit organization formed to certify and promote the compatibility and interoperability of broadband wireless products based upon the harmonized IEEE 802.16/ETSI HiperMAN standard [15]. A WiMAX Forum goal is to accelerate the introduction of these systems into the marketplace. WiMAX Forum Certified products are fully interoperable and support broadband fixed, portable, and mobile services. Along these lines, the WiMAX Forum works closely with service providers and regulators to ensure that WiMAX Forum Certified systems meet customer and government requirements.

The WiMAX Forum organizes its technical activities through a number of working groups coordinated by a Technical Steering Committee. The primary function of the technical activities is to develop the technical specifications underlying WiMAX Forum Certified products (Figure 52.3).

52.5 Integration with Other Networks

While WiMAX may seem like competing technology, it is really offering means for integration of various choices for Internet access or backhauling. Some of those options, namely the integration with DSL and 3GPP, will be briefly addressed here.

52.5.1 WiMAX-DSL Integration

Cable or DSL represent inexpensive broadband technologies, unlikely to disappear even in developing countries. While DSL or cable may be available within city limits, low-cost WiMAX offers a technology of choice for the rural and remote areas that lack broadband options. WiMAX, as robust mobile, portable, or fixed broadband access opens the door to a whole new world of location-variable broadband Internet access.

In the scenario of WiMAX-DSL integration, an incumbent digital subscriber line (DSL) operator would deploy a WiMAX access network (AN) and would integrate it with the existing DSL backend infrastructure, enabling maximal reuse of the existing switching, service provisioning, and billing infrastructures. The subscriber would perform first network entry and authentication in the WiMAX network, and then enable traffic to the DSL network. In a typical deployment scenario, a DSL deployment

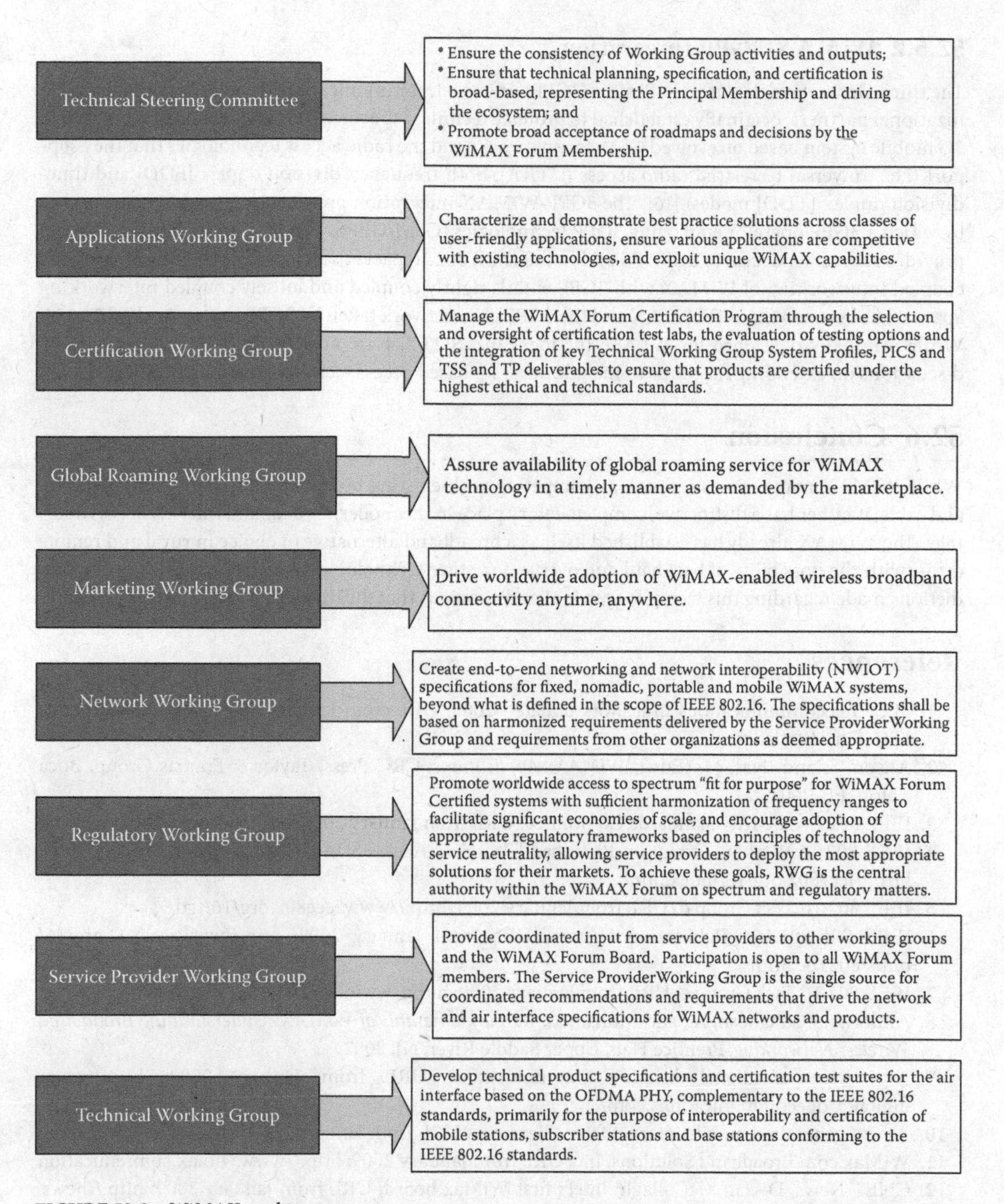

FIGURE 52.3 WiMAX working groups.

with Ethernet switching between the digital subscriber line access multiplexer (DSLAM) and the broadband remote access server (BRAS) can be reused by replacing the DSL link with a wireless WiMAX link with enabled Ethernet convergence sublayer. The higher layer network architecture remains unchanged. In this scenario, DSL reference architecture interface typically exists between the AN and the service providers. This interface connects either an application service provider (ASP) to the network service provider (NSP) owning the AN, or, in roaming scenarios, the NSP with the AN. The *T* interface is defined between the terminal equipment (TE) and the DSL modem in the customer premises network (CPN).

52.5.2 WiMAX-3GPP Integration

The third Generation Partnership Project (3GPP) unites telecommunications standards bodies as organizational partners, originally established to produce technical specifications and technical reports for a 3G mobile system based on evolved GSM core networks and the radio access technologies that they support (i.e., universal terrestrial radio access [UTRA], both frequency-division duplex [FDD] and time-division duplex [TDD] modes) [16]. The 3GPP-WiMAX integration presents interworking approaches by which a 3GPP operator employing 3GPP technologies (GSM/GPRS, EDGE, UMTS, and/or HSDPA) provides subscriber access to 3GPP PS and to the IEEE 802.16 packet service plane. Tightly and loosely coupled interworking of WiMAX with 3GPP entails tightly coupled and loosely coupled interworking approaches between the WiMAX AN and the 3GPP core network taken by 3GPP in standard TS 23.234. While originally developed for WLAN, the ideas in TS 23.234 are now the basis for ongoing 3GPP discussions on extending the scope to include other IP-based access technologies, including WiMAX.

52.6 Conclusion

While WiMAX may seem as technology competing with existing broadband alternatives such as DSL and cable, it rather has a distinctive, complementary place in the modern broadband Internet access offerings. The WiMAX already has established itself as a broadband alternative of choice in rural and remote areas, with the possibility of backhaul and overlay for other technologies as well. Regardless of the predictions made regarding this technology, it is already obvious that the times of WiMAX are yet to come.

References

1. ITU. Telecommunications indicators update—2004. URL from January 2009: http://www.itu.int/ITU-D/ict/statistics/
2. Ahson, S. and Ilyas, M. (Eds.), *WiMAX: Applications*, CRC Press, Taylor & Francis Group, Boca Raton, FL, 2008.
3. IEEE 802 LAN/MAN Standards Committee, URL from January 2009: http://www.ieee802.org/
4. IEEE 802.16 Working Group on Broadband Wireless Access Standards, URL from January 2009: http://ieee802.org/16/index.html
5. IEEE 802.16 Task Group d, URL from January 2009: http://www.ieee802.org/16/tgd/
6. IEEE Std 802.16e: Publication History, URL from January 2009: http://www.ieee802.org/16/pubs/80216e.html
7. IEEE 802.16 Task Group e, URL from January 2009: http://www.ieee802.org/16/tge/
8. Andrews, J.G., Ghosh, A., and Muhamed, R., *Fundamentals of WiMAX: Understanding Broadband Wireless Networking*, Prentice Hall, Upper Saddle River, NJ, 2007.
9. A service of digital bridge communications, URL from January 2009: http://www.digitalbridgecommunications.com/
10. A next generation mobile internet from Clearwire, URL from January 2009: http://www.clear.com
11. WiMax.com Broadband Solutions, Inc, URL from January 2009: http://www.wimax.com/education
12. CNET News, Declan McCullagh: Intel's first WiMax board, URL from January 2009: http://news.cnet.com/Photos-Welcoming-the-WiMax-chip/2009–1006_3–5675441.html
13. Katz, M.D. and Fitzek, F.H.P., *WiMAX Evolution, Emerging Technologies and Applications*, John Wiley & Sons, Chichester, U.K., 2009.
14. Kumar, A., *Mobile Broadcasting with WiMAX: Principles, Technology, and Applications*, Elsevier, Amsterdam, the Netherlands, 2008.
15. WiMAX Forum, URL from January 2009: http://www.wimaxforum.org/; http://www.ieee802.org/16/pubs/80216e.html
16. The mobile broadband standards, URL from January 2009: http://www.3gpp.org/

53

WirelessHART, ISA100.11a, and OCARI

53.1 Introduction 53-1
53.2 WirelessHART 53-2
The WirelessHART Physical Layer • The WirelessHART Data Link Layer • Time Keeping • The WirelessHART Network Layer and Topologies • The WirelessHART Upper Layers
53.3 ISA100.11a 53-8
The ISA100.11a Physical Layer • The ISA100.11a Data Link Layer • Time Keeping • The ISA100.11a Network Layer and Topologies • The ISA100.11a Upper Layers
53.4 OCARI 53-10
Physical Layer • MAC Layer • Network Layer and Topologies • Application Layer
53.5 Coexistence of the Three Protocols 53-13
53.6 Example of Platform Providers 53-15
53.7 Conclusions 53-16
References 53-16

Tuan Dang
EDF Research and Development

Emiliano Sisinni
University of Brescia

53.1 Introduction

The world of industrial communications shows increasing interest toward wireless fieldbuses, i.e., the use of wireless communications to interconnect devices at field level sensors; actuators; instruments; controllers; and so on [W08,FFM07]. In particular, the advent of hybrid wired/wireless networks promises to greatly improve efficiency and scalability. However, the success of this kind of solutions will depend on the standardization process that ensures multivendor compatibility and interoperability.

A big step toward the development of a truly open and standard solution has been done thanks to the introduction of the IEEE 802.15.4 specifications [I06]. In particular, this standard body created a physical layer (PHY) description that maintains good performances in presence of noise and interferences, but it is simply enough to allow system designer to produce low power consumption and low-cost devices. In fact, a wireless sensor for industrial applications must work in a noisy environment on one hand and must ensure an overall device lifespan and cost comparable with traditional wired systems, on the other one. The adoption of a direct sequence spread spectrum (DSSS) strategy mitigates interferences and offset quadrature phase shift keying with half-sine shaping (OQPSK-HSS) modulation ensures good spectral occupancy. However, the adoption of carrier sense multiple access with collision avoidance (CSMA/CA) at the medium access control (MAC) level poses severe limits on the respect of real-time deadlines, a typical requirement of industrial applications.

In order to overcome such limitations, a lot of solutions employing radios compliant with IEEE 802.15.4-PHY but using time division multiple access (TDMA) protocol at the MAC layer have appeared recently. In particular, in this chapter are described the proposal of the Hart Communication Foundation (HCF), also known as WirelessHART [SHM07,KHP08], the proposal of the International Society of Automation (ISA), also known as ISA100 [ISA09], and the Optimization of Ad hoc Communications in Industrial networks (OCARI [D08]) project, funded by the French national research council—ANR.

53.2 WirelessHART

The HART™ (Highway Addressable Remote Transducer) Communication Protocol is one of the earliest fieldbus to appear on the market. It was originally designed to preserve compatibility and improve communication over legacy 4–20 mA analog instrumentation wiring, sharing the pair of wires used by the older system. According to some, considering the huge number of 4–20 mA devices installed throughout the world, the HART protocol is probably the most diffused protocols for industrial communications today.

The protocol was proposed by Rosemount Inc. in the mid-1980s for their smart field instruments. It was based on the Bell 202 communications standard that AT&T developed for a modem operating in the audio frequency range. Starting from 1986, it was made an open protocol, thus speeding up its enhancement, as proved by several successive revisions to the specification.

As regards performances, it has been developed having in mind process automation as the target scenario, and when used in multichannel I/O system, it allows for about one transaction per second (if frequency shift keying (FSK) modulation is adopted), with latencies on the same order of magnitude (depending on the number of nodes). In addition, it is able to work on distances up to 1500 m.

WirelessHART [H09] is an extension of wired HART; it has been developed with the aim of preserving the same performance of the wired counterpart (or even improve them), allowing the implementation of a truly autonomous sensor. In fact, as suggested by its name, it uses radio frequencies (RF) as the communication medium with a nominal coverage area of about 100 m (line of sight) and devices may be powered either by wire, battery, solar, or a combination of these sources. Communicating devices are described in terms according to the well-known Open Systems Interconnection (OSI) Seven-layer communication model [H88]; a comparison with the traditional OSI model communications layers is given in Figure 53.1.

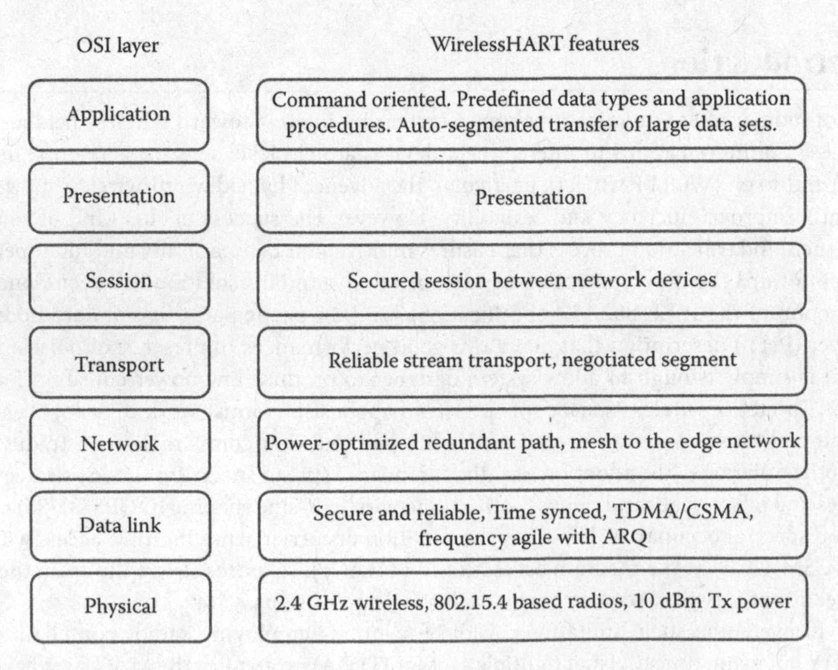

OSI layer
WirelessHART features
Application
Command oriented. Predefined data types and application procedures. Auto-segmented transfer of large data sets.
Presentation
Presentation
Session
Secured session between network devices
Transport
Reliable stream transport, negotiated segment
Network
Power-optimized redundant path, mesh to the edge network
Data link
Secure and reliable, Time synced, TDMA/CSMA, frequency agile with ARQ
Physical
2.4 GHz wireless, 802.15.4 based radios, 10 dBm Tx power

FIGURE 53.1 OSI Seven-layer model and WirelessHART comparison.

53.2.1 The WirelessHART Physical Layer

The lowest layer on the OSI model is the PHY; it is responsible for sending the bits across the network media. It enables the transmission and reception of the protocol data across the physical radio channel. In particular, the WirelessHART PHY is based on a tailored adoption of the IEEE STD 802.15.4-2006 [KHP08], in order to reduce the overall device cost. This standard is intended to conform to regulations in Europe, Canada, Japan, China, and the United States. Only a simplified set of services are defined and only the 2.4 GHz band, which uses OQPSK-HSS with DSSS modulation, is supported. In order to ensure worldwide availability, channel 26 (the one center around 2480 MHz) is not supported since it is not legal in many locales. In this way, transceivers readily available "off the shelf" or custom radio transceivers that are compliant to IEEE 802.15.4 can be adopted. Transceivers must only support some additional restrictions on channel switching time and power up sequence to satisfy the medium access strategy requirements and lower the power consumption. The radio must be capable to set the transmitting power (from −10 up to 10 dBm) and must provide some mechanisms to perform the so-called clear channel assessment (in particular, only the CCA mode 2, carrier sense, of IEEE 802.15.4 is supported). Finally, the radio can operate in two different states: awake, during normal operation; and sleep, when it is inactive.

53.2.2 The WirelessHART Data Link Layer

According to the OSI model, the data link layer (DLL) is the lowest information-centric layer, which exploits services offered by the PHY entities and provides basic low-level messaging among different peer entities. It provides only local addressing; it can forward messages to neighbors, i.e., devices within the local addressing domain, but cannot modify message addresses. Addresses at the DLL have only local scope and can be duplicated in other local links. The MAC layer is part of the DLL and sits directly on top of the PHY; since it controls the radio, it has a large impact on the overall energy consumption, and hence, the lifetime of a node. MAC protocols decide when and how nodes (devices) may access the shared medium, i.e., the radio channel, and try to ensure that no two nodes are interfering with each other's transmissions. WirelessHART uses TDMA and channel hopping to control accesses to the network. It is well known that TDMA is a technique that provides collision-free, deterministic communications and it is the preferable solution when time deadlines must be satisfied. When the TDMA approach is used, communications between devices can occur in time slots. Time slots are grouped together in structures called superframes, which are repeated continuously. Each device must be capable to manage several superframes. Both the slot size and the superframe length (in number of slots) are fixed and form a network cycle with a fixed repetition rate. A schematization of this mechanism is proposed in Figure 53.2.

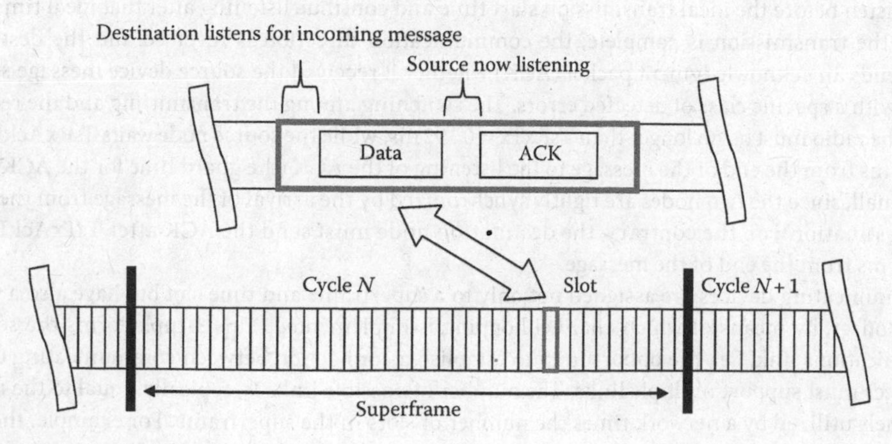

FIGURE 53.2 Basics of WirelessHART TDMA mechanism.

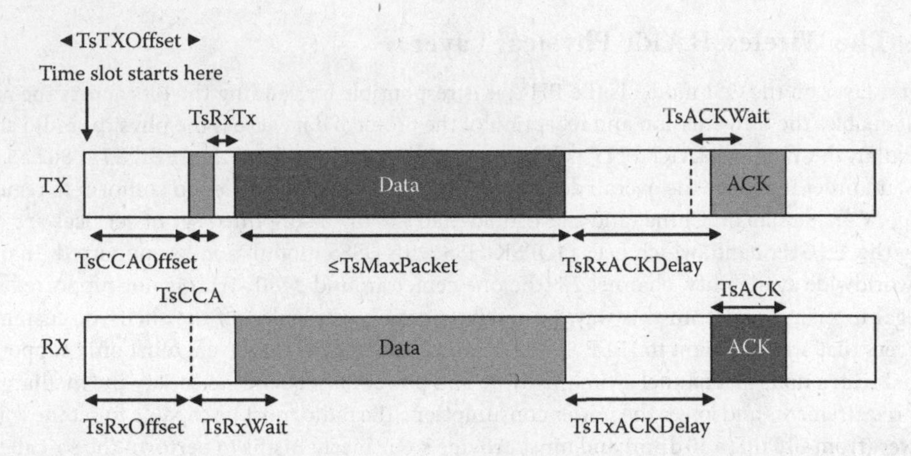

FIGURE 53.3 Packet timing and time slot format. Brighter box is for reception state and darker box is for transmitting state of devices.

For successful and efficient TDMA communications, all nodes within the same network must share a common sense of time; in other words, synchronization of clocks between devices is critical. Common sources of clock drift include temperature and aging. Consequently, tolerances on timekeeping and time synchronization mechanisms are specified to ensure that devices "exactly" know when the start of a slot occurs. The format of a timeslot is shown in Figure 53.3.

In this figure, it is shown how a unicast message exchange occurs between a transmitting source device (call it "A") that is transmitting to a destination device (call it "B"), which is listening. Assuming that both nodes are already synchronized, then "B" knows when to expect the first bit of the preamble of the message coming from "A" (TsMaxPacket is the duration of the longest packet allowed by the IEEE 802.15.4 standard, i.e., 4.256 ms).

There is an initial CCA (which starts with a TsCCAOffset = 1.8 ± 0.1 ms delay and lasts TsCCA = 0.128 ms) that is not strictly needed but has been taken into account in order to improve coexistence with other wireless networks. Within the slot, the actual transmission of the source message has a delay with respect to the beginning of a slot (TsTxOffset = 2.12 ± 0.1 ms in the Figure 53.3). This short time delay allows the source and destination to set their frequency channel and allows the receiver to begin listening on the correct channel. Also, the receiver has a delay with respect to time slot beginning (TsRxOffset = 1.12 ± 0.1 ms in Figure 53.3), needed to set up the radio. In addition, since there is a tolerance on clocks, the receiver has a guard interval (TsRxWait = 2.2 ± 0.1 ms in Figure 53.3), i.e., it must start to listen before the ideal transmission start time and continue listening after that ideal time.

Once the transmission is complete, the communication direction is reversed and the destination device sends an acknowledgment packet (ACK) whether it received the source device message successfully or with a specific class of detected errors. The switching among the transmitting and the receiving state of the radio must be no longer than TsRxTx = 0.192 ms, while the source node waits TsRxAckDelay = 0.8 ± 0.1 ms from the end of the message to the listening of the ACK (the guard time for the ACK arrival can be small, since the two nodes are tightly synchronized by the arrival of the message from the source to the destination); on the contrary, the destination node must send the ACK after TsTxAckDelay = 1.0 ± 0.1 ms from the end of the message.

Communicating devices are assigned not only to a superframe and time slot but have also a specific channel offset, by means of which channel hopping is implemented. This 3-tuple forms the so-called communications link, i.e., the opportunity to establish a connection between communicating devices. All devices must support multiple links. The number of possible links is, typically, equal to the number of channels utilized by a network times the number of slots in the superframe. For example, the use of 15 channels and 9,000 slots per superframe results in 135,000 possible links.

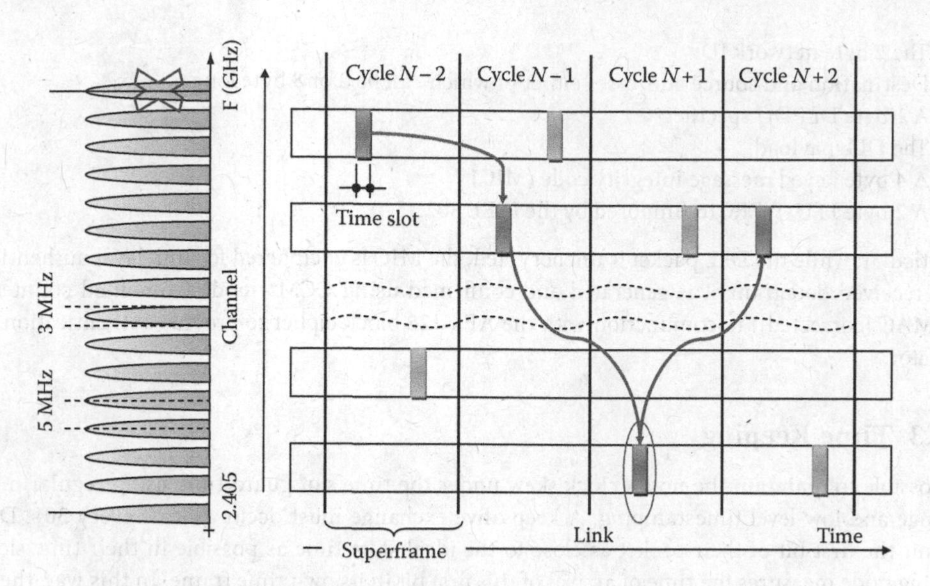

FIGURE 53.4 The frequency diversity mechanism of WirelessHART.

Usually, only two devices are assigned to a given slot (source and sink of the transaction). However, shared time slots are also considered; since collisions may occur, the traditional CSMA/CA MAC strategy is adopted.

Channel hopping (see Figure 53.4) is combined with TDMA in order to enhance reliability. It provides frequency diversity, which can avoid interferers and reduce multipath fading effects.

Channel hopping means also channel diversity, which is each slot may be used on multiple channels at the same time by different nodes. This can be achieved by creating links on the same slot, but with different channel offsets. The assignment of a link occurs in a centralized way, by the network manager described in the following subsection. In addition, the channel blacklisting mechanism, which allows the network administrator to restrict the channel hopping of network devices network-wide to selected channels in the RF band, has also been implemented. This feature could be useful in a crowded RF environment, increasing coexistence. However, it must be noticed that WirelessHART communication (like WiFi, Bluetooth, and other wireless communication) is very random and bursty in nature; consequently, blacklisting seldom provides tangible benefits.

53.2.2.1 The Data Link Layer Datagram

This subsection specifies the format of the data link datagram (DLPDU, see Figure 53.5). Each DLPDU consists of the following fields:

- A single byte set to 0x41
- A 1 byte address specifier
- The 1 byte sequence number

Data link layer
Data link payload

0x41	Address specifier	Sequence number	Network ID	Destination address	Source address	DLPDU specifier		MIC	CRC

Physical layer

Preamble	Delimiter	Length	

FIGURE 53.5 The datagram at the data link layer.

- The 2 byte network ID
- Destination and source addresses either of which can be 2 or 8 byte long
- A 1 byte DLPDU specifier
- The DLL payload
- A 4 byte keyed message integrity code (MIC)
- A 2 byte ITU-T CRC16 (imposed by the IEEE 802.15.4)

In particular, while the DLL packet is unencrypted, the MIC is enciphered for link layer authentication of the received datagram; it is generated and confirmed using CCM* mode (combined counter with CBC-MAC [corrected]) in conjunction with the AES-128 block cipher to provide authentication of the originator.

53.2.3 Time Keeping

It is possible to maintain the nodes clock skew under the time slot guard time using regular message exchange and low-level timestamping. A keep-alive exchange must occur at least every 30 s. Devices transmit the first bit of their packet as close to the ideal start time as possible in their time slot. The receiving node measures the time of arrival of this first bit in its own time frame. In this way, the destination knows the difference (offset) between its time slot and the transmitter's time slot. This information is then sent back to the source in the ACK packet in order to compensate for it.

53.2.4 The WirelessHART Network Layer and Topologies

The WirelessHART protocol allows to implement a sensor and actuator mesh communication system (see Figure 53.6).

Every WirelessHART network contains different kind of devices (logical and/or physical):

- One and only one security manager that distributes encryption keys to the network manager of each network
- One and only one active network manager, whose aim is to form the network, schedule resources, configure routing paths, etc.
- At least one gateway, whose aim is to interconnect field devices with the plant automation system by means of access points
- Several field devices that are connected with the process, i.e., devices with sensors and actuators

All communications occur, e.g., by moving data from the gateway, through the intermediate devices, to the packet's destination. Each movement of a packet from one device to another along the route to the

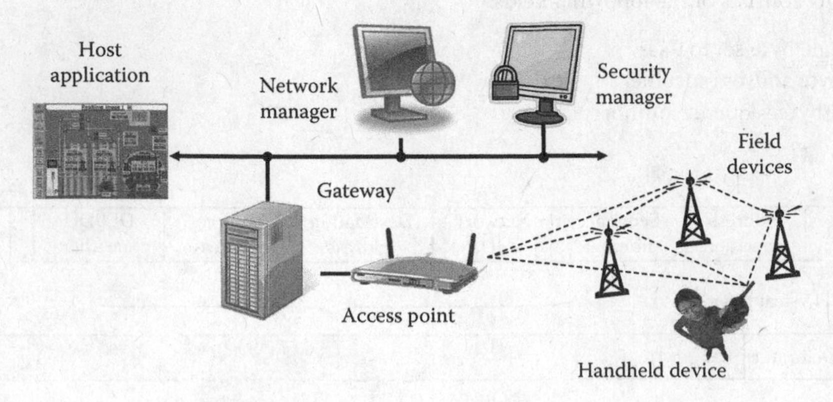

FIGURE 53.6 WirelessHART network topology.

packet's final destination is called a hop. WirelessHART networks can be implemented in a wide variety of topologies. In a high-performance scenario, a star topology is probably preferable (i.e., all devices are one hop to the gateway), while a multi-hop mesh topology is useful for a less demanding (e.g., monitoring) application. Obviously, any topology in between can also be realized.

There can be also devices that lack the connection with the process but have only communication facilities; they are routers, handheld devices (used for commissioning and/or maintenance purposes), and adapters (used to connect legacy hardware with the wireless network).

Routing of data packets is based on graph routing rather than on the (optional) source address routing. Each pair of nodes is interconnected by several graphs, i.e., a directed list of paths that connect them. Both upstream (toward the gateway) and downstream graphs are used. Only the network manager, who is responsible for correctly configuring each graph, knows the entire route; the graph information within a node only indicates the next hop destinations. Consequently, devices in a graph route must be configured prior to its use; i.e., it must contain a list of all the links that can be used to forward a packet along the graph. Graph routing is redundant, highly reliable, and should be used for normal communications.

On the contrary, a source route is a single, directed route between a source and a destination device. The source route is statically specified in the packet itself; for this reason, it is vulnerable and it is used only during the commissioning phase.

53.2.4.1 The Network Layer Datagram

The network layer datagram consists of the following fields (see Figure 53.7):

- A 1 byte control field
- The 1 byte time-to-live (TTL) hop counter
- The least-significant 2 byte of the absolute slot number (latency count)
- A 2 byte graph ID
- The (final) destination and (original) source addresses
- Optional routing fields
- A security layer is encapsulated within this datagram together with the enciphered payload
- Enciphered network protocol data unit (NPDU) payload

Communication between a pair of network addresses is organized in sessions for security purposes. Four sessions are generally set up as soon as any device joins the network, even if additional sessions may be added:

- Unicast session between the device and the network manager—it is used by the network manager to manage the device.
- Broadcast session between the device and the network manager—it is used to globally manage devices. For example, this can be used to roll a new network key out to all network devices. All devices in the network have the same key for this session.

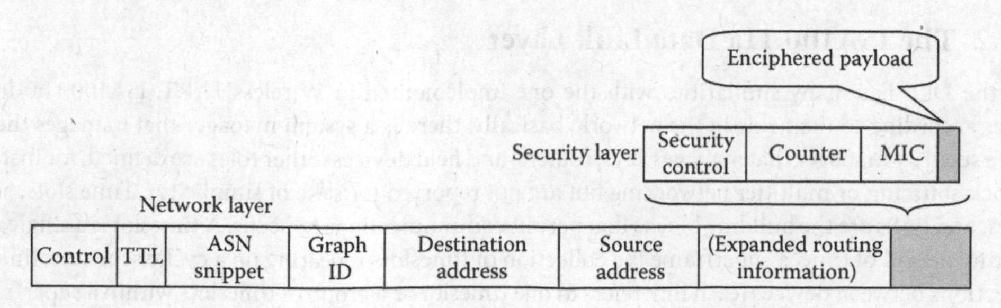

FIGURE 53.7 The network layer datagram.

- Unicast between the device and the gateway—it carries normal communications (e.g., process data) between the gateway and the device.
- Broadcast session between the device and the gateway—it is used by the gateway to send the identical application data to all devices.

In addition, each device always has a join session that cannot be deleted. The join key is unique; it is the only key that can be written by directly connecting to maintenance port physically implemented on the device itself. It can also be written by network manager. In fact, only the network manager may create or modify sessions and their corresponding keys.

53.2.5 The WirelessHART Upper Layers

A simple transport layer is provided in WirelessHART, which ensures data exchange reliability. It implements a master–slave transaction, where a "master" issues a request packet and one or more "slaves" reply with a response packet; or a slave cyclically publishes a response packet. The application layer (APL), instead, is the same of the wired counterpart. HART application layer is command based with standard data types and procedures. Universal, Common Practice, Device Family, and Wireless commands are specified. Extensive standard and device-specific status are available, including quality assessment and status for all process variables. A detailed discussion of these commands is outside the scope of this chapter.

53.3 ISA100.11a

The ISA100 [ISA09] committee was formed by International Society of Automation (ISA) to satisfy the need of a wireless manufacturing and control systems. In particular, the ISA10.11a is devoted to the development of a reliable and secure communication system for both noncritical monitoring and process control applications, which can tolerate latencies on the order of 100 ms. ISA100.11a networks can have a very rich structure. Several multi-hop networks having a mesh topology, each one built among devices with routing capabilities to which simple non-routing devices are associated, can be interconnected via a backbone. Routing support for inter-mesh communications is provided and each mesh can have one or more gateways to other types of plant networks. However, it must be underlined that the ISA100.11a is a very recent standard (it has been officially approved by the ISA association during September 2009), and only preliminary implementations are available.

53.3.1 The ISA100.11a Physical Layer

Similarly to WirelessHART, ISA100.11a also utilizes radios operating in the 2.4 GHz ISM band and based on the IEEE 802.15.4-2006 standard. Also in this case, the last channel is not used to be compliant with regulations in some countries and some requirements on switching time of transceivers are provided in order to implement a frequency agility mechanism.

53.3.2 The ISA100.11a Data Link Layer

Also, the DLL has many similarities with the one implemented in WirelessHART. ISA100.11a defines devices according to their role in the network; basically, there is a system manager that manages the network, a security manager that manages keys, routers, and field devices (other roles are defined, for instance, for clock sourcing or multitier networking but are not reported for sake of simplicity). Time slots, superframes, and links are the building blocks that permit communications to occur. A timeslot is a single, nonrepeating period of time; a superframe is a collection of timeslots repeating on a cyclic schedule; links are connections between devices (each link refers to one timeslot or a group of timeslots within a superframe). However, in the ISA specs, time slots do not have a fixed length but must be aligned to a 250 ms boundary.

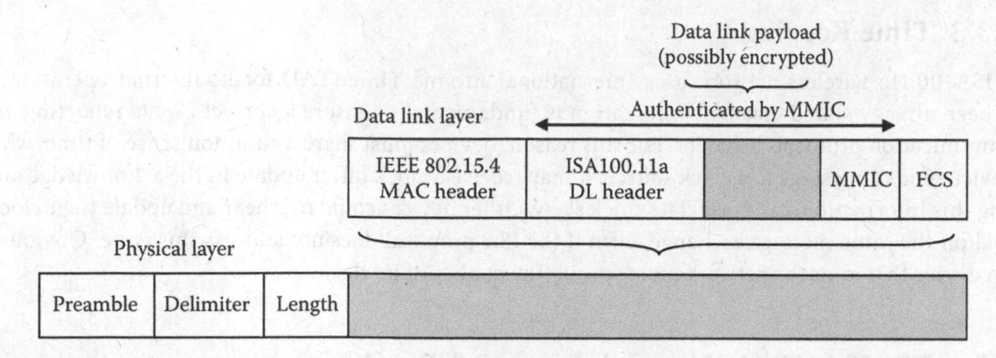

FIGURE 53.8 ISA100.11a datagram structure (DL Header: additional ISA100 Data Link layer header).

In addition, the compatibility with the IEEE 802.15-4-2006 has been preserved; in other words, the packet at the DLL preserves the header and footer specified in the IEEE standard; see Figure 53.8. Nevertheless, it must be highlighted that protocol itself is different; there are no IEEE MAC full function devices (FFDs), the superframe is defined in a different way; even if CSMA/CA is used in both solutions, details are not the same and the IEEE MAC backoff and retry mechanism is not used. Instead, the ISA standard implements its own retries, involving spatial diversity (retries to multiple devices) and frequency diversity (retries on multiple radio channels). Neighbor discovery and joining are also implemented differently.

Three general modes of operation are supported: the slotted channel hopping; the slow channel hopping; and hybrid combinations of slotted and slow hopping. In slotted hopping, each timeslot uses a different radio channel in a hopping pattern and it is intended to accommodate one message and its acknowledgment (see Figure 53.9). On the contrary, in slow hopping, a collection of contiguous timeslots is grouped on a single radio channel. Timeslot (bandwidth) allocation to devices may be centralized, i.e., handled by the system manager, or delegated. For instance, a simple form of delegation is based on channel hopping offsets; a frequency offset is assigned to each infrastructure device and it is free to create superframes using that offset.

According to the specs, packets forwarding within the wireless network is performed at the DLL, using the same graph routing adopted also by the WirelessHART solution. Routes are configured by the system manager, based on reports from devices that indicate instantaneous and historical quality of wireless connectivity to their immediate neighbors. The system manager accumulates these reports of link quality to make routing decisions. These reports are standardized, but not the routing decision process within the system manager.

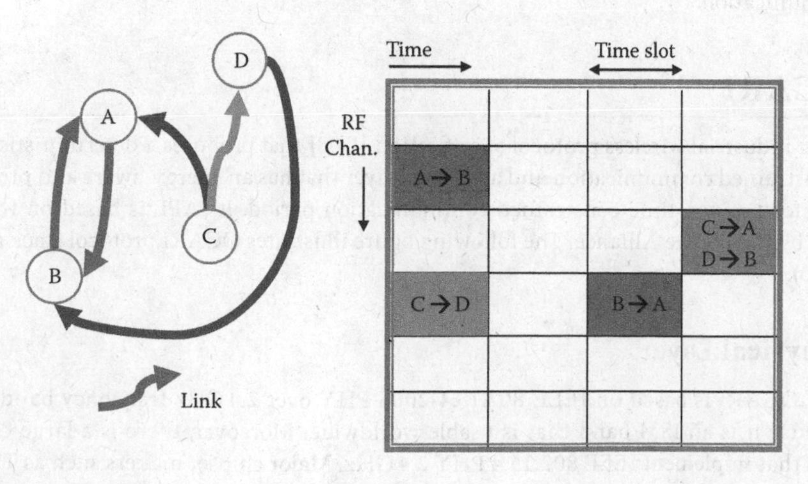

FIGURE 53.9 ISA100.11a link and channel hopping concept.

53.3.3 Time Keeping

An ISA100.11a wireless network uses International Atomic Time (TAI) for its internal operation. It has been already stated that synchronization is fundamental to ensure a correct events reporting and communication protocols behavior. For this reason, devices must share a common sense of time; when a device sends a message to a clock source, it may receive a clock offset update in the acknowledge and, using this information, can correct its clock skew. Other devices could overhear and update their clocks based on the same message exchange, even if the ISA proposal does not address this issue. Obviously, each device that joins the network must specify the quality of its clock.

53.3.4 The ISA100.11a Network Layer and Topologies

The network layer header fields have been influenced by 6LoWPAN proposal [MG07,MKH07], and an attempt is made to facilitate future compatibility. In the ISA proposal, the concept of a network made up of ISA100 devices only (called DL subnet in the following) with respect to a backbone network has been separated; a backbone device is defined as a device interfaced to a non-ISA100 network, e.g., an industrial Ethernet or any other network within the facility interfacing to the plant's network. The aim is to take into account power and bandwidth requirements on one side and allow for a huge variety of topologies on the other one. For this reason, several datagram formats have been considered for the network layer (e.g., using short addresses, 16 bit, for DL subnet devices and long addresses, 64 bit identifier for the joining process and 128 bit addresses for backbones). In particular, three formats have been considered: the basic header is only used in a DL subnet (it is expected to be the most common format and minimizes the overhead); also the contract-enabled must be used only over a DL subnet but allows to include more information; the full header is the IPv6 header and is intended for use over the backbone. It is a network layer task to translate addresses and to manage packet fragmentation. It must be remembered that at the network layer, any device within a DL subnet is a single network hop away.

53.3.5 The ISA100.11a Upper Layers

A transport layer that provides connectionless services with optional security has been defined in the standard; it extends the UDP over IP version 6 and 6LoWPAN. The APL defines software objects that mimic real-world objects and also defines the communication services necessary to enable object-to-object communication.

53.4 OCARI

OCARI is an industrial wireless protocol specification [D08] that proposes a deterministic MAC layer for time-constrained communication and a network layer that has an energy-aware and proactive routing strategy for the non-time-constrained communication period. Its APL is based on the specification defined by the ZigBee Alliance. The following figure illustrates OCARI protocol stack architecture (Figure 53.10).

53.4.1 Physical Layer

The PHY of OCARI is based on IEEE 802.15.4-2006 PHY over 2.4 GHz frequency band. This RF is chosen because it is an ISM band that is usable worldwide. Moreover, there is a large choice of RF transceivers that implement IEEE 802.15.4 PHY 2.4 GHz. Major chipset makers such as TI, Freescale, Atmel, etc. propose optimized RF transceivers over this frequency band.

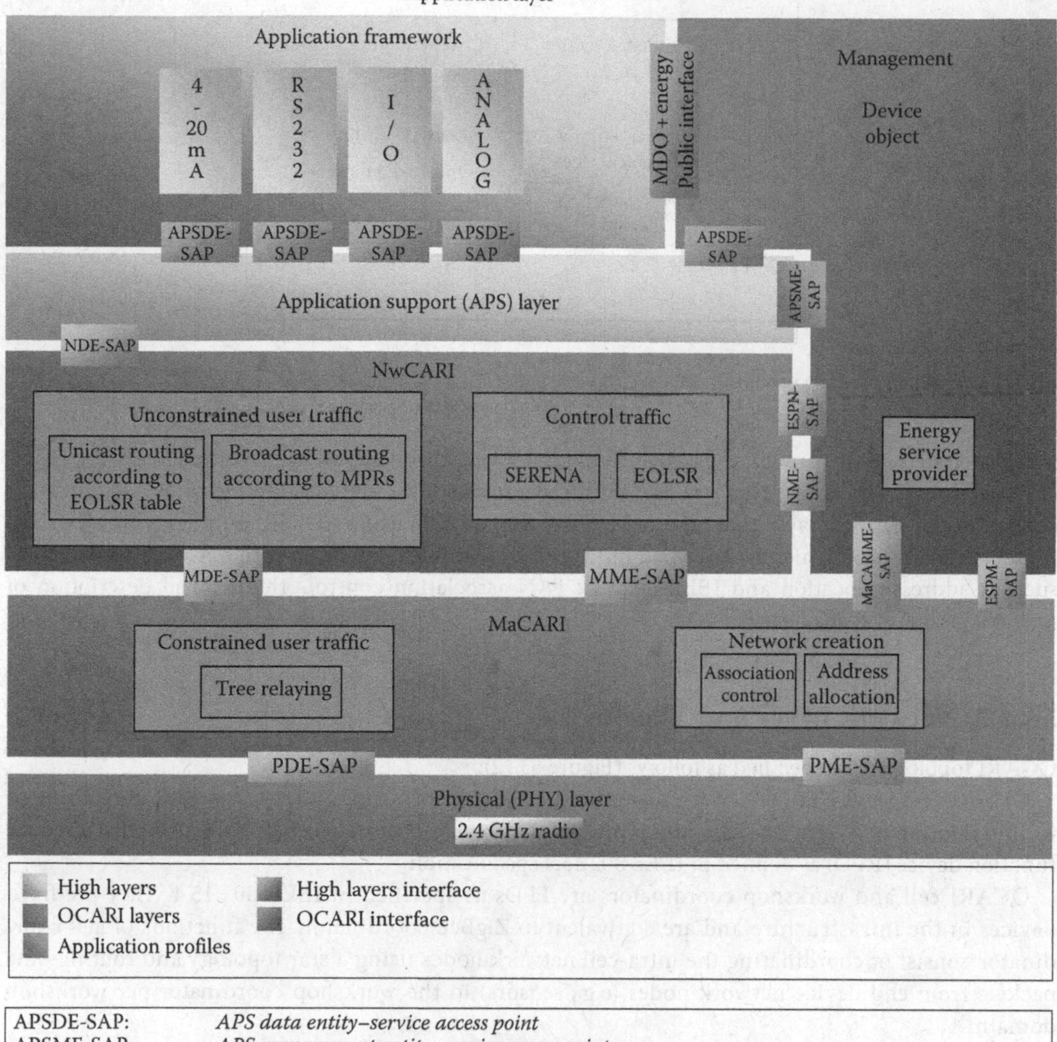

APSDE-SAP:	*APS data entity–service access point*
APSME-SAP:	*APS management entity–service access point*
NDE-SAP:	*NwCARI data entity–service access point*
NME-SAP:	*NwCARI management entity–service access point*
ESPN-SAP:	*Energy service provider NwCARI–service access point*
ESPM-SAP:	*Energy service provider MaCARI–service access point*
MME-SAP:	*MaCARI management entity between NwCARI and MaCARI–service access point*
MaCARIME-SAP:	*MaCARI management entity between MDO and MaCARI–service access point*
MDE-SAP:	*MaCARI data entity–service access point*
PME-SAP:	*Physical layer management entity–service access point*
PDE-SAP:	*Physical layer data entity–service access point*

FIGURE 53.10 OCARI stack architecture.

53.4.2 MAC Layer

MaCARI [CGM08,LVV08] is the MAC layer specified in OCARI. It supports both time-constrained communication and slotted CSMA/CA communication as depicted by Figure 53.11.

The design of MaCARI is based on a cross-layering approach. It proposes a tree-based strategy of packets relaying with deterministic activities schedule to reduce possible collision. In MaCARI, the PAN coordinator schedules the cells activity and fixes the beacons broadcasting sequence. The beacon

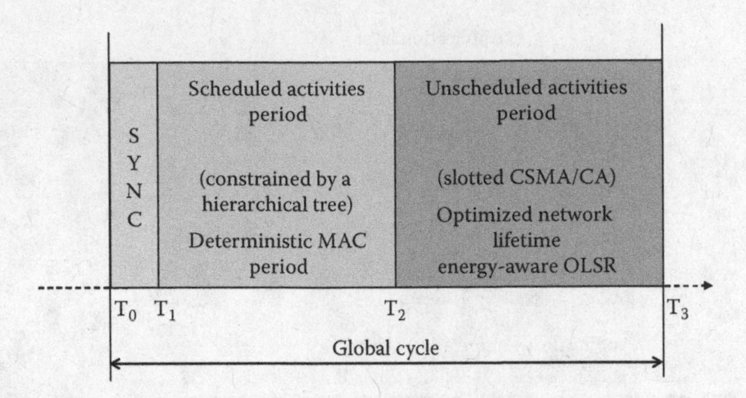

FIGURE 53.11 MaCARI concept.

is repeated in cascade through the tree and contains all synchronization information. Every node synchronizes on T_1 and knows the next T_0. Within the unscheduled activities period, MaCARI gives the control to the network layer that takes care of routing packets using EOLSR, an Energy-aware OLSR protocol (RFC 3626), which is described in [MM081]. MaCARI also deals with network initialization such as address allocation and IEEE 802.15.4 PAN association control. The detailed description of MaCARI is out of scope here.

53.4.3 Network Layer and Topologies

OCARI topologies are specified as follows (Figure 53.12):

OCARI End Device is a "radio fixed" network node (i.e., its position varies very little comparing to its initial location, so that its radio link is always managed by the same cell coordinator). It is a reduce function device (RFD) as defined in IEEE 802.15.4 specification.

OCARI cell and workshop coordinators are FFDs as specified in IEEE 802.15.4. They are fixed devices in the infrastructure and are equivalent to ZigBee coordinator. The functions of cell coordinator consist of coordinating the intra-cell network nodes using a star topology and routing data packets from end device network nodes (e.g., sensors) to the workshop coordinator per workshop domain.

Workshop domain is a permissive volume (delimited by a threshold of the RSSI) to electromagnetic wave that is covered by a unique workshop coordinator network node.

Workshop coordinator is a gateway between the wireless sensors network resided in a workshop domain and the industrial facility backbone.

OCARI network layer is based on a cross-layering approach. It manages packet routing for unscheduled activities period according to EOLSR proactive protocol (refer to [MM081] for a detailed description). It also offers a Scheduling RoutEr Node Activity (SERENA) strategy, which uses a three-hops neighbors decentralized coloring algorithm to schedule the router activities, so that energy consumption can be reduced, and the bandwidth efficiency is improved by avoiding the interferences between routers (a timeslot is assigned to a node according to its color and there is a spatial reuse of color code). The principle of SERENA is as follows [MM082].

A router node is woken up in the timeslots where

- It transmits
- One of its one-hop neighbors transmits
- It sleeps otherwise

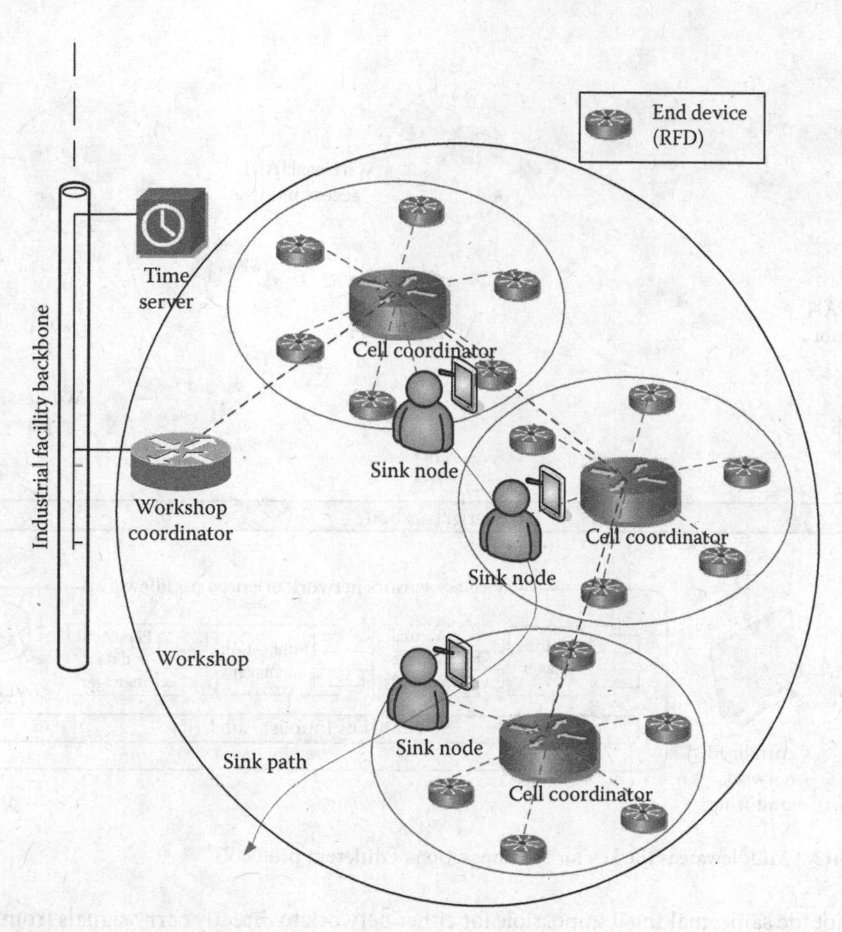

FIGURE 53.12 OCARI network topology.

53.4.4 Application Layer

OCARI APL is based on the specification released by the ZigBee Alliance. It defines some additional profiles such as

- "Sensor 4-20" that deals with the well-known 4–20 mA
- "Sensor binary" for binary sensor including electronic PNP commutation transistor
- "Accelerometer" for vibration analysis (up to 5 kHz) used in condition-based maintenance application

It also supports the well-known communication models:

- Request/reply
- Publish/subscribe (event-based notification)
- Periodic notification

53.5 Coexistence of the Three Protocols

From the users' point of view, a single wireless standard for the industrial automation world would be the most appealing solution. Even if protocols discussed here are quite similar, e.g., all of them are based on the same PHY, they are not identical. In particular, protocols used for the upper communications

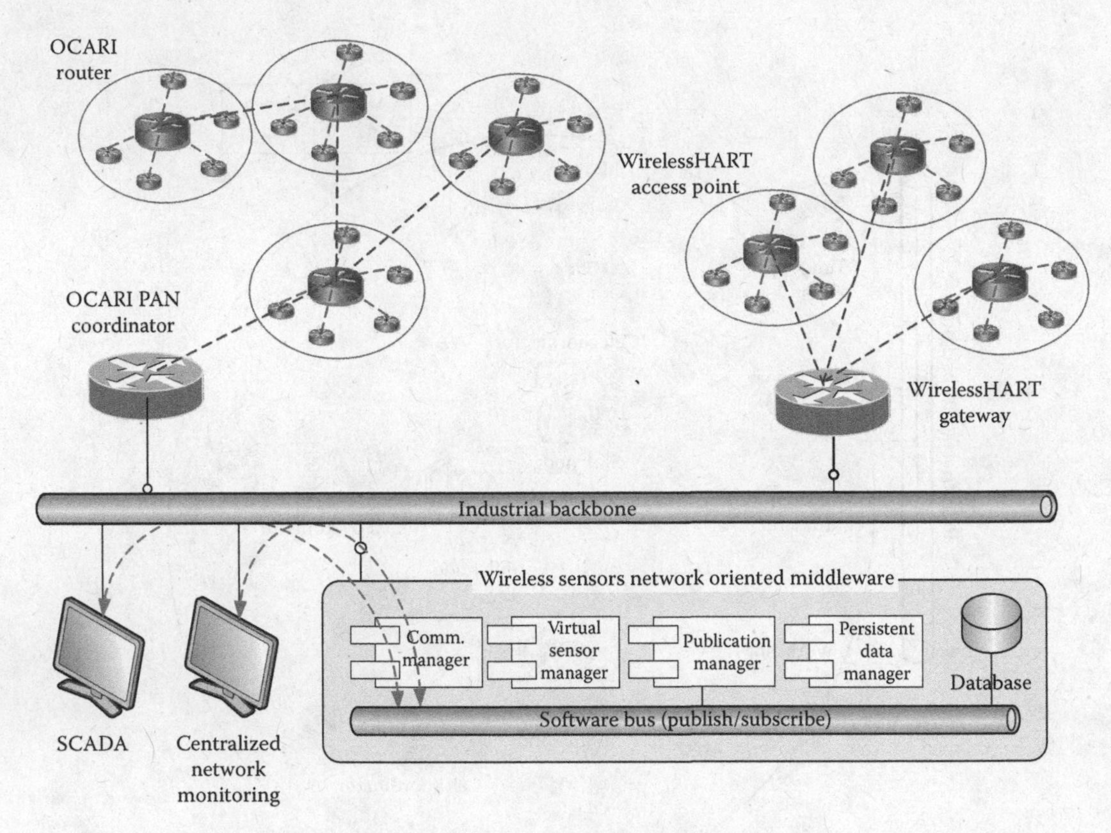

FIGURE 53.13 Middleware is the key for the integration of different protocols.

layers are not the same, making it impossible for either network to directly carry signals from the other. This means that gateways are needed to ensure interoperability and extensive research must be done in the development of new middleware paradigms. Middleware is understood as an abstraction layer that can be used to program applications in a network-transparent way. The high level of abstraction that is provided lowers the cost of development of control applications. Middleware is a generic name used to refer to a class of software whose purpose is to serve as glue between separately built systems (Figure 53.13).

However, the ISA100 committee is intended to be a single "universal" network for the wireless transport of information from all types of industrial wired protocols and a new subcommittee for the ISA100.11a and the WirelessHART convergence, called ISA100.12, has been created. The aim is not only to ensure coexistence of nodes of both kinds but also to make possible the cooperation between the two networks. Proposals like dual-mode gateways and a dual-mode node are under discussion. From the physical point of view, collisions will arise when different networks operate in the same area, especially in a crowded environment like the ISM band around 2.4 GHz. However, data delivery model of industrial communications must be considered, i.e., a cyclical exchange of a small quantity of information (i.e., determinism and often isochronisms is required but the average bandwidth is low). Both ISA100.11a and WirelessHART utilizes CCA to avoid collisions with other nonsynchronized systems, while frequency agility together with a short time-synchronized communications minimizes the bandwidth occupation and ensures successful data exchange since messages can be resent on other (free) channels exploiting automatic repeat query (ARQ) error control strategy.

Another important aspect is security. Both ISA and HART solutions include a security manager to manage and authenticate cryptographic keys. In particular, during normal operations, secret symmetric keys known to both the sender and the receiver are used. During the commissioning phase,

WirelessHART obliges to use a maintenance port for the key exchange while ISA100.11a supports asymmetric keys, i.e., credentials of new devices may be verified using public keys shared openly and a corresponding asymmetric secret private key kept inside the new device itself.

53.6 Example of Platform Providers

Even if all the solutions discussed in this chapter are recently proposed, the market already offers some development platforms and conformance testing tools. In fact, the main advantage deriving from the adoption of standards is probably manufacturers' interoperability.

For this reason, HART consortium has worked hard on the development of test specifications; it has recently released a "sniffer" to be used to verify messages over the air and has announced the so-called Wi-Htest [HSZ09], a linux box that automates the execution of single device testing, recreating the packet exchange of a whole virtual network around the device under test, including corrupted packets.

Besides, ISA organization has created an industry group within the Automation Standards Compliance Institute called the ISA100 Wireless Compliance Institute, in order to facilitate the proper use and application of automation standards through the development and implementation of conformance assessment programs and related activities. During the last ISA expo (November 2008), devices from 14 different instrumentation vendors successfully formed a wireless network based on the ISA100.11a draft technology.

As an example of platform providers, consider Nivis, a company launched in 1998, which is specialized in sensor networks, supporting an array of standard sensing and control protocols including WirelessHART and ISA100.11a, and most recently 6LoWPAN. The company supplies wireless sensors, edge routers, a management appliance and application, and a gateway to plant or other enterprise networks.

Most of difficulties in realizing an error-free wireless network are, obviously, in the development of protocol stack; to speed up the process, it has been founded the Wireless Industrial Technology Konsortium—WiTECK—http://witeck.org/. It is an open, nonprofit membership organization whose mission is to provide a reliable, cost-effective, high-quality portfolio of core enabling system software for industrial wireless sensing applications, under a company- and platform-neutral umbrella. They are now working on the realization of a WirelessHART stack.

Emerson Process Management has already announced some devices compliant with WirelessHART, as a core part of its Smart Wireless plant automation solutions portfolio. Other manufacturers that are working on wireless devices (or wireless adapters for legacy devices) are Honeywell (with their onewireless solution), Yokogawa, Elpro (now part of MTL), Pepperl + Fuchs (with WirelessHART nodes), and ABB, just to mention few of them (Table 53.1).

TABLE 53.1 Example of Platform Provider

	Environment		
Technology	Hardware	Stack (SW)	Debug Tools
WirelessHART	Emerson	WiTECK	NIVIS
	Honeywell	NIVIS	HCF consortium
	Yokogawa		
	NIVIS		
	MTL		
	P + F		
	ABB		
ISA100.11a	NIVIS	NIVIS	NIVIS
	ABB		
OCARI	Telit communication	Telit communication	Telit communication

53.7 Conclusions

In this chapter, a brief review of three emerging solutions implementing wireless sensor networks for industrial applications—primarily for process control—has been reported. Such networks are in general designed to carry traffic that is mainly constituted by exchanges of sensor readings (and actuator commands) toward (from) centralized controllers. Important characteristics of industrial traffic are the presence of deadlines, high-reliability requirements, and the predominance of short packets. All the proposed solutions address this problems in a very similar manner. They adopt a TDMA approach to wireless shared medium, while mesh or cluster topologies allow covering large areas with low-power RF signals. External interferences due to other transmitters operating in the same or in neighbored frequency bands—a problem especially pronounced in the license-free industrial, scientific, and medical (ISM) bands—are faced with frequency agile approaches. Security is ensured by means of cryptography, which allows to answer to questions like "who sent this message?" (authentication) and "is this the message originally sent?" (message integrity).

However, there is still a broad potential for future research in wireless industrial communication systems. Up to now, very few plants/demonstrations have been implemented to prove the viability of such solutions.

References

[CGM08] G. Chalhoub, A. Guitton, and M. Misson, MAC specifications for a WPAN allowing both energy saving and guaranteed delay—Part A: MaCARI: A synchronized tree-based MAC protocol, in *IFIP 2008 Conference on Wireless Sensor and Actor Networks*, Ottawa, Ontario, Canada.

[D08] T. Dang et al., OCARI: Optimization of communication for ad hoc reliable industrial networks, in *IEEE-INDIN'08 Conference*, Daejeon, Korea, July 13–16, 2008.

[FFM07] A. Flammini, P. Ferrari, D. Marioli, E. Sisinni, and A. Taroni, Sensor networks for industrial applications, in *Proceedings of Second International Workshop on Advances in Sensors and Interface, 2007. IWASI 2007*, Bari, Italy, June 26–27, 2007, pp. 1–15.

[H09] HART—The Logical Wireless Solution, HCF, http://www.hartcomm2.org/hart_protocol/wireless_hart/hart_the_logical_solution.html, February 2009.

[H88] Hayes, J. P., *Computer Architecture and Organization, McGraw-Hill Series in Computer Organization and Architecture*, 2nd en, McGraw Hill, New York, 1988.

[HSZ09] S. Han, J. Song, X. Zhu, A.K. Mok, D. Chen, M. Nixon, W. Pratt, and V. Gondhalekar, Wi-HTest: Compliance test suite for diagnosing devices in real-time WirelessHART network, *15th IEEE Real-Time and Embedded Technology and Applications Symposium (RTAS'09)*, San Francisco, CA pp. 327–336, 2009.

[I06] IEEE Standard for Information Technology—Telecommunications and Information Exchange between Systems—Local and Metropolitan Area Networks—Specific Requirements—Part 15.4: Wireless Medium Access Control (MAC) and Physical Layer (PHY) Specifications for Low Rate Wireless Personal Area Networks (LR-WPANs), 2006.

[ISA09] ISA100.11a, http://www.isa.org/MSTemplate.cfm?MicrositeID=1134&CommitteeID=6891, February 2009.

[KHP08] A. N. Kim, F. Hekland, S. Petersen, and P. Doyle, When HART goes wireless: Understanding and implementing the WH standard, in *Proceedings of ETFA2008*, Hamburg, Germany, September 15–18, 2008, pp. 899–907.

[LVV08] E. Livolant, A Van den Bossche, and T. Val, MAC specifications for a WPAN allowing both energy saving and guaranteed delay—Part B: Optimisation of the intra cellular exchanges for MaCARI, in *IFIP 2008 Conference on Wireless Sensor and Actor Networks*, Ottawa, Ontario, Canada.

[MG07] G. Mulligan and L.W. Group, The 6LoWPAN architecture, in *Proceedings of the EmNets*, Cork, Ireland, June 25–26, 2007.

[MKH07] G. Montenegro, N. Kushalnagar, J. Hui, and D. Culler, Transmission of IPv6 packets over IEEE 802.15.4 networks, RFC 4944, 2007.

[MM081] S. Mahfoudh and P. Minet, An energy efficient routing based on OLSR in wireless ad hoc and sensor networks, in *PAEWN 2008, IEEE International Workshop on Performance Analysis and Enhancement of Wireless Networks*, Ginowan, Japan, March 2008.

[MM082] S. Mahfoudh and P. Minet, Performance evaluation of the SERENA algorithm to SchEdule RoutEr nodes activity in wireless ad hoc and sensor networks, in *AINA 2008, IEEE 22nd International Conference on Advanced Information Networking and Applications*, Ginowan, Japan, March 2008.

[SHM07] J. Song, S. Han, A, K, Mok, D. Chen, M. Lucas, M. Nixon, and W. Pratt, WirelessHART: Applying wireless technology in real-time industrial process control, in *Proceedings of 14th IEEE Real-Time and Embedded Technology and Applications Symposium (RTAS)*, Bellevue, WA, April 3–6, 2007.

[W08] A. Willig, Recent and emerging topics in wireless industrial communications: A selection, *IEEE Transactions on Industrial Informatics*, May 2008, 4(2), 102–124.

54

Wireless Communication Standards

54.1 Introduction 54-1
54.2 A Wireless Standards Taxonomy 54-2
54.3 Regulations and EMC 54-2
54.4 Conclusion 54-6
References 54-6

Tuan Dang
EDF Research and Development

54.1 Introduction

There are a lot of books dealing with wireless communication principles and practice that one can easily find their editors on the Internet. The purpose of this chapter is not to explain again and again the wireless communication theories but to give the engineers a precise overview of the current wireless communication standards so that they can choose the right technology for the right application. Successful implementation of wireless communication in industrial environments requires the correct understanding of the strengths and the weaknesses of each wireless technology. Industrial wireless implementation is not quite as easy as one might hear. Electromagnetic compatibility (EMC) issues, bandwidth performance and cyber security requirements, battery autonomy, and interoperability considerations are among the constraints that one has to take into account when implementing industrial wireless solutions. Site assessment is often needed in order to understand the radio propagation conditions so that a correct network topology can be deployed to ensure the optimal communication coverage of the target facility. Be aware that the site environment may vary by time of day, depending on what is happening in and around where the signals propagate. Wireless communication is not always easy as it seems to be. The challenges concern the following topics:

- Robust physical layer that can guarantee a low bit errors rate (BER) without much processing power.
- Medium access method that can offer an efficient use of the communication channel capability while allowing deterministic access and low latency peer-to-peer transmission.
- Energy-aware route discovery strategy that helps to improve power conservation of the whole network while being proactive to maintain permanent route between two network nodes.
- Cyber security without much processing power to preserve battery lifetime.
- Reliability that guarantees the message delivery within a predefined time interval. Adaptability, frequency agility/hopping, and topology redundancy to avoid single points of failure are the important features that need to be studied when deploying in some application.
- Scalability and density (number of network nodes inside an acceptable transmission range of a transmitter node) are also important characteristics. How large and dense can the network get before interferences or loss of performance?

- Data rate, packet size, and fragmentation support are the features that should not be forgotten.
- Support of nomadism or mobility of network node may be important for some applications.

54.2 A Wireless Standards Taxonomy

Wireless technologies can be classified according to their communication range and their equivalent isotropically radiated power (EIRP), their data rate, their power consumption, and their "complexity." The frequency band and the EIRP are important parameters in terms of EMC. The EIRP has direct impact on the power consumption of the radio chipset. Figure 54.1 proposes a wireless standards taxonomy.

Industrial automation applications of wireless technologies concern mainly RFID, WPAN, and WLAN classes. WWAN and WMAN classes concern much more telecom operators except TErrestrial Trunked RAdio (TETRA). It is a digital trunked mobile radio standard (from 300 to 1000 MHz) developed to meet the needs of traditional professional mobile radio (PMR) user organizations such as

- Public safety
- Transportation
- Utilities
- Government
- Military
- Public access mobile radio (PAMR) operator
- Commercial and industry
- Oil and gas

Table 54.1 lists the characteristics of some technology belonging to RFID (see also reference [1]), WPAN, and WLAN classes that are important to know when choosing a wireless technology. For example, for DECT technology, it is worth checking regularly reference [2].

It is worth mentioning IEEE 802.15.4 standard which provides a foundation for *low-duty cycle, low power consumption, and low-bandwidth* mesh networking technology. ZigBee/ZigBee Pro, WirelessHART, ISA100.11a [3–5], and OCARI (see also Chapter 44 for more descriptions) are wireless communication standards that are aimed for wireless sensor and actuator networks. ZigBee Pro offers, for example, a *low-power active router* which is particularly useful for power conservation. OCARI offers support for mobility of network router with an advanced activity scheduling algorithm that helps to preserve the battery lifetime.

54.3 Regulations and EMC

Wireless classes that belong to RFID, WPAN, and WLAN categories mostly use license-free frequency bands. These frequency bands are specified by different international and national organizations such as

- The International Telecommunication Union (ITU)
- The Federal Communications Commission (FCC) in the United States
- The European Conference of Postal and Telecommunication Administrations (CEPT)
- The European Union, Radio, and Telecommunications Terminal Equipment (R&TTE)

It is important to check the recommendations issued by these organizations in order to avoid interferences with existing or future electronic equipments. For example, ITU-K series provide recommendations concerning protection against interference. IUT-K.76 defines EMC requirements for telecommunication network equipment in 9–150 kHz frequency band, and IUT-K.80 specifies EMC requirements for telecommunication network equipment in the band from 1 to 6 GHz.

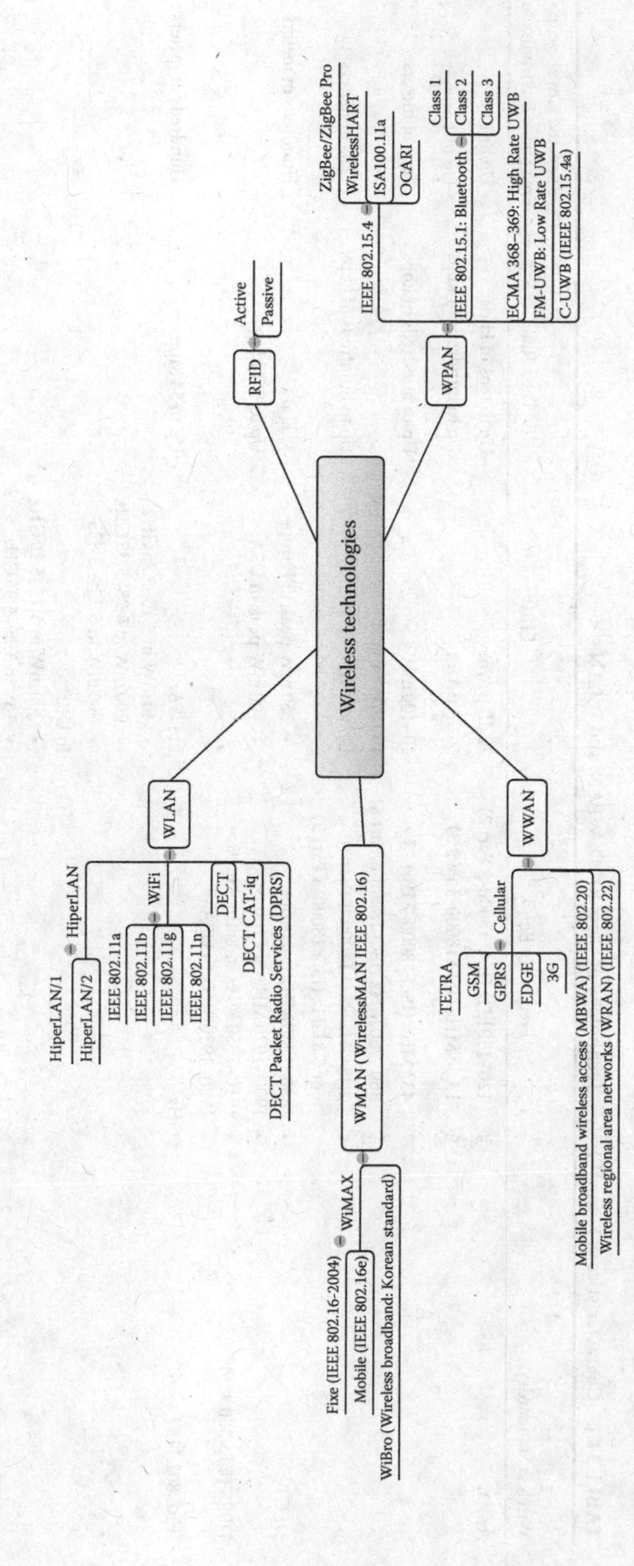

FIGURE 54.1 Wireless standards taxonomy.

TABLE 54.1 Characteristics of Some Wireless Technology of Class RFID, WPAN, and WLAN

Wireless Standards		Frequency Band	EIRP	Data Rate	Expected Coverage in Free Space Environment
RFID	Passive	120–150 kHz (ISO 18000-2 Part 2)	72 dB μA/m	Up to tens of kbps	Up to 1 m
	Passive	13.56 MHz (ISO 18000-3 Part 3)	60 dB μA/m	Up to hundreds of kbps	Up to 1 m
	Active	433 MHz (ISO 18000-7 Part 7)	10–100 mW	Up to hundreds of kbps	Tens of meters
	Active	860–960 MHz (ISO 18000-6 Part 6)	100 mW–4 W		
	Active	2.45 GHz (ISO 18000-4 Part 4) or 5.8 GHz (ISO 18000-5 Part 5)	Up to 4 W	Up to hundreds of kbps	Tens of meters
DECT DPRS (ETS 300 435)		• 1880–1900 MHz in Europe • 1900–1920 MHz in China • 1910–1930 MHz in Latin America • 1920–1930 MHz in USA	• 250 mW peak in Europe • 100 mW peak in USA	32 kbps Up to 552 kbps	Hundreds of meters
IEEE 802.11a		5 GHz	In USA: • 40 mW in 5.15–5.25 GHz • 200 mW in 5.25–5.35 GHz • 800 mW in 5.725–5.825 In UK: • 200 mW in 5.15–5.25 GHz • 1 W in 5.25–5.35 GHz • 1 W in 5.725–5.825 GHz In France: • 200 mW in 5.15–5.25 GHz • 100 mW in 5.25–5.35 GHz • 500 mW in 5.470–5.725 GHz	Up to 54 Mbps	Hundreds of meters

IEEE 802.11 g	2.4 GHz (ISM)	Europe standard: • 100 mW in 2.40–2.4835 GHz In USA: • 4 W in 2.40–2.4835 GHz	Up to 54 Mbps	Hundreds of meters
IEEE 802.11n (a.k.a WWiSE: World-Wide Spectrum Efficiency)	2.4 and 5 GHz	Regulations on 2.4 and 5 GHz bands are applied for total MIMO beam	Up to 600 Mbps	Hundreds of meters
IEEE 802.15.4 based technology: ZigBee/ZigBee Pro, WirelessHART, ISA100.11a, OCARI	• 16 channels in the 2450 MHz band • 30 channels in the 915 MHz band • 3 channels in the 868 MHz band	• 0.5–10 mW • 1 mW typically	Up to 250 kbit/s	Tens of meters
FM-UWB (Swiss standard) low rate UWB	• 3.1–10.6 GHz	• 0.05 mW	1–10 kbit/s and 100–1000 kbit/s	Up to 10 m
ECMA-368/369 high rate UWB (European association for standardizing information and communication systems)	• 3.1–10.6 GHz	• 1–15 mW (0–12 dBm)	53.3–480 Mbps	Up to 10 m

There are also dedicated standard bodies dealing with EMC. These include:

- The International Electrotechnical Commission (IEC), which has different working groups: "Comité International Spécial des Perturbations Radioélectriques" (CISPR) or International Special Committee on Radio Interference and Technical Committee 77 (TC77).
- The "Comité Européen de Normalisation Electrotechniques" (CENELEC) or European Committee for Electrotechnical Standardization.

CISPR works on EMC standards from 9 kHz upward (see different subcommittees from reference [8]). These offer protection of radio reception from interference sources such as electrical appliances of all types; the electricity supply system; industrial, scientific, and electromedical RF; broadcasting receivers (sound and TV); and, increasingly, IT equipment (ITE). TC77 works on immunity standards for the emission below 9 kHz.

54.4 Conclusion

This chapter proposes a brief overview of wireless communication standards that are mostly used for short-range communications. The potential applications of these wireless communication standards in industry, manufacturing, healthcare, and home automation are tremendous. Having knowledge of these wireless communication standards and regulation issues is vital for engineers.

References

1. RFID. http://www.skyrfid.com; http://www.hightechaid.com/tech/rfid/what_is_rfid.htm, March 2009.
2. DECT Forum. http://www.dect.org
3. HART Protocol Revision 7.0, August 23, 2007.
4. ZigBee Specification-Document 053474r17. http://www.zigbee.org
5. ISA100.11a. http://www.isa.org
6. ECMA-368/369. http://www.ecma-international.org, March 2009.
7. FM-UWB. http://www.fmuwb.ch, March 2009.
8. http://www.iec.ch/zone/emc/emc_cis.htm, March 2009.

55

Communication Aspects of IEC 61499 Architecture

55.1 Introduction 55-1
55.2 Illustrative Example 55-3
55.3 Logic Encapsulated in Basic FB 55-7
55.4 Extension 55-7
55.5 Distribution 55-9
55.6 Communication FBs 55-10
55.7 Communication Using Services of Internet Protocol Suite 55-10
55.8 Adding Distribution and Communication to the Sample System 55-13
55.9 Internals of Communication FBs: Modbus 55-15
55.10 Communication via the CIP 55-17
55.11 Impact of Communication Semantics on Application Behavior 55-19
55.12 Failures in Distributed Applications 55-20
55.13 Conclusion 55-20
References 55-21

Valeriy Vyatkin
University of Auckland

Mário de Sousa
University of Porto

Alois Zoitl
Vienna University of Technology

55.1 Introduction

The IEC 61499 standard [1] defines a system architecture for industrial measurement and control systems based on the concept of *function block* (FB). A FB is a software unit (or, more generally, an *intellectual property* capsule) that encapsulates some behavior and facilitates its re-use.

The International Electrotechnical Commission (IEC) initiated the IEC 61499 project to address the limitations of the legacy PLC* programming languages looking toward the realities of implementing real-time multiagent systems. Unlike previous standardization efforts, this is not a retrospective recognition of practices, but an attempt to guide future developments toward an open standard that allows genuine vendor interoperability.

FBs provide a pathway to integrate established automation programming languages, such as ladder logic, instruction list, or structured text, into modern component architecture. However, their application extends past simple replacement of legacy systems because of the inherent support for distributed applications and ability to provide a platform for modeling and simulating with well-defined interfaces. IEC 61499 defines three classes of FBs—basic FBs, composite FBs, and service interface FBs—for

* Programmable Logic Controller

capturing and hiding platform-dependent functions of devices. Each FB has a set of input and output variables. The input variables are read by the internal algorithm when it is executed, while the results from the algorithm are written to the outputs.

The use of FBs makes the control device openly programmable and easily reconfigurable. IEC 61499-compliant devices can easily interface each another, thus providing for seamless distribution of different tasks across different devices. Users may create their own program using standard FB types. Thus, the IEC 61499 architecture enables encapsulation, portability, interoperability, and configurability. Portability means that software tools and hardware devices can accept and correctly interpret software components and system configurations produced by other software tools. With interoperability, hardware devices can operate together to perform the cooperative functions specified by one or more distributed applications. With configurability, devices and their software components can be configured (selected, assigned locations, interconnected, and parameterized) by multiple software tools.

Being an architecture for distributed systems, IEC 61499 needs to address the issue of communication between distributed devices. It needs to be mentioned that particular networking mechanisms, such as protocols and middleware, are beyond the standard's scope. Interfaces to particular networking mechanisms can be implemented in IEC 61499 by communication interface function blocks (CIFBs).

There is a small but growing toolset for FB design. The Function Block Development Kit (FBDK) [2] remains the most widely used, because it is the oldest and is free for educational use. The new version of the ISaGRAF industrial control design software with support for IEC 61499 FBs is described in [3,4]. The nxtControl tool [5] is another example of a professionally developed design environment that can generate code for a range of BECKHOFF and WAGO control devices. A number of developments at academic and research organizations can be represented by CORFU [6] and the open-source projects 4DIAC IDE [7] and FBench [8,9].

In order for FBs to become executable on a variety of hardware, hardware vendors must provide support for the standard. The options remain limited, but are on the increase. There are currently several options for executing FBs. First, any platform that can execute standard Java byte code can run the FBRT that is a part of FBDK. This includes desktop computers running any major operating system. Embedded execution option includes the Elsist Netmaster II, which runs a cut-down version of Java standard edition (J2SE). Tait Control Systems MO'Intelligence units run Java micro edition (J2ME) and are supplied with a port of the FB run-time and vendor-supplied service interface FBs for hardware access. These units are available in several formats with support for DeviceNet and an integrated motor drive option.

The 4DIAC consortium has developed an open-source run-time for FB execution FORTE [10]. It has been ported to a variety of hardware platforms and was chosen by nxtControl as a primary run-time target. FB compiler based on synchronous semantics is reported in [11]. Other academic developments include Java-based FUBER run-time [12], real-time Java [13], and Linux [14]-based run-times, as well as the Java-based run-time implementing the cyclic execution model [15].

This chapter provides an introduction to IEC 61499 and especially to its communication-related features by discussing an example of a simple distributed control system. For a more fundamental introduction to IEC 61499, the reader is referred to [16,17] and for implementation ideas to [18].

The chapter is structured as follows. First, the example is introduced. Then implementation of its control in terms of IEC 61499 FB application is presented and discussed. This allows to see the simulation capability of IEC 61499 as well. Next, the two-stage design concept of IEC 61499 is presented followed by a detailed description of abstract communication models of IEC 61499. Then the example is extended functionally. Also, the hardware architecture is changed to a distributed one, where the connectivity is provided by two different types of networks. Explicit communication FBs are added to the application. The underlying mechanisms for implementation of those FBs using services of Modbus and control and information protocol (CIP) protocols are described in detail. These are followed by a discussion on the impact of communication semantics on the distributed application behavior and failure models for distributed applications. The chapter is concluded with a summary and a list of references.

55.2 Illustrative Example

We illustrate the ideas and problems of IEC 61499 using an example of an imaginary system that, nevertheless, represents many properties of real distributed measurement and control systems.

Let us imagine that our company is producing a kind of programmable running lights system as shown in Figure 55.1, left side. The system consists of four lamps blinking according to a required mode of operation. The operation modes include "blink all lamps together" or "chase the light to the left or to the right," and so on. The lamps are "controlled" by a controller box with four output control signals, each of which is connected to a lamp and serves as a switch, turning the lamp on and off. The control panel of the system consists of a start/stop switch and a selector knob having three fixed positions (from 0 to 2) corresponding to three pre-programmed modes of operation. It could also include a potentiometer knob giving a continuous range of values from 0 to 32,767 to set the frequency of blinking (time interval in milliseconds between two consecutive flashes). As seen from Figure 55.1, all the input/output devices are wired to the interface modules of the controller box.

Now, let us imagine that the R&D division of our company is working on the new generation of this system, with the main innovation being flexibility. As also shown in Figure 55.1, it is planned to extend the lights section on customer's demand with multiple 4-LED sections and allow them to be physically distributed along large areas. Frequency of blinking may be displayed using the numeric display. The input part of the system can be extended by the buttons increasing or decreasing the blinking frequency. These buttons can be placed anywhere, say near the 4-LED sections, and pressing them shall have global effect on the frequency of blinking across the whole system. Possible extensions can also include operating the system via a portable device or cellular phone.

Obviously, in this case, the direct connection of all the peripherals to the controller may require a lot of wires. The use of a network bus instead of direct wires allows saving on wiring as illustrated in Figure 55.2. In such a configuration, new peripheral devices can be easily plugged into the system. However, the hardware requires some modifications: the controller and peripherals need to be equipped with network interfaces.

The question is how to organize the control logic in such a way as to allow the same easy manipulation with functions within the control program. In the following, we show how that can be achieved in the IEC 61499 architecture.

The functionality of our system (simplest initial version) is programmed in Figure 55.2 as an IEC 61499 application, with the correspondence shown between peripheral devices and FBs. As can be seen from Figure 55.2, FBs are represented as strange shapes with a head and a body having inputs (on the left) and outputs (on the right). The blocks are connected to one another using connection arcs. Thus, the FB diagram defines not only the data flow between the blocks, but also the control flow. To indicate that all peripheral devices are physically connected to the control unit, we have used the single bus architecture, but at this stage the details of this physical network connection are not important.

The inputs and outputs located at the block's head are called *event* inputs and outputs. When an event output is connected to the event input of some other block, the event emitted on the output side activates execution of the recipient FB. Thus, the control logic of the application is determined by that of each FB and by event connections between the blocks.

The roles of core FBs in our example are as follows:

RUNSTOP — Implements interfacing with the start/stop switch. Upon every switching generates an event at the IND output. The switch status is delivered via the OUT logic output

DT — Interface to the time-setting knob. The initial position of the knob is set to 250 ms. Upon any change of the knob's position emits event via the IND output. The current value is available via the OUT data output

MODE — Interface to the mode-setting switch. Any change of the switch's position results in the IND event. The current status is available as a 0, 1, or 2 number at the I data output

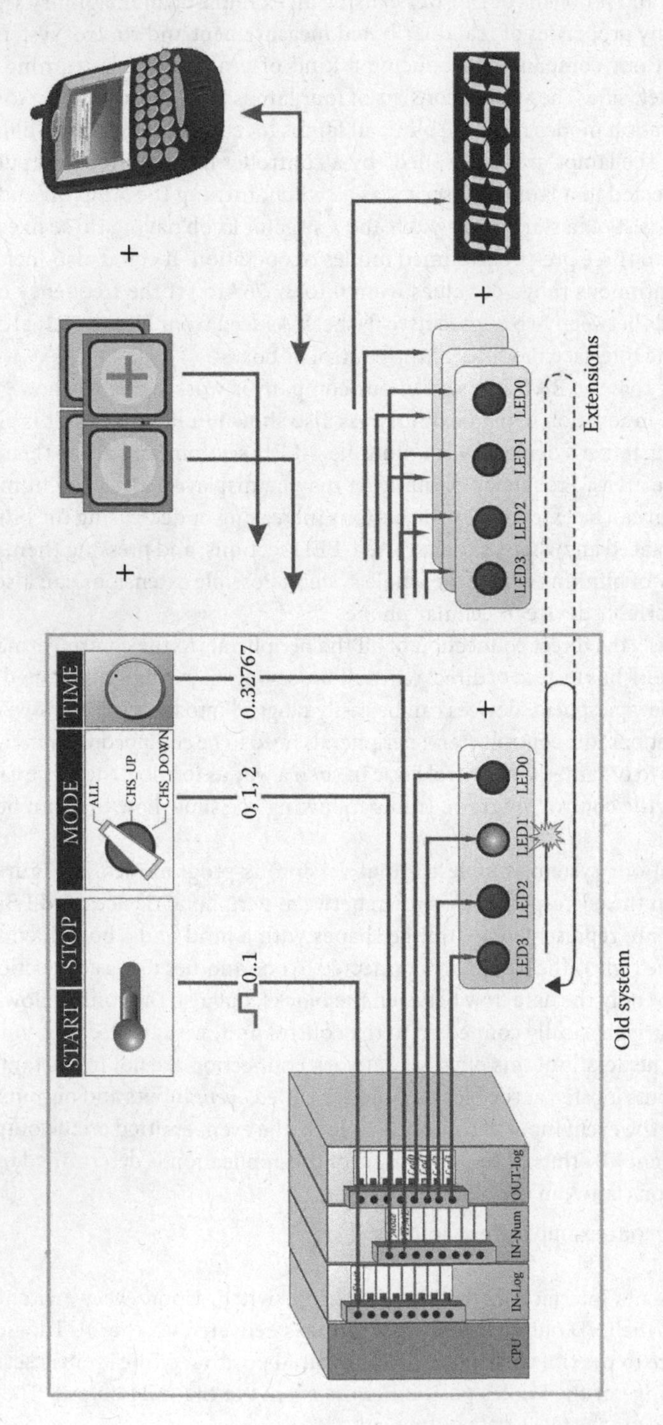

FIGURE 55.1 Programmable running lights system and its potential extensions.

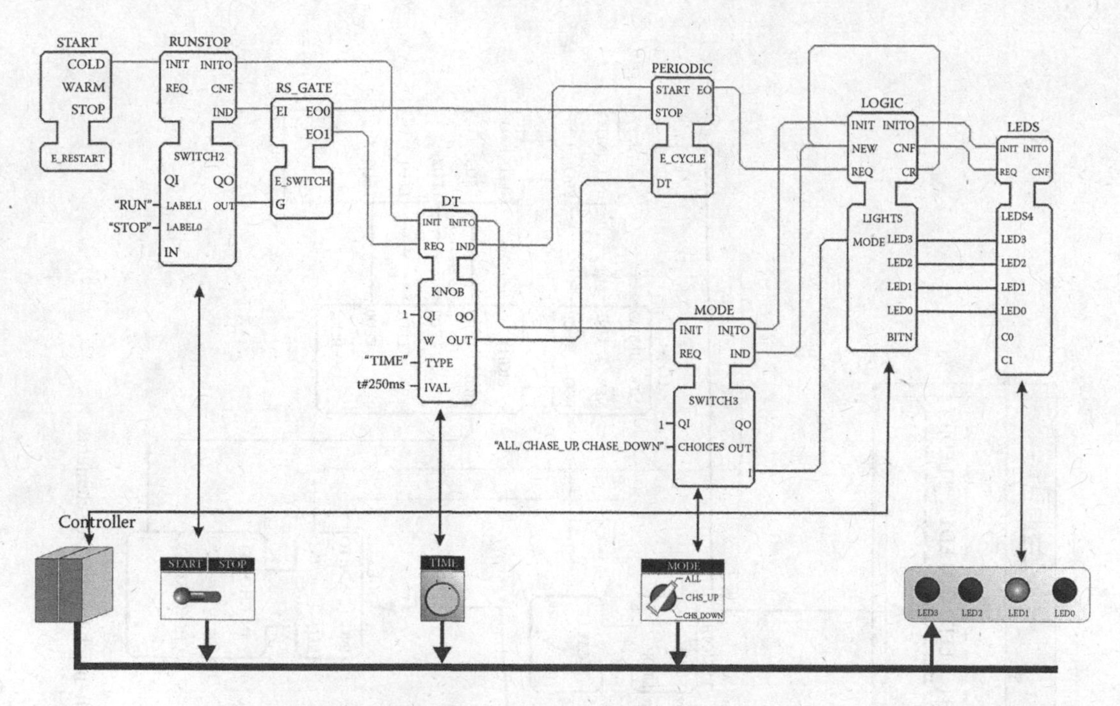

FIGURE 55.2 FB diagram shows the distribution of functionalities between blocks and event and data flow between them.

LOGIC — This block contains the control logic of the system. Given the current operation mode (as a number at the MODE data input), it produces a new combination of four values at the LED0 … LED3 data outputs upon the input event REQ

LEDS — Interface to the LEDs. Upon receiving the REQ input event refreshes the LEDs with the values of the LED0 … LED3 data inputs

Besides, there are three more service FBs:

START — Emits an event at its COLD event output upon physical restart of the device executing this application

RS_GATE — Trigger, emitting events corresponding to START and STOP commands of the start/stop switch

PERIODIC — Upon activation by the start event generates events periodically with the frequency determined by its DT data input. In our application, this event is connected to the LOGIC FB causing the next lights pattern generation

The IEC 61499 FBs are great for fast prototyping and simulation. Thus, in order to test the control logic, the developer can substitute interfaces to physical peripherals by instances of FB types that emulate those peripherals using human–machine interface elements, as illustrated in Figure 55.3. As long as the interfaces of two FB types are the same, this operation is very simple and preserves the structure of interblock connections.

Thus, the RUNSTOP FB is now implemented using the RADIO_BOOL FB type that creates and listens to the radio button element with two values: RUN and STOP. As a result of the FB-type substitution, the application logic can be initially tested without having the peripherals connected. The reverse substitution will produce the fully deployable application.

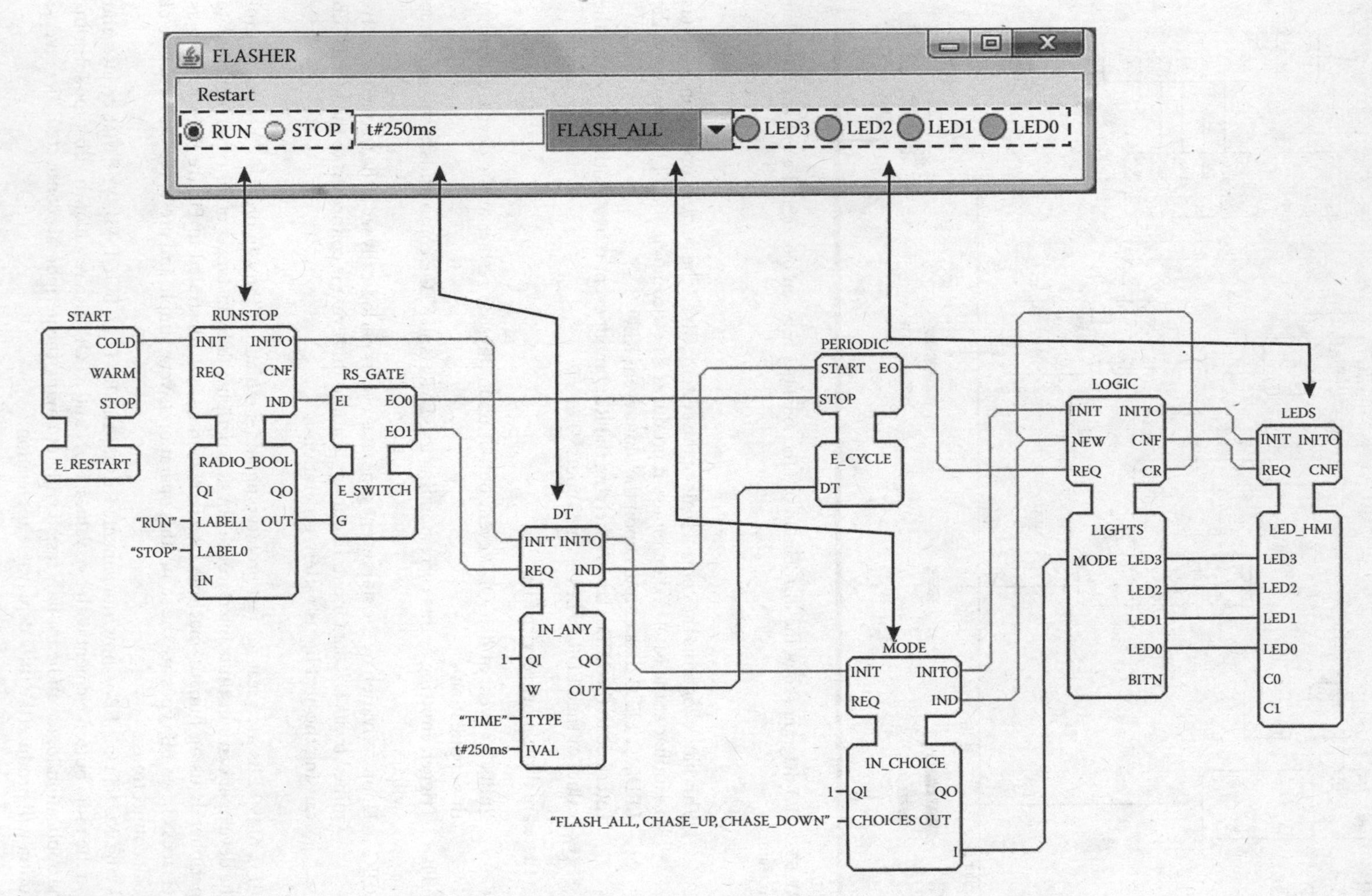

FIGURE 55.3 Simulation application with interfaces to physical peripherals substituted by human–machine interface elements.

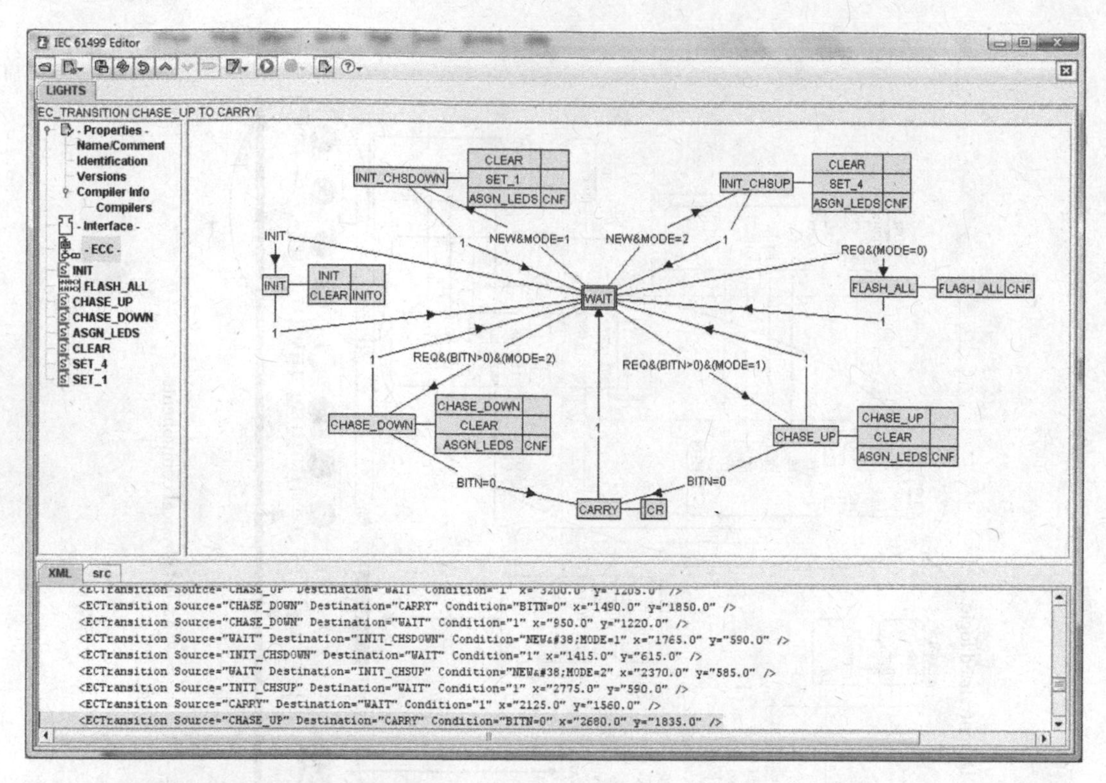

FIGURE 55.4 ECC of the FB-type LIGHTS.

55.3 Logic Encapsulated in Basic FB

Let us consider in more detail how the logic is programmed in FBs. We use as example the basic FB-type LIGHTS that encapsulates the behavioral logic of a single four-lamp LED section.

According to IEC 61499, basic FBs are defined by their interface (i.e., input and output event and data variables), internal variables, algorithms, and a state machine called execution control chart (ECC). Figure 55.4 shows the ECC of the LIGHTS FB as seen in FBDK. In the left pane, one sees also the list of eight algorithms (INIT, …, SET_4, SET_1).

The algorithms are invoked in the actions associated with the states of ECC. Initially, ECC is in a specifically designated start state; in our case this is WAIT state in the middle. For example, in the case of INIT input event, the ECC will "jump" into state INIT, execute the first action, consisting in calling algorithm with name INIT, then execute the second action, consisting in calling algorithm CLEAR and emitting of output event INITO, and then will return to the WAIT state.

Algorithms can be programmed in different programming languages. For example, it is expected that the PLC programming languages of IEC 61331-3 standard [19] may be used, among others.

55.4 Extension

Now let us consider one possible functional extension of our sample system that justifies the distribution of the application across several computing devices. Assume that in addition to the previously considered configuration in Figure 55.2, we add two buttons for increasing/decreasing the speed of the moving light, a numeric display indicating the current frequency of the light update, and, most importantly, one more section of LEDs, so the lights will "jump" from one LED section to another, as shown in Figure 55.5.

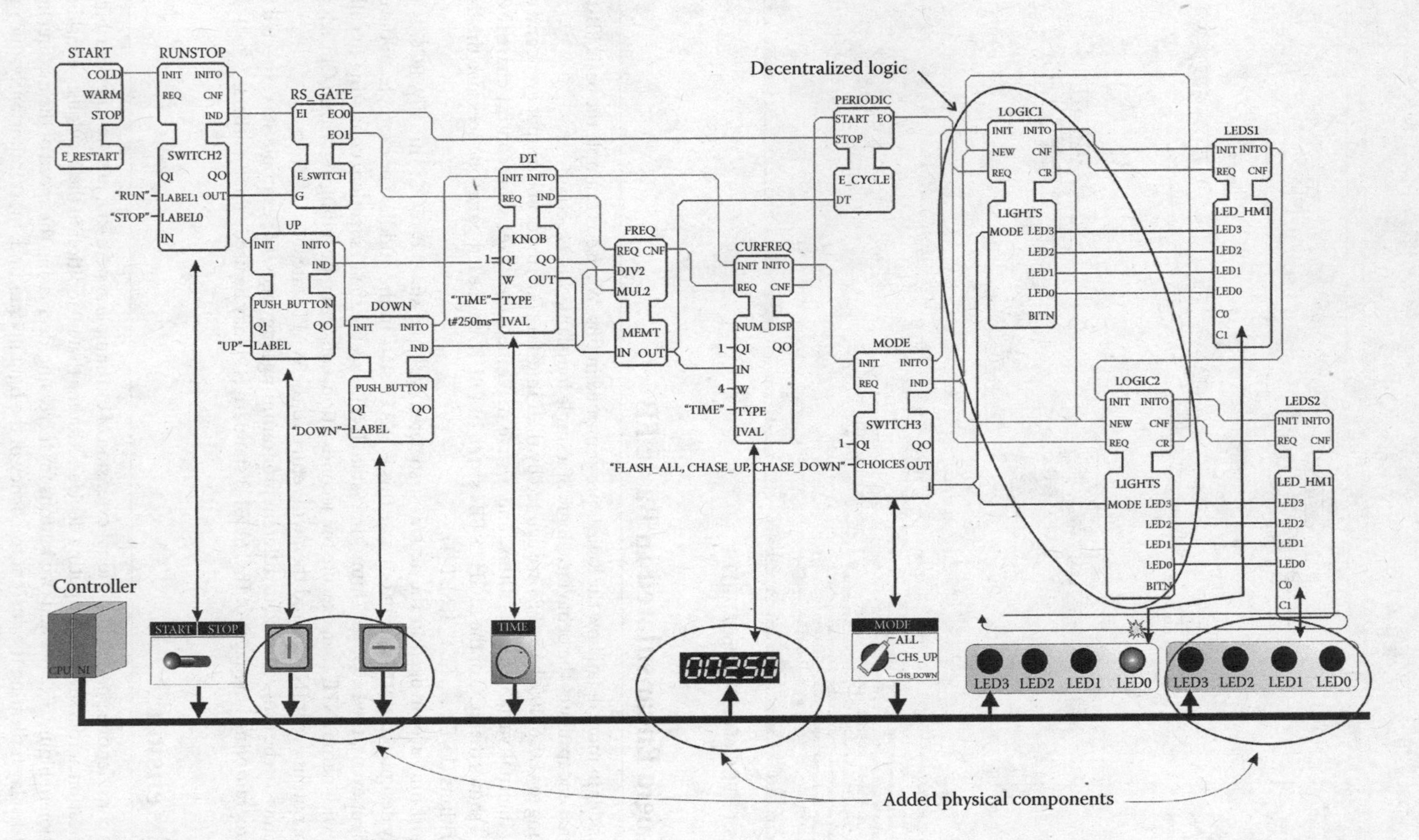

FIGURE 55.5 New system with added peripherals.

The logic and modes of operation of the new system shall remain the same. The changes to the FB model will closely follow the changes introduced into the physical system. For each new peripheral element, we have added a new FB. There is only one extra FB FREQ not corresponding to any physical component. This block implements a global variable storing the current flashing frequency value.

The frequency is set by the time-setting knob and can be changed (increased or decreased) as a result of pressing buttons "+" and "−." The current flashing frequency value is passed to the numeric display and to the "pulse" generator in the FB PERIODIC.

It is interesting that we did not need to change the logic at all! We just added one more FB LOGIC2 and "cascaded" it after LOGIC1. When the running light reaches the rightmost or leftmost position at the LED block, it "jumps" to the next LED block. For that, the event output CR of one block is connected to the event input NEW of the other.

Similar to the first version of the system, the extended system can be easily simulated by substituting FBs interfacing physical peripheral devices by the FBs imitating this on screen.

55.5 Distribution

IEC 61499 implies a two-stage design process. First, an application is created as a network of FBs. At this stage, communication between FBs is abstract, implemented via event and data flow arcs. Second, the application can be distributed across several devices as illustrated in Figure 55.6. The connections

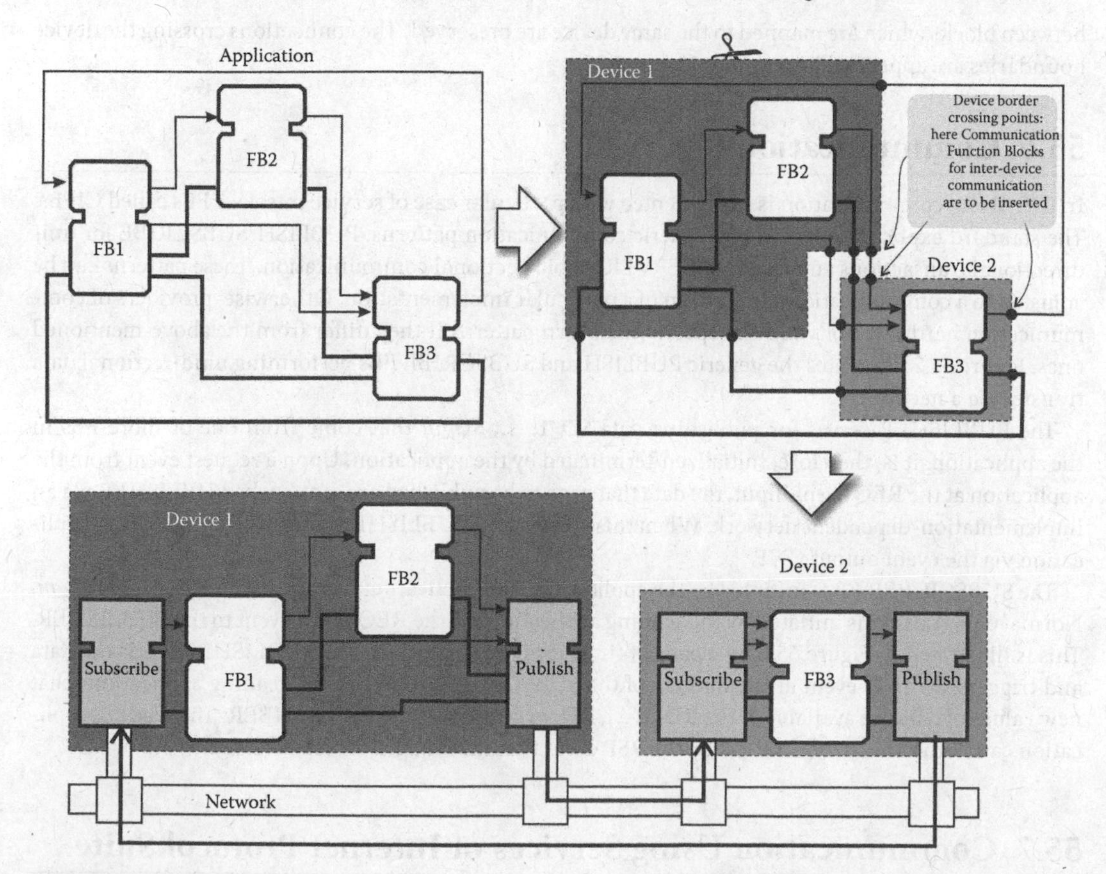

FIGURE 55.6 Distribution of the application across two devices: the connections between blocks which are mapped to the same device are preserved. The connections crossing the device boundaries are appended by communication FBs. (From Vyatkin, V., *IEC 61499 Function Blocks for Embedded Control Systems Design*, Instrumentation Society of America, Research Triangle Park, NC, 2007.)

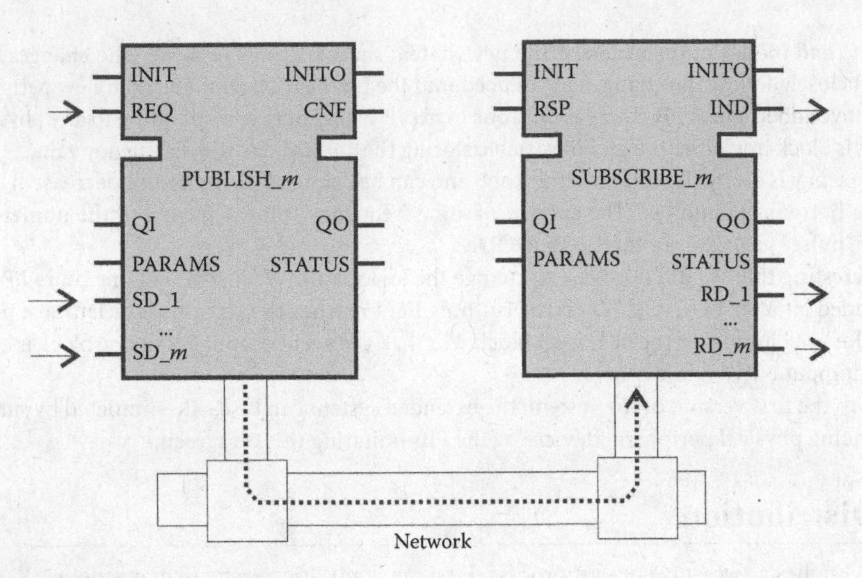

FIGURE 55.7 Publish and subscribe CIFBs.

between blocks which are mapped to the same device are preserved. The connections crossing the device boundaries are appended by communication FBs.

55.6 Communication FBs

In IEC 61499, communication is implemented via a particular case of service interface FBs called CIFBs. The standard explicitly refers to two generic communication patterns: PUBLISH/SUBSCRIBE for unidirectional transactions and CLIENT/SERVER for bidirectional communication. These patterns can be adjusted to a communication mechanism of a particular implementation. Otherwise, providers of communication hardware/software can specify their own patterns if they differ from the above-mentioned ones. Figure 55.7 illustrates the generic PUBLISH and SUBSCRIBE FBs performing unidirectional data transfer via a network.

The PUBLISHER serves for publishing data SD_1, …, SD_*m* that come from one or more FBs in the application. It is, therefore, initialized/terminated by the application. Upon a request event from the application at the REQ event input, the data that need to be published are sent by the PUBLISHER via an implementation-dependent network. When this is done, the PUBLISHER informs the publishing application via the event output CNF.

The SUBSCRIBER FB is initiated by the application. The application reads the data RD_1, …, RD_*m*. Normal data transfer is initiated by the sending application via the REQ input event to the PUBLISHER. This is illustrated in Figure 55.8 by means of time-sequence diagrams. The PUBLISHER sends the data and triggers the IND event at the outputs of the SUBSCRIBER to notify the reading applications that new values of data are available at the RD_1, …, RD_*m* outputs of the SUBSCRIBER. The reading application can notify the SUBSCRIBER by the RSP event that the data have been received.

55.7 Communication Using Services of Internet Protocol Suite

The generic communication patterns of the standard were further specified in the "IEC 61499 Compliance Profile for Feasibility Demonstrations" [1] for implementations using services of the Internet protocol suite.

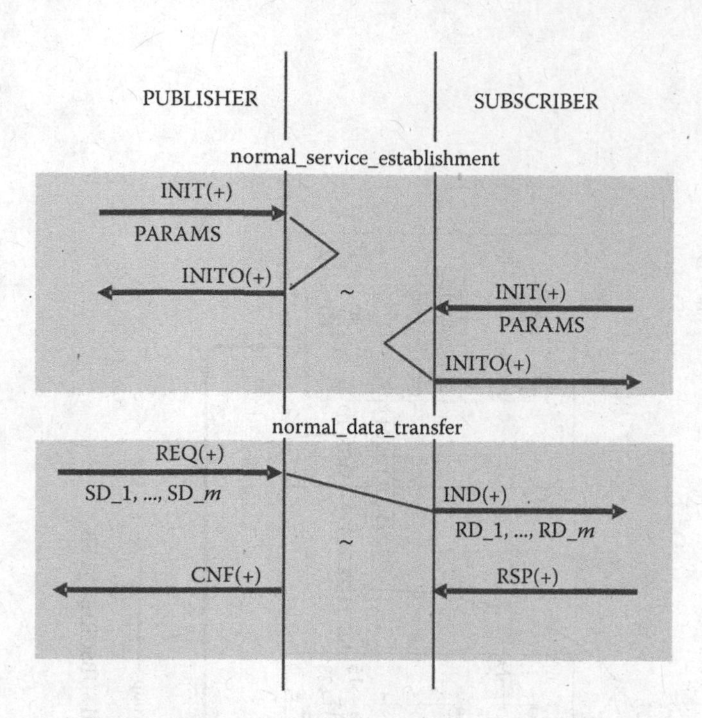

FIGURE 55.8 Communication establishment and normal data transfer sequence.

Provisions of the compliance profile were implemented, for example, in FBDK [2] as the pairs of PUBLISH_*n*/SUBSCRIBE_*n* and CLIENT_*n*_*m*/SERVER_*n*_*m* FBs. The PUBLISH/SUBSCRIBE communication is implemented using services of the universal datagram protocol (UDP). The communication is essentially unreliable and multicast. The CLIENT/SERVER communication is implemented via the transfer control protocol (TCP). It is point-to-point and reliable.

Using this library of communication FBs, a user can develop distributed applications communicating via local networks (Ethernet) and even via the Internet. The specifics of the implementation imply the addressing schema. The communication FBs have input parameter IDs, which are set as follows.

The PUBLISH/SUBSCRIBE communication pattern assumes that there could be many subscribers for one (or even several) publisher(s) as illustrated in Figure 55.9. This communication pattern is implemented in FBDK using the multicast services of the UDP.

The PUBLISH block has to have an ID that follows the rules for Internet multicast addresses. In general, an Internet address consists of four numbers in the interval 0-255; for example, 124.11.2.1. (The multicast addresses are in the interval 224.0.0.0 through 239.255.255.255, but the range of addresses between 224.0.0.0 and 224.0.0.255, inclusive, is reserved for system purposes.) In the example, the ID of the PUBLISHER (located in Device 1) is 225.0.0.1:1204 (where 225.0.0.1 is an IP address for that multicast group and 1204 is a unique port number selected in Device 1). The ID of the publisher completely identifies the multicast group.

Any SUBSCRIBE block located in a device on the same network segment as the PUBLISH can receive the published data if it has the same ID as the PUBLISH. The published data will be received by both subscribers located in Device 2 and Device 3.

The CLIENT_SERVER communication is point-to-point (i.e., only one client can send and receive data to/from one server). The addressing is illustrated in Figure 55.10. The SERVER, which is located in Device 2, has ID that refers to the address of the host device (localhost) and adds there a unique (for that device) port number. The CLIENT, located in Device 1, has an ID that is composed of two parts: IP address of the server's device and the same port number as used in the SERVER's ID.

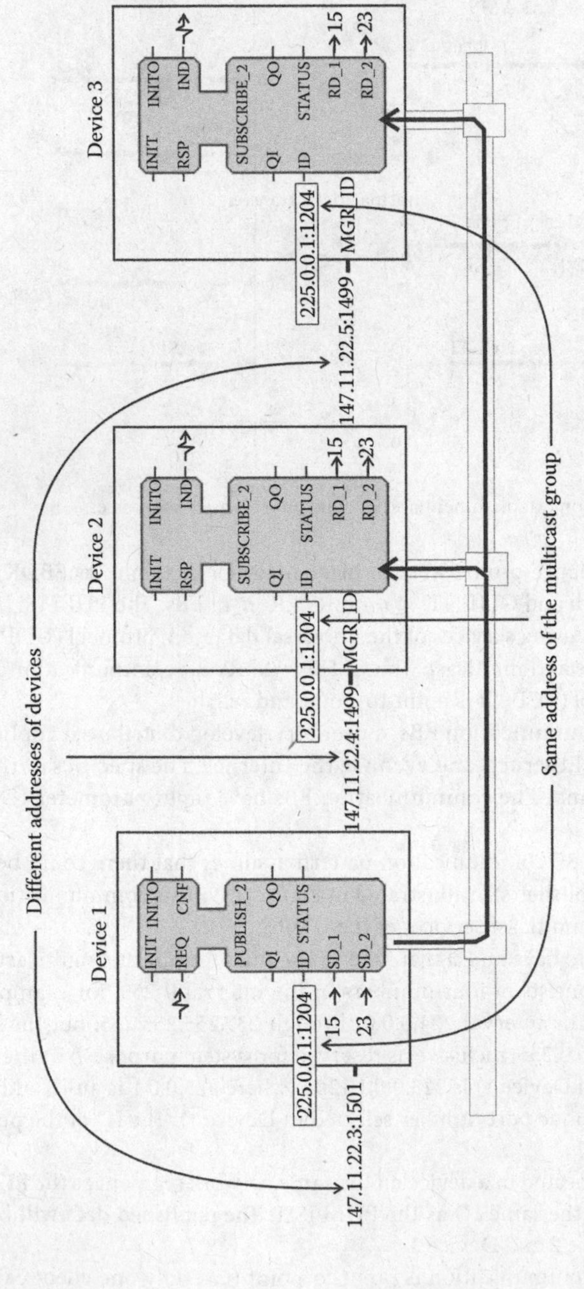

FIGURE 55.9 Multicast communication over a local network implemented using PUBLISH_2 and SUBSCRIBE_2 FBs.

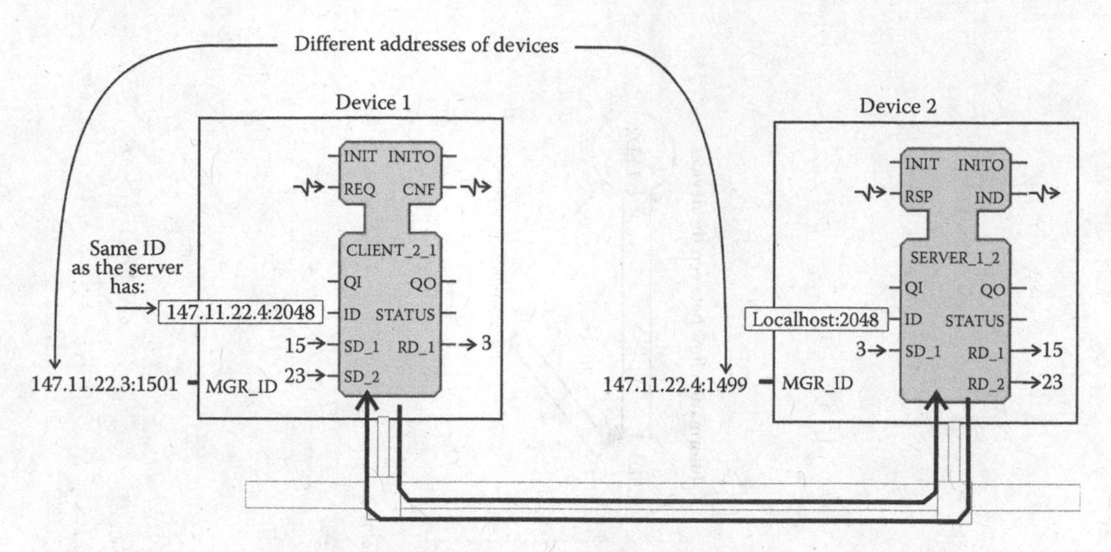

FIGURE 55.10 Point-to-point bidirectional communication implemented using SERVER_1_2 and CLIENT_2_1 FBs.

55.8 Adding Distribution and Communication to the Sample System

In this section, we see how distribution and communication can be added to our extended example from Figure 55.5. Assume that the peripherals are grouped into three networking devices as shown in Figure 55.11. There are two LED panels (LED1 and LED2), one of which (LED1) has increase–decrease speed buttons and control panel (device panel) for setting parameters of the running lights process. The devices are connected by two different networks: LED1 and LED2 with a DeviceNet segment, while Panel and LED1 with Ethernet.

These physical connections need to implement the logical connection between the devices that can be different. Thus, LED2 and panel have no direct physical connection, but information flow from Panel to LED2 is required.

In terms of IEC 61499, network topology of this system can be described within a simulation system configuration as shown in Figure 55.12. The system configuration is supposed to produce on-screen simulation of our system; that is why three devices have type "FRAME_DEVICE."

The FB application capturing the behavior logic of this system is exactly the same as was seen in Figure 55.8. Applying the mapping technique for distribution of our application across these three devices, we obtain the system configuration with explicit communication as shown in Figure 55.13.

The communication FBs SEND/RCV used in the distributed version of our system implement the point-to-point communication model. They are similar to PUBLISH/SUBSCRIBE, with the only difference that multicast is not supported. We also assume that if the ID parameter is left blank, then the compiler inserts the instance name as the ID, and each ID uniquely belongs to a pair of SEND and RCV FBs. (This way of addressing is implemented for the local communication FBs PUBL/SUBL in FBDK/FBRT.)

There are two modifications of the SEND/RCV FBs, one for communication over MODBUS protocol and the other for communication in DeviceNet using the CIP. The latter FBs are called CIP_SEND_*n*/CIP_RCV_*n*.

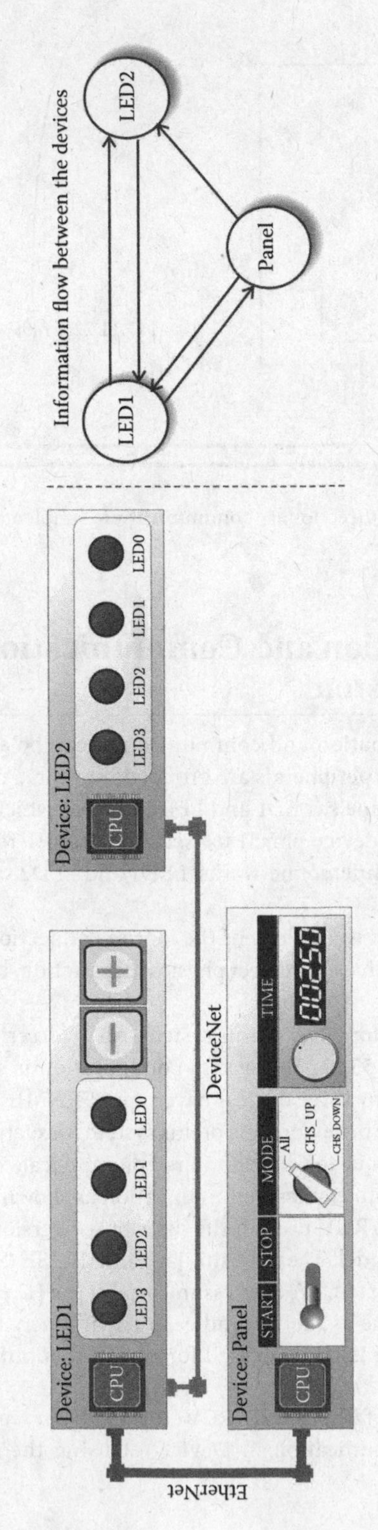

FIGURE 55.11 Distributed configuration of the "running lights" system (left) and information flow between the devices.

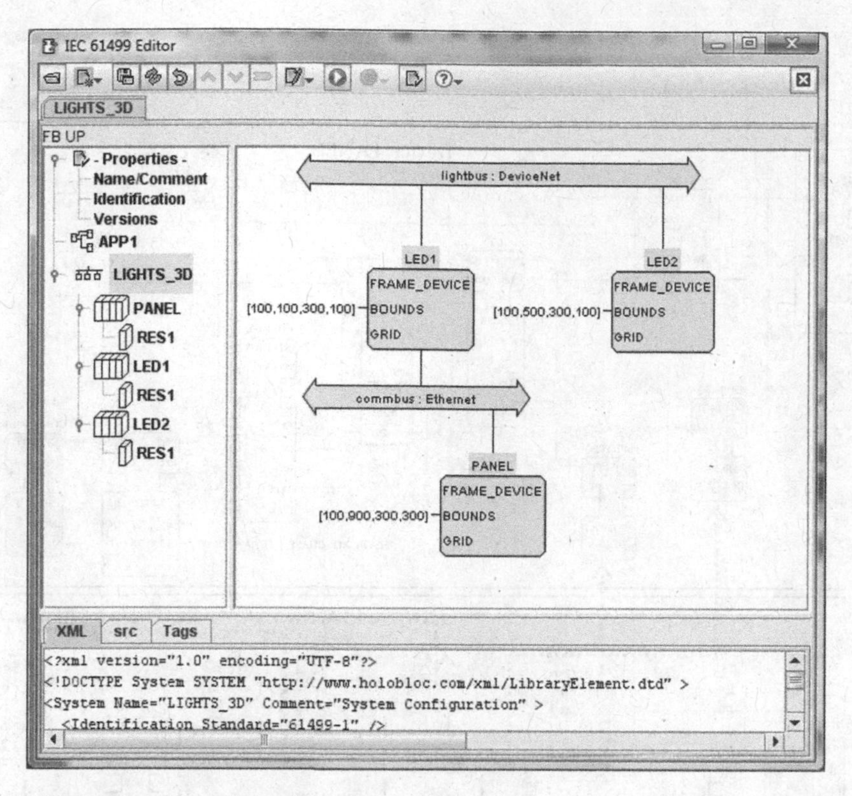

FIGURE 55.12 The network topology of our system described as a part of distributed system configuration of IEC 61499.

55.9 Internals of Communication FBs: Modbus

Modbus (described in Chapter 4.12 of this Handbook and in [20–24]) is one of the many communication protocols that may be used for transferring events and data over a network. As this protocol is based on a remote memory data model which is accessed and updated using the client–server interaction model, it is naturally implemented in IEC 61499 as CLIENT and SERVER communication service interface FBs illustrated in Figure 55.14.

The Modbus protocol includes three versions, namely RTU, ASCII, and TCP. The first two use a different method of byte encoding on the same physical media (RS232 or RS485), while the last transmits the messages over TCP/IP connections. Due to these variations of the protocol, one SERVER and one CLIENT FB type are needed for each version.

A description of a typical data and event transfer sequence is as follows:

- The device containing the RCV FB (the server device) initializes this server (INIT+ event, i.e., INIT event with QI input set to TRUE) while simultaneously providing configuration parameters. For the TCP version, these consist of the local IP addresses and port numbers on which it will listen awaiting for the arrival of connection requests from clients. For the RTU and ASCII versions, the parameters are the identification of the hardware interface (e.g., the serial port) to use as well as the serial line parameters (baud rate, number of start and stop bits, number of bits, and parity). All versions also require an additional parameter, the Modbus device ID.
- The device containing the SEND (client device) initializes this client (INIT+ event). Required parameters include the serial line interface details (for RTU and ASCII versions) and the Modbus device ID of the remote server with which it will communicate. For the TCP version, the IP address

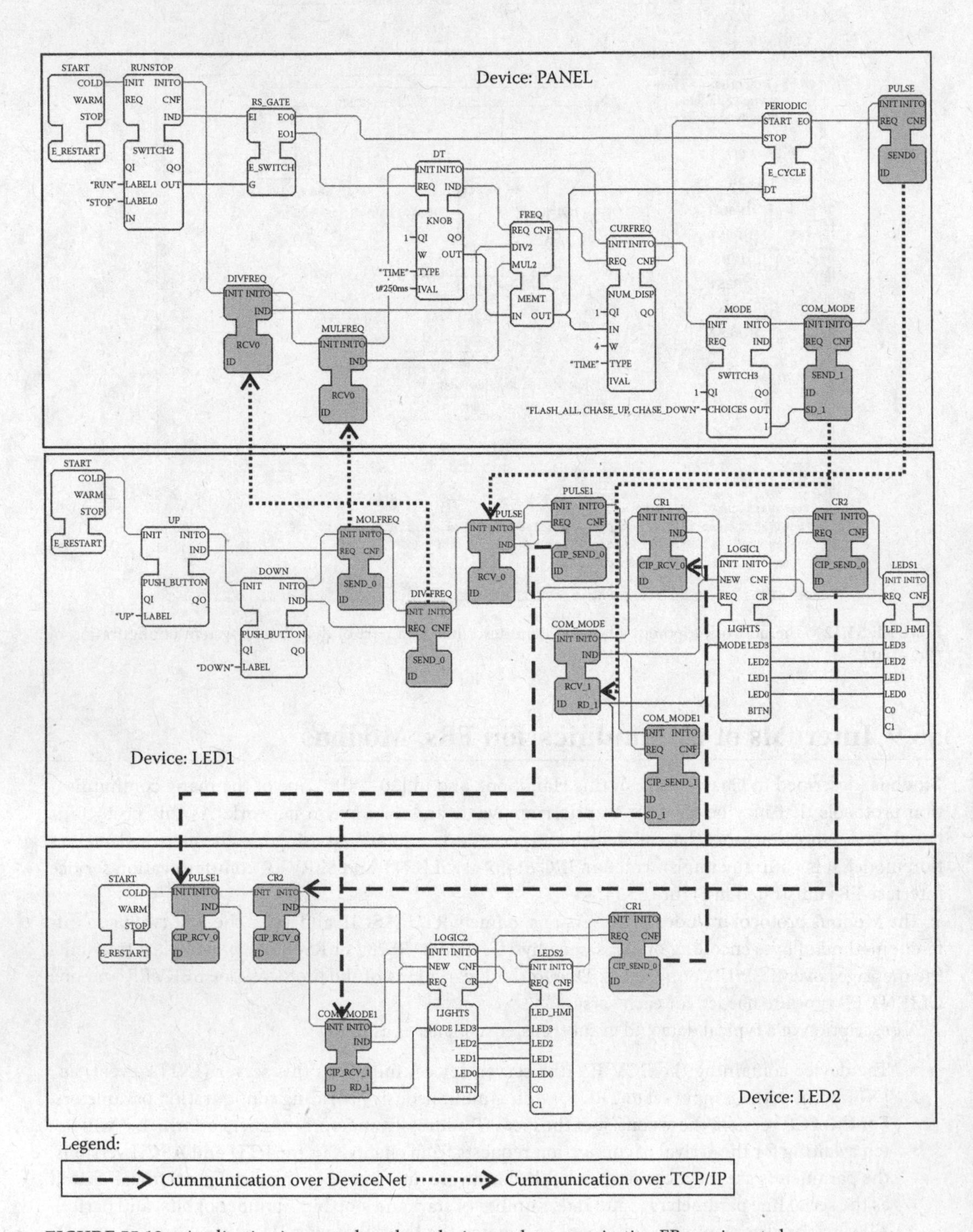

FIGURE 55.13 Application is mapped on three devices and communication FBs are inserted.

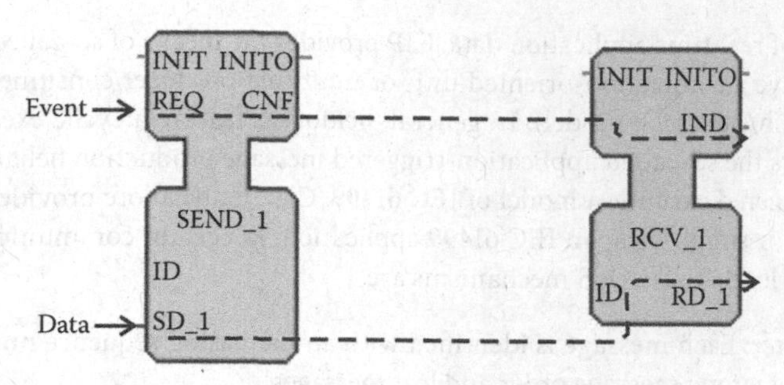

FIGURE 55.14 Producer–consumer pair implemented in MODBUS.

and port number of the remote server is also required, and this INIT+ event will result in the client establishing a TCP connection with the RCV FB. The SEND FB, upon receiving an REQ+ event, sends the data (16-bit words) available on its SD_x input ports and simultaneously asks for the return data, using the "read/write multiple registers" Modbus function (function code 23-17 hex).

- The RCV FB, upon receiving the message sent by the client, places the received data on its RD_x data output ports and emits an IND+ event.
- The server device must then make the return data available on the RCV SD_x input ports and emit the RSP+ event. Upon receiving this event, the RCV FB then replies to the "read/write multiple registers" request. The SEND FB, upon receiving the reply message, places the received data on its RD_x output ports and emits the CNF+ event.

If the RSP+ event is delayed beyond a certain timeout, the SEND FB will emit an RSP– event.

Since the IEC 61499 FBs do not use the same remote memory data model as the Modbus protocol, with the mapping proposed above it is not necessary to specify the Modbus register addresses. These are assumed to always start at 0, and the number of registers to read (write) is the same as the number of data outputs (inputs) on the SEND FB. Note that the number of input words of the SEND FB must match the number of output words in the RCV FB, and vice versa.

Similar SEND and RCV FBs are used for the transmission of Boolean data. However, since the Modbus protocol does not include a function that allows simultaneous reading and writing of Boolean data, these must only send data either from the SEND FB to the RCV FB (using the Modbus function "Write Multiple Coils," code 15-0F hex) or from the RCV FB to the SEND FB (using the Modbus function "Read Coils," code 01-01 hex).

The above communication assumes that the server device also supports IEC 61499. However, currently many remote terminal units (RTUs) support the Modbus protocol, but do not implement IEC 61499. To allow a controller device to communicate with these legacy RTUs, SEND FBs are used that connect directly to these legacy Modbus server devices. These RCV FBs require additional parameters when the function is initialized in order to specify the Modbus function, data address, and, when reading data, the number of registers to use when transferring data to/from the legacy Modbus server.

55.10 Communication via the CIP

In this section, we show how another communication network, based on CIP, can be used in our example. CIP has originally been developed as application layer for Rockwell Automation's CAN-based DeviceNet and their controller network ControlNet. With the development for the industrial Ethernet EtherNet/IP and the sensor actuator bus CompoNet, the CIP specification was further unified and provides a common application layer for these four networks. CIP is a multimaster system and therefore well-suited for the application in a distributed control system.

Interface of IEC FBs for CIP was defined in the compliance profile [25]. More implementation details can be found in [26,27].

For the exchange of real-time application data, CIP provides the means of so-called implicit messages. Implicit messages have a connection-oriented uni- or multicast producer/consumer messaging model (i.e., similar to publish/subscribe model). In general, fieldbuses feature a cyclic execution behavior. In difference, CIP allows the so-called application-triggered message production behavior, which fits very well to the event-triggered execution model of IEC 61499. CIP furthermore provides QoS mechanisms that can be utilized for simplifying an IEC 61499 application, as certain communication errors can be handled at CIP stack level. These QoS mechanisms are

1. *Sequence counter*: Each message is identified with an increasing sequence number, which allows the detection of wrong message order and lost messages.
2. *Production inhibit timer*: With this value, the minimum time between two messages can be specified on the producer's side. This allows to bound the network traffic.
3. *Expected packet rate*: This rate gives the consumer an estimation of how often messages will arrive. This can be used to determine if the producer was lost.

As CIP producers do not keep track of and do not request an acknowledgment from consumers, the producer has no CIP means for detecting the loss of consumers.

With these basic assumptions, the behavior of the CIP_SEND_1 and the CIP_RCV_1 communication FBs (Figure 55.15) is as follows:

- The device containing the CIP_SEND_1 FB initializes this FB with an INIT+ event. The FB will create a CIP assembly object with the given ID. To this assembly object, the CIP_RCV_1 can now connect. The size of the data and the meaning of the data are defined by the number of inputs and their data types. As CIP uses the same data types as IEC 61499, namely the data types of IEC 61131-3, the CIP data encoding mechanisms can be directly used.
- After the CIP_SEND_1 set up its side of the communication, the CIP_RCV_1 can establish the connection. This is again triggered with an INIT+ event. The ID input has to specify the address of the device and the assembly object to connect to (i.e., the assembly object created by the CIP_SEND_1). As a result, the CIP stack is configured for a consumer, and a unicast CIP connection between the two devices is established. As the application cannot guarantee that the CIP_SEND_1 is initialized before the CIP_RCV_1, the implementation has to be robust enough to handle this case (e.g., automatic retry to establish the connection).
- On triggering the REQ event of the CIP_SEND_1, the input data are encoded into the CIP network format, and the message is sent. With a CNF event, the IEC 61499 application is informed on the success of this procedure. No information on the consumer side, such as if the data have been received and processed is given.
- On receiving a message, the CIP stack will inform the CIP_RCV_1 FB. There the data is decoded and placed on the outputs. The application is then notified with an IND event.
- Through INIT events, the connection can be closed from both sides. If the connection is closed from the CIP_SEND_1 side, the CIP_RCV_1 will be informed, otherwise not.

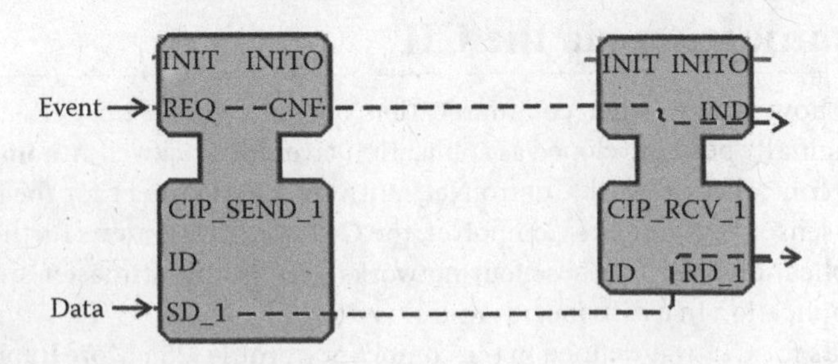

FIGURE 55.15 Producer–consumer pair implemented with CIP.

This investigation shows that CIP fits very well to the execution and object model of IEC 61499.

Our example made the assumption that all devices on the CIP network are IEC 61499-enabled control devices. However, several advantages can be gained from combining legacy CIP devices with IEC 61499-enabled devices. Two different application scenarios should be considered: first IEC 61499-enabled CIP slave device (i.e., programmable I/O devices) and second use standard CIP slave devices from IEC 61499 devices. In the first case, the CIP interface of the IEC 61499 device would be specified by utilizing the CIP FBs, and with this mechanism the interface can be adapted to the specific IEC 61499 application running in the device. On the other end of the communication, a CIP master device (e.g., a PLC) would interact with the IEC 61499 device in the same way as with any other CIP slave device. However, the use of IEC 61499 allows to pre-process I/O data and to provide higher level information to the PLC. This use case is an important step toward a fully distributed control system as described by IEC 61499. In the second use case, the IEC 61499 control device will take the place of the PLC, and the CIP communication FBs will mimic the behavior of a CIP master. However, in contrast to the existing PLC interface where the network data are presented to the application as a flat memory region, IEC 61499 allows a structured access to the data. This is achieved by representing each slave device as a separate FB. This allows to leverage existing investments in plants and I/O device development, or to extend dumb I/O modules by adding own CPUs. Therefore this will be an important part of an IEC 61499-based control environment.

Both use cases are not specific to the CIP environment. They can be applied to any fieldbus system used in industrial control systems.

55.11 Impact of Communication Semantics on Application Behavior

One word of caution however. Traditional control applications running on a single controller periodically or cyclically read from, and write to, all the remote devices connected to the network bus (even when no changes to the state of these devices is required); so if the network suffers a transient error during one period/cycle, the application will naturally recover in the next period/cycle, when a new network message is generated. In contrast, an IEC 61499 application is event oriented, so it will typically only generate a network message when the PUBLISH or SEND FBs receive an REQ event, or when an RCV FB receives an RSP event. While this may be considered an advantage due to the lower traffic imposed on the network, it also has the drawback that the application becomes much more susceptible to transient communication errors that may occur on this same network.

Consider, for example, the system in Figure 55.13. Now imagine that a CR event generated by FB Logic1 does not reach the NEW event input of Logic2 due to a transient communication error on the network used by the publish and subscribe CR2 FBs. The result will be that all the lights connected to Logic1 will be correctly switched off, and no light connected to Logic2 will be lit to continue the chasing sequence.

We may conclude that due to the asynchronous event-based architecture used by IEC 61499 applications, these are more susceptible to network failures. Nevertheless, this does not mean that the application will fail at the smallest network interference. In fact, many communication protocols include error detection and correction algorithms. If the communication protocol used between the CR2 PUBLISH and SUBSCRIBE FBs includes error detection and correction mechanisms, the IEC 61499 application will not be affected by the transient network failure. However, note that in the presence of failures, the exact semantics of the communication service interface FBs is not specified by the IEC 61499 standard, as well as the choice of which specific implementation of the communication service interface FBs will affect how the IEC 61499 application will react when in the presence of transient network failures.

The message delivery semantics in the presence of failures may generally be classified into three classes: "at most once," "at least once," and "exactly once." Although the "exactly once" semantics are usually the preferred semantics, due to the higher number of messages required, sometimes semantics with lower guarantees are sufficient. For example, when switching on a siren to warn of the occurrence of an alarm, the "at least once" semantics would be sufficient.

Another issue is that of event ordering. If two events are generated inside one device in the order EV1, EV2, these are not guaranteed to arrive at another device in the same order. If an application relies on the ordering of events to correctly execute its algorithms, care should be taken to choose communication service interface FBs whose implementation guarantees message delivery in the same order.

Some implementations of the communication service interface FBs may allow quality of service (QoS) parameters to be specified. Others may have fixed QoS parameters. In conclusion, one should be careful of the exact semantics implemented by the service interface FBs as these may affect the outcome of the IEC 61499 control application.

55.12 Failures in Distributed Applications

Although distributed applications may have many advantages, they are not without their drawbacks. One of these, as was discussed previously, is the fact that the communication network may exhibit failures that will affect the outcome of the application execution. Another is the failure of the devices executing the control logic itself.

When writing an industrial strength control application, usually all failure modes of the controlled process must be taken into account, so that the physical machinery under control always remains in a safe state. Likewise, the failure of the controlling devices, i.e., the PLCs executing the logic and the communication networks linking them, must also be taken into account.

For applications that run on a single PLC, usually either the control application fails completely when the executing PLC fails or it does not fail at all. Complete failures of the control application are usually handled by external safety devices.

However, when writing a distributed application, one must be careful of how this application may fail, as it may have partial failures when one or more, but not all, of the PLCs that are executing the distributed control application fail. Even more likely is the possibility of failure of the communication network that connects the distributed PLCs.

Now, to overcome partial failures of a distributed application, the application itself must be aware of this scenario and be prepared to take appropriate action when it occurs. However, in order to do so, it must first be able to identify the fact that a partial failure occurred (be it on the communication network or another executing device). This identification may be done either at the application level or by the communication service interface function blocks (SIFBs).

IEC 61499 communication SIFBs typically indicate the presence of failures by emitting local events while their QO output is set to false. However, as was stated previously, the capacity of a communication SIFB to detect the failure and consequently emit the event depends on the underlying communication protocol. For example, some protocols include "keep alive" messages that are exchanged even when no data need to be transmitted so that the lack of these messages may be safely assumed to be the result of a failure.

When communication SIFBs are not able to identify failures, then these will need to be identified at the application level. One simple means of doing this is to configure the control applications to periodically call the PUBLISH and/or CLIENT (or SEND, as in our example) SIFB, even when no new data need to be transmitted. Another possibility is the implementation of handshakes and timeouts at the application level.

55.13 Conclusion

The FBs of IEC 61499 are a powerful architectural model for abstract yet executable specification of distributed control systems. It can be further extended by adding communication over any particular network protocol. Interfaces to the protocol need to be implemented in libraries of CIFBs. Advanced software tools can make adding the explicit communication seamless and hidden from the user.

References

1. IEC 61499-1: Function Blocks—Part 1 Architecture, 1st edn., International Electrotechnical Commission, Geneva, Switzerland, 2005.
2. FBDK—Function Block Development Kit, Online: www.holobloc.com, 2008.
3. ICS-Triplex. ISaGRAF—IEC 61131 and IEC 61499 software, cited 2008. Available from: http://www.isagraf.com/
4. V. Vyatkin and J. Chouinard, On comparisons of the ISaGRAF implementation of IEC 61499 with FBDK and other implementations, in *Sixth IEEE International Conference on Industrial Informatics, 2008. INDIN 2008*, Daejeon, Korea, July 2008.
5. nxtControl website: http://www.nxtcontrol.com/
6. K. Thramboulidis, Development of distributed industrial control applications: The CORFU framework, in *Fourth IEEE International Workshop on Factory Communication Systems*, Vasteras, Sweden. August 2002.
7. T. Strasser et al., Framework for distributed industrial automation and control (4DIAC), in *Sixth IEEE International Conference on Industrial Informatics, 2008 (INDIN'08)*, Daejeon, Korea, July 13–16, 2008.
8. FBench website: http://www.ece.auckland.ac.nz/~vyatkin/fbench/
9. W. Dai, A. Shih, and V. Vyatkin, Development of distributed industrial automation systems and debugging functionality based on the Open Source OOONEIDA Workbench, in *Australasian Conference on Robotics and Industrial Automation*, Auckland, New Zealand, 2006.
10. PROFACTOR Produktionsforschungs GmbH, 4DIAC-RTE (FORTE): IEC 61499 Compliant Runtime Environment, 30/10/2007, 2007; http://www.fordiac.org
11. L.H. Yoong, P. Roop, V. Vyatkin, and Z. Salcic, Synchronous execution for IEC 61499 function blocks, in *Proceedings Fifth IEEE Conference on Industrial Informatics (INDIN'07)*, Vienna, Austria, July 23–27, 2007, pp. 1189–1194.
12. G. Cengic and K. Akesson, Definition of the execution model used in the FUBER IEC 61499 runtime environment, in *Sixth IEEE International Conference on Industrial Informatics, 2008 (INDIN 2008)*, Daejeon, Korea, July 13–16, 2008.
13. K. Thramboulidis and A. Zoupas, Real-time Java in control and automation: A model driven development approach, in *Tenth IEEE International Conference on Emerging Technologies and Factory Automation*, Catania, Italy, September 19–22, 2005.
14. G. Doukas and K. Thramboulidis, A real-time Linux execution environment for function-block based distributed control applications, in *Third IEEE International Conference on Industrial Informatics (INDIN'05)*, Perth, Australia, August 10–12, 2005.
15. P. Tata and V. Vyatkin, Proposing a novel IEC61499 Runtime Framework implementing the cyclic execution semantics, in *INDIN 2009*, Cardiff, U.K., June, 2009.
16. V. Vyatkin, *IEC 61499 Function Blocks for Embedded Control Systems Design*, Instrumentation Society of America, Research Triangle Park, NC, 2007.
17. H.-M. Hanisch and V. Vyatkin, Achieving reconfigurability of automation systems by using the new international standard IEC 61499: A developer's view, in *Integration Technologies for Industrial Automated Systems*, R. Zurawski (ed.), CRC Press/Taylor & Francis, Boca Raton, FL, July 2006.
18. A. Zoitl, *Real-Time Execution for IEC 61499*, ISA, Research Triangle Park, NC, 2008.
19. IEC 61131-3 International Standard, Programmable Controllers—Part 1: General information, 2.0 ed., International Electrotechnical Commission, Geneva, Switzerland, 2003.
20. IEC 61499 Compliance Profile for feasibility demonstrations, http://www.holobloc.com/doc/ita/index.htm
21. Modbus Application Protocol Specification, v1.1b, Modbus-IDA, December 28, 2006, available from: http://www.Modbus-IDA.org
22. A. Swales, Open Modbus/TCP Specification, Schneider Electric, March 29, 1999.

23. Modbus Messaging on TCP/IP Implementation Guide v1.0b, Modbus-IDA, October 24, 2006, available from http://www.Modbus-IDA.org
24. MODBUS over Serial Line Specification and Implementation Guide V1.02, Modbus-IDA, 2006, available from: http://www.Modbus-IDA.org
25. IEC 61499 Compliance Profile for CIP based communication function blocks, http://www-ist.massey.ac.nz/functionblocks/CIPCompliance.htm
26. F. Weehuizen, A. Brown, C.K. Sünder, and O. Hummer-Koppendorfer, Implementing IEC 61499 communication with the CIP protocol, in *Proceedings 12th IEEE International Conference on Emerging Technologies and Factory Automation (ETFA'07)*, Patras, Greece, September 25–28, 2007.
27. F. Weehuizen and A. Zoitl, Using the CIP protocol with IEC 61499 communication function blocks, in *Proceedings of the Fifth IEEE International Conference on Industrial Informatics (INDIN'07)*, Vienna, Austria, June 23–27, 2007, Vol. 1, pp. 261–265, ISBN: 1-4244-0864-4.

56

Industrial Internet

Martin Wollschlaeger
Dresden University of Technology

Thilo Sauter
Austrian Academy of Sciences

56.1 Introduction **56**-1
56.2 Application of Internet Technologies in Industry **56**-2
56.3 Technologies **56**-3
Transport and Communication Related Technologies • Technologies for Information Description and Presentation • Technologies for Server-Side and Client-Side Functions
56.4 Application Examples **56**-8
Description Technologies • Browser-Based Applications • Machine–Machine Communication Using Web Services
56.5 Conclusions and Outlook **56**-12
Acronyms **56**-12
References **56**-13

56.1 Introduction

The Internet and its applications can be seen as an integral part of the actual social development stage. Without any doubt, the Internet has changed communication and information retrieval. This is enabled by a rapidly developing set of technologies and easy-to-use applications, suitable for nearly all situations of every day life. On the other hand, the broad and increasing application range has in turn advanced technology development. The common and stable basis for Internet Applications is the Transmission Control Protocol/Internet Protocol (TCP/IP) as the general, always and everywhere available—in this sense ubiquitous—communication platform.

Internet Technologies is a common term for those technologies that represent the application-related layers 5–7 of the Open System Interconnection (OSI) communication model [HZ80], relying on IP protocol on layer 3 and the TCP or User Datagram Protocol (UDP) protocols at layer 4. Although it would be technically possible, in most cases a clear separation of these layers inside a specific application is not easy to achieve. However, this might be an additional benefit, since the slim application structures make application development for Internet use rather easy.

The application of Internet technologies is tremendously influenced by the development of the World Wide Web. Especially, Web browsers promise to be able to display nearly every type of data without any additional software. They are available for every computing platform, are easy to install or are already an integral part of operating systems. They are easy to use and provide access not only to static content like the traditional Web pages, but also to dynamic content including multimedia. The actual trends like Web 2.0 offer advanced technologies as well as platforms for social interconnections up to complete virtual communities like “Second Life.” Of course, the trends sketched above also create drawbacks. The problem that is mentioned most often is security. But without any doubt, Internet technologies have changed human life.

56.2 Application of Internet Technologies in Industry

The trend of applying Internet technologies in different application areas can hardly be stopped. The promising features like instant and ubiquitous information access without any additional efforts, no costs for software installation and maintenance, and finally the manifold de facto standards lead to a spreading application of Internet technologies also in industrial application, characterized by the term "Industrial Internet."

The application of Internet technologies seems to solve most major problems in the industrial automation domain. Exchange and presentation of information is greatly simplified by applying those methods and technologies that are well known from application in general IT systems. Because of communication system independence, Internet technologies seem to perfectly fit the requirements for vertical integration in automation. Especially, the trend to use Ethernet-based systems at the factory floor will simplify the steps needed for seamless integration of automation systems and company IT systems. This will reduce the efforts for data communication, conversion, and presentation, thus finally reducing complexity and costs for such systems. In addition, the adoption of Internet technologies introduces new concepts and offers new chances for tasks in different stages of an automation system's life cycle. Especially, Web technologies are state of the art and have developed into de facto standards (see Figure 56.1).

Browsers and servers exist for nearly every platform; they are widely accepted by different classes of users. The browser as a platform-independent application framework should allow re-using software components. Finally, location independence, multiple client access, and easy installation of software components are in the users' focus. On the other hand, this "promised land" has some drawbacks, which mostly encompass dealing with the heterogeneity of the different technologies with typically very short life cycles.

However, the application in industrial automation produces some additional requirements for the effective use of Internet technologies. Besides commonly mentioned requirements like real-time aspects, communication structures, and security aspects, in addition the availability and maturity of the technologies, as well as the acceptance and the knowledge of users and developers have to be considered. Finally, the selection of suitable technologies or technology combinations appears to be difficult. It is mainly based on the use cases to be addressed by an Industrial Internet application, but also on the automation and IT systems' design, and on considerations regarding the total cost of ownership (TCO).

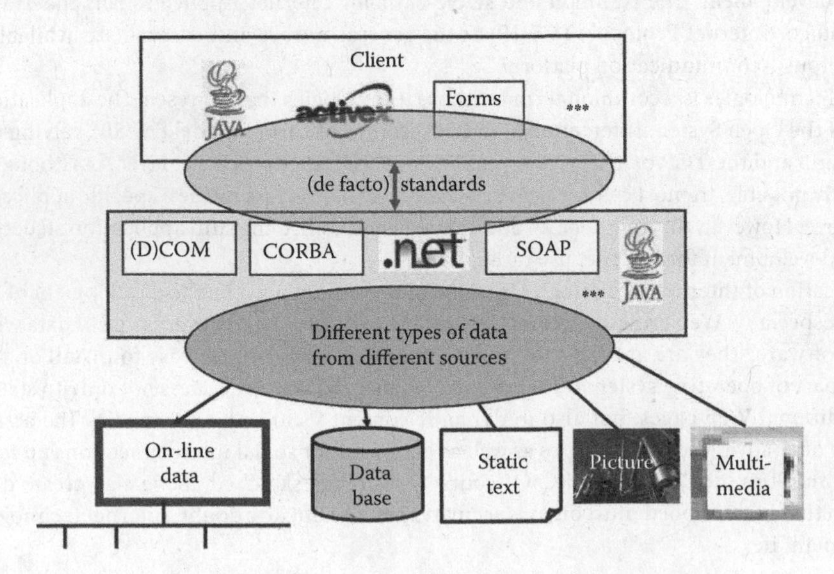

FIGURE 56.1 Internet and Web technologies as de facto standards.

A systematic solution for this problem should develop specific application scenarios, clearly identify technical requirements for the intended application functions, the features of the system component where the function will be implemented, and the location where the component is implemented inside the system. This may lead to different combinations. For all relevant combinations, the technical details of the information flows have to be specified and the requirements for the technologies have to be refined. This establishes a list of criteria by which the technologies have to be evaluated.

56.3 Technologies

It is obvious that an evaluation process as the one mentioned above is not possible without knowing communication paradigms and technologies in detail. The most common communication paradigm used by Internet applications is client–server communication. The server provides the data, while the client retrieves the data and displays it. Typically, confirmed services are used. The sequence can be described by the well-known service primitives (see Figure 56.2). The client issues the request by invoking the service at an appropriate service access point (SAP)—service primitive—. req—the server is informed by an indication (.ind), executes the service, and prepares the answer. The server passes the answer to the client as response (.rsp). Finally, the client gets the answer together with the confirmation (.cnf) from the service, and can now use the data. Thus, from an architectural point of view, a client layer and a server layer can be distinguished in Industrial Internet systems.

A further specialization of the functional components involved in the communication scenarios leads to more sophisticated architectures. The separation of application and data leads to the well-known three-tier-architecture (Figure 56.3), which can be further subdivided concerning application logic, data source, and communication server component (Figure 56.4). All types of architectural examples can be found in industrial applications. Thus, for the different functions the components of the architecture have to provide, different technologies exist—all assembled within the term Internet Technologies.

Internet technologies can be grouped by function. The following functional clusters can be distinguished:

- Transport and communication related technologies
- Technologies for information description and presentation
- Technologies for server-side and client-side functions

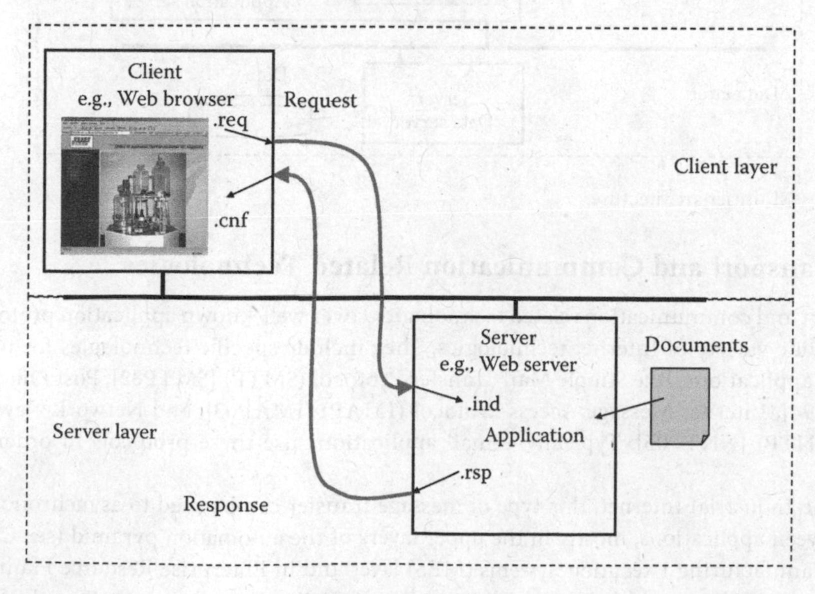

FIGURE 56.2 Basic architecture and communication paradigm of Industrial Internet applications.

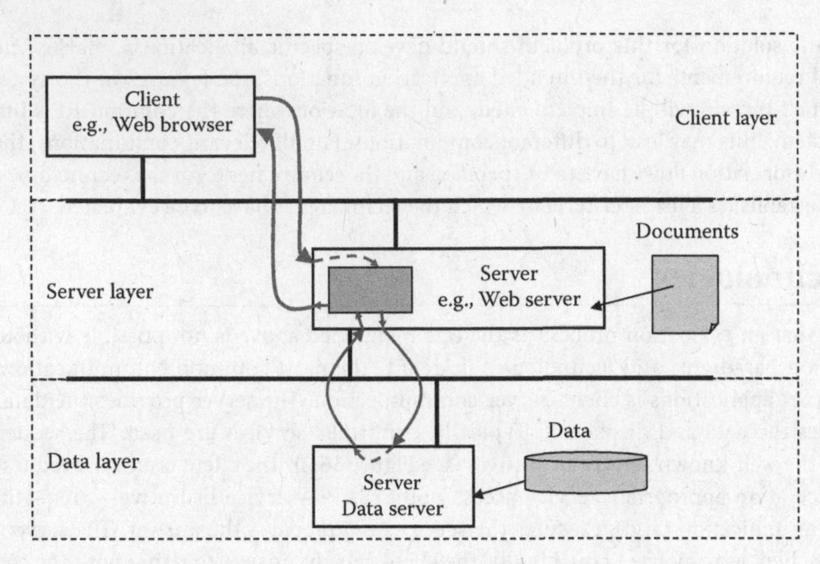

FIGURE 56.3 Three-tier architecture.

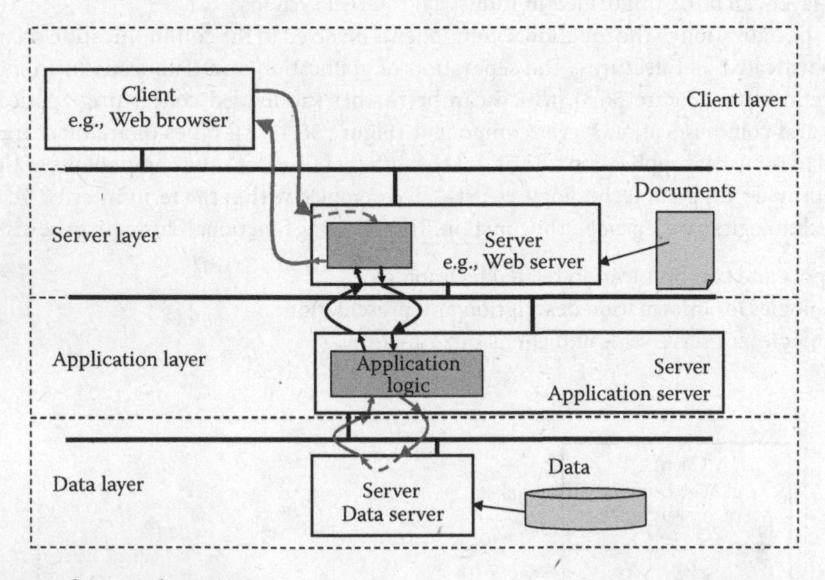

FIGURE 56.4 Multitier architecture.

56.3.1 Transport and Communication Related Technologies

The transport and communication related technologies cover well-known application protocol definitions for a wide variety of Internet technologies. They include specific technologies for use in message-related applications, like Simple Mail Transfer Protocol (SMTP) [SMTP82], Post Office Protocol (POP) [POP94], Internet Message Access Protocol (IMAP) [IMAP03], and Network News Transfer Protocol (NNTP) [NNTP06]. Typically, e-mail applications use these protocols in order to access messages.

Within the Industrial Internet, this type of message transfer can be used to asynchronously transfer data between applications, mostly in the upper layers of the automation pyramid (see Chapter 13), mainly in Manufacturing Execution Systems (MES) layer and in Enterprise Resource Planning (ERP) layer. It is, for example, possible to use such protocols and application to inform MES or ERP functions about events generated in the automation layer. Typically, event sources in the field layer or the sensor/actuator layer are using different techniques, addressing the real-time constraints, and the implementation effort (e.g., mail servers need a lot of resources).

In addition, other protocols like File Transfer Protocol (FTP) [FTP85] and its derivates are quite often used to transfer binary data between client and server. In Industrial Internet, such applications are useful in the upper layers down to the automation layer or even the field layer. For example, huge data sets in batch systems, large archives, programs, or parameter data are transferred using FTP-like protocols. Especially in systems where Industrial Ethernet is used at the field layer, FTP-based solutions are easy to implement.

56.3.1.1 HyperText Transfer Protocol

HyperText Transfer Protocol (HTTP) [HTTP99] is the most commonly known transport related protocol, used in any Web application to connect between Web browser and Web server. HTTP is a stateless protocol. The browser acts as a client and invokes the request (typically via HTTP-GET method), the server responds with the document requested or with an error message. Besides HTTP-GET, other methods can be used to transfer the header of the document only (HTTP-HEAD) or—if implemented at the server—to upload documents (HTTP-PUT). The method HTTP-POST is widely used for the transfer of form data from client to server and for invoking application logic at the server. The HTTPS protocol [HTTPS0] is a secured version of HTTP, often combined with other security methods at server and/or client.

The application of HTTP in Industrial Internet requires a Web server at the component acting as the server. The great variety of available Web server implementations for different purpose—ranging from high-end, high-load server software like Apache Web Server or Microsoft Internet Information Server (IIS) down to embedded Web servers with memory footprint of only a few kilobytes—enables a wide range of applications. However, it has to be considered that the typical communication paradigm is pure client–server. This means that transfer always starts upon client initiative. HTTP in its basic form is problematic to transfer events, since a client would have to poll the server all the time.

56.3.1.2 Simple Network Management Protocol

Developed for network management functions, SNMP [SNMP90] realizes the communication model and the information model part of OSI network management. Designed to be simple to use and to implement, SNMP has become the de facto standard for network management. The communication model part of SNMP can also be used as a transport protocol. It relies on the UDP of the TCP/IP protocol suite. Besides methods for reading attributes of managed objects like SNMP-GET, SNMP-GET-NEXT, or SNMP-GET-BULK, the protocol also provides SNMP-SET to set values of managed objects. Additionally, SNMP provides TRAP methods to transfer events occurring in managed objects of an agent to the corresponding manager. In contrast to client–server paradigms, this method can be used to propagate events within a system without client initiative.

Especially, the TRAP method makes SNMP suitable as a transport protocol for industrial applications. Automation systems using Industrial Ethernet typically provide SNMP implementations for the management of the network-related information. However, the full integration of industrial solutions into network management application requires using of the SNMP information model—the management information base (MIB) for description of components—as well as the integration of the functional model into SNMP applications. This is rather difficult, since both follow approaches different from those currently available in industrial applications. The use of SNMP as a transport layer only, for example, for industrial Web applications, is possible with limited effort.

56.3.1.3 Web Services Using Simple Object Access Protocol

In order to invoke functions of a server via the network, different systems have been developed. They range from open, complex, fully featured solutions like Common Object Request Broker Architectures (CORBA) [CORBA] to proprietary specifications like Microsoft's Distributed Component Object

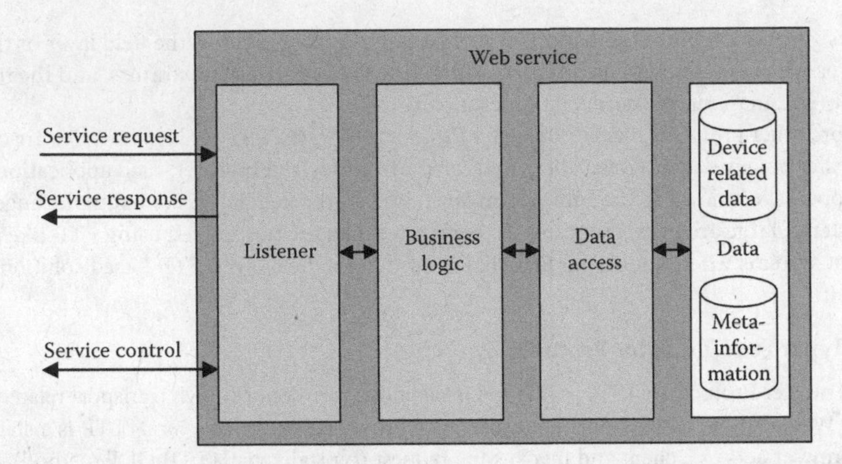

FIGURE 56.5 Principle structure of a Web service server.

Model (DCOM) [RGR97] or the transport layer of the.NET specification. However, due to complexity, management effort, and security reasons, a much simpler solution was requested. The Simple Object Access Protocol (SOAP) [DBX00] was defined to overcome the limitations and to provide a "firewall-friendly" way of invoking functions. SOAP is a protocol that can be bound to several underlying transport layers, including HTTP and SMTP. The functions calls, the responses, and possible error messages are serialized in coded format using eXtended Markup Language (XML) and are then embedded into the transport protocol.

Depending on the implementation (see Figure 56.5), a listener function is invoked by the transport layer, a Web server or a stand-alone application, e.g., the service request is passed to the listener, the function called and the arguments provided by the call are passed to business logic components. They finally provide the data for response and issue its transfer to the client. The services provided by a Web service server are described using a service description following the definitions in Web Service Description Language (WSDL), an XML-based format [WSD01].

Application of Web services in industrial applications is steadily growing. Because of the publicly available transport protocols Web services can be bound to, Web services provide an easy and commonly accepted way of handling communication between applications on different components and perhaps different operating systems.

56.3.2 Technologies for Information Description and Presentation

56.3.2.1 HyperText Markup Language

The content transferred by Internet technologies is manifold. Besides binary data representations for different formats and coding, technologies for information descriptions have been developed. The most important technology in this area is without any doubt the HyperText Markup Language (HTML) [HTM00], the main description technology of the World Wide Web. A client, a Web browser, interpreting the HTML statements of a hypertext document and thus rendering and presenting the content described in HTML documents is available on nearly every platform. Together with presentation-related technologies like Cascading Style Sheets (CSS) [CSS98], complex solutions for display of different types of data have become state of the art.

HTML documents can be used for any functions in automation domain that need to display data—especially data from different sources in different formats. In any case, the data are provided by the server upon client request. This might be difficult for some data sources (events).

56.3.2.2 eXtensible Markup Language

HTML has limitations concerning its mixture of presentation-related and content-related markup. In order to separate the content from its presentation, the XML has been specified [TBR98]. XML was developed and established as a markup language for generic purpose. XML is often treated as a Meta language suitable for development of customized, specific languages.

XML expands the description language HTML with user-defined tags, data types, and structures. Furthermore, it addresses one of the limitations in HTML—it introduces a clear separation between the data descriptions, the data themselves, and their representation (Figure 56.6). For declaring syntactical information, a separate definition file XML Schema Definition Language (XSD) [XSD01] or Document Type Definition (DTD) is used. This allows reusing the description structure in different contexts. This provides a number of benefits when using the same XML description file for different tasks. In combination with Namespace definitions semantics can be defined. Namespaces allow establishing a clear relation between tags defined in a schema and their usage context.

The availability of a wide range of software tools allows creating effective XML-based solutions. Publicly available solutions for parsers and development frameworks supporting different platforms and different operating systems contribute to the actual hype of using XML as a general-purpose description language. The XML data can be filtered and associated to software. The selection of the necessary information and the definition of their presentation details can be performed by means of scripts and style sheets. The style sheets [FBO98] are part of the development of XML. In most cases, they are implemented using the eXtensible Style Language for Transformation (XSLT). An XML document can be related to different style sheets for different presentation or conversion functions, so that only the XSLT references have to be changed. Since style sheets in XSLT can contain script elements, a Document Object Model (DOM) [DOM98] has been developed to support scripting. This DOM is used to access the XML file and its elements from a script. The DOM programming objects provide interfaces to the original XML file, representing its structure as a tree of objects. The output data of the DOM functions can be used to generate any textual or binary files, including new XML files with a different structure. The major benefit of this solution is a unique, reusable description with an excellent consistency and reduced efforts of the description process.

Applications of XML in the industrial domain have developed with increasing speed. They cover static descriptions for multiple purposes, for example, device descriptions, data exchange descriptions, and protocol descriptions using SOAP and WSDL (e.g., OPC and XML specification [OPC04]).

A special importance for visualization tasks is assigned to Scalable Vector Graphics (SVG) [SVG01], an XML format for presentation of two-dimensional vector-based graphics including functions for

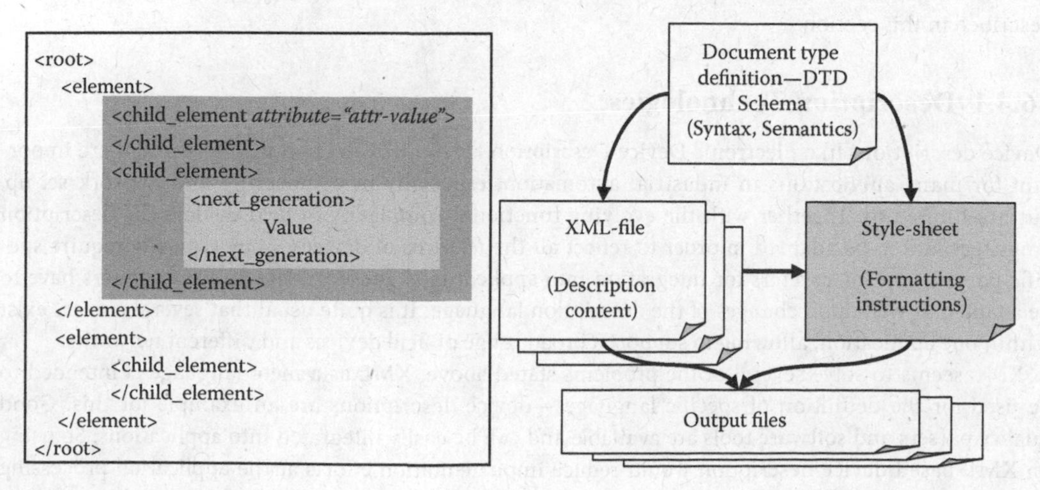

FIGURE 56.6 Structure of an XML file (left) and the XML environment.

scripting of the elements. Browsers can use plug-ins for rendering and display. Other XML-based formats are available for use in multimedia applications. They may be used in industrial applications whenever they become standards and the support by tools is stable.

56.3.3 Technologies for Server-Side and Client-Side Functions

The functionality of combining content description, application-depending interpretation, and transport is provided at both the server side and at the client. Different technologies have been developed. Popular client-based solutions are—without making any claim to be complete—JavaScript, Java-Applets, ActiveX-Controls, and Flash. They require client-based runtime environments, like Java Virtual Machine (JVM), Flash Players, and alike. They might be available bundled with the client software (e.g., a browser), or may require installation of plug-ins. An interesting and rapidly used approach is Asynchronous Javascript and XML (AJAX), which allows highly interactive applications with automatically, dynamically loading and rendering of information fragments.

Well-known server-based technologies include PHP, Java-Servlets, Active Server Pages (ASP), and Common Gateway Interface (CGI). In this case, the server component requires the runtime environment, for example, scripting engines or servlet containers. The latter are often bundled with Web servers; Apache Tomcat is an example.

Applications of client- and server-side technologies in the industrial domain can be compared to standard IT-based solutions. They are as manifold as the technologies supported. Depending on the use cases, a combination of different technologies may be required. Additional constraints may be derived concerning real-time aspects, data consistency, and data and interface management. Typically, gateways to underlying levels of the automation pyramid (see Chapter 13) are implemented at the server components.

An important feature for selection of a server- or client-side technology in the industrial application area is its maturity, often treated as a combination of time of availability, standardization, tool support, etc. Taking into account the typical life cycles of automation systems ranging from 3 to 5 years in factory automation to more than 40 years in process automation and power plants, newly introduced technologies are hesitantly applied.

56.4 Application Examples

The application of Internet technologies in Industrial automation depends on the specific use cases. These use cases also define the criteria for technology selection. Thus, the application range becomes broader and broader. Without claiming completeness, some of the typical application examples are described in this section.

56.4.1 Description Technologies

Device descriptions like Electronic Device Description (EDD) [IEC07] became more and more important for many applications in industrial automation, especially in engineering and network set-up, but not limited to. Together with the evolving functional complexity of field devices the description languages had to be adapted in order to reflect all the features of devices. They typically require specific parsers and interpreters for integration into applications. These parsers and interpreters have to be adapted as well upon changes of the description language. It is quite usual that several parsers exist within one application, allowing to support a broad range of field devices and different networks.

XML seems to solve several of the problems stated above. XML as a meta language is intended to be used for the definition of specific languages—device descriptions are an example for this. Good quality parsers and software tools are available and can be easily integrated into applications. So using an XML-based device description would reduce implementation efforts at the application processing device descriptions. Thus, several activities have been started to define XML-based device descriptions.

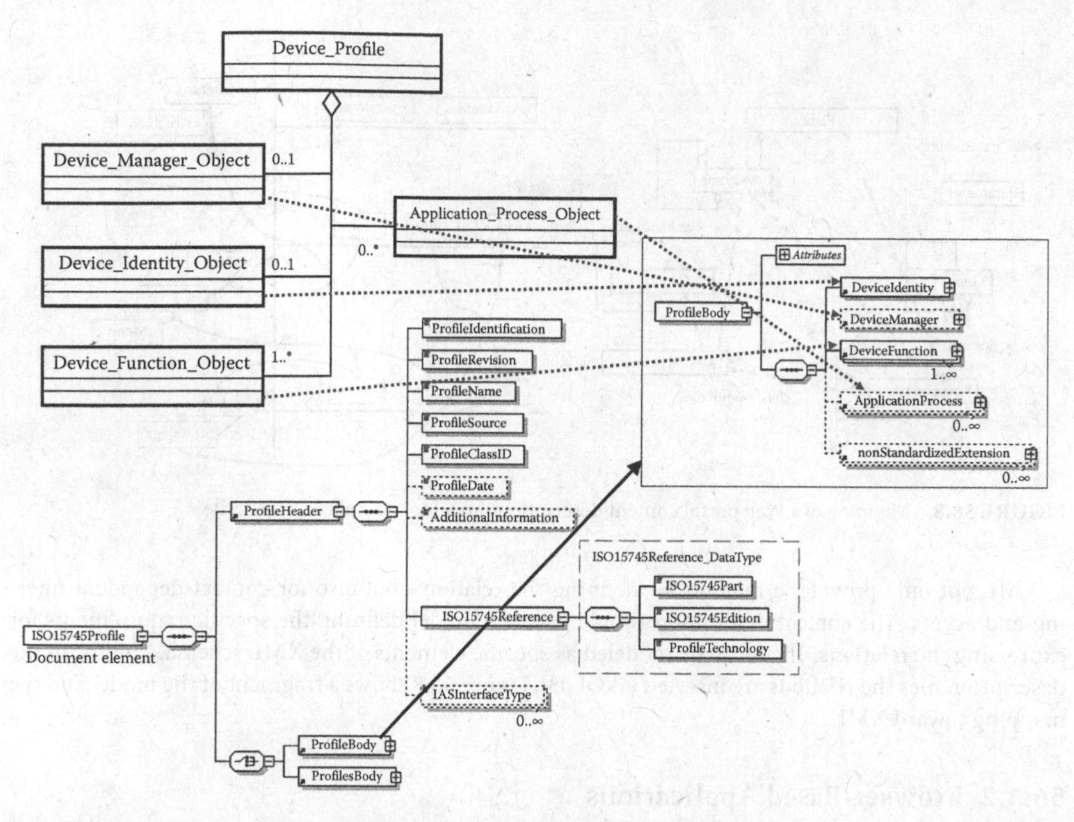

FIGURE 56.7 Mapping of device model acc. ISO 15745 to the FDCML XML schema.

Although with ISO 15745 [ISO03] a generic approach exists, the different user organizations provided their own schemas and mappings of the traditional description statements to XML formats (See Figure 56.7). With Field Device Configuration Markup Language (FDCML) [FDC01], a suitable description exists not only for InterBus-S, but also for Industrial Ethernet-based devices; it is applicable by extension for other systems, too (for example CANopen). The Electronic Data Sheet (EDS) for DeviceNet [EDS00] is based on XML. The Generic Station Description Markup Language (GSDML) [GSD06] is supporting PROFINET IO devices; XML-based formats for EDD are in discussion.

The benefits of XML concerning software tool support, flexibility, and extensibility have been proven by several applications. It is relatively easy to generate documents like catalog data sheets or user interfaces for visualization from existing device descriptions. However, XML also has some drawbacks. Because of the mostly long XML tags, the descriptions become quite large. This may lead to storage problems. The complexity of the XML schemas may lead to time-consuming parsing of descriptions. The number of existing description formats will produce integration efforts again—even if the same parsers can be used. Since this is not a technology problem, it cannot be expected that XML can solve this—it has to be initiated by the end users.

Another typical application of XML is description of different content that will be processed and/or displayed in applications. The flexibility of XML is a prerequisite for such type of modeling, especially when other Internet technologies are implemented, too. For example, a Web portal for maintenance of machines may provide different types of data—dynamic data from the machine components, static data like description, handbooks, plans, etc. The data need to be associated together depending on the functions the portal offers. It is not challenging to collect the data, but the assignment has to be performed at a higher level. This assignment depends on context, for example, the user role, the communication features, or specific phases of the life cycle.

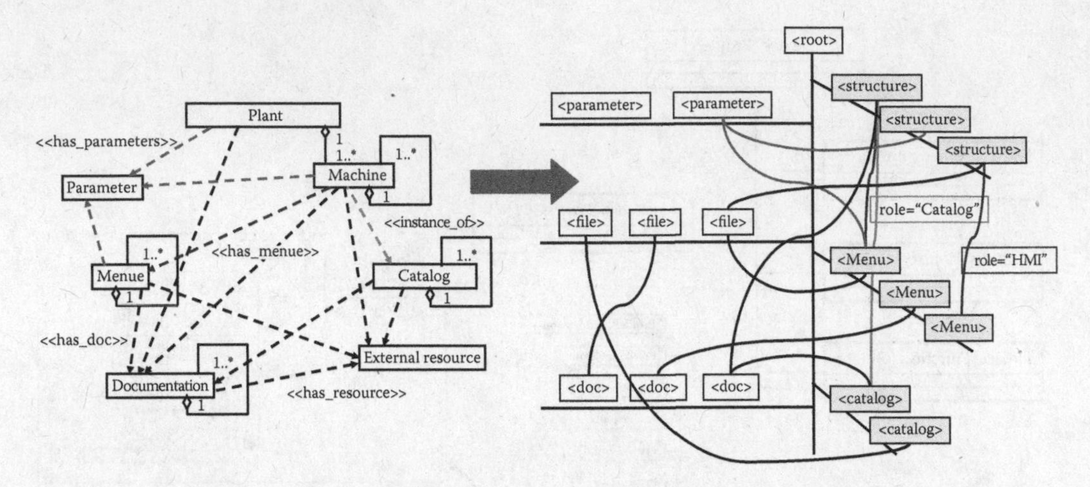

FIGURE 56.8 Mapping of a Web portal content description model to a (symbolic) XML file.

XML not only provides a method for defining the relations, but also for context-dependent filtering and access. The content description should follow a model defining the specific requirements for expressing the relations. These can be modeled as specific elements of the XML schema. Finally, in the description files the relations are inserted [WOL03]. Figure 56.8 shows a fragment of the model and the mapping toward XML.

56.4.2 Browser-Based Applications

Browser-based applications are typically used for visualization purposes. Respective solutions can be found in nearly every visualization-related product in automation. The ability to cover large distances makes such solutions especially interesting for all tasks that require remote access to the automation system, for example, remote maintenance. Instead of requiring specific protocol installation, only TCP/IP and the above mentioned transport-related protocols like HTTP are necessary, a Web browser on the client is typically also available. This simplifies installation and maintenance efforts dramatically. However, the browser alone is not always sufficient, since often complex, animated graphics have to be shown. In addition, the information presented has to be enriched with dynamically changing data from the process.

Several solutions are available to solve the problem sketched above. The data are provided by gateways to the automation system's components, by controllers, databases, or by using standard interfaces like OPC [OPC98]. The transfer of the data can be done using HTTP or SNMP. The dynamic parts of the solutions can be realized with client-side or server-side functions using technologies mentioned above. However, the specific solution has to reflect the requirements adopted from the use cases. For example, a requirement could be to allow access from clients with restricted capabilities in communication bandwidth and graphical resolution of the display like mobile phones or handhelds. This requires personalization of the data to be displayed. Another topic is the necessity for browser plug-ins. For example, displaying a flash animation or SVG content requires the prior installation of the appropriate plug-ins. This limits the general applicability of such a solution. Furthermore, an installation on-the-fly might not be always possible because of access restrictions and insufficient user rights.

Another important topic is the protection of intellectual properties. If client-side scripting is used, for example, JavaScript, the scripts need to be downloaded to the client's browser. Since the functions are executed within the browser's scripting engine, they are transferred and thus are available at source code level. So it is relatively easy to modify or to abuse them. A server-side solution, for example, using

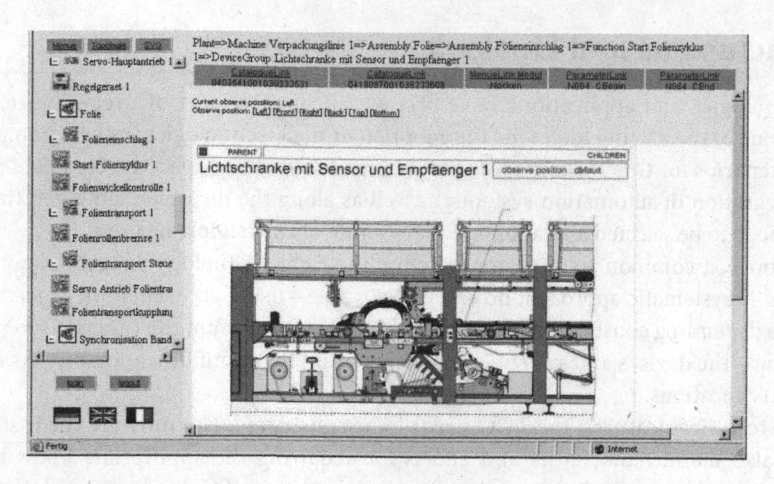

FIGURE 56.9 Screenshot of a browser-based visualization using SVG.

PHP, solves this problem. However, it requires frequent updates of the content, since every change is made at the server. Java applets seem to solve parts of the problem, since they are transferred as binary data (class files) from prespecified servers, reside in a sandbox within the browser's virtual machine, but can use several underlying transport mechanisms. But again, a general solution for this problem cannot be given.

Figure 56.9 shows an example for a browser-based application using SVG graphics with animation. The navigation areas are created on the server with PHP, the content model uses XML, and the SVG graphic is created from a Computer-aided design (CAD)-drawing of the machine [WOL03]. This shows the typical combination of different Internet technologies as mentioned above.

56.4.3 Machine–Machine Communication Using Web Services

Besides the application with user interfaces, Internet technologies may also be used effectively between distributed applications. This is often called machine–machine communication. The applications need to transfer data or to invoke functions. Thus, Web services using SOAP as the transport protocol with service descriptions in WSDL are the most promising technology in this scenario. Since this concept follows the actual trend of service-oriented architectures (SOA) with loosely coupled, distributed functions, many software development environments support the process of defining and implementing the appropriate software components. For example, Microsoft's Visual Studio and Eclipse provide functions to define WSDL files to generate service components or clients. They handle serialization of the function call and the parameters to SOAP messages and vice versa.

In the automation domain, Web services are increasingly used. The most important problem is not the tool support, but the common definition of services. This leads to the typical standardization topics, which are always hard to solve. However, the success of OPC enabled a definition of Web services to access existing OPC servers for reading and writing of OPC items. This solution is known as OPC&XML [OPCxml] and demonstrates the potential of Internet technologies. Using Web services reduces the complex efforts of providing the appropriate DCOM configuration dramatically, which was necessary for traditional OPC servers. The access can now easily pass firewalls; the service analysis at the server still provides security features. Actual developments like OPC-UA (see Chapter 56) also use Web services. However, a drawback is introduced by the enormous overhead of SOAP messages to be exchanged. So for OPC-UA, a specific binary protocol has been developed to reduce communication effort.

56.5 Conclusions and Outlook

Internet technologies and applications have become an integral part of everyday life. The overall acceptance is one of the driving forces for the adoption of such technologies in the automation domain. Application scenarios for this "Industrial Internet" are manifold and span the complete range of tasks in vertical integration of automation systems, as well as along the life cycle. However, the appropriate technologies need to be selected and adopted; they cannot always simply be used.

As stated above, a common solution for selecting the right technology or technology combination is not possible. A systematic approach, however, is possible—using an evaluation of features against a list of use-case depending constraints. This approach takes into account the functions to perform, their implementation—the devices and system components—and the communication aspects depending on the components' locations.

Besides the technical features, the TCO needs to be considered. Not only are the installation costs relevant, but also maintenance costs and efforts for acquiring the appropriate know-how. Another important aspect is the maturity of the technologies, since the short technology development cycles are in contrast to life cycle durations in industrial automation.

Application domain-related adoption of Internet technologies leads to standardization. This is problematic, since the influence of automation-related standardization bodies like ISO or IEC is rather limited in Internet standardization. Compared with the IT domain, industrial application is a small domain.

It is obvious that the actual developments of Internet and Web technologies, for example, Web 2.0, will find their applications in the industry domain, extending the role of Industrial Internet.

Acronyms

AJAX	Asynchronous Javascript and XML
ASP	Active Server Pages
CGI	Common Gateway Interface
CORBA	Common Object Request Broker Architectures
CSS	Cascading Style Sheets
DCOM	Distributed Component Object Model
DOM	Document Object Model
DTD	Document Type Definition
EDD	Electronic Device Description
EDS	Electronic Data Sheet
ERP	Enterprise Resource Planning
FDCML	Field Device Configuration Markup Language
FTP	File Transfer Protocol
GSDML	Generic Station Description Markup Language
HTML	HyperText Markup Language
HTTP	Hyper Text Transfer Protocol
HTTPS	HTTP-Secure
IMAP	Internet Message Access Protocol
IP	Internet Protocol
JVM	Java Virtual Machine
MES	Manufacturing Execution Systems
MIB	Management Information Base
NNTP	Network News Transfer Protocol
OPC	OLE for Process Control
OPC-UA	OPC Unified Architecture
POP	Post Office Protocol

SAP	Service Access Point
SMTP	Simple Mail Transfer Protocol
SNMP	Simple Network Management Protocol
SoA	Service-oriented architecture
SOAP	Simple Object Access Protocol
SVG	Scalable Vector Graphics
TCO	Total cost of ownership
TCP	Transmission Control Protocol
UDP	User Datagram Protocol
WSDL	Web Service Description Language
XML	eXtended Markup Language
XSD	XML Schema Definition Language
XSLT	eXtensible Style Language for Transformation

References

[CORBA] The Common Object Request Broker: Architecture and Specification, Object Management Group, ftp://ftp.omg.org/pub/docs/

[CSS98] RFC2318: The text/css Media Type, Internet Engineering Task Force (IETF), 1998, http://tools.ietf.org/html/rfc2318

[DBX00] D. Box et al., *SOAP: Simple Object Access Protocol*. Microsoft Corp., 2000. http://msdn.microsoft.com/xml/general/soapspec.asp

[DOM98] Document Object Model (DOM) Level 1 Specification, Version 1.0, W3C Recommendation, October 1, 1998, http://www.w3.org/TR/REC-DOM-Level-1

[EDS00] W.H. Moss, Report on ISO TC184/SC5/WG5 Open Systems Application Frameworks based on ISO 11898, In: *Proceedings "Industrial Automation" of 5th International CAN Conference (iCC'98)*, San Jose, CA, November 3–5, 1998, pp. 07-02–07-04.

[FBO98] F. Boumphrey, Professional Style Sheets for HTML and XML, Wrox Press, Birmingham, U.K., 1998.

[FDC01] Field Device Configuration Markup Language. FDCML 2.0 Specification, http://www.fdcml.org

[FTP85] STD 9: File Transfer Protocol, RFC 959, Internet Engineering Task Force (IETF), 1985, http://www.apps.ietf.org/rfc/rfc959.html

[GSD06] GSDML Specification for PROFINET IO, PROFIBUS Nutzerorganisation e.V, Version 2.1, August 2006.

[HTM00] ISO/IEC 15445:2000-Information technology—Document description and processing languages—HyperText Markup Language (HTML), International Organization for Standardization, 2000, https://www.cs.tcd.ie/15445/15445.HTML

[HTTP99] RFC2616: Hypertext Transfer Protocol—HTTP/1.1, Internet Engineering Task Force (IETF), 1999, http://tools.ietf.org/html/rfc2616

[HTTPS0] RFC2817: Upgrading to TLS within HTTP/1.1, Internet Engineering Task Force (IETF), 2000, http://tools.ietf.org/html/rfc2817

[HZ80] H. Zimmermann, OSI Reference Model—The ISO Model of Architecture for Open Systems Interconnection, *IEEE Transactions on Communications*, 28(4), April 1980, 425–432.

[IEC07] IEC 61804-3 (2007). Function Blocks for Process Control, Part 3: Electronic Device Description Language, International Electrotechnical Commission, June 2007.

[IMAP03] RFC3501: Internet Message Access Protocol-Version 4rev1, Internet Engineering Task Force (IETF), 2003, http://tools.ietf.org/html/rfc3501

[ISO03] ISO 15745: Industrial automation systems and integration—Open systems application integration frame-work. Part 1: Generic Reference Description. International Organization for Standardization, March 2003.

[NNTP06] RFC3977: Network News Transfer Protocol (NNTP), Internet Engineering Task Force (IETF), 2006, http://tools.ietf.org/html/rfc3977

[OPC04] OPC XML-DA Specification, Version 1.01. OPC Foundation, 2004, http://www.opcfoundation.org

[OPC98] OPC Data Access Automation Specification, Version 2.0, OPC Foundation, October 14, 1998.

[POP94] 1725 Post Office Protocol-Version 3, Internet Engineering Task Force (IETF), 1994, http://www.apps.ietf.org/rfc/rfc1725.html

[RGR97] R. Grimes, *Professional DCOM Programming*, Wrox Press, Birmingham, Canada, 1997.

[SMTP82] STD 10: Simple Mail Transfer Protocol, RFC 821, Internet Engineering Task Force (IETF), 1982, http://www.apps.ietf.org/rfc/rfc821.html

[SNMP90] STD 15 Simple Network Management Protocol (SNMP), RFC1157, Internet Engineering Task Force (IETF), 1990, http://www.apps.ietf.org/rfc/rfc1157.html

[SVG01] Ferraiolo, J. (editor), Scalable Vector Graphics (SVG) 1.0 Specification, W3C Recommendation, September 4, 2001, W3C Consortium, http://www.w3.org/TR/SVG/

[TBR98] T. Bray, J. Paoli, and C. M. Sperberg-McQueen, Extensible Markup Language (XML) 1.0. 1998, W3C Consortium, http://www.w3.org/TR/REC-xml

[WOL03] Wollschlaeger, M., Geyer, F., Krumsiek, D., and Wilzeck, R., XML-based description model of a Web portal for maintenance of machines and systems, *Ninth IEEE International Conference on Emerging Technologies and Factory Automation ETFA2003*, Lissabon, Portugal, October 16–19, 2003. Proceedings Vol. 1, IEEE 03TH8696, ISBN 0-7803-7937-3, S. 333-340.ETFA-paper 2003.

[WSD01] Christensen, E., Curbera, F., Meredith, G., and Weerawarana, S., W3C Note: Web Services Description Language (WSDL) 1.1. 2001, W3C Consortium, http://www.w3.org/TR/2001/NOTE-wsdl-20010315

[XSD01] XML Schema Part 1: Structures, W3C Recommendation, May 2, 2001, W3C Consortium, http://www.w3.org/TR/xmlschema-1/

57

OPC UA

57.1 Introduction 57-1
57.2 Overview of OPC UA System Architecture.................................... 57-2
UA Client Application Architecture • UA Server Architecture
57.3 Overview of UA AddressSpace... 57-5
57.4 Overview of UA Services.. 57-7
General Services • Discovery Service Set • SecureChannel Service Set • Session Service Set • NodeManagement Service Set • View Service Set • Query Service Set • Attribute Service Set • Method Service Set • MonitoredItem Service Set • Subscription Service Set
57.5 Implementations and Products 57-10
57.6 Conclusion 57-10
References.. 57-10

Tuan Dang
EDF Research and Development

Renaud Aubin
EDF Research and Development

57.1 Introduction

OPC Unified Architecture (OPC UA) is the new standard of the OPC Foundation [1] providing *interoperability* (e.g., Microsoft DCOM technology is not used anymore) in process automation and beyond. By defining abstract services, OPC UA provides a service-oriented architecture (SOA) for industrial applications from factory-floor devices to enterprise applications. Since the first introduction of OPC in 1996, OPC UA represents a tremendous achievement, which started at the end of 2003, that proposes to integrate the different flavors of the former OPC specifications (DA, HDA, DX, XML-DA…) into a *unified address space accessible with a single set of services including IT security.* This chapter gives an overview over the architecture of OPC UA, its address space model and its services, as well as different implementation flavors. For an in-depth study of OPC UA, it is advised to get the specifications from the OPC Foundation Web site [2], which offers a lot of information about implementations and product availability. It is interesting to know that the OPC Foundation is working with bodies such as MIMOSA [3], ISA [4], and OMAC [5] to extend the UA standard to batch, asset management, and other areas. Another important thing to know is that OPC UA provides support for the electronic device description language (EDDL) [6] so that it is possible to make device data parameters available to a large domain of applications through OPC UA. In this way, it gives an alternative to the Web Services for Devices (WS4D) initiative [7] that proposes to specify and implement WS4D using the Devices Profile for Web Services (DPWS) specification [8] that is the successor of the Universal Plug and Play (UPnP) protocol.

Typical use of OPC UA concerns the middleware services that offer an abstraction layer between devices (e.g., programmable logic controller [PLC], sensors, actuators,…) and plant-floor applications (SCADA, MES,…), as shown in Figure 57.1.

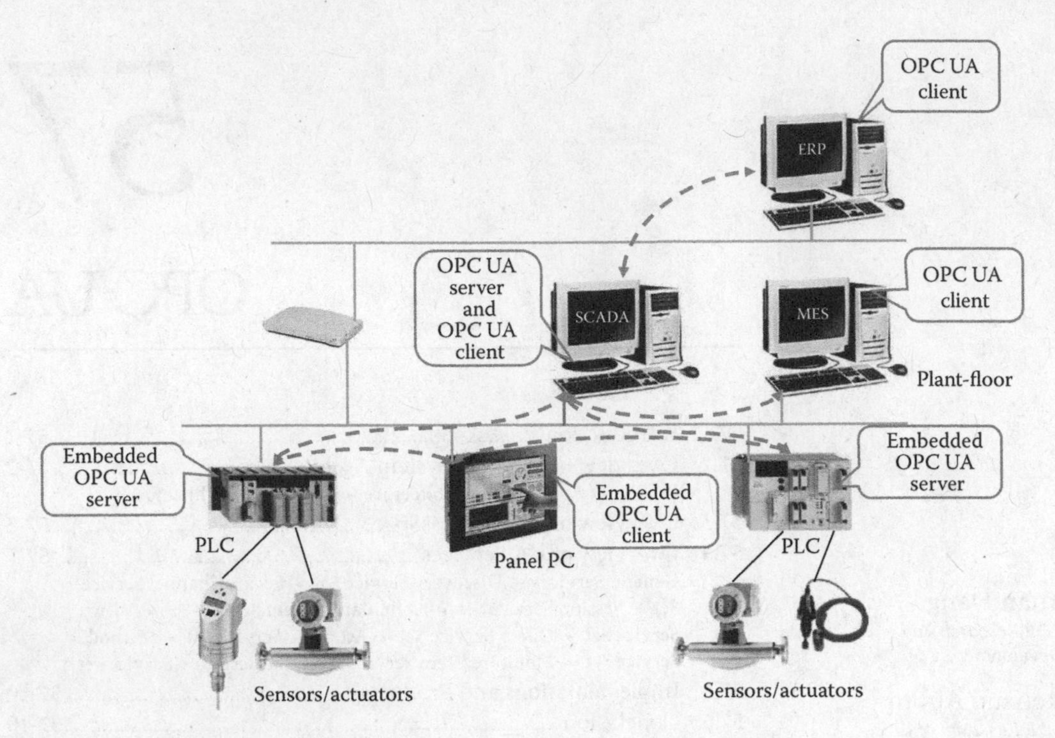

FIGURE 57.1 Example of OPC UA application.

57.2 Overview of OPC UA System Architecture

OPC UA software architecture is designed using the five-layer model, as shown in Figure 57.2. Business information models (e.g., EDDL, MIMOSA, ISA…) and vendor information model are typically built on top of OPC UA Base Services, which also provide the foundation for rebuilding existing OPC information models (DA, A&E, HDA…).

The release 1.01 of OPC UA addresses a wide range of target applications:

- Supervisory control and data acquisition (SCADA) and human machine interface (HMI) or more generally human system interface (HSI).
- Advanced control in distributed control system (DCS): Commands may be sent to devices and programmable logic controller (PLC).

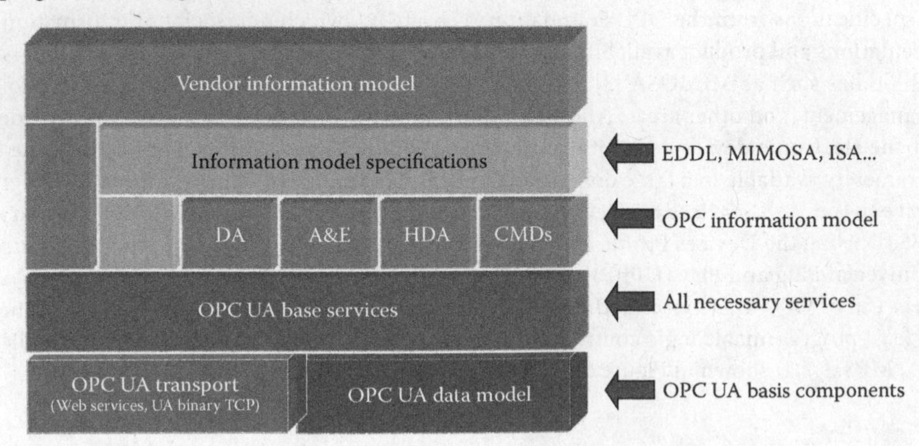

FIGURE 57.2 OPC UA software architecture.

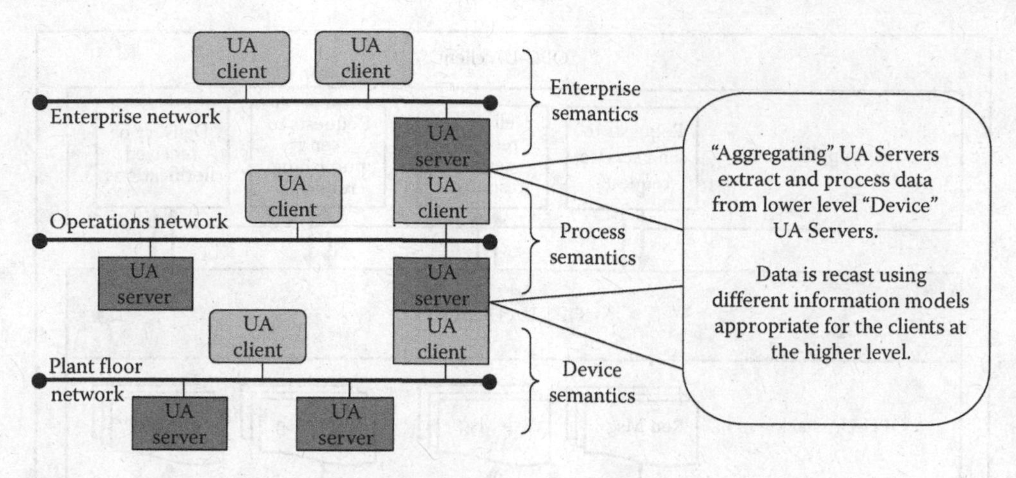

FIGURE 57.3 Chaining of UA servers and UA clients.

- Manufacturing execution system (MES) that is usually deployed in the supervisory, simulation, and scheduling of batch processes (e.g., pharmaceutical process).
- Middleware that serves intra-plant applications and extra-plant (enterprise wide) applications (Figure 57.3).

Middleware is a piece of software that serves to "glue together" or mediate between two separate and usually already existing programs. Middleware is sometimes called plumbing because it connects two sides of an application and passes data between them. *Middleware plays a crucial role in enterprise applications* by integrating system components, allowing them to interoperate correctly and reliably, and facilitating system management and evolution. This is a very important domain of applications that OPC UA is aimed at, bridging the industrial information systems with the enterprise one.

The system architecture of OPC UA is based on the well-known concept of client–server interaction. Each system may contain multiple clients and servers. *Each client may interact concurrently with one or more servers*, and *each server may interact concurrently with one or more clients*. An application may combine server and client components to allow interaction with other servers and clients, as shown in Figure 57.3. *Peer-to-peer interactions between servers are also possible*, as shown in Figure 57.4:

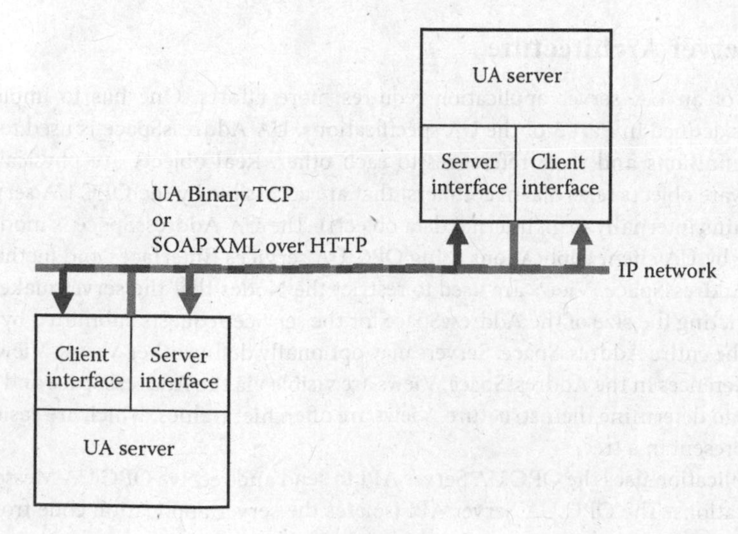

FIGURE 57.4 Peer-to-peer interactions between UA servers.

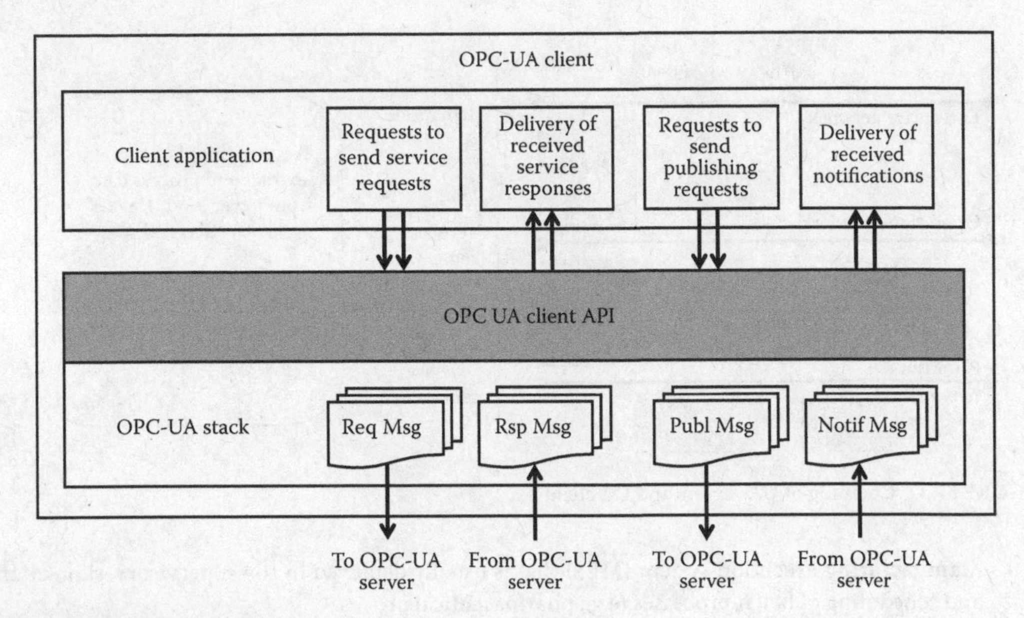

FIGURE 57.5 UA client architecture.

57.2.1 UA Client Application Architecture

To write an OPC UA client application, one typically needs to use the OPC UA client application programming interface (API), which exposes standard services that should be used to send and receive OPC UA service requests and responses to the OPC UA server, as shown in Figure 57.5. The *API isolates the client application code from an OPC UA communication stack* that may be chosen between different flavors: *UA Binary over TCP* and *SOAP XML/Text over HTTP*. The OPC UA communication stack converts OPC UA Client API calls into messages and sends them through the underlying communications entity to the server at the request of the client application. The OPC UA communication stack also receives responses and notification messages from the underlying communications entity and delivers them to the client application through the OPC UA Client API, as shown in Figure 57.5.

57.2.2 UA Server Architecture

The realization of an UA server application requires more efforts. One has to implement the *UA AddressSpace*, as defined in Part 3 of the UA specifications. UA AddressSpace is used to represent real objects, their definitions and their references to each other. Real objects are physical (i.e., physical devices) or software objects (e.g., alarm counters) that are accessible by the OPC UA server application or that it maintains internally (e.g., internal data objects). The UA AddressSpace is modeled as a set of *Nodes* accessible by UA client applications using OPC UA services (interfaces and methods). A *View* is a subset of the AddressSpace. Views are used to restrict the Nodes that the server makes visible to the client, thus restricting the size of the AddressSpace for the service requests submitted by the client. The *default View* is the entire AddressSpace. Servers may optionally define other Views. Views hide some of the Nodes or References in the AddressSpace. Views are visible via the AddressSpace and clients are able to browse Views to determine their structure. Views are often hierarchies, which are easier for clients to navigate and represent in a tree.

UA server application uses the OPC UA Server API to send and receive OPC UA Messages from OPC UA client applications. The OPC UA Server API isolates the server application code from an OPC UA communication stack, as shown in Figure 57.6.

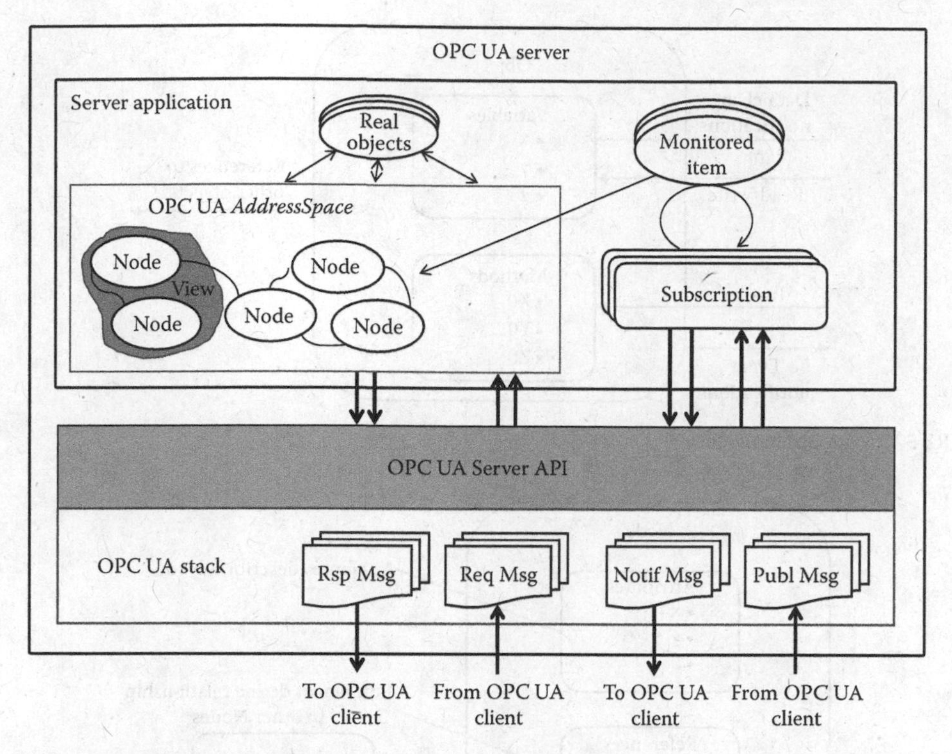

FIGURE 57.6 UA server architecture.

To finish the implementation, one has to realize publisher/subscriber entities, which offer a communication mechanism for informing a data change or an event/alarm occurrence that is modeled by *MonitoredItems*. MonitoredItems are entities in the server created by the client that monitors AddressSpace Nodes and their real-world counterparts. When they detect a change or an alarm occurrence, they generate a Notification that is transferred to the client by a Subscription. A Subscription is an endpoint in the server that publishes Notifications to clients. Clients control the rate at which publishing occurs by sending Publish Messages.

57.3 Overview of UA AddressSpace

AddressSpace is an important concept in OPC UA. It provides a standard way for UA servers to represent objects to UA clients. The OPC UA object model (Figure 57.7) defines objects in terms of Variables and Methods. It also allows relationships to other objects to be expressed, as shown in Figure 57.7.

The elements of this model are represented in the AddressSpace as *Nodes*. Each Node is assigned to a *NodeClass* and each NodeClass represents a different element of the object model. The Node Model that is used to define a NodeClass is illustrated by Figure 57.8:

Attributes are data elements that describe Nodes. Clients can access Attribute values using Read, Write, Query, and *Subscription/MonitoredItem* services. These services are detailed in Part 4 of the UA specifications. Attributes are elementary components of NodeClasses. Each Attribute definition consists of an attribute id1, a name, a description, a data type, and a mandatory/optional indicator. The set of Attributes defined for each NodeClass shall not be extended by UA clients or servers.

When a Node is instantiated in the AddressSpace, the values of the NodeClass Attributes are provided. The mandatory/optional indicator for the Attribute indicates whether the Attribute has to be instantiated.

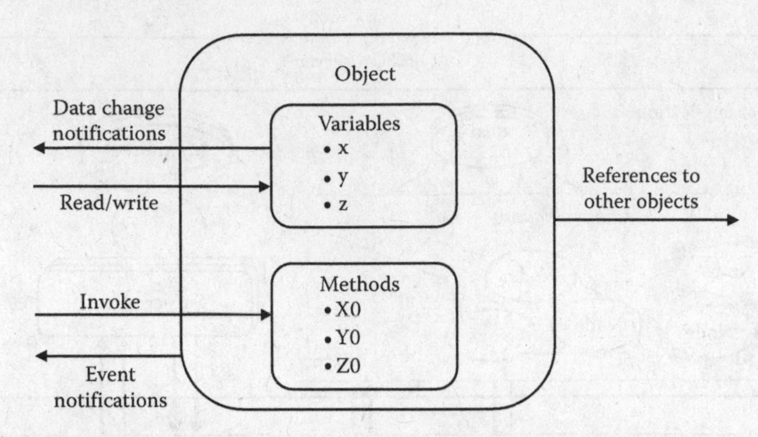

FIGURE 57.7 UA object model.

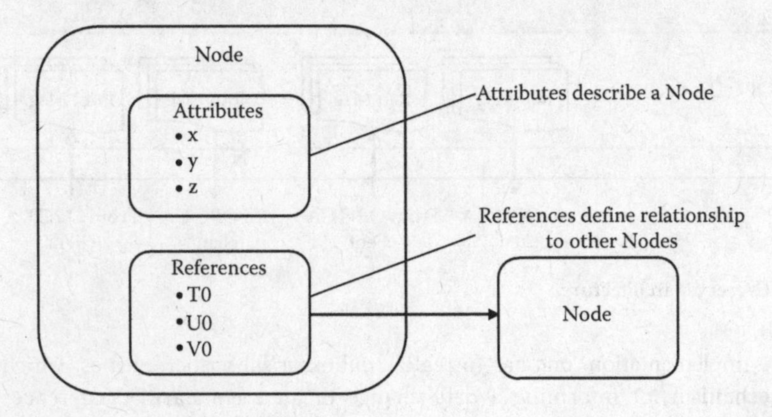

FIGURE 57.8 UA AddressSpace node model.

References are used to relate Nodes to each other. They can be accessed using the browsing and querying services. Like Attributes, they are defined as fundamental components of Nodes. Unlike Attributes, References are defined as instances of *ReferenceType Nodes*. ReferenceType Nodes are visible in the AddressSpace and are defined using the *ReferenceType NodeClass*.

The Node that contains the Reference is referred to as the *SourceNode* and the node that is referenced is referred to as the *TargetNode* (see Figure 57.9). The combination of the *SourceNode*, the *ReferenceType*,

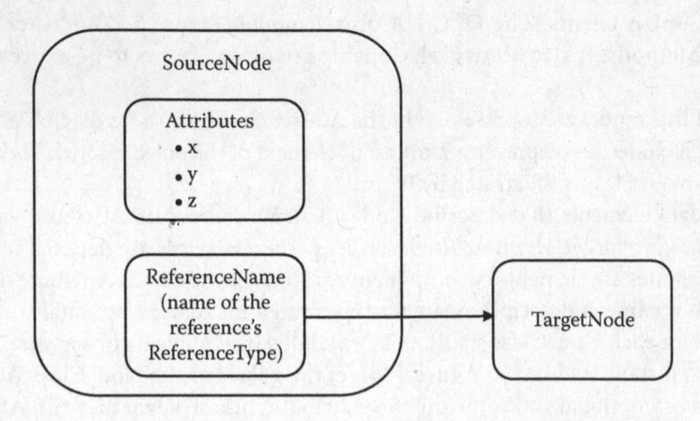

FIGURE 57.9 UA reference model.

and the *TargetNode* are used in OPC UA *services* to *uniquely* identify *References*. Thus, each Node can reference another Node with the same ReferenceType *only once*.

The *TargetNode* of a *reference* may be in the same AddressSpace or in the AddressSpace of another OPC UA server. TargetNodes located in other servers are identified in OPC UA *Services* using a combination of the *remote server name* and the *identifier* assigned to the Node by the remote server. OPC UA does not require that the *TargetNode* exists, thus References may point to a Node that does not exist!

57.4 Overview of UA Services

This section contains a summary of the description of UA services presented in Part 1 of the UA specifications. For a detailed description of UA Services, it is advised to read Part 4 of the UA specifications. Table 57.1 lists different parts of the UA specifications.

57.4.1 General Services

Request/response services are services invoked by the UA client application through the OPC UA service interface to perform a specific task on one or more Nodes in the AddressSpace and to return a response.

Publisher Services are services invoked through the OPC UA service interface for the purpose of *periodically* sending Notifications to UA clients. Notifications include Events, Alarms, Data Changes, and Program Outputs.

57.4.2 Discovery Service Set

This service set defines services used to discover OPC UA servers that are available in a system. It also provides a manner in which clients can read the *security configuration* required for connection to the Server. The Discovery services are implemented by *individual UA servers* and by *dedicated Discovery servers*. Well-known dedicated Discovery servers provide a way for clients to discover all registered OPC UA servers.

57.4.3 SecureChannel Service Set

This service set defines services used to open a communication channel that ensures the *confidentiality* and *integrity* of all messages exchanged with the UA server. More details about the base concepts for UA security are defined in Part 2 of the UA specifications.

TABLE 57.1 List of the UA Specifications

OPC UA Part N°	Description	Status
Part 1	Overview and concepts 1.01 specification	Released
Part 2	Security model 1.01 specification	Released
Part 3	Address space model 1.01 specification	Released
Part 4	Services 1.01 specification	Released
Part 5	Information model 1.01 specification	Released
Part 6	Mappings 1.00 specification	Released
Part 7	Profiles 1.00 specification	Released
Part 8	Data access 1.01 specification	Released
Part 9	Alarms draft 0.62 specification	Draft
Part 10	Programs 1.00 specification	Released
Part 11	Historical access 1.00 specification	Released

The SecureChannel services are unlike other services because they are typically not implemented by the UA application directly. Instead, they are *provided by the communication stack* that the UA application is built on. For example, a UA server may be built on a SOAP stack that allows applications to establish a SecureChannel using the *WS-SecureConversation* specification. In these cases, the UA application simply needs to verify that a *WS-SecureConversation* is active whenever it receives a message. Part 6 of the UA specifications describes how the SecureChannel services are implemented with different types of communication stacks.

57.4.4 Session Service Set

This service set defines services used to establish an application-layer connection in the context of a session on behalf of a specific user.

57.4.5 NodeManagement Service Set

The NodeManagement service set allows clients to add, modify, and delete Nodes in the AddressSpace. These services provide an interface for the configuration of UA Servers.

57.4.6 View Service Set

Views are publicly defined, UA server-created subsets of the AddressSpace. The entire AddressSpace is the default View, and therefore, the View services are capable of operating on the entire AddressSpace. Services to create client-defined Views may be specified in future versions of OPC UA.

The View service set allows clients to discover Nodes in a View by browsing. Browsing allows clients to navigate up and down the hierarchy, or to follow References between Nodes contained in the View. In this manner, browsing also allows clients to discover the structure of the View.

57.4.7 Query Service Set

The Query service set allows users to access the UA AddressSpace without browsing and without knowledge of the logical schema used for internal storage of the data.

Querying allows UA Clients to select a subset of the Nodes in a View based on some client-provided filter criteria. The Nodes selected from the View by the query statement are called a *result set*.

Notice: UA servers may find it difficult to process queries that require access to runtime data, such as device data, that involves resource-intensive operations or significant delays. In these cases, the UA server may find it necessary to reject the query.

57.4.8 Attribute Service Set

The Attribute Service Set is used to read and write Attribute values. Attributes are primitive characteristics of Nodes that are defined by OPC UA. They may not be defined by UA Clients or UA Servers. Attributes are the only elements in the AddressSpace permitted to have data values. A special Attribute, the Value Attribute is used to define the value of Variables.

57.4.9 Method Service Set

Methods represent the function calls of Objects. They are invoked and return after completion, whether *successful* or *unsuccessful*. Execution times for Methods may vary, depending on the function they are performing. A Method is *always* a component of an Object. Discovery is provided through the browse and query services. UA clients discover the Methods supported by an UA server by browsing for the owning Objects that identify their supported Methods.

Notice: Because Methods may control some aspect of plant operations, Method invocation may depend on use case conditions. For example, one may attempt to re-invoke a Method immediately after it has completed execution. Conditions that are required to invoke the Method may not yet have returned to the state that permits the Method to start again. In addition, some Methods may be capable of supporting concurrent invocations, while others may have a single invocation executing at a given time.

57.4.10 MonitoredItem Service Set

The MonitoredItem service set is used by the UA client to create and maintain *MonitoredItems. MonitoredItems* monitor *Variables, Attributes,* and *EventNotifiers.* They generate Notifications when they detect certain conditions. They monitor Variables for a change in value or status; Attributes for a change in value; and EventNotifiers for newly generated Alarm and Event reports.

Each *MonitoredItem* identifies the item to monitor and the *Subscription* to use to *periodically publish Notifications* to the UA client. Each *MonitoredItem* also specifies the *rate* at which the item is to be monitored (sampled) and, for Variables and EventNotifiers, the filter criteria used to determine when a Notification is to be generated. More details on filter criteria for Attributes are described in Part 4 of the UA specifications.

The *sample rate* defined for a *MonitoredItem* may be *faster* than the *publishing rate* of the *Subscription*. For this reason, the *MonitoredItem* may be configured to either queue all Notifications or to queue only the latest Notification for transfer by the *Subscription*. In this latter case, the queue size is one.

MonitoredItem services also define a *monitoring mode*. The monitoring mode is configured to disable *sampling* and *reporting*, to enable sampling only, or to enable both sampling and reporting. When sampling is enabled, the UA server samples the item. In addition, each sample is evaluated to determine if a Notification should be generated. If so, the Notification is queued. If reporting is enabled, the queue is made available to the *Subscription* for transfer.

MonitoredItems can be configured to *trigger* the reporting of other *MonitoredItems*. In this case, the monitoring mode of the items to report is typically set to sampling only, and when the triggering item generates a Notification, any queued Notifications of the items to report are made available to the *Subscription* for transfer.

57.4.11 Subscription Service Set

The Subscription service set is used by the UA client to create and maintain *Subscriptions. Subscriptions* are entities that *periodically* publish *NotificationMessages* for the *MonitoredItem* assigned to them. The *NotificationMessage* contains a common header followed by a series of Notifications. The format of Notifications is specific to the type of item being monitored (i.e., *Variables, Attributes*, and *EventNotifiers*).

Once created, the existence of a *Subscription* is independent of the UA client's session with the UA server. This allows one UA client to create a *Subscription*, and a second, possibly a *redundant* UA client, to receive *NotificationMessages* from it.

To protect against nonuse by UA Clients, *Subscriptions* have a configured *lifetime* that UA clients periodically renew. If any UA client fails to renew the lifetime, the lifetime *expires* and the *Subscription* is closed by the server. When a *Subscription* is closed, all *MonitoredItems* assigned to the Subscription are *deleted*.

Subscriptions include features that support *detection* and *recovery* of lost messages. Each *NotificationMessage* contains a *sequence number* that allows UA clients to detect missed messages. When there are no Notifications to send within the *keep-alive* time interval, the UA server sends a *keep-alive* message that contains the *sequence number* of the next *NotificationMessage* sent. If a UA client fails to receive a message after the *keep-alive* interval has expired, or if it determines that it has missed a message, it can request the UA server to resend one or more messages.

57.5 Implementations and Products

The OPC Foundation has released the OPC UA SDK (software development kit) and Java SDK for concept verification test. The UA SDK proposes a set of interfaces, libraries, and executables that allow developers to quickly create UA applications using C/C# languages with the .NET programming environment. The Java SDK is targeted for the development of UA applications with the Java environment.

Companies like Advosol, Unified Automation GmbH, and Technosoftware AG, are among the software vendors that have announced OPC UA products availability.

57.6 Conclusion

OPC UA is the next generation of the well-known OPC technologies from the OPC Foundation. It is a platform-independent standard through which various kinds of systems and devices can communicate by sending messages between clients and servers over various types of networks. The OPC UA specifications are layered to isolate the core design from the underlying computing technology and network transport. This allows OPC UA to be mapped to future technologies as necessary, without negating the core design.

References

1. W. Mahnke, S.-H. Leitner, and M. Damm, *OPC Unified Architecture*, Springer Verlag, Berlin/Heidelberg, Germany, 2009, ISBN: 978-3-540-68898-3.
2. About OPC page, OPC Foundation, August 2010, http://www.opcfoundation.org
3. What is MIMOSA?, Welcome page, August 2010, http://www.mimosa.org
4. About ISA page, August 2010, http://www.isa.org
5. About OMAC page, August 2010, http://www.omac.org
6. About EDDL page, August 2010, http://www.eddl.org
7. About WS4D page, August 2010, http://www.ws4d.org
8. S. Chan et al., Devices Profile for Web Services, May 2005, http://specs.xmlsoap.org/ws/2005/05/devprof/devicesprofile.pdf

58

DNP3 and IEC 60870-5

Andrew C. West
Invensys Operations Management

58.1 Requirements for SCADA Data Collection in Electric Power and Other Industries 58-1
58.2 Features Common to IEC 60870-5 and DNP3: Data Typing, Report by Exception, Error Recovery 58-3
58.3 Differentiation between IEC 60870-5 and DNP3 Operating Philosophy, Message Formatting, Efficiency, TCP/IP Transport 58-5
References 58-9

58.1 Requirements for SCADA Data Collection in Electric Power and Other Industries

Electric power networks consist of transmission systems and distribution systems. The transmission system consists of high-voltage power lines and a relatively small number of large substations. It delivers energy to terminal stations where transformers step the voltage down and feed energy into the distribution network. Distribution networks include large numbers of smaller substations and transformers that further convert the power to mains voltages and reticulate this to domestic and commercial consumers. Large generation stations connect directly to the transmission system and smaller generation stations can connect to the transmission or distribution networks.

Modern substations include equipment to control and protect the power system. These systems monitor many quantities such as voltages, currents, power flows, and the status of switches and other equipment. Some devices in substations perform functions to automatically disconnect power on the occurrence of problems such as overloads, short circuits, or faults such as those induced by electrical storms. The high-voltage transmission substations are the most heavily monitored and protected sites with lower voltage substations and distribution system equipment typically receiving less rigorous monitoring per site but having a much larger number of plant items that can be monitored.

Substations are usually monitored by supervisory control and data acquisition (SCADA) systems. These systems communicate their data to a control center where the information is used to allow operating staff to remotely monitor and control the power network. The operators perform routine switching of plant to bring it into and out of service and to quickly restore power after faults.

The dynamic characteristics of the electric power network can result in extremely rapid changes of the network behavior or operation in response to any changes in load, the availability of generation, or disturbances. The SCADA system must therefore provide rapid update of the monitored data and support minimal delay in issuing control command requests from the control center to the field equipment. The required responsiveness (being the time from a change in the field occurring until it is visible to the operator, or time for a request from the operator to cause an action in the field) in transmission networks is typically less than 5 s. Because of this rapid responsiveness, electric power SCADA systems are often considered to be real-time or near-real-time systems.

Power system control requires high integrity of both data reporting and control activation. The operator must be able to trust the accuracy of the information reported to the control center in order to quickly deduce any required responses. The system must correctly respond to any operator command and only activate exactly the equipment that is requested by the operator: Inadvertent operation of other devices cannot be tolerated and constitutes a safety risk to personnel and equipment. If any aspect of a control command fails a validation step, the command is discarded rather than allowing the possibility of incorrect execution of a command request.

When faults occur in electric power systems, automated protection equipment installed in substations typically act to identify the fault and take measures to minimize damage to the system. They take actions such as automatically disconnecting or isolating the faulty section of the network. The actions of this protection equipment are rapid and are tuned by carefully engineering a complex set of parameters that affect their operation. To verify the correct operation of this equipment and to facilitate the correct tuning of the parameters, the exact timing of changes of the data being monitored by the protection equipment and the responses of the protection equipment are important to facilitate the postmortem analysis of any fault and the corresponding operation of the equipment. This data gathering requires accurate collection of the time of the changes with resolution in the range from milliseconds down to microseconds and can involve capture of the power system waveforms for voltage and current at various points in the system. The correlation of data collected from different places in the power network requires close synchronization of the clocks used to time-stamp the data. The reporting of the time-stamped data to a system that archives this information is sometimes called "sequence of events" reporting or SOE.

Electric power SCADA systems are among the earliest applications for wide-scale distributed telemetry. The primary plant in generation stations and substations has a long service life (typically 30 or more years). The associated control equipment and communications systems typically have commensurate, though shorter, life spans in the order of 10 years. Thus, there are many SCADA systems operating over relatively old communication systems or systems with old architectures. These communication systems often have limited bandwidth and may have relatively high error rates when compared with those used in other applications such as Information Technology or factory automation. Many installed systems operate over serial data links with data rates in the range 300–9600 baud. Even where the equipment can be updated, it is not always feasible to significantly modify the available bandwidth without replacing the entire communication system. When upgrades occur, modern networking technologies are sometimes, but not always, adopted. For some applications, traditional serial data systems are more efficient or more cost-effective. This can be especially the case where it is necessary to provide access to a number of field devices over a shared media such as a limited-bandwidth radio channel.

The requirements for rapid data update and limited bandwidth availability dictate a need for efficient utilization of the available bandwidth. The data and command integrity requirements are satisfied by mechanisms that verify and validate data and commands. The accurate time-stamping of data requires that SOE information must be supported and that there is a mechanism to accurately synchronize the time across multiple widely dispersed devices. Some of these requirements are inherently contradictory (e.g., short latency and low bandwidth) and consequently a compromise must be achieved when addressing them.

Traditionally, these needs were met by vendors providing a variety of proprietary protocols and communication interfaces that specifically address these requirements. In more recent times, a number of open standards have been developed and widely adopted for this purpose.

Modern SCADA systems almost universally adopt standard interfaces from the telecommunication industries such as V.24/V.28 (RS-232) or V.11 (RS-422/RS-485) interfaces for serial or modem interfaces and any of the IEEE 802 interfaces for Ethernet (usually TCP/IP), with 100BaseT and 100BaseFX now being common. These interfaces are used with SCADA-specific protocols: In the electric power industry, the most commonly used SCADA protocols are IEC 60870-5-101 or IEC 60870-5-104 (dominant in Europe) and DNP3 (common in English-speaking countries). These two families of protocols are described further below.

Some other industries (e.g., rail transportation) have similar requirements for SCADA to those found in the electric power industry (e.g., traction power control) and adopt the same SCADA protocol standards for those applications, but will adopt different standards for other applications (e.g., for safety interfaces such as signaling).

Other major industries with large-scale widely distributed SCADA networks include oil and gas pipeline monitoring and water and wastewater monitoring. These industries have some requirements in common with electric power SCADA and some differences.

The oil and gas industries and water and wastewater industries typically have a relatively long time domain associated with their monitoring requirements: Pressures in gas pipelines and rates of flow and levels for liquids generally do not change rapidly or need prompt attention. This generally allows for a slower data collection strategy with a period of data collection in the range from minutes to hours often being acceptable. One notable exception to this is the prompt reporting of "urgent" alarm conditions that might necessitate operator intervention or maintenance call out.

For these pipeline industries, there are few additional requirements that are not covered by power system applications and therefore the electric power protocols and also Modbus (where no time tagging or SOE data is required) are applicable. Proprietary protocols are also still very common in the oil and gas industries. In recent years, the water industries in the United Kingdom and Australia have standardized on DNP3 for SCADA interfaces and are making use of the SOE functionality to allow accurate recording of field actions while allowing infrequent polling. DNP3 supports an unsolicited reporting mechanism where changes can be reported from the field without needing a request from the master station. The use of this feature allows prompt indication of urgent alarm conditions without requiring frequent polling of each remote site.

The IEC 60870 series profiles have not been widely adopted in the oil and gas and water industries, possibly because of a perception that they are power-system specific.

58.2 Features Common to IEC 60870-5 and DNP3: Data Typing, Report by Exception, Error Recovery

The International Electrotechnical Commission's 60870-5 series of standards were developed by the IEC Technical Committee 57 over the period 1988–2000. Some parts have been updated with subsequent revisions and extensions. This series includes the IEC 60870-5-101 [1] and 60870-5-104 [2] profiles for basic telecontrol tasks over serial and TCP/IP links. The IEC 60870-5 series are recognized as the international standards for electric power SCADA transmission protocols. Their use for that purpose is mandated in many countries by government legislation and they are widely adopted in Europe and in countries where European companies and conventions have significant influence.

IEC 60870 parts -5-1 through to -5-5 define a "cookbook" of rules for specifying SCADA Telecontrol protocols. Part -5-101 (sometimes also known as "Telecontrol 101" or "T101") specifies a fully defined profile for basic Telecontrol tasks based on these rules. T101 includes the definition of general-purpose and power-system-specific data objects and functions directly applicable to the substation to control center SCADA link. Part -5-104 ("T104") defines the transmission of T101 over TCP/IP with some minor functional extensions such as the addition of time tags to control objects.

DNP3 [3] is a general-purpose SCADA protocol based on the design rules in the early parts of IEC 60870 that adds some concepts that are quite different from those defined in later parts of IEC 60870-5. It defines data types and functions for the transmission of generic SCADA data. While it was designed for power system application, it does not include power-system-specific data objects.

DNP3 was initially developed by the Canadian SCADA company Westronic and was placed in the public domain in 1993. It is now maintained by an independent body called the DNP Users Group and is considered a de facto standard for electric power SCADA communication in North America where it is used by over 80% of power utilities [4]. DNP3 is also the de facto standard for electric power SCADA communications in most other English-speaking countries and places where North American vendors have significant market influence. The creation of DNP3 appears to have been influenced by a pragmatic response to utility market demands for conformance to the IEC 60870 standard in the years prior to the publication of IEC 60870-5-101. Its placement in the market as an open standard promoted its adoption and it has now been ratified as IEEE Standard 1815.

The heritage of the protocols and the dependencies in the documentation are illustrated in Figure 58.1.

The elements that are common to both IEC 60870-5-101/-104 and DNP3 include the following:

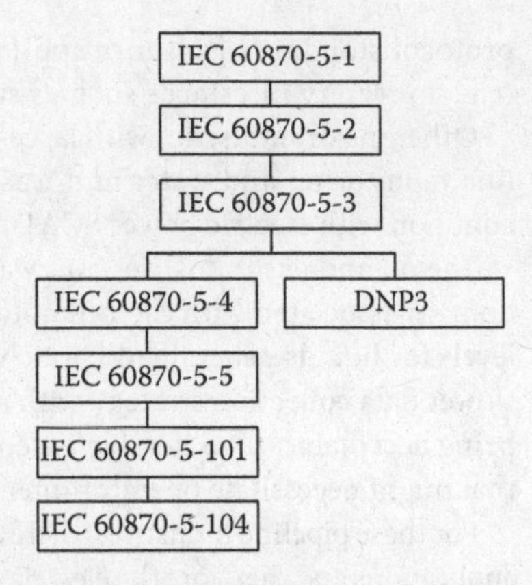

FIGURE 58.1 Dependencies of the specifications.

- T101/T104 and DNP3 adopt the "Enhanced Performance Architecture" (EPA) reference model described in IEC 60870-5-3 that uses only the Physical, Data Link, and Application layers of the ISO 7-Layer reference model and dispenses with the Network, Transport, Session and Presentation layers. DNP3 then extends the EPA model with a transport "function" (a pseudo-layer) that permits assembly of messages larger than a single data link frame.
- T101 and DNP3 use data link frame formats and data link frame handling rules defined in IEC 60870-5-1 and IEC 60870-5-2. These provide for good message efficiency with good error detection and rejection, meeting the basic electric power SCADA requirements for message integrity and efficiency.
- T101/T104 and DNP3 are based on the "Report by Exception" (or RBE) paradigm where, in normal operation, only changes are reported and data that has not changed is not reported. This meets the basic electric power SCADA requirement for efficiency of message reporting (by eliminating the unnecessary reporting of unchanged data) and also addresses the requirements to return SOE information (changes sent in sequence and with time-stamping). The RBE mechanism requires the initial collection of all data from a field device through an "integrity poll" and the subsequent reporting of all changes in the correct chronological time sequence. If this is done correctly with no loss of SOE data, then the resulting data image in the receiving station is always correct.
- Control command procedures include a "Select Before Operate" or "Select/Execute" mechanism also called a "two-pass control." This mechanism provides control commands with significantly enhanced security against random data errors introduced, for example, by induced noise interference in the communication system. This supports the electric power SCADA requirement for high integrity validation of control commands.
- Data objects include quality flags that provide extra information to qualify or validate the data that is being reported.
- Time stamps on change data that indicate the time of measurement of the reported data.
- Data and function types supported by both IEC 60870-5-101/-104 and DNP3 include
  - Reporting of single-bit binary status objects
  - Reporting of double-bit binary status objects
  - Reporting of integer analog measurands
  - Reporting of floating-point analog measurands
  - Reporting of counters (also called accumulators or integrated totals)
  - Control commands to binary outputs (on/off or trip/close commands)
  - Control commands to integer or floating-point setpoint outputs
  - Control commands to freeze counters/accumulators
  - Ability to read and write files to or from field devices
  - Ability to set the time and date (clock and calendar) in the field devices, including correcting for the communication system propagation delay
  - Ability to reset remote devices
  - Commands to perform an "integrity poll" (collect a refreshed data image of all data from a field device)
  - Commands to collect specific subsets of data from a field device

- Optional modes of operation where the field device can report change data without being polled by the master station.
- Test procedures to verify the correct operation of a protocol implementation in a device.
- To support SOE functionality, devices incorporate event data buffers. The protocols define mechanisms to verify the correct reporting of the data from these buffers: This data can be repeated if necessary to ensure that it is correctly reported. Once verified, it can be cleared from the event buffers.
- T104 and DNP3 have standardized XML file definitions for mapping between their standard data objects and data attributes of the Logical Node objects defined in the substation automation standard IEC 61850. T101 can also make use of the T104 mapping process.
- DNP3 has a secure authentication procedure that is based on IEC 62351-5. The equivalent functionality for T101 and T104 is under development at the time of writing.

Any device that implements these protocols can choose to implement only those parts of the protocol (data objects and functions) needed to support the operation of that device. Each protocol has some mandatory "housekeeping" functionality that must be implemented.

58.3 Differentiation between IEC 60870-5 and DNP3 Operating Philosophy, Message Formatting, Efficiency, TCP/IP Transport

While T101, T104, and DNP3 share many common operational characteristics such as the use of RBE, there are some detail differences in operating philosophy between the IEC 60870-5 protocols and DNP3. These differences are evident in the data object and message formats and in the command sequences used to manipulate them.

In some cases, the IEC 60870 and DNP3 protocol descriptions employ differing terminology to describe the same thing and similar terminology to describe different things. This easily leads to confusion, especially for people conversant with the details of one protocol who have relatively superficial contact with the other. In such cases, the existing familiarity with one protocol readily leads to misunderstanding or incorrect implementation of the other. Careful reading of both sets of specifications can be necessary in order to clearly understand the similarities and distinctions of the protocols. Sometimes, the end users also need to be aware of these details in order to correctly select configuration options, etc.

An example of differing terminology: The master station equipment typically found in a SCADA control center is called a "controlling station" in IEC 60870 and a "master" in DNP3; the substation equipment such as a remote terminal unit (RTU) that reports substation data to the control centre is called a "controlled station" in IEC 60870 and an "outstation" in DNP3.

An example of the same terms meaning somewhat different things exists in the use of data classes. In IEC 60870-5, there are two data classes used for polling data when using the unbalanced data link procedures, loosely corresponding to high priority (typically event data) in Class 1 and low priority (typically cyclic analog measurements) in Class 2. In DNP3, there are four classes: Class 0 for reporting "Static Data," meaning the current value of all kinds of data; and Classes 1, 2, and 3 that are three separate priority groups for event data. The mechanisms for deciding which class to read are different in the two protocols.

The principal tenet of the IEC 60870-5 series protocols is that the controlled station in the substation should determine what data is to be sent to the controlling station (master station). In the balanced mode of operation, the controlled station simply transmits whatever data it wishes at whatever time it wishes. The controlling station simply acknowledges receipt of the data and issues control commands when required. In the unbalanced mode of operation, the controlled station acts in a similar manner, but must wait for a poll request (effectively an "invitation to transmit") from the controlling station, to which it responds by sending a single message. The controlling station has little control over what the controlled station will transmit, other than being able to request "Class 1" or "Class 2" data. Normally the controlling station requests Class 2 (low priority, usually cyclic data) unless the previous response indicates that there is high-priority data available (by setting the access demand flag), in which case the controlling station requests Class 1 data. In this manner, the controlled station is fundamentally in control of the correct sequence of reporting of data to the controlled station. In DNP3, the outstation always replies with exactly the data that the master has requested: The master has full control over what will be sent. In order for the correct data reporting to be preserved in DNP3, the master must only request static data (the current value of inputs) if it is also requesting all buffered events for those same inputs in the same request. This is critical for correct data reporting.

Note: In all the protocols, an understanding of the underlying philosophy is required when implementing the protocol. It is possible to poorly implement T101 unbalanced in a controlled station so that the critical requirement to report all measurements in the order that they are measured is violated by arcane configuration settings. Similarly, it is possible to violate this same requirement in a DNP3 master by not requesting buffered events when static data is requested.

Differentiators between IEC 60870-5-101/-104 and DNP3 are

- The T101 profile is strictly for use over serial links, T104 is strictly for use over TCP/IP transports. The data link frame (IEC 60870-5-1 FT1.2) defined for T101 is replaced by a cut-down frame for T104 together with different message confirmation rules. T101 is permitted to use unbalanced data link services (required for use on multi-drop channels) or balanced data link services (requiring a dedicated full-duplex point-to-point link between pairs of devices). The operation of T104 is basically the same as the operation of T101 over balanced data links with the differentiation that a configurable number of T104 information messages (I-frames) can be sent before a confirmation is received, whereas the T101 rules require a confirmation of each individual frame. DNP3 always uses the serial data link frame (based on IEC 60870-5-1 FT3) and "balanced transmission" rules modified to allow multi-drop operation on half-duplex channels. When DNP3 is used with TCP/IP or UDP/IP, the serial data link frame is encapsulated in a TCP or UDP packet. This process simplifies the software for frame handling (same for serial or IP) and also simplifies the use of DNP3 with terminal server type interfaces that simply wrap or unwrap the serial message within the TCP/IP frames and do not need to apply a new serial data link frame.
- T101/T104 messages can only contain a single type of data (e.g., single-bit binary inputs or integer analog measurands). DNP3 messages can contain multiple types of data in one message, as long as the same "function" (e.g., read, control command, report response, etc.) applies to all data in the message. Because of this, the response to a command such as an integrity poll can consist of several smaller messages in T101 or T104 and a small number of large messages in DNP3. This can affect the efficiency of reporting the different protocols, especially over data links with special properties such as data radios (turn-around time may be long compared to transmit time) and satellite links (pay per byte: More packets = more overhead = more cost).
- T101/T104 messages are limited to the maximum size of a single data link frame. This typically allows about 250 bytes of data per message. DNP3 messages have no logical size limit: DNP3 uses a transport function and application layer fragmentation to build messages of any length and transmit them as a series of application fragments (typically 2048 bytes in size) split into many transport segments in order to fit the data link frames.
- DNP3 uses an explicit application confirmation mechanism to verify reporting of application data. T101 deduces correct application reporting by reliance on data link confirmation of each link frame or toggling of a single sequence number bit (known as the Frame Count Bit or FCB) in a subsequent request. T104 relies on the confirmation of a number of transmitted messages. The maximum window size in T104 is configurable (parameter k) but is typically set to 12. DNP3 also permits the use of the IEC data link confirm (except on TCP/IP links), however, the use of this provides no functional benefit and is not recommended.
- Within a device, T101 and T104 model every kind of data as being a member of a single set of objects, identified by the information object address (IOA) of the data element. The IOA values

may be sparsely assigned anywhere in the allowed range (which can vary depending on system-wide parameters that specify the IOA size and how it is formatted). DNP3 models all data within a device as a set of one-dimensional arrays with a separate array for each type of data. Each data element is identified in its array by an index number, starting at zero for the first element. In each protocol, maximum message reporting efficiency is achieved if the IOA range or index number range for any single data type is a single contiguous series with no gaps. For T101/T104, this means assigning IOA values X to $X + n - 1$ for n objects of the same type and for DNP3 it means assigning indices 0 to $n - 1$ for n objects of the same type. In this regard, the IEC data model somewhat resembles the register mapping concept used in Modbus.

- The identification of data in a T101 or T104 system is by the unique combination of the device identity (called the common address of ASDU or CAA) and IOA of each data element in the device. This identification is independent of the data link address (in T101) or the IP address (in T104) of the device reporting the data. Each different data object in a T101/T104 device (the device having one CAA value) must have a different IOA. In DNP3, the combination of the serial channel (or IP address) of the device, the DNP address of the device, the data type, and the object index together uniquely identify the data object.
- Because the IEC messages can only report a single kind of data or function in a single message, the response to some application functions consist of a series of messages starting with a "begin command sequence" or activation confirmation message (ACTCON) followed by one or more data messages and terminating with an "end command sequence" or activation termination message (ACTTERM). In DNP3, the same functionality typically only requires a single response message.
- The T101 and T104 messages include a "cause of transmission" (COT) value that serves to assist with the control of the multiple steps of the response to a command and is partly useful for indicating why data is being sent. For example, the COT can differentiate between a change that has occurred spontaneously (e.g., circuit breaker trip due to protection action) or because an operator has issued a control command to cause the change. This information can be added to the relevant event log entries. DNP3 does not include this kind of information and does not indicate why a change occurred, merely that a change has occurred.
- DNP3 defines generic SCADA data objects (e.g., binary input, setpoint output) without assigning specific meaning to those objects. In addition to these basic types, T101 and T104 define a number of power-system-specific objects such as "step position information" representing a transformer tap number, protection events, packed protection start events, and packed protection circuit information.
- DNP3 defines some complex generic data objects such as "Octet String" objects and "Data Set" objects that can be defined to provide functions such as multi-byte variable values or complex data structures. DNP3 also supports reporting of device data attributes (nameplate information). T101 and T104 support a generic 32 bit bitstring object whose purpose is left open for implementers to define.
- T101 and T104 define "normalized," "scaled," and "floating-point" data objects as different kinds of analog objects. Each of the three types also has separate setpoint command types. Normalized and scaled are both 16 bit 2's complement integers and floating-point uses the 32 bit "short" format defined in IEEE 754. For the integer formats, normalized data represents the values -1.0 to $+1.0-2^{-15}$ while scaled data represents the values -32768 to $+32767$. Both must be "scaled" before transmission and after reception, but by different scaling factors. There is an implication in the standard that the scaling factors used for the scaled values are always a factor of 10 (e.g., 0.001, 0.01, 0.1, 1, 10, 100, etc.) and this simplistic approach seems to be commonly used. Devices typically only support either the normalized or the scaled format, leading to some interoperability issues. The normalized format seems to be slightly more commonly used than the scaled format. DNP3 treats 16 bit integer, 32 bit integer, 32 bit floating-point, and 64 bit floating-point as alternate message formats for reporting the same object. The format that is used should be chosen to suit the data. The integer formats can be scaled as needed, with linear scaling ($Y = AX + B$) being common.

- DNP3 permits the use of "unsolicited reporting" on any channel where the devices are able to detect channel activity in order to avoid collisions. In this mode of reporting, an outstation can send an event notification to a master station without being polled. The approximately equivalent reporting mode for T101 requires the use of the balanced data link rules and needs a dedicated point-to-point full-duplex link between the master station and each outstation so that either device is permitted to transmit at any time with no possibility of message collision. As will be noted in Figure 58.2, the number of messages is the same in balanced and unbalanced modes, with the same application layer data being transferred in both modes. The order and content of the "link layer" messages differs slightly between the two modes. In DNP3, the use of unsolicited reporting removes the periodic polling for events, reducing the number of messages as shown in Figure 58.3.

Due to the limitation of a single data type per message and a single COT per message, most command sequences in T101 and T104 require many message transactions. In T101, each transaction requires that a message be sent in each direction. For example, the two-pass control command on a T101 link shown in Figure 58.2 typically involves 12 messages.

The equivalent transaction for DNP3 is shown in Figure 58.3, requiring that seven messages are sent in systems where the master polls the outstations, or six messages if unsolicited reporting is used. Similar distinctions between the protocols occur for most kinds of transactions.

When operating over TCP/IP, the difference between T104 and DNP3 is less pronounced and more difficult to evaluate. For some transactions, T104 will be more efficient when it is able to send a stream of information frames and receive a single confirmation for the set of frames (equivalent to removing

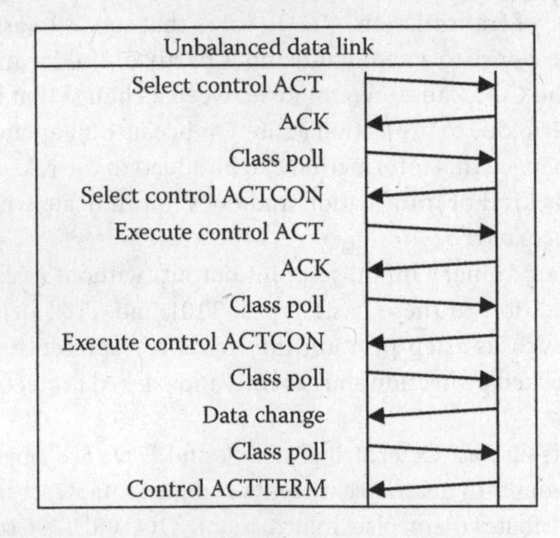

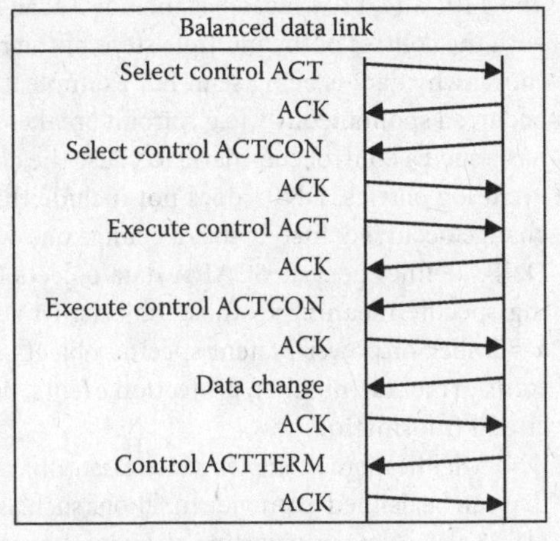

FIGURE 58.2 T101 Two-pass command sequence.

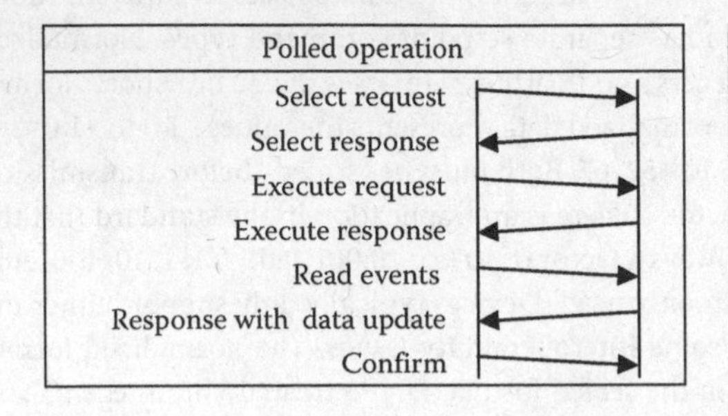

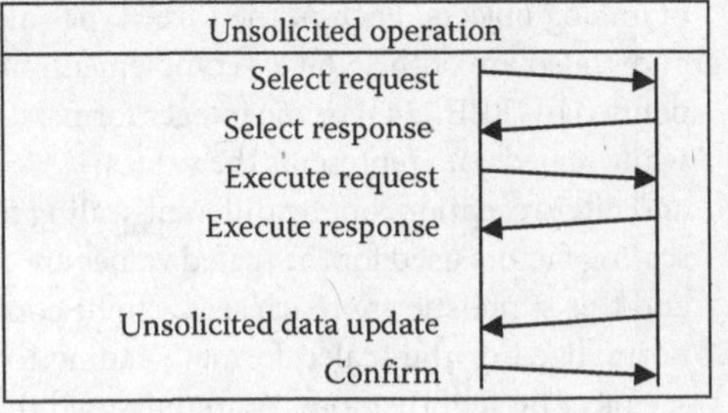

FIGURE 58.3 DNP3 Two-pass command sequence.

all except the last ACK from the right-hand column of Figure 58.2). Associated with this sequence, there will also be two or maybe more TCP/IP confirmation frames. In DNP3, the same function might be performed with two or three DNP3 messages (depending if unsolicited reporting is used) and two or three TCP/IP confirmation frames. If larger amounts of data are involved, the number of DNP3 and T104 messages will increase.

References

1. IEC 60870-5-101 Ed. 2, Telecontrol Equipment and System—Part 5: Transmission protocols—Section 101: Companion standard for basic Telecontrol tasks, International Electrotechnical Commission, 2002; www.iec.ch
2. IEC 60870-5-104 Ed. 2, Telecontrol equipment and systems—Part 5: Transmission protocols—Section 104: Network access for IEC 60870-5-101 using standard transport profiles, International Electrotechnical Commission, 2005, www.iec.ch
3. DNP3 Specification, Ver. 2, vv. 1–8, DNP Users Group 2007–2009; www.dnp.org
4. Leivo, E. and Newton, C., More than 80% of global utility respondents claim to have substation automation and integration programs underway, *Newton-Evans Research Company Market Trends Digest*, 28, 2–4, 2nd Quarter 2006.

59

IEC 61850 for Distributed Energy Resources

Sidonia Mesentean
Reinhold-Würth-University

Heinz Frank
Reinhold-Würth-University

Karlheinz Schwarz
Schwarz Consultancy Company

59.1 Introduction 59-1
59.2 Basic Concept 59-1
59.3 Modeling the Automation Functions 59-3
59.4 Communication Services 59-4
Client/Server Communication • GOOSE • Transmission of Sampled Analog Values • Clock Synchronization
59.5 Modeling with System Configuration Language 59-7
59.6 Different Types of DER 59-8
Wind Power Plants • Hydropower Plants • Other Specific Types of Distributed Energy Resources
References 59-10

59.1 Introduction

A dramatic change is going on currently in the supply of electrical energy, generally called power supply. In former times, this power supply was mainly provided by few big power stations. The technology of the automation systems in such power stations was designated by a few big enterprises. These days more and more small power supplies (DER—distributed energy resources) like wind power plants, combined heat and power, hydropower plants, photovoltaic systems, fuel cells, and others are being implemented everywhere.

The devices for the automation, protection, and monitoring of the distributed energy resources are developed and manufactured by many small and medium enterprises. So the technology is designated by many different participants. This causes a wide variety of technical approaches. In the application all distributed energy resources, however, all energy resources have to be integrated into one common electrical power grid. For the management of the energy flow in this power grid, for the maintenance of the systems, and for the electricity billing, most of the energy resources have to be integrated also into a common communication network (Figure 59.1). A required standard for this communication network is provided in the IEC 61850.

59.2 Basic Concept

The IEC 61850 standard was originally developed for substation automation [Sta1,Sch,Bra]. Afterwards it was used for distributed energy resources [Win,Sta2,Sta3]. It consists of the following parts:

- Modeling of the information that has to be exchanged
- Selection of abstract communication services from IEC 61850-7-2
- Selection of real protocols for the communication services

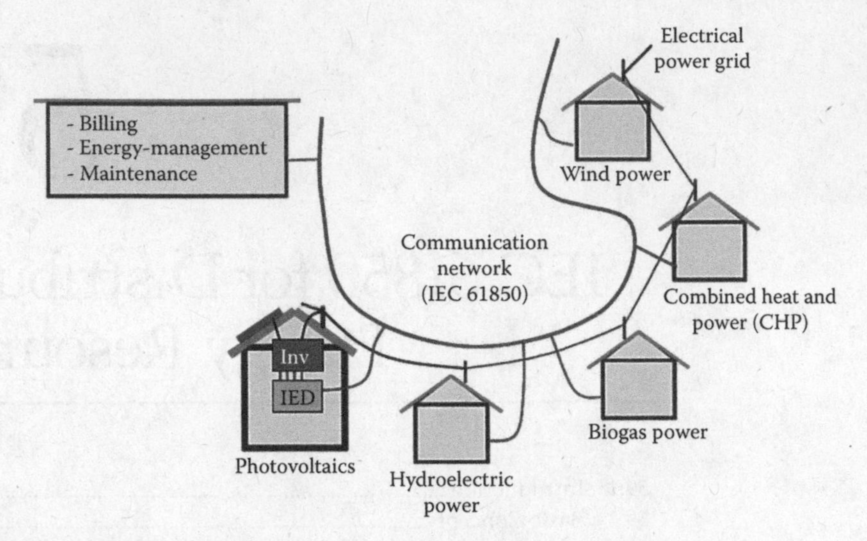

FIGURE 59.1 Communication network for distributed energy resources.

The basic concept of the IEC 61850 standard for its application in distributed energy resources will be described in the following example of a small photovoltaic system (Figure 59.2). This system consists of the following components:

- The photovoltaic array
- A DC-circuit breaker
- An inverter

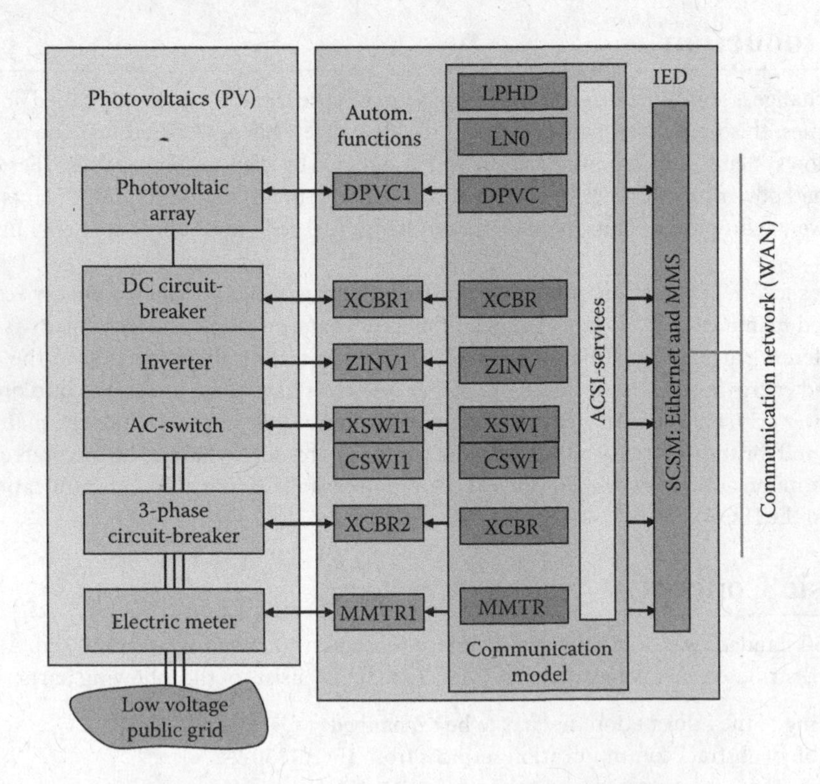

FIGURE 59.2 Example for the application of IEC 61850 on a photovoltaic system.

TABLE 59.1 Description of the Logical Nodes Classes

Logical Node	Description
LPHD	Describes the object for the control of a physical device, which is the microcontroller itself
LLNO	Describes the object for a logical device
DPVC	Describes the object for the photovoltaic array
XCBR	Describes the object for the DC circuit breaker
ZINV	Describes the object for the inverter
XSWI	Describes the object for the AC switch
CSWI	Describes the controller for the operation of the AC switch
MMTR	Describes the object for the electric meter

- An AC-switch
- A three-phase circuit breaker and
- An electric meter

For the integration of such a system into a communication network, a microcontroller is required. According to the IEC 61850 this computer is named as Intelligent Electronic Device (IED). The IEC 61850 provides standardized object classes to describe the information produced and consumed by the automation functions of such an IED. The most important object classes are the logical node (LN) classes. For the example shown in Figure 59.2 the Table 59.1 presents the logical node classes that were used.

The communication services required in an IED is described in a communication model by the so-called ACSI-services (ACSI—abstract communications service interface) defined in IEC 61850-7-2. These abstract services must be implemented using existing concrete communication protocols (SCSM—specific communication service mappings). At this stage of the standard, this mapping may be realized with Ethernet, TCP/IP, and MMS protocols.

59.3 Modeling the Automation Functions

In general, the automation functions in electrical power grids can be classified into three categories:

- Protection functions prevent hazard to people, damage to power network components, and breakdown of the power grid. Such functions have to be performed within 10 ms.
- Monitoring functions supervise the status of devices and primary equipment (circuit breakers, transformers, etc.) within the power grid.
- Control functions allow a local or remote operation of the devices.

These automation functions can be divided into subfunctions and functional elements. For modeling the automation functions in the communication model, objects are standardized on the level of the functional elements. They are named as LN. The LNs can be grouped to logical devices (LD).

An example of how LNs are built is shown in Figure 59.2. The LN CSWI1 is used in the example presented in Figure 59.2 in conjunction with XSWI1 for the control of the AC-switch. The corresponding LN-class CSWI consists of data objects (DO) that have standardized common data classes (CDC). As an example, the data object Pos (shown circled in Figure 59.3), has the CDC controllable double point (DPC). Each CDC contains a series of data attributes. As an example, the data attributes corresponding to data class DPC are shown in Figure 59.4. Some of this data attributes are mandatory, some are optional, and some can be mandatory or optional if certain conditions are fulfilled.

In our example, the data attributes corresponding to the LN CSWI1 are used to operate the switch while the LN XSWI1 is used to model the information received from the switch. The IED accesses the switch through binary inputs/outputs. Firstly, an operate command is sent from a client to the

LN	InClass	DataObject	CDC	M/O	
...					
CSWI1	CSWI	Mod	INC	M	Common LN information
		Beh	INS	M	
		Health	INS	M	
		NamPlt	LPL	M	
		Loc	SPS	O	
		OpCntRs	INS	O	
		Pos	DPC	M	Controls
		PosA	DPC	O	
		PosB	DPC	O	
		PosC	DPC	O	
		OpOpn	ACT	O	
		OpCls	ACT	O	
...					

Pos-This data is accessed when performing a switch command or to verify the switch status or position

FIGURE 59.3 Data objects of the logical node CSWI1 data (the mandatory and the optional data attributes are shown).

InClass	Data	CDC	M/O	
CSWI	Pos	DPC	M	Controls
	PosA	DPC	O	
	....			

Attr. Name	Value range	M/O
ctlVal	Off/On	AC_CO_M
stVal	Interm.-state/off/on/bad-state	M
q		M
t		M
ctlModel		M

ctlVal-Determines the control activity.
stVal-Level of the supervised value, which starts a dedicated action of the related function.
q-Quality of the attribute(s) representing the value of the data.
t-Timestamp of the last change in one of the attribute(s) representing the value of the data or in the q attribute.
ctlmodel-Specifies the control model of IEC 61850-7-2 that corresponds to the behavior of the data.

FIGURE 59.4 The common data class DPC and its attributes.

CSWI1-node within the IED. If the operation is permitted, then the AC-switch is activated through an output signal. The IED then supervises the return information from the switch. If the operation terminates with success, a command termination message will be then sent. The sequence of commands and responses is standardized in IEC 61850-7-2.

59.4 Communication Services

The communication services for an IED are described in the communication model by the so-called abstract communication services (Figure 59.5). These services must be implemented in an IED by using standardized protocols, (SCSM).

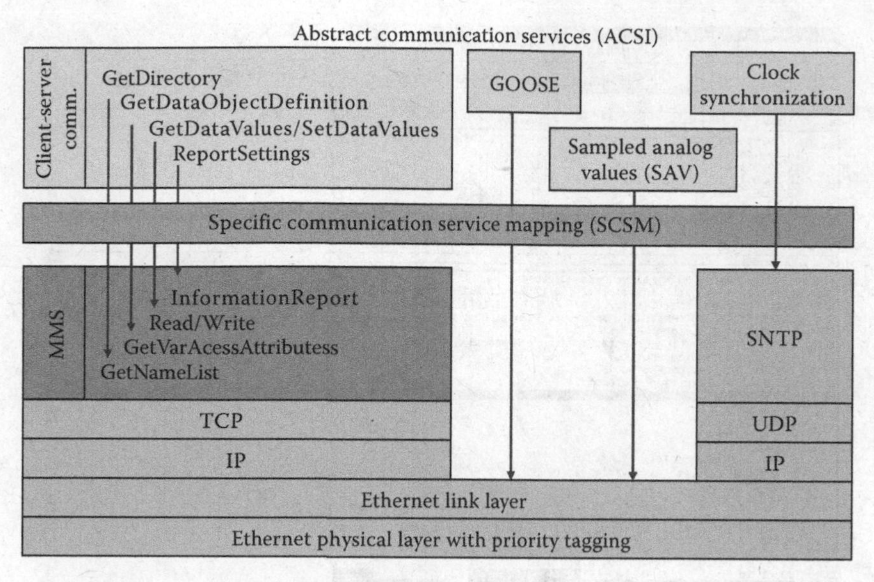

FIGURE 59.5 Comparison between ACSI and SCSM.

59.4.1 Client/Server Communication

The most important ACSI services for distributed energy resources are the client-server communication services. They include services such as the following:

- *GetDirectory*: When a client wants to communicate with another IED it can retrieve a complete list of the names of all accessible objects in an IED, which are LDs, LNs, DOs, and data attributes.
- *GetDataObjectDefinition*: A client can retrieve the data types for all objects.
- *GetDataValues, SetDataValues*: The values of the objects can be read and written.
- *Reporting*: It allows an event driven data exchange of packed values between a server and a client. The values can be transmitted by an IED either immediately or after some buffer time.
- *Logging*: The server provides the possibility to archive data for later retrieval by clients.

According to the actual state of the IEC 61850 standard, the client/server communication services have to be mapped to the Manufacturing Message Specification (MMS)-protocol that is based on Ethernet and TCP/IP (Figure 59.6). The MMS standard originally was developed for the communication between manufacturing devices like numerically controlled machine tools, transport systems and supervisory control and data acquisition (SCADA) systems. MMS defines a structure for the messages required to control and monitor the devices. It is not concerned with the way the messages are transferred between devices over a network. For that Ethernet and TCP/IP is used. The data objects in the IEC 61850 are mapped to MMS as presented in Table 59.2.

The mapping of some ACSI-services to MMS is already shown in Figure 59.5. Figure 59.6 shows a detailed example for the concrete message encoding of a GetDirectory ACSI-service to a GetNameList message of the MMS protocol.

59.4.2 GOOSE

Another ACSI service specified by the IEC 61850-standard is the so called Generic Object Oriented Substation Event (GOOSE). It is a mechanism for the fast transmission of events, such as commands, alarms, and indications. A single GOOSE broadcast-message sent by an IED (publisher) can be received

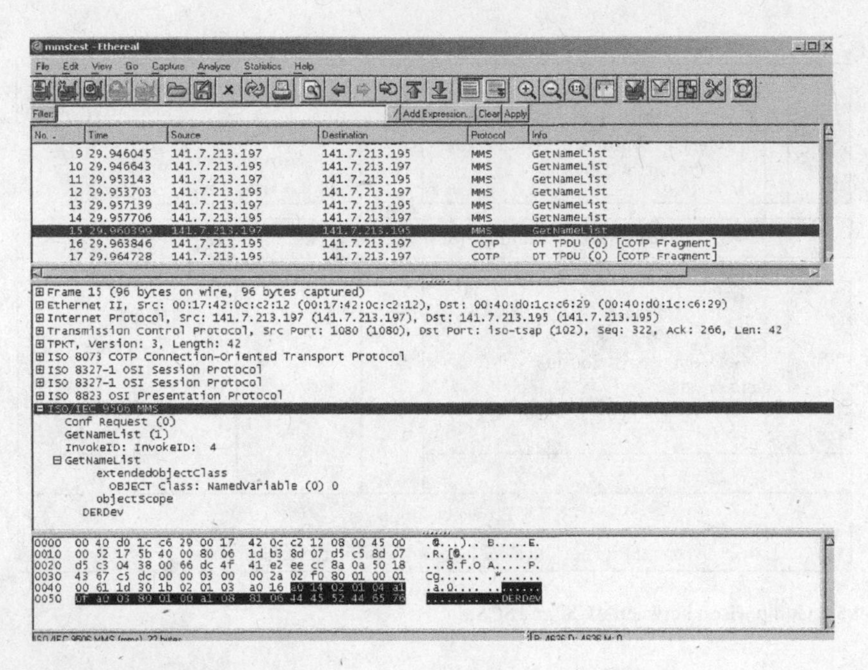

FIGURE 59.6 Example for a MMS message.

TABLE 59.2 Mapping of the Data Models in MMS Services

Objects in IEC 61850	Objects in MMS
Logical device	Domain
Logical node	NamedVariable
Data object	NamedVariable (structured components)
Data attribute	NamedVariable (structured components)
Data set	NamedVariableList

and used by several receivers (subscribers) located in the same subnetwork. It is implemented directly on the Ethernet Link Layer and supports real-time behavior. It is used, for example, for

- Tripping of switchgears
- Starting of disturbance recorders
- Providing position indications for interlocking

59.4.3 Transmission of Sampled Analog Values

The ACSI service called "transmission of sampled analog values" (SAV, defined in IEC 61850-9-2) also supports a time critical communication, mainly used for the exchange of current and voltage measurements from high voltage sensors. The characteristics of sampled process values can be transmitted as broadcast messages in real time from a sending IED (publisher) to receiving IEDs (subscribers). The publisher writes values together with a time stamp in a sending buffer. After the transmission the subscribers can read the values from their receiving buffers. This service has the following additional features:

- The transmission over the Ethernet can be in a point-to-point (unicast)-mode or in a multicast-mode.
- The transmission of sampled values is done in an organized and time-controlled way so that the combined jitter of sampling and transmission is minimized.
- The transmission of the values is controlled by a standardized control-block (SVC—sampled value control).

- Similar to the GOOSE model, the SAV model applies to the exchange of values of a DATA-SET. The difference in this case is that the data of the data set are of the common data class SAV (sampled analog value).

59.4.4 Clock Synchronization

For the clock-synchronization of an IED with a central time-server, the standard network time protocol (SNTP) is specified for nontime-critical applications. Separate synchronization is required for the SAV process. In future this will be implemented according to IEC 61588. The implementation of the use of a time-dissemination-service, like for instance Global Position System (GPS), is also allowed.

59.5 Modeling with System Configuration Language

An IED can control big systems like a wind-park or small systems like a simple sensor used for collecting data (e.g., temperature sensor or electric meter). The information produced and consumed by automation functions and the communication interface in such IEDs are modeled with the System Configuration Language (SCL) in a SCL file. For the SCL language, the standard IEC 61850-6 is still valid (developed for the substation automation) [Sta1] where SCL is named as Substation Configuration Language. A SCL file applied in distributed energy resources includes in a hierarchical structure the following information (Figure 59.7):

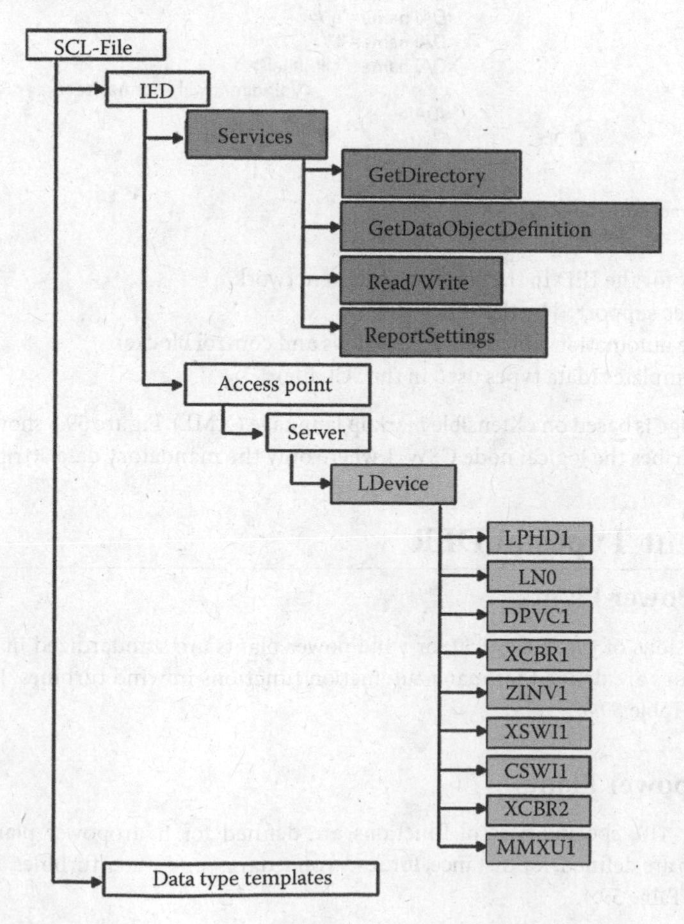

FIGURE 59.7 Example of how the SCL file is organized.

```
<LN lnType="CSWI1" lnClass="CSWI" inst="1" desc="Switch Controller">
        <DOI name="Mod" desc="Mode">
                <DAI name="ctlModel">
                        <Val>direct-with-normal-security</Val>
                </DAI>
        </DOI>
        <DOI name="Beh" desc="Behaviour">
                <DAI name="stVal"/>
                <DAI name="q"/>
                <DAI name="t"/>
        </DOI>
        <DOI name="Health" desc="Health">
                <DAI name="stVal"/>
                <DAI name="q"/>
                <DAI name="t"/>
        </DOI>
        <DOI name="NamPlt" desc="Name plate">
                <DAI name="vendor">
                        <Val>Maschinenfabrik Reinhausen</Val>
                </DAI>
                <DAI name="swRev">
                        <Val>1.0</Val>
                </DAI>
                <DAI name="d"/>
        </DOI>
        <DOI name="Pos" desc="Tap Position">
                <DAI name="stVal"/>
                <DAI name="q"/>
                <DAI name="t"/>
                <DAI name="ctlModel">
                        <Val>direct-with-normal-security</Val>
                </DAI>
        </DOI>
</LN>
```

FIGURE 59.8 Extract from an SCL file.

- Access point for the IED in the communication network
- ACSI-services supported by the IED
- Model of the automation functions of LDs (LNs and control blocks)
- Data type templates (data types used in the SCL file)

The SCL-language is based on eXtensible markup language (XML). Figure 59.8 shows an extract from an SCL file. It describes the logical node CSWI1 where only the mandatory data attributes are shown.

59.6 Different Types of DER

59.6.1 Wind Power Plants

The specific extensions of the IEC 61850 for wind power plants are standardized in the IEC 61400-25 [Win]. Logical nodes are defined for many automation functions in wind turbines. Examples for such LNs are shown in Table 59.3.

59.6.2 Hydropower Plants

In the IEC 61850-410, specific control functions are defined for hydropower plant systems [Sta2]. Specific LN classes are defined, for instance, for reservoirs, dams, and water turbines. Examples for such LNs are shown in Table 59.4.

TABLE 59.3 Logical Nodes for Wind Power Plants (Examples)

LN-Class	Description
WTUR	General information on a wind power plant
WROT	Information on the wind turbine rotor: Status information: turbine status, operation time, Control information: operation mode, setpoints,
WTRM	Information on the transmission: operation mode, the status of the generator cooling system, the generator speed
WGEN1	Information on the generator: generator speed, voltages, currents, power factor, temperatures
WMET	Wind power plant meteorological information: wind speed, ambient temperature, humidity

TABLE 59.4 Logical Nodes for Hydropower Plants (Examples)

LN-Class	Description
HDAM	Information on a hydropower reservoir: type of construction, calculated volumetric content.
HGTE	Information on a dam gate: Status information: upper/lower end position reached, gate is blocked Measured value: calculated water flow through the gate (m^3/s) Control information: gate position setpoint, change gate position
HUNT	Information on the hydropower unit (turbine and generator): Status information: Operating mode Settings: Rated maximum water flow, temporary limitation of power output

59.6.3 Other Specific Types of Distributed Energy Resources

The IEC 61850-420 standard includes LNs for the following types of distributed energy resources [Sta3]:

- *Photovoltaic systems*: An example for a PV system is already described above.
- *Diesel engines* (*reciprocating engine*): The most important LN for a diesel engine is the Reciprocating engine (DCIP) LN. It includes information such as the engine status, settings for the minimum and the maximum speed, the actual engine speed, and the actual engine torque. Examples for controls are engine on/off, crank relay driver on/off, and diagnostic mode enable.
- *Fuel cells*: The most important LN for fuel cells is the Fuel Cell Controller (DFCL) LN for the fuel cell controller. It includes status information (count of system starts, time until next maintenance, etc.), settings (maximum fuel consumption rate, system output voltage rating, etc.), controls (open close fuel valve, emergency stop), and measured values (system run energy, input fuel consumption, water level remaining, etc.).
- *Combined heat and power system* (*CHP*): Specific LNs are defined for CHP-system-controllers (DCHC), thermal storages (DCTS) and CHP boilers (DCHB).

References

[Bra] K.P. Brand, The Standard IEC 61850 as Prerequisite for Intelligent Applications in Substations, Power Engineering Society, *IEEE General Meeting*, Adelaide, Australia, 2004, pp. 714–718.

[Sch] K.H. Schwarz, An introduction to IEC 61850. *Basics and User-Oriented Project Examples for the IEC 61850 Series for Substation Automation*, Vogel Verlag, Würzburg, Germany, 2005.

[Sta1] IEC 61850, Communication networks and systems in substations, 10 parts, 2004.

[Sta2] IEC 61850, Communication networks and systems in substations, part 7-410, 2006.

[Sta3] IEC 61850, Communication networks and systems in substations, part 7-420, 2006.

[Win] IEC 61400-25, Communications for monitoring and control of wind power plants, 2006.

IV

Internet Programming

60 User Datagram Protocol—UDP *Aleksander Malinowski and Bogdan M. Wilamowski* **60**-1
Introduction • Protocol Operation • Programming Samples • References

61 Transmission Control Protocol—TCP *Aleksander Malinowski and Bogdan M. Wilamowski* **61**-1
Introduction • Protocol Operation • State Diagram • Programming Samples • References

62 Development of Interactive Web Pages *Pradeep Dandamudi* **62**-1
Introduction • Installations • Introduction to PHP • MySQL • Creating Dynamic Web Sites Using PHP and MySQL • References

63 Interactive Web Site Design Using Python Script *Hao Yu and Michael Carroll* **63**-1
Introduction • Software Installation • Database-Driven Web Site Design • Conclusion • References

64 Running Software over Internet *Nam Pham, Bogdan M. Wilamowski, and Aleksander Malinowski* **64**-1
Introduction • Most Commonly Used Network Programming Tools • Examples • Summary and Conclusion • References

65 Semantic Web Services for Manufacturing Industry *Chen Wu and Tharam S. Dillon* **65**-1
Background • Aims • Approach • Conclusion • References

66 Automatic Data Mining on Internet by Using PERL Scripting Language *Nam Pham and Bogdan M. Wilamowski* **66**-1
Introduction • Examples • Summary and Conclusion • References

60

User Datagram Protocol—UDP

Aleksander Malinowski
Bradley University

Bogdan M. Wilamowski
Auburn University

60.1 Introduction **60**-1
60.2 Protocol Operation **60**-1
UDP Datagram • Port Number Assignments • Connectionless Service—Flow and Error Control
60.3 Programming Samples **60**-8
References **60**-10

60.1 Introduction

User datagram protocol (UDP) is a part of TCP/IP suite [STD6,C02-1,F10,GW03,PD07]. It provides full transport layer services to applications. It belongs to the transport layer in the TCP/IP suite model, as shown in Figure 60.1. UDP provides a connection between two processes at both ends of transmission. This connection is provided with minimal overhead, without flow control or acknowledgment of received data. The minimal error control is provided by ignoring (dropping) received packets that fail the checksum test.

The UDP is standardized by *Internet Society*, which is an independent international not-for-profit organization. A complete information about UDP-related standards is published by *RFC Editor* [RFC]. Several relevant RFC articles as well as further reading materials are listed at the end of this article.

60.2 Protocol Operation

Underlying IP provides data transport through the network between two hosts. UDP allows addressing particular processes at each host so that incoming data is handled by a particular application. While underlying IP provides logical addressing and is responsible for data transmission through networks [STD5], UDP provides additional addressing that allows setting up connections for multiple applications and services on the same end. Addressing the processes is implemented by assigning a number to each connection to the network, which is called a port number. Figure 60.2 illustrates the concept of port numbers and shows the difference between IP and UDP addressing. UDP packets are also called *datagrams*. Each datagram is characterized by four parameters—the IP address and the port number at each end of the connection [STD6]. Certain class of destination IP addresses may not describe individual computers but a group of processes on multiple hosts that are to receive the same datagram [STD2,RFC3171]. Computers that need to receive multicast messages must "join" the multicast group associated with a particular multicast IP address.

Although the term connection is used here, unlike in case of transmission control protocol (TCP), there is no persistent connection between the two processes. The data exchange may follow certain

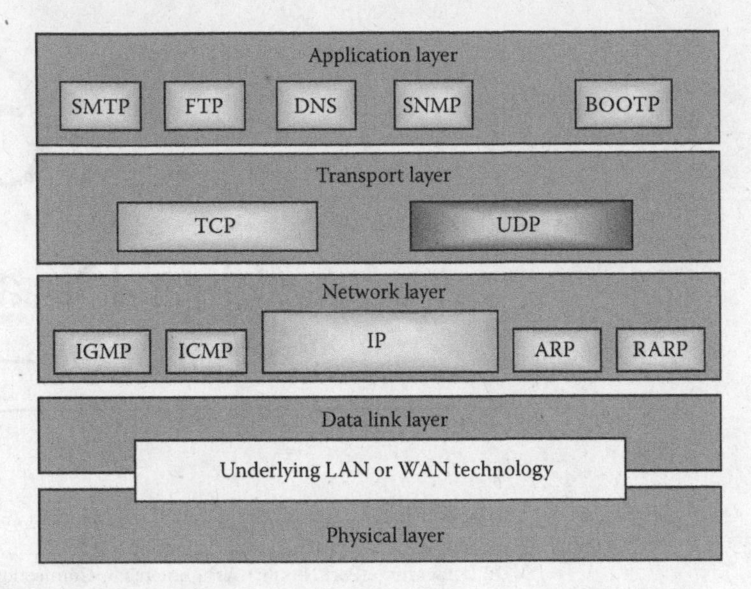

FIGURE 60.1 UDP and TCP/IP in the TCP/IP suite model.

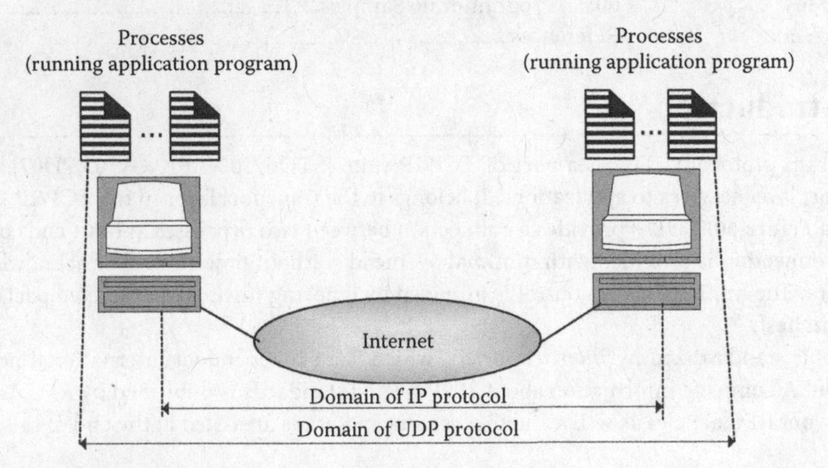

FIGURE 60.2 Comparison of IP and UDP addressing.

pattern or protocol but there is no guarantee that all data reach the destination and that the order of their receiving would be the same as the order they were sent. The client–server approach, which is usually used to describe interaction between two processes, still can be used. The process that may be identified as waiting for the data exchange is called the *server*. The process that may be identified as one that initializes that exchange is called the *client*.

60.2.1 UDP Datagram

Datagram have a constant size 8 byte header prepended to the transmitted data, as shown in Figure 60.3 [STD6]. The meaning of each header field is described below:

- *Source port address*: This is a 16 bit field that contains a port number of the process that sends options or data in this segment.
- *Destination port address*: This is a 16 bit field that contains a port number of the process that is supposed to receive options or data carried by this segment.

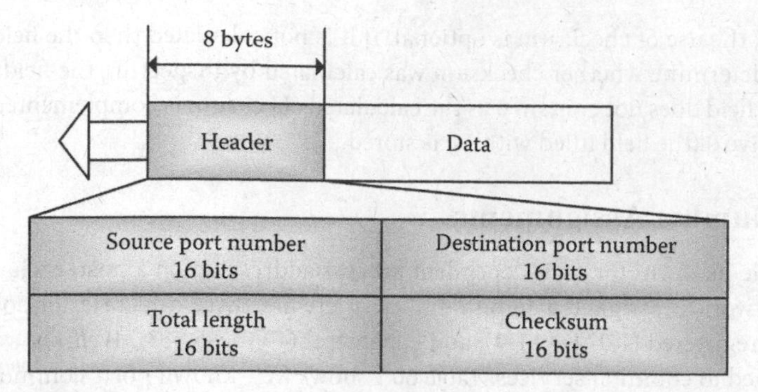

FIGURE 60.3 User datagram format.

- *Total length*: This is a 16 bit field that contains the total length of the packet. Although the number could be in the range from 0 to 65,535, the minimum length is 8 bytes that correspond to the packet with the header and no data. The maximum length is 65,507 because 20 bytes are used by the IP header and 8 bytes by the UDP header. Thus, this information is redundant to the packet length stored in the IP header. An extension to UDP that allows for transmission of larger datagrams over IPv6 packets has been standardized [RFC2675,RFC2147]. In such a case, the UDP header specified total length is ignored.
- *Checksum*: This 16 bit field contains the checksum. The checksum is calculated by
  - Initially filling it with 0s.
  - Adding a pseudoheader with information from IP, as illustrated in Figure 60.4.
  - Treating the whole segment with the pseudoheader prepended as a stream of 16 bit numbers. If the number of bytes is odd, 0 is appended at the end.
  - Adding all 16 bit numbers using 1s complement binary arithmetic.
  - Complementing the result. This complemented result is inserted into the checksum field.
- *Checksum verification*: The receiver calculates the new checksum for the received packet that includes the original checksum, after adding the so-called pseudoheader (see Figure 60.4). If the new checksum is nonzero, then the datagram is corrupt and is discarded.

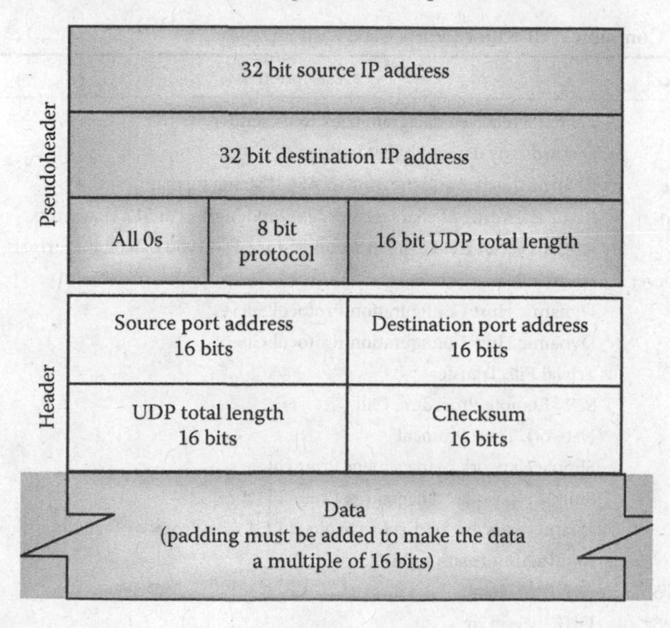

FIGURE 60.4 Pseudoheader added to the UDP datagram.

In case of UDP, the use of checksum is optional. If it is not calculated then the field is filled with 0s. The receiver can determine whether checksum was calculated by inspecting the field. Even in case the checksum is 0 the field does not contain 0 as the calculated checksum is complemented at the end of the process and negative 0 (the field filled with 1s) is stored.

60.2.2 Port Number Assignments

In order to provide platform for an independent process addressing on a host, each connection of the process to the network is assigned a 16 bit number. There are three categories of port numbers: well-known (0-1,023), registered (1,024-49,151), and ephemeral (49,152-6,553). *Well-known ports* have been historically assigned to common services. Table 60.1 shows well-known ports commonly used by UDP. Some operating systems require that the process that utilizes these ports must have administrative privileges. This requirement was historically created to avoid hackers running server imposters on multiuser systems. Well-known ports are registered and controlled by Internet Assigned Numbers Authority (IANA) [IANA] [STD2]. *Registered ports* are also registered by IANA to avoid possible conflicts among different applications attempting to use the same port for listening to incoming connections. *Ephemeral* ports are also called dynamic ports. These ports are used by outgoing connections that are typically assigned to the first available port above 49,151. Some operating systems may not follow IANA recommendations and treat the registered ports range as ephemeral. For example, BSD uses ports 1,024 through 4,999 as ephemeral ports, many Linux kernels use 32,768–61,000, Windows use 1,025–5,000, while FreeBSD follows IANA port range.

60.2.3 Connectionless Service—Flow and Error Control

As it was already outlined, UDP provides connectionless communication. Each datagram is not related to another datagrams coming earlier or later from the same source. The datagrams are not numbered. There is no need for establishing or tearing down the connection. In the extreme case, just one datagram might be sent in the process of data exchange between the two processes.

UDP does not provide any flow control. The receiving side may be overflowed with incoming data. Unlike in TCP, which has a windowing mechanism, there is no way to control the sender. There is no

TABLE 60.1 Common Well-Known Ports Used with UDP

Port	Protocol	Description	RFC/STD #
7	Echo	Echoes a received datagram back to its sender	STD20
9	Discard	Discards any datagram that is received	STD21
13	Daytime	Returns the date and the time in ASCII string format	STD25
19	Chargen	Returns a string of characters of random length 0 to 512 characters	STD22
37	Time	Returns the current time in seconds since 1/1/1900 in a 32 bit format	STD26
53	Nameserver	Domain Name Service	STD13
67	DHCP	Dynamic Host Configuration Protocol Server	RFC2131
68	DHCP	Dynamic Host Configuration Protocol Client	RFC2131
69	TFTP	Trivial File Transfer	STD33
111	RPC	SUN Remote Procedure Call	RFC5531
123	NTP	Network Time Protocol	STD12
161	SNMP	Simple Network Management Protocol	STD62
162	SNMP	Simple Network Management Protocol (trap)	STD62
213	IPX	Internetwork Packet Exchange Network Layer Protocol over IP	RFC1234
520	RIP	Routing Information Protocol	SDT56
546	DHCPv6	DHCPv6 client	RFC3315
547	DHCPv6	DHCPv6 server	RFC3315

```
/* * server.c - code for example server program that uses UDP * */
#ifndef unix
#include <winsock2.h>
/* also include Ws2_32.lib library in linking options */
#else
#define closesocket close
#define SOCKET int
#include <sys/types.h>
#include <sys/socket.h>
#include <netinet/in.h>
#include <arpa/inet.h>
#include <netdb.h>
#include <unistd.h>
/* also include xnet library for linking; on command line add: -lxnet */
#endif

#include <stdio.h>
#include <string.h>

#define PORT 1200                   /* server TCP port number          */
int main()
{
    SOCKET  sd;                   /* socket descriptor - (integer)     */
    struct  protoent *ptrp;  /* pointer to a protocol table entry   */
    struct  sockaddr_in sad; /* structure to hold server's address */
    struct  sockaddr_in cad; /* structure to hold client's address */
    int     alen;                 /* length of address                 */
    int     port;                 /* protocol port number              */

#ifdef WIN32
    WSADATA wsaData;
    if(WSAStartup(0x0101, &wsaData)!=0)
    { fprintf(stderr, "Windows Socket Init failed: %d\n", GetLastError()); exit(1); }
#endif

    memset((char *)&sad,0,sizeof(sad)); /* clear sockaddr structure */
    port = PROTOPORT;                    /* use predefined port number */
    sad.sin_port = htons((u_short)port); /* set server port number */
    sad.sin_family = AF_INET;            /* set family to Internet */
    sad.sin_addr.s_addr = INADDR_ANY;    /* set the local IP address */

    /* Map UDP transport protocol name to protocol number */
    ptrp = getprotobyname("udp");
    if ( ptrp == 0) { fprintf(stderr, "cannot map \"udp\" to number"); exit(1); }

    /* Create a socket */
    sd = socket(PF_INET, SOCK_DGRAM, ptrp->p_proto);
    if (sd < 0) { fprintf(stderr, "socket creation failed"); exit(1); }

    /* Bind a local address to the socket */
    if (bind(sd, (struct sockaddr *)&sad, sizeof(sad)) < 0)
    { fprintf(stderr, "bind failed"); exit(1); }

    {
        char buf[1000];         /* buffer for string the server sends  */
        int  visits = 0;        /* counts client connections           */
        int  n;                 /* number of characters received       */
        int  m;                 /* number of characters sent back      */

        while (1) { /* Main server loop - accept and handle requests */
            alen = sizeof(cad);
            n = recvfrom(sd,buf,sizeof(buf),0,(struct sockaddr*)&cad,&alen);
            if (n<0) {
                fprintf(stderr,"Error in receiving\n"); continue;
            } else if(n>=0) { /* We could recevie a useful emtpy packet */
                visits++;
                sprintf(buf,"This server has been contacted %d time%s\n",
                    visits,visits==1?".":"s.");
                m = sendto(sd,buf,strlen(buf)+1,0,(struct sockaddr*)&cad,alen);
                if(m<0) { fprintf(stderr,"Error in sending"); continue; }
            }
        }
    }

#ifdef WIN32
    WSACleanup();                        /* release use of winsock.dll */
#endif
    return(0);
}
```

FIGURE 60.5 Sample C program that illustrates a UDP server.

```
/* * code for example client program that uses UDP * */
#ifndef unix
#include <winsock2.h>
/* also include Ws2_32.lib library in linking options */
#else
#define closesocket close
#define SOCKET int
#include <sys/types.h>
#include <sys/socket.h>
#include <netinet/in.h>
#include <arpa/inet.h>
#include <netdb.h>
#include <unistd.h>
/* also include xnet library for linking; on command line add: -lxnet */
#endif

#include <stdio.h>
#include <string.h>

#define PORT 1200                 /* server TCP port number              */
char    hostname[] = "localhost";   /* server host name                  */
int main()
{
    SOCKET  sd;                   /* socket descriptor - (integer)       */
    struct  hostent  *ptrh;       /* pointer to a host table entry       */
    struct  protoent *ptrp;       /* pointer to a protocol table entry   */
    struct  sockaddr_in sad;      /* structure to hold an IP address     */
    int     alen;                 /* length of address                   */
    char    host[256];            /* pointer to host name                */

#ifdef WIN32
    WSADATA wsaData;
    if(WSAStartup(0x0101, &wsaData)!=0)
    { fprintf(stderr, "Windows Socket Init failed: %d\n", GetLastError()); exit(1); }
#endif

    /* prepare the client socket information */
    memset((char *)&sad,0,sizeof(sad)); /* clear sockaddr structure */
    sad.sin_family = AF_INET;           /* set family to Internet     */

    /* Convert host name to equivalent IP address and copy to sad. */
    strcpy(host, hostname);
    ptrh = gethostbyname(host);
    if (((char *)ptrh)==NULL) { fprintf(stderr, "invalid host: %s\n", host); exit(1); }
    memcpy(&sad.sin_addr, ptrh->h_addr, ptrh->h_length);

    sad.sin_port = htons((u_short)PORT); /* use the defined port number */
    alen = sizeof(sad);

    /* Map TCP transport protocol name to protocol number. */
    ptrp = getprotobyname("udp");
    if ( ptrp == 0) { fprintf(stderr, "cannot map \"udp\" to number\n"); exit(1); }

    /* Create a socket. */
    sd = socket(PF_INET, SOCK_DGRAM, ptrp->p_proto);
    if (sd < 0) { fprintf(stderr, "soclet creation failed\n"); exit(1); }

    {   int  n;                   /* number of characters received       */
        int  m;                   /* number of characters sent back      */
        char buf[1000];           /* buffer for data from the server     */

        /* Send data to socket in order to request reply. */
        m = 1;              /* sned zero bytes */
        m = sendto(sd,buf,m,0,(struct sockaddr*)&sad,alen);
        if(m<0) {
            fprintf(stderr,"Error in sending");
        } else {
            /* Read data from socket and write to user's screen. */
            n = recvfrom(sd,buf,sizeof(buf),0,(struct sockaddr*)&sad,&alen);
            if (n > 0) { buf[n]='\0'; printf("%s", buf); }
        }
    }
    closesocket(sd);

#ifdef WIN32
    WSACleanup();                          /* release use of winsock.dll */
#endif
    return(0);
}
```

FIGURE 60.6 Sample C program that illustrates a UDP client.

```
/* * Multicasting listener (server) * */
#ifndef unix
#include <winsock2.h>
#include <ws2tcpip.h>
/* also include Ws2_32.lib library in linking options */
#else
#define closesocket close
#define SOCKET int
#include <sys/types.h>
#include <sys/socket.h>
#include <netinet/in.h>
#include <arpa/inet.h>
#include <netdb.h>
#include <unistd.h>
/* also include xnet library for linking; on command line add: -lxnet */
#endif

#include <stdio.h>
#include <stdlib.h>

#define DEFAULT_IP "224.2.2.2"
#define DEFAULT_PORT 60001
#define MAX_MSG 100

int main(int argc, char *argv[]) {
    SOCKET          sd;
    int             rc, n;
    struct sockaddr_in cliAddr, servAddr;
    int             cliLen;
    struct in_addr  mcastAddr;
    struct ip_mreq  mreq;
    struct hostent *h;
    unsigned short  p;
    int             ttl;
    int             one;
    char            msg[MAX_MSG+1];

#ifdef WIN32
    WSADATA wsaData;
    if(WSAStartup(0x0101, &wsaData)!=0)
    { fprintf(stderr, "Windows Socket Init failed: %d\n", GetLastError()); exit(1); }
#endif

    /* get mcast address to listen to */
    if(argc>3) {
        printf("usage : %s [mcast address=%s] [port number=%d]\n",
          argv[0], DEFAULT_IP, DEFAULT_PORT); exit(0);
    } else if (argc>2) {
        h=gethostbyname(argv[1]); p =(unsigned short)atol(argv[2]);
    } else if (argc>=2) {
        h=gethostbyname(argv[1]);
    } else {
        h=gethostbyname(DEFAULT_IP); p=DEFAULT_PORT;
    }
    if(h==NULL) { printf("%s : unknown group %s\n", argv[0], argv[1]); exit(1); }
    if(p==0)    { printf("%s : invalid port %s\n", argv[0], argv[2]); exit(1); }
    memcpy((char *) &mcastAddr, h->h_addr_list[0], h->h_length);

    /* check given address is multicast */
    if(!IN_MULTICAST(ntohl(mcastAddr.s_addr)))
    { printf("%s : given address '%s' is not multicast\n",
        argv[0], inet_ntoa(mcastAddr)); exit(1); }

    /* create socket */
    sd = socket(AF_INET,SOCK_DGRAM,0);
    if(sd<0) { printf("%s : cannot create socket\n", argv[0]); exit(1); }

    /* allow multiple bind port - multiple sockets listening simultaneously */
    one = 1; setsockopt(sd, SOL_SOCKET, SO_REUSEADDR, (char *)&one, sizeof(one));
    /* bind port */
    servAddr.sin_family=AF_INET;
    servAddr.sin_addr.s_addr=htonl(INADDR_ANY);
    servAddr.sin_port=htons(p);
    if(bind(sd,(struct sockaddr *) &servAddr, sizeof(servAddr))<0)
    { printf("%s : cannot bind port %d\n", argv[0], p); exit(1); }
```

FIGURE 60.7 Sample C program that illustrates a UDP server utilizing multicasting.

(continued)

```
    /* set multicast TTL - range */
    ttl = 1; /* this allows to confine the range of transmission to a network segment */
    rc = setsockopt(sd, IPPROTO_IP, IP_MULTICAST_TTL, (char *)&ttl, sizeof(ttl));
    if (rc<0) { printf("%s : cannot set ttl = %d\n", argv[0], ttl); exit(1); }

    /* join multicast group */
    mreq.imr_multiaddr.s_addr=mcastAddr.s_addr;
    mreq.imr_interface.s_addr=htonl(INADDR_ANY);
    rc = setsockopt(sd, IPPROTO_IP, IP_ADD_MEMBERSHIP, (char *)&mreq, sizeof(mreq));
    if(rc<0) { printf("%s : cannot join multicast group %s\n", argv[0],
inet_ntoa(mcastAddr)); exit(1); }
    printf("%s : listening to mgroup %s:%d\n", argv[0], inet_ntoa(mcastAddr), p);

    /* infinite server loop */
    while(1) {
        cliLen=sizeof(cliAddr);
        n = recvfrom(sd, msg, MAX_MSG, 0, (struct sockaddr *)&cliAddr, &cliLen);
        if(n<0) { printf("%s : cannot receive data\n", argv[0]); continue; }
        msg[n]='\0';

        printf("%s : received from %s:%d: %s\n", argv[0], inet_ntoa(cliAddr.sin_addr),
ntohs(cliAddr.sin_port), msg);
    }/* end of infinite server loop */

    /* leave the group after you are done - actually never happens in this example */
    rc = setsockopt(sd,IPPROTO_IP,IP_DROP_MEMBERSHIP, (char *)&mreq, sizeof(mreq));
    if(rc<0) { printf("%s : cannot leave multicast group %s\n", argv[0],
inet_ntoa(mcastAddr)); exit(1); }

#ifdef WIN32
    WSACleanup();                       /* release use of winsock.dll */
#endif
    return 0;
}
```

FIGURE 60.7 (continued)

other error control than the checksum discussed above. The sender cannot be requested to resend any datagram. However, an upper level protocol that utilizes UDP may implement some kind of control. In that case, unlike in TCP, no data is repeated by resending the same datagram. Instead, the communicating process sends a request to send some information again. Trivial file transfer protocol (TFTP) is a very good example of that situation [STD33]. Since it has its own higher level flow and error control, it can use UDP as a transport layer protocol instead of TCP. Other examples of UDP utilization are simple network management protocol (SNMP) [STD62] or any other protocol that requires only simple short request-response communication.

60.3 Programming Samples

Figures 60.5 and 60.6 contain two C programs that demonstrate use of UDP for communication using client–server paradigm [C02-3]. Socket application programming interface (API) is a de facto standard. Programs use conditional compilation directives that include minor variations in the library headers needed for operation under Unix/Linux and Windows. Function calls should be modified, if desired, to run under other operating systems or with proprietary operating systems for embedded devices. Minor modifications need to be performed when utilizing underlying IPv6 API, as described in [RFC3493]. Figure 60.5 shows the server program that waits for incoming datagram from the client program listed in Figure 60.6. Because UDP is connectionless, the server is capable of servicing data from several clients almost simultaneously. In case a distinction is needed among clients, this must be implemented in the code of the server program. Programs show all steps necessary to set up, use, and tear down a UDP-based exchange of data.

Figures 60.7 and 60.8 contain another pair of C programs that demonstrate the use of UDP for communication using multicasting. Several server programs listed in Figure 60.7 receive one message sent by the client program listed in Figure 60.8. The server must subscribe to the multicast IP address, i.e., "join the multicast group" in order to receive data. In case of a client that only sends multicast data this step may be omitted.

Internet socket is an end point of a bidirectional interprocess communication flow across an IP-based computer network. Interface to the socket library is standardized only on Unix/Linux and Windows. Other operating systems including various embedded systems may employ different functions to carry out the same functionality. Programs of the same functionality implemented in other languages may combine several steps into calling a single function. This is especially true for the socket initialization and connection set up.

```
/* * Multicasting sender (client) * */
#ifndef unix
#include <winsock2.h>
#include <ws2tcpip.h>
/* also include Ws2_32.lib library in linking options */
#else
#define closesocket close
#define SOCKET int
#include <sys/types.h>
#include <sys/socket.h>
#include <netinet/in.h>
#include <arpa/inet.h>
#include <netdb.h>
#include <unistd.h>
/* also include xnet library for linking; on command line add: -lxnet */
#endif

#include <stdio.h>
#include <stdlib.h>

/* #define DEFAULT_IP "224.2.2.2" */
/* #define DEFAULT_PORT 60001 */

int main(int argc, char *argv[]) {

    SOCKET          sd;
    int             rc, i;
    struct sockaddr_in cliAddr, servAddr;
    struct in_addr  mcastAddr;
    struct ip_mreq  mreq;
    struct hostent *h;
    unsigned short  p;
    unsigned long   addr;
    unsigned char   ttl;
    int             one;

#ifdef WIN32
    WSADATA wsaData;
    if(WSAStartup(0x0101, &wsaData)!=0)
    { fprintf(stderr, "Windows Socket Init failed: %d\n", GetLastError()); exit(1); }
#endif

    if(argc<4) { printf("usage %s <maddress> <port> <data1> <data2> ... <dataN>\n",
argv[0]); exit(1); }
    h = gethostbyname(argv[1]);
    if(h==NULL) { printf("%s : unknown host %s\n", argv[0], argv[1]); exit(1); }
    p =(unsigned short)atol(argv[2]);
    if(p==0) { printf("%s : invalid port %s\n", argv[0], argv[2]); exit(1); }
    memcpy((char *) &mcastAddr, h->h_addr_list[0], h->h_length);

    /* check given address is multicast */
    if(!IN_MULTICAST(ntohl(mcastAddr.s_addr))){
        printf("%s : given address '%s' is not multicast\n", argv[0],
               inet_ntoa(mcastAddr)); exit(1); }

    /* create socket */
    sd = socket(AF_INET,SOCK_DGRAM,0);
    if(sd<0) { printf("%s : cannot create socket\n", argv[0]); exit(1); }

    /* enabling multicasting on a network interface is required only if sending data */

    /* allow multiple bind port to allow multiple instances on the same computer */
    one = 1; setsockopt(sd, SOL_SOCKET, SO_REUSEADDR, (char *)&one, sizeof(one));

    /* bind port */
    cliAddr.sin_family=AF_INET;
    cliAddr.sin_addr.s_addr=htonl(INADDR_ANY);
    cliAddr.sin_port=htons(p);
    if(bind(sd,(struct sockaddr *) &cliAddr, sizeof(cliAddr))<0)
    { printf("%s : cannot bind port %d\n", argv[0], p); exit(1); }
```

FIGURE 60.8 Sample C program that illustrates a UDP client utilizing multicasting.

(continued)

```
    /* set multicast TTL - range */
    ttl = 1;
    rc = setsockopt(sd, IPPROTO_IP, IP_MULTICAST_TTL, (char *)&ttl, sizeof(ttl));
    if (rc<0) { printf("%s : cannot set ttl = %d\n", argv[0], ttl); exit(1); }

    /* join multicast group */
    mreq.imr_multiaddr.s_addr=mcastAddr.s_addr;
    mreq.imr_interface.s_addr=htonl(INADDR_ANY);
    rc = setsockopt(sd, IPPROTO_IP, IP_ADD_MEMBERSHIP, (char *)&mreq, sizeof(mreq));
    if(rc<0) {
        printf("%s : cannot join multicast group %s\n",
                argv[0], inet_ntoa(mcastAddr)); exit(1); }

    servAddr.sin_family = h->h_addrtype;
    memcpy((char *) &servAddr.sin_addr.s_addr, h->h_addr, h->h_length);
    servAddr.sin_port = htons(p);

    printf("%s : sending data on multicast group %s (%s) port %d\n", argv[0],
        h->h_name, inet_ntoa(*(struct in_addr *) h->h_addr_list[0]),
        ntohs(servAddr.sin_port) );

    /* send data */
    for(i=3;i<argc;i++) {
        rc = sendto(sd, argv[i], strlen(argv[i])+1, 0,
                    (struct sockaddr *) &servAddr, sizeof(servAddr));

        if (rc<0) {
          printf("%s : cannot send data number %d\n", argv[0], i-2); break;
        } else {
          printf("%s : sent %d bytes in message %d\n", argv[0], strlen(argv[i])+1, i-2);
        }

    }/* end for */

    closesocket(sd);
#ifdef WIN32
    WSACleanup();                           /* release use of winsock.dll */
#endif
    exit(0);
}
```

FIGURE 60.8 (continued)

References

[C02-1] D. Comer, *Internetworking with TCP/IP Vol. 1: Principles, Protocols, and Architecture*, 5th edn., Prentice Hall, Upper Saddle River, NJ, 2005.

[C02-3] D. Comer et al., *Internetworking with TCP/IP*, Vol. III: *Client-Server Programming and Applications, Linux/Posix Sockets Version*, Prentice Hall, Upper Saddle River, NJ, 2000.

[F10] B.A. Forouzan, *TCP/IP Protocol Suite*, 4th edn., McGraw-Hill, New York, 2010.

[GW03] A. Leon-Garcia, I. Widjaja, *Communication Networks*, 2nd edn., McGraw-Hill, New York, 2003.

[IANA] Web Site for *IANA - Internet Assigned Numbers Authority*, http://www.iana.org/

[PD07] L.L. Peterson, B.S. Davie, *Computer Networks: A System Approach*, 4th edn., Morgan Kaufmann Publishers, San Francisco, CA, 2007.

[RFC] Web Site for *RFC Editor*, http://www.rfc-editor.org/

[RFC2147] D. Borman et al., TCP and UDP over IPv6 Jumbograms, May 1997. http://www.rfc-editor.org/

[RFC2675] D. Borman et al., IPv6 Jumbograms, August 1999. http://www.rfc-editor.org/

[RFC3171] Z. Albana, K. Almeroth, D. Meyer, M. Schipper, IANA Guidelines for IPv4 Multicast Address Assignments, August 2001. http://www.rfc-editor.org/

[RFC3493] R. Gilligan et al., Basic Socket Interface Extensions for IPv6, February 2003. http://www.rfc-editor.org/

[STD2] J. Reynolds, J. Postel, Assigned Numbers, October 1994. http://www.rfc-editor.org/

[STD5] J. Postel, Internet Protocol, September 1981. http://www.rfc-editor.org/

[STD6] RFC0768/STD0006, J. Postel, User Datagram Protocol, August 1980. http://www.rfc-editor.org/

[STD33] K. Sollins, The Tftp Protocol (Revision 2), July 1992. http://www.rfc-editor.org/

[STD62] D. Harrington, R. Presuhn, B. Wijnen, An Architecture for Describing Simple Network Management Protocol (SNMP) Management Frameworks, December 2002. http://www.rfc-editor.org/

61

Transmission Control Protocol—TCP

Aleksander Malinowski
Bradley University

Bogdan M. Wilamowski
Auburn University

61.1 Introduction 61-1
61.2 Protocol Operation 61-3
TCP Segment • Port Number Assignments • Connection Establishment • Maintaining the Open Connection • Flow Control and Sliding Window Protocol • Improving Flow Control • Error Control • Congestion Control • Connection Termination
61.3 State Diagram 61-14
61.4 Programming Samples 61-14
References 61-17

61.1 Introduction

Transmission control protocol (TCP) is a part of TCP/Internet protocol (IP) suite [STD7,C02-1,F10, GW03,PD07]. It provides full transport layer services to applications. It belongs to the transport layer in the TCP/IP suite model as shown in Figure 61.1. TCP is more complicated than user datagram protocol (UDP) and provides a reliable connection between both ends of transmission. This connection needs to be set up, maintained, and torn down. The protocol divides a stream of data into units of limited size called segments. These segments are reassembled in a reliable way at the other end. Reliability is achieved by assembling segments by preserving their original sequence, and arranging for retransmission of lost or corrupted data. While the underlying IP provides logical addressing of segments and is responsible for their transmission through networks [STD5], TCP provides additional addressing that allows setting up connections for multiple applications and services on the same end. Each application seeking TCP connection is assigned one or more numbers (one number per connection) that are called port numbers. Figure 61.2 illustrates the concept of port numbers.

TCP provides a logical full duplex connection between two application layer processes. The connection is perceived as a connection-oriented, reliable, in-sequence byte-stream service. The protocol also allows the recipient of data to control the rate at which the sender transmits data to avoid buffer overflow.

- *Connection-oriented* shows that the connection needs to be established and its parameters negotiated between the two processes. This negotiation is performed by sending transmission control blocks (TCB). At the end of the transmission, the connection must be torn down. TCP terminates each direction of the connection separately, which allows the data flow to continue even after the other direction has been closed.
- *Reliable* indicates that either all data are delivered or the connection is broken. TCP is designed to work over the IP and thus it does not assume that the underlying network services are reliable. To ensure reliability, each chunk of data are numbered, and a selective automatic repeat request (ARQ) is employed to request retransmission of lost or corrupt data chunk.

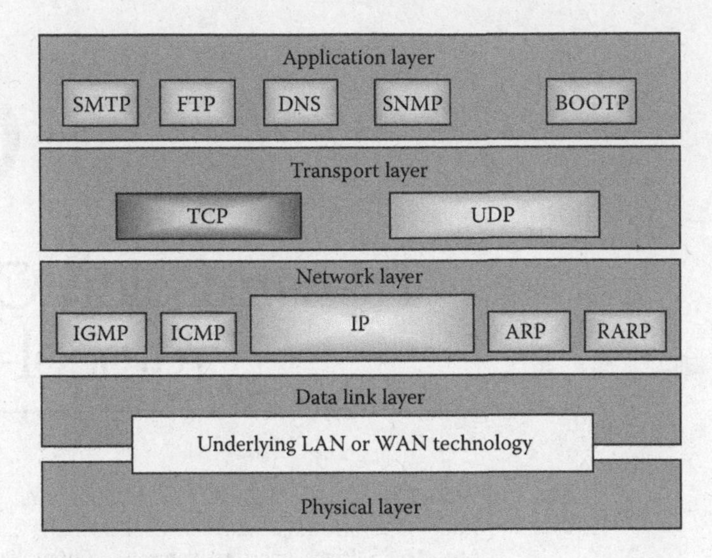

FIGURE 61.1 UDP and TCP/IP in the TCP/IP protocol suite model.

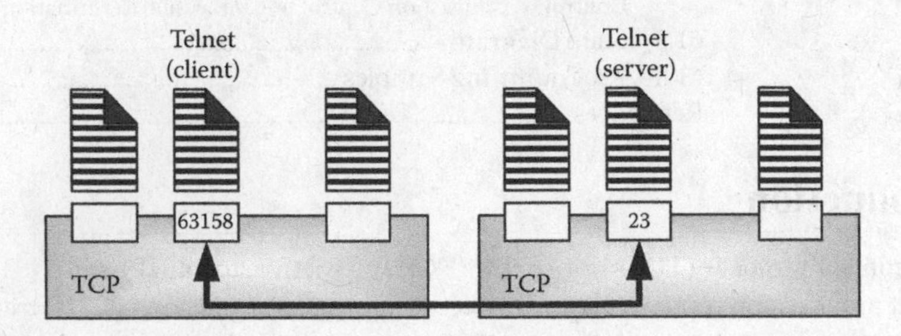

FIGURE 61.2 Process to process communication.

- *In-sequence* shows that all data are guaranteed to be received by the end receiving process in the same order that they were sent. Sequencing is implemented by the same mechanism that is used for missing data chunk detection. All received data are placed in a receiving buffer in location corresponding to the data chunk number. The process that receives data reads it from that buffer in a correct sequence.
- *Byte-stream* oriented implies that the connection can be perceived as continuous stream of data regardless of the fact that the underlying network services may transmit data in chunks. Those data chunks may be of various sizes and may have maximum size limit. The sending process, however, may send data either continuously or in bursts. All data to be sent are placed in the sending buffer. Data from the buffer are partitioned into chunks, which are called segments according to algorithms of TCP, and then they are sent over the network. On the receiving side, the data are placed in the receiving buffer in an appropriate sequence as mentioned earlier. The receiving process may read data either as chunks or as byte after byte.
- *Full duplex* indicates that each party involved in the transmission has control over it. For example, any side can send data at any time or terminate the connection.

The TCP is standardized by *Internet Society*, which is an independent international nonprofit organization. A complete information about TCP-related standards is published by *RFC Editor* [RFC]. Several relevant RFC articles as well as further reading materials are listed at the end of this article.

61.2 Protocol Operation

Before data transfer begins, TCP establishes a connection between two application processes by setting up the protocol parameters. These parameters are stored in a structure variable called TCB. Once the connection is established, the data transfer may begin. During the transfer, parameters may be modified to improve efficiency of transmission and prevent congestion. In the end, the connection is closed. Under certain circumstances, connection may be terminated without formal closing.

61.2.1 TCP Segment

Although TCP is byte-stream oriented, connection and buffering take care of the disparities among the speed of preparing data, sending data, and consuming of data at the other end, one more step is necessary before data are sent. The underneath IP layer that takes care of sending data through the entire network sends data in packets, not as a stream of bytes. Sending one byte in a packet would be very inefficient due to the large overhead caused by headers. Therefore, data are grouped into larger chunks called segments. Each segment is encapsulated in an IP datagram and then transmitted. Each TCP segment follows the same format that is shown in Figure 61.3 [STD7]. Each segment consists of a header that is followed by options and data. The meaning of each header field will be discussed now, and its importance is illustrated in the following sections, which show phases of a typical TCP connection.

- *Source port address*: This is a 16 bit field that contains a port number of the process that sends options or data in this segment.
- *Destination port address*: This is a 16 bit field that contains a port number of the process that is supposed to receive options or data carried by this segment.

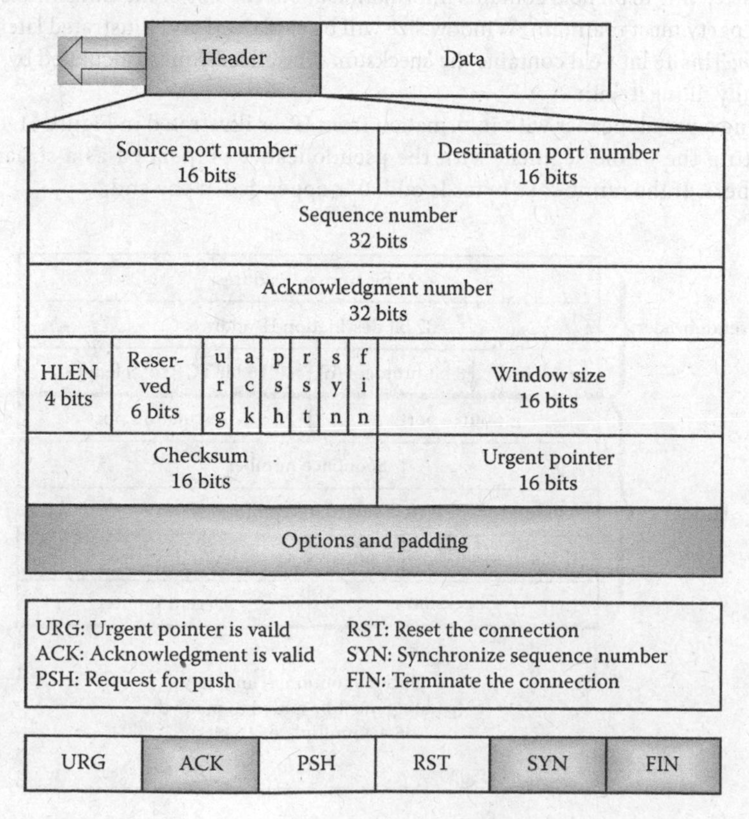

FIGURE 61.3 TCP segment format.

- *Sequence number*: This is a 32 bit field that contains the number assigned to the first byte of data carried by this segment. During the establishment of connection each side generates its own random initial sequence number (ISN). The sequence number is then increased by one with every 1 byte of data sent. For example, if a previous segment had a sequence number 700 and had 200 bytes of data then sequence number of the next segment is 900.
- *Acknowledgment number*: This is a 32 bit field that contains the sequence number for the next segment that is expected to be received from the other party. For example, if the last received segment had a sequence number 700 and had 200 bytes of data, then the acknowledgment number is 900.
- *Header length*: This 4 bit field contains the total size of the segment header divided by 4. Since headers may be 20–60 bytes long, the value is in the range 5–15.
- *Reserved*: This 6 bit field is reserved for future use.
- *Control*: This 6 bit field contains transmission control flags.
- *URG*: Urgent. If the bit is set, then the urgent pointer is valid.
- *ACK*: Acknowledgment. If the bit is set then the acknowledgment number is valid.
- *PSH*: Push. If the bit is set, then the other party should not buffer the outgoing data but send it immediately, as it becomes available, without buffering.
- *RST*: Reset. If the bit is set, then the connection should be aborted immediately due to abnormal condition.
- *SYN*: Synchronize. The bit is set in the connection request segment when the connection is being established.
- *FIN*: Finalize. If the bit is set, it means that the sender does not have anything more to send, and that the connection may be closed. The process will still accept data from the other party until the other party also sends FIN.
- *Window size*: This 16 bit field contains information about the size of the transmission window that the other party must maintain. Window size will be explained and illustrated later.
- *Checksum*: This 16 bit field contains the checksum. The checksum is calculated by
  - Initially filling it with 0s
  - Adding a pseudoheader with information from IP, as illustrated in Figure 61.4.
  - Treating the whole segment with the pseudoheader prepended as a stream of 16 bit numbers. If the number of bytes is odd, 0 is appended at the end.

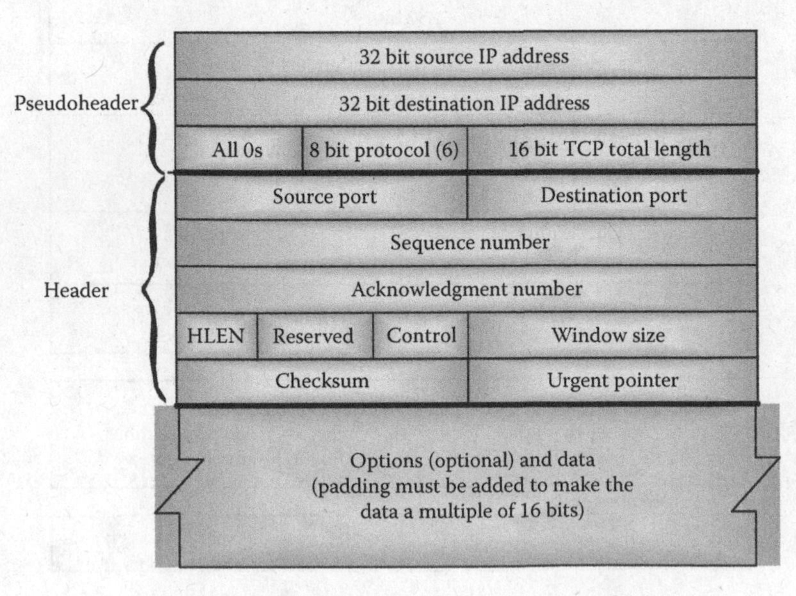

FIGURE 61.4 Pseudoheader added to the TCP datagram.

- Adding all 16 bit numbers using 1s complement binary arithmetic.
- Complementing the result. This complemented result is inserted into the checksum field.

- *Checksum verification*: The receiver calculates the new checksum for the received packet, which includes the original checksum, after adding the pseudoheader (see Figure 61.4). If the new checksum is nonzero then the packet is corrupt and is discarded.
- *Urgent pointer*: This 16 bit field is used only if the URG flag is set. It defines the location of the last urgent byte in the data section of the segment. Urgent data are those that are sent out-of-sequence. For example, telnet client sends a break character without queuing it in the sending buffer.
- *Options*: There can be up to 40 bytes of options appended to the header. The options are padded inside with 0s so that the total number of bytes in the header is divisible by 4. The possible options are listed below and are illustrated in Figure 61.5:

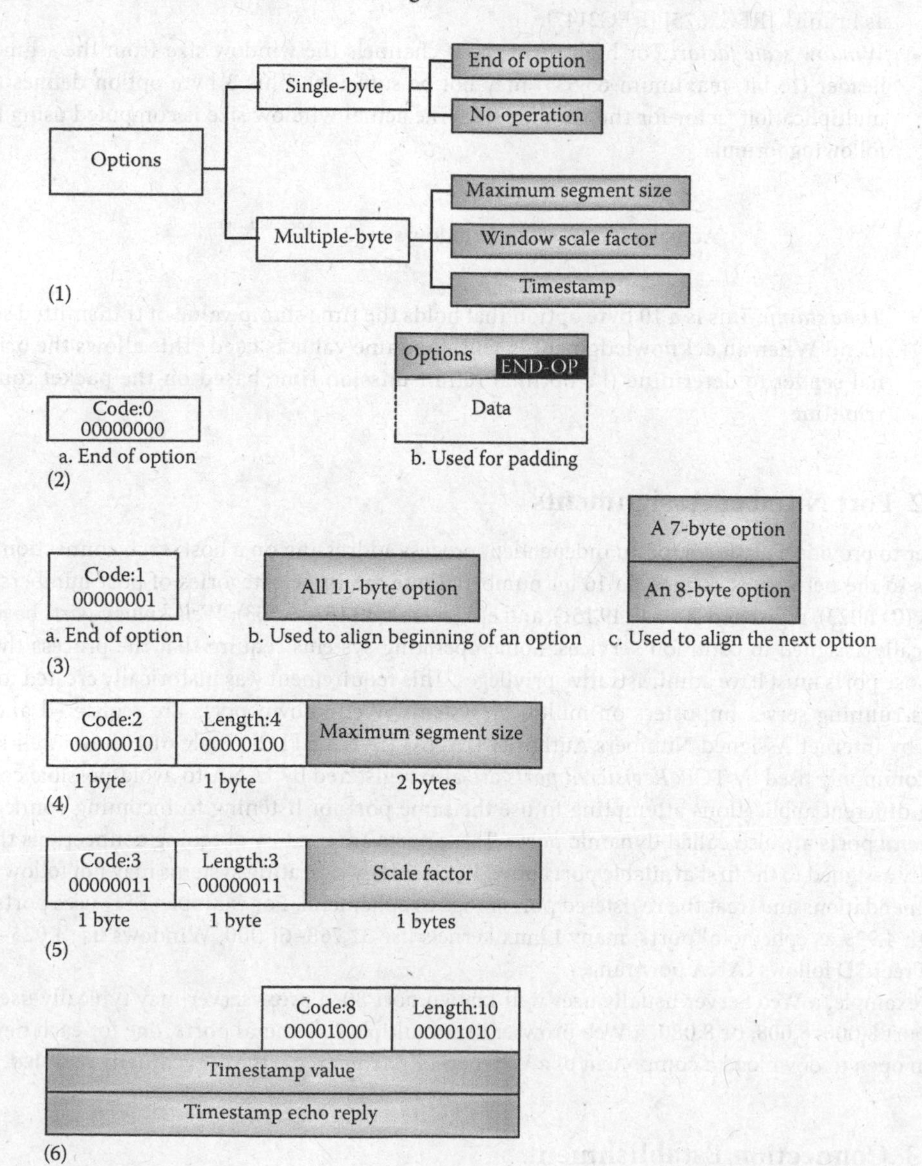

FIGURE 61.5 TCP options: 1. classification; 2. end of options; 3. no operation (used for padding); 4. setting maximum segment size; 5. setting window scale factor; 6. time stamp.

- *No operation*: 1 byte option used as filler between the options and as padding.
- *End of options*: 1 byte option used as padding at the end of options filed. Data portion of the segment follows immediately.
- *Maximum segment size (MSS)*: Option defines the maximum size of a chunk of data that can be received. It refers to the size of data not the whole segment. It is used only during the establishment of the connection. The receiving party defines the maximum size of the datagram sent by the sender. If not specified, the default maximum is 536. Because the length of the datagram is not specified in the TCP header, there is no length limitation for an individual TCP packet. However, the MSS value that is negotiated at the beginning of the connection limits the largest TCP packet that can be sent, and the Urgent Pointer cannot reference data beyond 65,535 bytes. MSS value of 65,535 is treated as infinity [RFC2675] [RFC2147].
- *Window scale factor*: For high throughput channels the window size from the segment header (16 bit, maximum 65,535) may not be sufficient. This 3 byte option defines the multiplication factor for the window size. The actual window size is computed using the following formula

$$\text{Actual window size} = \text{Window size} \cdot 2^{\text{window scale factor}}$$

- *Time stamp*: This is a 10 byte option that holds the time stamp value of transmitted segment. When an acknowledgment is sent, the same value is used. This allows the original sender to determine the optimal retransmission time based on the packet round trip time.

61.2.2 Port Number Assignments

In order to provide a platform for an independent process addressing on a host, each connection of the process to the network is assigned a 16 bit number. There are three categories of port numbers: well-known (0–1,023), registered (1,024–49,151), and ephemeral (49,152–6,553). *Well-known ports* have been historically assigned to common services. Some operating systems require that the process that utilizes those ports must have administrative privileges. This requirement was historically created to avoid hackers running server imposters on multiuser systems. Well-known ports are registered and controlled by Internet Assigned Numbers Authority (IANA) [IANA,STD2]. Table 61.1 shows well-known ports commonly used by TCP. *Registered ports* are also registered by IANA to avoid possible conflicts among different applications attempting to use the same port for listening to incoming connections. *Ephemeral* ports are also called dynamic ports. These ports are used by outgoing connections that are typically assigned to the first available port above 49,151. Many operating systems may not follow IANA recommendations and treat the registered ports range as ephemeral. For example, BSD uses ports 1,024 through 4,999 as ephemeral ports, many Linux kernels use 32,768–61,000, Windows use 1,025–5,000, while FreeBSD follows IANA port range.

For example, a Web Server usually uses well-known port 80, a proxyserver may typically use registered port 8,000, 8,008, or 8,080, a Web browser uses multiple ephemeral ports, one for each new connection open to download a component of a Web page. Ephemeral ports are eventually recycled.

61.2.3 Connection Establishment

Before data transfer begins, TCP establishes a connection between two application processes by setting up the protocol parameters. These parameters are stored in a structure variable called TCB.

TABLE 61.1 Common Well-Known Ports Used with TCP

Port	Protocol	Description	RFC/STD #
7	Echo	Echoes a received data back to its sender	STD20
9	Discard	Discards any data that are received	STD21
13	Daytime	Returns the date and the time in ASCII string format	STD25
19	Chargen	Sends back arbitrary characters until connection closed	STD22
20	FTP Data	File Transfer Protocol (data transmission connection, unencrypted)	STD9
21	FTP Control	File Transfer protocol (control connection, unencrypted)	STD9
22	SSH	Secure Shell (SSH)—used for secure logins, file transfers (scp, sftp), and port forwarding	RFC4250-4256
23	Telnet	Telnet (unencrypted)	STD8, STD27-32
25	SMTP	Simple Mail Transfer Protocol (sending email)	STD10
37	Time	Returns the current time in seconds since 1/1/1900 in a 32 bit format	STD26
53	Nameserver	Domain Name Service	STD13
80	HTTP	Hypertext Transfer Protocol	RFC2616
110	POP3	Post Office Protocol 3 (email download)	STD53
111	RPC	SUN Remote Procedure Call	RFC5531
143	IMAP	Interim Mail Access Protocol v2 (email access)	RFC2060
443	SHTTP	Secure Hypertext Transfer Protocol	RFC2660
995	POP3	POP3 over TLS/SSL	RFC2595

Before the two parties—called here as host A and host B—can start sending data, four actions must be performed:

1. Host A sends a segment announcing that it wants to establish connection. This segment includes initial information about traffic from A to B.
2. Host B sends a segment acknowledging the request received from A.
3. Host B sends a segment with information about traffic from B to A.
4. Host A sends a segment acknowledging the request received from B.

The second and the third steps can be combined into one segment. Therefore, the connection establishment described is also called three-way handshaking. Figure 61.6 illustrates an example of segment

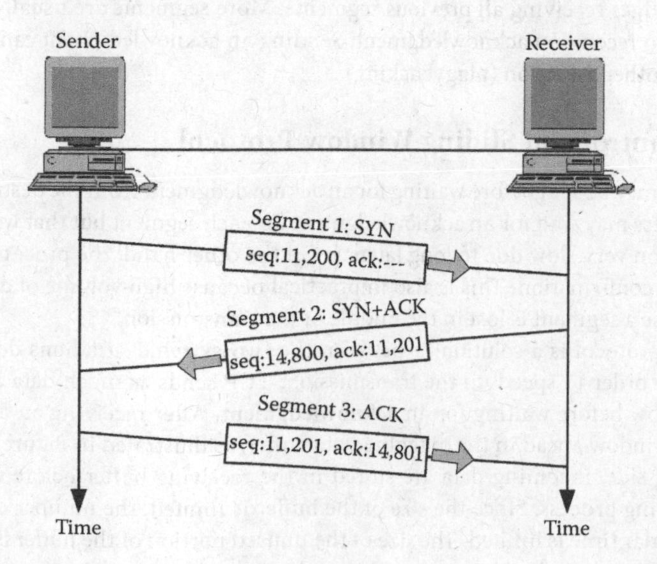

FIGURE 61.6 Three-way handshaking during connection establishment.

exchange during connection establishment. Among the two programs that are involved in the connection, the server program is the one that passively waits for another party to connect. It will be called shortly as the server. It tells the underlying TCP library in its system that it is ready to accept a connection. The program that actively seeks connection is called a client program. It will be called shortly as the client. The client tells the underlying library in its system that it needs to connect to a particular server that is defined by the computer IP address and the destination port number.

1. The client sends the first segment that includes the source and destination port with SYN bit set. The segment also contains the initial value of Sequence Number, which is also called ISN that will be used by the client in this connection. The client may also send some options as necessary, for example, to alter the window size, or maximum segment size in order to change their values from default.
2. The server sends the second segment that has both ACK and SYN bit sets. This is done in order to both acknowledge receiving the first segment from the client (ACK) and to initialize the sequence number in the direction from the server to the client (SYN). The Acknowledgment number is set to the client's ISN + 1. The server ISN is set to a random number. The option to define client window size must also be sent.
3. In reply to the segment from the server, the client sends as segment with ACK bit set and the segment number set to its ISN + 1. The Acknowledgment number is set to the server ISN + 1.
4. In very rare situation, the two processes may want to start connection at the same time. In that case, both will issue an active open segment with SYN and ISN, and then both would send the ACK segment. A situation when one of the segments is lost during the transmission and needs to be resent after time-out may also be analyzed.

61.2.4 Maintaining the Open Connection

After connection is established, each process can send data. Each segment has its sequence number increased by the number of bytes sent in the previous segment, with the previous sequence number. The number rolls over at 2^{32}. The process expects to receive segments with ACK bit set and acknowledgment numbers that confirm that data are received. Acknowledgment number is not merely the repetition of the received sequence number. It indicates the next expected sequence number of a segment to be received by the acknowledging process. Acknowledgments must be received within certain time frame, not every segment that is send needs to be acknowledged. Sending acknowledgment of a certain segment also acknowledges receiving all previous segments. More segments are usually sent ahead before the sender expects to receive an acknowledgment. Sending an acknowledgment can be combined with sending data in the other direction (piggybacking).

61.2.5 Flow Control and Sliding Window Protocol

The amount of data may be sent before waiting for an acknowledgment from the destination. In extreme situations, the process may wait for an acknowledgment for each segment but that would make the long distance transmission very slow due to long latency. On the other hand, the process may send all data without waiting for confirmation. This is also impractical because high volume of data may need to be retransmitted in case a segment is lost in the middle of the transmission.

Sliding window protocol is a solution in-between the two extreme situations described above that TCP implements in order to speed up the transmission. TCP sends as much data as allowed by a so-called sliding window before waiting for an acknowledgment. After receiving an acknowledgment, it moves (slides) the window ahead in the outgoing data buffer, as illustrated in Figure 61.7.

On the receiving side, incoming data are stored in the receiving buffer before data are consumed by the communicating process. Since the size of the buffer is limited, the number of bytes that can be received at a particular time is limited. The size of the unused portion of the buffer is called the *receiver window*. In order to maintain the flow control, the sender must be prevented from sending the excessive

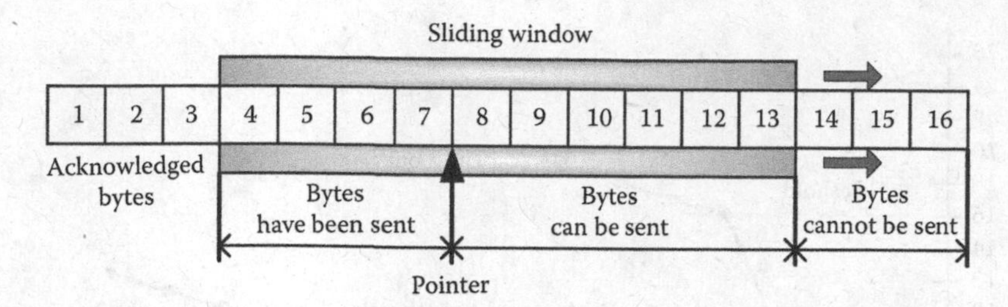

FIGURE 61.7 Sender buffer and sliding window protocol.

amount of data and overflowing the incoming data buffer. This is implemented by controlling the timing when the acknowledgments are sent.

If the receiving process consumes data at a faster rate than they arrive, then the receiver window expands. This information is passed to the sender as an option included in a segment so that the sender may synchronize the size if its *sender window*. This allows more data to be sent without receiving an acknowledgment. If the receiving process consumes data slower than they arrive, then the receiver buffer shrinks. In that case, the receiver must inform the sender to shrink the sender window as well. This causes transmission to wait for more frequent acknowledgments and prevents data from being sent without possibility of storing. If the receiving buffer is full, then the receiver may close (set to zero) the receiver window. The sender cannot send any more data until the receiver reopens the window.

61.2.6 Improving Flow Control

In case when one or both communicating processes process the data slowly, then a problem leading to significant bandwidth waste may arise. In case of slow transmission caused by the receiving processes, the sliding window may be reduced even as low as to 1 byte. In case of slow transmission caused by the sender, data may be sent as soon as it is produced even with a large window, one byte at a time, as well. This causes use of segments that carry small data payload, and lowers the ration of carried data size to the total segment size, and to the total packet size in the underlying protocols. Extensive overview of these and other problems that may develop is provided by [RFC2525] and [RFC2757]. Due to size limitation of this article only two most commonly known problems are discussed and their classic solutions are provided here.

To prevent this situation, data should not be sent immediately as and when it becomes available, but combined in larger chunks, provided that this approach is suitable for a particular application. Nagle found a very good and simple solution for the sender, which is called *Nagle's Algorithm* [RFC896]. The algorithm can be described by the following three steps:

- The first piece of data are sent immediately even if it is only 1 byte.
- After sending the first segment, the sender waits for the acknowledgment segment or until enough data are accumulated to fill in the segment of the maximum segment size, whichever happens first. Then, it sends the next segment with data.
- The previous step is repeated until the end of transmission.

On the receiving side, two different solutions are frequently used. Clark proposed to send the acknowledgment as soon as the incoming segment arrives, but to set the window to zero if the receiving buffer is full (Figure 61.8b) [RFC813]. Then, to increase the buffer only after a maximum size of a segment can be received or half of the buffer becomes empty. The other solution is to delay the acknowledgment and thus prevent the sender from sending more segments with data. The latter of the two solutions creates less network traffic but it may cause unnecessary retransmission of segment, if the delay is too long.

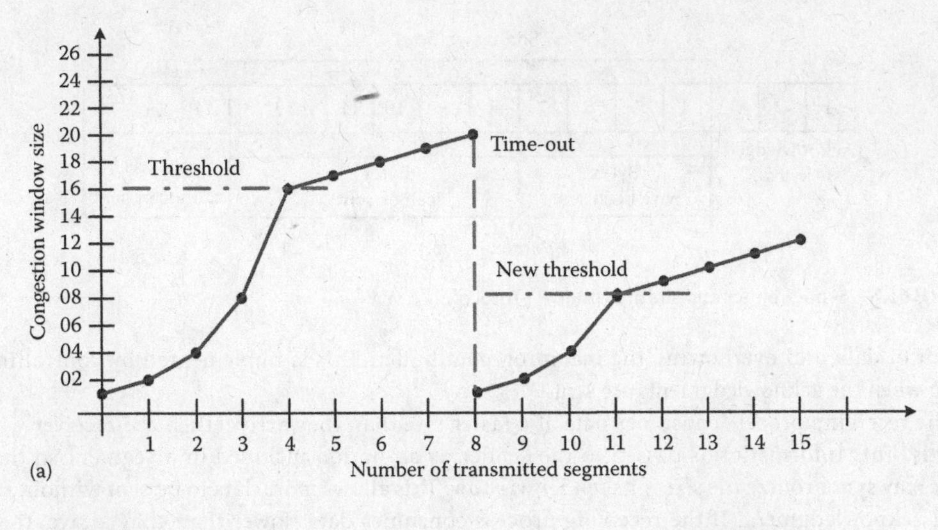

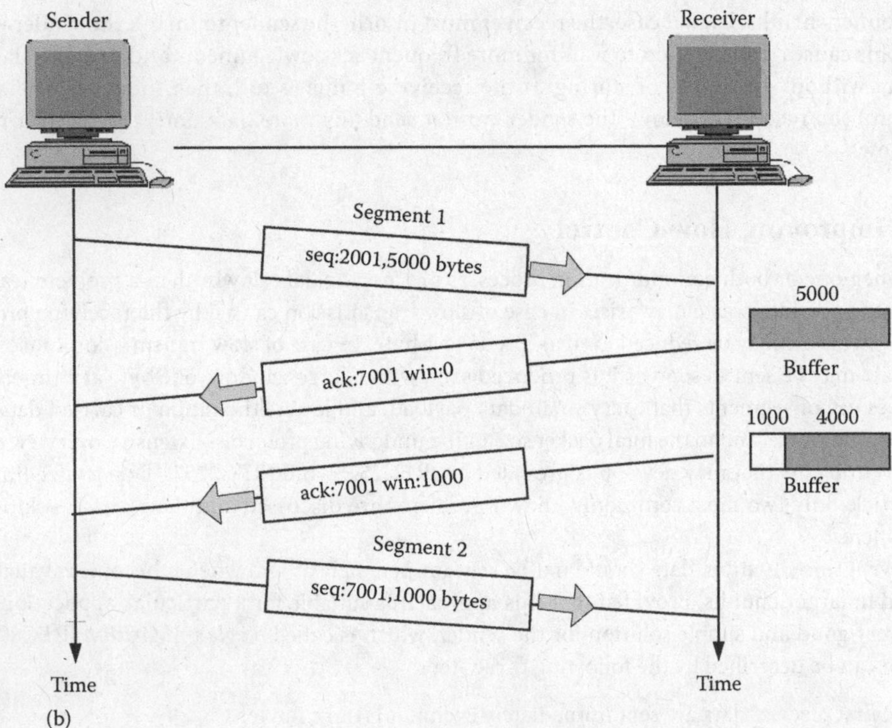

FIGURE 61.8 Congestion Control Window Size. Sender side (a) (upper/left), receiver side control (b) (lower/right).

61.2.7 Error Control

TCP as a reliable protocol that uses unreliable underlying IP must implement error control. This error control includes detecting lost segments, out-of-order segments, corrupted segments, and duplicated segments. After detection, the error must be corrected. The error detection is implemented by using three features: segment checksum, segment acknowledgment, and time-out. Segments with *incorrect checksum* (Figure 61.9a) are discarded. Since TCP uses only positive acknowledgment, no negative acknowledgment can be sent, and both the corrupted segment error and the *lost segment* (Figure 61.9b) error are detected by time-out for receiving an acknowledgment by the sender.

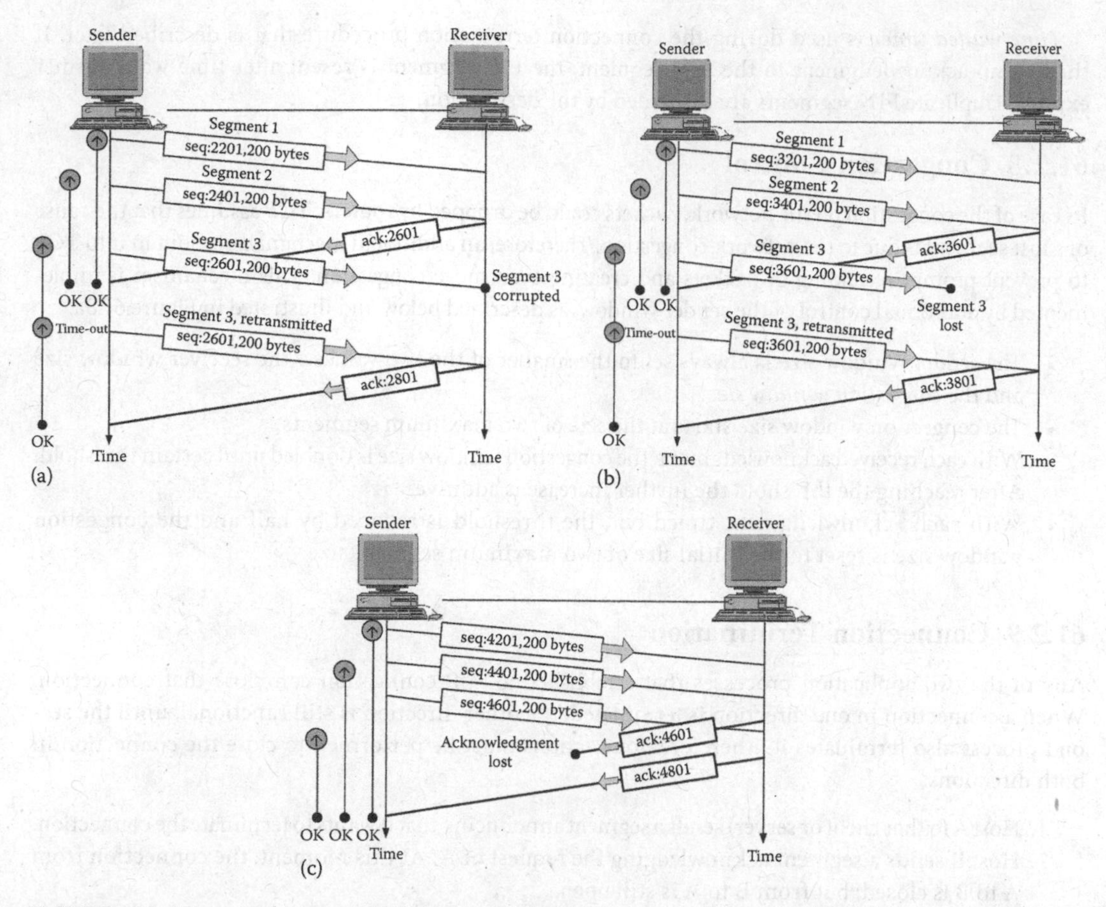

FIGURE 61.9 Corrupted segment (a), lost segment (b), and lost acknowledgment (c).

The error correction is implemented by establishing a *time-out counter* for each segment sent. If the acknowledgment is received, then the particular and all preceding counters and the segments are disposed. If the counter reaches time-out, then the segment is resent and the timer is reset.

A *duplicate segment* can be created if the acknowledgment segment is lost or is delayed too much. Since data are reassembled into a byte-stream at the destination, data from the duplicate segment is already in the buffer. The duplicate segment is discarded. No acknowledgment is sent.

An *out-of-order segment* is stored but not acknowledged before all segments that precede it are received. This may cause resending the segments. The duplicates are discarded.

In case of *lost acknowledgment* (Figure 61.9c), the error is corrected when the acknowledgment for the next segment is sent, since an acknowledgment to the next segment also acknowledges all previous segments. Therefore, a lost acknowledgment may not be noticed at all. This could generate a deadlock in case if the acknowledgment is sent on the last segment and for a time when there is no more segments to send. This deadlock is corrected by a sender that uses a so-called *persistence timer.* If no acknowledgment comes within a certain time frame, then the sender sends a 1 byte segment called probe and resets the persistence counter. With each attempt, the persistence timer time-out is doubled until 60 s time is reached. After receiving an acknowledgment, the persistence timer time-out is reset to the initial value.

Keep-alive timer is used to prevent a long idle connection. If no segments are exchanged for a given period of time, usually 2 h, the server send a probe. If there is no response after 10 probes the connection is reset and thus terminated.

Time-waited timer is used during the connection termination procedure that is described later. If there is no acknowledgment to the FIN segment, the FIN segment is resent after time-waited timer expires. Duplicate FIN segments are discarded by the destination.

61.2.8 Congestion Control

In case of the congestion on the network, packets could be dropped by routers. TCP assumes that the cause of a lost segment is due to the network congestion. Therefore, an additional mechanism is built in into TCP to prevent prompt resending of packets and creating even more congestion. This mechanism is implemented by additional control of the sender window, as described below and illustrated in Figure 61.8a.

- The sender window size is always set to the smaller of the two values—the receiver window size and the *congestion window size.*
- The congestion window size starts at the size of two maximum segments.
- With each received acknowledgment, the congestion window size is doubled until certain threshold. After reaching the threshold the further increase is additive.
- With each acknowledgment timed out, the threshold is reduced by half and the congestion window size is reset to the initial size of two maximum segment size.

61.2.9 Connection Termination

Any of the two application processes that established a TCP connection can close that connection. When a connection in one direction is terminated, the other direction is still functional, until the second process also terminates it. Therefore, four actions must be performed to close the connection in both directions:

1. Host A (either client or server) sends a segment announcing that it wants to terminate the connection.
2. Host B sends a segment acknowledging the request of A. At this moment, the connection from A to B is closed, but from B to A is still open.
3. When host B is ready to terminate the connection, for example, after sending the remaining data, it sends a segment announcing that it wants to terminate the connection as well.
4. Host A sends a segment acknowledging the request of B. At this moment, the connection in both directions is closed.

Unlike in case of opening the connection, none of the steps can be combined into one segment. Therefore, the connection termination described is also called *four-way handshaking.* Figure 61.10 illustrates an example of segment exchange during connection termination. Because it does not matter which host is a server and which is a client, they are denoted only as A and B.

1. Host A sends a segment with FIN bit set.
2. Host B sends a segment with ACK bit set to confirm receiving the FIN segment from host A.
3. Host B may continue to send data and expect to receive acknowledgment messages with no data from host A. Eventually, when there is nothing more to send, it sends a segment with FIN bit set.
4. Host A sends the last segment in the transmission with ACK bit set to confirm receiving the FIN segment from host B.

The connection termination procedure described above is so-called as graceful termination. In certain cases, the connection may be terminated abruptly by resetting. This may happen in the following cases:

- The client process requested a connection to a nonexistent port. The TCP library, on the other side, may send a segment with the RST bit set to indicate connection refusal.

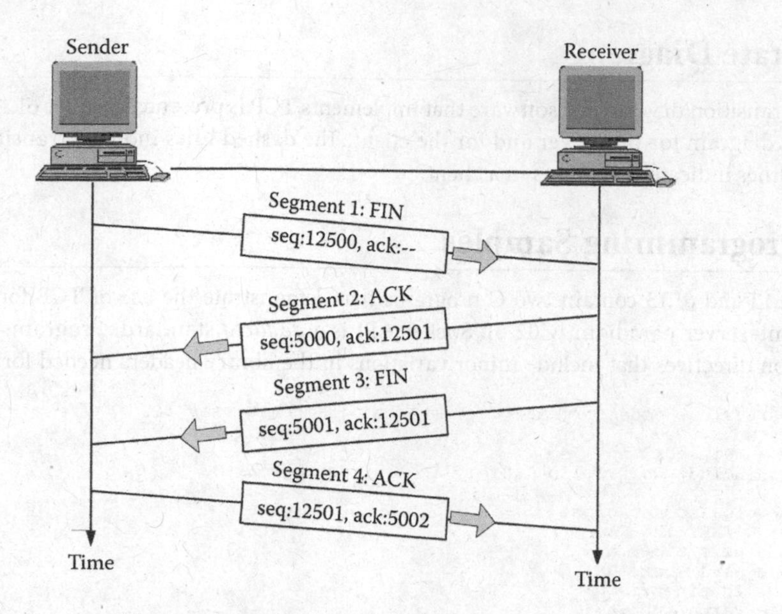

FIGURE 61.10 Four-way handshaking during connection termination.

- One of the processes wants to abandon the connection because of an abnormal situation. It can send the RST segment to destroy the connection.
- One of the processes determines that the other side is idle for a long time. The other system is considered idle if it does not send any data, any acknowledgments, or keep-alive segments. The awaiting process may send the RST segment to destroy the connection.

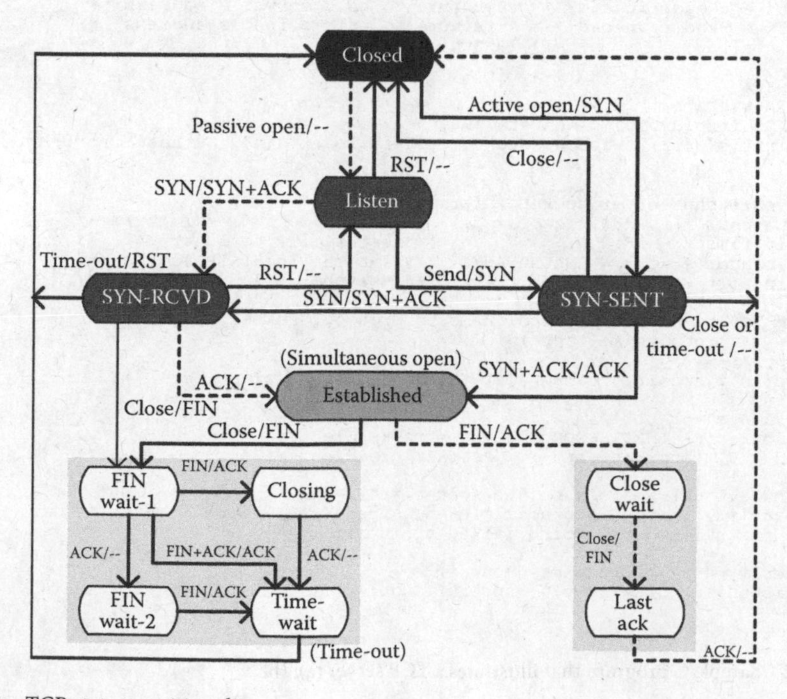

FIGURE 61.11 TCP state transition diagram.

61.3 State Diagram

The state transition diagram for software that implements TCP is presented. Figure 61.11 shows the state transition diagram for the server and for the client. The dashed lines indicate transitions in a server. The solid lines indicate transitions in a client.

61.4 Programming Samples

Figures 61.12 and 61.13 contain two C programs that demonstrate the use of TCP for communication using client–server paradigm [C02-3]. Socket API is a *de facto* standard. Programs use conditional compilation directives that include minor variations in the library headers needed for operation under

```
/* * server.c - code for example server program that uses TCP * */
#ifndef unix
#include <winsock2.h>
/* also include Ws2_32.lib library in linking options */
#else
#define closesocket close
#define SOCKET int
#include <sys/types.h>
#include <sys/socket.h>
#include <netinet/in.h>
#include <arpa/inet.h>
#include <netdb.h>
#include <unistd.h>
/* also include xnet library for linking; on command line add: -lxnet */
#endif

#include <stdio.h>
#include <string.h>

#define PORT 1200                /* server TCP port number         */
#define QLEN 3                   /* size of connection request queue  */

int main()
{
    SOCKET  sd, sd2;          /* socket descriptors - (integers)      */
    struct  protoent *ptrp;   /* pointer to a protocol table entry    */
    struct  sockaddr_in sad;  /* structure to hold server's address   */
    struct  sockaddr_in cad;  /* structure to hold client's address   */
    int     alen;             /* length of address                    */

#ifdef WIN32
    WSADATA wsaData;
    if(WSAStartup(0x0101, &wsaData)!=0)
    { fprintf(stderr, "Windows Socket Init failed: %d\n", GetLastError()); exit(1); }
#endif

    /* prepare the server socket information */
    memset((char *)&sad,0,sizeof(sad)); /* clear sockaddr structure */
    sad.sin_family = AF_INET;           /* set family to Internet    */
    sad.sin_addr.s_addr = INADDR_ANY;   /* set the local IP address */
    sad.sin_port = htons((u_short)PORT);/* use the defined port num */

    /* Map TCP transport protocol name to protocol number */
    ptrp = getprotobyname("tcp");
    if ( ptrp == 0) {
        fprintf(stderr, "cannot map \"tcp\" to protocol number\n"); exit(1); }

    /* Create a socket */
    sd = socket(PF_INET, SOCK_STREAM, ptrp->p_proto);
    if (sd < 0) { fprintf(stderr, "socket creation failed\n"); exit(1); }

    /* Bind a local address to the socket */
    if (bind(sd, (struct sockaddr *)&sad, sizeof(sad)) < 0)
        { fprintf(stderr, "bind failed\n"); exit(1); }

    /* Specify size of request queue */
    if (listen(sd, QLEN) < 0) { fprintf(stderr, "listen failed\n"); exit(1); }
```

(a)

FIGURE 61.12 Sample C program that illustrates a TCP server (a), (b).

```
    { /* Run the sample server application */
        char buf[1000];         /* buffer for string the server sends  */
        char name[80];          /* small buffer for received user data */
        int  visits = 0;        /* counts client connections           */

        while (1) { /* Main server loop - accept and handle requests */
            alen = sizeof(cad);
            sd2=accept(sd, (struct sockaddr *)&cad, &alen);
            if ( sd2<0) { fprintf(stderr, "accept failed\n"); continue; }

            /* This block could be run in a separate thread
               to allow for servicing concurrent clients */
            sprintf(buf,"Welcome to the server.\r\nWhat is your name please?\r\n");
            send(sd2,buf,strlen(buf),0);

            { /* collect one line of data from the user */
                int len=0; char bch;
                while (1) {
                    if (recv(sd2,&bch,1,0) < 0) break; /* client disconnected */
                    else if (bch=='\n') break;       /* done, received a line of text */
                    else if (bch=='\r')              /* ignore \r */ ;
                    else { name[len]=bch; len++; }   /* received another cahracter */
                    if (len+1>=sizeof(name)) break; // exit on exhausted data buffer
                }
                name[len]='\0';
            }
            fprintf(stdout, "%s has just contacted us.\n", name);

            visits++;
            sprintf(buf,"Nice to meet you %s.\r\nYou are the %dth connection today\r\n",
                name, visits);
            send(sd2,buf,strlen(buf),0);
            closesocket(sd2);
            /* End of the block that could be run in a separate thread */
        }
    }
#ifdef WIN32
    WSACleanup();                           /* release use of winsock.dll */
#endif
    return(0);
}
```

(b)

FIGURE 61.12 (continued)

Unix/Linux and Windows. Function calls should be modified if desired to run under other operating systems or with proprietary operating systems for embedded devices. Minor modifications need to be performed when utilizing underlying IPv6 API, as described in [RFC3493]. Figure 61.12 shows the server program that waits for the incoming connection from the client program listed in Figure 61.13. To allow maximum compatibility, the sample server does not use multithreading when servicing client connections, thus allowing only one connection at a time. However, such modification can be easily made by relocating code for servicing a client (marked by comments) into a function run in a separate thread, each time a new client is connected. Programs show all steps necessary to set-up, use, and tear down a TCP connection.

Internet socket is an end point of a bidirectional interprocess communication flow across an IP-based computer network. Interface to the socket library is standardized only on Unix/Linux and Windows. Other operating systems, including various embedded systems, may employ different functions to carry out the same functionality. Programs of the same functionality implemented in other languages may combine several steps into calling a single function. This is especially true for the socket initialization and connection set up.

```
/* * code for example client program that uses TCP * */
#ifndef unix
#include <winsock2.h>
/* also include Ws2_32.lib library in linking options */
#else
#define closesocket close
#define SOCKET int
#include <sys/types.h>
#include <sys/socket.h>
#include <netinet/in.h>
#include <arpa/inet.h>
#include <netdb.h>
#include <unistd.h>
/* also include xnet library for linking; on command line add: -lxnet */
#endif

#include <stdio.h>
#include <string.h>

#define PORT 1200              /* server TCP port number              */
char    hostname[] = "localhost";    /* server host name              */

int main()
{
    SOCKET  sd;                 /* socket descriptor - (integer)       */
    struct  hostent  *ptrh;     /* pointer to a host table entry       */
    struct  protoent *ptrp;     /* pointer to a protocol table entry   */
    struct  sockaddr_in sad;    /* structure to hold an IP address     */
    char    host[256];          /* pointer to host name                */

#ifdef WIN32
    WSADATA wsaData;
    if(WSAStartup(0x0101, &wsaData)!=0)
    { fprintf(stderr, "Windows Socket Init failed: %d\n", GetLastError()); exit(1); }
#endif

    /* prepare the client socket information */
    memset((char *)&sad,0,sizeof(sad)); /* clear sockaddr structure */
    sad.sin_family = AF_INET;          /* set family to Internet     */
    strcpy(host,hostname);

    /* Convert host name to equivalent IP address and copy to sad. */
    ptrh = gethostbyname(host);
    if ( ((char *)ptrh) == NULL )
        { fprintf(stderr, "invalid host: %s\n", host); exit(1); }
    memcpy(&sad.sin_addr, ptrh->h_addr, ptrh->h_length);
    sad.sin_port = htons((u_short)PORT); /* server port number on remote host */

    /* Map TCP transport protocol name to protocol number. */
    ptrp = getprotobyname("tcp");
    if ( ptrp == 0) { fprintf(stderr, "cannot map \"tcp\" to number\n"); exit(1); }

    /* Create a socket. */
    sd = socket(PF_INET, SOCK_STREAM, ptrp->p_proto);
    if (sd < 0) { fprintf(stderr, "soclet creation failed\n"); exit(1); }

    /* Connect the socket to the specified server. */
    if (connect(sd, (struct sockaddr *)&sad, sizeof(sad)) < 0)
        { fprintf(stderr, "connection failed\n"); exit(1); }

    /* Note: May want/need to read the initial greeting sent to us by the server */
    {   char buf[1000];          /* buffer for data sent to the server  */
        sprintf(buf,"Automated Client\r\n"); send(sd, buf, strlen(buf),0);
    }

    {   int  n;                   /* number of characters read          */
        char buf[1000];           /* buffer for data from the server    */
        /* Repeatedly read data from socket and write to user's screen. */
        n = recv(sd, buf, sizeof(buf), 0);
        while (n > 0) { fwrite(buf, n, 1, stdout); n = recv(sd, buf, sizeof(buf), 0); }
  /* Note: data may come in chunks of different size than chunks sent by the server */
    }
    closesocket(sd);            /* close as soon as no longer needed     */

#ifdef WIN32
    WSACleanup();                          /* release use of winsock.dll */
#endif
    return(0);
}
```

FIGURE 61.13 Sample C program that illustrates a TCP client.

References

[C02–1] D. Comer, *Internetworking with TCP/IP*, Vol. 1: *Principles, Protocols, and Architecture*, 5th edn., Prentice Hall, Upper Saddle River, NJ, 2005.

[C02-3] D. Comer et al., *Internetworking with TCP/IP*, Vol. III: *Client-Server Programming and Applications, Linux/Posix Sockets Version*, Prentice Hall, Upper Saddle River, NJ, 2000.

[F10] B.A. Forouzan, *TCP/IP Protocol Suite*, 4th edn., McGraw-Hill, New York, 2010.

[GW03] A. Leon-Garcia, I. Widjaja, *Communication Networks*, 2nd edn., McGraw-Hill, New York, 2003.

[IANA] Web Site for IANA—Internet Assigned Numbers Authority, http://www.iana.org/

[PD07] L.L. Peterson, B.S. Davie, *Computer Networks: A System Approach*, 4th edn., Morgan Kaufmann Publishers, San Francisco, CA, 2007.

[RFC] Web Site for *RFC Editor*, http://www.rfc-editor.org/

[RFC813] D. Clark, Window and Acknowledgement Strategy in TCP, July 1982, http://www.rfc-editor.org/

[RFC896] J. Nagle, Congestion Control in IP/TCP Internetworks, January 1984, http://www.rfc-editor.org/

[RFC2147] D. Borman et al., TCP and UDP over IPv6 Jumbograms, May 1997, http://www.rfc-editor.org/

[RFC2525] V. Paxson et al., Known TCP Implementation Problems, March 1999, http://www.rfc-editor.org/

[RFC2675] D. Borman et al., IPv6 Jumbograms, August 1999, http://www.rfc-editor.org/

[RFC2757] G. Montenegro et al., Long Thin Networks, January 2000, http://www.rfc-editor.org/

[RFC3493] R. Gilligan et al., Basic Socket Interface Extensions for IPv6, February 2003, http://www.rfc-editor.org/

[STD2] J. Reynolds, J. Postel, Assigned Numbers, October 1994, http://www.rfc-editor.org/

[STD5] J. Postel, Internet Protocol, September 1981, http://www.rfc-editor.org/

[STD7] RFC0793/STD0007, J. Postel (ed.), Transmission Control Protocol DARPA Internet Program Protocol Specification, September 1981, http://www.rfc-editor.org/

62

Development of Interactive Web Pages

62.1 Introduction 62-1
62.2 Installations 62-2
WAMP Server
62.3 Introduction to PHP 62-3
Variables • Conditional Statements • Loops • Functions • Include Function • $_GET and $_POST function
62.4 MySQL 62-7
Creating a Database • Creating a Table • Modifying Tables
62.5 Creating Dynamic Web Sites Using PHP and MySQL 62-10
Connecting to Database • SQL Queries with PHP • Creating a Database • Creating a Table • Insert, Update and Delete queries
References 62-14

Pradeep Dandamudi
Auburn University

62.1 Introduction

A static Web site, which is generally written in HTML, does not provide the functionality to dynamically extract the information because static Web sites do not store information in a database, lacking the functionality needed for dynamic Web sites. In most cases, because static Web pages do not carry large amount of information, all the information is embedded in the static Web site.

Dynamic Web sites [3] interact dynamically with the client by taking requests from the client, sending them to the server for processing, and displaying the results on the Web page. Meanwhile, the server interacts with the database to extract the information required for the user, formats the information so that the user can read it, and sends it back to the client browser. HTML alone does not provide the functionality needed for a dynamic, interactive environment. The most common scripting languages used for developing dynamic Web sites are PHP, Perl, and Python. Figure 62.1 shows the data flow in a dynamic, interactive client-server environment.

PHP, a server-side scripting language, is the most popular technology that is especially suited for developing dynamic, interactive Web sites. PHP, which originally stood for "personal home page," and is an open-source software initially created to replace a small set of Perl scripts that had been used in Web sites. Gradually PHP became a general purpose scripting language that is used especially for Web development and now called "hypertext preprocessor." PHP is embedded into HTML and interpreted on a Web server configured to operate PHP scripts. PHP is a server-side scripting language similar to other server-side scripting languages like Microsoft's Active Server Pages and Sun Microsystems' JavaServer Pages [1]. PHP resembles it a syntax to C and PERL. The main difference between PHP and PERL lays in the set of standards built in libraries that support the generations of HTML code, processing data from and to the Web server, and handling cookies [4]. PHP is used in conjunction with database systems, such

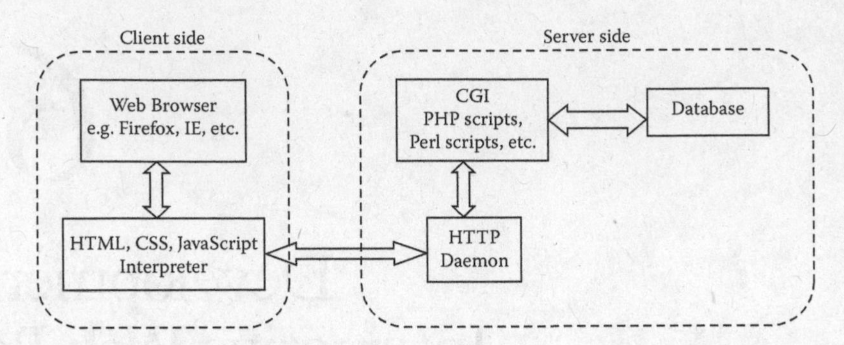

FIGURE 62.1 Data flow in a client–server partition.

as, PostgreSQL, Oracle, MySQL, to create powerful and dynamic server-side applications. Significant Web sites that are written in PHP include Joomla, Drupal, WordPress, Facebook, etc.

Database comes to picture when it comes to storing huge amount of information and retrieving the information back when needed by the Web site. The database can contain information collected from the user or from the administrator. MySQL (written in C and C++) is an open-source relational database management system (RDBMS) that is based on the structure query language (SQL) for processing the data in the database and manages multiuser access to a number of databases. In a relational database, there are tables that store data. The columns define what kind of information will be stored in the table and a row contains the actual values for these specified columns. The language provides a standard syntax for accessing relational databases. MySQL works on many different system platforms, including Linux, Mac OS X, Microsoft Windows, OpenSolaris and many others.

PHP (hypertext preprocessor) and MySQL (a portable SQL server) together have made the task of building and accessing relational databases much easier [2]. This chapter introduces you to PHP and MySQL and explains how to use these two together in order to build dynamic Web sites [7].

62.2 Installations

Before you begin to design a Web site, you need to set up an environment that facilitates your development. You can set up a test site on your computer by locally installing the following softwares.

1. Apache—an open-source Web server
   Download from: http://tomcat.apache.org/index.html
2. PHP—an open-source general purpose scripting language designed specifically for the Web
   Download from: www.php.net/
3. MySQL—an open source database management system
   Download from: www.mysql.com/

After the above softwares are downloaded and installed, you need to configure the Apache server to process the PHP code. You also need to create an MySQL account in order to use it in your PHP code. Sometimes you may find it difficult to install and configure these softwares but the Web sites where you download the software will provide you with general instructions to install and configure them. If you find it difficult to install these softwares will it would be easier for you to find software packages (often called software bundles) that can install and configure them automatically. A few software packages of independently created programs are WAMP server (for Windows environment), LAMP server (for Linux users), MAMP server (for Mac OS) and SAMP server (for Solaris OS).

62.2.1 WAMP Server

WAMP is an open-source software used to run dynamic Web sites on the Windows environment. The WAMP package principal components are: Apache Web server, MySQL database management system, PHP (or Perl/Python) scripting languages and PhpMyAdmin.

You can download the latest WAMP server software from the following Web site: http://www.wampserver.com/en/download.php.

After you download the package, double click on the executable file and choose a destination path (by default c:\wamp) for the software to be installed on your computer. Follow the instructions provided on the Web site and everything will be easily installed. Once the software is installed, you can add other releases of PHP, MySQL, and Apache by downloading them from their Web sites, and you will be able to easily switch between releases. After WampServer is installed, a tray icon to manage your server is shown at right end of the taskbar. By left clicking on the icon, you will see a menu to manage your server settings. Once you install WampServer, a "www" directory is created (c:\wamp\www) and you will see an "index.php" file in it. Do not modify this file because the software recognizes it as its default homepage. Create a directory inside it and store all your PHP or HTML files in it. To create a database, you can follow the traditional method by opening the SQL console and typing the commands that will create your database. Another way of creating databases and handling the administration of one or more MySQl servers is through phpMyAdmin. PhpMyAdmin is an open-source tool intended to create, modify, or delete databases, tables, fields, or rows; executing SQL statements; or managing users and permissions.

62.3 Introduction to PHP

A PHP script can be easily embedded into HTML by enclosing the code with PHP tags, < ?php and ? >, which is similar to HTML tags.

A PHP scripting block can be placed anywhere inside the HTML document.

Example:

```
<html>
<body>
<?php
echo "Hello-this is my first PHP program\n";
/*echo statement display's the text,Hello-this is my first PHP
program*/
?>
</body>
</html>
```

Save the HTML file as "test.php" or you can open a notepad, copy the code, and save it as "test.php."

62.3.1 Variables

All variables in PHP start with the "$" sign followed by the name of the variable. PHP variables are used to store information and can be used wherever they are needed in the script; like text strings, numbers, or arrays.

Example:

```
<?php
$n=20.5;
$string="Welcome";
echo $string;
echo "<br>";
echo $n;
?>
```

Output of above code: Welcome 20.5

PHP also provides "print_r" function to display the variable contents.

62.3.2 Conditional Statements

Conditional statements such as, if, if...else, if...elseif...else, switch statements are used to perform different actions based on different decisions.

a. An if statement is used to execute the code only if a specified condition is true.

```
<?php
$np=1;
//if statement.
if ($np==5)
  echo "Five";
?>
```

b. An if....else statement is used to execute some code if a condition is true and another code if condition is false.

```
<?php
$p=10.55;
//if....else statement.
if ($p==3)
   echo "Three";
else
   echo "Try different number";
?>
```

c. An if...elseif...else statement is used to selectively execute some code when a condition is true.

```
<?php
$i=10;
//if....elseif...else statement.
if ($n==1)
   echo "One";
elseif ($i==2)
   echo "Two";
if ($i==3)
   echo "Three";
else
   echo "Try different number";
?>
```

d. A switch statement is used to execute one of many blocks of code.

Example:

```
//Switch Statement
<?php
$k=10;
switch ($k)
```

```
{
case 1:
  echo "One";
  break;
case 2:
  echo "Two";
  break;
case 3:
  echo "Three";
  break;
default:
  echo "Try between one and three";
}
?>
```

62.3.3 Loops

Loops are used to repetitively execute a block of code until or when a specific condition is true. Loops are used to perform the same task again and again so that we do not have to write several, almost equal statements.

In PHP, loop statements such as while, do....while, and for are used differently depending on the condition.

Example:

```
//while statement.
<?php
$j=1;                       //initializing variable
while ($j<=5){
echo $j;
$j++;
/*incrementing the value of j each time until the condition is not
met.*/
} ?>
Output of above code: 1 2 3 4 5

//do.....while statement
<?php
$k=1;
do  {
  $k++;
  echo  $k ;
  }while ($k<=5);
?>
Output: 2 3 4 5 6

//for loop
Syntax:
for (init; condition; increment){
code to.be executed;
}
```

Example:

```
<?php
for ($i=1; $i<=5; $i++)   {
  echo $i;
  }
?>
Output: 1 2 3 4 5
```

62.3.4 Functions

Functions contain scripts that are called to execute whenever a page loads or is called from anywhere within a page. There are around 700 built-in functions available in PHP such as date function, string function, calendar function, array functions, etc. You can create your own function to achieve a specific goal.

Syntax:

```
function functionName(){
code to be executed;
}
```

Example 1: Without parameters

```
<?php
function demo() {                //user defined function
$string1="Function string";
echo "function code is executed and displayed";
}
demo();                          //calling a function, by using function name
?>
Output: function code is executed and displayed
```

Example 2: Parameters and return values

Parameters are declared inside the parenthesis. Parameters are just like variables that store information within the function.

```
<?php
function add($x,$y){    //$x and $y are two parameters.
$t=$x+$y;
return $t;      /* function returns value $t using 'return'
                statement.*/
}
echo "Sum is". add(1,5);
/*send values 1 and 5 to the function which in turn returns sum of
  1 and 5 and final result is displayed by the echo statement.*/
?>
Output: Sum is 6
```

62.3.5 Include Function

If you want to include one PHP file in another PHP file, include() or require() functions is used to take the content of one file and include it in the current file.

Both include() and require() functions serve the same purpose but handle errors differently. In the case of the include() function, it generates a warning and continues to execute the script whenever an error occurs. Whereas require() function generates a fatal error and terminates execution.

The include() function is used especially if a block of codes are reused on multiple pages. For example, menus, headers, footers, advertisements that are reused on different pages.

Example:

```
<html>
<body>
<?php include("footer.php"); ?>          //include("filename.php")
<h3>@Copyright 2009</h1>
</body>
</html>
```

62.3.6 $_GET and $_POST function

The built-in $_GET and $_POST functions are used to collect values from an HTML/PHP form. However, both functions handle information sent from a form in different ways. Information sent from a form using the GET method is visible to everyone (all variables are displayed in the URL) and is limited to send up to 100 characters. Whereas information sent from the POST method is not visible to everyone (all variables are not displayed in the URL) and the characters are not limited.

Example:

```
<form action="Login.php" method="get">
Username: <input type="text" name="username" />
Password: <input type="text" name="pwd" />
<input type="submit" />
</form>
```

When the user clicks the "submit" button, the form sends information attached to the URL to the server, which looks like http://www.companyname.com/Login.php?Username=Tom&Password=123

And "Login.php" can use $_GET function to collect data from the form:

```
<p> Your username: <?php echo $_GET["username"]; ?></p>
```

For the same example, replace "get" with "post" under method option and the URL will look like http://www.companyname.com/Login.php

And "Login.php" can use $_POST function to collect data from the form:

```
<p> Your username: <?php echo $_POST["username"]; ?> </p>
```

To collect data that is sent with both the GET and POST methods, $_REQUEST function can be used i.e.,

```
<p> Welcome <?php echo $_REQUEST["username"]; ?> </p>
```

62.4 MySQL

MySQL is a database system, which stores data in the form of tables. Each table is identified by a name and contains rows of data. See Table 62.1 for an example.

TABLE 62.1 SQL Database Table

Last Name	First Name	Employee Department	Employee ID
Rives	John	Sales	2
Tent	Tom	Accounts	1
Hold	Tim	Technical	5

After installing MySQL server, open command line to run the MySQL client program. To connect to your MySQL server, here's what you should type:

```
mysql -h <hostname> -u <username> -p <password>
```

If you installed and typed everything properly, the MySQL client program will introduce itself and then take you to the MySQL command line:

```
mysql>
```

If at any time you want to exit the MySQL client program, just type "quit" or "exit."

62.4.1 Creating a Database

To create a database and use it, type the following command (don't forget the semicolon), then ENTER.

```
mysql> CREATE DATABASE databasename;
mysql> USE databasename;
```

Ex:

```
mysql> CREATE DATABASE employee;
mysql> USE employee;
```

To see a list of databases available, type the following command, then ENTER.

```
mysql> SHOW DATABASES;
```

62.4.2 Creating a Table

To create a table, use the "CREATE TABLE" statement.

Syntax:

```
CREATE TABLE <table name> (
column_name1 data_type,
column_name2 data_type,
);
```

Example:

```
mysql> CREATE TABLE employeeinfo (
      LastName varchar(15),
      FirstName varchar(15),
      Department varchar(15),
      Employee_ID int
      );
```

To view tables, type the following command.

```
mysql> SHOW TABLES;
```

62.4.3 Modifying Tables

To delete a table, type

```
mysql> DROP TABLE <tableName>;
```

To insert data into a table, type

```
mysql> INSERT INTO <table name>
  (column1, column2, …)
  VALUES (value1, value2, …);
```

Example:

```
mysql> INSERT INTO employeeinfo
  (LastName, FirstName, Department, Employee_ID)
  VALUES (Rives, John, Sales, 2);
```

To update data stored in a database table, type

```
mysql> UPDATE <tableName> SET
  <col_name>=<new_value>, …
  WHERE <where clause>;
```

Example:

```
mysql> UPDATE employeeinfo SET FirstName='Johnny' WHERE
LastName='Rives' AND Employee_ID=2;
```

To delete a record from the table

```
mysql> DELETE FROM table_name WHERE some_column=some_value;
```

Example:

```
mysql> DELETE FROM employeeinfo WHERE LastName='Tent';
```

To delete all rows

```
DELETE * FROM employeeinfo;
```

62.4.3.1 To View Data

The SELECT command is used for viewing data stored in your database tables.

Syntax:

```
SELECT column_name(s) FROM table_name;
```

Example 1: To display data from columns LastName and Department.

```
mysql> SELECT LastName, Department from employeeinfo;
```

Example 2: To display all data that is in the employeeinfo database table,

```
mysql> SELECT * FROM employeeinfo;
```

62.4.3.2 WHERE Clause

To extract those records that fulfill specific criteria

```
mysql> SELECT LastName, Department FROM employeeinfo WHERE
Employee_ID=2;
```

62.4.3.3 ORDER BY Keyword

To sort records in the table

```
mysql> SELECT * FROM employeeinfo ORDER BY Employee_ID DESC;
```

To sort in ascending order, use ASC instead of DESC.

62.5 Creating Dynamic Web Sites Using PHP and MySQL

We are going to discuss some simple ways to create a dynamic Web site. These include connecting to the database through PHP, extracting data that sits in MySQL database, and presenting it as a part of the Web page.

62.5.1 Connecting to Database

To connect to a database using PHP, you do not need any other program. PHP features some built-in functions that can connect, manage and disconnect to MySQL database.

Syntax:

```
mysql_connect("localhost","username","password")
```

Since the connection returns an identifier, we can store it in a variable for later use.

Example:

```
<?php
$contn=mysql_connect("localhost","username","password");
?>
```

To display the error message when connection fails, "die" function serves that purpose.
To close the connection

Ex: `mysql_close($con);`

62.5.2 SQL Queries with PHP

To send SQL queries (commands) to the database and view the results of those queries

Syntax:

```
mysql_query("query", connection_id);
```

62.5.3 Creating a Database

To create a database on MySQL, use the CREATE DATABASE statement.

Example:

```
<?php
$contn=mysql_connect("localhost","username","password");
mysql_query("CREATE DATABASE employeeinfo",$contn);
?>
```

62.5.4 Creating a Table

To create a Table, use the CREATE TABLE statement.

Example:

```
<?php
//Establish connection to MySQL database.
$contn=mysql_connect("localhost","username","password");
//Create your database
mysql_query("CREATE DATABASE employee",$contn);
/*Select your database or for the matter, you can select any
database that you have already created.*/
mysql_select_db("employee", $contn);
// Create table
$tbl = "CREATE TABLE employeeinfo
(
LastName varchar(15),
FirstName varchar(15),
Department varchar(15),
Employee_ID int
)";
// Execute query
mysql_query($tbl,$contn);
//Close connection
mysql_close($contn);
?>
```

To see the last error message generated by MySQL server, mysql_error()returns a text describing the last error message.

Example:

```
if (mysql_query($tbl)) {
echo("Employee Table created</P>");
} else {
echo("Error creating employee table:". mysql_error());
}
```

62.5.5 Insert, Update and Delete queries

To insert data into a database table, use INSERT INTO Statement.

Example:

```
mysql_query("INSERT INTO employeeinfo (LastName, FirstName,
Department, Employee_ID)
VALUES ('Rives', 'John', 'sales', '2')");
```

$_POST is used to collect data from a form and if you substitute it in VALUES, parameters of a form can be inserted into a table.

To update existing records in a table, use UPDATE statement.

Example:

```
mysql_query("UPDATE employeeinfo SET LastName= 'Rives' WHERE
FirstName = 'John' AND Employee_ID = '2'");
or
mysql_query("UPDATE employeeinfo SET LastName= 'Rives'".
"WHERE FirstName = 'John' AND Employee_ID = '2'");
```

To delete existing records, use DELETE statement,

```
mysql_query("DELETE FROM employeeinfo WHERE LastName='Tent'");
```

62.5.5.1 Select Query

To select records from database, we use the SELECT statement.

Example:

To select all the records from the table employeeinfo

```
$result = mysql_query("SELECT * FROM employeeinfo");
```

The data returned by mysql_query is stored in the variable $result.

Now we use mysql_fetch_array() function to return to the first row of the table. And each time a call to mysql_fetch_array() returns the next row in the table.

Example:

```
while($row = mysql_fetch_array($result))
  {
  echo $row['LastName']. "". $row['FirstName']. "".
$row['Department']. "". $row['Employee_ID'];
  echo "<br />";
  }
```

```
Output:
  Rives John sales 2
  Tent Tom accounts 1
  Hold Tim technical 5
```

Example: Designing an online conference paper collection and evaluation Web site using PHP and MySQL: In this example, you will learn how to design a basic paper submission and evaluation system [5–7]. Three main code modules for this application are as follows:

a. Paper submission
b. Paper evaluation
c. Administration

Before writing the PHP code, make sure you have created a database with the following tables:

Loginaccnt_table for user, reviewer and administrator with lastname, firstname, userid, password (encrypted) fields; papers_submitted_table with title, author, date, abstract, file_size, file, file_type, and review_ok fields; comment_table with subject, name, date, status, and detail fields.

a. *Paper submission*: For the convenience of designing, the following design steps can be adapted:
   - The user has to login before submitting a paper (Figure 62.2). Required files: index.php, Login.php.
   - After user authentication, the user is directed to his homepage where he can submit paper, view reviewed papers, view comments, etc. For user authentication, login_check.php, session.php files are required.
   - If user ID and password are not found in the database, a message is generated to login again or a request to create an account. Required file: login_check.php, user_home.php, signup.php.
   - If user requests to create an account, change password. Required files: signup.php, signup_ok.php, password_change.php, password_updated.php.
   - After user logins to his homepage, user can submit a paper, view reviewed papers, download other papers, post comments for papers, view papers that are not yet reviewed, update his profile.

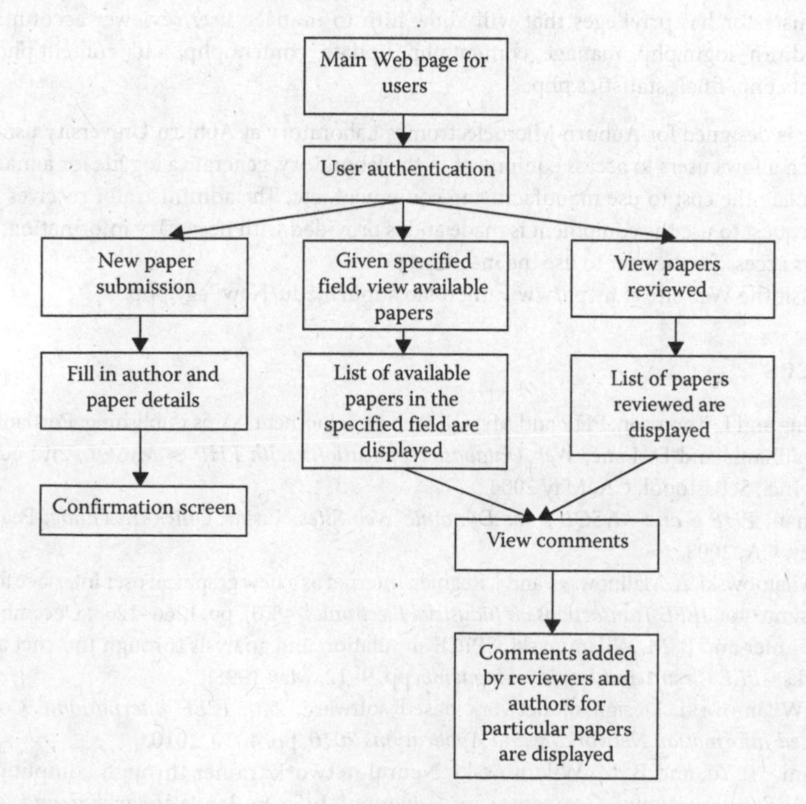

FIGURE 62.2 The user view of the paper evaluation system for user/authors.

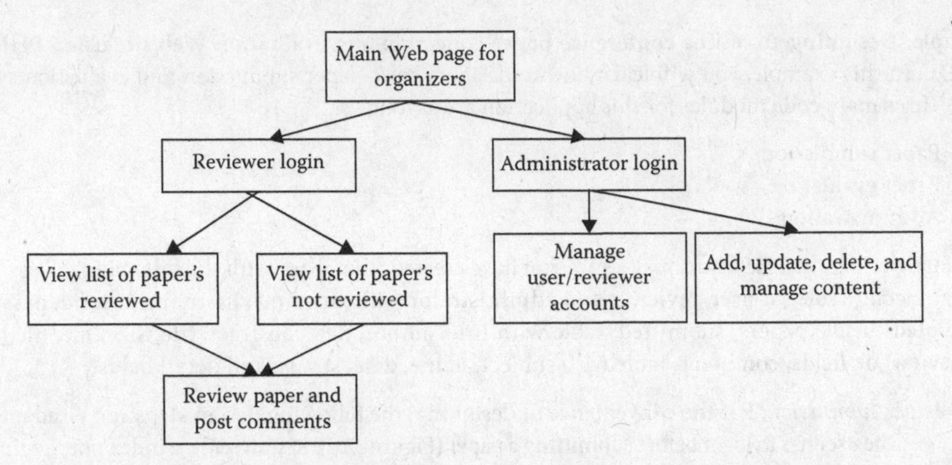

FIGURE 62.3 The organizer view of the paper evaluation system for reviewer/administrator.

Required files, submit_paper.php, submit_paper_check.php, submit_paper_ok.php, update_profile.php, profile_updated.php, paper_reviewed.php, paper_abstract.php, paper_download.php.

b. *Paper evaluation*: After the paper has been submitted by the user, the reviewer receives an email notification with the user information and the paper information Figure 62.3.
   - Reviewer access is provided only by the administrator. After reviewer logs in, he can view papers that were reviewed, review a paper. Required files: reviewer_login.php, reviewer_login_check.php, paper_reviewed.php, review_paper.php, paper_abstract.php.

c. *Administrator*: Administration interface lets you add, delete, and update from the database. Administrator has privileges that will allow him to manage user/reviewer accounts. Required files: admin_login.php, manage_content.php, update_content.php, add_content.php, manage_accounts.php, final_statistics.php.

A Web site is designed for Auburn Microelectronics Laboratory at Auburn University using PHP and MySQL, which allows users to access equipment in the laboratory, generate a log file for a manufacturing process, calculate the cost to use manufacturing equipment, etc. The administrator receives notification whenever a request to use the equipment is made and is provided with necessary information, the administrator allows access for the user to use the machinery.

You can visit the Web site at http://www.microlab.auburn.edu/NewPage.php.

References

1. L. Welling and L. Thomson, PHP and MySQL Web Development, Sams Publishing, Portland, OR, 2003.
2. H. E. Williams and D. Lane, *Web Database Applications with PHP & MySQL*, 2nd edn., O'Reilly Media, Inc., Sebastopol, CA, May 2004.
3. L. Ullman, *PHP 6 and MySQL 5 for Dynamic Web Sites: Visual QuickPro Guide*, Peachpit Press, Berkeley, CA, 2008.
4. B. M. Wilamowski, A. Malinowski, and J. Regnier, Internet as a new graphical user interface for the SPICE circuit simulator, *IEEE Transactions on Industrial Electronics*, 48(6), pp. 1266–1268, December 2001.
5. J. W. Regnier and B. M. Wilamowski, SPICE simulation and analysis through Internet and Intranet networks, *IEEE Circuit and Devices Magazine*, pp. 9–12, May 1998.
6. B. M. Wilamowski, Design of network based software, *24th IEEE International Conference on Advanced Information Networking and Applications 2010*, pp. 4–10, 2010.
7. N. Pham, H. Yu, and B. M. Wilamowski, Neural network trainer through computer networks, *24th IEEE International Conference on Advanced Information Networking and Applications 2010*, pp. 1203–1209, 2010.

63

Interactive Web Site Design Using Python Script

63.1 Introduction 63-1
Python • Django
63.2 Software Installation 63-2
63.3 Database-Driven Web Site Design 63-2
Create a Project • Run the Project • Setup Database • Create Model • Activate Model • Activate the Admin Site • Add the Application • Customize the Admin Site
63.4 Conclusion 63-7
References 63-8

Hao Yu
Auburn University

Michael Carroll
Auburn University

63.1 Introduction

With the development of Internet technology, more and more Web design languages such as Javascript, PHP, and CGI are developed to satisfy various online services. In this chapter, an open-source Web application framework, Django, which is developed in Python, will be introduced for database-driven Web site design.

63.1.1 Python

Python supports multiple programming paradigms, such as object oriented, imperative, and functional [F02]. Python features a fully dynamic-type system and automatic memory management. From the point of functions, python has very similar syntax to Perl, but it is simplified and more readable. Python is also flexible for programming, including both command line style, such as in MATAB®, and script style, such as in Perl.

As a high-level language, Python can be operated on many different operating systems, such as Windows, MacOS, Unix, and OS/2. The Python software can be freely downloaded and used, or can be included in applications. Python can also be freely modified and redistributed, since the language is copyrighted under an open-source license.

63.1.2 Django

Django is a high-level Python Web framework that encourages rapid development and clean, pragmatic design [HM09]. It was originally developed to ease the creation of complex database-driven Web sites in 2005. Using Python as a script language, Django focuses on automation as much as possible.

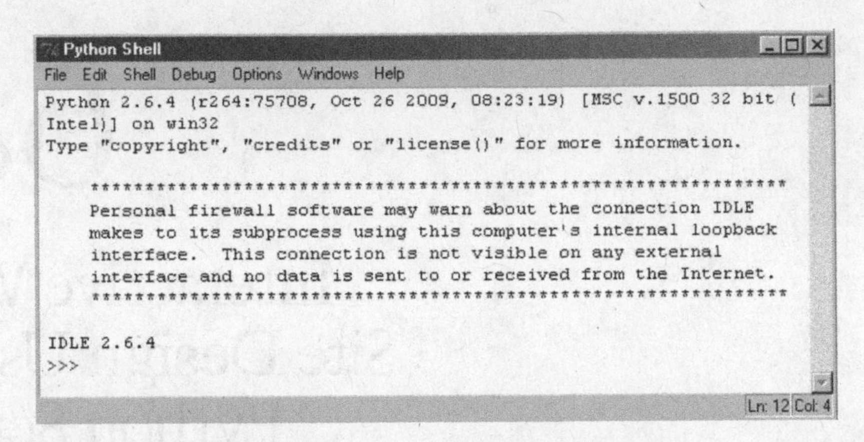

FIGURE 63.1 Shell interface of Python.

It adheres to the DRY principle, which means "don't repeat yourself." Django also provides optional administrative "create, read, update, and delete" (CRUD) interface templates, which are generated dynamically through introspection and configured via admin models.

63.2 Software Installation

Both Python and Django are required for setting up the programming environment. Python can be freely downloaded from http://www.python.org. Versions 2.4 to 2.6 are recommended (Django is not properly compatible with Python 3.0). The shell interface of Python after installation is shown in Figure 63.1.

Django can be downloaded from http://www.djangoproject.com/download/ as a packed file, like "Django-[version number].tar.gz." After unpacking the file, in command prompt model, the current directory should be set as where the Django files were released (by `cd [directory]`). Then the command `setup.py install` can be entered for installation. After installation, enter `import django` in the shell interface of Python; if there is no error information, it means Django is successfully installed and bonded with Python (Figure 63.2).

For database development, at least one database server, PostgreSQL, MySQL, or Orade, is required to be installed in the server. If Django is used on a production site, Apache with mod_wsgi is required for server configuration [A08].

63.3 Database-Driven Web Site Design

In order to introduce how to use Django for database-driven Web site design, let us have a simple example. The purpose is to build an online database with basic "CRUD" operations.

63.3.1 Create a Project

The project can be auto-generated by Django, and the important part is to set a proper instance, including database configuration, Django-specific options and application-specific settings [B08].

The whole process is shown in Figure 63.3. First of all, copy the file `django-admin.py` to the place where the project will be built, then locate into the directory by `cd` in command prompt model. After that,

```
>>> import django
>>>
```

FIGURE 63.2 Successful installation of Django.

```
C:\>copy c:\Python26\lib\site-packages\django\bin\django-admin.py c:\
        1 file(s) copied.

C:\>cd c:\

C:\>django-admin.py startproject paper_system

C:\>cd paper_system

C:\paper_system>dir
 Volume in drive C has no label.
 Volume Serial Number is 8476-550F

 Directory of C:\paper_system

02/24/2010  12:07 PM    <DIR>          .
02/24/2010  12:07 PM    <DIR>          ..
02/24/2010  12:07 PM               557 manage.py
02/24/2010  12:07 PM             2,864 settings.py
02/24/2010  12:07 PM               571 urls.py
02/24/2010  12:07 PM                 0 __init__.py
               4 File(s)          3,992 bytes
```

FIGURE 63.3 The process of creating a project.

execute the command `django-admin.py startproject paper_system`. Four files will be automatically generated and they are: `manage.py`, `settings.py`, `urls.py`, and `__init__.py`, each of which has specified functions:

- `manage.py`: This is a command line that can help to interact with the Django project in various ways in format: `manage.py <subcommand> [options]`. For example, `manage.py runserver` is used to start running the server.
- `settings.py`: This file is used to set/configure the Django project information, such as user name, password, and locations of the template resources.
- `urls.py`: The URL declarations of the Django project.
- `__init__.py`: An empty file specifying that the current directory should be considered as a Python package.

63.3.2 Run the Project

The work above can be verified by running the project: (1) `cd` to the directory of the project and (2) run command `manage.py runserver`. If the project is successfully created, the result is shown as in Figure 63.4.

Now the server is running and it can be locally visited by http://localhost:8000/ with the browser. Figure 63.5 shows the Web site information.

The link address and port can be modified using commands such as `manage.py runserver 131.204.222.54:80`.

```
C:\paper_system>manage.py runserver
Validating models...
0 errors found

Django version 1.1.1, using settings 'paper_system.settings'
Development server is running at http://127.0.0.1:8000/
Quit the server with CTRL-BREAK.
```

FIGURE 63.4 Result of running the project.

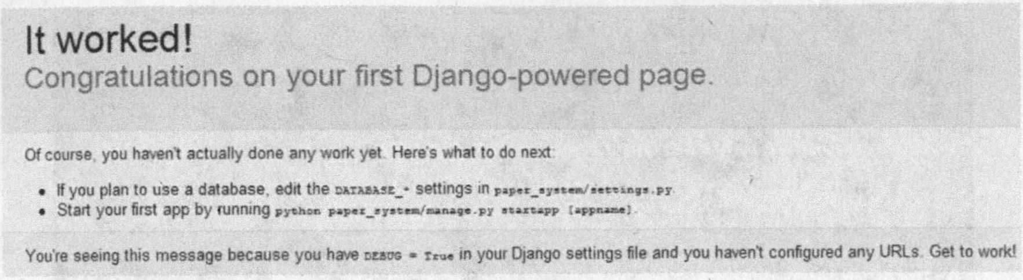

It worked!
Congratulations on your first Django-powered page.

Of course, you haven't actually done any work yet. Here's what to do next:

- If you plan to use a database, edit the DATABASE_* settings in paper_system/settings.py.
- Start your first app by running python paper_system/manage.py startapp [appname].

You're seeing this message because you have DEBUG = True in your Django settings file and you haven't configured any URLs. Get to work!

FIGURE 63.5 Web site of the server.

63.3.3 Setup Database

The database information is configured in the file `setting.py`. There are very detailed instructions about each parameter in the file. In this example, let us fulfill the basic settings as follows:

```
DATABASE_ENGINE = 'sqlite3'
DATABASE_NAME = 'C:/paper_system/sqlite3.db'
```

The first parameter indicates that SQLite database (included in the standard Python library) stores all the information in a single file on the computer, while the second parameter specifies the location of SQLite database file.

After the configuration of file `setting.py`, enter the command `manage.py syncdb` to create the database and register a super-user for the auth system. See Figure 63.6.

63.3.4 Create Model

So far, a project has been set up. In the following, the document database is going to be built based on the project. Notice that Django supports multiple applications, which means a project can have multiple applications and an application can be in multiple projects.

By entering the command `manage.py startapp documents`, a directory named `documents` is created with four files inside: `__init__.py`, `models.py`, `tests.py`, and `views.py`.

The file `models.py` is used to build the model. In this example, a `Document` model will be described in `documents/models.py` file as

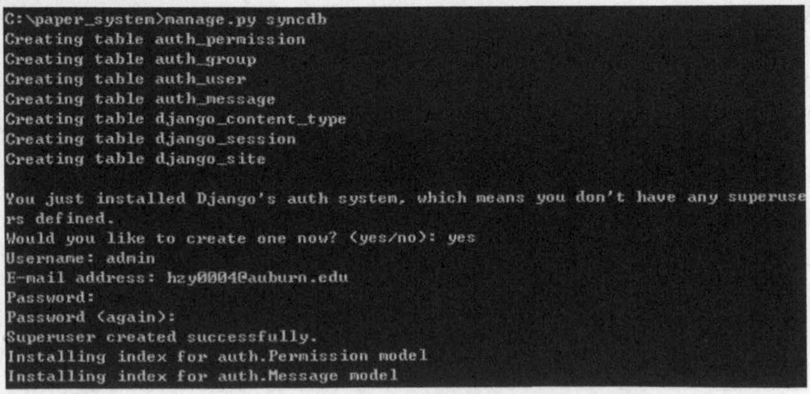

```
C:\paper_system>manage.py syncdb
Creating table auth_permission
Creating table auth_group
Creating table auth_user
Creating table auth_message
Creating table django_content_type
Creating table django_session
Creating table django_site

You just installed Django's auth system, which means you don't have any superuse
rs defined.
Would you like to create one now? (yes/no): yes
Username: admin
E-mail address: hzy0004@auburn.edu
Password:
Password (again):
Superuser created successfully.
Installing index for auth.Permission model
Installing index for auth.Message model
```

FIGURE 63.6 Database setup.

```
C:\paper_system>manage.py startapp documents

C:\paper_system>manage.py sql documents
BEGIN;
CREATE TABLE "documents_document" (
    "id" integer NOT NULL PRIMARY KEY,
    "document_name" varchar(500) NOT NULL,
    "document_auth" varchar(200) NOT NULL,
    "sub_date" datetime NOT NULL
)
;
COMMIT;
```

FIGURE 63.7 Model creation and activation.

```
from django.db import models
class Document(models.Model):
  document_name = models.CharField(max_length=500);
  document_auth = models.CharField(max_length=200);
  sub_date = models.DateTimeField('date submitted');
```

In the code, `.CharField` is used to create a character field; `.DateTimeField` is used to create a time field. This `Document` model consists of three parameters: the name of the document (with maximum length 500), the name of the author (with maximum length 200), and the date of submitting the document.

63.3.5 Activate Model

Models need to be activated for application. In this example, the steps are

1. Add an element `'paper_system.documents'` to list `INSTALLED_APPS` in `setting.py` file.
2. Enter command `manage.py sql documents`, and the result is as shown in Figure 63.7.

After activating the models, model tables (containing model information) need to be updated in the database, using command `manage.py syncdb`.

63.3.6 Activate the Admin Site

The Django admin site is not activated using default settings; it can be activated in three steps:

1. Add `'django.contrib.admin'` to the `INSTALLED_APPS` setting in `setting.py` file.
2. Run `manage.py syncdb` to update the database tables.
3. Edit the file `paper_system/urls.py` like this

```
from django.conf.urls.defaults import *
from django.contrib import admin
admin.autodiscover()
urlpatterns=patterns(' ',(r'^admin/',include(admin.site.urls)))
```

After activating the admin site, the link http://localhost:8000/admin/ becomes available, see Figure 63.8.

Using the administrator user name and password, which were configured in Section 63.3.3, one can log in, and follow with the administration management site, see Figure 63.9.

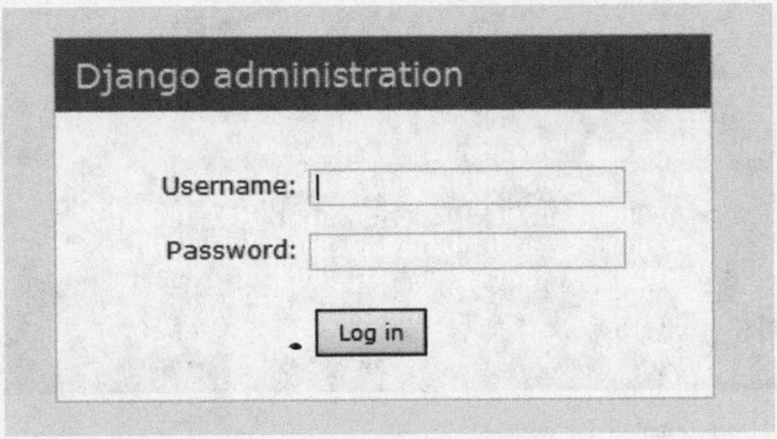

FIGURE 63.8 Activated admin site.

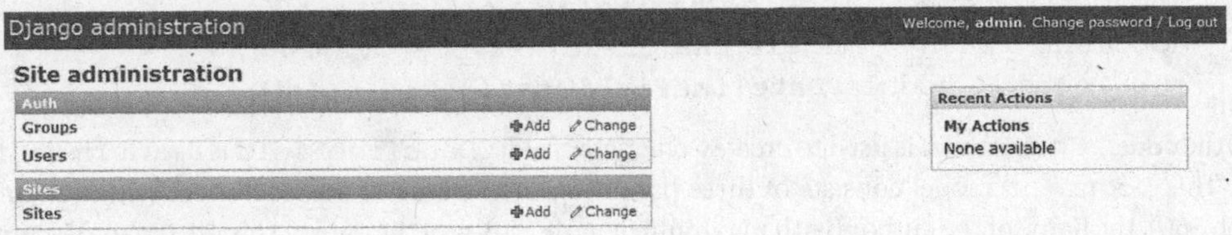

FIGURE 63.9 Default administration management site.

63.3.7 Add the Application

The previous step activated the default administration site, but without the required application. In order to add the `Document` module into the application, a file named `admin.py` needs to be created in `\paper _ system\documents` directory and the code inside could be

```
from paper _ system.documents.models import Document
from django.contrib import admin
admin.site.register(Document)
```

The updated Web site is shown in Figure 63.10.

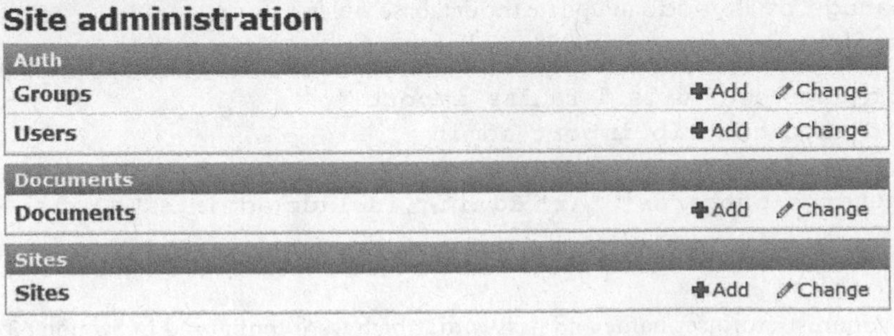

FIGURE 63.10 Administration management site with `Document` module.

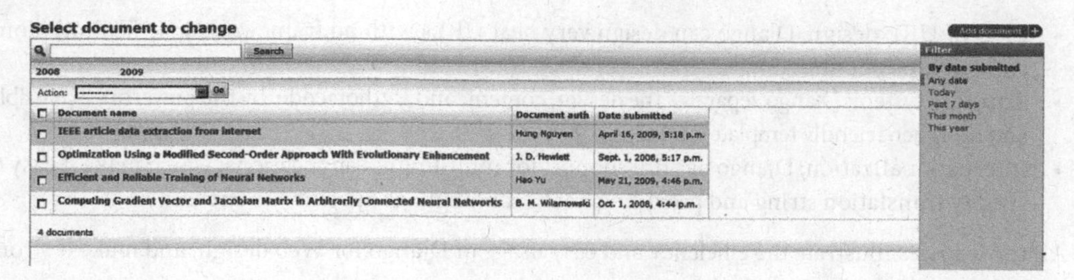

FIGURE 63.11 Customized administration Web site.

63.3.8 Customize the Admin Site

The application already has the basic functions of a database, such as add, change, and delete. But it is not good enough for application. The following steps will make the database much more powerful:

1. Add the Python code below into the `models.py` file. This will use the document names as the links to visit the related documents.

```
def __unicode__(self):
    return self.document_name
```

2. Edit the `admin.py` file using the code below. This will list all the document information row by row. In the table, all the document information can be ordered by either document name, author name, or submission date.

```
from paper_system.documents.models import Document
from django.contrib import admin
class DocumentAdmin(admin.ModelAdmin):
list_display = ('document_name','document_auth','sub_date')
admin.site.register(Document, DocumentAdmin)
```

3. Add code below to `admin.py` file. This will create sidebar used to filter the listed document information by specified date.

```
list_filter = ['sub_date']
```

4. Add code below to `admin.py` file. This will add a search box at the top of the document list, which can be used to search document information by author name.

```
search_fields = ['document_auth']
```

5. Add code below to `admin.py` file. This will add hierarchical navigation by submission date.

```
date_hierarchy = 'sub_date'
```

With all the above customization, the new administration Web site is as shown in Figure 63.11.

63.4 Conclusion

The above example built a framework for a database-driven Web site server. It includes interface design and basic SQL operations, such as create, search, add, and delete. As one may notice, the entire project is done in a considerably limited time and with very few codes.

By conclusion, the major features of Django can be described as follows [B08]:

- Object-relational mapper: By defining the data models in Python, Django can offer an efficient and dynamic database access APIs for free.
- Automatic admin interface: Django does all the work of creating interfaces automatically, avoiding tedious and duplicated coding.

- Elegant URL design: Django can design very neat URLs with no framework-specific limitations, as flexible as possible.
- Template system: Django separates the design, content, and Python code, by the powerful, extensible, and designer-friendly template language.
- Internationalization: Django has full support for multilingual applications, which make it easy to specify translation string and providing hooks for language-specific functionality.

All the features illustrate the efficiency and easy usage of Django for Web design, and make it strong in competing with various other Web programming tools.

References

[A08] M. Alchin, *Pro Django*. Apress, Lansing, MI, 2008.
[B08] J. Bennett, *Practical Django Projects*. Apress, Berkeley, CA, 2008.
[F02] C. Fehily, *Python*. Peachpit press, Berkeley, CA, 2002.
[HM09] Adrian Holovaty, Jacob Kaplan-Moss, *The Definitive Guide to Django: Web Development Done Right* (2nd version). Springer press, New York, 2009.

64

Running Software over Internet

Nam Pham
Auburn University

Bogdan M. Wilamowski
Auburn University

Aleksander Malinowski
Bradley University

64.1 Introduction **64**-1
64.2 Most Commonly Used Network Programming Tools **64**-2
Hypertext Markup Language • JavaScript • Java • ActiveX • CORBA and DCOM • Common Gateway Interface • PERL • PHP
64.3 Examples **64**-5
Neural Network Trainer through Computer Networks • Web-Based C++ Compiler • SPICE-Based Circuit Analysis Using Web Pages
64.4 Summary and Conclusion **64**-11
References **64**-11

64.1 Introduction

A static Web site, which is generally written in HTML, does not provide functionality to dynamically extract information and to store it in a database [WT08]. Dynamic Web sites interact dynamically with clients by taking requests from clients. When receiving requests from a client, a server interacts with a database to extract required information, formats information, and sends it back to the client. HTML alone does not provide the functionality needed for a dynamic, interactive environment. Additional technologies are used to implement the dynamic behavior both on the client side (inside a Web browser) and on the server side to render a web page dynamically. Most common network programming tools used for developing dynamic Web sites on the client side are JavaScript, and Java or ActiveX applets. Most common tools on the server side are PHP, Sun Microsystems' Java Server Pages, Java Servlets, Microsoft Active Server Pages (APS) technology, and Common Gateway Interface (CGI) scripts using scripting languages such as PERL and JavaScript, ActiveX and Python, or precompiled binary programs.

Communication through computer networks has become a popular and efficient means of computing and simulation. Most companies and research institutions use networks to some extent on a regular basis [W10]. Computer networks provide ability to access all kinds of information and made available from all around the world; and intranet networks provide connectivity for a smaller, more isolated domain like a company or a school. Users can run software through computer networks by interacting with user interface. Simulation through computer networks has several benefits:

- *Universal user interface on every system*: Every system can run software simulation with a web browser through user interface.
- *Portability*: Software located only on the center machine can be accessed at any time and everywhere.
- *Software protection*: Users can run software through computer networks but cannot own it.

- *Limitation*: Software can interact with any platform that is independent of the operation systems, users do not have to set up or configure software unless it is implemented on the server in the form of stored user profile.
- *Legacy software*: Old software can be run in a dedicated environment on the server while its new user interface runs through a web browser on new systems for which the particular application is not available.
- *Remote control*: Computer networks are used to control objects remotely.

The Internet bandwidth is already adequate for many software applications if their data flow is carefully designed. Furthermore, the bandwidth limitation will significantly improve with time. The key issue is to solve problems associated with a new way of software development so that application of software will be possible through the Internet and Intranet. It is therefore important to develop methods that take advantage of networks and then platform independent browsers. This would require solving several issues such as

- Minimization of the amount of data that must be sent through the network
- Task partitioning between the server and client
- Selection of programming tools used for various tasks
- Development of special user interfaces
- Use of multiple servers distributed around the world and job sharing among them
- Security and account handling
- Portability of software used on servers and clients
- Distributing and installing network packages on several servers
- Others

64.2 Most Commonly Used Network Programming Tools

For implementation, several different languages must be used simultaneously. The following sections review these languages and scopes of their application [W10].

64.2.1 Hypertext Markup Language

Hypertext Markup Language (HTML) was originally designed to describe a document layout regardless of the displaying device, its size, and other properties. It can be incorporated into networked application front-end development either to create form-based dialog boxes or as a tool for defining the layout of an interface, or wraparound for Java applets or ActiveX components. In a way, HTML can be classified as a programming language because the document is displayed as a result of the execution of its code. In addition, scripting language can be used to define simple interactions between a user and HTML components. Several improvements to the standard language are available: cascading style sheets (CSS) allow very precise description of the graphical view of the user interface; compressed HTML allows bandwidth conservation but can only be used by Microsoft Internet Explorer. HTML is also used directly as it was originally intended—as a publishing tool for instruction and help files that are bundled with the software [W10].

64.2.2 JavaScript

HTML itself lacks even basic programming construction such as conditional statements or loops. A few scripting interpretive languages were developed to allow for use of programming in HTML. They can be classified as extensions of HTML and are used to manipulate or dynamically create portions of HTML code. One of the most popular among them is JavaScript. The only drawback is that although JavaScript is already well developed, still there is no one uniform standard. Different web browsers may vary a little in the available functions. JavaScript is an interpretative language and the scripts are run as the web page is downloaded and displayed. There is no strong data typing or function prototyping. Yet the language includes support for object oriented programming with dynamically changing member functions. JavaScript programs can also communicate with Java applets that are embedded into an HTML page.

JavaScript is part of the HTML code. It can be placed in both the header and body of a web page. The script starts with `<script language="JavaScript">` line. One of the most useful applications of JavaScript is verification of the filled form before it is submitted online. That allows for immediate feedback and preserves the Internet bandwidth as well as lowers the web server load [W10].

JavaScript has certain limitations due to the security model of its implementation by a Web browser. One of those limitations is inability to retrieve data on demand dynamically from the server. This was changed by adding a new library that is typically referred to by the name of Ajax Technology. Ajax technology allows a JavaScript program embedded inside a web page to retrieve additional documents or web pages from the server, store them as local variables, and parse them in order to retrieve data and use it for dynamic alteration of the web page where the JavaScript is embedded. The additional data is typically generated on the server by means of ASP or CGI used to interface the Ajax query to the database on the server. Data is then sent back typically in the format of an XML formatted web page. Typically, everyday application of this technology is an auto-complete suggestion list in a web page form, for example auto-complete suggestions in a search engine web page before a user clicks the search button.

64.2.3 Java

Java is an object-oriented programming language compiled in two stages. The first stage of compilation, to so-called byte-code, is performed during the code development. Byte-code can be compared to machine code instructions for a microprocessor. Because no processor understands directly byte-code instructions, interpreters, called Java Virtual Machines (JVM), were developed for various microprocessors and operating systems. At some point, JVM were improved so that instead of interpreting the code they do perform the second stage of compilation, directly to the machine language. However, to cut down the initial time to run the program, the compilation is done only as necessary (just in time (JIT)), and there is no time for extensive code optimization. At current state of the art of JIT technology, programs written in Java run about two to five times slower than their C++ counterparts. Adding a JVM to a web browser allowed embedding software components that could be run on different platforms [W10].

64.2.4 ActiveX

Microsoft developed ActiveX is another technology allowing for the automatic transfer of software over the network. ActiveX, however, can be executed presently only on a PC with a Windows operating system, thus making the application platform dependent. Although this technology is very popular already, it does not allow for the development of applications running on multiple platforms. ActiveX components can be developed in Microsoft Visual Basic or Microsoft Visual C++. There is the only choice in cases when Java is too slow, or when some access to the operating system functionality or devices supported only by Windows OS is necessary. The easy access to the operating system form an ActiveX component makes it impossible to provide additional security by limiting the features or resources available to the components [W10].

64.2.5 CORBA and DCOM

Common Object Request Broker Architecture (CORBA) is a technology developed in the early 1990s for network distributed applications. It is a protocol for handling distributed data, which has to be exchanged among multiple platforms. A CORBA server or servers must be installed to access distributed data. CORBA in a way can be considered as a very high-level application programming interface (API). It allows sending data over the network, sharing local data that are registered with the CORBA server among multiple programs. Microsoft developed its own proprietary API that works only in the Windows operating system. It is called DCOM and can be used only in ActiveX technology [W10].

64.2.6 Common Gateway Interface

Common Gateway Interface (CGI) can be used for the dynamic creation of web pages. Such dynamically created pages are an excellent interface between a user and an application run on the server. CGI program is executed when a form embedded in HTML is submitted or when a program is referred directly via a web page link. The web server that receives a request is capable of distinguishing whether it should return a web page that is already provided on the hard drive or run a program that creates one. Any such program can be called a CGI script. CGI describes a variety of programming tools and strategies. All data processing can be done by one program, or one or more other programs can be called from a CGI script. The name CGI script does not denote that a scripting language must be used. However, developers in fact prefer scripting languages, and PERL is the most popular one [W10].

Because of the nature of the protocol that allows for transfer of web pages and execution of CGI scripts, there is a unique challenge that must be faced by a software developer. Although users working with CGI-based programs have the same expectations as in case of local user interface, the interface must be designed internally in an entirely different way. The web transfer is a stateless process. It means that no information is sent by web browsers to the web servers that identify each user. Each time the new user interface is sent as a web page, it must contain all information about the current state of the program. That state is recreated each time a new CGI script is sent and increases the network traffic and time latency caused by limited bandwidth and time necessary to process data once again.

In addition, the server-side software must be prepared for inconsistent data streams. For example, a user can back off through one or more web pages and give a different response to a particular dialog box, and execute the same CGI script. At the time of the second execution of the same script, the data sent back with the request may already be out of synchronization from the data kept on the server. Therefore, additional validation mechanisms must be implemented in the software, which is not necessary in case of a single program.

64.2.7 PERL

Practical Extraction Report Language (PERL) is an interpretive language dedicated for text processing. It is primarily used as a very advanced scripting language for batch programming and for text data processing. PERL interpreters have been developed for most of the existing computer platforms and operating systems. Modern PERL interpreters are in fact not interpreters but compilers that pre-compile the whole script before running it. PERL was originally developed for Unix as a scripting language that would allow for automation of administrative tasks. It has many very efficient strings, data streams, and file processing functions. Those functions make it especially attractive for CGI processing that deals with reading data from the networked streams, executing external programs, organizing data, and in the end producing the feedback to the user in the form of a text based HTML document that is sent back as an update of the user interface. Support of almost any possible computing platform and OS, and existence of many program libraries make it a platform independent tool [W10].

64.2.8 PHP

PHP, a server-side scripting language, is the most popular technology that is especially suited for developing the dynamic, interactive Web sites. PHP, which originally stood for "Personal Home Page," is the open-source software initially created to replace a small set of PERL scripts that had been used in Web sites. Gradually, PHP became a general purpose scripting language that is used especially for web development and now called "Hypertext Preprocessor." PHP is embedded into HTML and interpreted on a web server configured to operate PHP scripts. PHP is a server-side scripting language similar to other server-side scripting languages like Microsoft's Active Server Pages and Sun Microsystems' Java Server Pages or Java Servlets. PHP resembles its syntax to C and PERL. The main difference between PHP and PERL lays in the set of standard built-in libraries that support the generation of HTML code, processing data from and to the web server, and handling cookies [W10]. PHP is used in conjunction with database systems, such as PostgreSQL, Oracle, and MySQL, to create the powerful and dynamic server-side applications.

The database comes to picture when it comes to storing a huge amount of information and retrieving back when it is needed by the web site. The database can contain information collected from the user or from the administrator. MySQL (written in C and C++) is an open source relational database management system (RDBMS) that is based on the structure query language (SQL) for processing data in the database and manages multi-user access to a number of databases. In a relational database, there are tables that store data. The columns define what kind of information will be stored in the table, and a row contains the actual values for these specified columns. MySQL works on many different system platforms, including Linux, Mac OS X, Microsoft Windows, OpenSolaris, and many others. PHP (Hypertext Preprocessor) and MySQL (a portable SQL server) together have made the task of building and accessing the relational databases much easier [WL04,U07].

64.3 Examples

With the increase of Internet bandwidth, the World Wide Web (WWW) could revolutionize design processes by ushering in an area of pay-per-use tools. With this approach, very sophisticated tools will become accessible for engineers in large and small businesses and for educational and research processes in academia. Currently, such sophisticated systems are available only for specialized companies with large financial resources.

64.3.1 Neural Network Trainer through Computer Networks

Several neural network trainer tools are available on the market. One of the freeware available tools is "Stuttgart Neural Network Simulator" based on widely C platform and distributed in both executable and source code version. However, the installation of this tool requires certain knowledge of compiling and setting up the application. Also, it is based on XGUI that is not freeware and still single type architecture—Unix architecture [MWM02].

During software development, it is important to justify which part of the software should run on the client machine and which part should run on the server. CGI is quite different from writing Java applets. Applets are transferred though a network when requested and the execution is performed entirely on the client machine that made a request. In CGI, much less information has to be passed to the server, and the server executes instructions based on the given information and sends the results back to the local machine that makes the request. In case of neural network trainer, it only makes sense to use CGI for the training process. To send the trainer software through computer networks for every requesting time is not a wise choice because this makes the training process slower and software is not protected. Therefore, it is important to develop methods that take advantage of networks.

This training tool currently incorporates CGI, PHP, HTML, and Java-Script. A CGI program is executed on a server when it receives a request to process information from a web browser. A server then decides if a request should be granted. If the authorization is secured, a server executes a CGI program and sends the results back to a web browser that requested it. The trainer NBN 2.0 is developed based on Visual Studio 6.0 using C++ language hosting on a server and interacting with clients through PHP scripts. Its main interface is shown in Figure 64.1.

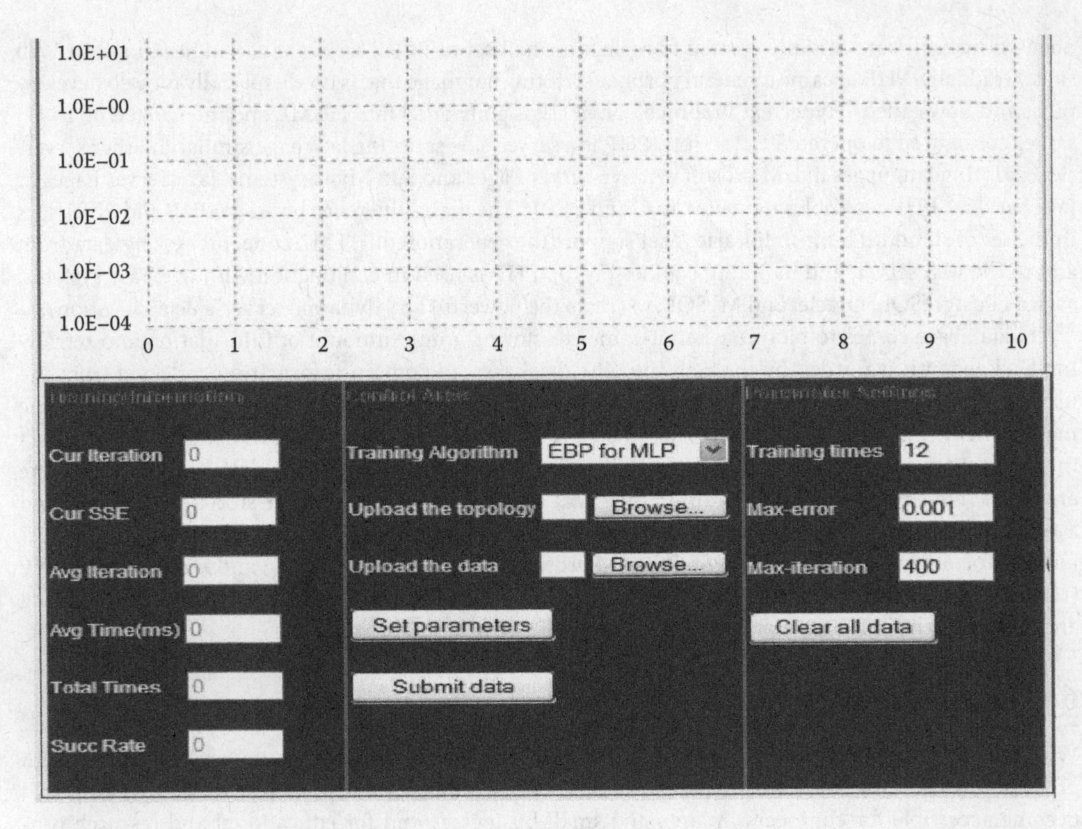

FIGURE 64.1 User interface of NBN 2.0.

The neural network interface will receive requests from users with uploading files and input parameters to generate data files and then send a command to the training software on the server machine. When all data requirements are set up properly, the training process will start. Otherwise, the training tool will send error warnings back to clients. If the training process is successful, the training result will be generated. The training result file is used to store training information and results such as training algorithm, training pattern file, topology, parameters, initial weights, and resultant weights (Figure 64.2).

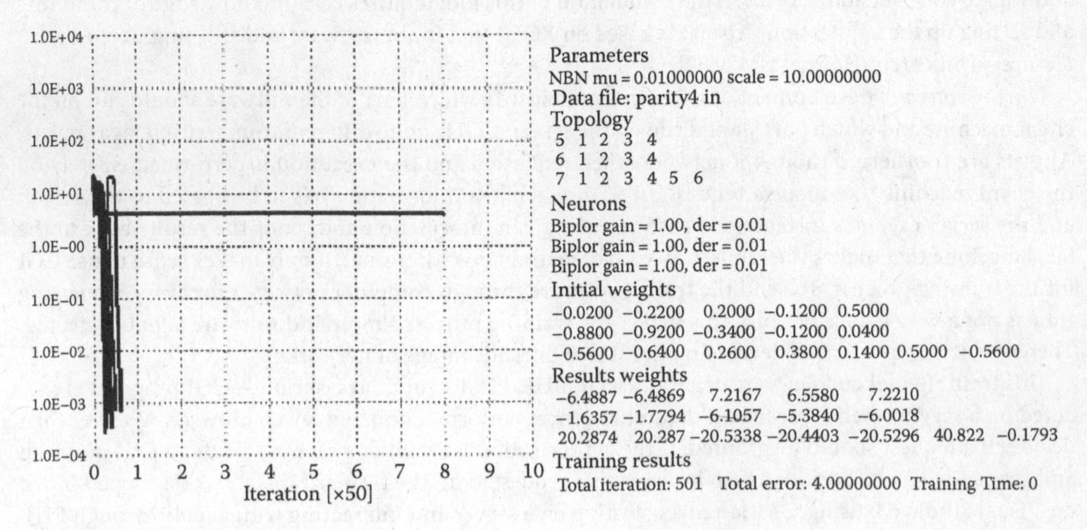

FIGURE 64.2 Training results.

The neural network trainer can be used remotely through any network connection or any operating system can be used to access it, making the application operating system independent. Also, much less installation time and configuration time is required because the training tool locates only on central machine. Many users can access at the same time. Users can train and see the training results directly through networks. And the most important thing is that the software developers can protect their intellectual property when network browsers are used as user interfaces.

64.3.2 Web-Based C++ Compiler

During the process of software development, more than one compiler package is frequently required. Some products are known to be very useful for locating errors or debugging, while others perform extremely well when a program or library is in the final stage of development and should be optimized as much as possible. Also, when facing obscure error messages, which may result in a time-consuming search for the error, a different error message from the second compiler frequently cuts that time dramatically [MW00].

Therefore, students should be to some extent exposed to different compilers at some point in their software courses curriculum. Although all necessary software is installed in the computer laboratories, most students prefer to work on their computers at home or dormitory and connect to the university network. That situation creates an unnecessary burden either for the network administrators who have to install additional software on many machines of non-standard configuration, or on students who must purchase and install on their own several software packages along with their full course of study.

In order to solve the problem at least partially in the area of programming, a software package was developed that allows for web-based interfacing of various compilers. Web-page based front end allows them to access without any restrictions regarding the computer system requirements, thus allowing for their use on different operating system platforms and also on older machines with lesser performance.

A common front end is used for all compilers and is presented in Figure 64.3. This HTML page allows for selecting a vendor and a language, and for setting a few basic compilation options. User uses copy and paste commands to enter the source code into the compiler front end. There are three menus located under the source code area as shown. The middle menu is used to select the programming language

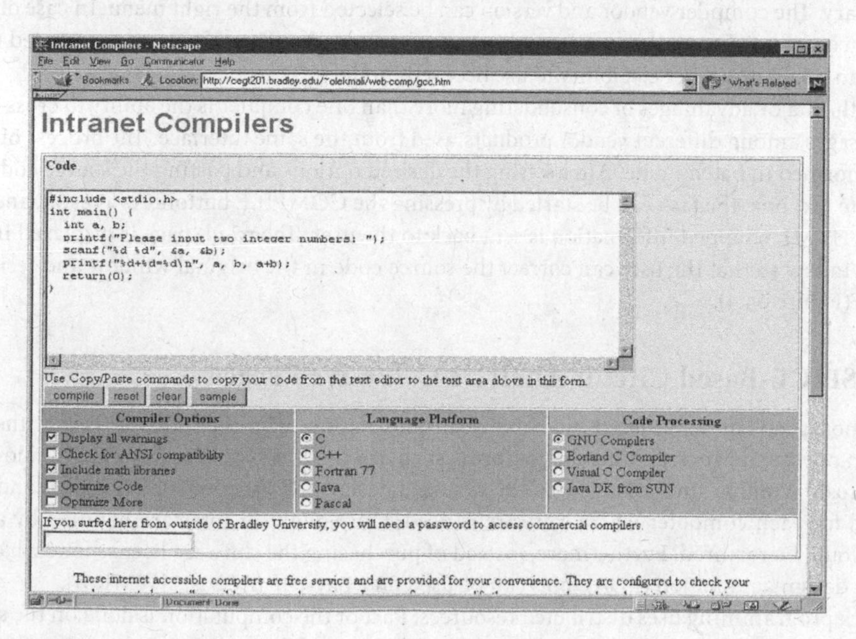

FIGURE 64.3 The common Web-based front-end to C++ compilers.

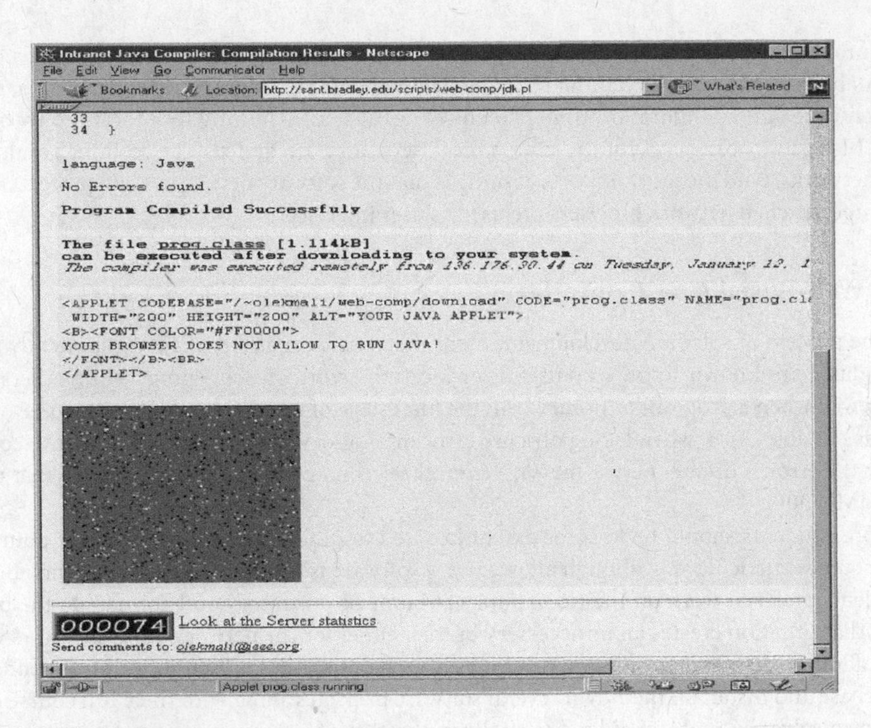

FIGURE 64.4 The result page.

while the right menu is used to select the compiler vendor. Currently the Intranet Compilers package supports C, C++, Ada, Fortran, Modula, Pascal, and Java languages. It utilizes MinGW (Minimalistic GNU for Windows ver. 3.4), Borland (ver. 5.0), and Microsoft (VS 2005) compilers for C and C++, compiler for Fortran and Pascal, and Sun's JDK (ver. 2.6) for Java.

One of the preset compiling configurations can be selected from the left menu. The user can decide whether aggressive binary code optimization or strict error checking and ASNI style violation checking are necessary. The compiler vendor and version can be selected from the right menu. In case of selecting one of the commercial compilers while working at off-campus location, the user is requested to input a password to verify his or her elegibility to use licensed products.

One of the major advantages of consolidating more than one compiler is the ability to cross-reference error messages among different vendor products used from the same interface. The process of compilation is performed in batch mode. After setting the desired options and pasting the source code into the appropriate text box, the task can be started by pressing the COMPILE button. As a result, another web page with HTML wrapped information is sent back to the user. The result page is displayed in another browser window so that the user can correct the source code in the original window and resubmit it if necessary (Figure 64.4).

64.3.3 SPICE-Based Circuit Analysis Using Web Pages

The common problem being faced by many electronic engineers in the industry is that their design tools often operate on several different platforms such as UNIX, DOS, Windows 95, Windows NT, or on Macintosh. Another limitation is that the required design software must be installed and a license purchased for each computer where software is used. Only one user interface handled by a network browser would be required. Furthermore, instead of purchasing the software license for each computer, electronic design automation (EDA) tools can be used on a pay-per-use basis [WMR01].

Network programming uses distributed resources. Part of the computation is done on the server and another part on the client machine. Certain information must be frequently sent both ways between the client and server. It would be nice to follow the JAVA applet concept and have most of the computation done on the client machine. This approach, however, is not visible for three major reasons:

- EDA programs are usually very large and thus not practical to be sent entirely via network as applets.
- Software developers are giving away their software without the ability of controlling its usage.
- JAVA applets used on-line and on demand are slower than regular software.

The Spice implementation used in this presentation is just one example of a networked application. An application called the Spice Internet Package has been developed for use through the Internet and Intranet networks. The SIP provides an operating system independent interface, which allows Spice simulation and analysis to be performed from any computer that has a web browser on the Internet or Intranet. The SIP has a user-friendly GUI (Graphical User Interface) and features include password protection and user accounts, local or remote files for simulation, editing of circuit files while viewing simulation results, and analysis of simulated data in the form of images or formatted text.

In the case of the Spice Internet it would be impossible to send the Spice engine through the network every time it was requested and this would be extremely slow. Java technology could also be used for functions like generating and manipulating graphs and implementing the GUI on the client side. The SIP program currently incorporates CGI, PERL, HTML, and JavaScript. A unique feature of the SIP versus other Spice simulators is that it is operating system independent. Anyone that has access to the Internet and a web browser, such as Mozilla Firefox or MS Internet Explorer, can run a Spice simulation and view the results graphically from anywhere in the world using any operating system (Figure 64.5).

The server is configured to accept requests from web browsers through network connections. The server processes the request for Spice simulation or analysis and returns the results to the requesting web browser as an HTML document. The heart of the software is a server-located PERL script. This script is executed first when the user logs in to SIP. Then each time the user selects any activity, a new dialog box in the form of an on-the-fly generated JavaScript enhanced HTML web page is sent back. Such pages may contain a text editor, a simulation report, a graphic postprocessor menu, or graphic image of plotted results. To complete some tasks, the PERL script may run additional programs installed only for the server such as the main CAD program, i.e., Berkeley Spice; GnuPlot, which generates plots; and some utility programs (netpbmp) to convert plotter files into standard images recognized by all graphical web browsers.

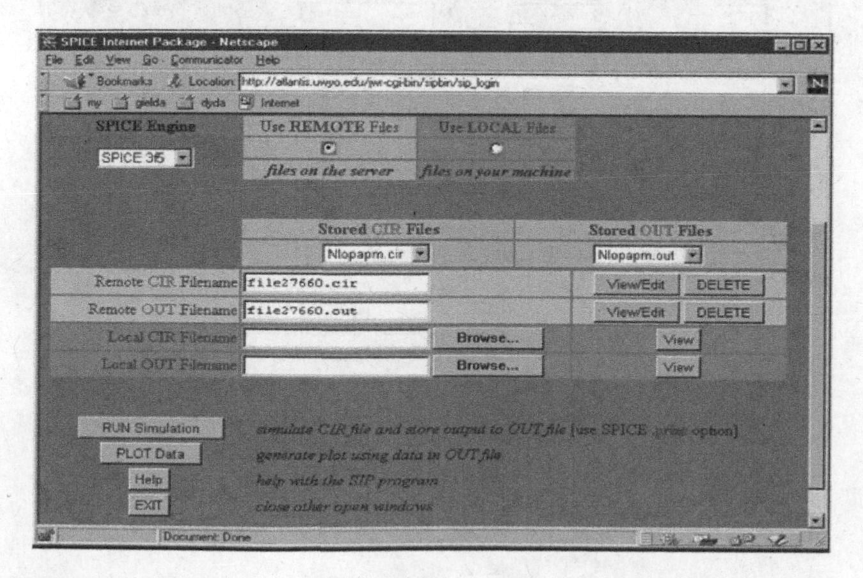

FIGURE 64.5 Graphical user interface for SIP package.

SIP is a very good example for network traffic considerations. The amount of data produced by a single simulation may differ significantly from a few hundred bytes to a few hundred KB, depending on the number of simulation steps requested. In case of large data files, it is better to generate graphical images of plots and send them to the user. Users frequently inspect the obtained results a few times, for example, by changing the range or variables to display. In the case when there may be many requests for different plots of the same data, it could be better to send the data once together with a custom Java applet that could display the same information in many different forms without further communicating with the server.

Several features make the Spice Internet Package a desirable program for computer-aided engineering and design. Only one copy of the Spice engine needs to be installed and configured. One machine acts as the server and other machines can simultaneously access the Spice engine through network connections. Remote access to SIP allows users to run Spice simulations from any computer on the network, and that might be from home or another office in another building or town. Also, the current Spice engine being used is Spice3f5 from Berkeley, which allows an unlimited number of transistors, unlike various "student versions" of Spice programs that are available (Figure 64.6).

Computer networks are also used in the systems for controlling objects such as robots, database management, etc. [MW01,WM01,PYW10]. A web server is used to provide the client application to the operator. The client uses custom TCP/IP protocol to connect to the server, which provides an interface to the specific robotic manipulators. Sensors and video cameras provide feedback to the client. Many robotic manipulators may be connected at a time to the server through either serial or parallel ports of the computer. In case of autonomous robots, the servers pass the commands addressed to the robot. In case of a simple robot, the server runs a separate process that interfaces to the robot. Data monitoring and control through a computer network or some other proprietary network is no longer prohibitively

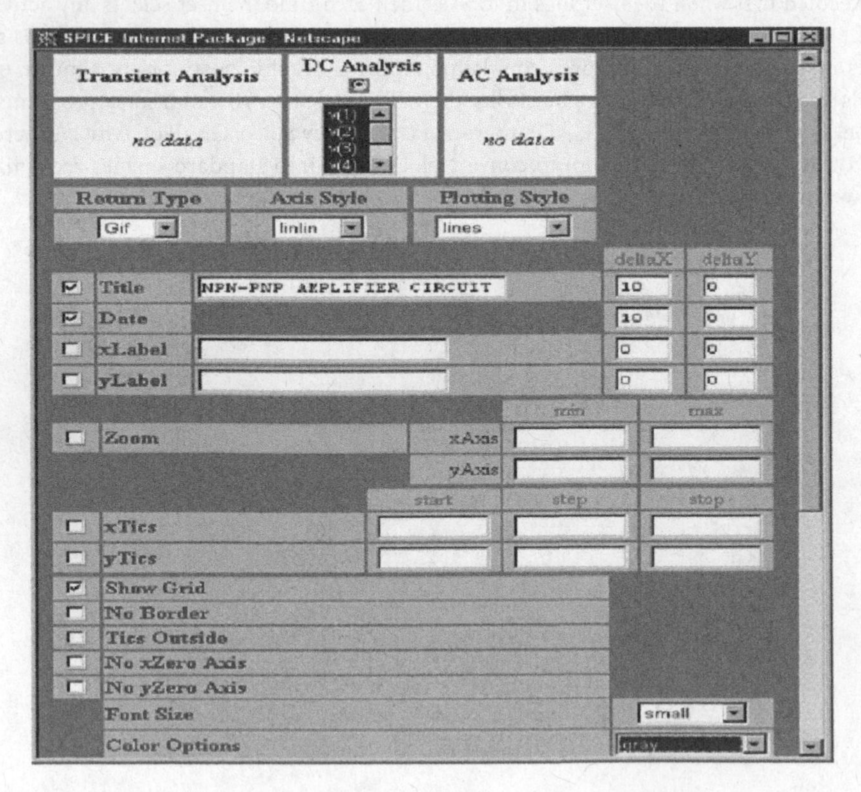

FIGURE 64.6 Upper part of the plotting configuration screen.

expensive, and thus restricted only to industrial or luxurious products. Continuously decreasing cost of microprocessors and network interfaces have opened additional possibilities in the home automation applications.

64.4 Summary and Conclusion

This chapter shows how to use the new opportunity created by Internet technologies for efficient and platform independent usage of software applications. The presented examples are just examples, but they show a way in which the technology can be implemented. They also illustrate how significant is the impact of the particular data flow on the programming tools and approaches taken to implement the Internet access to the server installed software. The authors are convinced that such approach will spread and it will revolutionize the general approach toward software development.

This approach will also have a synergetic effect on the development of better software applications since the market will increase significantly. Availability of software application via the Internet will boost design process in many new communities and improve our education processes at universities by allowing students to use the same sophisticated software as is used by leading industries.

Several features make the software run over Internet a desirable program. Only one copy of the software engine needs to be installed and configured. One machine acts as the server and other machines can simultaneously access the server engine through network connections. Remote access to server allows users to run simulations from any computer on the network. This may be from home, or another office in another building, or town.

References

[MW00] A. Malinowski and B. M. Wilamowski, Web-based C++ compilers, *ASEE 2000 Annual Conference*, St. Louis, MO, CD-ROM session 2532, June 18–21, 2000.

[MW01] A. Malinowski and B. M. Wilamowski, Controlling robots via internet, *Proceedings of the First International Conference on Information Technology in Mechatronics*, Istanbul, Turkey, pp. 101–107, October 1–3, 2001.

[MWM02] M. Manic, B. M. Wilamowski, and A. Malinowski, Internet based neural network online simulation tool, *Proceedings of the 28th Annual Conference of the IEEE Industrial Electronics Society*, Sevilla, Spain, pp. 2870–2874, Nov. 5–8, 2002.

[PYW10] N. Pham, H. Yu, and B. M. Wilamowski, Neural network trainer through computer networks, *24th IEEE International Conference on Advanced Information Networking and Applications 2010*, pp. 1203–1209, 2010.

[U07] L. Ullman, *PHP 6 and MySQL 5 for Dynamic Web Sites: Visual QuickPro Guide*, Peachpit Press, Berkeley, CA, Dec. 29, 2007.

[W10] B. M. Wilamowski, Design of network based software, *24th IEEE International Conference on Advanced Information Networking and Application 2010*, pp. 4–10, 2010.

[WL04] H. E. Williams and D. Lane, *Web Database Applications with PHP & MySQL*, 2nd edn., O'Reilly Media, Sebastopol, CA, May 16, 2004.

[WM01] B. M. Wilamowski and A. Malinowski, Paper collection and evaluation through the Internet, *Proceedings of the 27th Annual Conference of the IEEE Industrial Electronics Society*, Denver CO, pp. 1868–1873, Nov. 29–Dec. 2, 2001.

[WMR01] B. M. Wilamowski, A. Malinowski, and J. Regnier, Internet as a new graphical user interface for the SPICE circuit simulator, *IEEE Transactions on Industrial Electronics*, 48(6), 1266–1268, Dec. 2001.

[WT08] L. Welling and L. Thomson, *PHP and MySQL Web Development*, 4th edn., Addison-Wesley Professional, Boston, MA, October 11, 2008.

65

Semantic Web Services for Manufacturing Industry

Chen Wu
Curtin University of Technology

Tharam S. Dillon
Curtin University of Technology

65.1 Background....................65-1
65.2 Aims....................65-3
65.3 Approach....................65-3
Ontology-Driven Architecture • Multisite Issues and Rapid Reconfigurability • Customizations of Product and Materials Flow Monitoring • Optimization • Data, Process and Timing Consistency and Conformance • Customer-Centered Design through the Concept of Customer Value
65.4 Conclusion....................65-9
References....................65-10

65.1 Background

Many industries have introduced a new set of requirements on the manufacturing in the fabrication and assembly of products and/or project management of large projects. These are characterized by several features: they require the ability to support work carried out at sites remote from the final assembly site where primary materials and appropriate numbers of suitable skilled workers are available, transportation of large components to the remote fabrication sites, and final assembly of these at the site. The assembly site also needs rapid construction of these services. The manufacturing companies are often required to meet strict deadlines on construction so that they do not experience any delays, and a high degree of customization of the products to suit different levels of employees and different terrains and environmental conditions, which can sometimes be very unfriendly. These requirements have led to the manufacturing industry seeking to develop new modes of operation, which must allow for (1) multisite production of components with final assembly at a remote site on every occasion, (2) a high degree of reconfigurability suitable for a combined construction/manufacturing industrial setting capable of supporting the production of highly customized units or products, (3) an ability to carry out manufacturing and assembly to varying levels of granularity at multiple sites, (4) creation of major components through assembly of subcomponents sourced from different producers or factories, (5) the ability to monitor and control the flow of materials and partially assembled components within timing and resource constraints, (6) the ability to track and trace the production of a particular customized unit and to inform the relevant staff working on that unit at a subsequent stage in the process, (7) automated data collection in widely varying situations, and (8) ensuring dependency and timing constraints are fulfilled for feasible and on-time delivery. These requirements can only be met through new cutting edge IT and Web-based methodologies and technologies.

Previous methods for information-based production engineering methods such as client/server, blackboard methods, etc., have not addressed the multisite distributed production development environment with changing final assembly locations and how to manage such development efficiently. They, among other things, require (1) the site from which the data is entered is relatively fixed, (2) the production team members are collocated sufficiently to be able to communicate face to face to resolve issues and have a similar understanding of the underlying terminology and concepts and relationships within the domain, and (3) the production team members are sufficiently computer literate in the software systems deployed. The following scenarios demonstrate the need for a multisite distributed product development methodology:

- As the development project proceeds through the specification, design, and implementation phases, the project and contract developer who produced the project, buildings, and works specification and proposal for solution may be physically far removed from the design and component manufacturing teams, on either a regional or international basis. This physical distance could become a crucial issue if the specifications are not complete, are ambiguous, or are continually evolving.
- When the development is decomposed based on specialization, each implementation group could reside at different sites. The different components have to be integrated, and differences in the interpretation of specifications can lead to incompatible components.
- Since such development often uses a component-based approach, components may be developed by multiple remote teams, and the integration team who assemble the final product will be at a remote site; although working on the same product, different expert groups have different terminologies and often find it hard to overcome communication difficulties. This is exacerbated by the fact that the domain experts will use somewhat different terminology to the developers.
- Quality assurors and/or human factor evaluators involved in large project development, quality measurement, component-based testing, standards procedures auditing, user acceptance and usability evaluation may be located at sites remote to the designers and manufacturers of the customized component developers. Usability is best performed with a group of people who resemble the target users (e.g., employees of the company for whom the development is carried out).
- A recent trend for large project contracts is for the project managers or project directors to be in one place, and the team leaders and development teams to be spread across several countries or cities. It is crucial that smooth communication can take place to allow the project managers to monitor the progress and quality of production, so that the whole project can be delivered on time and within budget.
- The clients and the developers may also not be from the same region or city or the same country. This could arise from the organization not being able to identify the best quality team in the local area, or the system contract being let through a tender process. The challenge is to get the right model of development, specification, and solution and to ensure the end product meets the customer's needs.

The need for this multisite production is therefore of considerable significance. However, most of the production engineering process models that are in existence assume a centralized approach to development. As shown in the above examples, there are many situations where such an assumption is inappropriate. This assumption is also present in many of the current production engineering methodologies, which do not address multisite production. Most project management approaches do not consider multisite production. Finally, we see that many technologies have not yet become mature to facilitate multisite production. Those process models, methodologies, technologies, and approaches cannot be followed linearly for a multisite situation. Therefore, it is important to identify the key issues and solutions that are the subject of this chapter, which moves away from this "central" assumption to allow for different parts of the development team to be at physically different sites.

65.2 Aims

The main aims of this chapter are to describe a highly reconfigurable multisite production system that

1. Appropriately addresses the issues of communication, coordination, situation awareness for multisite manufacturing, and construction.
2. Provides a high degree of reconfigurability of process flows and/or manufacturing workstations to allow for production of highly customized products as well as incorporation of new materials, subcomponents, tools, and methods of production.
3. Allows information to be carried on the requirements of a particular product during its development.
4. Provides material flow monitoring and control across multiple sites.
5. Enables optimization of the process flows.
6. Allows the enforcement of constraints reflecting dependencies and inconsistencies for a given time or between different intervals of time; these should also be able to take into account limitations of resources, skills available at a particular site at a given time.
7. Allows the assessment and tracking of timing information related to particular events and processes including end-to-end timing.
8. Determines the conformance of the progression of a product's development against the proposed plan of production with the ability to reconfigure the production plan if necessary to meet timing and quality parameters.
9. Incorporation of the notion of customer centered design based on principles of customer value.
10. The composition and function of the various PEMFC parts are further discussed in the following subsections.

65.3 Approach

65.3.1 Ontology-Driven Architecture

To resolve the above issues, we use the development of a base ontology, which serves as a structure for an underlying knowledge base, allowing for full interaction, comprehension, and customization of products between remote teams. There are two main reasons that justify the need for an ontology-driven architecture.

The first is the fact that we need to have a conceptual framework that enforces an agreement on how information should be organized, without losing any of the flexibility of allowing people to express and view parts in their own familiar expression language (which is a very different concept compared to most areas of expertise). An ontology, which is defined as a shared conceptualization of some domain [Gruber 93], will permit coherent communication between the remote teams. Understanding the meaning of shared information on the web can substantially be enhanced if the information is mapped onto ontology. The base ontology, which encompasses diverse, complex, domain knowledge, technology and skills, will ensure a common ground for distributed collaboration and interactions.

The second reason why ontology is needed in this environment is to allow Web security and confidentiality to be defined more conveniently at the ontology level. One way to restrict user access to relevant information only within the base ontology is by creating a sub-ontology appropriate for each user profile. This is a notion currently used more frequently in ontology research although the naming conventions vary (e.g., Ontology Commitment [Jarrar 02], [Spyns 02], Ontology Version [Klein 02], Materialized Ontology View [Wouters 02]). Thus, the level of security and access can be defined at the conceptual level (sub-ontology level) where the semantics of related information and documents are kept, instead of hardwiring them at the instances or document level.

In general, four aspects of multisite production can be distinguished:

- *Communications* refers to the basic ability to exchange information of any required form in the collaboration process between the parties involved.
- *Integratability* relates to the ability of components from different sources to mesh relatively seamlessly and the ability to integrate these components to build the final product.
- *Coordination* focuses on scheduling and ordering tasks performed by these parties.
- *Awareness* and *cooperation* refer to the knowledge of the production process of work being performed, progress achieved, and decisions made by others.

The ontology-driven architecture proposed reduces conceptual and terminological confusion by providing a unifying framework within an organization through

- The normative model, which creates a semantic for the system and an extendible model that can later be refined and, which allows semantic transformations between different contexts from different sites.
- The network of relationships that can keep track of progress and monitoring quality and productivity from development teams in different locations.
- Creating consistency, it provides unambiguous definitions for terms in the production system.
- Allowing integration by providing a normative model and integration of components developed in different places can be achieved.

The integration of a number of (existing) ontologies for other domains can be used to augment the functionality of any framework using the base ontology. This base ontology would in general terms overlap with the external ontologies. A distributed workflow technique across multiple sites with appropriate shared repositories, an ability to enforce preconditions and post-conditions on development activities and the artefacts developed, plus task tracking will assist with overall coordination. As there are several aims that must be addressed, in what follows, we will discuss the approach taken for each of these aims in turn.

65.3.2 Multisite Issues and Rapid Reconfigurability

The multisite issues will be addressed through the use of an ontology-based approach. The three ontologies that have been identified as necessary are as follows:

1. A process ontology
2. A products ontology
3. A components ontology

The process ontology develops a representation of the different types of processes involved in manufacturing/construction. Previously taxonomies for manufacturing processes have been developed by [Holland 00] and refined by [Lastra 04]. These ideas were extended to an ontology by [Delamar 06]. These were largely relevant for mechanical production processes at a single site. Specifically they did not include construction activities nor did they allow specification of activities at multiple different sites. We intend to use part of the taxonomies previously developed and extend these to incorporate joint manufacturing/construction processes as well as allow for multiple sites. The product ontology will represent the different types of products or units that will result from the manufacturing/construction activity. They will involve classification of the products by features as well as capturing the geometrical layout, which is particularly important for construction. Here, the relationships involved and the layout configurations will also have to be captured by the product ontology.

The component ontology will provide a representation of all the components involved in the manufacture/construction activity at the different levels of granularity. As each large granularity component is likely to consist of subcomponents itself and so on, it is important these are all represented within the ontology. Here, again, it will be important to specify the functionality, geometrical aspects and the source of the component whether in-house fabricated or externally sourced. The first step is to understand the concepts and requirements of an ontology for a multisite production development environment, or stated another way, the ontological content.

First, we begin by properly defining the concept of a valid ontology in the specified domain. We need to ensure that the defined ontology conceptually represents the perceived domain knowledge through its concepts, attributes, taxonomies, relationships, and instances. However, in addition to this, other previously developed taxonomies for each of the manufacturing will also provide important inputs. Next, we also need to define the restrictions that need to be applied to the above definitions in order to ensure that it is practically usable. This is a refinement process of the high level abstractions defined earlier.

Secondly, the most appropriate representation to model the ontologies is needed. OWL will be useful at the implementation level; however, it will be necessary to use a graphical representation of ontologies at a higher level of abstraction to make it easier for the domain experts to understand and critique, and importantly, this model should be able to capture the semantic richness of the defined ontology In our previous work, we have developed such a notation [Wongthongtham 06] and utilized it successfully for the development of several ontologies including the Trust and Reputation Domain [Chang 06,07], the Software Engineering Domain [Pornpit 05,06], the Disease Ontology [Hadzic 05], and the Protein Domain [Sidhu 05,08]. We intend to utilize this for our representation language for the modeling phase. The implementation will then be carried out using OWL classes.

Next, we define the ontology-based architecture. The ontology-based architecture is grounded on the notion of a base ontology sub-ontologies [Wouters 02,03] or commitments [Jarrar 02, Spyns 02]. These sub-ontologies are used as independent systems (in functionality) for the various decentralized development teams. This would allow for a very versatile and scalable base for multisite production.

As our intention is to work with numerous sub-ontologies that provide custom portals for different expert groups to access information and communicate with other groups (at different locations), a solid representation becomes crucial as each sub-ontology can be viewed as an independent functioning ontology [Wouters 02, Jarrar 02], all of them have their own distinct representation, which is related or similar to the original but not necessarily the same and possibly not even a straight subset of elements of the general base ontology. Our previous work has provided a rigorous basis for doing this [Wouters 02]. A number of key issues are given here and some small examples of how extensions are added to the base ontology, whether new elements or existing external ontologies. These sub-ontologies define the essential elements of the ontology-based architecture.

Awareness of what work is being done by others becomes a problem in a multisite environment. Different teams might not be aware of what tasks are carried out by others potentially leading to issues such as two groups performing the same tasks, or tasks not performed at all, or incompatibilities between tasks (ignorance of who to contact to get proper details about a certain task). If everyone working on a certain project is located in the same area then situational awareness is relatively straightforward but the overheads in communication to find answers to the same question in a multisite environment can become very large. The proposed base ontology will deal with this on a conceptual level. Not only modeling the development process, but also the development methodology is the key. In this case, it would mean that the ontology not only provides the lower level concepts (such as a component that needs to be developed) but it also models concepts and semantics on the level of the interaction of the development team such as Task Responsibility. In practice, this allows teams to always be aware of who is performing each task.

Secure access control, which is another issue, considers the appropriateness of groups to access data. This is also a major issue in the single location development, but an issue that is easier to control, as the physical closeness automatically allows for a tight monitoring. In a multisite environment, it is easy to lose track of what the exact responsibilities are, and who should or should not access certain parts. The monitoring that is often taken for granted in a single site environment can become a huge problem.

Even if the environment manager is still aware of who should not have access, ensuring that it happens is another issue. Again, ontologies and, more specifically, the use of a base and several sub-ontologies, provide a solution to this problem. By integrating an access concept (related to all types of information/tasks/components in the project) in the base and sub-ontologies, it can be specified that certain groups do not have access (read, write, none or all) to specific parts of the project.

The second issue relates to rapid reconfigurability (aim2) and this has two dimensions, namely, (1) reconfigurability to bring about rapid production of a customized set of products demanded by the customer for delivery in a compressed timeframe and (2) introduction of new processes and/or components of a different level of granularity or made from new materials.

Previously, in order to address dimension one, a considerable amount of work has been done on flexible manufacturing systems (FMS). However, these systems are inappropriate for the rapid production of customized products because they have only a fixed number of processes that are in place and these will probably be inadequate for the range of customization of products that might be required in product construction. In addition, FMSs frequently absorb considerably more resources than are needed to produce the specific product currently being produced. For these reasons, there has been a tendency in recent research to move to other means of achieving reconfigurability. In this chapter, we explore the two different approaches, namely, (1) use a combination of the ontologies and multi-agents [Hadzic 07] and (2) use the semantic Web paradigm through a combination of ontologies and Web services [Wang 04], [Dillon 07a,07b]. In approach (1), ontologies will be used as common knowledge base by agents working at different sites. The agents themselves will be goal oriented and will collectively or individually seek to achieve these, updating relevant instances of ontology classes when appropriate to allow monitoring of the status of various processes through this. Our previous work on Software Engineering Ontology [Wongthongtham 06] and Project Management Ontologies [Cheah 07] will help provide important inputs on the way this can be achieved. The approach adopted for the design of Semantic Web services will use further extensions of the RESTFUL and Web-based approaches and makes use of ontologies to help in the identification of Web services that will be specified through the use of the triple computing paradigm [Dillon 07a,07b]. Note this will not use the WSDL-based description but the use of a tuple space to locate the required Web service. Previous studies [Pilioura 04], [Banaei-Kashani 04] suggested that the existing Web services architecture based on remote procedure call (RPC) is not indeed "Web oriented" because RPC is more suitable for a closed local network and raised concern that "Web services do not have much in common with the Web" [Buschmann 96], and lead to potential weakness such as scalability, performance, flexibility, and implementability [Piers 03].

Good manufacturing system architecture is crucial for multisite production. But how to define "good" or "bad?" In the keynote paper [Dillon 07a,07b] at *IFIP NPC 2007 Conference*, we reviewed four main architectural styles for SOA applications, namely, matchmaker, broker, peer-to-peer, Web-oriented. The Representational State Transfer (REST) [Fielding 02] uses a resource identifier (URI) to provide an unambiguous and unique label for one particular Web resource. In the RESTful architectural style, all resources are accessed with a generic interface resulting in a dramatic decrease in the complexity of the semantics of the service interface during the service interaction W3C has recognized this REST style as an alternative to WSDL. Triple Space Computing [Bussler 05] is built on top of several technologies: Tuple Space [Gelernter 85], Publish/Subscribe paradigm [Eugster 03], Semantic Web, and RDF [Vinoski 02]. [Klyne 04] Tuple Space employs the "persistently publish and read" paradigm by leveraging the Tuple Space architecture and APIs. From an architecture perspective, Triple Space Computing is, in effect, based on the natural confluence of Asynchronous Broker and RESTful styles. The basic interactions between service provider and requester are rather straightforward: the service provider can "write" one or more triples in a concrete identified Triple Space. The service requester is able to "subscribe" triples that match with a template specified against its interests in a particular concrete Triple Space. Whenever there is an update in the spaces, the Triple Space will "notify" related service requesters indicating that there are triples available that match the template specified in its preceding subscription, as shown in Figure 65.1.

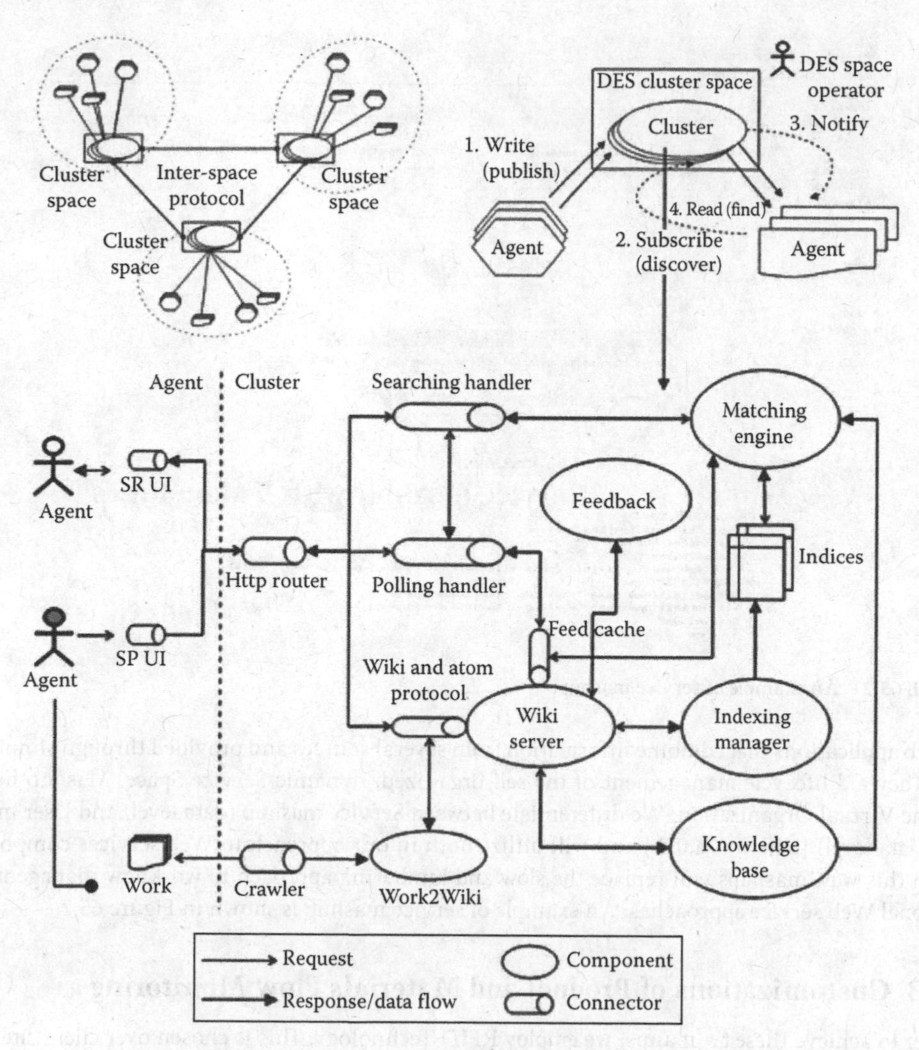

FIGURE 65.1 SOA-based system architecture for multisite production.

The notified service requesters "read" triples that match with the template within a particular transaction or the entire concrete space, and further process the triples accordingly. It provides intelligent middleware (broker like), to manage the spaces without requesting each service provider and requester to either download or search through the entire space. Moreover, it needs to provide security and trust while keeping the system scalable and the usage simple. [Krummenacher 05] proposed a minimal architecture for such provider middleware. [Bussler 05] identified a number of requirements for Triple Spaces (providers): autonomy (including four basic forms of autonomy, viz., time, location, reference, and data schema), simplicity, efficiency, scalability, decentralized architecture, security and trust mechanisms, persistent communications, and history. In order to overcome the lack of support for semantics-aware matching, Triple Space utilizes RDF to represent and match the machine-processable semantics. It is a promising, if immature, Web services architectural style and may represent the future paradigm for designing and implementing a truly service-oriented architecture. Based on our analysis and the criticism from the literature, we develop an SOA architectural design with the extended "Web-Oriented" style. [Dillon 07b], based on the extension of the triple space paradigm. We think of the Web as a platform of services, and applications are built by composing services. We utilize Web 2.0 ideas in developing these compositions. In current Web2.0 settings, service composition comes as the Mashup,

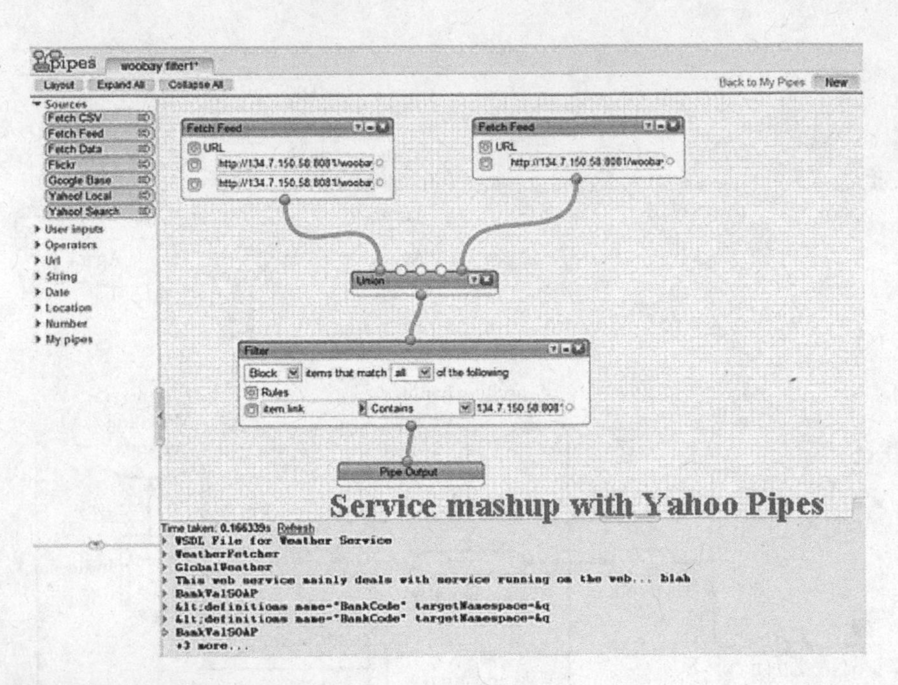

FIGURE 65.2 An example of service mashup.

i.e., Web applications that combine information from several sources and provided through simple Web APIs. They aid lifecycle management of the self-organized, dynamic Service Space. Mashup helps to form the Virtual Organization. We differentiate between Service mashup (data level) and User mashup (meta-data level) [Dillon 07a,07b] and will utilize both in our approach to Web services composition. Note in this way, mashups will replace the slow and lumbering approach to workflow management in traditional Web service approaches. An example of service mashup is shown in Figure 65.2.

65.3.3 Customizations of Product and Materials Flow Monitoring

In order to achieve these twin aims, we employ RFID Technology. This is chosen over client site workstations and barcode readers as it can be flexibly deployed at widely varying sites or new sites and is less dependent on an exact orientation of presentation of the barcode to the reader and can be employed in the presence of employees with relatively poor computer literacy. The use of RFID technology essentially requires the user of RFID devices and antennas. The RFID devices will be embedded in a particular customized product that is being produced and it could carry information and rules on the next step in the process that need to be carried out with the product itself. Such an approach has already been trialed with the automotive industry and in the production of electronic boards by IBM. In addition, it can be used to monitor the movement, flow, storage, and consumption of materials.

Also RFID technology will allow the determination of the precise quantities of certain materials and components at a given site effectively. This last factor is of considerable importance as the assembly of the products will frequently require the movement of specialized staff to the remote site to help carryout certain tasks and it would be extremely costly if they were sitting around doing nothing because certain component necessary for some stage of the work are not present at the site. In addition, to the actual gathering of data and information, which can be done by RFID technology, it is necessary to utilize this information to build a picture of the state of the overall state of production. Here we intend to explore two approaches creating a virtual representation of an RFID tag within the framework of a Web service. Such an approach has been attempted recently by [Chen 07] for the simpler problem of warehouse inventory management.

65.3.4 Optimization

The process flows are represented as network flow graphs representing all the data and process dependencies. A cost will be associated with each branch of the graph representing the cost of the process in the real world. All constraints including resource constraints and dependencies will then be represented. Given that some of these could be linear or even perhaps nonlinear, the network flow optimization methods used in [Sjèlvgren 83] can be used for the optimization of the problem. The methods have been shown to capable of dealing with problems with over 20,000 variables and limited sets of nonlinear constraints and currently constitute the most widely used methods worldwide by the power industry for medium term optimization of hydrothermal systems.

65.3.5 Data, Process and Timing Consistency and Conformance

The model proposed in this chapter utilizes Colored Petri Net (CPN) [Jensen 96], which will enable us to represent processes and subsystems involving complex preconditions, multiple threads of execution, and allow the use of guards to enforce security conditions. The CPNs also have clearly defined semantics. As they are a high level Net that allows folding of the Net, this helps deal with the problem of dimensionality and allows scaling up of the model. CPN is the high level Petri net, which allows for the representation of the multilevel behavioral abstractions through place/transition refinements. It allows the use of colored tokens, complex predicates associated with transitions, and the use of guards to control the enabling of transitions before firing. The advantages include avoiding cluttering the diagrammatic representation with excessive formalism.

This approach has four characteristics on the encoding: (1) events with associated pre/post-conditions, (2) individual methods that are activated to service events, (3) flow of control and data within the system, and (4) interactions between the controlled component processes and the interface, including the instance data and method. In contrast, the domain-specific behavioral aspects relate to the sequencing of events and the satisfaction of the pre- and post-conditions related to a specific service interface at the right point in the sequencing. The authors have previous experience at modeling FMS systems using high-level Petri nets [Zurawski 92] and Inter Object Interactions using CPNs [Hanish 97].

65.3.6 Customer-Centered Design through the Concept of Customer Value

It is important in the design of the systems and techniques that there is not only a manufacturer, constructor, and developer view embedded in the system, but also a customer-centered view. We intend to incorporate this customer-centered view through the use of the notion of customer value and also to carry out the setting up of the case studies incorporating these ideas of customer value.

The approach in this chapter is vital to the manufacturing and construction industries in which there are severe shortages, and in the IT industry in the area of ontology engineering. This is evidenced by the enormous number of unfilled academic positions and also the lack of quality project managers in industry. This is particularly the case where multisite production is being widely carried out in Australia and worldwide.

65.4 Conclusion

In this chapter, we proposed a reconfigurable multisite production system supported by a number of key emergent technologies: Semantic Web (i.e., ontology architecture), Web 2.0, and Web services technologies. The proposed solution aims to solve two key issues during the manufacturing processes: customization of products and multisite production. This enables a modern manufacturing system to carry out manufacturing and assembly to varying levels of granularity at multiple sites and to create major product components through the assembly of subcomponents sourced from different producers or factories in a highly flexible manner.

References

[Banaei-Kashani 04] F. Banaei-Kashani, C.-C. Chen, and C. Shahabi, WSPDS: Web services peer-to-peer discovery service, *International Symposium on Web Services and Applications*, Las Vegas, NV, 2004, pp. 733–743.

[Buschmann 96] F. Buschmann, R. Meunier, H. Rohnert, P. Sommerlad, and M. Stal, *Pattern-Oriented Software Architecture, a System of Patterns*, John Wiley & Sons, Chichester, U.K., 1996.

[Bussler 05] C. Bussler, E. Kilgarriff, R. Krummenacher et al., WSMX triple-space computing, WSMO Working Draft, http://www.wsmo.org/TR/d21/v0.1/20050613, 2005.

[Chang 07] E. Chang, T. S. Dillon, and F. Hussain, Trust ontologies for e-service environments, *International Journal of Intelligent Systems*, 22(5), 519–545, 2007.

[Chang 06] E. Chang, T. S. Dillon, and F. Hussain, *Trust and Reputation for Service-Oriented Environments: Technology for Building Business Intelligence and Consumer Confidence*, John Wiley & Sons, Chichester, U.K., 2006.

[Cheah 07] C. Cheah, Ontological methodologies—From open standards software development to open standards organizational project governance, *International Journal of Computer Science and Network Security*, 7(3), 283–289, 2007.

[Chen 07] H. Chen, P. B. Chow, N. H. Cohen, and S. Duri, Extending SOA/MDD to sensors and actuators for sense and respond business processes, *IEEE International Conference on E-Business Engineering*, Hong Kong, China, 2007, pp. 54–61.

[Delamar 06] J. Lastra and I. Delamer, Semantic web services in factory automation: Fundamental insights and research roadmap, 2006.

[Dillon 07a] T. S. Dillon, C. Wu, and E. Chang, Reference architectural styles for service-oriented computing (keynote), *IFIP International Conference on Network and Parallel Computing*, Vol. 4672, LNCS, Dalian, China, 2007, pp. 543–555, Springer.

[Dillon 07b] T. S. Dillon, C. Wu, and E. Chang, GRIDSpace: Semantic grid services on the web (keynote), *Third IEEE International Conference on Semantics, Knowledge and Grid (SKG 2007)*, Xian, China, 2007, pp. 7–13.

[Eugster 03] P. T. Eugster, P. A. Felber, R. Guerraoui, and A.-M. Kermarrec, The many faces of publish/subscribe, *ACM Computing Surveys*, 35, 114–131, 2003.

[Fielding 02] R. Fielding, and R. Taylor, Principled design of the modern Web architecture, *ACM Transactions on Internet Technology (TOIT)*, *2*, 115–150, 2002.

[Gelernter 85] D. Gelernter, Generative communication in Linda, *ACM Transactions on Programming Language and Systems*, 7(1), 80–112, 1985.

[Gruber 93] T. Gruber, A translation approach to portable ontology specifications, *Knowledge Acquisition*, *5*, 199–220, 1993.

[Hadzic 05] M. Hadzic and E. Chang, Ontology-based support for human disease study, *Proceedings of the Hawaii International Conference on System Sciences (HICSS-38)*, Big Island, HI, 2005.

[Hadzic 07] M. Hadzic and R. Cowan, Three fold system (3FS) for mental health domain, in *Proceedings of the 2007 OTM Confederated International Conference on On the Move to Meaningful Internet Systems—Volume Part II*, 2007, Springer-Verlag, Berlin, pp. 1355—1364.

[Hanish 97] A. A. Hanish and T. S. Dillon, Communication protocol design to facilitate re-use based on the object-oriented paradigm, *Mobile Networks and Applications Journal*, 2(3), 285–301, 1997 (Sp. Issue).

[Holland 00] W. Holland and W. Bronsvoort, Assembly features in modeling and planning, *Robotics and Computer-Integrated Manufacturing*, *16*, 277–294, 2000.

[Jarrar 02] M. Jarrar and R. Meersman, Formal ontology engineering in the DOGMA approach, *International Conference on Ontologies, Databases, and Applications of Semantics (ODBase 2002)*, Vol. 2519, LNCS, Irvine, CA, 2002, pp. 1238–1254, Springer.

[Jensen 96] K. Jensen, *Coloured Petri Nets: Basic Concepts, Analysis Methods and Practical Use*, Vol. 1, Springer-Verlag, London, U.K., 1996.

[Klein 02] M. Klein, D. Fensel, A. Kiryakov, and D. Ognyanov, Ontology versioning and change detection on the web, *Proceedings of the 13th International Conference on Knowledge Engineering and Knowledge Management* (*EKAW 02*), Siguenza, Spain, 2002, pp. 197–212.

[Klyne 04] G. Klyne and J. J. Carroll, Resource description framework (RDF): Concepts and ab-stract syntax, W3C Recommendation, http://www.w3.org/TR/rdf-concepts/, 2004.

[Krummenacher 05] R. Krummenacher, M. Hepp, A. Polleres, C. Bussler, and D. Fensel, WWW or what is wrong with web services, *Proceedings of the Third IEEE European Conference on Web Services*, Vaxjo, Sweden, 2005, pp. 235–243.

[Lastra 04] J. L. Martinez Lastra, Reference mechatronic architecture for actor-based assembly systems, Doctoral thesis, Tampere University of Technology, Tampere, Finland, 2004.

[Piers 03] P. Piers, M. Benevides, and M. Mattoso, Mediating heterogeneous web services, *Symposium on Applications and the Internet* (*SAINT 2003*), Orlando, FL, 2003.

[Pilioura 04] T. Pilioura, G. Kapos, and A. Tsalgatidou, PYRAMID-S: A scalable infrastructure for semantic web service publication and discovery, *Proceedings of the 14th International Workshop Research. Issues on Data Engineering*, Boston, MA, 2004.

[Pornpit 05] P. Wongthongtham, E. Chang, T. Dillon, and I. Sommerville, Software engineering ontologies and their implementation, in Kobol, P. (Ed.), *IASTED International Conference on Software Engineering* (*SE*), Innsbruck, Austria, pp. 208–213, February 15, 2005, IASTED, Austria.

[Pornpit 06] P. Wongthongtham, E. Chang, T. Dillon, and I. Sommerville, Ontology-based multi-site software development methodology and tools. *Journal of Systems Architecture, 52*, 640–653, 2006.

[Sidhu 08] A. S. Sidhu, T. S. Dillon, and E. Chang, *Protein Ontology*, Springer, New York, 2008.

[Sidhu 05] A. S. Sidhu, T. S. Dillon, E. Chang, and B. S. Sidhu, Protein ontology: Vocabulary for protein data, *Proceedings of the Third International IEEE Conference on Information Technology and Application* (*IEEE ICITA 2005*), Sydney, Australia, 2005, pp. 465–469, IEEE.

[Sjelvgren 83] D. Sjelvgren, S. Andersson, T. Andersson, U. Nyberg, and T. S. Dillon, Optimal operations planning in a large hydro-thermal power system, *IEEE Transactions on Power Apparatus and Systems*, 102(11), 3644–3651.

[Spyns 02] P. Spyns, R. Meersman, and M. Jarrar, Data modelling versus ontology engineering, *SIGMOD Record*, 31(4):12–17, March 2002.

[Vinoski 02] S. Vinoski, Putting the "Web" into Web services—Web services interaction models, part 2, *IEEE Internet Computing*, 6(4), 90–92, July 2002.

[Wang 04] J. Wang, B. Jin, and J. Li, An ontology-based publish/subscribe system, *Middleware 2004*, Vol. 3231, Lecture Notes in Computer Science, Toronto, Canada, 2004, pp. 232–253.

[Wongthongtham 06] P. Wongthongtham, E. Chang, T. S. Dillon, and I. Sommerville, Ontology-based multi-site software development methodology and tools, *Journal of Systems Architecture*, 52, 640–653.

[Wouters 03] C. Wouters, T. S. Dillon, J. W. Rahayu, E. Chang, and R. Meersman, Ontologies on the MOVE, *Ninth International Conference on Database Systems for Advanced Applications* (*DASFAA '04*), Jeju Island, Korea, 2004, pp. 812–823, Springer.

[Wouters 02] C. Wouters, T. S. Dillon, J. W. Rahayu, and E. Chang, A Practical walkthrough of the ontology derivation rules, *13th International Conference on Database and Expert Systems Applications*, Aix-en-Provence, France, 2002, pp. 259–268, Springer.

[Zurawski 92] R. Z. Zurawski and T. S. Dillon, Automatic synthesis of FMS system design models from system requirements, *IEEE International Conference on Emerging Technology and Factory Automation*, Melbourne, Australia, 1992, pp. 62–67.

[illegible]

66

Automatic Data Mining on Internet by Using PERL Scripting Language

Nam Pham
Auburn University

Bogdan M. Wilamowski
Auburn University

66.1 Introduction 66-1
PERL Scripting Language • Regular Expressions • Web Browser
66.2 Examples 66-4
Extract E-Mail Addresses from Excel Files • Extract Data from PDF Files • Extract Paper Information from IEEE XPLORE • List Papers Only in a Specific Subject from TIE Webpage • Using PERL with Google Scholar in Searching Data
66.3 Summary and Conclusion..66-8
References..66-9

66.1 Introduction

With the tremendous growth in available information to the masses, the question is how users can search the useful information in the shortest time [PW09,NGW08,PYW10]. In other words, making use of consolidated information requires such substantial efforts since the web pages are generated for visualization and not for data exchange [KCSG07,W10]. To reach this goal requires methods developed to optimize a user's searching process. This chapter introduces a method known as the data mining robot (DMR) to extract and process data by using PERL scripting language. The DMR can be understood quickly as a software program that serves for mining data automatically. Particularly, with data mining from servers, this method does not use any browser to handle the Web, but does so directly by using PERL modules (software programs are written for a specific function) such as LWP. The use of these modules turns the DMR into an effective solution to extract data with an accelerated speed.

The procedure of execution of the DMR can be divided into three steps. (1) Data collection: to extract all information from the data source. (2) Data Filtering: to extract the useful information built in step 1. (3) Data processing: to process and sort the extracted information in a format that is effective for users.

DMR can be written in scripting languages such as PHP, PERL, or C. PERL is one of the most powerful tools in data manipulation with the powerful regular expression. To run DMR is simple. Programmers only need to install PERL scripting language, which is an interpreted language, it is parsed and executed at runtime instead of being compiled into binary form and then run. Moreover, this is an open-source software for the users with the standard modules, which also come with PERL. Like the built-in functions, these modules provide users with hundreds of prewritten resources. The modules are made up of PERL code written in a way to conform to certain conventions so users can access that code from their program. The PERL modules provide users with a great deal of the prewritten code and are stored in files with extension ".pm." Users can load such the modules into their code by using the "use" statement.

66.1.1 PERL Scripting Language

PERL, which stands for "Practical Extraction and Report Language," was written by Larry Wall, a linguist working as a systems administrator for NASA in the late 1980s, as a way to make report processing easier. It is a powerful language for doing data manipulation tasks. It is portable, accessible, and excellent at handling textual information. It has been used with good success for system administration tasks on Unix systems, acting as glue between different computer systems, World Wide Web CGI programming, as well as customizing the output from other programs.

PERL is a prominent Web programming language because of its text processing features. Handling HTML forms is made simple by the CGI.pm module, a part of PERL's standard distribution.

66.1.2 Regular Expressions

PERL is especially good at handling text, and, in fact, that is what it was originally developed for. Regular expressions are a big part of text handling in PERL. Regular expressions let users work with pattern matching (that is, comparing strings to a test string—or a pattern—that may contain wildcards and other special characters) and text substitution, providing a very powerful way to manipulate text under programmatic control [H99]. Unfortunately, using regular expressions in PERL is one of the areas that programmers find the most daunting. Even relatively straightforward regular expressions can take some time to comprehend.

```
String1=~ m/String2/
```

"`=~ m//`" operator means that `string1` will be compared with the pattern that is `string2`. If the match is true, `string2` will be a part of `string1`, but not vice versa. In other words, the pattern matching operator is always used to compare the pattern on the right hand side with the pattern on the left hand side to see if it fits or not.

Assuming that there are 100 different names in a list and we want to know if John is in this list or not.

```
$a="Peter, Jack and Tom";
$b="Peter";
if ($a=~m/$b/){print ("yes\n");}
elsif($a!~m/$b/) {print("no\n")};
Result:
Yes
```

Instead of using the standard string as a pattern, users can replace it by "wildcards." The wildcards are just special characters to generalize a string. For example, a user may want to know if there are any numbers in the string. In this case, a user does not know what the number is in the string, so they cannot use a specific number as the pattern to compare, but they can use `(\d+)` instead. `d` is an integer, and + sign means that the integer of length 1 or more.

```
$a="Peter, Jack and Tom";
if ($a=~m/(\d+)/){print ("yes\n");}
elsif($a!~m/(\d+)/) {print("no\n")};
Result:
no
```

Below is a list of some widely used wildcards:

. match any character
\w match word character
\W match non-word character

\s match whitespace character
\S match non-whitespace character
\d match digit character
\D match non-digit character
+ match one or more times

66.1.3 Web Browser

The interaction between PERL and the Internet is one of the most special attributes of this language. Using PERL, you can easily create Internet applications that use File Transfer Protocol (FTP), e-mail, Usenet, and Hypertext Transfer Protocol (HTTP), and even browse the Web. There are a lot of modules supporting programmers to do that. Much of this power comes from CPAN modules that can be downloaded and installed from the Web. Since users often download files and data from the Internet, the question is, can PERL emulate a web browser to copy all web sources into local files? The modules as LWP::Simple or LWP::UserAgent have enough capability to do that.

By using these modules, the DMR does not use the Web. These modules have capability to download a web page. In other words, they can emulate browsers. The following example illustrates the download of the main FAQ index at CPAN by using the "get" function from the module LWP::Simple and store that web page into file *name.html*.

```
use LWP::Simple;
$add="http://www.cpan.org/doc/FAQs/index.html";
$content=get("$add");
open FILEHANDLE,">name.htm";
print FILEHANDLE $content;
close FILEHANDLE;
```

An address is called into the subroutine and content. This subroutine uses "get" function from LWP::Simple to copy the web source into the variable $content, and then the content of this web page is stored into the file name.html. The final result will be a file having the same content as the web page http://www.cpan.org/doc/FAQs/index.html.

Instead of using LWP::Simple to emulate a browser, we may also use the wget system call. GNU wget is an open source software package for retrieving files using HTTP, HTTPS, and FTP, the most widely-used Internet protocols. Its main function is to link a certain web page with an address input and copy all web sources of that web page into a file. It allows users to log off from the system and let wget finish its work. This is extremely effective and time-saving when users have to transfer a lot of data from different Web links. In contrast, other web browsers always require users' constant presence. Generally, wget works almost the same as the browser built from LWP::Simple module. However, it gives users more options [WSC05].

Below is the basic structure to call wget from a PERL script. "–O" means that documents will not be written to separate files, but all will be concatenated together and written to one file while "–q" turns off wget output to the prompt screen.

```
$add= ="http://www.cpan.org/doc/FAQs/index.html";
$fname= "filename.htm";
system("wget.exe", "-q", "-O", $fname,$add);
```

By assigning values to the variables $add and $fname as an address of a web page and a name of a file and using the "system" command, the new html file will contain the same content as that web link. Once the new html file contains all data that users want, they can open and copy this data into an array. From here, users can extract all information they want by using PERL commands.

66.2 Examples

66.2.1 Extract E-Mail Addresses from Excel Files

An excel file has three columns: the first is the order number, the second contains the e-mail addresses of the authors, and the third displays the categories of the author's interest. In this example, there are seven authors who are interested in three categories: power electronic converters, signal processing, and neural networks. Assume that you want to call for papers on neural networks by sending an e-mail to these authors. Our work is to extract e-mail addresses of these authors from the table (Table 66.1).

There are many ways to get this job done. The traditional way is to copy this excel file into a text file and use regular expressions to extract the desired data. This work becomes difficult and complicated if the excel file has more columns. However, it can be much simpler with PERL by using the module "Win32::OLE." This package is not included in the standard PERL library but can be downloaded from CPAN. What it does is to open an excel file, but a file has many different sheets, users have to define which sheets need to be opened by modifying the sheet number or the sheet name. Once this sheet is opened, each column can be saved into different arrays. From these arrays, the e-mail addresses of authors interested in neural networks can be retrieved (Table 66.2).

```
use Win32::OLE qw(in with);
use Win32::OLE::Const 'Microsoft Excel';
$Win32::OLE::Warn = 3;# die on errors...
$mydir='C:/ALL BACK UP/Industrial Electronics on Trans/';
$filename=$mydir."IESdatabase08.xls";
if (-e "$filename")
     {
      # get active Excel application or open Excel application
      my $Excel = Win32::OLE->
      GetActiveObject('Excel.Application')
        || Win32::OLE->new('Excel.Application', 'Quit');
        $Excel->{'Visible'} = 0; #opened file is visible
        # open Excel file
        my $Book = $Excel->Workbooks->Open("$filename");
        # select worksheet number 2 (you can also select a
        worksheet by name)
        my $Sheet = $Book->Worksheets(2);
        # count the number of columns
        my $LastCol = $Sheet->UsedRange->Find({What=>"*",
                     SearchDirection=>xlPrevious,
                     SearchOrder=>xlByColumns})->{Column};
        # count the number of rows
        my $LastRow = $Sheet->UsedRange->Find({What=>"*",
                     SearchDirection=>xlPrevious,
                     SearchOrder=>xlByRows})->{Row};
        $totalrow=$LastRow;
        foreach my $row (1..$LastRow)
            {
              foreach my $col (1..$LastCol)
                  {
                    next unless defined
                    $Sheet-> Cells(1,$col)->{'Value'};
```

```
                    $title_row=$Sheet->Cells(1,$col)->{'Value'};
                    #extract manuscript and save in an array
                    if ($title_col=~ m/Manuscript/)
                    {$area[$row-1]=$Sheet->Cells($row,$col)
                      ->{'Value'};}
                    #extract email and save in an array
                      if ($title_col =~ m/Email Address/)
                    {$email[$row-1]=$Sheet->Cells($row,$col)
                      ->{'Value'};}
                }
            }
        $Excel->Quit();
    }
 for($n=0;$n<$LastRow;++$n) //search data from saved arrays
    {
        if ($area[$n]=~ m/Neural Networks/)
          {print ("$email[$n]\n");}
    }
 Result
```

66.2.2 Extract Data from PDF Files

In data mining, we encounter different types of data in different formats; data could be in PDF format, Doc format, etc. The issue here is that these types of data have to be converted into the general text format that can be readable and reusable. There are two solutions to handle this problem, one is done manually and the other is done automatically.

With the manual solution, users have to copy and paste from the PDF format to the text format. However, this solution is time-consuming and impractical as the number of PDF documentations is scaled up. Users can call wget to open the PDF file and copy its content into the text file as they often use to handle text file or html file, but the result file will be corrupted in this way. Because of this problem, the straight solution is not applicable.

With the automatic solution, the work can be done in a much simpler way. In order to do that, users have to use the module CAM::PDF in their code. This module allows users to convert the PDF file into

TABLE 66.1 Excel file

Email Address	Manuscript
yuebook@yahoo.cn	Power Electronic Converters
tomfangok@126.com	Signal Processing
xmsun.whu@163.com	Power Electronic Converters
Hubert.RAzik@green.uhp-nancy.fr	Signal Processing
eexma10@nottingham.ac.uk	Power Electronic Converters
xu.tao.dragon@gmail.com	Neural Networks
raky829@gmail.com	Neural Networks

TALE 66.2 Result file displayed in the table

Email Address	Manuscript
xu.tao.dragon@gmail.com	Neural Networks
raky829@gmail.com	Neural Networks

a text file without corruption. In this way, users can preserve the original data as given in the PDF. When data is converted, it is still reusable.

```
use CAM::PDF;
use CAM::PDF::PageText;
$filename = "C:/ALL BACK UP/Hung Database/filename.pdf";
my $pdf = CAM::PDF->new($filename);
my $pageone_tree = $pdf->getPageContentTree(1);
open TEST, ">", "test.txt" or die $!;
print TEST CAM::PDF::PageText->render($pageone_tree);
close TEST;
```

66.2.3 Extract Paper Information from IEEE XPLORE

PERL is used not only to mine data from files in different formats, but also to extract data directly from the web through a network connection. The name for this functionality is Internet robot (IR). Any IR can be connected to servers remotely by embedding modules such as LWP::Simple, LWP::UserAgent, or wget into it. The accuracy and speed of mining data make this method become special in processing and extracting data from the Internet.

For this particular application, the IR accesses XPLORE and downloads all the information about the papers stored in the server. Then it extracts all desired information about authors, titles, page number, etc., and puts them together in an html format that is more human readable than the raw data. To be able to extract all information under different links, wget has to search and extract all hyperlinks (Figures 66.1 and 66.2).

66.2.4 List Papers Only in a Specific Subject from TIE Webpage

In this example, the IR will browse the TIE web page and extract papers about "neural network." After downloading data from the server, the IR will save all information in an array, where each line of the web source will be an index of the array. By using regular expressions to modify the indexes containing the title and compare with the pattern "neural network," the IR will list all papers about "neural network" (Figure 66.3).

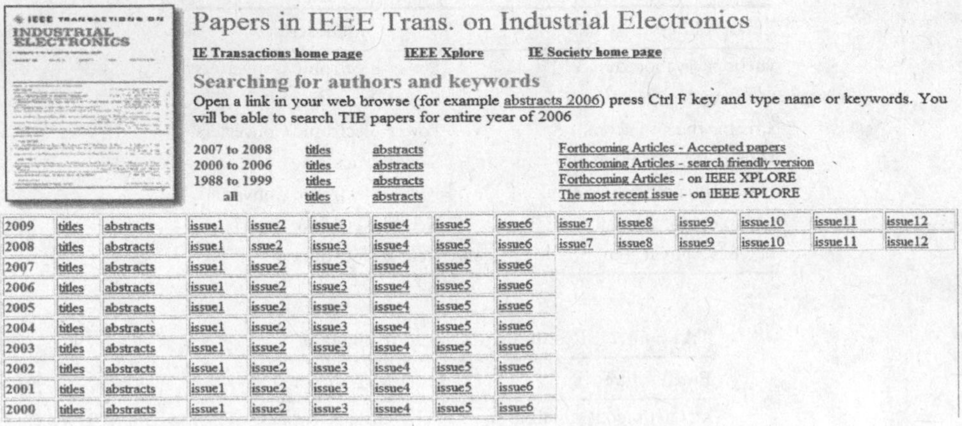

2009	titles	abstracts	issue1	issue2	issue3	issue4	issue5	issue6	issue7	issue8	issue9	issue10	issue11	issue12
2008	titles	abstracts	issue1	ssue2	issue3	issue4	issue5	issue6	issue7	issue8	issue9	issue10	issue11	issue12
2007	titles	abstracts	issue1	issue2	issue3	issue4	issue5	issue6						
2006	titles	abstracts	issue1	issue2	issue3	issue4	issue5	issue6						
2005	titles	abstracts	issue1	issue2	issue3	issue4	issue5	issue6						
2004	titles	abstracts	issue1	issue2	issue3	issue4	issue5	issue6						
2003	titles	abstracts	issue1	issue2	issue3	issue4	issue5	issue6						
2002	titles	abstracts	issue1	issue2	issue3	issue4	issue5	issue6						
2001	titles	abstracts	issue1	issue2	issue3	issue4	issue5	issue6						
2000	titles	abstracts	issue1	issue2	issue3	issue4	issue5	issue6						

FIGURE 66.1 A single link with all extracted information.

IEEE IEEE Transactions on Industrial Electronics ies

Volume 56, Number 5, May 2009 Access to the journal on IEEE XPLORE IE Transactions Home Page

56.5.1 T. Atsumi, "Feedforward Control Using Sampled-Data Polynomial for Track Seeking in Hard Disk Drives," *IEEE Trans. on Industrial Electronics,* vol. 56, no. 5, pp. 1338-1346, May 2009. Abstract Link Full Text

Abstract: To decrease the seek time of hard disk drives, a feedforward control method was developed by using a sampled-data polynomial. The sampled-data polynomial satisfies the boundary conditions that include the characteristics of the zero-order hold, and compensates for the discretization error caused by the zero-order hold without the need for complicated calculations. Therefore, the feedforward control using the sampled-data polynomial enables real-time calculation and does not require lookup tables which need a large amount of memory. The parameters of the sampled-data polynomial are designed by using shock-response-spectrum analysis to minimize the settling vibrations caused by the feedforward control inputs. When the proposed method was applied on a hard disk drive, it significantly reduced the amount of tracking error in the seek control and also reduced the seek time.

56.5.2 M. F. Heertjes, X. G. P. Schuurbiers, H. Nijmeijer, "Performance-Improved Design of N-PID Controlled Motion Systems With Applications to Wafer Stages," *IEEE Trans. on Industrial Electronics,* vol. 56, no. 5, pp. 1347-1355, May 2009. Abstract Link Full Text

Abstract: A nonlinear filter design is proposed to improve nanopositioning servo performances in high-speed (and generally linear) motion systems. The design offers a means to adapt fundamental control design tradeoffs—like disturbance suppression versus noise sensitivity—which are otherwise fixed. Typically performance-limiting oscillations in the feedback system that benefit from extra control are temporarily upscaled and subjected to nonlinear weighting. For sufficiently large amplitudes, this nonlinear filter operation induces extra controller gain. Oscillations that do not benefit from this extra control (typically because they represent noise contributions that should not be amplified) remain unscaled and, as such, do not induce extra controller gain. The combined usage of linear weighting filters with their exact inverses renders this part of the nonlinear filter design strictly performance based. The effective means to improve servo performance is demonstrated on a short-stroke wafer stage of an industrial wafer scanner. Since the nonlinear filter design is largely based on Lyapunov arguments, stability is guaranteed along the different design steps.

FIGURE 66.2 Web page with links generated by the IR.

56.1.24 F.-J. Lin, P.-H. Chou, "Adaptive Control of Two-Axis Motion Control System Using Interval Type-2 Fuzzy Neural Network," *IEEE Trans. on Industrial Electronics,* vol. 56, no. 1, pp. 178-193, Jan 2009.

56.3.26 E. Echenique, J. Dixon, R. Cardenas, R. Pena, "Sensorless Control for a Switched Reluctance Wind Generator, Based on Current Slopes and Neural Networks," *IEEE Trans. on Industrial Electronics,* vol. 56, no. 3, pp. 817-825, March 2009.

56.4.41 F.-J. Lin, Y.-C. Hung, S.-Y. Chen, "FPGA-Based Computed Force Control System Using Elman Neural Network for Linear Ultrasonic Motor," *IEEE Trans. on Industrial Electronics,* vol. 56 no. 4, pp. 1238-1253, April 2009.

56.5.16 M. A. M. Radzi, N. A. Rahim, "Neural Network and Bandless Hysteresis Approach to Control Switched Capacitor Active Power Filter for Reduction of Harmonics," *IEEE Trans. on Industrial Electronics,* vol. 56, no. 5, pp. 1477-1484, May 2009.

56.7.38 R.-J. Wai, C.-M. Liu, "Design of Dynamic Petri Recurrent Fuzzy Neural Network and Its Application to Path-Tracking Control of Nonholonomic Mobile Ro," *IEEE Trans. on Industrial Electronics,* vol. 56, no. 7, pp. 2667-2683, July 2009.

56.8.28 S. M. Gadoue, D. Giaouris, J. W. Finch, "Sensorless Control of Induction Motor Drives at Very Low and Zero Speeds Using Neural Network Flux Observ," *IEEE Trans. on Industrial Electronics,* vol. 56, no. 8, pp. 3029-3039, August 2009.

56.8.51 F. Moreno, J. Alarcon, R. Salvador, T. Riesgo, "Reconfigurable Hardware Architecture of a Shape Recognition System Based on Specialized Tiny Neural Networks With Online Train," *IEEE Trans. on Industrial Electronics,* vol. 56, no. 8, pp. 3253-3263, August 2009.

56.10.12 S. Cong, Y. Liang, "PID-Like Neural Network Nonlinear Adaptive Control for Uncertain Multivariable Motion Control Syst," *IEEE Trans. on Industrial Electronics,* vol. 56, no. 10, pp. 3872-3879, Oct 2009.

FIGURE 66.3 Papers about neural network.

```
$mydir="C:/ALL BACK UP/Hand Book/";
$add="http://tie.ieee-ies.org/tie/abs/56s.htm";
//copy website source into a local file
$fname= "handbook.htm";
system("wget.exe", "-q", "-O", $fname,$add);
$file=$mydir."handbook.htm";
open(r1,"<$file");
@lines = <r1>;
close(r1);
$N=@lines;
$j=0;
// search array indexes containing 'neural networks'string

for($i=0;$i<$N;++$i)
      {
         //transfer uppercase letters into lowercase letters
           $lines[$i] =~ tr/[A-Z]/[a-z]/;
```

```
    //search 'neural network'
      if(($lines[$i]=~ m/<td valign/)&&($lines[$i]=~ m/neural
        network/))
        { $a[$j]="<tr>".$lines[$i]."</td>"."</tr>";
            $j=$j+1;}
    }
Result:
```

66.2.5 Using PERL with Google Scholar in Searching Data

Google Scholar is as easy to use as the normal Google Web search can be, especially with the helpfulness of the "advanced search" option, which can automatically narrow search results to a specific journal or article. The most relevant results for searched keywords will be listed first, in order of the authors ranking, the amount of references that are linked to it and their relevance to other scholarly literature, and the ranking of the publication that the journal appears in and the citation index. The IR can be incorporated with this search engine to search for information about authors, citations, etc. With this type of searching, the IR can take advantage of the search engine Google Scholar, which is relatively quick and easy to use. The searching process can be modeled as following (Figure 66.4): input keywords from users, which can be journal name, year of published article, etc. When these keywords are defined, the IR will activate the search engine as Google Scholar, Web of Knowledge, etc., and search for selected information, then generate the output file. For example, if information about authors, citations of *Journal IEEE on Transactions Industrial Electronics* in the year 2006 is required, users can define the keywords by using this journal name and the given year to activate the search engine Google Scholar. The IR will copy all information about papers on this journal in this year through Google Scholar. When the extracting process is complete, the text file output will be generated (Figure 66.5). With this type of concept, there are many applications where users can benefit from a Robot. The searching process is optimized and time-saving.

66.3 Summary and Conclusion

Current tools that enable data extraction or data mining are both expensive to maintain and complex to design and use due to several potholes such as difference in data formats, varying attributes, and typographical errors in input documents. One such tool is an Extractor or Wrapper, which can perform the data extraction and processing tasks [CKGS06]. Wrapper induction based on inductive machine learning is the leading technique available nowadays. The user is asked to label or mark the target items in a set of training pages or a list of data records in one page. The system then learns extraction rules from these training pages. Inductive learning poses a major problem—the initial set of labeled training

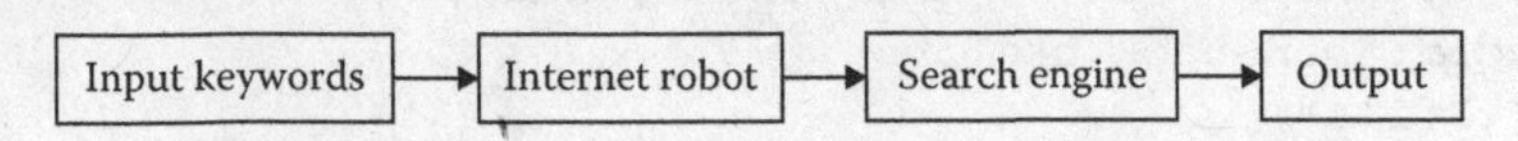

FIGURE 66.4 Searching process model.

```
Cites,Authors,
71,"J.M. Carrasco, L.G. Franquelo, J.T. Bialasiewicz,"
60,"F. Blaabjerg, R. Teodorescu, M. Liserre,"
48,"J. Holtz,"
38,"J. Moreno, M.E. Ortuzar, J.W. Dixon,"
35,"S. Katsura, Y. Matsumoto, K. Ohnishi,"
35,"P.P. Acarnley, J.F. Watson,"
33,"S. Alepuz, S. Busquets-Monge, J. Bordonau,"
32,"Y. Cheng, C. Qian, M.L. Crow,"
32,"J.H. Jung, J.J. Lee, B.H. Kwon,"
```

FIGURE 66.5 Output file.

pages may not be fully depictive of the templates of all other pages. Poor performance of learned rules is experienced for pages that follow templates uncovered by the labeled pages. This problem can be solved by labeling more pages, because more pages cover more templates. However, manual labeling requiring a large supply of labor and is time consuming with an unsatisfied coverage of all possible templates.

This method of data extraction, DMR, optimizes the data mining process and makes it become a popular tool in extracting data from web pages. PERL script with regular expressions and modules increases the speed of data extracting as well as the accuracy. The DMR is customized according to the required data and the format of data that users desire.

References

[CKGS06] C.H. Chang, M. Kayed, R. Girgis, and K.F. Shaalan, A survey of web information extraction systems, *IEEE Transactions on Knowledge and Data Engineering*, 18(10), 1411–1428, October 2006.

[H99] S. Holzner, *PERL Black Book*, CoriolisOpen Press, Scottsdale, AZ, Edition 2004.

[KCSG07] M. Keyed, C.-H. Chang, K. Shaalan, and M.R. Girgis, FiVa tech: Page-level data extraction from template pages, *Seventh IEEE International Conference on Data Mining Workshops 2007 (ICDM Workshops 2007)*, Omaha, NE, pp. 15–20, October 28–31, 2007.

[NGW08] S. Neeli, K. Govindasamy, B.M. Wilamowski, and A. Malinowski, Auto data mining from web servers using PERL script, *International Conference on Intelligent Engineering System 2008* (*INES 2008*), Miami, FL, pp 191–196, February 25–29, 2008.

[PW09] N. Pham and B.M. Wilamowski, IEEE article data extraction from Internet, *13th IEEE Intelligent Engineering Systems Conference* (*INES 2009*), Barbados, April 16–18, 2009.

[PYW10] N. Pham, H. Yu, and B.M. Wilamowski, Neural network trainer through computer networks, *24th IEEE International Conference on Advanced Information Networking and Applications 2010*, Perth, Australia, pp. 1203–1209, 2010.

[W10] B.M. Wilamowski, Design of network based software, *24th IEEE International Conference on Advanced Information Networking and Applications 2010*, Perth, Australia, pp. 4–10, 2010.

[WSC05] I.-C. Wu, J.-Y. Su, and L.-B. Chen, A web data extraction description language and its implementation, *29th Annual International Conference on Computer Software and Applications Conference 2005* (*COMPSAC 2005*), Vol. 2, Edinburgh, U.K., pp. 293–298, July 25–28, 2005.

V

Outlook

67 Trends and Challenges for Industrial Communication Systems *Peter Palensky* **67**-1
Introduction • Ubiquitous Global Connectivity and Digital Identity • Vertical Integration • Hybrid Local Networks and Quality of Service • M2M Communication • Scalability in Hardware and Software • References

68 Processing Data in Complex Communication Systems *Gerhard Zucker, Dietmar Bruckner, and Dietmar Dietrich* **68**-1
Introduction • An Archetype for Future Automation • Bottom-Up versus Top-Down Design: Behavioristic Model versus Functional Model • Automated Methods for Sensor and Actuator Systems • The Diagnostic System • Intelligent Surveillance Systems • The Human Mind as an Archetype for Cognitive Automation • References

67

Trends and Challenges for Industrial Communication Systems

67.1 Introduction 67-1
67.2 Ubiquitous Global Connectivity and Digital Identity 67-1
67.3 Vertical Integration 67-2
67.4 Hybrid Local Networks and Quality of Service 67-2
67.5 M2M Communication 67-3
67.6 Scalability in Hardware and Software 67-4
References 67-5

Peter Palensky
Austrian Institute of Technology

67.1 Introduction

In the era of ever-increasing connectivity, every possible industrial process is tapped with sensors, networks, and automation infrastructure to improve efficiency, safety, and transparency. This leads, depending on the domain-specific requirement, to specific technologies and methods, just as networks for mines are different from automotive networks. Some are still hoping for "the one," universal solution, but that seems a bit far-fetched. There are, however, several points where existing networks will experience improvements toward a more universal technology. This chapter lists major challenges that industrial communication systems (ICS) will face in the near future:

- Ubiquitous global connectivity and digital identity
- Vertical integration
- Machine-to-machine (M2M) communication
- Scalability of hardware and software

67.2 Ubiquitous Global Connectivity and Digital Identity

Personal experience tells us, Internet Protocol (IP) connectivity is ubiquitous. Cell phones and PDAs grab 3G or Wi-Fi signals and connect to the World Wide Web without much trouble. This delusive simplicity makes us believe that M2M communication enjoys the same luxury. This is, however, not the case. Global connectivity in this context shall mean mobile (but not necessarily wireless) connectivity over a certain region and maybe even over borders. It boils down to plugging a device into some network socket or registering it against some wireless network without much hassle.

One usual prerequisite for global connectivity is a subscription to a certain type of service plan. The service provider might be a public Internet provider, a telephone company, a corporate IT department, or some other IT operator. The subscriber's identity is typically defined via a subscriber identity module (SIM) card or via user name and password. Unfortunately, there is no universal concept of identity established. Every technology—and even providers—have their own flavor of the two mentioned methods.

An industrial communication system, hooked up to such a provider's infrastructure has to deal with a heterogeneous mix of technologies like IP-Proxy-Routers, dial-up protocols, digital certificates, authentication protocols, and passwords. Unfortunately, there is no single-sign-on (SSO) standard for ICS infrastructure that could be compared to SSO in the IT world.

Sun Java™ System Access Manager [SUN05] is an M2M-capable SSO solution that provides federated identity (authentication and authorization) support via Liberty Alliance Phase 2 and Security Assertion Markup Language (SAML), session management, delegated authority, and on-the-fly audits of authentication attempts.

Another approach, also out of the mobile communication world, comes from the European Telecommunications Standards Institute (ETSI). ETSI discusses future M2M architectures in its technical committee ETSI TC M2M#01 and calls SSO "Identity Federations Service," implemented by an M2M service gateway [ETSI09].

Traditional industrial networks still do not have anything like this, but it can be expected that IT, telecommunication, and industrial communication will further converge, so that IT solutions will be adopted for ICS.

67.3 Vertical Integration

ICS were and are traditionally used for automation, data acquisition, process monitoring, and other tasks that are "near the process." Recent trends in production show, however, an increasing link between the shop floor and administrative parts of an enterprise. Applications like supply chain management and digital ecosystems are first versions of totally integrated business processes.

This trend is reinforced by ecological constraints and the increased need for energy efficiency. In order to calculate the carbon footprint of a produced product, it is necessary to link data from a large variety of sources. This includes automation data, schedules, supplier information, and logistics data. Even if the various IT systems of these sources (enterprise resource planning software, manufacturing automation, logistics and fleet management, energy consumption equipment, etc.) are somehow interlinked, they usually lack a common management and a common language to exchange information.

Transmitting data "from I/O to CEO" is therefore not only a problem for the transport media but more a question of semantics and lossy translations between IT systems. The increased use of IT technologies like web services, XML, and service-oriented architecture (SOA) eases this goal. Organizations like OASIS [OASIS] are engaged in applying web services and XML-based formats to all kinds of businesses and processes. One example is oBIX [Ehrl03] for exchanging building data.

Also, the design and specification phase of industrial installations is more and more influenced by traditional IT methods. UML-PA [UMLPA] and OPC XML DA [OPCDA] are examples of the successful usage of IT methods for industrial systems.

True interworking and interoperability consists of a number of steps. The prerequisite is protocol conformance, i.e., using the right plugs, frames, addresses, coding, timing, etc. Above protocol conformance comes the application with its data types and functional profiles. One important, although often forgotten factor is interoperable network and application management and configuration. While protocol conformance usually leads to interoperable network management, the applications still might use some proprietary way of storing their configuration parameters. SOA and the use of XML-based data formats eases the "lower" parts (i.e., data representation and encoding) of application management; the semantics and usage of the individual parameters, however, must be defined in standardized high-level profiles.

67.4 Hybrid Local Networks and Quality of Service

Most ICS were designed with a particular application domain in mind. There are networks for network-based control, for drives, for data acquisition, multimedia streaming, or fire alarm systems. The application domain resulted in specific features like low latency, guaranteed time slots, good scalability, or low costs. "Universal" networks that can satisfy every possible connectivity need do not exist, but hybrid networks are getting close.

A hybrid network consists of more than one communication channel, ideally of very diverse channels. An example is a mobile device with GPRS and GPS. It combines a bidirectional best-effort channel with a unidirectional information source. Similar thoughts are behind energy management nodes that use Internet and electricity grid frequency as information sources [FK07].

Another possibility to move toward universal networks is newer developments where features of several domains were combined. Most notably, this happened on the media access control layer in the case of networks like IEEE 1394 or IEC 61580. Typically, one part of the bandwidth is reserved for deterministic traffic (i.e., isochronous slots), while the rest of it can be arbitrated in a CSMA way as many other networks do. With this hybrid design, two very different network qualities can be achieved: Easy scalability for non-real-time services and guaranteed service for real-time applications.

An important factor for hybrid designs is how quality of services (QoS) is implemented. As surely, the IP suite will play a continuously important role in ICS, it is good to know that version 6 of IP addresses QoS. Although one (especially an ICS) can still feel v4's problems in v6, it has at least (beside its often-mentioned better scalability) flow labeling, better priorities, and is prepared for extended headers that might contain additional QoS information. Surely, the designers of IPv6 QoS had multimedia services in mind, but ICS are also sensitive to (all four aspects of) QoS:

- Bandwidth
- Latency (packet delay)
- Latency variation (jitter)
- Packet loss

Bandwidth might be the least important aspect in 80% of all industrial communication (especially automation) cases, but robot vision and image processing is on the rise at the shop floor. Also, the size of installations and the number of nodes is dramatically increasing, so bandwidth will play a larger role than up to now where an IP network over 100 Mb Ethernet will practically never experience a bandwidth problem in an automation environment. Latency and jitter are most important for network-based control where guaranteed timely behavior is required. Especially, time synchronization—a very important service in distributed systems (think of system diagnostics in process control or multimedia)—needs accurately defined upper boundaries for latency. Packet loss is no problem for media streams but counted quantities during production or emergency shutdown messages demand high availability and reliability of the ICS. It is to be expected that QoS will play a more prominent and visible role in the future of ICS. The existing proposals and implementation of IPv6 QoS is certainly not sufficient for ICS and we have to work on suitable extensions, probably via its flexible header.

67.5 M2M Communication

Machines are—and will increasingly be—used to act on behalf of humans. In the past, this meant physical labor; nowadays, it is more and more decision making and "softer" tasks. Computer programs that act on behalf of certain individuals or roles are commonly called software agents. Just like humans, software agents need to communicate with other agents, the environment, and humans to be able to fulfill their tasks. Unlike humans, however, they are not legal and physical person with an associated identity and trust. This is a very important distinction when it comes to information security.

Most existing security measures rely on cryptographic methods to satisfy security needs like authenticity or confidentiality. The security problem is boiled down to protecting a certain cryptographic key during setup and operation. If the end points of the respective communication relation are humans, this key—which provides identity and confidentiality—might be protected with personal identification numbers (PINs) or other "side channels," which can only be provided by the intended person. Transferring these established procedures to M2M communication is not trivial. The only affordable method nowadays is using electronic chip cards (a.k.a. smart cards), trusted platform modules (TPM), or other chips that contain the key and algorithms to use the key. With this, the problem of identity and trust is—again—transferred further: to the administration of manufacturing and shipping the devices. Each one must get its own personal chip that defines its identity and helps with security.

The second big challenge for M2M communication is network management. As networks are getting larger, it is required that traditional network administration tasks are automated. Assigning addresses and baud rates is already past; most of these things happen in a plug-and-participate way. The next step is plug-and-work, where also the network higher layers and the application are configured automatically. The key ingredients for this are

- Service discovery and lookup tables
- Self-description of services and capabilities
- Application profiles

Naturally, not every application can be configured automatically without human involvement, but learning network management tools can ease that job dramatically.

67.6 Scalability in Hardware and Software

The trend for "total integration," i.e., the as-seamless-as-possible connection of different applications and networks, opens up the reason for the very existence of different networks. Their different features that made them suitable for specific application areas might be a barrier for integration. One of them is size. Size in this context partly means the physical size, but more the code size of protocol stacks, the silicon size for embedded functionality, and ultimately the price tag. Combining low-cost sensor networks with powerful multimedia or real-time networks does not work out-of-the-box. What is needed is a scalable architecture where—in terms of hardware, protocols, and software—very different nodes can interoperate, although they might only share a small set of common services [PAL04].

One first step toward such a scalable architecture is the introduction of 6LoPAN, a downsized version 6 IP over IEEE 802.15.4 (the popular wireless transport, also used by ZigBee and other technologies). Other wireless technologies like ISA-100.11a and WirelessHART show similar potential. But the idea should not be limited to wireless technologies. The concept of scalable architectures must be technology agnostic, ensuring that media access, addressing, management, and service discovery follow some minimum core requirements that allow for connecting a $0.5 node to a $500 node in order to share some functionality.

The Instrumentation, Systems, and Automation Society (ISA) defines six classes of industrial communication:

- Class 0: Emergency
- Class 1: Closed-loop regulatory control
- Class 2: Closed-loop supervisory control
- Class 3: Open-loop control (human in the loop)
- Class 4: Alerting
- Class 5: Logging, downloading

Class 0 falls into the category safety, the other classes are either control (1 and 2), monitoring (4 and 5), or something in between. Depending on its class, a certain service or application assumes certain QoS, and operating applications with different classes over one and the same transport is not easy. ISA-100.11a suggests using IEEE 802.15.4 as physical and TCP/UDP/IPv6 as network and transport layers. For Class 0, however, even wired networks are often not sufficient. It might still be a long way until a real universal and scalable network is available.

References

[Ehrl03] P. Ehrlich and K. Sinclair, What is oBIX?, AutomatedBuildings.com, December 2003.

[ETSI09] Sophia Antipolis, Initial proposed architecture for M2M networks, working document M2M01_026 of ETSI TC M2M#01, January 2009.

[FK07] F. Kupzog and P. Palensky, Wide-area control systems for balance-energy provision by energy consumers, *Proceedings of IFAC FET 2007*, Toulouse, France, 2007.

[OASIS] Organization for the Advancement of Structured Information Standards, Available at www.oasis-open.org, Accessed on August 2010.

[OPCDA] OPC Foundation, OPC XML-DA 1.00 specification, Version 1.01, OPC Foundation, July 2003.

[PAL04] P. Palensky, Requirements for the next generation of building networks, *Proceedings of International Conference on Cybernetics and Information Technologies, Systems and Applications (ISAS CITSA 2004)*, Orlando, FL, 2004.

[SUN05] M. B. Baikie and S. Gaede, The role of mobile operators and the JAVA™ platform in the machine-to-machine, White Paper, Sun Microsystems, Santa Clara, CA, June 2005.

[UMLPA] Specification of hard real-time industrial automation systems with UML-PA, White Paper, University of Stuttgart, Stuttgart, Germany, May 2005.

68 Processing Data in Complex Communication Systems

Gerhard Zucker
Vienna University of Technology

Dietmar Bruckner
Vienna University of Technology

Dietmar Dietrich
Vienna University of Technology

68.1 Introduction 68-1
68.2 An Archetype for Future Automation....................................68-2
68.3 Bottom-Up versus Top-Down Design: Behavioristic Model versus Functional Model..68-2
68.4 Automated Methods for Sensor and Actuator Systems............68-4
68.5 The Diagnostic System 68-4
Error Detection • Statistical Generative Models • Online Parameter Updates • Results
68.6 Intelligent Surveillance Systems...68-7
Architecture
68.7 The Human Mind as an Archetype for Cognitive Automation..68-9
Perception in Automation: A Historic Overview
References...68-10

68.1 Introduction

In automation, mechanical units are more and more replaced by electronic components. They contain sensors, actuators, and information computing systems. The goal is to achieve higher process and product quality. This means, e.g., to increase the performance by implementing more functionality. Systems become interconnected, which causes increased complexity—something that cannot easily be handled with traditional technological approaches. It is necessary to rethink existing approaches and find new solutions. These can be found in statistical models, hierarchical models, and bionic models. This chapter reviews recent research results about methods to reduce the effort for commission, maintenance, and operation of the system. Statistical models are used to automatically derive normal operation conditions. Hierarchical models help in abstracting information from the mere sensor values to semantic concepts that are also used in human thinking and conversation. In the bionic research direction, contemporary researchers like Mark Solms and Antonio R. Damasio build upon psychoanalysis, which could be such a new approach for engineers. The theories are widely unknown to engineers, still their methods, which originate from neurology, are partly similar to the methods of computer technology.

68.2 An Archetype for Future Automation

If we want to be able to handle complex control systems that we are facing today in automation—and which has been the driving force behind artificial intelligence for the last 50 years—then we should learn to understand how the human mind works, since it is the best control system that we have available. Especially the higher, more abstract levels are relevant; processing in the lower layers—the neurons—are in this context less interesting, since a technical solution will not ground on a biological base, it will instead use electronic computers.

Using a proper model of the human mind as an archetype allows us to design technical systems that are able to control complex processes: surveillance systems for airport security to detect dangerous scenarios; efficiently interpreting vast amounts of data from thousands of embedded systems in a building or in a production plant; providing support in a household and especially supporting the elderly in geriatric applications; controlling trains and airplanes. The principle behind all these applications is identical: a machine has to perceive its environment and interpret the perceived information on a high, abstract level and draw reasonable conclusions.

A model that can describe human behavior is not sufficient. A human being is much too complex that we could infer its function directly from its behavior. It will never reach a state that is identical to a state in the past—there are no two identical states in one life. The question for an appropriate functional model is also not answered in the description of synaptic connections between neural structures, since this description does not explain how a human "functions," i.e., it does not explain the functional modules that it contains. To give an analogy: if we have a description of the behavior of a word processing program, it will not suffice to examine the transistor structures of the CPU on which it runs; we will not be able to describe the functionality of a word processing program, too much complexity lies between the operation of a silicon chip and a software application. Still an electronic computer system is by far simpler than a biological brain.

If we look back at the development of artificial intelligence (AI), we see the great achievements that have been produced. Norbert Wiener was already aware that nature is the source of answers. We have to be ready to accept results from other scientific disciplines.

But we also realize that we have made some big mistakes in AI and, in our opinion, sometimes left the grounds of science. This should now be corrected by allowing alternative considerations beside the traditional approaches.

68.3 Bottom-Up versus Top-Down Design: Behavioristic Model versus Functional Model

According to [ENF07] and [Pal08] we can distinguish four generations of AI: First, symbolic AI, which works on symbolic coding, manipulation, and decoding of information. Symbolic processing methods are today an essential base for AI, which is also shown in the work of [Vel08].

Statistical AI employs networked structures and their couplings. The central topic is learning, which unfortunately always only leads to one or more optima, but does not go further. Learning from what has been learned cannot be modeled—the system is not able to change its learning structures based on what has been learned. This is also understandable from a different viewpoint, since only the ability to permanently store images and scenarios and integrate them into the learning process can yield the desired performance and diversity. We will discuss that a bit further below.

Next, we realize that the body with its needs and its requests for satisfaction is a necessary requirement for intelligence—this is the discipline of embodied intelligence. Intelligence does not end in itself; it serves the body to subsist efficiently, as Antonio R. Damasio states in [Dam94]. Seen from the philosophical viewpoint this was great progress, because this step had significant consequences on the discussions of the existence of a soul (Hegels body-mind problem). If we work scientifically, it becomes clear that thinking and body are the same thing, only seen from different viewpoints. This is exactly the statement of modern psychoanalysis and where neurology meets psychoanalysis and merges into neuropsychoanalysis [Sol02].

Finally, at the end of the 1980s, it was understood that the brain has to be more than just a system on which different programs run for different purposes [Pal08]. Emotions, language and consciousness are properties that are of great relevance for human beings. This is the domain of emotional AI. First, there was no separation between emotions and feelings and we simply assumed that these are evaluation mechanisms. Still, phenomena like *consciousness* or the sensation of *feeling something* could not be explained.

Analysis of this historical development makes two things clear. First, we see that engineers have chosen a bottom-up method for their designs, simply due to lack of knowledge. We put one piece on top of the other to work our way from neurons (which would be equivalent to hardware in computer models) to consciousness (which is the highest, most abstract level). But we should not do that; we should always work top-down like computer engineers are used to work. Second, we have selected certain structures, functions or behavior from the whole complex of human abilities, and tried to implement these in technical systems. We expected to understand the rest, which we left out, in good time. Especially, the last two generations of AI have used psychological knowledge, which was, however, not checked for consistency. This resulted in a patchwork—a mix of different, partly contradicting schools that were all merged together into one model. Such an example is shown in [Bre02], which violates scientific principles. If results of different psychological schools are used which have not been checked for consistency (or interoperability, to use a more technical term), the results are worthless.

This must have implications for future research, which, in our opinion, leads to the fifth generation of AI. The boundary conditions are clear:

1. Computer engineers must demand top-down design.
2. We need a *uniform* model, which is free of contradictions.
3. Engineers have to cooperate with scientists, who have always dealt with the brain and the mind.

Computer engineers successfully use the layering shown in Figure 68.1. Such a layer model is currently not available for the mind, but we have to try to apply the layer model anyway. If we work top-down, however, it becomes clear that an abstraction of the neural network [CR04] (the hardware, to compare it with the computer domain) is not necessary at the beginning; we will explain that a bit further below. We do not know how the brain is "programmed" on the neural level. And even if we knew, it would not be the right level of modeling (just think of aligning transistors until you get a word processing application); we have to focus on the psyche, and we have to remember that modeling is a multi-layer process (Figure 68.1). Therefore we cannot map specific behavior of a process (an "application" in the computer domain) directly to functional modules. In doing so we would reduce the complex system "human being" to a primitive system, which contradicts reality: the control unit of a human being does not execute the same behavior twice absolutely identical, since conditions always change [Jaa97]. Today's robots (which, in this context, are the same as puppets) have one thing in common: they are not human, but it is the human observer, who projects human-like properties into them. According to [Sol02], human beings consist physiologically and psychologically of many different control loops

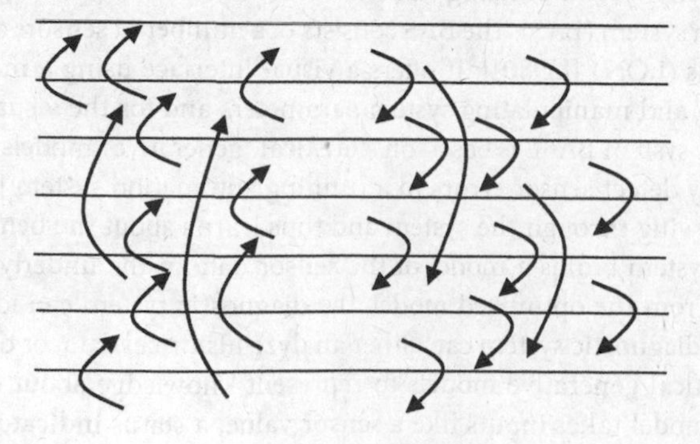

FIGURE 68.1 Layer model of the brain with reference to a computer.

that are hierarchical as well, but nevertheless strongly interconnected with each other (Figure 68.1). Observing the behavior within *one* experiment cannot lead to a complete description of the system. This is the wrong approach, even if it looks stunningly simple. To give an analogy: biologists do not explain flowers by visual features any more, but they look for functional units by considering the flower as a "process." Similarly [Dav97], trying to describe merely the behavior of a communication protocol under different conditions is a lost cause, since it will never cover the complete abilities of a communication protocol. But it is possible to develop a functional model of a process [HJ97] (where we use the word "process" to refer to both technical and biological occurrences) and understand its behavior based on this model. If we want to build robots or other systems that are intelligent in a more human-like way, we have to get away from behavioral description and employ functional development instead. While this is well understood in computer and communications engineering, some psychological schools like behaviorists have a slightly different view of the matter.

68.4 Automated Methods for Sensor and Actuator Systems

Today's building sensor and control systems are primarily based upon the processing of sensor information using predefined rules. The user or operator defines, e.g., the range of valid temperatures for a room by a rule—when the temperature value in that room is out of range (e.g. caused by a defect), the system reacts (e.g., with an error message). More complicated diagnostics require an experienced operator who can observe and interpret real-time sensor values. However, as systems become larger, are deployed in a wider variety of environments, and are targeted at technically less-sophisticated users, both possibilities (rule-based systems and expert users) become problematic. The control system would require comprehensive prior knowledge of possible operating conditions, ranges of values and error conditions. This knowledge may not be readily available, and will be difficult for an unsophisticated user to input. It is impractical for experienced operators to directly observe large systems, and naive users can not interpret sensor values.

A solution to this problem is to automatically recognize error conditions specific to a given sensor [HLS99], actuator, or system without the need of preprogrammed error conditions, user-entered parameters, or experienced operators. The system should observe sensor and actuator data over time [JGJS99], construct a model of "normality" [PL03], and issue error alerts when sensor or actuator values vary from normal. The result would be a system that can recognize sensor errors or abnormal sensor or actuator readings, with minimal manual configuration of the system [SH03]. Further, if sensor readings vary or drift over time, the system could automatically adapt itself to the new "normal" conditions, adjusting its error criteria accordingly.

68.5 The Diagnostic System

For illustration, a diagnostic system utilizing statistical methods (BASE [SBR05]) is compared to a standard building automation system (BAS). The BAS consists of a number of sensors and actuators connected by the LonWorks fieldbus (LON) [LDS01]. It offers a visual interface using a management information base (MIB) for retrieving and manipulating system parameters and for the visualization of system malfunctions. The diagnostic system BASE is based on statistical "generative" models (SGMs). The goal of the system is to automatically detect sensor errors in a running automation system [SBR05,FS94]. It does so by observing the data flowing through the system and thus learns about the behavior of the automation system. The diagnostic system builds a model of the sensor data in the underlying automation system, based on the data flow. From the optimized model, the diagnostic system can identify abnormal sensor and actuator values. The diagnostic system can either analyze historical data, or directly access live data.

We use a set of statistical generative models to represent knowledge about the automation system. A statistical generative model takes inputs like a sensor value, a status indicator, time of day, etc., and returns a probability between zero and one.

Using SGMs has several advantages. First, because the model encodes the probability of a sensor value, it provides a quantitative measure of "normality," which can be monitored to detect abnormal events. Second, the model can be queried as to what the "normal" state of the system would be, given an arbitrary subset of sensor readings. In other words, the model can "fill in" or predict sensor values, which can help to identify the source of abnormal system behavior. Third, the model can continuously be updated to adapt to sensor drift.

68.5.1 Error Detection

Given an SGM, implementation of this functionality is straightforward. The system assigns a probability to each newly observed data value. When this probability is high, the system returns that the new data value is a "normal" value. When the probability falls below a specific threshold, the system rates the value as "abnormal." The SGM system generates alarm events when it observes abnormal sensor values. This leaves open the question of how to assign the threshold for normality. In practice, the user sets the threshold using a graphical interface. Initially, before the system has learned normal system behavior, many alarms are generated, and the user may decide to set the threshold to a value near zero. As the system acquires a better model of the sensor system, the threshold can be raised. In any case, the threshold parameter tells us how improbable an event should be to raise an alarm. The system can also use a log-probability scale, so that the threshold can easily be set to only register extremely unlikely events.

68.5.2 Statistical Generative Models

The BASE system implements a number of SGMs (see Table 68.1).

The more complex models add additional capabilities, or relax assumptions in comparison to a simple Gaussian model [GLS+00].

Histogram: This is a very general model that is appropriate for discrete sensor values, as well as real-valued sensors with an arbitrary number of modes. One drawback is that a histogram requires a rather large quantity of data before it becomes usable or accurate.

Mixture of Gaussians: This model relaxes the Gaussian assumption that the distribution has only one mode. A mixture of Gaussians is composed of a number of Gaussian models, and each data value is attributed to the Gaussian modes with a weighting given by a "posterior probability." See, e.g., [Bis95].

Hidden Markov Model: This model is the equivalent of the mixture of Gaussians model or the histogram model, but with the addition that the current sensor value can be dependent on previous values [RJ86,Sal00].

The diagnostic system uses SGMs of automation data points. For any given data value x, model M assigns a probability to x: $P_M(x) \rightarrow [0,1]$.

Note that for discrete distributions such as a histogram, the value assigned to x by the model $P_M(x)$ is a well-defined probability, since the set of possible assignments to x is finite. For a probability density, such as a Gaussian or mixture of Gaussians, the probability value assigned to x by the model is the probability *density* at that value. In order to convert this density to a probability, the probability of generating a value within a neighborhood $\pm\delta$ around x is computed as $\int_{-\delta}^{\delta} P_M(x+\phi)\,d\phi$, and

TABLE 68.1 Statistical Generative Models

Model	Variable Type	Parameters
Gaussian	Real	μ, σ^2
Histogram	Discrete, Real	Bin counts
Mixture of Gaussians	Real	μ_i, σ_i^2, π_i
Hidden Markov model	Real	$T_{ij}\mu_i, \sigma_i^2$
Hidden Markov model	Discrete	T_{ij}, Bin counts

approximated as $2\delta P_M(x)$ for small δ. Alternatively, the probability under the model of equaling or exceeding the observed value can be computed:

$$P_M(x' \geq x) = \int_0^{\infty} P_M(x+\phi)\,d\phi \tag{68.1}$$

The data x can be a sensor reading such as an air pressure sensor, contact sensor, temperature sensor, and so on. Given a new data value, the system assigns a probability to this value. When the probability is above a given threshold, the system concludes that this data value is "normal."

Given a sequence of sensor readings $x = \{x_1, \ldots, x_T\}$ from times 1 to T, the system must create a model of "normal" sensor readings. The system uses an online version of the expectation maximization algorithm for maximum-likelihood parameter estimation. Given a model M with parameters θ, the log-likelihood of the model parameters given the data x is given by:

$$L(\theta) = \log P_M(x \mid \theta) \tag{68.2}$$

where the notation $P(x|\theta)$ denotes the conditional probability of x given the current values of the parameters θ.

The maximum-likelihood parameters are defined as the parameter values that maximize the log-likelihood over the observed data:

$$\theta_{ML} = \arg\max_{\theta} \left\{\log P_M(x \mid \theta)\right\} \tag{68.3}$$

68.5.3 Online Parameter Updates

In order for the system to continually adapt the model parameters, the parameter update algorithm must incrementally change the parameters based on newly observed sensor values. Such "online" updates have the advantage that there is no time during which the system is in an "optimization" phase, and unavailable for diagnostics [Kal60].

For the tests described here we used mixture of Gaussians models. For these models, the BASE system uses a simple stochastic estimation method, based on an expectation–maximization algorithm [NH98]. As each new data value x_i is observed, the parameters are adjusted in a two-step process. First, the posterior probability of each element of the mixture given the data value is computed. Second, the parameters are adjusted so as to increase the expected joint log-probability of the data and the Gaussian mixture component. See Section 2.6 of [Bis95] for details.

68.5.4 Results

The model log-likelihood, given by Equation 68.1, is a measure of model quality. As the parameter values are optimized online, the log-likelihood of the sensor models increases. The log-likelihood does not consistently increase, however, due to the online fitting of parameters, which happens simultaneously with reporting of abnormal sensor values. If sensors receive abnormal values, the log-likelihood decreases, until the values return to their normal range or the sensor model adapts to the new range of values. Figure 68.2 shows an example from a single sensor. The upper figure shows the sensor value; the lower figure shows the corresponding log-likelihood as a function of time. The large disturbance in the center (a power fluctuation) registers an alarm, just like the two small "spikes" near the end of the graph.

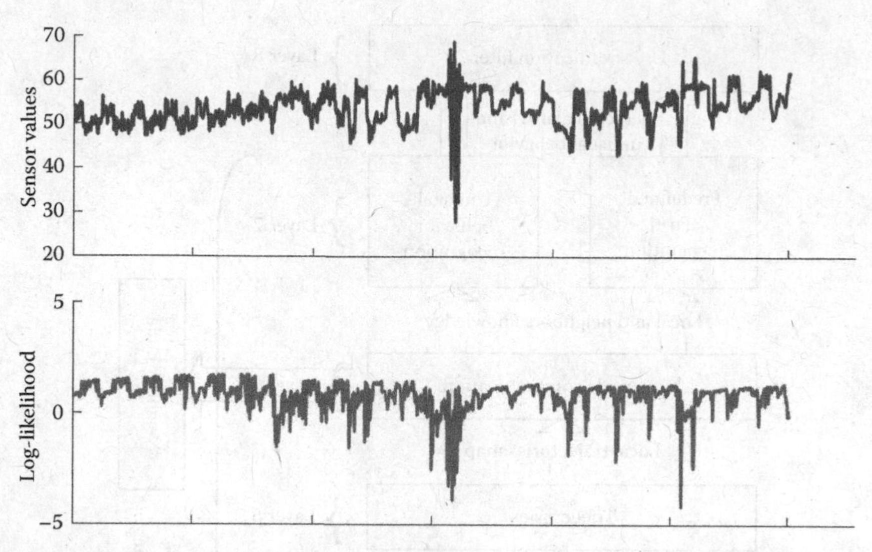

FIGURE 68.2 Sensor value and log-likelihood for a single sensor from the system. Unusual sensor values register as drops in the log-likelihood, causing alarms.

68.6 Intelligent Surveillance Systems

Today, surveillance has become a relevant means for protecting public and industrial areas against malicious subjects like burglars or vandals. For both keeping privacy of irreproachable citizens as well as enabling automated detection of potential threats, computer-based systems are needed that support human operators in recognizing unusual situations. A suitable approach for that purpose is to utilize a hierarchical architecture of semantic processing layers distributed in a network of nodes. The goal of these layers is to learn the "normality" in the environment of the network, in order to detect unusual situations and to inform the human operator in such cases. The SENSE project [WSe06,BKV+08] implements such an architecture. Therefore, we will briefly cover it here as an example of how hierarchical semantic processing can be used for surveillance systems.

SENSE consists of a network of communicating sensor nodes, each equipped with a camera and a microphone array. These sensor modalities observe their environment and deliver streams of mono-modal events to a reasoning unit, which derives fused high-level observations from this information. These observations are exchanged with neighbor nodes in order to establish a global view about the commonly observed environment. Detected potential threats are finally reported to the person(s) in charge.

Though most of the methods used in the particular layers are widely used in many applications, the benefit lies in the combination of them in order to let the messages of the system really appear meaningful to the user.

68.6.1 Architecture

In this case, an eight-layer data processing architecture is adopted, in which the lower layers are responsible for a stable and comprehensive world representation to be evaluated in the higher layers (Figure 68.3).

First, the visual low-level feature extraction (layer 0) processes frame by frame from the camera in 2D camera coordinates and extracts predefined visual objects. At the same time, the audio low-level extraction scans the acoustic signals from a linear eight-microphone array for trained sound patterns of predefined categories. Due to limited processing capabilities, this layer can deliver significantly unstable data in both modalities. In case of unfortunate conditions for the camera, detected symbols can change their label from one category to another and back for the same physical object within consecutive frames. The size of detected symbols can change from small elements to large ones covering tens of square

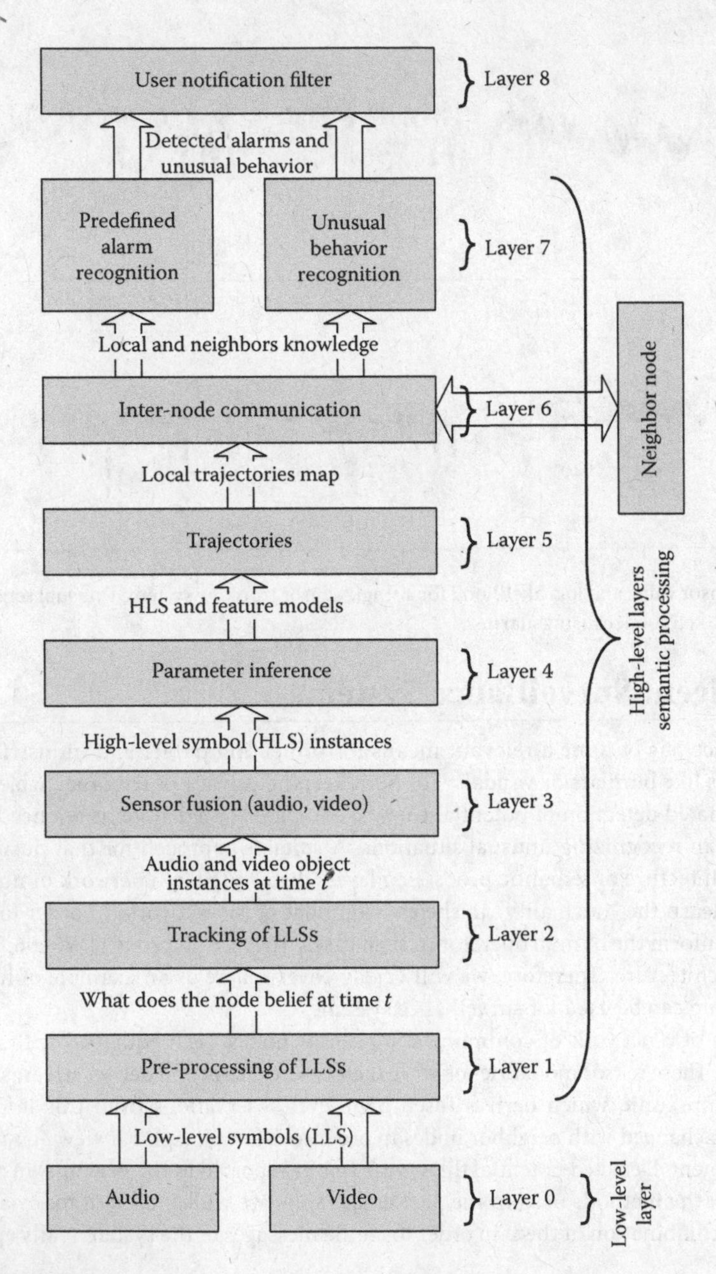

FIGURE 68.3 High level semantic processing software architecture.

meters and including previously detected and other objects. Consequently, the challenge for higher levels is to filter out significant information from very noisy data.

The modality events of layer 0 (low-level symbols in Figure 68.3) are checked at layer 1 for plausibility—e.g., the audio symbols with respect to position and intensity, the video symbols with respect to position and size. In the second layer, symbols that pass the first spatial check are subdued to a further check, regarding their temporal behavior. The output of this layer is a more stable and comprehensive world representation including unimodal symbols. Layer 3 is responsible for sensor fusion [BLS06] the unimodal symbols are fused here to form multi-modal symbols.

Layer 4 is the parameter inference machine in which probabilistic models for symbol parameters and events are optimized. The results of this layer are models of high-level symbols (HLSs) and features that describe their behavior. In layer 5, the system learns about trajectories of symbols. Typical paths through the view of a sensor node are stored. Layer 6 manages the communication to other nodes and establishes a global world view. The trajectories are also used to correlate observations between neighboring nodes. In layer 7, the recognition of unusual behavior and events takes place using two approaches: the first one compares current observations with learned models by calculating probabilities of occurrence of the observations with respect to their position, velocity, and direction. It also calculates the probabilities of the duration that symbols remain in an area, probabilities of the movement along trajectories, including trajectories across nodes. Observations with probabilities below defined thresholds raise "unusual behavior" alarms. The second part of this layer is concerned with the recognition of predefined scenarios and the creation of alarms in case predefined threat conditions are met. Finally, layer 8 is responsible for the communication to the user. It generates alarm or status messages and filters them if particular conditions would be announced too often or the same event is recognized by both methods in layer 7.

68.7 The Human Mind as an Archetype for Cognitive Automation

A long-awaited breakthrough in technology is the ability of machines to operate in an everyday human environment. Being able to sense the world in which they are immersed, robots may act in a way that is useful for human users. Decades ago, automation system designers started to follow the path that literature had drafted: intelligence and perception go hand in hand.

Thus, very simple devices can only carry out tasks that do not require sensing the real world. For instance, moving a certain piece 5 cm ahead in the conveyor belt. In a way, this is like driving blind a car: if the piece falls off the belt, the device won't know the reason and, therefore, won't be able to find a proper answer.

Obviously, more complex activities demand more complex devices and, in such cases, sophisticated perception is the turning point that allows automation systems to collect the information needed to control their actions and the consequences of them.

In the following section, we will give an overview on how machine perception has been addressed in the past, and what are promising approaches for perception in future automation systems in order to be able to fulfill useful tasks in more general environments—as humans can.

68.7.1 Perception in Automation: A Historic Overview

The term perception has been used in computer and automation systems from the 1950s onwards, since the foundation of AI. It means acquiring, interpreting, selecting, and organizing sensory information. The topic itself was not new to automation, but has gained a new quality from the moment information processing could be separated from energy flow and performed in completely new ways.

The foreseen timeframe for revolutionary, human-like applications in AI (sometimes referred to as artificial general intelligences) has been estimated to be roughly 10 years—for these last 60 years. Already in early phases of research, engineers, psychologists, and neuropsychiatrists cooperated in order to exploit synergies and achieve mutually fruitful results. Unfortunately, it soon turned out that loose couplings between scientists from different fields would not be enough to find comprehensive solutions. In 1961, E.E. David wrote in his introductory words for an issue of the *Transactions on Information Theory* [Dav61]:

> Not to say that cross-fertilization of engineering and the life-sciences should be scorned. But there must be more to these attempts than merely concocting a name, generating well-intentioned enthusiasm, speculating with the aid of brain-computer analogies, and holding symposia packed with "preliminary" results from inconclusive experiments. A bona fide "interdiscipline" draws its vitality from people of demonstrated achievement in the contributing disciplines, not from those who merely apply terminology of one field to another.

The authors of this text fully agree with this statement. It is of great importance to form teams with members from both sciences working together on a daily basis rather than meeting regularly presenting own theories about the other's profession. An indication that David's statement still remains true in many areas is that there are still no examples of human-like intelligence in automation.

The development in machine perception has taken two ways: one related with industrial process control, where machines are designed and built in order to increase productivity, reduce costs, as well as enhance quality and flexibility in the production process. These machines mostly need to perceive a well-known environment and therefore consist of a selected number of dedicated sensors. The sum of sensor views composes the machines' view of the world. Numerous publications have been written, which explain mathematical ways to cope with sensor data to fulfill the needs of automation in these respects [Ise84].

The second development path was and is concerned with the perception of humans [VLS+08] and human activities, on the one hand, and with implementing perception systems imitating human perception for broader application areas, on the other. Involved research fields are, among others, cognitive sciences, AI, image processing, audio data processing, natural language processing, user interfaces, and human machine interfaces.

Although the goal of perception systems always was to create human-like perception including all sensory modalities of humans [KBS01], from the early beginning of computer science until now, the user interface—the front end of the computer, through which computers can "perceive" humans—normally does not offer very human-like communication channels. User interfaces up to now, including today's so-called tangible interfaces are still unintuitive (batch interfaces in the 1960s, command-line interfaces in the 1970s, graphical user interfaces from the 1980s until today*).

The research field concerned with perceiving information about human users is called context aware systems. Context-aware systems are used to build devices in the fields of intelligent environments or ubiquitous computing. The common view in these communities is that computers will not only become cheaper, smaller, and more powerful; they will also more or less disappear and hide by becoming integrated in normal, everyday objects [Hai06,Mat04]. Technology will dissolve embedded into our surroundings. Smart objects will communicate, cooperate, and virtually amalgamate without explicit user interaction or commands to form consortia in order to offer or even fulfill tasks on behalf of a user. They will be capable of not only sensing values, but also of deriving context information about the reasons, intentions, desires, and beliefs of the user. This information may be shared over networks like the Internet and used to compare and classify activities, find connections to other people and/or devices, look up semantic databases, and much more.

One of the key issues in contemporary research toward this vision is scenario recognition [GJ07]. Scenario recognition tries to find sequences of particular behaviors in time, and groups it in a way humans would. This can range from very simple examples like "a person walking along a corridor" up to "there is a football match" in a stadium.

References

[Bis95] Bishop, C.M., *Neural Networks for Pattern Recognition*, New York, Oxford University Press, 1995.

[BKV+08] Bruckner, D., J. Kasbi, R. Velik, and W. Herzner, High-level hierarchical semantic processing framework for smart sensor networks, In: *Proceedings of the HSI*, Krakow, Poland, 2008.

[BLS06] Beyerer, J., F. Puente Leon, and K.-D. Sommer (eds.), Information fusion in der Mess- und Sensortechnik, Universitaetsverlag Karlsruhe, Karlsruhe, Germany, 2006.

[Bre02] Breazeal, C., *Designing Sociable Robots*, Cambridge, MA, MIT Press, 2002.

[CR04] Costello, M.C. and E.D. Reichle, LSDNet: A neural network for multisensory perception, In *Sixth International Conference on Cognitive Modeling*, Pittsburgh, PA, 2004, pp. 341–341.

* http://en. wikipedia. org/wiki/User_interface

[Dam94] Damasio, A., *Descartes' Error: Emotion, Reason, and the Human Brain*, New York, Penguine Books, 1994.

[Dav61] David, E., Bionics or electrology? An introduction to the sensory information issue, *IEEE Transactions on Information Theory*, 8(2), 74–77, 1961.

[Dav97] Davis, J., Biological sensor fusion inspires novel system design, In: *Joint Service Combat Identification Systems Conference*, San Diego, CA, 1997.

[ENF07] Dietrich, D., G. Fodor, G. Zucker, and D. Bruckner (eds.), *Simulating the Mind—A Technical Neuropsychoanalytical Approach*, Springer, Vienna, Austria, 2009.

[FS94] Fasolo, P. and D.E. Seborg, An SQC approach to monitoring and fault detection in HVAC control systems, In: *Proceedings of the American Control Conference*, Baltimore, MD, 1994.

[GJ07] George, D. and B. Jaros, The HTM learning algorithms, Numenta, Redwood City, CA, 2007.

[GLS+00] Goulding, P., B. Lennox, D. Sandoz, K. Smith, and O. Marjanovic, Fault detection in continuous processes using multivariate statistical methods, *International Journal of Systems Science*, 31(11), 1459–1471, 2000.

[Hai06] Hainich, R.R., *The End of Hardware, A Novel Approach to Augmented Reality*, Booksurge, Charleston, SC, 2006.

[HG06] Hawkins, J. and D. George, Hierarchical temporal memory—Concepts, theory, and terminology, Numenta, Redwood City, CA, 2006.

[HJ97] Hood, C.S. and C. Ji, Proactive network fault detection, In: *Proceedings of the INFOCOM*, Vol. 3, Kobe, Japan, 1997, pp. 1147–1155.

[HLS99] House, J., W.Y. Lee, and D.R. Shin, Classification techniques for fault detection and diagnosis of an air-handling unit, *ASHRAE Transactions*, 105(1), 1987–1997, 1999.

[Ise84] Isermann, R., Process fault detection based on modeling and estimation methods—A survey, *Automatica*, 20(4), 387–404, 1984.

[Jaa97] Jaakkola, T.S., Variational methods for inference and estimation in graphical models, PhD thesis, Department of Brain and Cognitive Sciences, MIT, Cambridge, MA, 1997.

[JGJS99] Jordan, M.I., Z. Ghahramani, T.S. Jaakkola, and L.K. Saul, An introduction to variational methods for graphical models, *Machine Learning*, 37, 183–233, 1999.

[Kal60] Kalman, R.E., A new approach to linear filtering and prediction problems, *Transactions of the ASME, Series D: Journal of Basis Engineering*, 82, 35–45, March 1960.

[KBS01] Kammermeier, P., M. Buss, and G. Schmidt, A systems theoretical model for human perception in multimodal presence systems, *IEEE/ASME Transactions on Mechatronics*, 6(3), 234–244, 2001.

[LDS01] Loy, D., D. Dietrich, and H. Schweinzer (eds.), *Open Control Networks: LonWorks/EIA 709 Technology*, Kluwer Academic Publishers, Boston, MA/Dordrecht, the Netherlands/London, U.K., 2001.

[Mat04] Mattern, F., Ubiquitous Computing: Schlaue Altagsgegenstnde—Die Vision von der Informatisierung des Alltags, *Bulletin des SEV/VSE*, 19, 9–13, 2004.

[NH98] Neal, R.M. and G.E. Hinton, A view of the EM algorithm that justifies incremental, sparse, and other variants, In: *Learning in Graphical Models*, M.I. Jordan (ed.), Kluwer Academic Publishers, Dordrecht, the Netherlands, 1998, pp. 355–368.

[Pal08] Palensky, B., From neuro-psychoanalysis to cognitive and affective automation systems, PhD thesis, Institute of Computer Technology, Vienna University of Technology, Vienna, Austria, 2008.

[PL03] Pontoppidan, N. and J. Larsen, Unsupervised condition change detection in large diesel engines, In: *IEEE Workshop on Neural Networks for Signal Processing*, Toulouse, France, C. Molina, T. Adali, J. Larsen, M.V. Hulle, S. Douoglas, and J. Rouat (eds.), IEEE Press, Piscataway, NJ, 2003, pp. 565–574.

[RJ86] Rabiner, L.R. and B.-H. Juang, An introduction to hidden Markov models, *IEEE ASSAP Magazine*, 3, 4–16, January 1986.

[Sal00] Sallans, B., Learning factored representations for partially observable Markov decision processes, In: *Advances in Neural Information Processing Systems*, S.A. Solla, T.K. Leen, and K.-R. Müller (eds.), Vol. 12, The MIT Press, Cambridge, MA, 2000, pp. 1050–1056.

[SBR05] Sallans, B., D. Bruckner, and G. Russ, Statistical model-based sensor diagnostic for automation systems, In: *Proceedings of the IFAC FET*, Puebla, Mexico, 2005.

[SH03] Schein, J. and J. House, Application of control charts for detecting faults in variable-air-volume boxes, *ASHRAE Transactions*, 109(2), 671–682, 2003.

[Sol02] Solms, M. and Turnbull, O. *The Brain and the Inner World: An Introduction to the Neuroscience of Subjective Experience*, Karnac/Other Press, Cathy Miller Foreign Rights Agency, London, U.K., 2002.

[Vel08] Velik, R., A bionic model for human-like machine perception, PhD thesis, Institute of Computer Technology, Vienna University of Technology, Vienna, Austria, 2008.

[VLS+08] Vadakkepat, P., P. Lim, L.C. De Silva, L. Jing, and L. L. Ling, Multimodal approach to human-face detection and tracking, *IEEE Transactions on Industrial Electronics*, 55(3), 1385–1393, 2008.

[WSe06] Tsahalis, D.T., SENSE-Smart Embedded Network of Sensing Entities, University of Patras, Available at www.sense-ist.org SENSE project web site

Index

A

Access control, RFID technology
  advantages, 9-8 thru 9-9
  babies, neonatal care, 9-9
  vehicle identification, 9-9
Acknowledged service, 41-6
Active managing node (AMN), 39-6
ActiveX technology, 64-3
Acyclic traffic, 39-10
Address autoconfiguration, 51-7
Addressing, EtherCAT
  logical, 38-6 thru 38-7
  physical, 38-6
Address space, 57-5 thru 57-7
Ad hoc networks
  applications, 7-2 thru 7-3
  characteristics, 7-3
  enabling technologies, 7-3 thru 7-4
  energy management, 7-9 thru 7-10
  MAC layer, 7-6 thru 7-7
  network layer, routing protocols
    flat and hierarchical, 7-6
    position- and nonposition-based, 7-6
    proactive, reactive, and hybrid, 7-5 thru 7-6
  performance evaluation, 7-8
  physical layer, 7-7 thru 7-8
  principles and benefits, 7-1 thru 7-2
  quality-of-service, 7-9
  security, 7-10
  topology and connectivity, 7-10
  transport layer, TCP
    classification, 7-4 thru 7-5
    suitability, 7-4
Adopted protocols, 49-7
Advanced audio distribution profile (A2DP), 49-8
Aloha mechanism, media access, 3-3
Antennas, 2-12 thru 2-13
Anticounterfeiting, RFID technology
  bank notes and automobile parts, 9-11
  electronic drug pedigree, 9-10 thru 9-11
  passports and visas, 9-11
Antiglare control, 41-8, 41-9
Apache server, 62-2
Aperiodic buffer transfers, 34-9 thru 34-10
Aperiodic busy interval, 34-16
Aperiodic traffic scheduling, 34-17
Application layer
  EtherCAT, 38-9 thru 38-10
  IEEE802.15.4, 50-7 thru 50-8
  modbus
    data access functions, 36-4 thru 36-6
    device classes, 36-7
    diagnostic functions, 36-6 thru 36-7
    error handling, 36-7
  OCARI, 53-13
  WorldFIPnetworks, 34-6 thru 34-7
Attributes, 57-5
Attribute service set, 57-8
Audio/video control transport protocol (AVRCP), 49-8
Authentication service, 41-6
Automatic data mining, *see* Practical extraction report language (PERL)
Automatic identification and data capture (AIDC) schemes, 8-2; *see also* Magnetic stripes
Automatic meter reading, 6-3
Automation systems
  archetype, future aspects, 68-2
  bottom-up *vs.* top-down design, 68-2 thru 68-4
  cognitive automation, perception
    artificial general intelligences, 68-9
    context-aware systems, 68-10
    human perception, 68-10
    machine perception, 68-10
    scenario recognition, 68-10
  sensor and actuator systems, 68-4
Automotive open system architecture (AUTOSAR), 44-6 thru 44-7
AUTOSAR, *see* Automotive open system architecture

B

Backbone couplers (BC), 42-2
Backdoor, 22-11

BACnet, building automation systems, 22-12
Bar code system
  hands-free technology, 8-16
  mentality of, 8-16 thru 8-17
  Morse and Code 128, 8-3 thru 8-4
Baseband resource manager, 49-5
BASE system
  online parameter updates, 68-6
  single sensor, sensor value and log-likelihood, 68-6 thru 68-7
  statistical generative models (SGMs)
    advantages, 68-5
    error detection, 68-5
    hidden Markov model, 68-5 thru 68-6
    histogram, 68-5
    maximum-likelihood parameters, 68-6
    mixture of Gaussians, 68-5
Basic imaging profile (BIP), 49-8
Basic printing profile (BPP), 49-9
BAT, *see* Bus arbitrator table
BioRobot, 30-11
Black channel principle, 46-3
Block acknowledgment (BlockACK), 48-8
Bluetooth
  Ad hoc networks, 7-4
  competitive technologies, 49-9 thru 49-10
  core architecture blocks
    baseband resource manager, 49-5
    channel manager, 49-3
    device manager, 49-4
    L2CAP resource manager, 49-3 thru 49-4
    link controller, 49-5
    link manager, 49-4
    radio frequency, 49-5
  future challenges, 49-10 thru 49-11
  history and technical background, 49-1 thru 49-2
  key features, 49-1
  low energy specification, 6-4
  medical applications, 30-7
  networks, 49-5 thru 49-6
  profiles
    four Bluetooth general profiles, 49-8
    general profiles, 49-8 thru 49-9
  protocol stack, 49-6 thru 49-7
  security, 49-6
  specifications, 49-2 thru 49-3
Boundary clock synchronization, 18-4 thru 18-5
Broadcast and multicast address mapping, 51-9
Broadcast mode, 36-8
Browser-based applications, 56-10 thru 56-11
Buffer transfer timings, 34-8
Building automation systems (BAS)
  applications
    climate control, 26-7
    different designs, 26-6
    SCADA systems, 26-8
    security and safety, 26-7 thru 26-8
    visual comfort, lighting, 26-7
  benefits of, 26-1 thru 26-2
  distributed functions
    building management system (BMS), 26-4
    data points, 26-3
    field layer, 26-3 thru 26-4
    management layer, 26-5
    process values, 26-5
  technologies and integration, 26-5 thru 26-6
Bus arbitrator table (BAT)
  HCF/LCM approach for, 34-8 thru 34-9
  periodic variables, 34-5
  possible schedule of variables, 34-5 thru 34-6
  WorldFIP, setting
    aperiodic busy interval, 34-16
    aperiodic traffic scheduling, 34-17
    earliest deadline approach, 34-12 thru 34-15
    rate monotonic approach, 34-10 thru 34-12
    response time analysis, 34-15
    upper bound for dead interval, 34-15 thru 34-16
    worst-case response time, 34-16 thru 34-17
Bus master, 33-2
Bus terminal module, 33-2
Byte-start sequence, 44-3

C

Cable data transmission, *see* Wired communication links
Cable replacement protocols, 49-7
Cables, 35-3 thru 35-4
CANopen device, 31-6 thru 31-7
CANopen over EtherCAT, 38-9
CAN technology, *see* Controller area network technology
Carrier sense multiple access (CSMA)
  collision avoidance, 3-4 thru 3-5
  collision detection, 3-4
  polling mechanisms, 3-5 thru 3-6
  real-time systems, 17-9
  token mechanisms, 3-5
CECED home appliances interoperating network (CHAIN), 42-10
CGI, *see* Common gateway interface
Challenge-response authentication method, 41-6
Channel hopping, 53-5, 53-9
Channel manager, 49-3
Client- and server-side technologies, 56-8
Client–server communication, 59-5
Client–server interaction model, 36-2
Clock synchronization
  distributed systems
    architecture, protocol, and algorithms, 18-8
    boundary clocks, 18-4 thru 18-5
    IEEE 1588 system model, 18-2 thru 18-3

network time protocol, 18-6 thru 18-10
ordinary clocks, 18-3 thru 18-4
precision time protocol, 18-1 thru 18-2
PTPv2, 18-5 thru 18-6
service access points, 18-3
TTCAN, 31-10
intelligent electronic device, 59-7
TTEthernet, 43-7
Code domain multiple access (CDMA), 2-16
Code 128 system, 8-3 thru 8-4
Coexistence, WSN and 2.4 GHz band, 12-8 thru 12-9
Colored Petri Net (CPN), 65-9
Combined heat and power system, 59-9
Common gateway interface (CGI), 56-8, 64-4, 64-5
Common ISDN access profile (CIP), 49-9
Common object request broker architecture (CORBA), 64-3 thru 64-4
Communication, PROFIsafe
code name, sender/receiver, 46-9
cyclic/acyclic, 46-4 thru 46-6
data consistency check, 46-9 thru 46-10
detected safety data failures, 46-10 thru 46-11
error-detection requirements, 46-4
error types and safeguards, 46-4
PDU, cyclic communication, 46-6 thru 46-7
time-out with receipt, 46-9
virtual consecutive number, 46-7 thru 46-9
Complex control systems; *see also* Automation systems
BASE system (*see* BASE system)
intelligent surveillance systems
architecture, 68-8 thru 68-9
SENSE project, 68-7
Compression, multimedia
compressors, 27-4 thru 27-5
image compressors, 27-6
spatial redundancy, 27-5
temporal redundancy, 27-5 thru 27-6
video compressors, 27-6 thru 27-7
Computer telephony integration (CTI), 14-2
Configuration and planning (CP) interface, 43-10
Congestion control window size, 61-9 thru 61-10
Controller area network (CAN) technology
data link layer, 31-3 thru 31-4
error detection and signaling, 31-4 thru 31-5
limitations
error detection capability, 31-12 thru 31-13
inconsistent communication, 31-12
node level faults, 31-13 thru 31-14
medical communication, 30-5
network topology, 31-6
physical layer, 31-2 thru 31-3
upper layer protocols
CANopen, 31-6 thru 31-7
DeviceNet, 31-8
FTT-CAN, 31-10 thru 31-12
TTCAN, 31-8 thru 31-10
CORBA, *see* Common object request broker architecture
Cordless telephony profile (CTP), 49-9
Cryptography, 22-6 thru 22-7
Cyclic/acyclic communication, 46-4 thru 46-6
Cyclical redundancy check (CRC) register, 33-4

D

Database-driven Web site design, *see* Django
Data consistency check, 46-9 thru 46-10
Data link layer (DLL)
asynchronous traffic, transmission of, 34-6
bus arbitrator table, 34-4 thru 34-6
cyclic traffic, transmission of, 34-3 thru 34-4
ISA 100.11a, 53-8 thru 53-9
LLC sublayer, 34-3
MAC, 34-2
WiMAX, 52-5
WirelessHART
datagram, 53-5 thru 53-6
frequency diversity mechanism, 53-5
packet timing and time slot format, 53-4
TDMA mechanism, 53-3
Data mining robot (DMR), 66-1; *see also* Practical extraction report language (PERL)
Data point types (DPTs), 42-9
DER, *see* Distributed energy resources
Destination-sequenced distance-vector (DSDV) protocol, 4-10 thru 4-11
Device attacks security
definition, 22-2
protected hardware and security token, 22-9 thru 22-10
secure software environments
malwares, 22-10 thru 22-11
prevention methods, 22-11 thru 22-12
Device manager, 49-4
DeviceNet communication, 31-8
Devices profile for web services (DPWS), 26-11
Diagnostic and maintenance (DM) interface, 43-9
Diagnostic frames, 45-10
Dial-Up network profile (DUN), 49-9
Diesel engines, 59-9
Differentiated services (DiffServ), IP networks QoS, 19-6 thru 19-8
Digital addressable lighting interface (DALI), 26-6
Digital living network association (DLNA), 26-11
DigitalSTROM, home automation, 26-9
Direct link protocol, 48-7
Direct link setup (DLS), 48-7
Discovery service set, 57-7
Distributed automation systems
application, profiles
IEC 61499, 5-4
IEC 61850 standard, 5-5 thru 5-6
LON, 5-4 thru 5-5

interoperating components
controller, 5-3
function blocks, 5-3
identical properties for, 5-2
nodes, 5-1
standardized system advantages, 5-3
type snvt_temp, 5-2
software interoperability, 5-6 thru 5-7
Distributed clock synchronization, 38-8
Distributed coordination function, 48-4 thru 48-5
Distributed energy resources (DER)
automation function, modeling, 59-3 thru 59-4
basic concept, 59-1 thru 59-3
communication services
client/server communication, 59-5
clock synchronization, 59-7
GOOSE, 59-5 thru 59-6
transmission of sampled analog values, 59-6 thru 59-7
system configuration language, modeling, 59-7 thru 59-8
types of
combined heat and power system, 59-9
diesel engines, 59-9
fuel cells, 59-9
hydropower plants, 59-8 thru 59-9
wind power plants, 59-8
Django
admin site activation
activated admin site, 63-6
default administration management site, 63-6
steps involved in, 63-5
automatic admin interface, 63-7
customized administration Web site, 63-7
database setup, 63-4
document module, 63-6
installation, 63-2
interactive web site design by, 63-1 thru 63-2
internationalization, 63-8
model creation and activation, 63-4 thru 63-5
object-relational mapper, 63-7
project creation, 63-2 thru 63-3
project running, 63-3 thru 63-4
template system, 63-8
URL design, 63-8
DLL, *see* Data link layer
DNP3
features, 58-3 thru 58-5
vs. IEC 60870-5
example, 58-5
explicit application confirmation mechanism, 58-6
normalized, scaled, and floating-point data objects, 58-7
TCP/IP, 58-6
T101 or T104 system, 58-7
two-pass command sequence, 58-8 thru 58-9
unsolicited reporting, 58-8
SCADA data collection
electric power networks, 58-1 thru 58-2
other major industries, 58-3
DoS prevention and detection, 22-8 thru 22-9
Drivers, 35-3
Dual authentication, RFID, 8-22
Duty-cycle, wireless communication, 10-4
Dynamic MANET on-demand routing protocol (DYMO) protocol, 4-11
Dynamic source routing protocol (DSR) protocol, 4-9 thru 4-10
Dynamic Web sites
Apache server, 62-2
client-server partition, data flow, 62-2
PHP and MySQL
database creation, 62-11
data insertion, 62-12
paper collection and evaluation, 62-13 thru 62-14
query deletion and selection, 62-12
record updation, 62-12
SQL queries, 62-10
table creation, 62-11
scripting languages, 62-1
WAMP server, 62-2 thru 62-3

E

Earliest deadline approach, 34-12 thru 34-15
EBUS, 38-2
Echonet, home automation, 26-11
Electrical commissioning, FF, 35-10
Electric power SCADA systems, 58-1 thru 58-2
Electronic cash, 8-7
Electronic commissioning, FF, 35-10
Electronic data sheet (EDS), 56-9
Electronic device description (EDD), 56-8 thru 56-10
Electronic device description language (EDDL), 35-3, 57-1
Electronic drug pedigree, anticounterfeiting, 9-10 thru 9-11
E-Mail address extraction, excel files, 66-4 thru 66-5
Energy management, wireless communication
application layer methods, 10-6 thru 10-8
communication protocol methods, 10-4 thru 10-6
composite materials, airplanes, 10-1
hardware methods, 10-2 thru 10-4
Enhanced distributed channel access (EDCA), 48-6, 48-9
EnOcean, home automation, 26-10
Ephemeral ports, 61-6
Equivalent isotropically radiated power (EIRP), 54-2
Error detection, 36-11
Error handling, 36-7

Ethernet for control automation technology (EtherCAT)
addressing
logical, 38-6 thru 38-7
physical, 38-6
application layer, 38-9 thru 38-10
communication protocol
commands, 38-3 thru 38-5
medium access mechanism, 38-2 thru 38-3
datagram structure, 38-4
definition, 38-1
distributed clock, 38-8
frame structure, 38-4
mailbox services, 38-9 thru 38-10
physical layer, 38-1 thru 38-2
SyncManager, 38-7 thru 38-8
Ethernet for plant automation (EPA), 37-7
Ethernet, industrial, *see* Industrial Ethernet
EtherNet/IP, 37-5 thru 37-7
Ethernet POWERLINK (EPL)
communication architecture, 39-1, 39-2
definition, 39-1
frame mapping, 39-3 thru 39-4
network configurations, 39-4 thru 39-5
parameters, 39-3
performance analysis
acyclic traffic, 39-10
cycle time, 39-9 thru 39-10
jitter, 39-8
turn-around time, 39-8 thru 39-9
protocol, 39-1 thru 39-3
redundancy
medium, 39-5, 39-6
MN, 39-5 thru 39-7
ring redundancy, 39-6
security, 39-7 thru 39-8
European Telecommunications Standards Institute (ETSI), 67-2
Event-triggered frame, 45-9
Exavera eShepherd™ system, 30-4
eXtensible markup language (XML), 56-7 thru 56-8
External interface files (XIF), 41-7 thru 41-8
External message interface (EMI), 42-5

F

Failure modes and effects analysis (FMEA), 21-6
Fast handover, WLANs
mechanisms on AP side, 48-10 thru 48-11
mechanisms on client side, 48-11
procedure, 48-9 thru 48-10
Fault handling, 43-5
Fault hypothesis, 43-5
Fault tolerance, 43-5
Fax profile (FAX), 49-9
FDMA, *see* Frequency division multiple access
FF, *see* Foundation fieldbus
FHSS, *see* Frequency hopping spread spectrum
Fiber-optic transmission, Profibus, 32-4; *see also* Optical communication systems (OCSs)
Fieldbus intrinsically safe concept (FISCO), 35-7 thru 35-8
Fieldbus memory management unit (FMMU), 38-2, 38-6 thru 38-7
Fieldbus power supplies, 35-5 thru 35-6
Fieldbus systems, vertical integration, 13-3
Field concentrators, 34-2
Field device configuration markup language (FDCML), 56-9
Field device tool/device-type manager (FDT/DTM), 35-3
File access over EtherCAT (FoE), 38-10
File transfer profile (FTP), 49-9
Firewalls, 22-5 thru 22-6
Fixed WiMAX, 52-3
Flexible time-triggered communication on CAN (FTT-CAN) protocol
asynchronous and synchronous windows, 31-11
elementary cycle (EC) concept, 31-10 thru 31-11
multislave access control, 31-11
system requirements database (SRDB), 31-12
FlexRay
protocol
communication cycles, 44-2 thru 44-3
framing, 44-3
physical layer, 44-3 thru 44-4
startup of cluster, 44-3
system architecture
node architecture, 44-5
star couplers, 44-5
topologies, 44-4
system design considerations
AUTOSAR, 44-6 thru 44-7
configuration, 44-5 thru 44-6
Foundation fieldbus (FF)
application, 35-2
cables, 35-3 thru 35-4
drivers, 35-3
fieldbus power supplies, 35-5 thru 35-6
high-speed Ethernet, 37-7
installations
in classified areas, 35-7 thru 35-8
and commissioning, 35-8 thru 35-9
in safe areas, 35-6
link active scheduler, 35-2
maintenance, 35-10
network topology, 35-4
project documentation, 35-8, 35-9
segment design, 35-4 thru 35-5
topology, 35-2 thru 35-3
Four-way handshaking, 61-12 thru 61-13
Frame end sequence, 44-3
Frame mapping, 39-3 thru 39-4
Frame start sequence, 44-3

Frequency division multiple access (FDMA), 2-16, 48-3
Frequency hopping spread spectrum (FHSS), 2-16, 6-4, 6-10, 7-4, 48-3, 49-2
FTT-CAN, *see* Flexible time-triggered communication on CAN protocol
Fuel cells, 59-9
Full-duplex media access, 3-2
Full Ethernet, 37-4
Functional safety
- definition, 21-1 thru 21-2
- fault tree analysis, 21-6 thru 21-7
- FMEA, 21-6
- generic lifecycle, 21-4 thru 21-5
- HAZOP, 21-5 thru 21-6
- industrial communication systems, 21-1
  - communication systems, 21-8 thru 21-10
  - failure mitigation, 21-12 thru 21-15
  - hazard and risk analysis, 21-10 thru 21-12
- safety cases, 21-7 thru 21-8
- standards
  - classification, 21-2
  - IEC61508, 21-3 thru 21-4
  - quagmire, 21-3

G

General audio/video distribution profile (GAVDP), 49-9
Generic message passing, 41-7
Generic object-exchange profile (GOEP), 49-8
Generic object oriented substation event (GOOSE), 59-5 thru 59-6
Generic object profile (GOEP), 49-9
Generic station description (GSD) file, 40-11
Generic station description markup language (GSDML), 56-9
Geographic adaptive fidelity (GAF), wireless networks, 4-13
Global positioning system (GPS), 8-14
GOOSE, *see* Generic object oriented substation event
Graphical human-machine interface (HMI), 36-2
Green interval, 40-9

H

Hands-free profile (HFP), 49-9
Hands-free technology, RFID, 8-8, 8-16
Hard copy cable replacement profile (HCRP), 49-9
Hard real-time system, 20-2
Hazard and operability study (HAZOP), 21-5 thru 21-6
HCF controlled channel access, 48-7
Header compression
- IPv6, 51-10
- UDP, 51-11
Headset profile (HSP), 49-9
Health device profile (HDP), 49-9
Hidden Markov Model, 68-5 thru 68-6
High-energy trunk–fieldbus barrier solution, 35-8
High-performance radio LAN (HIPERLAN), 49-10
High-speed Ethernet (HSE), 35-2 thru 35-3
Hollerith punch card, 8-2
Holonic multi-agent systems, 16-4 thru 16-5
Home automation systems
- integration of, 26-8
- privacy and security, 26-9
- technologies and integration
  - devices profile for web services (DPWS), 26-11
  - digital living network association (DLNA), 26-11
  - digitalSTROM, 26-9
  - Echonet and OSGi, 26-11
  - EnOcean and Z-Wave protocol, 26-10
  - HomePlug, 26-9
  - intelligent grouping and resource sharing (IGRS), 26-10 thru 26-11
  - universal plug and play (UPnP), 26-10
  - wireless protocols, 26-9 thru 26-10
- usability, 26-8 thru 26-9
HomePlug powerline alliance, 26-9
Horizontal communication, 13-4
HTML, *see* Hypertext markup language
HTTP, *see* Hypertext transfer protocol
Human body, wireless communication power, 10-3
Human interface device (HID) profile, 49-9
Hydropower plants, 59-8 thru 59-9
Hypertext markup language (HTML), 56-6, 64-2
Hypertext preprocessor (PHP), 62-1 thru 62-2; *see also* Dynamic Web sites
- conditional statements, 62-4 thru 62-5
- functions, 62-6
- $_GET and $_POST functions, 62-7
- include() function, 62-6 thru 62-7
- loops, 62-5 thru 62-6
- network programming tools, 64-4 thru 64-5
- variables, 62-3
Hypertext transfer protocol (HTTP), 56-5

I

IEC 60870-5
- *vs.* DNP3
  - example, 58-5
  - explicit application confirmation mechanism, 58-6
  - normalized, scaled, and floating-point data objects, 58-7
  - TCP/IP, 58-6
  - T101 or T104 system, 58-7
  - two-pass command sequence, 58-8 thru 58-9
  - unsolicited reporting, 58-8
- features, 58-3 thru 58-5
- SCADA data collection
  - electric power networks, 58-1 thru 58-2
  - other major industries, 58-3
IEC 61158-2, 35-3

IEC 61850
application of, 59-2
for DER (*see* Distributed energy resources)
logical nodes of, 5-5 thru 5-6
IEC 61158 and IEC 61784
EtherNet/IP, 37-5 thru 37-7
exotic solutions, 37-7, 37-9
foundation fieldbus high-speed Ethernet, 37-7
SERCOS III, 37-7
IEEE 802.11, 3-6, 3-7, 7-3, 7-7 thru 7-8, 28-2, 30-7 thru 30-8, 48-2, 48-4, 49-9 thru 49-10
IEEE 802.15.4
application layer, 50-7 thru 50-8
key features, 50-3 thru 50-4
6LoWPAN
defined PHY and MAC layers, 51-3
key features, 51-3
network topologies, 51-4
PHY and MAC frame format, 51-5
RFC4919, 51-5
specification, 51-6
MAC layer, 50-5 thru 50-6
network layer, 50-7
physical layer, 50-4 thru 50-5
protocol, 30-6
IEEE 802.11 protocol, 30-7 thru 30-8
IEEE 802.3 standard, *see* Industrial Ethernet
Image compressors, multimedia, 27-6
Industrial agent technology
application
aviation and space control industry, 16-10 thru 16-11
challenges, 16-10
comparison of, 16-9 thru 16-10
logistics, 16-11
order handling, 16-8 thru 16-9
PABADIS, 16-7
PROSA and AARIA, 16-6 thru 16-7
resource handling, 16-8
characteristics of, 16-6
emergence, 16-3
holonic multi-agent systems, 16-4 thru 16-5
holonic paradigm, 16-4
implementation, 16-5 thru 16-6
intelligent agents, 16-2
multi-agent systems, 16-2 thru 16-3
ontologies, 16-3
self-organizing system, 16-3
Industrial communication systems (ICS)
global connectivity and digital identity, 67-1 thru 67-2
hardware and software, scalability, 67-4
hybrid local networks and quality of services (QoS), 67-2 thru 67-3
M2M communication, 67-3 thru 67-4
security, 22-13 thru 22-14
automation systems, 22-12 thru 22-15
counteract device attacks, 22-9 thru 22-12
counteract network attacks, 22-4 thru 22-9
definitions, 22-1 thru 22-2
device attacks, 22-2
network attacks, 22-2
policy, setup and maintenance, 22-3 thru 22-4
vertical integration, 67-2
Industrial Ethernet
application domain profiles, 37-3 thru 37-4
classification of, 37-4 thru 37-5
definition, 37-2
features of major, 37-8, 37-9
IEC 61158 and IEC 61784
EtherNet/IP, 37-5 thru 37-7
exotic solutions, 37-7, 37-9
foundation fieldbus high-speed Ethernet, 37-7
SERCOS III, 37-7
network segmentation, 37-1 thru 37-2
properties, 37-3
redundancy, 37-3
synthesis, 37-9
Industrial wireless sensor networks (IWSNs)
advantages, 6-1
applications
automatic meter reading, 6-3
factory, building and industrial process automation, 6-2
inventory management, 6-2 thru 6-3
requirements design, 6-6 thru 6-7
utility automation, 6-3
hardware development
energy-harvesting techniques, 6-10 thru 6-11
low-power and low-cost sensor node, 6-7 thru 6-8
radio technologies, 6-8 thru 6-10
software development, 6-11
standardization, 6-3 thru 6-5
system architecture and protocol design, 6-11 thru 6-12
technical challenges, 6-5 thru 6-6
Information security, QoS
committees and standards, 19-4
policy, 19-3 thru 19-4
problems, 19-3
Infrared (irDA), 49-10
Instrumentation, Systems, and Automation Society (ISA), 67-4
Integrated Services (IntServ), IP networks QoS, 19-6
Intelligent agents, 16-2
Intelligent grouping and resource sharing (IGRS), 26-10 thru 26-11
Intelligent surveillance systems
architecture
inter-node communication, 68-9
parameter inference, 68-8 thru 68-9
sensor fusion, 68-8
spatial check, 68-8
trajectories of symbols, 68-9

unusual behavior and events recognition, 68-8, 69-9
user notification filter, 68-8, 69-9
visual low-level feature extraction layer, 68-7 thru 68-8
SENSE project, 68-7
Interactive voice response (IVR) telecom, 14-2 thru 14-3
INTERBUS
basic elements, 33-2
data transfer, 33-1 thru 33-2
diagnostic features, 33-7 thru 33-8
performance evaluation, 33-8 thru 33-9
protocol
acyclic parameter data transmission, 33-6
architecture, 33-5
CRC register, 33-4
layer 2 summation frame structure, 33-2 thru 33-3
logical structure, 33-4
slave node, 33-3
telegram formats, 33-5
VFD, 33-6 thru 33-7
topology, 33-1, 33-2
Intercom profile (ICP), 49-9
Interconnectable system, 1-3
Interface control unit information (ICI), ISO/OSI design, 1-8
Interface data unit (IDU), ISO/OSI design, 1-8
Interface file system (IFS), 43-8 thru 43-9
International Organization for Standardization (ISO)
advantages and disadvantages of, 1-9
definition, 1-1
horizontal communication, 1-5 thru 1-6
ISO/OSI model, 1-1 thru 1-2
management units, 1-9
open system
definition, 1-3
layer functionalities, 1-4 thru 1-5
origin, 1-1
services and protocols, dynamics
definition of, 1-6
interface process, 1-8
predefined primitives, 1-7 thru 1-8
service access points, 1-6 thru 1-7
specifications of, 1-6
specific description language, 1-1
vertical communication, 1-5 thru 1-6
Internet
applications
examples, 56-8 thru 56-11
industrial, 56-2 thru 56-3
technologies
basic architecture and communication paradigm, 56-3
for information description and presentation, 56-6 thru 56-8
multitier architecture, 56-3, 56-4
for server-side and client-side functions, 56-8
three-tier architecture, 56-3, 56-4
transport and communication related, 56-4 thru 56-6
Internet assigned numbers authority (IANA), 60-4
Internet protocol security (IPsec), 22-14
Interoperability, 5-6 thru 5-7; *see also* Distributed automation systems
IP-based networks
multimedia applications, 27-10 thru 27-11
quality of service
application supports, 19-5
classification and marking, 19-8 thru 19-9
DiffServ model, 19-6 thru 19-8
integrated Services (IntServ) model, 19-6
problems of, 19-4 thru 19-5
processes, 19-5 thru 19-6
queuing and congestion management, 19-9 thru 19-11
security, 22-14 thru 22-15
vertical integration, 13-5
IP-based transport layer, vertical integration, 13-5
IP telephony, *see* Voice over Internet protocol
IPv6 header compression, 51-10
IPv6 over low-power wireless personal area networks (6LoWPANs)
address autoconfiguration, 51-7
broadcast and multicast address mapping, 51-9
frame types and compression schemes, 51-11 thru 51-13
frame types and fragmentation, 51-7 thru 51-8
header compression, 51-9 thru 51-11
IEEE 802.15.4
defined PHY and MAC layers, 51-3
key features, 51-3
network topologies, 51-4
PHY and MAC frame format, 51-5
RFC4919, 51-5
specification, 51-6
mesh frame type, 51-8 thru 51-9
scopes, 51-11
security, 51-13
for smart cooperating objects, 51-1 thru 51-3
service-oriented architectures, 51-2
support in operating systems, 51-2
ISA 100.11a
data link layer, 53-8 thru 53-9
network layer and topologies, 53-10
physical layer, 53-8
time keeping, 53-10
upper layers, 53-10
ISA-100 standards, WSN, 12-7
ISO, *see* International Organization for Standardization
Isochronous real-time (IRT) communication
bus cycle, 40-9
conditions for, 40-8

cycle duration and constrains, 40-10
intervals, 40-9
orange interval, 40-9 thru 40-10
red interval, 40-10

J

Java, 64-4
Java agent development (JADE) framework, 16-5 thru 16-6
JavaScript
Ajax technology, 64-4
applications, 64-3
drawback, 64-2 thru 64-3
limitations, 64-3
Jitter, 39-8
Junction boxes, 34-2

K

KNX, building automation systems, 22-12
KNX network
association, 42-1
configuration, 42-12
devices, 42-11
interworking and application model, 42-2
medium-dependent layers, 42-5 thru 42-6
medium-independent layers
external message interface, 42-5
frame formats, 42-3 thru 42-4
network layer, 42-3
TL protocol data unit, 42-4
topology, 42-2 thru 42-3
net/IP tunneling, 42-6
powerline 110, 42-6
RF physical layer, 42-5
runtime interworking
CHAIN, 42-10
data point types (DPTs), 42-9
definition, 42-6
group objects, 42-6 thru 42-8
LTE, 42-10
user applications, 42-2

L

L2CAP resource manager, 49-3
Light emitting diodes (LEDs), optical links, 2-9
Link controller, 49-5
Link layer, 1-4, 41-4 thru 41-5
Link layer discovery protocol (LLDP), 40-12
Link manager, 49-4
Local bus, 33-2
Local interconnect network (LIN)-Bus
automotive communication architecture, 45-2
communication concept, 45-3, 45-4
configuration, 45-12
history and versions, 45-2 thru 45-3
message frames
break field, 45-6 thru 45-7
checksum, 45-8
data field, 45-8
diagnostic frame, 45-10
frame length, 45-8
frame types, 45-9
identifier, 45-7 thru 45-8
sync byte field, 45-7
time-triggered data transmission, 45-8
network and status management, 45-10 thru 45-11
physical layer
basic principle of, 45-5
signal specification, 45-5 thru 45-6
topology, 45-6
SAEJ2602andLIN2.0, relationship, 45-12
transport layer protocol, 45-11 thru 45-12
Local operating network (LON), *see* LonWorks
Logical addressing, 38-6 thru 38-7
Logical bomb, 22-11
Logical nodes classes, 59-3, 59-9
Logic-based monitoring, 47-9
LonWorks
application layer programming model, 41-7 thru 41-8
automatic design approach, 41-12
building automation systems, 22-12 thru 22-13
definition, 41-1
functional profiles in, 5-4 thru 5-5
function block-based design and system integration
antiglare control, 41-8, 41-9
standard network variable type, 41-11
sunblind controller, 41-10
network design tools, 41-11 thru 41-12
physical and logical view of, 41-3
protocol
application and presentation layer, 41-6 thru 41-7
link layer, 41-4 thru 41-5
network layer, 41-5
physical layer, 41-4
transport and session layer, 41-5 thru 41-6
routers, 41-3
system components
neuron chip structure, 41-2
physical and logical segmentation, 41-3
Low energy adaptive clustering hierarchy (LEACH) protocols, 4-12 thru 4-13
Luminance sensor, 41-8

M

MAC, *see* Medium access control
Machine-machine communication, 56-11
Machine-to-machine (M2M) communication, 67-3 thru 67-4

Magnetic stripes, 8-4
Manchester bus powered (MBP) transmissions, 32-4
Manual control, 41-8
Manufacturing industry
multisite distributed product development methodology, 65-2
multisite production system
Colored Petri Net (CPN), 65-9
customer-centered design, 65-9
ontology-driven architecture (*see* Ontology-driven architecture)
optimization, 65-9
product customization, 65-8
operation modes, requirements, 65-1
Manufacturing message specification (MMS), 33-6
Manufacturing message specification (MMS)-protocol, 59-5
Master/slave round, 43-11
Media
access methods
carrier sense mechanisms, 3-4 thru 3-6
full-duplex media access, 3-2
problem of, 3-6 thru 3-7
statistic access arbitration, 3-3
synchronous access arbitration, 3-2 thru 3-3
optical links, 2-7 thru 2-11
wired links, 2-1 thru 2-7
wireless links, 2-11 thru 2-17
Media access control (MAC)
processor, 41-2
Media redundancy protocol (MRP), 40-13
Medical applications
automation
healthcare robotics, 30-11 thru 30-12
smart homes, 30-9 thru 30-11
clinical monitoring
Bluetooth, 30-7
controller area network (CAN) technology, 30-5
IEEE 802.15.4, 30-6
IEEE 802.11 protocol, 30-7 thru 30-8
nonindustrial technologies, 30-8 thru 30-9
Profibus DP fieldbus, 30-5 thru 30-6
data communications, 30-2
localization, 30-4 thru 30-5
microcontroller-based devices, 30-1 thru 30-2
requirements, 30-2 thru 30-3
safety and reliability, 30-12
security and privacy, 30-12
standardization, 30-13
timeliness, 30-13
Medium access control (MAC)
architecture with HCF, 48-4
direct link protocol and block ACK, 48-7 thru 47-8
distributed coordination function, 48-4 thru 48-5
enhanced distributed channel access, 48-6
HCF controlled channel access, 48-7
point coordination function, 48-5 thru 48-6
protocol, 34-2
Medium access control (MAC) layer
IEEE802.15.4
slotted mode, 50-6
unslotted mode, 50-5 thru 50-6
OCARI, 53-11 thru 53-12
WiMAX, 52-5
Medium-dependent layers, 42-5 thru 42-6
Medium-independent layers
external message interface, 42-5
frame formats, 42-3 thru 42-4
network layer, 42-3
TL protocol data unit, 42-4
topology, 42-2 thru 42-3
Mesh frame type, 51-8 thru 51-9
Mesh networks, 50-1 thru 50-2
Message formatting, 58-5 thru 58-9
Message frames, LIN
break field, 45-6 thru 45-7
checksum, 45-8
data field, 45-8
diagnostic frame, 45-10
frame length, 45-8
frame types, 45-9
identifier, 45-7 thru 45-8
sync byte field, 45-7
time-triggered data transmission, 45-8
Method service set, 57-8 thru 57-9
Microsoft media server (MMS), 27-11
Middleware, 57-3
Mobile WiMAX, 52-3
Modbus
application layer
data access functions, 36-4 thru 36-6
device classes, 36-7
diagnostic functions, 36-6 thru 36-7
error handling, 36-7
interaction and data models
client–server interaction model, 36-2
memory areas, 36-2 thru 36-3
protocol architecture, 36-3
rope-making machine, automation of, 36-13 thru 36-15
serial network
ASCII mode, 36-10 thru 36-11
error detection, 36-11
frames, 36-8 thru 36-9
physical layer, 36-11
RTU mode, 36-9 thru 36-10
TCP network, 36-12
Modulation techniques, WLANs, 48-3
MonitoredItem service set, 57-9
Morse bar code system, 8-3 thru 8-4
Multi-agent systems (MAS), 16-2 thru 16-3
Multicast listener discovery (MLD), 27-11

Multimedia applications
compression
compressors, 27-4 thru 27-5
image compressors, 27-6
spatial redundancy, 27-5
temporal redundancy, 27-5 thru 27-6
video compressors, 27-6 thru 27-7
image transmission
IEEE 1394, 27-9 thru 27-10
IP-based networks, 27-10 thru 27-11
in industrial environment
communication network types, 27-2 thru 27-3
computer vision applications, 27-8 thru 27-9
monitoring applications, 27-8
pixels, resolution, 27-1
quality evaluation, 27-7
RGB format, 27-1 thru 27-2
security application, 27-3 thru 27-4
Multimedia service convergence
computer telephony integration (CTI), 14-2
definition of, 14-1 thru 14-2
enterprise communications, 14-1
interactive voice response telecom, 14-2 thru 14-3
MC_2 architecture
advantages of, 14-6
computer-aided interaction, 14-7
plugin architecture, 14-10
software PBX integration, 14-8
required features of, 14-2
service-oriented architecture
converged services, 14-5 thru 14-6
meta data model, 14-4 thru 14-5
software flexibility, 14-3
voice plugins, 14-9 thru 14-11
tailorability, 14-5
voice over Internet protocol (VoIP), 14-2 thru 14-3
Multipartner round, 43-11
Multiplexing
OFDM, 48-3
optical links, 2-10
wired links, 2-4 thru 2-5
wireless links, 2-16 thru 2-17
MySQL; *see also* Dynamic Web sites
database creation, 62-8
database table, 62-7 thru 62-8
interactive Web pages, development, 62-2
record sorting, 62-10
stored data view, 62-9 thru 62-10
table creation, 62-8 thru 62-9
table modification, 62-9
WHERE Clause, 62-10

N

Nagle's algorithm, 61-9
Net/IP tunneling, 42-6
Network access layer interface, 47-8
Network attacks security
cryptography, 22-6 thru 22-7
definition, 22-2
DoS prevention and detection, 22-8 thru 22-9
firewalls, 22-5 thru 22-6
guaranteed objectives, 22-4
virtual private networks, 22-5
Network-based control (NBC) system
architecture
basic capabilities, 20-3
defining feature, 20-2
remote control systems, 20-3 thru 20-6
shared-network control systems, 20-3
control loops, 20-1
designing of
constraints in, 20-7
network and control, 20-7
effects in control performance, 20-6
feedback control of, 20-1 thru 20-2
hard real-time system, 20-2
Network calculus (NC), real-time networks, 17-6
Network layer (NL)
IEEE802.15.4, 50-7
LonWorks, 41-5
medium-independent layers, KNX, 42-2
and topologies
ISA 100.11a, 53-10
OCARI, 53-12 thru 53-13
WirelessHART, 53-6
wirelessHART datagram, 53-7 thru 53-8
Network management and diagnostic messages, 41-7
Network programming tools
ActiveX technology, 64-3
autonomous robots, 64-10
CGI, 64-4
CORBA, 64-3 thru 64-4
data monitoring and control, 64-10 thru 64-11
HTML, 64-2
Java, 64-4
JavaScript, 64-2 thru 64-4
neural network trainer
CGI, 64-5
NBN 2.0. user interface, 64-5 thru 64-6
operating system, 64-7
training result file, 64-6
PERL, 64-4
PHP, 64-4 thru 64-5
simulation, benefits, 64-1 thru 64-2
software development, 64-2
Spice internet package
characteristics, 64-9
client machine, 64-8 thru 64-9
graphical user interface, 64-9
network traffic considerations, 64-10
server-located PERL script, 64-9
Web-based C_{++} compiler, 64-7 thru 64-8
Web server, client application, 64-10

Network time protocol (NTP), clock synchronization
 algorithms of, 18-9 thru 18-10
 clock discipline, 18-10
 components of, 18-8 thru 18-9
 hardware requirements for, 18-8
 phase and frequency prediction functions, 18-10
 standardization, 18-6
 stratum, 18-7
Network variable propagation, 41-6
Neuron Chip
 collision detections, 41-5
 neuron C program, 41-7
 structure of, 41-2
Node management service set, 57-8
Nuclear power plant automation systems, 29-3

O

Object access protocol, 56-5 thru 56-6
Object push profile (OPP), 49-9
OCARI
 application layer, 53-13
 MAC layer, 53-11 thru 53-12
 network layer and topologies, 53-12 thru 53-13
 physical layer, 53-10 thru 53-11
 stack architecture, 53-10, 53-11
Occupancy sensor, 41-8
Octopus card, *see* Smart card
Ontology-driven architecture
 awareness and cooperation, 65-4
 base ontology, 65-4
 communications, 65-4
 conceptual and terminological confusion reduction, 65-4
 conceptual framework, 65-3
 coordination, 65-4
 integratability, 65-4
 multisite issues
  components ontology, 65-4 thru 65-5
  ontology model representation, 65-5
  process ontology, 65-4
  products ontology, 65-4
  refinement process, 65-5
  secure access control, 65-5
  sub-ontologies, 65-5
  task responsibility, 65-5
 rapid reconfigurability, 65-6
 remote procedure call (RPC), 65-6
 representational state transfer (REST) approach, 65-6
 service mashup, 65-7 thru 65-8
 SOA-based system architecture, 65-6 thru 65-7
 triple space, 65-6
 Web security and confidentiality, 65-3
OPC unified architecture (OPC UA)
 address space, 57-5 thru 57-7
 application, 57-2
 implementations and products, 57-10
 services
  attribute service set, 57-8
  discovery service set, 57-7
  general, 57-7
  method service set, 57-8 thru 57-9
  MonitoredItem service set, 57-9
  NodeManagement service set, 57-8
  query service set, 57-8
  SecureChannel service set, 57-7 thru 57-8
  session service set, 57-8
  subscription service set, 57-9
 system architecture, 57-2 thru 57-4
  client application architecture, 57-4
  server architecture, 57-4 thru 57-5
Open standard system, 1-5 thru 1-6
OpenVPN (OVPN), 22-14 thru 22-15
Optical communication systems (OCSs)
 implementations and standards, 2-10 thru 2-11
 multiplexing, 2-10
 physical properties, 2-7 thru 2-8
 transmitters and receivers, 2-9 thru 2-10
 types and media access, 2-8
Optical time domain reflectometer (OTDR), 2-10 thru 2-11
Orange interval, 40-9 thru 40-10
Ordinary clocks synchronization, 18-3 thru 18-4
Orthogonal frequency-division multiplexing (OFDM), 48-3, 52-5
OSGi, home automation, 26-11
Overemitting, ultralow-power wireless communication, 10-5

P

PAN *vs.* WiMAX, 52-4
Passports anticounterfeiting, RFID technology, 9-11
Periodic communication, 33-6
Peripheral message specification (PMS), 33-6
Peripherals communication protocol (PCP), 33-6
PERL, *see* Practical extraction report language
Personal area networking profile (PAN), 49-9
Personal area network (PAN), security
 applications, 28-6
 security concerns
  authentication and location tracking, 28-6
  collision, 28-8
  configuration, 28-7
  eavesdropping, 28-6
  signal jamming, 28-8
Personal identity, RFID-ID card, 8-7 thru 8-8
Photomultipliers, optical links, 2-9
PHP, *see* Hypertext preprocessor
PhpMyAdmin tool, 62-3
Physical addressing, 38-6
Physical layer, 34-2
 EtherCAT, 38-1 thru 38-2
 FlexRay, 44-3 thru 44-4

IEEE802.15.4, 50-4 thru 50-5
ISA 100.11a, 53-8
LIN-Bus
basic principle of, 45-5
signal specification, 45-5 thru 45-6
topology, 45-6
LonWorks, 41-4
Modbus serial network, 36-11
OCARI, 53-10 thru 53-11
WiMAX architecture, 52-5
WirelessHART, 53-3
WLANs
frequency bands, 48-2 thru 48-3
modulation techniques, 48-3
Piconet, 49-5 thru 49-6
Pixels, multimedia resolution, 27-1
Plant automation based on distributed systems (PABADIS)
order handling, 16-8 thru 16-9
resource handling, 16-8
Platform providers, 53-15
PNET, 37-9
Point coordination function, 48-5 thru 48-6
Powerline 110, 42-6
Power line communication, 2-7
POWERLINK safety protocol, 39-7 thru 39-8
Power plant automation systems
combined cycle power plant information system, 29-1 thru 29-2
functional layers of, 29-2 thru 29-3
information systems
common information model, 29-4 thru 29-5
distributed energy resource (DER) model, 29-6
telecontrol protocols, 29-5
instrumentation and control functions, 29-4
nuclear power plant, 29-3
safety requirements, 29-3 thru 29-4
Practical extraction report language (PERL)
characteristics, 66-2
data extraction, PDF files, 66-5 thru 66-6
E-Mail address extraction, excel files, 66-4 thru 66-5
Google Scholar, 66-8
internet applications, 66-3
network programming tools, 64-4
neural network, 66-6 thru 66-8
output file, 66-8
paper information extraction, IR accesses XPLORE, 66-6
regular expressions, 66-2 thru 66-3
searching process model, 66-8
Process field bus (Profibus)
acyclic data exchange, 32-10 thru 32-12
application profiles, 32-12
cyclic data exchange, 32-9 thru 32-10
decentralized periphery (DP) system
application relations, 32-8
controllers, 32-7 thru 32-8
engineering stations, 32-8
field-devices, 32-8
definition, 32-1
fieldbus data link
framing, 32-5
medium access control, 32-6 thru 32-7
service, 32-5
physical transmissions
asynchronous (RS-485), 32-2 thru 32-3
fiber optics, 32-4
Manchester bus powered, 32-4
properties of, 32-2
Profibus DP fieldbus, medical applications, 30-5 thru 30-6
Profibus Nutzer Organisation (PNO), 32-1 thru 32-2
PROFINET component-based automation (CBA), 40-1 thru 40-2
PROFINET IO
basics
acyclic data traffic, 40-7
address resolution, 40-6
cyclic data traffic, 40-6 thru 40-7
device model, 40-5 thru 40-6
diagnostics, 40-7 thru 40-8
definition, 40-1
development of, 40-1 thru 40-2
device classes
conformance classes, 40-4
performance, 40-3 thru 40-4
prerequisites, 40-4 thru 40-5
engineering and commissioning
device addressing, 40-11
GSD file, 40-11
neighborhood and topology detection, 40-12
redundancy, 40-13 thru 40-14
system power-up, 40-11 thru 40-12
fieldbus systems integration and web applications, 40-14
IRT communication
bus cycle, 40-9
conditions for, 40-8
cycle duration and constrains, 40-10
intervals, 40-9
orange interval, 40-9 thru 40-10
red interval, 40-10
PROFIsafe
black channel principle, 46-3
communication
code name, sender/receiver, 46-9
cyclic/acyclic, 46-4 thru 46-6
data consistency check, 46-9 thru 46-10
detected safety data failures, 46-10 thru 46-11
error-detection requirements, 46-4
error types and safeguards, 46-4

PDU, cyclic communication, 46-6 thru 46-7
time-out with receipt, 46-9
virtual consecutive number, 46-7 thru 46-9
deployment
increased immunity, 46-11
installation guidelines, 46-11 thru 46-12
power supplies and electrical safety, 46-11
response time, 46-14
wireless transmission and security, 46-12 thru 46-13
principles, 46-1
standardization framework, 46-1 thru 46-3
Protocol control information (PCI), ISO/OSI design, 1-8
Protocol identifier, 36-12
Protocols, coexistence of three, 53-13 thru 53-15
Proximity card, 8-6
Proxy devices, vertical integration, 13-6
Python script
Django (*see* Django)
functions, 63-1
shell interface, 63-2

Q

Quality of service (QoS)
Ad hoc networks, 7-9
benefits, 19-2
congestion avoidance, 19-11 thru 19-13
definition, 19-2
high availability solutions, routers, 19-13 thru 19-4
information security
committees and standards, 19-4
policy, 19-3 thru 19-4
problems, 19-3
IP networks
application supports, 19-5
classification and marking, 19-8 thru 19-9
DiffServ model, 19-6 thru 19-8
integrated Services (IntServ) model, 19-6
problems of, 19-4 thru 19-5
processes, 19-5 thru 19-6
queuing and congestion management, 19-9 thru 19-11
professional knowledge, 19-1
virtual automation networks (VAN)
end-to-end communication, 15-9
monitoring modes, 15-10 thru 15-11
topology structure of, 15-10
warranty, 19-2
wireless multimedia sensor network (WMSN), 11-10
Query service set, 57-8

R

Radio frequency (RF), 49-5
Radio frequency identification (RFID) technology
active system, 8-8
affordable tag, 8-18
applications
access control, 9-8 thru 9-9
anticounterfeiting, 9-10 thru 9-11
cheating detection, 8-23
item tracking and tracing, 9-4 thru 9-8
lake utilization, 8-25 thru 8-26
in library, 8-23 thru 8-25
medicine dispense, 8-24 thru 8-26
architecture of
antennas, 9-2 thru 9-3
middleware, 9-3 thru 9-4
tags and readers, 9-2
artificial perception, 8-27 thru 8-28
bar code system
hands-free technology, 8-16
mentality of, 8-16 thru 8-17
Morse and Code 128, 8-3 thru 8-4
vs. Bluetooth, 49-10
dual authentication, 8-22
electronic cash, 8-7
EPC-TID, Kill command, 8-21 thru 8-22
frequency selection, 8-12 thru 8-13
hands-free technology, 8-8, 8-16
high frequency, 8-6
historical development, 8-20
Hollerith punch card, 8-2
international standard, 8-14
magnetic stripes, 8-4
national standards, 8-15 thru 8-16
nonionization radiation, 8-26 thru 8-27
personal identity, 8-7 thru 8-8
privacy infringement, 8-20 thru 8-21
prologue, 8-2
promiscuity, 8-14 thru 8-15
proximity card, 8-6
role reversal, 8-20
smart card, 8-5
supply chain management, 8-13 thru 8-14
trace-and-track, 8-23
ubiquity, 8-19 thru 8-20
vicinity card, 8-12
wake-up technology, 8-8 thru 8-9
Radio-frequency power, wireless communication, 10-3
Radio technology, hardware development, 6-8 thru 6-10
Rate monotonic (RM) approach, 34-10 thru 34-12
Real delivery transport (RDT), 27-11
Real-time service interface, 43-9
Real-time streaming protocol (RTSP), 27-11
Real-time systems
analytical assessment methods for, 17-6
application domains, 17-3
best effort *vs.* guaranteed service, 17-5
constraints characterization, 17-2
definition, 17-1

design paradigms
centralized access methods for, 17-8 thru 17-9
centralized *vs.* distributed architectures, 17-7
composability and scalability, 17-7
flexibility, 17-9 thru 17-10
security, 17-9
time-triggered *vs.* event-triggered systems, 17-7 thru 17-8
wireless communication support, 17-10
deterministic *vs.* statistical communication, 17-4 thru 17-5
elements of, 17-2
performance metrics, 17-5 thru 17-6
scheduling and metrics, 17-3 thru 17-4
Real-time transport protocol (RTP), 27-11
Red interval, 40-9, 40-10
Redundancy, 37-3
References, 57-6
Registered ports, 61-6
Remote bus, 33-2
Remote procedure call (RPC), 65-6
Remote terminal unit (RTU) mode, 36-8 thru 36-9
Repeaters, 34-2
Representational State Transfer (REST) approach, 65-6
Request/response services, 41-6, 57-7
Request to send/clear to send (RTS/CTS) mechanism, 48-5
Resource reservation protocol (RSVP), 27-10
Response-time analysis (RTA), 17-6
RFC4919, 51-5
Robotics, healthcare, 30-11 thru 30-12
Rootkit, 22-11
Rope-making machine, automation of, 36-13 thru 36-15
Routing methods, wireless networks
Ad Hoc Networks, 4-5 thru 4-6
AODV protocol, 4-10 thru 4-11
classification of, 4-4 thru 4-5
DSR protocol, 4-9 thru 4-10
DYMO protocol, 4-11
flat protocols, 4-7
GAF protocols, 4-13
hierarchical protocols, 4-7
LEACH protocols, 4-12 thru 4-13
location-based protocols, 4-7
optimized link-state routing protocol (OLSR), 4-8
sensor protocols, 4-12
TBRPF method, 4-8 thru 4-9
Runtime interworking
CHAIN, 42-10
data point types (DPTs), 42-9
definition, 42-6
group objects, 42-6 thru 42-8
LTE, 42-10

S

SAE J2602, 45-12
Safe failure fraction (SFF), 47-5
Safety function response time (SFRT), 46-13 thru 46-14
SafetyLon
application, 47-11
general concept, 47-1 thru 47-2
hardware
architecture, 47-4
safe failure fraction, 47-5
safety-related input, schematic of, 47-5
safety-related output, schematic of, 47-6
two-channel switch, 47-6 thru 47-7
library, 47-11 thru 47-12
safety-related firmware
logic-based monitoring, 47-9
network access layer interface, 47-8
node state machine, 47-10
primary function, 47-8
SafetyLon message structure, 47-9
software design, 47-7
software monitoring, 47-8 thru 47-9
safety-related lifecycle
characteristics, 47-2
parameters, 47-3
state diagram, 47-3
tools
application builder, 47-11
request message and response message, 47-12
structure, 47-11
types, 47-10
Sampled analog values (SAV), 59-6 thru 59-7
Scatternet, 49-6
Scheduling router node activity (SERENA) strategy, 53-12
Searching process model, 66-8
SecureChannel service set, 57-7 thru 57-8
Secure sockets layer (SSL), 22-14
Semantic Web, *see* Ontology-driven architecture
SERCOS III, 37-7
Serial network, Modbus
ASCII mode, 36-10 thru 36-11
error detection, 36-11
frames, 36-8 thru 36-9
physical layer, 36-11
RTU mode, 36-9 thru 36-10
Serial port profile (SPP), 49-8
Service access points (SAP), ISO/OSI design, 1-6 thru 1-7
Service data unit (SDU), ISO/OSI design, 1-8
Service discovery application profile (SDAP), 49-8, 49-9
Service port profile (SPP), 49-9
Servo drive over EtherCAT (SoE), 38-10
Session service set, 57-8
SFF, *see* Safe failure fraction
Shielding, cable data transmission, 2-2
SIM, *see* Subscriber identity module card
Simple network management protocol (SNMP), 56-5
Single-sign-on (SSO) standard, 67-1 thru 67-2

Sliding window protocol
flow control, 61-8 thru 61-9
receiver window, 61-8
sender window, 61-9
Smart card, 8-5
Smart transducer interface standard, 43-8
Smearing, optical links, 2-7
SNMP, *see* Simple network management protocol
Solar power, wireless communication, 10-3
Sparse time, 43-2
Spice internet package (SIP)
characteristics, 64-9
client machine, 64-8 thru 64-9
graphical user interface, 64-9
network traffic considerations, 64-10
server-located PERL script, 64-9
Sporadic frame, 45-9
Spyware, 22-11
Standard network variable types (SNVT), 41-10 thru 41-11
Stand-by managing node (SMN), 39-6
Star couplers, 44-5
Static Web site, 62-1
Statistic access arbitration, 3-3
Statistical generative models (SGMs), *see* BASE system
Status telegram, INTERBUS, 33-5
Stratum, NTP clock synchronization, 18-7
Subscriber identity module (SIM) card, 8-5
Subscription service set, 57-9
Sunblind actuator, 41-8, 41-10
Sun Java™ System Access Manager [SUN05], 67-2
Supervisory control and data acquisition (SCADA) systems, 26-8
in electric power industries, 58-1 thru 58-2
in oil and gas industries, 58-3
Sync byte field, 45-7
Synchronization profile (SYNC), 49-9
Synchronous access arbitration, 3-2 thru 3-3
SyncManager, 38-7 thru 38-8
System configuration language (SCL), 59-7 thru 59-8

T

Tailorable software, multimedia service convergence, 14-5
TargetNode, 57-7
Tcnet, 37-9
TCP network, 36-12
Technologies
CAN
data link layer, 31-3 thru 31-4
error detection and signaling, 31-4 thru 31-5
limitations, 31-12 thru 31-14
medical communication, 30-5
network topology, 31-6
physical layer, 31-2 thru 31-3
upper layer protocols, 31-6 thru 31-10
INTERBUS
basic elements, 33-2
data transfer, 33-1 thru 33-2
diagnostic features, 33-7 thru 33-8
performance evaluation, 33-8 thru 33-9
protocol, 33-3 thru 33-7
topology, 33-1, 33-2
internet
electronic device description, 56-8 thru 56-10
for information description and presentation, 56-6 thru 56-8
for server-side and client-side functions, 56-8
transport and communication related, 56-5 thru 56-6
Profibus
acyclic data exchange, 32-10 thru 32-12
application profiles, 32-12
cyclic data exchange, 32-9 thru 32-10
decentralized periphery (DP) system, 32-7 thru 32-8
definition, 32-1
fieldbus data link, 32-5 thru 32-7
physical transmissions, 32-2 thru 32-4
properties of, 32-2
Telecontrol profile, VAN
data transfer mechanisms, 15-12
definition and advantages, 15-11
method, 15-13
principle, 15-12
transfer buffer, 15-11
Telephony control protocols, 49-7
Temporal key integrity protocol (TKIP), 48-9
Thermalelectric power, wireless communication, 10-3
Third generation partnership project (3GPP), 52-8
Three-way handshaking, 61-7
Time division multiple access (TDMA), 2-16, 17-8, 31-9
Time keeping, 53-6, 53-10
Time-triggered architecture, protocols
clock synchronization, 43-7
concept, 43-1
fault-tolerant configuration, 43-7
flow control and temporal firewall, 43-2 thru 43-3
interface file system, 43-8 thru 43-9
principles of operation, 43-10 thru 43-11
smart transducer, three interfaces of, 43-9 thru 43-10
sparse time, 43-2
time-triggered communication, 43-3 thru 43-4
time-triggered Ethernet
periods, 43-6 thru 43-7
principles of operation, 43-6
time format, 43-6
time-triggered fieldbus, 43-2
time-triggered fieldbus TTP/A, 43-7 thru 43-8
time-triggered protocol (TTP)
advantages, 43-4
fault hypothesis and fault handling, 43-5
fault tolerance, 43-5

Time-triggered communication, 43-3 thru 43-4
Time-triggered communication on CAN (TTCAN)
  clock synchronization, 31-10
  levels, 31-8 thru 31-9
  TDMA bus access, 31-9
  time windows type, 31-10
Time-triggered Ethernet (TTEthernet)
  digital time format, 43-6
  features, 43-5
  periods, 43-6 thru 43-7
  principles of operation, 43-6
Time-triggered Ethernet (TTEthernet)
  fault-tolerant configuration, 43-7
Time-triggered paradigm
  flow control and temporal firewall, 43-2 thru 43-3
  sparse time, 43-2
Time-triggered protocol (TTP), 43-4 thru 43-5
Topology dissemination based on reverse-path forwarding (TBRPF), 4-8 thru 4-9
Tracking and tracing applications, RFID technology, 9-4 thru 9-8
  animal, 9-6
  baggage, 9-5
  children, 9-7 thru 9-8
  crowd control, 9-8
  golf ball, 9-8
  hospital equipment, 9-6 thru 9-7
  library book, 9-6
  newborn baby, 9-7
  patient, 9-7
Transaction identifier, 36-12
Transmission control protocol (TCP)
  byte-stream, 61-2
  congestion control, 61-12
  connection establishment
    client program, 61-8
    data transfer, 61-6 thru 61-7
    segment exchange, 61-7 thru 61-8
    three-way handshaking, 61-7
  connection-oriented, 61-1
  connection termination, 61-12 thru 61-13
  data sending process, acknowledgment number, 61-8
  error control
    duplicate segment, 61-11
    lost acknowledgment, 61-11 thru 61-12
    lost segment, 61-10, 61-11
    out-of-order segment, 61-11
    segment checksum, 61-10, 61-11
    time-out counter, 61-11
  flow control improvement, 61-9 thru 61-10
  full duplex, 61-2
  in-sequence data, 61-2
  port number assignments, 61-6
  port number concept, 61-1, 61-2
  programming samples, 61-14 thru 61-16
  reliability, 61-1
  sliding window protocol
    flow control, 61-8 thru 61-9
    receiver window, 61-8
    sender window, 61-9
  state transition, 61-13 thru 61-14
  TCP segment, 61-3 thru 61-6
  user datagram protocol (UDP), 61-1, 61-2
Transport and session layer, 41-5 thru 41-6
Transport layer protocol
  ad hoc networks, 7-4 thru 7-5
  LIN-Bus, 45-11 thru 45-12
Transport layer security (TLS), 22-14
Trapdoor, 22-11
Trojan horse, 22-11
TTCAN, *see* Time-triggered communication on CAN
TTP/A, 43-7 thru 43-8
Turn-around time, 39-8 thru 39-9

U

UDP, *see* User datagram protocol
UDP header compression, 51-11
Ultralow-power wireless communication
  application layer methods, 10-6 thru 10-8
  communication protocol methods
    duty-cycle, 10-4
    low duty-cycle, 10-5
    multi-hop method, 10-5
    single-hop star architecture, 10-4 thru 10-5
    topology control, 10-5
    wake-up method, 10-5 thru 10-6
  composite materials, airplanes, 10-1
  hardware methods
    basic and optional parts, 10-2 thru 10-3
    energy sources, 10-3 thru 10-4
Ultra-wideband (UWB), wireless communication, 6-4 thru 6-5
Unacknowledged repeated service, 41-6
Unacknowledged service, 41-5
Unconditional frame, 45-9
Unicast mode, 36-8
Unit identifier, 36-12
Universal plug and play (UPnP), home automation, 26-10
User datagram protocol (UDP)
  datagram, 60-1 thru 60-2
  flow and error control, 60-4 thru 60-8
  vs. IP, 60-2
  operation, 60-1 thru 60-2
  port number assignments, 60-4
  ports used with, 60-4
  programming samples, 60-8 thru 60-10
  in TCP/IP suite model, 60-1, 60-2
User-defined network variable type (UNVT), 41-11

V

Vertical integration
application, 13-7 thru 13-8
definition of, 13-1
history
computer-integrated manufacturing, 13-2 thru 13-3
fieldbus systems, 13-3
functional units of, 13-2
manufacturing automation protocol (MAP), 13-3
World Wide Web (WWW), 13-3 thru 13-4
holonic manufacturing systems, 13-10
network interconnections
cross-domain communication, 13-5
fieldbus media, 13-5 thru 13-6
flat network hierarchy, 13-5
horizontal communication, 13-4
IP-based transport layer, 13-5
multi-protocol gateways, 13-6 thru 13-7
protocol convergence, 13-5
proxy devices, 13-6
security, 13-8 thru 13-10
service-oriented architectures (SoA), 13-11
Vibration power, wireless communication, 10-3
Vicinity card, 8-12
Video compressors, multimedia, 27-6 thru 27-7
Video distribution profile (VDP), 49-9
View service set, 57-8
Virtual automation networks (VAN)
architecture, 15-6 thru 15-7
components, 15-5 thru 15-6
domains, 15-4
PROFINET network, 15-2
runtime tunnel establishment
name-based addressing and routing, 15-7 thru 15-8
quality-of-service monitoring, 15-9 thru 15-11
telecontrol profile, 15-11 thru 15-13
Virtual consecutive number, 46-7 thru 46-9
Virtual field device (VFD), 33-6 thru 33-7
Virtual router redundancy protocol (VRRP), QoS, 19-13 thru 19-14
Virus, 22-10
Visas anticounterfeiting, RFID technology, 9-11
Vnet/IP, 37-9
Voice over Internet protocol (VoIP), 14-2 thru 14-3
Voice plugins, 14-9 thru 14-11

W

Wake-up technology, RFID, 8-8 thru 8-9
WAMP server, 62-2 thru 62-3
Warranty, 19-2
Wavelength division multiplexing (WDM), 2-10
Wavelength division multiplexing (WDM) technology, optical links, 2-10
Web-based C_{++} compiler, 64-7 thru 64-8
Web integration, 40-14
Web service server, 56-6
Well-known ports, 61-6, 61-7
WiBree, 49-10
WiMAX
architecture
equipment, 52-6
MAC layer/data link layer, 52-5
physical layer, 52-5
broadband technology
backhaul/access network applications, 52-3 thru 52-4
mobile standards comparison, 52-4
example of, 52-2
fixed and mobile, 52-3
forum and working groups, 52-6
integration
WiMAX-DSL, 52-6 thru 52-7
WiMAX-3GPP, 52-8
statistics, 52-1
versions, 52-2
vs. Wi-Fi, 52-4
vs. WLAN and PAN, 52-4
WiMAX-DSL integration, 52-6 thru 52-7
Wind power plants, 59-8
Wind turbine power plant (WPP), 29-6
Wired communication links
bit encoding, 2-5 thru 2-6
cable types and operational characteristics, 2-2 thru 2-3
differential transmission, 2-4
multiplex communication, 2-4 thru 2-5
physical properties, 2-1 thru 2-2
power line communication, 2-7
simplex communication, 2-4
single-ended transmission, 2-3 thru 2-4
standards, 2-7
Wired equivalent privacy (WEP), 48-9
Wireless communication standards
challenges, 54-1 thru 54-2
characteristics of, 54-4 thru 54-5
IEEE 802.15.4, 54-2
professional mobile radio (PMR), 54-2
regulations and EMC, 54-2 thru 54-6
taxonomy, 54-2, 54-3
Wireless communication system; *see also* Industrial wireless sensor networks
bit coding, 2-16
classification
wireless mesh networks (WMN), 4-2 thru 4-3
WPAN and WLAN, 4-1 thru 4-2
media access, 2-15
modulation formats, 2-15 thru 2-16
multiplexing, 2-16 thru 2-17

physical properties
antennas, 2-12 thru 2-13
channel capacity, 2-11 thru 2-12
free-space path loss, 2-12
impairments, 2-13 thru 2-14
link budget, 2-14
thermal noise, 2-11
wavelength, 2-11
routing
Ad Hoc Networks, 4-5 thru 4-6
AODV protocol, 4-10 thru 4-11
classification of, 4-4 thru 4-5
DSR protocol, 4-9 thru 4-10
DYMO protocol, 4-11
flat protocols, 4-7
GAF protocols, 4-13
hierarchical protocols, 4-7
LEACH protocols, 4-12 thru 4-13
location-based protocols, 4-7
optimized link-state routing protocol (OLSR), 4-8
sensor protocols, 4-12
TBRPF method, 4-8 thru 4-9
security, 22-15
local area network (WLAN), 28-2 thru 28-6
personal area network (PAN), 28-6 thru 28-8
standards, 2-17
types, 2-14 thru 2-15
ultralow-power
application layer methods, 10-6 thru 10-8
communication protocol methods, 10-4 thru 10-6
composite materials, airplanes, 10-1
hardware methods, 10-2 thru 10-4
Wireless HART, 12-7
data link layer
datagram, 53-5 thru 53-6
frequency diversity mechanism, 53-5
packet timing and time slot format, 53-4
TDMA mechanism, 53-3
industrial wireless sensor networks, 6-3 thru 6-4
network layer datagram, 53-7 thru 53-8
network topologies, 53-6
vs. OSI seven-layer model, 53-2
physical layer, 53-3
timekeeping, 53-6
upperlayers, 53-8
Wireless local area networks (WLANs)
DCF and HCF, limitations of, 48-8 thru 47-9
802.11 family, 48-2
fast handover
mechanisms on AP side, 48-10 thru 48-11
mechanisms on client side, 48-11
procedure, 48-9 thru 48-10
future enhancements, 48-11
medium access control
architecture with HCF, 48-4
direct link protocol and block ACK, 48-7 thru 47-8
distributed coordination function, 48-4 thru 48-5
enhanced distributed channel access, 48-6
HCF controlled channel access, 48-7
point coordination function, 48-5 thru 48-6
physical layer
frequency bands, 48-2 thru 48-3
modulation techniques, 48-3
security mechanisms, 4-1, 48-9
deployment issues, 28-6
issues/attacks, WiFi, 28-2 thru 28-3
TKIP, CCMP, WAPI, 28-5 thru 28-6
WiFi protected access, 28-4 thru 28-5
wireless encryption protocol, 28-3 thru 28-4
Wireless mesh network (WMN)
applications, 4-2 thru 4-3
PAN, 28-7
types, 4-2
Wireless metropolitan area networks (WMAN), 4-2
Wireless multimedia sensor network (WMSN)
applications
habitat monitoring, 11-9
health care, 11-9
surveillance, 11-8
target tracking, 11-9
traffic monitoring, 11-8 thru 11-9
architecture
multitier architecture, 11-6
single-tier clustered architecture, 11-5 thru 11-6
single-tier flat architecture, 11-4 thru 11-5
energy-efficient design, 11-11
hardware
high-resolution motes, 11-7 thru 11-8
low-resolution motes, 11-6 thru 11-7
medium-resolution motes, 11-7
high bandwidth, 11-10
IP integration, 11-11 thru 11-12
localized processing and data fusion, 11-10 thru 11-11
multimedia coverage, 11-11
QoS requirements, 11-10
reliability and fault-tolerance, 11-11
scalable and flexible architectures, 11-10
vs. WSN, 11-3 thru 11-4
Wireless personal area networks (WPAN), 4-1
Wireless sensor network (WSN)
architecture and node, 11-2
characteristic system requirements, 11-3
industrial applications and requirements
battery lifetime, 12-5
coexistence with, 12-5
hazardous environments operation, 12-5 thru 12-6
reliable network performance, 12-4 thru 12-5
security, 12-5
standardized solutions, 12-4
motivation and drivers, 12-3 thru 12-4
node, 12-2

stacks, 12-2 thru 12-3
survey and evaluation of
2.4 GHz band coexistence, 12-8 thru 12-9
IEEE Std 802.15.4, 12-6
ISA-100 standards, 12-7
WirelessHART, 12-7
ZigBee specification, 12-6 thru 12-7
vs. WSN, 11-3 thru 11-4
Wireless traffic smoother (WTS), 48-8
Wireless wide area networks (WWAN), 4-2
WLAN *vs.* WiMAX, 52-4
WorldFIP
aperiodic buffer transfers, 34-9 thru 34-10
application layer, 34-6 thru 34-7
BAT, setting the
aperiodic busy interval, 34-16
aperiodic traffic scheduling, 34-17
earliest deadline approach, 34-12 thru 34-15
rate monotonic approach, 34-10 thru 34-12
response time analysis, aperiodic traffic, 34-15
upper bound for dead interval, 34-15 thru 34-16
worst-case response time, 34-16 thru 34-17
data link layer
asynchronous traffic, transmission of, 34-6
bus arbitrator table, 34-4 thru 34-6
cyclic traffic, transmission of, 34-3 thru 34-4
LLC sublayer, 34-3
MAC, 34-2
definition, 34-1
layered architecture, 34-1 thru 34-2
physical layer, 34-2
timing properties
aperiodic buffer transfers, 34-9 thru 34-10
buffer transfer timings, 34-8
bus arbitrator table, 34-8 thru 34-9
producer/distributor/consumer, 34-7
rate monotonic approach, 34-10 thru 34-17
World Wide Web (WWW), 13-3 thru 13-4
Worm, 22-10
Worst-case response time, 34-16 thru 34-17

X

XML, *see* eXtensible markup language
X10 protocol, medical application, 30-9

Y

Yellow interval, 40-9

Z

ZigBee
development and industrial applications, 50-8 thru 50-10
IEEE802.15.4
application layer, 50-7 thru 50-8
key features, 50-3 thru 50-4
MAC layer, 50-5 thru 50-6
network layer, 50-7
physical layer, 50-4 thru 50-5
industrial applications, 50-9 thru 50-10
integration with global intranet/Internet, 50-9
medical application, 30-10
and mesh networks, 50-1 thru 50-2
vs. other wireless networks, 50-2, 50-3
stack, 50-2, 50-4
standard, 49-10
standardization, 6-3
wireless sensor networks, 12-6 thru 12-7
Z-Wave protocol, home automation, 26-10

Printed in the United States
by Baker & Taylor Publisher Services